Problem Books in Mathematics

Series Editor

Peter Winkler
Department of Mathematics
Dartmouth College
Hanover, NH 03755
USA
peter.winkler@dartmouth.edu

D1363962

For other titles published in the series, go to
www.springer.com/series/714

Dušan Djukić • Vladimir Janković
Ivan Matić • Nikola Petrović

The IMO Compendium

A Collection of Problems Suggested
for The International Mathematical
Olympiads: 1959-2009

Second Edition

Springer

Dušan Djukić
Department of Mathematics
University of Toronto
Toronto Ontario, M5S3G3
Canada
dusan.djukic@utoronto.ca

Vladimir Janković
Department of Mathematics
University of Belgrade
Studentski Trg 16
11000 Belgrade
Serbia
vjankovic@matf.bg.ac.rs

Ivan Matić
Department of Mathematics
Duke University
Durham, North Carolina 27708
USA
matic@math.duke.edu

Nikola Petrović
Science Department
Texas A&M University
PO Box 23874
Doha
Qatar
nikola.petrovic@qatar.tamu.edu

ISSN 0941-3502
ISBN 978-1-4419-9853-8   e-ISBN 978-1-4419-9854-5
DOI 10.1007/978-1-4419-9854-5
Springer New York Dordrecht Heidelberg London

Library of Congress Control Number: 2011926996

Printed on acid-free paper

Springer is part of Springer Science+Business Media (www.springer.com)

Preface

The International Mathematical Olympiad (IMO) exists for more than 50 years and has already created a very rich legacy and firmly established itself as the most prestigious mathematical competition in which a high-school student could aspire to participate. Apart from the opportunity to tackle interesting and very challenging mathematical problems, the IMO represents a great opportunity for high-school students to see how they measure up against students from the rest of the world. Perhaps even more importantly, it is an opportunity to make friends and socialize with students who have similar interests, possibly even to become acquainted with their future colleagues on this first leg of their journey into the world of professional and scientific mathematics. Above all, however pleasing or disappointing the final score may be, preparing for an IMO and participating in one is an adventure that will undoubtedly linger in one's memory for the rest of one's life. It is to the high-school-aged aspiring mathematician and IMO participant that we devote this entire book.

The goal of this book is to include all problems ever shortlisted for the IMOs in a single volume. Up to this point, only scattered manuscripts traded among different teams have been available, and a number of manuscripts were lost for many years or unavailable to many.

In this book, all manuscripts have been collected into a single compendium of mathematics problems of the kind that usually appear on the IMOs. Therefore, we believe that this book will be the definitive and authoritative source for high-school students preparing for the IMO, and we suspect that it will be of particular benefit in countries lacking adequate preparation literature. A high-school student could spend an enjoyable year going through the numerous problems and novel ideas presented in the solutions and emerge ready to tackle even the most difficult problems on an IMO. In addition, the skill acquired in the process of successfully attacking difficult mathematics problems will prove to be invaluable in a serious and prosperous career in mathematics.

However, we must caution our aspiring IMO participant on the use of this book. Any book of problems, no matter how large, quickly depletes itself if the reader merely glances at a problem and then five minutes later, having determined that the problem seems unsolvable, glances at the solution.

The authors therefore propose the following plan for working through the book. Each problem is to be attempted at least half an hour before the reader looks at the solution. The reader is strongly encouraged to keep trying to solve the problem without looking at the solution as long as he or she is coming up with fresh ideas and possibilities for solving the problem. Only after all venues seem to have been exhausted is the reader to look at the solution, and then only in order to study it in close detail, carefully noting any previously unseen ideas or methods used. To condense the subject matter of this already very large book, most solutions have been streamlined, omitting obvious derivations and algebraic manipulations. Thus, reading the solutions requires a certain mathematical maturity, and in any case, the solutions, especially in geometry, are intended to be followed through with pencil and paper, the reader filling in all the omitted details. We highly recommend that the reader mark such unsolved problems and return to them in a few months to see whether they can be solved this time without looking at the solutions. We believe this to be the most efficient and systematic way (as with any book of problems) to raise one's level of skill and mathematical maturity.

We now leave our reader with final words of encouragement to persist in this journey even when the difficulties seem insurmountable and a sincere wish to the reader for all mathematical success one can hope to aspire to.

Belgrade,   *Dušan Djukić*
November 2010   *Vladimir Janković*
                                                    Ivan Matić
                                              Nikola Petrović

Over the previous years we have created the website: www.imomath.com. There you can find the most current information regarding the book, the list of detected errors with corrections, and the results from the previous olympiads. This site also contains problems from other competitions and olympiads, and a collection of training materials for students preparing for competitions.

We are aware that this book may still contain errors. If you find any, please notify us at imomath@gmail.com. If you have any questions, comments, or suggestions regarding both our book and our website, please do not hesitate to write to us at the above email address. We would be more than happy to hear from you.

Acknowledgements

The making of this book would have never been possible without the help of numerous individuals, whom we wish to thank.

First and foremost, obtaining manuscripts containing suggestions for IMOs was vital in order for us to provide the most complete listing of problems possible. We obtained manuscripts for many of the years from the former and current IMO team leaders of Yugoslavia / Serbia, who carefully preserved these valuable papers throughout the years. Special thanks are due to Prof. Vladimir Mićić, for some of the oldest manuscripts, and to Prof. Zoran Kadelburg. We also thank Prof. Djordje Dugošija and Prof. Pavle Mladenović. In collecting shortlisted and longlisted problems we were also assisted by Prof. Ioan Tomescu from Romania, Hà Duy Hưng from Vietnam, and Zhaoli from China.

A lot of work was invested in cleaning up our giant manuscript of errors. Special thanks in this respect go to David Kramer, our copy-editor, and to Prof. Titu Andreescu and his group for checking, in great detail, the validity of the solutions in this manuscript, and for their proposed corrections and alternative solutions to several problems. We also thank Prof. Abderrahim Ouardini from France for sending us the list of countries of origin for the shortlisted problems of 1998, Prof. Dorin Andrica for helping us compile the list of books for reference, and Prof. Ljubomir Čukić for proofreading part of the manuscript and helping us correct several errors.

We would also like to express our thanks to all anonymous authors of the IMO problems. Without them, the IMO would obviously not be what it is today. It is a pity that authors' names are not registered together with their proposed problems. In an attempt to change this, we have tried to trace down the authors of the problems, with partial success. We are thankful to all people who were so kind to help us in our investigation. The names we have found so far are listed in Appendix C. In many cases, the original solutions of the authors were used, and we duly acknowledge this immense contribution to our book, though once again, we regret that we cannot do this individually. In the same vein, we also thank all the students participating in the IMOs, since we have also included some of their original solutions in this book.

We thank the following individuals who discussed problems with us and helped us with correcting the mistakes from the previous edition of the book: Xiaomin Chen, Orlando Döhring, Marija Jelić, Rudolfs Kreicbergs, Stefan Mehner, Yasser Ahmady Phoulady, Dominic Shau Chin, Juan Ignacio Restrepo, Arkadii Slinko, Harun Šiljak, Josef Tkadlec, Ilan Vardi, Gerhard Woeginger, and Yufei Zhao.

The illustrations of geometry problems were done in WinGCLC, a program created by Prof. Predrag Janičić. This program is specifically designed for creating geometric pictures of unparalleled complexity quickly and efficiently. Even though it is still in its testing phase, its capabilities and utility are already remarkable and worthy of highest compliment.

Finally, we would like to thank our families for all their love and support during the making of this book.

Contents

1

Introduction

1.1 The International Mathematical Olympiad

The International Mathematical Olympiad (IMO) is the most important and prestigious mathematical competition for high-school students. It has played a significant role in generating wide interest in mathematics among high school students, as well as identifying talent.

In the beginning, the IMO was a much smaller competition than it is today. In 1959, the following seven countries gathered to compete in the first IMO: Bulgaria, Czechoslovakia, German Democratic Republic, Hungary, Poland, Romania, and the Soviet Union. Since then, the competition has been held annually. Gradually, other Eastern-block countries, countries from Western Europe, and ultimately numerous countries from around the world and every continent joined in. (The only year in which the IMO was not held was 1980, when for financial reasons no one stepped in to host it. Today this is hardly a problem, and hosts are lined up several years in advance.) In the 50th IMO, held in Bremen, no fewer than 104 countries took part.

The format of the competition quickly became stable and unchanging. Each country may send up to six contestants and each contestant competes individually (without any help or collaboration). The country also sends a team leader, who participates in problem selection and is thus isolated from the rest of the team until the end of the competition, and a deputy leader, who looks after the contestants.

The IMO competition lasts two days. On each day students are given four and a half hours to solve three problems, for a total of six problems. The first problem is usually the easiest on each day and the last problem the hardest, though there have been many notable exceptions. ((IMO96-5) is one of the most difficult problems from all the Olympiads, having been fully solved by only six students out of several hundred!) Each problem is worth 7 points, making 42 points the maximum possible score. The number of points obtained by a contestant on each problem is the result of intense negotiations and, ultimately, agreement among the problem coordinators, assigned by the host country, and the team leader and deputy, who defend the interests of their contestants. This system ensures a relatively objective grade that is seldom off by more than two or three points.

Though countries naturally compare each other's scores, only individual prizes, namely medals and honorable mentions, are awarded on the IMO. Fewer than one twelfth of participants are awarded the gold medal, fewer than one fourth are awarded the gold or silver medal, and fewer than one half are awarded the gold, silver or bronze medal. Among the students not awarded a medal, those who score 7 points on at least one problem are awarded an honorable mention. This system of determining awards works rather well. It ensures, on the one hand, strict criteria and appropriate recognition for each level of performance, giving every contestant something to strive for. On the other hand, it also ensures a good degree of generosity that does not greatly depend on the variable difficulty of the problems proposed.

According to the statistics, the hardest Olympiad was that in 1971, followed by those in 1996, 1993, and 1999. The Olympiad in which the winning team received the lowest score was that in 1977, followed by those in 1960 and 1999.

The selection of the problems consists of several steps. Participant countries send their proposals, which are supposed to be novel, to the IMO organizers. The organizing country does not propose problems. From the received proposals (the *longlisted* problems), the problem committee selects a shorter list (the *shortlisted* problems), which is presented to the IMO jury, consisting of all the team leaders. From the short-listed problems the jury chooses six problems for the IMO.

Apart from its mathematical and competitive side, the IMO is also a very large social event. After their work is done, the students have three days to enjoy events and excursions organized by the host country, as well as to interact and socialize with IMO participants from around the world. All this makes for a truly memorable experience.

1.2 The IMO Compendium

Olympiad problems have been published in many books [97]. However, the remaining shortlisted and longlisted problems have not been systematically collected and published, and therefore many of them are unknown to mathematicians interested in this subject. Some partial collections of shortlisted and longlisted problems can be found in the references, though usually only for one year. References [1], [39], [57], [88] contain problems from multiple years. In total, these books cover roughly 50% of the problems found in this book.

The goal of this book is to present, in a single volume, our comprehensive collection of problems proposed for the IMO. It consists of all problems selected for the IMO competitions, shortlisted problems from the 10th IMO and from the 12th through 50th IMOs, and longlisted problems from twenty IMOs. We do not have shortlisted problems from the 9th and the 11th IMOs, and we could not discover whether competition problems at those two IMOs were selected from the longlisted problems or whether there existed shortlisted problems that have not been preserved. Since IMO organizers usually do not distribute longlisted problems to the representatives of participant countries, our collection is incomplete. The practice of distributing these longlists effectively ended in 1989. A selection of problems from the first eight IMOs has been taken from [88].

The book is organized as follows. For each year, the problems that were given on the IMO contest are presented, along with the longlisted and/or shortlisted problems, if applicable. We present solutions to all shortlisted problems. The problems appearing on the IMOs are solved among the other shortlisted problems. The longlisted problems have not been provided with solutions, except for the two IMOs held in Yugoslavia (for patriotic reasons), since that would have made the book unreasonably long. This book has thus the added benefit for professors and team coaches of being a suitable book from which to assign problems. For each problem, we indicate the country that proposed it with a three-letter code. A complete list of country codes and the corresponding countries is given in the appendix. In all shortlists, we also indicate which problems were selected for the contest. We occasionally make references in our solutions to other problems in a straightforward way. After indicating with LL, SL, or IMO whether the problem is from a longlist, shortlist, or contest, we indicate the year of the IMO and then the number of the problem. For example, (SL89-15) refers to the fifteenth problem of the shortlist of 1989.

We also present a rough list of all formulas and theorems not obviously derivable that were called upon in our proofs. Since we were largely concerned with only the theorems used in proving the problems of this book, we believe that the list is a good compilation of the most useful theorems for IMO problem solving.

The gathering of such a large collection of problems into a book required a massive amount of editing. We reformulated the problems whose original formulations were not precise or clear. We translated the problems that were not in English. Some of the solutions are taken from the author of the problem or other sources, while others are original solutions of the authors of this book. Many of the non-original solutions were significantly edited before being included. We do not make any guarantee that the problems in this book fully correspond to the actual shortlisted or longlisted problems. However, we believe this book to be the closest possible approximation to such a list.

2

Basic Concepts and Facts

The following is a list of the most basic concepts and theorems frequently used in this book. We encourage the reader to become familiar with them and perhaps read up on them further in other literature.

2.1 Algebra

2.1.1 Polynomials

Theorem 2.1. *The quadratic equation* $ax^2 + bx + c = 0$ $(a, b, c \in \mathbb{R}, a \neq 0)$ *has solutions*

$$x_{1,2} = \frac{-b \pm \sqrt{b^2 - 4ac}}{2a}.$$

The discriminant D of a quadratic equation is defined as $D = b^2 - 4ac$. For $D < 0$ the solutions are complex and conjugate to each other, for $D = 0$ the solutions degenerate to one real solution, and for $D > 0$ the equation has two distinct real solutions.

Definition 2.2. *Binomial coefficients* $\binom{n}{k}$, $n, k \in \mathbb{N}_0$, $k \leq n$, *are defined as*

$$\binom{n}{i} = \frac{n!}{i!(n-i)!}.$$

They satisfy $\binom{n}{i} + \binom{n}{i-1} = \binom{n+1}{i}$ for $i > 0$ and also $\binom{n}{0} + \binom{n}{1} + \cdots + \binom{n}{n} = 2^n$, $\binom{n}{0} - \binom{n}{1} + \cdots + (-1)^n \binom{n}{n} = 0$, $\binom{n+m}{k} = \sum_{i=0}^{k} \binom{n}{i} \binom{m}{k-i}$, $\binom{n+r}{n} = \sum_{j=0}^{r} \binom{n+j-1}{n-1}$.

Theorem 2.3 ((Newton's) binomial formula). *For $x, y \in \mathbb{C}$ and $n \in \mathbb{N}$,*

$$(x+y)^n = \sum_{i=0}^{n} \binom{n}{i} x^{n-i} y^i.$$

Theorem 2.4 (Bézout's theorem). *A polynomial $P(x)$ is divisible by the binomial $x - a$ ($a \in \mathbb{C}$) if and only if $P(a) = 0$.*

Theorem 2.5 (The rational root theorem). *If $x = p/q$ is a rational zero of a polynomial $P(x) = a_n x^n + \cdots + a_0$ with integer coefficients and $(p,q) = 1$, then $p \mid a_0$ and $q \mid a_n$.*

Theorem 2.6 (The fundamental theorem of algebra). *Every nonconstant polynomial with coefficients in $\mathbb{C}$ has a complex root.*

Theorem 2.7 (Eisenstein's criterion (extended)). *Let $P(x) = a_n x^n + \cdots + a_1 x + a_0$ be a polynomial with integer coefficients. If there exist a prime p and an integer $k \in \{0, 1, \ldots, n-1\}$ such that $p \mid a_0, a_1, \ldots, a_k$, $p \nmid a_{k+1}$, and $p^2 \nmid a_0$, then there exists an irreducible factor $Q(x)$ of $P(x)$ whose degree is greater than k. In particular, if p can be chosen such that $k = n - 1$, then $P(x)$ is irreducible.*

Definition 2.8. *Symmetric polynomials* in $x_1, \ldots, x_n$ are polynomials that do not change on permuting the variables $x_1, \ldots, x_n$. *Elementary symmetric polynomials* are $\sigma_k(x_1, \ldots, x_n) = \sum x_{i_1} \cdots x_{i_k}$ (the sum is over all k-element subsets $\{i_1, \ldots, i_k\}$ of $\{1, 2, \ldots, n\}$).

Theorem 2.9. *Every symmetric polynomial in $x_1, \ldots, x_n$ can be expressed as a polynomial in the elementary symmetric polynomials $\sigma_1, \ldots, \sigma_n$.*

Theorem 2.10 (Viète's formulas). *Let $\alpha_1, \ldots, \alpha_n$ and $c_1, \ldots, c_n$ be complex numbers such that*

$$(x - \alpha_1)(x - \alpha_2) \cdots (x - \alpha_n) = x^n + c_1 x^{n-1} + c_2 x^{n-2} + \cdots + c_n .$$

Then $c_k = (-1)^k \sigma_k(\alpha_1, \ldots, \alpha_n)$ for $k = 1, 2, \ldots, n$.

Theorem 2.11 (Newton's formulas on symmetric polynomials). *Let $\sigma_k = \sigma_k(x_1, \ldots, x_n)$ and let $s_k = x_1^k + x_2^k + \cdots + x_n^k$, where $x_1, \ldots, x_n$ are arbitrary complex numbers. Then*

$$k\sigma_k = s_1 \sigma_{k-1} - s_2 \sigma_{k-2} + \cdots + (-1)^k s_{k-1} \sigma_1 + (-1)^{k-1} s_k .$$

2.1.2 Recurrence Relations

Definition 2.12. A *recurrence relation* is a relation that determines the elements of a sequence x_n, $n \in \mathbb{N}_0$, as a function of previous elements. A recurrence relation of the form

$$(\forall n \geq k) \quad x_n + a_1 x_{n-1} + \cdots + a_k x_{n-k} = 0$$

for constants $a_1, \ldots, a_k$ is called a *linear homogeneous recurrence relation of order* k. We define the *characteristic polynomial* of the relation as $P(x) = x^k + a_1 x^{k-1} + \cdots + a_k$.

Theorem 2.13. *Using the notation introduced in the above definition, let $P(x)$ factorize as $P(x) = (x - \alpha_1)^{k_1}(x - \alpha_2)^{k_2} \cdots (x - \alpha_r)^{k_r}$, where $\alpha_1, \ldots, \alpha_r$ are distinct complex* *numbers and $k_1, \ldots, k_r$ are positive integers. The general solution of this recurrence relation is in this case given by*

$$x_n = p_1(n)\alpha_1^n + p_2(n)\alpha_2^n + \cdots + p_r(n)\alpha_r^n,$$

where p_i is a polynomial of degree less than k_i. In particular, if $P(x)$ has k distinct roots, then all p_i are constant.

If $x_0, \ldots, x_{k-1}$ are set, then the coefficients of the polynomials are uniquely determined.

2.1.3 Inequalities

Theorem 2.14. *The squaring function is always positive; i.e., $(\forall x \in \mathbb{R})\ x^2 \geq 0$. By substituting different expressions for x, many of the inequalities below are obtained.*

Theorem 2.15 (Bernoulli's inequalities).

1. *If $n \geq 1$ is an integer and $x > -1$ a real number, then $(1+x)^n \geq 1 + nx$.*

2. *If $\alpha > 1$ or $\alpha < 0$, then for $x > -1$, the following inequality holds: $(1+x)^\alpha \geq 1 + \alpha x$.*

3. *If $\alpha \in (0,1)$ then for $x > -1$ the following inequality holds: $(1+x)^\alpha \leq 1 + \alpha x$.*

Theorem 2.16 (The mean inequalities). *For positive real numbers $x_1, x_2, \ldots, x_n$ it is always the case that $QM \geq AM \geq GM \geq HM$, where*

$$QM = \sqrt{\frac{x_1^2 + \cdots + x_n^2}{n}}, \qquad AM = \frac{x_1 + \cdots + x_n}{n},$$

$$GM = \sqrt[n]{x_1 \cdots x_n}, \qquad HM = \frac{n}{\frac{1}{x_1} + \cdots + \frac{1}{x_n}}.$$

Each of these inequalities becomes an equality if and only if $x_1 = x_2 = \cdots = x_n$. The numbers QM, AM, GM, and HM are respectively called the quadratic mean, *the* arithmetic mean, *the* geometric mean, *and the* harmonic mean *of $x_1, x_2, \ldots, x_n$.*

Theorem 2.17 (The general mean inequality). *Let $x_1, \ldots, x_n$ be positive real numbers. For each $p \in \mathbb{R}$ we define the* mean of order p *of $x_1, \ldots, x_n$ by*

$$M_p = \left(\frac{x_1^p + \cdots + x_n^p}{n} \right)^{1/p}$$

for $p \neq 0$, and $M_q = \lim_{p \to q} M_p$ for $q \in \{\pm\infty, 0\}$. Then

$$M_p \leq M_q \quad \text{whenever} \quad p \leq q.$$

Remark. In particular, $\max x_i$, QM, AM, GM, HM, and $\min x_i$ are M_∞, M_2, M_1, M_0, M_{-1}, and $M_{-\infty}$ respectively.

Theorem 2.18 (Cauchy–Schwarz inequality). *Let a_i, b_i, $i = 1, 2, \ldots, n$, be real numbers. Then*

$$\left(\sum_{i=1}^{n} a_i b_i \right)^2 \le \left(\sum_{i=1}^{n} a_i^2 \right) \left(\sum_{i=1}^{n} b_i^2 \right).$$

Equality occurs if and only if there exists $c \in \mathbb{R}$ such that $b_i = c a_i$ for $i = 1, \ldots, n$.

Theorem 2.19 (Hölder's inequality). *Let a_i, b_i, $i = 1, 2, \ldots, n$, be nonnegative real numbers, and let p, q be positive real numbers such that $1/p + 1/q = 1$. Then*

$$\sum_{i=1}^{n} a_i b_i \le \left(\sum_{i=1}^{n} a_i^p \right)^{1/p} \left(\sum_{i=1}^{n} b_i^q \right)^{1/q}.$$

Equality occurs if and only if there exists $c \in \mathbb{R}$ such that $b_i = c a_i$ for $i = 1, \ldots, n$. The Cauchy–Schwarz inequality is a special case of Hölder's inequality for $p = q = 2$.

Theorem 2.20 (Minkowski's inequality). *Let a_i, b_i ($i = 1, 2, \ldots, n$) be nonnegative real numbers and p any real number not smaller than 1. Then*

$$\left(\sum_{i=1}^{n} (a_i + b_i)^p \right)^{1/p} \le \left(\sum_{i=1}^{n} a_i^p \right)^{1/p} + \left(\sum_{i=1}^{n} b_i^p \right)^{1/p}.$$

For $p > 1$ equality occurs if and only if there exists $c \in \mathbb{R}$ such that $b_i = c a_i$ for $i = 1, \ldots, n$. For $p = 1$ equality occurs in all cases.

Theorem 2.21 (Chebyshev's inequality). *Let $a_1 \ge a_2 \ge \cdots \ge a_n$ and $b_1 \ge b_2 \ge \cdots \ge b_n$ be real numbers. Then*

$$n \sum_{i=1}^{n} a_i b_i \ge \left(\sum_{i=1}^{n} a_i \right) \left(\sum_{i=1}^{n} b_i \right) \ge n \sum_{i=1}^{n} a_i b_{n+1-i}.$$

The two inequalities become equalities at the same time when $a_1 = a_2 = \cdots = a_n$ or $b_1 = b_2 = \cdots = b_n$.

Definition 2.22. A real function f defined on an interval I is *convex* if $f(\alpha x + \beta y) \le \alpha f(x) + \beta f(y)$ for all $x, y \in I$ and all $\alpha, \beta > 0$ such that $\alpha + \beta = 1$. A function f is said to be *concave* if the opposite inequality holds, i.e., if $-f$ is convex.

Theorem 2.23. *If f is continuous on an interval I, then f is convex on that interval if and only if*

$$f\left(\frac{x+y}{2} \right) \le \frac{f(x) + f(y)}{2} \quad \text{for all } x, y \in I.$$

Theorem 2.24. *If f is differentiable, then it is convex if and only if the derivative f' is nondecreasing. Similarly, differentiable function f is concave if and only if f' is nonincreasing.*

Theorem 2.25 (Jensen's inequality). *If $f : I \to \mathbb{R}$ is a convex function, then the inequality*

$$f(\alpha_1 x_1 + \cdots + \alpha_n x_n) \leq \alpha_1 f(x_1) + \cdots + \alpha_n f(x_n)$$

holds for all $\alpha_i \geq 0$, $\alpha_1 + \cdots + \alpha_n = 1$, and $x_i \in I$. For a concave function the opposite inequality holds.

Theorem 2.26 (Muirhead's inequality). *Given $x_1, x_2, \ldots, x_n \in \mathbb{R}^+$ and an n-tuple $\mathbf{a} = (a_1, \ldots, a_n)$ of positive real numbers, we define*

$$T_{\mathbf{a}}(x_1, \ldots, x_n) = \sum y_1^{a_1} \cdots y_n^{a_n},$$

the sum being taken over all permutations $y_1, \ldots, y_n$ of $x_1, \ldots, x_n$. We say that an n-tuple $\mathbf{a}$ majorizes an n-tuple $\mathbf{b}$ if $a_1 + \cdots + a_n = b_1 + \cdots + b_n$ and $a_1 + \cdots + a_k \geq b_1 + \cdots + b_k$ for each $k = 1, \ldots, n-1$. If a nonincreasing n-tuple $\mathbf{a}$ majorizes a nonincreasing n-tuple $\mathbf{b}$, then the following inequality holds:

$$T_{\mathbf{a}}(x_1, \ldots, x_n) \geq T_{\mathbf{b}}(x_1, \ldots, x_n).$$

Equality occurs if and only if $x_1 = x_2 = \cdots = x_n$.

Theorem 2.27 (Schur's inequality). *Using the notation introduced for Muirhead's inequality,*

$$T_{\lambda + 2\mu, 0, 0}(x_1, x_2, x_3) + T_{\lambda, \mu, \mu}(x_1, x_2, x_3) \geq 2T_{\lambda + \mu, \mu, 0}(x_1, x_2, x_3),$$

where $\lambda \in \mathbb{R}$, $\mu > 0$. Equality occurs if and only if $x_1 = x_2 = x_3$ or $x_1 = x_2$, $x_3 = 0$ (and in analogous cases). An equivalent form of the Schur's inequality is

$$x^{\lambda}(x^{\mu} - y^{\mu})(x^{\mu} - z^{\mu}) + y^{\lambda}(y^{\mu} - x^{\mu})(y^{\mu} - z^{\mu}) + z^{\lambda}(z^{\mu} - x^{\mu})(z^{\mu} - y^{\mu}) \geq 0.$$

2.1.4 Groups and Fields

Definition 2.28. A *group* is a nonempty set G equipped with a binary operation $*$ satisfying the following conditions:

(i) $a * (b * c) = (a * b) * c$ for all $a, b, c \in G$.

(ii) There exists a (unique) *identity* $e \in G$ such that $e * a = a * e = a$ for all $a \in G$.

(iii) For each $a \in G$ there exists a (unique) *inverse* $a^{-1} = b \in G$ such that $a * b = b * a = e$.

If $n \in \mathbb{Z}$, we define a^n as $a * a * \cdots * a$ (n times) if $n \geq 0$, and as $(a^{-1})^{-n}$ otherwise.

Definition 2.29. A group $\mathscr{G} = (G, *)$ is *commutative* or *abelian* if $a * b = b * a$ for all $a, b \in G$.

Definition 2.30. A set A *generates* a group $(G, *)$ if every element of G can be obtained using powers of the elements of A and the operation $*$. In other words, if A is the generator of a group G, then every element $g \in G$ can be written as $a_1^{i_1} * \cdots * a_n^{i_n}$, where $a_j \in A$ and $i_j \in \mathbb{Z}$ for every $j = 1, 2, \ldots, n$.

Definition 2.31. The *order* of an element $a \in G$ is the smallest $n \in \mathbb{N}$, if it exists such that $a^n = e$. If no such n exists then the element a is said to be of infinite order. The *order* of a group is the number of its elements, if it is finite. Each element of a finite group has finite order.

Theorem 2.32 (Lagrange's theorem). *In a finite group, the order of an element divides the order of the group.*

Definition 2.33. A *ring* is a nonempty set R equipped with two operations $+$ and $\cdot$ such that $(R, +)$ is an abelian group and for any $a, b, c \in R$,

(i) $(a \cdot b) \cdot c = a \cdot (b \cdot c)$;

(ii) $(a + b) \cdot c = a \cdot c + b \cdot c$ and $c \cdot (a + b) = c \cdot a + c \cdot b$.

A ring is *commutative* if $a \cdot b = b \cdot a$ for any $a, b \in R$ and *with identity* if there exists a *multiplicative identity* $i \in R$ such that $i \cdot a = a \cdot i = a$ for all $a \in R$.

Definition 2.34. A *field* is a commutative ring with identity in which every element a other than the additive identity has a *multiplicative inverse* a^{-1} such that $a \cdot a^{-1} = a^{-1} \cdot a = i$.

Theorem 2.35. *The following are common examples of groups, rings, and fields:*

Groups: $(\mathbb{Z}_n, +)$, $(\mathbb{Z}_p \setminus \{0\}, \cdot)$, $(\mathbb{Q}, +)$, $(\mathbb{R}, +)$, $(\mathbb{R} \setminus \{0\}, \cdot)$.

Rings: $(\mathbb{Z}_n, +, \cdot)$, $(\mathbb{Z}, +, \cdot)$, $(\mathbb{Z}[x], +, \cdot)$, $(\mathbb{R}[x], +, \cdot)$.

Fields: $(\mathbb{Z}_p, +, \cdot)$, $(\mathbb{Q}, +, \cdot)$, $(\mathbb{Q}(\sqrt{2}), +, \cdot)$, $(\mathbb{R}, +, \cdot)$, $(\mathbb{C}, +, \cdot)$.

2.2 Analysis

Definition 2.36. A sequence $\{a_n\}_{n=1}^{\infty}$ of real numbers has a *limit* $a = \lim_{n \to \infty} a_n$ (also denoted by $a_n \to a$) if

$$(\forall \varepsilon > 0)(\exists n_\varepsilon \in \mathbb{N})(\forall n \geq n_\varepsilon) \, |a_n - a| < \varepsilon.$$

A function $f : (a, b) \to \mathbb{R}$ has a limit $y = \lim_{x \to c} f(x)$ if

$$(\forall \varepsilon > 0)(\exists \delta > 0)(\forall x \in (a, b)) \, 0 < |x - c| < \delta \Rightarrow |f(x) - y| < \varepsilon.$$

Definition 2.37. A sequence $\{x_n\}$ *converges* to $x \in \mathbb{R}$ if $\lim_{n \to \infty} x_n = x$. A series $\sum_{n=1}^{\infty} x_n$ converges to $s \in \mathbb{R}$ if and only if $\lim_{m \to \infty} \sum_{n=1}^{m} x_n = s$. A sequence or series that does not converge is said to *diverge*.

Theorem 2.38. *A sequence $\{a_n\}$ of real numbers is convergent if it is monotonic and bounded.*

Definition 2.39. A function f is *continuous* on $[a,b]$ if the following three relations hold:

$$\lim_{x \to x_0} f(x) = f(x_0), \text{ for every } x_0 \in (a,b),$$

$$\lim_{x \to a+} f(x) = f(a),$$

$$\text{and } \lim_{x \to b-} f(x) = f(b).$$

Definition 2.40. A function $f : (a,b) \to \mathbb{R}$ is *differentiable* at a point $x_0 \in (a,b)$ if the following limit exists:

$$f'(x_0) = \lim_{x \to x_0} \frac{f(x) - f(x_0)}{x - x_0}.$$

A function is differentiable on (a,b) if it is differentiable at every $x_0 \in (a,b)$. The function f' is called the *derivative* of f. We similarly define the second derivative f'' as the derivative of f', and so on.

Theorem 2.41. *A differentiable function is also continuous. If f and g are differentiable, then fg, $\alpha f + \beta g$ $(\alpha, \beta \in \mathbb{R})$, $f \circ g$, $1/f$ (if $f \neq 0$), f^{-1} (if well defined) are also differentiable. It holds that $(\alpha f + \beta g)' = \alpha f' + \beta g'$, $(fg)' = f'g + fg'$, $(f \circ g)' = (f' \circ g) \cdot g'$, $(1/f)' = -f'/f^2$, $(f/g)' = (f'g - fg')/g^2$, $(f^{-1})' = 1/(f' \circ f^{-1})$.*

Theorem 2.42. *The following are derivatives of some elementary functions (a denotes a real constant): $(x^a)' = ax^{a-1}$, $(\ln x)' = 1/x$, $(a^x)' = a^x \ln a$, $(\sin x)' = \cos x$, $(\cos x)' = -\sin x$.*

Theorem 2.43 (Fermat's theorem). *Let $f : [a,b] \to \mathbb{R}$ be a continuous function that is differentiable at every point of (a,b). The function f attains its maximum and minimum in $[a,b]$. If $x_0 \in (a,b)$ is a number at which the extremum is attained (i.e., $f(x_0)$ is the maximum or minimum), then $f'(x_0) = 0$.*

Theorem 2.44 (Rolle's theorem). *Let $f(x)$ be a continuous function defined on $[a,b]$, where $a, b \in \mathbb{R}$, $a < b$, and $f(a) = f(b)$. If f is differentiable in (a,b), then there exists $c \in (a,b)$ such that $f'(c) = 0$.*

Definition 2.45. Differentiable functions $f_1, f_2, \ldots, f_k$ defined on an open subset D of $\mathbb{R}^n$ are *independent* if there is no nonzero differentiable function $F : \mathbb{R}^k \to \mathbb{R}$ such that $F(f_1, \ldots, f_k)$ is identically zero on some open subset of D.

Theorem 2.46. *Functions $f_1, \ldots, f_k : D \to \mathbb{R}$ are independent if and only if the $k \times n$ matrix $[\partial f_i / \partial x_j]_{i,j}$ is of rank k, i.e., when its k rows are linearly independent at some point.*

Theorem 2.47 (Lagrange multipliers). *Let D be an open subset of $\mathbb{R}^n$ and f, f_1, f_2, ..., $f_k : D \to \mathbb{R}$ independent differentiable functions. Assume that a point a in D is an extremum of the function f within the set of points in D for which $f_1 = f_2 = \cdots = f_k = 0$. Then there exist real numbers $\lambda_1, \ldots, \lambda_k$ (so-called Lagrange multipliers) such that a is a stationary point of the function $F = f + \lambda_1 f_1 + \cdots + \lambda_k f_k$, i.e., such that all partial derivatives of F at a are zero.*

Definition 2.48. Let f be a real function defined on $[a,b]$ and let $a = x_0 \le x_1 \le \cdots \le x_n = b$ and $\xi_k \in [x_{k-1}, x_k]$. The sum $S = \sum_{k=1}^{n}(x_k - x_{k-1})f(\xi_k)$ is called a *Darboux sum*. If $I = \lim_{\delta \to 0} S$ exists (where $\delta = \max_k(x_k - x_{k-1})$), we say that f is *integrable* and that I is its *integral*. Every continuous function is integrable on a finite interval.

2.3 Geometry

2.3.1 Triangle Geometry

Definition 2.49. The *orthocenter* of a triangle is the common point of its three altitudes.

Definition 2.50. The *circumcenter* of a triangle is the center of its circumscribed circle (i.e., *circumcircle*). It is the common point of the perpendicular bisectors of the sides of the triangle.

Definition 2.51. The *incenter* of a triangle is the center of its inscribed circle (i.e., *incircle*). It is the common point of the internal bisectors of its angles.

Definition 2.52. The *centroid* of a triangle (*median point*) is the common point of its medians.

Theorem 2.53. *The orthocenter, circumcenter, incenter, and centroid are well defined (and unique) for every nondegenerate triangle.*

Theorem 2.54 (Euler's line). *The orthocenter H, centroid G, and circumcenter O of an arbitrary triangle lie on a line and satisfy $\overrightarrow{HG} = 2\overrightarrow{GO}$.*

Theorem 2.55 (The nine-point circle). *The feet of the altitudes from A, B, C and the midpoints of AB, BC, CA, AH, BH, CH lie on a circle.*

Theorem 2.56 (Feuerbach's theorem). *The nine-point circle of a triangle is tangent to the incircle and all three excircles of the triangle.*

Theorem 2.57 (Torricelli's point). *Given a triangle $\triangle ABC$, let $\triangle ABC'$, $\triangle AB'C$, and $\triangle A'BC$ be equilateral triangles constructed outward. Then AA', BB', CC' intersect in one point.*

Definition 2.58. Let ABC be a triangle, P a point, and X, Y, Z respectively the feet of the perpendiculars from P to BC, AC, AB. Triangle XYZ is called the *pedal triangle* of $\triangle ABC$ corresponding to point P.

Theorem 2.59 (Simson's line). *The pedal triangle XYZ is degenerate, i.e., X, Y, Z are collinear, if and only if P lies on the circumcircle of ABC. Points X, Y, Z are in this case said to lie on* Simson's line.

Theorem 2.60. *If M is a point on the circumcircle of $\triangle ABC$ with orthocenter H, then the Simson's line corresponding to M bisects the segment MH.*

Theorem 2.61 (Carnot's theorem). *The perpendiculars from X, Y, Z to BC, CA, AB respectively are concurrent if and only if*

$$BX^2 - XC^2 + CY^2 - YA^2 + AZ^2 - ZB^2 = 0.$$

Theorem 2.62 (Desargues's theorem). *Let $A_1B_1C_1$ and $A_2B_2C_2$ be two triangles. The lines A_1A_2, B_1B_2, C_1C_2 are concurrent or mutually parallel if and only if the points $A = B_1C_1 \cap B_2C_2$, $B = C_1A_1 \cap C_2A_2$, and $C = A_1B_1 \cap A_2B_2$ are collinear.*

Definition 2.63. Given a point C in the plane and a real number r, a *homothety* with center C and coefficient r is a mapping of the plane that sends each point A to the point A' such that $\overrightarrow{CA'} = k\overrightarrow{CA}$.

Theorem 2.64. *Let k_1, k_2, and k_3 be three circles. Then the three external similitude centers of these three circles are collinear (the external similitude center is the center of the homothety with positive coefficient that maps one circle to the other). Similarly, two internal similitude centers are collinear with the third external similitude center.*

All variants of the previous theorem can be directly obtained from the Desargues's theorem applied to the following two triangles: the first triangle is determined by the centers of k_1, k_2, k_3, while the second triangle is determined by the points of tangency of an appropriately chosen circle that is tangent to all three of k_1, k_2, k_3.

2.3.2 Vectors in Geometry

Definition 2.65. For any two vectors $\overrightarrow{a}, \overrightarrow{b}$ in space, we define the *scalar product* (also known as *dot product*) of $\overrightarrow{a}$ and $\overrightarrow{b}$ as $\overrightarrow{a} \cdot \overrightarrow{b} = |\overrightarrow{a}||\overrightarrow{b}|\cos\varphi$, and the *vector product* (also known as *cross product*) as $\overrightarrow{a} \times \overrightarrow{b} = \overrightarrow{p}$, where $\varphi = \angle(\overrightarrow{a}, \overrightarrow{b})$ and $\overrightarrow{p}$ is the vector with $|\overrightarrow{p}| = |\overrightarrow{a}||\overrightarrow{b}||\sin\varphi|$ perpendicular to the plane determined by $\overrightarrow{a}$ and $\overrightarrow{b}$ such that the triple of vectors $\overrightarrow{a}, \overrightarrow{b}, \overrightarrow{p}$ is positively oriented (note that if $\overrightarrow{a}$ and $\overrightarrow{b}$ are collinear, then $\overrightarrow{a} \times \overrightarrow{b} = \overrightarrow{0}$). Both these products are linear with respect to both factors. The scalar product is commutative, while the vector product is anticommutative, i.e., $\overrightarrow{a} \times \overrightarrow{b} = -\overrightarrow{b} \times \overrightarrow{a}$. We also define the *mixed vector product* of three vectors $\overrightarrow{a}, \overrightarrow{b}, \overrightarrow{c}$ as $[\overrightarrow{a}, \overrightarrow{b}, \overrightarrow{c}] = (\overrightarrow{a} \times \overrightarrow{b}) \cdot \overrightarrow{c}$.
Remark. The scalar product of vectors $\overrightarrow{a}$ and $\overrightarrow{b}$ is often denoted by $\langle \overrightarrow{a}, \overrightarrow{b} \rangle$.

Theorem 2.66 (Thales' theorem). *Let lines AA' and BB' intersect in a point O, $A' \neq O \neq B'$. Then $AB \parallel A'B' \Leftrightarrow \frac{\overrightarrow{OA}}{\overrightarrow{OA'}} = \frac{\overrightarrow{OB}}{\overrightarrow{OB'}}$ (Here $\frac{\overrightarrow{a}}{\overrightarrow{b}}$ denotes the ratio of two nonzero collinear vectors).*

Theorem 2.67 (Ceva's theorem). *Let ABC be a triangle and X, Y, Z points on lines BC, CA, AB respectively, distinct from A,B,C. Then the lines AX, BY, CZ are concurrent if and only if*

$$\frac{\overrightarrow{BX}}{\overrightarrow{XC}} \cdot \frac{\overrightarrow{CY}}{\overrightarrow{YA}} \cdot \frac{\overrightarrow{AZ}}{\overrightarrow{ZB}} = 1, \text{ or equivalently, } \frac{\sin \angle BAX}{\sin \angle XAC} \frac{\sin \angle CBY}{\sin \angle YBA} \frac{\sin \angle ACZ}{\sin \angle ZCB} = 1$$

(the last expression being called the trigonometric form *of Ceva's theorem).*

Theorem 2.68 (Menelaus's theorem). *Using the notation introduced for Ceva's theorem, points X,Y,Z are collinear if and only if*

$$\frac{\overrightarrow{BX}}{\overrightarrow{XC}} \cdot \frac{\overrightarrow{CY}}{\overrightarrow{YA}} \cdot \frac{\overrightarrow{AZ}}{\overrightarrow{ZB}} = -1.$$

Theorem 2.69 (Stewart's theorem). *If D is an arbitrary point on the line BC, then*

$$AD^2 = \frac{\overrightarrow{DC}}{\overrightarrow{BC}} BD^2 + \frac{\overrightarrow{BD}}{\overrightarrow{BC}} CD^2 - \overrightarrow{BD} \cdot \overrightarrow{DC}.$$

Specifically, if D is the midpoint of BC, then $4AD^2 = 2AB^2 + 2AC^2 - BC^2$.

2.3.3 Barycenters

Definition 2.70. A *mass point* (A,m) is a point A that is assigned a *mass* $m > 0$.

Definition 2.71. The *center of mass (barycenter)* of the set of mass points (A_i, m_i), $i = 1, 2, \ldots, n$, is the point T such that $\sum_i m_i \overrightarrow{TA_i} = \overrightarrow{0}$.

Theorem 2.72 (Leibniz's theorem). *Let T be the mass center of the set of mass points $\{(A_i, m_i) \mid i = 1, 2, \ldots, n\}$ of total mass $m = m_1 + \cdots + m_n$, and let X be an arbitrary point. Then*

$$\sum_{i=1}^{n} m_i XA_i^2 = \sum_{i=1}^{n} m_i TA_i^2 + mXT^2.$$

Specifically, if T is the centroid of $\triangle ABC$ and X an arbitrary point, then

$$AX^2 + BX^2 + CX^2 = AT^2 + BT^2 + CT^2 + 3XT^2.$$

2.3.4 Quadrilaterals

Theorem 2.73. *A quadrilateral ABCD is cyclic (i.e., there exists a circumcircle of ABCD) if and only if $\angle ACB = \angle ADB$ and if and only if $\angle ADC + \angle ABC = 180°$.*

Theorem 2.74 (Ptolemy's theorem). *A convex quadrilateral ABCD is cyclic if and only if*

$$AC \cdot BD = AB \cdot CD + AD \cdot BC.$$

For an arbitrary quadrilateral ABCD we have Ptolemy's inequality *(see 2.3.7, Geometric Inequalities).*

Theorem 2.75 (Casey's theorem). *Let k_1, k_2, k_3, and k_4 be four circles that all touch a given circle k. Let t_{ij} be the length of a segment determined by an external common tangent of circles k_i and k_j ($i, j \in \{1,2,3,4\}$) if both k_i and k_j touch k internally, or both touch k externally. Otherwise, t_{ij} is set to be the internal common tangent. Then one of the products $t_{12}t_{34}$, $t_{13}t_{24}$, and $t_{14}t_{23}$ is the sum of the other two.*

    Some of the circles k_1, k_2, k_3, k_4 may be degenerate, i.e., of 0 radius, and thus reduced to being points. In particular, for three points A, B, C on a circle k and a circle k' touching k at a point on the arc of AC not containing B, we have $AC \cdot b = AB \cdot c + a \cdot BC$, where a, b, and c are the lengths of the tangent segments from points A, B, and C to k'. Ptolemy's theorem is a special case of Casey's theorem when all four circles are degenerate.

Theorem 2.76. *A convex quadrilateral ABCD is tangent (i.e., there exists an incircle of ABCD) if and only if*

$$AB + CD = BC + DA.$$

Theorem 2.77. *For arbitrary points A,B,C,D in space, $AC \perp BD$ if and only if*

$$AB^2 + CD^2 = BC^2 + DA^2.$$

Theorem 2.78 (Newton's theorem). *Let ABCD be a quadrilateral, $AD \cap BC = E$, and $AB \cap DC = F$ (such points A,B,C,D,E,F form a* complete quadrilateral*). Then the midpoints of AC, BD, and EF are collinear. If ABCD is tangent, then the incenter also lies on this line.*

Theorem 2.79 (Brocard's theorem). *Let ABCD be a quadrilateral inscribed in a circle with center O, and let $P = AB \cap CD$, $Q = AD \cap BC$, $R = AC \cap BD$. Then O is the orthocenter of $\triangle PQR$.*

2.3.5 Circle Geometry

Theorem 2.80 (Pascal's theorem). *If $A_1, A_2, A_3, B_1, B_2, B_3$ are distinct points on a conic γ (e.g., circle), then points $X_1 = A_2B_3 \cap A_3B_2$, $X_2 = A_1B_3 \cap A_3B_1$, and $X_3 = A_1B_2 \cap A_2B_1$ are collinear. The special result when γ consists of two lines is called* Pappus's theorem.

Theorem 2.81 (Brianchon's theorem). *Let ABCDEF be a convex hexagon. If a conic (e.g., circle) can be inscribed in ABCDEF, then AD, BE, and CF meet in a point.*

Theorem 2.82 (The butterfly theorem). *Let AB be a chord of a circle k and C its midpoint. Let p and q be two different lines through C that, respectively, intersect k on one side of AB in P and Q and on the other in P' and Q'. Let E and F respectively be the intersections of PQ' and P'Q with AB. Then it follows that CE = CF.*

Definition 2.83. The *power* of a point X with respect to a circle $k(O,r)$ is defined by $\mathscr{P}(X) = OX^2 - r^2$. For an arbitrary line l through X that intersects k at A and B ($A = B$ when l is a tangent), it follows that $\mathscr{P}(X) = \overrightarrow{XA} \cdot \overrightarrow{XB}$.

Definition 2.84. The *radical axis* of two circles is the locus of points that have equal powers with respect to both circles. The radical axis of circles $k_1(O_1, r_1)$ and $k_2(O_2, r_2)$ is a line perpendicular to $O_1 O_2$. The radical axes of three distinct circles are concurrent or mutually parallel. If concurrent, the intersection of the three axes is called the *radical center*.

Definition 2.85. The *pole* of a line $l \not\ni O$ with respect to a circle $k(O,r)$ is a point A on the other side of l from O such that $OA \perp l$ and $d(O,l) \cdot OA = r^2$. In particular, if l intersects k in two points, its pole will be the intersection of the tangents to k at these two points.

Definition 2.86. The *polar* of the point A from the previous definition is the line l. In particular, if A is a point outside k and AM, AN are tangents to k ($M, N \in k$), then MN is the polar of A.
Poles and polars are generally defined in a similar way with respect to arbitrary nondegenerate conics.

Theorem 2.87. *If A belongs to the polar of B, then B belongs to the polar of A.*

2.3.6 Inversion

Definition 2.88. An *inversion* of the plane π about the circle $k(O,r)$ (which belongs to π) is a transformation of the set $\pi \setminus \{O\}$ onto itself such that every point P is transformed into a point P' on the ray $(OP$ such that $OP \cdot OP' = r^2$. In the following statements we implicitly assume exclusion of O.

Theorem 2.89. *The fixed points of an inversion about a circle k are on the circle k. The inside of k is transformed into the outside and vice versa.*

Theorem 2.90. *If A, B transform into A', B' after an inversion about a circle k, then $\angle OAB = \angle OB'A'$, and also ABB'A' is cyclic and perpendicular to k. A circle perpendicular to k transforms into itself. Inversion preserves angles between continuous curves (which includes lines and circles).*

Theorem 2.91. *An inversion transforms lines not containing O into circles containing O, lines containing O into themselves, circles not containing O into circles not containing O, circles containing O into lines not containing O.*

2.3.7 Geometric Inequalities

Theorem 2.92 (The triangle inequality). *For any three points A, B, C, $AB + BC \geq AC$. Equality occurs when A, B, C are collinear and B is between A and C. In the sequel we will use $\mathscr{B}(A,B,C)$ to emphasize that B is between A and C.*

Theorem 2.93 (Ptolemy's inequality). *For any four points A, B, C, D,*

$$AC \cdot BD \leq AB \cdot CD + AD \cdot BC.$$

Theorem 2.94 (The parallelogram inequality). *For any four points A, B, C, D,*

$$AB^2 + BC^2 + CD^2 + DA^2 \geq AC^2 + BD^2.$$

Equality occurs if and only if ABCD is a parallelogram.

Theorem 2.95. *For a given triangle $\triangle ABC$ the point X for which $AX + BX + CX$ is minimal is Toricelli's point when all angles of $\triangle ABC$ are less than or equal to $120°$, and is the vertex of the obtuse angle otherwise. The point X_2 for which $AX_2^2 + BX_2^2 + CX_2^2$ is minimal is the centroid (see Leibniz's theorem).*

Theorem 2.96 (The Erdős–Mordell inequality). *Let P be a point in the interior of $\triangle ABC$ and X,Y,Z projections of P onto BC,AC,AB, respectively. Then*

$$PA + PB + PC \geq 2(PX + PY + PZ).$$

Equality holds if and only if $\triangle ABC$ is equilateral and P is its center.

2.3.8 Trigonometry

Definition 2.97. The *trigonometric circle* is the unit circle centered at the origin O of a coordinate plane. Let A be the point $(1,0)$ and $P(x,y)$ a point on the trigonometric circle such that $\angle AOP = \alpha$. We define $\sin \alpha = y$, $\cos \alpha = x$, $\tan \alpha = y/x$, and $\cot \alpha = x/y$.

Theorem 2.98. *The functions* sin *and* cos *are periodic with period 2π. The functions* tan *and* cot *are periodic with period π. The following simple identities hold: $\sin^2 x + \cos^2 x = 1$, $\sin 0 = \sin \pi = 0$, $\sin(-x) = -\sin x$, $\cos(-x) = \cos x$, $\sin(\pi/2) = 1$, $\sin(\pi/4) = 1/\sqrt{2}$, $\sin(\pi/6) = 1/2$, $\cos x = \sin(\pi/2 - x)$. From these identities other identities can be easily derived.*

Theorem 2.99. *Additive formulas for trigonometric functions:*

$$\sin(\alpha \pm \beta) = \sin \alpha \cos \beta \pm \cos \alpha \sin \beta, \quad \cos(\alpha \pm \beta) = \cos \alpha \cos \beta \mp \sin \alpha \sin \beta,$$
$$\tan(\alpha \pm \beta) = \frac{\tan \alpha \pm \tan \beta}{1 \mp \tan \alpha \tan \beta}, \qquad \cot(\alpha \pm \beta) = \frac{\cot \alpha \cot \beta \mp 1}{\cot \alpha \pm \cot \beta}.$$

Theorem 2.100. *Formulas for trigonometric functions of 2x and 3x:*

$$\sin 2x = 2\sin x \cos x, \qquad \sin 3x = 3\sin x - 4\sin^3 x,$$
$$\cos 2x = 2\cos^2 x - 1, \qquad \cos 3x = 4\cos^3 x - 3\cos x,$$
$$\tan 2x = \frac{2\tan x}{1-\tan^2 x}, \qquad \tan 3x = \frac{3\tan x - \tan^3 x}{1-3\tan^2 x}.$$

Theorem 2.101. *For any* $x \in \mathbb{R}$, $\sin x = \frac{2t}{1+t^2}$ *and* $\cos x = \frac{1-t^2}{1+t^2}$, *where* $t = \tan\frac{x}{2}$.

Theorem 2.102. *Transformations from product to sum:*

$$2\cos\alpha\cos\beta = \cos(\alpha+\beta) + \cos(\alpha-\beta),$$
$$2\sin\alpha\cos\beta = \sin(\alpha+\beta) + \sin(\alpha-\beta),$$
$$2\sin\alpha\sin\beta = \cos(\alpha-\beta) - \cos(\alpha+\beta).$$

Theorem 2.103. *The angles* α, β, γ *of a triangle satisfy*

$$\cos^2\alpha + \cos^2\beta + \cos^2\gamma + 2\cos\alpha\cos\beta\cos\gamma = 1,$$
$$\tan\alpha + \tan\beta + \tan\gamma = \tan\alpha\tan\beta\tan\gamma.$$

Theorem 2.104 (De Moivre's formula). *If* $i^2 = -1$, *then*

$$(\cos x + i\sin x)^n = \cos nx + i\sin nx.$$

2.3.9 Formulas in Geometry

Theorem 2.105 (Heron's formula). *The area of a triangle ABC with sides* a, b, c *and semiperimeter s is given by*

$$S = \sqrt{s(s-a)(s-b)(s-c)} = \frac{1}{4}\sqrt{2a^2b^2 + 2a^2c^2 + 2b^2c^2 - a^4 - b^4 - c^4}.$$

Theorem 2.106 (The law of sines). *The sides* a, b, c *and angles* α, β, γ *of a triangle ABC satisfy*

$$\frac{a}{\sin\alpha} = \frac{b}{\sin\beta} = \frac{c}{\sin\gamma} = 2R,$$

where R is the circumradius of $\triangle ABC$.

Theorem 2.107 (The law of cosines). *The sides and angles of* $\triangle ABC$ *satisfy*

$$c^2 = a^2 + b^2 - 2ab\cos\gamma.$$

Theorem 2.108. *The circumradius R and inradius r of a triangle ABC satisfy* $R = \frac{abc}{4S}$ *and* $r = \frac{2S}{a+b+c} = R(\cos\alpha + \cos\beta + \cos\gamma - 1)$. *If* x, y, z *denote the distances of the circumcenter in an acute triangle to the sides, then* $x + y + z = R + r$.

Theorem 2.109 (Euler's formula). *If O and I are the circumcenter and incenter of* $\triangle ABC$, *then* $OI^2 = R(R - 2r)$, *where R and r are respectively the circumradius and the inradius of* $\triangle ABC$. *Consequently,* $R \geq 2r$.

Theorem 2.110. *If a, b, c, d are lengths of the sides of a convex quadrilateral, p its semiperimeter, and α and γ two non-adjacent angles of the quadrilateral, then its area S is given by*

$$S = \sqrt{(p-a)(p-b)(p-c)(p-d) - abcd\cos^2\frac{\alpha+\gamma}{2}}.$$

If the quadrilateral is cyclic, the above formula reduces to

$$S = \sqrt{(p-a)(p-b)(p-c)(p-d)}.$$

Theorem 2.111 (Euler's theorem for pedal triangles). *Let X, Y, Z be the feet of the perpendiculars from a point P to the sides of a triangle ABC. Let O denote the circumcenter and R the circumradius of $\triangle ABC$. Then*

$$S_{XYZ} = \frac{1}{4}\left|1 - \frac{OP^2}{R^2}\right|S_{ABC}.$$

Moreover, $S_{XYZ} = 0$ if and only if P lies on the circumcircle of $\triangle ABC$ (see Simson's line*).*

Theorem 2.112. *If $\vec{a} = (a_1, a_2, a_3)$, $\vec{b} = (b_1, b_2, b_3)$, $\vec{c} = (c_1, c_2, c_3)$ are three vectors in coordinate space, then*

$$\vec{a} \cdot \vec{b} = a_1b_1 + a_2b_2 + a_3b_3, \quad \vec{a} \times \vec{b} = (a_1b_2 - a_2b_1, a_2b_3 - a_3b_2, a_3b_1 - a_1b_3),$$

$$[\vec{a}, \vec{b}, \vec{c}] = \det\begin{bmatrix} a_1 & a_2 & a_3 \\ b_1 & b_2 & b_3 \\ c_1 & c_2 & c_3 \end{bmatrix}.$$

Here $\det M$ *denotes the determinant of the square matrix M.*

Theorem 2.113. *The area of a triangle ABC and the volume of a tetrahedron ABCD are equal to $\frac{1}{2}|\vec{AB} \times \vec{AC}|$ and $\frac{1}{6}\left|[\vec{AB}, \vec{AC}, \vec{AD}]\right|$, respectively.*

Theorem 2.114 (Cavalieri's principle). *If the sections of two solids by the same plane always have equal area, then the volumes of the two solids are equal.*

2.4 Number Theory

2.4.1 Divisibility and Congruences

Definition 2.115. *The greatest common divisor* $(a,b) = \gcd(a,b)$ *of $a, b \in \mathbb{N}$ is the largest positive integer that divides both a and b. Positive integers a and b are* coprime *or* relatively prime *if $(a,b) = 1$. The least common multiple* $[a,b] = \text{lcm}(a,b)$ *of $a, b \in \mathbb{N}$ is the smallest positive integer that is divisible by both a and b.* It holds that $[a,b](a,b) = ab$. The above concepts are easily generalized to more than two numbers; i.e., we also define $(a_1, a_2, \ldots, a_n)$ and $[a_1, a_2, \ldots, a_n]$.

Theorem 2.116 (Euclidean algorithm). *Since* $(a,b) = (|a-b|,a) = (|a-b|,b)$, *it follows that starting from positive integers a and b one eventually obtains* (a,b) *by repeatedly replacing a and b with* $|a-b|$ *and* $\min\{a,b\}$ *until the two numbers are equal. The algorithm can be generalized to more than two numbers.*

Theorem 2.117 (Corollary to Euclidean algorithm). *For each* $a,b \in \mathbb{N}$ *there exist* $x,y \in \mathbb{Z}$ *such that* $ax + by = (a,b)$. *The number* (a,b) *is the smallest positive number for which such x and y can be found.*

Theorem 2.118 (Second corollary to Euclid's algorithm). *For* $a,m,n \in \mathbb{N}$ *and* $a > 1$ *it follows that* $(a^m - 1, a^n - 1) = a^{(m,n)} - 1$.

Theorem 2.119 (Fundamental theorem of arithmetic). *Every positive integer can be uniquely represented as a product of primes, up to their order.*

Theorem 2.120. *The fundamental theorem of arithmetic also holds in some other rings, such as* $\mathbb{Z}[i] = \{a + bi \mid a,b \in \mathbb{Z}\}$, $\mathbb{Z}[\sqrt{2}]$, $\mathbb{Z}[\sqrt{-2}]$, $\mathbb{Z}[\omega]$ *(where* ω *is a complex third root of* 1*). In these cases, the factorization into primes is unique up to the order and divisors of* 1.

Definition 2.121. *Integers* a,b *are congruent modulo* $n \in \mathbb{N}$ *if* $n \mid a - b$. *We then write* $a \equiv b \pmod{n}$.

Theorem 2.122 (Chinese remainder theorem). *If* $m_1, m_2, \ldots, m_k$ *are positive integers pairwise relatively prime and* $a_1, \ldots, a_k$, $c_1, \ldots, c_k$ *are integers such that* $(a_i, m_i) = 1$ $(i = 1, \ldots, k)$, *then the system of congruences*

$$a_i x \equiv c_i \pmod{m_i}, \qquad i = 1, 2, \ldots, k,$$

has a unique solution modulo $m_1 m_2 \cdots m_k$.

2.4.2 Exponential Congruences

Theorem 2.123 (Wilson's theorem). *If p is a prime, then* $p \mid (p-1)! + 1$.

Theorem 2.124 (Fermat's (little) theorem). *Let p be a prime number and a an integer with* $(a,p) = 1$. *Then* $a^{p-1} \equiv 1 \pmod{p}$. *This theorem is a special case of Euler's theorem.*

Definition 2.125. *Euler's function* $\varphi(n)$ *is defined for* $n \in \mathbb{N}$ *as the number of positive integers less than or equal to* n *and coprime to* n. *It holds that*

$$\varphi(n) = n \left(1 - \frac{1}{p_1}\right) \cdots \left(1 - \frac{1}{p_k}\right),$$

where $n = p_1^{\alpha_1} \cdots p_k^{\alpha_k}$ *is the factorization of* n *into primes.*

Theorem 2.126 (Euler's theorem). *Let n be a natural number and a an integer with* $(a,n) = 1$. *Then* $a^{\varphi(n)} \equiv 1 \pmod{n}$.

Theorem 2.127 (Existence of primitive roots). *Let p be a prime number. There exists $g \in \{1,2,\ldots,p-1\}$ (called a* primitive root *modulo p such that the set $\{1,g,g^2,\ldots,g^{p-2}\}$ is equal to $\{1, 2, \ldots, p-1\}$ modulo p.*

Definition 2.128. Let p be a prime and α a nonnegative integer. We say that p^α is the *exact power* of p that divides an integer a (and α the *exact exponent*) if $p^\alpha \mid a$ and $p^{\alpha+1} \nmid a$.

Theorem 2.129. *Let a and n be positive integers and p an odd prime. If p^α ($\alpha \in \mathbb{N}$) is the exact power of p that divides $a-1$, then for any integer $\beta \geq 0$, $p^{\alpha+\beta} \mid a^n - 1$ if and only if $p^\beta \mid n$. (See (SL97-14).)*

A similar statement holds for $p = 2$. If 2^α ($\alpha \in \mathbb{N}$) is the exact power of 2 that divides $a^2 - 1$, then for any integer $\beta \geq 0$, $2^{\alpha+\beta} \mid a^n - 1$ if and only if $2^{\beta+1} \mid n$. (See (SL89-27).)

2.4.3 Quadratic Diophantine Equations

Theorem 2.130. *The solutions of $a^2 + b^2 = c^2$ in integers are given by $a = t(m^2 - n^2)$, $b = 2tmn$, $c = t(m^2 + n^2)$ (provided that b is even), where $t,m,n \in \mathbb{Z}$. The triples (a,b,c) are called* Pythagorean *(or* primitive Pythagorean *if $\gcd(a,b,c) = 1$).*

Definition 2.131. Given $D \in \mathbb{N}$ that is not a perfect square, a *Pell's equation* is an equation of the form $x^2 - Dy^2 = 1$, where $x, y \in \mathbb{Z}$.

Theorem 2.132. *If (x_0, y_0) is the least (nontrivial) solution in $\mathbb{N}$ of the Pell's equation $x^2 - Dy^2 = 1$, then all the nontrivial integer solutions (x,y) are given by $x + y\sqrt{D} = \pm(x_0 + y_0\sqrt{D})^n$, where $n \in \mathbb{Z}$.*

Definition 2.133. An integer a is a *quadratic residue* modulo a prime p if there exists $x \in \mathbb{Z}$ such that $x^2 \equiv a \pmod{p}$. Otherwise, a is a *quadratic nonresidue* modulo p.

Definition 2.134. *The Legendre symbol for an integer a and a prime p is defined by*

$$\left(\frac{a}{p}\right) = \begin{cases} 1 & \text{if } a \text{ is a quadratic residue mod } p \text{ and } p \nmid a; \\ 0 & \text{if } p \mid a; \\ -1 & \text{otherwise.} \end{cases}$$

Clearly $\left(\frac{a}{p}\right) = \left(\frac{a+p}{p}\right)$ and $\left(\frac{a^2}{p}\right) = 1$ if $p \nmid a$. The Legendre symbol is multiplicative, i.e., $\left(\frac{a}{p}\right)\left(\frac{b}{p}\right) = \left(\frac{ab}{p}\right)$.

Theorem 2.135 (Euler's criterion). *For each odd prime p and integer a not divisible by p, $a^{\frac{p-1}{2}} \equiv \left(\frac{a}{p}\right) \pmod{p}$.*

Theorem 2.136. *For a prime $p > 3$, $\left(\frac{-1}{p}\right)$, $\left(\frac{2}{p}\right)$, and $\left(\frac{-3}{p}\right)$ are equal to 1 if and only if $p \equiv 1 \pmod 4$, $p \equiv \pm 1 \pmod 8$ and $p \equiv 1 \pmod 6$, respectively.*

Theorem 2.137 (Gauss's reciprocity law). *For any two distinct odd primes p and q, we have that*

$$\left(\frac{p}{q}\right)\left(\frac{q}{p}\right) = (-1)^{\frac{p-1}{2}\cdot\frac{q-1}{2}}.$$

Definition 2.138. *Jacobi symbol* for an integer a and an odd positive integer b is defined as

$$\left(\frac{a}{b}\right) = \left(\frac{a}{p_1}\right)^{\alpha_1}\cdots\left(\frac{a}{p_k}\right)^{\alpha_k},$$

where $b = p_1^{\alpha_1}\cdots p_k^{\alpha_k}$ is the factorization of b into primes.

Theorem 2.139. *If $\left(\frac{a}{b}\right) = -1$, then a is a quadratic nonresidue modulo b, but the converse is false. All the above identities for Legendre symbols except Euler's criterion remain true for Jacobi symbols.*

2.4.4 Farey Sequences

Definition 2.140. For any positive integer n, the *Farey sequence F_n* is the sequence of rational numbers a/b with $0 \le a \le b \le n$ and $(a,b) = 1$ arranged in increasing order. For instance, $F_3 = \{\frac{0}{1}, \frac{1}{3}, \frac{1}{2}, \frac{2}{3}, \frac{1}{1}\}$.

Theorem 2.141. *If p_1/q_1, p_2/q_2, and p_3/q_3 are three successive terms in a Farey sequence, then*

$$p_2 q_1 - p_1 q_2 = 1 \quad and \quad \frac{p_1 + p_3}{q_1 + q_3} = \frac{p_2}{q_2}.$$

2.5 Combinatorics

2.5.1 Counting of Objects

Many combinatorial problems involving the counting of objects satisfying a given set of properties can be properly reduced to an application of one of the following concepts.

Definition 2.142. A *variation* of order n over k is a 1–to–1 mapping of $\{1,2,\ldots,k\}$ into $\{1,2,\ldots,n\}$. For a given n and k, where $n \ge k$, the number of different variations is $V_n^k = \frac{n!}{(n-k)!}$.

Definition 2.143. A *variation with repetition* of order n over k is an arbitrary mapping of $\{1,2,\ldots,k\}$ into $\{1,2,\ldots,n\}$. For a given n and k the number of different variations with repetition is $\overline{V}_n^k = k^n$.

Definition 2.144. A *permutation* of order n is a bijection of $\{1,2,\ldots,n\}$ into itself (a special case of variation for $k = n$). For a given n the number of different permutations is $P_n = n!$.

Definition 2.145. A *combination* of order n over k is a k-element subset of $\{1, 2, \ldots, n\}$. For a given n and k the number of different combinations is $C_n^k = \binom{n}{k}$.

Definition 2.146. A *permutation with repetition* of order n is a bijection of $\{1, 2, \ldots, n\}$ into a *multiset* of n elements. A multiset is defined to be a set in which certain elements are deemed mutually indistinguishable (for example, as in $\{1, 1, 2, 3\}$).

If $\{k_1, k_2, \ldots, k_s\}$ denotes the set of distinct elements in a multiset and the element k_i appears α_i times in the multiset, then number of different permutations with repetition is $P_{n,\alpha_1,\ldots,\alpha_s} = \frac{n!}{\alpha_1! \cdot \alpha_2! \cdots \alpha_s!}$. A combination is a special case of permutation with repetition for a multiset with two different elements.

Theorem 2.147 (The pigeonhole principle). *If a set of $nk + 1$ different elements is partitioned into n mutually disjoint subsets, then at least one subset will contain at least $k + 1$ elements.*

Theorem 2.148 (The inclusion–exclusion principle). *Let $S_1, S_2, \ldots, S_n$ be a family of subsets of the set S. The number of elements of S contained in none of the subsets is given by the formula*

$$|S \setminus (S_1 \cup \cdots \cup S_n)| = |S| - \sum_{k=1}^{n} \sum_{1 \leq i_1 < \cdots < i_k \leq n} (-1)^{k-1} |S_{i_1} \cap \cdots \cap S_{i_k}|.$$

2.5.2 Graph Theory

Definition 2.149. A *graph* $G = (V, E)$ is a set of objects, i.e., *vertices*, V paired with the multiset E of some pairs of elements of V, i.e., *edges*. When $(x, y) \in E$, for $x, y \in V$, the vertices x and y are said to be *connected* by an edge; i.e., the vertices are the *endpoints* of the edge.

A graph for which the multiset E reduces to a proper set (i.e., each pair of vertices are connected by at most one edge) and for which no vertex is connected to itself is called a *simple graph*.

A *finite graph* is one in which $|E|$ and $|V|$ are finite.

Definition 2.150. An *oriented graph* is one in which the pairs in E are ordered.

Definition 2.151. The simple graph K_n consisting of n vertices and in which each pair of vertices is connected is called a *complete* graph.

Definition 2.152. A *k-partite graph* (*bipartite* for $k = 2$) $K_{i_1, i_2, \ldots, i_k}$ is a graph whose set of vertices V can be partitioned into k nonempty disjoint subsets of cardinalities $i_1, i_2, \ldots, i_k$ such that each vertex x in a subset W of V is connected only with the vertices not in W.

Definition 2.153. *Given a bipartite graph (V, E), let W and M be a partition of its set of vertices (you can think of W as a set of women and M a set of men). Assume that $|W| \leq |M|$. A marriage is an injective map $f : W \rightarrow M$ for which $(w, f(w)) \in E$ for every $w \in W$.*

Theorem 2.154 (Hall's marriage theorem). *Let W, M be a partition of the set of vertices of a bipartite graph. There exists a marriage $f : W \rightarrow M$ if and only if for every $U \subseteq W$ the number $|U|$ is not greater than the total number of neighbors of U inside M.*

Definition 2.155. The *degree* $d(x)$ of a vertex x is the number of times x is the endpoint of an edge (thus, self-connecting edges are counted twice for corresponding vertices). An *isolated vertex* is one with degree 0.

Theorem 2.156. *For a graph $G = (V,E)$ the following identity holds:*

$$\sum_{x \in V} d(x) = 2|E|.$$

As a consequence, the number of vertices of odd degree is even.

Definition 2.157. A *trajectory* (*path*) of a graph is a finite sequence of vertices, each connected to the previous one. The *length* of a trajectory is the number of edges through which it passes. A *circuit* is a path that ends in the starting vertex. A *cycle* is a circuit in which no vertex appears more than once (except the initial/final vertex).

A graph is *connected* if there exists a trajectory between any two vertices.

Definition 2.158. A *subgraph* $G' = (V',E')$ of a graph $G = (V,E)$ is a graph such that $V' \subseteq V$ and E' contains exactly the edges of E connecting points in V'. A *connected component* of a graph is a connected subgraph such that no vertex of the subgraph is connected with any vertex outside of the subgraph.

Definition 2.159. A *tree* is a connected graph that contains no cycles.

Theorem 2.160. *A tree with n vertices has exactly $n - 1$ edges and at least two vertices of degree 1.*

Definition 2.161. An *Euler path* is a path in which each edge appears exactly once. Likewise, an *Euler circuit* is an Euler path that is also a circuit.

Theorem 2.162. *The following conditions are necessary and sufficient for a finite connected graph G to have an Euler path:*

- *The graph contains an Euler circuit if and only if each vertex has even degree.*
- *The graph contains an Euler path if and only if the number of vertices of odd degree is either 0 or 2 (in the latter case the path starts and ends in the two odd vertices).*

Definition 2.163. A *Hamiltonian circuit* is a circuit that contains each vertex of G exactly once (trivially, it is also a cycle).

A simple rule to determine whether a graph contains a Hamiltonian circuit has not yet been discovered.

Theorem 2.164 (Ore's theorem). *Let G be a graph with n vertices. If the sum of the degrees of any two nonadjacent vertices in G is greater than or equal to n, then G has a Hamiltonian circuit.*

Theorem 2.165 (Ramsey's theorem). *Let $r \geq 1$ and $q_1, q_2, \ldots, q_s \geq r$. There exists a minimal positive integer $N(q_1, q_2, \ldots, q_s; r)$ such that for $n \geq N$, if all subgraphs K_r of K_n are partitioned into s different sets, labeled $A_1, A_2 \ldots, A_s$, then for some i there exists a complete subgraph K_{q_i} whose subgraphs K_r all belong to A_i. For $r = 2$ this corresponds to coloring the edges of K_n with s different colors and looking for a monochromatic subgraph K_{q_i} in color i.*

Theorem 2.166. $N(p, q; r) \leq N(N(p-1, q; r), N(p, q-1; r); r-1) + 1$, *and in particular,* $N(p, q; 2) \leq N(p-1, q; 2) + N(p, q-1; 2)$.

The following values of N are known: $N(p, q; 1) = p + q - 1$, $N(2, p; 2) = p$, $N(3, 3; 2) = 6$, $N(3, 4; 2) = 9$, $N(3, 5; 2) = 14$, $N(3, 6; 2) = 18$, $N(3, 7; 2) = 23$, $N(3, 8; 2) = 28$, $N(3, 9; 2) = 36$, $N(4, 4; 2) = 18$, $N(4, 5; 2) = 25$.

Theorem 2.167 (Turán's theorem). *If a simple graph on $n = t(p-1) + r$ vertices $(0 \leq r < p-1)$ has more than $f(n, p) = \frac{(p-2)n^2 - r(p-1-r)}{2(p-1)}$ edges, then it contains K_p as a subgraph. The graph containing $f(n, p)$ edges that does not contain K_p is the complete multipartite graph with r parts with $t+1$ vertices, and $p-1-r$ parts with t vertices.*

Definition 2.168. A *planar graph* is one that can be embedded in a plane such that its vertices are represented by points and its edges by lines (not necessarily straight) connecting the vertices such that no two edges intersect each other.

Theorem 2.169. *A planar graph with n vertices has at most $3n - 6$ edges.*

Theorem 2.170 (Kuratowski's theorem). *Graphs K_5 and $K_{3,3}$ are not planar. Every nonplanar graph contains a subgraph that can be obtained from one of these two graphs by a subdivison of its edges.*

Theorem 2.171 (Euler's formula). *For a given convex polyhedron let E be the number of its edges, F the number of faces, and V the number of vertices. Then $E + 2 = F + V$. The same formula holds for a connected planar graph (F is in this case equal to the number of planar regions).*

3

Problems

3.1 The First IMO
Bucharest–Brasov, Romania, July 23–31, 1959

3.1.1 Contest Problems

First Day

1. **(POL)** For every integer n prove that the fraction $\frac{21n+4}{14n+3}$ cannot be reduced any further.

2. **(ROU)** For which real numbers x do the following equations hold:

$$\text{(a) } \sqrt{x+\sqrt{2x-1}}+\sqrt{x-\sqrt{2x-1}}=\sqrt{2},$$
$$\text{(b) } \sqrt{x+\sqrt{2x-1}}+\sqrt{x-\sqrt{2x-1}}=1,$$
$$\text{(c) } \sqrt{x+\sqrt{2x-1}}+\sqrt{x-\sqrt{2x-1}}=2?$$

3. **(HUN)** Let x be an angle and let the real numbers a, b, c, $\cos x$ satisfy the following equation:

$$a\cos^2 x+b\cos x+c=0.$$

Write the analogous quadratic equation for a, b, c, $\cos 2x$. Compare the given and the obtained equality for $a=4$, $b=2$, $c=-1$.

Second Day

4. **(HUN)** Construct a right-angled triangle whose hypotenuse c is given if it is known that the median from the right angle equals the geometric mean of the remaining two sides of the triangle.

5. **(ROU)** A segment AB is given and on it a point M. On the same side of AB squares $AMKD$ and $BMFE$ are constructed. The circumcircles of the two squares, whose centers are P and Q, intersect in M and another point N.
   (a) Prove that lines FA and BC intersect at N.

(b) Prove that all such constructed lines MN pass through the same point S, regardless of the selection of M.

(c) Find the locus of the midpoints of all segments PQ, as M varies along the segment AB.

6. **(CZS)** Let α and β be two planes intersecting at a line p. In α a point A is given and in β a point C is given, neither of which lies on p. Construct B in α and D in β such that $ABCD$ is an equilateral trapezoid, $AB \parallel CD$, in which a circle can be inscribed.

3.2 The Second IMO
Bucharest–Sinaia, Romania, July 18–25, 1960

3.2.1 Contest Problems

First Day

1. **(BGR)** Find all the three-digit numbers for which one obtains, when dividing the number by 11, the sum of the squares of the digits of the initial number.

2. **(HUN)** For which real numbers x does the following inequality hold:

$$\frac{4x^2}{(1 - \sqrt{1 + 2x})^2} < 2x + 9\,?$$

3. **(ROU)** A right-angled triangle ABC is given for which the hypotenuse BC has length a and is divided into n equal segments, where n is odd. Let α be the angle with which the point A sees the segment containing the middle of the hypotenuse. Prove that

$$\tan \alpha = \frac{4nh}{(n^2 - 1)a},$$

where h is the height of the triangle.

Second Day

4. **(HUN)** Construct a triangle ABC whose lengths of heights h_a and h_b (from A and B, respectively) and length of median m_a (from A) are given.

5. **(CZS)** A cube $ABCDA'B'C'D'$ is given.
   (a) Find the locus of all midpoints of segments XY, where X is any point on segment AC and Y any point on segment $B'D'$.
   (b) Find the locus of all points Z on segments XY such that $\overrightarrow{ZY} = 2\overrightarrow{XZ}$.

6. **(BGR)** An isosceles trapezoid with bases a and b and height h is given.
   (a) On the line of symmetry construct the point P such that both (nonbase) sides are seen from P with an angle of $90°$.
   (b) Find the distance of P from one of the bases of the trapezoid.
   (c) Under what conditions for a, b, and h can the point P be constructed (analyze all possible cases)?

7. **(GDR)** A sphere is inscribed in a regular cone. Around the sphere a cylinder is circumscribed so that its base is in the same plane as the base of the cone. Let V_1 be the volume of the cone and V_2 the volume of the cylinder.
   (a) Prove that $V_1 = V_2$ is impossible.
   (b) Find the smallest k for which $V_1 = kV_2$, and in this case construct the angle at the vertex of the cone.

3.3 The Third IMO
Budapest–Veszprem, Hungary, July 6–16, 1961

3.3.1 Contest Problems

First Day

1. **(HUN)** Solve the following system of equations:

$$x+y+z = a,$$
$$x^2+y^2+z^2 = b^2,$$
$$xy = z^2,$$

where a and b are given real numbers. What conditions must hold on a and b for the solutions to be positive and distinct?

2. **(POL)** Let a, b, and c be the lengths of a triangle whose area is S. Prove that

$$a^2+b^2+c^2 \geq 4S\sqrt{3}.$$

In what case does equality hold?

3. **(BGR)** Solve the equation $\cos^n x - \sin^n x = 1$, where n is a given positive integer.

Second Day

4. **(GDR)** In the interior of $\triangle P_1P_2P_3$ a point P is given. Let Q_1, Q_2, and Q_3 respectively be the intersections of PP_1, PP_2, and PP_3 with the opposing edges of $\triangle P_1P_2P_3$. Prove that among the ratios PP_1/PQ_1, PP_2/PQ_2, and PP_3/PQ_3 there exists at least one not larger than 2 and at least one not smaller than 2.

5. **(CZS)** Construct a triangle ABC if the following elements are given: $AC = b$, $AB = c$, and $\angle AMB = \omega$ ($\omega < 90^o$), where M is the midpoint of BC. Prove that the construction has a solution if and only if

$$b\tan\frac{\omega}{2} \leq c < b.$$

In what case does equality hold?

6. **(ROU)** A plane ε is given and on one side of the plane three noncollinear points A, B, and C such that the plane determined by them is not parallel to ε. Three arbitrary points A', B', and C' in ε are selected. Let L, M, and N be the midpoints of AA', BB', and CC', and G the centroid of $\triangle LMN$. Find the locus of all points obtained for G as A', B', C' are varied (independently of each other) across ε.

3.4 The Fourth IMO
Prague–Hluboka, Czechoslovakia, July 7–15, 1962

3.4.1 Contest Problems

First Day

1. **(POL)** Find the smallest natural number n with the following properties:
   (a) In decimal representation it ends with 6.
   (b) If we move this digit to the front of the number, we get a number 4 times larger.

2. **(HUN)** Find all real numbers x for which

$$\sqrt{3-x} - \sqrt{x+1} > \frac{1}{2} \, .$$

3. **(CZS)** A cube $ABCDA'B'C'D'$ is given. The point X is moving at a constant speed along the square $ABCD$ in the direction from A to B. The point Y is moving with the same constant speed along the square $BCC'B'$ in the direction from B' to C'. Initially, X and Y start out from A and B' respectively. Find the locus of all the midpoints of XY.

Second Day

4. **(ROU)** Solve the equation

$$\cos^2 x + \cos^2 2x + \cos^2 3x = 1 \, .$$

5. **(BGR)** On the circle k three points A, B, and C are given. Construct the fourth point on the circle D such that one can inscribe a circle in $ABCD$.

6. **(GDR)** Let ABC be an isosceles triangle with circumradius r and inradius ρ. Prove that the distance d between the circumcenter and incenter is given by

$$d = \sqrt{r(r-2\rho)} \, .$$

7. **(USS)** Prove that a tetrahedron $SABC$ has five different spheres that touch all six lines determined by its edges if and only if it is regular.

3.5 The Fifth IMO
Wroclaw, Poland, July 5–13, 1963

3.5.1 Contest Problems

First Day

1. **(CZS)** Determine all real solutions of the equation $\sqrt{x^2 - p} + 2\sqrt{x^2 - 1} = x$, where p is a real number.

2. **(USS)** Find the locus of points in space that are vertices of right angles of which one ray passes through a given point and the other intersects a given segment.

3. **(HUN)** Prove that if all the angles of a convex n-gon are equal and the lengths of consecutive edges $a_1, \ldots, a_n$ satisfy $a_1 \geq a_2 \geq \cdots \geq a_n$, then $a_1 = a_2 = \cdots = a_n$.

Second Day

4. **(USS)** Find all solutions $x_1, \ldots, x_5$ to the system of equations

$$\begin{cases} x_5 + x_2 = yx_1, \\ x_1 + x_3 = yx_2, \\ x_2 + x_4 = yx_3, \\ x_3 + x_5 = yx_4, \\ x_4 + x_1 = yx_5, \end{cases}$$

where y is a real parameter.

5. **(GDR)** Prove that $\cos \frac{\pi}{7} - \cos \frac{2\pi}{7} + \cos \frac{3\pi}{7} = \frac{1}{2}$.

6. **(HUN)** Five students A, B, C, D, and E have taken part in a certain competition. Before the competition, two persons X and Y tried to guess the rankings. X thought that the ranking would be A, B, C, D, E; and Y thought that the ranking would be D, A, E, C, B. At the end, it was revealed that X didn't guess correctly any rankings of the participants, and moreover, didn't guess any of the orderings of pairs of consecutive participants. On the other hand, Y guessed the correct rankings of two participants and the correct ordering of two pairs of consecutive participants. Determine the rankings of the competition.

3.6 The Sixth IMO
Moscow, Soviet Union, June 30–July 10, 1964

3.6.1 Contest Problems

First Day

1. **(CZS)** (a) Find all natural numbers n such that the number $2^n - 1$ is divisible by 7.
    (b) Prove that for all natural numbers n the number $2^n + 1$ is not divisible by 7.

2. **(HUN)** Denote by a, b, c the lengths of the sides of a triangle. Prove that

$$a^2(b+c-a) + b^2(c+a-b) + c^2(a+b-c) \le 3abc.$$

3. **(YUG)** The incircle is inscribed in a triangle ABC with sides a, b, c. Three tangents to the incircle are drawn, each of which is parallel to one side of the triangle ABC. These tangents form three smaller triangles (internal to $\triangle ABC$) with the sides of $\triangle ABC$. In each of these triangles an incircle is inscribed. Determine the sum of areas of all four incircles.

Second Day

4. **(HUN)** Each of 17 students talked with every other student. They all talked about three different topics. Each pair of students talked about one topic. Prove that there are three students that talked about the same topic among themselves.

5. **(ROU)** Five points are given in the plane. Among the lines that connect these five points, no two coincide and no two are parallel or perpendicular. Through each point we construct an altitude to each of the other lines. What is the maximal number of intersection points of these altitudes (excluding the initial five points)?

6. **(POL)** Given a tetrahedron $ABCD$, let D_1 be the centroid of the triangle ABC and let A_1, B_1, C_1 be the intersection points of the lines parallel to DD_1 and passing through the points A, B, C with the opposite faces of the tetrahedron. Prove that the volume of the tetrahedron $ABCD$ is one-third the volume of the tetrahedron $A_1B_1C_1D_1$. Does the result remain true if the point D_1 is replaced with any point inside the triangle ABC?

3.7 The Seventh IMO
Berlin, DR Germany, July 3–13, 1965

3.7.1 Contest Problems

First Day

1. **(YUG)** Find all real numbers $x \in [0, 2\pi]$ such that

$$2\cos x \leq |\sqrt{1+\sin 2x} - \sqrt{1-\sin 2x}| \leq \sqrt{2}.$$

2. **(POL)** Consider the system of equations

$$\begin{cases} a_{11}x_1 + a_{12}x_2 + a_{13}x_3 = 0, \\ a_{21}x_1 + a_{22}x_2 + a_{23}x_3 = 0, \\ a_{31}x_1 + a_{32}x_2 + a_{33}x_3 = 0, \end{cases}$$

whose coefficients satisfy the following conditions:
   (a) a_{11}, a_{22}, a_{33} are positive real numbers;
   (b) all other coefficients are negative;
   (c) in each of the equations the sum of the coefficients is positive.
   Prove that $x_1 = x_2 = x_3 = 0$ is the only solution to the system.

3. **(CZS)** A tetrahedron $ABCD$ is given. The lengths of the edges AB and CD are a and b, respectively, the distance between the lines AB and CD is d, and the angle between them is equal to ω. The tetrahedron is divided into two parts by the plane π parallel to the lines AB and CD. Calculate the ratio of the volumes of the parts if the ratio between the distances of the plane π from AB and CD is equal to k.

Second Day

4. **(USS)** Find all sets of four real numbers x_1, x_2, x_3, x_4 such that the sum of any of the numbers and the product of the other three is equal to 2.

5. **(ROU)** Given a triangle OAB such that $\angle AOB = \alpha < 90°$, let M be an arbitrary point of the triangle different from O. Denote by P and Q the feet of the perpendiculars from M to OA and OB, respectively. Let H be the orthocenter of the triangle OPQ. Find the locus of points H when:
   (a) M belongs to the segment AB;
   (b) M belongs to the interior of $\triangle OAB$.

6. **(POL)** We are given $n \geq 3$ points in the plane. Let d be the maximal distance between two of the given points. Prove that the number of pairs of points whose distance is equal to d is less than or equal to n.

3.8 The Eighth IMO
Sofia, Bulgaria, July 3–13, 1966

3.8.1 Contest Problems

First Day

1. **(USS)** Three problems A, B, and C were given on a mathematics olympiad. All 25 students solved at least one of these problems. The number of students who solved B and not A is twice the number of students who solved C and not A. The number of students who solved only A is greater by 1 than the number of students who along with A solved at least one other problem. Among the students who solved only one problem, half solved A. How many students solved only B?

2. **(HUN)** If a, b, and c are the sides and α, β, and γ the respective angles of the triangle for which $a + b = \tan \frac{\gamma}{2}(a \tan \alpha + b \tan \beta)$, prove that the triangle is isosceles.

3. **(BGR)** Prove that the sum of distances from the center of the circumsphere of the regular tetrahedron to its four vertices is less than the sum of distances from any other point to the four vertices.

Second Day

4. **(YUG)** Prove the following equality:

$$\frac{1}{\sin 2x} + \frac{1}{\sin 4x} + \frac{1}{\sin 8x} + \cdots + \frac{1}{\sin 2^n x} = \cot x - \cot 2^n x,$$

where $n \in \mathbb{N}$ and $x \notin \frac{\pi}{2^k}\mathbb{Z}$ for every $k \in \mathbb{N}$.

5. **(CZS)** Solve the following system of equations:

$$|a_1 - a_2|x_2 + |a_1 - a_3|x_3 + |a_1 - a_4|x_4 = 1,$$
$$|a_2 - a_1|x_1 + |a_2 - a_3|x_3 + |a_2 - a_4|x_4 = 1,$$
$$|a_3 - a_1|x_1 + |a_3 - a_2|x_2 + |a_3 - a_4|x_4 = 1,$$
$$|a_4 - a_1|x_1 + |a_4 - a_2|x_2 + |a_4 - a_3|x_3 = 1,$$

where a_1, a_2, a_3, and a_4 are mutually distinct real numbers.

6. **(POL)** Let M, K, and L be points on (AB), (BC), and (CA), respectively. Prove that the area of at least one of the three triangles $\triangle MAL$, $\triangle KBM$, and $\triangle LCK$ is less than or equal to one-fourth the area of $\triangle ABC$.

3.8.2 Some Longlisted Problems 1959–1966

1. **(CZS)** We are given $n > 3$ points in the plane, no three of which lie on a line. Does there necessarily exist a circle that passes through at least three of the given points and contains none of the other given points in its interior?

2. **(GDR)** Given n positive real numbers $a_1, a_2, \ldots, a_n$ such that $a_1 a_2 \cdots a_n = 1$, prove that
$$(1+a_1)(1+a_2)\cdots(1+a_n) \geq 2^n.$$

3. **(BGR)** A regular triangular prism has height h and a base of side length a. Both bases have small holes in the centers, and the inside of the three vertical walls has a mirror surface. Light enters through the small hole in the top base, strikes each vertical wall once and leaves through the hole in the bottom. Find the angle at which the light enters and the length of its path inside the prism.

4. **(POL)** Five points in the plane are given, no three of which are collinear. Show that some four of them form a convex quadrilateral.

5. **(USS)** Prove the inequality
$$\tan\frac{\pi \sin x}{4 \sin \alpha} + \tan\frac{\pi \cos x}{4 \cos \alpha} > 1$$
for any x, α with $0 \leq x \leq \pi/2$ and $\pi/6 < \alpha < \pi/3$.

6. **(USS)** A convex planar polygon $\mathcal{M}$ with perimeter l and area S is given. Let $M(R)$ be the set of all points in space that lie a distance at most R from a point of $\mathcal{M}$. Show that the volume $V(R)$ of this set equals
$$V(R) = \frac{4}{3}\pi R^3 + \frac{\pi}{2}lR^2 + 2SR.$$

7. **(USS)** For which arrangements of two infinite circular cylinders does their intersection lie in a plane?

8. **(USS)** We are given a bag of sugar, a two-pan balance, and a weight of 1 gram. How do we obtain 1 kilogram of sugar in the smallest possible number of weighings?

9. **(ROU)** Find x such that
$$\frac{\sin 3x \cos(60° - 4x) + 1}{\sin(60° - 7x) - \cos(30° + x) + m} = 0,$$
where m is a fixed real number.

10. **(GDR)** How many real solutions are there to the equation $x = 1964 \sin x - 189$?

11. **(CZS)** Does there exist an integer z that can be written in two different ways as $z = x! + y!$, where x, y are natural numbers with $x \leq y$?

12. **(BGR)** Find digits x, y, z such that the equality
$$\sqrt{\underbrace{\overline{xx\cdots x}}_{2n} - \underbrace{\overline{yy\cdots y}}_{n}} = \underbrace{\overline{zz\cdots z}}_{n}$$
holds for at least two values of $n \in \mathbb{N}$, and in that case find all n for which this equality is true.

13. **(YUG)** Let $a_1, a_2, \ldots, a_n$ be positive real numbers. Prove the inequality

$$\binom{n}{2} \sum_{i<j} \frac{1}{a_i a_j} \geq 4 \left(\sum_{i<j} \frac{1}{a_i + a_j} \right)^2$$

and find the conditions on the numbers a_i for equality to hold.

14. **(POL)** Compute the largest number of regions into which one can divide a disk by joining n points on its circumference.

15. **(POL)** Points A, B, C, D lie on a circle such that AB is a diameter and CD is not. If the tangents at C and D meet at P while AC and BD meet at Q, show that PQ is perpendicular to AB.

16. **(CZS)** We are given a circle K with center S and radius 1 and a square Q with center M and side 2. Let XY be the hypotenuse of an isosceles right triangle XYZ. Describe the locus of points Z as X varies along K and Y varies along the boundary of Q.

17. **(ROU)** Suppose $ABCD$ and $A'B'C'D'$ are two parallelograms arbitrarily arranged in space, and let points M, N, P, Q divide the segments AA', BB', CC', DD' respectively in equal ratios.
    (a) Show that $MNPQ$ is a parallelogram;
    (b) Find the locus of $MNPQ$ as M varies along the segment AA'.

18. **(HUN)** Solve the equation $\frac{1}{\sin x} + \frac{1}{\cos x} = \frac{1}{p}$, where p is a real parameter. Discuss for which values of p the equation has at least one real solution and determine the number of solutions in $[0, 2\pi)$ for a given p.

19. **(HUN)** Construct a triangle given the three exradii.

20. **(HUN)** We are given three equal rectangles with the same center in three mutually perpendicular planes, with the long sides also mutually perpendicular. Consider the polyhedron with vertices at the vertices of these rectangles.
    (a) Find the volume of this polyhedron;
    (b) can this polyhedron be regular, and under what conditions?

21. **(BGR)** Prove that the volume V and the lateral area S of a right circular cone satisfy the inequality $\left(\frac{6V}{\pi} \right)^2 \leq \left(\frac{2S}{\pi\sqrt{3}} \right)^3$. When does equality occur?

22. **(BGR)** Assume that two parallelograms P, P' of equal areas have sides a, b and a', b' respectively such that $a' \leq a \leq b \leq b'$ and a segment of length b' can be placed inside P. Prove that P and P' can be partitioned into four pairwise congruent parts.

23. **(BGR)** Three faces of a tetrahedron are right triangles, while the fourth is not an obtuse triangle.
    (a) Prove that a necessary and sufficient condition for the fourth face to be a right triangle is that at some vertex exactly two angles are right.

(b) Prove that if all the faces are right triangles, then the volume of the tetrahedron equals one -sixth the product of the three smallest edges not belonging to the same face.

24. **(POL)** There are $n \geq 2$ people in a room. Prove that there exist two among them having equal numbers of friends in that room. (Friendship is always mutual.)

25. **(GDR)** Show that $\tan 7°30' = \sqrt{6} + \sqrt{2} - \sqrt{3} - 2$.

26. **(CZS)** (a) Prove that $(a_1 + a_2 + \cdots + a_k)^2 \leq k(a_1^2 + \cdots + a_k^2)$, where $k \geq 1$ is a natural number and $a_1, \ldots, a_k$ are arbitrary real numbers.
    (b) If real numbers $a_1, \ldots, a_n$ satisfy

$$a_1 + a_2 + \cdots + a_n \geq \sqrt{(n-1)(a_1^2 + \cdots + a_n^2)},$$

show that they are all nonnegative.

27. **(GDR)** We are given a circle K and a point P lying on a line g. Construct a circle that passes through P and touches K and g.

28. **(CZS)** Let there be given a circle with center S and radius 1 in the plane, and let ABC be an arbitrary triangle circumscribed about the circle such that $SA \leq SB \leq SC$. Find the loci of the vertices A, B, C.

29. **(ROU)** (a) Find the number of ways 500 can be represented as a sum of consecutive integers.
    (b) Find the number of such representations for $N = 2^\alpha 3^\beta 5^\gamma$, $\alpha, \beta, \gamma \in \mathbb{N}$. Which of these representations consist only of natural numbers?
    (c) Determine the number of such representations for an arbitrary natural number N.

30. **(ROU)** If n is a natural number, prove that
    (a) $\log_{10}(n+1) > \frac{3}{10n} + \log_{10} n$;
    (b) $\log n! > \frac{3n}{10} \left(\frac{1}{2} + \frac{1}{3} + \cdots + \frac{1}{n} - 1 \right)$.

31. **(ROU)** Solve the equation $|x^2 - 1| + |x^2 - 4| = mx$ as a function of the parameter m. Which pairs (x, m) of integers satisfy this equation?

32. **(BGR)** The sides a, b, c of a triangle ABC form an arithmetic progression; the sides of another triangle $A_1 B_1 C_1$ also form an arithmetic progression. Suppose that $\angle A = \angle A_1$. Prove that the triangles ABC and $A_1 B_1 C_1$ are similar.

33. **(BGR)** Two circles touch each other from inside, and an equilateral triangle is inscribed in the larger circle. From the vertices of the triangle one draws segments tangent to the smaller circle. Prove that the length of one of these segments equals the sum of the lengths of the other two.

34. **(BGR)** Determine all pairs of positive integers (x, y) satisfying the equation $2^x = 3^y + 5$.

35. **(POL)** If a, b, c, d are integers such that ad is odd and bc is even, prove that at least one root of the polynomial $ax^3 + bx^2 + cx + d$ is irrational.

36. **(POL)** Let $ABCD$ be a cyclic quadrilateral. Show that the centroids of the triangles ABC, CDA, BCD, DAB lie on a circle.

37. **(POL)** Prove that the perpendiculars drawn from the midpoints of the sides of a cyclic quadrilateral to the opposite sides meet at one point.

38. **(ROU)** Two concentric circles have radii R and r respectively. Determine the greatest possible number of circles that are tangent to both these circles and mutually nonintersecting. Prove that this number lies between $\frac{3}{2} \cdot \frac{\sqrt{R}+\sqrt{r}}{\sqrt{R}-\sqrt{r}} - 1$ and $\frac{63}{20} \cdot \frac{R+r}{R-r}$.

39. **(ROU)** In a plane, a circle with center O and radius R and two points A, B are given.
    (a) Draw a chord CD parallel to AB so that AC and BD intersect at a point P on the circle.
    (b) Prove that there are two possible positions of point P, say P_1, P_2, and find the distance between them if $OA = a$, $OB = b$, $AB = d$.

40. **(CZS)** For a positive real number p, find all real solutions to the equation
$$\sqrt{x^2 + 2px - p^2} - \sqrt{x^2 - 2px - p^2} = 1.$$

41. **(CZS)** If $A_1 A_2 \ldots A_n$ is a regular n-gon ($n \geq 3$), how many different obtuse triangles $A_i A_j A_k$ exist?

42. **(CZS)** Let $a_1, a_2, \ldots, a_n$ ($n \geq 2$) be a sequence of integers. Show that there is a subsequence $a_{k_1}, a_{k_2}, \ldots, a_{k_m}$, where $1 \leq k_1 < k_2 < \cdots < k_m \leq n$, such that $a_{k_1}^2 + a_{k_2}^2 + \cdots + a_{k_m}^2$ is divisible by n.

43. **(CZS)** Five points in a plane are given, no three of which are collinear. Every two of them are joined by a segment, colored either red or gray, so that no three segments form a triangle colored in one color.
    (a) Prove that (1) every point is a vertex of exactly two red and two gray segments, and (2) the red segments form a closed path that passes through each point.
    (b) Give an example of such a coloring.

44. **(YUG)** What is the greatest number of balls of radius $1/2$ that can be placed within a rectangular box of size $10 \times 10 \times 1$?

45. **(YUG)** An alphabet consists of n letters. What is the maximal length of a word, if
    (i) two neighboring letters in a word are always different, and
    (ii) no word $abab$ ($a \neq b$) can be obtained by omitting letters from the given word?

46. **(YUG)** Let
$$f(a,b,c) = \left| \frac{|b-a|}{|ab|} + \frac{b+a}{ab} - \frac{2}{c} \right| + \frac{|b-a|}{|ab|} + \frac{b+a}{ab} + \frac{2}{c}.$$

Prove that $f(a,b,c) = 4\max\{1/a, 1/b, 1/c\}$.

47. **(ROU)** Find the number of lines dividing a given triangle into two parts of equal area which determine the segment of minimum possible length inside the triangle. Compute this minimum length in terms of the sides a, b, c of the triangle.

48. **(USS)** Find all positive numbers p for which the equation $x^2 + px + 3p = 0$ has integral roots.

49. **(USS)** Two mirror walls are placed to form an angle of measure α. There is a candle inside the angle. How many reflections of the candle can an observer see?

50. **(USS)** Given a quadrangle of sides a, b, c, d and area S, show that $S \leq \frac{a+c}{2} \cdot \frac{b+d}{2}$.

51. **(USS)** In a school, n children numbered 1 to n are initially arranged in the order $1, 2, \ldots, n$. At a command, every child can either exchange its position with any other child or not move. Can they rearrange into the order $n, 1, 2, \ldots, n-1$ after two commands?

52. **(USS)** A figure of area 1 is cut out from a sheet of paper and divided into 10 parts, each of which is colored in one of 10 colors. Then the figure is turned to the other side and again divided into 10 parts (not necessarily in the same way). Show that it is possible to color these parts in the 10 colors so that the total area of the portions of the figure both of whose sides are of the same color is at least 0.1.

53. **(USS, 1966)** Prove that in every convex hexagon of area S one can draw a diagonal that cuts off a triangle of area not exceeding $\frac{1}{6}S$.

54. **(USS, 1966)** Find the last two digits of a sum of eighth powers of 100 consecutive integers.

55. **(USS, 1966)** Given the vertex A and the centroid M of a triangle ABC, find the locus of vertices B such that all the angles of the triangle lie in the interval $[40°, 70°]$.

56. **(USS, 1966)** Let $ABCD$ be a tetrahedron such that $AB \perp CD$, $AC \perp BD$, and $AD \perp BC$. Prove that the midpoints of the edges of the tetrahedron lie on a sphere.

57. **(USS, 1966)** Is it possible to choose a set of 100 (or 200) points on the boundary of a cube such that this set is fixed under each isometry of the cube into itself? Justify your answer.

3.9 The Ninth IMO
Cetinje, Yugoslavia, July 2–13, 1967

3.9.1 Contest Problems

First Day (July 5)

1. *ABCD* is a parallelogram; $AB = a$, $AD = 1$, α is the size of $\angle DAB$, and the three angles of the triangle *ABD* are acute. Prove that the four circles K_A, K_B, K_C, K_D, each of radius 1, whose centers are the vertices A, B, C, D, cover the parallelogram if and only if $a \leq \cos \alpha + \sqrt{3} \sin \alpha$.

2. Exactly one side of a tetrahedron is of length greater than 1. Show that its volume is less than or equal to $1/8$.

3. Let k, m, and n be positive integers such that $m + k + 1$ is a prime number greater than $n + 1$. Write c_s for $s(s + 1)$. Prove that the product $(c_{m+1} - c_k)(c_{m+2} - c_k) \cdots (c_{m+n} - c_k)$ is divisible by the product $c_1 c_2 \cdots c_n$.

Second Day (July 6)

4. The triangles $A_0 B_0 C_0$ and $A'B'C'$ have all their angles acute. Describe how to construct one of the triangles *ABC*, similar to $A'B'C'$ and circumscribing $A_0 B_0 C_0$ (so that A, B, C correspond to A', B', C', and *AB* passes through C_0, *BC* through A_0, and *CA* through B_0). Among these triangles *ABC* describe, and prove, how to construct the triangle with the maximum area.

5. Consider the sequence (c_n):

$$c_1 = a_1 + a_2 + \cdots + a_8,$$
$$c_2 = a_1^2 + a_2^2 + \cdots + a_8^2,$$
$$\cdots \qquad \cdots$$
$$c_n = a_1^n + a_2^n + \cdots + a_8^n,$$
$$\cdots \qquad \cdots$$

where $a_1, a_2, \ldots, a_8$ are real numbers, not all equal to zero. Given that among the numbers of the sequence (c_n) there are infinitely many equal to zero, determine all the values of n for which $c_n = 0$.

6. In a sports competition lasting n days there are m medals to be won. On the first day, one medal and $1/7$ of the remaining $m - 1$ medals are won. On the second day, 2 medals and $1/7$ of the remainder are won. And so on. On the nth day exactly n medals are won. How many days did the competition last and what was the total number of medals?

3.9.2 Longlisted Problems

1. **(BGR 1)** Prove that all numbers in the sequence

$$\frac{107811}{3}, \quad \frac{110778111}{3}, \quad \frac{111077781111}{3}, \quad \ldots$$

are perfect cubes.

2. **(BGR 2)** Prove that $\frac{1}{3}n^2 + \frac{1}{2}n + \frac{1}{6} \geq (n!)^{2/n}$ (n is a positive integer) and that equality is possible only in the case $n = 1$.

3. **(BGR 3)** Prove the trigonometric inequality $\cos x < 1 - \frac{x^2}{2} + \frac{x^4}{16}$, where $x \in (0, \pi/2)$.

4. **(BGR 4)** Suppose medians m_a and m_b of a triangle are orthogonal. Prove that:
   (a) The medians of that triangle correspond to the sides of a right-angled triangle.
   (b) The inequality
   $$5(a^2 + b^2 - c^2) \geq 8ab$$
   is valid, where a, b, and c are side lengths of the given triangle.

5. **(BGR 5)** Solve the system
   $$\begin{aligned}
   x^2 + x - 1 &= y, \\
   y^2 + y - 1 &= z, \\
   z^2 + z - 1 &= x.
   \end{aligned}$$

6. **(BGR 6)** Solve the system
   $$\begin{aligned}
   |x+y| + |1-x| &= 6, \\
   |x+y+1| + |1-y| &= 4.
   \end{aligned}$$

7. **(CZS 1)** Find all real solutions of the system of equations
   $$\begin{aligned}
   x_1 + x_2 + \cdots + x_n &= a, \\
   x_1^2 + x_2^2 + \cdots + x_n^2 &= a^2, \\
   &\cdots \\
   x_1^n + x_2^n + \cdots + x_n^n &= a^n.
   \end{aligned}$$

8. **(CZS 2)**[IMO1] $ABCD$ is a parallelogram; $AB = a$, $AD = 1$, α is the size of $\angle DAB$, and the three angles of the triangle ABD are acute. Prove that the four circles K_A, K_B, K_C, K_D, each of radius 1, whose centers are the vertices A, B, C, D, cover the parallelogram if and only if $a \leq \cos \alpha + \sqrt{3} \sin \alpha$.

9. **(CZS 3)** The circle k and its diameter AB are given. Find the locus of the centers of circles inscribed in the triangles having one vertex on AB and two other vertices on k.

10. **(CZS 4)** The square $ABCD$ is to be decomposed into n triangles (nonoverlapping) all of whose angles are acute. Find the smallest integer n for which there exists a solution to this problem and construct at least one decomposition for this n. Answer whether it is possible to ask additionally that (at least) one of these triangles has a perimeter less than an arbitrarily given positive number.

11. **(CZS 5)** Let n be a positive integer. Find the maximal number of noncongruent triangles whose side lengths are integers less than or equal to n.

12. **(CZS 6)** Given a segment AB of the length 1, define the set M of points in the following way: it contains the two points A, B, and also all points obtained from A, B by iterating the following rule: for every pair of points X, Y in M, the set M also contains the point Z of the segment XY for which $YZ = 3XZ$.

    (a) Prove that the set M consists of points X from the segment AB for which the distance from the point A is either

$$AX = \frac{3k}{4^n} \quad \text{or} \quad AX = \frac{3k-2}{4^n},$$

    where n, k are nonnegative integers.

    (b) Prove that the point X_0 for which $AX_0 = 1/2 = X_0B$ does not belong to the set M.

13. **(GDR 1)** Find whether among all quadrilaterals whose interiors lie inside a semicircle of radius r there exists one (or more) with maximal area. If so, determine their shape and area.

14. **(GDR 2)** Which fraction p/q, where p, q are positive integers less than 100, is closest to $\sqrt{2}$? Find all digits after the decimal point in the decimal representation of this fraction that coincide with digits in the decimal representation of $\sqrt{2}$ (without using any tables).

15. **(GDR 3)** Suppose $\tan \alpha = p/q$, where p and q are integers and $q \neq 0$. Prove that the number $\tan \beta$ for which $\tan 2\beta = \tan 3\alpha$ is rational only when $p^2 + q^2$ is the square of an integer.

16. **(GDR 4)** Prove the following statement: If r_1 and r_2 are real numbers whose quotient is irrational, then any real number x can be approximated arbitrarily well by numbers of the form $z_{k_1,k_2} = k_1 r_1 + k_2 r_2$, k_1, k_2 integers; i.e., for every real number x and every positive real number p two integers k_1 and k_2 can be found such that $|x - (k_1 r_1 + k_2 r_2)| < p$.

17. **(UNK 1)**[IMO3] Let k, m, and n be positive integers such that $m + k + 1$ is a prime number greater than $n + 1$. Write c_s for $s(s+1)$. Prove that the product $(c_{m+1} - c_k)(c_{m+2} - c_k) \cdots (c_{m+n} - c_k)$ is divisible by the product $c_1 c_2 \cdots c_n$.

18. **(UNK 5)** If x is a positive rational number, show that x can be uniquely expressed in the form

$$x = a_1 + \frac{a_2}{2!} + \frac{a_3}{3!} + \cdots,$$

where $a_1, a_2, \ldots$ are integers, $0 \leq a_n \leq n-1$ for $n > 1$, and the series terminates. Show also that x can be expressed as the sum of reciprocals of different integers, each of which is greater than 10^6.

19. **(UNK 6)** The n points $P_1, P_2, \ldots, P_n$ are placed inside or on the boundary of a disk of radius 1 in such a way that the minimum distance d_n between any two

of these points has its largest possible value D_n. Calculate D_n for $n = 2$ to 7 and justify your answer.

20. **(HUN 1)** In space, n points $(n \geq 3)$ are given. Every pair of points determines some distance. Suppose all distances are different. Connect every point with the nearest point. Prove that it is impossible to obtain a polygonal line in such a way. [1]

21. **(HUN 2)** Without using any tables, find the exact value of the product

$$P = \cos \frac{\pi}{15} \cos \frac{2\pi}{15} \cos \frac{3\pi}{15} \cos \frac{4\pi}{15} \cos \frac{5\pi}{15} \cos \frac{6\pi}{15} \cos \frac{7\pi}{15}.$$

22. **(HUN 3)** The distance between the centers of the circles k_1 and k_2 with radii r is equal to r. Points A and B are on the circle k_1, symmetric with respect to the line connecting the centers of the circles. Point P is an arbitrary point on k_2. Prove that

$$PA^2 + PB^2 \geq 2r^2.$$

When does equality hold?

23. **(HUN 4)** Prove that for an arbitrary pair of vectors f and g in the plane, the inequality

$$af^2 + bfg + cg^2 \geq 0$$

holds if and only if the following conditions are fulfilled: $a \geq 0,\, c \geq 0,\, 4ac \geq b^2$.

24. **(HUN 5)**[IMO6] Father has left to his children several identical gold coins. According to his will, the oldest child receives one coin and one-seventh of the remaining coins, the next child receives two coins and one-seventh of the remaining coins, the third child receives three coins and one-seventh of the remaining coins, and so on through the youngest child. If every child inherits an integer number of coins, find the number of children and the number of coins.

25. **(HUN 6)** Three disks of diameter d are touching a sphere at their centers. Moreover, each disk touches the other two disks. How do we choose the radius R of the sphere so that the axis of the whole figure makes an angle of $60°$ with the line connecting the center of the sphere with the point on the disks that is at the largest distance from the axis? (The axis of the figure is the line having the property that rotation of the figure through $120°$ about that line brings the figure to its initial position. The disks are all on one side of the plane, pass through the center of the sphere, and are orthogonal to the axes.)

26. **(ITA 1)** Let $ABCD$ be a regular tetrahedron. To an arbitrary point M on one edge, say CD, corresponds the point $P = P(M)$, which is the intersection of two lines AH and BK, drawn from A orthogonally to BM and from B orthogonally to AM. What is the locus of P as M varies?

---

[1] The statement so formulated is false. It would be trivially true under the additional assumption that the polygonal line is closed. However, from the offered solution, which is not clear, it does not seem that the proposer had this in mind.

27. **(ITA 2)** Which regular polygons can be obtained (and how) by cutting a cube with a plane?

28. **(ITA 3)** Find values of the parameter u for which the expression

$$y = \frac{\tan(x-u) + \tan x + \tan(x+u)}{\tan(x-u)\tan x\tan(x+u)}$$

does not depend on x.

29. **(ITA 4)**[IMO4] The triangles $A_0B_0C_0$ and $A'B'C'$ have all their angles acute. Describe how to construct one of the triangles ABC, similar to $A'B'C'$ and circumscribing $A_0B_0C_0$ (so that A, B, C correspond to A', B', C', and AB passes through C_0, BC through A_0, and CA through B_0). Among these triangles ABC, describe, and prove, how to construct the triangle with the maximum area.

30. **(MNG 1)** Given $m+n$ numbers a_i $(i=1,2,\ldots,m)$, b_j $(j=1,2,\ldots,n)$, determine the number of pairs (a_i, b_j) for which $|i-j| \geq k$, where k is a nonnegative integer.

31. **(MNG 2)** An urn contains balls of k different colors; there are n_i balls of the ith color. Balls are drawn at random from the urn, one by one, without replacement. Find the smallest number of draws necessary for getting m balls of the same color.

32. **(MNG 3)** Determine the volume of the body obtained by cutting the ball of radius R by the trihedron with vertex in the center of that ball if its dihedral angles are α, β, γ.

33. **(MNG 4)** In what case does the system

$$x + y + mz = a,$$
$$x + my + z = b,$$
$$mx + y + z = c,$$

have a solution? Find the conditions under which the unique solution of the above system is an arithmetic progression.

34. **(MNG 5)** The faces of a convex polyhedron are six squares and eight equilateral triangles, and each edge is a common side for one triangle and one square. All dihedral angles obtained from the triangle and square with a common edge are equal. Prove that it is possible to circumscribe a sphere around this polyhedron and compute the ratio of the squares of the volumes of the polyhedron and of the ball whose boundary is the circumscribed sphere.

35. **(MNG 6)** Prove the identity

$$\sum_{k=0}^{n} \binom{n}{k} \left(\tan \frac{x}{2}\right)^{2k} \left[1 + 2^k \frac{1}{(1-\tan^2(x/2))^k}\right] = \sec^{2n}\frac{x}{2} + \sec^n x.$$

36. **(POL 1)** Prove that the center of the sphere circumscribed around a tetrahedron $ABCD$ coincides with the center of a sphere inscribed in that tetrahedron if and only if $AB = CD$, $AC = BD$, and $AD = BC$.

37. **(POL 2)** Prove that for arbitrary positive numbers the following inequality holds:

$$\frac{1}{a}+\frac{1}{b}+\frac{1}{c} \le \frac{a^8+b^8+c^8}{a^3b^3c^3}.$$

38. **(POL 3)** Does there exist an integer such that its cube is equal to $3n^2+3n+7$, where n is integer?

39. **(POL 4)** Show that the triangle whose angles satisfy the equality

$$\frac{\sin^2 A + \sin^2 B + \sin^2 C}{\cos^2 A + \cos^2 B + \cos^2 C} = 2$$

is a right-angled triangle.

40. **(POL 5)**[IMO2] Exactly one side of a tetrahedron is of length greater than 1. Show that its volume is less than or equal to $1/8$.

41. **(POL 6)** A line l is drawn through the intersection point H of the altitudes of an acute-angled triangle. Prove that the symmetric images l_a, l_b, l_c of l with respect to sides BC, CA, AB have one point in common, which lies on the circumcircle of ABC.

42. **(ROU 1)** Decompose into real factors the expression $1 - \sin^5 x - \cos^5 x$.

43. **(ROU 2)** The equation

$$x^5 + 5\lambda x^4 - x^3 + (\lambda \alpha - 4)x^2 - (8\lambda + 3)x + \lambda \alpha - 2 = 0$$

is given.
   (a) Determine α such that the given equation has exactly one root independent of λ.
   (b) Determine α such that the given equation has exactly two roots independent of λ.

44. **(ROU 3)** Suppose p and q are two different positive integers and x is a real number. Form the product $(x+p)(x+q)$.
   (a) Find the sum $S(x,n) = \Sigma(x+p)(x+q)$, where p and q take values from 1 to n.
   (b) Do there exist integer values of x for which $S(x,n) = 0$?

45. **(ROU 4)** (a) Solve the equation

$$\sin^3 x + \sin^3 \left(\frac{2\pi}{3}+x\right) + \sin^3 \left(\frac{4\pi}{3}+x\right) + \frac{3}{4}\cos 2x = 0.$$

   (b) Suppose the solutions are in the form of arcs AB of the trigonometric circle (where A is the beginning of arcs of the trigonometric circle), and P is a regular n-gon inscribed in the circle with one vertex at A.
      (1) Find the subset of arcs with the endpoint B at a vertex of the regular dodecagon.

(2) Prove that the endpoint B cannot be at a vertex of P if $2,3 \nmid n$ or n is prime.

46. **(ROU 5)** If x, y, z are real numbers satisfying the relations $x+y+z=1$ and $\arctan x + \arctan y + \arctan z = \pi/4$, prove that

$$x^{2n+1} + y^{2n+1} + z^{2n+1} = 1$$

for all positive integers n.

47. **(ROU 6)** Prove the inequality

$$x_1 x_2 \cdots x_k \left(x_1^{n-1} + x_2^{n-1} + \cdots + x_k^{n-1}\right) \le x_1^{n+k-1} + x_2^{n+k-1} + \cdots + x_k^{n+k-1},$$

where $x_i > 0$ $(i = 1,2,\ldots,k)$, $k \in N$, $n \in N$.

48. **(SWE 1)** Determine all positive roots of the equation $x^x = 1/\sqrt{2}$.

49. **(SWE 2)** Let n and k be positive integers such that $1 \le n \le N+1$, $1 \le k \le N+1$. Show that

$$\min_{n \ne k} |\sin n - \sin k| < \frac{2}{N}.$$

50. **(SWE 3)** The function $\varphi(x,y,z)$, defined for all triples (x,y,z) of real numbers, is such that there are two functions f and g defined for all pairs of real numbers such that

$$\varphi(x,y,z) = f(x+y,z) = g(x,y+z)$$

for all real x, y, and z. Show that there is a function h of one real variable such that

$$\varphi(x,y,z) = h(x+y+z)$$

for all real x, y, and z.

51. **(SWE 4)** A subset S of the set of integers $0,\ldots,99$ is said to have property A if it is impossible to fill a crossword puzzle with 2 rows and 2 columns with numbers in S (0 is written as 00, 1 as 01, and so on). Determine the maximal number of elements in sets S with property A.

52. **(SWE 5)** In the plane a point O and a sequence of points $P_1, P_2, P_3, \ldots$ are given. The distances $OP_1, OP_2, OP_3, \ldots$ are $r_1, r_2, r_3, \ldots$, where $r_1 \le r_2 \le r_3 \le \cdots$. Let α satisfy $0 < \alpha < 1$. Suppose that for every n the distance from the point P_n to any other point of the sequence is greater than or equal to r_n^α. Determine the exponent β, as large as possible, such that for some C independent of n,[2]

$$r_n \ge C n^\beta, \quad n = 1,2,\ldots .$$

---

[2] This problem is not elementary. The solution offered by the proposer, which is not quite clear and complete, only shows that if such a β exists, then $\beta \ge \frac{1}{2(1-\alpha)}$.

53. **(SWE 6)** In making Euclidean constructions in geometry it is permitted to use a straightedge and compass. In the constructions considered in this question, no compasses are permitted, but the straightedge is assumed to have two parallel edges, which can be used for constructing two parallel lines through two given points whose distance is at least equal to the breadth of the ruler. Then the distance between the parallel lines is equal to the breadth of the straightedge. Carry through the following constructions with such a straightedge. Construct:
   (a) The bisector of a given angle.
   (b) The midpoint of a given rectilinear segment.
   (c) The center of a circle through three given noncollinear points.
   (d) A line through a given point parallel to a given line.

54. **(USS 1)** Is it possible to put 100 (or 200) points on a wooden cube such that by all rotations of the cube the points map into themselves? Justify your answer.

55. **(USS 2)** Find all x for which for all n,

$$\sin x + \sin 2x + \sin 3x + \cdots + \sin nx \le \frac{\sqrt{3}}{2}.$$

56. **(USS 3)** In a group of interpreters each one speaks one or several foreign languages; 24 of them speak Japanese, 24 Malay, 24 Farsi. Prove that it is possible to select a subgroup in which exactly 12 interpreters speak Japanese, exactly 12 speak Malay, and exactly 12 speak Farsi.

57. **(USS 4)**[IMO5] Consider the sequence (c_n):

$$c_1 = a_1 + a_2 + \cdots + a_8,$$
$$c_2 = a_1^2 + a_2^2 + \cdots + a_8^2,$$
$$\cdots \qquad \cdots\cdots\cdots$$
$$c_n = a_1^n + a_2^n + \cdots + a_8^n,$$
$$\cdots \qquad \cdots\cdots\cdots$$

where $a_1, a_2, \ldots, a_8$ are real numbers, not all equal to zero. Given that among the numbers of the sequence (c_n) there are infinitely many equal to zero, determine all the values of n for which $c_n = 0$.

58. **(USS 5)** A linear binomial $l(z) = Az + B$ with complex coefficients A and B is given. It is known that the maximal value of $|l(z)|$ on the segment $-1 \le x \le 1$ ($y = 0$) of the real line in the complex plane ($z = x + iy$) is equal to M. Prove that for every z

$$|l(z)| \le M\rho,$$

where ρ is the sum of distances from the point $P = z$ to the points $Q_1: z = 1$ and $Q_3: z = -1$.

59. **(USS 6)** On the circle with center O and radius 1 the point A_0 is fixed and points $A_1, A_2, \ldots, A_{999}, A_{1000}$ are distributed in such a way that $\angle A_0 O A_k = k$ (in radians). Cut the circle at points $A_0, A_1, \ldots, A_{1000}$. How many arcs with different lengths are obtained?

3.10 The Tenth IMO
Moscow–Leningrad, Soviet Union, July 5–18, 1968

3.10.1 Contest Problems

First Day

1. Prove that there exists a unique triangle whose side lengths are consecutive natural numbers and one of whose angles is twice the measure of one of the others.

2. Find all positive integers x for which $p(x) = x^2 - 10x - 22$, where $p(x)$ denotes the product of the digits of x.

3. Let a, b, c be real numbers. Prove that the system of equations

$$\begin{cases} ax_1^2 + bx_1 + c = x_2, \\ ax_2^2 + bx_2 + c = x_3, \\ \cdots\cdots\cdots \\ ax_{n-1}^2 + bx_{n-1} + c = x_n, \\ ax_n^2 + bx_n + c = x_1, \end{cases}$$

  (a) has no real solutions if $(b-1)^2 - 4ac < 0$;
  (b) has a unique real solution if $(b-1)^2 - 4ac = 0$;
  (c) has more than one real solution if $(b-1)^2 - 4ac > 0$.

Second Day

4. Prove that in any tetrahedron there is a vertex such that the lengths of its sides through that vertex are sides of a triangle.

5. Let $a > 0$ be a real number and $f(x)$ a real function defined on all of $\mathbb{R}$, satisfying for all $x \in \mathbb{R}$,
$$f(x+a) = \frac{1}{2} + \sqrt{f(x) - f(x)^2}.$$

  (a) Prove that the function f is periodic; i.e., there exists $b > 0$ such that for all x, $f(x+b) = f(x)$.
  (b) Give an example of such a nonconstant function for $a = 1$.

6. Let $[x]$ denote the integer part of x, i.e., the greatest integer not exceeding x. If n is a positive integer, express as a simple function of n the sum
$$\left[\frac{n+1}{2}\right] + \left[\frac{n+2}{4}\right] + \cdots + \left[\frac{n+2^i}{2^{i+1}}\right] + \cdots.$$

3.10.2 Shortlisted Problems

1. **(SWE 2)** Two ships sail on the sea with constant speeds and fixed directions. It is known that at 9:00 the distance between them was 20 miles; at 9:35, 15 miles; and at 9:55, 13 miles. At what moment were the ships the smallest distance from each other, and what was that distance?

2. **(ROU 5)**[IMO1] Prove that there exists a unique triangle whose side lengths are consecutive natural numbers and one of whose angles is twice the measure of one of the others.

3. **(POL 4)**[IMO4] Prove that in any tetrahedron there is a vertex such that the lengths of its sides through that vertex are sides of a triangle.

4. **(BGR 2)**[IMO3] Let a, b, c be real numbers. Prove that the system of equations

$$\begin{cases} ax_1^2 + bx_1 + c = x_2, \\ ax_2^2 + bx_2 + c = x_3, \\ \qquad \cdots\cdots\cdots \\ ax_{n-1}^2 + bx_{n-1} + c = x_n, \\ ax_n^2 + bx_n + c = x_1, \end{cases}$$

has a unique real solution if and only if $(b-1)^2 - 4ac = 0$.

Remark. It is assumed that $a \neq 0$.

5. **(BGR 5)** Let h_n be the apothem (distance from the center to one of the sides) of a regular n-gon $(n \geq 3)$ inscribed in a circle of radius r. Prove the inequality

$$(n+1)h_{n+1} - nh_n > r.$$

Also prove that if r on the right side is replaced with a greater number, the inequality will not remain true for all $n \geq 3$.

6. **(HUN 1)** If a_i $(i = 1, 2, \ldots, n)$ are distinct non-zero real numbers, prove that the equation

$$\frac{a_1}{a_1 - x} + \frac{a_2}{a_2 - x} + \cdots + \frac{a_n}{a_n - x} = n$$

has at least $n-1$ real roots.

7. **(HUN 5)** Prove that the product of the radii of three circles exscribed to a given triangle does not exceed $\frac{3\sqrt{3}}{8}$ times the product of the side lengths of the triangle. When does equality hold?

8. **(ROU 2)** Given an oriented line Δ and a fixed point A on it, consider all trapezoids $ABCD$ one of whose bases AB lies on Δ, in the positive direction. Let E, F be the midpoints of AB and CD respectively.
   Find the loci of vertices B, C, D of trapezoids that satisfy the following:
   (i) $|AB| \leq a$ (a fixed);
   (ii) $|EF| = l$ (l fixed);
   (iii) the sum of squares of the nonparallel sides of the trapezoid is constant.

   Remark. The constants are chosen so that such trapezoids exist.

9. **(ROU 3)** Let ABC be an arbitrary triangle and M a point inside it. Let d_a, d_b, d_c be the distances from M to sides BC, CA, AB; a, b, c the lengths of the sides respectively, and S the area of the triangle ABC. Prove the inequality

$$abd_ad_b + bcd_bd_c + cad_cd_a \leq \frac{4S^2}{3}.$$

Prove that the left-hand side attains its maximum when M is the centroid of the triangle.

10. **(ROU 4)** Consider two segments of length a, b $(a > b)$ and a segment of length $c = \sqrt{ab}$.
   (a) For what values of a/b can these segments be sides of a triangle?
   (b) For what values of a/b is this triangle right-angled, obtuse-angled, or acute-angled?

11. **(ROU 6)** Find all solutions $(x_1, x_2, \ldots, x_n)$ of the equation

$$1 + \frac{1}{x_1} + \frac{x_1 + 1}{x_1 x_2} + \frac{(x_1 + 1)(x_2 + 1)}{x_1 x_2 x_3} + \cdots + \frac{(x_1 + 1) \cdots (x_{n-1} + 1)}{x_1 x_2 \cdots x_n} = 0.$$

12. **(POL 1)** If a and b are arbitrary positive real numbers and m an integer, prove that

$$\left(1 + \frac{a}{b}\right)^m + \left(1 + \frac{b}{a}\right)^m \geq 2^{m+1}.$$

13. **(POL 5)** Given two congruent triangles $A_1 A_2 A_3$ and $B_1 B_2 B_3$ $(A_i A_k = B_i B_k)$, prove that there exists a plane such that the orthogonal projections of these triangles onto it are congruent and equally oriented.

14. **(BGR 5)** A line in the plane of a triangle ABC intersects the sides AB and AC respectively at points X and Y such that $BX = CY$. Find the locus of the center of the circumcircle of triangle XAY.

15. **(UNK 1)**[IMO6] Let $[x]$ denote the integer part of x, i.e., the greatest integer not exceeding x. If n is a positive integer, express as a simple function of n the sum

$$\left[\frac{n+1}{2}\right] + \left[\frac{n+2}{4}\right] + \cdots + \left[\frac{n+2^i}{2^{i+1}}\right] + \cdots .$$

16. **(UNK 3)** A polynomial $p(x) = a_0 x^k + a_1 x^{k-1} + \cdots + a_k$ with integer coefficients is said to be divisible by an integer m if $p(x)$ is divisible by m for all integers x. Prove that if $p(x)$ is divisible by m, then $k! a_0$ is also divisible by m. Also prove that if a_0, k, m are nonnegative integers for which $k! a_0$ is divisible by m, there exists a polynomial $p(x) = a_0 x^k + \cdots + a_k$ divisible by m.

17. **(UNK 4)** Given a point O and lengths x, y, z, prove that there exists an equilateral triangle ABC for which $OA = x$, $OB = y$, $OC = z$, if and only if $x + y \geq z, y + z \geq x$, $z + x \geq y$ (the points O, A, B, C are coplanar).

18. **(ITA 2)** If an acute-angled triangle ABC is given, construct an equilateral triangle $A'B'C'$ in space such that lines AA', BB', CC' pass through a given point.

19. **(ITA 5)** We are given a fixed point on the circle of radius 1, and going from this point along the circumference in the positive direction on curved distances

$0, 1, 2, \ldots$ from it we obtain points with abscissas $n = 0, 1, 2, \ldots$ respectively. How many points among them should we take to ensure that some two of them are less than the distance $1/5$ apart?

20. **(CZS 1)** Given n ($n \geq 3$) points in space such that every three of them form a triangle with one angle greater than or equal to $120°$, prove that these points can be denoted by $A_1, A_2, \ldots, A_n$ in such a way that for each i, j, k, $1 \leq i < j < k \leq n$, angle $A_i A_j A_k$ is greater than or equal to $120°$.

21. **(CZS 2)** Let $a_0, a_1, \ldots, a_k$ ($k \geq 1$) be positive integers. Find all positive integers y such that

$$a_0 \mid y; \quad (a_0 + a_1) \mid (y + a_1); \quad \ldots; \quad (a_0 + a_n) \mid (y + a_n).$$

22. **(CZS 3)**[IMO2] Find all positive integers x for which $p(x) = x^2 - 10x - 22$, where $p(x)$ denotes the product of the digits of x.

23. **(CZS 4)** Find all complex numbers m such that polynomial

$$x^3 + y^3 + z^3 + mxyz$$

can be represented as the product of three linear trinomials.

24. **(MNG 1)** Find the number of all n-digit numbers for which some fixed digit stands only in the ith ($1 < i < n$) place and the last j digits are distinct.[3]

25. **(MNG 2)** Given k parallel lines and a few points on each of them, find the number of all possible triangles with vertices at these given points.[4]

26. **(GDR)**[IMO5] Let $a > 0$ be a real number and $f(x)$ a real function defined on all of $\mathbb{R}$, satisfying for all $x \in \mathbb{R}$,

$$f(x + a) = \frac{1}{2} + \sqrt{f(x) - f(x)^2}.$$

(a) Prove that the function f is periodic; i.e., there exists $b > 0$ such that for all x, $f(x + b) = f(x)$.
(b) Give an example of such a nonconstant function for $a = 1$.

---

[3] The problem is unclear. Presumably n, i, j, and the ith digit are fixed.
[4] The problem is unclear. The correct formulation could be the following:
  Given k parallel lines $l_1, \ldots, l_k$ and n_i points on the line l_i, $i = 1, 2, \ldots, k$, find the maximum possible number of triangles with vertices at these points.

3.11 The Eleventh IMO
Bucharest, Romania, July 5–20, 1969

3.11.1 Contest Problems

First Day (July 10)

1. Prove that there exist infinitely many natural numbers a with the following property: the number $z = n^4 + a$ is not prime for any natural number n.

2. Let $a_1, a_2, \ldots, a_n$ be real constants and

$$y(x) = \cos(a_1 + x) + \frac{\cos(a_2 + x)}{2} + \frac{\cos(a_3 + x)}{2^2} + \cdots + \frac{\cos(a_n + x)}{2^{n-1}}.$$

   If x_1, x_2 are real and $y(x_1) = y(x_2) = 0$, prove that $x_1 - x_2 = m\pi$ for some integer m.

3. Find conditions on the positive real number a such that there exists a tetrahedron k of whose edges ($k = 1, 2, 3, 4, 5$) have length a, and the other $6 - k$ edges have length 1.

Second Day (July 11)

4. Let AB be a diameter of a circle γ. A point C different from A and B is on the circle γ. Let D be the projection of the point C onto the line AB. Consider three other circles γ_1, γ_2, and γ_3 with the common tangent AB: γ_1 inscribed in the triangle ABC, and γ_2 and γ_3 tangent to both (the segment) CD and γ. Prove that γ_1, γ_2, and γ_3 have two common tangents.

5. Given n points in the plane such that no three of them are collinear, prove that one can find at least $\binom{n-3}{2}$ convex quadrilaterals with their vertices at these points.

6. Under the conditions $x_1, x_2 > 0$, $x_1 y_1 > z_1^2$, and $x_2 y_2 > z_2^2$, prove the inequality

$$\frac{8}{(x_1 + x_2)(y_1 + y_2) - (z_1 + z_2)^2} \leq \frac{1}{x_1 y_1 - z_1^2} + \frac{1}{x_2 y_2 - z_2^2}.$$

3.11.2 Longlisted Problems

1. **(BEL 1)** A parabola P_1 with equation $x^2 - 2py = 0$ and parabola P_2 with equation $x^2 + 2py = 0$, $p > 0$, are given. A line t is tangent to P_2. Find the locus of pole M of the line t with respect to P_1.

2. **(BEL 2)** (a) Find the equations of regular hyperbolas passing through the points $A(\alpha, 0)$, $B(\beta, 0)$, and $C(0, \gamma)$.
   (b) Prove that all such hyperbolas pass through the orthocenter H of the triangle ABC.
   (c) Find the locus of the centers of these hyperbolas.

(d) Check whether this locus coincides with the nine-point circle of the triangle ABC.

3. **(BEL 3)** Construct the circle that is tangent to three given circles.

4. **(BEL 4)** Let O be a point on a nondegenerate conic. A right angle with vertex O intersects the conic at points A and B. Prove that the line AB passes through a fixed point located on the normal to the conic through the point O.

5. **(BEL 5)** Let G be the centroid of the triangle OAB.
    (a) Prove that all conics passing through the points O, A, B, G are hyperbolas.
    (b) Find the locus of the centers of these hyperbolas.

6. **(BEL 6)** Evaluate $(\cos(\pi/4) + i\sin(\pi/4))^{10}$ in two different ways and prove that

$$\binom{10}{1} - \binom{10}{3} + \frac{1}{2}\binom{10}{5} = 2^4.$$

7. **(BGR 1)** Prove that the equation $\sqrt{x^3 + y^3 + z^3} = 1969$ has no integral solutions.

8. **(BGR 2)** Find all functions f defined for all x that satisfy the condition

$$xf(y) + yf(x) = (x+y)f(x)f(y),$$

for all x and y. Prove that exactly two of them are continuous.

9. **(BGR 3)** One hundred convex polygons are placed on a square with edge of length 38 cm. The area of each of the polygons is smaller than π cm^2, and the perimeter of each of the polygons is smaller than 2π cm. Prove that there exists a disk with radius 1 in the square that does not intersect any of the polygons.

10. **(BGR 4)** Let M be the point inside the right-angled triangle ABC ($\angle C = 90°$) such that
$$\angle MAB = \angle MBC = \angle MCA = \varphi.$$
Let ψ be the acute angle between the medians of AC and BC. Prove that $\frac{\sin(\varphi + \psi)}{\sin(\varphi - \psi)} = 5$.

11. **(BGR 5)** Let Z be a set of points in the plane. Suppose that there exists a pair of points that cannot be joined by a polygonal line not passing through any point of Z. Let us call such a pair of points *unjoinable*. Prove that for each real $r > 0$ there exists an unjoinable pair of points in plane separated by distance r.

12. **(CZS 1)** Given a unit cube, find the locus of the centroids of all tetrahedra whose vertices lie on the sides of the cube.

13. **(CZS 2)** Let p be a prime odd number. Is it possible to find $p - 1$ natural numbers $n + 1, n + 2, \ldots, n + p - 1$ such that the sum of the squares of these numbers is divisible by the sum of these numbers?

14. **(CZS 3)** Let a and b be two positive real numbers. If x is a real solution of the equation $x^2 + px + q = 0$ with real coefficients p and q such that $|p| \leq a, |q| \leq b$, prove that

$$|x| \le \frac{1}{2}\left(a + \sqrt{a^2 + 4b}\right). \tag{1}$$

Conversely, if x satisfies (1), prove that there exist real numbers p and q with $|p| \le a$, $|q| \le b$ such that x is one of the roots of the equation $x^2 + px + q = 0$.

15. **(CZS 4)** Let $K_1, \ldots, K_n$ be nonnegative integers. Prove that

$$K_1! K_2! \cdots K_n! \ge [K/n]!^n,$$

where $K = K_1 + \cdots + K_n$.

16. **(CZS 5)** A convex quadrilateral $ABCD$ with sides $AB = a$, $BC = b$, $CD = c$, $DA = d$ and angles $\alpha = \angle DAB$, $\beta = \angle ABC$, $\gamma = \angle BCD$, and $\delta = \angle CDA$ is given. Let $s = (a + b + c + d)/2$ and P be the area of the quadrilateral. Prove that

$$P^2 = (s - a)(s - b)(s - c)(s - d) - abcd \cos^2 \frac{\alpha + \gamma}{2}.$$

17. **(CZS 6)** Let d and p be two real numbers. Find the first term of an arithmetic progression $a_1, a_2, a_3, \ldots$ with difference d such that $a_1 a_2 a_3 a_4 = p$. Find the number of solutions in terms of d and p.

18. **(FRA 1)** Let a and b be two nonnegative integers. Denote by $H(a, b)$ the set of numbers n of the form $n = pa + qb$, where p and q are positive integers. Determine $H(a) = H(a, a)$. Prove that if $a \ne b$, it is enough to know all the sets $H(a, b)$ for coprime numbers a, b in order to know all the sets $H(a, b)$. Prove that in the case of coprime numbers a and b, $H(a, b)$ contains all numbers greater than or equal to $\omega = (a - 1)(b - 1)$ and also $\omega/2$ numbers smaller than ω.

19. **(FRA 2)** Let n be an integer that is not divisible by any square greater than 1. Denote by x_m the last digit of the number x^n in the number system with base n. For which integers x is it possible for x_m to be 0? Prove that the sequence x_m is periodic with period t independent of x. For which x do we have $x_t = 1$. Prove that if m and x are relatively prime, then $0_m, 1_m, \ldots, (n - 1)_m$ are different numbers. Find the minimal period t in terms of n. If n does not meet the given condition, prove that it is possible to have $x_m = 0 \ne x_1$ and that the sequence is periodic starting only from some number $k > 1$.

20. **(FRA 3)** A polygon (not necessarily convex) with vertices in the lattice points of a rectangular grid is given. The area of the polygon is S. If I is the number of lattice points that are strictly in the interior of the polygon and B the number of lattice points on the border of the polygon, find the number $T = 2S - B - 2I + 2$.

21. **(FRA 4)** A right-angled triangle OAB has its right angle at the point B. An arbitrary circle with center on the line OB is tangent to the line OA. Let AT be the tangent to the circle different from OA (T is the point of tangency). Prove that the median from B of the triangle OAB intersects AT at a point M such that $MB = MT$.

22. **(FRA 5)** Let $\alpha(n)$ be the number of pairs (x, y) of integers such that $x + y = n$, $0 \le y \le x$, and let $\beta(n)$ be the number of triples (x, y, z) such that $x + y + z = n$

and $0 \le z \le y \le x$. Find a simple relation between $\alpha(n)$ and the integer part of the number $\frac{n+2}{2}$ and the relation among $\beta(n)$, $\beta(n-3)$ and $\alpha(n)$. Then evaluate $\beta(n)$ as a function of the residue of n modulo 6. What can be said about $\beta(n)$ and $1 + \frac{n(n+6)}{12}$? And what about $\frac{(n+3)^2}{6}$?

Find the number of triples (x,y,z) with the property $x+y+z \le n$, $0 \le z \le y \le x$ as a function of the residue of n modulo 6. What can be said about the relation between this number and the number $\frac{(n+6)(2n^2+9n+12)}{72}$?

23. **(FRA 6)** Consider the integer $d = \frac{a^b-1}{c}$, where a, b, and c are positive integers and $c \le a$. Prove that the set G of integers that are between 1 and d and relatively prime to d (the number of such integers is denoted by $\varphi(d)$) can be partitioned into n subsets, each of which consists of b elements. What can be said about the rational number $\frac{\varphi(d)}{b}$?

24. **(UNK 1)** The polynomial $P(x) = a_0x^k + a_1x^{k-1} + \cdots + a_k$, where $a_0, \ldots, a_k$ are integers, is said to be divisible by an integer m if $P(x)$ is a multiple of m for every integral value of x. Show that if $P(x)$ is divisible by m, then $a_0 \cdot k!$ is a multiple of m. Also prove that if a,k,m are positive integers such that $ak!$ is a multiple of m, then a polynomial $P(x)$ with leading term ax^k can be found that is divisible by m.

25. **(UNK 2)** Let a,b,x,y be positive integers such that a and b have no common divisor greater than 1. Prove that the largest number not expressible in the form $ax+by$ is $ab-a-b$. If $N(k)$ is the largest number not expressible in the form $ax+by$ in only k ways, find $N(k)$.

26. **(UNK 3)** A smooth solid consists of a right circular cylinder of height h and base-radius r, surmounted by a hemisphere of radius r and center O. The solid stands on a horizontal table. One end of a string is attached to a point on the base. The string is stretched (initially being kept in the vertical plane) over the highest point of the solid and held down at the point P on the hemisphere such that OP makes an angle α with the horizontal. Show that if α is small enough, the string will slacken if slightly displaced and no longer remain in a vertical plane. If then pulled tight through P, show that it will cross the common circular section of the hemisphere and cylinder at a point Q such that $\angle SOQ = \phi$, S being where it initially crossed this section, and $\sin \phi = \frac{r \tan \alpha}{h}$.

27. **(UNK 4)** The segment AB perpendicularly bisects CD at X. Show that, subject to restrictions, there is a right circular cone whose axis passes through X and on whose surface lie the points A,B,C,D. What are the restrictions?

28. **(UNK 5)** Let us define $u_0 = 0$, $u_1 = 1$ and for $n \ge 0$, $u_{n+2} = au_{n+1} + bu_n$, a and b being positive integers. Express u_n as a polynomial in a and b. Prove the result. Given that b is prime, prove that b divides $a(u_b - 1)$.

29. **(GDR 1)** Find all real numbers λ such that the equation

$$\sin^4 x - \cos^4 x = \lambda(\tan^4 x - \cot^4 x)$$

(a) has no solution,
(b) has exactly one solution,
(c) has exactly two solutions,
(d) has more than two solutions (in the interval $(0, \pi/4)$).

30. **(GDR 2)**[IMO1] Prove that there exist infinitely many natural numbers a with the following property: The number $z = n^4 + a$ is not prime for any natural number n.

31. **(GDR 3)** Find the number of permutations $a_1, \ldots, a_n$ of the set $\{1, 2, \ldots, n\}$ such that $|a_i - a_{i+1}| \neq 1$ for all $i = 1, 2, \ldots, n-1$. Find a recurrence formula and evaluate the number of such permutations for $n \leq 6$.

32. **(GDR 4)** Find the maximal number of regions into which a sphere can be partitioned by n circles.

33. **(GDR 5)** Given a ring G in the plane bounded by two concentric circles with radii R and $R/2$, prove that we can cover this region with 8 disks of radius $2R/5$. (A region is covered if each of its points is inside or on the border of some disk.)

34. **(HUN 1)** Let a and b be arbitrary integers. Prove that if k is an integer not divisible by 3, then $(a+b)^{2k} + a^{2k} + b^{2k}$ is divisible by $a^2 + ab + b^2$.

35. **(HUN 2)** Prove that
$$1 + \frac{1}{2^3} + \frac{1}{3^3} + \cdots + \frac{1}{n^3} < \frac{5}{4}.$$

36. **(HUN 3)** In the plane 4000 points are given such that each line passes through at most 2 of these points. Prove that there exist 1000 disjoint quadrilaterals in the plane with vertices at these points.

37. **(HUN 4)**[IMO2] If $a_1, a_2, \ldots, a_n$ are real constants, and if
$$y = \cos(a_1 + x) + 2\cos(a_2 + x) + \cdots + n\cos(a_n + x)$$
has two zeros x_1 and x_2 whose difference is not a multiple of π, prove that $y \equiv 0$.

38. **(HUN 5)** Let r and m ($r \leq m$) be natural numbers and $A_k = \frac{2k-1}{2m}\pi$. Evaluate
$$\frac{1}{m^2} \sum_{k=1}^{m} \sum_{l=1}^{m} \sin(rA_k)\sin(rA_l)\cos(rA_k - rA_l).$$

39. **(HUN 6)** Find the positions of three points A, B, C on the boundary of a unit cube such that $\min\{AB, AC, BC\}$ is the greatest possible.

40. **(MNG 1)** Find the number of five-digit numbers with the following properties: there are two pairs of digits such that digits from each pair are equal and are next to each other, digits from different pairs are different, and the remaining digit (which does not belong to any of the pairs) is different from the other digits.

41. **(MNG 2)** Given four real numbers x_0, x_1, α, β, find an expression for the solution of the system
$$x_{n+2} - \alpha x_{n+1} - \beta x_n = 0, \quad n = 0, 1, 2, \ldots .$$

42. **(MNG 3)** Let A_k $(1 \leq k \leq h)$ be n-element sets such that each two of them have a nonempty intersection. Let A be the union of all the sets A_k, and let B be a subset of A such that for each k $(1 \leq k \leq h)$ the intersection of A_k and B consists of exactly two different elements a_k and b_k. Find all subsets X of the set A with r elements satisfying the condition that for at least one index k, both elements a_k and b_k belong to X.

43. **(MNG 4)** Let p and q be two prime numbers greater than 3. Prove that if their difference is 2^n, then for any two integers m and n, the number $S = p^{2m+1} + q^{2m+1}$ is divisible by 3.

44. **(MNG 5)** Find the radius of the circle circumscribed about the isosceles triangle whose sides are the solutions of the equation $x^2 - ax + b = 0$.

45. **(MNG 6)**[IMO5] Given n points in the plane such that no three of them are collinear, prove that one can find at least $\binom{n-3}{2}$ convex quadrilaterals with their vertices at these points.

46. **(NLD 1)** The vertices of an $(n+1)$-gon are placed on the edges of a regular n-gon so that the perimeter of the n-gon is divided into equal parts. How does one choose these $n+1$ points in order to obtain the $(n+1)$gon with
   (a) maximal area;
   (b) minimal area?

47. **(NLD 2)**[IMO4] Let A and B be points on the circle γ. A point C, different from A and B, is on the circle γ. Let D be the projection of the point C onto the line AB. Consider three other circles γ_1, γ_2, and γ_3 with the common tangent AB: γ_1 inscribed in the triangle ABC, and γ_2 and γ_3 tangent to both (the segment) CD and γ. Prove that γ_1, γ_2, and γ_3 have two common tangents.

48. **(NLD 3)** Let x_1, x_2, x_3, x_4, and x_5 be positive integers satisfying

$$
\begin{aligned}
x_1 + x_2 + x_3 + x_4 + x_5 &= 1000, \\
x_1 - x_2 + x_3 - x_4 + x_5 &> 0, \\
x_1 + x_2 - x_3 + x_4 - x_5 &> 0, \\
-x_1 + x_2 + x_3 - x_4 + x_5 &> 0, \\
x_1 - x_2 + x_3 + x_4 - x_5 &> 0, \\
-x_1 + x_2 - x_3 + x_4 + x_5 &> 0.
\end{aligned}
$$

   (a) Find the maximum of $(x_1 + x_3)^{x_2 + x_4}$.
   (b) In how many different ways can we choose $x_1, \ldots, x_5$ to obtain the desired maximum?

49. **(NLD 4)** A boy has a set of trains and pieces of railroad track. Each piece is a quarter of circle, and by concatenating these pieces, the boy obtained a closed railway. The railway does not intersect itself. In passing through this railway, the train sometimes goes in the clockwise direction, and sometimes in the opposite direction. Prove that the train passes an even number of times through the pieces in the clockwise direction and an even number of times in the counterclockwise direction. Also, prove that the number of pieces is divisible by 4.

50. **(NLD 5)** The bisectors of the exterior angles of a pentagon $B_1B_2B_3B_4B_5$ form another pentagon $A_1A_2A_3A_4A_5$. Construct $B_1B_2B_3B_4B_5$ from the given pentagon $A_1A_2A_3A_4A_5$.

51. **(NLD 6)** A curve determined by

$$y = \sqrt{x^2 - 10x + 52}, \quad 0 \le x \le 100,$$

is constructed in a rectangular grid. Determine the number of squares cut by the curve.

52. **(POL 1)** Prove that a regular polygon with an odd number of edges cannot be partitioned into four pieces with equal areas by two lines that pass through the center of polygon.

53. **(POL 2)** Given two segments AB and CD not in the same plane, find the locus of points M such that

$$MA^2 + MB^2 = MC^2 + MD^2.$$

54. **(POL 3)** Given a polynomial $f(x)$ with integer coefficients whose value is divisible by 3 for three integers k, $k+1$, and $k+2$, prove that $f(m)$ is divisible by 3 for all integers m.

55. **(POL 4)**[IMO3] Find the conditions on the positive real number a such that there exists a tetrahedron k of whose edges $(k = 1,2,3,4,5)$ have length a, and the other $6 - k$ edges have length 1.

56. **(POL 5)** Let a and b be two natural numbers that have an equal number n of digits in their decimal expansions. The first m digits (from left to right) of the numbers a and b are equal. Prove that if $m > n/2$, then

$$a^{1/n} - b^{1/n} < \frac{1}{n}.$$

57. **(POL 6)** On the sides AB and AC of triangle ABC two points K and L are given such that $\frac{KB}{AK} + \frac{LC}{AL} = 1$. Prove that KL passes through the centroid of ABC.

58. **(SWE 1)** Six points $P_1,\ldots,P_6$ are given in 3-dimensional space such that no four of them lie in the same plane. Each of the line segments P_jP_k is colored black or white. Prove that there exists one triangle $P_jP_kP_l$ whose edges are of the same color.

59. **(SWE 2)** For each λ $(0 < \lambda < 1$ and $\lambda \ne 1/n$ for all $n = 1,2,3,\ldots)$ construct a continuous function f such that there do not exist x,y with $0 < \lambda < y = x + \lambda \le 1$ for which $f(x) = f(y)$.

60. **(SWE 3)** Find the natural number n with the following properties:
    (i) Let $S = \{p_1, p_2, \ldots\}$ be an arbitrary finite set of points in the plane, and r_j the distance from P_j to the origin O. We assign to each P_j the closed disk D_j with center P_j and radius r_j. Then some n of these disks contain all points of S.

(ii) n is the smallest integer with the above property.

61. **(SWE 4)** Let a_0, a_1, a_2 be determined with $a_0 = 0$, $a_{n+1} = 2a_n + 2^n$. Prove that if n is power of 2, then so is a_n.

62. **(SWE 5)** Which natural numbers can be expressed as the difference of squares of two integers?

63. **(SWE 6)** Prove that there are infinitely many positive integers that cannot be expressed as the sum of squares of three positive integers.

64. **(USS 1)** Prove that for a natural number $n > 2$,

$$(n!)! > n[(n-1)!]^{n!}.$$

65. **(USS 2)** Prove that for $a > b^2$,

$$\sqrt{a - b\sqrt{a + b\sqrt{a - b\sqrt{a + \cdots}}}} = \sqrt{a - \frac{3}{4}b^2} - \frac{1}{2}b.$$

66. **(USS 3)** (a) Prove that if $0 \le a_0 \le a_1 \le a_2$, then

$$(a_0 + a_1 x - a_2 x^2)^2 \le (a_0 + a_1 + a_2)^2 \left(1 + \frac{1}{2}x + \frac{1}{3}x^2 + \frac{1}{2}x^3 + x^4\right).$$

(b) Formulate and prove the analogous result for polynomials of third degree.

67. **(USS 4)**[IMO6] Under the conditions $x_1, x_2 > 0$, $x_1 y_1 > z_1^2$, and $x_2 y_2 > z_2^2$, prove the inequality

$$\frac{8}{(x_1 + x_2)(y_1 + y_2) - (z_1 + z_2)^2} \le \frac{1}{x_1 y_1 - z_1^2} + \frac{1}{x_2 y_2 - z_2^2}.$$

68. **(USS 5)** Given 5 points in the plane, no three of which are collinear, prove that we can choose 4 points among them that form a convex quadrilateral.

69. **(YUG 1)** Suppose that positive real numbers x_1, x_2, x_3 satisfy

$$x_1 x_2 x_3 > 1, \qquad x_1 + x_2 + x_3 < \frac{1}{x_1} + \frac{1}{x_2} + \frac{1}{x_3}.$$

Prove that:
(a) None of x_1, x_2, x_3 equals 1.
(b) Exactly one of these numbers is less than 1.

70. **(YUG 2)** A park has the shape of a convex pentagon of area $5\sqrt{3}$ ha ($= 50000\sqrt{3}$ m^2). A man standing at an interior point O of the park notices that he stands at a distance of at most 200 m from each vertex of the pentagon. Prove that he stands at a distance of at least 100 m from each side of the pentagon.

71. **(YUG 3)** Let four points A_i ($i = 1, 2, 3, 4$) in the plane determine four triangles. In each of these triangles we choose the smallest angle. The sum of these angles is denoted by S. What is the exact placement of the points A_i if $S = 180°$?

3.12 The Twelfth IMO
Budapest–Keszthely, Hungary, July 8–22, 1970

3.12.1 Contest Problems

First Day (July 13)

1. Given a point M on the side AB of the triangle ABC, let r_1 and r_2 be the radii of the inscribed circles of the triangles ACM and BCM respectively while ρ_1 and ρ_2 are the radii of the excircles of the triangles ACM and BCM at the sides AM and BM respectively. Let r and ρ denote the respective radii of the inscribed circle and the excircle at the side AB of the triangle ABC. Prove that

$$\frac{r_1}{\rho_1}\frac{r_2}{\rho_2} = \frac{r}{\rho}.$$

2. Let a and b be the bases of two number systems and let

$$A_n = \overline{x_1 x_2 \ldots x_n}^{(a)}, \qquad A_{n+1} = \overline{x_0 x_1 x_2 \ldots x_n}^{(a)},$$
$$B_n = \overline{x_1 x_2 \ldots x_n}^{(b)}, \qquad B_{n+1} = \overline{x_0 x_1 x_2 \ldots x_n}^{(b)},$$

be numbers in the number systems with respective bases a and b, so that $x_0, x_1, x_2, \ldots, x_n$ denote digits in the number system with base a as well as in the number system with base b. Suppose that neither x_0 nor x_1 is zero. Prove that $a > b$ if and only if

$$\frac{A_n}{A_{n+1}} < \frac{B_n}{B_{n+1}}.$$

3. Let $1 = a_0 \le a_1 \le a_2 \le \cdots \le a_n \le \cdots$ be a sequence of real numbers. Consider the sequence $b_1, b_2, \ldots$ defined by

$$b_n = \sum_{k=1}^{n}\left(1 - \frac{a_{k-1}}{a_k}\right)\frac{1}{\sqrt{a_k}}.$$

Prove that:
   (a) For all natural numbers n, $0 \le b_n < 2$.
   (b) Given an arbitrary $0 \le b < 2$, there is a sequence $a_0, a_1, \ldots, a_n, \ldots$ of the above type such that $b_n > b$ is true for an infinity of natural numbers n.

Second Day (July 14)

4. For what natural numbers n can the product of some of the numbers $n, n+1, n+2, n+3, n+4, n+5$ be equal to the product of the remaining ones?

5. In the tetrahedron $ABCD$, the edges BD and CD are mutually perpendicular, and the projection of the vertex D to the plane ABC is the intersection of the altitudes of the triangle ABC. Prove that

$$(AB + BC + CA)^2 \le 6(DA^2 + DB^2 + DC^2).$$

For which tetrahedra does equality hold?

6. Given 100 points in the plane, no three of which are on the same line, consider all triangles that have all their vertices chosen from the 100 given points. Prove that at most 70% of these triangles are acute-angled.

3.12.2 Longlisted Problems

1. **(AUT 1)** Prove that

$$\frac{bc}{b+c} + \frac{ca}{c+a} + \frac{ab}{a+b} \le \frac{1}{2}(a+b+c) \quad (a,b,c>0).$$

2. **(AUT 2)** Prove that the two last digits of 9^{9^9} and $9^{9^{9^9}}$ in decimal representation are equal.

3. **(AUT 3)** Prove that for $a,b \in \mathbb{N}$, $a!b!$ divides $(a+b)!$.

4. **(AUT 4)** Solve the system of equations

$$\begin{aligned} x^2 + xy &= a^2 + ab \\ y^2 + xy &= a^2 - ab, \end{aligned} \qquad a,b \text{ real}, a \ne 0.$$

5. **(AUT 5)** Prove that $\sqrt[n]{\frac{1}{n+1} + \frac{2}{n+1} + \cdots + \frac{n}{n+1}} \ge 1$ for $n \ge 2$.

6. **(BEL 1)** Prove that the equation in x

$$\sum_{i=1}^{n} \frac{b_i}{x - a_i} = c, \qquad b_i > 0, \quad a_1 < a_2 < a_3 < \cdots < a_n,$$

has $n-1$ roots $x_1, x_2, x_3, \ldots, x_{n-1}$ such that $a_1 < x_1 < a_2 < x_2 < a_3 < x_3 < \cdots < x_{n-1} < a_n$.

7. **(BEL 2)** Let $ABCD$ be any quadrilateral. A square is constructed on each side of the quadrilateral, all in the same manner (i.e., outward or inward). Denote the centers of the squares by M_1, M_2, M_3, and M_4. Prove:
   (a) $M_1 M_3 = M_2 M_4$;
   (b) $M_1 M_3$ is perpendicular to $M_2 M_4$.

8. **(BEL 3)** (SL70-1).

9. **(BEL 4)** If n is even, prove that

$$1 - \frac{1}{2} + \frac{1}{3} - \frac{1}{4} + \cdots - \frac{1}{n} = 2\left(\frac{1}{n+2} + \frac{1}{n+4} + \frac{1}{n+6} + \cdots + \frac{1}{2n}\right).$$

10. **(BEL 5)** Let A, B, C be angles of a triangle. Prove that

$$1 < \cos A + \cos B + \cos C \le \frac{3}{2}.$$

11. **(BEL 6)** Let $ABCD$ and $A'B'C'D'$ be two squares in the same plane and oriented in the same direction. Let A'', B'', C'', and D'' be the midpoints of AA', BB', CC', and DD'. Prove that $A''B''C''D''$ is also a square.

12. **(BGR 1)** Let $x_1, x_2, x_3, x_4, x_5, x_6$ be given integers, not divisible by 7. Prove that at least one of the expressions of the form

$$\pm x_1 \pm x_2 \pm x_3 \pm x_4 \pm x_5 \pm x_6$$

is divisible by 7, where the signs are selected in all possible ways. (Generalize the statement to every prime number!)

13. **(BGR 2)** A triangle ABC is given. Each side of ABC is divided into equal parts, and through each of the division points are drawn lines parallel to AB, BC, and CA, thus cutting ABC into small triangles. A number 1, 2, or 3 is assign to each of the vertices of these triangles in such a way that the following conditions are satisfied:
    (i) to A, B, C are assigned 1, 2 and 3 respectively;
    (ii) points on AB are marked by 1 or 2;
    (iii) points on BC are marked by 2 or 3;
    (iv) points on CA are marked by 3 or 1.
    Prove that there must exist a small triangle whose vertices are marked by 1, 2, and 3.

14. **(BGR 3)** Let $\alpha + \beta + \gamma = \pi$. Prove that

$$\sin 2\alpha + \sin 2\beta + \sin 2\gamma = 2(\sin\alpha + \sin\beta + \sin\gamma)(\cos\alpha + \cos\beta + \cos\gamma)$$
$$-2(\sin\alpha + \sin\beta + \sin\gamma).$$

15. **(BGR 4)** Given a triangle ABC, let R be the radius of its circumcircle, O_1, O_2, O_3 the centers of its exscribed circles, and q the perimeter of $\triangle O_1 O_2 O_3$. Prove that $q \leq 6\sqrt{3} R$.

16. **(BGR 5)** Show that the equation

$$\sqrt{2 - x^2} + \sqrt[3]{3 - x^3} = 0$$

has no real roots.

17. **(BGR 6)** (SL70-3).
    Original formulation. In a triangular pyramid $SABC$ one of the angles at S is right and the projection of S onto the base ABC is the orthocenter of ABC. Let r be the radius of the circle inscribed in the base, $SA = m$, $SB = n$, $SC = p$, H the height of the pyramid (through S), and r_1, r_2, r_3 the radii of the circles inscribed in the intersections of the pyramid with the planes determined by the altitude of the pyramid and the lines SA, SB, SC respectively. Prove that:
    (a) $m^2 + n^2 + p^2 \geq 18r^2$;
    (b) the ratios r_1/H, r_2/H, r_3/H lie in the interval $[0.4, 0.5]$.

18. **(CZS 1)** (SL70-4).

19. **(CZS 2)** Let $n > 1$ be a natural number, $a \geq 1$ a real number, and $x_1, x_2, \ldots, x_n$ numbers such that $x_1 = 1$, $\frac{x_{k+1}}{x_k} = a + \alpha_k$ for $k = 1, 2, \ldots, n-1$, where α_k are real numbers with $\alpha_k \leq \frac{1}{k(k+1)}$. Prove that

$$\sqrt[n-1]{x_n} < a + \frac{1}{n-1}.$$

20. **(CZS 3)** (SL70-5).

21. **(CZS 4)** Find necessary and sufficient conditions on given positive numbers u, v for the following claim to be valid: there exists a right-angled triangle $\triangle ABC$ with $CD = u$, $CE = v$, where D, E are points of the segments AB such that $AD = DE = EB = \frac{1}{3}AB$.

22. **(FRA 1)** (SL70-6).

23. **(FRA 2)** Let E be a finite set, $\mathcal{P}_E$ the family of its subsets, and f a mapping from $\mathcal{P}_E$ to the set of nonnegative real numbers such that for any two disjoint subsets A, B of E,

$$f(A \cup B) = f(A) + f(B).$$

Prove that there exists a subset F of E such that if with each $A \subset E$ we associate a subset A' consisting of elements of A that are not in F, then $f(A) = f(A')$, and $f(A)$ is zero if and only if A is a subset of F.

24. **(FRA 3)** Let n and p be two integers such that $2p \leq n$. Prove the inequality

$$\frac{(n-p)!}{p!} \leq \left(\frac{n+1}{2}\right)^{n-2p}.$$

For which values does equality hold?

25. **(FRA 4)** Suppose that f is a real function defined for $0 \leq x \leq 1$ having the first derivative f' for $0 \leq x \leq 1$ and the second derivative f'' for $0 < x < 1$. Prove that if

$$f(0) = f'(0) = f'(1) = f(1) - 1 = 0,$$

there exists a number $0 < y < 1$ such that $|f''(y)| \geq 4$.

26. **(FRA 5)** Consider a finite set of vectors in space $\{a_1, a_2, \ldots, a_n\}$ and the set E of all vectors of the form $x = \lambda_1 a_1 + \lambda_2 a_2 + \cdots + \lambda_n a_n$, where λ_i are nonnegative numbers. Let F be the set consisting of all the vectors in E and vectors parallel to a given plane P. Prove that there exists a set of vectors $\{b_1, b_2, \ldots, b_p\}$ such that F is the set of all vectors y of the form

$$y = \mu_1 b_1 + \mu_2 b_2 + \cdots + \mu_p b_p,$$

where the μ_j are nonnegative.

27. **(FRA 6)** Find a natural number n such that for all prime numbers p, n is divisible by p if and only if n is divisible by $p - 1$.

28. **(GDR 1)** A set G with elements $u, v, w, \ldots$ is a group if the following conditions are fulfilled:
    (i) There is a binary algebraic operation $\circ$ defined on G such that for all $u, v \in G$ there is a $w \in G$ with $u \circ v = w$.
    (ii) This operation is associative; i.e., for all $u, v, w \in G$, $(u \circ v) \circ w = u \circ (v \circ w)$.
    (iii) For any two elements $u, v \in G$ there exists an element $x \in G$ such that $u \circ x = v$, and an element $y \in G$ such that $y \circ u = v$.
    Let K be a set of all real numbers greater than 1. On K is defined an operation by
    $$a \circ b = ab - \sqrt{(a^2 - 1)(b^2 - 1)}.$$
    Prove that K is a group.

29. **(GDR 2)** Prove that the equation $4^x + 6^x = 9^x$ has no rational solutions.

30. **(GDR 3)** (SL70-9).

31. **(GDR 4)** Prove that for any triangle with sides a, b, c and area P the following inequality holds:
    $$P \leq \frac{\sqrt{3}}{4}(abc)^{2/3}.$$
    Find all triangles for which equality holds.

32. **(NLD 1)** Let there be given an acute angle $\angle AOB = 3\alpha$, where $\overline{OA} = \overline{OB}$. The point A is the center of a circle with radius $\overline{OA}$. A line s parallel to OA passes through B. Inside the given angle a variable line t is drawn through O. It meets the circle in O and C and the given line s in D, where $\angle AOC = x$. Starting from an arbitrarily chosen position t_0 of t, the series $t_0, t_1, t_2, \ldots$ is determined by defining $\overline{BD_{i+1}} = \overline{OC_i}$ for each i (in which C_i and D_i denote the positions of C and D, corresponding to t_i). Making use of the graphical representations of BD and OC as functions of x, determine the behavior of t_i for $i \to \infty$.

33. **(NLD 2)** The vertices of a given square are clockwise lettered A, B, C, D. On the side AB is situated a point E such that $AE = AB/3$.
    Starting from an arbitrarily chosen point P_0 on segment AE and going clockwise around the perimeter of the square, a series of points $P_0, P_1, P_2, \ldots$ is marked on the perimeter such that $P_i P_{i+1} = AB/3$ for each i. It will be clear that when P_0 is chosen in A or in E, then some P_i will coincide with P_0. Does this possibly also happen if P_0 is chosen otherwise?

34. **(NLD 3)** In connection with a convex pentagon $ABCDE$ we consider the set of ten circles, each of which contains three of the vertices of the pentagon on its circumference. Is it possible that none of these circles contains the pentagon? Prove your answer.

35. **(NLD 4)** Find for every value of n a set of numbers p for which the following statement is true: Any convex n-gon can be divided into p isosceles triangles.

Alternative version. The same about division into p polygons with axis of symmetry.

36. **(NLD 5)** Let x, y, z be nonnegative real numbers satisfying

$$x^2 + y^2 + z^2 = 5 \quad \text{and} \quad yz + zx + xy = 2.$$

Which values can the greatest of the numbers $x^2 - yz$, $y^2 - xz$, $z^2 - xy$ have?

37. **(NLD 6)** Solve the set of simultaneous equations

$$\begin{aligned}
v^2 + w^2 + x^2 + y^2 &= 6 - 2u, \\
u^2 + \quad w^2 + x^2 + y^2 &= 6 - 2v, \\
u^2 + v^2 + \quad x^2 + y^2 &= 6 - 2w, \\
u^2 + v^2 + w^2 + \quad y^2 &= 6 - 2x, \\
u^2 + v^2 + w^2 + x^2 \quad &= 6 - 2y.
\end{aligned}$$

38. **(POL 1)** Find the greatest integer A for which in any permutation of the numbers $1, \ldots, 100$ there exist ten consecutive numbers whose sum is at least A.

39. **(POL 2)** (SL70-8).

40. **(POL 5)** Let ABC be a triangle with angles α, β, γ commensurable with π. Starting from a point P interior to the triangle, a ball reflects on the sides of ABC, respecting the law of reflection that the angle of incidence is equal to the angle of reflection.

Prove that, supposing that the ball never reaches any of the vertices A, B, C, the set of all directions in which the ball will move through time is finite. In other words, its path from the moment 0 to infinity consists of segments parallel to a finite set of lines.

41. **(POL 6)** Let a cube of side 1 be given. Prove that there exists a point A on the surface S of the cube such that every point of S can be joined to A by a path on S of length not exceeding 2. Also prove that there is a point of S that cannot be joined with A by a path on S of length less than 2.

42. **(ROU 1)** (SL70-2).

43. **(ROU 2)** Prove that the equation

$$x^3 - 3\tan\frac{\pi}{12}x^2 - 3x + \tan\frac{\pi}{12} = 0$$

has one root $x_1 = \tan\frac{\pi}{36}$, and find the other roots.

44. **(ROU 3)** If a, b, c are side lengths of a triangle, prove that

$$(a+b)(b+c)(c+a) \geq 8(a+b-c)(b+c-a)(c+a-b).$$

45. **(ROU 4)** Let M be an interior point of tetrahedron $VABC$. Denote by A_1, B_1, C_1 the points of intersection of lines MA, MB, MC with the planes VBC, VCA, VAB, and by A_2, B_2, C_2 the points of intersection of lines VA_1, VB_1, VC_1 with the sides BC, CA, AB.
    (a) Prove that the volume of the tetrahedron $VA_2B_2C_2$ does not exceed one-fourth of the volume of $VABC$.
    (b) Calculate the volume of the tetrahedron $V_1A_1B_1C_1$ as a function of the volume of $VABC$, where V_1 is the point of intersection of the line VM with the plane ABC, and M is the barycenter of $VABC$.

46. **(ROU 5)** Given a triangle ABC and a plane π having no common points with the triangle, find a point M such that the triangle determined by the points of intersection of the lines MA, MB, MC with π is congruent to the triangle ABC.

47. **(ROU 6)** Given a polynomial
$$P(x) = ab(a-c)x^3 + (a^3 - a^2c + 2ab^2 - b^2c + abc)x^2$$
$$+ (2a^2b + b^2c + a^2c + b^3 - abc)x + ab(b+c),$$
where $a, b, c \neq 0$, prove that $P(x)$ is divisible by
$$Q(x) = abx^2 + (a^2 + b^2)x + ab$$
and conclude that $P(x_0)$ is divisible by $(a+b)^3$ for $x_0 = (a+b+1)^n$, $n \in \mathbb{N}$.

48. **(ROU 7)** Let a polynomial $p(x)$ with integer coefficients take the value 5 for five different integer values of x. Prove that $p(x)$ does not take the value 8 for any integer x.

49. **(SWE 1)** For $n \in \mathbb{N}$, let $f(n)$ be the number of positive integers $k \leq n$ that do not contain the digit 9. Does there exist a positive real number p such that $\frac{f(n)}{n} \geq p$ for all positive integers n?

50. **(SWE 2)** The area of a triangle is S and the sum of the lengths of its sides is L. Prove that $36S \leq L^2\sqrt{3}$ and give a necessary and sufficient condition for equality.

51. **(SWE 3)** Let p be a prime number. A rational number x, with $0 < x < 1$, is written in lowest terms. The rational number obtained from x by adding p to both the numerator and the denominator differs from x by $1/p^2$. Determine all rational numbers x with this property.

52. **(SWE 4)** (SL70-10).

53. **(SWE 5)** A square $ABCD$ is divided into $(n-1)^2$ congruent squares, with sides parallel to the sides of the given square. Consider the grid of all n^2 corners obtained in this manner. Determine all integers n for which it is possible to construct a nondegenerate parabola with its axis parallel to one side of the square and that passes through exactly n points of the grid.

54. **(SWE 6)** (SL70-11).

55. **(USS 1)** A turtle runs away from an UFO with a speed of 0.2 m/s. The UFO flies 5 meters above the ground, with a speed of 20 m/s. The UFO's path is a broken line, where after flying in a straight path of length ℓ (in meters) it may turn through for any acute angle α such that $\tan\alpha < \frac{\ell}{1000}$. When the UFO's center approaches within 13 meters of the turtle, it catches the turtle. Prove that for any initial position the UFO can catch the turtle.

56. **(USS 2)** A square hole of depth h whose base is of length a is given. A dog is tied to the center of the square at the bottom of the hole by a rope of length $L > \sqrt{2a^2 + h^2}$, and walks on the ground around the hole. The edges of the hole are smooth, so that the rope can freely slide along it. Find the shape and area of the territory accessible to the dog (whose size is neglected).

57. **(USS 3)** Let the numbers $1, 2, \ldots, n^2$ be written in the cells of an $n \times n$ square board so that the entries in each column are arranged increasingly. What are the smallest and greatest possible sums of the numbers in the kth row? (k a positive integer, $1 \le k \le n$.)

58. **(USS 4)** (SL70-12).

59. **(USS 5)** (SL70-7).

3.12.3 Shortlisted Problems

1. **(BEL 3)** Consider a regular $2n$-gon and the n diagonals of it that pass through its center. Let P be a point of the inscribed circle and let $a_1, a_2, \ldots, a_n$ be the angles in which the diagonals mentioned are visible from the point P. Prove that

$$\sum_{i=1}^{n} \tan^2 a_i = 2n \frac{\cos^2 \frac{\pi}{2n}}{\sin^4 \frac{\pi}{2n}}.$$

2. **(ROU 1)**[IMO2] Let a and b be the bases of two number systems and let

$$A_n = \overline{x_1 x_2 \ldots x_n}^{(a)}, \qquad A_{n+1} = \overline{x_0 x_1 x_2 \ldots x_n}^{(a)},$$
$$B_n = \overline{x_1 x_2 \ldots x_n}^{(b)}, \qquad B_{n+1} = \overline{x_0 x_1 x_2 \ldots x_n}^{(b)},$$

be numbers in the number systems with respective bases a and b, so that $x_0, x_1, x_2, \ldots, x_n$ denote digits in the number system with base a as well as in the number system with base b. Suppose that neither x_0 nor x_1 is zero. Prove that $a > b$ if and only if $\frac{A_n}{A_{n+1}} < \frac{B_n}{B_{n+1}}$.

3. **(BGR 6)**[IMO5] In the tetrahedron $SABC$ the angle BSC is a right angle, and the projection of the vertex S to the plane ABC is the intersection of the altitudes of the triangle ABC. Let z be the radius of the inscribed circle of the triangle ABC. Prove that $SA^2 + SB^2 + SC^2 \ge 18z^2$.

4. **(CZS 1)**[IMO4] For what natural numbers n can the product of some of the numbers $n, n+1, n+2, n+3, n+4, n+5$ be equal to the product of the remaining ones?

5. **(CZS 3)** Let M be an interior point of the tetrahedron $ABCD$. Prove that

$$\overrightarrow{MA}\,\text{vol}(MBCD) + \overrightarrow{MB}\,\text{vol}(MACD)$$
$$+ \overrightarrow{MC}\,\text{vol}(MABD) + \overrightarrow{MD}\,\text{vol}(MABC) = 0$$

(vol($PQRS$) denotes the volume of the tetrahedron $PQRS$).

6. **(FRA 1)** In the triangle ABC let B' and C' be the midpoints of the sides AC and AB respectively and H the foot of the altitude passing through the vertex A. Prove that the circumcircles of the triangles $AB'C', BC'H$, and $B'CH$ have a common point I and that the line HI passes through the midpoint of the segment $B'C'$.

7. **(USS 5)** For which digits a do exist integers $n \geq 4$ such that each digit of $\frac{n(n+1)}{2}$ equals a?

8. **(POL 2)**[IMO1] Given a point M on the side AB of the triangle ABC, let r_1 and r_2 be the radii of the inscribed circles of the triangles ACM and BCM respectively and let ρ_1 and ρ_2 be the radii of the excircles of the triangles ACM and BCM at the sides AM and BM respectively. Let r and ρ denote the radii of the inscribed circle and the excircle at the side AB of the triangle ABC respectively. Prove that

$$\frac{r_1}{\rho_1}\frac{r_2}{\rho_2} = \frac{r}{\rho}.$$

9. **(GDR 3)** Let $u_1, u_2, \ldots, u_n, v_1, v_2, \ldots, v_n$ be real numbers. Prove that

$$1 + \sum_{i=1}^{n}(u_i + v_i)^2 \leq \frac{4}{3}\left(1 + \sum_{i=1}^{n}u_i^2\right)\left(1 + \sum_{i=1}^{n}v_i^2\right).$$

In what case does equality hold?

10. **(SWE 4)**[IMO3] Let $1 = a_0 \leq a_1 \leq a_2 \leq \cdots \leq a_n \leq \cdots$ be a sequence of real numbers. Consider the sequence $b_1, b_2, \ldots$ defined by:

$$b_n = \sum_{k=1}^{n}\left(1 - \frac{a_{k-1}}{a_k}\right)\frac{1}{\sqrt{a_k}}.$$

Prove that:
(a) For all natural numbers n, $0 \leq b_n < 2$.
(b) Given an arbitrary $0 \leq b < 2$, there is a sequence $a_0, a_1, \ldots, a_n, \ldots$ of the above type such that $b_n > b$ is true for infinitely many natural numbers n.

11. **(SWE 6)** Let P, Q, R be polynomials and let $S(x) = P(x^3) + xQ(x^3) + x^2R(x^3)$ be a polynomial of degree n whose roots $x_1, \ldots, x_n$ are distinct. Construct with the aid of the polynomials P, Q, R a polynomial T of degree n that has the roots $x_1^3, x_2^3, \ldots, x_n^3$.

12. **(USS 4)**[IMO6] We are given 100 points in the plane, no three of which are on the same line. Consider all triangles that have all vertices chosen from the 100 given points. Prove that at most 70% of these triangles are acute angled.

3.13 The Thirteenth IMO
Bratislava–Zilina, Czechoslovakia, July 10–21, 1971

3.13.1 Contest Problems

First Day (July 13)

1. Prove that the following statement is true for $n = 3$ and for $n = 5$, and false for all other $n > 2$:
   For any real numbers $a_1, a_2, \ldots, a_n$,

$$(a_1 - a_2)(a_1 - a_3) \cdots (a_1 - a_n) + (a_2 - a_1)(a_2 - a_3) \cdots (a_2 - a_n) + \ldots$$
$$+ (a_n - a_1)(a_n - a_2) \cdots (a_n - a_{n-1}) \geq 0.$$

2. Given a convex polyhedron P_1 with 9 vertices $A_1, \ldots, A_9$, let us denote by $P_2, P_3, \ldots, P_9$ the images of P_1 under the translations mapping the vertex A_1 to $A_2, A_3, \ldots, A_9$, respectively. Prove that among the polyhedra $P_1, \ldots, P_9$ at least two have a common interior point.

3. Prove that the sequence $2^n - 3$ $(n > 1)$ contains a subsequence of numbers relatively prime in pairs.

Second Day (July 14)

4. Given a tetrahedron $ABCD$ all of whose faces are acute-angled triangles, set

$$\sigma = \angle DAB + \angle BCD - \angle ABC - \angle CDA.$$

Consider all closed broken lines $XYZTX$ whose vertices X, Y, Z, T lie in the interior of segments AB, BC, CD, DA respectively. Prove that:
   (a) if $\sigma \neq 0$, then there is no broken line $XYZT$ of minimal length;
   (b) if $\sigma = 0$, then there are infinitely many such broken lines of minimal length. That length equals $2AC \sin(\alpha/2)$, where

$$\alpha = \angle BAC + \angle CAD + \angle DAB.$$

5. Prove that for every natural number $m \geq 1$ there exists a finite set S_m of points in the plane satisfying the following condition: If A is any point in S_m, then there are exactly m points in S_m whose distance to A equals 1.

6. Consider the $n \times n$ array of nonnegative integers

$$\begin{pmatrix} a_{11} & a_{12} & \cdots & a_{1n} \\ a_{21} & a_{22} & \cdots & a_{2n} \\ \vdots & \vdots & & \vdots \\ a_{n1} & a_{n2} & \cdots & a_{nn} \end{pmatrix},$$

with the following property: If an element a_{ij} is zero, then the sum of the elements of the ith row and the jth column is greater than or equal to n. Prove that the sum of all the elements is greater than or equal to $\frac{1}{2}n^2$.

3.13.2 Longlisted Problems

1. **(AUT 1)** The points $S(i,j)$ with integer Cartesian coordinates $0 < i \leq n, 0 < j \leq m$, $m \leq n$, form a lattice. Find the number of:
   (a) rectangles with vertices on the lattice and sides parallel to the coordinate axes;
   (b) squares with vertices on the lattice and sides parallel to the coordinate axes;
   (c) squares in total, with vertices on the lattice.

2. **(AUT 2)** Let us denote by $s(n) = \sum_{d|n} d$ the sum of divisors of a natural number n (1 and n included). If n has at most 5 distinct prime divisors, prove that $s(n) < \frac{77}{16}n$. Also prove that there exists a natural number n for which $s(n) > \frac{76}{16}n$ holds.

3. **(AUT 3)** Let a, b, c be positive real numbers, $0 < a \leq b \leq c$. Prove that for any positive real numbers x, y, z the following inequality holds:

$$(ax + by + cz)\left(\frac{x}{a} + \frac{y}{b} + \frac{z}{c}\right) \leq (x + y + z)^2 \frac{(a+c)^2}{4ac}.$$

4. **(BGR 1)** Let $x_n = 2^{2^n} + 1$ and let m be the least common multiple of $x_2, x_3, \ldots, x_{1971}$. Find the last digit of m.

5. **(BGR 2)** (SL71-1).
   Original formulation. Consider a sequence of polynomials $X_0(x), X_1(x), X_2(x), \ldots, X_n(x), \ldots$, where $X_0(x) = 2$, $X_1(x) = x$, and for every $n \geq 1$ the following equality holds:

$$X_n(x) = \frac{1}{x}(X_{n+1}(x) + X_{n-1}(x)).$$

   Prove that $(x^2 - 4)[X_n^2(x) - 4]$ is a square of a polynomial for all $n \geq 0$.

6. **(BGR 3)** Let squares be constructed on the sides BC, CA, AB of a triangle ABC, all to the outside of the triangle, and let A_1, B_1, C_1 be their centers. Starting from the triangle $A_1B_1C_1$ one analogously obtains a triangle $A_2B_2C_2$. If S, S_1, S_2 denote the areas of triangles $ABC, A_1B_1C_1, A_2B_2C_2$, respectively, prove that $S = 8S_1 - 4S_2$.

7. **(BGR 4)** In a triangle ABC, let H be its orthocenter, O its circumcenter, and R its circumradius. Prove that:
   (a) $|OH| = R\sqrt{1 - 8\cos\alpha\cos\beta\cos\gamma}$, where α, β, γ are angles of the triangle ABC;
   (b) $O \equiv H$ if and only if ABC is equilateral.

8. **(BGR 5)** (SL71-2).
   Original formulation. Prove that for every natural number $n \geq 1$ there exists an infinite sequence $M_1, M_2, \ldots, M_k, \ldots$ of distinct points in the plane such that for all i, exactly n among these points are at distance 1 from M_i.

9. **(BGR 6)** The base of an inclined prism is a triangle ABC. The perpendicular projection of B_1, one of the top vertices, is the midpoint of BC. The dihedral

angle between the lateral faces through BC and AB is α, and the lateral edges of the prism make an angle β with the base. If r_1, r_2, r_3 are exradii of a perpendicular section of the prism, assuming that in ABC, $\cos^2 A + \cos^2 B + \cos^2 C = 1$, $\angle A < \angle B < \angle C$, and $BC = a$, calculate $r_1 r_2 + r_1 r_3 + r_2 r_3$.

10. **(CUB 1)** In how many different ways can three knights be placed on a chessboard so that the number of squares attacked would be maximal?

11. **(CUB 2)** Prove that $n!$ cannot be the square of any natural number.

12. **(CUB 3)** A system of n numbers $x_1, x_2, \ldots, x_n$ is given such that

$$x_1 = \log_{x_{n-1}} x_n, \qquad x_2 = \log_{x_n} x_1, \qquad \ldots \qquad , \qquad x_n = \log_{x_{n-2}} x_{n-1}.$$

Prove that $\prod_{k=1}^{n} x_k = 1$.

13. **(CUB 4)** One Martian, one Venusian, and one Human reside on Pluto. One day they make the following conversation:
   Martian : I have spent $1/12$ of my life on Pluto.
   Human : I also have.
   Venusian : Me too.
   Martian : But Venusian and I have spend much more time here than
               you, Human.
   Human : That is true. However, Venusian and I are of the same age.
   Venusian : Yes, I have lived 300 Earth years.
   Martian : Venusian and I have been on Pluto for the past 13 years.
   It is known that Human and Martian together have lived 104 Earth years. Find the ages of Martian, Venusian, and Human.[5]

14. **(UNK 1)** Note that $8^3 - 7^3 = 169 = 13^2$ and $13 = 2^2 + 3^2$. Prove that if the difference between two consecutive cubes is a square, then it is the square of the sum of two consecutive squares.

15. **(UNK 2)** Let $ABCD$ be a convex quadrilateral whose diagonals intersect at O at an angle θ. Let us set $OA = a$, $OB = b$, $OC = c$, and $OD = d$, $c > a > 0$, and $d > b > 0$.
   Show that if there exists a right circular cone with vertex V, with the properties:
   (1) its axis passes through O, and
   (2) its curved surface passes through A, B, C and D, then

$$OV^2 = \frac{d^2 b^2 (c+a)^2 - c^2 a^2 (d+b)^2}{ca(d-b)^2 - db(c-a)^2}.$$

   Show also that if $\frac{c+a}{d+b}$ lies between $\frac{ca}{db}$ and $\sqrt{\frac{ca}{db}}$, and $\frac{c-a}{d-b} = \frac{ca}{db}$, then for a suitable choice of θ, a right circular cone exists with properties (1) and (2).

16. **(UNK 3)** (SL71-4).
   Original formulation. Two (intersecting) circles are given and a point P through

---

[5] The numbers in the problem are not necessarily in base 10.

which it is possible to draw a straight line on which the circles intercept two equal chords. Describe a construction by straightedge and compass for the straight line and prove the validity of your construction.

17. **(GDR 1)** (SL71-3).
Original formulation. Find all solutions of the system

$$x+y+z = 3,$$
$$x^3 + y^3 + z^3 = 15,$$
$$x^5 + y^5 + z^5 = 83.$$

18. **(GDR 2)** Let $a_1, a_2, \ldots, a_n$ be positive numbers, $m_g = (a_1 a_2 \cdots a_n)^{1/n}$ their geometric mean, and $m_a = (a_1 + a_2 + \cdots + a_n)/n$ their arithmetic mean. Prove that

$$(1+m_g)^n \le (1+a_1)\cdots(1+a_n) \le (1+m_a)^n.$$

19. **(GDR 3)** In a triangle $P_1 P_2 P_3$ let $P_i Q_i$ be the altitude from P_i for $i = 1,2,3$ (Q_i being the foot of the altitude). The circle with diameter $P_i Q_i$ meets the two corresponding sides at two points different from P_i. Denote the length of the segment whose endpoints are these two points by l_i. Prove that $l_1 = l_2 = l_3$.

20. **(GDR 4)** Let M be the circumcenter of a triangle ABC. The line through M perpendicular to CM meets the lines CA and CB at Q and P respectively. Prove that

$$\frac{\overline{CP}}{\overline{CM}} \frac{\overline{CQ}}{\overline{CM}} \frac{\overline{AB}}{\overline{PQ}} = 2.$$

21. **(HUN 1)** (SL71-5).

22. **(HUN 2)** We are given an $n \times n$ board, where n is an odd number. In each cell of the board either $+1$ or -1 is written. Let a_k and b_k denote the products of numbers in the kth row and in the kth column respectively. Prove that the sum $a_1 + a_2 + \cdots + a_n + b_1 + b_2 + \cdots + b_n$ cannot be equal to zero.

23. **(HUN 3)** Find all integer solutions of the equation

$$x^2 + y^2 = (x-y)^3.$$

24. **(HUN 4)** Let A, B, and C denote the angles of a triangle. If $\sin^2 A + \sin^2 B + \sin^2 C = 2$, prove that the triangle is right-angled.

25. **(HUN 5)** Let $ABC, AA_1 A_2, BB_1 B_2, CC_1 C_2$ be four equilateral triangles in the plane satisfying only that they are all positively oriented (i.e., in the counterclockwise direction). Denote the midpoints of the segments $A_2 B_1, B_2 C_1, C_2 A_1$ by P, Q, R in this order. Prove that the triangle PQR is equilateral.

26. **(HUN 6)** An infinite set of rectangles in the Cartesian coordinate plane is given. The vertices of each of these rectangles have coordinates $(0,0)$, $(p,0)$, (p,q), $(0,q)$ for some positive integers p,q. Show that there must exist two among them one of which is entirely contained in the other.

27. **(HUN 7)** (SL71-6).

28. **(NLD 1)** (SL71-7).
   Original formulation. A tetrahedron $ABCD$ is given. The sum of angles of the tetrahedron at the vertex A (namely $\angle BAC, \angle CAD, \angle DAB$) is denoted by α, and β, γ, δ are defined analogously. Let P, Q, R, S be variable points on edges of the tetrahedron: P on AD, Q on BD, R on BC, and S on AC, none of them at some vertex of $ABCD$. Prove that:
   (a) if $\alpha + \beta \neq 2\pi$, then $PQ + QR + RS + SP$ attains no minimal value;
   (b) if $\alpha + \beta = 2\pi$, then

$$AB\sin\frac{\alpha}{2} = CD\sin\frac{\gamma}{2} \quad \text{and} \quad PQ + QR + RS + SP \geq 2AB\sin\frac{\alpha}{2}.$$

29. **(NLD 2)** A rhombus with its incircle is given. At each vertex of the rhombus a circle is constructed that touches the incircle and two edges of the rhombus. These circles have radii r_1, r_2, while the incircle has radius r. Given that r_1 and r_2 are natural numbers and that $r_1 r_2 = r$, find r_1, r_2, and r.

30. **(NLD 3)** Prove that the system of equations

$$2yz + x - y - z = a,$$
$$2xz - x + y - z = a,$$
$$2xy - x - y + z = a,$$

a being a parameter, cannot have five distinct solutions. For what values of a does this system have four distinct integer solutions?

31. **(NLD 4)** (SL71-8).

32. **(NLD 5)** Two half-lines a and b, with the common endpoint O, make an acute angle α. Let A on a and B on b be points such that $OA = OB$, and let b' be the line through A parallel to b. Let β be the circle with center B and radius BO. We construct a sequence of half-lines $c_1, c_2, c_3, \ldots$, all lying inside the angle α, in the following manner:
   (i) c_1 is given arbitrarily;
   (ii) for every natural number k, the circle β intercepts on c_k a segment that is of the same length as the segment cut on b' by a and c_{k+1}.
   Prove that the angle determined by the lines c_k and b has a limit as k tends to infinity and find that limit.

33. **(NLD 6)** A square $2n \times 2n$ grid is given. Let us consider all possible paths along grid lines, going from the center of the grid to the border, such that (1) no point of the grid is reached more than once, and (2) each of the squares homothetic to the grid having its center at the grid center is passed through only once.
   (a) Prove that the number of all such paths is equal to $4\prod_{i=2}^{n}(16i - 9)$.
   (b) Find the number of pairs of such paths that divide the grid into two congruent figures.
   (c) How many quadruples of such paths are there that divide the grid into four congruent parts?

34. **(POL 1)** (SL71-9).

35. **(POL 2)** (SL71-10).

36. **(POL 3)** (SL71-11).

37. **(POL 4)** Let S be a circle, and $\alpha = \{A_1,\ldots,A_n\}$ a family of open arcs in S. Let $N(\alpha) = n$ denote the number of elements in α. We say that α is a covering of S if $\bigcup_{k=1}^{n} A_k \supset S$.
Let $\alpha = \{A_1,\ldots,A_n\}$ and $\beta = \{B_1,\ldots,B_m\}$ be two coverings of S. Show that we can choose from the family of all sets $A_i \cap B_j$, $i = 1,2,\ldots,n$, $j = 1,2,\ldots,m$, a covering γ of S such that $N(\gamma) \le N(\alpha) + N(\beta)$.

38. **(POL 5)** Let A,B,C be three points with integer coordinates in the plane and K a circle with radius R passing through A,B,C. Show that $AB \cdot BC \cdot CA \ge 2R$, and if the center of K is in the origin of the coordinates, show that $AB \cdot BC \cdot CA \ge 4R$.

39. **(POL 6)** (SL71-12).

40. **(SWE 1)** Prove that

$$\left(1 - \frac{1}{2^3}\right)\left(1 - \frac{1}{3^3}\right)\left(1 - \frac{1}{4^3}\right)\cdots\left(1 - \frac{1}{n^3}\right) > \frac{1}{2}, \quad n = 2,3,\ldots.$$

41. **(SWE 2)** Consider the set of grid points (m,n) in the plane, m,n integers. Let σ be a finite subset and define

$$S(\sigma) = \sum_{(m,n)\in\sigma} (100 - |m| - |n|).$$

Find the maximum of S, taken over the set of all such subsets σ.

42. **(SWE 3)** Let L_i, $i = 1,2,3$, be line segments on the sides of an equilateral triangle, one segment on each side, with lengths l_i, $i = 1,2,3$. By L_i^* we denote the segment of length l_i with its midpoint on the midpoint of the corresponding side of the triangle. Let $M(L)$ be the set of points in the plane whose orthogonal projections on the sides of the triangle are in L_1, L_2, and L_3, respectively; $M(L^*)$ is defined correspondingly. Prove that if $l_1 \ge l_2 + l_3$, we have that the area of $M(L)$ is less than or equal to the area of $M(L^*)$.

43. **(SWE 4)** Show that for nonnegative real numbers a,b and integers $n \ge 2$,

$$\frac{a^n + b^n}{2} \ge \left(\frac{a+b}{2}\right)^n.$$

When does equality hold?

44. **(SWE 5)** (SL71-13).

45. **(SWE 6)** Let m and n denote integers greater than 1, and let $v(n)$ be the number of primes less than or equal to n. Show that if the equation $\frac{n}{v(n)} = m$ has a solution (in n), then so does the equation $\frac{n}{v(n)} = m - 1$.

46. **(USS 1)** (SL71-14).

47. **(USS 2)** (SL71-15).

48. **(USS 3)** A sequence of real numbers $x_1, x_2, \ldots, x_n$ is given such that $x_{i+1} = x_i + \frac{1}{30000}\sqrt{1 - x_i^2}$, $i = 1, 2, \ldots$, and $x_1 = 0$. Can n be equal to 50000 if $x_n < 1$?

49. **(USS 4)** Diagonals of a convex quadrilateral $ABCD$ intersect at a point O. Find all angles of this quadrilateral if $\angle OBA = 30°$, $\angle OCB = 45°$, $\angle ODC = 45°$, and $\angle OAD = 30°$.

50. **(USS 5)** (SL71-16).

51. **(USS 6)** Suppose that the sides AB and DC of a convex quadrilateral $ABCD$ are not parallel. On the sides BC and AD, pairs of points (M, N) and (K, L) are chosen such that $BM = MN = NC$ and $AK = KL = LD$. Prove that the areas of triangles OKM and OLN are different, where O is the intersection point of AB and CD.

52. **(YUG 1)** (SL71-17).

53. **(YUG 2)** Denote by $x_n(p)$ the multiplicity of the prime p in the canonical representation of the number $n!$ as a product of primes. Prove that $\frac{x_n(p)}{n} < \frac{1}{p-1}$ and $\lim_{n \to \infty} \frac{x_n(p)}{n} = \frac{1}{p-1}$.

54. **(YUG 3)** A set M is formed of $\binom{2n}{n}$ men, $n = 1, 2, \ldots$. Prove that we can choose a subset P of the set M consisting of $n + 1$ men such that one of the following conditions is satisfied:
    (1) every member of the set P knows every other member of the set P;
    (2) no member of the set P knows any other member of the set P.

55. **(YUG 4)** Prove that the polynomial $x^4 + \lambda x^3 + \mu x^2 + vx + 1$ has no real roots if λ, μ, v are real numbers satisfying $|\lambda| + |\mu| + |v| \le \sqrt{2}$.

3.13.3 Shortlisted Problems

1. **(BGR 2)** Consider a sequence of polynomials $P_0(x), P_1(x), P_2(x), \ldots, P_n(x), \ldots$, where $P_0(x) = 2$, $P_1(x) = x$ and for every $n \ge 1$ the following equality holds: $P_{n+1}(x) + P_{n-1}(x) = xP_n(x)$. Prove that there exist three real numbers a, b, c such that for all $n \ge 1$,

$$(x^2 - 4)[P_n^2(x) - 4] = [aP_{n+1}(x) + bP_n(x) + cP_{n-1}(x)]^2. \tag{1}$$

2. **(BGR 5)**[IMO5] Prove that for every natural number $m \ge 1$ there exists a finite set S_m of points in the plane satisfying the following condition: If A is any point in S_m, then there are exactly m points in S_m whose distance to A equals 1.

3. **(GDR 1)** Knowing that the system

$$x+y+z = 3,$$
$$x^3+y^3+z^3 = 15,$$
$$x^4+y^4+z^4 = 35,$$

has a real solution x,y,z for which $x^2+y^2+z^2 < 10$, find the value of $x^5+y^5+z^5$ for that solution.

4. **(UNK 3)** We are given two mutually tangent circles in the plane, with radii r_1,r_2. A line intersects these circles in four points, determining three segments of equal length. Find this length as a function of r_1 and r_2 and the condition for the solvability of the problem.

5. **(HUN 1)**[IMO1] Let a, b, c, d, e be real numbers. Prove that the expression

$$(a-b)(a-c)(a-d)(a-e)+(b-a)(b-c)(b-d)(b-e)$$
$$+(c-a)(c-b)(c-d)(c-e)+(d-a)(d-b)(d-c)(d-e)$$
$$+(e-a)(e-b)(e-c)(e-d).$$

is nonnegative.

6. **(HUN 7)** Let $n \geq 2$ be a natural number. Find a way to assign natural numbers to the vertices of a regular 2^n-gon such that the following conditions are satisfied:
   (i) only digits 1 and 2 are used;
   (ii) each number consists of exactly n digits;
   (iii) different numbers are assigned to different vertices;
   (iv) the numbers assigned to two neighboring vertices differ at exactly one digit.

7. **(NLD 1)**[IMO4] Given a tetrahedron $ABCD$ all of whose faces are acute-angled triangles, set

$$\sigma = \angle DAB + \angle BCD - \angle ABC - \angle CDA.$$

Consider all closed broken lines $XYZTX$ whose vertices X,Y,Z,T lie in the interior of segments AB,BC,CD,DA respectively. Prove that:
   (a) if $\sigma \neq 0$, then there is no broken line $XYZT$ of minimal length;
   (b) if $\sigma = 0$, then there are infinitely many such broken lines of minimal length. That length equals $2AC\sin(\alpha/2)$, where

$$\alpha = \angle BAC + \angle CAD + \angle DAB.$$

8. **(NLD 4)** Determine whether there exist distinct real numbers a,b,c,t for which:
   (i) the equation $ax^2 + btx + c = 0$ has two distinct real roots x_1,x_2,
   (ii) the equation $bx^2 + ctx + a = 0$ has two distinct real roots x_2,x_3,
   (iii) the equation $cx^2 + atx + b = 0$ has two distinct real roots x_3,x_1.

9. **(POL 1)** Let $T_k = k-1$ for $k = 1,2,3,4$ and

$$T_{2k-1} = T_{2k-2} + 2^{k-2}, \qquad T_{2k} = T_{2k-5} + 2^k \qquad (k \geq 3).$$

Show that for all k,

$$1 + T_{2n-1} = \left[\frac{12}{7}2^{n-1}\right] \quad \text{and} \quad 1 + T_{2n} = \left[\frac{17}{7}2^{n-1}\right],$$

where $[x]$ denotes the greatest integer not exceeding x.

10. **(POL 2)**[IMO3] Prove that the sequence $2^n - 3$ $(n > 1)$ contains a subsequence of numbers relatively prime in pairs.

11. **(POL 3)** The matrix

$$\begin{pmatrix} a_{11} & \cdots & a_{1n} \\ \vdots & \cdots & \vdots \\ a_{n1} & \cdots & a_{nn} \end{pmatrix}$$

satisfies the inequality $\sum_{j=1}^{n} |a_{j1}x_1 + \cdots + a_{jn}x_n| \leq M$ for each choice of numbers x_i equal to ± 1. Show that

$$|a_{11} + a_{22} + \cdots + a_{nn}| \leq M.$$

12. **(POL 6)** Two congruent equilateral triangles ABC and $A'B'C'$ in the plane are given. Show that the midpoints of the segments AA', BB', CC' either are collinear or form an equilateral triangle.

13. **(SWE 5)**[IMO6] Consider the $n \times n$ array of nonnegative integers

$$\begin{pmatrix} a_{11} & a_{12} & \cdots & a_{1n} \\ a_{21} & a_{22} & \cdots & a_{2n} \\ \vdots & \vdots & & \vdots \\ a_{n1} & a_{n2} & \cdots & a_{nn} \end{pmatrix},$$

with the following property: If an element a_{ij} is zero, then the sum of the elements of the ith row and the jth column is greater than or equal to n. Prove that the sum of all the elements is greater than or equal to $\frac{1}{2}n^2$.

14. **(USS 1)** A broken line $A_1A_2\ldots A_n$ is drawn in a 50×50 square, so that the distance from any point of the square to the broken line is less than 1. Prove that its total length is greater than 1248.

15. **(USS 2)** Natural numbers from 1 to 99 (not necessarily distinct) are written on 99 cards. It is given that the sum of the numbers on any subset of cards (including the set of all cards) is not divisible by 100. Show that all the cards contain the same number.

16. **(USS 5)**[IMO2] Given a convex polyhedron P_1 with 9 vertices $A_1,\ldots,A_9$, let us denote by $P_2, P_3,\ldots,P_9$ the images of P_1 under the translations mapping the vertex A_1 to $A_2, A_3,\ldots,A_9$ respectively. Prove that among the polyhedra $P_1,\ldots,P_9$ at least two have a common interior point.

17. **(YUG 1)** Prove the inequality

$$\frac{a_1+a_3}{a_1+a_2} + \frac{a_2+a_4}{a_2+a_3} + \frac{a_3+a_1}{a_3+a_4} + \frac{a_4+a_2}{a_4+a_1} \geq 4,$$

where $a_i > 0$, $i = 1,2,3,4$.

3.14 The Fourteenth IMO
Warsaw–Toruń, Poland, July 5–17, 1972

3.14.1 Contest Problems

First Day (July 10)

1. A set of 10 positive integers is given such that the decimal expansion of each of them has two digits. Prove that there are two disjoint subsets of the set with equal sums of their elements.

2. Prove that for each $n \geq 4$ every cyclic quadrilateral can be decomposed into n cyclic quadrilaterals.

3. Let m and n be nonnegative integers. Prove that $\frac{(2m)!(2n)!}{m!n!(m+n)!}$ is an integer $(0! = 1)$.

Second Day (July 11)

4. Find all solutions in positive real numbers x_i $(i = 1,2,3,4,5)$ of the following system of inequalities:

$$
\begin{array}{ll}
(x_1^2 - x_3 x_5)(x_2^2 - x_3 x_5) \leq 0 & \text{(i)} \\
(x_2^2 - x_4 x_1)(x_3^2 - x_4 x_1) \leq 0 & \text{(ii)} \\
(x_3^2 - x_5 x_2)(x_4^2 - x_5 x_2) \leq 0 & \text{(iii)} \\
(x_4^2 - x_1 x_3)(x_5^2 - x_1 x_3) \leq 0 & \text{(iv)} \\
(x_5^2 - x_2 x_4)(x_1^2 - x_2 x_4) \leq 0. & \text{(v)}
\end{array}
$$

5. Let f and φ be real functions defined in the interval $(-\infty, \infty)$ satisfying the functional equation

$$f(x+y) + f(x-y) = 2\varphi(y)f(x),$$

for arbitrary real x, y (give examples of such functions). Prove that if $f(x)$ is not identically 0 and $|f(x)| \leq 1$ for all x, then $|\varphi(x)| \leq 1$ for all x.

6. Given four distinct parallel planes, show that a regular tetrahedron exists with a vertex on each plane.

3.14.2 Longlisted Problems

1. **(BGR 1)** Find all integer solutions of the equation

$$1 + x + x^2 + x^3 + x^4 = y^4.$$

2. **(BGR 2)** Find all real values of the parameter a for which the system of equations

$$
\begin{aligned}
x^4 &= yz - x^2 + a, \\
y^4 &= zx - y^2 + a, \\
z^4 &= xy - z^2 + a,
\end{aligned}
$$

has at most one real solution.

3. **(BGR 3)** On a line a set of segments is given of total length less than 1. Prove that every set of n points of the line can be translated in some direction along the line for a distance smaller than $n/2$ so that none of the points remain on the segments.

4. **(BGR 4)** Given a triangle, prove that the points of intersection of three pairs of trisectors of the inner angles at the sides lying closest to those sides are vertices of an equilateral triangle.

5. **(BGR 5)** Given a pyramid whose base is an n-gon inscribable in a circle, let H be the projection of the top vertex of the pyramid to its base. Prove that the projections of H to the lateral edges of the pyramid lie on a circle.

6. **(BGR 6)** Prove the inequality

$$(n+1)\cos\frac{\pi}{n+1} - n\cos\frac{\pi}{n} > 1$$

for all natural numbers $n \geq 2$.

7. **(BGR 7)** (SL72-1).

8. **(CZS 1)** (SL72-2).

9. **(CZS 2)** Given natural numbers k and n, $k \leq n$, $n \geq 3$, find the set of all values in the interval $(0, \pi)$ that the kth-largest among the interior angles of a convex $n$gon can take.

10. **(CZS 3)** Given five points in the plane, no three of which are collinear, prove that there can be found at least two obtuse-angled triangles with vertices at the given points. Construct an example in which there are exactly two such triangles.

11. **(CZS 4)** (SL72-3).

12. **(CZS 5)** A circle $k = (S, r)$ is given and a hexagon $AA'BB'CC'$ inscribed in it. The lengths of sides of the hexagon satisfy $AA' = A'B$, $BB' = B'C$, $CC' = C'A$. Prove that the area P of triangle ABC is not greater than the area P' of triangle $A'B'C'$. When does $P = P'$ hold?

13. **(CZS 6)** Given a sphere K, determine the set of all points A that are vertices of some parallelograms $ABCD$ that satisfy $AC \leq BD$ and whose entire diagonal BD is contained in K.

14. **(UNK 1)** (SL72-7).

15. **(UNK 2)** (SL72-8).

16. **(UNK 3)** Consider the set S of all the different odd positive integers that are not multiples of 5 and that are less than $30m$, m being a positive integer. What is the smallest integer k such that in any subset of k integers from S there must be two integers one of which divides the other? Prove your result.

17. **(UNK 4)** A solid right circular cylinder with height h and base-radius r has a solid hemisphere of radius r resting upon it. The center of the hemisphere O is on

the axis of the cylinder. Let P be any point on the surface of the hemisphere and Q the point on the base circle of the cylinder that is furthest from P (measuring along the surface of the combined solid). A string is stretched over the surface from P to Q so as to be as short as possible. Show that if the string is not in a plane, the straight line PO when produced cuts the curved surface of the cylinder.

18. **(UNK 5)** We have p players participating in a tournament, each player playing against every other player exactly once. A point is scored for each victory, and there are no draws. A sequence of nonnegative integers $s_1 \le s_2 \le s_3 \le \cdots \le s_p$ is given. Show that it is possible for this sequence to be a set of final scores of the players in the tournament if and only if

$$\text{(i) } \sum_{i=1}^{p} s_i = \frac{1}{2}p(p-1) \quad \text{and} \quad \text{(ii) for all } k < p, \ \sum_{i=1}^{k} s_i \ge \frac{1}{2}k(k-1).$$

19. **(UNK 6)** Let S be a subset of the real numbers with the following properties:
    (i) If $x \in S$ and $y \in S$, then $x - y \in S$;
    (ii) If $x \in S$ and $y \in S$, then $xy \in S$;
    (iii) S contains an exceptional number x' such that there is no number y in S satisfying $x'y + x' + y = 0$;
    (iv) If $x \in S$ and $x \ne x'$, there is a number y in S such that $xy + x + y = 0$.
    Show that
    (a) S has more than one number in it;
    (b) $x' \ne -1$ leads to a contradiction;
    (c) $x \in S$ and $x \ne 0$ implies $1/x \in S$.

20. **(GDR 1)** (SL72-4).

21. **(GDR 2)** (SL72-5).

22. **(GDR 3)** (SL72-6).

23. **(MNG 1)** Does there exist a $2n$-digit number $\overline{a_{2n}a_{2n-1}\dots a_1}$ (for an arbitrary n) for which the following equality holds:

$$\overline{a_{2n}\dots a_1} = (\overline{a_n \dots a_1})^2?$$

24. **(MNG 2)** The diagonals of a convex 18-gon are colored in 5 different colors, each color appearing on an equal number of diagonals. The diagonals of one color are numbered $1, 2, \dots$. One randomly chooses one-fifth of all the diagonals. Find the number of possibilities for which among the chosen diagonals there exist exactly n pairs of diagonals of the same color and with fixed indices i, j.

25. **(NLD 1)** We consider n real variables x_i ($1 \le i \le n$), where n is an integer and $n \ge 2$. The product of these variables will be denoted by p, their sum by s, and the sum of their squares by S. Furthermore, let α be a positive constant. We now study the inequality $ps \le S^\alpha$. Prove that it holds for every n-tuple (x_i) if and only if $\alpha = \frac{n+1}{2}$.

26. **(NLD 2)** (SL72-9).

27. **(NLD 3)** (SL72-10).

28. **(NLD 4)** The lengths of the sides of a rectangle are given to be odd integers. Prove that there does not exist a point within that rectangle that has integer distances to each of its four vertices.

29. **(NLD 5)** Let A, B, C be points on the sides B_1C_1, C_1A_1, A_1B_1 of a triangle $A_1B_1C_1$ such that A_1A, B_1B, C_1C are the bisectors of angles of the triangle. We have that $AC = BC$ and $A_1C_1 \neq B_1C_1$.
    (a) Prove that C_1 lies on the circumcircle of the triangle ABC.
    (b) Suppose that $\angle BAC_1 = \pi/6$; find the form of triangle ABC.

30. **(NLD 6)** (SL72-11).

31. **(ROU 1)** Find values of $n \in \mathbb{N}$ for which the fraction $\frac{3^n-2}{2^n-3}$ is reducible.

32. **(ROU 2)** If $n_1, n_2, \ldots, n_k$ are natural numbers and $n_1 + n_2 + \cdots + n_k = n$, show that

$$\max_{n_1 + \cdots + n_k = n} n_1 n_2 \cdots n_k = (t+1)^r t^{k-r},$$

where $t = [n/k]$ and r is the remainder of n upon division by k; i.e., $n = tk + r$, $0 \leq r \leq k-1$.

33. **(ROU 3)** A rectangle $ABCD$ is given whose sides have lengths 3 and $2n$, where n is a natural number. Denote by $U(n)$ the number of ways in which one can cut the rectangle into rectangles of side lengths 1 and 2.
    (a) Prove that $U(n+1) + U(n-1) = 4U(n)$;
    (b) Prove that $U(n) = \frac{1}{2\sqrt{3}}[(\sqrt{3}+1)(2+\sqrt{3})^n + (\sqrt{3}-1)(2-\sqrt{3})^n]$.

34. **(ROU 4)** If p is a prime number greater than 2 and a, b, c integers not divisible by p, prove that the equation

$$ax^2 + by^2 = pz + c$$

has an integer solution.

35. **(ROU 5)** (a) Prove that for $a, b, c, d \in \mathbb{R}$, $m \in [1, +\infty)$ with $am + b = -cm + d = m$,
    (i) $\sqrt{a^2 + b^2} + \sqrt{c^2 + d^2} + \sqrt{(a-c)^2 + (b-d)^2} \geq \frac{4m^2}{1+m^2}$, and
    (ii) $2 \leq \frac{4m^2}{1+m^2} < 4$.
    (b) Express a, b, c, d as functions of m so that there is equality in (1).

36. **(ROU 6)** A finite number of parallel segments in the plane are given with the property that for any three of the segments there is a line intersecting each of them. Prove that there exists a line that intersects all the given segments.

37. **(SWE 1)** On a chessboard (8×8 squares with sides of length 1) two diagonally opposite corner squares are taken away. Can the board now be covered with nonoverlapping rectangles with sides of lengths 1 and 2?

38. **(SWE 2)** Congruent rectangles with sides m (cm) and n (cm) are given (m, n positive integers). Characterize the rectangles that can be constructed from these rectangles (in the fashion of a jigsaw puzzle). (The number of rectangles is unbounded.)

39. **(SWE 3)** How many tangents to the curve $y = x^3 - 3x$ ($y = x^3 + px$) can be drawn from different points in the plane?

40. **(SWE 4)** Prove the inequalities

$$\frac{u}{v} \le \frac{\sin u}{\sin v} \le \frac{\pi}{2}\frac{u}{v}, \qquad \text{for } 0 \le u < v \le \frac{\pi}{2}.$$

41. **(SWE 5)** The ternary expansion $x = 0.10101010\ldots$ is given. Give the binary expansion of x.
    Alternatively, transform the binary expansion $y = 0.110110110\ldots$ into a ternary expansion.

42. **(SWE 6)** The decimal number 13^{101} is given. It is instead written as a ternary number. What are the two last digits of this ternary number?

43. **(USS 1)** A fixed point A inside a circle is given. Consider all chords XY of the circle such that $\angle XAY$ is a right angle, and for all such chords construct the point M symmetric to A with respect to XY. Find the locus of points M.

44. **(USS 2)** (SL72-12).

45. **(USS 3)** Let $ABCD$ be a convex quadrilateral whose diagonals AC and BD intersect at point O. Let a line through O intersect segment AB at M and segment CD at N. Prove that the segment MN is not longer than at least one of the segments AC and BD.

46. **(USS 4)** Numbers $1, 2, \ldots, 16$ are written in a 4×4 square matrix so that the sum of the numbers in every row, every column, and every diagonal is the same and furthermore that the numbers 1 and 16 lie in opposite corners. Prove that the sum of any two numbers symmetric with respect to the center of the square equals 17.

3.14.3 Shortlisted Problems

1. **(BGR 7)**[IMO5] Let f and φ be real functions defined on the set $\mathbb{R}$ satisfying the functional equation

$$f(x+y) + f(x-y) = 2\varphi(y)f(x), \qquad (1)$$

for arbitrary real x, y (give examples of such functions). Prove that if $f(x)$ is not identically 0 and $|f(x)| \le 1$ for all x, then $|\varphi(x)| \le 1$ for all x.

2. **(CZS 1)** We are given $3n$ points $A_1, A_2, \ldots, A_{3n}$ in the plane, no three of them collinear. Prove that one can construct n disjoint triangles with vertices at the points A_i.

3. **(CZS 4)** Let $x_1, x_2, \ldots, x_n$ be real numbers satisfying $x_1 + x_2 + \cdots + x_n = 0$. Let m be the least and M the greatest among them. Prove that

$$x_1^2 + x_2^2 + \cdots + x_n^2 \leq -nmM.$$

4. **(GDR 1)** Let n_1, n_2 be positive integers. Consider in a plane E two disjoint sets of points M_1 and M_2 consisting of $2n_1$ and $2n_2$ points, respectively, and such that no three points of the union $M_1 \cup M_2$ are collinear. Prove that there exists a straight line g with the following property: Each of the two half-planes determined by g on E (g not being included in either) contains exactly half of the points of M_1 and exactly half of the points of M_2.

5. **(GDR 2)** Prove the following assertion: The four altitudes of a tetrahedron $ABCD$ intersect in a point if and only if

$$AB^2 + CD^2 = BC^2 + AD^2 = CA^2 + BD^2.$$

6. **(GDR 3)** Show that for any $n \not\equiv 0 \pmod{10}$ there exists a multiple of n not containing the digit 0 in its decimal expansion.

7. **(UNK 1)**[IMO6] (a) A plane π passes through the vertex O of the regular tetrahedron $OPQR$. We define p, q, r to be the signed distances of P, Q, R from π measured along a directed normal to π. Prove that

$$p^2 + q^2 + r^2 + (q-r)^2 + (r-p)^2 + (p-q)^2 = 2a^2,$$

where a is the length of an edge of a tetrahedron.
(b) Given four parallel planes not all of which are coincident, show that a regular tetrahedron exists with a vertex on each plane.

8. **(UNK 2)**[IMO3] Let m and n be nonnegative integers. Prove that $m!n!(m+n)!$ divides $(2m)!(2n)!$.

9. **(NLD 2)**[IMO4] Find all solutions in positive real numbers x_i ($i = 1, 2, 3, 4, 5$) of the following system of inequalities:

$$(x_1^2 - x_3 x_5)(x_2^2 - x_3 x_5) \leq 0, \tag{i}$$
$$(x_2^2 - x_4 x_1)(x_3^2 - x_4 x_1) \leq 0, \tag{ii}$$
$$(x_3^2 - x_5 x_2)(x_4^2 - x_5 x_2) \leq 0, \tag{iii}$$
$$(x_4^2 - x_1 x_3)(x_5^2 - x_1 x_3) \leq 0, \tag{iv}$$
$$(x_5^2 - x_2 x_4)(x_1^2 - x_2 x_4) \leq 0. \tag{v}$$

10. **(NLD 3)**[IMO2] Prove that for each $n \geq 4$ every cyclic quadrilateral can be decomposed into n cyclic quadrilaterals.

11. **(NLD 6)** Consider a sequence of circles $K_1, K_2, K_3, K_4, \ldots$ of radii $r_1, r_2, r_3, r_4, \ldots$, respectively, situated inside a triangle ABC. The circle K_1 is tangent to AB and AC; K_2 is tangent to K_1, BA, and BC; K_3 is tangent to K_2, CA, and CB; K_4 is tangent to K_3, AB, and AC; etc.

(a) Prove the relation

$$r_1 \cot \frac{1}{2}A + 2\sqrt{r_1 r_2} + r_2 \cot \frac{1}{2}B = r\left(\cot \frac{1}{2}A + \cot \frac{1}{2}B\right),$$

where r is the radius of the incircle of the triangle ABC. Deduce the existence of a t_1 such that

$$r_1 = r \cot \frac{1}{2}B \cot \frac{1}{2}C \sin^2 t_1.$$

(b) Prove that the sequence of circles $K_1, K_2, \ldots$ is periodic.

12. **(USS 2)**[IMO1] A set of 10 positive integers is given such that the decimal expansion of each of them has two digits. Prove that there are two disjoint subsets of the set with equal sums of their elements.

3.15 The Fifteenth IMO
Moscow, Soviet Union, July 5–16, 1973

3.15.1 Contest Problems

First Day (July 9)

1. Let O be a point on the line l and $\overrightarrow{OP_1}, \overrightarrow{OP_2}, \ldots, \overrightarrow{OP_n}$ unit vectors such that points $P_1, P_2, \ldots, P_n$ and line l lie in the same plane and all points P_i lie in the same half-plane determined by l. Prove that if n is odd, then

$$\left\| \overrightarrow{OP_1} + \overrightarrow{OP_2} + \cdots + \overrightarrow{OP_n} \right\| \geq 1.$$

($\left\| \overrightarrow{OM} \right\|$ is the length of vector $\overrightarrow{OM}$).

2. Does there exist a finite set M of points in space, not all in the same plane, such that for each two points $A, B \in M$ there exist two other points $C, D \in M$ such that lines AB and CD are parallel but not equal?

3. Determine the minimum of $a^2 + b^2$ if a and b are real numbers for which the equation
$$x^4 + ax^3 + bx^2 + ax + 1 = 0$$
has at least one real solution.

Second Day (July 10)

4. A soldier has to investigate whether there are mines in an area that has the form of equilateral triangle. The radius of his detector's range is equal to one-half the altitude of the triangle. The soldier starts from one vertex of the triangle. Determine the shortest path through which the soldier has to pass in order to check the entire region.

5. Let G be a set of functions $f : \mathbb{R} \to \mathbb{R}$ of the form $f(x) = ax + b$, where a and b are real numbers and $a \neq 0$. Suppose that G satisfies the following conditions:
    (1) If $f, g \in G$, then $g \circ f \in G$, where $(g \circ f)(x) = g[f(x)]$.
    (2) If $f \in G$ and $f(x) = ax + b$, then the inverse f^{-1} of f belongs to G $(f^{-1}(x) = (x - b)/a)$.
    (3) For each $f \in G$ there exists a number $x_f \in \mathbb{R}$ such that $f(x_f) = x_f$.
    Prove that there exists a number $k \in \mathbb{R}$ such that $f(k) = k$ for all $f \in G$.

6. Let $a_1, a_2, \ldots, a_n$ be positive numbers and q a given real number, $0 < q < 1$. Find n real numbers $b_1, b_2, \ldots, b_n$ that satisfy:
    (1) $a_k < b_k$ for all $k = 1, 2, \ldots, n$;
    (2) $q < \frac{b_{k+1}}{b_k} < \frac{1}{q}$ for all $k = 1, 2, \ldots, n - 1$;
    (3) $b_1 + b_2 + \cdots + b_n < \frac{1+q}{1-q}(a_1 + a_2 + \cdots + a_n)$.

3.15.2 Shortlisted Problems

1. **(BGR 6)** Let a tetrahedron $ABCD$ be inscribed in a sphere S. Find the locus of points P inside the sphere S for which the equality

$$\frac{AP}{PA_1} + \frac{BP}{PB_1} + \frac{CP}{PC_1} + \frac{DP}{PD_1} = 4$$

holds, where A_1, B_1, C_1, and D_1 are the intersection points of S with the lines AP, BP, CP, and DP, respectively.

2. **(CZS 1)** Given a circle K, find the locus of vertices A of parallelograms $ABCD$ with diagonals $AC \leq BD$, such that BD is inside K.

3. **(CZS 6)**[IMO1] Prove that the sum of an odd number of unit vectors passing through the same point O and lying in the same half-plane whose border passes through O has length greater than or equal to 1.

4. **(UNK 1)** Let P be a set of 7 different prime numbers and C a set of 28 different composite numbers each of which is a product of two (not necessarily different) numbers from P. The set C is divided into 7 disjoint four-element subsets such that each of the numbers in one set has a common prime divisor with at least two other numbers in that set. How many such partitions of C are there?

5. **(FRA 2)** A circle of radius 1 is located in a right-angled trihedron and touches all its faces. Find the locus of centers of such circles.

6. **(POL 2)**[IMO2] Does there exist a finite set M of points in space, not all in the same plane, such that for each two points $A, B \in M$ there exist two other points $C, D \in M$ such that lines AB and CD are parallel?

7. **(POL 3)** Given a tetrahedron $ABCD$, let $x = AB \cdot CD$, $y = AC \cdot BD$, and $z = AD \cdot BC$. Prove that there exists a triangle with edges x, y, z.

8. **(ROU 1)** Prove that there are exactly $\binom{k}{[k/2]}$ arrays $a_1, a_2, \ldots, a_{k+1}$ of nonnegative integers such that $a_1 = 0$ and $|a_i - a_{i+1}| = 1$ for $i = 1, 2, \ldots, k$.

9. **(ROU 2)** Let Ox, Oy, Oz be three rays, and G a point inside the trihedron $Oxyz$. Consider all planes passing through G and cutting Ox, Oy, Oz at points A, B, C, respectively. How is the plane to be placed in order to yield a tetrahedron $OABC$ with minimal perimeter?

10. **(SWE 3)**[IMO6] Let $a_1, a_2, \ldots, a_n$ be positive numbers and q a given real number, $0 < q < 1$. Find n real numbers $b_1, b_2, \ldots, b_n$ that satisfy:
    (1) $a_k < b_k$ for all $k = 1, 2, \ldots, n$;
    (2) $q < \frac{b_{k+1}}{b_k} < \frac{1}{q}$ for all $k = 1, 2, \ldots, n-1$;
    (3) $b_1 + b_2 + \cdots + b_n < \frac{1+q}{1-q}(a_1 + a_2 + \cdots + a_n)$.

11. **(SWE 4)**[IMO3] Determine the minimum of $a^2 + b^2$ if a and b are real numbers for which the equation
$$x^4 + ax^3 + bx^2 + ax + 1 = 0$$

has at least one real solution.

12. **(SWE 6)** Consider the two square matrices

$$A = \begin{bmatrix} 1 & 1 & 1 & 1 & 1 \\ 1 & 1 & 1 & -1 & -1 \\ 1 & -1 & -1 & 1 & 1 \\ 1 & -1 & -1 & -1 & 1 \\ 1 & 1 & -1 & 1 & -1 \end{bmatrix} \quad \text{and} \quad B = \begin{bmatrix} 1 & 1 & 1 & 1 & 1 \\ 1 & 1 & 1 & -1 & -1 \\ 1 & 1 & -1 & 1 & -1 \\ 1 & -1 & -1 & 1 & 1 \\ 1 & -1 & 1 & -1 & 1 \end{bmatrix}$$

with entries 1 and -1. The following operations will be called *elementary*:
   (1) Changing signs of all numbers in one row;
   (2) Changing signs of all numbers in one column;
   (3) Interchanging two rows (two rows exchange their positions);
   (4) Interchanging two columns.
Prove that the matrix B cannot be obtained from the matrix A using these operations.

13. **(YUG 4)** Find the sphere of maximal radius that can be placed inside every tetrahedron that has all altitudes of length greater than or equal to 1.

14. **(YUG 5)**[IMO4] A soldier has to investigate whether there are mines in an area that has the form of an equilateral triangle. The radius of his detector is equal to one-half of an altitude of the triangle. The soldier starts from one vertex of the triangle. Determine the shortest path that the soldier has to traverse in order to check the whole region.

15. **(CUB 1)** Prove that for all $n \in \mathbb{N}$ the following is true:

$$2^n \prod_{k=1}^{n} \sin \frac{k\pi}{2n+1} = \sqrt{2n+1}.$$

16. **(CUB 2)** Given $a, \theta \in \mathbb{R}$, $m \in \mathbb{N}$, and $P(x) = x^{2m} - 2|a|^m x^m \cos\theta + a^{2m}$, factorize $P(x)$ as a product of m real quadratic polynomials.

17. **(POL 1)**[IMO5] Let $\mathscr{F}$ be a nonempty set of functions $f : \mathbb{R} \to \mathbb{R}$ of the form $f(x) = ax + b$, where a and b are real numbers and $a \neq 0$. Suppose that $\mathscr{F}$ satisfies the following conditions:
   (1) If $f, g \in \mathscr{F}$, then $g \circ f \in \mathscr{F}$, where $(g \circ f)(x) = g[f(x)]$.
   (2) If $f \in \mathscr{F}$ and $f(x) = ax + b$, then the inverse f^{-1} of f belongs to $\mathscr{F}$ $(f^{-1}(x) = (x - b)/a)$.
   (3) None of the functions $f(x) = x + c$, for $c \neq 0$, belong to $\mathscr{F}$.
Prove that there exists $x_0 \in \mathbb{R}$ such that $f(x_0) = x_0$ for all $f \in \mathscr{F}$.

3.16 The Sixteenth IMO
Erfurt–Berlin, DR Germany, July 4–17, 1974

3.16.1 Contest Problems

First Day (July 8)

1. Alice, Betty, and Carol took the same series of examinations. There was one grade of A, one grade of B, and one grade of C for each examination, where A, B, C are different positive integers. The final test scores were

Alice	Betty	Carol
20	10	9

    If Betty placed first in the arithmetic examination, who placed second in the spelling examination?

2. Let $\triangle ABC$ be a triangle. Prove that there exists a point D on the side AB such that CD is the geometric mean of AD and BD if and only if

$$\sqrt{\sin A \sin B} \le \sin \frac{C}{2}.$$

3. Prove that there does not exist a natural number n for which the number

$$\sum_{k=0}^{n} \binom{2n+1}{2k+1} 2^{3k}$$

is divisible by 5.

Second Day (July 9)

4. Consider a partition of an 8×8 chessboard into p rectangles whose interiors are disjoint such that each rectangle contains an equal number of white and black cells. Assume that $a_1 < a_2 < \cdots < a_p$, where a_i denotes the number of white cells in the ith rectangle. Find the maximal p for which such a partition is possible and for that p determine all possible corresponding sequences $a_1, a_2, \ldots, a_p$.

5. If a, b, c, d are arbitrary positive real numbers, find all possible values of

$$S = \frac{a}{a+b+d} + \frac{b}{a+b+c} + \frac{c}{b+c+d} + \frac{d}{a+c+d}.$$

6. Let $P(x)$ be a polynomial with integer coefficients. If $n(P)$ is the number of (distinct) integers k such that $P^2(k) = 1$, prove that $n(P) - \deg(P) \le 2$, where $\deg(P)$ denotes the degree of the polynomial P.

3.16.2 Longlisted Problems

1. **(BGR 1)** (SL74-11).

2. **(BGR 2)** Let $\{u_n\}$ be the Fibonacci sequence, i.e., $u_0 = 0$, $u_1 = 1$, $u_n = u_{n-1} + u_{n-2}$ for $n > 1$. Prove that there exist infinitely many prime numbers p that divide u_{p-1}.

3. **(BGR 3)** Let $ABCD$ be an arbitrary quadrilateral. Let squares ABB_1A_2, BCC_1B_2, CDD_1C_2, DAA_1D_2 be constructed in the exterior of the quadrilateral. Furthermore, let AA_1PA_2 and CC_1QC_2 be parallelograms. For any arbitrary point P in the interior of $ABCD$, parallelograms $RASC$ and $RPTQ$ are constructed. Prove that these two parallelograms have two vertices in common.

4. **(BGR 4)** Let K_a, K_b, K_c with centers O_a, O_b, O_c be the excircles of a triangle ABC, touching the interiors of the sides BC, CA, AB at points T_a, T_b, T_c respectively.
   Prove that the lines $O_a T_a, O_b T_b, O_c T_c$ are concurrent in a point P for which $PO_a = PO_b = PO_c = 2R$ holds, where R denotes the circumradius of ABC. Also prove that the circumcenter O of ABC is the midpoint of the segment PJ, where J is the incenter of ABC.

5. **(BGR 5)** A straight cone is given inside a rectangular parallelepiped B, with the apex at one of the vertices, say T, of the parallelepiped, and the base touching the three faces opposite to T. Its axis lies at the long diagonal through T. If V_1 and V_2 are the volumes of the cone and the parallelepiped respectively, prove that
$$V_1 \le \frac{\sqrt{3}\pi V_2}{27}.$$

6. **(CUB 1)** Prove that the product of two natural numbers with their sum cannot be the third power of a natural number.

7. **(CUB 2)** Let p be a prime number and n a natural number. Prove that the product
$$N = \frac{1}{p^{n^2}} \prod_{i=1;\, 2\nmid i}^{2n-1} \left[((p-1)i)! \binom{p^2 i}{pi} \right]$$
is a natural number that is not divisible by p.

8. **(CUB 3)** (SL74-9).

9. **(CZS 1)** Solve the following system of linear equations with unknown $x_1, \ldots, x_n$ ($n \ge 2$) and parameters $c_1, \ldots, c_n$:

$$
\begin{array}{rcl}
2x_1 \;-x_2 & = & c_1; \\
-x_1 +2x_2 \;-x_3 & = & c_2; \\
-x_2 +2x_3 \;-x_4 & = & c_3; \\
\cdots \qquad \cdots & & \cdots \\
-x_{n-2} +2x_{n-1} \;-x_n & = & c_{n-1}; \\
-x_{n-1} \;+2x_n & = & c_n.
\end{array}
$$

10. **(CZS 2)** A regular octagon P is given whose incircle k has diameter 1. About k is circumscribed a regular 16-gon, which is also inscribed in P, cutting from P eight isosceles triangles. To the octagon P, three of these triangles are added so that exactly two of them are adjacent and no two of them are opposite to each other. Every 11-gon so obtained is said to be P'.

Prove the following statement: Given a finite set M of points lying in P such that every two points of this set have a distance not exceeding 1, one of the 11-gons P' contains all of M.

11. **(CZS 3)** Given a line p and a triangle $\triangle$ in the plane, construct an equilateral triangle one of whose vertices lies on the line p, while the other two halve the perimeter of $\triangle$.

12. **(CZS 4)** A circle K with radius r, a point D on K, and a convex angle with vertex S and rays a and b are given in the plane. Construct a parallelogram $ABCD$ such that A and B lie on a and b respectively, $SA + SB = r$, and C lies on K.

13. **(FIN 1)** Prove that $2^{147} - 1$ is divisible by 343.

14. **(FIN 2)** Let n and k be natural numbers and $a_1, a_2, \ldots, a_n$ positive real numbers satisfying $a_1 + a_2 + \cdots + a_n = 1$. Prove that

$$a_1^{-k} + a_2^{-k} + \cdots + a_n^{-k} \geq n^{k+1}.$$

15. **(FIN 3)** (SL74-10).

16. **(UNK 1)** A pack of $2n$ cards contains n different pairs of cards. Each pair consists of two identical cards, either of which is called the *twin* of the other. A game is played between two players A and B. A third person called the *dealer* shuffles the pack and deals the cards one by one face upward onto the table. One of the players, called the *receiver*, takes the card dealt, provided he does not have already its twin. If he does already have the twin, his opponent takes the dealt card and becomes the receiver. A is initially the receiver and takes the first card dealt. The player who first obtains a complete set of n different cards wins the game. What fraction of all possible arrangements of the pack lead to A winning? Prove the correctness of your answer.

17. **(UNK 2)** Show that there exists a set S of 15 distinct circles on the surface of a sphere, all having the same radius and such that 5 touch exactly 5 others, 5 touch exactly 4 others, and 5 touch exactly 3 others.

18. **(UNK 3)** (SL74-5).

19. **(UNK 4)** (Alternative to UNK 2) Prove that there exists, for $n \geq 4$, a set S of $3n$ equal circles in space that can be partitioned into three subsets s_5, s_4, and s_3, each containing n circles, such that each circle in s_r touches exactly r circles in S.

20. **(NLD 1)** For which natural numbers n do there exist n natural numbers a_i ($1 \leq i \leq n$) such that $\sum_{i=1}^{n} a_i^{-2} = 1$?

21. **(NLD 2)** Let M be a nonempty subset of $\mathbb{Z}^+$ such that for every element x in M, the numbers $4x$ and $\lfloor \sqrt{x} \rfloor$ also belong to M. Prove that $M = \mathbb{Z}^+$.

22. **(NLD 3)** (SL74-8).

23. **(POL 1)** (SL74-2).

24. **(POL 2)** (SL74-7).

25. **(POL 3)** Let $f : \mathbb{R} \to \mathbb{R}$ be of the form $f(x) = x + \varepsilon \sin x$, where $0 < |\varepsilon| \le 1$. Define for any $x \in \mathbb{R}$,

$$x_n = \underbrace{f \circ \cdots \circ f}_{n \text{ times}}(x).$$

Show that for every $x \in \mathbb{R}$ there exists an integer k such that $\lim_{n \to \infty} x_n = k\pi$.

26. **(POL 4)** Let $g(k)$ be the number of partitions of a k-element set M, i.e., the number of families $\{A_1, A_2, \ldots, A_s\}$ of nonempty subsets of M such that $A_i \cap A_j = \emptyset$ for $i \ne j$ and $\bigcup_{i=1}^n A_i = M$. Prove that

$$n^n \le g(2n) \le (2n)^{2n} \quad \text{for every } n.$$

27. **(ROU 1)** Let C_1 and C_2 be circles in the same plane, P_1 and P_2 arbitrary points on C_1 and C_2 respectively, and Q the midpoint of segment $P_1 P_2$. Find the locus of points Q as P_1 and P_2 go through all possible positions.

    Alternative version. Let C_1, C_2, C_3 be three circles in the same plane. Find the locus of the centroid of triangle $P_1 P_2 P_3$ as P_1, P_2, and P_3 go through all possible positions on C_1, C_2, and C_3 respectively.

28. **(ROU 2)** Let M be a finite set and $P = \{M_1, M_2, \ldots, M_k\}$ a partition of M (i.e., $\bigcup_{i=1}^k M_i = M$, $M_i \ne \emptyset$, $M_i \cap M_j = \emptyset$ for all $i, j \in \{1, 2, \ldots, k\}$, $i \ne j$). We define the following elementary operation on P:

    Choose $i, j \in \{1, 2, \ldots, k\}$, such that $i \ne j$ and M_i has a elements and M_j has b elements such that $a \ge b$. Then take b elements from M_i and place them into M_j, i.e., M_j becomes the union of itself unifies and a b-element subset of M_i, while the same subset is subtracted from M_i (if $a = b$, M_i is thus removed from the partition).

    Let a finite set M be given. Prove that the property "for every partition P of M there exists a sequence $P = P_1, P_2, \ldots, P_r$ such that P_{i+1} is obtained from P_i by an elementary operation and $P_r = \{M\}$" is equivalent to "the number of elements of M is a power of 2."

29. **(ROU 3)** Let A, B, C, D be points in space. If for every point M on the segment AB the sum

    area(AMC)+area(CMD)+area(DMB)

    is constant show that the points A, B, C, D lie in the same plane.

30. **(ROU 4)** (SL74-6).

31. **(ROU 5)** Let $y^\alpha = \sum_{i=1}^n x_i^\alpha$, where $\alpha \neq 0$, $y > 0$, $x_i > 0$ are real numbers, and let $\lambda \neq \alpha$ be a real number. Prove that $y^\lambda > \sum_{i=1}^n x_i^\lambda$ if $\alpha(\lambda - \alpha) > 0$, and $y^\lambda < \sum_{i=1}^n x_i^\lambda$ if $\alpha(\lambda - \alpha) < 0$.

32. **(SWE 1)** Let $a_1, a_2, \ldots, a_n$ be n real numbers such that $0 < a \leq a_k \leq b$ for $k = 1, 2, \ldots, n$. If

$$m_1 = \frac{1}{n}(a_1 + a_2 + \cdots + a_n) \quad \text{and} \quad m_2 = \frac{1}{n}(a_1^2 + a_2^2 + \cdots + a_n^2),$$

prove that $m_2 \leq \frac{(a+b)^2}{4ab} m_1^2$ and find a necessary and sufficient condition for equality.

33. **(SWE 2)** Let a be a real number such that $0 < a < 1$, and let n be a positive integer. Define the sequence $a_0, a_1, a_2, \ldots, a_n$ recursively by

$$a_0 = a; \quad a_{k+1} = a_k + \frac{1}{n}a_k^2 \quad \text{for } k = 0, 1, \ldots, n-1.$$

Prove that there exists a real number A, depending on a but independent of n, such that

$$0 < n(A - a_n) < A^3.$$

34. **(SWE 3)** (SL74-3).

35. **(SWE 4)** If p and q are distinct prime numbers, then there are integers x_0 and y_0 such that $1 = px_0 + qy_0$. Determine the maximum value of $b - a$, where a and b are positive integers with the following property: If $a \leq t \leq b$, and t is an integer, then there are integers x and y with $0 \leq x \leq q - 1$ and $0 \leq y \leq p - 1$ such that $t = px + qy$.

36. **(SWE 5)** Consider infinite diagrams

$$D = \begin{vmatrix} \vdots & \vdots & \vdots & \\ n_{20} & n_{21} & n_{22} & \cdots \\ n_{10} & n_{11} & n_{12} & \cdots \\ n_{00} & n_{01} & n_{02} & \cdots \end{vmatrix}$$

where all but a finite number of the integers n_{ij}, $i = 0, 1, 2, \ldots$, $j = 0, 1, 2, \ldots$, are equal to 0. Three elements of a diagram are called *adjacent* if there are integers i and j with $i \geq 0$ and $j \geq 0$ such that the three elements are

(i) n_{ij}, $n_{i,j+1}$, $n_{i,j+2}$, or

(ii) n_{ij}, $n_{i+1,j}$, $n_{i+2,j}$, or

(iii) $n_{i+2,j}$, $n_{i+1,j+1}$, $n_{i,j+2}$.

An elementary operation on a diagram is an operation by which three adjacent elements n_{ij} are changed into n'_{ij} in such a way that $|n_{ij} - n'_{ij}| = 1$. Two diagrams are called equivalent if one of them can be changed into the other by a finite sequence of elementary operations. How many inequivalent diagrams exist?

37. **(USA 1)** Let a, b, and c denote the three sides of a billiard table in the shape of an equilateral triangle. A ball is placed at the midpoint of side a and then propelled toward side b with direction defined by the angle θ. For what values of θ will the ball strike the sides b, c, a in that order?

38. **(USA 2)** Consider the binomial coefficients $\binom{n}{k} = \frac{n!}{k!(n-k)!}$ $(k = 1, 2, \ldots, n-1)$. Determine all positive integers n for which $\binom{n}{1}, \binom{n}{2}, \ldots, \binom{n}{n-1}$ are all even numbers.

39. **(USA 3)** Let n be a positive integer, $n \geq 2$, and consider the polynomial equation

$$x^n - x^{n-2} - x + 2 = 0.$$

For each n, determine all complex numbers x that satisfy the equation and have modulus $|x| = 1$.

40. **(USA 4)** (SL74-1).

41. **(USA 5)** Through the circumcenter O of an arbitrary acute-angled triangle, chords A_1A_2, B_1B_2, C_1C_2 are drawn parallel to the sides BC, CA, AB of the triangle respectively. If R is the radius of the circumcircle, prove that

$$A_1O \cdot OA_2 + B_1O \cdot OB_2 + C_1O \cdot OC_2 = R^2.$$

42. **(USS 1)** (SL74-12).

43. **(USS 2)** An $(n^2 + n + 1) \times (n^2 + n + 1)$ matrix of zeros and ones is given. If no four ones are vertices of a rectangle, prove that the number of ones does not exceed $(n+1)(n^2 + n + 1)$.

44. **(USS 3)** We are given n mass points of equal mass in space. We define a sequence of points $O_1, O_2, O_3, \ldots$ as follows: O_1 is an arbitrary point (within the unit distance of at least one of the n points); O_2 is the center of gravity of all the n given points that are inside the unit sphere centered at O_1; O_3 is the center of gravity of all of the n given points that are inside the unit sphere centered at O_2; etc. Prove that starting from some m, all points $O_m, O_{m+1}, O_{m+2}, \ldots$ coincide.

45. **(USS 4)** (SL74-4).

46. **(USS 5)** Outside an arbitrary triangle ABC, triangles ADB and BCE are constructed such that $\angle ADB = \angle BEC = 90°$ and $\angle DAB = \angle EBC = 30°$. On the segment AC the point F with $AF = 3FC$ is chosen. Prove that

$$\angle DFE = 90° \quad \text{and} \quad \angle FDE = 30°.$$

47. **(VNM 1)** Given two points A, B outside of a given plane P, find the positions of points M in the plane P for which the ratio $\frac{MA}{MB}$ takes a minimum or maximum.

48. **(VNM 2)** Let a be a number different from zero. For all integers n define $S_n = a^n + a^{-n}$. Prove that if for some integer k both S_k and S_{k+1} are integers, then for each integer n the number S_n is an integer.

49. **(VNM 3)** Determine an equation of third degree with integral coefficients having roots $\sin \frac{\pi}{14}$, $\sin \frac{5\pi}{14}$, and $\sin \frac{-3\pi}{14}$.

50. **(YUG 1)** Let m and n be natural numbers with $m > n$. Prove that

$$2(m-n)^2(m^2 - n^2 + 1) \geq 2m^2 - 2mn + 1.$$

51. **(YUG 2)** There are n points on a flat piece of paper, any two of them at a distance of at least 2 from each other. An inattentive pupil spills ink on a part of the paper such that the total area of the damaged part equals $3/2$. Prove that there exist two vectors of equal length less than 1 and with their sum having a given direction, such that after a translation by either of these two vectors no points of the given set remain in the damaged area.

52. **(YUG 3)** A fox stands in the center of the field which has the form of an equilateral triangle, and a rabbit stands at one of its vertices. The fox can move through the whole field, while the rabbit can move only along the border of the field. The maximal speeds of the fox and rabbit are equal to u and v, respectively. Prove that:

(a) If $2u > v$, the fox can catch the rabbit, no matter how the rabbit moves.

(b) If $2u \leq v$, the rabbit can always run away from the fox.

3.16.3 Shortlisted Problems

1. **I 1 (USA 4)**[IMO1] Alice, Betty, and Carol took the same series of examinations. There was one grade of A, one grade of B, and one grade of C for each examination, where A, B, C are different positive integers. The final test scores were

Alice	Betty	Carol
20	10	9

If Betty placed first in the arithmetic examination, who placed second in the spelling examination?

2. **I 2 (POL 1)** Prove that the squares with sides $1/1, 1/2, 1/3, \ldots$ may be put into the square with side $3/2$ in such a way that no two of them have any interior point in common.

3. **I 3 (SWE 3)**[IMO6] Let $P(x)$ be a polynomial with integer coefficients. If $n(P)$ is the number of (distinct) integers k such that $P^2(k) = 1$, prove that

$$n(P) - \deg(P) \leq 2,$$

where $\deg(P)$ denotes the degree of the polynomial P.

4. **I 4 (USS 4)** The sum of the squares of five real numbers a_1, a_2, a_3, a_4, a_5 equals 1. Prove that the least of the numbers $(a_i - a_j)^2$, where $i, j = 1, 2, 3, 4, 5$ and $i \neq j$, does not exceed $1/10$.

5. **I 5 (UNK 3)** Let A_r, B_r, C_r be points on the circumference of a given circle S. From the triangle $A_r B_r C_r$, called $\triangle_r$, the triangle $\triangle_{r+1}$ is obtained by constructing the points $A_{r+1}, B_{r+1}, C_{r+1}$ on S such that $A_{r+1} A_r$ is parallel to $B_r C_r$, $B_{r+1} B_r$ is parallel to $C_r A_r$, and $C_{r+1} C_r$ is parallel to $A_r B_r$. Each angle of $\triangle_1$ is an integer number of degrees and those integers are not multiples of 45. Prove that at least two of the triangles $\triangle_1, \triangle_2, \ldots, \triangle_{15}$ are congruent.

6. **I 6 (ROU 4)**[IMO3] Does there exist a natural number n for which the number

$$\sum_{k=0}^{n} \binom{2n+1}{2k+1} 2^{3k}$$

is divisible by 5?

7. **II 1 (POL 2)** Let a_i, b_i be coprime positive integers for $i = 1, 2, \ldots, k$, and m the least common multiple of $b_1, \ldots, b_k$. Prove that the greatest common divisor of $a_1 \frac{m}{b_1}, \ldots, a_k \frac{m}{b_k}$ equals the greatest common divisor of $a_1, \ldots, a_k$.

8. **II 2 (NLD 3)**[IMO5] If a, b, c, d are arbitrary positive real numbers, find all possible values of

$$S = \frac{a}{a+b+d} + \frac{b}{a+b+c} + \frac{c}{b+c+d} + \frac{d}{a+c+d}.$$

9. **II 3 (CUB 3)** Let x, y, z be real numbers each of whose absolute value is different from $1/\sqrt{3}$ such that $x + y + z = xyz$. Prove that

$$\frac{3x - x^3}{1 - 3x^2} + \frac{3y - y^3}{1 - 3y^2} + \frac{3z - z^3}{1 - 3z^2} = \frac{3x - x^3}{1 - 3x^2} \cdot \frac{3y - y^3}{1 - 3y^2} \cdot \frac{3z - z^3}{1 - 3z^2}.$$

10. **II 4 (FIN 3)**[IMO2] Let $\triangle ABC$ be a triangle. Prove that there exists a point D on the side AB such that CD is the geometric mean of AD and BD if and only if $\sqrt{\sin A \sin B} \leq \sin \frac{C}{2}$.

11. **II 5 (BGR 1)**[IMO4] Consider a partition of an 8×8 chessboard into p rectangles whose interiors are disjoint such that each of them has an equal number of white and black cells. Assume that $a_1 < a_2 < \cdots < a_p$, where a_i denotes the number of white cells in the ith rectangle. Find the maximal p for which such a partition is possible and for that p determine all possible corresponding sequences $a_1, a_2, \ldots, a_p$.

12. **II 6 (USS 1)** In a certain language words are formed using an alphabet of three letters. Some words of two or more letters are not allowed, and any two such distinct words are of different lengths. Prove that one can form a word of arbitrary length that does not contain any nonallowed word.

3.17 The Seventeenth IMO
Burgas–Sofia, Bulgaria, 1975

3.17.1 Contest Problems

First Day (July 7)

1. Let $x_1 \geq x_2 \geq \cdots \geq x_n$ and $y_1 \geq y_2 \geq \cdots \geq y_n$ be two n-tuples of numbers. Prove that

$$\sum_{i=1}^{n}(x_i - y_i)^2 \leq \sum_{i=1}^{n}(x_i - z_i)^2$$

   is true when $z_1, z_2, \ldots, z_n$ denote $y_1, y_2, \ldots, y_n$ taken in another order.

2. Let $a_1, a_2, a_3, \ldots$ be any infinite increasing sequence of positive integers. (For every integer $i > 0$, $a_{i+1} > a_i$.) Prove that there are infinitely many m for which positive integers x, y, h, k can be found such that $0 < h < k < m$ and $a_m = x a_h + y a_k$.

3. On the sides of an arbitrary triangle ABC, triangles BPC, CQA, and ARB are externally erected such that
   $$\angle PBC = \angle CAQ = 45°,$$
   $$\angle BCP = \angle QCA = 30°,$$
   $$\angle ABR = \angle BAR = 15°.$$
   Prove that $\angle QRP = 90°$ and $QR = RP$.

Second Day (July 8)

4. Let A be the sum of the digits of the number 4444^{4444} and B the sum of the digits of the number A. Find the sum of the digits of the number B.

5. Is it possible to plot 1975 points on a circle with radius 1 so that the distance between any two of them is a rational number (distances have to be measured by chords)?

6. The function $f(x,y)$ is a homogeneous polynomial of the nth degree in x and y. If $f(1,0) = 1$ and for all a,b,c,

   $$f(a+b,c) + f(b+c,a) + f(c+a,b) = 0,$$

   prove that $f(x,y) = (x - 2y)(x + y)^{n-1}$.

3.17.2 Shortlisted Problems

1. **(FRA)** There are six ports on a lake. Is it possible to organize a series of routes satisfying the following conditions:
   (i) Every route includes exactly three ports;
   (ii) No two routes contain the same three ports;

    (iii) The series offers exactly two routes to each tourist who desires to visit two different arbitrary ports?

2. **(CZS)**[IMO1] Let $x_1 \geq x_2 \geq \cdots \geq x_n$ and $y_1 \geq y_2 \geq \cdots \geq y_n$ be two n-tuples of numbers. Prove that

$$\sum_{i=1}^{n}(x_i - y_i)^2 \leq \sum_{i=1}^{n}(x_i - z_i)^2$$

is true when $z_1, z_2, \ldots, z_n$ denote $y_1, y_2, \ldots, y_n$ taken in another order.

3. **(USA)** Find the integer represented by $\left[\sum_{n=1}^{10^9} n^{-2/3} \right]$. Here $[x]$ denotes the greatest integer less than or equal to x (e.g. $[\sqrt{2}] = 1$).

4. **(SWE)** Let $a_1, a_2, \ldots, a_n, \ldots$ be a sequence of real numbers such that $0 \leq a_n \leq 1$ and $a_n - 2a_{n+1} + a_{n+2} \geq 0$ for $n = 1, 2, 3, \ldots$. Prove that

$$0 \leq (n+1)(a_n - a_{n+1}) \leq 2 \quad \text{for } n = 1, 2, 3, \ldots.$$

5. **(SWE)** Let M be the set of all positive integers that do not contain the digit 9 (base 10). If $x_1, \ldots, x_n$ are arbitrary but distinct elements in M, prove that

$$\sum_{j=1}^{n} \frac{1}{x_j} < 80.$$

6. **(USS)**[IMO4] Let A be the sum of the digits of the number 16^{16} and B the sum of the digits of the number A. Find the sum of the digits of the number B without calculating 16^{16}.

7. **(GDR)** Prove that from $x + y = 1$ $(x, y \in \mathbb{R})$ it follows that

$$x^{m+1} \sum_{j=0}^{n} \binom{m+j}{j} y^j + y^{n+1} \sum_{i=0}^{m} \binom{n+i}{i} x^i = 1 \quad (m, n = 0, 1, 2, \ldots).$$

8. **(NLD)**[IMO3] On the sides of an arbitrary triangle ABC, triangles BPC, CQA, and ARB are externally erected such that
$$\angle PBC = \angle CAQ = 45°,$$
$$\angle BCP = \angle QCA = 30°,$$
$$\angle ABR = \angle BAR = 15°.$$
Prove that $\angle QRP = 90°$ and $QR = RP$.

9. **(NLD)** Let $f(x)$ be a continuous function defined on the closed interval $0 \leq x \leq 1$. Let $G(f)$ denote the graph of $f(x)$: $G(f) = \{(x, y) \in \mathbb{R}^2 \mid 0 \leq x \leq 1, y = f(x)\}$. Let $G_a(f)$ denote the graph of the translated function $f(x - a)$ (translated over a distance a), defined by $G_a(f) = \{(x, y) \in \mathbb{R}^2 \mid a \leq x \leq a+1, y = f(x-a)\}$. Is it possible to find for every a, $0 < a < 1$, a continuous function $f(x)$, defined on $0 \leq x \leq 1$, such that $f(0) = f(1) = 0$ and $G(f)$ and $G_a(f)$ are disjoint point sets?

10. **(UNK)**[IMO6] The function $f(x,y)$ is a homogeneous polynomial of the nth degree in x and y. If $f(1,0) = 1$ and for all a,b,c,

$$f(a+b,c)+f(b+c,a)+f(c+a,b)=0,$$

prove that $f(x,y) = (x-2y)(x+y)^{n-1}$.

11. **(UNK)**[IMO2] Let $a_1,a_2,a_3,\dots$ be any infinite increasing sequence of positive integers. (For every integer $i > 0$, $a_{i+1} > a_i$.) Prove that there are infinitely many m for which positive integers x,y,h,k can be found such that $0 < h < k < m$ and $a_m = xa_h + ya_k$.

12. **(HEL)** Consider on the first quadrant of the trigonometric circle the arcs $AM_1 = x_1, AM_2 = x_2, AM_3 = x_3, \dots, AM_v = x_v$, such that $x_1 < x_2 < x_3 < \cdots < x_v$. Prove that

$$\sum_{i=0}^{v-1} \sin 2x_i - \sum_{i=0}^{v-1} \sin(x_i - x_{i+1}) < \frac{\pi}{2} + \sum_{i=0}^{v-1} \sin(x_i + x_{i+1}).$$

13. **(ROU)** Let $A_0, A_1, \dots, A_n$ be points in a plane such that
    (i) $A_0A_1 \le \frac{1}{2}A_1A_2 \le \cdots \le \frac{1}{2^{n-1}}A_{n-1}A_n$ and
    (ii) $0 < \angle A_0A_1A_2 < \angle A_1A_2A_3 < \cdots < \angle A_{n-2}A_{n-1}A_n < 180°$,
    where all these angles have the same orientation. Prove that the segments A_kA_{k+1}, A_mA_{m+1} do not intersect for each k and m such that $0 \le k \le m-2 < n-2$.

14. **(YUG)** Let $x_0 = 5$ and $x_{n+1} = x_n + \frac{1}{x_n}$ $(n = 0,1,2,\dots)$. Prove that $45 < x_{1000} < 45.1$.

15. **(USS)**[IMO5] Is it possible to plot 1975 points on a circle with radius 1 so that the distance between any two of them is a rational number (distances have to be measured by chords)?

3.18 The Eighteenth IMO
Vienna–Lienz, Austria, 1976

3.18.1 Contest Problems

First Day (July 12)

1. In a convex quadrangle with area $32\,\text{cm}^2$, the sum of the lengths of two nonadjacent edges and of the length of one diagonal is equal to $16\,\text{cm}$. What is the length of the other diagonal?

2. Let $P_1(x) = x^2 - 2$, $P_j(x) = P_1(P_{j-1}(x))$, $j = 2,3,\dots$. Show that for arbitrary n, the roots of the equation $P_n(x) = x$ are real and different from one another.

3. A rectangular box can be filled completely with unit cubes. If one places the maximal number of cubes with volume 2 in the box such that their edges are parallel to the edges of the box, one can fill exactly 40% of the box. Determine all possible (interior) sizes of the box.

Second Day (July 13)

4. Find the largest number obtainable as the product of positive integers whose sum is 1976.

5. Let a set of p equations be given,

$$a_{11}x_1 + \cdots + a_{1q}x_q = 0,$$
$$a_{21}x_1 + \cdots + a_{2q}x_q = 0,$$
$$\vdots$$
$$a_{p1}x_1 + \cdots + a_{pq}x_q = 0,$$

with coefficients a_{ij} satisfying $a_{ij} = -1$, 0, or $+1$ for all $i = 1,\dots,p$ and $j = 1,\dots,q$. Prove that if $q = 2p$, there exists a solution $x_1,\dots,x_q$ of this system such that all x_j $(j = 1,\dots,q)$ are integers satisfying $|x_j| \le q$ and $x_j \ne 0$ for at least one value of j.

6. For all positive integral n, $u_{n+1} = u_n(u_{n-1}^2 - 2) - u_1$, $u_0 = 2$, and $u_1 = 2\frac{1}{2}$. Prove that

$$3\log_2 [u_n] = 2^n - (-1)^n,$$

where $[x]$ is the integral part of x.

3.18.2 Longlisted Problems

1. **(BGR 1)** (SL76-1).

2. **(BGR 2)** Let P be a set of n points and S a set of l segments. It is known that:
   (i) No four points of P are coplanar.
   (ii) Any segment from S has its endpoints at P.

(iii) There is a point, say g, in P that is the endpoint of a maximal number of segments from S and that is not a vertex of a tetrahedron having all its edges in S.

Prove that $l \leq \frac{n^2}{3}$.

3. **(BGR 3)** (SL76-2).

4. **(BGR 4)** Find all pairs of natural numbers (m,n) for which $2^m \cdot 3^n + 1$ is the square of some integer.

5. **(BGR 5)** Let $ABCDS$ be a pyramid with four faces and with $ABCD$ as a base, and let a plane α through the vertex A meet its edges SB and SD at points M and N, respectively. Prove that if the intersection of the plane α with the pyramid $ABCDS$ is a parallelogram, then

$$SM \cdot SN > BM \cdot DN.$$

6. **(CZS 1)** For each point X of a given polytope, denote by $f(X)$ the sum of the distances of the point X from all the planes of the faces of the polytope.

Prove that if f attains its maximum at an interior point of the polytope, then f is constant.

7. **(CZS 2)** Let P be a fixed point and T a given triangle that contains the point P. Translate the triangle T by a given vector $\mathbf{v}$ and denote by T' this new triangle. Let r, R, respectively, be the radii of the smallest disks centered at P that contain the triangles T, T', respectively.

Prove that

$$r + |\mathbf{v}| \leq 3R$$

and find an example to show that equality can occur.

8. **(CZS 3)** (SL76-3).

9. **(CZS 4)** Find all (real) solutions of the system

$$
\begin{aligned}
3x_1 - x_2 - x_3 \quad - x_5 \quad\quad\quad\quad &= 0, \\
-x_1 + 3x_2 \quad - x_4 \quad - x_6 \quad\quad &= 0, \\
-x_1 \quad + 3x_3 - x_4 \quad\quad - x_7 \quad &= 0, \\
-x_2 - x_3 + 3x_4 \quad\quad\quad\quad - x_8 &= 0, \\
-x_1 \quad\quad\quad + 3x_5 - x_6 - x_7 \quad &= 0, \\
-x_2 \quad\quad - x_5 + 3x_6 \quad - x_8 &= 0, \\
-x_3 \quad - x_5 \quad + 3x_7 - x_8 &= 0, \\
-x_4 \quad - x_6 - x_7 + 3x_8 &= 0.
\end{aligned}
$$

10. **(FIN 1)** Show that the reciprocal of any number of the form $2(m^2 + m + 1)$, where m is a positive integer, can be represented as a sum of consecutive terms in the sequence $(a_j)_{j=1}^{\infty}$,

$$a_j = \frac{1}{j(j+1)(j+2)}.$$

11. **(FIN 2)** (SL76-9).

12. **(FIN 3)** Five points lie on the surface of a ball of unit radius. Find the maximum of the smallest distance between any two of them.

13. **(UNK 1a)** (SL76-4).

14. **(UNK 1b)** A sequence $\{u_n\}$ of integers is defined by

$$u_1 = 2, \quad u_2 = u_3 = 7,$$
$$u_{n+1} = u_n u_{n-1} - u_{n-2}, \quad \text{for } n \geq 3.$$

Prove that for each $n \geq 1$, u_n differs by 2 from an integral square.

15. **(UNK 2)** Let ABC and $A'B'C'$ be any two coplanar triangles. Let L be a point such that $AL\|BC$, $A'L\|B'C'$, and M,N similarly defined. The line BC meets $B'C'$ at P, and similarly defined are Q and R. Prove that PL, QM, RN are concurrent.

16. **(UNK 3)** Prove that there is a positive integer n such that the decimal representation of 7^n contains a block of at least m consecutive zeros, where m is any given positive integer.

17. **(UNK 4)** Show that there exists a convex polyhedron with all its vertices on the surface of a sphere and with all its faces congruent isosceles triangles whose ratio of sides are $\sqrt{3} : \sqrt{3} : 2$.

18. **(GDR 1)** Prove that the number $19^{1976} + 76^{1976}$:
    (a) is divisible by the (Fermat) prime number $F_4 = 2^{2^4} + 1$;
    (b) is divisible by at least four distinct primes other than F_4.

19. **(GDR 2)** For a positive integer n, let $6^{(n)}$ be the natural number whose decimal representation consists of n digits 6. Let us define, for all natural numbers m,k with $1 \leq k \leq m$,

$$\begin{bmatrix} m \\ k \end{bmatrix} = \frac{6^{(m)} \cdot 6^{(m-1)} \cdots 6^{(m-k+1)}}{6^{(1)} \cdot 6^{(2)} \cdots 6^{(k)}}.$$

Prove that for all m,k, $\begin{bmatrix} m \\ k \end{bmatrix}$ is a natural number whose decimal representation consists of exactly $k(m+k-1)-1$ digits.

20. **(GDR 3)** Let (a_n), $n = 0,1,\ldots$, be a sequence of real numbers such that $a_0 = 0$ and

$$a_{n+1}^3 = \frac{1}{2}a_n^2 - 1, \quad n = 0,1,\ldots.$$

Prove that there exists a positive number q, $q < 1$, such that for all $n = 1,2,\ldots$,

$$|a_{n+1} - a_n| \leq q|a_n - a_{n-1}|,$$

and give one such q explicitly.

21. **(GDR 4)** Find the largest positive real number p (if it exists) such that the inequality

$$x_1^2 + x_2^2 + \cdots + x_n^2 \geq p(x_1x_2 + x_2x_3 + \cdots + x_{n-1}x_n) \tag{1}$$

is satisfied for all real numbers x_i, and (a) $n = 2$; (b) $n = 5$.
Find the largest positive real number p (if it exists) such that the inequality (1) holds for all real numbers x_i and all natural numbers n, $n \geq 2$.

22. **(GDR 5)** A regular pentagon $A_1A_2A_3A_4A_5$ with side length s is given. At each point A_i a sphere K_i of radius $s/2$ is constructed. There are two spheres K_1' and K_2' eah of radius $s/2$ touching all the five spheres K_i. Decide whether K_1' and K_2' intersect each other, touch each other, or have no common points.

23. **(NLD 1)** Prove that in a Euclidean plane there are infinitely many concentric circles C such that all triangles inscribed in C have at least one irrational side.

24. **(NLD 2)** Let $0 \leq x_1 \leq x_2 \leq \cdots \leq x_n \leq 1$. Prove that for all $A \geq 1$ there exists an interval I of length $2\sqrt[n]{A}$ such that for all $x \in I$,

$$|(x - x_1)(x - x_2) \cdots (x - x_n)| \leq A.$$

25. **(NLD 3)** (SL76-5).

26. **(NLD 4)** (SL76-6).

27. **(NLD 5)** In a plane three points P, Q, R, not on a line, are given. Let k, l, m be positive numbers. Construct a triangle ABC whose sides pass through P, Q, and R such that
    P divides the segment AB in the ratio $1 : k$,
    Q divides the segment BC in the ratio $1 : l$, and
    R divides the segment CA in the ratio $1 : m$.

28. **(POL 1a)** Let Q be a unit square in the plane: $Q = [0, 1] \times [0, 1]$. Let $T : Q \rightarrow Q$ be defined as follows:

$$T(x, y) = \begin{cases} (2x, y/2) & \text{if } 0 \leq x \leq 1/2; \\ (2x - 1, y/2 + 1/2) & \text{if } 1/2 < x \leq 1. \end{cases}$$

Show that for every disk $D \subset Q$ there exists an integer $n > 0$ such that $T^n(D) \cap D \neq \emptyset$.

29. **(POL 1b)** (SL76-7).

30. **(POL 2)** Prove that if $P(x) = (x - a)^k Q(x)$, where k is a positive integer, a is a nonzero real number, $Q(x)$ is a nonzero polynomial, then $P(x)$ has at least $k + 1$ nonzero coefficients.

31. **(POL 3)** Into every lateral face of a quadrangular pyramid a circle is inscribed. The circles inscribed into adjacent faces are tangent (have one point in common). Prove that the points of contact of the circles with the base of the pyramid lie on a circle.

32. **(POL 4)** We consider the infinite chessboard covering the whole plane. In every field of the chessboard there is a nonnegative real number. Every number is the

arithmetic mean of the numbers in the four adjacent fields of the chessboard. Prove that the numbers occurring in the fields of the chessboard are all equal.

33. **(SWE 1)** A finite set of points P in the plane has the following property: Every line through two points in P contains at least one more point belonging to P. Prove that all points in P lie on a straight line.

34. **(SWE 2)** Let $\{a_n\}_0^\infty$ and $\{b_n\}_0^\infty$ be two sequences determined by the recursion formulas
$$a_{n+1} = a_n + b_n,$$
$$b_{n+1} = 3a_n + b_n, \qquad n = 0, 1, 2, \ldots,$$
and the initial values $a_0 = b_0 = 1$. Prove that there exists a uniquely determined constant c such that $n|ca_n - b_n| < 2$ for all nonnegative integers n.

35. **(SWE 3)** (SL76-8).

36. **(USA 1)** Three concentric circles with common center O are cut by a common chord in successive points A, B, C. Tangents drawn to the circles at the points A, B, C enclose a triangular region. If the distance from point O to the common chord is equal to p, prove that the area of the region enclosed by the tangents is equal to
$$\frac{AB \cdot BC \cdot CA}{2p}.$$

37. **(USA 2)** From a square board 11 squares long and 11 squares wide, the central square is removed. Prove that the remaining 120 squares cannot be covered by 15 strips each 8 units long and one unit wide.

38. **(USA 3)** Let $x = \sqrt{a} + \sqrt{b}$, where a and b are natural numbers, x is not an integer, and $x < 1976$. Prove that the fractional part of x exceeds $10^{-19.76}$.

39. **(USA 4)** In $\triangle ABC$, the inscribed circle is tangent to side BC at X. Segment AX is drawn. Prove that the line joining the midpoint of segment AX to the midpoint of side BC passes through the center I of the inscribed circle.

40. **(USA 5)** Let $g(x)$ be a fixed polynomial and define $f(x)$ by $f(x) = x^2 + xg(x^3)$. Show that $f(x)$ is not divisible by $x^2 - x + 1$.

41. **(USA 6)** (SL76-10).

42. **(USS 1)** For a point O inside a triangle ABC, denote by A_1, B_1, C_1 the respective intersection points of AO, BO, CO with the corresponding sides. Let $n_1 = \frac{AO}{A_1O}$, $n_2 = \frac{BO}{B_1O}$, $n_3 = \frac{CO}{C_1O}$. What possible values of n_1, n_2, n_3 can all be positive integers?

43. **(USS 2)** Prove that if for a polynomial $P(x, y)$ we have $P(x - 1, y - 2x + 1) = P(x, y)$, then there exists a polynomial $\Phi(x)$ with $P(x, y) = \Phi(y - x^2)$.

44. **(USS 3)** A circle of radius 1 rolls around a circle of radius $\sqrt{2}$. Initially, the tangent point is colored red. Afterwards, the red points map from one circle to another by contact. How many red points will be on the bigger circle when the center of the smaller one has made n circuits around the bigger one?

45. **(USS 4)** We are given n $(n \geq 5)$ circles in a plane. Suppose that every three of them have a common point. Prove that all n circles have a common point.

46. **(USS 5)** For $a \geq 0, b \geq 0, c \geq 0, d \geq 0$, prove the inequality

$$a^4 + b^4 + c^4 + d^4 + 2abcd \geq a^2b^2 + a^2c^2 + a^2d^2 + b^2c^2 + b^2d^2 + c^2d^2.$$

47. **(VNM 1)** (SL76-11).

48. **(VNM 2)** (SL76-12).

49. **(VNM 3)** Determine whether there exist 1976 nonsimilar triangles with angles α, β, γ, each of them satisfying the relations

$$\frac{\sin\alpha + \sin\beta + \sin\gamma}{\cos\alpha + \cos\beta + \cos\gamma} = \frac{12}{7} \quad \text{and} \quad \sin\alpha\sin\beta\sin\gamma = \frac{12}{25}.$$

50. **(VNM 4)** Find a function $f(x)$ defined for all real values of x such that for all x, $f(x+2) - f(x) = x^2 + 2x + 4$, and if $x \in [0, 2)$, then $f(x) = x^2$.

51. **(YUG 1)** Four swallows are catching a fly. At first, the swallows are at the four vertices of a tetrahedron, and the fly is in its interior. Their maximal speeds are equal. Prove that the swallows can catch the fly.

3.18.3 Shortlisted Problems

1. **(BGR 1)** Let ABC be a triangle with bisectors AA_1, BB_1, CC_1 $(A_1 \in BC$, etc.) and M their common point. Consider the triangles $MB_1A, MC_1A, MC_1B, MA_1B, MA_1C, MB_1C$, and their inscribed circles. Prove that if four of these six inscribed circles have equal radii, then $AB = BC = CA$.

2. **(BGR 3)** Let $a_0, a_1, \ldots, a_n, a_{n+1}$ be a sequence of real numbers satisfying the following conditions:

$$a_0 = a_{n+1} = 0,$$
$$|a_{k-1} - 2a_k + a_{k+1}| \leq 1 \quad (k = 1, 2, \ldots, n).$$

Prove that $|a_k| \leq \frac{k(n+1-k)}{2}$ $(k = 0, 1, \ldots, n+1)$.

3. **(CZS 3)**[IMO1] In a convex quadrangle with area $32\,\text{cm}^2$, the sum of the lengths of two nonadjacent edges and of the length of one diagonal is equal to $16\,\text{cm}$.
   (a) What is the length of the other diagonal?
   (b) What are the lengths of the edges of the quadrangle if the perimeter is a minimum?
   (c) Is it possible to choose the edges in such a way that the perimeter is a maximum?

4. **(UNK 1a)**[IMO6] For all positive integral n, $u_{n+1} = u_n(u_{n-1}^2 - 2) - u_1$, $u_0 = 2$, and $u_1 = 5/2$. Prove that $3\log_2[u_n] = 2^n - (-1)^n$, where $[x]$ is the integral part of x.

5. **(NLD 3)**[IMO5] Let a set of p equations be given,

$$a_{11}x_1 + \cdots + a_{1q}x_q = 0,$$
$$a_{21}x_1 + \cdots + a_{2q}x_q = 0,$$
$$\vdots$$
$$a_{p1}x_1 + \cdots + a_{pq}x_q = 0,$$

with coefficients a_{ij} satisfying $a_{ij} = -1, 0$, or $+1$ for all $i = 1,\ldots,p$ and $j = 1,\ldots,q$. Prove that if $q = 2p$, there exists a solution $x_1,\ldots,x_q$ of this system such that all x_j $(j = 1,\ldots,q)$ are integers satisfying $|x_j| \le q$ and $x_j \ne 0$ for at least one value of j.

6. **(NLD 4)**[IMO3] A rectangular box can be filled completely with unit cubes. If one places the maximal number of cubes with volume 2 in the box such that their edges are parallel to the edges of the box, one can fill exactly 40% of the box. Determine all possible (interior) sizes of the box.

7. **(POL 1b)** Let $I = (0,1]$ be the unit interval of the real line. For a given number $a \in (0,1)$ we define a map $T : I \to I$ by the formula

$$T(x,y) = \begin{cases} x + (1-a) & \text{if } 0 < x \le a, \\ x - a & \text{if } a < x \le 1. \end{cases}$$

Show that for every interval $J \subset I$ there exists an integer $n > 0$ such that $T^n(J) \cap J \ne \emptyset$.

8. **(SWE 3)** Let P be a polynomial with real coefficients such that $P(x) > 0$ if $x > 0$. Prove that there exist polynomials Q and R with nonnegative coefficients such that $P(x) = \frac{Q(x)}{R(x)}$ if $x > 0$.

9. **(FIN 2)**[IMO2] Let $P_1(x) = x^2 - 2$, $P_j(x) = P_1(P_{j-1}(x))$, $j = 2,3,\ldots$. Show that for arbitrary n the roots of the equation $P_n(x) = x$ are real and different from one another.

10. **(USA 6)**[IMO4] Find the largest number obtainable as the product of positive integers whose sum is 1976.

11. **(VNM 1)** Prove that there exist infinitely many positive integers n such that the decimal representation of 5^n contains a block of 1976 consecutive zeros.

12. **(VNM 2)** The polynomial $1976(x + x^2 + \cdots + x^n)$ is decomposed into a sum of polynomials of the form $a_1x + a_2x^2 + \ldots + a_nx^n$, where $a_1, a_2, \cdots, a_n$ are distinct positive integers not greater than n. Find all values of n for which such a decomposition is possible.

3.19 The Nineteenth IMO
Belgrade–Arandjelovac, Yugoslavia, July 1–13, 1977

3.19.1 Contest Problems

First Day (July 6)

1. Equilateral triangles ABK, BCL, CDM, DAN are constructed inside the square $ABCD$. Prove that the midpoints of the four segments KL, LM, MN, NK and the midpoints of the eight segments AK, BK, BL, CL, CM, DM, DN, AN are the twelve vertices of a regular dodecagon.

2. In a finite sequence of real numbers the sum of any seven successive terms is negative, and the sum of any eleven successive terms is positive. Determine the maximum number of terms in the sequence.

3. Let n be a given integer greater than 2, and let V_n be the set of integers $1 + kn$, where $k = 1, 2, \ldots$. A number $m \in V_n$ is called indecomposable in V_n if there do not exist numbers $p, q \in V_n$ such that $pq = m$. Prove that there exists a number $r \in V_n$ that can be expressed as the product of elements indecomposable in V_n in more than one way. (Expressions that differ only in order of the elements of V_n will be considered the same.)

Second Day (July 7)

4. Let a, b, A, B be given constant real numbers and

$$f(x) = 1 - a\cos x - b\sin x - A\cos 2x - B\sin 2x.$$

Prove that if $f(x) \geq 0$ for all real x, then

$$a^2 + b^2 \leq 2 \quad \text{and} \quad A^2 + B^2 \leq 1.$$

5. Let a and b be natural numbers and let q and r be the quotient and remainder respectively when $a^2 + b^2$ is divided by $a + b$. Determine the numbers a and b if $q^2 + r = 1977$.

6. Let $f : \mathbb{N} \to \mathbb{N}$ be a function that satisfies the inequality $f(n+1) > f(f(n))$ for all $n \in \mathbb{N}$. Prove that $f(n) = n$ for all natural numbers n.

3.19.2 Longlisted Problems

1. **(BGR 1)** A pentagon $ABCDE$ inscribed in a circle for which $BC < CD$ and $AB < DE$ is the base of a pyramid with vertex S. If AS is the longest edge starting from S, prove that $BS > CS$.

2. **(BGR 2)** (SL77-1).

3. **(BGR 3)** In a company of n persons, each person has no more than d acquaintances, and in that company there exists a group of k persons, $k \geq d$, who are not acquainted with each other. Prove that the number of acquainted pairs is not greater than $[n^2/4]$.

4. **(BGR 4)** We are given n points in space. Some pairs of these points are connected by line segments so that the number of segments equals $[n^2/4]$, and a connected triangle exists. Prove that any point from which the maximal number of segments starts is a vertex of a connected triangle.

5. **(CZS 1)** (SL77-2).

6. **(CZS 2)** Let $x_1, x_2, \ldots, x_n$ ($n \geq 1$) be real numbers such that $0 \leq x_j \leq \pi$, $j = 1, 2, \ldots, n$. Prove that if $\sum_{j=1}^{n}(\cos x_j + 1)$ is an odd integer, then $\sum_{j=1}^{n} \sin x_j \geq 1$.

7. **(CZS 3)** Prove the following assertion: If $c_1, c_2, \ldots, c_n$ ($n \geq 2$) are real numbers such that
$$(n-1)(c_1^2 + c_2^2 + \cdots + c_n^2) = (c_1 + c_2 + \cdots + c_n)^2,$$
then either all these numbers are nonnegative or all these numbers are nonpositive.

8. **(CZS 4)** A hexahedron $ABCDE$ is made of two regular congruent tetrahedra $ABCD$ and $ABCE$. Prove that there exists only one isometry $\mathscr{I}$ that maps points A, B, C, D, E onto B, C, A, E, D, respectively. Find all points X on the surface of hexahedron whose distance from $\mathscr{I}(X)$ is minimal.

9. **(CZS 5)** Let $ABCD$ be a regular tetrahedron and $\mathscr{I}$ an isometry mapping A, B, C, D into B, C, D, A, respectively. Find the set $\mathscr{M}$ of all points X of the face ABC whose distance from $\mathscr{I}(X)$ is equal to a given number t. Find necessary and sufficient conditions for the set $\mathscr{M}$ to be nonempty.

10. **(FRG 1)** (SL77-3).

11. **(FRG 2)** Let n and z be integers greater than 1 and $(n, z) = 1$. Prove:
    (a) At least one of the numbers $z_i = 1 + z + z^2 + \cdots + z^i$, $i = 0, 1, \ldots, n-1$, is divisible by n.
    (b) If $(z-1, n) = 1$, then at least one of the numbers z_i, $i = 0, 1, \ldots, n-2$, is divisible by n.

12. **(FRG 3)** Let z be an integer > 1 and let M be the set of all numbers of the form $z_k = 1 + z + \cdots + z^k$, $k = 0, 1, \ldots$. Determine the set T of divisors of at least one of the numbers z_k from M.

13. **(FRG 4)** (SL77-4).

14. **(FRG 5)** (SL77-5).

15. **(GDR 1)** Let n be an integer greater than 1. In the Cartesian coordinate system we consider all squares with integer vertices (x, y) such that $1 \leq x, y \leq n$. Denote by p_k ($k = 0, 1, 2, \ldots$) the number of pairs of points that are vertices of exactly k such squares. Prove that $\sum_k (k-1)p_k = 0$.

16. **(GDR 2)** (SL77-6).

17. **(GDR 3)** A ball K of radius r is touched from the outside by mutually equal balls of radius R. Two of these balls are tangent to each other. Moreover, for two balls K_1 and K_2 tangent to K and tangent to each other there exist two other balls tangent to K_1, K_2 and also to K. How many balls are tangent to K? For a given r determine R.

18. **(GDR 4)** Given an isosceles triangle ABC with a right angle at C, construct the center M and radius r of a circle cutting on segments AB, BC, CA the segments DE, FG, and HK, respectively, such that $\angle DME + \angle FMG + \angle HMK = 180°$ and $DE : FG : HK = AB : BC : CA$.

19. **(UNK 1)** Given any integer $m > 1$ prove that there exist infinitely many positive integers n such that the last m digits of 5^n are a sequence $a_m, a_{m-1}, \ldots, a_1 = 5$ ($0 \le a_j < 10$) in which each digit except the last is of opposite parity to its successor (i.e., if a_i is even, then a_{i-1} is odd, and if a_i is odd, then a_{i-1} is even).

20. **(UNK 2)** (SL77-7).

21. **(UNK 3)** Given that $x_1 + x_2 + x_3 = y_1 + y_2 + y_3 = x_1 y_1 + x_2 y_2 + x_3 y_3 = 0$, prove that

$$\frac{x_1^2}{x_1^2 + x_2^2 + x_3^2} + \frac{y_1^2}{y_1^2 + y_2^2 + y_3^2} = \frac{2}{3}.$$

22. **(UNK 4)** (SL77-8).

23. **(HUN 1)** (SL77-9).

24. **(HUN 2)** Determine all real functions $f(x)$ that are defined and continuous on the interval $(-1, 1)$ and that satisfy the functional equation

$$f(x+y) = \frac{f(x) + f(y)}{1 - f(x)f(y)} \quad (x, y, x+y \in (-1, 1)).$$

25. **(HUN 3)** Prove the identity

$$(z+a)^n = z^n + a \sum_{k=1}^{n} \binom{n}{k} (a - kb)^{k-1} (z + kb)^{n-k}.$$

26. **(NLD 1)** Let p be a prime number greater than 5. Let V be the collection of all positive integers n that can be written in the form $n = kp + 1$ or $n = kp - 1$ ($k = 1, 2, \ldots$). A number $n \in V$ is called *indecomposable* in V if it is impossible to find $k, l \in V$ such that $n = kl$. Prove that there exists a number $N \in V$ that can be factorized into indecomposable factors in V in more than one way.

27. **(NLD 2)** (SL77-10).

28. **(NLD 3)** (SL77-11).

29. **(NLD 4)** (SL77-12).

30. **(NLD 5)** A triangle ABC with $\angle A = 30°$ and $\angle C = 54°$ is given. On BC a point D is chosen such that $\angle CAD = 12°$. On AB a point E is chosen such that $\angle ACE = 6°$. Let S be the point of intersection of AD and CE. Prove that $BS = BC$.

31. **(POL 1)** Let f be a function defined on the set of pairs of nonzero rational numbers whose values are positive real numbers. Suppose that f satisfies the following conditions:
    (1) $f(ab,c) = f(a,c)f(b,c)$, $f(c,ab) = f(c,a)f(c,b)$;
    (2) $f(a,1-a) = 1$.
    Prove that $f(a,a) = f(a,-a) = 1$, $f(a,b)f(b,a) = 1$.

32. **(POL 2)** In a room there are nine men. Among every three of them there are two mutually acquainted. Prove that some four of them are mutually acquainted.

33. **(POL 3)** A circle K centered at $(0,0)$ is given. Prove that for every vector (a_1,a_2) there is a positive integer n such that the circle K translated by the vector $n(a_1,a_2)$ contains a lattice point (i.e., a point both of whose coordinates are integers).

34. **(POL 4)** (SL77-13).

35. **(ROU 1)** Find all numbers $N = \overline{a_1a_2\ldots a_n}$ for which $9 \times \overline{a_1a_2\ldots a_n} = \overline{a_n\ldots a_2a_1}$ such that at most one of the digits $a_1, a_2, \ldots, a_n$ is zero.

36. **(ROU 2)** Consider a sequence of numbers $(a_1,a_2,\ldots,a_{2^n})$. Define the operation

$$S((a_1,a_2,\ldots,a_{2^n})) = (a_1a_2,a_2a_3,\ldots,a_{2^n-1}a_{2^n},a_{2^n}a_1).$$

Prove that whatever the sequence $(a_1,a_2,\ldots,a_{2^n})$ is, with $a_i \in \{-1,1\}$ for $i = 1,2,\ldots,2^n$, after finitely many applications of the operation we get the sequence $(1,1,\ldots,1)$.

37. **(ROU 3)** Let $A_1, A_2, \ldots, A_{n+1}$ be positive integers such that $(A_i, A_{n+1}) = 1$ for every $i = 1, 2, \ldots, n$. Show that the equation

$$x_1^{A_1} + x_2^{A_2} + \cdots + x_n^{A_n} = x_{n+1}^{A_{n+1}}$$

has an infinite set of solutions $(x_1,x_2,\ldots,x_{n+1})$ in positive integers.

38. **(ROU 4)** Let $m_j > 0$ for $j = 1,2,\ldots,n$ and $a_1 \le \cdots \le a_n < b_1 \le \cdots \le b_n < c_1 \le \cdots \le c_n$ be real numbers. Prove:

$$\left[\sum_{j=1}^{n} m_j(a_j + b_j + c_j)\right]^2 > 3\left(\sum_{j=1}^{n} m_j\right)\left[\sum_{j=1}^{n} m_j(a_jb_j + b_jc_j + c_ja_j)\right].$$

39. **(ROU 5)** Consider 37 distinct points in space, all with integer coordinates. Prove that we may find among them three distinct points such that their barycenter has integer coordinates.

40. **(SWE 1)** The numbers $1,2,3,\ldots,64$ are placed on a chessboard, one number in each square. Consider all squares on the chessboard of size 2×2. Prove that

there are at least three such squares for which the sum of the 4 numbers contained exceeds 100.

41. **(SWE 2)** A wheel consists of a fixed circular disk and a mobile circular ring. On the disk the numbers $1, 2, 3, \ldots, N$ are marked, and on the ring N integers $a_1, a_2, \ldots, a_N$ of sum 1 are marked (see the figure). The ring can be turned into N different positions in which the numbers on the disk and on the ring match each other. Multiply every number on the ring with the corresponding number on the disk and form the sum of N products. In this way a sum is obtained for every position of the ring. Prove that the N sums are different.

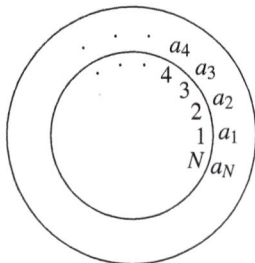

42. **(SWE 3)** The sequence $a_{n,k}$, $k = 1, 2, 3, \ldots, 2^n$, $n = 0, 1, 2, \ldots$, is defined by the following recurrence formula:

$$a_1 = 2, \quad a_{n,k} = 2a_{n-1,k}^3, \quad a_{n,k+2^{n-1}} = \frac{1}{2}a_{n-1,k}^3$$

$$\text{for } k = 1, 2, 3, \ldots, 2^{n-1}, \ n = 0, 1, 2, \ldots .$$

Prove that the numbers $a_{n,k}$ are all different.

43. **(FIN 1)** Evaluate

$$S = \sum_{k=1}^{n} k(k+1) \cdots (k+p),$$

where n and p are positive integers.

44. **(FIN 2)** Let E be a finite set of points in space such that E is not contained in a plane and no three points of E are collinear. Show that E contains the vertices of a tetrahedron $T = ABCD$ such that $T \cap E = \{A, B, C, D\}$ (including interior points of T) and such that the projection of A onto the plane BCD is inside a triangle that is similar to the triangle BCD and whose sides have midpoints B, C, D.

45. **(FIN 2′)** (SL77-14).

46. **(FIN 3)** Let f be a strictly increasing function defined on the set of real numbers. For x real and t positive, set

$$g(x,t) = \frac{f(x+t) - f(x)}{f(x) - f(x-t)}.$$

Assume that the inequalities

$$2^{-1} < g(x,t) < 2$$

hold for all positive t if $x = 0$, and for all $t \leq |x|$ otherwise.

Show that

$$14^{-1} < g(x,t) < 14$$

for all real x and positive t.

47. **(USS 1)** A square $ABCD$ is given. A line passing through A intersects CD at Q. Draw a line parallel to AQ that intersects the boundary of the square at points M and N such that the area of the quadrilateral $AMNQ$ is maximal.

48. **(USS 2)** The intersection of a plane with a regular tetrahedron with edge a is a quadrilateral with perimeter P. Prove that $2a \leq P \leq 3a$.

49. **(USS 3)** Find all pairs of integers (p,q) for which all roots of the trinomials $x^2 + px + q$ and $x^2 + qx + p$ are integers.

50. **(USS 4)** Determine all positive integers n for which there exists a polynomial $P_n(x)$ of degree n with integer coefficients that is equal to n at n different integer points and that equals zero at zero.

51. **(USS 5)** Several segments, which we shall call white, are given, and the sum of their lengths is 1. Several other segments, which we shall call black, are given, and the sum of their lengths is 1. Prove that every such system of segments can be distributed on the segment that is 1.51 long in the following way: Segments of the same color are disjoint, and segments of different colors are either disjoint or one is inside the other. Prove that there exists a system that cannot be distributed in that way on the segment that is 1.49 long.

52. **(USA 1)** Two perpendicular chords are drawn through a given interior point P of a circle with radius R. Determine, with proof, the maximum and the minimum of the sum of the lengths of these two chords if the distance from P to the center of the circle is kR.

53. **(USA 2)** Find all pairs of integers a and b for which

$$7a + 14b = 5a^2 + 5ab + 5b^2.$$

54. **(USA 3)** If $0 \leq a \leq b \leq c \leq d$, prove that

$$a^b b^c c^d d^a \geq b^a c^b d^c a^d.$$

55. **(USA 4)** Through a point O on the diagonal BD of a parallelogram $ABCD$, segments MN parallel to AB, and PQ parallel to AD, are drawn, with M on AD, and Q on AB. Prove that diagonals AO, BP, DN (extended if necessary) will be concurrent.

56. **(USA 5)** The four circumcircles of the four faces of a tetrahedron have equal radii. Prove that the four faces of the tetrahedron are congruent triangles.

57. **(VNM 1)** (SL77-15).

58. **(VNM 2)** Prove that for every triangle the following inequality holds:

$$\frac{ab + bc + ca}{4S} \geq \cot\frac{\pi}{6},$$

where a, b, c are lengths of the sides and S is the area of the triangle.

59. **(VNM 3)** (SL77-16).

60. **(VNM 4)** Suppose $x_0, x_1, \ldots, x_n$ are integers and $x_0 > x_1 > \cdots > x_n$. Prove that at least one of the numbers $|F(x_0)|, |F(x_1)|, |F(x_2)|, \ldots, |F(x_n)|$, where

$$F(x) = x^n + a_1 x^{n-1} + \cdots + a_n, \qquad a_i \in \mathbb{R}, \ i = 1, \ldots, n,$$

is greater than $\frac{n!}{2^n}$.

3.19.3 Shortlisted Problems

1. **(BGR 2)**[IMO6] Let $f : \mathbb{N} \to \mathbb{N}$ be a function that satisfies the inequality $f(n+1) > f(f(n))$ for all $n \in \mathbb{N}$. Prove that $f(n) = n$ for all natural numbers n.

2. **(CZS 1)** A lattice point in the plane is a point both of whose coordinates are integers. Each lattice point has four neighboring points: upper, lower, left, and right. Let k be a circle with radius $r \geq 2$, that does not pass through any lattice point. An interior boundary point is a lattice point lying inside the circle k that has a neighboring point lying outside k. Similarly, an exterior boundary point is a lattice point lying outside the circle k that has a neighboring point lying inside k. Prove that there are four more exterior boundary points than interior boundary points.

3. **(FRG 1)**[IMO5] Let a and b be natural numbers and let q and r be the quotient and remainder respectively when $a^2 + b^2$ is divided by $a + b$. Determine the numbers a and b if $q^2 + r = 1977$.

4. **(FRG 4)** Describe all closed bounded figures Φ in the plane any two points of which are connectable by a semicircle lying in Φ.

5. **(FRG 5)** There are 2^n words of length n over the alphabet $\{0, 1\}$. Prove that the following algorithm generates the sequence $w_0, w_1, \ldots, w_{2^n-1}$ of all these words such that any two consecutive words differ in exactly one digit.
   (1) $w_0 = 00\ldots0$ (n zeros).
   (2) Suppose $w_{m-1} = a_1 a_2 \ldots a_n$, $a_i \in \{0, 1\}$. Let $e(m)$ be the exponent of 2 in the representation of n as a product of primes, and let $j = 1 + e(m)$. Replace the digit a_j in the word w_{m-1} by $1 - a_j$. The obtained word is w_m.

6. **(GDR 2)** Let n be a positive integer. How many integer solutions (i, j, k, l), $1 \leq i, j, k, l \leq n$, does the following system of inequalities have:

$$\begin{array}{rl} 1 \leq & -j+k+l \leq n \\ 1 \leq & i-k+l \leq n \\ 1 \leq & i-j+l \leq n \\ 1 \leq & i+j-k \leq n \ ? \end{array}$$

7. **(UNK 2)**[IMO4] Let a, b, A, B be given constant real numbers and

$$f(x) = 1 - a\cos x - b\sin x - A\cos 2x - B\sin 2x.$$

Prove that if $f(x) \geq 0$ for all real x, then

$$a^2 + b^2 \leq 2 \quad \text{and} \quad A^2 + B^2 \leq 1.$$

8. **(UNK 4)** Let S be a convex quadrilateral $ABCD$ and O a point inside it. The feet of the perpendiculars from O to AB, BC, CD, DA are A_1, B_1, C_1, D_1 respectively. The feet of the perpendiculars from O to the sides of S_i, the quadrilateral $A_iB_iC_iD_i$, are $A_{i+1}B_{i+1}C_{i+1}D_{i+1}$, where $i = 1,2,3$. Prove that S_4 is similar to S.

9. **(HUN 1)** For which positive integers n do there exist two polynomials f and g with integer coefficients of n variables $x_1, x_2, \ldots, x_n$ such that the following equality is satisfied:

$$\left(\sum_{i=1}^{n} x_i \right) f(x_1, x_2, \ldots, x_n) = g(x_1^2, x_2^2, \ldots, x_n^2)?$$

10. **(NLD 2)**[IMO3] Let n be an integer greater than 2. Define $V = \{1 + kn \mid k = 1,2,\ldots\}$. A number $p \in V$ is called *indecomposable* in V if it is not possible to find numbers $q_1, q_2 \in V$ such that $q_1 q_2 = p$. Prove that there exists a number $N \in V$ that can be factorized into indecomposable factors in V in more than one way.

11. **(NLD 3)** Let n be an integer greater than 1. Define

$$x_1 = n, \ y_1 = 1, \quad x_{i+1} = \left[\frac{x_i + y_i}{2} \right], \ y_{i+1} = \left[\frac{n}{x_{i+1}} \right] \quad \text{for } i = 1, 2, \ldots,$$

where $[z]$ denotes the largest integer less than or equal to z. Prove that

$$\min\{x_1, x_2, \ldots x_n\} = [\sqrt{n}].$$

12. **(NLD 4)**[IMO1] On the sides of a square $ABCD$ one constructs inwardly equilateral triangles ABK, BCL, CDM, DAN. Prove that the midpoints of the four segments KL, LM, MN, NK, together with the midpoints of the eight segments AK, BK, BL, CL, CM, DM, DN, AN, are the 12 vertices of a regular dodecagon.

13. **(POL 4)** Let B be a set of k sequences each having n terms equal to 1 or -1. The product of two such sequences $(a_1, a_2, \ldots, a_n)$ and $(b_1, b_2, \ldots, b_n)$ is defined as $(a_1 b_1, a_2 b_2, \ldots, a_n b_n)$. Prove that there exists a sequence $(c_1, c_2, \ldots, c_n)$ such that the intersection of B and the set containing all sequences from B multiplied by $(c_1, c_2, \ldots, c_n)$ contains at most $k^2/2^n$ sequences.

14. **(FIN 2')** Let E be a finite set of points such that E is not contained in a plane and no three points of E are collinear. Show that at least one of the following alternatives holds:

   (i) E contains five points that are vertices of a convex pyramid having no other points in common with E;

(ii) some plane contains exactly three points from E.

15. **(VNM 1)**[IMO2] The length of a finite sequence is defined as the number of terms of this sequence. Determine the maximal possible length of a finite sequence that satisfies the following condition: The sum of each seven successive terms is negative, and the sum of each eleven successive terms is positive.

16. **(VNM 3)** Let E be a set of n points in the plane ($n \geq 3$) whose coordinates are integers such that any three points from E are vertices of a nondegenerate triangle whose centroid doesn't have both coordinates integers. Determine the maximal n.

3.20 The Twentieth IMO
Bucharest, Romania, 1978

3.20.1 Contest Problems

First Day (July 6)

1. Let $n > m \geq 1$ be natural numbers such that the groups of the last three digits in the decimal representation of $1978^m, 1978^n$ coincide. Find the ordered pair (m,n) of such m, n for which $m + n$ is minimal.

2. Given any point P in the interior of a sphere with radius R, three mutually perpendicular segments PA, PB, PC are drawn terminating on the sphere and having one common vertex in P. Consider the rectangular parallelepiped of which PA, PB, PC are coterminal edges. Find the locus of the point Q that is diagonally opposite P in the parallelepiped when P and the sphere are fixed.

3. Let $\{f(n)\}$ be a strictly increasing sequence of positive integers: $0 < f(1) < f(2) < f(3) < \dots$. Of the positive integers not belonging to the sequence, the nth in order of magnitude is $f(f(n)) + 1$. Determine $f(240)$.

Second day (July 7)

4. In a triangle ABC we have $AB = AC$. A circle is tangent internally to the circumcircle of ABC and also to the sides AB, AC, at P, Q respectively. Prove that the midpoint of PQ is the center of the incircle of ABC.

5. Let $\varphi : \{1,2,3,\dots\} \to \{1,2,3,\dots\}$ be injective. Prove that for all n,

$$\sum_{k=1}^{n} \frac{\varphi(k)}{k^2} \geq \sum_{k=1}^{n} \frac{1}{k}.$$

6. An international society has its members in 6 different countries. The list of members contains 1978 names, numbered $1, 2, \dots, 1978$. Prove that there is at least one member whose number is the sum of the numbers of two, not necessarily distinct, of his compatriots.

3.20.2 Longlisted Problems

1. **(BGR 1)** (SL78-1).

2. **(BGR 2)** If

$$f(x) = (x + 2x^2 + \dots + nx^n)^2 = a_2 x^2 + a_3 x^3 + \dots + a_{2n} x^{2n},$$

prove that

$$a_{n+1} + a_{n+2} + \dots + a_{2n} = \binom{n+1}{2} \frac{5n^2 + 5n + 2}{12}.$$

3. **(BGR 3)** Find all numbers α for which the equation

$$x^2 - 2x[x] + x - \alpha = 0$$

has two nonnegative roots. ($[x]$ denotes the largest integer less than or equal to x.)

4. **(BGR 4)** (SL78-2).

5. **(CUB 1)** Prove that for any triangle ABC there exists a point P in the plane of the triangle and three points A', B', and C' on the lines BC, AC, and AB respectively such that

$$AB \cdot PC' = AC \cdot PB' = BC \cdot PA' = 0.3M^2,$$

where $M = \max\{AB, AC, BC\}$.

6. **(CUB 2)** Prove that for all $X > 1$ there exists a triangle whose sides have lengths $P_1(X) = X^4 + X^3 + 2X^2 + X + 1$, $P_2(X) = 2X^3 + X^2 + 2X + 1$, and $P_3(X) = X^4 - 1$. Prove that all these triangles have the same greatest angle and calculate it.

7. **(CUB 3)** (SL78-3).

8. **(CZS 1)** For two given triangles $A_1A_2A_3$ and $B_1B_2B_3$ with areas Δ_A and Δ_B, respectively, $A_iA_k \geq B_iB_k$, $i, k = 1, 2, 3$. Prove that $\Delta_A \geq \Delta_B$ if the triangle $A_1A_2A_3$ is not obtuse-angled.

9. **(CZS 2)** (SL78-4).

10. **(CZS 3)** Show that for any natural number n there exist two prime numbers p and q, $p \neq q$, such that n divides their difference.

11. **(CZS 4)** Find all natural numbers $n < 1978$ with the following property:
    If m is a natural number, $1 < m < n$, and $(m, n) = 1$ (i.e., m and n are relatively prime), then m is a prime number.

12. **(FIN 1)** The equation $x^3 + ax^2 + bx + c = 0$ has three (not necessarily distinct) real roots t, u, v. For which a, b, c do the numbers t^3, u^3, v^3 satisfy the equation $x^3 + a^3x^2 + b^3x + c^3 = 0$?

13. **(FIN 2)** The satellites A and B circle the Earth in the equatorial plane at altitude h. They are separated by distance $2r$, where r is the radius of the Earth. For which h can they be seen in mutually perpendicular directions from some point on the equator?

14. **(FIN 3)** Let $p(x,y)$ and $q(x,y)$ be polynomials in two variables such that for $x \geq 0$, $y \geq 0$ the following conditions hold:
    (i) $p(x,y)$ and $q(x,y)$ are increasing functions of x for every fixed y.
    (ii) $p(x,y)$ is an increasing and $q(x)$ is a decreasing function of y for every fixed x.
    (iii) $p(x,0) = q(x,0)$ for every x and $p(0,0) = 0$.
    Show that the simultaneous equations $p(x,y) = a$, $q(x,y) = b$ have a unique solution in the set $x \geq 0$, $y \geq 0$ for all a, b satisfying $0 \leq b \leq a$ but lack a solution in the same set if $a < b$.

15. **(FRA 1)** Prove that for every positive integer n coprime to 10 there exists a multiple of n that does not contain the digit 1 in its decimal representation.

16. **(FRA 2)** (SL78-6).

17. **(FRA 3)** (SL78-17).

18. **(FRA 4)** Given a natural number n, prove that the number $M(n)$ of points with integer coordinates inside the circle $(O(0,0), \sqrt{n})$ satisfies

$$\pi n - 5\sqrt{n} + 1 < M(n) < \pi n + 4\sqrt{n} + 1.$$

19. **(FRA 5)** (SL78-7).

20. **(UNK 1)** Let O be the center of a circle. Let OU, OV be perpendicular radii of the circle. The chord PQ passes through the midpoint M of UV. Let W be a point such that $PM = PW$, where U, V, M, W are collinear. Let R be a point such that $PR = MQ$, where R lies on the line PW. Prove that $MR = UV$.

    Alternative version: A circle S is given with center O and radius r. Let M be a point whose distance from O is $\frac{r}{\sqrt{2}}$. Let PMQ be a chord of S. The point N is defined by $\overrightarrow{PN} = \overrightarrow{MQ}$. Let R be the reflection of N by the line through P that is parallel to OM. Prove that $MR = \sqrt{2}r$.

21. **(UNK 2)** A circle touches the sides AB, BC, CD, DA of a square at points K, L, M, N respectively, and BU, KV are parallel lines such that U is on DM and V on DN. Prove that UV touches the circle.

22. **(UNK 3)** Two nonzero integers x, y (not necessarily positive) are such that $x + y$ is a divisor of $x^2 + y^2$, and the quotient $\frac{x^2 + y^2}{x + y}$ is a divisor of 1978. Prove that $x = y$.

23. **(UNK 4)** (SL78-8).

24. **(UNK 5)** (SL78-9).

25. **(GDR 1)** Consider a polynomial $P(x) = ax^2 + bx + c$ with $a > 0$ that has two real roots x_1, x_2. Prove that the absolute values of both roots are less than or equal to 1 if and only if $a + b + c \geq 0$, $a - b + c \geq 0$, and $a - c \geq 0$.

26. **(GDR 2)** (SL78-5).

27. **(GDR 3)** Determine the sixth number after the decimal point in the number $\left(\sqrt{1978} + \left[\sqrt{1978}\right]\right)^{20}$.

28. **(GDR 4)** Let c, s be real functions defined on $\mathbb{R} \setminus \{0\}$ that are nonconstant on any interval and satisfy

$$c\left(\frac{x}{y}\right) = c(x)c(y) - s(x)s(y) \qquad \text{for any } x \neq 0, \ y \neq 0.$$

Prove that:

(a) $c(1/x) = c(x)$, $s(1/x) = -s(x)$ for any $x \neq 0$, and also $c(1) = 1$, $s(1) = s(-1) = 0$;

(b) c and s are either both even or both odd functions (a function f is even if $f(x) = f(-x)$ for all x, and odd if $f(x) = -f(-x)$ for all x).

Find functions c, s that also satisfy $c(x) + s(x) = x^n$ for all x, where n is a given positive integer.

29. **(GDR 5)** *(Variant of GDR 4)* Given a nonconstant function $f : \mathbb{R}^+ \to \mathbb{R}$ such that $f(xy) = f(x)f(y)$ for any $x, y > 0$, find functions $c, s : \mathbb{R}^+ \to \mathbb{R}$ that satisfy $c(x/y) = c(x)c(y) - s(x)s(y)$ for all $x, y > 0$ and $c(x) + s(x) = f(x)$ for all $x > 0$.

30. **(NLD 1)** (SL78-10).

31. **(NLD 2)** Let the polynomials

$$P(x) = x^n + a_{n-1}x^{n-1} + \cdots + a_1 x + a_0,$$
$$Q(x) = x^m + b_{m-1}x^{m-1} + \cdots + b_1 x + b_0,$$

be given satisfying the identity $P(x)^2 = (x^2 - 1)Q(x)^2 + 1$. Prove the identity

$$P'(x) = nQ(x).$$

32. **(NLD 3)** Let $\mathscr{C}$ be the circumcircle of the square with vertices $(0,0)$, $(0,1978)$, $(1978,0)$, $(1978,1978)$ in the Cartesian plane. Prove that $\mathscr{C}$ contains no other point for which both coordinates are integers.

33. **(SWE 1)** A sequence $(a_n)_0^\infty$ of real numbers is called *convex* if $2a_n \leq a_{n-1} + a_{n+1}$ for all positive integers n. Let $(b_n)_0^\infty$ be a sequence of positive numbers and assume that the sequence $(\alpha^n b_n)_0^\infty$ is convex for any choice of $\alpha > 0$. Prove that the sequence $(\log b_n)_0^\infty$ is convex.

34. **(SWE 2)** (SL78-11).

35. **(SWE 3)** A sequence $(a_n)_0^N$ of real numbers is called *concave* if $2a_n \geq a_{n-1} + a_{n+1}$ for all integers n, $1 \leq n \leq N-1$.
   (a) Prove that there exists a constant $C > 0$ such that

$$\left(\sum_{n=0}^N a_n \right)^2 \geq C(N-1) \sum_{n=0}^N a_n^2 \tag{1}$$

   for all concave positive sequences $(a_n)_0^N$.
   (b) Prove that (1) holds with $C = 3/4$ and that this constant is best possible.

36. **(TUR 1)** The integers 1 through 1000 are located on the circumference of a circle in natural order. Starting with 1, every fifteenth number (i.e., $1, 16, 31, \ldots$) is marked. The marking is continued until an already marked number is reached. How many of the numbers will be left unmarked?

37. **(TUR 2)** Simplify

$$\frac{1}{\log_a(abc)} + \frac{1}{\log_b(abc)} + \frac{1}{\log_c(abc)},$$

where a, b, c are positive real numbers.

38. **(TUR 3)** Given a circle, construct a chord that is trisected by two given non-collinear radii.

39. **(TUR 4)** A is a $2m$-digit positive integer each of whose digits is 1. B is an m-digit positive integer each of whose digits is 4. Prove that $A + B + 1$ is a perfect square.

40. **(TUR 5)** If $C_n^p = \frac{n!}{p!(n-p)!}$ $(p \geq 1)$, prove the identity

$$C_n^p = C_{n-1}^{p-1} + C_{n-2}^{p-1} + \cdots + C_p^{p-1} + C_{p-1}^{p-1}$$

and then evaluate the sum

$$S = 1 \cdot 2 \cdot 3 + 2 \cdot 3 \cdot 4 + \cdots + 97 \cdot 98 \cdot 99.$$

41. **(USA 1)** (SL78-12).

42. **(USA 2)** A, B, C, D, E are points on a circle O with radius equal to r. Chords AB and DE are parallel to each other and have length equal to x. Diagonals AC, AD, BE, CE are drawn. If segment XY on O meets AC at X and EC at Y, prove that lines BX and DY meet at Z on the circle.

43. **(USA 3)** If p is a prime greater than 3, show that at least one of the numbers $\frac{3}{p^2}, \frac{4}{p^2}, \ldots, \frac{p-2}{p^2}$ is expressible in the form $\frac{1}{x} + \frac{1}{y}$, where x and y are positive integers.

44. **(USA 4)** In $\triangle ABC$ with $\angle C = 60^\circ$, prove that $\frac{c}{a} + \frac{c}{b} \geq 2$.

45. **(USA 5)** If $r > s > 0$ and $a > b > c$, prove that

$$a^r b^s + b^r c^s + c^r a^s \geq a^s b^r + b^s c^r + c^s a^r.$$

46. **(USA 6)** (SL78-13).

47. **(VNM 1)** Given the expression

$$P_n(x) = \frac{1}{2^n} \left[\left(x + \sqrt{x^2 - 1} \right)^n + \left(x - \sqrt{x^2 - 1} \right)^n \right],$$

prove:
(a) $P_n(x)$ satisfies the identity $P_n(x) - x P_{n-1}(x) + \frac{1}{4} P_{n-2}(x) \equiv 0$.
(b) $P_n(x)$ is a polynomial in x of degree n.

48. **(VNM 2)** (SL78-14).

49. **(VNM 3)** Let A, B, C, D be four arbitrary distinct points in space.
(a) Prove that using the segments $AB + CD$, $AC + BD$ and $AD + BC$ it is always possible to construct a triangle T that is nondegenerate and has no obtuse angle.

(b) What should these four points satisfy in order for the triangle T to be right-angled?

50. **(VNM 4)** A variable tetrahedron $ABCD$ has the following properties: Its edge lengths can change as well as its vertices, but the opposite edges remain equal $(BC = DA, CA = DB, AB = DC)$; and the vertices A, B, C lie respectively on three fixed spheres with the same center P and radii $3, 4, 12$. What is the maximal length of PD?

51. **(VNM 5)** Find the relations among the angles of the triangle ABC whose altitude AH and median AM satisfy $\angle BAH = \angle CAM$.

52. **(YUG 1)** (SL78-15).

53. **(YUG 2)** (SL78-16).

54. **(YUG 3)** Let p, q and r be three lines in space such that there is no plane that is parallel to all three of them. Prove that there exist three planes α, β, and γ, containing p, q, and r respectively, that are perpendicular to each other ($\alpha \perp \beta$, $\beta \perp \gamma$, $\gamma \perp \alpha$).

3.20.3 Shortlisted Problems

1. **(BGR 1)** The set $M = \{1, 2, \ldots, 2n\}$ is partitioned into k nonintersecting subsets $M_1, M_2, \ldots, M_k$, where $n \geq k^3 + k$. Prove that there exist even numbers $2j_1, 2j_2, \ldots, 2j_{k+1}$ in M that are in one and the same subset M_i $(1 \leq i \leq k)$ such that the numbers $2j_1 - 1, 2j_2 - 1, \ldots, 2j_{k+1} - 1$ are also in one and the same subset M_j $(1 \leq j \leq k)$.

2. **(BGR 4)** Two identically oriented equilateral triangles, ABC with center S and $A'B'C$, are given in the plane. We also have $A' \neq S$ and $B' \neq S$. If M is the midpoint of $A'B$ and N the midpoint of AB', prove that the triangles $SB'M$ and $SA'N$ are similar.

3. **(CUB 3)**[IMO1] Let $n > m \geq 1$ be natural numbers such that the groups of the last three digits in the decimal representation of $1978^m, 1978^n$ coincide. Find the ordered pair (m, n) of such m, n for which $m + n$ is minimal.

4. **(CZS 2)** Let T_1 be a triangle having a, b, c as lengths of its sides and let T_2 be another triangle having u, v, w as lengths of its sides. If P, Q are the areas of the two triangles, prove that

$$16PQ \leq a^2(-u^2 + v^2 + w^2) + b^2(u^2 - v^2 + w^2) + c^2(u^2 + v^2 - w^2).$$

When does equality hold?

5. **(GDR 2)** For every integer $d \geq 1$, let M_d be the set of all positive integers that cannot be written as a sum of an arithmetic progression with difference d, having at least two terms and consisting of positive integers. Let $A = M_1$, $B = M_2 \setminus \{2\}$, $C = M_3$. Prove that every $c \in C$ may be written in a unique way as $c = ab$ with $a \in A, b \in B$.

6. **(FRA 2)**[IMO5] Let $\varphi : \{1,2,3,\dots\} \to \{1,2,3,\dots\}$ be injective. Prove that for all n,

$$\sum_{k=1}^{n} \frac{\varphi(k)}{k^2} \geq \sum_{k=1}^{n} \frac{1}{k}.$$

7. **(FRA 5)** We consider three distinct half-lines Ox, Oy, Oz in a plane. Prove the existence and uniqueness of three points $A \in Ox$, $B \in Oy$, $C \in Oz$ such that the perimeters of the triangles OAB, OBC, OCA are all equal to a given number $2p > 0$.

8. **(UNK 4)** Let S be the set of all the odd positive integers that are not multiples of 5 and that are less than $30m$, m being an arbitrary positive integer. What is the smallest integer k such that in any subset of k integers from S there must be two different integers, one of which divides the other?

9. **(UNK 5)**[IMO3] Let $\{f(n)\}$ be a strictly increasing sequence of positive integers: $0 < f(1) < f(2) < f(3) < \cdots$. Of the positive integers not belonging to the sequence, the nth in order of magnitude is $f(f(n)) + 1$. Determine $f(240)$.

10. **(NLD 1)**[IMO6] An international society has its members in 6 different countries. The list of members contains 1978 names, numbered $1, 2, \dots, 1978$. Prove that there is at least one member whose number is the sum of the numbers of two, not necessarily distinct, of his compatriots.

11. **(SWE 2)** A function $f : I \to \mathbb{R}$, defined on an interval I, is called *concave* if $f(\theta x + (1 - \theta)y) \geq \theta f(x) + (1 - \theta)f(y)$ for all $x, y \in I$ and $0 \leq \theta \leq 1$. Assume that the functions $f_1, \dots, f_n$, having all nonnegative values, are concave. Prove that the function $(f_1 f_2 \dots f_n)^{1/n}$ is concave.

12. **(USA 1)**[IMO4] In a triangle ABC we have $AB = AC$. A circle is tangent internally to the circumcircle of ABC and also to the sides AB, AC, at P, Q respectively. Prove that the midpoint of PQ is the center of the incircle of ABC.

13. **(USA 6)**[IMO2] Given any point P in the interior of a sphere with radius R, three mutually perpendicular segments PA, PB, PC are drawn terminating on the sphere and having one common vertex in P. Consider the rectangular parallelepiped of which PA, PB, PC are coterminal edges. Find the locus of the point Q that is diagonally opposite P in the parallelepiped when P and the sphere are fixed.

14. **(VNM 2)** Prove that it is possible to place $2n(2n + 1)$ parallelepipedic (rectangular) pieces of soap of dimensions $1 \times 2 \times (n + 1)$ in a cubic box with edge $2n + 1$ if and only if n is even or $n = 1$.

    Remark. It is assumed that the edges of the pieces of soap are parallel to the edges of the box.

15. **(YUG 1)** Let p be a prime and $A = \{a_1, \dots, a_{p-1}\}$ an arbitrary subset of the set of natural numbers such that none of its elements is divisible by p. Let us define a mapping f from $\mathscr{P}(A)$ (the set of all subsets of A) to the set $P = \{0, 1, \dots, p - 1\}$ in the following way:

(i) if $B = \{a_{i_1}, \ldots, a_{i_k}\} \subset A$ and $\sum_{j=1}^{k} a_{i_j} \equiv n \pmod{p}$, then $f(B) = n$,

(ii) $f(\emptyset) = 0$, $\emptyset$ being the empty set.

Prove that for each $n \in P$ there exists $B \subset A$ such that $f(B) = n$.

16. **(YUG 2)** Determine all the triples (a, b, c) of positive real numbers such that the system

$$ax + by - cz = 0,$$
$$a\sqrt{1-x^2} + b\sqrt{1-y^2} - c\sqrt{1-z^2} = 0,$$

is compatible in the set of real numbers, and then find all its real solutions.

17. **(FRA 3)** Prove that for any positive integers x, y, z with $xy - z^2 = 1$ one can find nonnegative integers a, b, c, d such that $x = a^2 + b^2$, $y = c^2 + d^2$, $z = ac + bd$. Set $z = (2q)!$ to deduce that for any prime number $p = 4q + 1$, p can be represented as the sum of squares of two integers.

3.21 The Twenty-First IMO
London, United Kingdom, 1979

3.21.1 Contest Problems

First Day (July 2)

1. Let p and q be natural numbers such that

$$1 - \frac{1}{2} + \frac{1}{3} - \frac{1}{4} + \cdots - \frac{1}{1318} + \frac{1}{1319} = \frac{p}{q}.$$

   Prove that p is divisible by 1979.

2. A pentagonal prism $A_1A_2 \ldots A_5B_1B_2 \ldots B_5$ is given. The edges, the diagonals of the lateral walls, and the internal diagonals of the prism are each colored either red or green in such a way that no triangle whose vertices are vertices of the prism has its three edges of the same color. Prove that all edges of the bases are of the same color.

3. Two circles in a plane intersect. Let A be one of the points of intersection. Starting simultaneously from A two points move with constant speeds, each point traveling along its own circle in the same sense. The two points return simultaneously after one revolution. Prove that there is a fixed point P in the plane such that, at any time, the distances from P to the moving points are equal.

Second Day (July 3)

4. Given a point P in a given plane π and a point Q not in π, determine all points R in π such that $\frac{QP+PR}{QR}$ is a maximum.

5. The nonnegative real numbers $x_1, x_2, x_3, x_4, x_5, a$ satisfy the following relations:

$$\sum_{i=1}^{5} i x_i = a, \qquad \sum_{i=1}^{5} i^3 x_i = a^2, \qquad \sum_{i=1}^{5} i^5 x_i = a^3.$$

   What are the possible values of a?

6. Let S and F be opposite vertices of a regular octagon. A frog starts jumping at vertex S. From any vertex of the octagon except F, it may jump to either of the two adjacent vertices. When it reaches vertex F, the frog stops and stays there. Let a_n be the number of distinct paths of exactly n jumps ending at F. Prove that for $n = 1, 2, 3, \ldots$,

$$a_{2n-1} = 0, \qquad a_{2n} = \frac{1}{\sqrt{2}}(x^{n-1} - y^{n-1}), \qquad \text{where } x = 2 + \sqrt{2}, y = 2 - \sqrt{2}.$$

3.21.2 Longlisted Problems

1. **(BEL 1)** (SL79-1).

2. **(BEL 2)** For a finite set E of cardinality $n \geq 3$, let $f(n)$ denote the maximum number of 3-element subsets of E, any two of them having exactly one common element. Calculate $f(n)$.

3. **(BEL 3)** Is it possible to partition 3-dimensional Euclidean space into 1979 mutually isometric subsets?

4. **(BEL 4)** (SL79-2).

5. **(BEL 5)** Describe which natural numbers do not belong to the set

$$E = \{[n + \sqrt{n} + 1/2] \mid n \in \mathbb{N}\}.$$

6. **(BEL 6)** Prove that $\frac{1}{2}\sqrt{4\sin^2 36° - 1} = \cos 72°$.

7. **(BRA 1)** $M = (a_{i,j})$, $i, j = 1, 2, 3, 4$, is a square matrix of order four. Given that:
   (i) for each $i = 1, 2, 3, 4$ and for each $k = 5, 6, 7$,

$$a_{i,k} = a_{i,k-4};$$
$$P_i = a_{1,i} + a_{2,i+1} + a_{3,i+2} + a_{4,i+3};$$
$$S_i = a_{4,i} + a_{3,i+1} + a_{2,i+2} + a_{1,i+3};$$
$$L_i = a_{i,1} + a_{i,2} + a_{i,3} + a_{i,4};$$
$$C_i = a_{1,i} + a_{2,i} + a_{3,i} + a_{4,i},$$

   (ii) for each $i, j = 1, 2, 3, 4$, $P_i = P_j$, $S_i = S_j$, $L_i = L_j$, $C_i = C_j$, and
   (iii) $a_{1,1} = 0$, $a_{1,2} = 7$, $a_{2,1} = 11$, $a_{2,3} = 2$, and $a_{3,3} = 15$;
   find the matrix M.

8. **(BRA 2)** The sequence (a_n) of real numbers is defined as follows:

$$a_1 = 1, \quad a_2 = 2 \quad \text{and} \quad a_n = 3a_{n-1} - a_{n-2}, \quad n \geq 3.$$

   Prove that for $n \geq 3$, $a_n = \left[\dfrac{a_{n-1}^2}{a_{n-2}}\right] + 1$, where $[x]$ denotes the integer p such that $p \leq x < p + 1$.

9. **(BRA 3)** The real numbers $\alpha_1, \alpha_2, \alpha_3, \ldots, \alpha_n$ are positive. Let us denote by $h = \dfrac{n}{1/\alpha_1 + 1/\alpha_2 + \cdots + 1/\alpha_n}$ the harmonic mean, $g = \sqrt[n]{\alpha_1 \alpha_2 \cdots \alpha_n}$ the geometric mean, $a = \dfrac{\alpha_1 + \alpha_2 + \cdots + \alpha_n}{n}$ the arithmetic mean. Prove that $h \leq g \leq a$, and that each of the equalities implies the other one.

10. **(BGR 1)** (SL79-3).

11. **(BGR 2)** Prove that a pyramid $A_1 A_2 \ldots A_{2k+1} S$ with equal lateral edges and equal space angles between adjacent lateral walls is regular.
    Variant. Prove that a pyramid $A_1 \ldots A_{2k+1} S$ with equal space angles between adjacent lateral walls is regular if there exists a sphere tangent to all its edges.

12. **(BGR 3)** (SL79-4).

13. **(BGR 4)** The plane is divided into equal squares by parallel lines; i.e., a square net is given. Let M be an arbitrary set of n squares of this net. Prove that it is possible to choose no fewer than $n/4$ squares of M in such a way that no two of them have a common point.

14. **(CZS 1)** Let S be a set of $n^2 + 1$ closed intervals (n is a positive integer). Prove that at least one of the following assertions holds:
   (i) There exists a subset S' of $n+1$ intervals from S such that the intersection of the intervals in S' is nonempty.
   (ii) There exists a subset S'' of $n+1$ intervals from S such that any two of the intervals in S'' are disjoint.

15. **(CZS 2)** (SL79-5).

16. **(CZS 3)** Let Q be a square with side length 6. Find the smallest integer n such that in Q there exists a set S of n points with the property that any square with side 1 completely contained in Q contains in its interior at least one point from S.

17. **(CZS 4)** (SL79-6).

18. **(FIN 1)** Show that for no integers $a \geq 1$, $n \geq 1$ is the sum

$$1 + \frac{1}{1+a} + \frac{1}{1+2a} + \cdots + \frac{1}{1+na}$$

an integer.

19. **(FIN 2)** For $k = 1, 2, \ldots$ consider the k-tuples $(a_1, a_2, \ldots, a_k)$ of positive integers such that

$$a_1 + 2a_2 + \cdots + ka_k = 1979.$$

Show that there are as many such k-tuples with odd k as there are with even k.

20. **(FIN 3)** (SL79-10).

21. **(FRA 1)** Let E be the set of all bijective mappings from $\mathbb{R}$ to $\mathbb{R}$ satisfying

$$(\forall t \in \mathbb{R}) \qquad f(t) + f^{-1}(t) = 2t,$$

where f^{-1} is the mapping inverse to f. Find all elements of E that are monotonic mappings.

22. **(FRA 2)** Consider two quadrilaterals $ABCD$ and $A'B'C'D'$ in an affine Euclidian plane such that $AB = A'B'$, $BC = B'C'$, $CD = C'D'$, and $DA = D'A'$. Prove that the following two statements are true:
   (a) If the diagonals BD and AC are mutually perpendicular, then the diagonals $B'D'$ and $A'C'$ are also mutually perpendicular.
   (b) If the perpendicular bisector of BD intersects AC at M, and that of $B'D'$ intersects $A'C'$ at M', then $\frac{\overline{MA}}{\overline{MC}} = \frac{\overline{M'A'}}{\overline{M'C'}}$ (if $\overline{MC} = 0$ then $\overline{M'C'} = 0$).

23. **(FRA 3)** Consider the set E consisting of pairs of integers (a,b), with $a \geq 1$ and $b \geq 1$, that satisfy in the decimal system the following properties:
    (i) b is written with three digits, as $\overline{\alpha_2 \alpha_1 \alpha_0}$, $\alpha_2 \neq 0$;
    (ii) a is written as $\overline{\beta_p \ldots \beta_1 \beta_0}$ for some p;
    (iii) $(a+b)^2$ is written as $\overline{\beta_p \ldots \beta_1 \beta_0 \alpha_2 \alpha_1 \alpha_0}$.
    Find the elements of E.

24. **(FRA 4)** Let a and b be coprime integers, greater than or equal to 1. Prove that all integers n greater than or equal to $(a-1)(b-1)$ can be written in the form:

$$n = ua + vb, \quad \text{with } (u,v) \in \mathbb{N} \times \mathbb{N}.$$

25. **(FRG 1)** (SL79-7).

26. **(FRG 2)** Let n be a natural number. If $4^n + 2^n + 1$ is a prime, prove that n is a power of three.

27. **(FRG 3)** (SL79-8).

28. **(FRG 4)** (SL79-9).

29. **(GDR 1)** (SL79-11).

30. **(GDR 2)** Let M be a set of points in a plane with at least two elements. Prove that if M has two axes of symmetry g_1 and g_2 intersecting at an angle $\alpha = q\pi$, where q is irrational, then M must be infinite.

31. **(GDR 3)** (SL79-12).

32. **(GDR 4)** Let $n,k \geq 1$ be natural numbers. Find the number $A(n,k)$ of solutions in integers of the equation

$$|x_1| + |x_2| + \cdots + |x_k| = n.$$

33. **(HEL 1)** (SL79-13).

34. **(HEL 2)** Notice that in the fraction $\frac{16}{64}$ we can perform a simplification as $\frac{16\!\!\!/}{6\!\!\!/4} = \frac{1}{4}$ obtaining a correct equality. Find all fractions whose numerators and denominators are two-digit positive integers for which such a simplification is correct.

35. **(HEL 3)** Given a sequence (a_n), with $a_1 = 4$ and $a_{n+1} = a_n^2 - 2$ $(\forall n \in \mathbb{N})$, prove that there is a triangle with side lengths $a_n - 1, a_n, a_n + 1$, and that its area is equal to an integer.

36. **(HEL 4)** A regular tetrahedron $A_1 B_1 C_1 D_1$ is inscribed in a regular tetrahedron $ABCD$, where A_1 lies in the plane BCD, B_1 in the plane ACD, etc. Prove that $A_1 B_1 \geq AB/3$.

37. **(HEL 5)** (SL79-14).

38. **(HUN 1)** Prove the following statement: If a polynomial $f(x)$ with real coefficients takes only nonnegative values, then there exists a positive integer n and polynomials $g_1(x), g_2(x), \ldots, g_n(x)$ such that

$$f(x) = g_1(x)^2 + g_2(x)^2 + \cdots + g_n(x)^2.$$

39. **(HUN 2)** A desert expedition camps at the border of the desert, and has to provide one liter of drinking water for another member of the expedition, residing
on the distance of n days of walking from the camp, under the following conditions:
    (i) Each member of the expedition can pick up at most 3 liters of water.
    (ii) Each member must drink one liter of water every day spent in the desert.
    (iii) All the members must return to the camp.
    How much water do they need (at least) in order to do that?

40. **(HUN 3)** A polynomial $P(x)$ has degree at most $2k$, where $k = 0, 1, 2, \ldots$. Given
that for an integer i, the inequality $-k \le i \le k$ implies $|P(i)| \le 1$, prove that for
all real numbers x, with $-k \le x \le k$, the following inequality holds:

$$|P(x)| < (2k+1)\binom{2k}{k}.$$

41. **(HUN 4)** Prove the following statement: There does not exist a pyramid with
square base and congruent lateral faces for which the measures of all edges, total
area, and volume are integers.

42. **(HUN 5)** Let a quadratic polynomial $g(x) = ax^2 + bx + c$ be given and an integer
$n \ge 1$. Prove that there exists at most one polynomial $f(x)$ of nth degree such
that $f(g(x)) = g(f(x))$.

43. **(ISR 1)** Let a, b, c denote the lengths of the sides BC, CA, AB, respectively, of a
triangle ABC. If P is any point on the circumference of the circle inscribed in the
triangle, show that $aPA^2 + bPB^2 + cPC^2$ is constant.

44. **(ISR 2)** (SL79-15).

45. **(ISR 3)** For any positive integer n we denote by $F(n)$ the number of ways in
which n can be expressed as the sum of three different positive integers, without
regard to order. Thus, since $10 = 7+2+1 = 6+3+1 = 5+4+1 = 5+3+2$,
we have $F(10) = 4$. Show that $F(n)$ is even if $n \equiv 2$ or 4 (mod 6), but odd if n is
divisible by 6.

46. **(ISR 4)** (SL79-16).

47. **(NLD 1)** (SL79-17).

48. **(NLD 2)** In the plane a circle C of unit radius is given. For any line l a number
$s(l)$ is defined in the following way: If l and C intersect in two points, $s(l)$ is
their distance; otherwise, $s(l) = 0$.
    Let P be a point at distance r from the center of C. One defines $M(r)$ to be the
maximum value of the sum $s(m) + s(n)$, where m and n are variable mutually
orthogonal lines through P. Determine the values of r for which $M(r) > 2$.

49. **(NLD 3)** Let there be given two sequences of integers $f_i(1), f_i(2), \ldots$ $(i = 1, 2)$
satisfying:

(i) $f_i(nm) = f_i(n)f_i(m)$ if $\gcd(n,m) = 1$;

(ii) for every prime P and all $k = 2,3,4,\dots$,

$$f_i(P^k) = f_i(P)f_i(P^{k-1}) - P^2 f(P^{k-2}).$$

Moreover, for every prime P:

(iii) $f_1(P) = 2P$,

(iv) $f_2(P) < 2P$.

Prove that $|f_2(n)| < f_1(n)$ for all n.

50. **(POL 1)** (SL79-18).

51. **(POL 2)** Let ABC be an arbitrary triangle and let $S_1, S_2, \dots, S_7$ be circles satisfying the following conditions:

S_1 is tangent to CA and AB,

S_2 is tangent to S_1, AB, and BC,

S_3 is tangent to S_2, BC, and CA,

....

S_7 is tangent to S_6, CA and AB.

Prove that the circles S_1 and S_7 coincide.

52. **(POL 3)** Let a real number $\lambda > 1$ be given and a sequence (n_k) of positive integers such that $\frac{n_{k+1}}{n_k} > \lambda$ for $k = 1,2,\dots$. Prove that there exists a positive integer c such that no positive integer n can be represented in more than c ways in the form $n = n_k + n_j$ or $n = n_r - n_s$.

53. **(POL 4)** An infinite increasing sequence of positive integers n_j $(j = 1,2,\dots)$ has the property that for a certain c, $\frac{1}{N}\sum_{n_j \leq N} n_j \leq c$, for every $N > 0$

Prove that there exist finitely many sequences $m_j^{(i)}$ $(i = 1,2,\dots,k)$ such that

$$\{n_1, n_2, \dots\} = \bigcup_{i=1}^{k}\{m_1^{(i)}, m_2^{(i)}, \dots\} \quad \text{and}$$
$$m_{j+1}^{(i)} > 2m_j^{(i)} \quad (1 \leq i \leq k,\, j = 1,2,\dots).$$

54. **(ROU 1)** (SL79-19).

55. **(ROU 2)** Let a,b be coprime integers. Show that the equation $ax^2 + by^2 = z^3$ has an infinite set of solutions (x,y,z) with $x,y,z \in \mathbb{Z}$ and x,y mutually coprime (in each solution).

56. **(ROU 3)** Show that for every natural number n, $n\sqrt{2} - [n\sqrt{2}] > \frac{1}{2n\sqrt{2}}$ and that for every $\varepsilon > 0$ there exists a natural number n with $n\sqrt{2} - [n\sqrt{2}] < \frac{1}{2n\sqrt{2}} + \varepsilon$.

57. **(ROU 4)** Let M be a set, and A,B,C given subsets of M. Find a necessary and sufficient condition for the existence of a set $X \subset M$ for which $(X \cup A) \setminus (X \cap B) = C$. Describe all such sets X.

58. **(ROU 5)** Prove that there exists a natural number k_0 such that for every natural number $k > k_0$ we may find a finite number of lines in the plane, not all parallel to one of them, that divide the plane exactly in k regions. Find k_0.

59. **(SWE 1)** Determine the maximum value of $x^2y^2z^2w$ when $x,y,z,w \geq 0$ and

$$2x + xy + z + yzw = 1.$$

60. **(SWE 2)** (SL79-20).

61. **(SWE 3)** Let $a_1 \leq a_2 \leq \cdots \leq a_n$ and $b_1 \leq b_2 \leq \cdots \leq b_n$ be two sequences such that $\sum_{k=1}^{m} a_k \geq \sum_{k=1}^{m} b_k$ for all $m \leq n$ with equality for $m = n$. Let f be a convex function defined on the real numbers. Prove that

$$\sum_{k=1}^{n} f(a_k) \leq \sum_{k=1}^{n} f(b_k).$$

62. **(SWE 4)** T is a given triangle with vertices P_1, P_2, P_3. Consider an arbitrary subdivision of T into finitely many subtriangles such that no vertex of a subtriangle lies strictly between two vertices of another subtriangle. To each vertex V of the subtriangles there is assigned a number $n(V)$ according to the following rules:
   (i) If $V = P_i$, then $n(V) = i$.
   (ii) If V lies on the side P_iP_j of T, then $n(V) = i$ or j.
   (iii) If V lies inside the triangle T, then $n(V)$ is any of the numbers 1,2,3.
   Prove that there exists at least one subtriangle whose vertices are numbered 1, 2, and 3.

63. **(USA 1)** If $a_1, a_2, \ldots, a_n$ denote the lengths of the sides of an arbitrary n-gon, prove that

$$2 \geq \frac{a_1}{s - a_1} + \frac{a_2}{s - a_2} + \cdots + \frac{a_n}{s - a_n} \geq \frac{n}{n - 1},$$

where $s = a_1 + a_2 + \cdots + a_n$.

64. **(USA 2)** From point P on arc BC of the circumcircle about triangle ABC, PX is constructed perpendicular to BC, PY is perpendicular to AC, and PZ perpendicular to AB (all extended if necessary). Prove that

$$\frac{BC}{PX} = \frac{AC}{PY} + \frac{AB}{PZ}.$$

65. **(USA 3)** Given $f(x) \leq x$ for all real x and

$$f(x + y) \leq f(x) + f(y) \qquad \text{for all real } x, y,$$

prove that $f(x) = x$ for all x.

66. **(USA 4)** (SL79-23).

67. **(USA 5)** (SL79-24).

68. **(USA 6)** (SL79-25).

69. **(USS 1)** (SL79-21).

70. **(USS 2)** There are 1979 equilateral triangles: $T_1, T_2, \ldots, T_{1979}$. A side of triangle T_k is equal to $1/k$, $k = 1, 2, \ldots, 1979$. At what values of a number a can one place all these triangles into the equilateral triangle with side length a so that they don't intersect (points of contact are allowed)?

71. **(USS 3)** (SL79-22).

72. **(VNM 1)** Let $f(x)$ be a polynomial with integer coefficients. Prove that if $f(x)$ equals 1979 for four different integer values of x, then $f(x)$ cannot be equal to 2×1979 for any integral value of x.

73. **(VNM 2)** In a plane a finite number of equal circles are given. These circles are mutually nonintersecting (they may be externally tangent). Prove that one can use at most four colors for coloring these circles so that two circles tangent to each other are of different colors. What is the smallest number of circles that requires four colors?

74. **(VNM 3)** Given an equilateral triangle ABC of side a in a plane, let M be a point on the circumcircle of the triangle. Prove that the sum $s = MA^4 + MB^4 + MC^4$ is independent of the position of the point M on the circle, and determine that constant value as a function of a.

75. **(VNM 4)** Given an equilateral triangle ABC, let M be an arbitrary point in space.
    (a) Prove that one can construct a triangle from the segments MA, MB, MC.
    (b) Suppose that P and Q are two points symmetric with respect to the center O of ABC. Prove that the two triangles constructed from the segments PA, PB, PC and QA, QB, QC are of equal area.

76. **(VNM 5)** Suppose that a triangle whose sides are of integer lengths is inscribed in a circle of diameter 6.25. Find the sides of the triangle.

77. **(YUG 1)** By $h(n)$, where n is an integer greater than 1, let us denote the greatest prime divisor of the number n. Are there infinitely many numbers n for which $h(n) < h(n+1) < h(n+2)$ holds?

78. **(YUG 2)** By $\omega(n)$, where n is an integer greater than 1, let us denote the number of different prime divisors of the number n. Prove that there exist infinitely many numbers n for which $\omega(n) < \omega(n+1) < \omega(n+2)$ holds.

79. **(YUG 3)** Let S be a unit circle and K a subset of S consisting of several closed arcs. Let K satisfy the following properties:
    (i) K contains three points A, B, C, that are the vertices of an acute-angled triangle;
    (ii) for every point A that belongs to K its diametrically opposite point A' and all points B on an arc of length $1/9$ with center A' do not belong to K.
    Prove that there are three points E, F, G on S that are vertices of an equilateral triangle and that do not belong to K.

80. **(YUG 4)** (SL79-26).

81. **(YUG 5)** Let $\mathscr{P}$ be the set of rectangular parallelepipeds that have at least one edge of integer length. If a rectangular parallelepiped P_0 can be decomposed into parallelepipeds $P_1, P_2, \ldots, P_n \in \mathscr{P}$, prove that $P_0 \in \mathscr{P}$.

3.21.3 Shortlisted Problems

1. **(BEL 1)** Prove that in the Euclidean plane every regular polygon having an even number of sides can be dissected into lozenges. (A lozenge is a quadrilateral whose four sides are all of equal length).

2. **(BEL 4)** From a bag containing 5 pairs of socks, each pair a different color, a random sample of 4 single socks is drawn. Any complete pairs in the sample are discarded and replaced by a new pair drawn from the bag. The process continues until the bag is empty or there are 4 socks of different colors held outside the bag. What is the probability of the latter alternative?

3. **(BGR 1)** Find all polynomials $f(x)$ with real coefficients for which

$$f(x)f(2x^2) = f(2x^3 + x).$$

4. **(BGR 3)**[IMO2] A pentagonal prism $A_1A_2 \ldots A_5B_1B_2 \ldots B_5$ is given. The edges, the diagonals of the lateral walls and the internal diagonals of the prism are each colored either red or green in such a way that no triangle whose vertices are vertices of the prism has its three edges of the same color. Prove that all edges of the bases are of the same color.

5. **(CZS 2)** Let $n \geq 2$ be an integer. Find the maximal cardinality of a set M of pairs (j, k) of integers, $1 \leq j < k \leq n$, with the following property: If $(j, k) \in M$, then $(k, m) \notin M$ for any m.

6. **(CZS 4)** Find the real values of p for which the equation

$$\sqrt{2p + 1 - x^2} + \sqrt{3x + p + 4} = \sqrt{x^2 + 9x + 3p + 9}$$

in x has exactly two real distinct roots ($\sqrt{t}$ means the positive square root of t).

7. **(FRG 1)**[IMO1] Given that $1 - \frac{1}{2} + \frac{1}{3} - \frac{1}{4} + \cdots - \frac{1}{1318} + \frac{1}{1319} = \frac{p}{q}$, where p and q are natural numbers having no common factor, prove that p is divisible by 1979.

8. **(FRG 3)** For all rational x satisfying $0 \leq x < 1$, f is defined by

$$f(x) = \begin{cases} f(2x)/4, & \text{for } 0 \leq x < 1/2, \\ 3/4 + f(2x-1)/4, & \text{for } 1/2 \leq x < 1. \end{cases}$$

Given that $x = 0.b_1b_2b_3 \ldots$ is the binary representation of x, find $f(x)$.

9. **(FRG 4)**[IMO6] Let S and F be two opposite vertices of a regular octagon. A counter starts at S and each second is moved to one of the two neighboring vertices of the octagon. The direction is determined by the toss of a coin. The

process ends when the counter reaches F. We define a_n to be the number of distinct paths of duration n seconds that the counter may take to reach F from S. Prove that for $n = 1, 2, 3, \ldots$,

$$a_{2n-1} = 0, \qquad a_{2n} = \frac{1}{\sqrt{2}}(x^{n-1} - y^{n-1}), \qquad \text{where } x = 2 + \sqrt{2}, \, y = 2 - \sqrt{2}.$$

10. **(FIN 3)** Show that for any vectors a, b in Euclidean space,

$$|a \times b|^3 \le \frac{3\sqrt{3}}{8}|a|^2|b|^2|a-b|^2.$$

Remark. Here $\times$ denotes the vector product.

11. **(GDR 1)** Given real numbers $x_1, x_2, \ldots, x_n$ $(n \ge 2)$, with $x_i \ge 1/n$ $(i = 1, 2, \ldots, n)$ and with $x_1^2 + x_2^2 + \cdots + x_n^2 = 1$, find whether the product $P = x_1 x_2 x_3 \cdots x_n$ has a greatest and/or least value and if so, give these values.

12. **(GDR 3)** Let R be a set of exactly 6 elements. A set F of subsets of R is called an *S-family over R* if and only if it satisfies the following three conditions:
    (i) For no two sets X, Y in F is $X \subseteq Y$;
    (ii) For any three sets X, Y, Z in F, $X \cup Y \cup Z \ne R$,
    (iii) $\bigcup_{X \in F} X = R$.
    We define $|F|$ to be the number of elements of F (i.e., the number of subsets of R belonging to F). Determine, if it exists, $h = \max |F|$, the maximum being taken over all S-families over R.

13. **(HEL 1)** Show that $\frac{20}{60} < \sin 20° < \frac{21}{60}$.

14. **(HEL 5)** Find all bases of logarithms in which a real positive number can be equal to its logarithm or prove that none exist.

15. **(ISR 2)**[IMO5] The nonnegative real numbers $x_1, x_2, x_3, x_4, x_5, a$ satisfy the following relations:

$$\sum_{i=1}^{5} i x_i = a, \qquad \sum_{i=1}^{5} i^3 x_i = a^2, \qquad \sum_{i=1}^{5} i^5 x_i = a^3.$$

What are the possible values of a?

16. **(ISR 4)** Let K denote the set $\{a, b, c, d, e\}$. F is a collection of 16 different subsets of K, and it is known that any three members of F have at least one element in common. Show that all 16 members of F have exactly one element in common.

17. **(NLD 1)** Inside an equilateral triangle ABC one constructs points P, Q and R such that
$$\angle QAB = \angle PBA = 15°,$$
$$\angle RBC = \angle QCB = 20°,$$
$$\angle PCA = \angle RAC = 25°.$$

Determine the angles of triangle PQR.

18. **(POL 1)** Let m positive integers $a_1,\ldots,a_m$ be given. Prove that there exist fewer than 2^m positive integers $b_1,\ldots,b_n$ such that all sums of distinct b_k's are distinct and all a_i $(i \le m)$ occur among them.

19. **(ROU 1)** Consider the sequences (a_n), (b_n) defined by

$$a_1 = 3, \quad b_1 = 100, \quad a_{n+1} = 3^{a_n}, \quad b_{n+1} = 100^{b_n}.$$

Find the smallest integer m for which $b_m > a_{100}$.

20. **(SWE 2)** Given the integer $n > 1$ and the real number $a > 0$ determine the maximum of $\sum_{i=1}^{n-1} x_i x_{i+1}$ taken over all nonnegative numbers x_i with sum a.

21. **(USS 1)** Let N be the number of integral solutions of the equation

$$x^2 - y^2 = z^3 - t^3$$

satisfying the condition $0 \le x, y, z, t \le 10^6$, and let M be the number of integral solutions of the equation

$$x^2 - y^2 = z^3 - t^3 + 1$$

satisfying the condition $0 \le x, y, z, t \le 10^6$. Prove that $N > M$.

22. **(USS 3)**[IMO3] There are two circles in the plane. Let a point A be one of the points of intersection of these circles. Two points begin moving simultaneously with constant speeds from the point A, each point along its own circle. The two points return to the point A at the same time. Prove that there is a point P in the plane such that at every moment of time the distances from the point P to the moving points are equal.

23. **(USA 4)** Find all natural numbers n for which $2^8 + 2^{11} + 2^n$ is a perfect square.

24. **(USA 5)** A circle O with center O on base BC of an isosceles triangle ABC is tangent to the equal sides AB, AC. If point P on AB and point Q on AC are selected such that $PB \times CQ = (BC/2)^2$, prove that line segment PQ is tangent to circle O, and prove the converse.

25. **(USA 6)**[IMO4] Given a point P in a given plane π and also a given point Q not in π, show how to determine a point R in π such that $\frac{QP+PR}{QR}$ is a maximum.

26. **(YUG 4)** Prove that the functional equations

$$f(x+y) = f(x)+f(y),$$
$$\text{and} \quad f(x+y+xy) = f(x)+f(y)+f(xy) \quad (x,y \in \mathbb{R})$$

are equivalent.

3.22 The Twenty-Second IMO
Washington DC, United States of America, July 8–20, 1981

3.22.1 Contest Problems

First Day (July 13)

1. Find the point P inside the triangle ABC for which

$$\frac{BC}{PD} + \frac{CA}{PE} + \frac{AB}{PF}$$

   is minimal, where PD, PE, PF are the perpendiculars from P to BC, CA, AB respectively.

2. Let $f(n,r)$ be the arithmetic mean of the minima of all r-subsets of the set $\{1,2,\ldots,n\}$. Prove that $f(n,r) = \frac{n+1}{r+1}$.

3. Determine the maximum value of $m^2 + n^2$ where m and n are integers satisfying

$$m,n \in \{1,2,\ldots,1981\} \quad \text{and} \quad (n^2 - mn - m^2)^2 = 1.$$

Second Day (July 14)

4. (a) For which values of $n > 2$ is there a set of n consecutive positive integers such that the largest number in the set in the set is a divisor of the least common multiple of the remaining $n-1$ numbers?
   (b) For which values of $n > 2$ is there a unique set having the stated property?

5. Three equal circles touch the sides of a triangle and have one common point O. Show that the center of the circle inscribed in and of the circle circumscribed about the triangle ABC and the point O are collinear.

6. Assume that $f(x,y)$ is defined for all positive integers x and y, and that the following equations are satisfied:

$$f(0,y) = y+1,$$
$$f(x+1,0) = f(x,1),$$
$$f(x+1,y+1) = f(x,f(x+1,y)).$$

   Determine $f(4,1981)$.

3.22.2 Shortlisted Problems

1. **(BEL)**[IMO4] (a) For which values of $n > 2$ is there a set of n consecutive positive integers such that the largest number in the set is a divisor of the least common multiple of the remaining $n-1$ numbers?
   (b) For which values of $n > 2$ is there a unique set having the stated property?

2. **(BGR)** A sphere S is tangent to the edges AB, BC, CD, DA of a tetrahedron $ABCD$ at the points E, F, G, H respectively. The points E, F, G, H are the vertices of a square. Prove that if the sphere is tangent to the edge AC, then it is also tangent to the edge BD.

3. **(CAN)** Find the minimum value of

$$\max(a+b+c, b+c+d, c+d+e, d+e+f, e+f+g)$$

subject to the constraints

(i) $a, b, c, d, e, f, g \geq 0$,   (ii) $a+b+c+d+e+f+g = 1$.

4. **(CAN)** Let $\{f_n\}$ be the Fibonacci sequence $\{1, 1, 2, 3, 5, \ldots\}$.
   (a) Find all pairs (a, b) of real numbers such that for each n, $af_n + bf_{n+1}$ is a member of the sequence.
   (b) Find all pairs (u, v) of positive real numbers such that for each n, $uf_n^2 + vf_{n+1}^2$ is a member of the sequence.

5. **(COL)** A cube is assembled with 27 white cubes. The larger cube is then painted black on the outside and disassembled. A blind man reassembles it. What is the probability that the cube is now completely black on the outside? Give an approximation of the size of your answer.

6. **(CUB)** Let $P(z)$ and $Q(z)$ be complex-variable polynomials, with degree not less than 1. Let

$$P_k = \{z \in \mathbb{C} \mid P(z) = k\}, \qquad Q_k = \{z \in \mathbb{C} \mid Q(z) = k\}.$$

Let also $P_0 = Q_0$ and $P_1 = Q_1$. Prove that $P(z) \equiv Q(z)$.

7. **(FIN)**[IMO6] Assume that $f(x, y)$ is defined for all positive integers x and y, and that the following equations are satisfied:

$$f(0, y) = y + 1,$$
$$f(x+1, 0) = f(x, 1),$$
$$f(x+1, y+1) = f(x, f(x+1, y)).$$

Determine $f(2, 2), f(3, 3)$ and $f(4, 4)$.
Alternative version: Determine $f(4, 1981)$.

8. **(FRG)**[IMO2] Let $f(n, r)$ be the arithmetic mean of the minima of all r-subsets of the set $\{1, 2, \ldots, n\}$. Prove that $f(n, r) = \frac{n+1}{r+1}$.

9. **(FRG)** A sequence (a_n) is defined by means of the recursion

$$a_1 = 1, \qquad a_{n+1} = \frac{1 + 4a_n + \sqrt{1 + 24a_n}}{16}.$$

Find an explicit formula for a_n.

10. **(FRA)** Determine the smallest natural number n having the following property: For every integer p, $p \geq n$, it is possible to subdivide (partition) a given square into p squares (not necessarily equal).

11. **(NLD)** On a semicircle with unit radius four consecutive chords AB, BC, CD, DE with lengths a, b, c, d, respectively, are given. Prove that

$$a^2 + b^2 + c^2 + d^2 + abc + bcd < 4.$$

12. **(NLD)**[IMO3] Determine the maximum value of $m^2 + n^2$ where m and n are integers satisfying

$$m, n \in \{1, 2, \ldots, 100\} \quad \text{and} \quad (n^2 - mn - m^2)^2 = 1.$$

13. **(ROU)** Let P be a polynomial of degree n satisfying

$$P(k) = \binom{n+1}{k}^{-1} \quad \text{for } k = 0, 1, \ldots, n.$$

Determine $P(n+1)$.

14. **(ROU)** Prove that a convex pentagon (a five-sided polygon) $ABCDE$ with equal sides and for which the interior angles satisfy the condition $\angle A \geq \angle B \geq \angle C \geq \angle D \geq \angle E$ is a regular pentagon.

15. **(UNK)**[IMO1] Find the point P inside the triangle ABC for which

$$\frac{BC}{PD} + \frac{CA}{PE} + \frac{AB}{PF}$$

is minimal, where PD, PE, PF are the perpendiculars from P to BC, CA, AB respectively.

16. **(UNK)** A sequence of real numbers $u_1, u_2, u_3, \ldots$ is determined by u_1 and the following recurrence relation for $n \geq 1$:

$$4u_{n+1} = \sqrt[3]{64u_n + 15}.$$

Describe, with proof, the behavior of u_n as $n \to \infty$.

17. **(USS)**[IMO5] Three equal circles touch the sides of a triangle and have one common point O. Show that the center of the circle inscribed in and of the circle circumscribed about the triangle ABC and the point O are collinear.

18. **(USS)** Several equal spherical planets are given in outer space. On the surface of each planet there is a set of points that is invisible from any of the remaining planets. Prove that the sum of the areas of all these sets is equal to the area of the surface of one planet.

19. **(YUG)** A finite set of unit circles is given in a plane such that the area of their union U is S. Prove that there exists a subset of mutually disjoint circles such that the area of their union is greater than $\frac{2S}{9}$.

3.23 The Twenty-Third IMO
Budapest, Hungary, July 5–14, 1982

3.23.1 Contest Problems

First Day (July 9)

1. The function $f(n)$ is defined for all positive integers n and takes on nonnegative integer values. Also, for all m, n,

$$f(m+n) - f(m) - f(n) = 0 \quad \text{or} \quad 1;$$

$$f(2) = 0, \quad f(3) > 0, \quad \text{and} \quad f(9999) = 3333.$$

Determine $f(1982)$.

2. A nonisosceles triangle $A_1A_2A_3$ is given with sides a_1, a_2, a_3 (a_i is the side opposite to A_i). For all $i = 1, 2, 3$, M_i is the midpoint of side a_i, T_i is the point where the incircle touches side a_i, and the reflection of T_i in the interior bisector of A_i yields the point S_i. Prove that the lines M_1S_1, M_2S_2, and M_3S_3 are concurrent.

3. Consider the infinite sequences $\{x_n\}$ of positive real numbers with the following properties:
$$x_0 = 1 \quad \text{and for all} \quad i \geq 0, \quad x_{i+1} \leq x_i.$$

   (a) Prove that for every such sequence there is an $n \geq 1$ such that

   $$\frac{x_0^2}{x_1} + \frac{x_1^2}{x_2} + \cdots + \frac{x_{n-1}^2}{x_n} \geq 3.999.$$

   (b) Find such a sequence for which $\frac{x_0^2}{x_1} + \frac{x_1^2}{x_2} + \cdots + \frac{x_{n-1}^2}{x_n} < 4$ for all n.

Second Day (July 10)

4. Prove that if n is a positive integer such that the equation $x^3 - 3xy^2 + y^3 = n$ has a solution in integers (x, y), then it has at least three such solutions. Show that the equation has no solution in integers when $n = 2891$.

5. The diagonals AC and CE of the regular hexagon $ABCDEF$ are divided by the inner points M and N, respectively, so that $\frac{AM}{AC} = \frac{CN}{CE} = r$. Determine r if B, M, and N are collinear.

6. Let S be a square with sides of length 100 and let L be a path within S that does not meet itself and that is composed of linear segments $A_0A_1, A_1A_2, \ldots, A_{n-1}A_n$ with $A_0 \neq A_n$. Suppose that for every point P of the boundary of S there is a point of L at a distance from P not greater than $\frac{1}{2}$. Prove that there are two points X and Y in L such that the distance between X and Y is not greater than 1 and the length of the part of L that lies between X and Y is not smaller than 198.

3.23.2 Longlisted Problems

1. **(AUS 1)** It is well known that the binomial coefficients $\binom{n}{k} = \frac{n!}{k!(n-k)!}$, $0 \le k \le n$, are positive integers. The factorial $n!$ is defined inductively by $0! = 1$, $n! = n \cdot (n-1)!$ for $n \ge 1$.
   (a) Prove that $\frac{1}{n+1}\binom{2n}{n}$ is an integer for $n \ge 0$.
   (b) Given a positive integer k, determine the smallest integer C_k with the property that $\frac{C_k}{n+k+1}\binom{2n}{n+k}$ is an integer for all $n \ge k$.

2. **(AUS 2)** Given a finite number of angular regions $A_1, \ldots, A_k$ in a plane, each A_i being bounded by two half-lines meeting at a vertex and provided with a $+$ or $-$ sign, we assign to each point P of the plane and not on a bounding half-line the number $k - l$, where k is the number of $+$ regions and l the number of $-$ regions that contain P. (Note that the boundary of A_i does not belong to A_i.) For instance, in the figure we have two $+$ regions QAP and RCQ, and one $-$ region RBP. Every point inside $\triangle ABC$ receives the number $+1$, while every point not inside $\triangle ABC$ and not on a boundary halfline the number 0. We say that the interior of $\triangle ABC$ is represented as a sum of the signed angular regions QAP, RBP, and RCQ.
   (a) Show how to represent the interior of any convex planar polygon as a sum of signed angular regions.
   (b) Show how to represent the interior of a tetrahedron as a sum of signed solid angular regions, that is, regions bounded by three planes intersecting at a vertex and provided with a $+$ or $-$ sign.

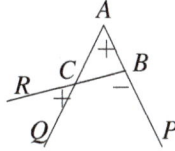

3. **(AUS 3)** Given n points $X_1, X_2, \ldots, X_n$ in the interval $0 \le X_i \le 1$, $i = 1, 2, \ldots, n$, show that there is a point y, $0 \le y \le 1$, such that

$$\frac{1}{n}\sum_{i=1}^{n} |y - X_i| = \frac{1}{2}.$$

4. **(AUS 4)** (SL82-14).
   Original formulation. Let $ABCD$ be a convex planar quadrilateral and let A_1 denote the circumcenter of $\triangle BCD$. Define B_1, C_1, D_1 in a corresponding way.
   (a) Prove that either all of A_1, B_1, C_1, D_1 coincide in one point, or they are all distinct. Assuming the latter case, show that A_1, C_1 are on opposite sides of the line B_1D_1, and similarly, B_1, D_1 are on opposite sides of the line A_1C_1. (This establishes the convexity of the quadrilateral $A_1B_1C_1D_1$.)
   (b) Denote by A_2 the circumcenter of $B_1C_1D_1$, and define B_2, C_2, D_2 in an analogous way. Show that the quadrilateral $A_2B_2C_2D_2$ is similar to the quadrilateral $ABCD$.
   (c) If the quadrilateral $A_1B_1C_1D_1$ was obtained from the quadrilateral $ABCD$ by the above process, what condition must be satisfied by the four points

A_1, B_1, C_1, D_1? Assuming that the four points A_1, B_1, C_1, D_1 satisfying this condition are given, describe a construction by straightedge and compass to obtain the original quadrilateral $ABCD$. (It is not necessary to actually perform the construction).

5. (**BEL 1**) Among all triangles with a given perimeter, find the one with the maximal radius of its incircle.

6. (**BEL 2**) On the three distinct lines a, b, and c three points A, B, and C are given, respectively. Construct three collinear points X, Y, Z on lines a, b, c, respectively, such that $\frac{BY}{AX} = 2$ and $\frac{CZ}{AX} = 3$.

7. (**BEL 3**) Find all solutions $(x, y) \in \mathbb{Z}^2$ of the equation

$$x^3 - y^3 = 2xy + 8.$$

8. (**BRA 1**) (SL82-10).

9. (**BRA 2**) Let n be a natural number, $n \geq 2$, and let ϕ be Euler's function; i.e., $\phi(n)$ is the number of positive integers not exceeding n and coprime to n. Given any two real numbers α and β, $0 \leq \alpha < \beta \leq 1$, prove that there exists a natural number m such that

$$\alpha < \frac{\phi(m)}{m} < \beta.$$

10. (**BRA 3**) Let $r_1, \ldots, r_n$ be the radii of n spheres. Call $S_1, S_2, \ldots, S_n$ the areas of the set of points of each sphere from which one cannot see any point of any other sphere. Prove that

$$\frac{S_1}{r_1^2} + \frac{S_2}{r_2^2} + \cdots + \frac{S_n}{r_n^2} = 4\pi.$$

11. (**BRA 4**) A rectangular pool table has a hole at each of three of its corners. The lengths of sides of the table are the real numbers a and b. A billiard ball is shot from the fourth corner along its angle bisector. The ball falls in one of the holes. What should the relation between a and b be for this to happen?

12. (**BRA 5**) Let there be 3399 numbers arbitrarily chosen among the first 6798 integers $1, 2, \ldots, 6798$ in such a way that none of them divides another. Prove that there are exactly 1982 numbers in $\{1, 2, \ldots, 6798\}$ that must end up being chosen.

13. (**BGR 1**) A regular n-gonal truncated pyramid is circumscribed around a sphere. Denote the areas of the base and the lateral surfaces of the pyramid by S_1, S_2, and S, respectively. Let σ be the area of the polygon whose vertices are the tangential points of the sphere and the lateral faces of the pyramid. Prove that

$$\sigma S = 4S_1 S_2 \cos^2 \frac{\pi}{n}.$$

14. (**BGR 2**) (SL82-4).

15. **(CAN 1)** Show that the set S of natural numbers n for which $3/n$ cannot be written as the sum of two reciprocals of natural numbers ($S = \{n \mid 3/n \neq 1/p + 1/q$ for any $p, q \in \mathbb{N}\}$) is not the union of finitely many arithmetic progressions.

16. **(CAN 2)** (SL82-7).

17. **(CAN 3)** (SL82-11).

18. **(CAN 4)** You are given an algebraic system admitting addition and multiplication for which all the laws of ordinary arithmetic are valid except commutativity of multiplication. Show that

$$(a + ab^{-1}a)^{-1} + (a+b)^{-1} = a^{-1},$$

where x^{-1} is the element for which $x^{-1}x = xx^{-1} = e$, where e is the element of the system such that for all a the equality $ea = ae = a$ holds.

19. **(CAN 5)** (SL82-15).

20. **(CZS 1)** Consider a cube C and two planes σ, τ, which divide Euclidean space into several regions. Prove that the interior of at least one of these regions meets at least three faces of the cube.

21. **(CZS 2)** All edges and all diagonals of regular hexagon $A_1A_2A_3A_4A_5A_6$ are colored blue or red such that each triangle $A_jA_kA_m$, $1 \leq j < k < m \leq 6$ has at least one red edge. Let R_k be the number of red segments A_kA_j, $(j \neq k)$. Prove the inequality

$$\sum_{k=1}^{6} (2R_k - 7)^2 \leq 54.$$

22. **(CZS 3)** (SL82-19).

23. **(FIN 1)** Determine the sum of all positive integers whose digits (in base ten) form either a strictly increasing or a strictly decreasing sequence.

24. **(FIN 2)** Prove that if a person a has infinitely many descendants (children, their children, etc.), then a has an infinite sequence $a_0, a_1, \ldots$ of descendants (i.e., $a = a_0$ and for all $n \geq 1$, a_{n+1} is always a child of a_n). It is assumed that no-one can have infinitely many children.
    Variant 1. Prove that if a has infinitely many ancestors, then a has an infinite descending sequence of ancestors (i.e., $a_0, a_1, \ldots$ where $a = a_0$ and a_n is always a child of a_{n+1}).
    Variant 2. Prove that if someone has infinitely many ancestors, then all people cannot descend from $A(dam)$ and $E(ve)$.

25. **(FIN 3)** (SL82-12).

26. **(FRA 1)** Let $(a_n)_{n \geq 0}$ and $(b_n)_{n \geq 0}$ be two sequences of natural numbers. Determine whether there exists a pair (p, q) of natural numbers that satisfy

$$p < q \quad \text{and} \quad a_p \leq a_q, \; b_p \leq b_q.$$

27. **(FRA 2)** (SL82-18).

28. **(FRA 3)** Let $(u_1, \ldots, u_n)$ be an ordered $n$tuple. For each k, $1 \leq k \leq n$, define $v_k = \sqrt[k]{u_1 u_2 \cdots u_k}$. Prove that

$$\sum_{k=1}^{n} v_k \leq e \cdot \sum_{k=1}^{n} u_k.$$

(e is the base of the natural logarithm).

29. **(FRA 4)** Let $f : \mathbb{R} \to \mathbb{R}$ be a continuous function. Suppose that the restriction of f to the set of irrational numbers is injective. What can we say about f? Answer the analogous question if f is restricted to rationals.

30. **(UNK 1)** (SL82-9).

31. **(UNK 2)** (SL82-16).

32. **(UNK 3)** (SL82-1).

33. **(UNK 4)** A sequence (u_n) of integers is defined for $n \geq 0$ by $u_0 = 0$, $u_1 = 1$, and $u_n - 2u_{n-1} + (1 - c)u_{n-2} = 0$ $(n \geq 2)$, where c is a fixed integer independent of n. Find the least value of c for which both of the following statements are true:
   (i) If p is a prime less than or equal to P, then p divides u_p.
   (ii) If p is a prime greater than P, then p does not divide u_p.

34. **(GDR 1)** Let M be the set of all functions f with the following properties:
   (i) f is defined for all real numbers and takes only real values.
   (ii) For all $x, y \in \mathbb{R}$ the following equality holds: $f(x)f(y) = f(x+y) + f(x-y)$.
   (iii) $f(0) \neq 0$.
   Determine all functions $f \in M$ such that
   (a) $f(1) = 5/2$;
   (b) $f(1) = \sqrt{3}$.

35. **(GDR 2)** If the inradius of a triangle is half of its circumradius, prove that the triangle is equilateral.

36. **(NLD 1)** (SL82-13).

37. **(NLD 2)** (SL82-5).

38. **(POL 1)** Numbers $u_{n,k}$ $(1 \leq k \leq n)$ are defined as follows:

$$u_{1,1} = 1, \quad u_{n,k} = \binom{n}{k} - \sum_{d|n,\, d|k,\, d>1} u_{n/d, k/d}$$

(the empty sum is defined to be equal to zero). Prove that $n \mid u_{n,k}$ for every natural number n and for every k $(1 \leq k \leq n)$.

39. **(POL 2)** Let S be the unit circle with center O and let $P_1, P_2, \ldots, P_n$ be points of S such that the sum of vectors $v_i = \overrightarrow{OP_i}$ is the zero vector. Prove that the inequality $\sum_{i=1}^{n} XP_i \geq n$ holds for every point X.

40. **(POL 3)** We consider a game on an infinite chessboard similar to that of solitaire: If two adjacent fields are occupied by pawns and the next field is empty (the three fields lie on a vertical or horizontal line), then we may remove these two pawns and put one of them on the third field. Prove that if in the initial position pawns fill a $3k \times n$ rectangle, then it is impossible to reach a position with only one pawn on the board.

41. **(POL 4)** (SL82-8).

42. **(POL 5)** Let $\mathscr{F}$ be the family of all k-element subsets of the set $\{1, 2, \ldots, 2k+1\}$. Prove that there exists a bijective function $f : \mathscr{F} \to \mathscr{F}$ such that for every $A \in \mathscr{F}$, the sets A and $f(A)$ are disjoint.

43. **(TUN 1)** (a) What is the maximal number of acute angles in a convex polygon?
    (b) Consider m points in the interior of a convex n-gon. The n-gon is partitioned into triangles whose vertices are among the $n+m$ given points (the vertices of the n-gon and the given points). Each of the m points in the interior is a vertex of at least one triangle. Find the number of triangles obtained.

44. **(TUN 2)** Let A and B be positions of two ships M and N, respectively, at the moment when N saw M moving with constant speed v following the line Ax. In search of help, N moves with speed kv ($k < 1$) along the line By in order to meet M as soon as possible. Denote by C the point of meeting of the two ships, and set
$$AB = d, \quad \angle BAC = \alpha, \quad 0 \le \alpha < \frac{\pi}{2}.$$
Determine the angle $\angle ABC = \beta$ and time t that N needs in order to meet M.

45. **(TUN 3)** (SL82-20).

46. **(USA 1)** Prove that if a diagonal is drawn in a quadrilateral inscribed in a circle, the sum of the radii of the circles inscribed in the two triangles thus formed is the same, no matter which diagonal is drawn.

47. **(USA 2)** Evaluate $\sec'' \frac{\pi}{4} + \sec'' \frac{3\pi}{4} + \sec'' \frac{5\pi}{4} + \sec'' \frac{7\pi}{4}$. (Here $\sec''$ means the second derivative of sec.)

48. **(USA 3)** Given a finite sequence of complex numbers $c_1, c_2, \ldots, c_n$, show that there exists an integer k ($1 \le k \le n$) such that for every finite sequence $a_1, a_2, \ldots, a_n$ of real numbers with $1 \ge a_1 \ge a_2 \ge \cdots \ge a_n \ge 0$, the following inequality holds:
$$\left| \sum_{m=1}^{n} a_m c_m \right| \le \left| \sum_{m=1}^{k} c_m \right|.$$

49. **(USA 4)** Simplify
$$\sum_{k=0}^{n} \frac{(2n)!}{(k!)^2 ((n-k)!)^2}.$$

50. **(USS 1)** Let O be the midpoint of the axis of a right circular cylinder. Let A and B be diametrically opposite points of one base, and C a point of the other

base circle that does not belong to the plane OAB. Prove that the sum of dihedral angles of the trihedral $OABC$ is equal to 2π.

51. **(USS 2)** Let n numbers $x_1, x_2, \ldots, x_n$ be chosen in such a way that $1 \geq x_1 \geq x_2 \geq \cdots \geq x_n \geq 0$. Prove that

$$(1 + x_1 + x_2 + \cdots + x_n)^\alpha \leq 1 + x_1^\alpha + 2^{\alpha-1} x_2^\alpha + \cdots + n^{\alpha-1} x_n^\alpha$$

if $0 \leq \alpha \leq 1$.

52. **(USS 3)** We are given $2n$ natural numbers

$$1, 1, 2, 2, 3, 3, \ldots, n-1, n-1, n, n.$$

Find all n for which these numbers can be arranged in a row such that for each $k \leq n$, there are exactly k numbers between the two numbers k.

53. **(USS 4)** (SL82-3).

54. **(USS 5)** (SL82-17).

55. **(VNM 1)** (SL82-6).

56. **(VNM 2)** Let $f(x) = ax^2 + bx + c$ and $g(x) = cx^2 + bx + a$. If $|f(0)| \leq 1$, $|f(1)| \leq 1$, $|f(-1)| \leq 1$, prove that for $|x| \leq 1$,
    (a) $|f(x)| \leq 5/4$,
    (b) $|g(x)| \leq 2$.

57. **(YUG 1)** (SL82-2).

3.23.3 Shortlisted Problems

1. **A1 (UNK 3)**[IMO1] The function $f(n)$ is defined for all positive integers n and takes on nonnegative integer values. Also, for all m, n,

$$f(m+n) - f(m) - f(n) = 0 \text{ or } 1;$$

$$f(2) = 0, \quad f(3) > 0, \quad \text{and} \quad f(9999) = 3333.$$

Determine $f(1982)$.

2. **A2 (YUG 1)** Let K be a convex polygon in the plane and suppose that K is positioned in the coordinate system in such a way that

$$\text{area } (K \cap Q_i) = \frac{1}{4} \text{ area } K \quad (i = 1, 2, 3, 4,),$$

where the Q_i denote the quadrants of the plane. Prove that if K contains no nonzero lattice point, then the area of K is less than 4.

3. **A3 (USS 4)**[IMO3] Consider the infinite sequences $\{x_n\}$ of positive real numbers with the following properties:

$$x_0 = 1 \text{ and for all } i \geq 0, \ x_{i+1} \leq x_i.$$

(a) Prove that for every such sequence there is an $n \geq 1$ such that $\frac{x_0^2}{x_1} + \frac{x_1^2}{x_2} + \cdots + \frac{x_{n-1}^2}{x_n} \geq 3.999$.

(b) Find such a sequence for which $\frac{x_0^2}{x_1} + \frac{x_1^2}{x_2} + \cdots + \frac{x_{n-1}^2}{x_n} < 4$ for all n.

4. **A4 (BGR 2)** Determine all real values of the parameter a for which the equation

$$16x^4 - ax^3 + (2a + 17)x^2 - ax + 16 = 0$$

has exactly four distinct real roots that form a geometric progression.

5. **A5 (NLD 2)**[IMO5] Let $A_1A_2A_3A_4A_5A_6$ be a regular hexagon. Each of its diagonals $A_{i-1}A_{i+1}$ is divided into the same ratio $\frac{\lambda}{1-\lambda}$, where $0 < \lambda < 1$, by a point B_i in such a way that A_i, B_i, and B_{i+2} are collinear ($i \equiv 1, \ldots, 6 \pmod 6$). Compute λ.

6. **A6 (VNM 1)**[IMO6] Let S be a square with sides of length 100 and let L be a path within S that does not meet itself and that is composed of linear segments $A_0A_1, A_1A_2, \ldots, A_{n-1}A_n$ with $A_0 \neq A_n$. Suppose that for every point P of the boundary of S there is a point of L at a distance from P not greater than $\frac{1}{2}$. Prove that there are two points X and Y in L such that the distance between X and Y is not greater than 1 and the length of that part of L that lies between X and Y is not smaller than 198.

7. **B1 (CAN 2)** Let $p(x)$ be a cubic polynomial with integer coefficients with leading coefficient 1 and with one of its roots equal to the product of the other two. Show that $2p(-1)$ is a multiple of $p(1) + p(-1) - 2(1 + p(0))$.

8. **B2 (POL 4)** A convex, closed figure lies inside a given circle. The figure is seen from every point of the circumference at a right angle (that is, the two rays drawn from the point and supporting the convex figure are perpendicular). Prove that the center of the circle is a center of symmetry of the figure.

9. **B3 (UNK 1)** Let ABC be a triangle, and let P be a point inside it such that $\angle PAC = \angle PBC$. The perpendiculars from P to BC and CA meet these lines at L and M, respectively, and D is the midpoint of AB. Prove that $DL = DM$.

10. **B4 (BRA 1)** A box contains p white balls and q black balls. Beside the box there is a pile of black balls. Two balls are taken out of the box. If they have the same color, a black ball from the pile is put into the box. If they have different colors, the white ball is put back into the box. This procedure is repeated until the last two balls are removed from the box and one last ball is put in. What is the probability that this last ball is white?

11. **B5 (CAN 3)** (a) Find the rearrangement $\{a_1, \ldots, a_n\}$ of $\{1, 2, \ldots, n\}$ that maximizes

$$a_1a_2 + a_2a_3 + \cdots + a_na_1 = Q.$$

(b) Find the rearrangement that minimizes Q.

12. **B6 (FIN 3)** Four distinct circles C, C_1, C_2, C_3 and a line L are given in the plane such that C and L are disjoint and each of the circles C_1, C_2, C_3 touches the other

two, as well as C and L. Assuming the radius of C to be 1, determine the distance between its center and L.

13. **C1 (NLD 1)**[IMO2] A scalene triangle $A_1A_2A_3$ is given with sides a_1, a_2, a_3 (a_i is the side opposite to A_i). For all $i = 1, 2, 3$, M_i is the midpoint of side a_i, T_i is the point where the incircle touches side a_i, and the reflection of T_i in the interior bisector of A_i yields the point S_i. Prove that the lines M_1S_1, M_2S_2, and M_3S_3 are concurrent.

14. **C2 (AUS 4)** Let $ABCD$ be a convex plane quadrilateral and let A_1 denote the circumcenter of $\triangle BCD$. Define B_1, C_1, D_1 in a corresponding way.
    (a) Prove that either all of A_1, B_1, C_1, D_1 coincide in one point, or they are all distinct. Assuming the latter case, show that A_1, C_1 are on opposite sides of the line B_1D_1, and similarly, B_1, D_1 are on opposite sides of the line A_1C_1. (This establishes the convexity of the quadrilateral $A_1B_1C_1D_1$.)
    (b) Denote by A_2 the circumcenter of $B_1C_1D_1$, and define B_2, C_2, D_2 in an analogous way. Show that the quadrilateral $A_2B_2C_2D_2$ is similar to the quadrilateral $ABCD$.

15. **C3 (CAN 5)** Show that
$$\frac{1-s^a}{1-s} \le (1+s)^{a-1}$$
holds for every $1 \ne s > 0$ real and $0 < a \le 1$ rational.

16. **C4 (UNK 2)**[IMO4] Prove that if n is a positive integer such that the equation $x^3 - 3xy^2 + y^3 = n$ has a solution in integers (x, y), then it has at least three such solutions. Show that the equation has no solution in integers when $n = 2891$.

17. **C5 (USS 5)** The right triangles ABC and AB_1C_1 are similar and have opposite orientation. The right angles are at C and C_1, and we also have $\angle CAB = \angle C_1AB_1$. Let M be the point of intersection of the lines BC_1 and B_1C. Prove that if the lines AM and CC_1 exist, they are perpendicular.

18. **C6 (FRA 2)** Let O be a point of three-dimensional space and let l_1, l_2, l_3 be mutually perpendicular straight lines passing through O. Let S denote the sphere with center O and radius R, and for every point M of S, let S_M denote the sphere with center M and radius R. We denote by P_1, P_2, P_3 the intersection of S_M with the straight lines l_1, l_2, l_3, respectively, where we put $P_i \ne O$ if l_i meets S_M at two distinct points and $P_i = O$ otherwise ($i = 1, 2, 3$). What is the set of centers of gravity of the (possibly degenerate) triangles $P_1P_2P_3$ as M runs through the points of S?

19. **C7 (CZS 3)** Let M be the set of real numbers of the form $\frac{m+n}{\sqrt{m^2+n^2}}$, where m and n are positive integers. Prove that for every pair $x \in M$, $y \in M$ with $x < y$, there exists an element $z \in M$ such that $x < z < y$.

20. **C8 (TUN 3)** Let $ABCD$ be a convex quadrilateral and draw regular triangles ABM, CDP, BCN, ADQ, the first two outward and the other two inward. Prove that $MN = AC$. What can be said about the quadrilateral $MNPQ$?

3.24 The Twenty-Fourth IMO
Paris, France, July 1–12, 1983

3.24.1 Contest Problems

First Day (July 6)

1. Find all functions f defined on the positive real numbers and taking positive real values that satisfy the following conditions:
   (i) $f(xf(y)) = yf(x)$ for all positive real x, y;
   (ii) $f(x) \to 0$ as $x \to +\infty$.

2. Let K be one of the two intersection points of the circles W_1 and W_2. Let O_1 and O_2 be the centers of W_1 and W_2. The two common tangents to the circles meet W_1 and W_2 respectively in P_1 and P_2, the first tangent, and Q_1 and Q_2 the second tangent. Let M_1 and M_2 be the midpoints of P_1Q_1 and P_2Q_2, respectively. Prove that $\angle O_1KO_2 = \angle M_1KM_2$.

3. Let a, b, c be positive integers satisfying $(a,b) = (b,c) = (c,a) = 1$. Show that $2abc - ab - bc - ca$ is the largest integer not representable as

$$xbc + yca + zab$$

with nonnegative integers x, y, z.

Second Day (July 7)

4. Let ABC be an equilateral triangle. Let E be the set of all points from segments AB, BC, and CA (including A, B, and C). Is it true that for any partition of the set E into two disjoint subsets, there exists a right-angled triangle all of whose vertices belong to the same subset in the partition?

5. Prove or disprove the following statement: In the set $\{1, 2, 3, \ldots, 10^5\}$ a subset of 1983 elements can be found that does not contain any three consecutive terms of an arithmetic progression.

6. If a, b, and c are sides of a triangle, prove that

$$a^2b(a-b) + b^2c(b-c) + c^2a(c-a) \geq 0$$

and determine when there is equality.

3.24.2 Longlisted Problems

1. **(AUS 1)** (SL83-1).

2. **(AUS 2)** Seventeen cities are served by four airlines. It is noted that there is direct service (without stops) between any two cities and that all airline schedules offer round-trip flights. Prove that at least one of the airlines can offer a round trip with an odd number of landings.

3. **(AUS 3)** (a) Given a tetrahedron $ABCD$ and its four altitudes (i.e., lines through each vertex, perpendicular to the opposite face), assume that the altitude dropped from D passes through the orthocenter H_4 of $\triangle ABC$. Prove that this altitude DH_4 intersects all the other three altitudes.

   (b) If we further know that a second altitude, say the one from vertex A to the face BCD, also passes through the orthocenter H_1 of $\triangle BCD$, then prove that all four altitudes are concurrent and each one passes through the orthocenter of the respective triangle.

4. **(BEL 1)** (SL83-2).

5. **(BEL 2)** Consider the set $\mathbb{Q}^2$ of points in $\mathbb{R}^2$, both of whose coordinates are rational.

   (a) Prove that the union of segments with vertices from $\mathbb{Q}^2$ is the entire set $\mathbb{R}^2$.

   (b) Is the convex hull of $\mathbb{Q}^2$ (i.e., the smallest convex set in $\mathbb{R}^2$ that contains $\mathbb{Q}^2$) equal to $\mathbb{R}^2$?[6]

6. **(BEL 3)** (SL83-3).

7. **(BEL 4)** Find all numbers $x \in \mathbb{Z}$ for which the number

$$x^4 + x^3 + x^2 + x + 1$$

is a perfect square.

8. **(BEL 5)** (SL83-4).

9. **(BRA 1)** (SL83-5).

10. **(BRA 2)** Which of the numbers $1, 2, \ldots, 1983$ has the largest number of divisors?

11. **(BRA 3)** A boy at point A wants to get water at a circular lake and carry it to point B. Find the point C on the lake such that the distance walked by the boy is the shortest possible given that the line AB and the lake are exterior to each other.

12. **(BRA 4)** The number 0 or 1 is to be assigned to each of the n vertices of a regular polygon. In how many different ways can this be done (if we consider two assignments that can be obtained one from the other through rotation in the plane of the polygon to be identical)?

13. **(BGR 1)** Let p be a prime number and $a_1, a_2, \ldots, a_{(p+1)/2}$ different natural numbers less than or equal to p. Prove that for each natural number r less than or equal to p, there exist two numbers (perhaps equal) a_i and a_j such that

$$p \equiv a_i a_j \pmod{r}.$$

14. **(BGR 2)** Let l be tangent to the circle k at B. Let A be a point on k and P the foot of perpendicular from A to l. Let M be symmetric to P with respect to AB. Find the set of all such points M.

---

[6] The problem is unclear. In this form, part (a) is false and part (b) is trivial.

15. **(CAN 1)** Find all possible finite sequences $\{n_0, n_1, n_2, \ldots, n_k\}$ of integers such that for each i, i appears in the sequence n_i times $(0 \le i \le k)$.

16. **(CAN 2)** (SL83-6).

17. **(CAN 3)** In how many ways can $1, 2, \ldots, 2n$ be arranged in a $2 \times n$ rectangular array $\begin{pmatrix} a_1 & a_2 & \cdots & a_n \\ b_1 & b_2 & \cdots & b_n \end{pmatrix}$ for which:
    (i) $a_1 < a_2 < \cdots < a_n$,
    (ii) $b_1 < b_2 < \cdots < b_n$,
    (iii) $a_1 < b_1, a_2 < b_2, \ldots, a_n < b_n$?

18. **(CAN 4)** Let $b \ge 2$ be a positive integer.
    (a) Show that for an integer N, written in base b, to be equal to the sum of the squares of its digits, it is necessary either that $N = 1$ or that N have only two digits.
    (b) Give a complete list of all integers not exceeding 50 that, relative to some base b, are equal to the sum of the squares of their digits.
    (c) Show that for any base b the number of two-digit integers that are equal to the sum of the squares of their digits is even.
    (d) Show that for any odd base b there is an integer other than 1 that is equal to the sum of the squares of its digits.

19. **(CAN 5)** (SL83-7).

20. **(COL 1)** Let f and g be functions from the set A to the same set A. We define f to be *a functional nth root of g* (n is a positive integer) if $f^n(x) = g(x)$, where $f^n(x) = f^{n-1}(f(x))$.
    (a) Prove that the function $g : \mathbb{R} \to \mathbb{R}$, $g(x) = 1/x$ has an infinite number of nth functional roots for each positive integer n.
    (b) Prove that there is a bijection from $\mathbb{R}$ onto $\mathbb{R}$ that has no nth functional root for each positive integer n.

21. **(COL 2)** Prove that there are infinitely many positive integers n for which it is possible for a knight, starting at one of the squares of an $n \times n$ chessboard, to go through each of the squares exactly once.

22. **(CUB 1)** Does there exist an infinite number of sets C consisting of 1983 consecutive natural numbers such that each of the numbers is divisible by some number of the form a^{1983}, with $a \in \mathbb{N}$, $a \ne 1$?

23. **(FIN 1)** (SL83-10).

24. **(FIN 2)** Every x, $0 \le x \le 1$, admits a unique representation $x = \sum_{j=0}^\infty a_j 2^{-j}$, where all the a_j belong to $\{0, 1\}$ and infinitely many of them are 0. If $b(0) = \frac{1+c}{2+c}$, $b(1) = \frac{1}{2+c}$, $c > 0$, and

$$f(x) = a_0 + \sum_{j=0}^\infty b(a_0) \cdots b(a_j) a_{j+1},$$

show that $0 < f(x) - x < c$ for every x, $0 < x < 1$.
(**FIN 2′**) (SL83-11).

25. (**FRG 1**) How many permutations $a_1, a_2, \ldots, a_n$ of $\{1, 2, \ldots, n\}$ are sorted into increasing order by at most three repetitions of the following operation: Move from left to right and interchange a_i and a_{i+1} whenever $a_i > a_{i+1}$ for i running from 1 up to $n - 1$?

26. (**FRG 2**) Let a, b, c be positive integers satisfying $(a, b) = (b, c) = (c, a) = 1$. Show that $2abc - ab - bc - ca$ cannot be represented as $bcx + cay + abz$ with nonnegative integers x, y, z.

27. (**FRG 3**) (SL83-18).

28. (**UNK 1**) Show that if the sides a, b, c of a triangle satisfy the equation

$$2(ab^2 + bc^2 + ca^2) = a^2b + b^2c + c^2a + 3abc,$$

then the triangle is equilateral. Show also that the equation can be satisfied by positive real numbers that are not the sides of a triangle.

29. (**UNK 2**) Let O be a point outside a given circle. Two lines OAB, OCD through O meet the circle at A, B, C, D, where A, C are the midpoints of OB, OD, respectively. Additionally, the acute angle θ between the lines is equal to the acute angle at which each line cuts the circle. Find $\cos \theta$ and show that the tangents at A, D to the circle meet on the line BC.

30. (**UNK 3**) Prove the existence of a unique sequence $\{u_n\}$ $(n = 0, 1, 2 \ldots)$ of positive integers such that

$$u_n^2 = \sum_{r=0}^{n} \binom{n+r}{r} u_{n-r} \qquad \text{for all } n \geq 0,$$

where $\binom{m}{r}$ is the usual binomial coefficient.

31. (**UNK 4**) (SL83-12).

32. (**UNK 5**) Let a, b, c be positive real numbers and let $[x]$ denote the greatest integer that does not exceed the real number x. Suppose that f is a function defined on the set of nonnegative integers n and taking real values such that $f(0) = 0$ and

$$f(n) \leq an + f([bn]) + f([cn]), \qquad \text{for all } n \geq 1.$$

Prove that if $b + c < 1$, there is a real number k such that

$$f(n) \leq kn \qquad \text{for all } n, \tag{1}$$

while if $b + c = 1$, there is a real number K such that $f(n) \leq Kn \log_2 n$ for all $n \geq 2$. Show that if $b + c = 1$, there may not be a real number k that satisfies (1).

33. (**GDR 1**) (SL83-16).

34. **(GDR 2)** In a plane are given n points P_i $(i = 1, 2, \ldots, n)$ and two angles α and β. Over each of the segments $P_i P_{i+1}$ $(P_{n+1} = P_1)$ a point Q_i is constructed such that for all i:
    (i) upon moving from P_i to P_{i+1}, Q_i is seen on the same side of $P_i P_{i+1}$,
    (ii) $\angle P_{i+1} P_i Q_i = \alpha$,
    (iii) $\angle P_i P_{i+1} Q_i = \beta$.
    Furthermore, let g be a line in the same plane with the property that all the points P_i, Q_i lie on the same side of g. Prove that

$$\sum_{i=1}^{n} d(P_i, g) = \sum_{i=1}^{n} d(Q_i, g),$$

    where $d(M, g)$ denotes the distance from point M to line g.

35. **(GDR 3)** (SL83-17).

36. **(ISR 1)** The set X has 1983 members. There exists a family of subsets $\{S_1, S_2, \ldots, S_k\}$ such that:
    (i) the union of any three of these subsets is the entire set X, while
    (ii) the union of any two of them contains at most 1979 members.
    What is the largest possible value of k?

37. **(ISR 2)** The points $A_1, A_2, \ldots, A_{1983}$ are set on the circumference of a circle and each is given one of the values ± 1. Show that if the number of points with the value $+1$ is greater than 1789, then at least 1207 of the points will have the property that the partial sums that can be formed by taking the numbers from them to any other point, in either direction, are strictly positive.

38. **(KWT 1)** Let $\{u_n\}$ be the sequence defined by its first two terms u_0, u_1 and the recursion formula

$$u_{n+2} = u_n - u_{n+1}.$$

    (a) Show that u_n can be written in the form $u_n = \alpha a^n + \beta b^n$, where a, b, α, β are constants independent of n that have to be determined.
    (b) If $S_n = u_0 + u_1 + \cdots + u_n$, prove that $S_n + u_{n-1}$ is a constant independent of n. Determine this constant.

39. **(KWT 2)** If α is the real root of the equation

$$E(x) = x^3 - 5x - 50 = 0$$

    such that $x_{n+1} = (5x_n + 50)^{1/3}$ and $x_1 = 5$, where n is a positive integer, prove that:
    (a) $x_{n+1}^3 - \alpha^3 = 5(x_n - \alpha)$
    (b) $\alpha < x_{n+1} < x_n$

40. **(LUX 1)** Four faces of tetrahedron $ABCD$ are congruent triangles whose angles form an arithmetic progression. If the lengths of the sides of the triangles are $a < b < c$, determine the radius of the sphere circumscribed about the tetrahedron as a function on a, b, and c. What is the ratio c/a if $R = a$?

41. **(LUX 2)** (SL83-13).

42. **(LUX 3)** Consider the square $ABCD$ in which a segment is drawn between each vertex and the midpoints of both opposite sides. Find the ratio of the area of the octagon determined by these segments and the area of the square $ABCD$.

43. **(LUX 4)** Given a square $ABCD$, let P, Q, R, and S be four variable points on the sides AB, BC, CD, and DA, respectively. Determine the positions of the points P, Q, R, and S for which the quadrilateral $PQRS$ is a parallelogram, a rectangle, a square, or a trapezoid.

44. **(LUX 5)** We are given twelve coins, one of which is a fake with a different mass from the other eleven. Determine that coin with three weighings and whether it is heavier or lighter than the others.

45. **(LUX 6)** Let two glasses, numbered 1 and 2, contain an equal quantity of liquid, milk in glass 1 and coffee in glass 2. One does the following: Take one spoon of mixture from glass 1 and pour it into glass 2, and then take the same spoon of the new mixture from glass 2 and pour it back into the first glass. What happens after this operation is repeated n times, and what as n tends to infinity?

46. **(LUX 7)** Let f be a real-valued function defined on $I = (0, +\infty)$ and having no zeros on I. Suppose that
$$\lim_{x \to +\infty} \frac{f'(x)}{f(x)} = +\infty.$$
For the sequence $u_n = \ln \left| \frac{f(n+1)}{f(n)} \right|$, prove that $u_n \to +\infty \ (n \to +\infty)$.

47. **(NLD 1)** In a plane, three pairwise intersecting circles C_1, C_2, C_3 with centers M_1, M_2, M_3 are given. For $i = 1, 2, 3$, let A_i be one of the points of intersection of C_j and C_k ($\{i, j, k\} = \{1, 2, 3\}$). Prove that if $\angle M_3 A_1 M_2 = \angle M_1 A_2 M_3 = \angle M_2 A_3 M_1 = \pi/3$ (directed angles), then $M_1 A_1$, $M_2 A_2$, and $M_3 A_3$ are concurrent.

48. **(NLD 2)** Prove that in any parallelepiped the sum of the lengths of the edges is less than or equal to twice the sum of the lengths of the four diagonals.

49. **(POL 1)** Given positive integers k, m, n with $km \le n$ and nonnegative real numbers $x_1, \ldots, x_k$, prove that
$$n \left(\prod_{i=1}^{k} x_i^m - 1 \right) \le m \sum_{i=1}^{k} (x_i^n - 1).$$

50. **(POL 2)** (SL83-14).

51. **(POL 3)** (SL83-15).

52. **(ROU 1)** (SL83-19).

53. **(ROU 2)** Let $a \in \mathbb{R}$ and let $z_1, z_2, \ldots, z_n$ be complex numbers of modulus 1 satisfying the relation

$$\sum_{k=1}^{n} z_k^3 = 4(a + (a-n)i) - 3\sum_{k=1}^{n} \overline{z_k}.$$

Prove that $a \in \{0, 1, \ldots, n\}$ and $z_k \in \{1, i\}$ for all k.

54. **(ROU 3)** (SL83-20).

55. **(ROU 4)** For every $a \in \mathbb{N}$ denote by $M(a)$ the number of elements of the set

$$\{b \in \mathbb{N} \mid a + b \text{ is a divisor of } ab\}.$$

Find $\max_{a \leq 1983} M(a)$.

56. **(ROU 5)** Consider the expansion

$$(1 + x + x^2 + x^3 + x^4)^{496} = a_0 + a_1 x + \cdots + a_{1984} x^{1984}.$$

   (a) Determine the greatest common divisor of the coefficients $a_3, a_8, a_{13}, \ldots,$ a_{1983}.
   (b) Prove that $10^{340} < a^{992} < 10^{347}$.

57. **(ESP 1)** In the system of base $n^2 + 1$ find a number N with n different digits such that:
   (i) N is a multiple of n. Let $N = nN'$.
   (ii) The number N and N' have the same number n of different digits in base $n^2 + 1$, none of them being zero.
   (iii) If $s(C)$ denotes the number in base $n^2 + 1$ obtained by applying the permutation s to the n digits of the number C, then for each permutation s, $s(N) = ns(N')$.

58. **(ESP 2)** (SL83-8).

59. **(ESP 3)** Solve the equation

$$\tan^2(2x) + 2\tan(2x) \cdot \tan(3x) - 1 = 0.$$

60. **(SWE 1)** (SL83-21).

61. **(SWE 2)** Let a and b be integers. Is it possible to find integers p and q such that the integers $p + na$ and $q + nb$ have no common prime factor no matter how the integer n is chosen.

62. **(SWE 3)** A circle γ is drawn and let AB be a diameter. The point C on γ is the midpoint of the line segment BD. The line segments AC and DO, where O is the center of γ, intersect at P. Prove that there is a point E on AB such that P is on the circle with diameter AE.

63. **(SWE 4)** (SL83-22).

64. **(USA 1)** The sum of all the face angles about all of the vertices except one of a given polyhedron is 5160. Find the sum of all of the face angles of the polyhedron.

65. **(USA 2)** Let $ABCD$ be a convex quadrilateral whose diagonals AC and BD intersect in a point P. Prove that

$$\frac{AP}{PC} = \frac{\cot \angle BAC + \cot \angle DAC}{\cot \angle BCA + \cot \angle DCA}.$$

66. **(USA 3)** (SL83-9).

67. **(USA 4)** The altitude from a vertex of a given tetrahedron intersects the opposite face in its orthocenter. Prove that all four altitudes of the tetrahedron are concurrent.

68. **(USA 5)** Three of the roots of the equation $x^4 - px^3 + qx^2 - rx + s = 0$ are $\tan A$, $\tan B$, and $\tan C$, where A, B, and C are angles of a triangle. Determine the fourth root as a function only of p, q, r, and s.

69. **(USS 1)** (SL83-23).

70. **(USS 2)** (SL83-24).

71. **(USS 3)** (SL83-25).

72. **(USS 4)** Prove that for all $x_1, x_2, \ldots, x_n \in \mathbb{R}$ the following inequality holds:

$$\sum_{n \geq i > j \geq 1} \cos^2(x_i - x_j) \geq \frac{n(n-2)}{4}.$$

73. **(VNM 1)** Let ABC be a nonequilateral triangle. Prove that there exist two points P and Q in the plane of the triangle, one in the interior and one in the exterior of the circumcircle of ABC, such that the orthogonal projections of any of these two points on the sides of the triangle are vertices of an equilateral triangle.

74. **(VNM 2)** In a plane we are given two distinct points A, B and two lines a, b passing through B and A respectively $(a \ni B, b \ni A)$ such that the line AB is equally inclined to a and b. Find the locus of points M in the plane such that the product of distances from M to A and a equals the product of distances from M to B and b (i.e., $MA \cdot MA' = MB \cdot MB'$, where A' and B' are the feet of the perpendiculars from M to a and b respectively).

75. **(VNM 3)** Find the sum of the fiftieth powers of all sides and diagonals of a regular 100-gon inscribed in a circle of radius R.

3.24.3 Shortlisted Problems

1. **(AUS 1)** The localities $P_1, P_2, \ldots, P_{1983}$ are served by ten international airlines $A_1, A_2, \ldots, A_{10}$. It is noticed that there is direct service (without stops) between any two of these localities and that all airline schedules offer round-trip flights. Prove that at least one of the airlines can offer a round trip with an odd number of landings.

2. **(BEL 1)** Let n be a positive integer. Let $\sigma(n)$ be the sum of the natural divisors d of n (including 1 and n). We say that an integer $m \geq 1$ is *superabundant* (P.Erdös, 1944) if $\forall k \in \{1, 2, \ldots, m-1\}$, $\frac{\sigma(m)}{m} > \frac{\sigma(k)}{k}$. Prove that there exists an infinity of superabundant numbers.

3. **(BEL 3)**[IMO4] We say that a set E of points of the Euclidian plane is "Pythagorean" if for any partition of E into two sets A and B, at least one of the sets contains the vertices of a right-angled triangle. Decide whether the following sets are Pythagorean:
   (a) a circle;
   (b) an equilateral triangle (that is, the set of three vertices and the points of the three edges).

4. **(BEL 5)** On the sides of the triangle ABC, three similar isosceles triangles ABP $(AP = PB)$, AQC $(AQ = QC)$, and BRC $(BR = RC)$ are constructed. The first two are constructed externally to the triangle ABC, but the third is placed in the same half-plane determined by the line BC as the triangle ABC. Prove that $APRQ$ is a parallelogram.

5. **(BRA 1)** Consider the set of all strictly decreasing sequences of n natural numbers having the property that in each sequence no term divides any other term of the sequence. Let $A = (a_j)$ and $B = (b_j)$ be any two such sequences. We say that A precedes B if for some k, $a_k < b_k$ and $a_i = b_i$ for $i < k$. Find the terms of the first sequence of the set under this ordering.

6. **(CAN 2)** Suppose that $\{x_1, x_2, \ldots, x_n\}$ are positive integers for which $x_1 + x_2 + \cdots + x_n = 2(n+1)$. Show that there exists an integer r with $0 \leq r \leq n-1$ for which the following $n-1$ inequalities hold:

$$x_{r+1} + \cdots + x_{r+i} \leq 2i+1 \qquad \forall i, \ 1 \leq i \leq n-r;$$
$$x_{r+1} + \cdots + x_n + x_1 + \cdots + x_i \leq 2(n-r+i)+1 \qquad \forall i, \ 1 \leq i \leq r-1.$$

Prove that if all the inequalities are strict, then r is unique and that otherwise there are exactly two such r.

7. **(CAN 5)** Let a be a positive integer and let $\{a_n\}$ be defined by $a_0 = 0$ and

$$a_{n+1} = (a_n+1)a + (a+1)a_n + 2\sqrt{a(a+1)a_n(a_n+1)} \qquad (n = 1, 2 \ldots).$$

Show that for each positive integer n, a_n is a positive integer.

8. **(ESP 2)** In a test, $3n$ students participate, who are located in three rows of n students in each. The students leave the test room one by one. If $N_1(t)$, $N_2(t)$, $N_3(t)$ denote the numbers of students in the first, second, and third row respectively at time t, find the probability that for each t during the test,

$$|N_i(t) - N_j(t)| < 2, \quad i \neq j, \quad i, j = 1, 2, \ldots .$$

9. **(USA 3)**[IMO6] If a, b, and c are sides of a triangle, prove that

$$a^2b(a-b)+b^2c(b-c)+c^2a(c-a) \geq 0.$$

Determine when there is equality.

10. **(FIN 1)** Let p and q be integers. Show that there exists an interval I of length $1/q$ and a polynomial P with integral coefficients such that

$$\left| P(x) - \frac{p}{q} \right| < \frac{1}{q^2}$$

for all $x \in I$.

11. **(FIN 2′)** Let $f : [0,1] \to \mathbb{R}$ be continuous and satisfy:

$$\begin{aligned} bf(2x) &= f(x), & 0 \leq x \leq 1/2; \\ f(x) &= b + (1-b)f(2x-1), & 1/2 \leq x \leq 1, \end{aligned}$$

where $b = \frac{1+c}{2+c}$, $c > 0$. Show that $0 < f(x) - x < c$ for every x, $0 < x < 1$.

12. **(UNK 4)**[IMO1] Find all functions f defined on the positive real numbers and taking positive real values that satisfy the following conditions:
    (i) $f(xf(y)) = yf(x)$ for all positive real x, y.
    (ii) $f(x) \to 0$ as $x \to +\infty$.

13. **(LUX 2)** Let E be the set of 1983^3 points of the space $\mathbb{R}^3$ all three of whose coordinates are integers between 0 and 1982 (including 0 and 1982). A coloring of E is a map from E to the set $\{red, blue\}$. How many colorings of E are there satisfying the following property: The number of red vertices among the 8 vertices of any right-angled parallelepiped is a multiple of 4?

14. **(POL 2)**[IMO5] Prove or disprove: From the interval $[1, \ldots, 30000]$ one can select a set of 1000 integers containing no arithmetic triple (three consecutive numbers of an arithmetic progression).

15. **(POL 3)** Decide whether there exists a set M of natural numbers satisfying the following conditions:
    (i) For any natural number $m > 1$ there are $a, b \in M$ such that $a + b = m$.
    (ii) If $a, b, c, d \in M$, $a, b, c, d > 10$ and $a + b = c + d$, then $a = c$ or $a = d$.

16. **(GDR 1)** Let $F(n)$ be the set of polynomials $P(x) = a_0 + a_1 x + \cdots + a_n x^n$, with $a_0, a_1, \ldots, a_n \in \mathbb{R}$ and $0 \leq a_0 = a_n \leq a_1 = a_{n-1} \leq \cdots \leq a_{[n/2]} = a_{[(n+1)/2]}$. Prove that if $f \in F(m)$ and $g \in F(n)$, then $fg \in F(m+n)$.

17. **(GDR 3)** Let $P_1, P_2, \ldots, P_n$ be distinct points of the plane, $n \geq 2$. Prove that

$$\max_{1 \leq i < j \leq n} P_i P_j > \frac{\sqrt{3}}{2}(\sqrt{n} - 1) \min_{1 \leq i < j \leq n} P_i P_j.$$

18. **(FRG 3)**[IMO3] Let a, b, c be positive integers satisfying $(a,b) = (b,c) = (c,a) = 1$. Show that $2abc - ab - bc - ca$ is the largest integer not representable as

$$xbc + yca + zab$$

with nonnegative integers x, y, z.

19. **(ROU 1)** Let $(F_n)_{n \geq 1}$ be the Fibonacci sequence $F_1 = F_2 = 1$, $F_{n+2} = F_{n+1} + F_n$ $(n \geq 1)$, and $P(x)$ the polynomial of degree 990 satisfying

$$P(k) = F_k, \quad \text{for } k = 992, \ldots, 1982.$$

Prove that $P(1983) = F_{1983} - 1$.

20. **(ROU 3)** Solve the system of equations

$$x_1|x_1| = x_2|x_2| + (x_1 - a)|x_1 - a|,$$
$$x_2|x_2| = x_3|x_3| + (x_2 - a)|x_2 - a|,$$
$$\cdots$$
$$x_n|x_n| = x_1|x_1| + (x_n - a)|x_n - a|,$$

in the set of real numbers, where $a > 0$.

21. **(SWE 1)** Find the greatest integer less than or equal to $\sum_{k=1}^{2^{1983}} k^{1/1983-1}$.

22. **(SWE 4)** Let n be a positive integer having at least two different prime factors. Show that there exists a permutation $a_1, a_2, \ldots, a_n$ of the integers $1, 2, \ldots, n$ such that

$$\sum_{k=1}^{n} k \cdot \cos \frac{2\pi a_k}{n} = 0.$$

23. **(USS 1)**[IMO2] Let K be one of the two intersection points of the circles W_1 and W_2. Let O_1 and O_2 be the centers of W_1 and W_2. The two common tangents to the circles meet W_1 and W_2 respectively in P_1 and P_2, the first tangent, and Q_1 and Q_2, the second tangent. Let M_1 and M_2 be the midpoints of P_1Q_1 and P_2Q_2, respectively. Prove that

$$\angle O_1 K O_2 = \angle M_1 K M_2.$$

24. **(USS 2)** Let d_n be the last nonzero digit of the decimal representation of $n!$. Prove that d_n is aperiodic; that is, there do not exist T and n_0 such that for all $n \geq n_0$, $d_{n+T} = d_n$.

25. **(USS 3)** Prove that every partition of 3-dimensional space into three disjoint subsets has the following property: One of these subsets contains all possible distances; i.e., for every $a \in \mathbb{R}_+$, there are points M and N inside that subset such that distance between M and N is exactly a.

3.25 The Twenty-Fifth IMO
Prague, Czechoslovakia, June 29–July 10, 1984

3.25.1 Contest Problems

First Day (July 4)

1. Let x, y, z be nonnegative real numbers with $x + y + z = 1$. Show that

$$0 \le xy + yz + zx - 2xyz \le \frac{7}{27}.$$

2. Find two positive integers a, b such that none of the numbers $a, b, a + b$ is divisible by 7 and $(a + b)^7 - a^7 - b^7$ is divisible by 7^7.

3. In a plane two different points O and A are given. For each point $X \ne O$ of the plane denote by $\alpha(X)$ the angle AOX measured in radians ($0 \le \alpha(X) < 2\pi$) and by $C(X)$ the circle with center O and radius $OX + \frac{\alpha(X)}{OX}$. Suppose each point of the plane is colored by one of a finite number of colors. Show that there exists a point X with $\alpha(X) > 0$ such that its color appears somewhere on the circle $C(X)$.

Second Day (July 5)

4. Let $ABCD$ be a convex quadrilateral for which the circle of diameter AB is tangent to the line CD. Show that the circle of diameter CD is tangent to the line AB if and only if the lines BC and AD are parallel.

5. Let d be the sum of the lengths of all diagonals of a convex polygon of n ($n > 3$) vertices, and let p be its perimeter. Prove that

$$\frac{n-3}{2} < \frac{d}{p} < \frac{1}{2} \left(\left[\frac{n}{2} \right] \left[\frac{n+1}{2} \right] - 2 \right).$$

6. Let a, b, c, d be odd positive integers such that $a < b < c < d$, $ad = bc$, and $a + d = 2^k$, $b + c = 2^m$ for some integers k and m. Prove that $a = 1$.

3.25.2 Longlisted Problems

1. **(AUS 1)** The fraction $\frac{3}{10}$ can be written as the sum of two positive fractions with numerator 1 as follows: $\frac{3}{10} = \frac{1}{5} + \frac{1}{10}$ and also $\frac{3}{10} = \frac{1}{4} + \frac{1}{20}$. There are the only two ways in which this can be done.
   In how many ways can $\frac{3}{1984}$ be written as the sum of two positive fractions with numerator 1?
   Is there a positive integer n, not divisible by 3, such that $\frac{3}{n}$ can be written as the sum of two positive fractions with numerator 1 in exactly 1984 ways?

2. **(AUS 2)** Given a regular convex $2m$-sided polygon P, show that there is a $2m$-sided polygon π with the same vertices as P (but in different order) such that π has exactly one pair of parallel sides.

3. **(AUS 3)** The opposite sides of the reentrant hexagon $AFBDCE$ intersect at the points K, L, M (as shown in the figure). It is given that $AL = AM = a$, $BM = BK = b$, $CK = CL = c$, $LD = DM = d$, $ME = EK = e$, $FK = FL = f$.
   (a) Given length a and the three angles α, β, and γ at the vertices A, B, and C, respectively, satisfying the condition $\alpha + \beta + \gamma < 180°$, show that all the angles and sides of the hexagon are thereby uniquely determined.
   (b) Prove that
   $$\frac{1}{a} + \frac{1}{e} = \frac{1}{b} + \frac{1}{d}.$$
   Easier version of (b). Prove that
   $$(a+f)(b+d)(c+e) = (a+e)(b+f)(c+d).$$

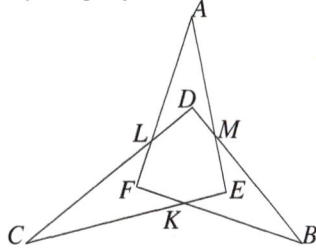

4. **(BEL 1)** Given a triangle ABC, three equilateral triangles AEB, BFC, and CGA are constructed in the exterior of ABC. Prove that:
   (a) $CE = AF = BG$;
   (b) CE, AF, and BG have a common point.

5. **(BEL 2)** For a real number x, let $[x]$ denote the greatest integer not exceeding x. If $m \geq 3$, prove that
   $$\left[\frac{m(m+1)}{2(2m-1)}\right] = \left[\frac{m+1}{4}\right].$$

6. **(BEL 3)** Let P, Q, R be the polynomials with real or complex coefficients such that at least one of them is not constant. If $P^n + Q^n + R^n = 0$, prove that $n < 3$.

7. **(BGR 1)** Prove that for any natural number n, the number $\binom{2n}{n}$ divides the least common multiple of the numbers $1, 2, \ldots, 2n-1, 2n$.

8. **(BGR 2)** In the plane of a given triangle $A_1A_2A_3$ determine (with proof) a straight line l such that the sum of the distances from A_1, A_2, and A_3 to l is the least possible.

9. **(BGR 3)** The circle inscribed in the triangle $A_1A_2A_3$ is tangent to its sides A_1A_2, A_2A_3, A_3A_1 at points T_1, T_2, T_3, respectively. Denote by M_1, M_2, M_3 the midpoints of the segments A_2A_3, A_3A_1, A_1A_2, respectively. Prove that the perpendiculars through the points M_1, M_2, M_3 to the lines T_2T_3, T_3T_1, T_1T_2 meet at one point.

10. **(BGR 4)** Assume that the bisecting plane of the dihedral angle at edge AB of the tetrahedron $ABCD$ meets the edge CD at point E. Denote by S_1, S_2, S_3, respectively the areas of the triangles ABC, ABE, and ABD. Prove that no tetrahedron exists for which S_1, S_2, S_3 (in this order) form an arithmetic or geometric progression.

11. **(BGR 5)** (SL84-13).

12. **(CAN 1)** (SL84-11).
    Original formulation. Suppose that $a_1, a_2, \ldots, a_{2n}$ are distinct integers such that

    $$(x-a_1)(x-a_2)\cdots(x-a_{2n})+(-1)^{n-1}(n!)^2 = 0$$

    has an integer solution r. Show that $r = \frac{a_1+a_2+\cdots+a_{2n}}{2n}$.

13. **(CAN 2)** (SL84-2).
    Original formulation. Let m,n be nonzero integers. Show that $4mn - m - n$ can be a square infinitely many times, but that this never happens when either m or n is positive.

    Alternative formulation. Let m,n be positive integers. Show that $4mn - m - n$ can be 1 less than a perfect square infinitely often, but can never be a square.

14. **(CAN 3)** (SL84-6).

15. **(CAN 4)** Consider all the sums of the form

    $$\sum_{k=1}^{1985} e_k k^5 = \pm 1^5 \pm 2^5 \pm \cdots \pm 1985^5,$$

    where $e_k = \pm 1$. What is the smallest nonnegative value attained by a sum of this type?

16. **(CAN 5)** (SL84-19).
    Original formulation. The triangular array $(a_{n,k})$ of numbers is given by $a_{n,1} = 1/n$, for $n = 1, 2, \ldots$, $a_{n,k+1} = a_{n-1,k} - a_{n,k}$, for $1 \le k \le n-1$. Find the harmonic mean of the 1985th row.

17. **(FRA 1)** (SL84-1).

18. **(FRA 2)** Let c be the inscribed circle of the triangle ABC, d a line tangent to c which does not pass through the vertices of triangle ABC. Prove the existence of points A_1, B_1, C_1, respectively, on the lines BC, CA, AB satisfying the following two properties:
    (i) Lines AA_1, BB_1, and CC_1 are parallel.
    (ii) Lines AA_1, BB_1, and CC_1 meet d respectively at points A', B', and C' such that

    $$\frac{\overline{A'A_1}}{\overline{A'A}} = \frac{\overline{B'B_1}}{\overline{B'B}} = \frac{\overline{C'C_1}}{\overline{C'C}}.$$

19. **(FRA 3)** Let ABC be an isosceles triangle with right angle at point A. Find the minimum of the function F given by

    $$F(M) = BM + CM - \sqrt{3}AM.$$

20. **(FRG 1)** (SL84-5).

21. **(FRG 2)**
    (1) Start with a white balls and b black balls.
    (2) Draw one ball at random.

(3) If the ball is white, then stop. Otherwise, add two black balls and go to step 2.

Let S be the number of draws before the process terminates. For the cases $a = b = 1$ and $a = b = 2$ only, find $a_n = P(S = n)$, $b_n = P(S \leq n)$, $\lim_{n \to \infty} b_n$, and the expectation value of the number of balls drawn: $E(S) = \sum_{n \geq 1} n a_n$.

22. **(FRG 3)** (SL84-17).
Original formulation. In a permutation $(x_1, x_2, \ldots, x_n)$ of the set $1, 2, \ldots, n$ we call a pair (x_i, x_j) *discordant* if $i < j$ and $x_i > x_j$. Let $d(n, k)$ be the number of such permutations with exactly k discordant pairs.
   (a) Find $d(n, 2)$.
   (b) Show that

$$d(n, k) = d(n, k-1) + d(n-1, k) - d(n-1, k-1)$$

with $d(n, k) = 0$ for $k < 0$ and $d(n, 0) = 1$ for $n \geq 1$. Compute with this recursion a table of $d(n, k)$ for $n = 1$ to 6.

23. **(FRG 4)** A $2 \times 2 \times 12$ box fixed in space is to be filled with twenty-four $1 \times 1 \times 2$ bricks. In how many ways can this be done?

24. **(FRG 5)** (SL84-7).
Original formulation. Consider several types of 4-cell figures:

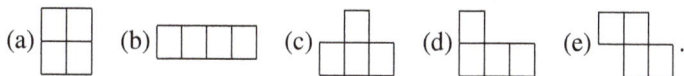

Find, with proof, for which of these types of figures it is not possible to number the fields of the 8×8 chessboard using the numbers $1, 2, \ldots, 64$ in such a way that the sum of the four numbers in each of its parts congruent to the given figure is divisible by 4.

25. **(UNK 1)** (SL84-10).

26. **(UNK 2)** A cylindrical container has height 6 cm and radius 4 cm. It rests on a circular hoop, also of radius 4 cm, fixed in a horizontal plane with its axis vertical and with each circular rim of the cylinder touching the hoop at two points.
The cylinder is now moved so that each of its circular rims still touches the hoop in two points. Find with proof the locus of one of the cylinder's vertical ends.

27. **(UNK 3)** The function $f(n)$ is defined on the nonnegative integers n by: $f(0) = 0$, $f(1) = 1$,

$$f(n) = f\left(n - \frac{1}{2}m(m-1)\right) - f\left(\frac{1}{2}m(m+1) - n\right),$$

for $\frac{1}{2}m(m-1) < n \leq \frac{1}{2}m(m+1)$, $m \geq 2$. Find the smallest integer n for which $f(n) = 5$.

28. **(UNK 4)** A "number triangle" (t_{nk}) $(0 \leq k \leq n)$ is defined by $t_{n,0} = t_{n,n} = 1$ $(n \geq 0)$,

$$t_{n+1,m} = \left(2-\sqrt{3}\right)^m t_{n,m} + \left(2+\sqrt{3}\right)^{n-m+1} t_{n,m-1} \quad (1 \le m \le n).$$

Prove that all $t_{n,m}$ are integers.

29. **(GDR 1)** Let $S_n = \{1,\dots,n\}$ and let f be a function that maps every subset of S_n into a positive real number and satisfies the following condition: For all $A \subseteq S_n$ and $x,y \in S_n$, $x \neq y$, $f(A \cup \{x\}) f(A \cup \{y\}) \le f(A \cup \{x,y\}) f(A)$. Prove that for all $A, B \subseteq S_n$ the following inequality holds:

$$f(A) \cdot f(B) \le f(A \cup B) \cdot f(A \cap B).$$

30. **(GDR 2)** Decide whether it is possible to color the 1984 natural numbers $1,2,3,\dots,1984$ using 15 colors so that no geometric sequence of length 3 of the same color exists.

31. **(LUX 1)** Let $f_1(x) = x^3 + a_1 x^2 + b_1 x + c_1 = 0$ be an equation with three positive roots $\alpha > \beta > \gamma > 0$. From the equation $f_1(x) = 0$ one constructs the equation $f_2(x) = x^3 + a_2 x^2 + b_2 x + c_2 = x(x+b_1)^2 - (a_1 x + c_1)^2 = 0$. Continuing this process, we get equations $f_3,\dots,f_n$. Prove that

$$\lim_{n\to\infty} \sqrt[2^{n-1}]{-a_n} = \alpha.$$

32. **(LUX 2)** (SL84-15).

33. **(MNG 1)** (SL84-4).

34. **(MNG 2)** One country has n cities and every two of them are linked by a railroad. A railway worker should travel by train exactly once through the entire railroad system (reaching each city exactly once). If it is impossible for worker to travel by train between two cities, he can travel by plane. What is the minimal number of flights that the worker will have to use?

35. **(MNG 3)** Prove that there exist distinct natural numbers $m_1, m_2, \dots, m_k$ satisfying the conditions

$$\pi^{-1984} < 25 - \left(\frac{1}{m_1} + \frac{1}{m_2} + \cdots + \frac{1}{m_k}\right) < \pi^{-1960}$$

where π is the ratio between circle and its diameter.

36. **(MNG 4)** The set $\{1,2,\dots,49\}$ is divided into three subsets. Prove that at least one of these subsets contains three different numbers a,b,c such that $a+b=c$.

37. **(MAR 1)** Denote by $[x]$ the greatest integer not exceeding x. For all real $k > 1$, define two sequences:

$$a_n(k) = [nk] \quad \text{and} \quad b_n(k) = \left[\frac{nk}{k-1}\right].$$

If $A(k) = \{a_n(k) : n \in \mathbb{N}\}$ and $B(k) = \{b_n(k) : n \in \mathbb{N}\}$, prove that $A(k)$ and $B(k)$ form a partition of $\mathbb{N}$ if and only if k is irrational.

38. **(MAR 2)** Determine all continuous functions f such that
$$(\forall (x,y) \in \mathbb{R}^2) \quad f(x+y)f(x-y) = (f(x)f(y))^2.$$

39. **(MAR 3)** Let ABC be an isosceles triangle, $AB = AC$, $\angle A = 20°$. Let D be a point on AB, and E a point on AC such that $\angle ACD = 20°$ and $\angle ABE = 30°$. What is the measure of the angle $\angle CDE$?

40. **(NLD 1)** (SL84-12).

41. **(NLD 2)** Determine positive integers p, q, and r such that the diagonal of a block consisting of $p \times q \times r$ unit cubes passes through exactly 1984 of the unit cubes, while its length is minimal. (The diagonal is said to pass through a unit cube if it has more than one point in common with the unit cube.)

42. **(NLD 3)** Triangle ABC is given for which $BC = AC + \frac{1}{2}AB$. The point P divides AB such that $RP : PA = 1 : 3$. Prove that $\angle CAP = 2\angle CPA$.

43. **(POL 1)** (SL84-16).

44. **(POL 2)** (SL84-9).

45. **(POL 3)** Let X be an arbitrary nonempty set contained in the plane and let sets $A_1, A_2, \ldots, A_m$ and $B_1, B_2, \ldots, B_n$ be its images under parallel translations. Let us suppose that
$$A_1 \cup A_2 \cup \cdots \cup A_m \subset B_1 \cup B_2 \cup \cdots \cup B_n$$
and that the sets $A_1, A_2, \ldots, A_m$ are disjoint. Prove that $m \leq n$.

46. **(ROU 1)** Let $(a_n)_{n \geq 1}$ and $(b_n)_{n \geq 1}$ be two sequences of natural numbers such that $a_{n+1} = na_n + 1$, $b_{n+1} = nb_n - 1$ for every $n \geq 1$. Show that these two sequences can have only a finite number of terms in common.

47. **(ROU 2)** (SL84-8).

48. **(ROU 3)** Let ABC be a triangle with interior angle bisectors AA_1, BB_1, CC_1 and incenter I. If $\sigma[IA_1B] + \sigma[IB_1C] + \sigma[IC_1A] = \frac{1}{2}\sigma[ABC]$, where $\sigma[ABC]$ denotes the area of ABC, show that ABC is isosceles.

49. **(ROU 4)** Let $n > 1$ and $x_i \in \mathbb{R}$ for $i = 1, \ldots, n$. Set $S_k = x_1^k + x_2^k + \cdots + x_n^k$ for $k \geq 1$. If $S_1 = S_2 = \cdots = S_{n+1}$, show that $x_i \in \{0, 1\}$ for every $i = 1, 2, \ldots, n$.

50. **(ROU 5)** (SL84-14).

51. **(ESP 1)** Two cyclists leave simultaneously a point P in a circular runway with constant velocities v_1, v_2 ($v_1 > v_2$) and in the same sense. A pedestrian leaves P at the same time, moving with velocity $v_3 = \frac{v_1 + v_2}{12}$. If the pedestrian and the cyclists move in opposite directions, the pedestrian meets the second cyclist 91 seconds after he meets the first. If the pedestrian moves in the same direction as the cyclists, the first cyclist overtakes him 187 seconds before the second does. Find the point where the first cyclist overtakes the second cyclist the first time.

52. **(ESP 2)** Construct a scalene triangle such that

$$a(\tan B - \tan C) = b(\tan A - \tan C).$$

53. **(ESP 3)** Find a sequence of natural numbers a_i such that $a_i = \sum_{r=1}^{i+4} d_r$, where $d_r \neq d_s$ for $r \neq s$ and d_r divides a_i.

54. **(ESP 4)** Let P be a convex planar polygon with equal angles. Let $l_1,\ldots,l_n$ be its sides. Show that a necessary and sufficient condition for P to be regular is that the sum of the ratios $\frac{l_i}{l_{i+1}}$ ($i = 1,\ldots,n$; $l_{n+1} = l_1$) equals the number of sides.

55. **(ESP 5)** Let a,b,c be natural numbers such that $a+b+c = 2pq(p^{30} - q^{30})$, $p > q$ being two given positive integers.
    (a) Prove that $k = a^3 + b^3 + c^3$ is not a prime number.
    (b) Prove that if $a \cdot b \cdot c$ is maximum, then 1984 divides k.

56. **(SWE 1)** Let a,b,c be nonnegative integers such that $a \leq b \leq c$, $2b \neq a+c$ and $\frac{a+b+c}{3}$ is an integer. Is it possible to find three nonnegative integers d, e, and f such that $d \leq e \leq f$, $f \neq c$, and such that $a^2 + b^2 + c^2 = d^2 + e^2 + f^2$?

57. **(SWE 2)** Let a,b,c,d be a permutation of the numbers $1,9,8,4$ and let $n = (10a+b)^{10c+d}$. Find the probability that 1984! is divisible by n.

58. **(SWE 3)** Let $(a_n)_1^{\infty}$ be a sequence such that $a_n \leq a_{n+m} \leq a_n + a_m$ for all positive integers n and m. Prove that $\frac{a_n}{n}$ has a limit as n approaches infinity.

59. **(USA 1)** Determine the smallest positive integer m such that $529^n + m \cdot 132^n$ is divisible by 262417 for all odd positive integers n.

60. **(USA 2)** (SL84-20).

61. **(USA 3)** A fair coin is tossed repeatedly until there is a run of an odd number of heads followed by a tail. Determine the expected number of tosses.

62. **(USA 4)** From a point P exterior to a circle K, two rays are drawn intersecting K in the respective pairs of points A,A' and B, B'. For any other pair of points C,C' on K, let D be the point of intersection of the circumcircles of triangles PAC and $PB'C'$ other than point P. Similarly, let D' be the point of intersection of the circumcircles of triangles $PA'C'$ and PBC other than point P. Prove that the points P, D, and D' are collinear.

63. **(USA 5)** (SL84-18).

64. **(USS 1)** For a matrix (p_{ij}) of the format $m \times n$ with real entries, set

$$a_i = \sum_{j=1}^{n} p_{ij} \text{ for } i = 1,\ldots,m \quad \text{and} \quad b_j = \sum_{i=1}^{m} p_{ij} \text{ for } j = 1,\ldots,n. \quad (1)$$

By *integering* a real number we mean replacing the number with the integer closest to it.

Prove that integering the numbers a_i, b_j, p_{ij} can be done in such a way that (1) still holds.

65. **(USS 2)** A tetrahedron is inscribed in a sphere of radius 1 such that the center of the sphere is inside the tetrahedron.
    Prove that the sum of lengths of all edges of the tetrahedron is greater than 6.

66. **(USS 3)** (SL84-3).
    Original formulation. All the divisors of a positive integer n arranged in increasing order are $x_1 < x_2 < \cdots < x_k$. Find all such numbers n for which $x_5^2 + x_6^2 - 1 = n$.

67. **(USS 4)** With the medians of an acute-angled triangle another triangle is constructed. If R and R_m are the radii of the circles circumscribed about the first and the second triangle, respectively, prove that

$$R_m > \frac{5}{6}R.$$

68. **(USS 5)** In the Martian language every finite sequence of letters of the Latin alphabet letters is a word. The publisher "Martian Words" makes a collection of all words in many volumes. In the first volume there are only one-letter words, in the second, two-letter words, etc., and the numeration of the words in each of the volumes continues the numeration of the previous volume. Find the word whose numeration is equal to the sum of numerations of the words *Prague, Olympiad, Mathematics*.

3.25.3 Shortlisted Problems

1. **(FRA 1)** Find all solutions of the following system of n equations in n variables:

$$x_1|x_1| - (x_1 - a)|x_1 - a| = x_2|x_2|,$$
$$x_2|x_2| - (x_2 - a)|x_2 - a| = x_3|x_3|,$$
$$\cdots$$
$$x_n|x_n| - (x_n - a)|x_n - a| = x_1|x_1|,$$

where a is a given number.

2. **(CAN 2)** Prove:
   (a) There are infinitely many triples of positive integers m, n, p such that $4mn - m - n = p^2 - 1$.
   (b) There are no positive integers m, n, p such that $4mn - m - n = p^2$.

3. **(USS 3)** Find all positive integers n such that

$$n = d_6^2 + d_7^2 - 1,$$

where $1 = d_1 < d_2 < \cdots < d_k = n$ are all positive divisors of the number n.

4. **(MNG 1)**[IMO5] Let d be the sum of the lengths of all diagonals of a convex polygon of n ($n > 3$) vertices and let p be its perimeter. Prove that

$$\frac{n-3}{2} < \frac{d}{p} < \frac{1}{2}\left(\left[\frac{n}{2}\right]\left[\frac{n+1}{2}\right] - 2\right).$$

5. **(FRG 1)**[IMO1] Let x,y,z be nonnegative real numbers with $x+y+z=1$. Show that

$$0 \le xy+yz+zx-2xyz \le \frac{7}{27}.$$

6. **(CAN 3)** Let c be a positive integer. The sequence $\{f_n\}$ is defined as follows:

$$f_1 = 1, \quad f_2 = c, \quad f_{n+1} = 2f_n - f_{n-1} + 2 \quad (n \ge 2).$$

Show that for each $k \in \mathbb{N}$ there exists $r \in \mathbb{N}$ such that $f_k f_{k+1} = f_r$.

7. **(FRG 5)**
   (a) Decide whether the fields of the 8×8 chessboard can be numbered by the numbers $1, 2, \ldots, 64$ in such a way that the sum of the four numbers in each of its parts of one of the forms

   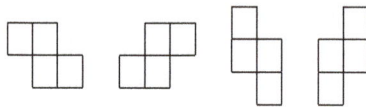

   is divisible by four.
   (b) Solve the analogous problem for

   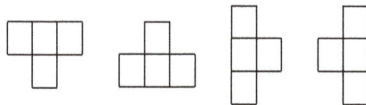

8. **(ROU 2)**[IMO3] In a plane two different points O and A are given. For each point $X \ne O$ of the plane denote by $\alpha(X)$ the angle AOX measured in radians ($0 \le \alpha(X) < 2\pi$) and by $C(X)$ the circle with center O and radius $OX + \frac{\alpha(X)}{OX}$. Suppose each point of the plane is colored by one of a finite number of colors. Show that there exists a point X with $\alpha(X) > 0$ such that its color appears somewhere on the circle $C(X)$.

9. **(POL 2)** Let a,b,c be positive numbers with $\sqrt{a}+\sqrt{b}+\sqrt{c} = \frac{\sqrt{3}}{2}$. Prove that the system of equations

$$\begin{aligned}
\sqrt{y-a}+\sqrt{z-a} &= 1, \\
\sqrt{z-b}+\sqrt{x-b} &= 1, \\
\sqrt{x-c}+\sqrt{y-c} &= 1,
\end{aligned}$$

   has exactly one solution (x,y,z) in real numbers.

10. **(UNK 1)** Prove that the product of five consecutive positive integers cannot be the square of an integer.

11. **(CAN 1)** Let n be a natural number and $a_1, a_2, \ldots, a_{2n}$ mutually distinct integers. Find all integers x satisfying

$$(x-a_1) \cdot (x-a_2) \cdots (x-a_{2n}) = (-1)^n (n!)^2.$$

12. **(NLD 1)**[IMO2] Find two positive integers a, b such that none of the numbers $a, b, a+b$ is divisible by 7 and $(a+b)^7 - a^7 - b^7$ is divisible by 7^7.

13. **(BGR 5)** Prove that the volume of a tetrahedron inscribed in a right circular cylinder of volume 1 does not exceed $\frac{2}{3\pi}$.

14. **(ROU 5)**[IMO4] Let $ABCD$ be a convex quadrilateral for which the circle with diameter AB is tangent to the line CD. Show that the circle with diameter CD is tangent to the line AB if and only if the lines BC and AD are parallel.

15. **(LUX 2)** Angles of a given triangle ABC are all smaller than $120°$. Equilateral triangles AFB, BDC and CEA are constructed in the exterior of $\triangle ABC$.
    (a) Prove that the lines AD, BE, and CF pass through one point S.
    (b) Prove that $SD + SE + SF = 2(SA + SB + SC)$.

16. **(POL 1)**[IMO6] Let a, b, c, d be odd positive integers such that $a < b < c < d$, $ad = bc$, and $a + d = 2^k$, $b + c = 2^m$ for some integers k and m. Prove that $a = 1$.

17. **(FRG 3)** In a permutation $(x_1, x_2, \ldots, x_n)$ of the set $1, 2, \ldots, n$ we call a pair (x_i, x_j) *discordant* if $i < j$ and $x_i > x_j$. Let $d(n, k)$ be the number of such permutations with exactly k discordant pairs. Find $d(n, 2)$ and $d(n, 3)$.

18. **(USA 5)** Inside triangle ABC there are three circles k_1, k_2, k_3 each of which is tangent to two sides of the triangle and to its incircle k. The radii of k_1, k_2, k_3 are 1, 4, and 9. Determine the radius of k.

19. **(CAN 5)** The triangular array $(a_{n,k})$ of numbers is given by $a_{n,1} = 1/n$, for $n = 1, 2, \ldots$, $a_{n,k+1} = a_{n-1,k} - a_{n,k}$, for $1 \le k \le n-1$. Prove that the geometric mean of the 1985th row is greater than 2^{-1984}.

20. **(USA 2)** Determine all pairs (a, b) of positive real numbers with $a \ne 1$ such that

$$\log_a b < \log_{a+1}(b+1).$$

3.26 The Twenty-Sixth IMO
Joutsa, Finland, June 29–July 11, 1985

3.26.1 Contest Problems

First Day (July 4)

1. A circle whose center is on the side ED of the cyclic quadrilateral $BCDE$ touches the other three sides. Prove that $EB + CD = ED$.

2. Each of the numbers in the set $N = \{1, 2, 3, \ldots, n-1\}$, where $n \geq 3$, is colored with one of two colors, say red or black, so that:
   (i) i and $n - i$ always receive the same color, and
   (ii) for some $j \in N$ relatively prime to n, i and $|j - i|$ receive the same color for all $i \in N$, $i \neq j$.
   Prove that all numbers in N must receive the same color.

3. The weight $w(p)$ of a polynomial p, $p(x) = \sum_{i=0}^{n} a_i x^i$, with integer coefficients a_i is defined as the number of its odd coefficients. For $i = 0, 1, 2, \ldots$, let $q_i(x) = (1+x)^i$. Prove that for any finite sequence $0 \leq i_1 < i_2 < \cdots < i_n$ the inequality

$$w(q_{i_1} + \cdots + q_{i_n}) \geq w(q_{i_1})$$

   holds.

Second Day (July 5)

4. Given a set M of 1985 positive integers, none of which has a prime divisor larger than 26, prove that M has four distinct elements whose geometric mean is an integer.

5. A circle with center O passes through points A and C and intersects the sides AB and BC of the triangle ABC at points K and N, respectively. The circumscribed circles of the triangles ABC and KBN intersect at two distinct points B and M. Prove that $\angle OMB = 90°$.

6. The sequence $f_1, f_2, \ldots, f_n, \ldots$ of functions is defined for $x > 0$ recursively by

$$f_1(x) = x, \qquad f_{n+1}(x) = f_n(x)\left(f_n(x) + \frac{1}{n}\right).$$

   Prove that there exists one and only one positive number a such that $0 < f_n(a) < f_{n+1}(a) < 1$ for all integers $n \geq 1$.

3.26.2 Longlisted Problems

1. **(AUS 1)** (SL85-4).

2. **(AUS 2)** We are given a triangle ABC and three rectangles R_1, R_2, R_3 with sides parallel to two fixed perpendicular directions and such that their union covers the sides AB, BC, and CA; i.e., each point on the perimeter of ABC is contained in or on at least one of the rectangles. Prove that all points inside the triangle are also covered by the union of R_1, R_2, R_3.

3. **(AUS 3)** A function f has the following property: If $k > 1$, $j > 1$, and $(k, j) = m$, then $f(kj) = f(m)(f(k/m) + f(j/m))$. What values can $f(1984)$ and $f(1985)$ take?

4. **(BEL 1)** Let x, y, and z be real numbers satisfying $x + y + z = xyz$. Prove that

$$x(1 - y^2)(1 - z^2) + y(1 - z^2)(1 - x^2) + z(1 - x^2)(1 - y^2) = 4xyz.$$

5. **(BEL 2)** (SL85-16).

6. **(BEL 3)** On a one-way street, an unending sequence of cars of width a, length b passes with velocity v. The cars are separated by the distance c. A pedestrian crosses the street perpendicularly with velocity w, without paying attention to the cars.

   (a) What is the probability that the pedestrian crosses the street uninjured?

   (b) Can he improve this probability by crossing the road in a direction other than perpendicular?

7. **(BRA 1)** A convex quadrilateral is inscribed in a circle of radius 1. Prove that the difference between its perimeter and the sum of the lengths of its diagonals is greater than zero and less than 2.

8. **(BRA 2)** Let K be a convex set in the xy-plane, symmetric with respect to the origin and having area greater than 4. Prove that there exists a point $(m, n) \neq (0, 0)$ in K such that m and n are integers.

9. **(BRA 3)** (SL85-2).

10. **(BGR 1)** (SL85-13).

11. **(BGR 2)** Let a and b be integers and n a positive integer. Prove that

$$\frac{b^{n-1}a(a+b)(a+2b)\cdots(a+(n-1)b)}{n!}$$

is an integer.

12. **(CAN 1)** Find the maximum value of

$$\sin^2 \theta_1 + \sin^2 \theta_2 + \cdots + \sin^2 \theta_n$$

subject to the restrictions $0 \leq \theta_i \leq \pi$, $\theta_1 + \theta_2 + \cdots + \theta_n = \pi$.

13. **(CAN 2)** Find the average of the quantity

$$(a_1 - a_2)^2 + (a_2 - a_3)^2 + \cdots + (a_{n-1} - a_n)^2$$

taken over all permutations $(a_1, a_2, \ldots, a_n)$ of $(1, 2, \ldots, n)$.

14. **(CAN 3)** Let k be a positive integer. Define $u_0 = 0$, $u_1 = 1$, and $u_n = ku_{n-1} - u_{n-2}$, $n \geq 2$. Show that for each integer n, the number $u_1^3 + u_2^3 + \cdots + u_n^3$ is a multiple of $u_1 + u_2 + \cdots + u_n$.

15. **(CAN 4)** *Superchess* is played on a 12×12 board, and it uses *superknights*, which move between opposite corner cells of any 3×4 subboard. Is it possible for a superknight to visit every other cell of a superchessboard exactly once and return to its starting cell?

16. **(CAN 5)** (SL85-18).

17. **(CUB 1)** Set

$$A_n = \sum_{k=1}^{n} \frac{k^6}{2^k}.$$

Find $\lim_{n \to \infty} A_n$.

18. **(CYP 1)** The circles (R, r) and (P, ρ), where $r > \rho$, touch externally at A. Their direct common tangent touches (R, r) at B and (P, ρ) at C. The line RP meets the circle (P, ρ) again at D and the line BC at E. If $|BC| = 6|DE|$, prove that:
   (a) the lengths of the sides of the triangle RBE are in an arithmetic progression, and
   (b) $|AB| = 2|AC|$.

19. **(CYP 2)** Solve the system of simultaneous equations

$$\begin{aligned}
\sqrt{x} - 1/y - 2w + 3z &= 1, \\
x + 1/y^2 - 4w^2 - 9z^2 &= 3, \\
x\sqrt{x} - 1/y^3 - 8w^3 + 27z^3 &= -5, \\
x^2 + 1/y^4 - 16w^4 - 81z^4 &= 15.
\end{aligned}$$

20. **(CZS 1)** Let T be the set of all lattice points (i.e., all points with integer coordinates) in three-dimensional space. Two such points (x, y, z) and (u, v, w) are called *neighbors* if $|x - u| + |y - v| + |z - w| = 1$. Show that there exists a subset S of T such that for each $p \in T$, there is exactly one point of S among p and its neighbors.

21. **(CZS 2)** Let A be a set of positive integers such that for any two elements x, y of A, $|x - y| \geq \frac{xy}{25}$. Prove that A contains at most nine elements. Give an example of such a set of nine elements.

22. **(CZS 3)** (SL85-7).

23. **(CZS 4)** Let $\mathbb{N} = \{1, 2, 3, \ldots\}$. For real x, y, set $S(x, y) = \{s \mid s = [nx + y], n \in \mathbb{N}\}$. Prove that if $r > 1$ is a rational number, there exist real numbers u and v such that

$$S(r, 0) \cap S(u, v) = \emptyset, \quad S(r, 0) \cup S(u, v) = \mathbb{N}.$$

24. **(FRA 1)** Let $d \geq 1$ be an integer that is not the square of an integer. Prove that for every integer $n \geq 1$,

$$(n\sqrt{d} + 1)|\sin(n\pi\sqrt{d})| \geq 1.$$

25. **(FRA 2)** Find eight positive integers $n_1, n_2, \ldots, n_8$ with the following property: For every integer k, $-1985 \leq k \leq 1985$, there are eight integers $\alpha_1, \alpha_2, \ldots, \alpha_8$, each belonging to the set $\{-1, 0, 1\}$, such that $k = \sum_{i=1}^{8} \alpha_i n_i$.

26. **(FRA 3)** (SL85-15).

27. **(FRA 4)** Let O be a point on the oriented Euclidean plane and $(\mathbf{i}, \mathbf{j})$ a directly oriented orthonormal basis. Let C be the circle of radius 1, centered at O. For every real number t and nonnegative integer n let M_n be the point on C for which $\langle \mathbf{i}, \overrightarrow{OM_n} \rangle = \cos 2^n t$ (or $\overrightarrow{OM_n} = \cos 2^n t \mathbf{i} + \sin 2^n t \mathbf{j}$). Let $k \geq 2$ be an integer. Find all real numbers $t \in [0, 2\pi)$ that satisfy
    (i) $M_0 = M_k$, and
    (ii) if one starts from M_0 and goes once around C in the positive direction, one meets successively the points $M_0, M_1, \ldots, M_{k-2}, M_{k-1}$, in this order.

28. **(FRG 1)** Let M be the set of the lengths of an octahedron whose sides are congruent quadrangles. Prove that M has at most three elements.
    (FRG 1a) Let an octahedron whose sides are congruent quadrangles be given. Prove that each of these quadrangles has two equal sides meeting at a common vertex.

29. **(FRG 2)** Call a four-digit number $(xyzt)_B$ in the number system with base B stable if $(xyzt)_B = (dcba)_B - (abcd)_B$, where $a \leq b \leq c \leq d$ are the digits of $(xyzt)_B$ in ascending order. Determine all stable numbers in the number system with base B.
    (FRG 2a) The same problem with $B = 1985$.
    (FRG 2b) With assumptions as in FRG 2, determine the number of bases $B \leq 1985$ such that there is a stable number with base B.

30. **(UNK 1)** A plane rectangular grid is given and a "rational point" is defined as a point (x, y) where x and y are both rational numbers. Let A, B, A', B' be four distinct rational points. Let P be a point such that $\frac{A'B'}{AB} = \frac{B'P}{BC} = \frac{PA'}{PA}$. In other words, the triangles ABP, $A'B'P$ are directly or oppositely similar. Prove that P is in general a rational point and find the exceptional positions of A' and B' relative to A and B such that there exists a P that is not a rational point.

31. **(UNK 2)** Let E_1, E_2, and E_3 be three mutually intersecting ellipses, all in the same plane. Their foci are respectively F_2, F_3; F_3, F_1; and F_1, F_2. The three foci are not on a straight line. Prove that the common chords of each pair of ellipses are concurrent.

32. **(UNK 3)** A collection of $2n$ letters contains 2 each of n different letters. The collection is partitioned into n pairs, each pair containing 2 letters, which may be the same or different. Denote the number of distinct partitions by u_n. (Partitions differing in the order of the pairs in the partition or in the order of the two letters in the pairs are not considered distinct.) Prove that $u_{n+1} = (n+1)u_n - \frac{n(n-1)}{2}u_{n-2}$.
    (UNK 3a) A pack of n cards contains n pairs of 2 identical cards. It is shuffled and 2 cards are dealt to each of n different players. Let p_n be the probability

that every one of the n players is dealt two identical cards. Prove that $\frac{1}{p_{n+1}} = \frac{n+1}{p_n} - \frac{n(n-1)}{2p_{n-2}}$.

33. **(UNK 4)** (SL85-12).

34. **(UNK 5)** (SL85-20).

35. **(GDR 1)** We call a coloring f of the elements in the set $M = \{(x,y) \mid x = 0,1,\ldots,kn-1; y = 0,1,\ldots,ln-1\}$ with n colors *allowable* if every color appears exactly k and l times in each row and column and there are no rectangles with sides parallel to the coordinate axes such that all the vertices in M have the same color. Prove that every allowable coloring f satisfies $kl \le n(n+1)$.

36. **(GDR 2)** Determine whether there exist 100 distinct lines in the plane having exactly 1985 distinct points of intersection.

37. **(GDR 3)** Prove that a triangle with angles α, β, γ, circumradius R, and area A satisfies

$$\tan\frac{\alpha}{2} + \tan\frac{\beta}{2} + \tan\frac{\gamma}{2} \le \frac{9R^2}{4A}.$$

38. **(IRL 1)** (SL85-21).

39. **(IRL 2)** Given a triangle ABC and external points X, Y, and Z such that $\angle BAZ = \angle CAY$, $\angle CBX = \angle ABZ$, and $\angle ACY = \angle BCX$, prove that AX, BY, and CZ are concurrent.

40. **(IRL 3)** Each of the numbers $x_1, x_2, \ldots, x_n$ equals 1 or -1 and

$$x_1 x_2 x_3 x_4 + x_2 x_3 x_4 x_5 + \cdots + x_{n-3} x_{n-2} x_{n-1} x_n$$
$$+ x_{n-2} x_{n-1} x_n x_1 + x_{n-1} x_n x_1 x_2 + x_n x_1 x_2 x_3 = 0.$$

Prove that n is divisible by 4.

41. **(IRL 4)** (SL85-14).

42. **(ISR 1)** Prove that the product of two sides of a triangle is always greater than the product of the diameters of the inscribed circle and the circumscribed circle.

43. **(ISR 2)** Suppose that 1985 points are given inside a unit cube. Show that one can always choose 32 of them in such a way that every (possibly degenerate) closed polygon with these points as vertices has a total length of less than $8\sqrt{3}$.

44. **(ISR 3)** (SL85-19).

45. **(ITA 1)** Two persons, X and Y, play with a die. X wins a game if the outcome is 1 or 2; Y wins in the other cases. A player wins a match if he wins two consecutive games. For each player determine the probability of winning a match within 5 games. Determine the probabilities of winning in an unlimited number of games. If X bets 1, how much must Y bet for the game to be fair?

46. **(ITA 2)** Let C be the curve determined by the equation $y = x^3$ in the rectangular coordinate system. Let t be the tangent to C at a point P of C; t intersects C at

another point Q. Find the equation of the set L of the midpoints M of PQ as P describes C. Is the correspondence associating P and M a bijection of C on L? Find a similarity that transforms C into L.

47. **(ITA 3)** Let F be the correspondence associating with every point $P = (x,y)$ the point $P' = (x',y')$ such that

$$x' = ax+b, \quad y' = ay+2b. \qquad (1)$$

Show that if $a \neq 1$, all lines PP' are concurrent. Find the equation of the set of points corresponding to $P = (1,1)$ for $b = a^2$. Show that the composition of two mappings of type (1) is of the same type.

48. **(ITA 4)** In a given country, all inhabitants are knights or knaves. A knight never lies; a knave always lies. We meet three persons, A, B, and C. Person A says, "If C is a knight, B is a knave." Person C says, "A and I are different; one is a knight and the other is a knave." Who are the knights, and who are the knaves?

49. **(MNG 1)** (SL85-1).

50. **(MNG 2)** From each of the vertices of a regular n-gon a car starts to move with constant speed along the perimeter of the n-gon in the same direction. Prove that if all the cars end up at a vertex A at the same time, then they never again meet at any other vertex of the n-gon. Can they meet again at A?

51. **(MNG 3)** Let $f_1 = (a_1, a_2, \ldots, a_n)$, $n > 2$, be a sequence of integers. From f_1 one constructs a sequence f_k of sequences as follows: if $f_k = (c_1, c_2, \ldots, c_n)$, then $f_{k+1} = (c_{i_1}, c_{i_2}, c_{i_3}+1, c_{i_4}+1, \ldots, c_{i_n}+1)$, where $(c_{i_1}, c_{i_2}, \ldots, c_{i_n})$ is a permutation of $(c_1, c_2, \ldots, c_n)$. Give a necessary and sufficient condition for f_1 under which it is possible for f_k to be a constant sequence $(b_1, b_2, \ldots, b_n)$, $b_1 = b_2 = \cdots = b_n$, for some k.

52. **(MNG 4)** In the triangle ABC, let B_1 be on AC, E on AB, G on BC, and let EG be parallel to AC. Furthermore, let EG be tangent to the inscribed circle of the triangle ABB_1 and intersect BB_1 at F. Let r, r_1, and r_2 be the inradii of the triangles ABC, ABB_1, and BFG, respectively. Prove that $r = r_1 + r_2$.

53. **(MNG 5)** For each P inside the triangle ABC, let $A(P)$, $B(P)$, and $C(P)$ be the points of intersection of the lines AP, BP, and CP with the sides opposite to A, B, and C, respectively. Determine P in such a way that the area of the triangle $A(P)B(P)C(P)$ is as large as possible.

54. **(MAR 1)** Set $S_n = \sum_{p=1}^{n}(p^5 + p^7)$. Determine the greatest common divisor of S_n and S_{3n}.

55. **(MAR 2)** The points A, B, C are in this order on line D, and $AB = 4BC$. Let M be a variable point on the perpendicular to D through C. Let MT_1 and MT_2 be tangents to the circle with center A and radius AB. Determine the locus of the orthocenter of the triangle MT_1T_2.

56. **(MAR 3)** Let $ABCD$ be a rhombus with angle $\angle A = 60°$. Let E be a point, different from D, on the line AD. The lines CE and AB intersect at F. The lines

DF and *BE* intersect at *M*. Determine the angle $\angle BMD$ as a function of the position of *E* on *AD*.

57. **(NLD 1)** The solid *S* is defined as the intersection of the six spheres with the six edges of a regular tetrahedron *T*, with edge length 1, as diameters. Prove that *S* contains two points at a distance $\frac{1}{\sqrt{6}}$.

   (NLD 1a) Using the same assumptions, prove that no pair of points in *S* has a distance larger than $\frac{1}{\sqrt{6}}$.

58. **(NLD 2)** Prove that there are infinitely many pairs (k,N) of positive integers such that $1+2+\cdots+k = (k+1)+(k+2)+\cdots+N$.

59. **(NLD 3)** (SL85-3).

60. **(NOR 1)** The sequence (s_n), where $s_n = \sum_{k=1}^{n} \sin k$, $n = 1,2,\ldots$, is bounded. Find an upper and a lower bound.

61. **(NOR 2)** Consider the set $A = \{0,1,2,\ldots,9\}$ and let $(B_1,B_2,\ldots,B_k)$ be a collection of nonempty subsets of *A* such that $B_i \cap B_j$ has at most two elements for $i \neq j$. What is the maximal value of *k*?

62. **(NOR 3)** A "large" circular disk is attached to a vertical wall. It rotates clockwise with one revolution per minute. An insect lands on the disk and immediately starts to climb vertically upward with constant speed $\frac{\pi}{3}$ cm per second (relative to the disk). Describe the path of the insect
   (a) relative to the disk;
   (b) relative to the wall.

63. **(POL 1)** (SL85-6).

64. **(POL 2)** Let *p* be a prime. For which *k* can the set $\{1,2,\ldots,k\}$ be partitioned into *p* subsets with equal sums of elements?

65. **(POL 3)** Define the functions f, $F : \mathbb{N} \to \mathbb{N}$, by

$$f(n) = \left[\frac{3-\sqrt{5}}{2}n\right], \quad F(k) = \min\{n \in \mathbb{N}|f^k(n) > 0\},$$

where $f^k = f \circ \cdots \circ f$ is f iterated *n* times. Prove that $F(k+2) = 3F(k+1) - F(k)$ for all $k \in \mathbb{N}$.

66. **(ROU 1)** (SL85-5).

67. **(ROU 2)** Let $k \geq 2$ and $n_1,n_2,\ldots,n_k \geq 1$ be natural numbers having the property $n_2 \mid 2^{n_1} - 1$, $n_3 \mid 2^{n_2} - 1,\ldots,n_k \mid 2^{n_{k-1}} - 1$, and $n_1 \mid 2^{n_k} - 1$. Show that $n_1 = n_2 = \cdots = n_k = 1$.

68. **(ROU 3)** Show that the sequence $\{a_n\}_{n\geq 1}$ defined by $a_n = [n\sqrt{2}]$ contains an infinite number of integer powers of 2. ($[x]$ is the integer part of *x*.)

69. **(ROU 4)** Let *A* and *B* be two finite disjoint sets of points in the plane such that no three distinct points in $A \cup B$ are collinear. Assume that at least one of the sets

A, B contains at least five points. Show that there exists a triangle all of whose vertices are contained in A or in B that does not contain in its interior any point from the other set.

70. **(ROU 5)** Let C be a class of functions $f : \mathbb{N} \to \mathbb{N}$ that contains the functions $S(x) = x+1$ and $E(x) = x - [\sqrt{x}]^2$ for every $x \in \mathbb{N}$. ($[x]$ is the integer part of x.) If C has the property that for every $f, g \in C$, $f+g, fg, f \circ g \in C$, show that the function $\max(f(x) - g(x), 0)$ is in C for all $f, g \in C$.

71. **(ROU 6)** For every integer $r > 1$ find the smallest integer $h(r) > 1$ having the following property: For any partition of the set $\{1, 2, \ldots, h(r)\}$ into r classes, there exist integers $a \geq 0$, $1 \leq x \leq y$ such that the numbers $a+x, a+y, a+x+y$ are contained in the same class of the partition.

72. **(ESP 1)** Construct a triangle ABC given the side AB and the distance OH from the circumcenter O to the orthocenter H, assuming that OH and AB are parallel.

73. **(ESP 2)** Let A_1A_2, B_1B_2, C_1C_2 be three equal segments on the three sides of an equilateral triangle. Prove that in the triangle formed by the lines B_2C_1, C_2A_1, A_2B_1, the segments B_2C_1, C_2A_1, A_2B_1 are proportional to the sides in which they are contained.

74. **(ESP 3)** Find the triples of positive integers x, y, z satisfying

$$\frac{1}{x} + \frac{1}{y} + \frac{1}{z} = \frac{4}{5}.$$

75. **(ESP 4)** Let $ABCD$ be a rectangle, $AB = a$, $BC = b$. Consider the family of parallel and equidistant straight lines (the distance between two consecutive lines being d) that are at an angle ϕ, $0 \leq \phi \leq 90°$, with respect to AB. Let L be the sum of the lengths of all the segments intersecting the rectangle. Find:
   (a) how L varies,
   (b) a necessary and sufficient condition for L to be a constant, and
   (c) the value of this constant.

76. **(SWE 1)** Are there integers m and n such that

$$5m^2 - 6mn + 7n^2 = 1985?$$

77. **(SWE 2)** Two equilateral triangles are inscribed in a circle with radius r. Let A be the area of the set consisting of all points interior to both triangles. Prove that $2A \geq r^2\sqrt{3}$.

78. **(SWE 3)** (SL85-17).

79. **(SWE 4)** Let a, b, and c be real numbers such that

$$\frac{1}{bc - a^2} + \frac{1}{ca - b^2} + \frac{1}{ab - c^2} = 0.$$

Prove that

$$\frac{a}{(bc - a^2)^2} + \frac{b}{(ca - b^2)^2} + \frac{c}{(ab - c^2)^2} = 0.$$

80. **(TUR 1)** Let $E = \{1, 2, \ldots, 16\}$ and let M be the collection of all 4×4 matrices whose entries are distinct members of E. If a matrix $A = (a_{ij})_{4 \times 4}$ is chosen randomly from M, compute the probability $p(k)$ of $\max_i \min_j a_{ij} = k$ for $k \in E$. Furthermore, determine $l \in E$ such that $p(l) = \max\{p(k) \mid k \in E\}$.

81. **(TUR 2)** Given the side a and the corresponding altitude h_a of a triangle ABC, find a relation between a and h_a such that it is possible to construct, with straight-edge and compass, triangle ABC such that the altitudes of ABC form a right triangle admitting h_a as hypotenuse.

82. **(TUR 3)** Find all cubic polynomials $x^3 + ax^2 + bx + c$ admitting the rational numbers a, b, and c as roots.

83. **(TUR 4)** Let Γ_i, $i = 0, 1, 2, \ldots$, be a circle of radius r_i inscribed in an angle of measure 2α such that each Γ_i is externally tangent to Γ_{i+1} and $r_{i+1} < r_i$. Show that the sum of the areas of the circles Γ_i is equal to the area of a circle of radius $r = \frac{1}{2} r_0 (\sqrt{\sin\alpha} + \sqrt{\csc\alpha})$.

84. **(TUR 5)** (SL85-8).

85. **(USA 1)** Let CD be a diameter of circle K. Let AB be a chord that is parallel to CD. The line segment AE, with E on K, is parallel to CB; F is the point of intersection of line segments AB and DE. The line segment FG, with G on DC, extended is parallel to CB. Is GA tangent to K at point A?

86. **(USA 2)** Let l denote the length of the smallest diagonal of all rectangles inscribed in a triangle T. (By inscribed, we mean that all four vertices of the rectangle lie on the boundary of T.) Determine the maximum value of $\frac{l^2}{S(T)}$ taken over all triangles ($S(T)$ denotes the area of triangle T).

87. **(USA 3)** (SL85-9).

88. **(USA 4)** Determine the range of $w(w+x)(w+y)(w+z)$, where x, y, z, and w are real numbers such that

$$x + y + z + w = x^7 + y^7 + z^7 + w^7 = 0.$$

89. **(USA 5)** Given that n elements a_1, a_2, $\ldots$, a_n are organized into n pairs P_1, P_2, $\ldots$, P_n in such a way that two pairs P_i, P_j share exactly one element when (a_i, a_j) is one of the pairs, prove that every element is in exactly two of the pairs.

90. **(USS 1)** Decompose the number $5^{1985} - 1$ into a product of three integers, each of which is larger than 5^{100}.

91. **(USS 2)** Thirty-four countries participated in a jury session of the IMO, each represented by the leader and the deputy leader of the team. Before the meeting, some participants exchanged handshakes, but no team leader shook hands with his deputy. After the meeting, the leader of the Illyrian team asked every other participant the number of people they had shaken hands with, and all the answers she got were different. How many people did the deputy leader of the Illyrian team greet?

92. **(USS 3)** (SL85-11).

    (USS 3a) Given six numbers, find a method of computing by using not more than 15 additions and 14 multiplications the following five numbers: the sum of the numbers, the sum of products of the numbers taken two at a time, and the sums of the products of the numbers taken three, four, and five at a time.

93. **(USS 4)** The sphere inscribed in tetrahedron $ABCD$ touches the sides ABD and DBC at points K and M, respectively. Prove that $\angle AKB = \angle DMC$.

94. **(USS 5)** (SL85-22).

95. **(VNM 1)** (SL85-10).

    (VNM 1a) Prove that for each point M on the edges of a regular tetrahedron there is one and only one point M' on the surface of the tetrahedron such that there are at least three curves joining M and M' on the surface of the tetrahedron of minimal length among all curves joining M and M' on the surface of the tetrahedron. Denote this minimal length by d_M. Determine the positions of M for which d_M attains an extremum.

96. **(VNM 2)** Determine all functions $f : \mathbb{R} \to \mathbb{R}$ satisfying the following two conditions:

    (a) $f(x+y) + f(x-y) = 2f(x)f(y)$ for all $x, y \in \mathbb{R}$,
    (b) $\lim_{x \to \infty} f(x) = 0$.

97. **(VNM 3)** In a plane a circle with radius R and center w and a line Λ are given. The distance between w and Λ is d, $d > R$. The points M and N are chosen on Λ in such a way that the circle with diameter MN is externally tangent to the given circle. Show that there exists a point A in the plane such that all the segments MN are seen in a constant angle from A.

3.26.3 Shortlisted Problems

Proposals of the Problem Selection Committee.

1. **(MNG 1)**[IMO4] Given a set M of 1985 positive integers, none of which has a prime divisor larger than 26, prove that the set has four distinct elements whose geometric mean is an integer.

2. **(BRA 3)** A polyhedron has 12 faces and is such that:
    (i) all faces are isosceles triangles,
    (ii) all edges have length either x or y,
    (iii) at each vertex either 3 or 6 edges meet, and
    (iv) all dihedral angles are equal.
    Find the ratio x/y.

3. **(NLD 3)**[IMO3] The *weight* $w(p)$ of a polynomial p, $p(x) = \sum_{i=0}^n a_i x^i$, with integer coefficients a_i is defined as the number of its odd coefficients. For $i = 0, 1, 2, \ldots$, let $q_i(x) = (1+x)^i$. Prove that for any finite sequence $0 \le i_1 < i_2 < \cdots < i_n$, the inequality

$$w(q_{i_1} + \cdots + q_{i_n}) \geq w(q_{i_1})$$

holds.

4. **(AUS 1)**[IMO2] Each of the numbers in the set $N = \{1, 2, 3, \ldots, n-1\}$, where $n \geq 3$, is colored with one of two colors, say red or black, so that:
   (i) i and $n - i$ always receive the same color, and
   (ii) for some $j \in N$, relatively prime to n, i and $|j - i|$ receive the same color for all $i \in N$, $i \neq j$.
   Prove that all numbers in N must receive the same color.

5. **(ROU 1)** Let D be the interior of the circle C and let $A \in C$. Show that the function $f : D \to \mathbb{R}$, $f(M) = \frac{|MA|}{|MM'|}$, where $M' = (AM \cap C)$, is strictly convex; i.e., $f(P) < \frac{f(M_1) + f(M_2)}{2}$, $\forall M_1, M_2 \in D$, $M_1 \neq M_2$, where P is the midpoint of the segment $M_1 M_2$.

6. **(POL 1)** Let $x_n = \sqrt[2]{2 + \sqrt[3]{3 + \ldots + \sqrt[n]{n}}}$. Prove that

$$x_{n+1} - x_n < \frac{1}{n!}, \qquad n = 2, 3, \ldots .$$

Alternatives

7. **1a.(CZS 3)** The positive integers $x_1, \ldots, x_n$, $n \geq 3$, satisfy $x_1 < x_2 < \cdots < x_n < 2x_1$. Set $P = x_1 x_2 \cdots x_n$. Prove that if p is a prime number, k a positive integer, and P is divisible by p^k, then $\frac{P}{p^k} \geq n!$.

8. **1b.(TUR 5)** Find the smallest positive integer n such that
   (i) n has exactly 144 distinct positive divisors, and
   (ii) there are ten consecutive integers among the positive divisors of n.

9. **2a.(USA 3)** Determine the radius of a sphere S that passes through the centroids of each face of a given tetrahedron T inscribed in a unit sphere with center O. Also, determine the distance from O to the center of S as a function of the edges of T.

10. **2b.(VNM 1)** Prove that for every point M on the surface of a regular tetrahedron there exists a point M' such that there are at least three different curves on the surface joining M to M' with the smallest possible length among all curves on the surface joining M to M'.

11. **3a.(USS 3)** Find a method by which one can compute the coefficients of $P(x) = x^6 + a_1 x^5 + \cdots + a_6$ from the roots of $P(x) = 0$ by performing not more than 15 additions and 15 multiplications.

12. **3b.(UNK 4)** A sequence of polynomials $P_m(x, y, z)$, $m = 0, 1, 2, \ldots$, in x, y, and z is defined by $P_0(x, y, z) = 1$ and by

$$P_m(x, y, z) = (x+z)(y+z)P_{m-1}(x, y, z+1) - z^2 P_{m-1}(x, y, z)$$

for $m > 0$. Prove that each $P_m(x,y,z)$ is symmetric, in other words, is unaltered by any permutation of x,y,z.

13. **4a.(BGR 1)** Let m boxes be given, with some balls in each box. Let $n < m$ be a given integer. The following operation is performed: choose n of the boxes and put 1 ball in each of them. Prove:
   (a) If m and n are relatively prime, then it is possible, by performing the operation a finite number of times, to arrive at the situation that all the boxes contain an equal number of balls.
   (b) If m and n are not relatively prime, there exist initial distributions of balls in the boxes such that an equal distribution is not possible to achieve.

14. **4b.(IRL 4)** A set of 1985 points is distributed around the circumference of a circle and each of the points is marked with 1 or -1. A point is called "good" if the partial sums that can be formed by starting at that point and proceeding around the circle for any distance in either direction are all strictly positive. Show that if the number of points marked with -1 is less than 662, there must be at least one good point.

15. **5a.(FRA 3)** Let K and K' be two squares in the same plane, their sides of equal length. Is it possible to decompose K into a finite number of triangles $T_1, T_2, \ldots, T_p$ with mutually disjoint interiors and find translations $t_1, t_2, \ldots, t_p$ such that
$$K' = \bigcup_{i=1}^{p} t_i(T_i)?$$

16. **5b.(BEL 2)** If possible, construct an equilateral triangle whose three vertices are on three given circles.

17. **6a.(SWE 3)**[IMO6] The sequence $f_1, f_2, \ldots, f_n, \ldots$ of functions is defined for $x > 0$ recursively by
$$f_1(x) = x, \qquad f_{n+1}(x) = f_n(x)\left(f_n(x) + \frac{1}{n}\right).$$

Prove that there exists one and only one positive number a such that $0 < f_n(a) < f_{n+1}(a) < 1$ for all integers $n \geq 1$.

18. **6b.(CAN 5)** Let $x_1, x_2, \ldots, x_n$ be positive numbers. Prove that
$$\frac{x_1^2}{x_1^2 + x_2 x_3} + \frac{x_2^2}{x_2^2 + x_3 x_4} + \cdots + \frac{x_{n-1}^2}{x_{n-1}^2 + x_n x_1} + \frac{x_n^2}{x_n^2 + x_1 x_2} \leq n - 1.$$

Supplementary Problems

19. **(ISR 3)** For which integers $n \geq 3$ does there exist a regular n-gon in the plane such that all its vertices have integer coordinates in a rectangular coordinate system?

20. **(UNK 5)**[IMO1] A circle whose center is on the side ED of the cyclic quadrilateral $BCDE$ touches the other three sides. Prove that $EB + CD = ED$.

21. **(IRL 1)** The tangents at B and C to the circumcircle of the acute-angled triangle ABC meet at X. Let M be the midpoint of BC. Prove that
    (a) $\angle BAM = \angle CAX$, and
    (b) $\frac{AM}{AX} = \cos \angle BAC$.

22. **(USS 5)**[IMO5] A circle with center O passes through points A and C and intersects the sides AB and BC of the triangle ABC at points K and N, respectively. The circumscribed circles of the triangles ABC and KBN intersect at two distinct points B and M. Prove that $\angle OMB = 90°$.

3.27 The Twenty-Seventh IMO
Warsaw, Poland, July 4–15, 1986

3.27.1 Contest Problems

First Day (July 9)

1. The set $S = \{2,5,13\}$ has the property that for every $a,b \in S$, $a \neq b$, the number $ab - 1$ is a perfect square. Show that for every positive integer d not in S, the set $S \cup \{d\}$ does not have the above property.

2. Let A,B,C be fixed points in the plane. A man starts from a certain point P_0 and walks directly to A. At A he turns his direction by $60°$ to the left and walks to P_1 such that $P_0 A = A P_1$. After he performs the same action 1986 times successively around the points $A,B,C,A,B,C,\ldots$, he returns to the starting point. Prove that ABC is an equilateral triangle, and that the vertices A,B,C are arranged counter-clockwise.

3. To each vertex P_i $(i = 1,\ldots,5)$ of a pentagon an integer x_i is assigned, the sum $s = \sum x_i$ being positive. The following operation is allowed, provided at least one of the x_i's is negative: Choose a negative x_i, replace it by $-x_i$, and add the former value of x_i to the integers assigned to the two neighboring vertices of P_i (the remaining two integers are left unchanged).
   This operation is to be performed repeatedly until all negative integers disappear. Decide whether this procedure must eventually terminate.

Second Day (July 10)

4. Let A,B be adjacent vertices of a regular n-gon in the plane and let O be its center. Now let the triangle ABO glide around the polygon in such a way that the points A and B move along the whole circumference of the polygon. Describe the figure traced by the vertex O.

5. Find, with proof, all functions f defined on the nonnegative real numbers and taking nonnegative real values such that
   (i) $f[xf(y)]f(y) = f(x+y)$,
   (ii) $f(2) = 0$ but $f(x) \neq 0$ for $0 \leq x < 2$.

6. Prove or disprove: Given a finite set of points with integer coefficients in the plane, it is possible to color some of these points red and the remaining ones white in such a way that for any straight line L parallel to one of the coordinate axes, the number of red colored points and the number of white colored points on L differ by at most 1.

3.27.2 Longlisted Problems

1. **(AUS 1)** Let k be one of the integers $2,3,4$ and let $n = 2^k - 1$. Prove the inequality
$$1 + b^k + b^{2k} + \cdots + b^{nk} \geq (1 + b^n)^k$$
for all real $b \geq 0$.

2. **(AUS 2)** Let $ABCD$ be a convex quadrilateral. DA and CB meet at F and AB and DC meet at E. The bisectors of the angles DFC and AED are perpendicular. Prove that these angle bisectors are parallel to the bisectors of the angles between the lines AC and BD.

3. **(AUS 3)** A line parallel to the side BC of a triangle ABC meets AB in F and AC in E. Prove that the circles on BE and CF as diameters intersect in a point lying on the altitude of the triangle ABC dropped from A to BC.

4. **(BEL 1)** Find the last eight digits of the binary development of 27^{1986}.

5. **(BEL 2)** Let ABC and DEF be acute-angled triangles. Write $d = EF$, $e = FD$, $f = DE$. Show that there exists a point P in the interior of ABC for which the value of the expression $d \cdot AP + e \cdot BP + f \cdot CP$ attains a minimum.

6. **(BEL 3)** In an urn there are one ball marked 1, two balls marked 2, and so on, up to n balls marked n. Two balls are randomly drawn without replacement. Find the probability that the two balls are assigned the same number.

7. **(BGR 1)** (SL86-11).

8. **(BGR 2)** (SL86-19).

9. **(CAN 1)** In a triangle ABC, $\angle BAC = 100°$, $AB = AC$. A point D is chosen on the side AC such that $\angle ABD = \angle CBD$. Prove that $AD + DB = BC$.

10. **(CAN 2)** A set of n standard dice are shaken and randomly placed in a straight line. If $n < 2r$ and $r < s$, then the probability that there will be a string of at least r, but not more than s, consecutive 1's can be written as $P/6^{s+2}$. Find an explicit expression for P.

11. **(CAN 3)** (SL86-20).

12. **(CHN 1)** Let O be an interior point of a tetrahedron $A_1A_2A_3A_4$. Let S_1, S_2, S_3, S_4 be spheres with centers A_1, A_2, A_3, A_4, respectively, and let U, V be spheres with centers at O. Suppose that for $i, j = 1, 2, 3, 4$, $i \neq j$, the spheres S_i and S_j are tangent to each other at a point B_{ij} lying on A_iA_j. Suppose also that U is tangent to all edges A_iA_j and V is tangent to the spheres S_1, S_2, S_3, S_4. Prove that $A_1A_2A_3A_4$ is a regular tetrahedron.

13. **(CHN 2)** Let $N = \{1, 2, \ldots, n\}$, $n \geq 3$. To each pair i, j of elements of N, $i \neq j$, there is assigned a number $f_{ij} \in \{0, 1\}$ such that $f_{ij} + f_{ji} = 1$. Let $r(i) =$

$\sum_{j\neq i} f_{ij}$ and write $M = \max_{i\in N} r(i)$, $m = \min_{i\in N} r(i)$. Prove that for any $w \in N$ with $r(w) = m$ there exist $u,v \in N$ such that $r(u) = M$ and $f_{uv} f_{vw} = 1$.

14. **(CHN 3)** (SL86-17).

15. **(CHN 4)** Let $\mathbb{N} = B_1 \cup \cdots \cup B_q$ be a partition of the set $\mathbb{N}$ of all positive integers and let an integer $l \in \mathbb{N}$ be given. Prove that there exist a set $X \subset \mathbb{N}$ of cardinality l, an infinite set $T \subset \mathbb{N}$, and an integer k with $1 \leq k \leq q$ such that for any $t \in T$ and any finite set $Y \subset X$, the sum $t + \sum_{y \in Y} y$ belongs to B_k.

16. **(CZS 1)** Given a positive integer k, find the least integer n_k for which there exist five sets S_1, S_2, S_3, S_4, S_5 with the following properties:

$$|S_j| = k \ \text{ for } j = 1,\ldots,5, \quad \left| \bigcup_{j=1}^{5} S_j \right| = n_k;$$

$$|S_i \cap S_{i+1}| = 0 = |S_5 \cap S_1|, \quad \text{for } i = 1,\ldots,4.$$

17. **(CZS 2)** We call a tetrahedron right-faced if each of its faces is a right-angled triangle.
    (a) Prove that every orthogonal parallelepiped can be partitioned into six right-faced tetrahedra.
    (b) Prove that a tetrahedron with vertices A_1, A_2, A_3, A_4 is right-faced if and only if there exist four distinct real numbers c_1, c_2, c_3, and c_4 such that the edges $A_j A_k$ have lengths $A_j A_k = \sqrt{|c_j - c_k|}$ for $1 \leq j < k \leq 4$.

18. **(CZS 3)** (SL86-4).

19. **(FIN 1)** Let $f : [0,1] \to [0,1]$ satisfy $f(0) = 0$, $f(1) = 1$ and

$$f(x+y) - f(x) = f(x) - f(x-y)$$

for all $x,y \geq 0$ with $x-y, x+y \in [0,1]$. Prove that $f(x) = x$ for all $x \in [0,1]$.

20. **(FIN 2)** For any angle α with $0 < \alpha < 180°$, we call a closed convex planar set an α-*set* if it is bounded by two circular arcs (or an arc and a line segment) whose angle of intersection is α. Given a (closed) triangle T, find the greatest α such that any two points in T are contained in an α-set $S \subset T$.

21. **(FRA 1)** Let AB be a segment of unit length and let C, D be variable points of this segment. Find the maximum value of the product of the lengths of the six distinct segments with endpoints in the set $\{A, B, C, D\}$.

22. **(FRA 2)** Let $(a_n)_{n\in\mathbb{N}}$ be the sequence of integers defined recursively by $a_0 = 0$, $a_1 = 1$, $a_{n+2} = 4a_{n+1} + a_n$ for $n \geq 0$. Find the common divisors of a_{1986} and a_{6891}.

23. **(FRA 3)** Let I and J be the centers of the incircle and the excircle in the angle BAC of the triangle ABC. For any point M in the plane of the triangle, not on the line BC, denote by I_M and J_M the centers of the incircle and the excircle (touching BC) of the triangle BCM. Find the locus of points M for which $II_M J J_M$ is a rectangle.

24. **(FRA 4)** Two families of parallel lines are given in the plane, consisting of 15 and 11 lines, respectively. In each family, any two neighboring lines are at a unit distance from one another; the lines of the first family are perpendicular to the lines of the second family. Let V be the set of 165 intersection points of the lines under consideration. Show that there exist not fewer than 1986 distinct squares with vertices in the set V.

25. **(FRA 5)** (SL86-7).

26. **(FRG 1)** (SL86-5).

27. **(FRG 2)** In an urn there are n balls numbered $1, 2, \ldots, n$. They are drawn at random one by one without replacement and the numbers are recorded. What is the probability that the resulting random permutation has only one local maximum? A term in a sequence is a local maximum if it is greater than all its neighbors.

28. **(FRG 3)** (SL86-13).

29. **(FRG 4)** We define a binary operation $\star$ in the plane as follows: Given two points A and B in the plane, $C = A \star B$ is the third vertex of the equilateral triangle ABC oriented positively. What is the relative position of three points I, M, O in the plane if $I \star (M \star O) = (O \star I) \star M$ holds?

30. **(FRG 5)** Prove that a convex polyhedron all of whose faces are equilateral triangles has at most 30 edges.

31. **(UNK 1)** Let P and Q be distinct points in the plane of a triangle ABC such that $AP : AQ = BP : BQ = CP : CQ$. Prove that the line PQ passes through the circumcenter of the triangle.

32. **(UNK 2)** Find, with proof, all solutions of the equation $\frac{1}{x} + \frac{2}{y} - \frac{3}{z} = 1$ in positive integers x, y, z.

33. **(UNK 3)** (SL86-1).

34. **(UNK 4)** For each nonnegative integer n, $F_n(x)$ is a polynomial in x of degree n. Prove that if the identity

$$F_n(2x) = \sum_{r=0}^{n} (-1)^{n-r} \binom{n}{r} 2^r F_r(x)$$

holds for each n, then

$$F_n(tx) = \sum_{r=0}^{n} \binom{n}{r} t^r (1-t)^{n-r} F_r(x)$$

for each n and all t.

35. **(UNK 5)** Establish the maximum and minimum values that the sum $|a| + |b| + |c|$ can have if a, b, c are real numbers such that the maximum value of $|ax^2 + bx + c|$ is 1 for $-1 \leq x \leq 1$.

36. **(GDR 1)** (SL86-9).

37. **(GDR 2)** Prove that the set $\{1, 2, \ldots, 1986\}$ can be partitioned into 27 disjoint sets so that none of these sets contains an arithmetic triple (i.e., three distinct numbers in an arithmetic progression).

38. **(GDR 3)** (SL86-12).

39. **(HEL 1)** Let S be a k-element set.
    (a) Find the number of mappings $f : S \to S$ such that

$$\text{(i) } f(x) \neq x \text{ for } x \in S, \quad \text{(ii) } f(f(x)) = x \text{ for } x \in S.$$

    (b) The same with the condition (i) left out.

40. **(HEL 2)** Find the maximum value that the quantity $2m + 7n$ can have such that there exist distinct positive integers x_i ($1 \leq i \leq m$), y_j ($1 \leq j \leq n$) such that the x_i's are even, the y_j's are odd, and $\sum_{i=1}^{m} x_i + \sum_{j=1}^{n} y_j = 1986$.

41. **(HEL 3)** Let M, N, P be the midpoints of the sides BC, CA, AB of a triangle ABC. The lines AM, BN, CP intersect the circumcircle of ABC at points A', B', C', respectively. Show that if $A'B'C'$ is an equilateral triangle, then so is ABC.

42. **(HUN 1)** The integers $1, 2, \ldots, n^2$ are placed on the fields of an $n \times n$ chessboard ($n > 2$) in such a way that any two fields that have a common edge or a vertex are assigned numbers differing by at most $n + 1$. What is the total number of such placements?

43. **(HUN 2)** (SL86-10).

44. **(IRL 1)** (SL86-14).

45. **(IRL 2)** Given n real numbers $a_1 \leq a_2 \leq \cdots \leq a_n$, define

$$M_1 = \frac{1}{n} \sum_{i=1}^{n} a_i, \quad M_2 = \frac{2}{n(n-1)} \sum_{1 \leq i < j \leq n} a_i a_j, \quad Q = \sqrt{M_1^2 - M_2}.$$

    Prove that
$$a_1 \leq M_1 - Q \leq M_1 + Q \leq a_n$$

    and that equality holds if and only if $a_1 = a_2 = \cdots = a_n$.

46. **(IRL 3)** We wish to construct a matrix with 19 rows and 86 columns, with entries $x_{ij} \in \{0, 1, 2\}$ ($1 \leq i \leq 19$, $1 \leq j \leq 86$), such that:
    (i) in each column there are exactly k terms equal to 0;
    (ii) for any distinct $j, k \in \{1, \ldots, 86\}$ there is $i \in \{1, \ldots, 19\}$ with $x_{ij} + x_{ik} = 3$.
    For what values of k is this possible?

47. **(ISL 1)** (SL86-16).

48. **(ISL 2)** Let P be a convex 1986-gon in the plane. Let A, D be interior points of two distinct sides of P and let B, C be two distinct interior points of the line segment AD. Starting with an arbitrary point Q_1 on the boundary of P, define recursively a sequence of points Q_n as follows: given Q_n extend the directed line segment $Q_n B$ to meet the boundary of P in a point R_n and then extend $R_n C$ to

meet the boundary of P again in a point, which is defined to be Q_{n+1}. Prove that for all n large enough the points Q_n are on one of the sides of P containing A or D.

49. **(ISL 3)** Let C_1, C_2 be circles of radius $1/2$ tangent to each other and both tangent internally to a circle C of radius 1. The circles C_1 and C_2 are the first two terms of an infinite sequence of distinct circles C_n defined as follows: C_{n+2} is tangent externally to C_n and C_{n+1} and internally to C. Show that the radius of each C_n is the reciprocal of an integer.

50. **(LUX 1)** Let D be the point on the side BC of the triangle ABC such that AD is the bisector of $\angle CAB$. Let I be the incenter of $\triangle ABC$.
    (a) Construct the points P and Q on the sides AB and AC, respectively, such that PQ is parallel to BC and the perimeter of the triangle APQ is equal to $k \cdot BC$, where k is a given rational number.
    (b) Let R be the intersection point of PQ and AD. For what value of k does the equality $AR = RI$ hold?
    (c) In which case do the equalities $AR = RI = ID$ hold?

51. **(MNG 1)** Let a, b, c, d be the lengths of the sides of a quadrilateral circumscribed about a circle and let S be its area. Prove that $S \le \sqrt{abcd}$ and find conditions for equality.

52. **(MNG 2)** Solve the system of equations

$$\tan x_1 + \cot x_1 = 3 \tan x_2,$$
$$\tan x_2 + \cot x_2 = 3 \tan x_3,$$
$$\cdots \cdots$$
$$\tan x_n + \cot x_n = 3 \tan x_1.$$

53. **(MNG 3)** For given positive integers r, v, n let $S(r, v, n)$ denote the number of n-tuples of nonnegative integers $(x_1, \ldots, x_n)$ satisfying the equation $x_1 + \cdots + x_n = r$ and such that $x_i \le v$ for $i = 1, \ldots, n$. Prove that

$$S(r, v, n) = \sum_{k=0}^{m} (-1)^k \binom{n}{k} \binom{r - (v+1)k + n - 1}{n - 1},$$

where $m = \min \left\{ n, \left[\frac{r}{v+1} \right] \right\}$.

54. **(MNG 4)** Find the least integer n with the following property: For any set V of 8 points in the plane, no three lying on a line, and for any set E of n line segments with endpoints in V, one can find a straight line intersecting at least 4 segments in E in interior points.

55. **(MNG 5)** Given an integer $n \ge 2$, determine all n-digit numbers $M_0 = \overline{a_1 a_2 \ldots a_n}$ ($a_i \ne 0$, $i = 1, 2, \ldots, n$) divisible by the numbers $M_1 = \overline{a_2 a_3 \ldots a_n a_1}$, $M_2 = \overline{a_3 a_4 \ldots a_n a_1 a_2}$, $\ldots$, $M_{n-1} = \overline{a_n a_1 a_2 \ldots a_{n-1}}$.

56. **(MAR 1)** Let $A_1 A_2 A_3 A_4 A_5 A_6$ be a hexagon inscribed into a circle with center O. Consider the circular arc with endpoints A_1, A_6 not containing A_2. For any point

M of that arc denote by h_i the distance from M to the line A_iA_{i+1} ($1 \leq i \leq 5$). Construct M such that the sum $h_1 + \cdots + h_5$ is maximal.

57. **(MAR 2)** In a triangle ABC, the incircle touches the sides BC, CA, AB in the points A', B', C', respectively; the excircle in the angle A touches the lines containing these sides in A_1, B_1, C_1, and similarly, the excircles in the angles B and C touch these lines in A_2, B_2, C_2 and A_3, B_3, C_3. Prove that the triangle ABC is right-angled if and only if one of the point triples (A', B_3, C'), (A_3, B', C_3), (A', B', C_2), (A_2, B_2, C'), (A_2, B_1, C_2), (A_3, B_3, C_1), (A_1, B_2, C_1), (A_1, B_1, C_3) is collinear.

58. **(NLD 1)** (SL86-6).

59. **(NLD 2)** (SL86-15).

60. **(NLD 3)** Prove the inequality

$$(-a+b+c)^2(a-b+c)^2(a+b-c)^2$$
$$\geq (-a^2+b^2+c^2)(a^2-b^2+c^2)(a^2+b^2-c^2)$$

for all real numbers a, b, c.

61. **(ROU 1)** Given a positive integer n, find the greatest integer p with the property that for any function $f : \mathbb{P}(X) \rightarrow C$, where X and C are sets of cardinality n and p, respectively, there exist two distinct sets $A, B \in \mathbb{P}(X)$ such that $f(A) = f(B) = f(A \cup B)$. ($\mathbb{P}(X)$ is the family of all subsets of X.)

62. **(ROU 2)** Determine all pairs of positive integers (x, y) satisfying the equation $p^x - y^3 = 1$, where p is a given prime number.

63. **(ROU 3)** Let AA', BB', CC' be the bisectors of the angles of a triangle ABC ($A' \in BC$, $B' \in CA$, $C' \in AB$). Prove that each of the lines $A'B'$, $B'C'$, $C'A'$ intersects the incircle in two points.

64. **(ROU 4)** Let $(a_n)_{n \in \mathbb{N}}$ be the sequence of integers defined recursively by $a_1 = a_2 = 1$, $a_{n+2} = 7a_{n+1} - a_n - 2$ for $n \geq 1$. Prove that a_n is a perfect square for every n.

65. **(ROU 5)** Let $A_1A_2A_3A_4$ be a quadrilateral inscribed in a circle C. Show that there is a point M on C such that $MA_1 - MA_2 + MA_3 - MA_4 = 0$.

66. **(SWE 1)** One hundred red points and one hundred blue points are chosen in the plane, no three of them lying on a line. Show that these points can be connected pairwise, red ones with blue ones, by disjoint line segments.

67. **(SWE 2)** (SL86-2).

68. **(SWE 3)** Consider the equation $x^4 + ax^3 + bx^2 + ax + 1 = 0$ with real coefficients a, b. Determine the number of distinct real roots and their multiplicities for various values of a and b. Display your result graphically in the (a, b) plane.

69. **(TUR 1)** (SL86-18).

70. **(TUR 2)** (SL86-21).

71. **(TUR 3)** Two straight lines perpendicular to each other meet each side of a triangle in points symmetric with respect to the midpoint of that side. Prove that these two lines intersect in a point on the nine-point circle.

72. **(TUR 4)** A one-person game with two possible outcomes is played as follows: After each play, the player receives either a or b points, where a and b are integers with $0 < b < a < 1986$. The game is played as many times as one wishes and the total score of the game is defined as the sum of points received after successive plays. It is observed that every integer $x \geq 1986$ can be obtained as the total score whereas 1985 and 663 cannot. Determine a and b.

73. **(TUR 5)** Let $(a_i)_{i\in\mathbb{N}}$ be a strictly increasing sequence of positive real numbers such that $\lim_{i\to\infty} a_i = +\infty$ and $a_{i+1}/a_i \leq 10$ for each i. Prove that for every positive integer k there are infinitely many pairs (i, j) with $10^k \leq a_i/a_j \leq 10^{k+1}$.

74. **(USA 1)** (SL86-8).
Alternative formulation. Let A be a set of n points in space. From the family of all segments with endpoints in A, q segments have been selected and colored yellow. Suppose that all yellow segments are of different length. Prove that there exists a polygonal line composed of m yellow segments, where $m \geq \frac{2q}{n}$, arranged in order of increasing length.

75. **(USA 2)** The incenter of a triangle is the midpoint of the line segment of length 4 joining the centroid and the orthocenter of the triangle. Determine the maximum possible area of the triangle.

76. **(USA 3)** (SL86-3).

77. **(USS 1)** Find all integers x, y, z that satisfy

$$x^3 + y^3 + z^3 = x + y + z = 8.$$

78. **(USS 2)** If T and T_1 are two triangles with angles x, y, z and x_1, y_1, z_1, respectively, prove the inequality

$$\frac{\cos x_1}{\sin x} + \frac{\cos y_1}{\sin y} + \frac{\cos z_1}{\sin z} \leq \cot x + \cot y + \cot z.$$

79. **(USS 3)** Let AA_1, BB_1, CC_1 be the altitudes in an acute-angled triangle ABC. K and M are points on the line segments A_1C_1 and B_1C_1 respectively. Prove that if the angles MAK and CAA_1 are equal, then the angle C_1KM is bisected by AK.

80. **(USS 4)** Let $ABCD$ be a tetrahedron and O its incenter, and let the line OD be perpendicular to AD. Find the angle between the planes DOB and DOC.

3.27.3 Shortlisted Problems

1. **(UNK 3)**[IMO5] Find, with proof, all functions f defined on the nonnegative real numbers and taking nonnegative real values such that
   (i) $f[xf(y)]f(y) = f(x+y)$,

(ii) $f(2) = 0$ but $f(x) \neq 0$ for $0 \leq x < 2$.

2. **(SWE 2)** Let $f(x) = x^n$ where n is a fixed positive integer and $x = 1, 2, \ldots$. Is the decimal expansion $a = 0.f(1)f(2)f(3)\ldots$ rational for any value of n? The decimal expansion of a is defined as follows: If $f(x) = d_1(x)d_2(x)\ldots \ldots d_{r(x)}(x)$ is the decimal expansion of $f(x)$, then

$$a = 0.1d_1(2)d_2(2)\ldots d_{r(2)}(2)d_1(3)\ldots d_{r(3)}(3)d_1(4)\ldots .$$

3. **(USA 3)** Let A, B, and C be three points on the edge of a circular chord such that B is due west of C and ABC is an equilateral triangle whose side is 86 meters long. A boy swam from A directly toward B. After covering a distance of x meters, he turned and swam westward, reaching the shore after covering a distance of y meters. If x and y are both positive integers, determine y.

4. **(CZS 3)** Let n be a positive integer and let p be a prime number, $p > 3$. Find at least $3(n+1)$ [easier version: $2(n+1)$] sequences of positive integers x, y, z satisfying

$$xyz = p^n(x+y+z)$$

that do not differ only by permutation.

5. **(FRG 1)**[IMO1] The set $S = \{2, 5, 13\}$ has the property that for every $a, b \in S$, $a \neq b$, the number $ab - 1$ is a perfect square. Show that for every positive integer d not in S, the set $S \cup \{d\}$ does not have the above property.

6. **(NLD 1)** Find four positive integers each not exceeding 70000 and each having more than 100 divisors.

7. **(FRA 5)** Let real numbers $x_1, x_2, \ldots, x_n$ satisfy $0 < x_1 < x_2 < \cdots < x_n < 1$ and set $x_0 = 0$, $x_{n+1} = 1$. Suppose that these numbers satisfy the following system of equations:

$$\sum_{j=0, j\neq i}^{n+1} \frac{1}{x_i - x_j} = 0 \quad \text{where } i = 1, 2, \ldots, n. \tag{1}$$

Prove that $x_{n+1-i} = 1 - x_i$ for $i = 1, 2, \ldots, n$.

8. **(USA 1)** From a collection of n persons q distinct two-member teams are selected and ranked $1, \ldots, q$ (no ties). Let m be the least integer larger than or equal to $2q/n$. Show that there are m distinct teams that may be listed so that (i) each pair of consecutive teams on the list have one member in common and (ii) the chain of teams on the list are in rank order.

Alternative formulation. Given a graph with n vertices and q edges numbered $1, \ldots, q$, show that there exists a chain of m edges, $m \geq \frac{2q}{n}$, each two consecutive edges having a common vertex, arranged monotonically with respect to the numbering.

9. **(GDR 1)**[IMO6] Prove or disprove: Given a finite set of points with integer coordinates in the plane, it is possible to color some of these points red and the remaining ones white in such a way that for any straight line L parallel to one of

the coordinate axes, the number of red colored points and the number of white colored points on L differ by at most 1.

10. **(HUN 2)** Three persons A,B,C, are playing the following game: A k-element subset of the set $\{1,\ldots,1986\}$ is randomly chosen, with an equal probability of each choice, where k is a fixed positive integer less than or equal to 1986. The winner is A,B or C, respectively, if the sum of the chosen numbers leaves a remainder of 0, 1, or 2 when divided by 3. For what values of k is this game a fair one? (A game is fair if the three outcomes are equally probable.)

11. **(BGR 1)** Let $f(n)$ be the least number of distinct points in the plane such that for each $k = 1,2,\ldots,n$ there exists a straight line containing exactly k of these points. Find an explicit expression for $f(n)$.
    Simplified version. Show that $f(n) = \left[\frac{n+1}{2}\right]\left[\frac{n+2}{2}\right]$ ($[x]$ denoting the greatest integer not exceeding x).

12. **(GDR 3)**[IMO3] To each vertex P_i $(i = 1,\ldots,5)$ of a pentagon an integer x_i is assigned, the sum $s = \sum x_i$ being positive. The following operation is allowed, provided at least one of the x_i's is negative: Choose a negative x_i, replace it by $-x_i$, and add the former value of x_i to the integers assigned to the two neighboring vertices of P_i (the remaining two integers are left unchanged).
    This operation is to be performed repeatedly until all negative integers disappear. Decide whether this procedure must eventually terminate.

13. **(FRG 3)** A particle moves from $(0,0)$ to (n,n) directed by a fair coin. For each head it moves one step east and for each tail it moves one step north. At (n,y), $y < n$, it stays there if a head comes up and at (x,n), $x < n$, it stays there if a tail comes up. Let k be a fixed positive integer. Find the probability that the particle needs exactly $2n + k$ tosses to reach (n,n).

14. **(IRL 1)** The circle inscribed in a triangle ABC touches the sides BC,CA,AB in D,E,F, respectively, and X,Y,Z are the midpoints of EF,FD,DE, respectively. Prove that the centers of the inscribed circle and of the circles around XYZ and ABC are collinear.

15. **(NLD 2)** Let $ABCD$ be a convex quadrilateral whose vertices do not lie on a circle. Let $A'B'C'D'$ be a quadrangle such that A',B',C',D' are the centers of the circumcircles of triangles BCD,ACD,ABD, and ABC. We write $T(ABCD) = A'B'C'D'$. Let us define $A''B''C''D'' = T(A'B'C'D') = T(T(ABCD))$.
    (a) Prove that $ABCD$ and $A''B''C''D''$ are similar.
    (b) The ratio of similitude depends on the size of the angles of $ABCD$. Determine this ratio.

16. **(ISL 1)**[IMO4] Let A,B be adjacent vertices of a regular n-gon in the plane and let O be its center. Now let the triangle ABO glide around the polygon in such a way that the points A and B move along the whole circumference of the polygon. Describe the figure traced by the vertex O.

17. **(CHN 3)**[IMO2] Let A, B, C be fixed points in the plane. A man starts from a certain point P_0 and walks directly to A. At A he turns his direction by $60°$ to the left and walks to P_1 such that $P_0 A = AP_1$. After he does the same action 1986 times successively around the points $A, B, C, A, B, C, \ldots$, he returns to the starting point. Prove that $\triangle ABC$ is equilateral and that the vertices A, B, C are arranged counterclockwise.

18. **(TUR 1)** Let AX, BY, CZ be three cevians concurrent at an interior point D of a triangle ABC. Prove that if two of the quadrangles $DYAZ, DZBX, DXCY$ are circumscribable, so is the third.

19. **(BGR 2)** A tetrahedron $ABCD$ is given such that $AD = BC = a$; $AC = BD = b$; $AB \cdot CD = c^2$. Let $f(P) = AP + BP + CP + DP$, where P is an arbitrary point in space. Compute the least value of $f(P)$.

20. **(CAN 3)** Prove that the sum of the face angles at each vertex of a tetrahedron is a straight angle if and only if the faces are congruent triangles.

21. **(TUR 2)** Let $ABCD$ be a tetrahedron having each sum of opposite sides equal to 1. Prove that

$$r_A + r_B + r_C + r_D \leq \frac{\sqrt{3}}{3},$$

where r_A, r_B, r_C, r_D are the inradii of the faces, equality holding only if $ABCD$ is regular.

3.28 The Twenty-Eighth IMO
Havana, Cuba, July 5–16, 1987

3.28.1 Contest Problems

First Day (July 10)

1. Let S be a set of n elements. We denote the number of all permutations of S that have exactly k fixed points by $p_n(k)$. Prove that

$$\sum_{k=0}^{n} k p_n(k) = n!.$$

2. The prolongation of the bisector AL ($L \in BC$) in the acute-angled triangle ABC intersects the circumscribed circle at point N. From point L to the sides AB and AC are drawn the perpendiculars LK and LM respectively. Prove that the area of the triangle ABC is equal to the area of the quadrilateral $AKNM$.

3. Suppose $x_1, x_2, \ldots, x_n$ are real numbers with $x_1^2 + x_2^2 + \cdots + x_n^2 = 1$. Prove that for any integer $k > 1$ there are integers e_i not all 0 and with $|e_i| < k$ such that

$$|e_1 x_1 + e_2 x_2 + \cdots + e_n x_n| \leq \frac{(k-1)\sqrt{n}}{k^n - 1}.$$

Second Day (July 11)

4. Does there exist a function $f : \mathbb{N} \to \mathbb{N}$, such that $f(f(n)) = n + 1987$ for every natural number n?

5. Prove that for every natural number $n \geq 3$ it is possible to put n points in the Euclidean plane such that the distance between each pair of points is irrational and each three points determine a nondegenerate triangle with rational area.

6. Let $f(x) = x^2 + x + p$, for $p \in \mathbb{N}$. Prove that if the numbers $f(0)$, $f(1)$, ..., $f([\sqrt{p/3}])$ are primes, then all the numbers $f(0)$, $f(1)$, ..., $f(p-2)$ are primes.

3.28.2 Longlisted Problems

1. **(AUS 1)** Let $x_1, x_2, \ldots, x_n$ be n integers. Let $n = p + q$, where p and q are positive integers. For $i = 1, 2, \ldots, n$, put

$$S_i = x_i + x_{i+1} + \cdots + x_{i+p-1} \quad \text{and} \quad T_i = x_{i+p} + x_{i+p+1} + \cdots + x_{i+n-1}$$

(it is assumed that $x_{i+n} = x_i$ for all i). Next, let $m(a,b)$ be the number of indices i for which S_i leaves the remainder a and T_i leaves the remainder b on division by 3, where $a, b \in \{0, 1, 2\}$. Show that $m(1,2)$ and $m(2,1)$ leave the same remainder when divided by 3.

2. **(AUS 2)** Suppose we have a pack of $2n$ cards, in the order $1, 2, \ldots, 2n$. A perfect shuffle of these cards changes the order to $n+1, 1, n+2, 2, \ldots, n-1, 2n, n$; i.e., the cards originally in the first n positions have been moved to the places $2, 4, \ldots, 2n$, while the remaining n cards, in their original order, fill the odd positions $1, 3, \ldots, 2n-1$.
Suppose we start with the cards in the above order $1, 2, \ldots, 2n$ and then successively apply perfect shuffles. What conditions on the number n are necessary for the cards eventually to return to their original order? Justify your answer.

Remark. This problem is trivial. Alternatively, it may be required to find the least number of shuffles after which the cards will return to the original order.

3. **(AUS 3)** A town has a road network that consists entirely of one-way streets that are used for bus routes. Along these routes, bus stops have been set up. If the one-way signs permit travel from bus stop X to bus stop $Y \neq X$, then we shall say *Y can be reached from X*.
We shall use the phrase *Y comes after X* when we wish to express that every bus stop from which the bus stop X can be reached is a bus stop from which the bus stop Y can be reached, and every bus stop that can be reached from Y can also be reached from X. A visitor to this town discovers that if X and Y are any two different bus stops, then the two sentences "Y can be reached from X" and "Y comes after X" have exactly the same meaning in this town.
Let A and B be two bus stops. Show that of the following two statements, exactly one is true: (i) B can be reached from A; (ii) A can be reached from B.

4. **(AUS 4)** Let $a_1, a_2, a_3, b_1, b_2, b_3$ be positive real numbers. Prove that

$$(a_1 b_2 + a_2 b_1 + a_1 b_3 + a_3 b_1 + a_2 b_3 + a_3 b_2)^2 \\ \geq 4(a_1 a_2 + a_2 a_3 + a_3 a_1)(b_1 b_2 + b_2 b_3 + b_3 b_1)$$

and show that the two sides of the inequality are equal if and only if $a_1/b_1 = a_2/b_2 = a_3/b_3$.

5. **(AUS 5)** Let there be given three circles K_1, K_2, K_3 with centers O_1, O_2, O_3 respectively, which meet at a common point P. Also, let $K_1 \cap K_2 = \{P, A\}$, $K_2 \cap K_3 = \{P, B\}$, $K_3 \cap K_1 = \{P, C\}$. Given an arbitrary point X on K_1, join X to A to meet K_2 again in Y, and join X to C to meet K_3 again in Z.
   (a) Show that the points Z, B, Y are collinear.
   (b) Show that the area of triangle XYZ is less than or equal to 4 times the area of triangle $O_1 O_2 O_3$.

6. **(AUS 6)** (SL87-1).

7. **(BEL 1)** Let $f : (0, +\infty) \to \mathbb{R}$ be a function having the property that $f(x) = f(1/x)$ for all $x > 0$. Prove that there exists a function $u : [1, +\infty) \to \mathbb{R}$ satisfying $u\left(\frac{x+1/x}{2}\right) = f(x)$ for all $x > 0$.

8. **(BEL 2)** Determine the least possible value of the natural number n such that $n!$ ends in exactly 1987 zeros.

9. **(BEL 3)** In the set of 20 elements $\{1, 2, 3, 4, 5, 6, 7, 8, 9, 0, A, B, C, D, J, K, L, U, X, Y, Z\}$ we have made a random sequence of 28 throws. What is the probability that the sequence *CUBA JULY* 1987 appears in this order in the sequence already thrown?

10. **(FIN 1)** In a Cartesian coordinate system, the circle C_1 has center $O_1(-2,0)$ and radius 3. Denote the point $(1,0)$ by A and the origin by O. Prove that there is a constant $c > 0$ such that for every X that is exterior to C_1,

$$OX - 1 \geq c\min\{AX, AX^2\}.$$

Find the largest possible c.

11. **(FIN 2)** Let $S \subset [0,1]$ be a set of 5 points with $\{0,1\} \subset S$. The graph of a real function $f : [0,1] \to [0,1]$ is continuous and increasing, and it is linear on every subinterval I in $[0,1]$ such that the endpoints but no interior points of I are in S. We want to compute, using a computer, the extreme values of $g(x,t) = \frac{f(x+t)-f(x)}{f(x)-f(x-t)}$ for $x - t, x + t \in [0,1]$. At how many points (x,t) is it necessary to compute $g(x,t)$ with the computer?

12. **(FIN 3)** (SL87-3).

13. **(FIN 4)** Let A be an infinite set of positive integers such that every $n \in A$ is the product of at most 1987 prime numbers. Prove that there are an infinite set $B \subset A$ and a number p such that the greatest common divisor of any two distinct numbers in B is p.

14. **(FRA 1)** Given n real numbers $0 < t_1 \leq t_2 \leq \cdots \leq t_n < 1$, prove that

$$(1 - t_n^2)\left(\frac{t_1}{(1-t_1^2)^2} + \frac{t_2^2}{(1-t_2^3)^2} + \cdots + \frac{t_n^n}{(1-t_n^{n+1})^2}\right) < 1.$$

15. **(FRA 2)** Let $a_1, a_2, a_3, b_1, b_2, b_3, c_1, c_2, c_3$ be nine strictly positive real numbers. We set

$$S_1 = a_1 b_2 c_3, \quad S_2 = a_2 b_3 c_1, \quad S_3 = a_3 b_1 c_2;$$
$$T_1 = a_1 b_3 c_2, \quad T_2 = a_2 b_1 c_3, \quad T_3 = a_3 b_2 c_1.$$

Suppose that the set $\{S_1, S_2, S_3, T_1, T_2, T_3\}$ has at most two elements. Prove that

$$S_1 + S_2 + S_3 = T_1 + T_2 + T_3.$$

16. **(FRA 3)** Let ABC be a triangle. For every point M belonging to segment BC we denote by B' and c' the orthogonal projections of M on the straight lines AC and BC. Find points M for which the length of segment $B'C'$ is a minimum.

17. **(FRA 4)** Consider the number α obtained by writing one after another the decimal representations of $1, 1987, 1987^2, \ldots$ to the right the decimal point. Show that α is irrational.

18. **(FRA 5)** (SL87-4).

19. **(FRG 1)** (SL87-14).

20. **(FRG 2)** (SL87-15).

21. **(FRG 3)** (SL87-16).

22. **(UNK 1)** (SL87-5).

23. **(UNK 2)** A lampshade is part of the surface of a right circular cone whose axis is vertical. Its upper and lower edges are two horizontal circles. Two points are selected on the upper smaller circle and four points on the lower larger circle. Each of these six points has three of the others that are its nearest neighbors at a distance d from it. By distance is meant the shortest distance measured over the curved survace of the lampshade.
Prove that the area of the lampshade if $d^2(2\theta + \sqrt{3})$, where $\cot \frac{\theta}{2} = \frac{3}{\theta}$.

24. **(UNK 3)** Prove that if the equation $x^4 + ax^3 + bx + c = 0$ has all its roots real, then $ab \leq 0$.

25. **(UNK 4)** Numbers $d(n,m)$, with m,n integers, $0 \leq m \leq n$, are defined by $d(n,0) = d(n,n) = 0$ for all $n \geq 0$ and

$$md(n,m) = md(n-1,m) + (2n-m)d(n-1,m-1) \quad \text{for all } 0 < m < n.$$

Prove that all the $d(n,m)$ are integers.

26. **(UNK 5)** Prove that if x,y,z are real numbers such that $x^2 + y^2 + z^2 = 2$, then

$$x + y + z \leq xyz + 2.$$

27. **(UNK 6)** Find, with proof, the smallest real number C with the following property: For every infinite sequence $\{x_i\}$ of positive real numbers such that $x_1 + x_2 + \cdots + x_n \leq x_{n+1}$ for $n = 1, 2, 3, \ldots$, we have

$$\sqrt{x_1} + \sqrt{x_2} + \cdots + \sqrt{x_n} \leq C\sqrt{x_1 + x_2 + \cdots + x_n} \quad \text{for } n = 1, 2, 3, \ldots.$$

28. **(GDR 1)** In a chess tournament there are $n \geq 5$ players, and they have already played $\left[\frac{n^2}{4}\right] + 2$ games (each pair have played each other at most once).
   (a) Prove that there are five players a,b,c,d,e for which the pairs ab, ac, bc, ad, ae, de have already played.
   (b) Is the statement also valid for the $\left[\frac{n^2}{4}\right] + 1$ games played?
   Make the proof by induction over n.

29. **(GDR 2)** (SL87-13).

30. **(HEL 1)** Consider the regular 1987-gon $A_1 A_2 \ldots A_{1987}$ with center O. Show that the sum of vectors belonging to any proper subset of $M = \{OA_j \mid j = 1, 2, \ldots, 1987\}$ is nonzero.

31. **(HEL 2)** Construct a triangle ABC given its side $a = BC$, its circumradius R ($2R \geq a$), and the difference $1/k = 1/c - 1/b$, where $c = AB$ and $b = AC$.

32. **(HEL 3)** Solve the equation $28^x = 19^y + 87^z$, where x, y, z are integers.

33. **(HEL 4)** (SL87-6).

34. **(HUN 1)** (SL87-8).

35. **(HUN 2)** (SL87-9).

36. **(ISL 1)** A game consists in pushing a flat stone along a sequence of squares $S_0, S_1, S_2, \ldots$ that are arranged in linear order. The stone is initially placed on square S_0. When the stone stops on a square S_k it is pushed again in the same direction and so on until it reaches S_{1987} or goes beyond it; then the game stops. Each time the stone is pushed, the probability that it will advance exactly n squares is $1/2^n$. Determine the probability that the stone will stop exactly on square S_{1987}.

37. **(ISL 2)** Five distinct numbers are drawn successively and at random from the set $\{1, \ldots, n\}$. Show that the probability of a draw in which the first three numbers as well as all five numbers can be arranged to form an arithmetic progression is greater than $\frac{6}{(n-2)^3}$.

38. **(ISL 3)** (SL87-10).

39. **(LUX 1)** Let A be a set of polynomials with real coefficients and let them satisfy the following conditions:
    (i) if $f \in A$ and $\deg f \le 1$, then $f(x) = x - 1$;
    (ii) if $f \in A$ and $\deg f \ge 2$, then either there exists $g \in A$ such that $f(x) = x^{2+\deg g} + xg(x) - 1$ or there exist $g, h \in A$ such that $f(x) = x^{1+\deg g}g(x) + h(x)$;
    (iii) for every $f, g \in A$, both $x^{2+\deg f} + xf(x) - 1$ and $x^{1+\deg f}f(x) + g(x)$ belong to A.
    Let $R_n(f)$ be the remainder of the Euclidean division of the polynomial $f(x)$ by x^n. Prove that for all $f \in A$ and for all natural numbers $n \ge 1$ we have
    $$R_n(f)(1) \le 0 \quad \text{and} \quad R_n(f)(1) = 0 \Rightarrow R_n(f) \in A.$$

40. **(MNG 1)** The perpendicular line issued from the center of the circumcircle to the bisector of angle C in a triangle ABC divides the segment of the bisector inside ABC into two segments with ratio of lengths λ. Given $b = AC$ and $a = BC$, find the length of side c.

41. **(MNG 2)** Let n points be given arbitrarily in the plane, no three of them collinear. Let us draw segments between pairs of these points. What is the minimum number of segments that can be colored red in such a way that among any four points, three of them are connected by segments that form a red triangle?

42. **(MNG 3)** Find the integer solutions of the equation
    $$\left[\sqrt{2}\,m\right] = \left[(2 + \sqrt{2})n\right].$$

43. **(MNG 4)** Let $2n+3$ points be given in the plane in such a way that no three lie on a line and no four lie on a circle. Prove that the number of circles that pass through three of these points and contain exactly n interior points is not less than $\frac{1}{3}\binom{2n+3}{2}$.

44. **(MAR 1)** Let $\theta_1, \theta_2, \ldots, \theta_n$ be real numbers such that $\sin\theta_1 + \cdots + \sin\theta_n = 0$. Prove that
$$|\sin\theta_1 + 2\sin\theta_2 + \cdots + n\sin\theta_n| \le \left[\frac{n^2}{4}\right].$$

45. **(MAR 2)** Let us consider a variable polygon with $2n$ sides ($n \in \mathbb{N}$) in a fixed circle such that $2n-1$ of its sides pass through $2n-1$ fixed points lying on a straight line Δ. Prove that the last side also passes through a fixed point lying on Δ.

46. **(NLD 1)** (SL87-7).

47. **(NLD 2)** Through a point P within a triangle ABC the lines l, m, and n perpendicular respectively to AP, BP, CP are drawn. Prove that if l intersects the line BC in Q, m intersects AC in R, and n intersects AB in S, then the points Q, R, and S are collinear.

48. **(POL 1)** (SL87-11).

49. **(POL 2)** In the coordinate system in the plane we consider a convex polygon W and lines given by equations $x = k$, $y = m$, where k and m are integers. The lines determine a tiling of the plane with unit squares. We say that the boundary of W intersects a square if the boundary contains an interior point of the square. Prove that the boundary of W intersects at most $4\lceil d \rceil$ unit squares, where d is the maximal distance of points belonging to W (i.e., the diameter of W) and $\lceil d \rceil$ is the least integer not less than d.

50. **(POL 3)** Let P, Q, R be polynomials with real coefficients, satisfying $P^4 + Q^4 = R^2$. Prove that there exist real numbers p, q, r and a polynomial S such that $P = pS$, $Q = qS$ and $R = rS^2$.
    Variants: (1) $P^4 + Q^4 = R^4$; (2) $\gcd(P, Q) = 1$; (3) $\pm P^4 + Q^4 = R^2$ or R^4.

51. **(POL 4)** The function F is a one-to-one transformation of the plane into itself that maps rectangles into rectangles (rectangles are closed; continuity is not assumed). Prove that F maps squares into squares.

52. **(POL 5)** (SL87-12).

53. **(ROU 1)** (SL87-17).

54. **(ROU 2)** Let n be a natural number. Solve in integers the equation
$$x^n + y^n = (x - y)^{n+1}.$$

55. **(ROU 3)** Two moving bodies M_1, M_2 are displaced uniformly on two coplanar straight lines. Describe the union of all straight lines $M_1 M_2$.

56. **(ROU 4)** (SL87-18).

57. **(ROU 5)** The bisectors of the angles B, C of a triangle ABC intersect the opposite sides in B', C' respectively. Prove that the straight line $B'C'$ intersects the inscribed circle in two different points.

58. **(ESP 1)** Find, with argument, the integer solutions of the equation

$$3z^2 = 2x^3 + 385x^2 + 256x - 58195.$$

59. **(ESP 2)** It is given that a_{11}, a_{22} are real numbers, that $x_1, x_2, a_{12}, b_1, b_2$ are complex numbers, and that $a_{11}a_{22} = a_{12}\overline{a_{12}}$ (where $\overline{a_{12}}$ is the conjugate of a_{12}). We consider the following system in x_1, x_2:

$$\overline{x_1}(a_{11}x_1 + a_{12}x_2) = b_1,$$
$$\overline{x_2}(a_{12}x_1 + a_{22}x_2) = b_2.$$

   (a) Give one condition to make the system consistent.
   (b) Give one condition to make $\arg x_1 - \arg x_2 = 98°$.

60. **(TUR 1)** It is given that $x = -2272$, $y = 10^3 + 10^2 c + 10b + a$, and $z = 1$ satisfy the equation $ax + by + cz = 1$, where a, b, c are positive integers with $a < b < c$. Find y.

61. **(TUR 2)** Let PQ be a line segment of constant length λ taken on the side BC of a triangle ABC with the order B, P, Q, C, and let the lines through P and Q parallel to the lateral sides meet AC at P_1 and Q_1 and AB at P_2 and Q_2 respectively. Prove that the sum of the areas of the trapezoids PQQ_1P_1 and PQQ_2P_2 is independent of the position of PQ on BC.

62. **(TUR 3)** Let l, l' be two lines in 3-space and let A, B, C be three points taken on l with B as midpoint of the segment AC. If a, b, c are the distances of A, B, C from l', respectively, show that $b \leq \sqrt{\frac{a^2 + c^2}{2}}$, equality holding if l, l' are parallel.

63. **(TUR 4)** Compute $\sum_{k=0}^{2n}(-1)^k a_k^2$, where a_k are the coefficients in the expansion

$$(1 - \sqrt{2}x + x^2)^n = \sum_{k=0}^{2n} a_k x^k.$$

64. **(USA 1)** Let $r > 1$ be a real number, and let n be the largest integer smaller than r. Consider an arbitrary real number x with $0 \leq x \leq \frac{n}{r-1}$. By a *base-r expansion* of x we mean a representation of x in the form

$$x = \frac{a_1}{r} + \frac{a_2}{r^2} + \frac{a_3}{r^3} + \cdots,$$

where the a_i are integers with $0 \leq a_i < r$.
You may assume without proof that every number x with $0 \leq x \leq \frac{n}{r-1}$ has at least one base-r expansion.
Prove that if r is not an integer, then there exists a number p, $0 \leq p \leq \frac{n}{r-1}$, which has infinitely many distinct base-r expansions.

65. **(USA 2)** The *runs* of a decimal number are its increasing or decreasing blocks of digits. Thus 024379 has three runs: 024, 43, and 379. Determine the average number of runs for a decimal number in the set $\{d_1d_2\ldots d_n \mid d_k \neq d_{k+1},\ k = 1,2,\ldots,n-1\}$, where $n \geq 2$.

66. **(USA 3)** (SL87-2).

67. **(USS 1)** If a,b,c,d are real numbers such that $a^2 + b^2 + c^2 + d^2 \leq 1$, find the maximum of the expression

$$(a+b)^4 + (a+c)^4 + (a+d)^4 + (b+c)^4 + (b+d)^4 + (c+d)^4.$$

68. **(USS 2)** (SL87-19).
Original formulation. Let there be given positive real numbers α, β, γ such that $\alpha + \beta + \gamma < \pi$, $\alpha + \beta > \gamma$, $\beta + \gamma > \alpha$, $\gamma + \alpha > \beta$. Prove that it is possible to draw a triangle with the lengths of the sides $\sin\alpha$, $\sin\beta$, $\sin\gamma$. Moreover, prove that its area is less than

$$\frac{1}{8}(\sin 2\alpha + \sin 2\beta + \sin 2\gamma).$$

69. **(USS 3)** (SL87-20).

70. **(USS 4)** (SL87-21).

71. **(USS 5)** To every natural number k, $k \geq 2$, there corresponds a sequence $a_n(k)$ according to the following rule:

$$a_0 = k, \qquad a_n = \tau(a_{n-1}) \ \text{ for } n \geq 1,$$

in which $\tau(a)$ is the number of different divisors of a. Find all k for which the sequence $a_n(k)$ does not contain the square of an integer.

72. **(VNM 1)** Is it possible to cover a rectangle of dimensions $m \times n$ with bricks that have the tromino angular shape (an arrangement of three unit squares forming the letter L) if:
    (a) $m \times n = 1985 \times 1987$;
    (b) $m \times n = 1987 \times 1989$?

73. **(VNM 2)** Let $f(x)$ be a periodic function of period $T > 0$ defined over $\mathbb{R}$. Its first derivative is continuous on $\mathbb{R}$. Prove that there exist $x,y \in [0,T)$ such that $x \neq y$ and

$$f(x)f'(y) = f(y)f'(x).$$

74. **(VNM 3)** (SL87-22).

75. **(VNM 4)** Let a_k be positive numbers such that $a_1 \geq 1$ and $a_{k+1} - a_k \geq 1$ ($k = 1,2,\ldots$). Prove that for every $n \in \mathbb{N}$,

$$\sum_{k=1}^{n} \frac{1}{a_{k+1} \sqrt[1987]{a_k}} < 1987.$$

76. **(VNM 5)** Given two sequences of positive numbers $\{a_k\}$ and $\{b_k\}$ $(k \in \mathbb{N})$ such that

  (i) $a_k < b_k$,
  (ii) $\cos a_k x + \cos b_k x \geq -\frac{1}{k}$ for all $k \in \mathbb{N}$ and $x \in \mathbb{R}$,
  prove the existence of $\lim_{k \to \infty} \frac{a_k}{b_k}$ and find this limit.

77. **(YUG 1)** Find the least natural number k such that for any $a \in [0,1]$ and any natural number n,

$$a^k(1-a)^n < \frac{1}{(n+1)^3}.$$

78. **(YUG 2)** (SL87-23).

3.28.3 Shortlisted Problems

1. **(AUS 6)** Let f be a function that satisfies the following conditions:
   (i) If $x > y$ and $f(y) - y \geq v \geq f(x) - x$, then $f(z) = v + z$, for some number z between x and y.
   (ii) The equation $f(x) = 0$ has at least one solution, and among the solutions of this equation, there is one that is not smaller than all the other solutions;
   (iii) $f(0) = 1$.
   (iv) $f(1987) \leq 1988$.
   (v) $f(x)f(y) = f(xf(y) + yf(x) - xy)$.
   Find $f(1987)$.

2. **(USA 3)** At a party attended by n married couples, each person talks to everyone else at the party except his or her spouse. The conversations involve sets of persons or cliques $C_1, C_2, \ldots, C_k$ with the following property: no couple are members of the same clique, but for every other pair of persons there is exactly one clique to which both members belong. Prove that if $n \geq 4$, then $k \geq 2n$.

3. **(FIN 3)** Does there exist a second-degree polynomial $p(x,y)$ in two variables such that every nonnegative integer n equals $p(k,m)$ for one and only one ordered pair (k,m) of nonnegative integers?

4. **(FRA 5)** Let $ABCDEFGH$ be a parallelepiped with $AE \| BF \| CG \| DH$. Prove the inequality
$$AF + AH + AC \leq AB + AD + AE + AG.$$

   In what cases does equality hold?

5. **(UNK 1)** Find, with proof, the point P in the interior of an acute-angled triangle ABC for which $BL^2 + CM^2 + AN^2$ is a minimum, where L, M, N are the feet of the perpendiculars from P to BC, CA, AB respectively.

6. **(HEL 4)** Show that if a, b, c are the lengths of the sides of a triangle and if $2S = a + b + c$, then
$$\frac{a^n}{b+c} + \frac{b^n}{c+a} + \frac{c^n}{a+b} \geq \left(\frac{2}{3}\right)^{n-2} S^{n-1}, \quad n \geq 1.$$

7. **(NLD 1)** Given five real numbers u_0, u_1, u_2, u_3, u_4, prove that it is always possible to find five real numbers v_0, v_1, v_2, v_3, v_4 that satisfy the following conditions:
   (i) $u_i - v_i \in \mathbb{N}$.
   (ii) $\sum_{0 \leq i < j \leq 4}(v_i - v_j)^2 < 4$.

8. **(HUN 1)** (a) Let $(m, k) = 1$. Prove that there exist integers $a_1, a_2, \ldots, a_m$ and $b_1, b_2, \ldots, b_k$ such that each product $a_i b_j$ ($i = 1, 2, \ldots, m$; $j = 1, 2, \ldots, k$) gives a different residue when divided by mk.
   (b) Let $(m, k) > 1$. Prove that for any integers $a_1, a_2, \ldots, a_m$ and $b_1, b_2, \ldots, b_k$ there must be two products $a_i b_j$ and $a_s b_t$ ($(i, j) \neq (s, t)$) that give the same residue when divided by mk.

9. **(HUN 2)** Does there exist a set M in usual Euclidean space such that for every plane λ the intersection $M \cap \lambda$ is finite and nonempty?

10. **(ISL 3)** Let S_1 and S_2 be two spheres with distinct radii that touch externally. The spheres lie inside a cone C, and each sphere touches the cone in a full circle. Inside the cone there are n additional solid spheres arranged in a ring in such a way that each solid sphere touches the cone C, both of the spheres S_1 and S_2 externally, as well as the two neighboring solid spheres. What are the possible values of n?

11. **(POL 1)** Find the number of partitions of the set $\{1, 2, \ldots, n\}$ into three subsets A_1, A_2, A_3, some of which may be empty, such that the following conditions are satisfied:
    (i) After the elements of every subset have been put in ascending order, every two consecutive elements of any subset have different parity.
    (ii) If A_1, A_2, A_3 are all nonempty, then in exactly one of them the minimal number is even.

12. **(POL 5)** Given a nonequilateral triangle ABC, the vertices listed counterclockwise, find the locus of the centroids of the equilateral triangles $A'B'C'$ (the vertices listed counterclockwise) for which the triples of points A, B', C'; A', B, C'; and A', B', C are collinear.

13. **(GDR 2)**[IMO5] Is it possible to put 1987 points in the Euclidean plane such that the distance between each pair of points is irrational and each three points determine a nondegenerate triangle with rational area?

14. **(FRG 1)** How many words with n digits can be formed from the alphabet $\{0, 1, 2, 3, 4\}$, if neighboring digits must differ by exactly one?

15. **(FRG 2)**[IMO3] Suppose $x_1, x_2, \ldots, x_n$ are real numbers with $x_1^2 + x_2^2 + \cdots + x_n^2 = 1$. Prove that for any integer $k > 1$ there are integers e_i not all 0 and with $|e_i| < k$ such that
$$|e_1 x_1 + e_2 x_2 + \cdots + e_n x_n| \leq \frac{(k-1)\sqrt{n}}{k^n - 1}.$$

16. **(FRG 3)**[IMO1] Let S be a set of n elements. We denote the number of all permutations of S that have exactly k fixed points by $p_n(k)$. Prove:

(a) $\sum_{k=0}^{n} k p_n(k) = n!$;

(b) $\sum_{k=0}^{n} (k-1)^2 p_n(k) = n!$.

17. **(ROU 1)** Prove that there exists a four-coloring of the set $M = \{1, 2, \ldots, 1987\}$ such that any arithmetic progression with 10 terms in the set M is not monochromatic.

    Alternative formulation. Let $M = \{1, 2, \ldots, 1987\}$. Prove that there is a function $f : M \rightarrow \{1, 2, 3, 4\}$ that is not constant on every set of 10 terms from M that form an arithmetic progression.

18. **(ROU 4)** For any integer $r \geq 1$, determine the smallest integer $h(r) \geq 1$ such that for any partition of the set $\{1, 2, \ldots, h(r)\}$ into r classes, there are integers $a \geq 0, 1 \leq x \leq y$, such that $a+x, a+y, a+x+y$ belong to the same class.

19. **(USS 2)** Let α, β, γ be positive real numbers such that $\alpha + \beta + \gamma < \pi$, $\alpha + \beta > \gamma$, $\beta + \gamma > \alpha$, $\gamma + \alpha > \beta$. Prove that with the segments of lengths $\sin \alpha$, $\sin \beta$, $\sin \gamma$ we can construct a triangle and that its area is not greater than

$$\frac{1}{8}(\sin 2\alpha + \sin 2\beta + \sin 2\gamma).$$

20. **(USS 3)**[IMO6] Let $f(x) = x^2 + x + p$, $p \in \mathbb{N}$. Prove that if the numbers $f(0), f(1)$, $\ldots, f([\sqrt{p/3}\,])$ are primes, then all the numbers $f(0), f(1), \ldots, f(p-2)$ are primes.

21. **(USS 4)**[IMO2] The prolongation of the bisector AL ($L \in BC$) in the acute-angled triangle ABC intersects the circumscribed circle at point N. From point L to the sides AB and AC are drawn the perpendiculars LK and LM respectively. Prove that the area of the triangle ABC is equal to the area of the quadrilateral $AKNM$.

22. **(VNM 3)**[IMO4] Does there exist a function $f : \mathbb{N} \rightarrow \mathbb{N}$, such that $f(f(n)) = n + 1987$ for every natural number n?

23. **(YUG 2)** Prove that for every natural number k ($k \geq 2$) there exists an irrational number r such that for every natural number m,

$$[r^m] \equiv -1 \pmod{k}.$$

Remark. An easier variant: Find r as a root of a polynomial of second degree with integer coefficients.

3.29 The Twenty-Ninth IMO
Canberra, Australia, July 9–21, 1988

3.29.1 Contest Problems

First Day (July 15)

1. Consider two concentric circles of radii R and r $(R > r)$ with center O. Fix P on the small circle and consider the variable chord PA of the small circle. Points B and C lie on the large circle; B, P, C are collinear and BC is perpendicular to AP.
   (a) For which value(s) of $\angle OPA$ is the sum $BC^2 + CA^2 + AB^2$ extremal?
   (b) What are the possible positions of the midpoints U of BA and V of AC as $\angle OPA$ varies?

2. Let n be an even positive integer. Let $A_1, A_2, \ldots, A_{n+1}$ be sets having n elements each such that any two of them have exactly one element in common, while every element of their union belongs to at least two of the given sets. For which n can one assign to every element of the union one of the numbers 0 and 1 in such a manner that each of the sets has exactly $n/2$ zeros?

3. A function f defined on the positive integers (and taking positive integer values) is given by
$$f(1) = 1, \quad f(3) = 3,$$
$$f(2n) = f(n),$$
$$f(4n+1) = 2f(2n+1) - f(n),$$
$$f(4n+3) = 3f(2n+1) - 2f(n),$$

   for all positive integers n. Determine with proof the number of positive integers less than or equal to 1988 for which $f(n) = n$.

Second Day (July 16)

4. Show that the solution set of the inequality
$$\sum_{k=1}^{70} \frac{k}{x-k} \geq \frac{5}{4}$$
   is the union of disjoint half-open intervals with the sum of lengths 1988.

5. In a right-angled triangle ABC let AD be the altitude drawn to the hypotenuse and let the straight line joining the incenters of the triangles ABD, ACD intersect the sides AB, AC at the points K, L respectively. If E and E_1 denote the areas of the triangles ABC and AKL respectively, show that $\frac{E}{E_1} \geq 2$.

6. Let a and b be two positive integers such that $ab+1$ divides $a^2 + b^2$. Show that $\frac{a^2+b^2}{ab+1}$ is a perfect square.

3.29.2 Longlisted Problems

1. **(BGR 1)** (SL88-1).

2. **(BGR 2)** Let $a_n = \left[\sqrt{(n+1)^2 + n^2} \right]$, $n = 1, 2, \ldots$, where $[x]$ denotes the integer part of x. Prove that
   (a) there are infinitely many positive integers m such that $a_{m+1} - a_m > 1$;
   (b) there are infinitely many positive integers m such that $a_{m+1} - a_m = 1$.

3. **(BGR 3)** (SL88-2).

4. **(CAN 1)** (SL88-3).

5. **(CUB 1)** Let k be a positive integer and M_k the set of all the integers that are between $2k^2 + k$ and $2k^2 + 3k$, both included. Is it possible to partition M_k into two subsets A and B such that

$$\sum_{x \in A} x^2 = \sum_{x \in B} x^2?$$

6. **(CZS 1)** (SL88-4).

7. **(CZS 2)** (SL88-5).

8. **(CZS 3)** (SL88-6).

9. **(FRA 1)** If a_0 is a positive real number, consider the sequence $\{a_n\}$ defined by

$$a_{n+1} = \frac{a_n^2 - 1}{n+1} \qquad \text{for } n \geq 0.$$

   Show that there exists a real number $a > 0$ such that:
   (i) for all real $a_0 \geq a$, the sequence $\{a_n\} \to +\infty \ (n \to \infty)$;
   (ii) for all real $a_0 < a$, the sequence $\{a_n\} \to 0$.

10. **(FRA 2)** (SL88-7).

11. **(FRA 3)** (SL88-8).

12. **(FRA 4)** Show that there do not exist more than 27 half-lines (or rays) emanating from the origin in 3-dimensional space such that the angle between each pair of rays is greater than or equal to $\pi/4$.

13. **(FRA 5)** Let T be a triangle with inscribed circle C. A square with sides of length a is circumscribed about the same circle C. Show that the total length of the parts of the edges of the square interior to the triangle T is at least $2a$.

14. **(FRG 1)** (SL88-9).

15. **(FRG 2)** Let $1 \leq k < n$. Consider all finite sequences of positive integers with sum n. Find $T(n, k)$, the total number of terms of size k in all of these sequences.

16. **(FRG 3)** Show that if n runs through all positive integers, then

$$f(n) = \left[n + \sqrt{\frac{n}{3}} + \frac{1}{2} \right]$$

runs through all positive integers skipping the terms of the sequence $a_n = 3n^2 - 2n$.

17. (**FRG 4**) Show that if n runs through all positive integers, then

$$f(n) = \left[n + \sqrt{3n} + \frac{1}{2} \right]$$

runs through all positive integers skipping the terms of the sequence $a_n = \left[\frac{n^2 + 2n}{3} \right]$.

18. (**UNK 1**) (SL88-25).

19. (**UNK 2**) (SL88-26).

20. (**UNK 3**) It is proposed to partition the set of positive integers into two disjoint subsets A and B subject to the following conditions:
    (i) 1 is in A;
    (ii) no two distinct members of A have a sum of the form $2^k + 2$ ($k = 0, 1, 2, \ldots$); and
    (iii) no two distinct members of B have a sum of that form.
    Show that this partitioning can be carried out in a unique manner and determine the subsets to which 1987, 1988, and 1989 belong.

21. (**UNK 4**) (SL88-27).

22. (**UNK 5**) (SL88-28).

23. (**GDR 1**) (SL88-10).

24. (**GDR 2**) Let $Z_{m,n}$ be the set of all ordered pairs (i, j) with $i \in \{1, \ldots, m\}$ and $j \in \{1, \ldots, n\}$. Also let $a_{m,n}$ be the number of all those subsets of $Z_{m,n}$ that contain no two ordered pairs $(i_1, j_1), (i_2, j_2)$ with $|i_1 - i_2| + |j_1 - j_2| = 1$. Show that for all positive integers m and k,

$$a_{m,2k}^2 \leq a_{m,2k-1} a_{m,2k+1}.$$

25. (**GDR 3**) (SL88-11).

26. (**HEL 1**) Let AB and CD be two perpendicular chords of a circle with center O and radius r, and let X, Y, Z, W denote in cyclical order the four parts into which the disk is thus divided. Find the maximum and minimum of the quantity $\frac{A(Z)}{A(Y) + A(W)}$, where $A(U)$ denotes the area of U.

27. (**HEL 2**) (SL88-12).

28. (**HEL 3**) (SL88-13).

29. (**HEL 4**) Find positive integers $x_1, x_2, \ldots, x_{29}$, at least one of which is greater than 1988, such that

$$x_1^2 + x_2^2 + \cdots + x_{29}^2 = 29x_1x_2\ldots x_{29}.$$

30. **(HKG 1)** Find the total number of different integers that the function

$$f(x) = [x] + [2x] + \left[\frac{5x}{3}\right] + [3x] + [4x]$$

takes for $0 \le x \le 100$.

31. **(HKG 2)** The circle $x^2 + y^2 = r^2$ meets the coordinate axes at $A = (r,0)$, $B = (-r,0)$, $C = (0,r)$, and $D = (0,-r)$. Let $P = (u,v)$ and $Q = (-u,v)$ be two points on the circumference of the circle. Let N be the point of intersection of PQ and the y-axis, and let M be the foot of the perpendicular drawn from P to the x-axis. If r^2 is odd, $u = p^m > q^n = v$, where p and q are prime numbers, and m and n are natural numbers, show that

$$|AM| = 1, \quad |BM| = 9, \quad |DN| = 8, \quad |PQ| = 8.$$

32. **(HKG 3)** Assuming that the roots of $x^3 + px^2 + qx + r = 0$ are all real and positive, find a relation between p, q, and r that gives a necessary condition for the roots to be exactly the cosines of three angles of a triangle.

33. **(HKG 4)** Find a necessary and sufficient condition on the natural number n for the equation $x^n + (2+x)^n + (2-x)^n = 0$ to have a real root.

34. **(HKG 5)** Express the number 1988 as the sum of some positive integers in such a way that the product of these positive integers is maximal.

35. **(HKG 6)** In the triangle ABC, let D, E, and F be the midpoints of the three sides, X, Y, and Z the feet of the three altitudes, H the orthocenter, and P, Q, and R the midpoints of the line segments joining H to the three vertices. Show that the nine points $D, E, F, P, Q, R, X, Y, Z$ lie on a circle.

36. **(HUN 1)** (SL88-14).

37. **(HUN 2)** Let n points be given on the surface of a sphere. Show that the surface can be divided into n congruent regions such that each of them contains exactly one of the given points.

38. **(HUN 3)** In a multiple choice test there were 4 questions and 3 possible answers for each question. A group of students was tested and it turned out that for any 3 of them there was a question that the three students answered differently. What is the maximal possible number of students tested?

39. **(ISL 1)** (SL88-15).

40. **(ISL 2)** A sequence of numbers a_n, $n = 1, 2, \ldots$, is defined as follows: $a_1 = 1/2$, and for each $n \ge 2$,

$$a_n = \left(\frac{2n-3}{2n}\right) a_{n-1}.$$

Prove that $\sum_{k=1}^{n} a_k < 1$ for all $n \ge 1$.

41. (IDN 1)

(a) Let ABC be a triangle with $AB = 12$ and $AC = 16$. Suppose M is the midpoint of side BC and points E and F are chosen on sides AC and AB respectively, and suppose that the lines EF and AM intersect at G. If $AE = 2AF$ then find the ratio EG/GF.

(b) Let E be a point external to a circle and suppose that two chords EAB and ECD meet at an angle of $40°$. If $AB = BC = CD$, find the size of $\angle ACD$.

42. (IDN 2)

(a) Four balls of radius 1 are mutually tangent, three resting on the floor and the fourth resting on the others. A tetrahedron, each of whose edges has length s, is circumscribed around the balls. Find the value of s.

(b) Suppose that $ABCD$ and $EFGH$ are opposite faces of a rectangular solid, with $\angle DHC = 45°$ and $\angle FHB = 60°$. Find the cosine of $\angle BHD$.

43. (IDN 3)

(a) The polynomial $x^{2k} + 1 + (x+1)^{2k}$ is not divisible by $x^2 + x + 1$. Find the value of k.

(b) If p, q, and r are distinct roots of $x^3 - x^2 + x - 2 = 0$, find the value of $p^3 + q^3 + r^3$.

(c) If r is the remainder when each of the numbers 1059, 1417, and 2312 is divided by d, where d is an integer greater than one, find the value of $d - r$.

(d) What is the smallest positive odd integer n such that the product of $2^{1/7}$, $2^{3/7}, \ldots, 2^{(2n+1)/7}$ is greater than 1000?

44. (IDN 4)

(a) Let $g(x) = x^5 + x^4 + x^3 + x^2 + x + 1$. What is the remainder when the polynomial $g(x^{12})$ is divided by the polynomial $g(x)$?

(b) If k is a positive integer and f is a function such that for every positive number x, $f(x^2 + 1)^{\sqrt{x}} = k$, find the value of $f\left(\frac{9+y^2}{y^2}\right)^{\sqrt{12/y}}$ for every positive number y.

(c) The function f satisfies the functional equation $f(x) + f(y) = f(x+y) - xy - 1$ for every pair x, y of real numbers. If $f(1) = 1$, find the number of integers n for which $f(n) = n$.

45. (IDN 5)

(a) Consider a circle K with diameter AB, a circle L tangent to AB and to K, and a circle M tangent to circle K, circle L, and AB. Calculate the ratio of the area of circle K to the area of circle M.

(b) In triangle ABC, $AB = AC$ and $\angle CAB = 80°$. If points D, E, and F lie on sides BC, AC, and AB, respectively, and $CE = CD$ and $BF = BD$, find the measure of $\angle EDF$.

46. (IDN 6)

(a) Calculate $x = \dfrac{(11+6\sqrt{2})\sqrt{11-6\sqrt{2}} - (11-6\sqrt{2})\sqrt{11+6\sqrt{2}}}{(\sqrt{\sqrt{5}+2} + \sqrt{\sqrt{5}-2}) - (\sqrt{\sqrt{5}+1})}$.

(b) For each positive number x, let $k = \frac{(x+1/x)^6-(x^6+1/x^6)-2}{(x+1/x)^3+(x^3+1/x^3)}$. Calculate the minimum value of k.

47. **(IRL 1)** (SL88-16).

48. **(IRL 2)** Find all plane triangles whose sides have integer length and whose incircles have unit radius.

49. **(IRL 3)** Let $-1 < x < 1$. Show that

$$\sum_{k=0}^{6} \frac{1-x^2}{1-2x\cos(2\pi k/7)+x^2} = \frac{7(1+x^7)}{1-x^7}.$$

Deduce that

$$\csc^2 \frac{\pi}{7} + \csc^2 \frac{2\pi}{7} + \csc^2 \frac{3\pi}{7} = 8.$$

50. **(IRL 4)** Let $g(n)$ be defined as follows:

$$g(1) = 0, \quad g(2) = 1,$$
$$g(n+2) = g(n) + g(n+1) + 1 \quad (n \geq 1).$$

Prove that if $n > 5$ is a prime, then n divides $g(n)(g(n)+1)$.

51. **(ISR 1)** Let $A_1, A_2, \ldots, A_{29}$ be 29 different sequences of positive integers. For $1 \leq i < j \leq 29$ and any natural number x, we define $N_i(x)$ to be the number of elements of the sequence A_i that are less than or equal to x, and $N_{ij}(x)$ to be the number of elements of the intersection $A_i \cap A_j$ that are less than or equal to x. It is given that for all $1 \leq i \leq 29$ and every natural number x,

$$N_i(x) \geq \frac{x}{e}, \quad \text{where } e = 2.71828\ldots.$$

Prove that there exists at least one pair i, j ($1 \leq i < j \leq 29$) such that $N_{ij}(1988) > 200$.

52. **(ISR 2)** (SL88-17).

53. **(KOR 1)** Let $x = p$, $y = q$, $z = r$, $w = s$ be the unique solution of the system of linear equations

$$x + a_i y + a_i^2 z + a_i^3 w = a_i^4, \quad i = 1, 2, 3, 4.$$

Express the solution of the following system in terms of p, q, r, and s:

$$x + a_i^2 y + a_i^4 z + a_i^6 w = a_i^8, \quad i = 1, 2, 3, 4.$$

Assume the uniqueness of the solution.

54. **(KOR 2)** (SL88-22).

55. **(KOR 3)** Find all positive integers x such that the product of all digits of x is given by $x^2 - 10x - 22$.

56. **(KOR 4)** The Fibonacci sequence is defined by

$$a_{n+1} = a_n + a_{n-1} \quad (n \geq 1), \qquad a_0 = 0, \, a_1 = a_2 = 1.$$

Find the greatest common divisor of the 1960th and 1988th terms of the Fibonacci sequence.

57. **(KOR 5)** Let C be a cube with edges of length 2. Construct a solid with fourteen faces by cutting off all eight corners of C, keeping the new faces perpendicular to the diagonals of the cube and keeping the newly formed faces identical. If at the conclusion of this process the fourteen faces so formed have the same area, find the area of each face of the new solid.

58. **(KOR 6)** For each pair of positive integers k and n, let $S_k(n)$ be the base-k digit sum of n. Prove that there are at most two primes p less than 20,000 for which $S_{31}(p)$ is a composite number.

59. **(LUX 1)** (SL88-18).

60. **(MEX 1)** (SL88-19).

61. **(MEX 2)** Prove that the numbers A, B, and C are equal, where we define A as the number of ways that we can cover a $2 \times n$ rectangle with 2×1 rectangles, B as the number of sequences of ones and twos that add up to n, and C as

$$\begin{cases} \binom{m}{0} + \binom{m+1}{2} + \cdots + \binom{2m}{2m} & \text{if } n = 2m, \\ \binom{m+1}{1} + \binom{m+2}{3} + \cdots + \binom{2m+1}{2m+1} & \text{if } n = 2m+1. \end{cases}$$

62. **(MNG 1)** The positive integer n has the property that in any set of n integers chosen from the integers $1, 2, \ldots, 1988$, twenty-nine of them form an arithmetic progression. Prove that $n > 1788$.

63. **(MNG 2)** Let $ABCD$ be a quadrilateral. Let $A'BCD'$ be the reflection of $ABCD$ in BC, while $A''B'CD'$ is the reflection of $A'BCD'$ in CD' and $A''B''C'D'$ is the reflection of $A''B'CD'$ in $D'A''$. Show that if the lines AA'' and BB'' are parallel, then $ABCD$ is a cyclic quadrilateral.

64. **(MNG 3)** Given n points $A_1, A_2, \ldots, A_n$, no three collinear, show that the n-gon $A_1 A_2 \ldots A_n$ can be inscribed in a circle if and only if

$$A_1 A_2 \cdot A_3 A_n \cdots A_{n-1} A_n + A_2 A_3 \cdot A_4 A_n \cdots A_{n-1} A_n \cdot A_1 A_n + \cdots$$
$$+ A_{n-1} A_{n-2} \cdot A_1 A_n \cdots A_{n-3} A_n = A_1 A_{n-1} \cdot A_2 A_n \cdots A_{n-2} A_n.$$

65. **(MNG 4)** (SL88-20).

66. **(MNG 5)** Suppose $\alpha_i > 0$, $\beta_i > 0$ for $1 \leq i \leq n$ $(n > 1)$ and that $\sum_{i=1}^{n} \alpha_i = \sum_{i=1}^{n} \beta_i = \pi$. Prove that

$$\sum_{i=1}^{n} \frac{\cos \beta_i}{\sin \alpha_i} \leq \sum_{i=1}^{n} \cot \alpha_i.$$

67. **(NLD 1)** Given a set of 1988 points in the plane, no three points of the set collinear, the points of a subset with 1788 points are colored blue, and the remaining 200 are colored red. Prove that there exists a line in the plane such that each of the two parts into which the line divides the plane contains 894 blue points and 100 red points.

68. **(NLD 2)** Let S be the set of all sequences $\{a_i \mid 1 \leq i \leq 7, a_i = 0 \text{ or } 1\}$. The distance between two elements $\{a_i\}$ and $\{b_i\}$ of S is defined as $\sum_{i=1}^{7} |a_i - b_i|$. Let T be a subset of S in which any two elements have a distance apart greater than or equal to 3. Prove that T contains at most 16 elements. Give an example of such a subset with 16 elements.

69. **(POL 1)** For a convex polygon P in the plane let P' denote the convex polygon with vertices at the midpoints of the sides of P. Given an integer $n \geq 3$, determine sharp bounds for the ratio $\dfrac{\text{area}(P')}{\text{area}(P)}$ over all convex n-gons P.

70. **(POL 2)** In 3-dimensional space a point O is given and a finite set A of segments with the sum of the lengths equal to 1988. Prove that there exists a plane disjoint from A such that the distance from it to O does not exceed 574.

71. **(POL 3)** Given integers $a_1, \ldots, a_{10}$, prove that there exists a nonzero sequence $(x_1, \ldots, x_{10})$ such that all x_i belong to $\{-1, 0, 1\}$ and the number $\sum_{i=1}^{10} x_i a_i$ is divisible by 1001.

72. **(POL 4)** (SL88-21).

73. **(SGP 1)** In a group of n people each one knows exactly three others. They are seated around a table. We say that the seating is *perfect* if everyone knows the two sitting by their sides. Show that if there is a perfect seating S for the group, then there is always another perfect seating that cannot be obtained from S by rotation or reflection.

74. **(SGP 2)** (SL88-23).

75. **(ESP 1)** Let ABC be a triangle with inradius r and circumradius R. Show that

$$\sin\frac{A}{2}\sin\frac{B}{2} + \sin\frac{B}{2}\sin\frac{C}{2} + \sin\frac{C}{2}\sin\frac{A}{2} \leq \frac{5}{8} + \frac{r}{4R}.$$

76. **(ESP 2)** The quadrilateral $A_1A_2A_3A_4$ is cyclic and its sides are $a_1 = A_1A_2$, $a_2 = A_2A_3$, $a_3 = A_3A_4$, and $a_4 = A_4A_1$. The respective circles with centers I_i and radii ρ_i are tangent externally to each side a_i and to the sides a_{i+1} and a_{i-1} extended $(a_0 = a_4)$. Show that

$$\prod_{i=1}^{4} \frac{a_i}{\rho_i} = 4(\csc A_1 + \csc A_2)^2.$$

77. **(ESP 3)** Consider $h + 1$ chessboards. Number the squares of each board from 1 to 64 in such a way that when the perimeters of any two boards of the collection are brought into coincidence in any possible manner, no two squares in the same position have the same number. What is the maximum value of h?

78. **(SWE 1)** A two-person game is played with nine boxes arranged in a 3×3 square, initially empty, and with white and black stones. At each move a player puts three stones, not necessarily of the same color, in three boxes in either a horizontal or a vertical row. No box can contain stones of different colors: If, for instance, a player puts a white stone in a box containing black stones, the white stone and one of the black stones are removed from the box. The game is over when the center box and the corner boxes each contain one black stone and the other boxes are empty. At one stage of the game x boxes contained one black stone each and the other boxes were empty. Determine all possible values of x.

79. **(SWE 2)** (SL88-24).

80. **(SWE 3)** Let S be an infinite set of integers containing zero and such that the distance between successive numbers never exceeds a given fixed number. Consider the following procedure: Given a set X of integers, we construct a new set consisting of all numbers $x \pm s$, where x belongs to X and s belongs to S.
    Starting from $S_0 = \{0\}$ we successively construct sets $S_1, S_2, S_3, \ldots$ using this procedure. Show that after a finite number of steps we do not obtain any new sets; i.e., $S_k = S_{k_0}$ for $k \geq k_0$.

81. **(USA 1)** There are $n \geq 3$ job openings at a factory, ranked 1 to n in order of increasing pay. There are n job applicants, ranked 1 to n in order of increasing ability. Applicant i is qualified for job j if and only if $i \geq j$.
    The applicants arrive one at a time in random order. Each in turn is hired to the highest-ranking job for which he or she is qualified and that is lower in rank than any job already filled. (Under these rules, job 1 is always filled and hiring terminates thereafter.)
    Show that applicants n and $n-1$ have the same probability of being hired.

82. **(USA 2)** The triangle ABC has a right angle at C. The point P is located on segment AC such that triangles PBA and PBC have congruent inscribed circles. Express the length $x = PC$ in terms of $a = BC$, $b = CA$, and $c = AB$.

83. **(USA 3)** (SL88-29).

84. **(USS 1)** (SL88-30).

85. **(USS 2)** (SL88-31).

86. **(USS 3)** Let a, b, c be integers different from zero. It is known that the equation $ax^2 + by^2 + cz^2 = 0$ has a solution (x, y, z) in integers different from the solution $x = y = z = 0$. Prove that the equation $ax^2 + by^2 + cz^2 = 1$ has a solution in rational numbers.

87. **(USS 4)** All the irreducible positive rational numbers such that the product of the numerator and the denominator is less than 1988 are written in increasing order. Prove that any two adjacent fractions a/b and c/d, $a/b < c/d$, satisfy the equation $bc - ad = 1$.

88. **(USS 5)** There are six circles inside a fixed circle, each tangent to the fixed circle and tangent to the two adjacent smaller circles. If the points of contact between

the six circles and the larger circle are, in order, A_1, A_2, A_3, A_4, A_5, and A_6, prove that

$$A_1A_2 \cdot A_3A_4 \cdot A_5A_6 = A_2A_3 \cdot A_4A_5 \cdot A_6A_1.$$

89. **(VNM 1)** We match sets $\mathcal{M}$ of points in the coordinate plane to sets $\mathcal{M}^*$ according to the rule that (x^*, y^*) belongs to $\mathcal{M}^*$ if and only if $xx^* + yy^* \leq 1$ whenever $(x, y) \in \mathcal{M}$. Find all triangles $\mathcal{Y}$ such that $\mathcal{Y}^*$ is the reflection of $\mathcal{Y}$ at the origin.

90. **(VNM 2)** Does there exist a number α $(0 < \alpha < 1)$ such that there is an infinite sequence $\{a_n\}$ of positive numbers satisfying

$$1 + a_{n+1} \leq a_n + \frac{\alpha}{n} a_n, \quad n = 1, 2, \ldots?$$

91. **(VNM 3)** A regular 14-gon with side length a is inscribed in a circle of radius one. Prove that

$$\frac{2 - a}{2a} > \sqrt{3 \cos \frac{\pi}{7}}.$$

92. **(VNM 4)** Let $p \geq 2$ be a natural number. Prove that there exists an integer n_0 such that

$$\sum_{i=1}^{n_0} \frac{1}{i \sqrt[i]{i+1}} > p.$$

93. **(VNM 5)** Given a natural number n, find all polynomials $P(x)$ of degree less than n satisfying the following condition:

$$\sum_{i=0}^{n} P(i)(-1)^i \binom{n}{i} = 0.$$

94. **(VNM 6)** Let $n + 1$ $(n \geq 1)$ positive integers be given such that for each integer, the set of all prime numbers dividing this integer is a subset of the set of n given prime numbers. Prove that among these $n + 1$ integers one can find numbers (possibly one number) whose product is a perfect square.

3.29.3 Shortlisted Problems

1. **(BGR 1)** An integer sequence is defined by

$$a_n = 2a_{n-1} + a_{n-2} \quad (n > 1), \qquad a_0 = 0, \ a_1 = 1.$$

Prove that 2^k divides a_n if and only if 2^k divides n.

2. **(BGR 3)** Let n be a positive integer. Find the number of odd coefficients of the polynomial

$$u_n(x) = (x^2 + x + 1)^n.$$

3. **(CAN 1)** The triangle ABC is inscribed in a circle. The interior bisectors of the angles A, B, and C meet the circle again at A', B', and C' respectively. Prove that the area of triangle $A'B'C'$ is greater than or equal to the area of triangle ABC.

4. **(CZS 1)** An $n \times n$ chessboard ($n \geq 2$) is numbered by the numbers $1, 2, \ldots, n^2$ (every number occurs once). Prove that there exist two neighboring (which share a common edge) squares such that their numbers differ by at least n.

5. **(CZS 2)**[IMO2] Let n be an even positive integer. Let $A_1, A_2, \ldots, A_{n+1}$ be sets having n elements each such that any two of them have exactly one element in common while every element of their union belongs to at least two of the given sets. For which n can one assign to every element of the union one of the numbers 0 and 1 in such a manner that each of the sets has exactly $n/2$ zeros?

6. **(CZS 3)** In a given tetrahedron $ABCD$ let K and L be the centers of edges AB and CD respectively. Prove that every plane that contains the line KL divides the tetrahedron into two parts of equal volume.

7. **(FRA 2)** Let a be the greatest positive root of the equation $x^3 - 3x^2 + 1 = 0$. Show that $[a^{1788}]$ and $[a^{1988}]$ are both divisible by 17. ($[x]$ denotes the integer part of x.)

8. **(FRA 3)** Let $u_1, u_2, \ldots, u_m$ be m vectors in the plane, each of length less than or equal to 1, which add up to zero. Show that one can rearrange $u_1, u_2, \ldots, u_m$ as a sequence $v_1, v_2, \ldots, v_m$ such that each partial sum $v_1, v_1 + v_2, v_1 + v_2 + v_3, \ldots, v_1 + v_2 + \cdots + v_m$ has length less than or equal to $\sqrt{5}$.

9. **(FRG 1)**[IMO6] Let a and b be two positive integers such that $ab + 1$ divides $a^2 + b^2$. Show that $\frac{a^2+b^2}{ab+1}$ is a perfect square.

10. **(GDR 1)** Let $N = \{1, 2, \ldots, n\}$, $n \geq 2$. A collection $F = \{A_1, \ldots, A_t\}$ of subsets $A_i \subseteq N$, $i = 1, \ldots, t$, is said to be *separating* if for every pair $\{x, y\} \subseteq N$, there is a set $A_i \in F$ such that $A_i \cap \{x, y\}$ contains just one element. A collection F is said to be *covering* if every element of N is contained in at least one set $A_i \in F$. What is the smallest value $f(n)$ of t such that there is a set $F = \{A_1, \ldots, A_t\}$ that is simultaneously separating and covering?

11. **(GDR 3)** The lock on a safe consists of three wheels, each of which may be set in eight different positions. Due to a defect in the safe mechanism the door will open if any two of the three wheels are in the correct position. What is the smallest number of combinations that must be tried if one is to guarantee being able to open the safe (assuming that the "right combination" is not known)?

12. **(HEL 2)** In a triangle ABC, choose any points $K \in BC$, $L \in AC$, $M \in AB$, $N \in LM$, $R \in MK$, and $F \in KL$. If E_1, E_2, E_3, E_4, E_5, E_6, and E denote the areas of the triangles AMR, CKR, BKF, ALF, BNM, CLN, and ABC respectively, show that

$$E \geq 8 \sqrt[6]{E_1 E_2 E_3 E_4 E_5 E_6}.$$

Remark. Points K, L, M, N, R, F lie on segments BC, AC, AB, LM, MK, KL respectively.

13. **(HEL 3)**[IMO5] In a right-angled triangle ABC, let AD be the altitude drawn to the hypotenuse and let the straight line joining the incenters of the triangles

ABD, ACD intersect the sides *AB, AC* at the points *K, L* respectively. If *E* and E_1 denote the areas of the triangles *ABC* and *AKL* respectively, show that $\frac{E}{E_1} \geq 2$.

14. **(HUN 1)** For what values of *n* does there exist an $n \times n$ array of entries $-1, 0$, or 1 such that the $2n$ sums obtained by summing the elements of the rows and the columns are all different?

15. **(ISL 1)** Let *ABC* be an acute-angled triangle. Three lines L_A, L_B, and L_C are constructed through the vertices *A*, *B*, and *C* respectively according to the following prescription: Let *H* be the foot of the altitude drawn from the vertex *A* to the side *BC*; let S_A be the circle with diameter *AH*; let S_A meet the sides *AB* and *AC* at *M* and *N* respectively, where *M* and *N* are distinct from *A*; then L_A is the line through *A* perpendicular to *MN*. The lines L_B and L_C are constructed similarly. Prove that L_A, L_B, and L_C are concurrent.

16. **(IRL 1)**[IMO4] Show that the solution set of the inequality

$$\sum_{k=1}^{70} \frac{k}{x-k} \geq \frac{5}{4}$$

is a union of disjoint intervals the sum of whose lengths is 1988.

17. **(ISR 2)** In the convex pentagon *ABCDE*, the sides *BC, CD, DE* have the same length. Moreover, each diagonal of the pentagon is parallel to a side (*AC* is parallel to *DE*, *BD* is parallel to *AE*, etc.). Prove that *ABCDE* is a regular pentagon.

18. **(LUX 1)**[IMO1] Consider two concentric circles of radii *R* and *r* $(R > r)$ with center *O*. Fix *P* on the small circle and consider the variable chord *PA* of the small circle. Points *B* and *C* lie on the large circle; *B, P, C* are collinear and *BC* is perpendicular to *AP*.
   (a) For what value(s) of ∠*OPA* is the sum $BC^2 + CA^2 + AB^2$ extremal?
   (b) What are the possible positions of the midpoints *U* of *BA* and *V* of *AC* as ∠*OPA* varies?

19. **(MEX 1)** Let $f(n)$ be a function defined on the set of all positive integers and having its values in the same set. Suppose that $f(f(m) + f(n)) = m + n$ for all positive integers n, m. Find all possible values for $f(1988)$.

20. **(MNG 4)** Find the least natural number *n* such that if the set $\{1, 2, \ldots, n\}$ is arbitrarily divided into two nonintersecting subsets, then one of the subsets contains three distinct numbers such that the product of two of them equals the third.

21. **(POL 4)** Forty-nine students solve a set of three problems. The score for each problem is a whole number of points from 0 to 7. Prove that there exist two students *A* and *B* such that for each problem, *A* will score at least as many points as *B*.

22. **(KOR 2)** Let *p* be the product of two consecutive integers greater than 2. Show that there are no integers $x_1, x_2, \ldots, x_p$ satisfying the equation

$$\sum_{i=1}^{p} x_i^2 - \frac{4}{4p+1} \left(\sum_{i=1}^{p} x_i \right)^2 = 1.$$

Alternative formulation. Show that there are only two values of p for which there are integers $x_1, x_2, \ldots, x_p$ satisfying the above inequality.

23. **(SGP 2)** Let Q be the center of the inscribed circle of a triangle ABC. Prove that for any point P,

$$a(PA)^2 + b(PB)^2 + c(PC)^2 = a(QA)^2 + b(QB)^2 + c(QC)^2 + (a+b+c)(QP)^2,$$

where $a = BC$, $b = CA$, and $c = AB$.

24. **(SWE 2)** Let $\{a_k\}_1^{\infty}$ be a sequence of nonnegative real numbers such that $a_k - 2a_{k+1} + a_{k+2} \geq 0$ and $\sum_{j=1}^{k} a_j \leq 1$ for all $k = 1, 2, \ldots$. Prove that $0 \leq (a_k - a_{k+1}) < \frac{2}{k^2}$ for all $k = 1, 2, \ldots$.

25. **(UNK 1)** A positive integer is called a *double number* if its decimal representation consists of a block of digits, not commencing with 0, followed immediately by an identical block. For instance, 360360 is a double number, but 36036 is not. Show that there are infinitely many double numbers that are perfect squares.

26. **(UNK 2)**[IMO3] A function f defined on the positive integers (and taking positive integer values) is given by

$$\begin{aligned}
f(1) &= 1, \quad f(3) = 3, \\
f(2n) &= f(n), \\
f(4n+1) &= 2f(2n+1) - f(n), \\
f(4n+3) &= 3f(2n+1) - 2f(n),
\end{aligned}$$

for all positive integers n. Determine with proof the number of positive integers less than or equal to 1988 for which $f(n) = n$.

27. **(UNK 4)** The triangle ABC is acute-angled. Let L be any line in the plane of the triangle and let u, v, w be the lengths of the perpendiculars from A, B, C respectively to L. Prove that

$$u^2 \tan A + v^2 \tan B + w^2 \tan C \geq 2\Delta,$$

where Δ is the area of the triangle, and determine the lines L for which equality holds.

28. **(UNK 5)** The sequence $\{a_n\}$ of integers is defined by $a_1 = 2$, $a_2 = 7$, and

$$-\frac{1}{2} < a_{n+1} - \frac{a_n^2}{a_{n-1}} \leq \frac{1}{2}, \quad \text{for } n \geq 2.$$

Prove that a_n is odd for all $n > 1$.

29. **(USA 3)** A number of signal lights are equally spaced along a one-way railroad track, labeled in order $1, 2, \ldots, N$ ($N \geq 2$). As a safety rule, a train is not allowed to pass a signal if any other train is in motion on the length of track between it and the following signal. However, there is no limit to the number of trains that can be parked motionless at a signal, one behind the other. (Assume that the trains have zero length.)

A series of K freight trains must be driven from Signal 1 to Signal N. Each train travels at a distinct but constant speed (i.e., the speed is fixed and different from that of each of the other trains) at all times when it is not blocked by the safety rule. Show that regardless of the order in which the trains are arranged, the same time will elapse between the first train's departure from Signal 1 and the last train's arrival at Signal N.

30. **(USS 1)** A point M is chosen on the side AC of the triangle ABC in such a way that the radii of the circles inscribed in the triangles ABM and BMC are equal. Prove that

$$BM^2 = \Delta \cot \frac{B}{2},$$

where Δ is the area of the triangle ABC.

31. **(USS 2)** Around a circular table an even number of persons have a discussion. After a break they sit again around the circular table in a different order. Prove that there are at least two people such that the number of participants sitting between them before and after the break is the same.

3.30 The Thirtieth IMO
Braunschweig–Niedersachen, FR Germany, July 13–24, 1989

3.30.1 Contest Problems

First Day (July 18)

1. Prove that the set $\{1,2,\ldots,1989\}$ can be expressed as the disjoint union of 17 subsets $A_1, A_2, \ldots, A_{17}$ such that:
   (i) each A_i contains the same number of elements;
   (ii) the sum of all elements of each A_i is the same for $i = 1, 2, \ldots, 17$.

2. Let ABC be a triangle. The bisector of angle A meets the circumcircle of triangle ABC in A_1. Points B_1 and C_1 are defined similarly. Let AA_1 meet the lines that bisect the two external angles at B and C in point A^0. Define B^0 and C^0 similarly. If $S_{X_1 X_2 \ldots X_n}$ denotes the area of the polygon $X_1 X_2 \ldots X_n$, prove that

$$S_{A^0 B^0 C^0} = 2 S_{AC_1 BA_1 CB_1} \geq 4 S_{ABC}.$$

3. Given a set S in the plane containing n points and satisfying the conditions
   (i) no three points of S are collinear,
   (ii) for every point P of S there exist at least k points in S that have the same distance to P,
   prove that the following inequality holds:

$$k < \frac{1}{2} + \sqrt{2n}.$$

Second Day (July 19)

4. The quadrilateral $ABCD$ has the following properties:
   (i) $AB = AD + BC$;
   (ii) there is a point P inside it at a distance x from the side CD such that $AP = x + AD$ and $BP = x + BC$.
   Show that

$$\frac{1}{\sqrt{x}} \geq \frac{1}{\sqrt{AD}} + \frac{1}{\sqrt{BC}}.$$

5. For which positive integers n does there exist a positive integer N such that none of the integers $1+N, 2+N, \ldots, n+N$ is the power of a prime number?

6. We consider permutations $(x_1, \ldots, x_{2n})$ of the set $\{1, \ldots, 2n\}$ such that $|x_i - x_{i+1}| = n$ for at least one $i \in \{1, \ldots, 2n-1\}$. For every natural number n, find out whether permutations with this property are more or less numerous than the remaining permutations of $\{1, \ldots, 2n\}$.

3.30.2 Longlisted Problems

1. **(AUS 1)** In the set $S_n = \{1, 2, \ldots, n\}$ a new multiplication $a * b$ is defined with the following properties:
    (i) $c = a * b$ is in S_n for any $a \in S_n, b \in S_n$.
    (ii) If the ordinary product $a \cdot b$ is less than or equal to n, then $a * b = a \cdot b$.
    (iii) The ordinary rules of multiplication hold for $*$, i.e.,
        (1) $a * b = b * a$ (commutativity)
        (2) $(a * b) * c = a * (b * c)$ (associativity)
        (3) If $a * b = a * c$ then $b = c$ (cancellation law).
    Find a suitable multiplication table for the new product for $n = 11$ and $n = 12$.

2. **(AUS 2)** (SL89-1).

3. **(AUS 3)** (SL89-2).

4. **(AUS 4)** (SL89-3).

5. **(BGR 1)** The sequences $a_0, a_1, \ldots$ and $b_0, b_1, \ldots$ are defined by the equalities

$$a_0 = \frac{\sqrt{2}}{2}, \qquad a_{n+1} = \frac{\sqrt{2}}{2}\sqrt{1 - \sqrt{1 - a_n^2}}, \qquad n = 0, 1, 2, \ldots$$

and

$$b_0 = 1, \qquad b_{n+1} = \frac{\sqrt{1 + b_n^2} - 1}{b_n}, \qquad n = 0, 1, 2, \ldots.$$

Prove the inequalities

$$2^{n+2} a_n < \pi < 2^{n+2} b_n, \qquad \text{for every } n = 0, 1, 2, \ldots.$$

6. **(BGR 2)** The circles c_1 and c_2 are tangent at the point A. A straight line l through A intersects c_1 and c_2 at points C_1 and C_2 respectively. A circle c, which contains C_1 and C_2, meets c_1 and c_2 at points B_1 and B_2 respectively. Let κ be the circle circumscribed around triangle AB_1B_2. The circle k tangent to κ at the point A meets c_1 and c_2 at the points D_1 and D_2 respectively. Prove that
    (a) the points C_1, C_2, D_1, D_2 are concyclic or collinear;
    (b) the points B_1, B_2, D_1, D_2 are concyclic if and only if AC_1 and AC_2 are diameters of c_1 and c_2.

7. **(BGR 3)** (SL89-4).

8. **(COL 1)** (SL89-5).

9. **(COL 2)** Let m be a positive integer and define $f(m)$ to be the number of factors of 2 in $m!$ (that is, the greatest positive integer k such that $2^k \mid m!$). Prove that there are infinitely many positive integers m such that $m - f(m) = 1989$.

10. **(CUB 1)** Given the equation

$$4x^3 + 4x^2 y - 15xy^2 - 18y^3 - 12x^2 + 6xy + 36y^2 + 5x - 10y = 0,$$

find all positive integer solutions.

11. (**CUB 2**) Given the equation

$$y^4 + 4y^2x - 11y^2 + 4xy - 8y + 8x^2 - 40x + 52 = 0,$$

find all real solutions.

12. (**CUB 3**) Let $P(x)$ be a polynomial such that the following inequalities are satisfied:

$$P(0) > 0;$$
$$P(1) > P(0);$$
$$P(2) > 2P(1) - P(0);$$
$$P(3) > 3P(2) - 3P(1) + P(0);$$

and also for every natural number n, $P(n+4) > 4P(n+3) - 6P(n+2) + 4P(n+1) - P(n)$. Prove that for every positive natural number n, $P(n)$ is positive.

13. (**CUB 4**) Let n be a natural number not greater than 44. Prove that for any function f defined over $\mathbb{N}^2$ whose images are in the set $\{1, 2, \ldots, n\}$, there are four ordered pairs $(i, j), (i, k), (l, j)$, and (l, k) such that $f(i, j) = f(i, k) = f(l, j) = f(l, k)$, where i, j, k, l are chosen in such a way that there are natural numbers n, p that satisfy

$$1989m \le i < l < 1989 + 1989m, \qquad 1989p \le j < k < 1989 + 1989p.$$

14. (**CZS 1**) (SL89-6).

15. (**CZS 2**) A sequence $a_1, a_2, a_3, \ldots$ is defined recursively by $a_1 = 1$ and $a_{2^k+j} = -a_j$ ($j = 1, 2, \ldots, 2^k$). Prove that this sequence is not periodic.

16. (**FIN 1**) (SL89-7).

17. (**FIN 2**) Let a, $0 < a < 1$, be a real number and f a continuous function on $[0, 1]$ satisfying $f(0) = 0$, $f(1) = 1$, and

$$f\left(\frac{x+y}{2}\right) = (1-a)f(x) + af(y)$$

for all $x, y \in [0, 1]$ with $x \le y$. Determine $f(1/7)$.

18. (**FIN 3**) There are some boys and girls sitting in an $n \times n$ quadratic array. We know the number of girls in every column and row and every line parallel to the diagonals of the array. For which n is this information sufficient to determine the exact positions of the girls in the array? For which seats can we say for sure that a girl sits there or not?

19. (**FRA 1**) Let $a_1, \ldots, a_n$ be distinct positive integers that do not contain a 9 in their decimal representations. Prove that

$$\frac{1}{a_1} + \cdots + \frac{1}{a_n} \le 30.$$

20. (**FRA 2**) (SL89-8).

21. **(FRA 2b)** Same problem as previous, but with a rectangular parallelepiped having at least one integral side.

22. **(FRA 3)** Let ABC be an equilateral triangle with side length equal to a natural number N. Consider the set S of all points M inside the triangle ABC such that $\overrightarrow{AM} = \frac{1}{N}(n\overrightarrow{AB} + m\overrightarrow{AC})$, where m, n are integers and $0 \leq m, n, m+n \leq N$. Every point of S is colored in one of the three colors blue, white, red such that no point on AB is colored blue, no point on AC is colored white, and no point on BC is colored red. Prove that there exists an equilateral triangle with vertices in S and side length 1 whose three vertices are colored blue, white, and red.

23. **(FRA 3b)** Like the previous problem, but with a regular tetrahedron and four different colors used.

24. **(FRA 4)** (SL89-9).

25. **(UNK 1)** Let ABC be a triangle. Prove that there is a unique point U in the plane of ABC such that there exist real numbers $\lambda, \mu, \nu, \kappa$, not all zero, such that

$$\lambda PL^2 + \mu PM^2 + \nu PN^2 - \kappa UP^2$$

is constant for all points P of the plane, where L, M, N are the feet of the perpendiculars from P to BC, CA, AB respectively.

26. **(UNK 2)** Let a, b, c, d be positive integers such that $ab = cd$ and $a+b = c-d$. Prove that there exists a right-angled triangle the measures of whose sides (in some unit) are integers and whose area measure is ab square units.

27. **(UNK 3)** Integers $c_{m,n}$ $(m \geq 0, n \geq 0)$ are defined by $c_{m,0} = 1$ for all $m \geq 0$, $c_{0,n} = 1$ for all $n \geq 0$, and $c_{m,n} = c_{m-1,n} - nc_{m-1,n-1}$ for all $m > 0, n > 0$. Prove that $c_{m,n} = c_{n,m}$ for all $m \geq 0, n \geq 0$.

28. **(UNK 4)** Let $b_1, b_2, \ldots, b_{1989}$ be positive real numbers such that the equations
$$x_{r-1} - 2x_r + x_{r+1} + b_r x_r = 0 \quad (1 \leq r \leq 1989)$$
have a solution with $x_0 = x_{1990} = 0$ but not all of $x_1, \ldots, x_{1989}$ are equal to zero. Prove that
$$b_1 + b_2 + \cdots + b_{1989} \geq \frac{2}{995}.$$

29. **(HEL 1)** Let L denote the set of all lattice points of the plane (points with integral coordinates). Show that for any three points A, B, C of L there is a fourth point D, different from A, B, C, such that the interiors of the segments AD, BD, CD contain no points of L. Is the statement true if one considers four points of L instead of three?

30. **(HEL 2)** In a triangle ABC for which $6(a+b+c)r^2 = abc$, we consider a point M on the inscribed circle and the projections D, E, F of M on the sides BC, AC, and AB respectively. Let S, S_1 denote the areas of the triangles ABC and DEF respectively. Find the maximum and minimum values of the quotient $\frac{S}{S_1}$ (here r denotes the inradius of ABC and, as usual, $a = BC$, $b = AC$, $c = AB$).

31. **(HEL 3)** (SL89-10).

32. **(HKG 1)** Let ABC be an equilateral triangle. Let D, E, F, M, N, and P be the midpoints of BC, CA, AB, FD, FB, and DC respectively.
    (a) Show that the line segments AM, EN, and FP are concurrent.
    (b) Let O be the point of intersection of AM, EN, and FP. Find $OM : OF : ON : OE : OP : OA$.

33. **(HKG 2)** Let n be a positive integer. Show that $(\sqrt{2}+1)^n = \sqrt{m} + \sqrt{m-1}$ for some positive integer m.

34. **(HKG 3)** Given an acute triangle find a point inside the triangle such that the sum of the distances from this point to the three vertices is the least.

35. **(HKG 4)** Find all square numbers S_1 and S_2 such that $S_1 - S_2 = 1989$.

36. **(HKG 5)** Prove the identity

$$1 + \frac{1}{2} - \frac{2}{3} + \frac{1}{4} + \frac{1}{5} - \frac{2}{6} + \cdots + \frac{1}{478} + \frac{1}{479} - \frac{2}{480} = 2 \sum_{k=0}^{159} \frac{641}{(161+k)(480-k)}.$$

37. **(HUN 1)** (SL89-11).

38. **(HUN 2)** Connecting the vertices of a regular n-gon we obtain a closed (not necessarily convex) n-gon. Show that if n is even, then there are two parallel segments among the connecting segments and if n is odd then there cannot be exactly two parallel segments.

39. **(HUN 3)** (SL89-12).

40. **(ISL 1)** A sequence of real numbers $x_0, x_1, x_2, \ldots$ is defined as follows: $x_0 = 1989$ and for each $n \geq 1$

$$x_n = -\frac{1989}{n} \sum_{k=0}^{n-1} x_k.$$

Calculate the value of $\sum_{n=0}^{1989} 2^n x_n$.

41. **(ISL 2)** Alice has two urns. Each urn contains four balls and on each ball a natural number is written. She draws one ball from each urn at random, notes the sum of the numbers written on them, and replaces the balls in the urns from which she took them. This she repeats a large number of times. Bill, on examining the numbers recorded, notices that the frequency with which each sum occurs is the same as if it were the sum of two natural numbers drawn at random from the range 1 to 4. What can he deduce about the numbers on the balls?

42. **(ISL 3)** (SL89-13).

43. **(IDN 1)** Let $f(x) = a\sin^2 x + b\sin x + c$, where a, b, and c are real numbers. Find all values of a, b, and c such that the following three conditions are satisfied simultaneously:
    (i) $f(x) = 381$ if $\sin x = 1/2$.
    (ii) The absolute maximum of $f(x)$ is 444.

(iii) The absolute minimum of $f(x)$ is 364.

44. **(IDN 2)** Let A and B be fixed distinct points on the X axis, none of which coincides with the origin $O(0,0)$, and let C be a point on the Y axis of an orthogonal Cartesian coordinate system. Let g be a line through the origin $O(0,0)$ and perpendicular to the line AC. Find the locus of the point of intersection of the lines g and BC as C varies along the Y axis. (Give an equation and a description of the locus.)

45. **(IDN 3)** The expressions $a+b+c$, $ab+ac+bc$, and abc are called the elementary symmetric expressions on the three letters a,b,c; symmetric because if we interchange any two letters, say a and c, the expressions remain algebraically the same. The common degree of its terms is called the order of the expression.
Let $S_k(n)$ denote the elementary expression on k different letters of order n; for example $S_4(3) = abc + abd + acd + bcd$. There are four terms in $S_4(3)$. How many terms are there in $S_{9891}(1989)$? (Assume that we have 9891 different letters.)

46. **(IDN 4)** Given two distinct numbers b_1 and b_2, their product can be formed in two ways: $b_1 \times b_2$ and $b_2 \times b_1$. Given three distinct numbers, b_1, b_2, b_3, their product can be formed in twelve ways: $b_1 \times (b_2 \times b_3)$; $(b_1 \times b_2) \times b_3$; $b_1 \times (b_3 \times b_2)$; $(b_1 \times b_3) \times b_2$; $b_2 \times (b_1 \times b_3)$; $(b_2 \times b_1) \times b_3$; $b_2 \times (b_3 \times b_1)$; $(b_2 \times b_3) \times b_1$; $b_3 \times (b_1 \times b_2)$; $(b_3 \times b_1) \times b_2$; $b_3 \times (b_2 \times b_1)$; $(b_3 \times b_2) \times b_1$. In how many ways can the product of n distinct letters be formed?

47. **(IDN 5)** Let $\log_2^2 x - 4\log_2 x - m^2 - 2m - 13 = 0$ be an equation in x. Prove:
    (a) For any real value of m the equation has has two distinct solutions.
    (b) The product of the solutions of the equation does not depend on m.
    (c) One of the solutions of the equation is less than 1, while the other solution is greater than 1.
Find the minimum value of the larger solution and the maximum value of the smaller solution.

48. **(IDN 6)** Let S be the point of intersection of the two lines $l_1 : 7x - 5y + 8 = 0$ and $l_2 : 3x + 4y - 13 = 0$. Let $P = (3,7)$, $Q = (11,13)$, and let A and B be points on the line PQ such that P is between A and Q, and B is between P and Q, and such that $PA/AQ = PB/BQ = 2/3$. Without finding the coordinates of B find the equations of the lines SA and SB.

49. **(IND 1)** Let A,B denote two distinct fixed points in space. Let X,P denote variable points (in space), while K,N,n denote positive integers. Call (X,K,N,P) *admissible* if $(N-K)PA + K \cdot PB \geq N \cdot PX$. Call (X,K,N) admissible if (X,K,N,P) is admissible for all choices of P. Call (X,N) admissible if (X,K,N) is admissible for some choice of K in the interval $0 < K < N$. Finally, call X admissible if (X,N) is admissible for some choice of N ($N > 1$). Determine:
    (a) the set of admissible X;
    (b) the set of X for which $(X, 1989)$ is admissible but not (X,n), $n < 1989$.

50. **(IND 2)** (SL89-14).

51. **(IND 3)** Let $t(n)$, for $n = 3,4,5,\ldots$, represent the number of distinct, incongruent, integer-sided triangles whose perimeter is n; e.g., $t(3) = 1$. Prove that

$$t(2n-1) - t(2n) = \left[\frac{n}{6}\right] \text{ or } \left[\frac{n}{6}+1\right].$$

52. **(IRL 1)** (SL89-15).

53. **(IRL 2)** Let $f(x) = (x-a_1)(x-a_2)\cdots(x-a_n) - 2$, where $n \geq 3$ and $a_1, a_2, \ldots, a_n$ are distinct integers. Suppose that $f(x) = g(x)h(x)$, where $g(x), h(x)$ are both nonconstant polynomials with integer coefficients. Prove that $n = 3$.

54. **(IRL 3)** Let f be a function from the real numbers to the real numbers such that $f(1) = 1$, $f(a+b) = f(a) + f(b)$ for all a, b, and $f(x)f(1/x) = 1$ for all $x \neq 0$. Prove that $f(x) = x$ for all real numbers x.

55. **(IRL 4)** Let $[x]$ denote the greatest integer less than or equal to x. Let α be the positive root of the equation $x^2 - 1989x - 1 = 0$. Prove that there exist infinitely many natural numbers n that satisfy the equation

$$[\alpha n + 1989\alpha[\alpha n]] = 1989n + (1989^2+1)[\alpha n].$$

56. **(IRL 5)** Let $n = 2k - 1$, where $k \geq 6$ is an integer. Let T be the set of all n-tuples $(x_1, x_2, \ldots, x_n)$ where x_i is 0 or 1 $(i = 1, 2, \ldots, n)$. For $\mathbf{x} = (x_1, \ldots, x_n)$ and $\mathbf{y} = (y_1, \ldots, y_n)$ in T, let $d(\mathbf{x}, \mathbf{y})$ denote the number of integers j with $1 \leq j \leq n$ such that $x_j \neq y_j$. (In particular $d(\mathbf{x}, \mathbf{x}) = 0$.) Suppose that there exists a subset S of T with 2^k elements that has the following property: Given any element $\mathbf{x}$ in T, there is a unique element $\mathbf{y}$ in S with $d(\mathbf{x}, \mathbf{y}) \leq 3$. Prove that $n = 23$.

57. **(ISR 1)** (SL89-16).

58. **(ISR 2)** Let $P_1(x), P_2(x), \ldots, P_n(x)$ be polynomials with real coefficients. Show that there exist real polynomials $A_r(x), B_r(x)$ $(r = 1, 2, 3)$ such that

$$\begin{aligned}
\Sigma_{s=1}^{n}(P_s(x))^2 &= (A_1(x))^2 + (B_1(x))^2 \\
&= (A_2(x))^2 + x(B_2(x))^2 \\
&= (A_3(x))^2 - x(B_3(x))^2.
\end{aligned}$$

59. **(ISR 3)** Let $v_1, v_2, \ldots, v_{1989}$ be a set of coplanar vectors with $|v_r| \leq 1$ for $1 \leq r \leq 1989$. Show that it is possible to find ε_r $(1 \leq r \leq 1989)$, each equal to ± 1, such that

$$\left|\sum_{r=1}^{1989} \varepsilon_r v_r\right| \leq \sqrt{3}.$$

60. **(KOR 1)** A real-valued function f on $\mathbb{Q}$ satisfies the following conditions for arbitrary $\alpha, \beta \in \mathbb{Q}$:

(i) $f(0) = 0$,
(iii) $f(\alpha\beta) = f(\alpha)f(\beta)$,
(v) $f(m) \leq 1989$ for all $m \in \mathbb{Z}$.

(ii) $f(\alpha) > 0$ if $\alpha \neq 0$,
(iv) $f(\alpha+\beta) \leq f(\alpha) + f(\beta)$,

Prove that $f(\alpha+\beta) = \max\{f(\alpha), f(\beta)\}$ if $f(\alpha) \neq f(\beta)$.

Here, $\mathbb{Z}, \mathbb{Q}$ denote the sets of integers and rational numbers, respectively.

61. **(KOR 2)** Let A be a set of positive integers such that no positive integer greater than 1 divides all the elements of A. Prove that any sufficiently large positive integer can be written as a sum of elements of A. (Elements may occur several times in the sum.)

62. **(KOR 3)** (SL89-25).

63. **(KOR 4)** (SL89-26).

64. **(KOR 5)** Let a regular $(2n+1)$-gon be inscribed in a circle of radius r. We consider all the triangles whose vertices are from those of the regular $(2n+1)$-gon.

   (a) How many triangles among them contain the center of the circle in their interior?

   (b) Find the sum of the areas of all those triangles that contain the center of the circle in their interior.

65. **(LUX 1)** A regular n-gon $A_1 A_2 A_3 \ldots A_k \ldots A_n$ inscribed in a circle of radius R is given. If S is a point on the circle, calculate $T = SA_1^2 + SA_2^2 + \cdots + SA_n^2$.

66. **(MNG 1)** (SL89-17).

67. **(MNG 2)** A family of sets $A_1, A_2, \ldots, A_n$ has the following properties:

   (i) Each A_i contains 30 elements.

   (ii) $A_i \cap A_j$ contains exactly one element for all i, j, $1 \leq i < j \leq 30$.

Find the largest possible n if the intersection of all these sets is empty.

68. **(MNG 3)** If $0 < k \leq 1$ and a_i are positive real numbers, $i = 1, 2, \ldots, n$, prove that

$$\left(\frac{a_1}{a_2 + \cdots + a_n} \right)^k + \cdots + \left(\frac{a_n}{a_1 + \cdots + a_{n-1}} \right)^k \geq \frac{n}{(n-1)^k}.$$

69. **(MNG 4)** (SL89-18).

70. **(MNG 5)** Three mutually nonparallel lines l_i $(i = 1, 2, 3)$ are given in a plane. The lines l_i determine a triangle and reflections f_i with axes on lines l_i. Prove that for every point of the plane, there exists a finite composition of the reflections f_i that maps that point to a point interior to the triangle.

71. **(MNG 6)** (SL89-19).

72. **(MAR 1)** Let $ABCD$ be a quadrilateral inscribed in a circle with diameter AB such that $BC = a$, $CD = 2a$, $DA = \frac{3\sqrt{5}-1}{2}a$. For each point M on the semicircle AB not containing C and D, denote by h_1, h_2, h_3 the distances from M to the sides BC, CD, and DA. Find the maximum of $h_1 + h_2 + h_3$.

73. **(NLD 1)** (SL89-20).

74. **(NLD 2)** (SL89-21).

75. **(PHI 1)** (SL89-22).

76. **(PHI 2)** Let k and s be positive integers. For sets of real numbers $\{\alpha_1, \alpha_2, \ldots, \alpha_s\}$ and $\{\beta_1, \beta_2, \ldots, \beta_s\}$ that satisfy $\sum_{i=1}^{s} \alpha_i^j = \sum_{i=1}^{s} \beta_i^j$ for each $j = 1, 2, \ldots, k$, we write

$$\{\alpha_1, \alpha_2, \ldots, \alpha_s\} =_k \{\beta_1, \beta_2, \ldots, \beta_s\}.$$

Prove that if $\{\alpha_1, \alpha_2, \ldots, \alpha_s\} =_k \{\beta_1, \beta_2, \ldots, \beta_s\}$ and $s \leq k$, then there exists a permutation π of $\{1, 2, \ldots, s\}$ such that $\beta_i = \alpha_{\pi(i)}$ for $i = 1, 2, \ldots, s$.

77. **(POL 1)** Given that

$$\frac{\cos x + \cos y + \cos z}{\cos(x+y+z)} = \frac{\sin x + \sin y + \sin z}{\sin(x+y+z)} = a,$$

show that

$$\cos(y+z) + \cos(z+x) + \cos(x+y) = a.$$

78. **(POL 2)** (SL89-23).

   Alternative formulation. Two identical packs of n different cards are shuffled together; all arrangements are equiprobable. The cards are then laid face up, one at a time. For every natural number n, find out which is more probable, that at least one pair of identical cards will appear in immediate succession or that there will be no such pair.

79. **(POL 3)** To each pair (x, y) of distinct elements of a finite set X a number $f(x, y)$ equal to 0 or 1 is assigned in such a way that $f(x, y) \neq f(y, x)$ for all x, y $(x \neq y)$. Prove that exactly one of the following situations occurs:
   (i) X is the union of two disjoint nonempty subsets U, V such that $f(u, v) = 1$ for every $u \in U, v \in V$.
   (ii) The elements of X can be labeled $x_1, \ldots, x_n$ so that $f(x_1, x_2) = f(x_2, x_3) = \cdots = f(x_{n-1}, x_n) = f(x_n, x_1) = 1$.

   Alternative formulation. In a tournament of n participants, each pair plays one game (no ties). Prove that exactly one of the following situations occurs:
   (i) The league can be partitioned into two nonempty groups such that each player in one of these groups has won against each player of the other.
   (ii) All participants can be ranked 1 through n so that ith player wins the game against the $(i+1)$st and the nth player wins against the first.

80. **(POL 4)** We are given a finite collection of segments in the plane, of total length 1. Prove that there exists a line ℓ such that the sum of the lengths of the projections of the given segments to the line ℓ is less than $2/\pi$.

81. **(POL 5)** (SL89-24).

82. **(POR 1)** Solve in the set of real numbers the equation $3x^3 - [x] = 3$, where $[x]$ denotes the integer part of x.

83. **(POR 2)** Poldavia is a strange kingdom. Its currency unit is the bourbaki and there exist only two types of coins: gold ones and silver ones. Each gold coin

is worth n bourbakis and each silver coin is worth m bourbakis (n and m are positive integers). Using gold and silver coins, it is possible to obtain sums such as 10000 bourbakis, 1875 bourbakis, 3072 bourbakis, and so on. But Poldavia's monetary system is not as strange as it seems:

   (a) Prove that it is possible to buy anything that costs an integral number of bourbakis, as long as one can receive change.
   (b) Prove that any payment above $mn - 2$ bourbakis can be made without the need to receive change.

84. **(POR 3)** Let a, b, c, r, and s be real numbers. Show that if r is a root of $ax^2 + bx + c = 0$ and s is a root of $-ax^2 + bx + c = 0$, then $\frac{a}{2}x^2 + bx + c = 0$ has a root between r and s.

85. **(POR 4)** Let $P(x)$ be a polynomial with integer coefficients such that $P(m_1) = P(m_2) = P(m_3) = P(m_4) = 7$ for given distinct integers m_1, m_2, m_3, and m_4. Show that there is no integer m such that $P(m) = 14$.

86. **(POR 5)** Given two natural numbers w and n, the tower of n w's is the natural number $T_n(w)$ defined by

$$T_n(w) = w^{w^{\cdot^{\cdot^{\cdot^{w}}}}},$$

with n w's on the right side. More precisely, $T_1(w) = w$ and $T_{n+1}(w) = w^{T_n(w)}$. For example, $T_3(2) = 2^{2^2} = 16$, $T_4(2) = 2^{16} = 65536$, and $T_2(3) = 3^3 = 27$. Find the smallest tower of 3's that exceeds the tower of 1989 2's. In other words, find the smallest value of n such that $T_n(3) > T_{1989}(2)$. Justify your answer.

87. **(POR 6)** A balance has a left pan, a right pan, and a pointer that moves along a graduated ruler. Like many other grocer balances, this one works as follows: An object of weight L is placed in the left pan and another of weight R in the right pan, the pointer stops at the number $R - L$ on the graduated ruler. There are n (≥ 2) bags of coins, each containing $\frac{n(n-1)}{2} + 1$ coins. All coins look the same (shape, color, and so on). Of the bags, $n - 1$ contain genuine coins, all with the same weight. The remaining bag (we don't know which one it is) contains counterfeit coins. All counterfeit coins have the same weight, and this weight is different from the weight of the genuine coins. A legal weighing consists of placing a certain number of coins in one of the pans, putting a certain number of coins in the other pan, and reading the number given by the pointer in the graduated ruler. With just two legal weighings it is possible to identify the bag containing counterfeit coins. Find a way to do this and explain it.

88. **(ROU 1)** (SL89-27).

89. **(ROU 2)** (SL89-28).

90. **(ROU 3)** Prove that the sequence $(a_n)_{n \geq 0}$, $a_n = \lfloor n\sqrt{2} \rfloor$, contains an infinite number of perfect squares.

91. **(ROU 4)** (SL89-29).

92. **(ROU 5)** Find the set of all $a \in \mathbb{R}$ for which there is no infinite sequence $(x_n)_{n \geq 0} \subset \mathbb{R}$ satisfying $x_0 = a$,

$$x_{n+1} = \frac{x_n + \alpha}{\beta x_n + 1}, \quad n = 0, 1, \ldots, \quad \text{where } \alpha\beta > 0.$$

93. **(ROU 6)** For $\Phi : \mathbb{N} \to \mathbb{Z}$ let us define

$$M_\Phi = \{f : \mathbb{N} \to \mathbb{Z}; f(x) > f(\Phi(x)), \forall x \in \mathbb{N}\}.$$

(a) Prove that if $M_{\Phi_1} = M_{\Phi_2} \neq \emptyset$, then $\Phi_1 = \Phi_2$.
(b) Does this property remain true if

$$M_\Phi = \{f : \mathbb{N} \to \mathbb{N}; f(x) > f(\Phi(x)), \forall x \in \mathbb{N}\}?$$

94. **(SWE 1)** Prove that $a < b$ implies that $a^3 - 3a \leq b^3 - 3b + 4$. When does equality occur?

95. **(SWE 2)** (SL89-30).

96. **(SWE 3)** (SL89-31).

97. **(THA 1)** Let n be a positive integer, $X = \{1, 2, \ldots, n\}$, and k a positive integer such that $n/2 \leq k \leq n$. Determine, with proof, the number of all functions $f : X \to X$ that satisfy the following conditions:
   (i) $f^2 = f$;
   (ii) the number of elements in the image of f is k;
   (iii) for each y in the image of f, the number of all points x in X such that $f(x)=y$ is at most 2.

98. **(THA 2)** Let $f : \mathbb{N} \to \mathbb{N}$ be such that
   (i) f is strictly increasing;
   (ii) $f(mn) = f(m)f(n) \ \forall m, n \in \mathbb{N}$; and
   (iii) if $m \neq n$ and $m^n = n^m$, then $f(m) = n$ or $f(n) = m$.
   Determine $f(30)$.

99. **(THA 3)** An arithmetic function is a real-valued function whose domain is the set of positive integers. Define the convolution product of two arithmetic functions f and g to be the arithmetic function $f \star g$, where

$$(f \star g)(n) = \sum_{ij=n} f(i)g(j), \quad \text{and} \quad f^{\star k} = f \star f \star \cdots \star f \ (k \text{ times}).$$

We say that two arithmetic functions f and g are dependent if there exists a nontrivial polynomial of two variables $P(x, y) = \sum_{i,j} a_{ij} x^i y^j$ with real coefficients such that

$$P(f, g) = \sum_{i,j} a_{ij} f^{\star i} \star g^{\star j} = 0,$$

and say that they are independent if they are not dependent. Let p and q be two distinct primes and set

$$f_1(n) = \begin{cases} 1 & \text{if } n = p, \\ 0 & \text{otherwise}; \end{cases} \qquad f_2(n) = \begin{cases} 1 & \text{if } n = q, \\ 0 & \text{otherwise}. \end{cases}$$

Prove that f_1 and f_2 are independent.

100. **(THA 4)** Let A be an $n \times n$ matrix whose elements are nonnegative real numbers. Assume that A is a nonsingular matrix and all elements of A^{-1} are nonnegative real numbers. Prove that every row and every column of A has exactly one nonzero element.

101. **(TUR 1)** Let ABC be an equilateral triangle and Γ the semicircle drawn exteriorly to the triangle, having BC as diameter. Show that if a line passing through A trisects BC, it also trisects the arc Γ.

102. **(TUR 2)** If in a convex quadrilateral $ABCD$, E and F are the midpoints of the sides BC and DA respectively. Show that the sum of the areas of the triangles EDA and FBC is equal to the area of the quadrangle.

103. **(USA 1)** An accurate 12-hour analog clock has an hour hand, a minute hand, and a second hand that are aligned at 12:00 o'clock and make one revolution in 12 hours, 1 hour, and 1 minute, respectively. It is well known, and not difficult to prove, that there is no time when the three hands are equally spaced around the clock, with each separating angle $2\pi/3$. Let $f(t), g(t), h(t)$ be the respective absolute deviations of the separating angles from $2\pi/3$ at t hours after 12:00 o'clock. What is the minimum value of $\max\{f(t), g(t), h(t)\}$?

104. **(USA 2)** For each nonzero complex number z, let $\arg z$ be the unique real number t such that $-\pi < t \le \pi$ and $z = |z|(\cos t + \imath \sin t)$. Given a real number $c > 0$ and a complex number $z \ne 0$ with $\arg z \ne \pi$, define

$$B(c, z) = \{b \in \mathbb{R} \mid |w - z| < b \Rightarrow |\arg w - \arg z| < c\}.$$

Determine necessary and sufficient conditions, in terms of c and z, such that $B(c, z)$ has a maximum element, and determine what this maximum element is in this case.

105. **(USA 3)** (SL89-32).

106. **(USA 4)** Let $n > 1$ be a fixed integer. Define functions $f_0(x) = 0$, $f_1(x) = 1 - \cos x$, and for $k > 0$,

$$f_{k+1}(x) = 2 f_k(x) \cos x - f_{k-1}(x).$$

If $F(x) = f_1(x) + f_2(x) + \cdots + f_n(x)$, prove that
(a) $0 < F(x) < 1$ for $0 < x < \frac{\pi}{n+1}$, and
(b) $F(x) > 1$ for $\frac{\pi}{n+1} < x < \frac{\pi}{n}$.

107. **(VNM 1)** Let E be the set of all triangles whose only points with integer coordinates (in the Cartesian coordinate system in space), in its interior or on its sides, are its three vertices, and let f be the function of area of a triangle. Determine the set of values $f(E)$ of f.

108. **(VNM 2)** For every sequence $(x_1, x_2, \ldots, x_n)$ of the numbers $\{1, 2, \ldots, n\}$ arranged in any order, denote by $f(s)$ the sum of absolute values of the differences between two consecutive members of s. Find the maximum value of $f(s)$ (where s runs through the set of all such sequences).

109. **(VNM 3)** Let Ax, By be two noncoplanar rays with AB as a common perpendicular, and let M, N be two mobile points on Ax and By respectively such that $AM + BN = MN$.

First version. Prove that there exist infinitely many lines coplanar with each of the lines MN.

Second version. Prove that there exist infinitely many rotations around a fixed axis Δ mapping the line Ax onto a line coplanar with each of the lines MN.

110. **(VNM 4)** Do there exist two sequences of real numbers $\{a_i\}, \{b_i\}, i \in \mathbb{N} = \{1, 2, 3, \ldots\}$, satisfying the following conditions:

$$\frac{3\pi}{2} \le a_i \le b_i, \quad \cos a_i x + \cos b_i x \ge -\frac{1}{i}$$

for all $i \in \mathbb{N}$ and all x, $0 < x < 1$?

111. **(VNM 5)** Find the greatest number c such that for all natural numbers n, $\{n\sqrt{2}\} \ge \frac{c}{n}$ (where $\{n\sqrt{2}\} = n\sqrt{2} - [n\sqrt{2}]$; $[x]$ is the integer part of x). For this number c, find all natural numbers n for which $\{n\sqrt{2}\} = \frac{c}{n}$.

3.30.3 Shortlisted Problems

1. **(AUS 2)**[IMO2] Let ABC be a triangle. The bisector of angle A meets the circumcircle of triangle ABC in A_1. Points B_1 and C_1 are defined similarly. Let AA_1 meet the lines that bisect the two external angles at B and C in point A^0. Define B^0 and C^0 similarly. If $S_{X_1 X_2 \ldots X_n}$ denotes the area of the polygon $X_1 X_2 \ldots X_n$, prove that

$$S_{A^0 B^0 C^0} = 2 S_{AC_1 BA_1 CB_1} \ge 4 S_{ABC}.$$

2. **(AUS 3)** Ali Barber, the carpet merchant, has a rectangular piece of carpet whose dimensions are unknown. Unfortunately, his tape measure is broken and he has no other measuring instruments. However, he finds that if he lays it flat on the floor of either of his storerooms, then each corner of the carpet touches a different wall of that room. If the two rooms have dimensions of 38 feet by 55 feet and 50 feet by 55 feet, what are the carpet dimensions?

3. **(AUS 4)** Ali Barber, the carpet merchant, has a rectangular piece of carpet whose dimensions are unknown. Unfortunately, his tape measure is broken and he has no other measuring instruments. However, he finds that if he lays it flat on the floor of either of his storerooms, then each corner of the carpet touches a different wall of that room. He knows that the sides of the carpet are integral numbers of feet and that his two storerooms have the same (unknown) length, but widths of 38 feet and 50 feet respectively. What are the carpet dimensions?

4. **(BGR 3)** Prove that for every integer $n > 1$ the equation

$$\frac{x^n}{n!} + \frac{x^{n-1}}{(n-1)!} + \cdots + \frac{x^2}{2!} + \frac{x}{1!} + 1 = 0$$

has no rational roots.

5. **(COL 1)** Consider the polynomial $p(x) = x^n + nx^{n-1} + a_2x^{n-2} + \cdots + a_n$ having all real roots. If $r_1^{16} + r_2^{16} + \cdots + r_n^{16} = n$, where the r_j are the roots of $p(x)$, find all such roots.

6. **(CZS 1)** For a triangle ABC, let k be its circumcircle with radius r. The bisectors of the inner angles A, B, and C of the triangle intersect respectively the circle k again at points A', B', and C'. Prove the inequality

$$16Q^3 \geq 27r^4P,$$

where Q and P are the areas of the triangles $A'B'C'$ and ABC respectively.

7. **(FIN 1)** Show that any two points lying inside a regular n-gon E can be joined by two circular arcs lying inside E and meeting at an angle of at least $\left(1 - \frac{2}{n}\right)\pi$.

8. **(FRA 2)** Let R be a rectangle that is the union of a finite number of rectangles R_i, $1 \leq i \leq n$, satisfying the following conditions:
   (i) The sides of every rectangle R_i are parallel to the sides of R.
   (ii) The interiors of any two different R_i are disjoint.
   (iii) Every R_i has at least one side of integral length.
   Prove that R has at least one side of integral length.

9. **(FRA 4)** For all integers n, $n \geq 0$, there exist uniquely determined integers a_n, b_n, c_n such that

$$\left(1 + 4\sqrt[3]{2} - 4\sqrt[3]{4}\right)^n = a_n + b_n\sqrt[3]{2} + c_n\sqrt[3]{4}.$$

Prove that $c_n = 0$ implies $n = 0$.

10. **(HEL 3)** Let $g : \mathbb{C} \to \mathbb{C}$, $w \in \mathbb{C}$, $a \in \mathbb{C}$, $w^3 = 1$ ($w \neq 1$). Show that there is one and only one function $f : \mathbb{C} \to \mathbb{C}$ such that

$$f(z) + f(wz + a) = g(z), \quad z \in \mathbb{C}.$$

Find the function f.

11. **(HUN 1)** Define sequence a_n by $\sum_{d|n} a_d = 2^n$. Show that $n | a_n$.

12. **(HUN 3)** At n distinct points of a circular race course there are n cars ready to start. Each car moves at a constant speed and covers the circle in an hour. On hearing the initial signal, each of them selects a direction and starts moving immediately. If two cars meet, both of them change directions and go on without loss of speed.
    Show that at a certain moment each car will be at its starting point.

13. **(ISL 3)**[IMO4] The quadrilateral $ABCD$ has the following properties:
    (i) $AB = AD + BC$;
    (ii) there is a point P inside it at a distance x from the side CD such that $AP = x + AD$ and $BP = x + BC$.
    Show that
    $$\frac{1}{\sqrt{x}} \geq \frac{1}{\sqrt{AD}} + \frac{1}{\sqrt{BC}}.$$

14. **(IND 2)** A *bicentric* quadrilateral is one that is both inscribable in and circumscribable about a circle. Show that for such a quadrilateral, the centers of the two associated circles are collinear with the point of intersection of the diagonals.

15. **(IRL 1)** Let a, b, c, d, m, n be positive integers such that
    $$a^2 + b^2 + c^2 + d^2 = 1989, \quad a + b + c + d = m^2,$$
    and the largest of a, b, c, d is n^2. Determine, with proof, the values of m and n.

16. **(ISR 1)** The set $\{a_0, a_1, \ldots, a_n\}$ of real numbers satisfies the following conditions:
    (i) $a_0 = a_n = 0$;
    (ii) for $1 \leq k \leq n - 1$,
    $$a_k = c + \sum_{i=k}^{n-1} a_{i-k}(a_i + a_{i+1}).$$
    Prove that $c \leq \frac{1}{4n}$.

17. **(MNG 1)** Given seven points in the plane, some of them are connected by segments so that:
    (i) among any three of the given points, two are connected by a segment;
    (ii) the number of segments is minimal.
    How many segments does a figure satisfying (i) and (ii) contain? Give an example of such a figure.

18. **(MNG 4)** Given a convex polygon $A_1 A_2 \ldots A_n$ with area S and a point M in the same plane, determine the area of polygon $M_1 M_2 \ldots M_n$, where M_i is the image of M under rotation $\mathscr{R}_{A_i}^{\alpha}$ around A_i by α, $i = 1, 2, \ldots, n$.

19. **(MNG 6)** A positive integer is written in each square of an $m \times n$ board. The allowed move is to add an integer k to each of two adjacent numbers in such a way that no negative numbers are obtained. (Two squares are adjacent if they have a common side.) Find a necessary and sufficient condition for it to be possible for all the numbers to be zero by a finite sequence of moves.

20. **(NLD 1)**[IMO3] Given a set S in the plane containing n points and satisfying the conditions:
    (i) no three points of S are collinear,
    (ii) for every point P of S there exist at least k points in S that have the same distance to P,

prove that the following inequality holds:

$$k < \frac{1}{2} + \sqrt{2n}.$$

21. **(NLD 2)** Prove that the intersection of a plane and a regular tetrahedron can be an obtuse-angled triangle and that the obtuse angle in any such triangle is always smaller than $120°$.

22. **(PHI 1)**[IMO1] Prove that the set $\{1,2,\ldots,1989\}$ can be expressed as the disjoint union of 17 subsets $A_1, A_2, \ldots, A_{17}$ such that:
    (i) each A_i contains the same number of elements;
    (ii) the sum of all elements of each A_i is the same for $i = 1, 2, \ldots, 17$.

23. **(POL 2)**[IMO6] We consider permutations $(x_1, \ldots, x_{2n})$ of the set $\{1, \ldots, 2n\}$ such that $|x_i - x_{i+1}| = n$ for at least one $i \in \{1, \ldots, 2n-1\}$. For every natural number n, find out whether permutations with this property are more or less numerous than the remaining permutations of $\{1, \ldots, 2n\}$.

24. **(POL 5)** For points $A_1, \ldots, A_5$ on the sphere of radius 1, what is the maximum value that $\min_{1 \le i,j \le 5} A_i A_j$ can take? Determine all configurations for which this maximum is attained. (Or: determine the diameter of any set $\{A_1, \ldots, A_5\}$ for which this maximum is attained.)

25. **(KOR 3)** Let a, b be integers that are not perfect squares. Prove that if

$$x^2 - ay^2 - bz^2 + abw^2 = 0$$

has a nontrivial solution in integers, then so does

$$x^2 - ay^2 - bz^2 = 0.$$

26. **(KOR 4)** Let n be a positive integer and let a, b be given real numbers. Determine the range of x_0 for which

$$\sum_{i=0}^{n} x_i = a \quad \text{and} \quad \sum_{i=0}^{n} x_i^2 = b,$$

where $x_0, x_1, \ldots, x_n$ are real variables.

27. **(ROU 1)** Let m be a positive odd integer, $m \ge 2$. Find the smallest positive integer n such that 2^{1989} divides $m^n - 1$.

28. **(ROU 2)** Consider in a plane Π the points O, A_1, A_2, A_3, A_4 such that $\sigma(OA_iA_j) \ge 1$ for all $i, j = 1, 2, 3, 4$, $i \ne j$. Prove that there is at least one pair $i_0, j_0 \in \{1, 2, 3, 4\}$ such that $\sigma(OA_{i_0}A_{j_0}) \ge \sqrt{2}$.
    (We have denoted by $\sigma(OA_iA_j)$ the area of triangle OA_iA_j.)

29. **(ROU 4)** A flock of 155 birds sit down on a circle C. Two birds P_i, P_j are mutually visible if $m(P_iP_j) \le 10°$. Find the smallest number of mutually visible pairs of birds. (One assumes that a position (point) on C can be occupied simultaneously by several birds.)

30. **(SWE 2)**[IMO5] For which positive integers n does there exist a positive integer N such that none of the integers $1+N, 2+N, \ldots, n+N$ is the power of a prime number?

31. **(SWE 3)** Let $a_1 \geq a_2 \geq a_3$ be given positive integers and let $N(a_1, a_2, a_3)$ be the number of solutions (x_1, x_2, x_3) of the equation

$$\frac{a_1}{x_1} + \frac{a_2}{x_2} + \frac{a_3}{x_3} = 1,$$

where x_1, x_2, and x_3 are positive integers. Show that

$$N(a_1, a_2, a_3) \leq 6a_1 a_2 (3 + \ln(2a_1)).$$

32. **(USA 3)** The vertex A of the acute triangle ABC is equidistant from the circumcenter O and the orthocenter H. Determine all possible values for the measure of angle A.

3.31 The Thirty-First IMO
Beijing, China, July 8–19, 1990

3.31.1 Contest Problems

First Day (July 12)

1. Given a circle with two chords AB, CD that meet at E, let M be a point of chord AB other than E. Draw the circle through D, E, and M. The tangent line to the circle DEM at E meets the lines BC, AC at F, G, respectively. Given $\frac{AM}{AB} = \lambda$, find $\frac{GE}{EF}$.

2. On a circle, $2n - 1$ $(n \geq 3)$ different points are given. Find the minimal natural number N with the property that whenever N of the given points are colored black, there exist two black points such that the interior of one of the corresponding arcs contains exactly n of the given $2n - 1$ points.

3. Find all positive integers n having the property that $\frac{2^n+1}{n^2}$ is an integer.

Second Day (July 13)

4. Let $\mathbb{Q}^+$ be the set of positive rational numbers. Construct a function $f : \mathbb{Q}^+ \to \mathbb{Q}^+$ such that
$$f(xf(y)) = \frac{f(x)}{y}, \qquad \text{for all } x, y \text{ in } \mathbb{Q}^+.$$

5. Two players A and B play a game in which they choose numbers alternately according to the following rule: At the beginning, an initial natural number $n_0 > 1$ is given. Knowing n_{2k}, player A may choose any $n_{2k+1} \in \mathbb{N}$ such that
$$n_{2k} \leq n_{2k+1} \leq n_{2k}^2.$$
Then player B chooses a number $n_{2k+2} \in \mathbb{N}$ such that
$$\frac{n_{2k+1}}{n_{2k+2}} = p^r,$$
where p is a prime number and $r \in \mathbb{N}$.
It is stipulated that player A wins the game if he (she) succeeds in choosing the number 1990, and player B wins if he (she) succeeds in choosing 1. For which natural numbers n_0 can player A manage to win the game, for which n_0 can player B manage to win, and for which n_0 can players A and B each force a tie?

6. Is there a 1990-gon with the following properties (i) and (ii)?
   (i) All angles are equal;
   (ii) The lengths of the 1990 sides form a permutation of the numbers 1^2, 2^2, ..., 1989^2, 1990^2.

3.31.2 Longlisted Problems

1. **(AUS 1)** In triangle ABC, point O is the circumcenter, H is the orthocenter. Denote by A_1, B_1, and C_1 the circumcenters of the triangles CHB, CHA, and AHB respectively. Prove that the triangles ABC and $A_1B_1C_1$ are congruent and that their nine-point circles coincide.

2. **(AUS 2)** Prove that

$$1990 \cdot \left(\frac{1}{1990}\binom{1990}{0} - \frac{1}{1989}\binom{1989}{1} + \frac{1}{1988}\binom{1988}{2} \right.$$
$$- \cdots + \frac{(-1)^m}{1990-m}\binom{1990-m}{m} + \cdots - \left. \frac{1}{995}\binom{995}{995} \right) + 1 = 0.$$

3. **(AUS 3)** (SL90-1)

4. **(CAN 1)** (SL90-2)

5. **(COL 1)** Let b be a positive integer. Assume that there exist exactly 1990 triangles ABC with integral side-lengths satisfying the following conditions:
   (i) $\angle ABC = \frac{1}{2}\angle BAC$;
   (ii) $AC = b$.
   Find the minimal value for b.

6. **(COL 2)** Assume that the function $f : (\mathbb{Z}^+)^3 \to \mathbb{N}$ satisfy the following conditions:
   (i) $f(0,0,0) = 1$;
   (ii) $f(x,y,z) = f(x-1,y,z) + f(x,y-1,z) + f(x,y,z-1)$;
   (iii) When applying the above relation iteratively, if any of x', y', z' is negative, then $f(x',y',z') = 0$.
   Prove that if x, y, z are the side lengths of a triangle, then $\frac{(f(x,y,z))^k}{f(mx,my,mz)}$ is not an integers for any integers $k, m > 1$.

7. **(CUB 1)** Let A and B be two points in the plane α, and let r be the line passing through A and B. There are n distinct points P_1, P_2, ..., P_n in one of the half-planes divided by the line r. Prove that there are at least $\sqrt{n}$ distinct values among the distances AP_1, AP_2, ..., AP_n, BP_1, BP_2, ..., BP_n.

8. **(CZS 1)** (SL90-3)

9. **(CZS 2)** (SL90-4)

10. **(CZS 3)** Let $p, k,$ and x be positive integers such that $p > k$ and $x < \left\lceil \frac{p(p-k+1)}{2(k-1)} \right\rceil$, where $\lfloor q \rfloor$ denotes the largest integer not greater than q. Prove that when x balls are put into p boxes arbitrarily, there exist k boxes with the same number of balls.

11. **(CZS 4)** In a group of mathematicians, every mathematician has some friends (friendship is symmetrical relation). Prove that there exists a mathematician, such that the average of the numbers of friends of all his friends is not less than the average of the number of friends of all the mathematicians.

12. **(CZS 5)** For any permutation p of the set $\{1,2,\ldots,n\}$ define $d(p) = |p(1)-1|+|p(2)-2|+\cdots+|p(n)-n|$. Denote by $i(p)$ the number of integer pairs (i,j) in the permutation p such that $1 \le i < j \le n$ and $p(i) > p(j)$. Find all real numbers c such that the inequality $i(p) \le cd(p)$ holds for any positive integer n and any permutation p.

13. **(FIN 1)** Six cities A, B, C, D, E, and F are located at the vertices of a regular hexagon in that order. Let G be the center of the hexagon. The sides of the hexagon are the roads connecting these cities. Furthermore, there are roads connecting the cities B, C, E, F, and G. Because of raining, one or more of the roads may be destroyed. The probability of each road remaining undestroyed is equal to p. Determine the probability that it is possible to travel between the cities A and D.

14. **(FIN 2)** We call a set $S \subseteq \mathbb{R}$ *superinvariant*, if for any stretching A of the set S by the transformation taking x to $A(x) = x_0 + a(x-x_0)$ (here $a > 0$ is a real number), there exists a transformation B, $B(x) = x+b$ such that the images of S under A and B agree: i.e. for any $x \in S$, there exists $y \in S$ such that $A(x) = B(y)$ and for any $t \in S$ there is $u \in S$ such that $B(t) = A(u)$. Determine all *superinvariant* sets.

15. **(FRA 1)** (SL90-5)

16. **(FRA 2)** We say that an integer $k \ge 1$ has property P, if there exists at least one integer $m \ge 1$ which cannot be expressed in the form $m = \varepsilon_1 z_1 + \varepsilon_2 z_2^k + \varepsilon_{2k} z_{2k}^k$, where z_i is nonnegative integer and $\varepsilon_i = \pm 1$, $i = 1,2,\ldots,2k$. Prove that there are infinitely many integers k having the property P.

17. **(FRA 3)** 1990 mathematicians attend a meeting. Every mathematician has at least 1327 friends (friendship is symmetric relation). Prove that it is possible to find four mathematicians such that any two of them are friends.

18. **(FRG 1)** Find, with proof, the least positive integer n having the following property: All the binary representations of 1, 2, ..., 1990 appear after the decimal point in the binary expression of $1/n$.

19. **(FRG 2)** (SL90-6)

20. **(FRG 3)** Is it possible to express the three-dimensional space as a union of disjoint circles?

21. **(HEL 1)** Point O is in the interior of $\triangle ABC$. Three lines through O parallel to BC, CA, and AB intersect the sides AB and AC at D and E; the sides BC and BA at F and G; and the sides CA and CB at H and I, respectively.

22. **(HEL 2)** (SL90-7)

23. **(HUN 1)** (SL90-8)

24. **(HUN 2)** Find the real number t such that the following system of equations has a unique real number solution (x,y,z,v):

$$x+y+z+v=0$$
$$(xy+yz+zv)+t(xz+xv+yv)=0.$$

25. **(HUN 3)** (SL90-9)

26. **(ISL 1)** Prove that there exist infinitely many positive integers n such that $\frac{1^2+2^2+\cdots+n^2}{n}$ is a perfect square. Obviously, 1 is the least integer having this property. Find the next two least integers with this property.

27. **(ISL 2)** (SL90-10)

28. **(IND 1)** Let ABC be an acute-angled triangle. Assume that the circle Γ satisfies the following two conditions:
    (i) Γ intersects all three sides of $\triangle ABC$.
    (ii) These points form a hexagon whose three pairs of opposite sides are parallel. (The hexagon may be degenerate if two or more vertices coincide. In this case opposite sides being parallel is defined through limit behavior.)
    Construct the locus of the centers of such circles Γ.

29. **(IND 2)** Function $f(n)$, $n \in \mathbb{N}$ is defined as follows: Let $A(n)$ and $B(n)$ be coprime positive integers such that

$$\frac{A(n)}{B(n)} = \frac{(2n)!}{n!(n+1000)!}.$$

    If $B(n) = 1$ then $f(n) = 1$; if $B(n) \neq 1$ then $f(n)$ is the largest prime factor of $B(n)$. Prove that the set of values of $f(n)$ is finite and find the maximum value for $f(n)$.

30. **(IND 3)** (SL90-11)

31. **(IND 4)** Let $S = \{1,2,\ldots,1990\}$. A 31-element subset of S is called *good* if the sum of its elements is divisible by 5. Find the number of good subsets of S.

32. **(IRN 1)** Using the following five figures is it possible to construct a parallelepiped whose side lengths are all integers greater than 1 and whose volume is 1990? In the following figure, every square represents a unit cube.

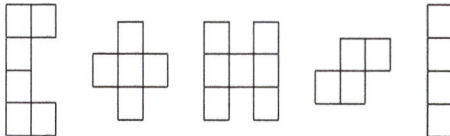

33. **(IRN 2)** Let S be a set with 1990 elements. Let P be a set of ordered sequences of 100 elements from S. If $x = (\ldots,a,\ldots,b,\ldots) \in P$, for $a,b \in S$, then we say that the ordered pair (a,b) *appears* in x. Assume that any ordered pair from S *appears* in at most one element of P. Prove that P has at most 800 elements.

34. **(IRN 3)** There are n non-coplanar points in the space. Prove that there exists a circle that passes through exactly three of those points.

35. **(IRN 4)** Prove that if $|x| < 1$, then

$$\frac{x}{(1-x)^2} + \frac{x^2}{(1+x^2)^2} + \frac{x^3}{(1-x^3)^2} + \cdots = \frac{x}{1-x} + \frac{2x^2}{1+x^2} + \frac{3x^3}{1-x^3} + \cdots$$

36. **(IRL 1)** (SL90-12)

37. **(IRL 2)** (SL90-13)

38. **(IRL 3)** Let α be a positive solution of the quadratic equation $x^2 = 1990x + 1$. For every $m, n \in \mathbb{N}$ define the operation $m \star n = mn + [\alpha m][\alpha n]$, where $[x]$ denotes the largest integer not exceeding x. Prove that $(p \star q) \star r = p \star (q \star r)$ holds for all $p, q, r \in \mathbb{N}$.

39. **(IRL 4)** Let a, b, c be integers. Prove that there are integers p_1, q_1, r_1, p_2, q_2, and r_2 satisfying $a = q_1 r_2 - q_2 r_1$, $b = r_1 p_2 - r_2 p_1$, and $c = p_1 q_2 - p_2 q_1$.

40. **(ISR 1)** Given three letters X, Y, Z, we can construct letter sequences such as XZ, $ZZYXYY$, $XXYZXX$, etc. For any given sequence, one can perform the following operations:

   T_1 If the right-most letter is Y, we add YZ after it, for example: $T_1(XYZZXY) = XYZZXYYZ$;

   T_2 If the sequence contains YYY, this can be replaced by Z as in the following example: $T_2(XXYYZYYYX) = XXYYZZX$;

   T_3 Xp can be replaced by XpX where p is any subsequence of letters: Example: $T_3(XXYZ) = XXYZX$;

   T_4 In a sequence that contains one or more letters Z, we can replace the first Z by XY. Example: $T_4(XXYYZZX) = XXYYXYZX$;

   T_5 We can replace any of XX, YY, ZZ by X, for example: $T_5(ZZYXYY) = XYXX$, or $T_5(ZZYXYY) = XYXYY$, or $T_5(ZZYXYY) = ZZYXX$.

   Using the above operations is it possible to obtain $XYZZ$ starting from XYZ?

41. **(ISR 2)** Given a positive integer n, calculate $S_n = \sum_{r=0}^{n} 2^{r-2n} \cdot \binom{2n-r}{n}$.

42. **(ITA 1)** Find n points $P_1, \ldots, P_n$ on the circumference of a unit circle such that $\sum_{1 \le i < j \le n} P_i P_j$ is maximal.

43. **(ITA 2)** Let V be a finite set of points in the three-dimensional space. Let S_1, S_2, S_3 be the sets consisting of the orthogonal projections of the points of V onto the planes Oyz, Ozx, and Oxy respectively. Prove that $|V|^2 \le |S_1| \cdot |S_2| \cdot |S_3|$, where $|A|$ denotes the number of elements in the set A.

44. **(ITA 3)** Prove that for any positive integer n, the number of odd integers among the binomial coefficients $\binom{n}{k}$ ($0 \le k \le n$) is a power of 2.

45. **(ITA 4)** A tourist is looking for a treasure on an island. The treasure is hidden behind the series of doors each of which is colored with one of n possible colors. The tourist has n keys, all of different colors. Each key can open any door, however, a key gets destroyed when it opens the door of the same color as the key itself (if it opens a door of some other color, it remains intact). Once the tourist

starts using a particular key, she must continue using only that key until it gets destroyed.

Find the least number of doors to ensure that no tourist can get the treasure, no matter how he chooses the order of keys.

46. **(JPN 1)** Let P be an interior point of triangle ABC. Let Q, R, S be the intersections of AP, BP, CP with sides BC, CA, AB respectively. Prove that $S_{QRS} \leq \frac{1}{4} S_{ABC}$.

47. **(JPN 2)** (SL90-14)

48. **(JPN 3)** Prove that $\sqrt{2} + \sqrt{5} + \sqrt{1990}$ is irrational.

49. **(LUX 1)** Let AB and AC be two chords of a circle with center O. The diameter perpendicular to BC intersects AB and AC at F and G respectively (F is inside the circle). Let T be the point on tangency of the circle and the tangent from G. Prove that F is the projection of T on OG.

50. **(MEX 1)** During the duration of the class, n children sit in a circle and play the following game: The teacher goes around the children in the clockwise direction and hands out candies according to the following rules: The teacher selects a child, gives him/her a candy as well to the child child next to him; then the teacher skips one child and gives a candy to the next one; then the teacher skips two children, gives a candy to the next; then skips over three children, ... Find the value of n such that the teacher ends up giving at least one candy to each of the children after finitely many steps.

51. **(MEX 2)** (SL90-15)

52. **(MNG 1)** Let $a > 0$ be a real number. Assume that real numbers $a_1, \ldots, a_n$ satisfy $0 < a_i \leq a$ for $i = 1, 2, \ldots, n$. Prove that:
   (a) If $n = 4$, then

$$\frac{1}{a} \sum_{i=1}^{4} a_i - \frac{a_1 a_2 + a_2 a_3 + a_3 a_4 + a_4 a_1}{a^2} \leq 2.$$

   (b) If $n = 6$, then

$$\frac{1}{a} \sum_{i=1}^{6} a_i - \frac{a_1 a_2 + a_2 a_3 + \cdots + a_5 a_6 + a_6 a_1}{a^2} \leq 3.$$

53. **(MNG 2)** Find all real solutions to the system of equations:

$$x^3 + y^3 = 1,$$
$$x^5 + y^5 = 1.$$

54. **(MNG 3)** Given a set $M = \{1, 2, \ldots, n\}$, let $\phi : M \to M$ be a bijection.
   (a) Prove that there are bijections $\phi_1, \phi_2 : M \to M$ such that $\phi_2 \circ \phi_1 = \phi$ and $\phi_1^2 = \phi_2^2 = \text{id}$, where id is the identity mapping.

(b) Prove that the conclusion in (a) still holds if M is the set of all positive integers.

55. **(MNG 4)** Given points A, M, M_1, and a rational number $\lambda \neq -1$, construct a triangle ABC such that: $M \in BC$, $M_1 \in B_1C_1$, where B_1 and C_1 are the projections of B, C to AC and AB respectively, and

$$\frac{BM}{MC} = \frac{B_1M_1}{M_1C_1} = \lambda.$$

56. **(MAR 1)** For positive integers n, p, $n \geq p$, define real number $K_{n,p}$ as follows: $K_{n,0} = \frac{1}{n+1}$, $K_{n,p} = K_{n-1,p-1} - K_{n,p-1}$ for $1 \leq p \leq n$.
    (a) If $S_n = \sum_{p=0}^{n} K_{n,p}$, $n = 0,1,2,\ldots$, find $\lim_{n\to\infty} S_n$.
    (b) Find $T_n = \sum_{p=0}^{n}(-1)^p K_{n,p}$, $n = 0,1,2,\ldots$.

57. **(MAR 2)** The sequence $\{u_n\}$ is defined by $u_1 = 1$, $u_2 = 1$, $u_n = u_{n-1} + 2u_{n-2}$, for $n \geq 3$.
    (a) Prove that for any positive integers n, p $(p > 1)$, $u_{n+p} = u_{n+1}u_p + 2u_n u_{p-1}$.
    (b) Find the greatest common divisor of u_n and u_{n+3}.

58. **(NLD 1)** (SL90-16)

59. **(NLD 2)** Given eight real numbers $a_1 \leq a_2 \leq \cdots \leq a_7 \leq a_8$, let $x = \frac{a_1 + \cdots + a_8}{8}$, $y = \frac{a_1^2 + \cdots + a_8^2}{8}$. Prove that

$$2\sqrt{y - x^2} \leq a_8 - a_1 \leq 4\sqrt{y - x^2}.$$

60. **(NLD 3)** (SL90-17)

61. **(NLD 4)** Prove that we can fill in the three dimensional space with regular tetrahedrons and regular octahedrons all of which have the same edge-lengths. Find the ratio of the number of regular tetrahedrons used to the number of regular octahedrons.

62. **(NOR 1)** (SL90-18)

63. **(POL 1)** (SL90-19)

64. **(POL 2)** Given an m-element set M and its k-element subset $K \subseteq M$, we say that a function $f : K \to M$ has a path, if there exists an element $x_0 \in K$ such that $f(x_0) = x_0$, or there exists a chain $x_0, x_1, \ldots, x_j = x_0 \in K$ such that $x_i = f(x_{i-1})$, for $i = 1, 2, \ldots, j$. Find the number of functions $f : K \to M$ that have paths.

65. **(POL 3)** (SL90-20)

66. **(POL 4)** Find all continuous bounded functions $f : \mathbb{R} \to \mathbb{R}$ such that

$$(f(x))^2 - (f(y))^2 = f(x+y)f(x-y), \quad \text{for all } x,y, \in \mathbb{R}.$$

67. **(PRK 1)** Let $a + bi$ and $c + di$ be two roots of the equation $x^n = 1990$ ($n \geq 3$ is an integer). Assume that $f(2,1) = (1,2)$ where f is the linear transformation:

$$f = \begin{bmatrix} a & c \\ b & d \end{bmatrix}.$$

Denote by r the distance between the image of $(2,2)$ and the origin. Find the range for the values of r.

68. **(PRK 2)** A mobile point M starts from the origin $O(0,0)$ and moves along the line l with slope k, where k is an irrational number.
    (a) Prove that the point $O(0,0)$ is the only rational point (i.e. with both rational coordinates) on the line l.
    (b) Prove that for any number $\varepsilon > 0$ there are integers m and n such that the distance between l and the point (m,n) is less than ε.

69. **(PRK 3)** Consider the set of *cuboids*: three edges a, b, c from a common vertex satisfy the condition: $\frac{a}{b} = \frac{a^2}{c^3}$.
    (a) Prove that there are 100 pairs of cuboids in this set with equal volumes in each pair.
    (b) For each pair of the above cuboids, find the ratio of the sum of their edges.

70. **(PRK 4)** Let M be a point on the side BC of a triangle ABC.
    (a) Prove that if M is the midpoint of BC, then $AB^2 + AC^2 = 2(AM^2 + BM^2)$.
    (b) If there exists a point $N \in BC$ different than M satisfying $AB^2 + AC^2 = 2(AN^2 + BN^2)$, find the region that the point A might occupy.

71. **(PRK 5)** Given a point $P = (p_1, \ldots, p_n)$, find the point $X = (x_1, \ldots, x_n)$ satisfying $x_1 \leq x_2 \leq \cdots \leq x_n$ such that X minimizes the expression

$$\sqrt{(x_1 - p_1)^2 + \cdots + (x_n - p_n)^2}.$$

72. **(KOR 1)** Let $n \geq 5$ be a positive integer. Let $a_1, b_1, a_2, b_2, \ldots, a_n, b_n$ be integers such that the ordered (a_i, b_i) are distinct for $i = 1, 2, \ldots, n$ and $|a_1 b_2 - a_2 b_1| = |a_2 b_3 - a_3 b_2| = \cdots = |a_{n-1} b_n - a_n b_{n-1}| = 1$. Prove that there exists a pair of indices i and j that satisfy $2 \leq |i - j| \leq n - 2$ and $|a_i b_j - a_j b_i| = 1$.
    Alternative formulation. Let $n \geq 5$ be a positive integers and let $P_1, \ldots, P_n$ be the points with integral coordinates in the coordinate system with the origin O. The areas of the triangles $OP_1 P_2, OP_2 P_3, \ldots, OP_{n-1} P_n$ are equal to $\frac{1}{2}$. Prove that there exists a pair of integers i, j, such that $2 \leq |i - j| \leq n - 2$ for which the area of $\triangle OP_i P_j$ is equal to $\frac{1}{2}$.

73. **(KOR 2)** A function $f : \mathbb{Q} \to \mathbb{R}$ satisfies the following conditions:
    (i) $f(0) = 0$ and for every nonzero $a \in \mathbb{Q}$, $f(a) > 0$;
    (ii) $f(\alpha + \beta) = f(\alpha)f(\beta)$;
    (iii) $f(\alpha + \beta) \leq \max\{f(\alpha), f(\beta)\}$.
    Let x be an integer for which $f(x) \neq 1$. Prove that $f(1 + x + \cdots + x^n) = 1$ for every positive integer n.

74. **(KOR 3)** Let L be a subset of the plane defined by $L = \{(41x + 2y, 59x + 15y) : x, y \in \mathbb{Z}\}$. Prove that every parallelogram with center at the origin and area of 1990 contains at least two points of L.

75. **(ROU 1)** (SL90-21)

76. **(ROU 2)** Prove that there are at least two non-congruent cyclic quadrilaterals with equal areas and perimeters.

77. **(ROU 3)** Let $a, b, c \in \mathbb{R}$. Prove that

$$(a^2 + ab + b^2)(b^2 + bc + c^2)(c^2 + ca + a^2) \geq (ab + bc + ca)^3.$$

When does the equality hold?

78. **(ROU 4)** (SL90-22)

79. **(ROU 5)** (SL90-23)

80. **(ESP 1)** Function $f : \mathbb{N} \times \mathbb{N} \to \mathbb{Q}$ satisfies the following conditions:
    (i) $f(1,1) = 1$;
    (ii) $f(p+1,q) + f(p,q+1) = f(p,q)$ for all $p, q \in \mathbb{N}$;
    (iii) $qf(p+1,q) = pf(p,q+1)$ for all $p, q \in \mathbb{N}$.
    Find $f(1990, 31)$.

81. **(ESP 2)** Circle $k(K, \rho)$ tangents the sides AB and BC of $\triangle ABC$ and intersects the side BC at points D and E. Let p be the distance from K to the side BC.
    (a) Prove that $a(p - \rho) = 2s(r - \rho)$, where r is the inradius, s the semi-perimeter of $\triangle ABC$ and a the length of the side BC.
    (b) Prove that

$$DE = \frac{\sqrt{rr_1(\rho - r)(r_1 - \rho)}}{r_1 - r},$$

    where r_1 is the radius of the excircle of $\triangle ABC$ opposite to A.

82. **(ESP 3)** Define the *symmedian* S_a of triangle ABC as the line symmetric to the median from A with respect to the bisector of $\angle CAB$. Assume that the median m_a intersects BC at A' and the circumcircle at A_1. Assume that the symmedian S_a intersects BC at M and the circumcircle at A_2. Denote by O the circumcenter of $\triangle ABC$. If A_1, O, and A_2 are collinear, prove that:
    (a) $\frac{AA'}{AM} = \frac{b^2 + c^2}{2bc}$;
    (b) $(b^2 + c^2)^2 + 4b^2c^2 = a^2(b^2 + c^2)$.

83. **(SWE 1)** Point D lies on the hypothenuse BC of the right triangle ABC. The inradii of the triangles ADB and ADC are equal. Prove that $S_{ABC} = AD^2$.

84. **(SWE 2)** Let $a_1, a_2, \ldots, a_n \in (0, 2n)$ be n distinct integers $(n \geq 4)$. Prove that there exists a subset of the set $\{a_1, \ldots, a_n\}$ whose sum of elements is divisible by $2n$.

85. **(SWE 3)** Let $A_1, A_2, \ldots, A_n$ $(n \geq 4)$ be n convex sets in plane. Given that every three of these sets have a common point, prove that there exists a point belonging to all the sets.

86. **(SWE 4)** Given a function $f(x) = \sin x + \sin(\pi x)$ and a positive number d, prove that for every $n \in \mathbb{N}$ there exists a real number p such that $p > n$ and $|f(x+p) - f(x)| < d$ holds for all real numbers x.

87. **(THA 1)** Let m be a positive odd integer not divisible by 3. Prove that

$$112 \mid \left[4^m + \left(2 + \sqrt{2}\right)^m \right].$$

88. **(THA 2)** (SL90-24)

89. **(THA 3)** Let n be a positive integer. Let $S_1, \ldots, S_n$ be pairwise non-intersecting sets such that S_k has exactly k elements $(k = 1, 2, \ldots, n)$. Denote $S = S_1 \cup S_2 \cup \cdots \cup S_n$. A function $f : S \to S$ maps all elements of S_k to a fixed element of S_k for $k = 1, 2, \ldots, n$. Find the number of functions $g : S \to S$ satisfying $f \circ g \circ f = f$ (i.e. $f(g(f(x))) = f(x)$ for all x).

90. **(TUR 1)** Let P be a variable point on the circumference of a quarter-circle with radii OA, OB ($\angle AOB = 90°$). Let H be the projection of P on OA. Find the locus of the incenters of $\triangle HPO$.

91. **(TUR 2)** Quadrilateral $ABCD$ is circumscribed around the circle with center O. If $AB = CD$ and M and K are the midpoints of BC and AD respectively prove that $OM = OK$.

92. **(TUR 3)** Let n be a positive integer and $m = \frac{(n+1)(n+2)}{2}$. There are n distinct lines $L_1, L_2, \ldots, L_n$ in coordinate plane and m distinct points $A_1, A_2, \ldots, A_m$ satisfying the following two conditions:
    (i) Any two of the lines are non-parallel.
    (ii) Any three lines are non-concurrent.
    (iii) Only A_1 does not line on any line L_k and there are exactly $k + 1$ among the points $A_1, \ldots, A_m$ that lie on line L_k $(k = 1, 2, \ldots, n)$.
    Prove that there exists a unique polynomial $p(x, y)$ of degree n satisfying $p(A_1) = 1$ and $p(A_j) = 0$ for $j = 2, 3, \ldots, m$.

93. **(TUR 4)** (SL90-25)

94. **(USA 1)** Given an integer $n > 1$ and a real number $t \geq 1$ let P be a parallelogram with vertices $(0,0)$, $(0,t)$, (tF_{2n+1}, tF_{2n}), $(tF_{2n+1}, tF_{2n} + t)$, where F_n is the n-th term of the Fibonacci sequence defined by $F_0 = 1$, $F_1 = 1$, and $F_{m+1} = F_m + F_{m-1}$ for $m \geq 1$. Let L be the number of integral points (i.e. points whose all coordinates are integers) in the interior of P, and let M be the area of P.
    (a) Prove that for any integral point (a, b) there exists a unique pair of integers (j, k) such that $j(F_{n+1}, F_n) + k(F_n, F_{n-1}) = (a, b)$.
    (b) Prove that $\left| \sqrt{L} - \sqrt{M} \right| \leq \sqrt{2}$.

95. **(USA 2)** (SL90-26)

96. **(USA 3)** Given a triangle ABC, points X, Y, Z are on the sides BC, CA, AB, respectively, such that $\triangle XYZ \sim \triangle ABC$ with $\angle X = \angle A$, $\angle Y = \angle B$, and $\angle Z = \angle C$. Prove that the orthocenter of triangle XYZ is the circumcenter of the triangle ABC.

97. **(USA 4)** Given a convex hexagon $ABCDEF$, assume that $\angle BCA = \angle DEC = \angle AFB = \angle CBD = \angle EDG$. Prove that $AB = CD = EF$.

98. **(USS 1)** (SL90-27)

99. **(USS 2)** Given a 10×10 chessboard colored in a standard way, prove that for any 46 unit squares without common edges one can choose 30 squares of the same color.

100. **(USS 3)** (SL90-28)

101. **(USS 4)** The side lengths of two equilateral triangles ABC and KLM are 1 and $\frac{1}{4}$ respectively. The triangle KLM is located in the interior of the triangle ABC. Denote by Σ the sum of the distances from A to the lines KL, LM, and MK. Find the position of KLM that maximizes Σ.

102. **(USS 5)** We call a point (x, y) a *lattice point* of the coordinate plane if both x and y are integers. Knowing that the vertices of triangle ABC are all lattice points, and that there exists exactly one lattice point in the interior of $\triangle ABC$ (excluding the sides), prove that $S_{ABC} \leq \frac{9}{2}$.

103. **(VNM 1)** Find the minimal value of the function

$$f(x) = \sqrt{15 - 12\cos x} + \sqrt{4 - 2\sqrt{3}\sin x}$$
$$+ \sqrt{7 - 4\sqrt{3}\sin x} + \sqrt{10 - 4\sqrt{3}\sin x - 6\cos x}.$$

104. **(VNM 2)** Let $x, y, z \in \mathbb{R}$ such that $x \geq y \geq z > 0$. Prove that

$$\frac{x^2 y}{z} + \frac{y^2 z}{x} + \frac{z^2 x}{y} \geq x^2 + y^2 + z^2.$$

105. **(YUG 1)** Let S and T respectively be the circumcenter and the centroid of the triangle ABC. If M is a point in the plane of $\triangle ABC$ such that $90° \leq \angle SMT < 180°$, denote by A_1, B_1, and C_1 the intersections of AM, BM, CM with the circumcircle of the triangle ABC respectively. Prove that $MA_1 + MB_1 + MC_1 \geq MA + MB + MC$.

106. **(YUG 2)** Let S be the incenter of the triangle ABC. Let A_1, B_1, C_1 be the intersections of AS, BS, and CS respectively with the circumcircle of the triangle ABC. Prove that $SA_1 + SB_1 + SC_1 \geq SA + SB + SC$.

107. **(YUG 3)** Let a, b, c, and P be the side lengths and the area of a triangle, respectively. Prove that

$$\left(a^2 + b^2 + c^2 - 4\sqrt{3}P\right) \cdot \left(a^2 + b^2 + c^2\right)$$
$$\geq 2\left[a^2(b - c)^2 + b^2(c - a)^2 + c^2(a - b)^2\right].$$

108. **(YUG 4)** Let $(a_1, a_2, \ldots a_n)$ be a permutation of the set $\{1, 2, \ldots, n\}$. Prove that

$$\frac{1}{2} + \frac{2}{3} + \cdots + \frac{n-1}{n} \leq \frac{a_1}{a_2} + \frac{a_2}{a_3} + \cdots + \frac{a_{n-1}}{a_n}.$$

3.31.3 Shortlisted Problems

1. **(AUS 3)** The integer 9 can be written as a sum of two consecutive integers: $9 = 4 + 5$. Moreover, it can be written as a sum of (more than one) consecutive positive integers in exactly two ways: $9 = 4 + 5 = 2 + 3 + 4$. Is there an integer that can be written as a sum of 1990 consecutive integers and that can be written as a sum of (more than one) consecutive positive integers in exactly 1990 ways?

2. **(CAN 1)** Given n countries with three representatives each, m committees $A(1), A(2), \ldots A(m)$ are called *a cycle* if
   (i) each committee has n members, one from each country;
   (ii) no two committees have the same membership;
   (iii) for $i = 1, 2, \ldots, m$, committee $A(i)$ and committee $A(i+1)$ have no member in common, where $A(m+1)$ denotes $A(1)$;
   (iv) if $1 < |i - j| < m - 1$, then committees $A(i)$ and $A(j)$ have at least one member in common.
   Is it possible to have a cycle of 1990 committees with 11 countries?

3. **(CZS 1)**[IMO2] On a circle, $2n - 1$ $(n \geq 3)$ different points are given. Find the minimal natural number N with the property that whenever N of the given points are colored black, there exist two black points such that the interior of one of the corresponding arcs contains exactly n of the given $2n - 1$ points.

4. **(CZS 2)** Assume that the set of all positive integers is decomposed into r (disjoint) subsets $A_1 \cup A_2 \cup \cdots A_r = \mathbb{N}$. Prove that one of them, say A_i, has the following property: There exists a positive m such that for any k one can find numbers $a_1, a_2, \ldots, a_k$ in A_i with $0 < a_{j+1} - a_j \leq m$ $(1 \leq j \leq k - 1)$.

5. **(FRA 1)** Given $\triangle ABC$ with no side equal to another side, let G, K, and H be its centroid, incenter, and orthocenter, respectively. Prove that $\angle GKH > 90°$.

6. **(FRG 2)**[IMO5] Two players A and B play a game in which they choose numbers alternately according to the following rule: At the beginning, an initial natural number $n_0 > 1$ is given. Knowing n_{2k}, player A may choose any $n_{2k+1} \in \mathbb{N}$ such that
   $$n_{2k} \leq n_{2k+1} \leq n_{2k}^2.$$
   Then player B chooses a number $n_{2k+2} \in \mathbb{N}$ such that
   $$\frac{n_{2k+1}}{n_{2k+2}} = p^r,$$
   where p is a prime number and $r \in \mathbb{N}$.
   It is stipulated that player A wins the game if he (she) succeeds in choosing the number 1990, and player B wins if he (she) succeeds in choosing 1. For which natural numbers n_0 can player A manage to win the game, for which n_0 can player B manage to win, and for which n_0 can players A and B each force a tie?

7. **(HEL 2)** Let $f(0) = f(1) = 0$ and

$$f(n+2) = 4^{n+2}f(n+1) - 16^{n+1}f(n) + n \cdot 2^{n^2}, \quad n = 0, 1, 2, 3, \ldots.$$

Show that the numbers $f(1989), f(1990), f(1991)$ are divisible by 13.

8. **(HUN 1)** For a given positive integer k denote the square of the sum of its digits by $f_1(k)$ and let $f_{n+1}(k) = f_1(f_n(k))$.
Determine the value of $f_{1991}(2^{1990})$.

9. **(HUN 3)** The incenter of the triangle ABC is K. The midpoint of AB is C_1 and that of AC is B_1. The lines $C_1 K$ and AC meet at B_2, the lines $B_1 K$ and AB at C_2. If the areas of the triangles AB_2C_2 and ABC are equal, what is the measure of angle $\angle CAB$?

10. **(ISL 2)** A plane cuts a right circular cone into two parts. The plane is tangent to the circumference of the base of the cone and passes through the midpoint of the altitude. Find the ratio of the volume of the smaller part to the volume of the whole cone.

11. **(IND 3′)**[IMO1] Given a circle with two chords AB, CD that meet at E, let M be a point of chord AB other than E. Draw the circle through D, E, and M. The tangent line to the circle DEM at E meets the lines BC, AC at F, G, respectively. Given $\frac{AM}{AB} = \lambda$, find $\frac{GE}{EF}$.

12. **(IRL 1)** Let ABC be a triangle and L the line through C parallel to the side AB. Let the internal bisector of the angle at A meet the side BC at D and the line L at E and let the internal bisector of the angle at B meet the side AC at F and the line L at G. If $GF = DE$, prove that $AC = BC$.

13. **(IRL 2)** An eccentric mathematician has a ladder with n rungs that he always ascends and descends in the following way: When he ascends, each step he takes covers a rungs of the ladder, and when he descends, each step he takes covers b rungs of the ladder, where a and b are fixed positive integers. By a sequence of ascending and descending steps he can climb from ground level to the top rung of the ladder and come back down to ground level again. Find, with proof, the minimum value of n, expressed in terms of a and b.

14. **(JPN 2)** In the coordinate plane a rectangle with vertices $(0,0), (m,0), (0,n)$, (m,n) is given where both m and n are odd integers. The rectangle is partitioned into triangles in such a way that
   (i) each triangle in the partition has at least one side (to be called a "good" side) that lies on a line of the form $x = j$ or $y = k$, where j and k are integers, and the altitude on this side has length 1;
   (ii) each "bad" side (i.e., a side of any triangle in the partition that is not a "good" one) is a common side of two triangles in the partition.
Prove that there exist at least two triangles in the partition each of which has two good sides.

15. **(MEX 2)** Determine for which positive integers k the set

$$X = \{1990, 1990+1, 1990+2, \ldots, 1990+k\}$$

can be partitioned into two disjoint subsets A and B such that the sum of the elements of A is equal to the sum of the elements of B.

16. **(NLD 1)**[IMO6] Is there a 1990-gon with the following properties (i) and (ii)?
   (i) All angles are equal;
   (ii) The lengths of the 1990 sides form a permutation of the numbers 1^2, 2^2, ..., 1989^2, 1990^2.

17. **(NLD 3)** Unit cubes are made into beads by drilling a hole through them along a diagonal. The beads are put on a string in such a way that they can move freely in space under the restriction that the vertices of two neighboring cubes are touching. Let A be the beginning vertex and B be the end vertex. Let there be $p \times q \times r$ cubes on the string $(p,q,r \geq 1)$.
   (a) Determine for which values of p, q, and r it is possible to build a block with dimensions p, q, and r. Give reasons for your answers.
   (b) The same question as (a) with the extra condition that $A = B$.

18. **(NOR)** Let a,b be natural numbers with $1 \leq a \leq b$, and $M = \left[\frac{a+b}{2}\right]$. Define the function $f : \mathbb{Z} \to \mathbb{Z}$ by

$$f(n) = \begin{cases} n+a, & \text{if } n < M, \\ n-b, & \text{if } n \geq M. \end{cases}$$

Let $f^1(n) = f(n)$, $f^{i+1}(n) = f(f^i(n))$, $i = 1,2,\ldots$. Find the smallest natural number k such that $f^k(0) = 0$.

19. **(POL 1)** Let P be a point inside a regular tetrahedron T of unit volume. The four planes passing through P and parallel to the faces of T partition T into 14 pieces. Let $f(P)$ be the joint volume of those pieces that are neither a tetrahedron nor a parallelepiped (i.e., pieces adjacent to an edge but not to a vertex). Find the exact bounds for $f(P)$ as P varies over T.

20. **(POL 3)** Prove that every integer k greater than 1 has a multiple that is less than k^4 and can be written in the decimal system with at most four different digits.

21. **(ROU 1')** Let n be a composite natural number and p a proper divisor of n. Find the binary representation of the smallest natural number N such that $\frac{(1+2^p+2^{n-p})N-1}{2^n}$ is an integer.

22. **(ROU 4)** Ten localities are served by two international airlines such that there exists a direct service (without stops) between any two of these localities and all airline schedules offer round-trip service between the cities they serve. Prove that at least one of the airlines can offer two disjoint round trips each containing an odd number of landings.

23. **(ROU 5)**[IMO3] Find all positive integers n having the property that $\frac{2^n+1}{n^2}$ is an integer.

24. **(THA 2)** Let a,b,c,d be nonnegative real numbers such that $ab+bc+cd+da = 1$. Show that

$$\frac{a^3}{b+c+d} + \frac{b^3}{a+c+d} + \frac{c^3}{a+b+d} + \frac{d^3}{a+b+c} \geq \frac{1}{3}.$$

25. **(TUR 4)**[IMO4] Let $\mathbb{Q}^+$ be the set of positive rational numbers. Construct a function $f : \mathbb{Q}^+ \to \mathbb{Q}^+$ such that

$$f(xf(y)) = \frac{f(x)}{y}, \qquad \text{for all } x, y \text{ in } \mathbb{Q}^+.$$

26. **(USA 2)** Let P be a cubic polynomial with rational coefficients, and let $q_1, q_2, q_3, \ldots$ be a sequence of rational numbers such that $q_n = P(q_{n+1})$ for all $n \geq 1$. Prove that there exists $k \geq 1$ such that for all $n \geq 1$, $q_{n+k} = q_n$.

27. **(USS 1)** Find all natural numbers n for which every natural number whose decimal representation has $n - 1$ digits 1 and one digit 7 is prime.

28. **(USS 3)** Prove that in the coordinate plane it is impossible to draw a closed broken line satisfying the following conditions:
    (i) the coordinates of each vertex are rational;
    (ii) the length of each of its edges is 1;
    (iii) the line has an odd number of vertices.

3.32 The Thirty-Second IMO
Sigtuna, Sweden, July 12–23, 1991

3.32.1 Contest Problems

First Day (July 17)

1. Prove for each triangle ABC the inequality

$$\frac{1}{4} < \frac{IA \cdot IB \cdot IC}{l_A l_B l_C} \leq \frac{8}{27},$$

where I is the incenter and l_A, l_B, l_C are the lengths of the angle bisectors of ABC.

2. Let $n > 6$ and let $a_1 < a_2 < \ldots < a_k$ be all natural numbers that are less than n and relatively prime to n. Show that if $a_1, a_2, \ldots, a_k$ is an arithmetic progression, then n is a prime number or a natural power of two.

3. Let $S = \{1, 2, 3, \ldots, 280\}$. Find the minimal natural number n such that in any n-element subset of S there are five numbers that are pairwise relatively prime.

Second Day (July 18)

4. Suppose G is a connected graph with n edges. Prove that it is possible to label the edges of G from 1 to n in such a way that in every vertex v of G with two or more incident edges, the set of numbers labeling those edges has no common divisor greater than 1.

5. Let ABC be a triangle and M an interior point in ABC. Show that at least one of the angles $\angle MAB, \angle MBC$, and $\angle MCA$ is less than or equal to $30°$.

6. Given a real number $a > 1$, construct an infinite and bounded sequence $x_0, x_1,$ $x_2, \ldots$ such that for all natural numbers i and j, $i \neq j$, the following inequality holds:

$$|x_i - x_j||i - j|^a \geq 1.$$

3.32.2 Shortlisted Problems

1. (**PHI 3**) Let ABC be any triangle and P any point in its interior. Let P_1, P_2 be the feet of the perpendiculars from P to the two sides AC and BC. Draw AP and BP, and from C drop perpendiculars to AP and BP. Let Q_1 and Q_2 be the feet of these perpendiculars. Prove that the lines $Q_1 P_2, Q_2 P_1$, and AB are concurrent.

2. (**JPN 5**) For an acute triangle ABC, M is the midpoint of the segment BC, P is a point on the segment AM such that $PM = BM$, H is the foot of the perpendicular line from P to BC, Q is the point of intersection of segment AB and the line passing through H that is perpendicular to PB, and finally, R is the point of intersection of the segment AC and the line passing through H that is perpendicular to PC.
Show that the circumcircle of $\triangle QHR$ is tangent to the side BC at point H.

3. **(PRK 1)** Let S be any point on the circumscribed circle of $\triangle PQR$. Then the feet of the perpendiculars from S to the three sides of the triangle lie on the same straight line. Denote this line by $l(S, PQR)$. Suppose that the hexagon $ABCDEF$ is inscribed in a circle. Show that the four lines $l(A, BDF), l(B, ACE), l(D, ABF)$, and $l(E, ABC)$ intersect at one point if and only if $CDEF$ is a rectangle.

4. **(FRA 2)**[IMO5] Let ABC be a triangle and M an interior point in ABC. Show that at least one of the angles $\angle MAB, \angle MBC$, and $\angle MCA$ is less than or equal to $30°$.

5. **(ESP 4)** In the triangle ABC, with $\angle A = 60°$, a parallel IF to AC is drawn through the incenter I of the triangle, where F lies on the side AB. The point P on the side BC is such that $3BP = BC$. Show that $\angle BFP = \angle B/2$.

6. **(USS 4)**[IMO1] Prove for each triangle ABC the inequality

$$\frac{1}{4} < \frac{IA \cdot IB \cdot IC}{l_A l_B l_C} \le \frac{8}{27},$$

where I is the incenter and l_A, l_B, l_C are the lengths of the angle bisectors of ABC.

7. **(CHN 2)** Let O be the center of the circumsphere of a tetrahedron $ABCD$. Let L, M, N be the midpoints of BC, CA, AB respectively, and assume that $AB + BC = AD + CD, BC + CA = BD + AD$, and $CA + AB = CD + BD$. Prove that $\angle LOM = \angle MNG = \angle NOL$.

8. **(NLD 1)** Let S be a set of n points in the plane. No three points of S are collinear. Prove that there exists a set P containing $2n - 5$ points satisfying the following condition: In the interior of every triangle whose three vertices are elements of S lies a point that is an element of P.

9. **(FRA 3)** In the plane we are given a set E of 1991 points, and certain pairs of these points are joined with a path. We suppose that for every point of E, there exist at least 1593 other points of E to which it is joined by a path. Show that there exist six points of E every pair of which are joined by a path.

   Alternative version. Is it possible to find a set E of 1991 points in the plane and paths joining certain pairs of the points in E such that every point of E is joined with a path to at least 1592 other points of E, and in every subset of six points of E there exist at least two points that are not joined?

10. **(USA 5)**[IMO4] Suppose G is a connected graph with n edges. Prove that it is possible to label the edges of G from 1 to n in such a way that in every vertex v of G with two or more incident edges, the set of numbers labeling those edges has no common divisor greater than 1.

11. **(AUS 4)** Prove that

$$\sum_{m=0}^{995} \frac{(-1)^m}{1991 - m} \binom{1991 - m}{m} = \frac{1}{1991}.$$

12. **(CHN 3)**[IMO3] Let $S = \{1,2,3,\ldots,280\}$. Find the minimal natural number n such that in any n-element subset of S there are five numbers that are pairwise relatively prime.

13. **(POL 4)** Given any integer $n \geq 2$, assume that the integers $a_1, a_2, \ldots, a_n$ are not divisible by n and, moreover, that n does not divide $a_1 + a_2 + \cdots + a_n$. Prove that there exist at least n different sequences $(e_1, e_2, \cdots, e_n)$ consisting of zeros or ones such that $e_1 a_1 + e_2 a_2 + \cdots + e_n a_n$ is divisible by n.

14. **(POL 3)** Let a, b, c be integers and p an odd prime number. Prove that if $f(x) = ax^2 + bx + c$ is a perfect square for $2p - 1$ consecutive integer values of x, then p divides $b^2 - 4ac$.

15. **(USS 2)** Let a_n be the last nonzero digit in the decimal representation of the number $n!$. Does the sequence $a_1, a_2, \ldots, a_n, \ldots$ become periodic after a finite number of terms?

16. **(ROU 1)**[IMO2] Let $n > 6$ and $a_1 < a_2 < \cdots < a_k$ be all natural numbers that are less than n and relatively prime to n. Show that if $a_1, a_2, \ldots, a_k$ is an arithmetic progression, then n is a prime number or a natural power of two.

17. **(HKG 4)** Find all positive integer solutions x, y, z of the equation $3^x + 4^y = 5^z$.

18. **(BGR 1)** Find the highest degree k of 1991 for which 1991^k divides the number

$$1990^{1991^{1992}} + 1992^{1991^{1990}}.$$

19. **(IRL 5)** Let a be a rational number with $0 < a < 1$ and suppose that

$$\cos 3\pi a + 2\cos 2\pi a = 0.$$

(Angle measurements are in radians.) Prove that $a = 2/3$.

20. **(IRL 3)** Let α be the positive root of the equation $x^2 = 1991x + 1$. For natural numbers m, n define

$$m * n = mn + [\alpha m][\alpha n],$$

where $[x]$ is the greatest integer not exceeding x. Prove that for all natural numbers p, q, r,

$$(p * q) * r = p * (q * r).$$

21. **(HKG 6)** Let $f(x)$ be a monic polynomial of degree 1991 with integer coefficients. Define $g(x) = f^2(x) - 9$. Show that the number of distinct integer solutions of $g(x) = 0$ cannot exceed 1995.

22. **(USA 4)** Real constants a, b, c are such that there is exactly one square all of whose vertices lie on the cubic curve $y = x^3 + ax^2 + bx + c$. Prove that the square has sides of length $\sqrt[4]{72}$.

23. **(IND 2)** Let f and g be two integer-valued functions defined on the set of all integers such that
    (i) $f(m + f(f(n))) = -f(f(m+1)) - n$ for all integers m and n;

(ii) g is a polynomial function with integer coefficients and $g(n) = g(f(n))$ for all integers n.

Determine $f(1991)$ and the most general form of g.

24. **(IND 1)** An odd integer $n \geq 3$ is said to be "nice" if there is at least one permutation $a_1, a_2, \ldots, a_n$ of $1, 2, \ldots, n$ such that the n sums $a_1 - a_2 + a_3 - \cdots - a_{n-1} + a_n, a_2 - a_3 + a_4 - \cdots - a_n + a_1, a_3 - a_4 + a_5 - \cdots - a_1 + a_2, \ldots, a_n - a_1 + a_2 - \cdots - a_{n-2} + a_{n-1}$ are all positive. Determine the set of all "nice" integers.

25. **(USA 1)** Suppose that $n \geq 2$ and $x_1, x_2, \ldots, x_n$ are real numbers between 0 and 1 (inclusive). Prove that for some index i between 1 and $n-1$ the inequality

$$x_i(1 - x_{i+1}) \geq \frac{1}{4}x_1(1 - x_n)$$

holds.

26. **(CZS 1)** Let $n \geq 2$ be a natural number and let the real numbers p, a_1, a_2, $\ldots$, a_n, b_1, b_2, $\ldots$, b_n satisfy $1/2 \leq p \leq 1$, $0 \leq a_i$, $0 \leq b_i \leq p$, $i = 1, \ldots, n$, and $\sum_{i=1}^{n} a_i = \sum_{i=1}^{n} b_i = 1$. Prove the inequality

$$\sum_{i=1}^{n} b_i \prod_{\substack{j=1 \\ j \neq i}}^{n} a_j \leq \frac{p}{(n-1)^{n-1}}.$$

27. **(POL 2)** Determine the maximum value of the sum $\sum_{i<j} x_i x_j (x_i + x_j)$ over all n-tuples $(x_1, \ldots, x_n)$, satisfying $x_i \geq 0$ and $\sum_{i=1}^{n} x_i = 1$.

28. **(NLD 2)**[IMO6] Given a real number $a > 1$, construct an infinite and bounded sequence $x_0, x_1, x_2, \ldots$ such that for all natural numbers i and j, $i \neq j$, the following inequality holds: $|x_i - x_j||i - j|^a \geq 1$.

29. **(FIN 2)** We call a set S on the real line $\mathbb{R}$ *superinvariant* if for any stretching A of the set by the transformation taking x to $A(x) = x_0 + a(x - x_0)$ there exists a translation B, $B(x) = x + b$, such that the images of S under A and B agree; i.e., for any $x \in S$ there is a $y \in S$ such that $A(x) = B(y)$ and for any $t \in S$ there is a $u \in S$ such that $B(t) = A(u)$. Determine all superinvariant sets.

Remark. It is assumed that $a > 0$.

30. **(BGR 3)** Two students A and B are playing the following game: Each of them writes down on a sheet of paper a positive integer and gives the sheet to the referee. The referee writes down on a blackboard two integers, one of which is the sum of the integers written by the players. After that, the referee asks student A: "Can you tell the integer written by the other student?" If A answers "no," the referee puts the same question to student B. If B answers "no," the referee puts the question back to A, and so on. Assume that both students are intelligent and truthful. Prove that after a finite number of questions, one of the students will answer "yes."

3.33 The Thirty-Third IMO
Moscow, Russia, July 10–21, 1992

3.33.1 Contest Problems

First Day (July 15)

1. Find all integer triples (p,q,r) such that $1 < p < q < r$ and $(p-1)(q-1)(r-1)$ is a divisor of $(pqr-1)$.

2. Find all functions $f : \mathbb{R} \to \mathbb{R}$ such that

$$f(x^2 + f(y)) = y + f(x)^2 \text{ for all } x,y \text{ in } \mathbb{R}.$$

3. Given nine points in space, no four of which are coplanar, find the minimal natural number n such that for any coloring with red or blue of n edges drawn between these nine points there always exists a triangle having all edges of the same color.

Second Day (July 16)

4. In the plane, let there be given a circle C, a line l tangent to C, and a point M on l. Find the locus of points P that has the following property: There exist two points Q and R on l such that M is the midpoint of QR and C is the incircle of PQR.

5. Let V be a finite subset of Euclidean space consisting of points (x,y,z) with integer coordinates. Let S_1, S_2, S_3 be the projections of V onto the yz, xz, xy planes, respectively. Prove that

$$|V|^2 \le |S_1||S_2||S_3|$$

($|X|$ denotes the number of elements of X).

6. For each positive integer n, denote by $s(n)$ the greatest integer such that for all positive integer $k \le s(n)$, n^2 can be expressed as a sum of squares of k positive integers.
   (a) Prove that $s(n) \le n^2 - 14$ for all $n \ge 4$.
   (b) Find a number n such that $s(n) = n^2 - 14$.
   (c) Prove that there exist infinitely many positive integers n such that $s(n) = n^2 - 14$.

3.33.2 Longlisted Problems

1. (**AUS 1**) Points D and E are chosen on the sides AB and AC of the triangle ABC in such a way that if F is the intersection point of BE and CD, then $AE + EF = AD + DF$. Prove that $AC + CF = AB + BF$.

2. **(AUS 2)** (SL92-1).

   Original formulation. Let m be a positive integer and x_0, y_0 integers such that x_0, y_0 are relatively prime, y_0 divides $x_0^2 + m$, and x_0 divides $y_0^2 + m$.
   Prove that there exist positive integers x and y such that x and y are relatively prime, y divides $x^2 + m$, x divides $y^2 + m$, and $x + y \leq m + 1$.

3. **(AUS 3)** Let ABC be a triangle, O its circumcenter, S its centroid, and H its orthocenter. Denote by A_1, B_1, and C_1 the centers of the circles circumscribed about the triangles CHB, CHA, and AHB, respectively. Prove that the triangle ABC is congruent to the triangle $A_1B_1C_1$ and that the nine-point circle of $\triangle ABC$ is also the nine-point circle of $\triangle A_1B_1C_1$.

4. **(CAN 1)** Let p, q, and r be the angles of a triangle, and let $a = \sin 2p$, $b = \sin 2q$, and $c = \sin 2r$. If $s = (a+b+c)/2$, show that

$$s(s-a)(s-b)(s-c) \geq 0.$$

   When does equality hold?

5. **(CAN 2)** Let I, H, O be the incenter, centroid, and circumcenter of the non-isosceles triangle ABC. Prove that $AI \| HO$ if and only if $\angle BAC = 120°$.

6. **(CAN 3)** Suppose that n numbers $x_1, x_2, \ldots, x_n$ are chosen randomly from the set $\{1, 2, 3, 4, 5\}$. Prove that the probability that $x_1^2 + x_2^2 + \cdots + x_n^2 \equiv 0 \pmod 5$ is at least $1/5$.

7. **(CAN 4)** Let X be a bounded, nonempty set of points in the Cartesian plane. Let $f(X)$ be the set of all points that are at a distance of at most 1 from some point in X. Let $f^n(X) = f(f(\ldots(f(X))\ldots))$ (n times). Show that $f^n(X)$ becomes "more circular" as n gets larger. In other words, if $r_n = \sup\{$radii of circles contained in $f^n(X)\}$ and $R_n = \inf\{$radii of circles containing $f^n(X)\}$, then show that R_n/r_n gets arbitrarily close to 1 as n becomes arbitrarily large.

8. **(CHN 1)** (SL92-2).

9. **(CHN 2)** (SL92-3).

10. **(CHN 3)** (SL92-4).

11. **(COL 1)** Let $\phi(n, m)$, $m \neq 1$, be the number of positive integers less than or equal to n that are coprime with m. Clearly, $\phi(m, m) = \phi(m)$, where $\phi(m)$ is Euler's phi function. Find all integers m that satisfy the following inequality:

$$\frac{\phi(n, m)}{n} \geq \frac{\phi(m)}{m}$$

   for every positive integer n.

12. **(COL 2)** Given a triangle ABC such that the circumcenter is in the interior of the incircle, prove that the triangle ABC is acute-angled.

13. **(COL 3)** (SL92-5).

14. **(FIN 1)** Integers $a_1, a_2, \ldots, a_n$ satisfy $|a_k| = 1$ and

$$\sum_{k=1}^{n} a_k a_{k+1} a_{k+2} a_{k+3} = 2,$$

where $a_{n+j} = a_j$. Prove that $n \neq 1992$.

15. **(FIN 2)** Prove that there exist 78 lines in the plane such that they have exactly 1992 points of intersection.

16. **(FIN 3)** Find all triples (x, y, z) of integers such that

$$\frac{1}{x^2} + \frac{2}{y^2} + \frac{3}{z^2} = \frac{2}{3}.$$

17. **(FRA 1)** (SL92-20).

18. **(FRG 1)** Fibonacci numbers are defined as follows: $F_1 = F_2 = 1$, $F_{n+2} = F_{n+1} + F_n$, $n \geq 1$. Let a_n be the number of words that consist of n letters 0 or 1 and contain no two letters 1 at distance two from each other. Express a_n in terms of Fibonacci numbers.

19. **(FRG 2)** Denote by a_n the greatest number that is not divisible by 3 and that divides n. Consider the sequence $s_0 = 0$, $s_n = a_1 + a_2 + \cdots + a_n$, $n \in \mathbb{N}$. Denote by $A(n)$ the number of all sums s_k ($0 \leq k \leq 3^n$, $k \in \mathbb{N}_0$) that are divisible by 3. Prove the formula

$$A(n) = 3^{n-1} + 2 \cdot 3^{(n/2)-1} \cos(n\pi/6), \quad n \in \mathbb{N}_0.$$

20. **(FRG 3)** Let X and Y be two sets of points in the plane and M be a set of segments connecting points from X and Y. Let k be a natural number. Prove that the segments from M can be painted using k colors in such a way that for any point $x \in X \cup Y$ and two colors α and β ($\alpha \neq \beta$), the difference between the number of α-colored segments and the number of β-colored segments originating in X is less than or equal to 1.

21. **(UNK 1)** Prove that if $x, y, z > 1$ and $\frac{1}{x} + \frac{1}{y} + \frac{1}{z} = 2$, then

$$\sqrt{x+y+z} \geq \sqrt{x-1} + \sqrt{y-1} + \sqrt{z-1}.$$

22. **(UNK 2)** (SL92-21).

23. **(HKG 1)** An *Egyptian number* is a positive integer that can be expressed as a sum of positive integers, not necessarily distinct, such that the sum of their reciprocals is 1. For example, $32 = 2 + 3 + 9 + 18$ is Egyptian because $\frac{1}{2} + \frac{1}{3} + \frac{1}{9} + \frac{1}{18} = 1$. Prove that all integers greater than 23 are Egyptian.

24. **(ISL 1)** Let $\mathbb{Q}^+$ denote the set of nonnegative rational numbers. Show that there exists exactly one function $f : \mathbb{Q}^+ \rightarrow \mathbb{Q}^+$ satisfying the following conditions:

   (i) if $0 < q < \frac{1}{2}$, then $f(q) = 1 + f\left(\frac{q}{1-2q}\right)$;

(ii) if $1 < q \leq 2$, then $f(q) = 1 + f(q+1)$;

(iii) $f(q)f(1/q) = 1$ for all $q \in \mathbb{Q}^+$.

Find the smallest rational number $q \in \mathbb{Q}^+$ such that $f(q) = 19/92$.

25. **(IND 1)** (a) Show that the set $\mathbb{N}$ of all natural numbers can be partitioned into three disjoint subsets A, B, and C satisfying the following conditions:

$$A^2 = A, \quad B^2 = C, \quad C^2 = B,$$
$$AB = B, \quad AC = C, \quad BC = A,$$

where HK stands for $\{hk \mid h \in H, k \in K\}$ for any two subsets H, K of $\mathbb{N}$, and H^2 denotes HH.

(b) Show that for every such partition of $\mathbb{N}$, $\min\{n \in \mathbb{N} \mid n \in A \text{ and } n+1 \in A\}$ is less than or equal to 77.

26. **(IND 2)** (SL92-6).

27. **(IND 3)** Let ABC be an arbitrary scalene triangle. Define Σ to be the set of all circles y that have the following properties:

(i) y meets each side of $\triangle ABC$ in two (possibly coincident) points;

(ii) if the points of intersection of y with the sides of the triangle are labeled by P, Q, R, S, T, U, with the points occurring on the sides in orders $\mathcal{B}(B,P,Q,C)$, $\mathcal{B}(C,R,S,A)$, $\mathcal{B}(A,T,U,B)$, then the following relations of parallelism hold: $TS \| BC$; $PU \| CA$; $RQ \| AB$. (In the limiting cases, some of the conditions of parallelism will hold vacuously; e.g., if A lies on the circle y, then T, S both coincide with A and the relation $TS \| BC$ holds vacuously.)

(a) Under what circumstances is Σ nonempty?

(b) Assuming that Σ is nonempty, show how to construct the locus of centers of the circles in the set Σ.

(c) Given that the set Σ has just one element, deduce the size of the largest angle of $\triangle ABC$.

(d) Show how to construct the circles in Σ that have, respectively, the largest and the smallest radii.

28. **(IND 4)** (SL92-7).

Alternative formulation. Two circles G_1 and G_2 are inscribed in a segment of a circle G and touch each other externally at a point W. Let A be a point of intersection of a common internal tangent to G_1 and G_2 with the arc of the segment, and let B and C be the endpoints of the chord. Prove that W is the incenter of the triangle ABC.

29. **(IND 5)** (SL92-8).

30. **(IND 6)** Let $P_n = (19 + 92)(19^2 + 92^2) \cdots (19^n + 92^n)$ for each positive integer n. Determine, with proof, the least positive integer m, if it exists, for which P_m is divisible by 33^{33}.

31. **(IRL 1)** (SL92-19).

32. **(IRL 2)** Let $S_n = \{1, 2, \ldots, n\}$ and $f_n : S_n \to S_n$ be defined inductively as follows: $f_1(1) = 1$, $f_n(2j) = j$ $(j = 1, 2, \ldots, [n/2])$ and
    (i) if $n = 2k$ $(k \geq 1)$, then $f_n(2j-1) = f_k(j) + k$ $(j = 1, 2, \ldots, k)$;
    (ii) if $n = 2k+1$ $(k \geq 1)$, then $f_n(2k+1) = k + f_{k+1}(1)$, $f_n(2j-1) = k + f_{k+1}(j+1)$ $(j = 1, 2, \ldots, k)$.
    Prove that $f_n(x) = x$ if and only if x is an integer of the form

$$\frac{(2n+1)(2^d - 1)}{2^{d+1} - 1}$$

    for some positive integer d.

33. **(IRL 3)** Let a, b, c be positive real numbers and p, q, r complex numbers. Let S be the set of all solutions (x, y, z) in $\mathbb{C}$ of the system of simultaneous equations

$$ax + by + cz = p,$$
$$ax^2 + by^2 + cz^2 = q,$$
$$ax^3 + bx^3 + cx^3 = r.$$

    Prove that S has at most six elements.

34. **(IRL 4)** Let a, b, c be integers. Prove that there are integers $p_1, q_1, r_1, p_2, q_2, r_2$ such that

$$a = q_1 r_2 - q_2 r_1, \quad b = r_1 p_2 - r_2 p_1, \quad c = p_1 q_2 - p_2 q_1.$$

35. **(IRN 1)** (SL92-9).

36. **(IRN 2)** Find all rational solutions of

$$a^2 + c^2 + 17(b^2 + d^2) = 21,$$
$$ab + cd = 2.$$

37. **(IRN 3)** Let the circles C_1, C_2, and C_3 be orthogonal to the circle C and intersect each other inside C forming acute angles of measures A, B, and C. Show that $A + B + C < \pi$.

38. **(ITA 1)** (SL92-10).

39. **(ITA 2)** Let $n \geq 2$ be an integer. Find the minimum k for which there exists a partition of $\{1, 2, \ldots, k\}$ into n subsets $X_1, X_2, \ldots, X_n$ such that the following condition holds: for any i, j, $1 \leq i < j \leq n$, there exist $x_1 \in X_1$, $x_2 \in X_2$ such that $|x_i - x_j| = 1$.

40. **(ITA 3)** The colonizers of a spherical planet have decided to build N towns, each having area $1/1000$ of the total area of the planet. They also decided that any two points belonging to different towns will have different latitude and different longitude. What is the maximal value of N?

41. **(JPN 1)** Let S be a set of positive integers $n_1, n_2, \ldots, n_6$ and let $n(f)$ denote the number $n_1 n_{f(1)} + n_2 n_{f(2)} + \cdots + n_6 n_{f(6)}$, where f is a permutation of $\{1, 2, \ldots, 6\}$. Let

$$\Omega = \{n(f) \mid f \text{ is a permutation of } \{1,2,\dots,6\}\}.$$

Give an example of positive integers $n_1,\dots,n_6$ such that Ω contains as many elements as possible and determine the number of elements of Ω.

42. **(JPN 2)** (SL92-11).

43. **(KOR 1)** Find the number of positive integers n satisfying $\phi(n) \mid n$ such that

$$\sum_{m=1}^{\infty} \left(\frac{n}{m} - \frac{n-1}{m} \right) = 1992.$$

What is the largest number among them? As usual, $\phi(n)$ is the number of positive integers less than or equal to n and relatively prime to n.[7]

44. **(KOR 2)** (SL92-16).

45. **(KOR 3)** Let n be a positive integer. Prove that the number of ways to express n as a sum of distinct positive integers (up to order) and the number of ways to express n as a sum of odd positive integers (up to order) are the same.

46. **(KOR 4)** Prove that the sequence $5,12,19,26,33,\dots$ contains no term of the form $2^n - 1$.

47. **(KOR 5)** Find the largest integer not exceeding $\prod_{n=1}^{1992} \frac{3n+2}{3n+1}$.

48. **(MNG 1)** Find all the functions $f : \mathbb{R}^+ \to \mathbb{R}$ satisfying the identity

$$f(x)f(y) = y^{\alpha} \cdot f\left(\frac{x}{2}\right) + x^{\beta} \cdot f\left(\frac{y}{2}\right), \quad x,y \in \mathbb{R}^+,$$

where α, β are given real numbers.

49. **(MNG 2)** Given real numbers x_i ($i = 1,2,\dots,4x+2$) such that

$$\sum_{i=1}^{4x+2} (-1)^{i+1} x_i x_{i+1} = 4m \quad (x_1 = x_{4k+3}),$$

prove that it is possible to choose numbers $x_{k_1},\dots,x_{k_6}$ such that

$$\sum_{i=1}^{6} (-1)^i x_{k_i} x_{k_{i+1}} > m \quad (x_{k_1} = x_{k_7}).$$

50. **(MNG 3)** Let N be a point inside the triangle ABC. Through the midpoints of the segments AN, BN, and CN the lines parallel to the opposite sides of $\triangle ABC$ are constructed. Let A_N, B_N, and C_N be the intersection points of these lines. If N is the orthocenter of the triangle ABC, prove that the nine-point circles of $\triangle ABC$ and $\triangle A_N B_N C_N$ coincide.

Remark. The statement of the original problem was that the nine-point circles of the triangles $A_N B_N C_N$ and $A_M B_M C_M$ coincide, where N and M are the orthocenter and the centroid of $\triangle ABC$. This statement is false.

---

[7] The problem in this formulation is senseless. The correct formulation could be, "Find ...
such that $\sum_{m=1}^{\infty} \left(\left[\frac{n}{m}\right] - \left[\frac{n-1}{m}\right] \right) = 1992 \dots$."

51. **(NLD 1)** (SL92-12).

52. **(NLD 2)** Let n be an integer > 1. In a circular arrangement of n lamps $L_0, \ldots, L_{n-1}$, each one of which can be either ON or OFF, we start with the situation that all lamps are ON, and then carry out a sequence of steps, $Step_0, Step_1, \ldots$. If L_{j-1} (j is taken mod n) is ON, then $Step_j$ changes the status of L_j (it goes from ON to OFF or from OFF to ON) but does not change the status of any of the other lamps. If L_{j-1} is OFF, then $Step_j$ does not change anything at all. Show that:

   (a) There is a positive integer $M(n)$ such that after $M(n)$ steps all lamps are ON again.
   (b) If n has the form 2^k, then all lamps are ON after $n^2 - 1$ steps.
   (c) If n has the form $2^k + 1$, then all lamps are ON after $n^2 - n + 1$ steps.

53. **(NZL 1)** (SL92-13).

54. **(POL 1)** Suppose that $n > m \geq 1$ are integers such that the string of digits 143 occurs somewhere in the decimal representation of the fraction m/n. Prove that $n > 125$

55. **(POL 2)** (SL92-14).

56. **(POL 3)** A directed graph (any two distinct vertices joined by at most one directed line) has the following property: If x, u, and v are three distinct vertices such that $x \to u$ and $x \to v$, then $u \to w$ and $v \to w$ for some vertex w. Suppose that $x \to u \to y \to \cdots \to z$ is a path of length n, that cannot be extended to the right (no arrow goes away from z). Prove that every path beginning at x arrives after n steps at z.

57. **(POL 4)** For positive numbers a, b, c define $A = (a+b+c)/3$, $G = (abc)^{1/3}$, $H = 3/(a^{-1} + b^{-1} + c^{-1})$. Prove that

$$\left(\frac{A}{G}\right)^3 \geq \frac{1}{4} + \frac{3}{4} \cdot \frac{A}{H},$$

for every $a, b, c > 0$.

58. **(POR 1)** Let ABC be a triangle. Denote by a, b, and c the lengths of the sides opposite to the angles A, B, and C, respectively. Prove that[8]

$$\frac{bc}{a+b+c} = \frac{\sin A + \sin B + \sin C}{\cos(A/2)\sin(B/2)\sin(C/2)}.$$

59. **(PRK 1)** Let a regular 7-gon $A_0A_1A_2A_3A_4A_5A_6$ be inscribed in a circle. Prove that for any two points P, Q on the arc A_0A_6 the following equality holds:

$$\sum_{i=0}^{6}(-1)^i PA_i = \sum_{i=0}^{6}(-1)^i QA_i.$$

---

[8] The statement of the problem is obviously wrong, and the authors couldn't determine a suitable alteration of the formulation which would make the problem correct. We put it here only for completeness of the problem set.

60. **(PRK 2)** (SL92-15).

61. **(PRK 3)** There are a board with $2n \cdot 2n$ $(= 4n^2)$ squares and $4n^2 - 1$ cards numbered with different natural numbers. These cards are put one by one on each of the squares. One square is empty. We can move a card to an empty square from one of the adjacent squares (two squares are adjacent if they have a common edge). Is it possible to exchange two cards on two adjacent squares of a column (or a row) in a finite number of movements?

62. **(ROU 1)** Let $c_1, \ldots, c_n$ $(n \geq 2)$ be real numbers such that $0 \leq \sum c_i \leq n$. Prove that there exist integers $k_1, \ldots, k_n$ such that $\sum k_i = 0$ and $1 - n \leq c_i + n k_i \leq n$ for every $i = 1, \ldots, n$.

63. **(ROU 2)** Let a and b be integers. Prove that $\frac{2a^2 - 1}{b^2 + 2}$ is not an integer.

64. **(ROU 3)** For any positive integer n consider all representations $n = a_1 + \cdots + a_k$, where $a_1 > a_2 > \cdots > a_k > 0$ are integers such that for all $i \in \{1, 2, \ldots, k - 1\}$, the number a_i is divisible by a_{i+1}. Find the longest such representation of the number 1992.

65. **(SAF 1)** If A, B, C, and D are four distinct points in space, prove that there is a plane P on which the orthogonal projections of A, B, C, and D form a parallelogram (possibly degenerate).

66. **(ESP 1)** A circle of radius ρ is tangent to the sides AB and AC of the triangle ABC, and its center K is at a distance p from BC.
    (a) Prove that $a(p - \rho) = 2s(r - \rho)$, where r is the inradius and $2s$ the perimeter of ABC.
    (b) Prove that if the circle intersect BC at D and E, then

$$DE = \frac{4\sqrt{rr_1(\rho - r)(r_1 - \rho)}}{(r_1 - r)},$$

where r_1 is the exradius corresponding to the vertex A.

67. **(ESP 2)** In a triangle, a symmedian is a line through a vertex that is symmetric to the median with respect to the internal bisector (all relative to the same vertex). In the triangle ABC, the median m_a meets BC at A' and the circumcircle again at A_1. The symmedian s_a meets BC at M and the circumcircle again at A_2. Given that the line $A_1 A_2$ contains the circumcenter O of the triangle, prove that:
    (a) $\dfrac{AA'}{AM} = \dfrac{b^2 + c^2}{2bc}$;
    (b) $1 + 4b^2 c^2 = a^2(b^2 + c^2)$.

68. **(ESP 3)** Show that the numbers $\tan(r\pi/15)$, where r is a positive integer less than 15 and relatively prime to 15, satisfy

$$x^8 - 92x^6 + 134x^4 - 28x^2 + 1 = 0.$$

69. **(SWE 1)** (SL92-17).

70. **(THA 1)** Let two circles A and B with unequal radii r and R, respectively, be tangent internally at the point A_0. If there exists a sequence of distinct circles (C_n) such that each circle is tangent to both A and B, and each circle C_{n+1} touches circle C_n at the point A_n, prove that

$$\sum_{n=1}^{\infty} |A_{n+1}A_n| < \frac{4\pi Rr}{R+r}.$$

71. **(THA 2)** Let $P_1(x,y)$ and $P_2(x,y)$ be two relatively prime polynomials with complex coefficients. Let $Q(x,y)$ and $R(x,y)$ be polynomials with complex coefficients and each of degree not exceeding d. Prove that there exist two integers A_1, A_2 not simultaneously zero with $|A_i| \le d+1$ $(i=1,2)$ and such that the polynomial $A_1 P_1(x,y) + A_2 P_2(x,y)$ is coprime to $Q(x,y)$ and $R(x,y)$.

72. **(TUR 1)** In a school six different courses are taught: mathematics, physics, biology, music, history, geography. The students were required to rank these courses according to their preferences, where equal preferences were allowed. It turned out that:

   (i) mathematics was ranked among the most preferred courses by all students;
   (ii) no student ranked music among the least preferred ones;
   (iii) all students preferred history to geography and physics to biology; and
   (iv) no two rankings were the same.

   Find the greatest possible value for the number of students in this school.

73. **(TUR 2)** Let $\{A_n \mid n=1,2,\dots\}$ be a set of points in the plane such that for each n, the disk with center A_n and radius 2^n contains no other point A_j. For any given positive real numbers $a < b$ and R, show that there is a subset G of the plane satisfying:

   (i) the area of G is greater than or equal to R;
   (ii) for each point P in G, $a < \sum_{n=1}^{\infty} \frac{1}{|A_n P|} < b$.

74. **(TUR 3)** Let $S = \left\{ \frac{\pi^n}{1992^m} \mid n,m \in \mathbb{Z} \right\}$. Show that every real number $x \ge 0$ is an accumulation point of S.

75. **(TWN 1)** A sequence $\{a_n\}$ of positive integers is defined by

$$a_n = \left[n + \sqrt{n} + \frac{1}{2} \right], \quad n \in \mathbb{N}.$$

   Determine the positive integers that occur in the sequence.

76. **(TWN 2)** Given any triangle ABC and any positive integer n, we say that n is a *decomposable* number for triangle ABC if there exists a decomposition of the triangle ABC into n subtriangles with each subtriangle similar to $\triangle ABC$. Determine the positive integers that are decomposable numbers for every triangle.

77. **(TWN 3)** Show that if 994 integers are chosen from $1,2,\dots,1992$ and one of the chosen integers is less than 64, then there exist two among the chosen integers such that one of them is a factor of the other.

78. **(USA 1)** Let F_n be the nth Fibonacci number, defined by $F_1 = F_2 = 1$ and $F_n = F_{n-1} + F_{n-2}$ for $n > 2$. Let $A_0, A_1, A_2, \ldots$ be a sequence of points on a circle of radius 1 such that the minor arc from A_{k-1} to A_k runs clockwise and such that

$$\mu(A_{k-1}A_k) = \frac{4F_{2k+1}}{F_{2k+1}^2 + 1}$$

for $k \geq 1$, where $\mu(XY)$ denotes the radian measure of the arc XY in the clockwise direction. What is the limit of the radian measure of arc $A_0 A_n$ as n approaches infinity?

79. **(USA 2)** (SL92-18).

80. **(USA 3)** Given a graph with n vertices and a positive integer m that is less than n, prove that the graph contains a set of $m+1$ vertices in which the difference between the largest degree of any vertex in the set and the smallest degree of any vertex in the set is at most $m - 1$.

81. **(USA 4)** Suppose that points X, Y, Z are located on sides BC, CA, and AB, respectively, of $\triangle ABC$ in such a way that $\triangle XYZ$ is similar to $\triangle ABC$. Prove that the orthocenter of $\triangle XYZ$ is the circumcenter of $\triangle ABC$.

82. **(VNM 1)** Let $f(x) = x^m + a_1 x^{m-1} + \cdots + a_{m-1} x + a_m$ and $g(x) = x^n + b_1 x^{n-1} + \cdots + b_{n-1} + b_n$ be two polynomials with real coefficients such that for each real number x, $f(x)$ is the square of an integer if and only if so is $g(x)$. Prove that if $n + m > 0$, then there exists a polynomial $h(x)$ with real coefficients such that $f(x) \cdot g(x) = (h(x))^2$.

3.33.3 Shortlisted Problems

1. **(AUS 2)** Prove that for any positive integer m there exists an infinite number of pairs of integers (x, y) such that (i) x and y are relatively prime; (ii) y divides $x^2 + m$; (iii) x divides $y^2 + m$.

2. **(CHN 1)** Let $\mathbb{R}^+$ be the set of all nonnegative real numbers. Given two positive real numbers a and b, suppose that a mapping $f : \mathbb{R}^+ \to \mathbb{R}^+$ satisfies the functional equation

$$f(f(x)) + af(x) = b(a + b)x.$$

Prove that there exists a unique solution of this equation.

3. **(CHN 2)** The diagonals of a quadrilateral $ABCD$ are perpendicular: $AC \perp BD$. Four squares, $ABEF, BCGH, CDIJ, DAKL$, are erected externally on its sides. The intersection points of the pairs of straight lines CL, DF; DF, AH; AH, BJ; BJ, CL are denoted by P_1, Q_1, R_1, S_1, respectively, and the intersection points of the pairs of straight lines AI, BK; BK, CE; CE, DG; DG, AI are denoted by P_2, Q_2, R_2, S_2, respectively. Prove that $P_1 Q_1 R_1 S_1 \cong P_2 Q_2 R_2 S_2$.

4. **(CHN 3)**[IMO3] Given nine points in space, no four of which are coplanar, find the minimal natural number n such that for any coloring with red or blue of n edges

drawn between these nine points there always exists a triangle having all edges of the same color.

5. **(COL 3)** Let $ABCD$ be a convex quadrilateral such that $AC = BD$. Equilateral triangles are constructed on the sides of the quadrilateral. Let O_1, O_2, O_3, O_4 be the centers of the triangles constructed on AB, BC, CD, DA respectively. Show that $O_1 O_3$ is perpendicular to $O_2 O_4$.

6. **(IND 2)**[IMO2] Find all functions $f : \mathbb{R} \to \mathbb{R}$ such that

$$f(x^2 + f(y)) = y + f(x)^2 \quad \text{for all } x, y \text{ in } \mathbb{R}.$$

7. **(IND 4)** Circles G, G_1, G_2 are three circles related to each other as follows: Circles G_1 and G_2 are externally tangent to one another at a point W and both these circles are internally tangent to the circle G. Points A, B, C are located on the circle G as follows: Line BC is a direct common tangent to the pair of circles G_1 and G_2, and line WA is the transverse common tangent at W to G_1 and G_2, with W and A lying on the same side of the line BC. Prove that W is the incenter of the triangle ABC.

8. **(IND 5)** Show that in the plane there exists a convex polygon of 1992 sides satisfying the following conditions:
   (i) its side lengths are $1, 2, 3, \ldots, 1992$ in some order;
   (ii) the polygon is circumscribable about a circle.

   Alternative formulation. Does there exist a 1992-gon with side lengths 1, 2, 3, ..., 1992 circumscribed about a circle? Answer the same question for a 1990-gon.

9. **(IRN 1)** Let $f(x)$ be a polynomial with rational coefficients and α be a real number such that $\alpha^3 - \alpha = f(\alpha)^3 - f(\alpha) = 33^{1992}$. Prove that for each $n \geq 1$,

$$(f^{(n)}(\alpha))^3 - f^{(n)}(\alpha) = 33^{1992},$$

   where $f^{(n)}(x) = f(f(\ldots f(x)))$, and n is a positive integer.

10. **(ITA 1)**[IMO5] Let V be a finite subset of Euclidean space consisting of points (x, y, z) with integer coordinates. Let S_1, S_2, S_3 be the projections of V onto the yz, xz, xy planes, respectively. Prove that

$$|V|^2 \leq |S_1||S_2||S_3|$$

   ($|X|$ denotes the number of elements of X).

11. **(JPN 2)** In a triangle ABC, let D and E be the intersections of the bisectors of $\angle ABC$ and $\angle ACB$ with the sides AC, AB, respectively. Determine the angles $\angle A, \angle B, \angle C$ if

$$\angle BDE = 24°, \qquad \angle CED = 18°.$$

12. **(NLD 1)** Let f, g, and a be polynomials with real coefficients, f and g in one variable and a in two variables. Suppose

$$f(x) - f(y) = a(x,y)(g(x) - g(y)) \quad \text{for all } x, y \in \mathbb{R}.$$

Prove that there exists a polynomial h with $f(x) = h(g(x))$ for all $x \in \mathbb{R}$.

13. **(NZL 1)**[IMO1] Find all integer triples (p, q, r) such that $1 < p < q < r$ and $(p - 1)(q - 1)(r - 1)$ is a divisor of $(pqr - 1)$.

14. **(POL 2)** For any positive integer x define

$$g(x) = \text{greatest odd divisor of } x,$$

$$f(x) = \begin{cases} x/2 + x/g(x), & \text{if } x \text{ is even}; \\ 2^{(x+1)/2}, & \text{if } x \text{ is odd}. \end{cases}$$

Construct the sequence $x_1 = 1, x_{n+1} = f(x_n)$. Show that the number 1992 appears in this sequence, determine the least n such that $x_n = 1992$, and determine whether n is unique.

15. **(PRK 2)** Does there exist a set M with the following properties?
    (i) The set M consists of 1992 natural numbers.
    (ii) Every element in M and the sum of any number of elements have the form m^k $(m, k \in \mathbb{N}, k \geq 2)$.

16. **(KOR 2)** Prove that $N = \frac{5^{125} - 1}{5^{25} - 1}$ is a composite number.

17. **(SWE 1)** Let $\alpha(n)$ be the number of digits equal to one in the binary representation of a positive integer n. Prove that:
    (a) the inequality $\alpha(n^2) \leq \frac{1}{2}\alpha(n)(\alpha(n) + 1)$ holds;
    (b) the above inequality is an equality for infinitely many positive integers;
    (c) there exists a sequence $(n_i)_1^\infty$ such that $\alpha(n_i^2)/\alpha(n_i) \to 0$ as $i \to \infty$.

    Alternative parts: Prove that there exists a sequence $(n_i)_1^\infty$ such that $\alpha(n_i^2)/\alpha(n_i)$ tends to
    (d) ∞;
    (e) an arbitrary real number $\gamma \in (0, 1)$;
    (f) an arbitrary real number $\gamma \geq 0$.

18. **(USA 2)** Let $[x]$ denote the greatest integer less than or equal to x. Pick any x_1 in $[0, 1)$ and define the sequence $x_1, x_2, x_3, \ldots$ by $x_{n+1} = 0$ if $x_n = 0$ and $x_{n+1} = 1/x_n - [1/x_n]$ otherwise. Prove that

$$x_1 + x_2 + \cdots + x_n < \frac{F_1}{F_2} + \frac{F_2}{F_3} + \cdots + \frac{F_n}{F_{n+1}},$$

where $F_1 = F_2 = 1$ and $F_{n+2} = F_{n+1} + F_n$ for $n \geq 1$.

19. **(IRL 1)** Let $f(x) = x^8 + 4x^6 + 2x^4 + 28x^2 + 1$. Let $p > 3$ be a prime and suppose there exists an integer z such that p divides $f(z)$. Prove that there exist integers $z_1, z_2, \ldots, z_8$ such that if

$$g(x) = (x - z_1)(x - z_2) \cdots (x - z_8),$$

then all coefficients of $f(x) - g(x)$ are divisible by p.

20. **(FRA 1)**[IMO4] In the plane, let there be given a circle C, a line l tangent to C, and a point M on l. Find the locus of points P that have the following property: There exist two points Q and R on l such that M is the midpoint of QR and C is the incircle of PQR.

21. **(UNK 2)**[IMO6] For each positive integer n, denote by $s(n)$ the greatest integer such that for all positive integers $k \leq s(n)$, n^2 can be expressed as a sum of squares of k positive integers.
    - (a) Prove that $s(n) \leq n^2 - 14$ for all $n \geq 4$.
    - (b) Find a number n such that $s(n) = n^2 - 14$.
    - (c) Prove that there exist infinitely many positive integers n such that $s(n) = n^2 - 14$.

3.34 The Thirty-Fourth IMO
Istanbul, Turkey, July 13–24, 1993

3.34.1 Contest Problems

First Day (July 18)

1. Let $n > 1$ be an integer and let $f(x) = x^n + 5x^{n-1} + 3$. Prove that there do not exist polynomials $g(x), h(x)$, each having integer coefficients and degree at least one, such that $f(x) = g(x)h(x)$.

2. A, B, C, D are four points in the plane, with C, D on the same side of the line AB, such that $AC \cdot BD = AD \cdot BC$ and $\angle ADB = 90° + \angle ACB$. Find the ratio

$$\frac{AB \cdot CD}{AC \cdot BD},$$

and prove that circles ACD, BCD are orthogonal. (Intersecting circles are said to be orthogonal if at either common point their tangents are perpendicular.)

3. On an infinite chessboard, a solitaire game is played as follows: At the start, we have n^2 pieces occupying n^2 squares that form a square of side n. The only allowed move is a jump horizontally or vertically over an occupied square to an unoccupied one, and the piece that has been jumped over is removed. For what positive integers n can the game end with only one piece remaining on the board?

Second Day (July 19)

4. For three points A, B, C in the plane we define $m(ABC)$ to be the smallest length of the three altitudes of the triangle ABC, where in the case of A, B, C collinear, $m(ABC) = 0$. Let A, B, C be given points in the plane. Prove that for any point X in the plane,
$$m(ABC) \leq m(ABX) + m(AXC) + m(XBC).$$

5. Let $\mathbb{N} = \{1, 2, 3, \ldots\}$. Determine whether there exists a strictly increasing function $f : \mathbb{N} \to \mathbb{N}$ with the following properties:

$$f(1) = 2; \tag{1}$$
$$f(f(n)) = f(n) + n \quad (n \in \mathbb{N}). \tag{2}$$

6. Let n be an integer greater than 1. In a circular arrangement of n lamps $L_0, \ldots, L_{n-1}$, each one of that can be either on or off, we start with the situation where all lamps are on, and then carry out a sequence of steps, $Step_0, Step_1, \ldots$. If L_{j-1} (indices are taken modulo n) is on, then $Step_j$ changes the status of L_j (it goes from on to off or from off to on) but does not change the status of any of the other lamps. If L_{j-1} is off, then $Step_j$ does not change anything at all. Show that:

(a) There is a positive integer $M(n)$ such that after $M(n)$ steps all lamps are on again.

(b) If n has the form 2^k, then all lamps are on after $n^2 - 1$ steps.

(c) If n has the form $2^k + 1$, then all lamps are on after $n^2 - n + 1$ steps.

3.34.2 Shortlisted Problems

1. **(BRA 1)** Show that there exists a finite set $A \subset \mathbb{R}^2$ such that for every $X \in A$ there are points $Y_1, Y_2, \ldots, Y_{1993}$ in A such that the distance between X and Y_i is equal to 1, for every i.

2. **(CAN 2)** Let triangle ABC be such that its circumradius R is equal to 1. Let r be the inradius of ABC and let p be the inradius of the orthic triangle $A'B'C'$ of triangle ABC.
Prove that $p \leq 1 - \frac{1}{3}(1 + r)^2$.

   Remark. The *orthic* triangle is the triangle whose vertices are the feet of the altitudes of ABC.

3. **(ESP 1)** Consider the triangle ABC, its circumcircle k with center O and radius R, and its incircle with center I and radius r. Another circle k_c is tangent to the sides CA, CB at D, E, respectively, and it is internally tangent to k.
Show that the incenter I is the midpoint of DE.

4. **(ESP 2)** In the triangle ABC, let D, E be points on the side BC such that $\angle BAD = \angle CAE$. If M, N are, respectively, the points of tangency with BC of the incircles of the triangles ABD and ACE, show that

$$\frac{1}{MB} + \frac{1}{MD} = \frac{1}{NC} + \frac{1}{NE}.$$

5. **(FIN 3)**[IMO3] On an infinite chessboard, a solitaire game is played as follows: At the start, we have n^2 pieces occupying n^2 squares that form a square of side n. The only allowed move is a jump horizontally or vertically over an occupied square to an unoccupied one, and the piece that has been jumped over is removed. For what positive integers n can the game end with only one piece remaining on the board?

6. **(GER 1)**[IMO5] Let $\mathbb{N} = \{1, 2, 3, \ldots\}$. Determine whether there exists a strictly increasing function $f : \mathbb{N} \to \mathbb{N}$ with the following properties:

$$f(1) = 2; \tag{1}$$
$$f(f(n)) = f(n) + n \quad (n \in \mathbb{N}). \tag{2}$$

7. **(GEO 3)** Let a, b, c be given integers $a > 0$, $ac - b^2 = P = P_1 \cdots P_m$ where $P_1, \ldots, P_m$ are (distinct) prime numbers. Let $M(n)$ denote the number of pairs of integers (x, y) for which

$$ax^2 + 2bxy + cy^2 = n.$$

Prove that $M(n)$ is finite and $M(n) = M(P^k \cdot n)$ for every integer $k \geq 0$.

8. **(IND 1)** Define a sequence $\langle f(n)\rangle_{n=1}^{\infty}$ of positive integers by $f(1) = 1$ and

$$f(n) = \begin{cases} f(n-1) - n, & \text{if } f(n-1) > n; \\ f(n-1) + n, & \text{if } f(n-1) \leq n, \end{cases}$$

for $n \geq 2$. Let $S = \{n \in \mathbb{N} \mid f(n) = 1993\}$.
   (a) Prove that S is an infinite set.
   (b) Find the least positive integer in S.
   (c) If all the elements of S are written in ascending order as $n_1 < n_2 < n_3 < \cdots$, show that

$$\lim_{i \to \infty} \frac{n_{i+1}}{n_i} = 3.$$

9. **(IND 4)**
   (a) Show that the set $\mathbb{Q}^+$ of all positive rational numbers can be partitioned into three disjoint subsets A, B, C satisfying the following conditions:

$$BA = B, \qquad B^2 = C, \qquad BC = A,$$

where HK stands for the set $\{hk \mid h \in H, k \in K\}$ for any two subsets H, K of $\mathbb{Q}^+$ and H^2 stands for HH.
   (b) Show that all positive rational cubes are in A for such a partition of $\mathbb{Q}^+$.
   (c) Find such a partition $\mathbb{Q}^+ = A \cup B \cup C$ with the property that for no positive integer $n \leq 34$ are both n and $n+1$ in A; that is,

$$\min\{n \in \mathbb{N} \mid n \in A, n+1 \in A\} > 34.$$

10. **(IND 5)** A natural number n is said to have the property P if whenever n divides $a^n - 1$ for some integer a, n^2 also necessarily divides $a^n - 1$.
   (a) Show that every prime number has property P.
   (b) Show that there are infinitely many composite numbers n that possess property P.

11. **(IRL 1)**[IMO1] Let $n > 1$ be an integer and let $f(x) = x^n + 5x^{n-1} + 3$. Prove that there do not exist polynomials $g(x), h(x)$, each having integer coefficients and degree at least one, such that $f(x) = g(x)h(x)$.

12. **(IRL 2)** Let n, k be positive integers with $k \leq n$ and let S be a set containing n distinct real numbers. Let T be the set of all real numbers of the form $x_1 + x_2 + \cdots + x_k$, where $x_1, x_2, \ldots, x_k$ are distinct elements of S. Prove that T contains at least $k(n-k) + 1$ distinct elements.

13. **(IRL 3)** Let S be the set of all pairs (m, n) of relatively prime positive integers m, n with n even and $m < n$. For $s = (m, n) \in S$ write $n = 2^k n_0$, where k, n_0 are positive integers with n_0 odd and define $f(s) = (n_0, m + n - n_0)$. Prove that f is a function from S to S and that for each $s = (m, n) \in S$, there exists a positive integer $t \leq \frac{m+n+1}{4}$ such that $f^t(s) = s$, where

$$f^t(s) = \underbrace{(f \circ f \circ \cdots \circ f)}_{t \text{ times}}(s).$$

If $m+n$ is a prime number that does not divide $2^k - 1$ for $k = 1, 2, \ldots, m+n-2$, prove that the smallest value of t that satisfies the above conditions is $\left[\frac{m+n+1}{4}\right]$, where $[x]$ denotes the greatest integer less than or equal to x.

14. **(ISR 1)** The vertices D, E, F of an equilateral triangle lie on the sides BC, CA, AB respectively of a triangle ABC. If a, b, c are the respective lengths of these sides, and S the area of ABC, prove that

$$DE \geq \frac{2\sqrt{2}S}{\sqrt{a^2 + b^2 + c^2 + 4\sqrt{3}S}}.$$

15. **(MKD 1)**[IMO4] For three points A, B, C in the plane we define $m(ABC)$ to be the smallest length of the three altitudes of the triangle ABC, where in the case of A, B, C collinear, $m(ABC) = 0$. Let A, B, C be given points in the plane. Prove that for any point X in the plane,

$$m(ABC) \leq m(ABX) + m(AXC) + m(XBC).$$

16. **(MKD 3)** Let $n \in \mathbb{N}$, $n \geq 2$, and $A_0 = (a_{01}, a_{02}, \ldots, a_{0n})$ be any n-tuple of natural numbers such that $0 \leq a_{0i} \leq i - 1$, for $i = 1, \ldots, n$. The n-tuples $A_1 = (a_{11}, a_{12}, \ldots, a_{1n})$, $A_2 = (a_{21}, a_{22}, \ldots, a_{2n}), \ldots$ are defined by

$$a_{i+1,j} = \mathrm{Card}\{a_{i,l} \mid 1 \leq l \leq j-1, a_{i,l} \geq a_{i,j}\}, \quad \text{for } i \in \mathbb{N} \text{ and } j = 1, \ldots, n.$$

Prove that there exists $k \in \mathbb{N}$, such that $A_{k+2} = A_k$.

17. **(NLD 2)**[IMO6] Let n be an integer greater than 1. In a circular arrangement of n lamps $L_0, \ldots, L_{n-1}$, each one of that can be either on or off, we start with the situation where all lamps are on, and then carry out a sequence of steps, Step_0, $\mathrm{Step}_1, \ldots$. If L_{j-1} (indices are taken modulo n) is on, then Step_j changes the status of L_j (it goes from on to off or from off to on) but does not change the status of any of the other lamps. If L_{j-1} is off, then Step_j does not change anything at all. Show that:
    (a) There is a positive integer $M(n)$ such that after $M(n)$ steps all lamps are on again.
    (b) If n has the form 2^k, then all lamps are on after $n^2 - 1$ steps.
    (c) If n has the form $2^k + 1$, then all lamps are on after $n^2 - n + 1$ steps.

18. **(POL 1)** Let S_n be the number of sequences $(a_1, a_2, \ldots, a_n)$, where $a_i \in \{0, 1\}$, in which no six consecutive blocks are equal. Prove that $S_n \to \infty$ as $n \to \infty$.

19. **(ROU 2)** Let a, b, n be positive integers, $b > 1$ and $b^n - 1 \mid a$. Show that the representation of the number a in the base b contains at least n digits different from zero.

20. **(ROU 3)** Let $c_1, \ldots, c_n \in \mathbb{R}$ $(n \geq 2)$ such that $0 \leq \sum_{i=1}^n c_i \leq n$. Show that we can find integers $k_1, \ldots, k_n$ such that $\sum_{i=1}^n k_i = 0$ and

$$1 - n \leq c_i + nk_i \leq n \quad \text{for every } i = 1, \ldots, n.$$

21. **(UNK 1)** A circle S is said to cut a circle Σ *diametrally* if their common chord is a diameter of Σ.

    Let S_A, S_B, S_C be three circles with distinct centers A, B, C respectively. Prove that A, B, C are collinear if and only if there is no unique circle S that cuts each of S_A, S_B, S_C diametrally. Prove further that if there exists more than one circle S that cuts each of S_A, S_B, S_C diametrally, then all such circles pass through two fixed points. Locate these points in relation to the circles S_A, S_B, S_C.

22. **(UNK 2)**[IMO2] A, B, C, D are four points in the plane, with C, D on the same side of the line AB, such that $AC \cdot BD = AD \cdot BC$ and $\angle ADB = 90° + \angle ACB$. Find the ratio

$$\frac{AB \cdot CD}{AC \cdot BD},$$

    and prove that circles ACD, BCD are orthogonal. (Intersecting circles are said to be orthogonal if at either common point their tangents are perpendicular.)

23. **(UNK 3)** A finite set of (distinct) positive integers is called a "*DS*-set" if each of the integers divides the sum of them all. Prove that every finite set of positive integers is a subset of some *DS*-set.

24. **(USA 3)** Prove that

$$\frac{a}{b+2c+3d} + \frac{b}{c+2d+3a} + \frac{c}{d+2a+3b} + \frac{d}{a+2b+3c} \geq \frac{2}{3}$$

    for all positive real numbers a, b, c, d.

25. **(VNM 1)** Solve the following system of equations, in which a is a given number satisfying $|a| > 1$:

$$\begin{aligned}
x_1^2 &= ax_2 + 1, \\
x_2^2 &= ax_3 + 1, \\
&\cdots \quad \cdots \\
x_{999}^2 &= ax_{1000} + 1, \\
x_{1000}^2 &= ax_1 + 1.
\end{aligned}$$

26. **(VNM 2)** Let a, b, c, d be four nonnegative numbers satisfying $a+b+c+d = 1$. Prove the inequality

$$abc + bcd + cda + dab \leq \frac{1}{27} + \frac{176}{27}abcd.$$

3.35 The Thirty-Fifth IMO
Hong Kong, July 9–22, 1994

3.35.1 Contest Problems

First Day (July 13)

1. Let m and n be positive integers. The set $A = \{a_1, a_2, \ldots, a_m\}$ is a subset of $1, 2, \ldots, n$. Whenever $a_i + a_j \leq n$, $1 \leq i \leq j \leq m$, $a_i + a_j$ also belongs to A. Prove that
$$\frac{a_1 + a_2 + \cdots + a_m}{m} \geq \frac{n+1}{2}.$$

2. N is an arbitrary point on the bisector of $\angle BAC$. P and O are points on the lines AB and AN, respectively, such that $\angle ANP = 90° = \angle APO$. Q is an arbitrary point on NP, and an arbitrary line through Q meets the lines AB and AC at E and F respectively. Prove that $\angle OQE = 90°$ if and only if $QE = QF$.

3. For any positive integer k, A_k is the subset of $\{k+1, k+2, \ldots, 2k\}$ consisting of all elements whose digits in base 2 contain exactly three 1's. Let $f(k)$ denote the number of elements in A_k.
   (a) Prove that for any positive integer m, $f(k) = m$ has at least one solution.
   (b) Determine all positive integers m for which $f(k) = m$ has a unique solution.

Second Day (July 14)

4. Determine all pairs (m, n) of positive integers such that $\frac{n^3+1}{mn-1}$ is an integer.

5. Let S be the set of real numbers greater than -1. Find all functions $f : S \to S$ such that
$$f(x + f(y) + xf(y)) = y + f(x) + yf(x) \qquad \text{for all } x \text{ and } y \text{ in } S,$$
and $f(x)/x$ is strictly increasing for $-1 < x < 0$ and for $0 < x$.

6. Find a set A of positive integers such that for any infinite set P of prime numbers, there exist positive integers $m \in A$ and $n \notin A$, both the product of the same number (at least two) of distinct elements of P.

3.35.2 Shortlisted Problems

1. **A1 (USA)** Let $a_0 = 1994$ and $a_{n+1} = \frac{a_n^2}{a_n+1}$ for each nonnegative integer n. Prove that $1994 - n$ is the greatest integer less than or equal to a_n, $0 \leq n \leq 998$.

2. **A2 (FRA)**[IMO1] Let m and n be positive integers. The set $A = \{a_1, a_2, \ldots, a_m\}$ is a subset of $\{1, 2, \ldots, n\}$. Whenever $a_i + a_j \leq n$, $1 \leq i \leq j \leq m$, $a_i + a_j$ also belongs to A. Prove that
$$\frac{a_1 + a_2 + \cdots + a_m}{m} \geq \frac{n+1}{2}.$$

3. **A3 (UNK)**[IMO5] Let S be the set of real numbers greater than -1. Find all functions $f : S \to S$ such that

$$f(x + f(y) + xf(y)) = y + f(x) + yf(x) \quad \text{for all } x \text{ and } y \text{ in } S,$$

and $f(x)/x$ is strictly increasing for $-1 < x < 0$ and for $0 < x$.

4. **A4 (MNG)** Let $\mathbb{R}$ denote the set of all real numbers and $\mathbb{R}^+$ the subset of all positive ones. Let α and β be given elements in $\mathbb{R}$, not necessarily distinct. Find all functions $f : \mathbb{R}^+ \to \mathbb{R}$ such that

$$f(x)f(y) = y^\alpha f\left(\frac{x}{2}\right) + x^\beta f\left(\frac{y}{2}\right) \quad \text{for all } x \text{ and } y \text{ in } \mathbb{R}^+.$$

5. **A5 (POL)** Let $f(x) = \frac{x^2+1}{2x}$ for $x \neq 0$. Define $f^{(0)}(x) = x$ and $f^{(n)}(x) = f(f^{(n-1)}(x))$ for all positive integers n and $x \neq 0$. Prove that for all nonnegative integers n and $x \neq -1, 0,$ or 1,

$$\frac{f^{(n)}(x)}{f^{(n+1)}(x)} = 1 + \frac{1}{f\left(\left(\frac{x+1}{x-1}\right)^{2^n}\right)}.$$

6. **C1 (UKR)** On a 5×5 board, two players alternately mark numbers on empty cells. The first player always marks 1's, the second 0's. One number is marked per turn, until the board is filled. For each of the nine 3×3 squares the sum of the nine numbers on its cells is computed. Denote by A the maximum of these sums. How large can the first player make A, regardless of the responses of the second player?

7. **C2 (COL)** In a certain city, age is reckoned in terms of real numbers rather than integers. Every two citizens x and x' either know each other or do not know each other. Moreover, if they do not, then there exists a chain of citizens $x = x_0, x_1, \ldots, x_n = x'$ for some integer $n \geq 2$ such that x_{i-1} and x_i know each other. In a census, all male citizens declare their ages, and there is at least one male citizen. Each female citizen provides only the information that her age is the average of the ages of all the citizens she knows. Prove that this is enough to determine uniquely the ages of all the female citizens.

8. **C3 (MKD)** Peter has three accounts in a bank, each with an integral number of dollars. He is only allowed to transfer money from one account to another so that the amount of money in the latter is doubled.
   (a) Prove that Peter can always transfer all his money into two accounts.
   (b) Can Peter always transfer all his money into one account?

9. **C4 (EST)** There are $n + 1$ fixed positions in a row, labeled 0 to n in increasing order from right to left. Cards numbered 0 to n are shuffled and dealt, one in each position. The object of the game is to have card i in the ith position for $0 \leq i \leq n$. If this has not been achieved, the following move is executed. Determine the smallest k such that the kth position is occupied by a card $l > k$. Remove this card, slide all cards from the $(k+1)$st to the lth position one place to the right, and replace the card l in the lth position.

(a) Prove that the game lasts at most $2^n - 1$ moves.

(b) Prove that there exists a unique initial configuration for which the game lasts exactly $2^n - 1$ moves.

10. **C5 (SWE)** At a round table are 1994 girls, playing a game with a deck of n cards. Initially, one girl holds all the cards. In each turn, if at least one girl holds at least two cards, one of these girls must pass a card to each of her two neighbors. The game ends when and only when each girl is holding at most one card.

(a) Prove that if $n \geq 1994$, then the game cannot end.

(b) Prove that if $n < 1994$, then the game must end.

11. **C6 (FIN)** On an infinite square grid, two players alternately mark symbols on empty cells. The first player always marks X's, the second O's. One symbol is marked per turn. The first player wins if there are 11 consecutive X's in a row, column, or diagonal. Prove that the second player can prevent the first from winning.

12. **C7 (BRA)** Prove that for any integer $n \geq 2$, there exists a set of 2^{n-1} points in the plane such that no 3 lie on a line and no $2n$ are the vertices of a convex $2n$-gon.

13. **G1 (FRA)** A semicircle Γ is drawn on one side of a straight line l. C and D are points on Γ. The tangents to Γ at C and D meet l at B and A respectively, with the center of the semicircle between them. Let E be the point of intersection of AC and BD, and F the point on l such that EF is perpendicular to l. Prove that EF bisects $\angle CFD$.

14. **G2 (UKR)** $ABCD$ is a quadrilateral with BC parallel to AD. M is the midpoint of CD, P that of MA and Q that of MB. The lines DP and CQ meet at N. Prove that N is not outside triangle ABM.[9]

15. **G3 (RUS)** A circle ω is tangent to two parallel lines l_1 and l_2. A second circle ω_1 is tangent to l_1 at A and to ω externally at C. A third circle ω_2 is tangent to l_2 at B, to ω externally at D, and to ω_1 externally at E. AD intersects BC at Q. Prove that Q is the circumcenter of triangle CDE.

16. **G4 (AUS-ARM)**[IMO2] N is an arbitrary point on the bisector of $\angle BAC$. P and O are points on the lines AB and AN, respectively, such that $\angle ANP = 90° = \angle APO$. Q is an arbitrary point on NP, and an arbitrary line through Q meets the lines AB and AC at E and F respectively. Prove that $\angle OQE = 90°$ if and only if $QE = QF$.

17. **G5 (CYP)** A line l does not meet a circle ω with center O. E is the point on l such that OE is perpendicular to l. M is any point on l other than E. The tangents from M to ω touch it at A and B. C is the point on MA such that EC is perpendicular to MA. D is the point on MB such that ED is perpendicular to MB.

---

[9] This problem is false. However, it is true if "not outside ABM" is replaced by "not outside $ABCD$".

The line CD cuts OE at F. Prove that the location of F is independent of that of M.

18. **N1 (BGR)** M is a subset of $\{1,2,3,\ldots,15\}$ such that the product of any three distinct elements of M is not a square. Determine the maximum number of elements in M.

19. **N2 (AUS)**[IMO4] Determine all pairs (m,n) of positive integers such that $\frac{n^3+1}{mn-1}$ is an integer.

20. **N3 (FIN)**[IMO6] Find a set A of positive integers such that for any infinite set P of prime numbers, there exist positive integers $m \in A$ and $n \notin A$, both the product of the same number of distinct elements of P.

21. **N4 (FRA)** For any positive integer x_0, three sequences $\{x_n\}, \{y_n\}$, and $\{z_n\}$ are defined as follows:
    (i) $y_0 = 4$ and $z_0 = 1$;
    (ii) if x_n is even for $n \geq 0$, $x_{n+1} = \frac{x_n}{2}$, $y_{n+1} = 2y_n$, and $z_{n+1} = z_n$;
    (iii) if x_n is odd for $n \geq 0$, $x_{n+1} = x_n - \frac{y_n}{2} - z_n$, $y_{n+1} = y_n$, and $z_{n+1} = y_n + z_n$.
    The integer x_0 is said to be *good* if $x_n = 0$ for some $n \geq 1$. Find the number of good integers less than or equal to 1994.

22. **N5 (ROU)**[IMO3] For any positive integer k, A_k is the subset of $\{k+1, k+2, \ldots, 2k\}$ consisting of all elements whose digits in base 2 contain exactly three 1's. Let $f(k)$ denote the number of elements in A_k.
    (a) Prove that for any positive integer m, $f(k) = m$ has at least one solution.
    (b) Determine all positive integers m for which $f(k) = m$ has a unique solution.

23. **N6 (LVA)** Let x_1 and x_2 be relatively prime positive integers. For $n \geq 2$, define $x_{n+1} = x_n x_{n-1} + 1$.
    (a) Prove that for every $i > 1$, there exists $j > i$ such that x_i^i divides x_j^j.
    (b) Is it true that x_1 must divide x_j^j for some $j > 1$?

24. **N7 (UNK)** A *wobbly* number is a positive integer whose digits in base 10 are alternately nonzero and zero, the units digit being nonzero. Determine all positive integers that do not divide any wobbly number.

3.36 The Thirty-Sixth IMO
Toronto, Canada, July 13–25, 1995

3.36.1 Contest Problems

First Day (July 19)

1. Let A, B, C, and D be distinct points on a line, in that order. The circles with diameters AC and BD intersect at X and Y. O is an arbitrary point on the line XY but not on AD. CO intersects the circle with diameter AC again at M, and BO intersects the other circle again at N. Prove that the lines AM, DN, and XY are concurrent.

2. Let a, b, and c be positive real numbers such that $abc = 1$. Prove that

$$\frac{1}{a^3(b+c)} + \frac{1}{b^3(a+c)} + \frac{1}{c^3(a+b)} \geq \frac{3}{2}.$$

3. Determine all integers $n > 3$ such that there are n points $A_1, A_2, \ldots, A_n$ in the plane that satisfy the following two conditions simultaneously:
   (a) No three lie on the same line.
   (b) There exist real numbers $p_1, p_2, \ldots, p_n$ such that the area of $\triangle A_i A_j A_k$ is equal to $p_i + p_j + p_k$, for $1 \leq i < j < k \leq n$.

Second Day (July 20)

4. The positive real numbers $x_0, x_1, \ldots, x_{1995}$ satisfy $x_0 = x_{1995}$ and

$$x_{i-1} + \frac{2}{x_{i-1}} = 2x_i + \frac{1}{x_i}$$

   for $i = 1, 2, \ldots, 1995$. Find the maximum value that x_0 can have.

5. Let $ABCDEF$ be a convex hexagon with $AB = BC = CD$, $DE = EF = FA$, and $\angle BCD = \angle EFA = \pi/3$ (that is, $60°$). Let G and H be two points interior to the hexagon, such that angles AGB and DHE are both $2\pi/3$ (that is, $120°$). Prove that $AG + GB + GH + DH + HE \geq CF$.

6. Let p be an odd prime. Find the number of p-element subsets A of $\{1, 2, \ldots, 2p\}$ such that the sum of all elements of A is divisible by p.

3.36.2 Shortlisted Problems

1. **A1 (RUS)**[IMO2] Let a, b, and c be positive real numbers such that $abc = 1$. Prove that

$$\frac{1}{a^3(b+c)} + \frac{1}{b^3(a+c)} + \frac{1}{c^3(a+b)} \geq \frac{3}{2}.$$

2. **A2 (SWE)** Let a and b be nonnegative integers such that $ab \geq c^2$, where c is an integer. Prove that there is a number n and integers $x_1, x_2, \ldots, x_n, y_1, y_2, \ldots, y_n$ such that

$$\sum_{i=1}^{n} x_i^2 = a, \quad \sum_{i=1}^{n} y_i^2 = b, \quad \text{and} \quad \sum_{i=1}^{n} x_i y_i = c.$$

3. **A3 (UKR)** Let n be an integer, $n \geq 3$. Let $a_1, a_2, \ldots, a_n$ be real numbers such that $2 \leq a_i \leq 3$ for $i = 1, 2, \ldots, n$. If $s = a_1 + a_2 + \cdots + a_n$, prove that

$$\frac{a_1^2 + a_2^2 - a_3^2}{a_1 + a_2 - a_3} + \frac{a_2^2 + a_3^2 - a_4^2}{a_2 + a_3 - a_4} + \cdots + \frac{a_n^2 + a_1^2 - a_2^2}{a_n + a_1 - a_2} \leq 2s - 2n.$$

4. **A4 (USA)** Let a, b, and c be given positive real numbers. Determine all positive real numbers x, y, and z such that

$$x + y + z = a + b + c$$

and

$$4xyz - (a^2 x + b^2 y + c^2 z) = abc.$$

5. **A5 (UKR)** Let $\mathbb{R}$ be the set of real numbers. Does there exist a function $f : \mathbb{R} \to \mathbb{R}$ that simultaneously satisfies the following three conditions?
   (a) There is a positive number M such that $-M \leq f(x) \leq M$ for all x.
   (b) $f(1) = 1$.
   (c) If $x \neq 0$, then

$$f\left(x + \frac{1}{x^2}\right) = f(x) + \left[f\left(\frac{1}{x}\right)\right]^2.$$

6. **A6 (JPN)** Let n be an integer, $n \geq 3$. Let $x_1, x_2, \ldots, x_n$ be real numbers such that $x_i < x_{i+1}$ for $1 \leq i \leq n - 1$. Prove that

$$\frac{n(n-1)}{2} \sum_{i<j} x_i x_j > \left(\sum_{i=1}^{n-1} (n-i) x_i\right) \left(\sum_{j=2}^{n} (j-1) x_j\right).$$

7. **G1 (BGR)**[IMO1] Let A, B, C, and D be distinct points on a line, in that order. The circles with diameters AC and BD intersect at X and Y. O is an arbitrary point on the line XY but not on AD. CO intersects the circle with diameter AC again at M, and BO intersects the other circle again at N. Prove that the lines AM, DN, and XY are concurrent.

8. **G2 (GER)** Let A, B, and C be noncollinear points. Prove that there is a unique point X in the plane of ABC such that $XA^2 + XB^2 + AB^2 = XB^2 + XC^2 + BC^2 = XC^2 + XA^2 + CA^2$.

9. **G3 (TUR)** The incircle of ABC touches BC, CA, and AB at D, E, and F respectively. X is a point inside ABC such that the incircle of XBC touches BC at D also, and touches CX and XB at Y and Z, respectively. Prove that $EFZY$ is a cyclic quadrilateral.

10. **G4 (UKR)** An acute triangle ABC is given. Points A_1 and A_2 are taken on the side BC (with A_2 between A_1 and C), B_1 and B_2 on the side AC (with B_2 between B_1 and A), and C_1 and C_2 on the side AB (with C_2 between C_1 and B) such that

$$\angle AA_1A_2 = \angle AA_2A_1 = \angle BB_1B_2 = \angle BB_2B_1 = \angle CC_1C_2 = \angle CC_2C_1.$$

The lines AA_1, BB_1, and CC_1 form a triangle, and the lines AA_2, BB_2, and CC_2 form a second triangle. Prove that all six vertices of these two triangles lie on a single circle.

11. **G5 (NZL)**[IMO5] Let $ABCDEF$ be a convex hexagon with $AB = BC = CD$, $DE = EF = FA$, and $\angle BCD = \angle EFA = \pi/3$ (that is, $60°$). Let G and H be two points interior to the hexagon such that angles AGB and DHE are both $2\pi/3$ (that is, $120°$). Prove that $AG + GB + GH + DH + HE \geq CF$.

12. **G6 (USA)** Let $A_1A_2A_3A_4$ be a tetrahedron, G its centroid, and A'_1, A'_2, A'_3, and A'_4 the points where the circumsphere of $A_1A_2A_3A_4$ intersects GA_1, GA_2, GA_3, and GA_4, respectively. Prove that

$$GA_1 \cdot GA_2 \cdot GA_3 \cdot GA_4 \leq GA'_1 \cdot GA'_2 \cdot GA'_3 \cdot GA'_4$$

and

$$\frac{1}{GA'_1} + \frac{1}{GA'_2} + \frac{1}{GA'_3} + \frac{1}{GA'_4} \leq \frac{1}{GA_1} + \frac{1}{GA_2} + \frac{1}{GA_3} + \frac{1}{GA_4}.$$

13. **G7 (LVA)** O is a point inside a convex quadrilateral $ABCD$ of area S. K, L, M, and N are interior points of the sides AB, BC, CD, and DA respectively. If $OKBL$ and $OMDN$ are parallelograms, prove that $\sqrt{S} \geq \sqrt{S_1} + \sqrt{S_2}$, where S_1 and S_2 are the areas of $ONAK$ and $OLCM$ respectively.

14. **G8 (COL)** Let ABC be a triangle. A circle passing through B and C intersects the sides AB and AC again at C' and B', respectively. Prove that BB', CC', and HH' are concurrent, where H and H' are the orthocenters of triangles ABC and $AB'C'$ respectively.

15. **N1 (ROU)** Let k be a positive integer. Prove that there are infinitely many perfect squares of the form $n2^k - 7$, where n is a positive integer.

16. **N2 (RUS)** Let $\mathbb{Z}$ denote the set of all integers. Prove that for any integers A and B, one can find an integer C for which $M_1 = \{x^2 + Ax + B : x \in \mathbb{Z}\}$ and $M_2 = \{2x^2 + 2x + C : x \in \mathbb{Z}\}$ do not intersect.

17. **N3 (CZE)**[IMO3] Determine all integers $n > 3$ such that there are n points A_1, A_2, $\ldots, A_n$ in the plane that satisfy the following two conditions simultaneously:
    (a) No three lie on the same line.
    (b) There exist real numbers $p_1, p_2, \ldots, p_n$ such that the area of $\triangle A_iA_jA_k$ is equal to $p_i + p_j + p_k$, for $1 \leq i < j < k \leq n$.

18. **N4 (BGR)** Find all positive integers x and y such that $x + y^2 + z^3 = xyz$, where z is the greatest common divisor of x and y.

19. **N5 (IRL)** At a meeting of $12k$ people, each person exchanges greetings with exactly $3k + 6$ others. For any two people, the number who exchange greetings with both is the same. How many people are at the meeting?

20. **N6 (POL)**[IMO6] Let p be an odd prime. Find the number of p-element subsets A of $\{1, 2, \ldots, 2p\}$ such that the sum of all elements of A is divisible by p.

21. **N7 (BLR)** Does there exist an integer $n > 1$ that satisfies the following condition?

The set of positive integers can be partitioned into n nonempty subsets such that an arbitrary sum of $n - 1$ integers, one taken from each of any $n - 1$ of the subsets, lies in the remaining subset.

22. **N8 (GER)** Let p be an odd prime. Determine positive integers x and y for which $x \leq y$ and $\sqrt{2p} - \sqrt{x} - \sqrt{y}$ is nonnegative and as small as possible.

23. **S1 (UKR)** Does there exist a sequence $F(1), F(2), F(3), \ldots$ of nonnegative integers that simultaneously satisfies the following three conditions?
    (a) Each of the integers $0, 1, 2, \ldots$ occurs in the sequence.
    (b) Each positive integer occurs in the sequence infinitely often.
    (c) For any $n \geq 2$,

$$F\left(F\left(n^{163}\right)\right) = F(F(n)) + F(F(361)).$$

24. **S2 (POL)**[IMO4] The positive real numbers $x_0, x_1, \ldots, x_{1995}$ satisfy $x_0 = x_{1995}$ and

$$x_{i-1} + \frac{2}{x_{i-1}} = 2x_i + \frac{1}{x_i}$$

for $i = 1, 2, \ldots, 1995$. Find the maximum value that x_0 can have.

25. **S3 (POL)** For an integer $x \geq 1$, let $p(x)$ be the least prime that does not divide x, and define $q(x)$ to be the product of all primes less than $p(x)$. In particular, $p(1) = 2$. For x such that $p(x) = 2$, define $q(x) = 1$. Consider the sequence $x_0, x_1, x_2, \ldots$ defined by $x_0 = 1$ and

$$x_{n+1} = \frac{x_n p(x_n)}{q(x_n)}$$

for $n \geq 0$. Find all n such that $x_n = 1995$.

26. **S4 (NZL)** Suppose that $x_1, x_2, x_3, \ldots$ are positive real numbers for which

$$x_n^n = \sum_{j=0}^{n-1} x_n^j$$

for $n = 1, 2, 3, \ldots$. Prove that for all n,

$$2 - \frac{1}{2^{n-1}} \leq x_n < 2 - \frac{1}{2^n}.$$

27. **S5 (FIN)** For positive integers n, the numbers $f(n)$ are defined inductively as follows: $f(1) = 1$, and for every positive integer n, $f(n+1)$ is the greatest integer m such that there is an arithmetic progression of positive integers $a_1 < a_2 < \cdots < a_m = n$ for which

$$f(a_1) = f(a_2) = \cdots = f(a_m).$$

Prove that there are positive integers a and b such that $f(an+b) = n+2$ for every positive integer n.

28. **S6 (IND)** Let $\mathbb{N}$ denote the set of all positive integers. Prove that there exists a unique function $f : \mathbb{N} \to \mathbb{N}$ satisfying

$$f(m + f(n)) = n + f(m + 95)$$

for all m and n in $\mathbb{N}$. What is the value of $\sum_{k=1}^{19} f(k)$?

3.37 The Third-Seventh IMO
Mumbai, India, July 5–17, 1996

3.37.1 Contest Problems

First Day (July 10)

1. We are given a positive integer r and a rectangular board $ABCD$ with dimensions $|AB| = 20$, $|BC| = 12$. The rectangle is divided into a grid of 20×12 unit squares. The following moves are permitted on the board: One can move from one square to another only if the distance between the centers of the two squares is $\sqrt{r}$. The task is to find a sequence of moves leading from the square corresponding to vertex A to the square corresponding to vertex B.
   (a) Show that the task cannot be done if r is divisible by 2 or 3.
   (b) Prove that the task is possible when $r = 73$.
   (c) Is there a solution when $r = 97$?

2. Let P be a point inside $\triangle ABC$ such that

$$\angle APB - \angle C = \angle APC - \angle B.$$

   Let D, E be the incenters of $\triangle APB, \triangle APC$ respectively. Show that AP, BD, and CE meet in a point.

3. Let $\mathbb{N}_0$ denote the set of nonnegative integers. Find all functions f from $\mathbb{N}_0$ into itself such that

$$f(m + f(n)) = f(f(m)) + f(n), \quad \forall m, n \in \mathbb{N}_0.$$

Second Day (July 11)

4. The positive integers a and b are such that the numbers $15a + 16b$ and $16a - 15b$ are both squares of positive integers. What is the least possible value that can be taken on by the smaller of these two squares?

5. Let $ABCDEF$ be a convex hexagon such that AB is parallel to DE, BC is parallel to EF, and CD is parallel to AF. Let R_A, R_C, R_E be the circumradii of triangles FAB, BCD, DEF respectively, and let P denote the perimeter of the hexagon. Prove that

$$R_A + R_C + R_E \geq \frac{P}{2}.$$

6. Let p, q, n be three positive integers with $p + q < n$. Let $(x_0, x_1, \ldots, x_n)$ be an $(n+1)$-tuple of integers satisfying the following conditions:
   (i) $x_0 = x_n = 0$.
   (ii) For each i with $1 \leq i \leq n$, either $x_i - x_{i-1} = p$ or $x_i - x_{i-1} = -q$.
   Show that there exists a pair (i, j) of distinct indices with $(i, j) \neq (0, n)$ such that $x_i = x_j$.

3.37.2 Shortlisted Problems

1. **A1 (SVN)** Let a, b, and c be positive real numbers such that $abc = 1$. Prove that

$$\frac{ab}{a^5 + b^5 + ab} + \frac{bc}{b^5 + c^5 + bc} + \frac{ca}{c^5 + a^5 + ca} \leq 1.$$

When does equality hold?

2. **A2 (IRL)** Let $a_1 \geq a_2 \geq \cdots \geq a_n$ be real numbers such that

$$a_1^k + a_2^k + \cdots + a_n^k \geq 0$$

for all integers $k > 0$. Let $p = \max\{|a_1|, \ldots, |a_n|\}$. Prove that $p = a_1$ and that

$$(x - a_1)(x - a_2) \cdots (x - a_n) \leq x^n - a_1^n$$

for all $x > a_1$.

3. **A3 (HEL)** Let $a > 2$ be given, and define recursively

$$a_0 = 1, \quad a_1 = a, \quad a_{n+1} = \left(\frac{a_n^2}{a_{n-1}^2} - 2\right) a_n.$$

Show that for all $k \in \mathbb{N}$, we have

$$\frac{1}{a_0} + \frac{1}{a_1} + \frac{1}{a_2} + \cdots + \frac{1}{a_k} < \frac{1}{2}\left(2 + a - \sqrt{a^2 - 4}\right).$$

4. **A4 (KOR)** Let $a_1, a_2, \ldots, a_n$ be nonnegative real numbers, not all zero.
   (a) Prove that $x^n - a_1 x^{n-1} - \cdots - a_{n-1} x - a_n = 0$ has precisely one positive real root.
   (b) Let $A = \sum_{j=1}^n a_j$, $B = \sum_{j=1}^n j a_j$, and let R be the positive real root of the equation in part (a). Prove that

$$A^A \leq R^B.$$

5. **A5 (ROU)** Let $P(x)$ be the real polynomial function $P(x) = ax^3 + bx^2 + cx + d$. Prove that if $|P(x)| \leq 1$ for all x such that $|x| \leq 1$, then

$$|a| + |b| + |c| + |d| \leq 7.$$

6. **A6 (IRL)** Let n be an even positive integer. Prove that there exists a positive integer k such that
$$k = f(x)(x + 1)^n + g(x)(x^n + 1)$$
for some polynomials $f(x), g(x)$ having integer coefficients. If k_0 denotes the least such k, determine k_0 as a function of n.
   A6′ Let n be an even positive integer. Prove that there exists a positive integer k such that

$$k = f(x)(x+1)^n + g(x)(x^n+1)$$

for some polynomials $f(x), g(x)$ having integer coefficients. If k_0 denotes the least such k, show that $k_0 = 2^q$, where q is the odd integer determined by $n = q2^r$, $r \in \mathbb{N}$.

A6″ Prove that for each positive integer n, there exist polynomials $f(x), g(x)$ having integer coefficients such that

$$f(x)(x+1)^{2^n} + g(x)(x^{2^n}+1) = 2.$$

7. **A7 (ARM)** Let f be a function from the set of real numbers $\mathbb{R}$ into itself such that for all $x \in \mathbb{R}$, we have $|f(x)| \le 1$ and

$$f\left(x+\frac{13}{42}\right) + f(x) = f\left(x+\frac{1}{6}\right) + f\left(x+\frac{1}{7}\right).$$

Prove that f is a periodic function (that is, there exists a nonzero real number c such that $f(x+c) = f(x)$ for all $x \in \mathbb{R}$).

8. **A8 (ROU)**[IMO3] Let $\mathbb{N}_0$ denote the set of nonnegative integers. Find all functions f from $\mathbb{N}_0$ into itself such that

$$f(m+f(n)) = f(f(m)) + f(n), \qquad \forall m, n \in \mathbb{N}_0.$$

9. **A9 (POL)** Let the sequence $a(n), n = 1, 2, 3, \ldots$, be generated as follows: $a(1) = 0$, and for $n > 1$,

$$a(n) = a([n/2]) + (-1)^{\frac{n(n+1)}{2}}. \qquad \text{(Here } [t] = \text{the greatest integer} \le t.)$$

(a) Determine the maximum and minimum value of $a(n)$ over $n \le 1996$ and find all $n \le 1996$ for which these extreme values are attained.
(b) How many terms $a(n), n \le 1996$, are equal to 0?

10. **G1 (UNK)** Let triangle ABC have orthocenter H, and let P be a point on its circumcircle, distinct from A, B, C. Let E be the foot of the altitude BH, let $PAQB$ and $PARC$ be parallelograms, and let AQ meet HR in X. Prove that EX is parallel to AP.

11. **G2 (CAN)**[IMO2] Let P be a point inside $\triangle ABC$ such that

$$\angle APB - \angle C = \angle APC - \angle B.$$

Let D, E be the incenters of $\triangle APB, \triangle APC$ respectively. Show that AP, BD and CE meet in a point.

12. **G3 (UNK)** Let ABC be an acute-angled triangle with $BC > CA$. Let O be the circumcenter, H its orthocenter, and F the foot of its altitude CH. Let the perpendicular to OF at F meet the side CA at P. Prove that $\angle FHP = \angle BAC$.
Possible second part: What happens if $|BC| \le |CA|$ (the triangle still being acute-angled)?

13. **G4 (USA)** Let $\triangle ABC$ be an equilateral triangle and let P be a point in its interior. Let the lines AP, BP, CP meet the sides BC, CA, AB in the points A_1, B_1, C_1 respectively. Prove that

$$A_1B_1 \cdot B_1C_1 \cdot C_1A_1 \geq A_1B \cdot B_1C \cdot C_1A.$$

14. **G5 (ARM)**[IMO5] Let $ABCDEF$ be a convex hexagon such that AB is parallel to DE, BC is parallel to EF, and CD is parallel to AF. Let R_A, R_C, R_E be the circumradii of triangles FAB, BCD, DEF respectively, and let P denote the perimeter of the hexagon. Prove that

$$R_A + R_C + R_E \geq \frac{P}{2}.$$

15. **G6 (ARM)** Let the sides of two rectangles be $\{a, b\}$ and $\{c, d\}$ with $a < c \leq d < b$ and $ab < cd$. Prove that the first rectangle can be placed within the second one if and only if

$$(b^2 - a^2)^2 \leq (bd - ac)^2 + (bc - ad)^2.$$

16. **G7 (UNK)** Let ABC be an acute-angled triangle with circumcenter O and circumradius R. Let AO meet the circle BOC again in A', let BO meet the circle COA again in B', and let CO meet the circle AOB again in C'. Prove that

$$OA' \cdot OB' \cdot OC' \geq 8R^3.$$

When does equality hold?

17. **G8 (RUS)** Let $ABCD$ be a convex quadrilateral, and let R_A, R_B, R_C, and R_D denote the circumradii of the triangles DAB, ABC, BCD, and CDA respectively. Prove that $R_A + R_C > R_B + R_D$ if and only if

$$\angle A + \angle C > \angle B + \angle D.$$

18. **G9 (UKR)** In the plane are given a point O and a polygon $\mathscr{F}$ (not necessarily convex). Let P denote the perimeter of $\mathscr{F}$, D the sum of the distances from O to the vertices of $\mathscr{F}$, and H the sum of the distances from O to the lines containing the sides of $\mathscr{F}$. Prove that

$$D^2 - H^2 \geq \frac{P^2}{4}.$$

19. **N1 (UKR)** Four integers are marked on a circle. At each step we simultaneously replace each number by the difference between this number and the next number on the circle, in a given direction (that is, the numbers a, b, c, d are replaced by $a - b, b - c, c - d, d - a$). Is it possible after 1996 such steps to have numbers a, b, c, d such that the numbers $|bc - ad|, |ac - bd|, |ab - cd|$ are primes?

20. **N2 (RUS)**[IMO4] The positive integers a and b are such that the numbers $15a + 16b$ and $16a - 15b$ are both squares of positive integers. What is the least possible value that can be taken on by the smaller of these two squares?

21. **N3 (BGR)** A finite sequence of integers $a_0, a_1, \ldots, a_n$ is called *quadratic* if for each $i \in \{1, 2, \ldots, n\}$ we have the equality $|a_i - a_{i-1}| = i^2$.
    (a) Prove that for any two integers b and c, there exist a natural number n and a quadratic sequence with $a_0 = b$ and $a_n = c$.
    (b) Find the smallest natural number n for which there exists a quadratic sequence with $a_0 = 0$ and $a_n = 1996$.

22. **N4 (BGR)** Find all positive integers a and b for which

$$\left[\frac{a^2}{b}\right] + \left[\frac{b^2}{a}\right] = \left[\frac{a^2 + b^2}{ab}\right] + ab$$

where as usual, $[t]$ refers to greatest integer that is less than or equal to t.

23. **N5 (ROU)** Let $\mathbb{N}_0$ denote the set of nonnegative integers. Find a bijective function f from $\mathbb{N}_0$ into $\mathbb{N}_0$ such that for all $m, n \in \mathbb{N}_0$,

$$f(3mn + m + n) = 4f(m)f(n) + f(m) + f(n).$$

24. **C1 (FIN)**[IMO1] We are given a positive integer r and a rectangular board $ABCD$ with dimensions $|AB| = 20$, $|BC| = 12$. The rectangle is divided into a grid of 20×12 unit squares. The following moves are permitted on the board: One can move from one square to another only if the distance between the centers of the two squares is $\sqrt{r}$. The task is to find a sequence of moves leading from the square corresponding to vertex A to the square corresponding to vertex B.
    (a) Show that the task cannot be done if r is divisible by 2 or 3.
    (b) Prove that the task is possible when $r = 73$.
    (c) Is there a solution when $r = 97$?

25. **C2 (UKR)** An $(n-1) \times (n-1)$ square is divided into $(n-1)^2$ unit squares in the usual manner. Each of the n^2 vertices of these squares is to be colored red or blue. Find the number of different colorings such that each unit square has exactly two red vertices. (Two coloring schemes are regarded as different if at least one vertex is colored differently in the two schemes.)

26. **C3 (USA)** Let k, m, n be integers such that $1 < n \leq m - 1 \leq k$. Determine the maximum size of a subset S of the set $\{1, 2, 3, \ldots, k\}$ such that no n distinct elements of S add up to m.

27. **C4 (FIN)** Determine whether or not there exist two disjoint infinite sets $\mathscr{A}$ and $\mathscr{B}$ of points in the plane satisfying the following conditions:
    (i) No three points in $\mathscr{A} \cup \mathscr{B}$ are collinear, and the distance between any two points in $\mathscr{A} \cup \mathscr{B}$ is at least 1.
    (ii) There is a point of $\mathscr{A}$ in any triangle whose vertices are in $\mathscr{B}$, and there is a point of $\mathscr{B}$ in any triangle whose vertices are in $\mathscr{A}$.

28. **C5 (FRA)**[IMO6] Let p, q, n be three positive integers with $p + q < n$. Let $(x_0, x_1, \ldots, x_n)$ be an $(n+1)$-tuple of integers satisfying the following conditions:
    (i) $x_0 = x_n = 0$.

(ii) For each i with $1 \le i \le n$, either $x_i - x_{i-1} = p$ or $x_i - x_{i-1} = -q$.
Show that there exists a pair (i, j) of distinct indices with $(i, j) \ne (0, n)$ such that
$x_i = x_j$.

29. **C6 (CAN)** A finite number of beans are placed on an infinite row of squares. A sequence of moves is performed as follows: At each stage a square containing more than one bean is chosen. Two beans are taken from this square; one of them is placed on the square immediately to the left, and the other is placed on the square immediately to the right of the chosen square. The sequence terminates if at some point there is at most one bean on each square. Given some initial configuration, show that any legal sequence of moves will terminate after the same number of steps and with the same final configuration.

30. **C7 (IRL)** Let U be a finite set and let f, g be bijective functions from U onto itself. Let

$$S = \{w \in U : f(f(w)) = g(g(w))\}, \qquad T = \{w \in U : f(g(w)) = g(f(w))\},$$

and suppose that $U = S \cup T$. Prove that for every $w \in U$, $f(w) \in S$ if and only if $g(w) \in S$.

3.38 The Thirty-Eighth IMO
Mar del Plata, Argentina, July 18–31, 1997

3.38.1 Contest Problems

First Day (July 24)

1. An infinite square grid is colored in the chessboard pattern. For any pair of positive integers m, n consider a right-angled triangle whose vertices are grid points and whose legs, of lengths m and n, run along the lines of the grid. Let S_b be the total area of the black part of the triangle and S_w the total area of its white part. Define the function $f(m,n) = |S_b - S_w|$.
   (a) Calculate $f(m,n)$ for all m, n that have the same parity.
   (b) Prove that $f(m,n) \leq \frac{1}{2} \max(m,n)$.
   (c) Show that $f(m,n)$ is not bounded from above.

2. In triangle ABC the angle at A is the smallest. A line through A meets the circumcircle again at the point U lying on the arc BC opposite to A. The perpendicular bisectors of CA and AB meet AU at V and W, respectively, and the lines CV, BW meet at T. Show that $AU = TB + TC$.

3. Let $x_1, x_2, \ldots, x_n$ be real numbers satisfying the conditions
$$|x_1 + x_2 + \cdots + x_n| = 1 \quad \text{and} \quad |x_i| \leq \frac{n+1}{2} \quad \text{for} \quad i = 1, 2, \ldots, n.$$
   Show that there exists a permutation $y_1, \ldots, y_n$ of the sequence $x_1, \ldots, x_n$ such that
$$|y_1 + 2y_2 + \cdots + ny_n| \leq \frac{n+1}{2}.$$

Second Day (July 25)

4. An $n \times n$ matrix with entries from $\{1, 2, \ldots, 2n-1\}$ is called a *silver matrix* if for each i the union of the ith row and the ith column contains $2n-1$ distinct entries. Show that:
   (a) There exist no silver matrices for $n = 1997$.
   (b) Silver matrices exist for infinitely many values of n.

5. Find all pairs of integers $x, y \geq 1$ satisfying the equation $x^{y^2} = y^x$.

6. For a positive integer n, let $f(n)$ denote the number of ways to represent n as the sum of powers of 2 with nonnegative integer exponents. Representations that differ only in the ordering in their summands are not considered to be distinct. (For instance, $f(4) = 4$ because the number 4 can be represented in the following four ways: 4; 2+2; 2+1+1; 1+1+1+1.) Prove that the inequality
$$2^{n^2/4} < f(2^n) < 2^{n^2/2}$$
   holds for any integer $n \geq 3$.

3.38.2 Shortlisted Problems

1. **(BLR)**[IMO1] An infinite square grid is colored in the chessboard pattern. For any pair of positive integers m, n consider a right-angled triangle whose vertices are grid points and whose legs, of lengths m and n, run along the lines of the grid. Let S_b be the total area of the black part of the triangle and S_w the total area of its white part. Define the function $f(m, n) = |S_b - S_w|$.
   (a) Calculate $f(m, n)$ for all m, n that have the same parity.
   (b) Prove that $f(m, n) \leq \frac{1}{2} \max(m, n)$.
   (c) Show that $f(m, n)$ is not bounded from above.

2. **(CAN)** Let $R_1, R_2, \ldots$ be the family of finite sequences of positive integers defined by the following rules: $R_1 = (1)$, and if $R_{n-1} = (x_1, \ldots, x_s)$, then

$$R_n = (1, 2, \ldots, x_1, 1, 2, \ldots, x_2, \ldots, 1, 2, \ldots, x_s, n).$$

   For example, $R_2 = (1, 2), R_3 = (1, 1, 2, 3), R_4 = (1, 1, 1, 2, 1, 2, 3, 4)$.
   Prove that if $n > 1$, then the kth term from the left in R_n is equal to 1 if and only if the kth term from the right in R_n is different from 1.

3. **(GER)** For each finite set U of nonzero vectors in the plane we define $l(U)$ to be the length of the vector that is the sum of all vectors in U. Given a finite set V of nonzero vectors in the plane, a subset B of V is said to be maximal if $l(B)$ is greater than or equal to $l(A)$ for each nonempty subset A of V.
   (a) Construct sets of 4 and 5 vectors that have 8 and 10 maximal subsets respectively.
   (b) Show that for any set V consisting of $n \geq 1$ vectors, the number of maximal subsets is less than or equal to $2n$.

4. **(IRN)**[IMO4] An $n \times n$ matrix with entries from $\{1, 2, \ldots, 2n - 1\}$ is called a *coveralls matrix* if for each i the union of the ith row and the ith column contains $2n - 1$ distinct entries. Show that:
   (a) There exist no coveralls matrices for $n = 1997$.
   (b) Coveralls matrices exist for infinitely many values of n.

5. **(ROU)** Let $ABCD$ be a regular tetrahedron and M, N distinct points in the planes ABC and ADC respectively. Show that the segments MN, BN, MD are the sides of a triangle.

6. **(IRL)** (a) Let n be a positive integer. Prove that there exist distinct positive integers x, y, z such that
$$x^{n-1} + y^n = z^{n+1}.$$

   (b) Let a, b, c be positive integers such that a and b are relatively prime and c is relatively prime either to a or to b. Prove that there exist infinitely many triples (x, y, z) of distinct positive integers x, y, z such that

$$x^a + y^b = z^c.$$

Original formulation: Let a,b,c,n be positive integers such that n is odd and ac is relatively prime to $2b$. Prove that there exist distinct positive integers x,y,z such that

(i) $x^a + y^b = z^c$, and

(ii) xyz is relatively prime to n.

7. **(RUS)** Let $ABCDEF$ be a convex hexagon such that $AB = BC, CD = DE, EF = FA$. Prove that

$$\frac{BC}{BE} + \frac{DE}{DA} + \frac{FA}{FC} \geq \frac{3}{2}.$$

When does equality occur?

8. **(UNK)**[IMO2] In triangle ABC the angle at A is the smallest. A line through A meets the circumcircle again at the point U lying on the arc BC opposite to A. The perpendicular bisectors of CA and AB meet AU at V and W, respectively, and the lines CV, BW meet at T. Show that $AU = TB + TC$.

   Original formulation. Four different points A, B, C, D are chosen on a circle Γ such that the triangle BCD is not right-angled. Prove that:

   (a) The perpendicular bisectors of AB and AC meet the line AD at certain points W and V, respectively, and that the lines CV and BW meet at a certain point T.

   (b) The length of one of the line segments AD, BT, and CT is the sum of the lengths of the other two.

9. **(USA)** Let $A_1A_2A_3$ be a nonisosceles triangle with incenter I. Let C_i, $i = 1, 2, 3$, be the smaller circle through I tangent to A_iA_{i+1} and A_iA_{i+2} (the addition of indices being mod 3). Let B_i, $i = 1, 2, 3$, be the second point of intersection of C_{i+1} and C_{i+2}. Prove that the circumcenters of the triangles $A_1B_1I, A_2B_2I, A_3B_3I$ are collinear.

10. **(CZE)** Find all positive integers k for which the following statement is true: If $F(x)$ is a polynomial with integer coefficients satisfying the condition

$$0 \leq F(c) \leq k \quad \text{for each } c \in \{0, 1, \ldots, k+1\},$$

then $F(0) = F(1) = \cdots = F(k+1)$.

11. **(NLD)** Let $P(x)$ be a polynomial with real coefficients such that $P(x) > 0$ for all $x \geq 0$. Prove that there exists a positive integer n such that $(1+x)^n P(x)$ is a polynomial with nonnegative coefficients.

12. **(ITA)** Let p be a prime number and let $f(x)$ be a polynomial of degree d with integer coefficients such that:

   (i) $f(0) = 0, f(1) = 1$;

   (ii) for every positive integer n, the remainder of the division of $f(n)$ by p is either 0 or 1.

   Prove that $d \geq p - 1$.

13. **(IND)** In town A, there are n girls and n boys, and each girl knows each boy. In town B, there are n girls $g_1, g_2, \ldots, g_n$ and $2n - 1$ boys $b_1, b_2, \ldots, b_{2n-1}$. The

girl g_i, $i = 1,2,\ldots,n$, knows the boys $b_1,b_2,\ldots,b_{2i-1}$, and no others. For all $r = 1,2,\ldots,n$, denote by $A(r),B(r)$ the number of different ways in which r girls from town A, respectively town B, can dance with r boys from their own town, forming r pairs, each girl with a boy she knows. Prove that $A(r) = B(r)$ for each $r = 1,2,\ldots,n$.

14. **(IND)** Let b,m,n be positive integers such that $b > 1$ and $m \neq n$. Prove that if $b^m - 1$ and $b^n - 1$ have the same prime divisors, then $b + 1$ is a power of 2.

15. **(RUS)** An infinite arithmetic progression whose terms are positive integers contains the square of an integer and the cube of an integer. Show that it contains the sixth power of an integer.

16. **(BLR)** In an acute-angled triangle ABC, let AD,BE be altitudes and AP,BQ internal bisectors. Denote by I and O the incenter and the circumcenter of the triangle, respectively. Prove that the points D, E, and I are collinear if and only if the points P, Q, and O are collinear.

17. **(CZE)**[IMO5] Find all pairs of integers $x,y \geq 1$ satisfying the equation $x^{y^2} = y^x$.

18. **(UNK)** The altitudes through the vertices A,B,C of an acute-angled triangle ABC meet the opposite sides at D,E,F, respectively. The line through D parallel to EF meets the lines AC and AB at Q and R, respectively. The line EF meets BC at P. Prove that the circumcircle of the triangle PQR passes through the midpoint of BC.

19. **(IRL)** Let $a_1 \geq \cdots \geq a_n \geq a_{n+1} = 0$ be a sequence of real numbers. Prove that

$$\sqrt{\sum_{k=1}^{n} a_k} \leq \sum_{k=1}^{n} \sqrt{k}(\sqrt{a_k} - \sqrt{a_{k+1}}).$$

20. **(IRL)** Let D be an internal point on the side BC of a triangle ABC. The line AD meets the circumcircle of ABC again at X. Let P and Q be the feet of the perpendiculars from X to AB and AC, respectively, and let γ be the circle with diameter XD. Prove that the line PQ is tangent to γ if and only if $AB = AC$.

21. **(RUS)**[IMO3] Let $x_1,x_2,\ldots,x_n$ be real numbers satisfying the conditions

$$|x_1 + x_2 + \cdots + x_n| = 1 \quad \text{and} \quad |x_i| \leq \frac{n+1}{2} \quad \text{for} \quad i = 1,2,\ldots,n.$$

Show that there exists a permutation $y_1,\ldots,y_n$ of the sequence $x_1,\ldots,x_n$ such that

$$|y_1 + 2y_2 + \cdots + ny_n| \leq \frac{n+1}{2}.$$

22. **(UKR)** (a) Do there exist functions $f : \mathbb{R} \to \mathbb{R}$ and $g : \mathbb{R} \to \mathbb{R}$ such that

$$f(g(x)) = x^2 \quad \text{and} \quad g(f(x)) = x^3 \quad \text{for all } x \in \mathbb{R}?$$

(b) Do there exist functions $f : \mathbb{R} \to \mathbb{R}$ and $g : \mathbb{R} \to \mathbb{R}$ such that

$$f(g(x)) = x^2 \quad \text{and} \quad g(f(x)) = x^4 \quad \text{for all } x \in \mathbb{R}?$$

23. **(UNK)** Let $ABCD$ be a convex quadrilateral and O the intersection of its diagonals AC and BD. If

$$OA \sin \angle A + OC \sin \angle C = OB \sin \angle B + OD \sin \angle D,$$

prove that $ABCD$ is cyclic.

24. **(LTU)**[IMO6] For a positive integer n, let $f(n)$ denote the number of ways to represent n as the sum of powers of 2 with nonnegative integer exponents. Representations that differ only in the ordering in their summands are not considered to be distinct. (For instance, $f(4) = 4$ because the number 4 can be represented in the following four ways: 4; $2+2$; $2+1+1$; $1+1+1+1$.) Prove that the inequality

$$2^{n^2/4} < f(2^n) < 2^{n^2/2}$$

holds for any integer $n \geq 3$.

25. **(POL)** The bisectors of angles A, B, C of a triangle ABC meet its circumcircle again at the points K, L, M, respectively. Let R be an internal point on the side AB. The points P and Q are defined by the following conditions: RP is parallel to AK, and BP is perpendicular to BL; RQ is parallel to BL, and AQ is perpendicular to AK. Show that the lines KP, LQ, MR have a point in common.

26. **(ITA)** For every integer $n \geq 2$ determine the minimum value that the sum $a_0 + a_1 + \cdots + a_n$ can take for nonnegative numbers $a_0, a_1, \ldots, a_n$ satisfying the condition

$$a_0 = 1, \quad a_i \leq a_{i+1} + a_{i+2} \quad \text{for } i = 0, \ldots, n-2.$$

3.39 The Thirty-Ninth IMO
Taipei, Taiwan, July 10–21, 1998

3.39.1 Contest Problems

First Day (July 15)

1. A convex quadrilateral $ABCD$ has perpendicular diagonals. The perpendicular bisectors of AB and CD meet at a unique point P inside $ABCD$. Prove that $ABCD$ is cyclic if and only if triangles ABP and CDP have equal areas.

2. In a contest, there are m candidates and n judges, where $n \geq 3$ is an odd integer. Each candidate is evaluated by each judge as either pass or fail. Suppose that each pair of judges agrees on at most k candidates. Prove that

$$\frac{k}{m} \geq \frac{n-1}{2n}.$$

3. For any positive integer n, let $\tau(n)$ denote the number of its positive divisors (including 1 and itself). Determine all positive integers m for which there exists a positive integer n such that $\frac{\tau(n^2)}{\tau(n)} = m$.

Second Day (July 16)

4. Determine all pairs (x, y) of positive integers such that $x^2 y + x + y$ is divisible by $xy^2 + y + 7$.

5. Let I be the incenter of triangle ABC. Let K, L, and M be the points of tangency of the incircle of ABC with AB, BC, and CA, respectively. The line t passes through B and is parallel to KL. The lines MK and ML intersect t at the points R and S. Prove that $\angle RIS$ is acute.

6. Determine the least possible value of $f(1998)$, where f is a function from the set $\mathbb{N}$ of positive integers into itself such that for all $m, n \in \mathbb{N}$,

$$f(n^2 f(m)) = m[f(n)]^2.$$

3.39.2 Shortlisted Problems

1. (**LUX**)[IMO1] A convex quadrilateral $ABCD$ has perpendicular diagonals. The perpendicular bisectors of AB and CD meet at a unique point P inside $ABCD$. Prove that $ABCD$ is cyclic if and only if triangles ABP and CDP have equal areas.

2. (**POL**) Let $ABCD$ be a cyclic quadrilateral. Let E and F be variable points on the sides AB and CD, respectively, such that $AE : EB = CF : FD$. Let P be the point on the segment EF such that $PE : PF = AB : CD$. Prove that the ratio between the areas of triangles APD and BPC does not depend on the choice of E and F.

3. **(UKR)^{IMO5}** Let I be the incenter of triangle ABC. Let K, L, and M be the points of tangency of the incircle of ABC with AB, BC, and CA, respectively. The line t passes through B and is parallel to KL. The lines MK and ML intersect t at the points R and S. Prove that $\angle RIS$ is acute.

4. **(ARM)** Let M and N be points inside triangle ABC such that

$$\angle MAB = \angle NAC \quad \text{and} \quad \angle MBA = \angle NBC.$$

Prove that

$$\frac{AM \cdot AN}{AB \cdot AC} + \frac{BM \cdot BN}{BA \cdot BC} + \frac{CM \cdot CN}{CA \cdot CB} = 1.$$

5. **(FRA)** Let ABC be a triangle, H its orthocenter, O its circumcenter, and R its circumradius. Let D be the reflection of A across BC, E that of B across CA, and F that of C across AB. Prove that D, E, and F are collinear if and only if $OH = 2R$.

6. **(POL)** Let $ABCDEF$ be a convex hexagon such that $\angle B + \angle D + \angle F = 360°$ and

$$\frac{AB}{BC} \cdot \frac{CD}{DE} \cdot \frac{EF}{FA} = 1.$$

Prove that

$$\frac{BC}{CA} \cdot \frac{AE}{EF} \cdot \frac{FD}{DB} = 1.$$

7. **(UNK)** Let ABC be a triangle such that $\angle ACB = 2\angle ABC$. Let D be the point on the side BC such that $CD = 2BD$. The segment AD is extended to E so that $AD = DE$. Prove that

$$\angle ECB + 180° = 2\angle EBC.$$

8. **(IND)** Let ABC be a triangle such that $\angle A = 90°$ and $\angle B < \angle C$. The tangent at A to its circumcircle ω meets the line BC at D. Let E be the reflection of A across BC, X the foot of the perpendicular from A to BE, and Y the midpoint of AX. Let the line BY meet ω again at Z. Prove that the line BD is tangent to the circumcircle of triangle ADZ.

9. **(MNG)** Let $a_1, a_2, \ldots, a_n$ be positive real numbers such that $a_1 + a_2 + \cdots + a_n < 1$. Prove that

$$\frac{a_1 a_2 \cdots a_n [1 - (a_1 + a_2 + \cdots + a_n)]}{(a_1 + a_2 + \cdots + a_n)(1 - a_1)(1 - a_2) \cdots (1 - a_n)} \leq \frac{1}{n^{n+1}}.$$

10. **(AUS)** Let $r_1, r_2, \ldots, r_n$ be real numbers greater than or equal to 1. Prove that

$$\frac{1}{r_1 + 1} + \frac{1}{r_2 + 1} + \cdots + \frac{1}{r_n + 1} \geq \frac{n}{\sqrt[n]{r_1 r_2 \cdots r_n} + 1}.$$

11. **(RUS)** Let x, y, and z be positive real numbers such that $xyz = 1$. Prove that

$$\frac{x^3}{(1+y)(1+z)} + \frac{y^3}{(1+z)(1+x)} + \frac{z^3}{(1+x)(1+y)} \geq \frac{3}{4}.$$

12. **(POL)** Let $n \geq k \geq 0$ be integers. The numbers $c(n,k)$ are defined as follows:

$$c(n,0) = c(n,n) = 1 \qquad \text{for all } n \geq 0;$$
$$c(n+1,k) = 2^k c(n,k) + c(n,k-1) \qquad \text{for } n \geq k \geq 1.$$

Prove that $c(n,k) = c(n, n-k)$ for all $n \geq k \geq 0$.

13. **(BGR)**[IMO6] Determine the least possible value of $f(1998)$, where f is a function from the set $\mathbb{N}$ of positive integers into itself such that for all $m, n \in \mathbb{N}$,

$$f(n^2 f(m)) = m[f(n)]^2.$$

14. **(UNK)**[IMO4] Determine all pairs (x,y) of positive integers such that $x^2 y + x + y$ is divisible by $xy^2 + y + 7$.

15. **(AUS)** Determine all pairs (a,b) of real numbers such that $a\lfloor bn \rfloor = b\lfloor an \rfloor$ for all positive integers n. (Note that $\lfloor x \rfloor$ denotes the greatest integer less than or equal to x.)

16. **(UKR)** Determine the smallest integer $n \geq 4$ for which one can choose four different numbers a, b, c, and d from any n distinct integers such that $a + b - c - d$ is divisible by 20.

17. **(UNK)** A sequence of integers $a_1, a_2, a_3, \ldots$ is defined as follows: $a_1 = 1$, and for $n \geq 1$, a_{n+1} is the smallest integer greater than a_n such that $a_i + a_j \neq 3a_k$ for any i, j, k in $\{1, 2, \ldots, n+1\}$, not necessarily distinct. Determine a_{1998}.

18. **(BGR)** Determine all positive integers n for which there exists an integer m such that $2^n - 1$ is a divisor of $m^2 + 9$.

19. **(BLR)**[IMO3] For any positive integer n, let $\tau(n)$ denote the number of its positive divisors (including 1 and itself). Determine all positive integers m for which there exists a positive integer n such that $\frac{\tau(n^2)}{\tau(n)} = m$.

20. **(ARG)** Prove that for each positive integer n, there exists a positive integer with the following properties:
   (i) It has exactly n digits.
   (ii) None of the digits is 0.
   (iii) It is divisible by the sum of its digits.

21. **(CAN)** Let $a_0, a_1, a_2, \ldots$ be an increasing sequence of nonnegative integers such that every nonnegative integer can be expressed uniquely in the form $a_i + 2a_j + 4a_k$, where i, j, k are not necessarily distinct. Determine a_{1998}.

22. **(UKR)** A rectangular array of numbers is given. In each row and each column, the sum of all numbers is an integer. Prove that each nonintegral number x in the array can be changed into either $\lceil x \rceil$ or $\lfloor x \rfloor$ so that the row sums and column sums remain unchanged. (Note that $\lceil x \rceil$ is the least integer greater than or equal to x, while $\lfloor x \rfloor$ is the greatest integer less than or equal to x.)

23. **(BLR)** Let n be an integer greater than 2. A positive integer is said to be *attainable* if it is 1 or can be obtained from 1 by a sequence of operations with the following properties:
    (i) The first operation is either addition or multiplication.
    (ii) Thereafter, additions and multiplications are used alternately.
    (iii) In each addition one can choose independently whether to add 2 or n.
    (iv) In each multiplication, one can choose independently whether to multiply by 2 or by n.
    A positive integer that cannot be so obtained is said to be *unattainable*.
    (a) Prove that if $n \geq 9$, there are infinitely many unattainable positive integers.
    (b) Prove that if $n = 3$, all positive integers except 7 are attainable.

24. **(SWE)** Cards numbered 1 to 9 are arranged at random in a row. In a move, one may choose any block of consecutive cards whose numbers are in ascending or descending order, and switch the block around. For example, 916532748 may be changed to 913562748. Prove that in at most 12 moves, one can arrange the 9 cards so that their numbers are in ascending or descending order.

25. **(NZL)** Let $U = \{1, 2, \ldots, n\}$, where $n \geq 3$. A subset S of U is said to be *split* by an arrangement of the elements of U if an element not in S occurs in the arrangement somewhere between two elements of S. For example, 13542 splits $\{1, 2, 3\}$ but not $\{3, 4, 5\}$. Prove that for any $n - 2$ subsets of U, each containing at least 2 and at most $n - 1$ elements, there is an arrangement of the elements of U that splits all of them.

26. **(IND)**[IMO2] In a contest, there are m candidates and n judges, where $n \geq 3$ is an odd integer. Each candidate is evaluated by each judge as either pass or fail. Suppose that each pair of judges agrees on at most k candidates. Prove that $\frac{k}{m} \geq \frac{n-1}{2n}$.

27. **(BLR)** Ten points such that no three of them lie on a line are marked in the plane. Each pair of points is connected with a segment. Each of these segments is painted with one of k colors in such a way that for any k of the ten points, there are k segments each joining two of them with no two being painted the same color. Determine all integers k, $1 \leq k \leq 10$, for which this is possible.

28. **(IRN)** A solitaire game is played on an $m \times n$ rectangular board, using mn markers that are white on one side and black on the other. Initially, each square of the board contains a marker with its white side up, except for one corner square, which contains a marker with its black side up. In each move, one can take away one marker with its black side up, but must then turn over all markers that are in squares having an edge in common with the square of the removed marker. Determine all pairs (m, n) of positive integers such that all markers can be removed from the board.

3.40 The Fortieth IMO
Bucharest, Romania, July 10–22, 1999

3.40.1 Contest Problems

First Day (July 16)

1. A set S of points in the plane will be called *completely symmetric* if it has at least three elements and satisfies the following condition: For every two distinct points A, B from S the perpendicular bisector of the segment AB is an axis of symmetry for S.
   Prove that if a completely symmetric set is finite, then it consists of the vertices of a regular polygon.

2. Let $n \geq 2$ be a fixed integer. Find the least constant C such that the inequality

$$\sum_{i<j} x_i x_j (x_i^2 + x_j^2) \leq C \left(\sum_i x_i \right)^4$$

   holds for every $x_1, \ldots, x_n \geq 0$ (the sum on the left consists of $\binom{n}{2}$ summands). For this constant C, characterize the instances of equality.

3. Let n be an even positive integer. We say that two different cells of an $n \times n$ board are *neighboring* if they have a common side. Find the minimal number of cells on the $n \times n$ board that must be marked so that every cell (marked or not marked) has a marked neighboring cell.

Second Day (July 17)

4. Find all pairs of positive integers (x, p) such that p is a prime, $x \leq 2p$, and x^{p-1} is a divisor of $(p-1)^x + 1$.

5. Two circles Ω_1 and Ω_2 touch internally the circle Ω in M and N, and the center of Ω_2 is on Ω_1. The common chord of the circles Ω_1 and Ω_2 intersects Ω in A and B. MA and MB intersect Ω_1 in C and D. Prove that Ω_2 is tangent to CD.

6. Find all the functions $f : \mathbb{R} \to \mathbb{R}$ that satisfy

$$f(x - f(y)) = f(f(y)) + xf(y) + f(x) - 1$$

   for all $x, y \in \mathbb{R}$.

3.40.2 Shortlisted Problems

1. **N1 (TWN)**[IMO4] Find all pairs of positive integers (x, p) such that p is a prime, $x \leq 2p$, and x^{p-1} is a divisor of $(p-1)^x + 1$.

2. **N2 (ARM)** Prove that every positive rational number can be represented in the form $\dfrac{a^3 + b^3}{c^3 + d^3}$, where a, b, c, d are positive integers.

3. **N3 (RUS)** Prove that there exist two strictly increasing sequences (a_n) and (b_n) such that $a_n(a_n + 1)$ divides $b_n^2 + 1$ for every natural number n.

4. **N4 (FRA)** Denote by S the set of all primes p such that the decimal representation of $\frac{1}{p}$ has its fundamental period divisible by 3. For every $p \in S$ such that $\frac{1}{p}$ has its fundamental period $3r$ one may write $\frac{1}{p} = 0.a_1 a_2 \ldots a_{3r} a_1 a_2 \ldots a_{3r}$ $\ldots$, where $r = r(p)$; for every $p \in S$ and every integer $k \geq 1$ define $f(k, p)$ by

$$f(k, p) = a_k + a_{k+r(p)} + a_{k+2r(p)}.$$

   (a) Prove that S is infinite.
   (b) Find the highest value of $f(k, p)$ for $k \geq 1$ and $p \in S$.

5. **N5 (ARM)** Let n, k be positive integers such that n is not divisible by 3 and $k \geq n$. Prove that there exists a positive integer m that is divisible by n and the sum of whose digits in decimal representation is k.

6. **N6 (BLR)** Prove that for every real number M there exists an infinite arithmetic progression such that:
   (i) each term is a positive integer and the common difference is not divisible by 10;
   (ii) the sum of the digits of each term (in decimal representation) exceeds M.

7. **G1 (ARM)** Let ABC be a triangle and M an interior point. Prove that

$$\min\{MA, MB, MC\} + MA + MB + MC < AB + AC + BC.$$

8. **G2 (JPN)** A circle is called a *separator* for a set of five points in a plane if it passes through three of these points, it contains a fourth point in its interior, and the fifth point is outside the circle.
   Prove that every set of five points such that no three are collinear and no four are concyclic has exactly four separators.

9. **G3 (EST)**[IMO1] A set S of points in space will be called *completely symmetric* if it has at least three elements and satisfies the following condition: For every two distinct points A, B from S the perpendicular bisector of the segment AB is an axis of symmetry for S.
   Prove that if a completely symmetric set is finite, then it consists of the vertices of either a regular polygon, a regular tetrahedron, or a regular octahedron.

10. **G4 (UNK)** For a triangle $T = ABC$ we take the point X on the side (AB) such that $AX/XB = 4/5$, the point Y on the segment (CX) such that $CY = 2YX$, and, if possible, the point Z on the ray $(CA$ such that $\angle CXZ = 180° - \angle ABC$. We denote by Σ the set of all triangles T for which $\angle XYZ = 45°$.
   Prove that all the triangles from Σ are similar and find the measure of their smallest angle.

11. **G5 (FRA)** Let ABC be a triangle, Ω its incircle and $\Omega_a, \Omega_b, \Omega_c$ three circles orthogonal to Ω passing through B and C, A and C, and A and B respectively.

The circles Ω_a, Ω_b meet again in C'; in the same way we obtain the points B' and A'. Prove that the radius of the circumcircle of $A'B'C'$ is half the radius of Ω.

12. **G6 (RUS)**[IMO5] Two circles Ω_1 and Ω_2 touch internally the circle Ω in M and N, and the center of Ω_2 is on Ω_1. The common chord of the circles Ω_1 and Ω_2 intersects Ω in A and B. MA and MB intersect Ω_1 in C and D. Prove that Ω_2 is tangent to CD.

13. **G7 (ARM)** The point M inside the convex quadrilateral $ABCD$ is such that

$$MA = MC, \quad \angle AMB = \angle MAD + \angle MKD, \quad \angle CMD = \angle MCB + \angle MAB.$$

Prove that $AB \cdot CM = BC \cdot MD$ and $BM \cdot AD = MA \cdot CD$.

14. **G8 (RUS)** Points A, B, C divide the circumcircle Ω of the triangle ABC into three arcs. Let X be a variable point on the arc AB, and let O_1, O_2 be the incenters of the triangles CAX and CBX. Prove that the circumcircle of the triangle XO_1O_2 intersects Ω in a fixed point.

15. **A1 (POL)**[IMO2] Let $n \geq 2$ be a fixed integer. Find the least constant C such that the inequality

$$\sum_{i<j} x_i x_j (x_i^2 + x_j^2) \leq C \left(\sum_i x_i \right)^4$$

holds for every $x_1, \ldots, x_n \geq 0$ (the sum on the left consists of $\binom{n}{2}$ summands). For this constant C, characterize the instances of equality.

16. **A2 (RUS)** The numbers from 1 to n^2 are randomly arranged in the cells of an $n \times n$ square ($n \geq 2$). For any pair of numbers situated in the same row or in the same column, the ratio of the greater number to the smaller one is calculated. Let us call the *characteristic* of the arrangement the smallest of these $n^2(n-1)$ fractions. What is the highest possible value of the characteristic?

17. **A3 (FIN)** A game is played by n girls ($n \geq 2$), everybody having a ball. Each of the $\binom{n}{2}$ pairs of players, in an arbitrary order, exchange the balls they have at that moment. The game is called *nice* if at the end nobody has her own ball, and it is called *tiresome* if at the end everybody has her initial ball. Determine the values of n for which there exists a nice game and those for which there exists a tiresome game.

18. **A4 (BLR)** Prove that the set of positive integers cannot be partitioned into three nonempty subsets such that for any two integers x, y taken from two different subsets, the number $x^2 - xy + y^2$ belongs to the third subset.

19. **A5 (JPN)**[IMO6] Find all the functions $f : \mathbb{R} \to \mathbb{R}$ that satisfy

$$f(x - f(y)) = f(f(y)) + xf(y) + f(x) - 1$$

for all $x, y \in \mathbb{R}$.

20. **A6 (SWE)** For $n \geq 3$ and $a_1 \leq a_2 \leq \cdots \leq a_n$ given real numbers we have the following instructions:

(1) place the numbers in some order in a circle;
(2) delete one of the numbers from the circle;
(3) if just two numbers are remaining in the circle, let S be the sum of these two numbers. Otherwise, if there are more than two numbers in the circle, replace $(x_1, x_2, x_3, \ldots, x_{p-1}, x_p)$ with $(x_1 + x_2, x_2 + x_3, \ldots, x_{p-1} + x_p, x_p + x_1)$. Afterwards, start again with step (2).

Show that the largest sum S that can result in this way is given by the formula

$$S_{\max} = \sum_{k=2}^{n} \binom{n-2}{\left[\frac{k}{2}\right]-1} a_k.$$

21. **C1 (IND)** Let $n \geq 1$ be an integer. A *path* from $(0,0)$ to (n,n) in the xy plane is a chain of consecutive unit moves either to the right (move denoted by E) or upwards (move denoted by N), all the moves being made inside the half-plane $x \geq y$. A *step* in a path is the occurrence of two consecutive moves of the form EN.

Show that the number of paths from $(0,0)$ to (n,n) that contain exactly s steps $(n \geq s \geq 1)$ is

$$\frac{1}{s}\binom{n-1}{s-1}\binom{n}{s-1}.$$

22. **C2 (CAN)** (a) If a $5 \times n$ rectangle can be tiled using n pieces like those shown in the diagram, prove that n is even.

(b) Show that there are more than $2 \cdot 3^{k-1}$ ways to tile a fixed $5 \times 2k$ rectangle $(k \geq 3)$ with $2k$ pieces. (Symmetric constructions are considered to be different.)

23. **C3 (UNK)** A biologist watches a chameleon. The chameleon catches flies and rests after each catch. The biologist notices that:
   (i) the first fly is caught after a resting period of one minute;
   (ii) the resting period before catching the $2m$th fly is the same as the resting period before catching the mth fly and one minute shorter than the resting period before catching the $(2m+1)$th fly;
   (iii) when the chameleon stops resting, he catches a fly instantly.
   (a) How many flies were caught by the chameleon before his first resting period of 9 minutes?
   (b) After how many minutes will the chameleon catch his 98th fly?
   (c) How many flies were caught by the chameleon after 1999 minutes passed?

24. **C4 (UNK)** Let A be a set of N residues (mod N^2). Prove that there exists a set B of N residues (mod N^2) such that the set $A + B = \{a + b \mid a \in A, b \in B\}$ contains at least half of all residues (mod N^2).

25. **C5 (BLR)**[IMO3] Let n be an even positive integer. We say that two different cells of an $n \times n$ board are *neighboring* if they have a common side. Find the minimal number of cells on the $n \times n$ board that must be marked so that every cell (marked or not marked) has a marked neighboring cell.

26. **C6 (UNK)** Suppose that every integer has been given one of the colors red, blue, green, yellow. Let x and y be odd integers such that $|x| \neq |y|$. Show that there are two integers of the same color whose difference has one of the following values: $x, y, x+y, x-y$.

27. **C7 (IRL)** Let $p > 3$ be a prime number. For each nonempty subset T of $\{0, 1, 2, 3, \ldots, p-1\}$ let $E(T)$ be the set of all $(p-1)$-tuples $(x_1, \ldots, x_{p-1})$, where each $x_i \in T$ and $x_1 + 2x_2 + \cdots + (p-1)x_{p-1}$ is divisible by p and let $|E(T)|$ denote the number of elements in $E(T)$.
Prove that
$$|E(\{0, 1, 3\})| \geq |E(\{0, 1, 2\})|,$$
with equality if and only if $p = 5$.

3.41 The Forty-First IMO
Taejon, South Korea, July 13–25, 2000

3.41.1 Contest Problems

First day (July 18)

1. Two circles G_1 and G_2 intersect at M and N. Let AB be the line tangent to these circles at A and B, respectively, such that M lies closer to AB than N. Let CD be the line parallel to AB and passing through M, with C on G_1 and D on G_2. Lines AC and BD meet at E; lines AN and CD meet at P; lines BN and CD meet at Q. Show that $EP = EQ$.

2. Let a, b, c be positive real numbers with product 1. Prove that

$$\left(a - 1 + \frac{1}{b}\right)\left(b - 1 + \frac{1}{c}\right)\left(c - 1 + \frac{1}{a}\right) \le 1.$$

3. Let $n \ge 2$ be a positive integer and λ a positive real number. Initially there are n fleas on a horizontal line, not all at the same point. We define a move of choosing two fleas at some points A and B, with A to the left of B, and letting the flea from A jump over the flea from B to the point C such that $BC/AB = \lambda$.
Determine all values of λ such that for any point M on the line and for any initial position of the n fleas, there exists a sequence of moves that will take them all to a position right of M.

Second Day (July 19)

4. A magician has one hundred cards numbered 1 to 100. He puts them into three boxes, a red one, a white one, and a blue one, so that each box contains at least one card. A member of the audience draws two cards from two different boxes and announces the sum of numbers on those cards. Given this information, the magician locates the box from which no card has been drawn. How many ways are there to put the cards in the three boxes so that the trick works?

5. Does there exist a positive integer n such that n has exactly 2000 prime divisors and $2^n + 1$ is divisible by n?

6. $A_1 A_2 A_3$ is an acute-angled triangle. The foot of the altitude from A_i is K_i, and the incircle touches the side opposite A_i at L_i. The line $K_1 K_2$ is reflected in the line $L_1 L_2$. Similarly, the line $K_2 K_3$ is reflected in $L_2 L_3$ and $K_3 K_1$ is reflected in $L_3 L_1$. Show that the three new lines form a triangle with vertices on the incircle.

3.41.2 Shortlisted Problems

1. **C1 (HUN)**[IMO4] A magician has one hundred cards numbered 1 to 100. He puts them into three boxes, a red one, a white one, and a blue one, so that each box

contains at least one card. A member of the audience draws two cards from two different boxes and announces the sum of numbers on those cards. Given this information, the magician locates the box from which no card has been drawn. How many ways are there to put the cards in the three boxes so that the trick works?

2. **C2 (ITA)** A brick staircase with three steps of width 2 is made of twelve unit cubes. Determine all integers n for which it is possible to build a cube of side n using such bricks.

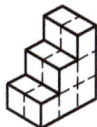

3. **C3 (COL)** Let $n \geq 4$ be a fixed positive integer. Given a set $S = \{P_1, P_2, \ldots, P_n\}$ of points in the plane such that no three are collinear and no four concyclic, let a_t, $1 \leq t \leq n$, be the number of circles $P_i P_j P_k$ that contain P_t in their interior, and let

$$m(S) = a_1 + a_2 + \cdots + a_n.$$

Prove that there exists a positive integer $f(n)$, depending only on n, such that the points of S are the vertices of a convex polygon if and only if $m(S) = f(n)$.

4. **C4 (CZE)** Let n and k be positive integers such that $n/2 < k \leq 2n/3$. Find the least number m for which it is possible to place m pawns on m squares of an $n \times n$ chessboard so that no column or row contains a block of k adjacent unoccupied squares.

5. **C5 (RUS)** In the plane we have n rectangles with parallel sides. The sides of distinct rectangles lie on distinct lines. The boundaries of the rectangles cut the plane into connected regions. A region is *nice* if it has at least one of the vertices of the n rectangles on its boundary. Prove that the sum of the numbers of the vertices of all nice regions is less than $40n$. (There can be nonconvex regions as well as regions with more than one boundary curve.)

6. **C6 (FRA)** Let p and q be relatively prime positive integers. A subset S of $\{0, 1, 2, \ldots\}$ is called *ideal* if $0 \in S$ and for each element $n \in S$, the integers $n + p$ and $n + q$ belong to S. Determine the number of ideal subsets of $\{0, 1, 2 \ldots\}$.

7. **A1 (USA)**[IMO2] Let a, b, c be positive real numbers with product 1. Prove that

$$\left(a - 1 + \frac{1}{b}\right)\left(b - 1 + \frac{1}{c}\right)\left(c - 1 + \frac{1}{a}\right) \leq 1.$$

8. **A2 (UNK)** Let a, b, c be positive integers satisfying the conditions $b > 2a$ and $c > 2b$. Show that there exists a real number t with the property that all the three numbers ta, tb, tc have their fractional parts lying in the interval $(1/3, 2/3]$.

9. **A3 (BLR)** Find all pairs of functions $f : \mathbb{R} \to \mathbb{R}$, $g : \mathbb{R} \to \mathbb{R}$ such that

$$f(x + g(y)) = xf(y) - yf(x) + g(x) \qquad \text{for all } x, y \in \mathbb{R}.$$

10. **A4 (UNK)** The function F is defined on the set of nonnegative integers and takes nonnegative integer values satisfying the following conditions: For every $n \geq 0$,

(i) $F(4n) = F(2n) + F(n)$;
(ii) $F(4n+2) = F(4n) + 1$;
(iii) $F(2n+1) = F(2n) + 1$.

Prove that for each positive integer m, the number of integers n with $0 \leq n < 2^m$ and $F(4n) = F(3n)$ is $F(2^{m+1})$.

11. **A5 (BLR)**[IMO3] Let $n \geq 2$ be a positive integer and λ a positive real number. Initially there are n fleas on a horizontal line, not all at the same point. We define a move of choosing two fleas at some points A and B, with A to the left of B, and letting the flea from A jump over the flea from B to the point C such that $BC/AB = \lambda$.

Determine all values of λ such that for any point M on the line and for any initial position of the n fleas, there exists a sequence of moves that will take them all to a position right of M.

12. **A6 (IRL)** A nonempty set A of real numbers is called a B_3-set if the conditions $a_1, a_2, a_3, a_4, a_5, a_6 \in A$ and $a_1 + a_2 + a_3 = a_4 + a_5 + a_6$ imply that the sequences (a_1, a_2, a_3) and (a_4, a_5, a_6) are identical up to a permutation. Let $A = \{a_0 = 0 < a_1 < a_2 < \cdots\}$, $B = \{b_0 = 0 < b_1 < b_2 < \cdots\}$ be infinite sequences of real numbers with $D(A) = D(B)$, where, for a set X of real numbers, $D(X)$ denotes the difference set $\{|x - y| \mid x, y \in X\}$. Prove that if A is a B_3-set, then $A = B$.

13. **A7 (RUS)** For a polynomial P of degree 2000 with distinct real coefficients let $M(P)$ be the set of all polynomials that can be produced from P by permutation of its coefficients. A polynomial P will be called n-independent if $P(n) = 0$ and we can get from any Q in $M(P)$ a polynomial Q_1 such that $Q_1(n) = 0$ by interchanging at most one pair of coefficients of Q. Find all integers n for which n-independent polynomials exist.

14. **N1 (JPN)** Determine all positive integers $n \geq 2$ that satisfy the following condition: For all integers a, b relatively prime to n,

$$a \equiv b \pmod{n} \quad \text{if and only if} \quad ab \equiv 1 \pmod{n}.$$

15. **N2 (FRA)** For a positive integer n, let $d(n)$ be the number of all positive divisors of n. Find all positive integers n such that $d(n)^3 = 4n$.

16. **N3 (RUS)**[IMO5] Does there exist a positive integer n such that n has exactly 2000 prime divisors and $2^n + 1$ is divisible by n?

17. **N4 (BRA)** Determine all triples of positive integers (a, m, n) such that $a^m + 1$ divides $(a+1)^n$.

18. **N5 (BGR)** Prove that there exist infinitely many positive integers n such that $p = nr$, where p and r are respectively the semiperimeter and the inradius of a triangle with integer side lengths.

19. **N6 (ROU)** Show that the set of positive integers that cannot be represented as a sum of distinct perfect squares is finite.

20. **G1 (NLD)** In the plane we are given two circles intersecting at X and Y. Prove that there exist four points A, B, C, D with the following property:
    For every circle touching the two given circles at A and B, and meeting the line XY at C and D, each of the lines AC, AD, BC, BD passes through one of these points.

21. **G2 (RUS)**[IMO1] Two circles G_1 and G_2 intersect at M and N. Let AB be the line tangent to these circles at A and B, respectively, such that M lies closer to AB than N. Let CD be the line parallel to AB and passing through M, with C on G_1 and D on G_2. Lines AC and BD meet at E; lines AN and CD meet at P; lines BN and CD meet at Q. Show that $EP = EQ$.

22. **G3 (IND)** Let O be the circumcenter and H the orthocenter of an acute triangle ABC. Show that there exist points D, E, and F on sides BC, CA, and AB respectively such that $OD + DH = OE + EH = OF + FH$ and the lines AD, BE, and CF are concurrent.

23. **G4 (RUS)** Let $A_1A_2 \ldots A_n$ be a convex polygon, $n \geq 4$. Prove that $A_1A_2 \ldots A_n$ is cyclic if and only if to each vertex A_j one can assign a pair (b_j, c_j) of real numbers, $j = 1, 2, \ldots n$, such that

    $$A_iA_j = b_jc_i - b_ic_j \quad \text{for all } i, j \text{ with } 1 \leq i \leq j \leq n.$$

24. **G5 (UNK)** The tangents at B and A to the circumcircle of an acute-angled triangle ABC meet the tangent at C at T and U respectively. AT meets BC at P, and Q is the midpoint of AP; BU meets CA at R, and S is the midpoint of BR. Prove that $\angle ABQ = \angle BAS$. Determine, in terms of ratios of side lengths, the triangles for which this angle is a maximum.

25. **G6 (ARG)** Let $ABCD$ be a convex quadrilateral with AB not parallel to CD, let X be a point inside $ABCD$ such that $\angle ADX = \angle BCX < 90°$ and $\angle DAX = \angle CBX < 90°$. If Y is the point of intersection of the perpendicular bisectors of AB and CD, prove that $\angle AYB = 2\angle ADX$.

26. **G7 (IRN)** Ten gangsters are standing on a flat surface, and the distances between them are all distinct. At twelve o'clock, when the church bells start chiming, each of them fatally shoots the one among the other nine gangsters who is the nearest. At least how many gangsters will be killed?

27. **G8 (RUS)**[IMO6] $A_1A_2A_3$ is an acute-angled triangle. The foot of the altitude from A_i is K_i, and the incircle touches the side opposite A_i at L_i. The line K_1K_2 is reflected in the line L_1L_2. Similarly, the line K_2K_3 is reflected in L_2L_3, and K_3K_1 is reflected in L_3L_1. Show that the three new lines form a triangle with vertices on the incircle.

3.42 The Forty-Second IMO
Washington DC, United States of America, July 1–14, 2001

3.42.1 Contest Problems

First Day (July 8)

1. In acute triangle ABC with circumcenter O and altitude AP, $\angle C \geq \angle B + 30°$. Prove that $\angle A + \angle COP < 90°$.

2. Prove that for all positive real numbers a, b, c,
$$\frac{a}{\sqrt{a^2 + 8bc}} + \frac{b}{\sqrt{b^2 + 8ca}} + \frac{c}{\sqrt{c^2 + 8ab}} \geq 1.$$

3. Twenty-one girls and twenty-one boys took part in a mathematical competition. It turned out that
   (i) each contestant solved at most six problems, and
   (ii) for each pair of a girl and a boy, there was at least one problem that was solved by both the girl and the boy.
   Show that there is a problem that was solved by at least three girls and at least three boys.

Second Day (July 9)

4. Let n be an odd integer greater than 1 and let $c_1, c_2, \ldots, c_n$ be integers. For each permutation $a = (a_1, a_2, \ldots, a_n)$ of $\{1, 2, \ldots, n\}$, define $S(a) = \sum_{i=1}^{n} c_i a_i$. Prove that there exist permutations $a \neq b$ of $\{1, 2, \ldots, n\}$ such that $n!$ is a divisor of $S(a) - S(b)$.

5. Let ABC be a triangle with $\angle BAC = 60°$. Let AP bisect $\angle BAC$ and let BQ bisect $\angle ABC$, with P on BC and Q on AC. If $AB + BP = AQ + QB$, what are the angles of the triangle?

6. Let $a > b > c > d$ be positive integers and suppose
$$ac + bd = (b + d + a - c)(b + d - a + c).$$
Prove that $ab + cd$ is not prime.

3.42.2 Shortlisted Problems

1. **A1 (IND)** Let T denote the set of all ordered triples (p, q, r) of nonnegative integers. Find all functions $f : T \rightarrow \mathbb{R}$ such that
$$f(p,q,r) = \begin{cases} 0 & \text{if } pqr = 0, \\ 1 + \frac{1}{6}\,(f(p+1,q-1,r) + f(p-1,q+1,r) \\ \quad + f(p-1,q,r+1) + f(p+1,q,r-1) \\ \quad + f(p,q+1,r-1) + f(p,q-1,r+1)) & \text{otherwise.} \end{cases}$$

2. **A2 (POL)** Let $a_0, a_1, a_2, \ldots$ be an arbitrary infinite sequence of positive numbers. Show that the inequality $1 + a_n > a_{n-1}\sqrt[n]{2}$ holds for infinitely many positive integers n.

3. **A3 (ROU)** Let $x_1, x_2, \ldots, x_n$ be arbitrary real numbers. Prove the inequality

$$\frac{x_1}{1+x_1^2} + \frac{x_2}{1+x_1^2+x_2^2} + \cdots + \frac{x_n}{1+x_1^2+\cdots+x_n^2} < \sqrt{n}.$$

4. **A4 (LTU)** Find all functions $f : \mathbb{R} \to \mathbb{R}$ satisfying

$$f(xy)(f(x) - f(y)) = (x-y)f(x)f(y)$$

for all x, y.

5. **A5 (BGR)** Find all positive integers $a_1, a_2, \ldots, a_n$ such that

$$\frac{99}{100} = \frac{a_0}{a_1} + \frac{a_1}{a_2} + \cdots + \frac{a_{n-1}}{a_n},$$

where $a_0 = 1$ and $(a_{k+1} - 1)a_{k-1} \geq a_k^2(a_k - 1)$ for $k = 1, 2, \ldots, n-1$.

6. **A6 (KOR)**[IMO2] Prove that for all positive real numbers a, b, c,

$$\frac{a}{\sqrt{a^2+8bc}} + \frac{b}{\sqrt{b^2+8ca}} + \frac{c}{\sqrt{c^2+8ab}} \geq 1.$$

7. **C1 (COL)** Let $A = (a_1, a_2, \ldots, a_{2001})$ be a sequence of positive integers. Let m be the number of 3-element subsequences (a_i, a_j, a_k) with $1 \leq i < j < k \leq 2001$ such that $a_j = a_i + 1$ and $a_k = a_j + 1$. Considering all such sequences A, find the greatest value of m.

8. **C2 (CAN)**[IMO4] Let n be an odd integer greater than 1 and let $c_1, c_2, \ldots, c_n$ be integers. For each permutation $a = (a_1, a_2, \ldots, a_n)$ of $\{1, 2, \ldots, n\}$, define $S(a) = \sum_{i=1}^{n} c_i a_i$. Prove that there exist permutations $a \neq b$ of $\{1, 2, \ldots, n\}$ such that $n!$ is a divisor of $S(a) - S(b)$.

9. **C3 (RUS)** Define a *k-clique* to be a set of k people such that every pair of them are acquainted with each other. At a certain party, every pair of 3-cliques has at least one person in common, and there are no 5-cliques. Prove that there are two or fewer people at the party whose departure leaves no 3-clique remaining.

10. **C4 (NZL)** A set of three nonnegative integers $\{x, y, z\}$ with $x < y < z$ is called *historic* if $\{z - y, y - x\} = \{1776, 2001\}$. Show that the set of all nonnegative integers can be written as the union of disjoint historic sets.

11. **C5 (FIN)** Find all finite sequences $(x_0, x_1, \ldots, x_n)$ such that for every $j, 0 \leq j \leq n$, x_j equals the number of times j appears in the sequence.

12. **C6 (CAN)** For a positive integer n define a sequence of zeros and ones to be *balanced* if it contains n zeros and n ones. Two balanced sequences a and b are *neighbors* if you can move one of the $2n$ symbols of a to another position to form

b. For instance, when $n = 4$, the balanced sequences 01101001 and 00110101 are neighbors because the third (or fourth) zero in the first sequence can be moved to the first or second position to form the second sequence. Prove that there is a set S of at most $\frac{1}{n+1}\binom{2n}{n}$ balanced sequences such that every balanced sequence is equal to or is a neighbor of at least one sequence in S.

13. **C7 (FRA)** A pile of n pebbles is placed in a vertical column. This configuration is modified according to the following rules. A pebble can be moved if it is at the top of a column that contains at least two more pebbles than the column immediately to its right. (If there are no pebbles to the right, think of this as a column with 0 pebbles.) At each stage, choose a pebble from among those that can be moved (if there are any) and place it at the top of the column to its right. If no pebbles can be moved, the configuration is called a *final configuration*. For each n, show that no matter what choices are made at each stage, the final configuration is unique. Describe that configuration in terms of n.

14. **C8 (GER)**[IMO3] Twenty-one girls and twenty-one boys took part in a mathematical competition. It turned out that
    (i) each contestant solved at most six problems, and
    (ii) for each pair of a girl and a boy, there was at least one problem that was solved by both the girl and the boy.
    Show that there is a problem that was solved by at least three girls and at least three boys.

15. **G1 (UKR)** Let A_1 be the center of the square inscribed in acute triangle ABC with two vertices of the square on side BC. Thus one of the two remaining vertices of the square is on side AB and the other is on AC. Points B_1, C_1 are defined in a similar way for inscribed squares with two vertices on sides AC and AB, respectively. Prove that lines AA_1, BB_1, CC_1 are concurrent.

16. **G2 (KOR)**[IMO1] In acute triangle ABC with circumcenter O and altitude AP, $\angle C \geq \angle B + 30°$. Prove that $\angle A + \angle COP < 90°$.

17. **G3 (UNK)** Let ABC be a triangle with centroid G. Determine, with proof, the position of the point P in the plane of ABC such that

$$AP \cdot AG + BP \cdot BG + CP \cdot CG$$

is a minimum, and express this minimum value in terms of the side lengths of ABC.

18. **G4 (FRA)** Let M be a point in the interior of triangle ABC. Let A' lie on BC with MA' perpendicular to BC. Define B' on CA and C' on AB similarly. Define

$$p(M) = \frac{MA' \cdot MB' \cdot MC'}{MA \cdot MB \cdot MC}.$$

Determine, with proof, the location of M such that $p(M)$ is maximal. Let $\mu(ABC)$ denote the maximum value. For which triangles ABC is the value of $\mu(ABC)$ maximal?

19. **G5 (HEL)** Let ABC be an acute triangle. Let DAC, EAB, and FBC be isosceles triangles exterior to ABC, with $DA = DC$, $EA = EB$, and $FB = FC$ such that

$$\angle ADC = 2\angle BAC, \qquad \angle BEA = 2\angle ABC, \qquad \angle CFB = 2\angle ACB.$$

Let D' be the intersection of lines DB and EF, let E' be the intersection of EC and DF, and let F' be the intersection of FA and DE. Find, with proof, the value of the sum

$$\frac{DB}{DD'} + \frac{EC}{EE'} + \frac{FA}{FF'}.$$

20. **G6 (IND)** Let ABC be a triangle and P an exterior point in the plane of the triangle. Suppose AP, BP, CP meet the sides BC, CA, AB (or extensions thereof) in D, E, F, respectively. Suppose further that the areas of triangles PBD, PCE, PAF are all equal. Prove that each of these areas is equal to the area of triangle ABC itself.

21. **G7 (BGR)** Let O be an interior point of acute triangle ABC. Let A_1 lie on BC with OA_1 perpendicular to BC. Define B_1 on CA and C_1 on AB similarly. Prove that O is the circumcenter of ABC if and only if the perimeter of $A_1 B_1 C_1$ is not less than any one of the perimeters of AB_1C_1, BC_1A_1, and CA_1B_1.

22. **G8 (ISR)**[IMO5] Let ABC be a triangle with $\angle BAC = 60°$. Let AP bisect $\angle BAC$ and let BQ bisect $\angle ABC$, with P on BC and Q on AC. If $AB + BP = AQ + QB$, what are the angles of the triangle?

23. **N1 (AUS)** Prove that there is no positive integer n such that for $k = 1, 2, \ldots, 9$, the leftmost digit (in decimal notation) of $(n + k)!$ equals k.

24. **N2 (COL)** Consider the system

$$x + y = z + u,$$
$$2xy = zu.$$

Find the greatest value of the real constant m such that $m \le x/y$ for every positive integer solution x, y, z, u of the system with $x \ge y$.

25. **N3 (UNK)** Let $a_1 = 11^{11}$, $a_2 = 12^{12}$, $a_3 = 13^{13}$, and

$$a_n = |a_{n-1} - a_{n-2}| + |a_{n-2} - a_{n-3}|, \qquad n \ge 4.$$

Determine $a_{14^{14}}$.

26. **N4 (VNM)** Let $p \ge 5$ be a prime number. Prove that there exists an integer a with $1 \le a \le p - 2$ such that neither $a^{p-1} - 1$ nor $(a + 1)^{p-1} - 1$ is divisible by p^2.

27. **N5 (BGR)**[IMO6] Let $a > b > c > d$ be positive integers and suppose

$$ac + bd = (b + d + a - c)(b + d - a + c).$$

Prove that $ab + cd$ is not prime.

28. **N6 (RUS)** Is it possible to find 100 positive integers not exceeding 25,000 such that all pairwise sums of them are different?

3.43 The Forty-Third IMO
Glasgow, United Kingdom, July 19–30, 2002

3.43.1 Contest Problems

First Day (July 24)

1. Let n be a positive integer. Each point (x,y) in the plane, where x and y are nonnegative integers with $x+y=n$, is colored red or blue, subject to the following condition: If a point (x,y) is red, then so are all points (x',y') with $x' \le x$ and $y' \le y$. Let A be the number of ways to choose n blue points with distinct x-coordinates, and let B be the number of ways to choose n blue points with distinct y-coordinates. Prove that $A = B$.

2. The circle S has center O, and BC is a diameter of S. Let A be a point of S such that $\angle AOB < 120°$. Let D be the midpoint of the arc AB that does not contain C. The line through O parallel to DA meets the line AC at I. The perpendicular bisector of OA meets S at E and at F. Prove that I is the incenter of the triangle CEF.

3. Find all pairs of positive integers $m, n \ge 3$ for which there exist infinitely many positive integers a such that
$$\frac{a^m + a - 1}{a^n + a^2 - 1}$$
is itself an integer.

Second Day (July 25)

4. Let $n \ge 2$ be a positive integer, with divisors $1 = d_1 < d_2 < \cdots < d_k = n$. Prove that $d_1 d_2 + d_2 d_3 + \cdots + d_{k-1} d_k$ is always less than n^2, and determine when it is a divisor of n^2.

5. Find all functions f from the reals to the reals such that
$$(f(x) + f(z))(f(y) + f(t)) = f(xy - zt) + f(xt + yz)$$
for all real x, y, z, t.

6. Let $n \ge 3$ be a positive integer. Let $C_1, C_2, C_3, \ldots, C_n$ be unit circles in the plane, with centers $O_1, O_2, O_3, \ldots, O_n$ respectively. If no line meets more than two of the circles, prove that
$$\sum_{1 \le i < j \le n} \frac{1}{O_i O_j} \le \frac{(n-1)\pi}{4}.$$

3.43.2 Shortlisted Problems

1. **N1 (UZB)** What is the smallest positive integer t such that there exist integers $x_1, x_2, \ldots, x_t$ with
$$x_1^3 + x_2^3 + \cdots + x_t^3 = 2002^{2002}?$$

2. **N2 (ROU)**[IMO4] Let $n \geq 2$ be a positive integer, with divisors $1 = d_1 < d_2 < \cdots < d_k = n$. Prove that $d_1 d_2 + d_2 d_3 + \cdots + d_{k-1} d_k$ is always less than n^2, and determine when it is a divisor of n^2.

3. **N3 (MNG)** Let $p_1, p_2, \ldots, p_n$ be distinct primes greater than 3. Show that $2^{p_1 p_2 \cdots p_n} + 1$ has at least 4^n divisors.

4. **N4 (GER)** Is there a positive integer m such that the equation

$$\frac{1}{a} + \frac{1}{b} + \frac{1}{c} + \frac{1}{abc} = \frac{m}{a+b+c}$$

has infinitely many solutions in positive integers a, b, c?

5. **N5 (IRN)** Let $m, n \geq 2$ be positive integers, and let $a_1, a_2, \ldots, a_n$ be integers, none of which is a multiple of m^{n-1}. Show that there exist integers $e_1, e_2, \ldots, e_n$, not all zero, with $|e_i| < m$ for all i, such that $e_1 a_1 + e_2 a_2 + \cdots + e_n a_n$ is a multiple of m^n.

6. **N6 (ROU)**[IMO3] Find all pairs of positive integers $m, n \geq 3$ for which there exist infinitely many positive integers a such that

$$\frac{a^m + a - 1}{a^n + a^2 - 1}$$

is itself an integer.

7. **G1 (FRA)** Let B be a point on a circle S_1, and let A be a point distinct from B on the tangent at B to S_1. Let C be a point not on S_1 such that the line segment AC meets S_1 at two distinct points. Let S_2 be the circle touching AC at C and touching S_1 at a point D on the opposite side of AC from B. Prove that the circumcenter of triangle BCD lies on the circumcircle of triangle ABC.

8. **G2 (KOR)** Let ABC be a triangle for which there exists an interior point F such that $\angle AFB = \angle BFC = \angle CFA$. Let the lines BF and CF meet the sides AC and AB at D and E respectively. Prove that

$$AB + AC \geq 4DE.$$

9. **G3 (KOR)**[IMO2] The circle S has center O, and BC is a diameter of S. Let A be a point of S such that $\angle AOB < 120°$. Let D be the midpoint of the arc AB that does not contain C. The line through O parallel to DA meets the line AC at I. The perpendicular bisector of OA meets S at E and at F. Prove that I is the incenter of the triangle CEF.

10. **G4 (RUS)** Circles S_1 and S_2 intersect at points P and Q. Distinct points A_1 and B_1 (not at P or Q) are selected on S_1. The lines $A_1 P$ and $B_1 P$ meet S_2 again at A_2 and B_2 respectively, and the lines $A_1 B_1$ and $A_2 B_2$ meet at C. Prove that as A_1 and B_1 vary, the circumcenters of triangles $A_1 A_2 C$ all lie on one fixed circle.

11. **G5 (AUS)** For any set S of five points in the plane, no three of which are collinear, let $M(S)$ and $m(S)$ denote the greatest and smallest areas, respectively,

of triangles determined by three points from S. What is the minimum possible value of $M(S)/m(S)$?

12. **G6 (UKR)**[IMO6] Let $n \geq 3$ be a positive integer. Let $C_1, C_2, C_3, \ldots, C_n$ be unit circles in the plane, with centers $O_1, O_2, O_3, \ldots, O_n$ respectively. If no line meets more than two of the circles, prove that

$$\sum_{1 \leq i < j \leq n} \frac{1}{O_i O_j} \leq \frac{(n-1)\pi}{4}.$$

13. **G7 (BGR)** The incircle Ω of the acute-angled triangle ABC is tangent to BC at K. Let AD be an altitude of triangle ABC and let M be the midpoint of AD. If N is the other common point of Ω and KM, prove that Ω and the circumcircle of triangle BCN are tangent at N.

14. **G8 (ARM)** Let S_1 and S_2 be circles meeting at the points A and B. A line through A meets S_1 at C and S_2 at D. Points M, N, K lie on the line segments CD, BC, BD respectively, with MN parallel to BD and MK parallel to BC. Let E and F be points on those arcs BC of S_1 and BD of S_2 respectively that do not contain A. Given that EN is perpendicular to BC and FK is perpendicular to BD, prove that $\angle EMF = 90°$.

15. **A1 (CZE)** Find all functions f from the reals to the reals such that

$$f(f(x) + y) = 2x + f(f(y) - x)$$

for all real x, y.

16. **A2 (YUG)** Let $a_1, a_2, \ldots$ be an infinite sequence of real numbers for which there exists a real number c with $0 \leq a_i \leq c$ for all i such that

$$|a_i - a_j| \geq \frac{1}{i+j} \qquad \text{for all } i, j \text{ with } i \neq j.$$

Prove that $c \geq 1$.

17. **A3 (POL)** Let P be a cubic polynomial given by $P(x) = ax^3 + bx^2 + cx + d$, where a, b, c, d are integers and $a \neq 0$. Suppose that $xP(x) = yP(y)$ for infinitely many pairs x, y of integers with $x \neq y$. Prove that the equation $P(x) = 0$ has an integer root.

18. **A4 (IND)**[IMO5] Find all functions f from the reals to the reals such that

$$(f(x) + f(z))(f(y) + f(t)) = f(xy - zt) + f(xt + yz)$$

for all real x, y, z, t.

19. **A5 (IND)** Let n be a positive integer that is not a perfect cube. Define real numbers a, b, c by

$$a = \sqrt[3]{n}, \qquad b = \frac{1}{a - [a]}, \qquad c = \frac{1}{b - [b]},$$

where $[x]$ denotes the integer part of x. Prove that there are infinitely many such integers n with the property that there exist integers r, s, t, not all zero, such that $ra + sb + tc = 0$.

20. **A6 (IRN)** Let A be a nonempty set of positive integers. Suppose that there are positive integers $b_1, \ldots, b_n$ and $c_1, \ldots, c_n$ such that
    (i) for each i the set $b_i A + c_i = \{b_i a + c_i \mid a \in A\}$ is a subset of A, and
    (ii) the sets $b_i A + c_i$ and $b_j A + c_j$ are disjoint whenever $i \neq j$.
    Prove that
    $$\frac{1}{b_1} + \cdots + \frac{1}{b_n} \leq 1.$$

21. **C1 (COL)$^{\text{IMO1}}$** Let n be a positive integer. Each point (x, y) in the plane, where x and y are nonnegative integers with $x + y \leq n$, is colored red or blue, subject to the following condition: If a point (x, y) is red, then so are all points (x', y') with $x' \leq x$ and $y' \leq y$. Let A be the number of ways to choose n blue points with distinct x-coordinates, and let B be the number of ways to choose n blue points with distinct y-coordinates. Prove that $A = B$.

22. **C2 (ARM)** For n an odd positive integer, the unit squares of an $n \times n$ chessboard are colored alternately black and white, with the four corners colored black. A *tromino* is an L-shape formed by three connected unit squares. For which values of n is it possible to cover all the black squares with nonoverlapping trominos? When it is possible, what is the minimum number of trominos needed?

23. **C3 (COL)** Let n be a positive integer. A sequence of n positive integers (not necessarily distinct) is called *full* if it satisfies the following condition: For each positive integer $k \geq 2$, if the number k appears in the sequence, then so does the number $k - 1$, and moreover, the first occurrence of $k - 1$ comes before the last occurrence of k. For each n, how many full sequences are there?

24. **C4 (BGR)** Let T be the set of ordered triples (x, y, z), where x, y, z are integers with $0 \leq x, y, z \leq 9$. Players A and B play the following guessing game: Player A chooses a triple (x, y, z) in T, and Player B has to discover A's triple in as few moves as possible. A *move* consists of the following: B gives A a triple (a, b, c) in T, and A replies by giving B the number $|x + y - a - b| + |y + z - b - c| + |z + x - c - a|$. Find the minimum number of moves that B needs to be sure of determining A's triple.

25. **C5 (BRA)** Let $r \geq 2$ be a fixed positive integer, and let $\mathscr{F}$ be an infinite family of sets, each of size r, no two of which are disjoint. Prove that there exists a set of size $r - 1$ that meets each set in $\mathscr{F}$.

26. **C6 (POL)** Let n be an even positive integer. Show that there is a permutation $x_1, x_2, \ldots, x_n$ of $1, 2, \ldots, n$ such that for every $1 \leq i \leq n$ the number x_{i+1} is one of $2x_i, 2x_i - 1, 2x_i - n, 2x_i - n - 1$ (where we take $x_{n+1} = x_1$).

27. **C7 (NZL)** Among a group of 120 people, some pairs are friends. A *weak quartet* is a set of four people containing exactly one pair of friends. What is the maximum possible number of weak quartets?

3.44 The Forty-Fourth IMO
Tokyo, Japan, July 7–19, 2003

3.44.1 Contest Problems

First Day (July 13)

1. Let A be a 101-element subset of the set $S = \{1,2,\ldots,1000000\}$. Prove that there exist numbers $t_1, t_2, \ldots, t_{100}$ in S such that the sets

$$A_j = \{x + t_j | x \in A\}, \quad j = 1,2,\ldots,100,$$

are pairwise disjoint.

2. Determine all pairs (a,b) of positive integers such that

$$\frac{a^2}{2ab^2 - b^3 + 1}$$

is a positive integer.

3. Each pair of opposite sides of a convex hexagon has the following property: The distance between their midpoints is equal to $\sqrt{3}/2$ times the sum of their lengths. Prove that all the angles of the hexagon are equal.

Second Day (July 14)

4. Let $ABCD$ be a cyclic quadrilateral. Let P, Q, R be the feet of the perpendiculars from D to the lines BC, CA, AB, respectively. Show that $PQ = QR$ if and only if the bisectors of $\angle ABC$ and $\angle ADC$ are concurrent with AC.

5. Let n be a positive integer and let $x_1 \leq x_2 \leq \cdots \leq x_n$ be real numbers.
   (a) Prove that

$$\left(\sum_{i,j=1}^{n} |x_i - x_j| \right)^2 \leq \frac{2(n^2 - 1)}{3} \sum_{i,j=1}^{n} (x_i - x_j)^2.$$

   (b) Show that equality holds if and only if $x_1, \ldots, x_n$ is an arithmetic progression.

6. Let p be a prime number. Prove that there exists a prime number q such that for every integer n, the number $n^p - p$ is not divisible by q.

3.44.2 Shortlisted Problems

1. **A1 (USA)** Let a_{ij}, $i = 1,2,3$, $j = 1,2,3$, be real numbers such that a_{ij} is positive for $i = j$ and negative for $i \neq j$.
   Prove that there exist positive real numbers c_1, c_2, c_3 such that the numbers

$$a_{11}c_1 + a_{12}c_2 + a_{13}c_3, \quad a_{21}c_1 + a_{22}c_2 + a_{23}c_3, \quad a_{31}c_1 + a_{32}c_2 + a_{33}c_3$$

   are all negative, all positive, or all zero.

2. **A2 (AUS)** Find all nondecreasing functions $f : \mathbb{R} \to \mathbb{R}$ such that
   (i) $f(0) = 0$, $f(1) = 1$;
   (ii) $f(a) + f(b) = f(a)f(b) + f(a+b-ab)$ for all real numbers a, b such that $a < 1 < b$.

3. **A3 (GEO)** Consider pairs of sequences of positive real numbers $a_1 \geq a_2 \geq a_3 \geq \cdots$, $b_1 \geq b_2 \geq b_3 \geq \cdots$ and the sums $A_n = a_1 + \cdots + a_n$, $B_n = b_1 + \cdots + b_n$, $n = 1, 2, \ldots$. For any pair define $c_i = \min\{a_i, b_i\}$ and $C_n = c_1 + \cdots + c_n$, $n = 1, 2, \ldots$.
   (a) Does there exist a pair $(a_i)_{i \geq 1}, (b_i)_{i \geq 1}$ such that the sequences $(A_n)_{n \geq 1}$ and $(B_n)_{n \geq 1}$ are unbounded while the sequence $(C_n)_{n \geq 1}$ is bounded?
   (b) Does the answer to question (1) change by assuming additionally that $b_i = 1/i$, $i = 1, 2, \ldots$?
   Justify your answer.

4. **A4 (IRL)**[IMO5] Let n be a positive integer and let $x_1 \leq x_2 \leq \cdots \leq x_n$ be real numbers.
   (a) Prove that
   $$\left(\sum_{i,j=1}^{n} |x_i - x_j| \right)^2 \leq \frac{2(n^2 - 1)}{3} \sum_{i,j=1}^{n} (x_i - x_j)^2.$$
   (b) Show that equality holds if and only if $x_1, \ldots, x_n$ is an arithmetic progression.

5. **A5 (KOR)** Let $\mathbb{R}^+$ be the set of all positive real numbers. Find all functions $f : \mathbb{R}^+ \to \mathbb{R}^+$ that satisfy the following conditions:
   (i) $f(xyz) + f(x) + f(y) + f(z) = f(\sqrt{xy})f(\sqrt{yz})f(\sqrt{zx})$ for all $x, y, z \in \mathbb{R}^+$.
   (ii) $f(x) < f(y)$ for all $1 \leq x < y$.

6. **A6 (USA)** Let n be a positive integer and let $(x_1, \ldots, x_n)$, $(y_1, \ldots, y_n)$ be two sequences of positive real numbers. Suppose $(z_2, z_3, \ldots, z_{2n})$ is a sequence of positive real numbers such that
   $$z_{i+j}^2 \geq x_i y_j \quad \text{for all } 1 \leq i, j \leq n.$$
   Let $M = \max\{z_2, \ldots, z_{2n}\}$. Prove that
   $$\left(\frac{M + z_2 + \cdots + z_{2n}}{2n} \right)^2 \geq \left(\frac{x_1 + \cdots + x_n}{n} \right) \left(\frac{y_1 + \cdots + y_n}{n} \right).$$

7. **C1 (BRA)**[IMO1] Let A be a 101-element subset of the set $S = \{1, 2, \ldots, 1000000\}$. Prove that there exist numbers $t_1, t_2, \ldots, t_{100}$ in S such that the sets
   $$A_j = \{x + t_j \mid x \in A\}, \quad j = 1, 2, \ldots, 100,$$
   are pairwise disjoint.

8. **C2 (GEO)** Let $D_1, \ldots, D_n$ be closed disks in the plane. (A closed disk is a region bounded by a circle, taken jointly with this circle.) Suppose that every point in the plane is contained in at most 2003 disks D_i. Prove that there exists a disk D_k that intersects at most $7 \cdot 2003 - 1$ other disks D_i.

9. **C3 (LTU)** Let $n \geq 5$ be a given integer. Determine the largest integer k for which there exists a polygon with n vertices (convex or not, with non-self-intersecting boundary) having k internal right angles.

10. **C4 (IRN)** Let $x_1, \ldots, x_n$ and $y_1, \ldots, y_n$ be real numbers. Let $A = (a_{ij})_{1 \leq i,j \leq n}$ be the matrix with entries

$$a_{ij} = \begin{cases} 1, & \text{if } x_i + y_j \geq 0; \\ 0, & \text{if } x_i + y_j < 0. \end{cases}$$

Suppose that B is an $n \times n$ matrix whose entries are 0, 1 such that the sum of the elements in each row and each column of B is equal to the corresponding sum for the matrix A. Prove that $A = B$.

11. **C5 (ROU)** Every point with integer coordinates in the plane is the center of a disk with radius $1/1000$.
    (a) Prove that there exists an equilateral triangle whose vertices lie in different disks.
    (b) Prove that every equilateral triangle with vertices in different disks has side length greater than 96.

12. **C6 (SAF)** Let $f(k)$ be the number of integers n that satisfy the following conditions:
    (i) $0 \leq n < 10^k$, so n has exactly k digits (in decimal notation), with leading zeros allowed;
    (ii) the digits of n can be permuted in such a way that they yield an integer divisible by 11.
    Prove that $f(2m) = 10f(2m-1)$ for every positive integer m.

13. **G1 (FIN)**[IMO4] Let $ABCD$ be a cyclic quadrilateral. Let P, Q, R be the feet of the perpendiculars from D to the lines BC, CA, AB, respectively. Show that $PQ = QR$ if and only if the bisectors of $\angle ABC$ and $\angle ADC$ are concurrent with AC.

14. **G2 (HEL)** Three distinct points A, B, C are fixed on a line in this order. Let Γ be a circle passing through A and C whose center does not lie on the line AC. Denote by P the intersection of the tangents to Γ at A and C. Suppose Γ meets the segment PB at Q. Prove that the intersection of the bisector of $\angle AQC$ and the line AC does not depend on the choice of Γ.

15. **G3 (IND)** Let ABC be a triangle and let P be a point in its interior. Denote by D, E, F the feet of the perpendiculars from P to the lines BC, CA, and AB, respectively. Suppose that

$$AP^2 + PD^2 = BP^2 + PE^2 = CP^2 + PF^2.$$

Denote by I_A, I_B, I_C the excenters of the triangle ABC. Prove that P is the circumcenter of the triangle $I_A I_B I_C$.

16. **G4 (ARM)** Let $\Gamma_1, \Gamma_2, \Gamma_3, \Gamma_4$ be distinct circles such that Γ_1, Γ_3 are externally tangent at P, and Γ_2, Γ_4 are externally tangent at the same point P. Suppose that

Γ_1 and Γ_2; Γ_2 and Γ_3; Γ_3 and Γ_4; Γ_4 and Γ_1 meet at A, B, C, D, respectively, and that all these points are different from P. Prove that

$$\frac{AB \cdot BC}{AD \cdot DC} = \frac{PB^2}{PD^2}.$$

17. **G5 (KOR)** Let ABC be an isosceles triangle with $AC = BC$, whose incenter is I. Let P be a point on the circumcircle of the triangle AIB lying inside the triangle ABC. The lines through P parallel to CA and CB meet AB at D and E, respectively. The line through P parallel to AB meets CA and CB at F and G, respectively. Prove that the lines DF and EG intersect on the circumcircle of the triangle ABC.

18. **G6 (POL)**[IMO3] Each pair of opposite sides of a convex hexagon has the following property: The distance between their midpoints is equal to $\sqrt{3}/2$ times the sum of their lengths.
   Prove that all the angles of the hexagon are equal.

19. **G7 (SAF)** Let ABC be a triangle with semiperimeter s and inradius r. The semicircles with diameters BC, CA, AB are drawn outside of the triangle ABC. The circle tangent to all three semicircles has radius t. Prove that

$$\frac{s}{2} < t \le \frac{s}{2} + \left(1 - \frac{\sqrt{3}}{2}\right) r.$$

20. **N1 (POL)** Let m be a fixed integer greater than 1. The sequence $x_0, x_1, x_2, \ldots$ is defined as follows:

$$x_i = \begin{cases} 2^i, & \text{if } 0 \le i \le m - 1; \\ \sum_{j=1}^{m} x_{i-j}, & \text{if } i \ge m. \end{cases}$$

Find the greatest k for which the sequence contains k consecutive terms divisible by m.

21. **N2 (USA)** Each positive integer a undergoes the following procedure in order to obtain the number $d = d(a)$:
   (1) move the last digit of a to the first position to obtain the number b;
   (2) square b to obtain the number c;
   (3) move the first digit of c to the end to obtain the number d.
   (All the numbers in the problem are considered to be represented in base 10.) For example, for $a = 2003$, we have $b = 3200$, $c = 10240000$, and $d = 02400001 = 2400001 = d(2003)$.
   Find all numbers a for which $d(a) = a^2$.

22. **N3 (BGR)**[IMO2] Determine all pairs (a, b) of positive integers such that

$$\frac{a^2}{2ab^2 - b^3 + 1}$$

is a positive integer.

23. **N4 (ROU)** Let b be an integer greater than 5. For each positive integer n, consider the number

$$x_n = \underbrace{11\ldots1}_{n-1}\underbrace{22\ldots2}_{n}5,$$

written in base b. Prove that the following condition holds if and only if $b = 10$: There exists a positive integer M such that for every integer n greater than M, the number x_n is a perfect square.

24. **N5 (KOR)** An integer n is said to be *good* if $|n|$ is not the square of an integer. Determine all integers m with the following property: m can be represented in infinitely many ways as a sum of three distinct good integers whose product is the square of an odd integer.

25. **N6 (FRA)**[IMO6] Let p be a prime number. Prove that there exists a prime number q such that for every integer n, the number $n^p - p$ is not divisible by q.

26. **N7 (BRA)** The sequence $a_0, a_1, a_2, \ldots$ is defined as follows:

$$a_0 = 2, \qquad a_{k+1} = 2a_k^2 - 1 \qquad \text{for } k \geq 0.$$

Prove that if an odd prime p divides a_n, then 2^{n+3} divides $p^2 - 1$.

27. **N8 (IRN)** Let p be a prime number and let A be a set of positive integers that satisfies the following conditions:
    (i) the set of prime divisors of the elements in A consists of $p - 1$ elements;
    (ii) for any nonempty subset of A, the product of its elements is not a perfect pth power.
    What is the largest possible number of elements in A?

3.45 The Forty-Fifth IMO
Athens, Greece, July 7–19, 2004

3.45.1 Contest Problems

First Day (July 12)

1. Let ABC be an acute-angled triangle with $AB \neq AC$. The circle with diameter BC intersects the sides AB and AC at M and N, respectively. Denote by O the midpoint of BC. The bisectors of the angles BAC and MNG intersect at R. Prove that the circumcircles of the triangles BMR and CNR have a common point lying on the line segment BC.

2. Find all polynomials $P(x)$ with real coefficients that satisfy the equality

$$P(a-b) + P(b-c) + P(c-a) = 2P(a+b+c)$$

   for all triples a, b, c of real numbers such that $ab + bc + ca = 0$.

3. Determine all $m \times n$ rectangles that can be covered with *hooks* made up of 6 unit squares, as in the figure:

   Rotations and reflections of hooks are allowed. The rectangle must be covered without gaps and overlaps. No part of a hook may cover area outside the rectangle.

Second Day (July 13)

4. Let $n \geq 3$ be an integer and $t_1, t_2, \ldots, t_n$ positive real numbers such that

$$n^2 + 1 > (t_1 + t_2 + \cdots + t_n)\left(\frac{1}{t_1} + \frac{1}{t_2} + \cdots + \frac{1}{t_n}\right).$$

   Show that t_i, t_j, t_k are the side lengths of a triangle for all i, j, k that satisfy $1 \leq i < j < k \leq n$.

5. In a convex quadrilateral $ABCD$ the diagonal BD does not bisect the angles ABC and CDA. The point P lies inside $ABCD$ and satisfies

$$\angle PBC = \angle DBA \quad \text{and} \quad \angle PDC = \angle BDA.$$

   Prove that $ABCD$ is a cyclic quadrilateral if and only if $AP = CP$.

6. We call a positive integer *alternate* if its decimal digits are alternately odd and even. Find all positive integers n such that n has an alternate multiple.

3.45.2 Shortlisted Problems

1. **A1 (KOR)**[IMO4] Let $n \geq 3$ be an integer and $t_1, t_2, \ldots, t_n$ positive real numbers such that

$$n^2 + 1 > (t_1 + t_2 + \cdots + t_n) \left(\frac{1}{t_1} + \frac{1}{t_2} + \cdots + \frac{1}{t_n} \right).$$

Show that t_i, t_j, t_k are the side lengths of a triangle for all i, j, k that satisfy $1 \leq i < j < k \leq n$.

2. **A2 (ROU)** An infinite sequence $a_0, a_1, a_2, \ldots$ of real numbers satisfies the condition

$$a_n = |a_{n+1} - a_{n+2}| \quad \text{for every } n \geq 0$$

with a_0 and a_1 positive and distinct. Can this sequence be bounded?

3. **A3 (CAN)** Does there exist a function $s : \mathbb{Q} \to \{-1, 1\}$ such that if x and y are distinct rational numbers satisfying $xy = 1$ or $x + y \in \{0, 1\}$, then $s(x)s(y) = -1$? Justify your answer.

4. **A4 (KOR)**[IMO2] Find all polynomials $P(x)$ with real coefficients that satisfy the equality

$$P(a - b) + P(b - c) + P(c - a) = 2P(a + b + c)$$

for all triples a, b, c of real numbers such that $ab + bc + ca = 0$.

5. **A5 (THA)** Let $a, b, c > 0$ and $ab + bc + ca = 1$. Prove the inequality

$$\sqrt[3]{\frac{1}{a} + 6b} + \sqrt[3]{\frac{1}{b} + 6c} + \sqrt[3]{\frac{1}{c} + 6a} \leq \frac{1}{abc}.$$

6. **A6 (RUS)** Find all functions $f : \mathbb{R} \to \mathbb{R}$ satisfying the equation

$$f\left(x^2 + y^2 + 2f(xy)\right) = (f(x+y))^2 \quad \text{for all } x, y \in \mathbb{R}.$$

7. **A7 (IRL)** Let $a_1, a_2, \ldots, a_n$ be positive real numbers, $n > 1$. Denote by g_n their geometric mean, and by $A_1, A_2, \ldots, A_n$ the sequence of arithmetic means defined by $A_k = \frac{a_1 + a_2 + \cdots + a_k}{k}$, $k = 1, 2, \ldots, n$. Let G_n be the geometric mean of $A_1, A_2, \ldots, A_n$. Prove the inequality

$$n \sqrt[n]{\frac{G_n}{A_n}} + \frac{g_n}{G_n} \leq n + 1$$

and establish the cases of equality.

8. **C1 (PRI)** There are 10001 students at a university. Some students join together to form several clubs (a student may belong to different clubs). Some clubs join together to form several societies (a club may belong to different societies). There are a total of k societies. Suppose that the following conditions hold:
   (i) Each pair of students are in exactly one club.
   (ii) For each student and each society, the student is in exactly one club of the society.

(iii) Each club has an odd number of students. In addition, a club with $2m+1$ students (m is a positive integer) is in exactly m societies.
Find all possible values of k.

9. **C2 (GER)** Let n and k be positive integers. There are given n circles in the plane. Every two of them intersect at two distinct points, and all points of intersection they determine are distinct. Each intersection point must be colored with one of n distinct colors so that each color is used at least once, and exactly k distinct colors occur on each circle. Find all values of $n \geq 2$ and k for which such a coloring is possible.

10. **C3 (AUS)** The following operation is allowed on a finite graph: Choose an arbitrary cycle of length 4 (if there is any), choose an arbitrary edge in that cycle, and delete it from the graph. For a fixed integer $n \geq 4$, find the least number of edges of a graph that can be obtained by repeated applications of this operation from the complete graph on n vertices (where each pair of vertices are joined by an edge).

11. **C4 (POL)** Consider a matrix of size $n \times n$ whose entries are real numbers of absolute value not exceeding 1, and the sum of all entries is 0. Let n be an even positive integer. Determine the least number C such that every such matrix necessarily has a row or a column with the sum of its entries not exceeding C in absolute value.

12. **C5 (NZL)** Let N be a positive integer. Two players A and B, taking turns, write numbers from the set $\{1,\ldots,N\}$ on a blackboard. A begins the game by writing 1 on his first move. Then, if a player has written n on a certain move, his adversary is allowed to write $n+1$ or $2n$ (provided the number he writes does not exceed N). The player who writes N wins. We say that N is of type A or of type B according as A or B has a winning strategy.
   (a) Determine whether $N = 2004$ is of type A or of type B.
   (b) Find the least $N > 2004$ whose type is different from that of 2004.

13. **C6 (IRN)** For an $n \times n$ matrix A, let X_i be the set of entries in row i, and Y_j the set of entries in column j, $1 \leq i, j \leq n$. We say that A is *golden* if $X_1,\ldots,X_n,Y_1,\ldots,Y_n$ are distinct sets. Find the least integer n such that there exists a 2004×2004 golden matrix with entries in the set $\{1,2,\ldots,n\}$.

14. **C7 (EST)**[IMO3] Determine all $m \times n$ rectangles that can be covered with *hooks* made up of 6 unit squares, as in the figure:

Rotations and reflections of hooks are allowed. The rectangle must be covered without gaps and overlaps. No part of a hook may cover area outside the rectangle.

15. **C8 (POL)** For a finite graph G, let $f(G)$ be the number of triangles and $g(G)$ the number of tetrahedra formed by edges of G. Find the least constant c such that

$$g(G)^3 \le c \cdot f(G)^4 \text{ for every graph } G.$$

16. **G1 (ROU)**[IMO1] Let ABC be an acute-angled triangle with $AB \ne AC$. The circle with diameter BC intersects the sides AB and AC at M and N, respectively. Denote by O the midpoint of BC. The bisectors of the angles BAC and MNG intersect at R. Prove that the circumcircles of the triangles BMR and CNR have a common point lying on the line segment BC.

17. **G2 (KAZ)** The circle Γ and the line ℓ do not intersect. Let AB be the diameter of Γ perpendicular to ℓ, with B closer to ℓ than A. An arbitrary point $C \ne A, B$ is chosen on Γ. The line AC intersects ℓ at D. The line DE is tangent to Γ at E, with B and E on the same side of AC. Let BE intersect ℓ at F, and let AF intersect Γ at $G \ne A$. Prove that the reflection of G in AB lies on the line CF.

18. **G3 (KOR)** Let O be the circumcenter of an acute-angled triangle ABC with $\angle B < \angle C$. The line AO meets the side BC at D. The circumcenters of the triangles ABD and ACD are E and F, respectively. Extend the sides BA and CA beyond A, and choose on the respective extension points G and H such that $AG = AC$ and $AH = AB$. Prove that the quadrilateral $EFGH$ is a rectangle if and only if $\angle ACB - \angle ABC = 60°$.

19. **G4 (POL)**[IMO5] In a convex quadrilateral $ABCD$ the diagonal BD does not bisect the angles ABC and CDA. The point P lies inside $ABCD$ and satisfies

$$\angle PBC = \angle DBA \quad \text{and} \quad \angle PDC = \angle BDA.$$

Prove that $ABCD$ is a cyclic quadrilateral if and only if $AP = CP$.

20. **G5 (SCG)** Let $A_1 A_2 \ldots A_n$ be a regular n-gon. The points $B_1, \ldots, B_{n-1}$ are defined as follows:
  (i) If $i = 1$ or $i = n - 1$, then B_i is the midpoint of the side $A_i A_{i+1}$.
  (ii) If $i \ne 1$, $i \ne n - 1$, and S is the intersection point of $A_1 A_{i+1}$ and $A_n A_i$, then B_i is the intersection point of the bisector of the angle $A_i S A_{i+1}$ with $A_i A_{i+1}$.
  Prove the equality

$$\angle A_1 B_1 A_n + \angle A_1 B_2 A_n + \cdots + \angle A_1 B_{n-1} A_n = 180°.$$

21. **G6 (UNK)** Let $\mathscr{P}$ be a convex polygon. Prove that there is a convex hexagon that is contained in $\mathscr{P}$ and that occupies at least 75 percent of the area of $\mathscr{P}$.

22. **G7 (RUS)** For a given triangle ABC, let X be a variable point on the line BC such that C lies between B and X and the incircles of the triangles ABX and ACX intersect at two distinct points P and Q. Prove that the line PQ passes through a point independent of X.

23. **G8 (SCG)** A cyclic quadrilateral $ABCD$ is given. The lines AD and BC intersect at E, with C between B and E; the diagonals AC and BD intersect at F. Let M

be the midpoint of the side CD, and let $N \neq M$ be a point on the circumcircle of the triangle ABM such that $AN/BN = AM/BM$. Prove that the points E, F, and N are collinear.

24. **N1 (BLR)** Let $\tau(n)$ denote the number of positive divisors of the positive integer n. Prove that there exist infinitely many positive integers a such that the equation

$$\tau(an) = n$$

does not have a positive integer solution n.

25. **N2 (RUS)** The function ψ from the set $\mathbb{N}$ of positive integers into itself is defined by the equality

$$\psi(n) = \sum_{k=1}^{n} (k,n), \qquad n \in \mathbb{N},$$

where (k,n) denotes the greatest common divisor of k and n.
   (a) Prove that $\psi(mn) = \psi(m)\psi(n)$ for every two relatively prime $m,n \in \mathbb{N}$.
   (b) Prove that for each $a \in \mathbb{N}$ the equation $\psi(x) = ax$ has a solution.
   (c) Find all $a \in \mathbb{N}$ such that the equation $\psi(x) = ax$ has a unique solution.

26. **N3 (IRN)** A function f from the set of positive integers $\mathbb{N}$ into itself is such that for all $m,n \in \mathbb{N}$ the number $(m^2 + n)^2$ is divisible by $f^2(m) + f(n)$. Prove that $f(n) = n$ for each $n \in \mathbb{N}$.

27. **N4 (POL)** Let k be a fixed integer greater than 1, and let $m = 4k^2 - 5$. Show that there exist positive integers a and b such that the sequence (x_n) defined by

$$x_0 = a, \quad x_1 = b, \quad x_{n+2} = x_{n+1} + x_n \quad \text{for} \quad n = 0,1,2,\ldots$$

has all of its terms relatively prime to m.

28. **N5 (IRN)**[IMO6] We call a positive integer *alternate* if its decimal digits are alternately odd and even. Find all positive integers n such that n has an alternate multiple.

29. **N6 (IRL)** Given an integer $n > 1$, denote by P_n the product of all positive integers x less than n and such that n divides $x^2 - 1$. For each $n > 1$, find the remainder of P_n on division by n.

30. **N7 (BGR)** Let p be an odd prime and n a positive integer. In the coordinate plane, eight distinct points with integer coordinates lie on a circle with diameter of length p^n. Prove that there exists a triangle with vertices at three of the given points such that the squares of its side lengths are integers divisible by p^{n+1}.

3.46 The Forty-Sixth IMO
Mérida, Mexico, July 8–19, 2005

3.46.1 Contest Problems

First Day (July 13)

1. Six points are chosen on the sides of an equilateral triangle ABC: A_1, A_2 on BC; B_1, B_2 on CA; C_1, C_2 on AB. These points are vertices of a convex hexagon $A_1A_2B_1B_2C_1C_2$ with equal side lengths. Prove that the lines A_1B_2, B_1C_2 and C_1A_2 are concurrent.

2. Let $a_1, a_2, \ldots$ be a sequence of integers with infinitely many positive terms and infinitely many negative terms. Suppose that for each positive integer n, the numbers $a_1, a_2, \ldots, a_n$ leave n different remainders on division by n. Prove that each integer occurs exactly once in the sequence.

3. Let x, y, and z be positive real numbers such that $xyz \geq 1$. Prove that

$$\frac{x^5 - x^2}{x^5 + y^2 + z^2} + \frac{y^5 - y^2}{y^5 + z^2 + x^2} + \frac{z^5 - z^2}{z^5 + x^2 + y^2} \geq 0.$$

Second Day (July 14)

4. Consider the sequence $a_1, a_2, \ldots$ defined by

$$a_n = 2^n + 3^n + 6^n - 1 \quad (n = 1, 2, \ldots).$$

Determine all positive integers that are relatively prime to every term of the sequence.

5. Let $ABCD$ be a given convex quadrilateral with sides BC and AD equal in length and not parallel. Let E and F be interior points of the sides BC and AD respectively such that $BE = DF$. The lines AC and BD meet at P; the lines BD and EF meet at Q; the lines EF and AC meet at R. Consider all the triangles PQR as E and F vary. Show that the circumcircles of these triangles have a common point other than P.

6. In a mathematical competition, six problems were posed to the contestants. Each pair of problems was solved by more than $2/5$ of the contestants. Nobody solved all six problems. Show that there were at least two contestants who each solved exactly five problems.

3.46.2 Shortlisted Problems

1. **A1 (ROU)** Find all monic polynomials $p(x)$ with integer coefficients of degree two for which there exists a polynomial $q(x)$ with integer coefficients such that $p(x)q(x)$ is a polynomial having all coefficients ± 1.

2. **A2 (BGR)** Let $\mathbb{R}^+$ denote the set of positive real numbers. Determine all functions $f : \mathbb{R}^+ \to \mathbb{R}^+$ such that

$$f(x)f(y) = 2f(x+yf(x))$$

for all positive real numbers x and y.

3. **A3 (CZE)** Four real numbers p,q,r,s satisfy

$$p+q+r+s = 9 \quad \text{and} \quad p^2+q^2+r^2+s^2 = 21.$$

Prove that $ab - cd \geq 2$ holds for some permutation (a,b,c,d) of (p,q,r,s).

4. **A4 (IND)** Find all functions $f : \mathbb{R} \to \mathbb{R}$ satisfying the equation

$$f(x+y) + f(x)f(y) = f(xy) + 2xy + 1$$

for all real x and y.

5. **A5 (KOR)**[IMO3] Let x, y and z be positive real numbers such that $xyz \geq 1$. Prove that

$$\frac{x^5 - x^2}{x^5 + y^2 + z^2} + \frac{y^5 - y^2}{y^5 + z^2 + x^2} + \frac{z^5 - z^2}{z^5 + x^2 + y^2} \geq 0.$$

6. **C1 (AUS)** A house has an even number of lamps distributed among its rooms in such a way that there are at least three lamps in every room. Each lamp shares a switch with exactly one other lamp, not necessarily from the same room. Each change in the switch shared by two lamps changes their states simultaneously. Prove that for every initial state of the lamps there exists a sequence of changes in some of the switches at the end of which each room contains lamps that are on as well as lamps that are off.

7. **C2 (IRN)** Let k be a fixed positive integer. A company has a special method to sell sombreros. Each customer can convince two persons to buy a sombrero after he/she buys one; convincing someone already convinced does not count. Each of these new customers can convince two others and so on. If each of the two customers convinced by someone makes at least k persons buy sombreros (directly or indirectly), then that someone wins a free instructional video. Prove that if n persons bought sombreros, then at most $n/(k+2)$ of them got videos.

8. **C3 (IRN)** In an $m \times n$ rectangular board of mn unit squares, *adjacent* squares are ones with a common edge, and a *path* is a sequence of squares in which any two consecutive squares are adjacent. Each square of the board can be colored black or white. Let N denote the number of colorings of the board such that there exists at least one black path from the left edge of the board to its right edge, and let M denote the number of colorings in which there exist at least two nonintersecting black paths from the left edge to the right edge. Prove that $N^2 \geq 2^{mn}M$.

9. **C4 (COL)** Let $n \geq 3$ be a given positive integer. We wish to label each side and each diagonal of a regular n-gon $P_1 \ldots P_n$ with a positive integer less than or equal to r so that:

  (i) every integer between 1 and r occurs as a label;
  (ii) in each triangle $P_iP_jP_k$ two of the labels are equal and greater than the third.
Given these conditions:
  (a) Determine the largest positive integer r for which this can be done.
  (b) For that value of r, how many such labelings are there?

10. **C5 (SCG)** There are n markers, each with one side white and the other side black, aligned in a row so that their white sides are up. In each step, if possible, we choose a marker with the white side up (but not one of the outermost markers), remove it, and reverse the closest marker to the left and the closest marker to the right of it. Prove that one can achieve the state with only two markers remaining if and only if $n - 1$ is not divisible by 3.

11. **C6 (ROU)**[IMO6] In a mathematical competition, six problems were posed to the contestants. Each pair of problems was solved by more than $2/5$ of the contestants. Nobody solved all six problems. Show that there were at least two contestants who each solved exactly five problems.

12. **C7 (USA)** Let $n \geq 1$ be a given integer, and let $a_1, \ldots, a_n$ be a sequence of integers such that n divides the sum $a_1 + \cdots + a_n$. Show that there exist permutations σ and τ of $1, 2, \ldots, n$ such that $\sigma(i) + \tau(i) \equiv a_i \pmod{n}$ for all $i = 1, \ldots, n$.

13. **C8 (BGR)** Let M be a convex n-gon, $n \geq 4$. Some $n - 3$ of its diagonals are colored green and some other $n - 3$ diagonals are colored red, so that no two diagonals of the same color meet inside M. Find the maximum possible number of intersection points of green and red diagonals inside M.

14. **G1 (HEL)** In a triangle ABC satisfying $AB + BC = 3AC$ the incircle has center I and touches the sides AB and BC at D and E, respectively. Let K and L be the symmetric points of D and E with respect to I. Prove that the quadrilateral $ACKL$ is cyclic.

15. **G2 (ROU)**[IMO1] Six points are chosen on the sides of an equilateral triangle ABC: A_1, A_2 on BC; B_1, B_2 on CA; C_1, C_2 on AB. These points are vertices of a convex hexagon $A_1A_2B_1B_2C_1C_2$ with equal side lengths. Prove that the lines A_1B_2, B_1C_2 and C_1A_2 are concurrent.

16. **G3 (UKR)** Let $ABCD$ be a parallelogram. A variable line l passing through the point A intersects the rays BC and DC at points X and Y, respectively. Let K and L be the centers of the excircles of triangles ABX and ADY, touching the sides BX and DY, respectively. Prove that the size of angle KCL does not depend on the choice of the line l.

17. **G4 (POL)**[IMO5] Let $ABCD$ be a given convex quadrilateral with sides BC and AD equal in length and not parallel. Let E and F be interior points of the sides BC and AD respectively such that $BE = DF$. The lines AC and BD meet at P; the lines BD and EF meet at Q; the lines EF and AC meet at R. Consider all the triangles PQR as E and F vary. Show that the circumcircles of these triangles have a common point other than P.

18. **G5 (ROU)** Let ABC be an acute-angled triangle with $AB \neq AC$; let H be its orthocenter and M the midpoint of BC. Points D on AB and E on AC are such that $AE = AD$ and D, H, E are collinear. Prove that HM is orthogonal to the common chord of the circumcircles of triangles ABC and ADE.

19. **G6 (RUS)** The median AM of a triangle ABC intersects its incircle ω at K and L. The lines through K and L parallel to BC intersect ω again at X and Y. The lines AX and AY intersect BC at P and Q. Prove that $BP = CQ$.

20. **G7 (KOR)** In an acute triangle ABC, let D, E, F, P, Q, R be the feet of perpendiculars from A, B, C, A, B, C to BC, CA, AB, EF, FD, DE, respectively. Prove that $p(ABC)p(PQR) \geq p(DEF)^2$, where $p(T)$ denotes the perimeter of triangle T.

21. **N1 (POL)**[IMO4] Consider the sequence $a_1, a_2, \ldots$ defined by

$$a_n = 2^n + 3^n + 6^n - 1 \quad (n = 1, 2, \ldots).$$

Determine all positive integers that are relatively prime to every term of the sequence.

22. **N2 (NLD)**[IMO2] Let $a_1, a_2, \ldots$ be a sequence of integers with infinitely many positive terms and infinitely many negative terms. Suppose that for each positive integer n, the numbers $a_1, a_2, \ldots, a_n$ leave n different remainders on division by n. Prove that each integer occurs exactly once in the sequence.

23. **N3 (MNG)** Let a, b, c, d, e, and f be positive integers. Suppose that the sum $S = a + b + c + d + e + f$ divides both $abc + def$ and $ab + bc + ca - de - ef - fd$. Prove that S is composite.

24. **N4 (COL)** Find all positive integers $n > 1$ for which there exists a unique integer a with $0 < a \leq n!$ such that $a^n + 1$ is divisible by $n!$.

25. **N5 (NLD)** Denote by $d(n)$ the number of divisors of the positive integer n. A positive integer n is called *highly divisible* if $d(n) > d(m)$ for all positive integers $m < n$. Two highly divisible integers m and n with $m < n$ are called *consecutive* if there exists no highly divisible integer s satisfying $m < s < n$.
    (a) Show that there are only finitely many pairs of consecutive highly divisible integers of the form (a, b) with $a \mid b$.
    (b) Show that for every prime number p there exist infinitely many positive highly divisible integers r such that pr is also highly divisible.

26. **N6 (IRN)** Let a and b be positive integers such that $a^n + n$ divides $b^n + n$ for every positive integer n. Show that $a = b$.

27. **N7 (RUS)** Let $P(x) = a_n x^n + a_{n-1} x^{n-1} + \cdots + a_0$, where $a_0, \ldots, a_n$ are integers, $a_n > 0$, $n \geq 2$. Prove that there exists a positive integer m such that $P(m!)$ is a composite number.

3.47 The Forty-Seventh IMO
Ljubljana, Slovenia, July 6–18, 2006

3.47.1 Contest Problems

First Day (July 12)

1. Let ABC be a triangle with incenter I. A point P in the interior of the triangle satisfies

$$\angle PBA + \angle PCA = \angle PBC + \angle PCB.$$

   Show that $AP \geq AI$, and that equality holds if and only if $P = I$.

2. Let $\mathscr{P}$ be a regular 2006-gon. A diagonal of $\mathscr{P}$ is called *good* if its endpoints divide the boundary of $\mathscr{P}$ into two parts, each composed of an odd number of sides of $\mathscr{P}$. The sides of $\mathscr{P}$ are also called good.
   Suppose $\mathscr{P}$ has been dissected into triangles by 2003 diagonals, no two of which have a common point in the interior of $\mathscr{P}$. Find the maximum number of isosceles triangles having two good sides that could appear in such a configuration.

3. Determine the least real number M such that the inequality

$$\left| ab(a^2 - b^2) + bc(b^2 - c^2) + ca(c^2 - a^2) \right| \leq M(a^2 + b^2 + c^2)^2$$

   holds for all real numbers a, b, and c.

Second Day (July 13)

4. Determine all pairs (x, y) of integers such that

$$1 + 2^x + 2^{2x+1} = y^2.$$

5. Let $P(x)$ be a polynomial of degree $n > 1$ with integer coefficients and let k be a positive integer. Consider the polynomial

$$Q(x) = P(P(\ldots P(P(x))\ldots)),$$

   where P occurs k times. Prove that there are at most n integers t that satisfy the equality $Q(t) = t$.

6. Assign to each side b of a convex polygon $\mathscr{P}$ the maximum area of a triangle that has b as a side and is contained in $\mathscr{P}$. Show that the sum of the areas assigned to the sides of $\mathscr{P}$ is at least twice the area of $\mathscr{P}$.

3.47.2 Shortlisted Problems

1. **A1 (EST)** A sequence of real numbers $a_0, a_1, a_2, \ldots$ is defined by the formula

$$a_{i+1} = [a_i] \cdot \{a_i\}, \quad \text{for } i \geq 0;$$

   here a_0 is an arbitrary number, $[a_i]$ denotes the greatest integer not exceeding a_i, and $\{a_i\} = a_i - [a_i]$. Prove that $a_i = a_{i+2}$ for i sufficiently large.

2. **A2 (POL)** The sequence of real numbers $a_0, a_1, a_2, \ldots$ is defined recursively by $a_0 = -1$ and

$$\sum_{k=0}^{n} \frac{a_{n-k}}{k+1} = 0, \quad \text{for } n \geq 1.$$

Show that $a_n > 0$ for $n \geq 1$.

3. **A3 (RUS)** The sequence $c_0, c_1, \ldots, c_n, \ldots$ is defined by $c_0 = 1$, $c_1 = 0$, and $c_{n+2} = c_{n+1} + c_n$ for $n \geq 0$. Consider the set S of ordered pairs (x, y) for which there is a finite set J of positive integers such that $x = \sum_{j \in J} c_j$, $y = \sum_{j \in J} c_{j-1}$. Prove that there exist real numbers α, β, and M with the following property: an ordered pair of nonnegative integers (x, y) satisfies the inequality $m < \alpha x + \beta y < M$ if and only if $(x, y) \in S$.
Remark: A sum over the elements of the empty set is assumed to be 0.

4. **A4 (SRB)** Prove the inequality

$$\sum_{i<j} \frac{a_i a_j}{a_i + a_j} \leq \frac{n}{2(a_1 + a_2 + \cdots + a_n)} \sum_{i<j} a_i a_j$$

for positive real numbers $a_1, a_2, \ldots, a_n$.

5. **A5 (KOR)** Let a, b, c be the sides of a triangle. Prove that

$$\frac{\sqrt{b+c-a}}{\sqrt{b}+\sqrt{c}-\sqrt{a}} + \frac{\sqrt{c+a-b}}{\sqrt{c}+\sqrt{a}-\sqrt{b}} + \frac{\sqrt{a+b-c}}{\sqrt{a}+\sqrt{b}-\sqrt{c}} \leq 3.$$

6. **A6 (IRL)**[IMO3] Determine the smallest number M such that the inequality

$$|ab(a^2 - b^2) + bc(b^2 - c^2) + ca(c^2 - a^2)| \leq M(a^2 + b^2 + c^2)^2$$

holds for all real numbers a, b, c

7. **C1 (FRA)** We have $n \geq 2$ lamps $L_1, \ldots, L_n$ in a row, each of them being either *on* or *off*. Every second we simultaneously modify the state of each lamp as follows: if the lamp L_i and its neighbors (only one neighbor for $i = 1$ or $i = n$, two neighbors for other i) are in the same state, then L_i is switched off; otherwise, L_i is switched on.
Initially all the lamps are off except the leftmost one which is on.
   (a) Prove that there are infinitely many integers n for which all the lamps will eventually be off.
   (b) Prove that there are infinitely many integers n for which the lamps will never be all off.

8. **C2 (SRB)**[IMO2] A diagonal of a regular 2006-gon is called *odd* if its endpoints divide the boundary into two parts, each composed of an odd number of sides. Sides are also regarded as odd diagonals. Suppose the 2006-gon has been dissected into triangles by 2003 nonintersecting diagonals. Find the maximum possible number of isosceles triangles with two odd sides.

9. **C3 (COL)** Let S be a finite set of points in the plane such that no three of them are on a line. For each convex polygon P whose vertices are in S, let $a(P)$ be the number of vertices of P, and let $b(P)$ be the number of points of S that are outside P. Prove that for every real number x

$$\sum_P x^{a(P)}(1-x)^{b(P)} = 1,$$

where the sum is taken over all convex polygons with vertices in S.
Remark. A line segment, a point, and the empty set are considered convex polygons of 2, 1, and 0 vertices respectively.

10. **C4 (TWN)** A cake has the form of an $n \times n$ square composed of n^2 unit squares. Strawberries lie on some of the unit squares so that each row and each column contains exactly one strawberry; call this arrangement $\mathscr{A}$.
Let $\mathscr{B}$ be another such arrangement. Suppose that every grid rectangle with one vertex at the top left corner of the cake contains no fewer strawberries of arrangement $\mathscr{B}$ than of arrangement $\mathscr{A}$. Prove that arrangement $\mathscr{B}$ can be obtained from $\mathscr{A}$ by performing a number of *switches*, defined as follows:
A *switch* consists in selecting a grid rectangle with only two strawberries, situated at its top right corner and bottom left corner, and moving these two strawberries to the other two corners of that rectangle.

11. **C5 (ARG)** An (n,k)-tournament is a contest with n players held in k rounds such that:
    (i) Each player plays in each round, and every two players meet at most once.
    (ii) If player A meets player B in round i, player C meets player D in round i, and player A meets player C in round j, then player B meets player D in round j.
Determine all pairs (n,k) for which there exists an (n,k)-tournament.

12. **C6 (COL)** A *holey triangle* is an upward equilateral triangle of side length n with n upward unit triangular holes cut out. A *diamond* is a $60°$–$120°$ unit rhombus. Prove that a holey triangle T can be tiled with diamonds if and only if the following condition holds: every upward equilateral triangle of side length k in T contains at most k holes, for $1 \le k \le n$.

13. **C7 (JPN)** Consider a convex polyhedron without parallel edges and without an edge parallel to any face other than the two faces adjacent to it. Call a pair of points of the polyhedron *antipodal* if there exist two parallel planes passing through these points and such that the polyhedron is contained between these planes.
Let A be the number of antipodal pairs of vertices, and let B be the number of antipodal pairs of midpoint edges. Determine the difference $A - B$ in terms of the numbers of vertices, edges, and faces.

14. **G1 (KOR)**[IMO1] Let ABC be a triangle with incenter I. A point P in the interior of the triangle satisfies $\angle PBA + \angle PCA = \angle PBC + \angle PCB$. Show that $AP \ge AI$ and that equality holds if and only if P coincides with I.

15. **G2 (UKR)** Let ABC be a trapezoid with parallel sides $AB > CD$. Points K and L lie on the line segments AB and CD, respectively, so that $AK/KB = DL/LC$. Suppose that there are points P and Q on the line segment KL satisfying $\angle APB = \angle BCD$ and $\angle CQD = \angle ABC$. Prove that the points P, Q, B, and C are concyclic.

16. **G3 (USA)** Let $ABCDE$ be a convex pentagon such that $\angle BAC = \angle CAD = \angle DAE$ and $\angle ABC = \angle ACD = \angle ADE$. The diagonals BD and CE meet at P. Prove that the line AP bisects the side CD.

17. **G4 (RUS)** A point D is chosen on the side AC of a triangle ABC with $\angle C < \angle A < 90°$ in such a way that $BD = BA$. The incircle of ABC is tangent to AB and AC at points K and L, respectively. Let J be the incenter of triangle BCD. Prove that the line KL intersects the line segment AJ at its midpoint.

18. **G5 (HEL)** In triangle ABC, let J be the center of the excircle tangent to side BC at A_1 and to the extensions of sides AC and AB at B_1 and C_1, respectively. Suppose that the lines A_1B_1 and AB are perpendicular and intersect at D. Let E be the foot of the perpendicular from C_1 to line DJ. Determine the angles $\angle BEA_1$ and $\angle AEB_1$.

19. **G6 (BRA)** Circles ω_1 and ω_2 with centers O_1 and O_2 are externally tangent at point D and internally tangent to a circle ω at points E and F, respectively. Line t is the common tangent of ω_1 and ω_2 at D. Let AB be the diameter of ω perpendicular to t, so that A, E, and O_1 are on the same side of t. Prove that the lines AO_1, BO_2, EF, and t are concurrent.

20. **G7 (SVK)** In a triangle ABC, let M_a, M_b, M_c, be respectively the midpoints of the sides BC, CA, AB, and let T_a, T_b, T_c be the midpoints of the arcs BC, CA, AB of the circumcircle of ABC, not counting the opposite vertices. For $i \in \{a,b,c\}$ let ω_i be the circle with M_iT_i as diameter. Let p_i be the common external tangent to ω_j, ω_k ($\{i,j,k\} = \{a,b,c\}$) such that ω_i lies on the opposite side of p_i from ω_j, ω_k. Prove that the lines p_a, p_b, p_c form a triangle similar to ABC and find the ratio of similitude.

21. **G8 (POL)** Let $ABCD$ be a convex quadrilateral. A circle passing through the points A and D and a circle passing through the points B and C are externally tangent at a point P inside the quadrilateral. Suppose that $\angle PAB + \angle PDC \leq 90°$ and $\angle PBA + \angle PCD \leq 90°$. Prove that $AB + CD \geq BC + AD$.

22. **G9 (RUS)** Points A_1, B_1, C_1 are chosen on the sides BC, CA, AB of a triangle ABC respectively. The circumcircles of triangles AB_1C_1, BC_1A_1, CA_1B_1 intersect the circumcircle of triangle ABC again at points A_2, B_2, C_2 respectively ($A_2 \neq A$, $B_2 \neq B$, $C_2 \neq C$). Points A_3, B_3, C_3 are symmetric to A_1, B_1, C_1 with respect to the midpoints of the sides BC, CA, AB, respectively. Prove that the triangles $A_2B_2C_2$ and $A_3B_3C_3$ are similar.

23. **G10 (SRB)**[IMO6] Assign to each side b of a convex polygon $\mathscr{P}$ the maximum area of a triangle that has b as a side and is contained in $\mathscr{P}$. Show that the sum of the areas assigned to the sides of $\mathscr{P}$ is at least twice the area of $\mathscr{P}$.

24. **N1 (USA)**[IMO4] Determine all pairs (x, y) of integers satisfying the equation $1 + 2^x + 2^{2x+1} = y^2$.

25. **N2 (CAN)** For $x \in (0, 1)$ let $y \in (0, 1)$ be the number whose nth digit after the decimal point is the 2^nth digit after the decimal point of x. Show that if x is rational then so is y.

26. **N3 (SAF)** The sequence $f(1), f(2), f(3), \ldots$ is defined by

$$f(n) = \frac{1}{n} \left(\left[\frac{n}{1} \right] + \left[\frac{n}{2} \right] + \cdots + \left[\frac{n}{n} \right] \right),$$

where $[x]$ denotes the integral part of x.
   (a) Prove that $f(n+1) > f(n)$ infinitely often.
   (b) Prove that $f(n+1) < f(n)$ infinitely often.

27. **N4 (ROU)**[IMO5] Let $P(x)$ be a polynomial of degree $n > 1$ with integer coefficients and let k be a positive integer. Consider the polynomial $Q(x) = P(P(\ldots P(P(x))\ldots))$, where P occurs k times. Prove that there are at most n integers t such that $Q(t) = t$.

28. **N5 (RUS)** Find all integer solutions of the equation

$$\frac{x^7 - 1}{x - 1} = y^5 - 1.$$

29. **N6 (USA)** Let $a > b > 1$ be relatively prime positive integers. Define the *weight* of an integer c, denoted by $w(c)$, to be the minimal possible value of $|x| + |y|$ taken over all pairs of integers x and y such that $ax + by = c$. An integer c is called a *local champion* if $w(c) \geq w(c \pm a)$ and $w(c) \geq w(c \pm b)$. Find all local champions and determine their number.

30. **N7 (EST)** Prove that for every positive integer n there exists an integer m such that $2^m + m$ is divisible by n.

3.48 The Forty-Eighth IMO
Hanoi, Vietnam, July 19–31, 2007

3.48.1 Contest Problems

First Day (July 25)

1. Real numbers $a_1, a_2, \ldots, a_n$ are given. For each i ($1 \leq i \leq n$) define

$$d_i = \max\{a_j \mid 1 \leq j \leq i\} - \min\{a_j \mid i \leq j \leq n\}$$

and let $d = \max\{d_i \mid 1 \leq i \leq n\}$.
   (a) Prove that for any real numbers $x_1 \leq x_2 \leq \cdots \leq x_n$,

$$\max\{|x_i - a_i| \mid 1 \leq i \leq n\} \geq \frac{d}{2}. \qquad (1)$$

   (b) Show that there are real numbers $x_1 \leq x_2 \leq \cdots \leq x_n$ such that equality holds in (1).

2. Consider five points A, B, C, D, and E such that $ABCD$ is a parallelogram and $BCED$ is a cyclic quadrilateral. Let ℓ be a line passing through A. Suppose that ℓ intersects the interior of the segment DC at F and intersects line BC at G. Suppose also that $EF = EG = EC$. Prove that ℓ is the bisector of angle DAB.

3. In a mathematical competition some competitors are friends. Friendship is always mutual. Call a group of competitors a *clique* if each two of them are friends. (In particular, any group of fewer than two competitors is a clique.) The number of members of a clique is called its *size*.
   Given that in this competition, the largest size of a clique is even, prove that the competitors can be arranged in two rooms such that the largest size of a clique contained in one room is the same as the largest size of a clique contained in the other room.

Second Day (July 26)

4. In triangle ABC the bisector of angle BCA intersects the circumcircle again at R, the perpendicular bisector of BC at P, and the perpendicular bisector of AC at Q. The midpoint of BC is K and the midpoint of AC is L. Prove that the triangles RPK and RQL have the same area.

5. Let a and b be positive integers. Show that if $4ab - 1$ divides $(4a^2 - 1)^2$, then $a = b$.

6. Let n be a positive integer. Consider

$$S = \big\{(x,y,z) \mid x,y,z \in \{0,1,\ldots,n\},\ x+y+z > 0\big\}$$

as a set of $(n+1)^3 - 1$ points in three-dimensional space. Determine the smallest possible number of planes, the union of which contains S but does not include $(0,0,0)$.

3.48.2 Shortlisted Problems

1. **A1 (NZL)** [IMO1] Given a sequence $a_1, a_2, \ldots, a_n$ of real numbers, for each i ($1 \le i \le n$) define

$$d_i = \max\{a_j : 1 \le j \le i\} - \min\{a_j : i \le j \le n\}$$

and let $d = \max\{d_i : 1 \le i \le n\}$.
   (a) Prove that for arbitrary real numbers $x_1 \le x_2 \le \cdots \le x_n$,

$$\max\{|x_i - a_i| : 1 \le i \le n\} \ge \frac{d}{2}. \tag{1}$$

   (b) Show that there exists a sequence $x_1 \le x_2 \le \cdots \le x_n$ of real numbers such that we have equality in (1).

2. **A2 (BGR)** Consider those functions $f : \mathbb{N} \to \mathbb{N}$ that satisfy the condition

$$f(m+n) \ge f(m) + f(f(n)) - 1, \text{ for all } m, n \in \mathbb{N}.$$

   Find all possible values of $f(2007)$.

3. **A3 (EST)** Let n be a positive integer, and let x and y be positive real numbers such that $x^n + y^n = 1$. Prove that

$$\left(\sum_{k=1}^{n} \frac{1 + x^{2k}}{1 + x^{4k}} \right) \left(\sum_{k=1}^{n} \frac{1 + y^{2k}}{1 + y^{4k}} \right) < \frac{1}{(1-x)(1-y)}.$$

4. **A4 (THA)** Find all functions $f : \mathbb{R}^+ \to \mathbb{R}^+$ such that

$$f(x + f(y)) = f(x + y) + f(y)$$

   for all $x, y \in \mathbb{R}^+$.

5. **A5 (HRV)** Let $c > 2$, and let $a(1), a(2), \ldots$ be a sequence of nonnegative real numbers such that

$$a(m+n) \le 2a(m) + 2a(n) \text{ for all } m, n \ge 1, \text{ and}$$
$$a(2^k) \le \frac{1}{(k+1)^c} \text{ for all } k \ge 0.$$

   Prove that the sequence $a(n)$ is bounded.

6. **A6 (POL)** Let $a_1, a_2, \ldots, a_{100}$ be nonnegative real numbers such that $a_1^2 + a_2^2 + \cdots + a_{100}^2 = 1$. Prove that

$$a_1^2 a_2 + a_2^2 a_3 + \cdots + a_{100}^2 a_1 < \frac{12}{25}.$$

7. **A7 (NLD)**[IMO6] Let $n > 1$ be an integer. Consider the following subset of space:

$$S = \{(x,y,z) | x,y,z \in \{0,1,\dots,n\}, x+y+z > 0\}.$$

Find the smallest number of planes that jointly contain all $(n+1)^3 - 1$ points of S but none of them passes through the origin.

8. **C1 (SRB)** Let n be an integer. Find all sequences $a_1, a_2, \dots, a_{n^2+n}$ satisfying the following conditions:
   (i) $a_i \in \{0,1\}$ for all $1 \le i \le n^2 + n$;
   (ii) $a_{i+1} + a_{i+2} + \cdots + a_{i+n} < a_{i+n+1} + a_{i+n+2} + \cdots + a_{i+2n}$ for all $0 \le i \le n^2 - n$.

9. **C2 (JPN)** A unit square is dissected into $n > 1$ rectangles such that their sides are parallel to the sides of the square. Any line parallel to a side of the square and intersecting its interior also intersects the interior of some rectangle. Prove that in this dissection, there exists a rectangle having no point on the boundary of the square.

10. **C3 (NLD)** Find all positive integers n for which the numbers in the set $S = \{1,2,\dots,n\}$ can be colored red and blue, with the following condition being satisfied: the set $S \times S \times S$ contains exactly 2007 ordered triples (x,y,z) such that
    (i) x,y,z are of the same color, and
    (ii) $x+y+z$ is divisible by n.

11. **C4 (IRN)** Let $A_0 = \{a_1, \dots, a_n\}$ be a finite sequence of real numbers. For each $k \ge 0$, from the sequence $A_k = (x_1, \dots, x_n)$ we construct a new sequence A_{k+1} in the following way:
    (i) We choose a partition $\{1, \dots, n\} = I \cup J$, where I and J are two disjoint sets, such that the expression

$$\left| \sum_{i \in I} x_i - \sum_{j \in J} x_j \right|$$

attains the smallest possible value. (We allow the set I or J to be empty; in this case the corresponding sum is 0.) If there are several such partitions, one is chosen arbitrarily.
    (ii) We set $A_{k+1} = (y_1, \dots, y_n)$, where $y_i = x_i + 1$ if $i \in I$, and $y_i = x_i - 1$ if $i \in J$.
    Prove that for some k, the sequence A_k contains an element x such that $|x| \ge n/2$.

12. **C5 (ROU)** In the Cartesian coordinate plane define the strip

$$S_n = \{(x,y) : n \le x < n+1\}$$

for every integer n. Assume that each strip S_n is colored either red or blue, and let a and b be two distinct positive integers. Prove that there exists a rectangle with side lengths a and b such that its vertices have the same color.

13. **C6 (RUS)**[IMO3] In a mathematical competition some competitors are friends; friendship is always mutual. Call a group of competitors a *clique* if each two of them are friends. The number of members in a clique is called its *size*.

It is known that the largest size of a clique is even. Prove that the competitors can be arranged in two rooms such that the largest size of a clique in one room is the same as the largest size of a clique in the other room.

14. **C7 (AUT)** Let $\alpha < \frac{3-\sqrt{5}}{2}$ be a positive real number. Prove that there exist positive integers n and p such that $p > \alpha \cdot 2^n$ and for which one can select $2p$ distinct subsets $S_1, \ldots, S_p, T_1, \ldots, T_p$ of the set $\{1, 2, \ldots, n\}$ such that $S_i \cap T_j \neq \emptyset$ for all $1 \leq i, j \leq p$.

15. **C8 (UKR)** Given a convex n-gon P in the plane, for every three vertices of P, consider the triangle determined by them. Call such a triangle *good* if all its sides are of unit length. Prove that there are not more than $2n/3$ good triangles.

16. **G1 (CZE)** [IMO4] In a triangle ABC the bisector of angle BCA intersects the circumcircle again at R, the perpendicular bisector of BC at P, and the perpendicular bisector of AC at Q. The midpoint of BC is K and the midpoint of AC is L. Prove that the triangles RPK and RQL have the same area.

17. **G2 (CAN)** Given an isosceles triangle ABC, assume that $AB = AC$. The midpoint of the side BC is denoted by M. Let X be a variable point on the shorter arc MA of the circumcircle of triangle ABM. Let T be the point in the angle domain BMA for which $\angle TMX = 90°$ and $TX = BX$. Prove that $\angle MTB - \angle CTM$ does not depend on X.

18. **G3 (UKR)** The diagonals of a trapezoid $ABCD$ intersect at point P. Point Q lies between the parallel lines BC and AD such that $\angle AQD = \angle CQB$, and the line CD separates the points P and Q. Prove that $\angle BQP = \angle DAQ$.

19. **G4 (LUX)** [IMO2] Consider five points A, B, C, D, and E such that $ABCD$ is a parallelogram and $BCED$ is a cyclic quadrilateral. Let ℓ be a line passing through A. Suppose that ℓ intersects the interior of the segment DC at F and intersects line BC at G. Suppose also that $EF = EG = EC$. Prove that ℓ is the bisector of angle DAB.

20. **G5 (UNK)** Let ABC be a fixed triangle, and let A_1, B_1, C_1 be the midpoints of sides BC, CA, AB respectively. Let P be a variable point on the circumcircle. Let lines PA_1, PB_1, PC_1 meet the circumcircle again at A', B', C' respectively. Assume that the points A, B, C, A', B', C' are distinct, and lines AA', BB', CC' form a triangle. Prove that the area of this triangle does not depend on P.

21. **G6 (USA)** Let $ABCD$ be a convex quadrilateral, and let points A_1, B_1, C_1, and D_1 lie on sides AB, BC, CD, and DA respectively. Consider the areas of triangles AA_1D_1, BB_1A_1, CC_1B_1, and DD_1C_1; let S be the sum of the two smallest ones, and let S_1 be the area of the quadrilateral $A_1B_1C_1D_1$.
Find the smallest positive real number k such that $kS_1 \geq S$ holds for every convex quadrilateral $ABCD$.

22. **G7 (IRN)** Given an acute triangle ABC with angles α, β, and γ at vertices A, B, and C respectively such that $\beta > \gamma$, let I be its incenter, and R the circumradius. Point D is the foot of the altitude from vertex A. Point K lies on line AD such

that $AK = 2R$, and D separates A and K. Finally, lines DI and KI meet sides AC and BC at E and F respectively. Prove that if $IE = IF$, then $\beta \leq 3\gamma$.

23. **G8 (POL)** A point P lies on the side AB of a convex quadrilateral $ABCD$. Let ω be the incircle of the triangle CPD, and let I be its incenter. Suppose that ω is tangent to the incircles of triangles APD and BPC at points K and L, respectively. Let the lines AC and BD meet at E, and let the lines AK and BL meet at F. Prove that the points E, I, and F are collinear.

24. **N1 (AUT)** Find all pairs (k,n) of positive integers for which $7^k - 3^n$ divides $k^4 + n^2$.

25. **N2 (CAN)** Let $b,n > 1$ be integers. Suppose that for each $k > 1$ there exists an integer a_k such that $b - a_k^n$ is divisible by k. Prove that $b = A^n$ for some integer A.

26. **N3 (NLD)** Let X be a set of 10000 integers, none of which is divisible by 47. Prove that there exists a 2007-element subset Y of X such that $a - b + c - d + e$ is not divisible by 47 for any $a,b,c,d,e \in Y$.

27. **N4 (POL)** For every integer $k \geq 2$, prove that 2^{3k} divides the number

$$\binom{2^{k+1}}{2^k} - \binom{2^k}{2^{k-1}}$$

but 2^{3k+1} does not.

28. **N5 (IRN)** Find all surjective functions $f : \mathbb{N} \to \mathbb{N}$ such that for every $m,n \in \mathbb{N}$ and every prime p, the number $f(m+n)$ is divisible by p if and only if $f(m) + f(n)$ is divisible by p.

29. **N6 (UNK)** [IMO5] Let k be a positive integer. Prove that the number $(4k^2 - 1)^2$ has a positive divisor of the form $8kn - 1$ if and only if k is even.

30. **N7 (IND)** For a prime p and a positive integer n, denote by $v_p(n)$ the exponent of p in the prime factorization of $n!$. Given a positive integer d and a finite set $\{p_1, \ldots, p_k\}$ of primes, show that there are infinitely many positive integers n such that $d \mid v_{p_i}(n)$ for all $1 \leq i \leq k$.

3.49 The Forty-Ninth IMO
Madrid, Spain, July 10–22, 2008

3.49.1 Contest Problems

First Day (July 16)

1. An acute-angled triangle ABC has orthocenter H. The circle passing through H with center the midpoint of BC intersects the line BC at A_1 and A_2. Similarly, the circle passing through H with center the midpoint of CA intersects the line CA at B_1 and B_2, and the circle passing through H with center the midpoint of AB intersects the line AB at C_1 and C_2. Show that $A_1, A_2, B_1, B_2, C_1, C_2$ lie on a circle.

2. (a) Prove that
$$\frac{x^2}{(x-1)^2} + \frac{y^2}{(y-1)^2} + \frac{z^2}{(z-1)^2} \geq 1$$
   for all real numbers x, y, z each different from 1 and satisfying $xyz = 1$.
   (b) Prove that equality holds above for infinitely many triples of rational numbers x, y, z each different from 1 and satisfying $xyz = 1$.

3. Prove that there exist infinitely many positive integers n such that $n^2 + 1$ has a prime divisor that is greater than $2n + \sqrt{2n}$.

Second Day (July 17)

4. Find all functions $f : (0, +\infty) \rightarrow (0, +\infty)$ (so f is a function from the positive real numbers to the positive real numbers) such that
$$\frac{(f(w))^2 + (f(x))^2}{f(y^2) + f(z^2)} = \frac{w^2 + x^2}{y^2 + z^2}$$
for all positive real numbers w, x, y, z satisfying $wx = yz$.

5. Let n and k be positive integers with $k \geq n$ and $k - n$ an even number. Let $2n$ lamps labeled $1, 2, \ldots, 2n$ be given, each of which can be either *on* or *off*. Initially all the lamps are off. We consider sequence of *steps*: at each step one of the lamps is switched (from on to off or from off to on).
   Let N be the number of such sequences consisting of k steps and resulting in the state in which lamps 1 through n are all on, and lamps $n + 1$ through $2n$ are all off.
   Let M be the number of such sequences consisting of k steps and resulting in the state in which lamps 1 through n are all on, and lamps $n + 1$ through $2n$ are all off, but where none of the lamps $n + 1$ through $2n$ is ever switched on.
   Determine the ratio N/M.

6. Let $ABCD$ be a convex quadrilateral with $|BA| \neq |BC|$. Denote the incircles of triangles ABC and ADC by ω_1 and ω_2 respectively. Suppose that there exists a circle ω tangent to the ray BA beyond A and to the ray BC beyond C that is also tangent to the lines AD and CD. Prove that the common external tangents of ω_1 and ω_2 intersect on ω.

3.49.2 Shortlisted Problems

1. **A1 (KOR)** [IMO4] Find all functions $f : (0,+\infty) \to (0,+\infty)$ (so f is a function from the positive real numbers to the positive real numbers) such that

$$\frac{(f(w))^2 + (f(x))^2}{f(y^2) + f(z^2)} = \frac{w^2 + x^2}{y^2 + z^2}$$

for all positive real numbers w, x, y, z, satisfying $wx = yz$.

2. **A2 (AUT)** [IMO2]
   (a) Prove that

   $$\frac{x^2}{(x-1)^2} + \frac{y^2}{(y-1)^2} + \frac{z^2}{(z-1)^2} \geq 1$$

   for all real numbers x, y, z each different from 1 and satisfying $xyz = 1$.
   (b) Prove that equality holds above for infinitely many triples of rational numbers x, y, z each different from 1 and satisfying $xyz = 1$.

3. **A3 (NLD)** Let $S \subseteq \mathbb{R}$ be a set of real numbers. We say that a pair (f,g) of functions from S to S is a *Spanish couple* on S if they satisfy the following conditions:
   (i) Both functions are strictly increasing, i.e., $f(x) < f(y)$ and $g(x) < g(y)$ for all $x, y \in S$ with $x < y$;
   (ii) The inequality $f(g(g(x))) < g(f(x))$ holds for all $x \in S$.
   Decide whether there exists a Spanish couple
   (a) on the set $S = \mathbb{N}$ of positive integers;
   (b) on the set $S = \{a - 1/b : a,b \in \mathbb{N}\}$.

4. **A4 (AUT)** For an integer m, denote by $t(m)$ the unique number in $\{1,2,3\}$ such that $m + t(m)$ is a multiple of 3. A function $f : \mathbb{Z} \to \mathbb{Z}$ satisfies $f(-1) = 0$, $f(0) = 1$, $f(1) = -1$, and

   $$f(2^n + m) = f(2^n - t(m)) - f(m) \quad \text{for all integers } m, n \geq 0 \text{ with } 2^n > m.$$

   Prove that $f(3p) \geq 0$ holds for all integers $p \geq 0$.

5. **A5 (SVK)** Let a, b, c, d be positive real numbers such that

   $$abcd = 1 \quad \text{and} \quad a+b+c+d > \frac{a}{b} + \frac{b}{c} + \frac{c}{d} + \frac{d}{a}.$$

   Prove that

   $$a+b+c+d < \frac{b}{a} + \frac{c}{b} + \frac{d}{c} + \frac{a}{d}.$$

6. **A6 (LTU)** Let $f : \mathbb{R} \to \mathbb{N}$ be a function that satisfies

$$f\left(x + \frac{1}{f(y)}\right) = f\left(y + \frac{1}{f(x)}\right), \qquad \text{for all } x, y \in \mathbb{R}.$$

Prove that there is a positive integer that is not a value of f.

7. **A7 (GER)** Prove that for any four positive real numbers a, b, c, d, the inequality

$$\frac{(a-b)(a-c)}{a+b+c} + \frac{(b-c)(b-d)}{b+c+d} + \frac{(c-d)(c-a)}{c+d+a} + \frac{(d-a)(d-b)}{d+a+b} \geq 0$$

holds. Determine all cases of equality.

8. **C1 (NLD)** A *box* is a rectangle in the plane whose sides are parallel to the coordinate axes and have positive lengths. Two boxes *intersect* if they have a common point in their interior or on their boundary.
Find the largest n for which there exist n boxes $B_1, \ldots, B_n$ such that B_i and B_j intersect if and only if $i \not\equiv j \pm 1 \pmod{n}$.

9. **C2 (SRB)** For every positive integer n determine the number of permutations $(a_1, \ldots, a_n)$ of the set $\{1, 2, \ldots, n\}$ with the following property:

$$2(a_1 + a_2 + \cdots + a_k) \text{ is divisible by } k \text{ for } k = 1, 2, \ldots, n.$$

10. **C3 (PER)** Consider the set S of all points with integer coordinates in the coordinate plane. For a positive integer k, two distinct points $A, B \in S$ will be called k-*friends* if there is a point $C \in S$ such that the area of the triangle ABC is equal to k. A set $T \subseteq S$ will be called a k-*clique* if every two points in T are k-friends. Find the least positive integer k for which there exists a k-clique with more than 200 elements.

11. **C4 (FRA)** [IMO5] Let n and k be positive integers with $k \geq n$ and $k - n$ an even number. Let $2n$ lamps labeled 1, 2, $\ldots$, $2n$ be given, each of which can be either *on* or *off*. Initially all the lamps are off. We consider sequence of *steps*: at each step one of the lamps is switched (from on to off or from off to on).
Let N be the number of such sequences consisting of k steps and resulting in the state in which lamps 1 through n are all on, and lamps $n + 1$ through $2n$ are all off.
Let M be the number of such sequences consisting of k steps and resulting in the state in which lamps 1 through n are all on, and lamps $n + 1$ through $2n$ are all off, but where none of the lamps $n + 1$ through $2n$ is ever switched on.
Determine the ratio N/M.

12. **C5 (RUS)** Let $S = \{x_1, x_2, \ldots, x_{k+l}\}$ be a $(k + l)$-element set of real numbers contained in the interval $[0, 1]$; k and l are positive integers. A k-element subset $A \subseteq S$ is called *nice* if

$$\left| \frac{1}{k} \sum_{x_i \in A} x_i - \frac{1}{l} \sum_{x_j \in S \setminus A} x_j \right| \leq \frac{k+l}{2kl}.$$

Prove that the number of nice subsets is at least $\frac{2}{k+l} \cdot \binom{k+l}{k}$.

13. **C6 (NLD)** For $n \geq 2$, let $S_1, S_2, \ldots, S_{2^n}$ be 2^n subsets of $A = \{1, 2, 3, \ldots, 2^{n+1}\}$ that satisfy the following property: There do not exist indices a and b with $a < b$ and elements $x, y, z \in A$ with $x < y < z$ such that $y, z \in S_a$ and $x, z \in S_b$. Prove that at least one of the sets $S_1, S_2, \ldots, S_{2^n}$ contains no more than $4n$ elements.

14. **G1 (RUS)** [IMO1] An acute-angled triangle ABC has orthocenter H. The circle passing through H with center the midpoint of BC intersects the line BC at A_1 and A_2. Similarly, the circle passing through H with center the midpoint of CA intersects the line CA at B_1 and B_2, and the circle passing through H with center the midpoint of AB intersects the line AB at C_1 and C_2. Show that $A_1, A_2, B_1, B_2, C_1, C_2$ lie on a circle.

15. **G2 (LUX)** Given a trapezoid $ABCD$ with parallel sides AB and CD, assume that there exist points E on line BC outside the segment BC, and F inside the segment AD, such that $\angle DAE = \angle CBF$. Denote by I the intersection point of CD and EF, and by J the intersection point of AB and EF. Let K be the midpoint of the segment EF. Assume that K does not lie on the lines AB and CD.
Prove that I belongs to the circumcircle of $\triangle ABK$ if and only if K belongs to the circumcircle of $\triangle CDJ$.

16. **G3 (PER)** Let $ABCD$ be a convex quadrilateral and let P and Q be the points such that $PQDA$ and $QPBC$ are cyclic quadrilaterals. Suppose that there exists a point E on the line segment PQ such that $\angle PAE = \angle QDE$ and $\angle PBE = \angle QCE$. Show that the quadrilateral $ABCD$ is cyclic.

17. **G4 (IRN)** Let BE and CF be altitudes in an acute triangle ABC. Two circles passing through the points A and F are tangent to the line BC at the points P and Q so that B lies between C and Q. Prove that the lines PE and QF intersect on the circumcircle of $\triangle AEF$.

18. **G5 (NLD)** Let k and n be integers with $0 \leq k \leq n - 2$. Consider a set L of n lines in the plane such that no two of them are parallel and no three have a common point. Denote by I the set of intersection points of lines in L. Let O be a point in the plane not lying on any line of L.
A point $X \in I$ is colored red if the open line segment (OX) intersects at most k lines from L. Prove that I contains at least $\frac{1}{2}(k+1)(k+2)$ red points.

19. **G6 (SRB)** Let $ABCD$ be a convex quadrilateral. Prove that there exists a point P inside the quadrilateral such that

$$\angle PAB + \angle PDC = \angle PBC + \angle PAD = \angle PCD + \angle PBA = \angle PDA + \angle PCB = 90°$$

if and only if the diagonals AC and BD are perpendicular.

20. **G7 (RUS)** [IMO6] Let $ABCD$ be a convex quadrilateral with $|BA| \neq |BC|$. Denote the incircles of triangles ABC and ADC by ω_1 and ω_2 respectively. Suppose that there exists a circle ω tangent to the ray BA beyond A and to the ray BC beyond

C that is also tangent to the lines AD and CD. Prove that the common external tangents of ω_1 and ω_2 intersect on ω.

21. **N1 (AUS)** Let n be a positive integer and let p be a prime number. Prove that if a, b, c are integers (not necessarily positive) satisfying the equations

$$a^n + pb = b^n + pc = c^n + pa,$$

then $a = b = c$.

22. **N2 (IRN)** Let $a_1, a_2, \ldots, a_n$ be distinct positive integers, $n \geq 3$. Prove that there exist distinct indices i and j such that $a_i + a_j$ does not divide any of the numbers $3a_1, 3a_2, \ldots, 3a_n$.

23. **N3 (IRN)** Let a_0, a_1, a_2 be a sequence of positive integers such that the greatest common divisor of any two consecutive terms is greater than the preceding term, i.e. $(a_i, a_{i+1}) > a_{i-1}$ for all $i \geq 1$. Prove that $a_n \geq 2^n$ for all $n \geq 0$.

24. **N4 (SRB)** Let n be a positive integer. Show that the numbers

$$\binom{2^n - 1}{0}, \binom{2^n - 1}{1}, \binom{2^n - 1}{2}, \ldots, \binom{2^n - 1}{2^{n-1} - 1}$$

are congruent modulo 2^n to $1, 3, 5, \ldots, 2^n - 1$ in some order.

25. **N5 (FRA)** For every $n \in \mathbb{N}$ let $d(n)$ denote the number of (positive) divisors of n. Find all functions $f : \mathbb{N} \to \mathbb{N}$ with the following properties:
    (i) $d(f(x)) = x$ for all $x \in \mathbb{N}$;
    (ii) $f(xy)$ divides $(x-1)y^{xy-1}f(x)$ for all $x, y \in \mathbb{N}$.

26. **N6 (LTU)** [IMO3] Prove that there exist infinitely many positive integers n such that $n^2 + 1$ has a prime divisor that is greater than $2n + \sqrt{2n}$.

3.50 The Fiftieth IMO
Bremen, Germany, July 10–22, 2009

3.50.1 Contest Problems

First Day (July 15)

1. Let n be a positive integer and let $a_1, \ldots, a_k$ ($k \geq 2$) be distinct integers in the set $\{1,\ldots,n\}$ such that n divides $a_i(a_{i+1} - 1)$ for $i = 1,\ldots,k - 1$. Prove that n does not divide $a_k(a_1 - 1)$.

2. Let ABC be a triangle with circumcenter O. The points P and Q are interior points of the sides CA and AB, respectively. Let K, L and M be the midpoints of the segments BP, CQ, and PQ, respectively, and let Γ be the circle passing through K, L, and M. Suppose that the line PQ is tangent to the circle Γ. Prove that $OP = OQ$.

3. Suppose that $s_1, s_2, s_3, \ldots$ is a strictly increasing sequence of positive integers such that the subsequences

$$s_{s_1}, s_{s_2}, s_{s_3}, \ldots \quad \text{and} \quad s_{s_1+1}, s_{s_2+1}, s_{s_3+1}, \ldots$$

are both arithmetic progressions. Prove that the sequence $s_1, s_2, s_3, \ldots$ is itself an arithmetic progression.

Second Day (July 16)

4. Let ABC be a triangle with $AB = AC$. The angle bisectors of $\angle CAB$ and $\angle ABC$ meet the sides BC and CA at D and E, respectively. Let K be the incenter of triangle ADC. Suppose that $\angle BEK = 45°$. Find all possible values of $\angle CAB$.

5. Determine all functions f from the set of positive integers to the set of positive integers such that, for all positive integers a and b, there exists a nondegenerate triangle with sides of lengths

$$a, \quad f(b), \quad \text{and} \quad f(b + f(a) - 1).$$

(A triangle is *nondegenerate* if its vertices are not collinear.)

6. Let $a_1, a_2, \ldots, a_n$ be distinct positive integers and let M be a set of $n - 1$ positive integers not containing $s = a_1 + a_2 + \cdots + a_n$. A grasshopper is to jump along the real axis, starting at the point 0 and making n jumps to the right with lengths $a_1, a_2, \ldots, a_n$ in some order. Prove that the order can be chosen in such a way that the grasshopper never lands on any point in M.

3.50.2 Shortlisted Problems

1. **A1 (CZE)** Find the largest possible integer k such that the following statement is true:

Let 2009 arbitrary nondegenerate triangles be given. In every triangle the three sides are colored, such that one is blue, one is red, and one is white. Now, for every color separately, let us sort the lengths of the sides. We obtain

$$b_1 \leq b_2 \leq \cdots \leq b_{2009} \quad \text{the lengths of the blue sides,}$$
$$r_1 \leq r_2 \leq \cdots \leq r_{2009} \quad \text{the lengths of the red sides,}$$
$$\text{and} \quad w_1 \leq w_2 \leq \cdots \leq w_{2009} \quad \text{the lengths of the white sides.}$$

Then there exist k indices j such that we can form a nondegenerate triangle with side lengths b_j, r_j, w_j.

2. **A2 (EST)** Let a, b, c be positive real numbers such that $\frac{1}{a} + \frac{1}{b} + \frac{1}{c} = a + b + c$. prove that

$$\frac{1}{(2a+b+c)^2} + \frac{1}{(2b+c+a)^2} + \frac{1}{(2c+a+b)^2} \leq \frac{3}{16}.$$

3. **A3 (FRA)** [IMO5] Determine all functions f from the set of positive integers to the set of positive integers such that for all positive integers a and b, there exists a nondegenerate triangle with sides of lengths

$$a, \ f(b), \ \text{and} \ f(b + f(a) - 1).$$

(A triangle is *nondegenerate* if its vertices are not collinear.)

4. **A4 (BLR)** Let a, b, c be positive real numbers such that $ab + bc + ca \leq 3abc$. Prove that

$$\sqrt{\frac{a^2 + b^2}{a + b}} + \sqrt{\frac{b^2 + c^2}{b + c}} + \sqrt{\frac{c^2 + a^2}{c + a}} + 3 \leq \sqrt{2}\left(\sqrt{a + b} + \sqrt{b + c} + \sqrt{c + a}\right).$$

5. **A5 (BLR)** Let f be any function that maps the set of real numbers into the set of real numbers. Prove that there exist real numbers x and y such that

$$f(x - f(y)) > yf(x) + x.$$

6. **A6 (USA)** [IMO3] Suppose that $s_1, s_2, s_3, \ldots$ is a strictly increasing sequence of positive integers such that the subsequences

$$s_{s_1}, s_{s_2}, s_{s_3}, \ldots \quad \text{and} \quad s_{s_1+1}, s_{s_2+1}, s_{s_3+1}, \ldots$$

are both arithmetic progressions. Prove that the sequence $s_1, s_2, s_3, \ldots$ is itself an arithmetic progression.

7. **A7 (JPN)** Find all functions f from the set of real numbers into the set of real numbers that satisfy for all real x, y the identity

$$f(xf(x + y)) = f(yf(x)) + x^2.$$

8. **C1 (NZL)** Consider 2009 cards, each having one gold side and one black side, lying in parallel on a long table. Initially all cards show their gold sides. Two players, standing by the same long side of the table, play a game with alternating moves. Each move consists in choosing a block of 50 consecutive cards, the leftmost of which is showing gold, and turning them all over, so those that showed gold now show black and vice versa. The last player who can make a legal move wins.
   (a) Does the game necessarily end?
   (b) Does there exist a winning strategy for the starting player?

9. **C2 (ROU)** For any integer $n \geq 2$, let $N(n)$ be the maximal number of triples (a_i, b_i, c_i), $i = 1, \ldots, N(n)$, consisting of nonnegative integers a_i, b_i, and c_i such that the following two conditions are satisfied:
   (i) $a_i + b_i + c_i = n$ for all $i = 1, \ldots, N(n)$,
   (ii) If $i \neq j$, then $a_i \neq a_j$, $b_i \neq b_j$, and $c_i \neq c_j$.
   Determine $N(n)$ for all $n \geq 2$.

10. **C3 (RUS)** Let n be a positive integer. Given a sequence $\varepsilon_1, \ldots, \varepsilon_{n-1}$ with $\varepsilon_i = 0$ or $\varepsilon_i = 1$ for each $i = 1, \ldots, n-1$, the sequences $a_0, \ldots, a_n$ and $b_0, \ldots, b_n$ are constructed by the following rules:

$$a_0 = b_0 = 1, \quad a_1 = b_1 = 7,$$

$$a_{i+1} = \begin{cases} 2a_{i-1} + 3a_i, & \text{if } \varepsilon_i = 0, \\ 3a_{i-1} + a_i, & \text{if } \varepsilon_i = 1, \end{cases} \quad \text{for } i = 1, \ldots, n-1,$$

$$b_{i+1} = \begin{cases} 2b_{i-1} + 3b_i, & \text{if } \varepsilon_{n-i} = 0, \\ 3b_{i-1} + b_i, & \text{if } \varepsilon_{n-i} = 1, \end{cases} \quad \text{for } i = 1, \ldots, n-1.$$

Prove that $a_n = b_n$.

11. **C4 (NLD)** For an integer $m \geq 1$ we consider partitions of a $2^m \times 2^m$ chessboard into rectangles consisting of cells of the chessboard in which each of the 2^m cells along one diagonal forms a separate rectangle of side length 1. Determine the smallest possible sum of rectangle perimeters in such a partition.

12. **C5 (NLD)** Five identical empty buckets of 2-liter capacity stand at the vertices of a regular pentagon. Cinderella and her wicked stepmother go through a sequence of rounds: At the beginning of every round the stepmother takes one liter of water from the nearby river and distributes it arbitrarily among the five buckets. Then Cinderella chooses a pair of neighboring buckets, empties them into the river, and puts them back. Then the next round begins. The stepmother's goal is to make one of the buckets overflow. Cinderella's goal is to prevent this. Can the wicked stepmother enforce a bucket overflow?

13. **C6 (BGR)** On a 999×999 board a *limp rook* can move in the following way: From any square it can move to any of its adjacent squares, i.e., a square having a common side with it, and every move must be a turn: i.e., the directions of any two consecutive moves must be perpendicular. A *nonintersecting route* of the limp rook consists of a sequence of distinct squares that the limp rook can visit

in that order by an admissible sequence of moves. Such a nonintersecting route is called *cyclic* if the limp rook can, after reaching the last square of the route, move directly to the first square of the route and start over.

How many squares does the longest possible cyclic, nonintersecting route of a limp rook visit?

14. **C7 (RUS)** [IMO6] Let $a_1, a_2, \dots, a_n$ be distinct positive integers and let M be a set of $n - 1$ positive integers not containing $s = a_1 + a_2 + \cdots + a_n$. A grasshopper is to jump along the real axis, starting at the point 0 and making n jumps to the right with lengths $a_1, a_2, \dots, a_n$ in some order. Prove that the order can be chosen in such a way that the grasshopper never lands on any point in M.

15. **C8 (AUT)** For any integer $n \geq 2$ we compute the integer $h(n)$ by applying the following procedure to its decimal representation. Denote by r the rightmost digit of n.

   1° If $r = 0$, then the decimal representation of $h(n)$ results from the decimal representation of n by removing this rightmost digit 0.

   2° If $1 \leq r \leq 9$ we split the decimal representation of n into a maximal right part R that consists solely of digits not less than r and into the left part L that either is empty or ends with a digit strictly smaller than r. Then the decimal representation of $h(n)$ consists of the decimal representation of L, followed by two copies of the decimal representation of $R - 1$. For instance, for the number $n = 17{,}151{,}345{,}543$ we will have $L = 17{,}151$, $R = 345{,}543$, and $h(n) = 17{,}151{,}345{,}542{,}345{,}542$.

   Prove that, starting with an arbitrary integer $n \geq 2$, iterated application of h produces the integer 1 after finitely many steps.

16. **G1 (BEL)** [IMO4] Let ABC be a triangle with $AB = AC$. The angle bisectors of $\angle CAB$ and $\angle ABC$ meet the sides BC and CA at D and E, respectively. Let K be the incenter of triangle ADC. Suppose that $\angle BEK = 45°$. Find all possible values of $\angle CAB$.

17. **G2 (RUS)** [IMO2] Let ABC be a triangle with circumcenter O. The points P and Q are interior points of the sides CA and AB, respectively. Let K, L, and M be the midpoints of the segments BP, CQ, and PQ, respectively, and let Γ be the circle passing through K, L, and M. Suppose that the line PQ is tangent to the circle Γ. Prove that $OP = OQ$.

18. **G3 (IRN)** Let ABC be a triangle. The incircle of ABC touches the sides AB and AC at the points Z and Y, respectively. Let G be the point where the lines BY and CZ meet, and let R and S be the points such that the two quadrilaterals $BCYR$ and $BCSZ$ are parallelograms. Prove that $GR = GS$.

19. **G4 (UNK)** Given a cyclic quadrilateral $ABCD$, let the diagonals AC and BD meet at E and the lines AD and BC meet at F. The midpoints of AB and CD are G and H, respectively. Show that EF is tangent at E to the circle through the points E, G, and H.

20. **G5 (POL)** Let P be a polygon that is convex and symmetric with respect to some point O. Prove that for some parallelogram R satisfying $P \subseteq R$ we have

$$\frac{|S_R|}{|S_P|} \leq \sqrt{2}.$$

21. **G6 (UKR)** Let the sides AD and BC of the quadrilateral $ABCD$ (such that AB is not parallel to CD) intersect at point P. Points O_1 and O_2 are the circumcenters and points H_1 and H_2 are the orthocenters of the triangles ABP and DCP, respectively. Denote the midpoints of segments $O_1 H_1$ and $O_2 H_2$ by E_1 and E_2, respectively. Prove that the perpendicular from E_1 on CD, the perpendicular from E_2 on AB, and the line $H_1 H_2$ are concurrent.

22. **G7 (IRN)** Let ABC be a triangle with incenter I and let X, Y, and Z be the incenters of the triangles BIC, CIA, and AIB respectively. Let the triangle XYZ be equilateral. Prove that ABC is equilateral too.

23. **G8 (BGR)** Let $ABCD$ be a circumscribed quadrilateral. Let g be a line through A that meets the segment BC in M and the line CD in N. Denote by I_1, I_2, and I_3 the incenters of $\triangle ABM$, $\triangle MNC$, and $\triangle NDA$, respectively. Show that the orthocenter of $\triangle I_1 I_2 I_3$ lies on g.

24. **N1 (AUS)** [IMO1] Let n be a positive integer and let $a_1, \ldots, a_k$ ($k \geq 2$) be distinct integers in the set $\{1, \ldots, n\}$ such that n divides $a_i(a_{i+1} - 1)$ for $i = 1, \ldots, k-1$. Prove that n does not divide $a_k(a_1 - 1)$.

    Original formulation: A social club has n members. They have the membership numbers 1, 2, ..., n, respectively. From time to time members send presents to other members including items they have already received as presents from other members. In order to avoid the embarrassing situation that a member might receive a present that he or she has sent to other members, the club adds the following rule to its statutes at one of its annual general meetings: "A member with membership number a is permitted to send a present to a member with membership number b if and only if $a(b-1)$ is a multiple of n." Prove that if each member follows this rule, none will receive a present from another member that he or she has already sent to other members.

    Alternative formulation: Let G be a directed graph with n vertices $v_1, v_2, \ldots, v_n$ such that there is an edge going from v_a to v_b if and only if a and b are distinct and $a(b-1)$ is a multiple of n. Prove that this graph does not contain a directed cycle.

25. **N2 (PER)** A positive integer N is called *balanced* if $N = 1$ or if N can be written as a product of an even number of not necessarily distinct primes. Given positive integers a and b, consider the polynomial P defined by $P(x) = (x+a)(x+b)$.
    (a) Prove that there exist distinct positive integers a and b such that all the numbers $P(1)$, $P(2)$, ..., $P(50)$ are balanced.
    (b) Prove that if $P(n)$ is balanced for all positive integers n, then $a = b$.

26. **N3 (EST)** Let f be a nonconstant function from the set of positive integers into the set of positive integers, such that $a - b$ divides $f(a) - f(b)$ for all distinct

positive integers a and b. Prove that there exist infinitely many primes p such that p divides $f(c)$ for some positive integer c.

27. **N4 (PRK)** Find all positive integers n such that there exists a sequence of positive integers $a_1, a_2, \ldots, a_n$ satisfying

$$a_{k+1} = \frac{a_k^2 + 1}{a_{k-1} + 1} - 1$$

for every k with $2 \le k \le n-1$.

28. **N5 (HUN)** Let $P(x)$ be a nonconstant polynomial with integer coefficients. Prove that there is no function T from the set of integers into the set of integers such that the number of integers x with $T^n(x) = x$ is equal to $P(n)$ for every $n \ge 1$, where T^n denotes the n-fold application of T.

29. **N6 (TUR)** Let k be a positive integer. Show that if there exists a sequence $a_0, a_1, \ldots$ of integers satisfying the condition

$$a_n = \frac{a_{n-1} + n^k}{n} \quad \text{for all } n \ge 1,$$

then $k - 2$ is divisible by 3.

30. **N7 (MNG)** Let a and b be distinct integers greater than 1. Prove that there exists a positive integer n such that $(a^n - 1)(b^n - 1)$ is not a perfect square.

4

Solutions

4.1 Solutions to the Contest Problems of IMO 1959

1. The desired result $(14n+3, 21n+4) = 1$ follows from

$$3(14n+3) - 2(21n+4) = 1.$$

2. For the square roots to be real we must have $2x - 1 \geq 0 \Rightarrow x \geq 1/2$ and $x \geq \sqrt{2x-1} \Rightarrow x^2 \geq 2x - 1 \Rightarrow (x-1)^2 \geq 0$, which always holds. Then we have $\sqrt{x + \sqrt{2x-1}} + \sqrt{x - \sqrt{2x-1}} = c \iff$

$$c^2 = 2x + 2\sqrt{x^2 - \sqrt{2x-1}^2} = 2x + 2|x-1| = \begin{cases} 2, & 1/2 \leq x \leq 1, \\ 4x - 2, & x \geq 1. \end{cases}$$

   (a) $c^2 = 2$. The equation holds for $1/2 \leq x \leq 1$.
   (b) $c^2 = 1$. The equation has no solution.
   (c) $c^2 = 4$. The equation holds for $4x - 2 = 4 \Rightarrow x = 3/2$.

3. Multiplying the equality by $4(a\cos^2 x - b\cos x + c)$, we obtain $4a^2 \cos^4 x + 2(4ac - 2b^2)\cos^2 x + 4c^2 = 0$. Plugging in $2\cos^2 x = 1 + \cos 2x$ we obtain (after quite a bit of manipulation):

$$a^2 \cos^2 2x + (2a^2 + 4ac - 2b^2)\cos 2x + (a^2 + 4ac - 2b^2 + 4c^2) = 0.$$

   For $a = 4$, $b = 2$, and $c = -1$ we get $4\cos^2 x + 2\cos x - 1 = 0$ and $16\cos^2 2x + 8\cos 2x - 4 = 0 \Rightarrow 4\cos^2 2x + 2\cos 2x - 1 = 0$.

4. *Analysis.* Let a and b be the other two sides of the triangle. From the conditions of the problem we have $c^2 = a^2 + b^2$ and $c/2 = \sqrt{ab} \iff 3/2c^2 = a^2 + b^2 + 2ab = (a+b)^2 \iff \sqrt{3/2}c = a + b$. Given a desired $\triangle ABC$ let D be a point on $(AC$ such that $CD = CB$. In that case, $AD = a + b = \sqrt{3/2}c$, and also, since $BC = CD$, it follows that $\angle ADB = 45°$.

   Construction. From a segment of length c we elementarily construct a segment AD of length $\sqrt{3/2}\,c$. We then construct a ray $(DX$ such that $\angle ADX = 45°$

357

and a circle $k(A,c)$ that intersects the ray at point B. Finally, we construct the perpendicular from B to AD; point C is the foot of that perpendicular.

Proof. It holds that $AB = c$, and, since $CB = CD$, it also holds that $AC + CB = AC + CD = AD = \sqrt{3/2}\,c$. From this it follows that $\sqrt{AC \cdot CB} = c/2$. Since BC is perpendicular to AD, it follows that $\angle BCA = 90°$. Thus ABC is the desired triangle.

Discussion. Since $AB\sqrt{2} = \sqrt{2}c > \sqrt{3/2}\,c = AD > AB$, the circle k intersects the ray DX in exactly two points, which correspond to two symmetric solutions.

5. (a) It suffices to prove that $AF \perp BC$, since then for the intersection point X we have $\angle AXC = \angle BXF = 90°$, implying that X belongs to the circumcircles of both squares and thus that $X = N$. The relation $AF \perp BC$ holds because from $MA = MC$, $MF = MB$, and $\angle AMC = \angle FMB$ it follows that $\triangle AMF$ is obtained by rotating $\triangle BMC$ by $90°$ around M.

    (b) Since N is on the circumcircle of $BMFE$, it follows that $\angle ANM = \angle MNB = 45°$. Hence MN is the bisector of $\angle ANB$. It follows that MN passes through the midpoint of the arc $\overset{\frown}{AB}$ of the circle with diameter AB (i.e., the circumcircle of $\triangle ABN$) not containing N.

    (c) Let us introduce a coordinate system such that $A = (0,0)$, $B = (b,0)$, and $M = (m,0)$. Setting in general $W = (x_W, y_W)$ for an arbitrary point W and denoting by R the midpoint of PQ, we have $y_R = (y_P + y_Q)/2 = (m + b - m)/4 = b/4$ and $x_R = (x_P + x_Q)/2 = (m + m + b)/4 = (2m + b)/4$, the parameter m varying from 0 to b. Thus the locus of all points R is the closed segment $R_1 R_2$ where $R_1 = (b/4, b/4)$ and $R_2 = (b/4, 3b/4)$.

6. *Analysis.* For $AB \parallel CD$ to hold evidently neither must intersect p and hence constructing lines r in α through A and s in β through C, both being parallel to p, we get that $B \in r$ and $D \in s$. Hence the problem reduces to a planar problem in γ, determined by r and s. Denote by A' the foot of the perpendicular from A to s. Since $ABCD$ is isosceles and has an incircle, it follows that $AD = BC = (AB + CD)/2 = A'C$. The remaining parts of the problem are now obvious.

4.2 Solutions to the Contest Problems of IMO 1960

1. Given the number $\overline{acb}$, since $11 \mid \overline{acb}$, it follows that $c = a + b$ or $c = a + b - 11$. In the first case, $a^2 + b^2 + (a+b)^2 = 10a + b$, and in the second case, $a^2 + b^2 + (a+b-11)^2 = 10(a-1) + b$. In the first case the LHS is even, and hence $b \in \{0, 2, 4, 6, 8\}$, while in the second case it is odd, and hence $b \in \{1, 3, 5, 7, 9\}$. Analyzing the 10 quadratic equations for a we obtain that the only valid solutions are 550 and 803.

2. The LHS term is well-defined for $x \geq -1/2$ and $x \neq 0$. Furthermore, $4x^2/(1 - \sqrt{1+2x})^2 = (1 + \sqrt{1+2x})^2$. Since

$$f(x) = \left(1 + \sqrt{1+2x}\right)^2 - 2x - 9 = 2\sqrt{1+2x} - 7$$

is increasing and since $f(45/8) = 0$, it follows that the inequality holds precisely for $-1/2 \leq x < 45/8$ and $x \neq 0$.

3. Let $B'C'$ be the middle of the $n = 2k+1$ segments and let D be the foot of the perpendicular from A to the hypotenuse. Let us assume $\mathscr{B}(C, D, C', B', B)$. Then from $CD < BD$, $CD + BD = a$, and $CD \cdot BD = h^2$ we have $CD^2 - a \cdot CD + h^2 = 0 \implies CD = (a - \sqrt{a^2 - 4h^2})/2$. Let us define $\angle DAC' = \gamma$ and $\angle DAB' = \beta$; then $\tan\beta = DB'/h$ and $\tan\gamma = DC'/h$. Since $DB' = CB' - CD = (k+1)a/(2k+1) - (c - \sqrt{c^2 - 4h^2})/2$ and $DC' = ka/(2k+1) - (c - \sqrt{c^2 - 4h^2})/2$, we have

$$\tan\alpha = \tan(\beta - \gamma) = \frac{\tan\beta - \tan\gamma}{1 + \tan\beta \cdot \tan\gamma} = \frac{\frac{a}{(2k+1)h}}{1 + \frac{a^2 - 4h^2}{4h^2} - \frac{a^2}{4h^2(2k+1)^2}}$$

$$= \frac{4h(2k+1)}{4ak(k+1)} = \frac{4nh}{(n^2-1)a}.$$

The case $\mathscr{B}(C, C', D, B', B)$ is similar.

4. *Analysis.* Let A' and B' be the feet of the perpendiculars from A and B, respectively, to the opposite sides, A_1 the midpoint of BC, and let D' be the foot of the perpendicular from A_1 to AC. We then have $AA_1 = m_a$, $AA' = h_a$, $\angle AA'A_1 = 90°$, $A_1D' = h_b/2$, and $\angle AD'A_1 = 90°$.

Construction. We construct the quadrilateral $AD'A_1A'$ (starting from the circle with diameter AA_1). Then C is the intersection of $A'A_1$ and AD', and B is on the line A_1C such that $CA_1 = A_1B$ and $\mathscr{B}(B, A_1, C)$.

Discussion. We must have $m_a \geq h_a$ and $m_a \geq h_b/2$. The number of solutions is 0 if $m_a = h_a = h_b/2$, 1 if two of $m_a, h_a, h_b/2$ are equal, and 2 otherwise.

5. (a) The locus of the points is the square $EFGH$ where these four points are the centers of the faces $ABB'A'$, $BCC'B'$, $CDD'C'$ and $DAA'D'$.

   (b) The locus of the points is the rectangle $IJKL$ where these points are on AB', CB', CD', and AD' at a distance of $AA'/3$ with respect to the plane $ABCD$.

6. Let E, F respectively be the midpoints of the bases AB, CD of the isosceles trapezoid $ABCD$.

   (a) The point P is on the intersection of EF and the circle with diameter BC.
   (b) Let $x = EP$. Since $\triangle BEP \sim \triangle PFC$, we have $x(h-x) = ab/4 \Rightarrow x_{1,2} = (h \pm \sqrt{h^2 - ab})/2$.
   (c) If $h^2 > ab$ there are two solutions, if $h^2 = ab$ there is only one solution, and if $h^2 < ab$ there are no solutions.

7. Let A be the vertex of the cone, O the center of the sphere, S the center of the base of the cone, B a point on the base circle, and r the radius of the sphere. Let $\angle SAB = \alpha$. We easily obtain $AS = r(1 + \sin\alpha)/\sin\alpha$ and $SB = r(1 + \sin\alpha)\tan\alpha/\sin\alpha$ and hence $V_1 = \pi SB^2 \cdot SA/3 = \pi r^3(1 + \sin\alpha)^2/[3\sin\alpha(1 - \sin\alpha)]$. We also have $V_2 = 2\pi r^3$ and hence

$$k = \frac{(1 + \sin\alpha)^2}{6\sin\alpha(1 - \sin\alpha)} \Rightarrow (1 + 6k)\sin^2\alpha + 2(1 - 3k)\sin\alpha + 1 = 0.$$

The discriminant of this quadratic must be nonnegative: $(1 - 3k)^2 - (1 + 6k) \geq 0 \Rightarrow k \geq 4/3$. Hence we cannot have $k = 1$. For $k = 4/3$ we have $\sin\alpha = 1/3$, whose construction is elementary.

4.3 Solutions to the Contest Problems of IMO 1961

1. This is a problem solvable using elementary manipulations, so we shall state only the final solutions. For $a = 0$ we get $(x,y,z) = (0,0,0)$. For $a \neq 0$ we get $(x,y,z) \in \{(t_1,t_2,z_0),(t_2,t_1,z_0)\}$, where

$$z_0 = \frac{a^2 - b^2}{2a} \quad \text{and} \quad t_{1,2} = \frac{a^2 + b^2 \pm \sqrt{(3a^2 - b^2)(3b^2 - a^2)}}{4a}.$$

For the solutions to be positive and distinct the following conditions are necessary and sufficient: $3b^2 > a^2 > b^2$ and $a > 0$.

2. Using $S = bc\sin\alpha/2$, $a^2 = b^2 + c^2 - 2bc\cos\alpha$ and $(\sqrt{3}\sin\alpha + \cos\alpha)/2 = \cos(\alpha - 60°)$ we have

$$a^2 + b^2 + c^2 \geq 4S\sqrt{3} \Leftrightarrow b^2 + c^2 \geq bc(\sqrt{3}\sin\alpha + \cos\alpha) \Leftrightarrow$$
$$\Leftrightarrow (b - c)^2 + 2bc(1 - \cos(\alpha - 60°)) \geq 0,$$

where equality holds if and only if $b = c$ and $\alpha = 60°$, i.e., if the triangle is equilateral.

3. For $n \geq 2$ we have

$$1 = \cos^n x - \sin^n x \leq |\cos^n x - \sin^n x|$$
$$\leq |\cos^n x| + |\sin^n x| \leq \cos^2 x + \sin^2 x = 1.$$

Hence $\sin^2 x = |\sin^n x|$ and $\cos^2 x = |\cos^n x|$, from which it follows that $\sin x$, $\cos x \in \{1,0,-1\} \Rightarrow x \in \pi\mathbb{Z}/2$. By inspection one obtains the set of solutions

$$\{m\pi \mid m \in \mathbb{Z}\} \quad \text{for even } n \quad \text{and} \quad \{2m\pi, 2m\pi - \pi/2 \mid m \in \mathbb{Z}\} \quad \text{for odd } n.$$

For $n = 1$ we have $1 = \cos x - \sin x = -\sqrt{2}\sin(x - \pi/4)$, which yields the set of solutions

$$\{2m\pi, 2m\pi - \pi/2 \mid m \in \mathbb{Z}\}.$$

4. Let $x_i = PP_i/PQ_i$ for $i = 1,2,3$. For all i we have

$$\frac{1}{x_i + 1} = \frac{PQ_i}{P_iQ_i} = \frac{S_{PP_jP_k}}{S_{P_1P_2P_3}},$$

where the indices j and k are distinct and different from i. Hence we have

$$f(x_1,x_2,x_3) = \frac{1}{x_1 + 1} + \frac{1}{x_2 + 1} + \frac{1}{x_3 + 1}$$
$$= \frac{S(PP_2P_3) + S(PP_1P_3) + S(PP_2P_3)}{S(P_1P_2P_3)} = 1.$$

It follows that $1/(x_i + 1) \geq 1/3$ for some i and $1/(x_j + 1) \leq 1/3$ for some j. Consequently, $x_i \leq 2$ and $x_j \geq 2$.

5. *Analysis.* Let C_1 be the midpoint of AB. In $\triangle AMB$ we have $MC_1 = b/2$, $AB = c$, and $\angle AMB = \omega$. Thus, given $AB = c$, the point M is at the intersection of the circle $k(C', b/2)$ and the set of points e that view AB at an angle of ω. The construction of ABC is now obvious.

Discussion. It suffices to establish the conditions for which k and e intersect. Let E be the midpoint of one of the arcs that make up e. A necessary and sufficient condition for k to intersect e is

$$\frac{c}{2} = C'A \le \frac{b}{2} \le C'E = \frac{c}{2}\cot\frac{\omega}{2} \Leftrightarrow b\tan\frac{\omega}{2} \le c < b.$$

6. Let $h(X)$ denote the distance of a point X from ε, X restricted to being on the same side of ε as A, B, and C. Let G_1 be the (fixed) centroid of $\triangle ABC$ and G'_1 the centroid of $\triangle A'B'C'$. It is trivial to prove that G is the midpoint of $G_1 G'_1$. Hence varying G'_1 across ε, we get that the locus of G is the plane α parallel to ε such that

$$X \in \alpha \Leftrightarrow h(X) = \frac{h(G_1)}{2} = \frac{h(A) + h(B) + h(C)}{6}.$$

4.4 Solutions to the Contest Problems of IMO 1962

1. From the conditions of the problem we have $n = 10x + 6$ and $4n = 6 \cdot 10^m + x$ for some integer x. Eliminating x from these two equations, we get $40n = 6 \cdot 10^{m+1} + n - 6 \Rightarrow n = 2(10^{m+1} - 1)/13$. Hence we must find the smallest m such that this fraction is an integer. By inspection, this happens for $m = 6$, and for this m we obtain $n = 153846$, which indeed satisfies the conditions of the problem.

2. We note that $f(x) = \sqrt{3 - x} - \sqrt{x + 1}$ is well-defined only for $-1 \leq x \leq 3$ and is decreasing (and obviously continuous) on this interval. We also note that $f(-1) = 2 > 1/2$ and

$$ f\left(1 - \frac{\sqrt{31}}{8}\right) = \sqrt{\left(\frac{1}{4} + \frac{\sqrt{31}}{4}\right)^2} - \sqrt{\left(\frac{1}{4} - \frac{\sqrt{31}}{4}\right)^2} = \frac{1}{2}. $$

Hence the inequality is satisfied for $-1 \leq x < 1 - \sqrt{31}/8$.

3. By inspecting the four different stages of this periodic motion we easily obtain that the locus of the midpoints of XY is the edges of $MNCQ$, where M, N, and Q are the centers of $ABB'A'$, $BCC'B'$, and $ABCD$, respectively.

4. Since $\cos 2x = 1 + \cos^2 x$ and $\cos \alpha + \cos \beta = 2\cos\left(\frac{\alpha+\beta}{2}\right)\cos\left(\frac{\alpha-\beta}{2}\right)$, we have $\cos^2 x + \cos^2 2x + \cos^2 3x = 1 \Leftrightarrow \cos 2x + \cos 4x + 2\cos^2 3x = 2\cos 3x(\cos x + \cos 3x) = 0 \Leftrightarrow 4\cos 3x \cos 2x \cos x = 0$. Hence the solutions are $x \in \{\pi/2 + m\pi, \ \pi/4 + m\pi/2, \ \pi/6 + m\pi/3 \mid m \in \mathbb{Z}\}$.

5. *Analysis.* Let $ABCD$ be the desired quadrilateral. Let us assume w.l.o.g. that $AB > BC$ (for $AB = BC$ the construction is trivial). For a tangent quadrilateral we have $AD - DC = AB - BC$. Let X be a point on AD such that $DX = DC$. We then have $AX = AB - BC$ and $\angle AXC = \angle ADC + \angle CDX = 180° - \angle ABC/2$. Constructing X and hence D is now obvious.

6. This problem is a special case, when the triangle is isosceles, of Euler's formula, which holds for all triangles.

7. The spheres are arranged in a similar manner as in the planar case where we have one incircle and three excircles. Here we have one "insphere" and four "exspheres" corresponding to each of the four sides. Each vertex of the tetrahedron effectively has three tangent lines drawn from it to each of the five spheres. Repeatedly using the equality of the three tangent segments from a vertex (in the same vein as for tangent planar quadrilaterals) we obtain $SA + BC = SB + CA = SC + AB$ from the insphere. From the exsphere opposite of S we obtain $SA - BC = SB - CA = SC - AB$, hence $SA = SB = SC$ and $AB = BC = CA$. By symmetry, we also have $AB = AC = AS$. Hence indeed, all the edges of the tetrahedron are equal in length and thus we have shown that the tetrahedron is regular.

4.5 Solutions to the Contest Problems of IMO 1963

1. Obviously, $x \geq 0$ and $p \geq 0$; hence squaring the given equation yields an equivalent equation $5x^2 - p - 4 + 4\sqrt{(x^2-1)(x^2-p)} = x^2$, i.e., $4\sqrt{(x^2-1)(x^2-p)} = (p+4) - 4x^2$. If $4x^2 \leq (p+4)$, we may square the equation once again to get $-16(p+1)x^2 + 16p = -8(p+4)x^2 + (p+4)^2$, which is equivalent to $x^2 = (4-p)^2/[4(4-2p)]$. We immediately get $p < 2$ hence

$$x = \frac{4-p}{2\sqrt{4-2p}}.$$

Placing this x in the original equation gives $|3p - 4| = 4 - 3p$ hence $0 \leq p \leq \frac{4}{3}$. Thus for $0 \leq p \leq \frac{4}{3}$ we have $x = (4-p)^2/[4(4-2p)]$, otherwise the solution doesn't exist.

2. Let A be the given point, BC the given segment, and $\mathscr{B}_1, \mathscr{B}_2$ the closed balls with the diameters AB and AC respectively. Consider one right angle $\angle AOK$ with $K \in [BC]$. If B', C' are the feet of the perpendiculars from B, C to AO respectively, then O lies on the segment $B'C'$, which implies that it lies on exactly one of the segments AB', AC'. Hence O belongs to exactly one of the balls $\mathscr{B}_1, \mathscr{B}_2$; i.e., $O \in \mathscr{B}_1 \triangle \mathscr{B}_2$. This is obviously the required locus.

3. Let $\overrightarrow{OA_1}, \overrightarrow{OA_2}, \ldots, \overrightarrow{OA_n}$ be the vectors corresponding respectively to the edges $a_1, a_2, \ldots, a_n$ of the polygon. By the conditions of the problem, these vectors satisfy $\overrightarrow{OA_1} + \cdots + \overrightarrow{OA_n} = \overrightarrow{0}$, $\angle A_1 OA_2 = \angle A_2 OA_3 = \cdots = \angle A_n OA_1 = 2\pi/n$ and $OA_1 \geq OA_2 \geq \cdots \geq OA_n$. Our task is to prove that $OA_1 = \cdots = OA_n$.
   Let l be the line through O perpendicular to OA_n, and $B_1, \ldots, B_{n-1}$ the projections of $A_1, \ldots, A_{n-1}$ onto l respectively. By the assumptions, the sum of the $\overrightarrow{OB_i}$'s is $\overrightarrow{0}$. On the other hand, since $OB_i \leq OB_{n-i}$ for all $i \leq n/2$, all the sums $\overrightarrow{OB_i} + \overrightarrow{OB_{n-i}}$ lie on the same side of the point O. Hence all these sums must be equal to $\overrightarrow{0}$. Consequently, $OA_i = OA_{n-i}$, from which the result immediately follows.

4. Summing up all the equations yields $2(x_1 + x_2 + x_3 + x_4 + x_5) = y(x_1 + x_2 + x_3 + x_4 + x_5)$. If $y = 2$, then the given equations imply $x_1 - x_2 = x_2 - x_3 = \cdots = x_5 - x_1$; hence $x_1 = x_2 = \cdots = x_5$, which is clearly a solution. If $y \neq 2$, then $x_1 + \cdots + x_5 = 0$, and summing the first three equalities gives $x_2 = y(x_1 + x_2 + x_3)$. Using that $x_1 + x_3 = yx_2$ we obtain $x_2 = (y^2 + y)x_2$, i.e., $(y^2 + y - 1)x_2 = 0$. If $y^2 + y - 1 \neq 0$, then $x_2 = 0$, and similarly $x_1 = \cdots = x_5 = 0$. If $y^2 + y - 1 = 0$, it is easy to prove that the last two equations are the consequence of the first three. Thus choosing any values for x_1 and x_5 will give exactly one solution for x_2, x_3, x_4.

5. The LHS of the desired identity equals $S = \cos(\pi/7) + \cos(3\pi/7) + \cos(5\pi/7)$. Now

$$S\sin\frac{\pi}{7} = \frac{\sin\frac{2\pi}{7}}{2} + \frac{\sin\frac{4\pi}{7} - \sin\frac{2\pi}{7}}{2} + \frac{\sin\frac{6\pi}{7} - \sin\frac{4\pi}{7}}{2} = \frac{\sin\frac{6\pi}{7}}{2} \Rightarrow S = \frac{1}{2}.$$

6. The result is $EDACB$.

4.6 Solutions to the Contest Problems of IMO 1964

1. Let $n = 3k + r$, where $0 \le r < 2$. Then $2^n = 2^{3k+r} = 8^k \cdot 2^r \equiv 2^r \pmod{7}$. Thus the remainder of 2^n modulo 7 is $1, 2, 4$ if $n \equiv 0, 1, 2 \pmod 3$. Hence $2^n - 1$ is divisible by 7 if and only if $3 \mid n$, while $2^n + 1$ is never divisible by 7.

2. By substituting $a = x+y$, $b = y+z$, and $c = z+x$ $(x, y, z > 0)$ the given inequality becomes
$$6xyz \le x^2 y + xy^2 + y^2 z + yz^2 + z^2 x + zx^2,$$
which follows immediately by the AM–GM inequality applied to $x^2 y$, xy^2, $x^2 z$, xz^2, $y^2 z$, yz^2.

3. Let r be the radius of the incircle of $\triangle ABC$, r_a, r_b, r_c the radii of the smaller circles corresponding to A, B, C, and h_a, h_b, h_c the altitudes from A, B, C respectively. The coefficient of similarity between the smaller triangle at A and the triangle ABC is $1 - 2r/h_a$, from which we easily obtain $r_a = (h_a - 2r)r/h_a = (s-a)r/s$. Similarly, $r_b = (s-b)r/s$ and $r_c = (s-c)r/s$. Now a straightforward computation gives that the sum of areas of the four circles is given by
$$\Sigma = \frac{(b+c-a)(c+a-b)(a+b-c)(a^2+b^2+c^2)\pi}{(a+b+c)^3}.$$

4. Let us call the topics T_1, T_2, T_3. Consider an arbitrary student A. By the pigeonhole principle there is a topic, say T_3, he discussed with at least 6 other students. If two of these 6 students discussed T_3, then we are done.
Suppose now that the 6 students discussed only T_1 and T_2 and choose one of them, say B. By the pigeonhole principle he discussed one of the topics, say T_2, with three of these students. If two of these three students also discussed T_2, then we are done. Otherwise, all the three students discussed only T_1, which completes the task.

5. Let us first compute the number of intersection points of the perpendiculars passing through two distinct points B and C. The perpendiculars from B to the lines through C other than BC meet all perpendiculars from C, which counts to $3 \cdot 6 = 18$ intersection points. Each perpendicular from B to the 3 lines not containing C can intersect at most 5 of the perpendiculars passing through C, which counts to another $3 \cdot 5 = 15$ intersection points. Thus there are $18 + 15 = 33$ intersection points corresponding to B, C.
It follows that the required total number is at most $10 \cdot 33 = 330$. But some of these points, namely the orthocenters of the triangles with vertices at the given points, are counted thrice. There are 10 such points. Hence the maximal number of intersection points is $330 - 2 \cdot 10 = 310$.

Remark. The jury considered only the combinatorial part of the problem and didn't require an example in which 310 points appear. However, it is "easily" verified that, for instance, the set of points $A(1,1)$, $B(e, \pi)$, $C(e^2, \pi^2)$, $D(e^3, \pi^3)$, $E(e^4, \pi^4)$ works.

6. We shall prove that the statement is valid in the general case, for an arbitrary point D_1 inside $\triangle ABC$. Since D_1 belongs to the plane ABC, there are real numbers a, b, c such that $(a+b+c)\overrightarrow{DD_1} = a\overrightarrow{DA} + b\overrightarrow{DB} + c\overrightarrow{DC}$. Since $AA_1 \parallel DD_1$, it holds that $\overrightarrow{AA_1} = k\overrightarrow{DD_1}$ for some $k \in \mathbb{R}$. Now it is easy to get $\overrightarrow{DA_1} = -(b\overrightarrow{DB}+c\overrightarrow{DC})/a$, $\overrightarrow{DB_1} = -(a\overrightarrow{DA}+c\overrightarrow{DC})/b$, and $\overrightarrow{DC_1} = -(a\overrightarrow{DA}+b\overrightarrow{DB})/c$. This implies

$$\overrightarrow{D_1A_1} = -\frac{a^2\overrightarrow{DA} + b(a+2b+c)\overrightarrow{DB} + c(a+b+2c)\overrightarrow{DC}}{a(a+b+c)},$$

$$\overrightarrow{D_1B_1} = -\frac{a(2a+b+c)\overrightarrow{DA} + b^2\overrightarrow{DB} + c(a+b+2c)\overrightarrow{DC}}{b(a+b+c)}, \text{ and}$$

$$\overrightarrow{D_1C_1} = -\frac{a(2a+b+c)\overrightarrow{DA} + b(a+2b+c)\overrightarrow{DB} + c^2\overrightarrow{DC}}{c(a+b+c)}.$$

By using

$$6V_{D_1A_1B_1C_1} = \left\| \left[\overrightarrow{D_1A_1}, \overrightarrow{D_1B_1}, \overrightarrow{D_1C_1} \right] \right\| \text{ and } 6V_{DABC} = \left\| \left[\overrightarrow{DA}, \overrightarrow{DB}, \overrightarrow{DC} \right] \right\|$$

we get

$$V_{D_1A_1B_1C_1} = \frac{\begin{vmatrix} a^2 & b(a+2b+c) & c(a+b+2c) \\ a(2a+b+c) & b^2 & c(a+b+2c) \\ a(2a+b+c) & b(a+2b+c) & c^2 \end{vmatrix}}{6abc(a+b+c)^3}$$

$$\cdot \left\| \left[\overrightarrow{DA}, \overrightarrow{DB}, \overrightarrow{DC} \right] \right\|$$

$$= 3V_{DABC}.$$

4.7 Solutions to the Contest Problems of IMO 1965

1. Let us set $S = \left| \sqrt{1 + \sin 2x} - \sqrt{1 - \sin 2x} \right|$. Notice that $S^2 = 2 - 2\sqrt{1 - \sin^2 2x} = 2 - 2|\cos 2x| \leq 2$, implying $S \leq \sqrt{2}$. Thus the righthand inequality holds for all x.

   It remains to investigate the left-hand inequality. If $\pi/2 \leq x \leq 3\pi/2$, then $\cos x \leq 0$ and the inequality trivially holds. Assume now that $\cos x > 0$. Then the inequality is equivalent to $2 + 2\cos 2x = 4\cos^2 x \leq S^2 = 2 - 2|\cos 2x|$, which is equivalent to $\cos 2x \leq 0$, i.e., to $x \in [\pi/4, \pi/2] \cup [3\pi/2, 7\pi/4]$. Hence the solution set is $\pi/4 \leq x \leq 7\pi/4$.

2. Suppose that (x_1, x_2, x_3) is a solution. We may assume w.l.o.g. that $|x_1| \geq |x_2| \geq |x_3|$. Suppose that $|x_1| > 0$. From the first equation we obtain that

$$0 = |x_1| \cdot \left| a_{11} + a_{12}\frac{x_2}{x_1} + a_{13}\frac{x_3}{x_1} \right| \geq |x_1| \cdot (a_{11} - |a_{12}| - |a_{13}|) > 0,$$

   which is a contradiction. Hence $|x_1| = 0$ and consequently $x_1 = x_2 = x_3 = 0$.

3. Let d denote the distance between the lines AB and CD. Being parallel to AB and CD, the plane π intersects the faces of the tetrahedron in a parallelogram $EFGH$. Let $X \in AB$ be a point such that $HX \parallel DB$.

   Clearly $V_{AEHBFG} = V_{AXEH} + V_{XEHBFG}$. Let MN be the common perpendicular to lines AB and CD ($M \in AB$, $N \in CD$) and let MN, BN meet the plane π at Q and R respectively. Then it holds that $BR/RN = MQ/QN = k$ and consequently $AX/XB = AE/EC = AH/HD = BF/FC = BG/GD = k$. Now we have $V_{AXEH}/V_{ABCD} = k^3/(k+1)^3$. Furthermore, if $h = 3V_{ABCD}/S_{ABC}$ is the height of $ABCD$ from D, then

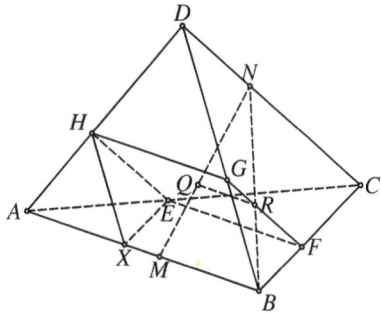

$$V_{XEHBFG} = \frac{1}{2}S_{XBFE}\frac{k}{k+1}h \text{ and}$$

$$S_{XBFE} = S_{ABC} - S_{AXE} - S_{EFC} = \frac{(k+1)^2 - 1 - k^2}{(k+1)^2} = \frac{2k}{(1+k)^2}.$$

These relations give us $V_{XEHBFG}/V_{ABCD} = 3k^2/(1+k)^3$. Finally,

$$\frac{V_{AEHBFG}}{V_{ABCD}} = \frac{k^3 + 3k^2}{(k+1)^3}.$$

Similarly, $V_{CEFDHG}/V_{ABCD} = (3k+1)/(k+1)^3$, and hence the required ratio is $(k^3 + 3k^2)/(3k+1)$.

4. It is easy to see that all x_i are nonzero. Let $x_1 x_2 x_3 x_4 = p$. The given system of equations can be rewritten as $x_i + p/x_i = 2$, $i = 1, 2, 3, 4$. The equation $x + p/x = 2$ has at most two real solutions, say y and z. Then each x_i is equal either to y or to z. There are three cases:

  (i) $x_1 = x_2 = x_3 = x_4 = y$. Then $y + y^3 = 2$ and hence $y = 1$.
  (ii) $x_1 = x_2 = x_3 = y$, $x_4 = z$. Then $z + y^3 = y + y^2 z = 2$. It is easy to obtain that the only possibilities for (y, z) are $(-1, 3)$ and $(1, 1)$.
  (iii) $x_1 = x_2 = y$, $x_3 = x_4$. In this case the only possibility is $y = z = 1$.
  Hence the solutions for (x_1, x_2, x_3, x_4) are $(1, 1, 1, 1)$, $(-1, -1, -1, 3)$, and the cyclic permutations.

5. (a) Let A' and B' denote the feet of the perpendiculars from A and B to OB and OA respectively. We claim that $H \in A'B'$. Indeed, since $MPHQ$ is a parallelogram, we have $B'P/B'A = BM/BA = MQ/AA' = PH/AA'$, which implies by Thales's theorem that $H \in A'B'$. It is easy to see that the locus of H is the whole segment $A'B'$.
   (b) In this case the locus of points H is obviously the interior of the triangle $OA'B'$.

6. We recall the simple statement that every two diameters of a set must have a common point.
   Consider any point B that is an endpoint of $k \geq 2$ diameters $BC_1, BC_2, \ldots, BC_k$. We may assume w.l.o.g. that all the points $C_1, \ldots, C_k$ lie on the arc $C_1 C_k$, whose center is B and measure does not exceed $60°$. We observe that for $1 < i < k$ any diameter with the endpoint C_i has to intersect both the diameters $C_1 B$ and $C_k B$. Hence $C_i B$ is the only diameter with an endpoint at C_i if $i = 2, \ldots, k - 1$. In other words, with each point that is an endpoint of $k \geq 2$ we can associate $k - 2$ points that are endpoints of exactly one diameter.
   We now assume that each A_i is an endpoint of exactly $k_i \geq 0$ diameters, and that $k_1, \ldots, k_s \geq 2$, while $k_{s+1}, \ldots, k_n \leq 1$. The total number D of diameters satisfies the inequality $2D \leq k_1 + k_2 + \cdots + k_s + (n - s)$. On the other hand, by the above consideration we have $(k_1 - 2) + \cdots + (k_s - 2) \leq n - s$, i.e., $k_1 + \cdots + k_s \leq n + s$. Hence $2D \leq (n + s) + (n - s) = 2n$, which proves the result.

4.8 Solutions to the Contest Problems of IMO 1966

1. Let $N_a, N_b, N_c, N_{ab}, N_{ac}, N_{bc}, N_{abc}$ denote the number of students who solved exactly the problems whose letters are stated in the index of the variable. From the conditions of the problem we have

$$N_a + N_b + N_c + N_{ab} + N_{bc} + N_{ac} + N_{abc} = 25,$$

$$N_b + N_{bc} = 2(N_c + N_{bc}), \qquad N_a - 1 = N_{ac} + N_{abc} + N_{ab}, \qquad N_a = N_b + N_c.$$

From the first and third equations we get $2N_a + N_b + N_c + N_{bc} = 26$, and from the second and fourth we get $4N_b + N_c = 26$ and thus $N_b \leq 6$. On the other hand, we have from the second equation $N_b = 2N_c + N_{bc} \Rightarrow N_c \leq N_b/2 \Rightarrow 26 \leq 9N_b/2 \Rightarrow N_b \geq 6$; hence $N_b = 6$.

2. Angles α and β are less than $90°$, otherwise if w.l.o.g. $\alpha \geq 90°$ we have $\tan(\gamma/2) \cdot (a\tan\alpha + b\tan\beta) < b\tan(\gamma/2)\tan\beta \leq b\tan(\gamma/2)\cot(\gamma/2) = b < a + b$. Since $a \geq b \Leftrightarrow \tan a \geq \tan b$, Chebyshev's inequality gives $a\tan\alpha + b\tan\beta \geq (a+b)(\tan\alpha + \tan\beta)/2$. Due to the convexity of the tan function we also have $(\tan\alpha + \tan\beta)/2 \geq \tan[(\alpha+\beta)/2] = \cot(\gamma/2)$. Hence we have

$$\tan\frac{\gamma}{2}(a\tan\alpha + b\tan\beta) \geq \frac{1}{2}\tan\frac{\gamma}{2}(a+b)(\tan\alpha + \tan\beta)$$
$$\geq \tan\frac{\gamma}{2}(a+b)\cot\frac{\gamma}{2} = a+b.$$

The equalities can hold only if $a = b$. Thus the triangle is isosceles.

3. Consider a coordinate system in which the points of the regular tetrahedron are placed at $A(-a,-a,-a)$, $B(-a,a,a)$, $C(a,-a,a)$ and $D(a,a,-a)$. Then the center of the tetrahedron is at $O(0,0,0)$. For a point $X(x,y,z)$ we see that the sum $XA + XB + XC + XD$ by the QM–AM inequality does not exceed $2\sqrt{XA^2 + XB^2 + XC^2 + XD^2}$. Now, since $XA^2 = (x+a)^2 + (y+a)^2 + (z+a)^2$ etc., we easily obtain

$$XA^2 + XB^2 + XC^2 + XD^2 = 4(x^2 + y^2 + z^2) + 12a^2$$
$$\geq 12a^2 = OA^2 + OB^2 + OC^2 + OD^2.$$

Hence $XA + XB + XC + XD \geq 2\sqrt{OA^2 + OB^2 + OC^2 + OD^2} = OA + OB + OC + OD$.

4. It suffices to prove $1/\sin 2^k x = \cot 2^{k-1}x - \cot 2^k x$ for any integer k and real x, i.e., $1/\sin 2x = \cot x - \cot 2x$ for all real x. We indeed have

$$\cot x - \cot 2x = \cot x - \frac{\cot^2 x - 1}{2\cot x} = \frac{\left(\frac{\cos x}{\sin x}\right)^2 + 1}{2\frac{\cos x}{\sin x}} = \frac{1}{2\sin x\cos x} = \frac{1}{\sin 2x}.$$

5. We define $L_1 = |a_1 - a_2|x_2 + |a_1 - a_3|x_3 + |a_1 - a_4|x_4$ and analogously L_2, L_3, and L_4. Let us assume w.l.o.g. that $a_1 < a_2 < a_3 < a_4$. In that case,

$$2|a_1 - a_2||a_2 - a_3|x_2 = |a_3 - a_2|L_1 - |a_1 - a_3|L_2 + |a_1 - a_2|L_3$$
$$= |a_3 - a_2| - |a_1 - a_3| + |a_1 - a_2| = 0,$$
$$2|a_2 - a_3||a_3 - a_4|x_3 = |a_4 - a_3|L_2 - |a_2 - a_4|L_3 + |a_2 - a_3|L_4$$
$$= |a_4 - a_3| - |a_2 - a_4| + |a_2 - a_3| = 0.$$

Hence it follows that $x_2 = x_3 = 0$ and consequently $x_1 = x_4 = 1/|a_1 - a_4|$. This solution set indeed satisfies the starting equations. It is easy to generalize this result to any ordering of a_1, a_2, a_3, a_4.

6. Let S denote the area of $\triangle ABC$. Let A_1, B_1, C_1 be the midpoints of BC, AC, AB respectively. We note that $S_{A_1 B_1 C} = S_{A_1 BC_1} = S_{AB_1 C_1} = S_{A_1 B_1 C_1} = S/4$. Let us assume w.l.o.g. that $M \in [AC_1]$. We then must have $K \in [BA_1]$ and $L \in [CB_1]$. However, we then have $S(KLM) > S(KLC_1) > S(KB_1C_1) = S(A_1B_1C_1) = S/4$. Hence, by the pigeonhole principle one of the remaining three triangles $\triangle MAL$, $\triangle KBM$, and $\triangle LCK$ must have an area less than or equal to $S/4$. This completes the proof.

4.9 Solutions to the Longlisted Problems of IMO 1967

1. Let us denote the nth term of the given sequence by a_n. Then

$$a_n = \frac{1}{3}\left(\frac{10^{3n+3}-10^{2n+3}}{9} + 7\frac{10^{2n+2}-10^{n+1}}{9} + \frac{10^{n+2}-1}{9}\right)$$

$$= \frac{1}{27}(10^{3n+3} - 3\cdot 10^{2n+2} + 3\cdot 10^{n+1} - 1) = \left(\frac{10^{n+1}-1}{3}\right)^3.$$

2. $(n!)^{2/n} = ((1\cdot 2\cdots n)^{1/n})^2 \le \left(\frac{1+2+\cdots+n}{n}\right)^2 = \left(\frac{n+1}{2}\right)^2 \le \frac{1}{3}n^2 + \frac{1}{2}n + \frac{1}{6}.$

3. Consider the function $f : [0,\pi/2] \to \mathbb{R}$ defined by $f(x) = 1 - x^2/2 + x^4/16 - \cos x$.

   It is easy to calculate that $f'(0) = f''(0) = f'''(0) = 0$ and $f''''(x) = 3/2 - \cos x$. Since $f''''(x) > 0$, $f'''(x)$ is increasing. Together with $f'''(0) = 0$, this gives $f'''(x) > 0$ for $x > 0$; hence $f''(x)$ is increasing, etc. Continuing in the same way we easily conclude that $f(x) > 0$.

4. (a) Let $ABCD$ be a parallelogram, and K,L the midpoints of segments BC and CD respectively. The sides of $\triangle AKL$ are equal and parallel to the medians of $\triangle ABC$.

   (b) Using the formulas $4m_a^2 = 2b^2 + 2c^2 - a^2$ etc., it is easy to obtain that $m_a^2 + m_b^2 = m_c^2$ is equivalent to $a^2 + b^2 = 5c^2$. Then

   $$5(a^2 + b^2 - c^2) = 4(a^2 + b^2) \ge 8ab.$$

5. If one of x,y,z is equal to 1 or -1, then we obtain solutions $(-1,-1,-1)$ and $(1,1,1)$. We claim that these are the only solutions to the system.

   Let $f(t) = t^2 + t - 1$. If among x,y,z one is greater than 1, say $x > 1$, we have $x < f(x) = y < f(y) = z < f(z) = x$, which is impossible. It follows that $x,y,z \le 1$.

   Suppose now that one of x,y,z, say x, is less than -1. Since $\min_t f(t) = -5/4$, we have $x = f(z) \in [-5/4,-1)$. Also, since $f([-5/4,-1)) = (-1,-11/16) \subseteq (-1,0)$ and $f((-1,0)) = [-5/4,-1)$, it follows that $y = f(x) \in (-1,0)$, $z = f(y) \in [-5/4,-1)$, and $x = f(z) \in (-1,0)$, which is a contradiction. Therefore $-1 \le x,y,z \le 1$.

   If $-1 < x,y,z < 1$, then $x > f(x) = y > f(y) = z > f(z) = x$, a contradiction. This proves our claim.

6. The given system has two solutions: $(-2,-1)$ and $(-14/3, 13/3)$.

7. Let $S_k = x_1^k + x_2^k + \cdots + x_n^k$ and let σ_k, $k = 1,2,\ldots,n$ denote the kth elementary symmetric polynomial in $x_1,\ldots,x_n$. The given system can be written as $S_k = a^k$, $k = 1,\ldots,n$. Using Newton's formulas

$$k\sigma_k = S_1\sigma_{k-1} - S_2\sigma_{k-2} + \cdots + (-1)^k S_{k-1}\sigma_1 + (-1)^{k-1}S_k, \quad k = 1,2,\ldots,n,$$

the system easily leads to $\sigma_1 = a$ and $\sigma_k = 0$ for $k = 2, \ldots, n$. By Vieta's formulas, $x_1, x_2, \ldots, x_n$ are the roots of the polynomial $x^n - ax^{n-1}$, i.e., $a, 0, 0, \ldots, 0$ in some order.

Remark. This solution does not use the assumption that the x_j's are real.

8. The circles K_A, K_B, K_C, K_D cover the parallelogram if and only if for every point X inside the parallelogram, the length of one of the segments XA, XB, XC, XD does not exceed 1.

   Let O and r be the center and radius of the circumcircle of $\triangle ABD$. For every point X inside $\triangle ABD$, it holds that $XA \leq r$ or $XB \leq r$ or $XD \leq r$. Similarly, for X inside $\triangle BCD$, $XB \leq r$ or $XC \leq r$ or $XD \leq r$. Hence K_A, K_B, K_C, K_D cover the parallelogram if and only if $r \leq 1$, which is equivalent to $\angle ABD \geq 30°$. However, this last is exactly equivalent to $a = AB = 2r\sin\angle ADB \leq 2\sin(\alpha + 30°) = \sqrt{3}\sin\alpha + \cos\alpha$.

9. The incenter of any such triangle lies inside the circle k. We shall show that every point S interior to the circle S is the incenter of one such triangle. If S lies on the segment AB, then it is obviously the incenter of an isosceles triangle inscribed in k that has AB as an axis of symmetry. Let us now suppose S does not lie on AB. Let X and Y be the intersection points of lines AS and BS with k, and let Z be the foot of the perpendicular from S to AB. Since the quadrilateral $BZSX$ is cyclic, we have $\angle ZXS = \angle ABS = \angle SXY$ and analogously $\angle ZYS = \angle SYX$, which implies that S is the incenter of $\triangle XYZ$.

10. Let n be the number of triangles and let b and i be the numbers of vertices on the boundary and in the interior of the square, respectively.

    Since all the triangles are acute, each of the vertices of the square belongs to at least two triangles. Additionally, every vertex on the boundary belongs to at least three, and every vertex in the interior belongs to at least five triangles. Therefore

    $$3n \geq 8 + 3b + 5i. \tag{1}$$

Moreover, the sum of angles at any vertex that lies in the interior, on the boundary, or at a vertex of the square is equal to $2\pi, \pi, \pi/2$ respectively. The sum of all angles of the triangles equals $n\pi$, which gives us $n\pi = 4 \cdot \pi/2 + b\pi + 2i\pi$, i.e., $n = 2 + b + 2i$. This relation together with (1) easily yields that $i \geq 2$. Since each of the vertices inside the square belongs to at least five triangles, and at most two contain both, it follows that $n \geq 8$.

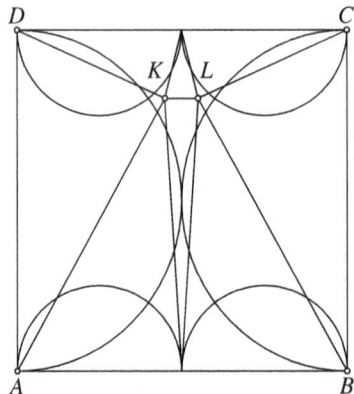

It is shown in the figure that the square can be decomposed into eight acute triangles. Obviously one of them can have an arbitrarily small perimeter.

11. We have to find the number p_n of triples of positive integers (a, b, c) satisfying $a \le b \le c \le n$ and $a + b > c$. Let us denote by $p_n(k)$ the number of such triples with $c = k$, $k = 1, 2, \ldots, n$. For k even, $p_n(k) = k + (k - 2) + (k - 4) + \cdots + 2 = (k^2 + 2k)/4$, and for k odd, $p_n(k) = (k^2 + 2k + 1)/4$. Hence

$$p_n = p_n(1) + p_n(2) + \cdots + p_n(n) = \begin{cases} n(n+2)(2n+5)/24, & \text{for } 2 \mid n, \\ (n+1)(n+3)(2n+1)/24, & \text{for } 2 \nmid n. \end{cases}$$

12. Let us denote by M_n the set of points of the segment AB obtained from A and B by not more than n iterations of $(*)$. It can be proved by induction that

$$M_n = \left\{ X \in AB \mid AX = \frac{3k}{4^n} \text{ or } \frac{3k-2}{4^n} \text{ for some } k \in \mathbb{N} \right\}.$$

Thus (a) immediately follows from $M = \bigcup M_n$. It also follows that if $a, b \in \mathbb{N}$ and $a/b \in M$, then $3 \mid a(b - a)$. Therefore $1/2 \notin M$.

13. The maximum area is $3\sqrt{3}r^2/4$ (where r is the radius of the semicircle) and is attained in the case of a trapezoid with two vertices at the endpoints of the diameter of the semicircle and the other two vertices dividing the semicircle into three equal arcs.

14. We have that

$$\left| \frac{p}{q} - \sqrt{2} \right| = \frac{|p - q\sqrt{2}|}{q} = \frac{|p^2 - 2q^2|}{q(p + q\sqrt{2})} \ge \frac{1}{q(p + q\sqrt{2})}, \tag{1}$$

because $|p^2 - 2q^2| \ge 1$.

The greatest $p, q \le 100$ satisfying the equation $|p^2 - 2q^2| = 1$ are $(p, q) = (99, 70)$. It is easy to verify using (1) that $\frac{99}{70}$ best approximates $\sqrt{2}$ among the fractions p/q with $p, q \le 100$. The numbers $\frac{99}{70} = 1.41428\ldots$ and $\sqrt{2}$ coincide up to the fourth decimal digit: indeed, (1) gives $7 \cdot 10^{-5} < \frac{p}{q} - \sqrt{2} < 8 \cdot 10^{-5}$.

Second solution. By using some basic facts about Farey sequences, one can find that $\frac{41}{29} < \sqrt{2} < \frac{99}{70}$ and that $\frac{41}{29} < \frac{p}{q} < \frac{99}{70}$ implies $p \ge 41 + 99 > 100$ because $99 \cdot 29 - 41 \cdot 70 = 1$. Of the two fractions $\frac{41}{29}$ and $\frac{99}{70}$, the latter is closer to $\sqrt{2}$.

15. Given that $\tan \alpha \in \mathbb{Q}$, we have that $\tan \beta$ is rational if and only if $\tan \gamma$ is rational, where $\gamma = \beta - \alpha$ and $2\gamma = \alpha$. Putting $t = \tan \gamma$ we obtain $\frac{p}{q} = \tan 2\gamma = \frac{2t}{1 - t^2}$, which leads to the quadratic equation $pt^2 + 2qt - p = 0$. This equation has rational solutions if and only if its discriminant $4(p^2 + q^2)$ is a perfect square, and the result follows.

16. First let us notice that all the numbers $z_{m_1, m_2} = m_1 r_1 + m_2 r_2$ $(m_1, m_2 \in \mathbb{Z})$ are distinct, since r_1/r_2 is irrational. Thus for any $n \in \mathbb{N}$ the interval $[-n(|r_1| + |r_2|), n(|r_1| + |r_2|)]$ contains $(2n + 1)^2$ numbers z_{m_1, m_2}, where $|m_1|, |m_2| \le n$. Therefore some two of these $(2n + 1)^2$ numbers, say $z_{m_1, m_2}, z_{n_1, n_2}$, differ by at most $\frac{2n(|r_1| + |r_2|)}{(2n+1)^2 - 1} = \frac{(|r_1| + |r_2|)}{2(n+1)}$. By taking n large enough we can achieve that $z_{q_1, q_2} = |z_{m_1, m_2} - z_{n_1, n_2}| \le p$. If now k is the integer such that $kz_{q_1, q_2} \le x < (k + 1)z_{q_1, q_2}$, then $z_{kq_1, kq_2} = kz_{q_1, q_2}$ differs from x by at most p, as desired.

17. Using $c_r - c_s = (r-s)(r+s+1)$ we can easily get

$$\frac{(c_{m+1} - c_k) \cdots (c_{m+n} - c_k)}{c_1 c_2 \cdots c_n} = \frac{(m-k+n)!}{(m-k)!n!} \cdot \frac{(m+k+n+1)!}{(m+k+1)!(n+1)!}.$$

The first factor $\frac{(m-k+n)!}{(m-k)!n!} = \binom{m-k+n}{n}$ is clearly an integer. The second factor is also an integer because by the assumption, $m+k+1$ and $(m+k)!(n+1)!$ are coprime, and $(m+k+n+1)!$ is divisible by both; hence it is also divisible by their product.

18. In the first part, it is sufficient to show that each rational number of the form $m/n!$, $m,n \in \mathbb{N}$, can be written uniquely in the required form. We prove this by induction on n.

The statement is trivial for $n = 1$. Let us assume it holds for $n-1$, and let there be given a rational number $m/n!$. Let us take $a_n \in \{0,\ldots,n-1\}$ such that $m - a_n = nm_1$ for some $m_1 \in \mathbb{N}$. By the inductive hypothesis, there are unique $a_1 \in \mathbb{N}_0$, $a_i \in \{0,\ldots,i-1\}$ $(i = 1,\ldots,n-1)$ such that $m_1/(n-1)! = \sum_{i=1}^{n-1} a_i/i!$, and then

$$\frac{m}{n!} = \frac{m_1}{(n-1)!} + \frac{a_n}{n!} = \sum_{i=1}^{n} \frac{a_i}{i!},$$

as desired. On the other hand, if $m/n! = \sum_{i=1}^{n} a_i/i!$, multiplying by $n!$ we see that $m - a_n$ must be a multiple of n, so the choice of a_n was unique and therefore the representation itself. This completes the induction.

In particular, since $a_i \mid i!$ and $i!/a_i > (i-1)! \geq (i-1)!/a_{i-1}$, we conclude that each rational q, $0 < q < 1$, can be written as the sum of different reciprocals.

Now we prove the second part. Let $x > 0$ be a rational number. For any integer $m > 10^6$, let $n > m$ be the greatest integer such that $y = x - \frac{1}{m} - \frac{1}{m+1} - \cdots - \frac{1}{n} > 0$. Then y can be written as the sum of reciprocals of different positive integers, which all must be greater than n. The result follows immediately.

19. Suppose $n \leq 6$. Let us decompose the disk by its radii into n congruent regions, so that one of the points P_j lies on the boundaries of two of these regions. Then one of these regions contains two of the n given points. Since the diameter of each of these regions is $2 \sin \frac{\pi}{n}$, we have $d_n \leq 2 \sin \frac{\pi}{n}$. This value is attained if P_i are the vertices of a regular n-gon inscribed in the boundary circle. Hence $D_n = 2 \sin \frac{\pi}{n}$.

For $n = 7$ we have $D_7 \leq D_6 = 1$. This value is attained if six of the seven points form a regular hexagon inscribed in the boundary circle and the seventh is at the center. Hence $D_7 = 1$.

20. The statement so formulated is false. It would be true under the additional assumption that the polygonal line is closed. However, from the offered solution, which is not clear, it does not seem that the proposer had this in mind.

21. Using the formula $\cos x \cos 2x \cos 4x \cdots \cos 2^{n-1}x = \frac{\sin 2^n x}{2^n \sin x}$, which is shown by simple induction, we obtain

$$\cos \frac{\pi}{15} \cos \frac{2\pi}{15} \cos \frac{4\pi}{15} \cos \frac{7\pi}{15} = -\cos \frac{\pi}{15} \cos \frac{2\pi}{15} \cos \frac{4\pi}{15} \cos \frac{8\pi}{15} = \frac{1}{16},$$

$$\cos\frac{3\pi}{15}\cos\frac{6\pi}{15}=\frac{1}{4}, \qquad \cos\frac{5\pi}{15}=\frac{1}{2}.$$

Multiplying these equalities, we get that the required product P equals $1/128$.

22. Let O_1 and O_2 be the centers of circles k_1 and k_2 and let C be the midpoint of the segment AB. Using the well-known relation for elements of a triangle, we obtain

$$PA^2+PB^2=2PC^2+2CA^2\geq 2O_1C^2+2CA^2=2O_1A^2=2r^2.$$

Equality holds if P coincides with O_1 or if A and B coincide with O_2.

23. Suppose that $a\geq 0$, $c\geq 0$, $4ac\geq b^2$. If $a=0$, then $b=0$, and the inequality reduces to the obvious $cg^2\geq 0$. Also, if $a>0$, then

$$af^2+bfg+cg^2=a\left(f+\frac{b}{2a}g\right)^2+\frac{4ac-b^2}{4a}g^2\geq 0.$$

Suppose now that $af^2+bfg+cg^2\geq 0$ holds for an arbitrary pair of vectors f,g. Substituting f by tg ($t\in\mathbb{R}$) we get that $(at^2+bt+c)g^2\geq 0$ holds for any real number t. Therefore $a\geq 0$, $c\geq 0$, $4ac\geq b^2$.

24. Let m be the total number of coins and suppose that the kth child receive x_k coins. By the condition of the problem, the number of coins that remain after him was $6(x_k-k)$. This gives us a recurrence relation

$$x_{k+1}=k+1+\frac{6(x_k-k)-k-1}{7}=\frac{6}{7}x_k+\frac{6}{7},$$

which, together with the condition $x_1=1+(m-1)/7$, yields

$$x_k=\frac{6^{k-1}}{7^k}(m-36)+6 \quad\text{for}\quad 1\leq k\leq n.$$

Since we are given $x_n=n$, we obtain $6^{n-1}(m-36)=7^n(n-6)$. It follows that $6^{n-1}\mid n-6$, which is possible only for $n=6$. Hence, $n=6$ and $m=36$.

25. The answer is $R=(4+\sqrt{3})d/6$.

26. Let L be the midpoint of the edge AB. Since P is the orthocenter of $\triangle ABM$ and ML is its altitude, P lies on ML and therefore belongs to the triangular area LCD. Moreover, from the similarity of triangles ALP and MLB we have

$$LP\cdot LM=LA\cdot LB=a^2/4,$$

where a is the side length of tetrahedron $ABCD$. It easily follows that the locus of P is the image of the segment CD under the inversion of the plane LCD with center L and radius $a/2$. This locus is the arc of a circle with center L and endpoints at the orthocenters of triangles ABC and ABD.

27. Regular polygons with 3, 4, and 6 sides can be obtained by cutting a cube with a plane, as shown in the figure. A polygon with more than 6 sides cannot be obtained in such a way, for a cube has 6 faces. Also, if a pentagon is obtained by cutting a cube with a plane, then its sides lying on opposite faces are parallel; hence it cannot be regular.

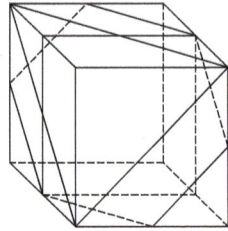

28. The given expression can be transformed into

$$y = \frac{4\cos 2u + 2}{\cos 2u - \cos 2x} - 3.$$

It does not depend on x if and only if $\cos 2u = -1/2$, i.e., $u = \pm\pi/3 + k\pi$ for some $k \in \mathbb{Z}$.

29. Let arc l_a be the locus of points A lying on the opposite side from A_0 with respect to the line B_0C_0 such that $\angle B_0AC_0 = \angle A'$. Let k_a be the circle containing l_a, and let S_a be the center of k_a. We similarly define l_b, l_c, k_b, k_c, S_b, S_c. It is easy to show that circles k_a, k_b, k_c have a common point S inside $\triangle ABC$. Let A_1, B_1, C_1 be the points on the arcs l_a, l_b, l_c diametrically opposite to S with respect to S_a, S_b, S_c respectively. Then $A_0 \in B_1C_1$ because $\angle B_1A_0S = \angle C_1A_0S = 90°$; similarly, $B_0 \in A_1C_1$ and $C_0 \in A_1B_1$. Hence the triangle $A_1B_1C_1$ is circumscribed about $\triangle A_0B_0C_0$ and similar to $\triangle A'B'C'$.

Moreover, we claim that $\triangle A_1B_1C_1$ is the triangle ABC with the desired properties having the maximum side BC and hence the maximum area. Indeed, if ABC is any other such triangle and S_b', S_c' are the projections of S_b and S_c onto the line BC, it holds that

$$BC = 2S_b'S_c' \le 2S_bS_c = B_1C_1,$$

which proves the maximality of B_1C_1.

30. We assume without loss of generality that $m \le n$. Let r and s be the numbers of pairs for which $i - j \ge k$ and of those for which $j - i \ge k$. The desired number is $r + s$. We easily find that

$$r = \begin{cases} (m-k)(m-k+1)/2, & k < m, \\ 0, & k \ge m, \end{cases}$$

$$s = \begin{cases} m(2n-2k-m+1)/2, & k < n - m, \\ (n-k)(n-k+1)/2, & n - m \le k < n, \\ 0, & k \ge n. \end{cases}$$

31. Suppose that $n_1 \le n_2 \le \cdots \le n_k$. If $n_k < m$, there is no solution. Otherwise, the solution is

$$1 + (m-1)(k-s+1) + \sum_{i<s} n_i,$$

where s is the smallest i for which $m \le n_i$ holds.

32. Let us denote by V the volume of the given body, and by V_a, V_b, V_c the volumes of the parts of the given ball that lie inside the dihedra of the given trihedron. It holds that $V_a = 2R^3\alpha/3$, $V_b = 2R^3\beta/3$, $V_c = 2R^3\gamma/3$. It is easy to see that $2(V_a + V_b + V_c) = 4V + 4\pi R^3/3$, from which it follows that

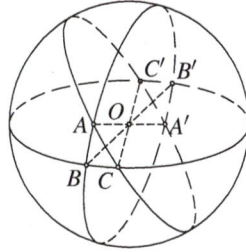

$$V = \frac{1}{3}R^3(\alpha + \beta + \gamma - \pi).$$

33. If $m \notin \{-2, 1\}$, the system has the unique solution

$$x = \frac{b+a-(1+m)c}{(2+m)(1-m)}, \quad y = \frac{a+c-(1+m)b}{(2+m)(1-m)}, \quad z = \frac{b+c-(1+m)a}{(2+m)(1-m)}.$$

The numbers x, y, z form an arithmetic progression if and only if a, b, c do so. For $m = 1$ the system has a solution if and only if $a = b = c$, while for $m = -2$ it has a solution if and only if $a + b + c = 0$. In both these cases it has infinitely many solutions.

34. Each vertex of the polyhedron is a vertex of exactly two squares and triangles (more than two is not possible; otherwise, the sum of angles at a vertex exceeds $360°$). By using the condition that the trihedral angles are equal it is easy to see that such a polyhedron is uniquely determined by its side length. The polyhedron obtained from a cube by "cutting" its vertices, as shown in the figure, satisfies the conditions.

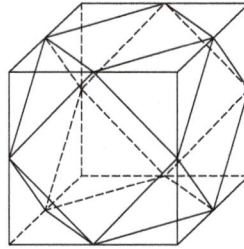

Now it is easy to calculate that the ratio of the squares of volumes of that polyhedron and of the ball whose boundary is the circumscribed sphere is equal to $25/(8\pi^2)$.

35. The given sum can be rewritten as

$$\sum_{k=0}^{n} \binom{n}{k} \left(\tan^2\frac{x}{2}\right)^k + \sum_{k=0}^{n} \binom{n}{k} \left(\frac{2\tan^2\frac{x}{2}}{1-\tan^2\frac{x}{2}}\right)^k.$$

Since $\frac{2\tan^2(x/2)}{1-\tan^2(x/2)} = \frac{1-\cos x}{\cos x}$, the above sum is transformed using the binomial formula into

$$\left(1 + \tan^2\frac{x}{2}\right)^n + \left(1 + \frac{1-\cos x}{\cos x}\right)^n = \sec^{2n}\frac{x}{2} + \sec^n x.$$

36. Suppose that the skew edges of the tetrahedron $ABCD$ are equal. Let K, L, M, P, Q, R be the midpoints of edges AB, AC, AD, CD, DB, BC respectively. Segments KP, LQ, MR have the common midpoint T.

We claim that the lines KP, LQ and MR are axes of symmetry of the tetrahedron $ABCD$. From $LM \parallel CD \parallel RQ$ and similarly $LR \parallel MQ$ and $LM = CD/2 = AB/2 = LR$ it follows that $LMQR$ is a rhombus and therefore $LQ \perp MR$. We similarly show that KP is perpendicular to LQ and MR, and thus it is perpendicular to the plane

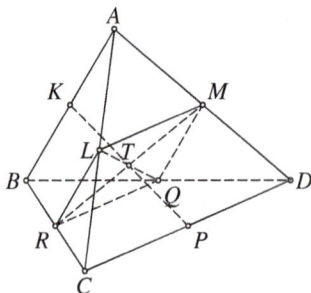

$LMQR$. Since the lines AB and CD are parallel to the plane $LMQR$, they are perpendicular to KP. Hence the points A and C are symmetric to B and D with respect to the line KP, which means that KP is an axis of symmetry of the tetrahedron $ABCD$. Similarly, so are the lines LQ and MR.

The centers of circumscribed and inscribed spheres of tetrahedron $ABCD$ must lie on every axis of symmetry of the tetrahedron, and hence both must coincide with T.

Conversely, suppose that the centers of circumscribed and inscribed spheres of the tetrahedron $ABCD$ coincide with some point T. Then the orthogonal projections of T onto the faces ABC and ABD are the circumcenters O_1 and O_2 of these two triangles, and moreover, $TO_1 = TO_2$. Pythagoras's theorem gives $AO_1 = AO_2$, which by the law of sines implies $\angle ACB = \angle ADB$. Now it easily follows that the sum of the angles at one vertex of the tetrahedron is equal to $180°$. Let D', D'', and D''' be the points in the plane ABC lying outside $\triangle ABC$ such that $\triangle D'BC \cong \triangle DBC$, $\triangle D''CA \cong \triangle DCA$, and $\triangle D'''AB \cong \triangle DAB$. The angle $D''AD'''$ is then straight, and hence A, B, C are midpoints of the segments $D''D''', D'''D', D'D''$ respectively. Hence $AD = D''D'''/2 = BC$, and analogously $AB = CD$ and $AC = BD$.

37. Using the A–G mean inequality we obtain

$$8a^2b^3c^3 \le 2a^8 + 3b^8 + 3c^8,$$
$$8a^3b^2c^3 \le 3a^8 + 2b^8 + 3c^8,$$
$$8a^3b^3c^2 \le 3a^8 + 3b^8 + 2c^8.$$

By adding these inequalities and dividing by $3a^3b^3c^3$ we obtain the desired one.

38. Suppose that there exist integers n and m such that $m^3 = 3n^2 + 3n + 7$. Then from $m^3 \equiv 1 \pmod 3$ it follows that $m = 3k + 1$ for some $k \in \mathbb{Z}$. Substituting into the initial equation we obtain $3k(3k^2 + 3k + 1) = n^2 + n + 2$. It is easy to check that $n^2 + n + 2$ cannot be divisible by 3, and so this equality cannot be true. Therefore our equation has no solutions in integers.

39. Since $\sin^2 A + \sin^2 B + \sin^2 C + \cos^2 A + \cos^2 B + \cos^2 C = 3$, the given equality is equivalent to $\cos^2 A + \cos^2 B + \cos^2 C = 1$, which by multiplying by 2 is transformed into

$$0 = \cos 2A + \cos 2B + 2\cos^2 C = 2\cos(A+B)\cos(A-B) + 2\cos^2 C$$
$$= 2\cos C(\cos(A-B) - \cos C).$$

It follows that either $\cos C = 0$ or $\cos(A - B) = \cos C$. In both cases the triangle is right-angled.

40. Suppose CD is the longest edge of the tetrahedron $ABCD$, $AB = a$, CK and DL are the altitudes of the triangles ABC and ABD respectively, and DM is the altitude of the tetrahedron $ABCD$. Then $CK^2 \le 1 - a^2/4$, since CK is a leg of the right triangle whose other leg has length not less than $a/2$ and whose hypotenuse has length not greater than 1 (AKC or BKC). In the similar way we can show that $DL^2 \le 1 - a^2/4$. Since $DM \le DL$, then $DM^2 \le 1 - a^2/4$. It follows that

$$V = \frac{1}{3}\left(\frac{a}{2}CK\right)DM \le \frac{1}{6}a\left(1 - \frac{a^2}{4}\right) = \frac{1}{24}a(2-a)(2+a)$$
$$= \frac{1}{24}[1 - (a-1)^2](2+a) \le \frac{1}{24}\cdot 1 \cdot 3 = \frac{1}{8}.$$

41. It is well known that the points K, L, M, symmetric to H with respect to BC, CA, AB respectively, lie on the circumcircle k of the triangle ABC. For K, this follows from an elementary calculation of angles of triangles HBC and noting that $\angle KBC = \angle HBC = \angle KAC$. For other points the proof is analogous.

Since the lines l_a, l_b pass through K and L and l_b is obtained from l_a by rotation about C for an angle $2\gamma = \angle LCK$, it follows that the intersection point P of l_a and l_b is at the circumcircle of KLC, that is, k. Similarly, l_b and l_c meet at a point on k; hence they must pass through the same point P.

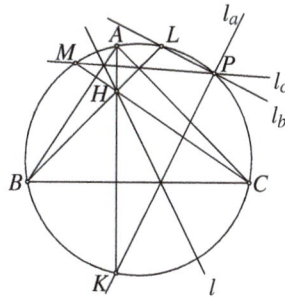

42. $E = (1 - \sin x)(1 - \cos x)[3 + 2(\sin x + \cos x) + 2\sin x \cos x + \sin x \cos x(\sin x + \cos x)]$.

43. We can write the given equation in the form

$$x^5 - x^3 - 4x^2 - 3x - 2 + \lambda(5x^4 + \alpha x^2 - 8x + \alpha) = 0.$$

A root of this equation is independent of λ if and only if it is a common root of the equations

$$x^5 - x^3 - 4x^2 - 3x - 2 = 0 \quad \text{and} \quad 5x^4 + \alpha x^2 - 8x + \alpha = 0.$$

The first of these two equations is equivalent to $(x - 2)(x^2 + x + 1)^2 = 0$ and has three different roots: $x_1 = 2$, $x_{2,3} = (-1 \pm i\sqrt{3})/2$.
   (a) For $\alpha = -64/5$, $x_1 = 2$ is the unique root independent of λ.
   (b) For $\alpha = -3$ there are two roots independent of λ: $x_1 = \omega$ and $x_2 = \omega^2$.

44. (a) $S(x,n) = n(n-1)\left[x^2 + (n+1)x + (n+1)(3n+2)/12\right]$.

(b) It is easy to see that the equation $S(x,n) = 0$ has two roots

$$x_{1,2} = \left(-(n+1) \pm \sqrt{(n+1)/3}\right)/2.$$

They are integers if and only if $n = 3k^2 - 1$ for some $k \in \mathbb{N}$.

45. (a) Using the formula $4\sin^3 x = 3\sin x - \sin 3x$ one can easily reduce the given equation to $\sin 3x = \cos 2x$. Its solutions are given by $x = (4k+1)\pi/10$, $k \in \mathbb{Z}$.

   (b) (1) The point B corresponding to the solution $x = (4k+1)\pi/10$ is a vertex of the regular dodecagon if and only if $(4k+1)\pi/10 = 2m\pi/12$, i.e., $3(4k+1) = 5m$ for some $m \in \mathbb{Z}$. This is possible if and only if $5 \mid 4k+1$, i.e., $k \equiv 1 \pmod 5$.

   (2) Similarly, if the point B corresponding to $x = (4k+1)\pi/10$ is a vertex of a polygon P, then $(4k+1)n = 20m$ for some $m \in \mathbb{N}$, which implies that $4 \mid n$.

46. Let us set $\arctan x = a$, $\arctan y = b$, $\arctan z = c$. Then $\tan(a+b) = \frac{x+y}{1-xy}$ and $\tan(a+b+c) = \frac{x+y+z-xyz}{1-yz-zx-xy} = 1$, which implies that

$$(x-1)(y-1)(z-1) = xyz - xy - yz - zx + x + y + z - 1 = 0.$$

One of x,y,z is equal to 1, say $z = 1$, and consequently $x + y = 0$. Therefore

$$x^{2n+1} + y^{2n+1} + z^{2n+1} = x^{2n+1} + (-x)^{2n+1} + 1^{2n+1} = 1.$$

47. Using the A–G mean inequality we get

$$(n+k-1)x_1^n x_2 \cdots x_k \le n x_1^{n+k-1} + x_2^{n+k-1} + \cdots + x_k^{n+k-1},$$
$$(n+k-1)x_1 x_2^n \cdots x_k \le x_1^{n+k-1} + n x_2^{n+k-1} + \cdots + x_k^{n+k-1},$$
$$\cdots\cdots$$
$$(n+k-1)x_1 x_2 \cdots x_k^n \le x_1^{n+k-1} + x_2^{n+k-1} + \cdots + n x_k^{n+k-1}.$$

By adding these inequalities and dividing by $n+k-1$ we obtain the desired one.

Remark. This is also an immediate consequence of Muirhead's inequality.

48. Put $f(x) = x\ln x$. The given equation is equivalent to $f(x) = f(1/2)$, which has the solutions $x_1 = 1/2$ and $x_2 = 1/4$. Since the function f is decreasing on $(0, 1/e)$, and increasing on $(1/e, +\infty)$, this equation has no other solutions.

49. Since $\sin 1, \sin 2, \ldots, \sin(N+1) \in (-1,1)$, two of these $N+1$ numbers have distance less than $2/N$. Therefore $|\sin n - \sin k| < 2/N$ for some integers $1 \le k, n \le N+1$, $n \ne k$.

50. Since $\varphi(x,y,z) = f(x+y,z) = \varphi(0,x+y,z) = g(0,x+y+z)$, it is enough to put $h(t) = g(0,t)$.

51. If there exist two numbers $\overline{ab}, \overline{bc} \in S$, then one can fill a crossword puzzle as $\begin{pmatrix} a & b \\ b & c \end{pmatrix}$. The converse is obvious. Hence the set S has property A if and only if

the set of first digits and the set of second digits of numbers in S are disjoint. Thus the maximum size of S is 25.

52. This problem is not elementary. The solution offered by the proposer was not quite clear and complete (the existence was not proved).

53. (a) We can construct two lines parallel to the rays of the angle, at equal distances from the rays. The intersection of these two lines lies on the bisector of the angle.

 (b) If the length of a segment AB exceeds the breadth of the ruler, we can construct parallel lines through A and B in two different ways. The diagonal in the resulting rhombus is the perpendicular bisector of the segment AB.
 If the segment AB is too short, we can construct a line l parallel to AB and centrally project AB onto l from a point C chosen sufficiently close to the segment, thus obtaining an arbitrarily long segment $A'B' \parallel AB$. Then we construct the midpoint D' of $A'B'$ as above. The line $D'C$ intersects the segment AB at its midpoint D. By means of lines parallel to DC the segment AB can be prolonged symmetrically, and then the perpendicular bisector can be found as above.

 (c) follows immediately from part (b).

 (d) Let there be given a point P and a line l. We draw an arbitrary line through P that intersects l at A, and two lines l_1 and l_2 parallel to AP, at equal distances from AP and on either side of AP. Line l_1 intersects l at B. We can construct the midpoint C of AP. If BC intersects l_2 at D, then PD is parallel to l.

54. Let S be the given set of points on the cube. Let x, y, z denote the numbers of points from S lying at a vertex, at the midpoint of an edge, at the midpoint of a face of the cube, respectively, and let u be the number of all other points from S. Either there are no points from S at the vertices of the cube, or there is a point from S at each vertex. Hence x is either 0 or 8. Similarly, y is either 0 or 12, and z is either 0 or 6. Any other point of S has 24 possible images under rotations of the cube. Hence u is divisible by 24. Since $n = x + y + z + u$ and $6 \mid y, z, u$, it follows that either $6 \mid n$ or $6 \mid n - 8$, i.e., $n \equiv 0$ or $n \equiv 2 \pmod 6$. Thus $n = 200$ is possible, while $n = 100$ is not, because $n \equiv 4 \pmod 6$.

55. It is enough to find all x from $(0, 2\pi]$ such that the given inequality holds for all integers n.
Suppose $0 < x < 2\pi/3$. If n is the maximum integer for which $nx \le 2\pi/3$, we have $\pi/3 < nx \le 2\pi/3$, and consequently $\sin nx \ge \sqrt{3}/2$. Thus $\sin x + \sin 2x + \cdots + \sin nx > \sqrt{3}/2$.
Suppose now that $2\pi/3 \le x < 2\pi$. We have

$$\sin x + \cdots + \sin nx = \frac{\cos \frac{x}{2} - \cos \frac{2n+1}{2}x}{2 \sin \frac{x}{2}} \le \frac{\cos \frac{x}{2} + 1}{2 \sin \frac{x}{2}} = \frac{\cot \frac{x}{4}}{2} \le \frac{\sqrt{3}}{2}.$$

For $x = 2\pi$ the given inequality clearly holds for all n. Hence, the inequality holds for all n if and only if $2\pi/3 + 2k\pi \le x \le 2\pi + 2k\pi$ for some integer k.

56. We shall prove by induction on n the following statement: If in some group of interpreters exactly n persons, $n \geq 2$, speak each of the three languages, then it is possible to select a subgroup in which each language is spoken by exactly two persons.

The statement of the problem easily follows from this: it suffices to select six such groups.

The case $n = 2$ is trivial. Let us assume $n \geq 2$, and let $N_j, N_m, N_f, N_{jm}, N_{jf}, N_{mf}, N_{jmf}$ be the sets of those interpreters who speak only Japanese, only Malay, only Farsi, only Japanese and Malay, only Japanese and Farsi, only Malay and Farsi, and all the three languages, respectively, and $n_j, n_m, n_f, n_{jm}, n_{jf}, n_{mf}, n_{jmf}$ the cardinalities of these sets, respectively. By the condition of the problem,
$$n_j + n_{jm} + n_{jf} + n_{jmf} = n_m + n_{jm} + n_{mf} + n_{jmf} = n_f + n_{jf} + n_{mf} + n_{jmf} = 24,$$
and consequently

$$n_j - n_{mf} = n_m - n_{jf} = n_f - n_{jm} = c.$$

Now if $c < 0$, then $n_{jm}, n_{jf}, n_{mf} > 0$, and it is enough to select one interpreter from each of the sets N_{jm}, N_{jf}, N_{mf}. If $c > 0$, then $n_j, n_m, n_f > 0$, and it is enough to select one interpreter from each of the sets N_j, N_m, N_f and then use the inductive assumption. Also, if $c = 0$, then w.l.o.g. $n_j = n_{mf} > 0$, and it is enough to select one interpreter from each of the sets N_j, N_{mf} and then use the inductive hypothesis. This completes the induction.

57. Obviously $c_n > 0$ for all even n. Thus $c_n = 0$ is possible only for an odd n. Let us assume $a_1 \leq a_2 \leq \cdots \leq a_8$: in particular, $a_1 \leq 0 \leq a_8$.

If $|a_1| < |a_8|$, then there exists n_0 such that for every odd $n > n_0$, $7|a_1|^n < a_8^n$ $\Rightarrow a_1^n + \cdots + a_7^n + a_8^n > 7a_1^n + a_8^n > 0$, contradicting the condition that $c_n = 0$ for infinitely many n. Similarly $|a_1| > |a_8|$ is impossible, and we conclude that $a_1 = -a_8$.

Continuing in the same manner we can show that $a_2 = -a_7$, $a_3 = -a_6$ and $a_4 = -a_5$. Hence $c_n = 0$ for every odd n.

58. The following sequence of equalities and inequalities gives an even stronger estimate than needed.

$$\begin{aligned}
|l(z)| = |Az + B| &= \frac{1}{2}|(z+1)(A+B) + (z-1)(A-B)| \\
&= \frac{1}{2}|(z+1)f(1) + (z-1)f(-1)| \\
&\leq \frac{1}{2}(|z+1| \cdot |f(1)| + |z-1| \cdot |f(-1)|) \\
&\leq \frac{1}{2}(|z+1| + |z-1|)M = \frac{1}{2}\rho M.
\end{aligned}$$

59. By the arc AB we shall always mean the positive arc AB. We denote by $|AB|$ the length of arc AB. Let a *basic* arc be one of the $n+1$ arcs into which the circle is partitioned by the points $A_0, A_1, \ldots, A_n$, where $n \in \mathbb{N}$.

Suppose that A_pA_0 and A_0A_q are the basic arcs with an endpoint at A_0, and that x_n, y_n are their lengths, respectively. We show by induction on n that for each n the length of a basic arc is equal to x_n, y_n, or $x_n + y_n$.

The statement is trivial for $n = 1$. Assume that it holds for n, and let A_iA_{n+1}, $A_{n+1}A_j$ be basic arcs. We shall prove that these two arcs have lengths x_n, y_n, or $x_n + y_n$. If i, j are both strictly positive, then $|A_iA_{n+1}| = |A_{i-1}A_n|$ and $|A_{n+1}A_j| = |A_nA_{j-1}|$ are equal to x_n, y_n, or $x_n + y_n$ by the inductive hypothesis.

Let us assume now that $i = 0$, i.e., that A_pA_{n+1} and $A_{n+1}A_0$ are basic arcs. Then $|A_pA_{n+1}| = |A_0A_{n+1-p}| \geq |A_0A_q| = y_n$ and similarly $|A_{n+1}A_q| \geq x_n$, but $|A_pA_q| = x_n + y_n$, from which it follows that $|A_pA_{n+1}| = |A_0A_q| = y_n$ and consequently $n + 1 = p + q$. Also, $x_{n+1} = |A_{n+1}A_0| = y_n - x_n$ and $y_{n+1} = y_n$. Now, all basic arcs have lengths $y_n - x_n$, x_n, y_n, $x_n + y_n$. A presence of a basic arc of length $x_n + y_n$ would spoil our inductive step. However, if any basic arc A_kA_l has length $x_n + y_n$, then we must have $l - q = k - p$ because 2π is irrational, and therefore the arc A_kA_l contains either the point A_{k-p} (if $k \geq p$) or the point A_{k+q} (if $k < p$), which is impossible; hence, the proof is complete for $i = 0$. The proof for $j = 0$ is analogous. This completes the induction.

It can be also seen from the above considerations that the basic arcs take only two distinct lengths if and only if $n = p + q - 1$. If we denote by n_k the sequence of n's for which this holds, and by p_k, q_k the sequences of the corresponding p, q, we have $p_1 = q_1 = 1$ and

$$(p_{k+1}, q_{k+1}) = \begin{cases} (p_k + q_k, q_k), & \text{if } \{p_k/(2\pi)\} + \{q_k/(2\pi)\} > 1, \\ (p_k, p_k + q_k), & \text{if } \{p_k/(2\pi)\} + \{q_k/(2\pi)\} < 1. \end{cases}$$

It is now "easy" to calculate that $p_{19} = p_{20} = 333$, $q_{19} = 377$, $q_{20} = 710$, and thus $n_{19} = 709 < 1000 < 1042 = n_{20}$. It follows that the lengths of the basic arcs for $n = 1000$ take exactly three different values.

4.10 Solutions to the Shortlisted Problems of IMO 1968

1. Since the ships are sailing with constant speeds and directions, the second ship is sailing at a constant speed and direction in reference to the first ship. Let A be the constant position of the first ship in this frame. Let B_1, B_2, B_3, and B on line b defining the trajectory of the ship be positions of the second ship with respect to the first ship at 9:00, 9:35, 9:55, and at the moment the two ships were closest. Then we have the following equations for distances (in miles):

$$AB_1 = 20, \quad AB_2 = 15, \quad AB_3 = 13,$$
$$B_1B_2 : B_2B_3 = 7 : 4, \quad AB_i^2 = AB^2 + BB_i^2.$$

Since $BB_1 > BB_2 > BB_3$, it follows that $\mathscr{B}(B_3, B, B_2, B_1)$ or $\mathscr{B}(B, B_3, B_2, B_1)$. We get a system of three quadratic equations with three unknowns: AB, BB_3 and B_3B_2 (BB_3 being negative if $\mathscr{B}(B_3, B, B_1, B_2)$, positive otherwise). This can be solved by eliminating AB and then BB_3. The unique solution ends up being

$$AB = 12, \quad BB_3 = 5, \quad B_3B_2 = 4,$$

and consequently, the two ships are closest at 10:20 when they are at a distance of 12 miles.

2. The sides a, b, c of a triangle ABC with $\angle ABC = 2\angle BAC$ satisfy $b^2 = a(a+c)$ (this statement is the lemma in (SL98-7)). Taking into account the remaining condition that a, b, c are consecutive integers with $a < b$, we obtain three cases:
   (i) $a = n, b = n+1, c = n+2$. We get the equation $(n+1)^2 = n(2n+2)$, giving us $(a, b, c) = (1, 2, 3)$, which is not a valid triangle.
   (ii) $a = n, b = n+2, c = n+1$. We get $(n+2)^2 = n(2n+1) \Rightarrow (n-4)(n+1) = 0$, giving us the triangle $(a, b, c) = (4, 6, 5)$.
   (iii) $a = n+1, b = n+2, c = n$. We get $(n+2)^2 = (n+1)(2n+1) \Rightarrow n^2 - n - 3 = 0$, which has no positive integer solutions for n.
   Hence, the only solution is the triangle with sides of lengths 4, 5, and 6.

3. A triangle cannot be formed out of three lengths if and only if one of them is larger than the sum of the other two. Let us assume this is the case for all triplets of edges out of each vertex in a tetrahedron $ABCD$. Let w.l.o.g. AB be the largest edge of the tetrahedron. Then $AB \geq AC + AD$ and $AB \geq BC + BD$, from which it follows that $2AB \geq AC + AD + BC + BD$. This implies that either $AB \geq AC + BC$ or $AB \geq AD + BD$, contradicting the triangle inequality. Hence the three edges coming out of at least one of the vertices A and B form a triangle.

   Remark. The proof can be generalized to prove that in a polyhedron with only triangular surfaces there is a vertex such that the edges coming out of this vertex form a triangle.

4. We will prove the equivalence in the two directions separately:
   ($\Rightarrow$) Suppose $\{x_1, \ldots, x_n\}$ is the unique solution of the equation. Since $\{x_n, x_1, x_2, \ldots, x_{n-1}\}$ is also a solution, it follows that $x_1 = x_2 = \cdots = x_n = x$ and

the system of equations reduces to a single equation $ax^2 + (b-1)x + c = 0$. For the solution for x to be unique the discriminant $(b-1)^2 - 4ac$ of this quadratic equation must be 0.

($\Leftarrow$) Assume $(b-1)^2 - 4ac = 0$. Adding up the equations, we get

$$\sum_{i=1}^{n} f(x_i) = 0, \quad \text{where} \quad f(x) = ax^2 + (b-1)x + c.$$

But by the assumed condition, $f(x) = a\left(x + \frac{b-1}{2a}\right)^2$. Hence we must have $f(x_i) = 0$ for all i, and $x_i = -\frac{b-1}{2a}$, which is indeed a solution.

5. We have $h_k = r\cos(\pi/k)$ for all $k \in \mathbb{N}$. Using $\cos x = 1 - 2\sin^2(x/2)$ and $\cos x = 2/(1 + \tan^2(x/2)) - 1$ and $\tan x > x > \sin x$ for all $0 < x < \pi/2$, it suffices to prove

$$(n+1)\left(1 - 2\frac{\pi^2}{4(n+1)^2}\right) - n\left(\frac{2}{1 + \pi^2/(4n^2)} - 1\right) > 1$$

$$\Leftrightarrow 1 + 2n\left(1 - \frac{1}{1 + \pi^2/(4n^2)}\right) - \frac{\pi^2}{2(n+1)} > 1$$

$$\Leftrightarrow 1 + \frac{\pi^2}{2}\left(\frac{1}{n + \pi^2/(4n)} - \frac{1}{n+1}\right) > 1,$$

where the last inequality holds because $\pi^2 < 4n$. It is also apparent that as n tends to infinity the term in parentheses tends to 0, and hence it is not possible to strengthen the bound. This completes the proof.

6. We define $f(x) = \frac{a_1}{a_1 - x} + \frac{a_2}{a_2 - x} + \cdots + \frac{a_n}{a_n - x}$. Let us assume w.l.o.g. $a_1 < a_2 < \cdots < a_n$. We note that for all $1 \le i < n$ the function f is continuous in the interval (a_i, a_{i+1}) and satisfies $\lim_{x \to a_i} f(x) = -\infty$ and $\lim_{x \to a_{i+1}} f(x) = \infty$. Hence the equation $f(x) = n$ will have a real solution in each of the $n-1$ intervals (a_i, a_{i+1}).

Remark. In fact, this equation has exactly n solutions, and hence they are all real. Moreover, the solutions are distinct if all a_i are of the same sign, since $x = 0$ is an evident solution.

7. Let r_a, r_b, r_c denote the radii of the exscribed circles corresponding to the sides of lengths a, b, c respectively, and R, p and S denote the circumradius, semiperimeter, and area of the given triangle. It is well-known that $r_a(p-a) = r_b(p-b) = r_c(p-c) = S = \sqrt{p(p-a)(p-b)(p-c)} = \frac{abc}{4R}$. Hence, the desired inequality $r_a r_b r_c \le \frac{3\sqrt{3}}{8} abc$ reduces to $p \le \frac{3\sqrt{3}}{2} R$, which is by the law of sines equivalent to

$$\sin \alpha + \sin \beta + \sin \gamma \le \frac{3\sqrt{3}}{2}.$$

This inequality immediately follows from Jensen's inequality, since the sine is concave on $[0, \pi]$. Equality holds if and only if the triangle is equilateral.

8. Let G be the point such that $BCDG$ is a parallelogram and let H be the midpoint of AG. Obviously $HEFD$ is also a parallelogram, and thus $DH = EF = l$. If $AD^2 + BC^2 = m^2$ is fixed, then from the Stewart theorem we have

$$DH^2 = \frac{2DA^2 + 2DG^2 - AG^2}{4} = \frac{2m^2 - AG^2}{4},$$

which is fixed.

Thus G and H are fixed points, and from here the locus of D is a circle with center H and radius l. The locus of B is the segment $(GI]$, where $I \in \Delta$ is a point in the positive direction such that $AI = a$. Finally, the locus of C is a region of the plane consisting of a rectangle sandwiched between two semicircles of radius l centered at points H and H', where H' is a point such that $\overrightarrow{IH'} = \overrightarrow{GH}$.

9. We note that $S_a = ad_a/2$, $S_b = bd_b/2$, and $S_c = cd_c/2$ are the areas of the triangles MBC, MCA, and MAB respectively. The desired inequality now follows from

$$S_aS_b + S_bS_c + S_cS_a \le \frac{1}{3}(S_a + S_b + S_c)^2 = \frac{S^2}{3}.$$

Equality holds if and only if $S_a = S_b = S_c$, which is equivalent to M being the centroid of the triangle.

10. (a) Let us set $k = a/b > 1$. Then $a = kb$ and $c = \sqrt{k}b$, and $a > c > b$. The segments a, b, c form a triangle if and only if $k < \sqrt{k} + 1$, which holds if and only if $1 < k < \frac{3+\sqrt{5}}{2}$.

   (b) The triangle is right-angled if and only if $a^2 = b^2 + c^2 \Leftrightarrow k^2 = k + 1 \Leftrightarrow k = \frac{1+\sqrt{5}}{2}$. Also, it is acute-angled if and only if $k^2 < k + 1 \Leftrightarrow 1 < k < \frac{1+\sqrt{5}}{2}$ and obtuse-angled if $\frac{1+\sqrt{5}}{2} < k < \frac{3+\sqrt{5}}{2}$.

11. Introducing $y_i = \frac{1}{x_i}$, we transform our equation to

$$\begin{aligned}
0 &= 1 + y_1 + (1 + y_1)y_2 + \cdots + (1 + y_1)\cdots(1 + y_{n-1})y_n \\
&= (1 + y_1)(1 + y_2)\cdots(1 + y_n).
\end{aligned}$$

The solutions are n-tuples $(y_1, \ldots, y_n)$ with $y_i \ne 0$ for all i and $y_j = -1$ for at least one index j. Returning to x_i, we conclude that the solutions are all the n-tuples $(x_1, \ldots, x_n)$ with $x_i \ne 0$ for all i, and $x_j = -1$ for at least one index j.

12. The given inequality is equivalent to $(a+b)^m/b^m + (a+b)^m/a^m \ge 2^{m+1}$, which can be rewritten as

$$\frac{1}{2}\left(\frac{1}{a^m} + \frac{1}{b^m}\right) \ge \left(\frac{2}{a+b}\right)^m.$$

Since $f(x) = 1/x^m$ is a convex function for every $m \in \mathbb{Z}$, the last inequality immediately follows from Jensen's inequality $(f(a) + f(b))/2 \ge f((a+b)/2)$.

13. Translating one of the triangles if necessary, we may assume w.l.o.g. that $B_1 \equiv A_1$. We also assume that $B_2 \ne A_2$ and $B_3 \ne A_3$, since the result is obvious otherwise.

There exists a plane π through A_1 that is parallel to both A_2B_2 and A_3B_3. Let A_2', A_3', B_2', B_3' denote the orthogonal projections of A_2, A_3, B_2, B_3 onto π, and let h_2, h_3 denote the distances of A_2, B_2 and of A_3, B_3 from π. By the Pythagorean theorem, $A_2'A_3'^2 = A_2A_3^2 - (h_2 + h_3)^2 = B_2B_3^2 - (h_2 + h_3)^2 = B_2'B_3'^2$, and similarly

$A_1A_2' = A_1B_2'$ and $A_1A_3' = A_1B_3'$; hence $\triangle A_1A_2'A_3'$ and $\triangle A_1B_2'B_3'$ are congruent. If these two triangles are equally oriented, then we have finished. Otherwise, they are symmetric with respect to some line a passing through A_1, and consequently the projections of the triangles $A_1A_2A_3$ and $A_1B_2B_3$ onto the plane through a perpendicular to π coincide.

14. Let O, D, E be the circumcenter of $\triangle ABC$ and the midpoints of AB and AC, and given arbitrary $X \in AB$ and $Y \in AC$ such that $BX = CY$, let O_1, D_1, E_1 be the circumcenter of $\triangle AXY$ and the midpoints of AX and AY, respectively. Since $AD = AB/2$ and $AD_1 = AX/2$, it follows that $DD_1 = BX/2$ and similarly $EE_1 = CY/2$. Hence O_1 is at the same distance $BX/2 = CY/2$ from the lines OD and OE and lies on the half-line bisector l of $\angle DOE$.

If we let X, Y vary along the segments AB and AC, we obtain that the locus of O_1 is the segment OP, where $P \in l$ is a point at distance $\min(AB, AC)/2$ from OD and OE.

15. Set
$$f(n) = \left[\frac{n+1}{2}\right] + \left[\frac{n+2}{4}\right] + \cdots + \left[\frac{n+2^i}{2^{i+1}}\right] + \ldots.$$

We prove by induction that $f(n) = n$. This obviously holds for $n = 1$. Let us assume that $f(n-1) = n-1$. Define
$$g(i,n) = \left[\frac{n+2^i}{2^{i+1}}\right] - \left[\frac{n-1+2^i}{2^{i+1}}\right].$$

We have that $f(n) - f(n+1) = \sum_{i=0}^{\infty} g(i,n)$. We also note that $g(i,n) = 1$ if and only if $2^{i+1} \mid n + 2^i$; otherwise, $g(i,n) = 0$. The divisibility $2^{i+1} \mid n + 2^i$ is equivalent to $2^i \mid n$ and $2^{i+1} \nmid n$, which for a given n holds for exactly one $i \in \mathbb{N}_0$. Thus it follows that $f(n) - f(n-1) = 1 \Rightarrow f(n) = n$. The proof by induction is now complete.

Second solution. It is easy to show that $[x + 1/2] = [2x] - [x]$ for $x \in \mathbb{R}$. Now $f(x) = ([x] - [x/2]) + ([x/2] - [x/4]) + \cdots = [x]$. Hence, $f(n) = n$ for all $n \in \mathbb{N}$.

16. We shall prove the result by induction on k. It trivially holds for $k = 0$. Assume that the statement is true for some $k - 1$, and let $p(x)$ be a polynomial of degree k. Let us set $p_1(x) = p(x+1) - p(x)$. Then $p_1(x)$ is a polynomial of degree $k - 1$ with leading coefficient ka_0. Also, $m \mid p_1(x)$ for all $x \in \mathbb{Z}$ and hence by the inductive assumption $m \mid (k-1)! \cdot ka_0 = k!a_0$, which completes the induction. On the other hand, for any a_0, k and $m \mid k!a_0$, $p(x) = k!a_0\binom{x}{k}$ is a polynomial with leading coefficient a_0 that is divisible by m.

17. Let there be given an equilateral triangle ABC and a point O such that $OA = x$, $OB = y$, $OC = z$. Let X be the point in the plane such that $\triangle CXB$ and $\triangle COA$ are congruent and equally oriented. Then $BX = x$ and the triangle XOC is equilateral, which implies $OX = z$. Thus we have a triangle OBX with $BX = x$, $BO = y$, and $OX = z$.

Conversely, given a triangle OBX with $BX = x$, $BO = y$ and $OX = z$ it is easy to construct the triangle ABC.

18. The required construction is not feasible. In fact, let us consider the special case $\angle BOC = 135°$, $\angle AOC = 120°$, $\angle AOB = 90°$, where $AA' \cap BB' \cap CC' = \{O\}$. Denoting OA', OB', OC' by a, b, c respectively we obtain the system of equations

$$a^2 + b^2 = a^2 + c^2 + ac = b^2 + c^2 + \sqrt{2}bc.$$

Assuming w.l.o.g. $c = 1$ we easily obtain $a^3 - a^2 - a - 1 = 0$, which is an irreducible equation of third degree. By a known theorem, its solution a is not constructible by ruler and compass.

19. We shall denote by d_n the shortest curved distance from the initial point to the nth point in the positive direction. The sequence d_n goes as follows: 0, 1, 2, 3, 4, 5, 6, 0.72, 1.72, ..., 5.72, 0.43, 1.43, ..., 5.43, 0.15 $= d_{19}$. Hence the required number of points is 20.

20. Let us denote the points $A_1, A_2, \ldots, A_n$ in such a manner that $A_1 A_n$ is a diameter of the set of given points, and $A_1 A_2 \leq A_1 A_3 \leq \cdots \leq A_1 A_n$.
Since for each $1 < i < n$ it holds that $A_1 A_i < A_1 A_n$, we have $\angle A_i A_1 A_n < 120°$ and hence $\angle A_i A_1 A_n < 60°$ (otherwise, all angles in $\triangle A_1 A_i A_n$ are less than $120°$). It follows that for all $1 < i < j \leq n$, $\angle A_i A_1 A_j < 120°$. Consequently, the angle in the triangle $A_1 A_i A_j$ that is at least $120°$ must be $\angle A_1 A_i A_j$. Moreover, for any $1 < i < j < k \leq n$ it holds that $\angle A_i A_j A_k \geq \angle A_1 A_j A_k - \angle A_1 A_j A_i > 120° - 60° = 60°$ (because $\angle A_1 A_j A_i < 60°$); hence $\angle A_i A_j A_k \geq 120°$. This proves that the denotation is correct.

Remark. It is easy to show that the diameter is unique. Hence the denotation is also unique.

21. The given conditions are equivalent to $y - a_0$ being divisible by $a_0, a_0 + a_1, a_0 + a_2, \ldots, a_0 + a_n$, i.e., to $y = k[a_0, a_0 + a_1, \ldots, a_0 + a_n] + a_0$, $k \in \mathbb{N}_0$.

22. It can be shown by induction on the number of digits of x that $p(x) \leq x$ for all $x \in \mathbb{N}$. It follows that $x^2 - 10x - 22 \leq x$, which implies $x \leq 12$. Since $0 < x^2 - 10x - 22 = (x - 12)(x + 2) + 2$, one easily obtains $x \geq 12$. Now one can directly check that $x = 12$ is indeed a solution, and thus the only one.

23. We may assume w.l.o.g. that in all the factors the coefficient of x is 1. Suppose that $x + ay + bz$ is one of the linear factors of $p(x, y, z) = x^3 + y^3 + z^3 + mxyz$. Then $p(x)$ is 0 at every point (x, y, z) with $z = -ax - by$. Hence $x^3 + y^3 + (-ax - by)^3 + mxy(-ax - by) = (1 - a^3)x^3 - (3ab + m)(ax + by)xy + (1 - b^3)y^3 \equiv 0$. This is obviously equivalent to $a^3 = b^3 = 1$ and $m = -3ab$, from which it follows that $m \in \{-3, -3\omega, -3\omega^2\}$, where $\omega = \frac{-1 + i\sqrt{3}}{2}$. Conversely, for each of the three possible values for m there are exactly three possibilities (a, b). Hence $-3, -3\omega, -3\omega^2$ are the desired values.

24. If the ith digit is 0, then the result is

$$\begin{cases} 9^{k-j} \frac{9!}{(10-j)!}, & \text{if } i > k - j, \\ 9^{k-j-1} \frac{9!}{(9-j)!}, & \text{otherwise} \end{cases}.$$

If the ith digit is not 0, then the above results are multiplied by 8.

25. The answer is

$$\sum_{1\leq p<q<r\leq k} n_p n_q n_r + \sum_{1\leq p<q\leq k}\left[n_p\binom{n_q}{2}+n_q\binom{n_p}{2}\right].$$

26. (a) We shall show that the period of f is $2a$. From $(f(x+a)-1/2)^2 = f(x)-f(x)^2$ we obtain

$$\left(f(x)-f(x)^2\right)+\left(f(x+a)-f(x+a)^2\right)=\frac{1}{4}.$$

Subtracting the above relation for $x+a$ in place of x we get $f(x)-f(x)^2 = f(x+2a)-f(x+2a)^2$, which implies $(f(x)-1/2)^2 = (f(x+2a)-1/2)^2$. Since $f(x)\geq 1/2$ holds for all x by the condition of the problem, we conclude that $f(x+2a)=f(x)$.

(b) The following function, as is directly verified, satisfies the conditions:

$$f(x) = \begin{cases} 1/2 & \text{if } 2n \leq x < 2n+1, \\ 1 & \text{if } 2n+1 \leq x < 2n+2, \end{cases} \text{ for } n = 0,1,2,\dots .$$

4.11 Solutions to the Contest Problems of IMO 1969

1. Set $a = 4m^4$, where $m \in \mathbb{N}$ and $m > 1$. We then have $z = n^4 + 4m^4 = (n^2 + 2m^2)^2 - (2mn)^2 = (n^2 + 2m^2 + 2mn)(n^2 + 2m^2 - 2mn)$. Since $n^2 + 2m^2 - 2mn = (n-m)^2 + m^2 \geq m^2 > 1$, it follows that z must be composite. Thus we have found infinitely many a that satisfy the condition of the problem.

2. Using $\cos(a + x) = \cos a \cos x - \sin a \sin x$, we obtain $y(x) = A \sin x + B \cos x$ where $A = -\sin a_1 - \sin a_2/2 - \cdots - \sin a_n/2^{n-1}$ and $B = \cos a_1 + \cos a_2/2 + \cdots + \cos a_n/2^{n-1}$. Numbers A and B cannot both be equal to 0, for otherwise y would be identically equal to 0, while on the other hand, we have $y(-a_1) = \cos(a_1 - a_1) + \cos(a_2 - a_1)/2 + \cdots + \cos(a_n - a_1)/2^{n-1} \geq 1 - 1/2 - \cdots - 1/2^{n-1} = 1/2^{n-1} > 0$. Setting $A = C \cos \phi$ and $B = C \sin \phi$, where $C \neq 0$ (such C and ϕ always exist), we get $y(x) = C \sin(x + \phi)$. It follows that the zeros of y are of the form $x_0 \in -\phi + \pi\mathbb{Z}$, from which $y(x_1) = y(x_2) \Rightarrow x_1 - x_2 = m\pi$ immediately follows.

3. We have several cases:
   $1°$ $k = 1$. W.l.o.g. let $AB = a$ and the remaining segments have length 1. Let M be the midpoint of CD. Then $AM = BM = \sqrt{3}/2$ ($\triangle CDA$ and $\triangle CDB$ are equilateral) and $0 < AB < AM + BM = \sqrt{3}$, i.e., $0 < a < \sqrt{3}$. It is evident that all values of a within this interval are realizable.
   $2°$ $k = 2$. We have two subcases.
   First, let $AC = AD = a$. Let M be the midpoint of CD. We have $CD = 1$, $AM = \sqrt{a^2 - 1/4}$, and $BM = \sqrt{3}/2$. Then we have $1 - \sqrt{3}/2 = AB - BM < AM < AB + BM = 1 + \sqrt{3}/2$, which gives us $\sqrt{2 - \sqrt{3}} < a < \sqrt{2 + \sqrt{3}}$.
   Second, let $AB = CD = a$. Let M be the midpoint of CD. From $\triangle MAB$ we get $a < \sqrt{2}$.
   Thus, from $\sqrt{2 - \sqrt{3}} < \sqrt{2} < \sqrt{2 + \sqrt{3}}$ it follows that the required condition in this case is $0 < a < \sqrt{2 + \sqrt{3}}$. All values for a in this range are realizable.
   $3°$ $k = 3$. We show that such a tetrahedron exists for all a. Assume $a > 1$. Assume $AB = AC = AD = a$. Varying A along the line perpendicular to the plane BCD and through the center of $\triangle BCD$ we achieve all values of $a > 1/\sqrt{3}$. For $a \leq 1/\sqrt{3}$ we can observe a similar tetrahedron with three edges of length $1/a$ and three of length 1 and proceed as before.
   $4°$ $k = 4$. By observing the similar tetrahedron we reduce this case to $k = 2$ with length $1/a$ instead of a. Thus we get $a > \sqrt{2 - \sqrt{3}}$.
   $5°$ $k = 5$. We reduce to $k = 1$ and get $a > 1/\sqrt{3}$.

4. Let O be the midpoint of AB, i.e., the center of γ. Let O_1, O_2, and O_3 respectively be the centers of γ_1, γ_2, and γ_3 and let r_1, r_2, r_3 respectively be the radii of γ_1, γ_2 and γ_3. Let C_1, C_2, and C_3 respectively be the points of tangency of γ_1, γ_2 and γ_3 with AB. Let D_2 and D_3 respectively be the points of tangency of γ_2 and γ_3 with CD. Finally, let G_2 and G_3 respectively be the points of tangency of γ_2 and γ_3 with γ. We have $\mathcal{B}(G_2, O_2, O)$, $G_2 O_2 = O_2 D_2$, and

$G_2O = OB$. Hence, G_2, D_2, B are collinear. Similarly, G_3, D_3, A are collinear. It follows that AG_2D_2D and BG_3D_3D are cyclic, since $\angle AG_2D_2 = \angle D_2DA = \angle D_3DB = \angle BG_3D_3 = 90°$. Hence $BC_2^2 = BD_2 \cdot BG_2 = BD \cdot BA = BC^2 \Rightarrow BC_2 = BC$ and hence $AC_2 = AB - BC$. Similarly, $AC_3 = AC$. We thus have $AC_1 = (AC + AB - BC)/2 = (AC_3 + AC_2)/2$. Hence, C_1 is the midpoint of C_2C_3. We also have $r_2 + r_3 = C_2C_3 = AC + BC - AB = 2r_1$, from which it follows that O_1, O_2, O_3 are collinear.

Second solution. We shall prove the statement for arbitrary points A, B, C on γ. Let us apply the inversion ψ with respect to the circle γ_1. We denote by $\widehat{X}$ the image of an object X under ψ. Also, ψ maps lines BC, CA, AB onto circles $\widehat{a}, \widehat{b}, \widehat{c}$, respectively. Circles $\widehat{a}, \widehat{b}, \widehat{c}$ pass through the center O_1 of γ_1 and have radii equal to the radius of $\widehat{\gamma}$. Let P, Q, R be the centers of $\widehat{a}, \widehat{b}, \widehat{c}$ respectively. The line CD maps onto a circle k through $\widehat{C}$ and O_1 that is perpendicular to $\widehat{c}$. Therefore its center K lies in the intersection of the tangent t to $\widehat{c}$ and the line PQ (which bisects $\widehat{C}O_1$). Let O be a point such that RO_1KO is a parallelogram and γ_2', γ_3' the circles centered at O tangent to k. It is easy to see that γ_2' and γ_3' are also tangent to $\widehat{c}$, since OR and OK have lengths equal to the radii of k and $\widehat{c}$. Hence γ_2' and γ_3' are the images of γ_2 and γ_3 under ψ. Moreover, since $Q\widehat{A}OK$ and $P\widehat{B}OK$ are parallelograms and Q, P, K are collinear, it follows that $\widehat{A}, \widehat{B}, O$ are also collinear. Hence the centers of $\gamma_1, \gamma_2, \gamma_3$ are collinear, lying on the line O_1O, and the statement follows.

Third solution. Moreover, the statement holds for an arbitrary point $D \in BC$. Let E, F, G, H be the points of tangency of γ_2 with AB, CD and of γ_3 with AB, CD, respectively. Let O_i be the center of γ_i, $i = 1, 2, 3$. As is shown in the third solution of (SL93-3), EF and GH meet at O_1. Hence the problem of proving the collinearity of O_1, O_2, O_3 reduces to the following simple problem:

> Let D, E, F, G, H be points such that $D \in EG$, $F \in DH$ and $DE = DF$, $DG = DH$. Let O_1, O_2, O_3 be points such that $\angle O_2ED = \angle O_2FD = 90°$, $\angle O_3GD = \angle O_3HD = 90°$, and $O_1 = EF \cap GH$. Then the points O_1, O_2, and O_3 are collinear.

Let $K_2 = DO_2 \cap EF$ and $K_3 = DO_3 \cap GH$. Then $O_2K_2/O_2D = DK_3/DO_3 = K_2O_1/DO_3$ and hence by Thales' theorem $O_1 \in O_2O_3$.

5. We first prove the following lemma.

   Lemma. If of five points in a plane no three belong to a single line, then there exist four that are the vertices of a convex quadrilateral.

   Proof. If the convex hull of the five points A, B, C, D, E is a pentagon or a quadrilateral, the statement automatically holds. If the convex hull is a triangle, then w.l.o.g. let $\triangle ABC$ be that triangle and D, E points in its interior. Let the line DE w.l.o.g. intersect $[AB]$ and $[AC]$. Then B, C, D, E form the desired quadrilateral.

   We now observe each quintuplet of points within the set. There are $\binom{n}{5}$ such quintuplets, and for each of them there is at least one quadruplet of points forming a convex quadrilateral. Each quadruplet, however, will be counted up to $n-4$ times. Hence we have found at least $\frac{1}{n-4}\binom{n}{5}$ quadruplets. Since $\frac{1}{n-4}\binom{n}{5} \geq \binom{n-3}{2} \Leftrightarrow (n-5)(n-6)(n+8) \geq 0$, which always holds, it follows that we have found at least $\binom{n-3}{2}$ desired quadruplets of points.

6. Define $u_1 = \sqrt{x_1 y_1} + z_1$, $u_2 = \sqrt{x_2 y_2} + z_2$, $v_1 = \sqrt{x_1 y_1} - z_1$, and $v_2 = \sqrt{x_2 y_2} - z_2$. By expanding both sides of the equation we can easily verify $(x_1 + x_2)(y_1 + y_2) - (z_1 + z_2)^2 = (u_1 + u_2)(v_1 + v_2) + (\sqrt{x_1 y_2} - \sqrt{x_2 y_1})^2 \geq (u_1 + u_2)(v_1 + v_2)$. Since $x_i y_i - z_i^2 = u_i v_i$ for $i = 1, 2$, it suffices to prove

$$\frac{8}{(u_1 + u_2)(v_1 + v_2)} \leq \frac{1}{u_1 v_1} + \frac{1}{u_2 v_2}$$
$$\Leftrightarrow 8 u_1 u_2 v_1 v_2 \leq (u_1 + u_2)(v_1 + v_2)(u_1 v_1 + u_2 v_2).$$

This follows from the AM–GM inequalities $2\sqrt{u_1 u_2} \leq u_1 + u_2$, $2\sqrt{v_1 v_2} \leq v_1 + v_2$ and $2\sqrt{u_1 v_1 u_2 v_2} \leq u_1 v_1 + u_2 v_2$.

Equality holds if and only if $x_1 y_2 = x_2 y_1$, $u_1 = u_2$ and $v_1 = v_2$, i.e. if and only if $x_1 = x_2$, $y_1 = y_2$ and $z_1 = z_2$.

Second solution. Let us define $f(x, y, z) = 1/(xy - z^2)$. The problem actually states that

$$2f\left(\frac{x_1 + x_2}{2}, \frac{y_1 + y_2}{2}, \frac{z_1 + z_2}{2}\right) \leq f(x_1, y_1, z_1) + f(x_2, y_2, z_2),$$

i.e., that the function f is convex on the set $D = \{(x, y, z) \in \mathbb{R}^2 \mid xy - z^2 > 0\}$. It is known that a twice continuously differentiable function $f(t_1, t_2, \ldots, t_n)$ is convex if its Hessian $[f''_{ij}]^n_{i,j=1}$ is positive definite, or equivalently (by Sylvester's criterion), if its principal minors $D_k = \det[f''_{ij}]^k_{i,j=1}$, $k = 1, 2, \ldots, n$, are positive. In the case of our f this is directly verified: $D_1 = 2y^2/(xy - z^2)^3$, $D_2 = 3xy + z^2/(xy - z^2)^5$, $D_3 = 6/(xy - z^2)^6$ are obviously positive.

4.12 Solutions to the Shortlisted Problems of IMO 1970

1. Denote respectively by R and r the radii of the circumcircle and incircle, by $A_1, \ldots, A_n, B_1, \ldots, B_n$,the vertices of the $2n$-gon and by O its center. Let P' be the point symmetric to P with respect to O. Then $A_i P' B_i P$ is a parallelogram, and applying cosine theorem on triangles $A_i B_i P$ and $PP'B_i$ yields

$$4R^2 = PA_i^2 + PB_i^2 - 2PA_i \cdot PB_i \cos a_i$$
$$4r^2 = PB_i^2 + P'B_i^2 - 2PB_i \cdot P'B_i \cos \angle PB_i P'.$$

Since $A_i P' B_i P$ is a parallelogram, we have that $P'B_i = PA_i$ and $\angle PB_i P' = \pi - a_i$. Subtracting the expression for $4r^2$ from the one for $4R^2$ yields $4(R^2 - r^2) = -4PA_i \cdot PB_i \cos a_i = -8S_{\triangle A_i B_i P} \cot a_i$, hence we conclude that

$$\tan^2 a_i = \frac{4S_{\triangle A_i B_i P}^2}{(R^2 - r^2)^2}. \tag{1}$$

Denote by M_i the foot of the perpendicular from P to $A_i B_i$ and let $m_i = PM_i$. Then $S_{\triangle A_i B_i P} = Rm_i$. Substituting this into (1) and adding up these relations for $i = 1, 2, \ldots, n$, we obtain

$$\sum_{i=1}^{n} \tan^2 a_i = \frac{4R^2}{(R^2 - r^2)^2} \left(\sum_{i=1}^{n} m_i^2 \right).$$

Note that all the points M_i lie on a circle with diameter OP and form a regular n-gon. Denote its center by F. We have that $m_i^2 = \|\overrightarrow{PM_i}\|^2 = \|\overrightarrow{FM_i} - \overrightarrow{FP}\|^2 = \|\overrightarrow{FM_i}^2\| + \|\overrightarrow{FP}^2\| - 2\langle \overrightarrow{FM_i}, \overrightarrow{FP} \rangle = r^2/2 - 2\langle \overrightarrow{FM_i}, \overrightarrow{FP} \rangle$. From this it follows that

$$\sum_{i=1}^{n} m_i^2 = 2n(r/2)^2 - 2\sum_{i=1}^{n} \langle \overrightarrow{FM_i}, \overrightarrow{FP} \rangle = 2n(r/2)^2 - 2\langle \sum_{i=1}^{n} \overrightarrow{FM_i}, \overrightarrow{FP} \rangle = 2n(r/2)^2,$$

because $\sum_{i=1}^{n} \overrightarrow{FM_i} = \overrightarrow{0}$. Thus

$$\sum_{i=1}^{n} \tan^2 a_i = \frac{4R^2}{(R^2 - r^2)^2} 2n \left(\frac{r}{2} \right)^2 = 2n \frac{(r/R)^2}{(1 - (r/R)^2)^2} = 2n \frac{\cos^2 \frac{\pi}{2n}}{\sin^4 \frac{\pi}{2n}}.$$

Remark. For $n = 1$ there is no regular 2-gon. However, if we think of a 2-gon as a line segment, the statement will remain true.

2. Suppose that $a > b$. Consider the polynomial $P(X) = x_1 X^{n-1} + x_2 X^{n-2} + \cdots + x_{n-1} X + x_n$. We have $A_n = P(a)$, $B_n = P(b)$, $A_{n+1} = x_0 a^n + P(a)$, and $B_{n+1} = x_0 b^n + P(b)$. The inequality $A_n / A_{n+1} < B_n / B_{n+1}$ becomes $P(a)/(x_0 a^n + P(a)) < P(b)/(x_0 b^n + P(b))$, i.e.,

$$b^n P(a) < a^n P(b).$$

Since $a > b$, we have that $a^i > b^i$ and hence $x_i a^n b^{n-i} \geq x_i b^n a^{n-i}$ (also, for $i \geq 1$ the inequality is strict). Summing up all these inequalities for $i = 1, \ldots, n$ we get $a^n P(b) > b^n P(a)$, which completes the proof for $a > b$.

On the other hand, for $a < b$ we analogously obtain the opposite inequality $A_n/A_{n+1} > B_n/B_{n+1}$, while for $a = b$ we have equality. Thus $A_n/A_{n+1} < B_n/B_{n+1} \Leftrightarrow a > b$.

3. We shall use the following lemma

   Lemma. If an altitude of a tetrahedron passes through the orthocenter of the opposite side, then each of the other altitudes possesses the same property.

   Proof. Denote the tetrahedron by $SABC$ and let $a = BC$, $b = CA$, $c = AB$, $m = SA$, $n = SB$, $p = SC$. It is enough to prove that an altitude passes through the orthocenter of the opposite side if and only if $a^2 + m^2 = b^2 + n^2 = c^2 + p^2$. Suppose that the foot S' of the altitude from S is the orthocenter of ABC. Then $SS' \perp ABC \Rightarrow SB^2 - SC^2 = S'B^2 - S'C^2$. But from $AS' \perp BC$ it follows that $AB^2 - AC^2 = S'B^2 - S'C^2$. From these two equalities it can be concluded that $n^2 - p^2 = c^2 - b^2$, or equivalently, $n^2 + b^2 = c^2 + p^2$. Analogously, $a^2 + m^2 = n^2 + b^2$, so we have proved the first part of the equivalence.

   Now suppose that $a^2 + m^2 = b^2 + n^2 = c^2 + p^2$. Defining S' as before, we get $n^2 - p^2 = S'B^2 - S'C^2$. From the condition $n^2 - p^2 = c^2 - b^2$ ($\Leftrightarrow b^2 + n^2 = c^2 + p^2$) we conclude that $AS' \perp BC$. In the same way $CS' \perp AB$, which proves that S' is the orthocenter of $\triangle ABC$. The lemma is thus proven.

   Now using the lemma it is easy to see that if one of the angles at S is right, than so are the others. Indeed, suppose that $\angle ASB = \pi/2$. From the lemma we have that the altitude from C passes through the orthocenter of $\triangle ASB$, which is S, so $CS \perp ASB$ and $\angle CSA = \angle CSB = \pi/2$.

   Therefore $m^2 + n^2 = c^2$, $n^2 + p^2 = a^2$, and $p^2 + m^2 = b^2$, so it follows that $m^2 + n^2 + p^2 = (a^2 + b^2 + c^2)/2$. By the inequality between the arithmetic and quadric means, we have that $(a^2 + b^2 + c^2)/2 \geq 2s^2/3$, where s denotes the semiperimeter of $\triangle ABC$. It remains to be shown that $2s^2/3 \geq 18r^2$. Since $S_{\triangle ABC} = sr$, this is equivalent to $2s^4/3 \geq 18S_{ABC}^2 = 18s(s - a)(s - b)(s - c)$ by Heron's formula. This reduces to $s^3 \geq 27(s - a)(s - b)(s - c)$, which is an obvious consequence of the AM–GM mean inequality.

   Remark. In the place of the lemma one could prove that the opposite edges of the tetrahedron are mutually perpendicular and proceed in the same way.

4. Suppose that n is such a natural number. If a prime number p divides any of the numbers $n, n + 1, \ldots, n + 5$, then it must divide another one of them, so the only possibilities are $p = 2, 3, 5$. Moreover, $n + 1, n + 2, n + 3, n + 4$ have no prime divisors other than 2 and 3 (if some prime number greater than 3 divides one of them, then none of the remaining numbers can have that divisor). Since two of these numbers are odd, they must be powers of 3 (greater than 1). However, there are no two powers of 3 whose difference is 2. Therefore there is no such natural number n.

   Second solution. Obviously, none of $n, n + 1, \ldots, n + 5$ is divisible by 7; hence they form a reduced system of residues. We deduce that $n(n + 1) \cdots (n + 5) \equiv 1 \cdot 2 \cdots 6 \equiv -1 \pmod 7$. If $\{n, \ldots, n + 5\}$ can be partitioned into two subsets with the same products, both congruent to, say, p modulo 7, then $p^2 \equiv -1 \pmod 7$, which is impossible.

Remark. Erdős has proved that a set $n, n+1, \ldots, n+m$ of consecutive natural numbers can never be partitioned into two subsets with equal products of elements.

5. Denote respectively by A_1, B_1, C_1 and D_1 the points of intersection of the lines AM, BM, CM, and DM with the opposite sides of the tetrahedron. Since $\mathrm{vol}(MBCD) = \mathrm{vol}(ABCD)\overrightarrow{MA_1}/\overrightarrow{AA_1}$, the relation we have to prove is equivalent to

$$\overrightarrow{MA} \cdot \frac{\overrightarrow{MA_1}}{\overrightarrow{AA_1}} + \overrightarrow{MB} \cdot \frac{\overrightarrow{MB_1}}{\overrightarrow{BB_1}} + \overrightarrow{MC} \cdot \frac{\overrightarrow{MC_1}}{\overrightarrow{CC_1}} + \overrightarrow{MD} \cdot \frac{\overrightarrow{MD_1}}{\overrightarrow{DD_1}} = 0. \tag{1}$$

There exist unique real numbers α, β, γ, and δ such that $\alpha + \beta + \gamma + \delta = 1$ and for every point O in space

$$\overrightarrow{OM} = \alpha\overrightarrow{OA} + \beta\overrightarrow{OB} + \gamma\overrightarrow{OC} + \delta\overrightarrow{OD}. \tag{2}$$

(This follows easily from $\overrightarrow{OM} = \overrightarrow{OA} + \overrightarrow{AM} = \overrightarrow{OA} + k\overrightarrow{AB} + l\overrightarrow{AC} + m\overrightarrow{AD} = \overrightarrow{AB} + k(\overrightarrow{OB} - \overrightarrow{OA}) + l(\overrightarrow{OC} - \overrightarrow{OA}) + m(\overrightarrow{OD} - \overrightarrow{OA})$ for some $k, l, m \in \mathbb{R}$.) Further, from the condition that A_1 belongs to the plane BCD we obtain for every O in space the following equality for some β', γ', δ':

$$\overrightarrow{OA_1} = \beta'\overrightarrow{OB} + \gamma'\overrightarrow{OC} + \delta'\overrightarrow{OD}. \tag{3}$$

However, for $\lambda = \overrightarrow{MA_1}/\overrightarrow{AA_1}$, $\overrightarrow{OM} = \lambda\overrightarrow{OA} + (1 - \lambda)\overrightarrow{OA_1}$; hence substituting (2) and (3) in this expression and equating coefficients for $\overrightarrow{OA}$ we obtain $\lambda = \overrightarrow{MA_1}/\overrightarrow{AA_1} = \alpha$. Analogously, $\beta = \overrightarrow{MB_1}/\overrightarrow{BB_1}$, $\gamma = \overrightarrow{MC_1}/\overrightarrow{CC_1}$, and $\delta = \overrightarrow{MD_1}/\overrightarrow{DD_1}$; hence (1) follows immediately for $O = M$.

Remark. The statement of the problem actually follows from the fact that M is the center of mass of the system with masses $\mathrm{vol}(MBCD)$, $\mathrm{vol}(MACD)$, $\mathrm{vol}(MABD)$, $\mathrm{vol}(MABC)$ at A, B, C, D respectively. Our proof is actually a formal verification of this fact.

6. Let F be the midpoint of $B'C'$, A' the midpoint of BC, and I the intersection point of the line HF and the circle circumscribed about $\triangle BHC'$. Denote by M the intersection point of the line AA' with the circumscribed circle about the triangle ABC. Triangles $HB'C'$ and ABC are similar. Since $\angle C'IF = \angle ABC = \angle A'MC$, $\angle C'FI = \angle AA'B = \angle MA'C$, $2C'F = C'B'$, and $2A'C = CB$, it follows that $\triangle C'IB' \sim \triangle CMB$, hence $\angle FIB' = \angle A'MB = \angle ACB$. Now one concludes that I belongs to the circumscribed circles of $\triangle AB'C'$ (since $\angle C'IB' = 180° - \angle C'AB'$) and $\triangle HCB'$.

Second Solution. We denote the angles of $\triangle ABC$ by α, β, γ. Evidently $\triangle ABC \sim \triangle HC'B'$. Within $\triangle HC'B'$ there exists a unique point I such that $\angle HIB' = 180° - \gamma$, $\angle HIC' = 180° - \beta$, and $\angle C'IB' = 180° - \alpha$, and all three circles must contain this point. Let HI and $B'C'$ intersect in F. It remains to show that $FB' = FC'$. From $\angle HIB' + \angle HB'F = 180°$ we obtain $\angle IHB' = \angle IB'F$. Similarly, $\angle IHC' = \angle IC'F$. Thus circles around $\triangle IHC'$ and $\triangle IHB'$ are both tangent to $B'C'$, giving us $FB'^2 = FI \cdot FH = FC'^2$.

7. For $a = 5$ one can take $n = 10$, while for $a = 6$ one takes $n = 11$. Now assume $a \notin \{5,6\}$.

If there exists an integer n such that each digit of $n(n+1)/2$ is equal to a, then there is an integer k such that $n(n+1)/2 = (10^k - 1)a/9$. After multiplying both sides of the equation by 72, one obtains $36n^2 + 36n = 8a \cdot 10^k - 8a$, which is equivalent to

$$9(2n+1)^2 = 8a \cdot 10^k - 8a + 9. \tag{1}$$

So $8a \cdot 10^k - 8a + 9$ is the square of some odd integer. This means that its last digit is 1, 5, or 9. Therefore $a \in \{1,3,5,6,8\}$.

If $a = 3$ or $a = 8$, the number on the RHS of (1) is divisible by 5, but not by 25 (for $k \geq 2$), and thus cannot be a square. It remains to check the case $a = 1$. In that case, (1) becomes $9(2n+1)^2 = 8 \cdot 10^k + 1$, or equivalently $[3(2n+1) - 1][3(2n+1) + 1] = 8 \cdot 10^k \Rightarrow (3n+1)(3n+2) = 2 \cdot 10^k$. Since the factors $3n+1, 3n+2$ are relatively prime, this implies that one of them is 2^{k+1} and the other one is 5^k. It is directly checked that their difference really equals 1 only for $k = 1$ and $n = 1$, which is excluded. Hence, the desired n exists only for $a \in \{5,6\}$.

8. Let $AC = b, BC = a, AM = x, BM = y, CM = l$. Denote by I_1 the incenter and by S_1 the center of the excircle of $\triangle AMC$. Suppose that P_1 and Q_1 are feet of perpendiculars from I_1 and S_1, respectively, to the line AC. Then $\triangle I_1CP_1 \sim \triangle S_1CQ_1$, hence $r_1/\rho_1 = CP_1/CQ_1$. We have $CP_1 = (AC + MC - AM)/2 = (b+l-x)/2$ and $CQ_1 = (AC + MC + AM)/2 = (b+l+x)/2$. Hence

$$\frac{r_1}{\rho_1} = \frac{b+l-x}{b+l+x}.$$

We similarly obtain

$$\frac{r_2}{\rho_2} = \frac{b+l-y}{b+l+y} \quad \text{and} \quad \frac{r}{\rho} = \frac{a+b-x-y}{a+b+x+y}.$$

What we have to prove is now equivalent to

$$\frac{(b+l-x)(a+l-y)}{(b+l+x)(a+l+y)} = \frac{a+b-x-y}{a+b+x+y}. \tag{1}$$

Multiplying both sides of (1) by $(a+l+y)(b+l+x)(a+b+x+y)$ we obtain an expression that reduces to $l^2x + l^2y + x^2y + xy^2 = b^2y + a^2x$. Dividing both sides by $c = x+y$, we get that (1) is equivalent to $l^2 = b^2y/(x+y) + a^2x/(x+y) - xy$, which is exactly Stewart's theorem for l. This finally proves the desired result.

9. Let us set $a = \sqrt{\sum_{i=1}^n u_i^2}$ and $b = \sqrt{\sum_{i=1}^n v_i^2}$. By Minkowski's inequality (for $p = 2$) we have $\sum_{i=1}^n (u_i + v_i)^2 \leq (a+b)^2$. Hence the LHS of the desired inequality

is not greater than $1 + (a+b)^2$, while the RHS is equal to $4(1+a^2)(1+b^2)/3$. Now it is sufficient to prove that

$$3 + 3(a+b)^2 \le 4(1+a^2)(1+b^2).$$

The last inequality can be reduced to the trivial $0 \le (a-b)^2 + (2ab-1)^2$. The equality in the initial inequality holds if and only if $u_i/v_i = c$ for some $c \in \mathbb{R}$ and $a = b = 1/\sqrt{2}$.

10. (a) Since $a_{n-1} < a_n$, we have

$$\left(1 - \frac{a_{k-1}}{a_k}\right)\frac{1}{\sqrt{a_k}} = \frac{a_k - a_{k-1}}{a_k^{3/2}}$$

$$\le \frac{2(\sqrt{a_k} - \sqrt{a_{k-1}})\sqrt{a_k}}{a_k\sqrt{a_{k-1}}} = 2\left(\frac{1}{\sqrt{a_{k-1}}} - \frac{1}{\sqrt{a_k}}\right).$$

Summing up all these inequalities for $k = 1, 2, \ldots, n$ we obtain

$$b_n \le 2\left(\frac{1}{\sqrt{a_0}} - \frac{1}{\sqrt{a_n}}\right) < 2.$$

(b) Choose a real number $q > 1$, and let $a_k = q^k$, $k = 1, 2, \ldots$. Then we deduce $(1 - a_{k-1}/a_k)/\sqrt{a_k} = (1 - 1/q)/q^{k/2}$, and consequently

$$b_n = \left(1 - \frac{1}{q}\right)\sum_{k=1}^n \frac{1}{q^{k/2}} = \frac{\sqrt{q}+1}{q}\left(1 - \frac{1}{q^{n/2}}\right).$$

Since $(\sqrt{q}+1)/q$ can be arbitrarily close to 2, one can set q such that $(\sqrt{q}+1)/q > b$. Then $b_n \ge b$ for all sufficiently large n.

Second solution.

(a) Note that

$$b_n = \sum_{k=1}^n \left(1 - \frac{a_{k-1}}{a_k}\right)\frac{1}{\sqrt{a_k}} = \sum_{k=1}^n (a_k - a_{k-1})\cdot\frac{1}{a_k^{3/2}};$$

hence b_n represents exactly the lower Darboux sum for the function $f(x) = x^{-3/2}$ on the interval $[a_0, a_n]$. Then $b_n \le \int_{a_0}^{a_n} x^{-3/2}dx < \int_1^{+\infty} x^{-3/2}dx = 2$.

(b) For each $b < 2$ there exists a number $\alpha > 1$ such that $\int_1^\alpha x^{-3/2}dx > b + (2-b)/2$. Now, by Darboux's theorem, there exists a sequence $1 = a_0 \le a_1 \le \cdots \le a_n = \alpha$ such that the corresponding Darboux sums are arbitrarily close to the value of the integral. In particular, there is a sequence $a_0, \ldots, a_n$ with $b_n > b$.

11. Let $S(x) = (x-x_1)(x-x_2)\cdots(x-x_n)$. We have $x^3 - x_i^3 = (x-x_i)(\omega x - x_i)(\omega^2 x - x_i)$, where ω is a primitive third root of 1. Multiplying these equalities for $i = 1, \ldots, n$ we obtain

$$T(x^3) = (x^3 - x_1^3)(x^3 - x_2^3)\cdots(x^3 - x_n^3) = S(x)S(\omega x)S(\omega^2 x).$$

Since $S(\omega x) = P(x^3) + \omega x Q(x^3) + \omega^2 x^2 R(x^3)$ and $S(\omega^2 x) = P(x^3) + \omega^2 x Q(x^3) + \omega x^2 R(x^3)$, the above expression reduces to

$$T(x^3) = P^3(x^3) + x^3 Q^3(x^3) + x^6 R^3(x^3) - 3P(x^3)Q(x^3)R(x^3).$$

Therefore the zeros of the polynomial

$$T(x) = P^3(x) + xQ^3(x) + x^2 R^3(x) - 3P(x)Q(x)R(x)$$

are exactly $x_1^3, \dots, x_n^3$. It is easily verified that $\deg T = \deg S = n$, and hence T is the desired polynomial.

12. *Lemma.* Five points are given in the plane such that no three of them are collinear. Then there are at least three triangles with vertices at these points that are not acute-angled.

   Proof. We consider three cases, according to whether the convex hull of these points is a triangle, quadrilateral, or pentagon.

   (i) Let a triangle ABC be the convex hull and two other points D and E lie inside the triangle. At least two of the triangles ADB, BDC and CDA have obtuse angles at the point D. Similarly, at least two of the triangles AEB, BEC and CEA are obtuse-angled. Thus there are at least four non-acute-angled triangles.

   (ii) Suppose that $ABCD$ is the convex hull and that E is a point of its interior. At least one angle of the quadrilateral is not acute, determining one non-acute-angled triangle. Also, the point E lies in the interior of either $\triangle ABC$ or $\triangle CDA$; hence, as in the previous case, it determines another two obtuse-angled triangles.

   (iii) It is easy to see that at least two of the angles of the pentagon are not acute. We may assume that these two angles are among the angles corresponding to vertices A, B, and C. Now consider the quadrilateral $ACDE$. At least one of its angles is not acute. Hence, there are at least three triangles that are not acute-angled.

Now we consider all combinations of 5 points chosen from the given 100. There are $\binom{100}{5}$ such combinations, and for each of them there are at least three non-acute-angled triangles with vertices in it. On the other hand, vertices of each of the triangles are counted $\binom{97}{2}$ times. Hence there are at least $3\binom{100}{5}/\binom{97}{2}$ non-acute-angled triangles with vertices in the given 100 points. Since the number of all triangles with vertices in the given points is $\binom{100}{3}$, the ratio between the number of acute-angled triangles and the number of all triangles cannot be greater than

$$1 - \frac{3\binom{100}{5}}{\binom{97}{2}\binom{100}{3}} = 0.7.$$

4.13 Solutions to the Shortlisted Problems of IMO 1971

1. Assuming that a,b,c in (1) exist, let us find what their values should be. Since $P_2(x) = x^2 - 2$, equation (1) for $n = 1$ becomes $(x^2 - 4)^2 = [a(x^2 - 2) + bx + 2c]^2$. Therefore, there are two possibilities for (a,b,c): $(1,0,-1)$ and $(-1,0,1)$. In both cases we must prove that

$$(x^2 - 4)[P_n(x)^2 - 4] = [P_{n+1}(x) - P_{n-1}(x)]^2. \tag{2}$$

It suffices to prove (2) for all x in the interval $[-2,2]$. In this interval we can set $x = 2\cos t$ for some real t. We prove by induction that

$$P_n(x) = 2\cos nt \quad \text{for all } n. \tag{3}$$

This is trivial for $n = 0, 1$. Assume (3) holds for some $n - 1$ and n. Then $P_{n+1}(x) = 4\cos t \cos nt - 2\cos(n-1)t = 2\cos(n+1)t$ by the additive formula for the cosine. This completes the induction.

Now (2) reduces to the obviously correct equality

$$16\sin^2 t \sin^2 nt = (2\cos(n+1)t - 2\cos(n-1)t)^2.$$

Second solution. If x is fixed, the linear recurrence relation $P_{n+1}(x) + P_{n-1}(x) = xP_n(x)$ can be solved in the standard way. The characteristic polynomial $t^2 - xt + 1$ has zeros $t_{1,2}$ with $t_1 + t_2 = x$ and $t_1 t_2 = 1$; hence, the general $P_n(x)$ has the form $at_1^n + bt_2^n$ for some constants a, b. From $P_0 = 2$ and $P_1 = x$ we obtain that

$$P_n(x) = t_1^n + t_2^n.$$

Plugging in these values and using $t_1 t_2 = 1$ one easily verifies (2).

2. We will construct such a set S_m of 2^m points.
Take vectors $u_1, \ldots, u_m$ in a given plane such that $|u_i| = 1/2$ and $0 \neq |c_1 u_1 + c_2 u_2 + \cdots + c_n u_n| \neq 1/2$ for any choice of numbers c_i equal to 0 or ± 1 (where two or more of the numbers c_i are nonzero). Such vectors are easily constructed by induction on m: For $u_1, \ldots, u_{m-1}$ fixed, there are only finitely many vector values u_m that violate the upper condition, and we may set u_m to be any other vector of length $1/2$.
Let S_m be the set of all points $M_0 + \varepsilon_1 u_1 + \varepsilon_2 u_2 + \cdots + \varepsilon_m u_m$, where M_0 is any fixed point in the plane and $\varepsilon_i = \pm 1$ for $i = 1, \ldots, m$. Then S_m obviously satisfies the condition of the problem.

3. Let x, y, z be a solution of the given system with $x^2 + y^2 + z^2 = \alpha < 10$. Then

$$xy + yz + zx = \frac{(x+y+z)^2 - (x^2+y^2+z^2)}{2} = \frac{9 - \alpha}{2}.$$

Furthermore, $3xyz = x^3 + y^3 + z^3 - (x+y+z)(x^2+y^2+z^2 - xy - yz - zx)$, which gives us $xyz = 3(9 - \alpha)/2 - 4$. We now have

$$35 = x^4 + y^4 + z^4 = (x^3 + y^3 + z^3)(x + y + z)$$
$$- (x^2 + y^2 + z^2)(xy + yz + zx) + xyz(x + y + z)$$
$$= 45 - \frac{\alpha(9 - \alpha)}{2} + \frac{9(9 - \alpha)}{2} - 12.$$

The solutions in α are $\alpha = 7$ and $\alpha = 11$. Therefore $\alpha = 7$, $xyz = -1$, $xy + xz + yz = 1$, and

$$x^5 + y^5 + z^5 = (x^4 + y^4 + z^4)(x + y + z)$$
$$- (x^3 + y^3 + z^3)(xy + xz + yz) + xyz(x^2 + y^2 + z^2)$$
$$= 35 \cdot 3 - 15 \cdot 1 + 7 \cdot (-1) = 83.$$

4. In the coordinate system in which the x-axis passes through the centers of the circles and the y-axis is their common tangent, the circles have equations

$$x^2 + y^2 + 2r_1 x = 0, \qquad x^2 + y^2 - 2r_2 x = 0.$$

Let p be the desired line with equation $y = ax + b$. The abscissas of points of intersection of p with both circles satisfy one of

$$(1 + a^2)x^2 + 2(ab + r_1)x + b^2 = 0, \qquad (1 + a^2)x^2 + 2(ab - r_2)x + b^2 = 0.$$

Let us denote the lengths of the chords and their projections onto the x-axis by d and d_1, respectively. From these equations it follows that

$$d_1^2 = \frac{4(ab + r_1)^2}{(1 + a^2)^2} - \frac{4b^2}{1 + a^2} = \frac{4(ab - r_2)^2}{(1 + a^2)^2} - \frac{4b^2}{1 + a^2}. \tag{1}$$

Consider the point of intersection of p with the y-axis. This point has equal powers with respect to both circles. Hence, if that point divides the segment determined on p by the two circles into two segments of lengths x and y, this power equals $x(x + d) = y(y + d)$, which implies $x = y = d/2$. Thus each of the equations in (1) has two roots, one of which is thrice the other. This fact gives us $(ab + r_1)^2 = 4(1 + a^2)b^2/3$. We can now use (1) to obtain

$$ab = \frac{r_2 - r_1}{2}, \qquad 4b^2 + a^2 b^2 = 3[(ab + r_1)^2 - a^2 b^2] = 3r_1 r_2;$$
$$a^2 = \frac{4(r_2 - r_1)^2}{14r_1 r_2 - r_1^2 - r_2^2}, \qquad b^2 = \frac{14r_1 r_2 - r_1^2 - r_2^2}{16};$$
$$d_1^2 = \frac{(14r_1 r_2 - r_1^2 - r_2^2)^2}{36(r_1 + r_2)^2}.$$

Finally, since $d^2 = d_1^2(1 + a^2)$, we conclude that

$$d^2 = \frac{1}{12}(14r_1 r_2 - r_1^2 - r_2^2),$$

and that the problem is solvable if and only if $7 - 4\sqrt{3} \le \frac{r_1}{r_2} \le 7 + 4\sqrt{3}$.

5. Without loss of generality, we may assume that $a \geq b \geq c \geq d \geq e$. Then $a - b = -(b - a) \geq 0$, $a - c \geq b - c \geq 0$, $a - d \geq b - d \geq 0$ and $a - e \geq b - e \geq 0$, and hence

$$(a - b)(a - c)(a - d)(a - e) + (b - a)(b - c)(b - d)(b - e) \geq 0.$$

Analogously, $(d - a)(d - b)(d - c)(d - e) + (e - a)(e - b)(e - c)(e - d) \geq 0$. Finally, $(c - a)(c - b)(c - d)(c - e) \geq 0$ as a product of two nonnegative numbers, from which the inequality stated in the problem follows.

Remark. The problem in an alternative formulation, accepted for the IMO, asked to prove that the analogous inequality

$$(a_1 - a_2)(a_1 - a_2) \cdots (a_1 - a_n) + (a_2 - a_1)(a_2 - a_3) \cdots (a_2 - a_n) + \cdots$$
$$+ (a_n - a_1)(a_n - a_2) \cdots (a_n - a_{n-1}) \geq 0$$

holds for arbitrary real numbers a_i if and only if $n = 3$ or $n = 5$. The case $n = 3$ is analogous to $n = 5$. For $n = 4$, a counterexample is $a_1 = 0$, $a_2 = a_3 = a_4 = 1$, while for $n > 5$ one can take $a_1 = a_2 = \cdots = a_{n-4} = 0$, $a_{n-3} = a_{n-2} = a_{n-1} = 2$, $a_n = 1$ as a counterexample.

6. The proof goes by induction on n. For $n = 2$, the following labeling satisfies the conditions (i)–(iv): $C_1 = 11, C_2 = 12, C_3 = 22, C_4 = 21$. Suppose that $n > 2$, and that the numeration $C_1, C_2, \ldots, C_{2^{n-1}}$ of a regular 2^{n-1}-gon, in cyclical order, satisfies (i)–(iv). Then one can assign to the vertices of a 2^n-gon cyclically the following numbers:

$$\overline{1C_1}, \overline{1C_2}, \ldots, \overline{1C_{2^{n-1}}}, \overline{2C_{2^{n-1}}}, \ldots, \overline{2C_2}, \overline{2C_1}.$$

The conditions (i), (ii) obviously hold, while (iii) and (iv) follow from the inductive assumption.

7. (a) Suppose that X, Y, Z are fixed on segments AB, BC, CD. It is proven in a standard way that if $\angle ATX \neq \angle ZTD$, then $ZT + TX$ can be reduced. It follows that if there exists a broken line $XYZTX$ of minimal length, then the following conditions hold:

$$\angle DAB = \pi - \angle ATX - \angle AXT,$$
$$\angle ABC = \pi - \angle BXY - \angle BYX = \pi - \angle AXT - \angle CYZ,$$
$$\angle BCD = \pi - \angle CYZ - \angle CZY,$$
$$\angle CDA = \pi - \angle DTZ - \angle DZT = \pi - \angle ATX - \angle CZY.$$

Thus $\sigma = 0$.

(b) Now let $\sigma = 0$. Let us cut the surface of the tetrahedron along the edges AC, CD, and DB and set it down into a plane. Consider the plane figure $\mathscr{S} = ACD'BD''C'$ thus obtained made up of triangles BCD', ABC, ABD'', and $AC'D''$, with Z', T', Z'' respectively on $CD', AD'', C'D''$ (here C' corresponds to C, etc.). Since $\angle C'D''A + \angle D''AB + \angle ABC + \angle BCD' = 0$ as an oriented angle (because $\sigma = 0$), the lines CD' and $C'D''$ are parallel and equally oriented; i.e., $CD'D''C'$ is a parallelogram.

The broken line $XYZTX$ has minimal length if and only if Z'', T', X, Y, Z' are collinear (where $Z'Z'' \parallel CC'$), and then this length equals $Z'Z'' = CC' = 2AC\sin(\alpha/2)$. There is an infinity of such lines, one for every line $Z'Z''$ parallel to CC' that meets the interiors of all the segments CB, BA, AD''. Such $Z'Z''$ exist. Indeed, the triangles CAB and $D''AB$ are acute-angled, and thus the segment AB has a common interior point with the parallelogram $CD'D''C'$. Therefore the desired result follows.

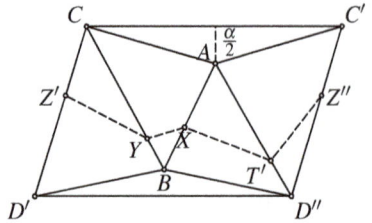

8. Suppose that a, b, c, t satisfy all the conditions. Then $abc \neq 0$ and

$$x_1 x_2 = \frac{c}{a}, \qquad x_2 x_3 = \frac{a}{b}, \qquad x_3 x_1 = \frac{b}{c}.$$

Multiplying these equations, we obtain $x_1^2 x_2^2 x_3^2 = 1$, and hence $x_1 x_2 x_3 = \varepsilon = \pm 1$. From the above equalities we get $x_1 = \varepsilon b/a$, $x_2 = \varepsilon c/b$, $x_3 = \varepsilon a/c$. Substituting x_1 in the first equation, we get $ab^2/a^2 + t\varepsilon b^2/a + c = 0$, which gives us

$$b^2(1 + t\varepsilon) = -ac. \tag{1}$$

Analogously, $c^2(1 + t\varepsilon) = -ab$ and $a^2(1 + t\varepsilon) = -bc$, and therefore $(1 + t\varepsilon)^3 = -1$; i.e., $1 + t\varepsilon = -1$, since it is real. This also implies together with (1) that $b^2 = ac$, $c^2 = ab$, and $a^2 = bc$, and consequently

$$a = b = c.$$

Thus the three equations in the problem are equal, which is impossible. Hence, such a, b, c, t do not exist.

9. We use induction. Since $T_1 = 0$, $T_2 = 1$, $T_3 = 2$, $T_4 = 3$, $T_5 = 5$, $T_6 = 8$, the statement is true for $n = 1, 2, 3$. Suppose that both formulas from the problem hold for some $n \geq 3$. Then

$$T_{2n+1} = 1 + T_{2n} + 2^{n-1} = \left[\frac{17}{7}2^{n-1} + 2^{n-1}\right] = \left[\frac{12}{7}2^n\right],$$

$$T_{2n+2} = 1 + T_{2n-3} + 2^{n+1} = \left[\frac{12}{7}2^{n-2} + 2^{n+1}\right] = \left[\frac{17}{7}2^n\right].$$

Therefore the formulas hold for $n + 1$, which completes the proof.

10. We use induction. Suppose that every two of the numbers $a_1 = 2^{n_1} - 3, a_2 = 2^{n_2} - 3, \ldots, a_k = 2^{n_k} - 3$, where $2 = n_1 < n_2 < \cdots < n_k$, are coprime. Then one can construct $a_{k+1} = 2^{n_{k+1}} - 3$ in the following way:

Set $s = a_1 a_2 \ldots a_k$. Among the numbers $2^0, 2^1, \ldots, 2^s$, two give the same residue upon division by s, say $s \mid 2^\alpha - 2^\beta$. Since s is odd, it can be assumed w.l.o.g. that $\beta = 0$ (this is actually a direct consequence of Euler's theorem). Let $2^\alpha - 1 = qs$, $q \in \mathbb{N}$. Since $2^{\alpha+2} - 3 = 4qs + 1$ is then coprime to s, it is enough to take $n_{k+1} = \alpha + 2$. We obviously have $n_{k+1} > n_k$.

11. We use induction. The statement for $n = 1$ is trivial. Suppose that it holds for $n = k$ and consider $n = k+1$. From the given condition, we have

$$\sum_{j=1}^{k} |a_{j,1}x_1 + \cdots + a_{j,k}x_k + a_{j,k+1}|$$
$$+ |a_{k+1,1}x_1 + \cdots + a_{k+1,k}x_k + a_{k+1,k+1}| \le M,$$
$$\sum_{j=1}^{k} |a_{j,1}x_1 + \cdots + a_{j,k}x_k - a_{j,k+1}|$$
$$+ |a_{k+1,1}x_1 + \cdots + a_{k+1,k}x_k - a_{k+1,k+1}| \le M$$

for each choice of $x_i = \pm 1$. Since $|a+b| + |a-b| \ge 2|a|$ for all a, b, we obtain

$$2\sum_{j=1}^{k} |a_{j1}x_1 + \cdots + a_{jk}x_k| + 2|a_{k+1,k+1}| \le 2M, \text{ that is,}$$
$$\sum_{j=1}^{k} |a_{j1}x_1 + \cdots + a_{jk}x_k| \le M - |a_{k+1,k+1}|.$$

Now by the inductive assumption $\sum_{j=1}^{k} |a_{jj}| \le M - |a_{k+1,k+1}|$, which is equivalent to the desired inequality.

12. Let us start with the case $A = A'$. If the triangles ABC and $A'B'C'$ are oppositely oriented, then they are symmetric with respect to some axis, and the statement is true. Suppose that they are equally oriented. There is a rotation around A by $60°$ that maps ABB' onto ACC'. This rotation also maps the midpoint B_0 of BB' onto the midpoint C_0 of CC', hence the triangle AB_0C_0 is equilateral.

In the general case, when $A \ne A'$, let us denote by T the translation that maps A onto A'. Let X' be the image of a point X under the (unique) isometry mapping ABC onto $A'B'C'$, and X'' the image of X under T. Furthermore, let X_0, X_0' be the midpoints of segments $XX', X'X''$. Then X_0 is the image of X_0' under the translation $-(1/2)T$. However, since it has already been proven that the triangle $A_0'B_0'C_0'$ is equilateral, its image $A_0B_0C_0$ under $(1/2)T$ is also equilateral. The statement of the problem is thus proven.

13. Let p be the least of all the sums of elements in one row or column. If $p \ge n/2$, then the sum of all elements of the array is $s \ge np \ge n^2/2$. Now suppose that $p < n/2$. Without loss of generality, one can assume that the sum of elements in the first row is p, and that exactly the first q elements of it are different from zero. Then the sum of elements in the last $n - q$ columns is greater than or equal to $(n-p)(n-q)$. Furthermore, the sum of elements in the first q columns is greater than or equal to pq. This implies that the sum of all elements in the array is

$$s \geq (n-p)(n-q) + pq = \frac{1}{2}n^2 + \frac{1}{2}(n-2p)(n-2q) \geq \frac{1}{2}n^2,$$

since $n \geq 2p \geq 2q$.

14. Denote by V the figure made by a circle of radius 1 whose center moves along the broken line. From the condition of the problem, V contains the whole 50×50 square, and thus the area $S(V)$ of V is not less than 2500.

Let L be the length of the broken line. We shall show that $S(V) \leq 2L + \pi$, from which it will follow that $L \geq 1250 - \pi/2 > 1248$. For each segment $l_i = A_i A_{i+1}$ of the broken line, consider the figure V_i obtained by a circle of radius 1 whose center moves along it, and let $\overline{V}_i$ be obtained by cutting off the circle of radius 1 with center at the starting point of l_i. The area of $\overline{V}_i$ is equal to $2A_i A_{i+1}$. It is clear that the union of all the figures $\overline{V}_i$ together with a semicircle with center in A_1 and a semicircle with center in A_n contains V completely. Therefore

$$S(V) \leq \pi + 2A_1 A_2 + 2A_2 A_3 + \cdots + 2A_{n-1} A_n = \pi + 2L.$$

This completes the proof.

15. Assume the opposite. Then one can numerate the cards 1 to 99, with a number n_i written on the card i, so that $n_{98} \neq n_{99}$. Denote by x_i the remainder of $n_1 + n_2 + \cdots + n_i$ upon division by 100, for $i = 1, 2, \ldots, 99$. All x_i must be distinct: Indeed, if $x_i = x_j$, $i < j$, then $n_{i+1} + \cdots + n_j$ is divisible by 100, which is impossible. Also, no x_i can be equal to 0. Thus, the numbers $x_1, x_2, \ldots, x_{99}$ take exactly the values $1, 2, \ldots, 99$ in some order.

Let x be the remainder of $n_1 + n_2 + \cdots + n_{97} + n_{99}$ upon division by 100. It is not zero; hence it must be equal to x_k for some $k \in \{1, 2, \ldots, 99\}$. There are three cases:

(i) $x = x_k$, $k \leq 97$. Then $n_{k+1} + n_{k+2} + \cdots + n_{97} + n_{99}$ is divisible by 100, a contradiction;

(ii) $x = x_{98}$. Then $n_{98} = n_{99}$, a contradiction;

(iii) $x = x_{99}$. Then n_{98} is divisible by 100, a contradiction.

Therefore, all the cards contain the same number.

16. Denote by P' the polyhedron defined as the image of P under the homothety with center at A_1 and coefficient of similarity 2. It is easy to see that all P_i, $i = 1, \ldots, 9$, are contained in P' (indeed, if $M \in P_k$, then

$$\frac{1}{2}\overrightarrow{A_1 M} = \frac{1}{2}(\overrightarrow{A_1 A_k} + \overrightarrow{A_1 M'})$$

for some $M' \in P$, and the claim follows from the convexity of P). But the volume of P' is exactly 8 times the volume of P, while the volumes of P_i add up to 9 times that volume. We conclude that not all P_i have disjoint interiors.

17. We use the following obvious consequences of $(a+b)^2 \geq 4ab$:

$$\frac{1}{(a_1 + a_2)(a_3 + a_4)} \geq \frac{4}{(a_1 + a_2 + a_3 + a_4)^2},$$

$$\frac{1}{(a_1+a_4)(a_2+a_3)} \geq \frac{4}{(a_1+a_2+a_3+a_4)^2}.$$

Now we have

$$\begin{aligned}
&\frac{a_1+a_3}{a_1+a_2} + \frac{a_2+a_4}{a_2+a_3} + \frac{a_3+a_1}{a_3+a_4} + \frac{a_4+a_2}{a_4+a_1}\\
&= \frac{(a_1+a_3)(a_1+a_2+a_3+a_4)}{(a_1+a_2)(a_3+a_4)} + \frac{(a_2+a_4)(a_1+a_2+a_3+a_4)}{(a_1+a_4)(a_2+a_3)}\\
&\geq \frac{4(a_1+a_3)}{a_1+a_2+a_3+a_4} + \frac{4(a_2+a_4)}{a_1+a_2+a_3+a_4} = 4.
\end{aligned}$$

4.14 Solutions to the Shortlisted Problems of IMO 1972

1. Suppose that $f(x_0) \neq 0$ and for a given y define the sequence x_k by the formula

$$x_{k+1} = \begin{cases} x_k + y, & \text{if } |f(x_k + y)| \geq |f(x_k - y)|; \\ x_k - y, & \text{otherwise.} \end{cases}$$

It follows from (1) that $|f(x_{k+1})| \geq |\varphi(y)||f(x_k)|$; hence by induction, $|f(x_k)| \geq |\varphi(y)|^k |f(x_0)|$. Since $|f(x_k)| \leq 1$ for all k, we obtain $|\varphi(y)| \leq 1$.

Second solution. Let $M = \sup |f(x)| \leq 1$, and x_k any sequence, possibly constant, such that $|f(x_k)| \to M$, $k \to \infty$. Then for all k,

$$|\varphi(y)| = \frac{|f(x_k + y) + f(x_k - y)|}{2|f(x_k)|} \leq \frac{2M}{2|f(x_k)|} \to 1, \quad k \to \infty.$$

2. We use induction. For $n = 1$ the assertion is obvious. Assume that it is true for a positive integer n. Let $A_1, A_2, \ldots, A_{3n+3}$ be given $3n+3$ points, and let w.l.o.g. $A_1 A_2 \ldots A_m$ be their convex hull.

   Among all the points A_i distinct from A_1, A_2, we choose the one, say A_k, for which the angle $\angle A_k A_1 A_2$ is minimal (this point is uniquely determined, since no three points are collinear). The line $A_1 A_k$ separates the plane into two half-planes, one of which contains A_2 only, and the other one all the remaining $3n$ points. By the inductive hypothesis, one can construct n disjoint triangles with vertices in these $3n$ points. Together with the triangle $A_1 A_2 A_k$, they form the required system of disjoint triangles.

3. We have for each $k = 1, 2, \ldots, n$ that $m \leq x_k \leq M$, which gives $(M - x_k)(m - x_k) \leq 0$. It follows directly that

$$0 \geq \sum_{k=1}^{n} (M - x_k)(m - x_k) = nmM - (m + M) \sum_{k=1}^{n} x_k + \sum_{k=1}^{n} x_k^2.$$

   But $\sum_{k=1}^{n} x_k = 0$, implying the required inequality.

4. Choose in E a half-line s beginning at a point O. For every α in the interval $[0, 180°]$, denote by $s(\alpha)$ the line obtained by rotation of s about O by α, and by $g(\alpha)$ the oriented line containing $s(\alpha)$ on which $s(\alpha)$ defines the positive direction. For each P in M_i, $i = 1, 2$, let $P(\alpha)$ be the foot of the perpendicular from P to $g(\alpha)$, and $l_P(\alpha)$ the oriented (positive, negative or zero) distance of $P(\alpha)$ from O. Then for $i = 1, 2$ one can arrange the $l_P(\alpha)$ $(P \in M_i)$ in ascending order, as $l_1(\alpha), l_2(\alpha), \ldots, l_{2n_i}(\alpha)$. Call $J_i(\alpha)$ the interval $[l_{n_i}(\alpha), l_{n_i+1}(\alpha)]$. It is easy to see that any line perpendicular to $g(\alpha)$ and passing through the point with the distance l in the interior of $J_i(\alpha)$ from O, will divide the set M_i into two subsets of equal cardinality.

   Therefore it remains to show that for some α, the interiors of intervals $J_1(\alpha)$ and $J_2(\alpha)$ have a common point. If this holds for $\alpha = 0$, then we have finished. Suppose w.l.o.g. that $J_1(0)$ lies on $g(0)$ to the left of $J_2(0)$; then $J_1(180°)$ lies to the right of $J_2(180°)$. Note that J_1 and J_2 cannot simultaneously degenerate to a point (otherwise, we would have four collinear points in $M_1 \cup M_2$); also, each of them degenerates to a point for only finitely many values of α. Since $J_1(\alpha)$ and $J_2(\alpha)$ move continuously, there exists a subinterval I of $[0, 180°]$ on which they are not disjoint. Thus, at some point of I, they are both nondegenerate and have a common interior point, as desired.

5. *Lemma.* If X, Y, Z, T are points in space, then the lines XZ and YT are perpendicular if and only if $XY^2 + ZT^2 = YZ^2 + TX^2$.

   Proof. Consider the plane π through XZ parallel to YT. If Y', T' are the feet of the perpendiculars to π from Y, T respectively, then

   $$XY^2 + ZT^2 = XY'^2 + ZT'^2 + 2YY'^2,$$
   $$\text{and} \quad YZ^2 + TX^2 = Y'Z^2 + T'X^2 + 2YY'^2.$$

   Since by the Pythagorean theorem $XY'^2 + ZT'^2 = Y'Z^2 + T'X^2$, i.e., $XY'^2 - Y'Z^2 = XT'^2 - T'Z^2$, if and only if $Y'T' \perp XZ$, the statement follows.

   Assume that the four altitudes intersect in a point P. Then we have $DP \perp ABC \Rightarrow DP \perp AB$ and $CP \perp ABD \Rightarrow CP \perp AB$, which implies that $CDP \perp AB$, and $CD \perp AB$. By the lemma, $AC^2 + BD^2 = AD^2 + BC^2$. Using the same procedure we obtain the relation $AD^2 + BC^2 = AB^2 + CD^2$.

   Conversely, assume that $AB^2 + CD^2 = AC^2 + BD^2 = AD^2 + BC^2$. The lemma implies that $AB \perp CD$, $AC \perp BD$, $AD \perp BC$. Let π be the plane containing CD that is perpendicular to AB, and let h_D be the altitude from D to ABC. Since $\pi \perp AB$, we have $\pi \perp ABC \Rightarrow h_D \subset \pi$ and $\pi \perp ABD \Rightarrow h_C \subset \pi$. The altitudes h_D and h_C are not parallel; thus they have an intersection point P_{CD}. Analogously, $h_B \cap h_C = \{P_{BC}\}$ and $h_B \cap h_D = \{P_{BD}\}$, where both these points belong to π. On the other hand, h_B doesn't belong to π; otherwise, it would be perpendicular to both ACD and $AB \subset \pi$, i.e. $AB \subset ACD$, which is impossible. Hence, h_B can have at most one common point with π, implying $P_{BD} = P_{CD}$. Analogously, $P_{AB} = P_{BD} = P_{CD} = P_{ABCD}$.

6. Let $n = 2^\alpha 5^\beta m$, where $\alpha = 0$ or $\beta = 0$. These two cases are analogous, and we treat only $\alpha = 0$, $n = 5^\beta m$. The case $m = 1$ is settled by the following lemma.

   Lemma. For any integer $\beta \geq 1$ there exists a multiple M_β of 5^β with β digits in decimal expansion, all different from 0.

   Proof. For $\beta = 1$, $M_1 = 5$ works. Assume that the lemma is true for $\beta = k$. There is a positive integer $C_k \leq 5$ such that $C_k 2^k + m_k \equiv 0 \pmod 5$, where $5^k m_k = M_k$, i.e. $C_k 10^k + M_k \equiv 0 \pmod{5^{k+1}}$. Then $M_{k+1} = C_k 10^k + M_k$ satisfies the conditions, and proves the lemma.

   In the general case, consider, the sequence $1, 10^\beta, 10^{2\beta}, \ldots$. It contains two numbers congruent modulo $(10^\beta - 1)m$, and therefore for some $k > 0$, $10^{k\beta} \equiv 1 \pmod{(10^\beta - 1)m}$ (this is in fact a consequence of Fermat's theorem). The number

   $$\frac{10^{k\beta} - 1}{10^\beta - 1} M_\beta = 10^{(k-1)\beta} M_\beta + 10^{(k-2)\beta} M_\beta + \cdots + M_\beta$$

is a multiple of $n = 5^\beta m$ with the required property.

7. (a) Consider the circumscribing cube $OQ_1PR_1O_1QP_1R$ (that is, the cube in which the edges of the tetrahedron are small diagonals), of side $b = a\sqrt{2}/2$. The left-hand side is the sum of squares of the projections of the edges of the tetrahedron onto a perpendicular l to π. On the other hand, if l

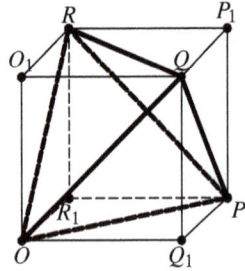

forms angles $\varphi_1, \varphi_2, \varphi_3$ with OO_1, OQ_1, OR_1 respectively, then the projections of OP and QR onto l have lengths $b(\cos\varphi_2 + \cos\varphi_3)$ and $b|\cos\varphi_2 - \cos\varphi_3|$. Summing up all these expressions, we obtain

$$4b^2(\cos^2\varphi_1 + \cos^2\varphi_2 + \cos^2\varphi_3) = 4b^2 = 2a^2.$$

(b) We construct a required tetrahedron of edge length a given in (a). Take O arbitrarily on π_0, and let p, q, r be the distances of O from π_1, π_2, π_3. Since $a > p, q, r, |p - q|$, we can choose P on π_1 anywhere at distance a from O, and Q at one of the two points on π_2 at distance a from both O and P. Consider the fourth vertex of the tetrahedron: its distance from π_0 will satisfy the equation from (a); i.e., there are two values for this distance; clearly, one of them is r, putting R on π_3.

8. Let $f(m,n) = \frac{(2m)!(2n)!}{m!n!(m+n)!}$. Then it is directly shown that

$$f(m,n) = 4f(m, n-1) - f(m+1, n-1),$$

and thus n may be successively reduced until one obtains $f(m,n) = \sum_r c_r f(r,0)$. Now $f(r,0)$ is a simple binomial coefficient, and the c_r's are integers.

Second solution. For each prime p, the greatest exponents of p that divide the numerator $(2m)!(2n)!$ and denominator $m!n!(m+n)!$ are respectively

$$\sum_{k>0}\left(\left[\frac{2m}{p^k}\right] + \left[\frac{2n}{p^k}\right]\right) \quad \text{and} \quad \sum_{k>0}\left(\left[\frac{m}{p^k}\right] + \left[\frac{n}{p^k}\right] + \left[\frac{m+n}{p^k}\right]\right);$$

hence it suffices to show that the first exponent is not less than the second one for every p. This follows from the fact that for each real x, $[2x] + [2y] \geq [x] + [y] + [x+y]$, which is straightforward to prove (for example, using $[2x] = [x] + [x+1/2]$).

9. Clearly $x_1 = x_2 = x_3 = x_4 = x_5$ is a solution. We shall show that this describes all solutions.

Suppose that not all x_i are equal. Then among x_3, x_5, x_2, x_4, x_1 two consecutive are distinct: Assume w.l.o.g. that $x_3 \neq x_5$. Moreover, since $(1/x_1, \ldots, 1/x_5)$ is a solution whenever $(x_1, \ldots, x_5)$ is, we may assume that $x_3 < x_5$.

Consider first the case $x_1 \leq x_2$. We infer from (i) that $x_1 \leq \sqrt{x_3 x_5} < x_5$ and $x_2 \geq \sqrt{x_3 x_5} > x_3$. Then $x_5^2 > x_1 x_3$, which together with (iv) gives $x_4^2 \leq x_1 x_3 < x_3 x_5$; but we also have $x_3^2 \leq x_5 x_2$; hence by (iii), $x_4^2 \geq x_5 x_2 > x_5 x_3$, a contradiction. Consider next the case $x_1 > x_2$. We infer from (i) that $x_1 \geq \sqrt{x_3 x_5} > x_3$ and $x_2 \leq \sqrt{x_3 x_5} < x_5$. Then by (ii) and (v),

$$x_1 x_4 \leq \max(x_2^2, x_3^2) \leq x_3 x_5 \quad \text{and} \quad x_2 x_4 \geq \min(x_1^2, x_5^2) \geq x_3 x_5,$$

which contradicts the assumption $x_1 > x_2$.

Second solution.

$$0 \geq L_1 = (x_1^2 - x_3 x_5)(x_2^2 - x_3 x_5) = x_1^2 x_2^2 + x_3^2 x_5^2 - (x_1^2 + x_2^2)x_3 x_5$$
$$\geq x_1^2 x_2^2 + x_3^2 x_5^2 - \frac{1}{2}(x_1^2 x_3^2 + x_1^2 x_5^2 + x_2^2 x_3^2 + x_2^2 x_5^2),$$

and analogously for $L_2, \ldots, L_5$. Therefore $L_1 + L_2 + L_3 + L_4 + L_5 \geq 0$, with the only case of equality $x_1 = x_2 = x_3 = x_4 = x_5$.

10. Consider first a triangle. It can be decomposed into $k = 3$ cyclic quadrilaterals by perpendiculars from some interior point of it to the sides; also, it can be decomposed into a cyclic quadrilateral and a triangle, and it follows by induction that this decomposition is possible for every k. Since every triangle can be cut into two triangles, the required decomposition is possible for each $n \geq 6$. It remains to treat the cases $n = 4$ and $n = 5$.

$n = 4$. If the center O of the circumcircle is inside a cyclic quadrilateral $ABCD$, then the required decomposition is effected by perpendiculars from O to the four sides. Otherwise, let C and D be the vertices of the obtuse angles of the quadrilateral. Draw the perpendiculars at C and D to the lines BC and AD respectively, and choose points P and Q on them such that $PQ \parallel AB$. Then the required decomposition is effected by CP, PQ, QD and the perpendiculars from P and Q to AB.

$n = 5$. If $ABCD$ is an isosceles trapezoid with $AB \parallel CD$ and $AD = BC$, then it is trivially decomposed by lines parallel to AB. Otherwise, $ABCD$ can be decomposed into a cyclic quadrilateral and a trapezoid; this trapezoid can be cut into an isosceles trapezoid and a triangle, which can further be cut into three cyclic quadrilaterals and an isosceles trapezoid.

Remark. It can be shown that the assertion is not true for $n = 2$ and $n = 3$.

11. Let $\angle A = 2x$, $\angle B = 2y$, $\angle C = 2z$.
    (a) Denote by M_i the center of K_i, $i = 1, 2, \ldots$. If N_1, N_2 are the projections of M_1, M_2 onto AB, we have $AN_1 = r_1 \cot x$, $N_2 B = r_2 \cot y$, and $N_1 N_2 = \sqrt{(r_1 + r_2)^2 - (r_1 - r_2)^2} = 2\sqrt{r_1 r_2}$. The required relation between r_1, r_2 follows from $AB = AN_1 + N_1 N_2 + N_2 B$.
    If this relation is further considered as a quadratic equation in $\sqrt{r_2}$, then its discriminant, which equals

$$\Delta = 4\left(r(\cot x + \cot y)\cot y - r_1(\cot x \cot y - 1)\right),$$

must be nonnegative, and therefore $r_1 \leq r \cot y \cot z$. Then $t_1, t_2, \ldots$ exist, and we can assume that $t_i \in [0, \pi/2]$.

(b) Substituting $r_1 = r \cot y \cot z \sin^2 t_1$, $r_2 = r \cot z \cot x \sin^2 t_2$ in the relation of (a) we obtain that $\sin^2 t_1 + \sin^2 t_2 + k^2 + 2k \sin t_1 \sin t_2 = 1$, where we set $k = \sqrt{\tan x \tan y}$. It follows that $(k + \sin t_1 \sin t_2)^2 = (1 - \sin^2 t_1)(1 - \sin^2 t_2) = \cos^2 t_1 \cos^2 t_2$, and hence

$$\cos(t_1 + t_2) = \cos t_1 \cos t_2 - \sin t_1 \sin t_2 = k = \sqrt{\tan x \tan y},$$

which is constant. Writing the analogous relations for each t_i, t_{i+1} we conclude that $t_1 + t_2 = t_4 + t_5$, $t_2 + t_3 = t_5 + t_6$, and $t_3 + t_4 = t_6 + t_7$. It follows that $t_1 = t_7$, i.e., $K_1 = K_7$.

12. First we observe that it is not essential to require the subsets to be disjoint (if they aren't, one simply excludes their intersection). There are $2^{10} - 1 = 1023$ different subsets and at most 990 different sums. By the pigeonhole principle there are two different subsets with equal sums.

4.15 Solutions to the Shortlisted Problems of IMO 1973

1. The condition of the point P can be written in the form $\frac{AP^2}{AP \cdot PA_1} + \frac{BP^2}{BP \cdot PB_1} + \frac{CP^2}{CP \cdot PC_1} + \frac{DP^2}{DP \cdot PD_1} = 4$. All the four denominators are equal to $R^2 - OP^2$, i.e., to the power of P with respect to S. Thus the condition becomes

$$AP^2 + BP^2 + CP^2 + DP^2 = 4(R^2 - OP^2). \tag{1}$$

Let M and N be the midpoints of segments AB and CD respectively, and G the midpoint of MN, or the centroid of $ABCD$. By Stewart's formula, an arbitrary point P satisfies

$$AP^2 + BP^2 + CP^2 + DP^2 = 2MP^2 + 2NP^2 + \frac{1}{2}AB^2 + \frac{1}{2}CD^2$$
$$= 4GP^2 + MN^2 + \frac{1}{2}(AB^2 + CD^2).$$

Particularly, for $P \equiv O$ we get $4R^2 = 4OG^2 + MN^2 + \frac{1}{2}(AB^2 + CD^2)$, and the above equality becomes

$$AP^2 + BP^2 + CP^2 + DP^2 = 4GP^2 + 4R^2 - 4OG^2.$$

Therefore (1) is equivalent to $OG^2 = OP^2 + GP^2 \Leftrightarrow \angle OPG = 90°$. Hence the locus of points P is the sphere with diameter OG. Now the converse is easy.

2. Let D' be the reflection of D across A. Since $BCAD'$ is then a parallelogram, the condition $BD \geq AC$ is equivalent to $BD \geq BD'$, which is in turn equivalent to $\angle BAD \geq \angle BAD'$, i.e. to $\angle BAD \geq 90°$. Thus the needed locus is actually the locus of points A for which there exist points B, D inside K with $\angle BAD = 90°$. Such points B, D exist if and only if the two tangents from A to K, say AP and AQ, determine an obtuse angle. Then if $P, Q \in K$, we have $\angle PAO = \angle QAO = \varphi > 45°$; hence $OA = \frac{OP}{\sin \varphi} < OP\sqrt{2}$. Therefore the locus of A is the interior of the circle K' with center O and radius $\sqrt{2}$ times the radius of K.

3. We use induction on odd numbers n. For $n = 1$ there is nothing to prove. Suppose that the result holds for $n - 2$ vectors, and let us be given vectors $v_1, v_2, \ldots, v_n$ arranged clockwise. Set $v' = v_2 + v_3 + \cdots + v_{n-1}$, $u = v_1 + v_n$, and $v = v_1 + v_2 + \cdots + v_n = v' + u$. By the inductive hypothesis we have $|v'| \geq 1$. Now if the angles between v' and the vectors v_1, v_n are α and β respectively, then the angle between u and v' is $|\alpha - \beta|/2 \leq 90°$. Hence $|v' + u| \geq |v'| \geq 1$.

Second solution. Again by induction, it can be easily shown that all possible values of the sum $v = v_1 + v_2 + \cdots + v_n$, for n vectors $v_1, \ldots, v_n$ in the upper half-plane (with $y \geq 0$), are those for which $|v| \leq n$ and $|v - ke| \geq 1$ for every integer k for which $n - k$ is odd, where e is the unit vector on the x axis.

4. Each of the subsets must be of the form $\{a^2, ab, ac, ad\}$ or $\{a^2, ab, ac, bc\}$. It is now easy to count up the partitions. The result is 26460.

5. Let O be the vertex of the trihedron, Z the center of a circle k inscribed in the trihedron, and A, B, C points in which the plane of the circle meets the edges of the trihedron. We claim that the distance OZ is constant.

Set $OA = x$, $OB = y$, $OC = z$, $BC = a$, $CA = b$, $AB = c$, and let S and $r = 1$ be the area and inradius of $\triangle ABC$. Since Z is the incenter of ABC, we have $(a+b+c)\overrightarrow{OZ} = a\overrightarrow{OA} + b\overrightarrow{OB} + c\overrightarrow{OC}$. Hence

$$(a+b+c)^2 OZ^2 = (a\overrightarrow{OA} + b\overrightarrow{OB} + c\overrightarrow{OC})^2 = a^2 x^2 + b^2 y^2 + c^2 z^2. \qquad (1)$$

But since $y^2 + z^2 = a^2$, $z^2 + x^2 = b^2$ and $x^2 + y^2 = c^2$, we obtain $x^2 = \frac{-a^2 + b^2 + c^2}{2}$, $y^2 = \frac{a^2 - b^2 + c^2}{2}$, $z^2 = \frac{a^2 + b^2 - c^2}{2}$. Substituting these values in (1) yields

$$(a+b+c)^2 OZ^2 = \frac{2a^2 b^2 + 2b^2 c^2 + 2c^2 a^2 - a^4 - b^4 - c^4}{2}$$

$$= 8S^2 = 2(a+b+c)^2 r^2.$$

Hence $OZ = r\sqrt{2} = \sqrt{2}$, and Z belongs to a sphere σ with center O and radius $\sqrt{2}$.

Moreover, the distances of Z from the faces of the trihedron do not exceed 1; hence Z belongs to a part of σ that lies inside the unit cube with three faces lying on the faces of the trihedron. It is easy to see that this part of σ is exactly the required locus.

6. Yes. Take for $\mathscr{M}$ the set of vertices of a cube $ABCDEFGH$ and two points I, J symmetric to the center O of the cube with respect to the laterals $ABCD$ and $EFGH$.

Remark. We prove a stronger result: Given an arbitrary finite set of points $\mathscr{S}$, then there is a finite set $\mathscr{M} \supset \mathscr{S}$ with the described property. Choose a point $A \in \mathscr{S}$ and any point O such that $AO \parallel BC$ for some two points $B, C \in \mathscr{S}$. Now let X' be the point symmetric to X with respect to O, and $\mathscr{S}' = \{X, X' \mid X \in S\}$. Finally, take $\mathscr{M} = \{X, \overline{X} \mid X \in S'\}$, where $\overline{X}$ denotes the point symmetric to X with respect to A. This $\mathscr{M}$ has the desired property: If $X, Y \in \mathscr{M}$ and $Y \neq \overline{X}$, then $XY \parallel \overline{X}\overline{Y}$; otherwise, $X\overline{X}$, i.e., XA is parallel to $X'A'$ if $X \neq A'$, or to BC otherwise.

7. The result follows immediately from Ptolemy's inequality.

8. Let f_n be the required total number, and let $f_n(k)$ denote the number of sequences $a_1, \ldots, a_n$ of nonnegative integers such that $a_1 = 0$, $a_n = k$, and $|a_i - a_{i+1}| = 1$ for $i = 1, \ldots, n-1$. In particular, $f_1(0) = 1$ and $f_n(k) = 0$ if $k < 0$ or $k \geq n$. Since a_{n-1} is either $k-1$ or $k+1$, we have

$$f_n(k) = f_{n-1}(k+1) + f_{n-1}(k-1) \quad \text{for } k \geq 1. \qquad (1)$$

By successive application of (1) we obtain

$$f_n(k) = \sum_{i=0}^{r} \left[\binom{r}{i} - \binom{r}{i-k-1} \right] f_{n-r}(k+r-2i). \qquad (2)$$

This can be verified by direct induction. Substituting $r = n-1$ in (2), we get at most one nonzero summand, namely the one for which $i = \frac{k+n-1}{2}$. Therefore $f_n(n-1-2j) = \binom{n-1}{j} - \binom{n-1}{j-1}$. Adding up these equalities for $j = 0, 1, \ldots, \left[\frac{n-1}{2}\right]$ we obtain $f_n = \binom{n-1}{\left[\frac{n-1}{2}\right]}$, as required.

9. Let a, b, c be vectors going along Ox, Oy, Oz, respectively, such that $\overrightarrow{OG} = a + b + c$. Now let $A \in Ox$, $B \in Oy$, $C \in Oz$ and let $\overrightarrow{OA} = \alpha a$, $\overrightarrow{OB} = \beta b$, $\overrightarrow{OC} = \gamma c$, where $\alpha, \beta, \gamma > 0$. Point G belongs to a plane ABC with $A \in Ox$, $B \in Oy$, $C \in Oz$ if and only if there exist positive real numbers λ, μ, ν with sum 1 such that $\lambda \overrightarrow{OA} + \mu \overrightarrow{OB} + \nu \overrightarrow{OC} = \overrightarrow{OG}$, which is equivalent to $\lambda \alpha = \mu \beta = \nu \gamma = 1$. Such λ, μ, ν exist if and only if

$$\alpha, \beta, \gamma > 0 \quad \text{and} \quad \frac{1}{\alpha} + \frac{1}{\beta} + \frac{1}{\gamma} = 1.$$

Since the volume of $OABC$ is proportional to the product $\alpha\beta\gamma$, it is minimized when $\frac{1}{\alpha} \cdot \frac{1}{\beta} \cdot \frac{1}{\gamma}$ is maximized, which occurs when $\alpha = \beta = \gamma = 3$ and G is the centroid of $\triangle ABC$.

10. Let

$$b_k = a_1 q^{k-1} + \cdots + a_{k-1}q + a_k + a_{k+1}q + \cdots + a_n q^{n-k}, \quad k = 1, 2, \ldots, n.$$

We show that these numbers satisfy the required conditions. Obviously $b_k > a_k$. Further,

$$b_{k+1} - q b_k = -[(q^2-1)a_{k+1} + \cdots + q^{n-k-1}(q^2-1)a_n] > 0;$$

we analogously obtain $q b_{k+1} - b_k < 0$. Finally,

$$\begin{aligned}
b_1 + b_2 + \cdots + b_n &= a_1(q^{n-1} + \cdots + q + 1) + \cdots \\
&\quad + a_k(q^{n-k} + \cdots + q + 1 + q + \cdots + q^{k-1}) + \cdots \\
&\leq (a_1 + a_2 + \cdots + a_n)(1 + 2q + 2q^2 + \cdots + 2q^{n-1}) \\
&< \frac{1+q}{1-q}(a_1 + \cdots + a_n).
\end{aligned}$$

11. Putting $x + \frac{1}{x} = t$ we also get $x^2 + \frac{1}{x^2} = t^2 - 2$, and the given equation reduces to $t^2 + at + b - 2 = 0$. Since $x = \frac{t \pm \sqrt{t^2-4}}{2}$, x will be real if and only if $|t| \geq 2$, $t \in \mathbb{R}$. Thus we need the minimum value of $a^2 + b^2$ under the condition $at + b = -(t^2 - 2)$, $|t| \geq 2$.

However, by the Cauchy–Schwarz inequality we have

$$(a^2 + b^2)(t^2 + 1) \geq (at + b)^2 = (t^2 - 2)^2.$$

It follows that $a^2 + b^2 \geq h(t) = \frac{(t^2-2)^2}{t^2+1}$. Since $h(t) = (t^2 + 1) + \frac{9}{t^2+1} - 6$ is increasing for $t \geq 2$, we conclude that $a^2 + b^2 \geq h(2) = \frac{4}{5}$.

The cases of equality are easy to examine: These are $a = \pm\frac{4}{5}$ and $b = -\frac{2}{5}$.

Second solution. In fact, there was no need for considering $x = t + 1/t$. By the Cauchy–Schwarz inequality we have $(a^2 + 2b^2 + a^2)(x^6 + x^4/2 + x^2) \geq (ax^3 + bx^2 + ax)^2 = (x^4 + 1)^2$. Hence

$$a^2 + b^2 \geq \frac{(x^4 + 1)^2}{2x^6 + x^4 + 2x^2} \geq \frac{4}{5},$$

with equality for $x = 1$.

12. Observe that the absolute values of the determinants of the given matrices are invariant under all the admitted operations. The statement follows from $\det A = 16 \neq \det B = 0$.

13. Let S_1, S_2, S_3, S_4 denote the areas of the faces of the tetrahedron, V its volume, h_1, h_2, h_3, h_4 its altitudes, and r the radius of its inscribed sphere. Since

$$3V = S_1 h_1 = S_2 h_2 = S_3 h_3 = S_4 h_4 = (S_1 + S_2 + S_3 + S_4)r,$$

it follows that

$$\frac{1}{h_1} + \frac{1}{h_2} + \frac{1}{h_3} + \frac{1}{h_4} = \frac{1}{r}.$$

In our case, $h_1, h_2, h_3, h_4 \geq 1$, hence $r \geq 1/4$. On the other hand, it is clear that a sphere of radius greater than $1/4$ cannot be inscribed in a tetrahedron all of whose altitudes have length equal to 1. Thus the answer is $1/4$.

14. Suppose that the soldier starts at the vertex A of the equilateral triangle ABC of side length a. Let φ, ψ be the arcs of circles with centers B and C and radii $a\sqrt{3}/4$ respectively, that lie inside the triangle. In order to check the vertices B, C, he must visit some points $D \in \varphi$ and $E \in \psi$. Thus his path cannot be shorter than the path ADE (or AED) itself.

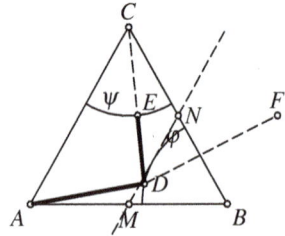

The length of the path ADE is $AD + DE \geq AD + DC - a\sqrt{3}/4$. Let F be the reflection of C across the line MN, where M, N are the midpoints of AB and BC. Then $DC \geq DF$ and hence $AD + DC \geq AD + DF \geq AF$. Consequently

$$AD + DE \geq AF - a\frac{\sqrt{3}}{4} = a\left(\frac{\sqrt{7}}{2} - \frac{\sqrt{3}}{4}\right),$$

with equality if and only if D is the midpoint of arc φ and $E = (CD) \cap \psi$. Moreover, it is easy to verify that, in following the path ADE, the soldier will check the whole region. Therefore this path (as well as the one symmetric to it) is the shortest possible path that the soldier can take in order to check the entire field.

15. If $z = \cos\theta + i\sin\theta$, then $z - z^{-1} = 2i\sin\theta$. Now put $z = \cos\frac{\pi}{2n+1} + i\sin\frac{\pi}{2n+1}$. Using de Moivre's formula we transform the required equality into

$$A = \prod_{k=1}^{n}(z^k - z^{-k}) = i^n\sqrt{2n+1}. \tag{1}$$

On the other hand, the complex numbers z^{2k} $(k = -n, -n+1, \ldots, n)$ are the roots of $x^{2n+1} - 1$, and hence

$$\prod_{k=1}^{n}(x - z^{2k})(x - z^{-2k}) = \frac{x^{2n+1} - 1}{x - 1} = x^{2n} + \cdots + x + 1. \tag{2}$$

Now we go back to proving (1). We have

$$(-1)^n z^{n(n+1)/2}A = \prod_{k=1}^{n}(1 - z^{2k}) \quad\text{and}\quad z^{-n(n+1)/2}A = \prod_{k=1}^{n}(1 - z^{-2k}).$$

Multiplying these two equalities, we get $(-1)^n A^2 = \prod_{k=1}^{n}(1 - z^{2k})(1 - z^{-2k}) = 2n+1$, by (2). Therefore $A = \pm i^{-n}\sqrt{2n+1}$. This actually implies that the required product is $\pm\sqrt{2n+1}$, but it must be positive, since all the sines are, and the result follows.

16. First, we have $P(x) = Q(x)R(x)$ for $Q(x) = x^m - |a|^m e^{i\theta}$ and $R(x) = x^m - |a|^m e^{-i\theta}$, where $e^{i\varphi}$ means of course $\cos\varphi + i\sin\varphi$. It remains to factor both Q and R. Suppose that $Q(x) = (x - q_1)\cdots(x - q_m)$ and $R(x) = (x - r_1)\cdots(x - r_m)$. Considering $Q(x)$, we see that $|q_k^m| = |a|^m$ and also $|q_k| = |a|$ for $k = 1, \ldots, m$. Thus we may put $q_k = |a|e^{i\beta_k}$ and obtain by de Moivre's formula $q_k^m = |a|^m e^{im\beta_k}$. It follows that $m\beta_k = \theta + 2j\pi$ for some $j \in \mathbb{Z}$, and we have exactly m possibilities for β_k modulo 2π: $\beta_k = \frac{\theta + 2(k-1)\pi}{m}$ for $k = 1, 2, \ldots, m$.

Thus $q_k = |a|e^{i\beta_k}$; analogously we obtain for $R(x)$ that $r_k = |a|e^{-i\beta_k}$. Consequently,

$$x^m - |a|^m e^{i\theta} = \prod_{k=1}^{m}(x - |a|e^{i\beta_k}) \quad\text{and}\quad x^m - |a|^m e^{-i\theta} = \prod_{k=1}^{m}(x - |a|e^{-i\beta_k}).$$

Finally, grouping the kth factors of both polynomials, we get

$$P(x) = \prod_{k=1}^{m}(x - |a|e^{i\beta_k})(x - |a|e^{-i\beta_k}) = \prod_{k=1}^{m}(x^2 - 2|a|x\cos\beta_k + a^2)$$

$$= \prod_{k=1}^{m}\left(x^2 - 2|a|x\cos\frac{\theta + 2(k-1)\pi}{m} + a^2\right).$$

17. Let $f_1(x) = ax + b$ and $f_2(x) = cx + d$ be two functions from $\mathscr{F}$. We define

$$g(x) = f_1 \circ f_2(x) = acx + (ad + b) \quad\text{and}\quad h(x) = f_2 \circ f_1(x) = acx + (bc + d).$$

By the condition for $\mathscr{F}$, both $g(x)$ and $h(x)$ belong to $\mathscr{F}$. Moreover, there exists $h^{-1}(x) = \frac{x - (bc + d)}{ac}$, and

$$h^{-1} \circ g(x) = \frac{acx + (ad+b) - (bc+d)}{ac} = x + \frac{(ad+b) - (bc+d)}{ac}$$

belongs to $\mathscr{F}$. Now it follows that we must have $ad + b = bc + d$ for every $f_1, f_2 \in \mathscr{F}$, which is equivalent to $\frac{b}{1-a} = \frac{d}{1-c} = k$. But these formulas exactly describe the fixed points of f_1 and f_2: $f_1(x) = ax + b = x \Rightarrow x = \frac{b}{1-a}$. Hence all the functions in $\mathscr{F}$ fix the point k.

4.16 Solutions to the Shortlisted Problems of IMO 1974

1. Denote by n the number of exams. We have $n(A+B+C) = 20+10+9 = 39$, and since A, B, C are distinct, their sum is at least 6; therefore $n = 3$ and $A+B+C = 13$.

   Assume w.l.o.g. that $A > B > C$. Since Betty gained A points in arithmetic, but fewer than 13 points in total, she had C points in both remaining exams (in spelling as well). Furthermore, Carol also gained fewer than 13 points, but with at least B points on two examinations (on which Betty scored C), including spelling. If she had A in spelling, then she would have at least $A+B+C = 13$ points in total, a contradiction. Hence, Carol scored B and placed second in spelling.

   Remark. Moreover, it follows that Alice, Betty, and Carol scored $B+A+A$, $A+C+C$, and $C+B+B$ respectively, and that $A = 8$, $B = 4$, $C = 1$.

2. We denote by q_i the square with side $\frac{1}{i}$. Let us divide the big square into rectangles r_i by parallel lines, where the size of r_i is $\frac{3}{2} \times \frac{1}{2^i}$ for $i = 2, 3, \ldots$ and $\frac{3}{2} \times 1$ for $i = 1$ (this can be done because $1 + \sum_{i=2}^{\infty} \frac{1}{2^i} = \frac{3}{2}$). In rectangle r_1, one can put the squares q_1, q_2, q_3, as is done on the figure. Also, since $\frac{1}{2^i} + \cdots + \frac{1}{2^{i+1}-1} < 2^i \cdot \frac{1}{2^i} = 1 < \frac{3}{2}$, in each r_i, $i \geq 2$, one can put $q_{2^i}, \ldots, q_{2^{i+1}-1}$. This completes the proof.

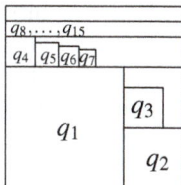

   Remark. It can be shown that the squares q_1, q_2 cannot fit in any square of side less than $\frac{3}{2}$.

3. For $\deg(P) \leq 2$ the statement is obvious, since $n(P) \leq \deg(P^2) = 2\deg(P) \leq \deg(P) + 2$.

   Suppose now that $\deg(P) \geq 3$ and $n(P) > \deg(P) + 2$. Then there is at least one integer b for which $P(b) = -1$, and at least one x with $P(x) = 1$. We may assume w.l.o.g. that $b = 0$ (if necessary, we consider the polynomial $P(x+b)$ instead). If $k_1, \ldots, k_m$ are all integers for which $P(k_i) = 1$, then $P(x) = Q(x)(x - k_1) \cdots (x - k_m) + 1$ for some polynomial $Q(x)$ with integer coefficients. Setting $x = 0$ we obtain $(-1)^m Q(0)k_1 \cdots k_m = 1 - P(0) = 2$. It follows that $k_1 \cdots k_m \mid 2$, and hence m is at most 3. The same holds for the polynomial $-P(x)$, and thus $P(x) = -1$ also has at most 3 integer solutions. This counts for 6 solutions of $P^2(x) = 1$ in total, implying the statement for $\deg(P) \geq 4$.

   It remains to verify the statement for $n = 3$. If $\deg(P) = 3$ and $n(P) = 6$, then it follows from the above consideration that $P(x)$ is either $-(x^2 - 1)(x - 2) + 1$ or $(x^2 - 1)(x+2) + 1$. It is directly checked that $n(P)$ equals only 4 in both cases.

4. Assume w.l.o.g. that $a_1 \leq a_2 \leq a_3 \leq a_4 \leq a_5$. If m is the least value of $|a_i - a_j|$, $i \neq j$, then $a_{i+1} - a_i \geq m$ for $i = 1, 2, \ldots, 5$, and consequently $a_i - a_j \geq (i - j)m$ for any $i, j \in \{1, \ldots, 5\}$, $i > j$. Then it follows that

$$\sum_{i>j}(a_i - a_j)^2 \geq m^2 \sum_{i>j}(i - j)^2 = 50m^2.$$

On the other hand, by the condition of the problem,

$$\sum_{i>j}(a_i - a_j)^2 = 5 \sum_{i=1}^{5} a_i^2 - (a_1 + \cdots + a_5)^2 \leq 5.$$

Therefore $50m^2 \leq 5$; i.e., $m^2 \leq \frac{1}{10}$.

5. All the angles are assumed to be oriented and measured modulo $180°$. Denote by $\alpha_i, \beta_i, \gamma_i$ the angles of triangle $\triangle_i$, at A_i, B_i, C_i respectively. Let us determine the angles of $\triangle_{i+1}$. If D_i is the intersection of lines $B_i B_{i+1}$ and $C_i C_{i+1}$, we have $\angle B_{i+1} A_{i+1} C_{i+1} = \angle D_i B_i C_{i+1} = \angle B_i D_i C_{i+1} + \angle D_i C_{i+1} B_i = \angle B_i D_i C_i - \angle B_i C_{i+1} C_i = -2\angle B_i A_i C_i$. We conclude that

$$\alpha_{i+1} = -2\alpha_i, \quad \text{and analogously} \quad \beta_{i+1} = -2\beta_i, \quad \gamma_{i+1} = -2\gamma_i.$$

Therefore $\alpha_{r+t} = (-2)^t \alpha_r$. However, since $(-2)^{12} \equiv 1 \pmod{45}$ and consequently $(-2)^{14} \equiv (-2)^2 \pmod{180}$, it follows that $\alpha_{15} = \alpha_3$, since all values are modulo $180°$. Analogously, $\beta_{15} = \beta_3$ and $\gamma_{15} = \gamma_3$, and moreover, $\triangle_3$ and $\triangle_{15}$ are inscribed in the same circle; hence $\triangle_3 \cong \triangle_{15}$.

6. We set

$$x = \sum_{k=0}^{n} \binom{2n+1}{2k+1} 2^{3k} = \frac{1}{\sqrt{8}} \sum_{k=0}^{n} \binom{2n+1}{2k+1} \sqrt{8}^{2k+1},$$

$$y = \sum_{k=0}^{n} \binom{2n+1}{2k} 2^{3k} = \sum_{k=0}^{n} \binom{2n+1}{2k} \sqrt{8}^{2k}.$$

Both x and y are positive integers. Also, from the binomial formula we obtain

$$y + x\sqrt{8} = \sum_{i=0}^{2n+1} \binom{2n+1}{i} \sqrt{8}^{i} = (1 + \sqrt{8})^{2n+1},$$

and similarly $y - x\sqrt{8} = (1 - \sqrt{8})^{2n+1}$.

Multiplying these equalities, we get $y^2 - 8x^2 = (1 + \sqrt{8})^{2n+1}(1 - \sqrt{8})^{2n+1} = -7^{2n+1}$. Reducing modulo 5 gives us

$$3x^2 - y^2 \equiv 2^{2n+1} \equiv 2 \cdot (-1)^n.$$

Now we see that if x is divisible by 5, then $y^2 \equiv \pm 2 \pmod{5}$, which is impossible. Therefore x is never divisible by 5.

Second solution. Another standard way is considering recurrent formulas. If we set

$$x_m = \sum_k \binom{m}{2k+1} 8^k, \qquad y_m = \sum_k \binom{m}{2k} 8^k,$$

then since $\binom{a}{b} = \binom{a-1}{b} + \binom{a-1}{b-1}$, it follows that $x_{m+1} = x_m + y_m$ and $y_{m+1} = 8x_m + y_m$; therefore $x_{m+1} = 2x_m + 7x_{m-1}$. We need to show that none of x_{2n+1} are divisible by 5. Considering the sequence $\{x_m\}$ modulo 5, we get that $x_m = 0, 1, 2, 1, 1, 4, 0, 3, 1, 3, 3, 2, 0, 4, 3, 4, 4, 1, \ldots$. Zeros occur in the initial position of blocks of length 6, where each subsequent block is obtained by multiplying the previous one by 3 (modulo 5). Consequently, x_m is divisible by 5 if and only if m is a multiple of 6, which cannot happen if $m = 2n + 1$.

7. Consider an arbitrary prime number p. If $p \mid m$, then there exists b_i that is divisible by the same power of p as m. Then p divides neither $a_i \frac{m}{b_i}$ nor a_i, because $(a_i, b_i) = 1$. If otherwise $p \nmid m$, then $\frac{m}{b_i}$ is not divisible by p for any i, hence p divides a_i and $a_i \frac{m}{b_i}$ to the same power. Therefore $(a_1, \ldots, a_k)$ and $\left(a_1 \frac{m}{b_1}, \ldots, a_k \frac{m}{b_k}\right)$ have the same factorization; hence they are equal.

Second solution. For $k = 2$ we can easily verify the formula

$$\left(m\frac{a_1}{b_1}, m\frac{a_2}{b_2}\right) = \frac{m}{b_1 b_2}(a_1 b_2, a_2 b_1) = \frac{1}{b_1 b_2}[b_1, b_2](a_1, a_2)(b_1, b_2) = (a_1, a_2),$$

since $[b_1, b_2] \cdot (b_1, b_2) = b_1 b_2$. We proceed by induction:

$$\left(a_1 \frac{m}{b_1}, \ldots, a_k \frac{m}{b_k}, a_{k+1} \frac{m}{b_{k+1}}\right) = \left(\frac{m}{[b_1, \ldots, b_k]}(a_1, \ldots, a_k), a_{k+1} \frac{m}{b_{k+1}}\right)$$
$$= (a_1, \ldots, a_k, a_{k+1}).$$

8. It is clear that

$$\frac{a}{a+b+c+d} + \frac{b}{a+b+c+d} + \frac{c}{a+b+c+d} + \frac{d}{a+b+c+d} < S$$

$$\text{and } S < \frac{a}{a+b} + \frac{b}{a+b} + \frac{c}{c+d} + \frac{d}{c+d},$$

or equivalently, $1 < S < 2$.

On the other hand, all values from $(1, 2)$ are attained. Since $S = 1$ for $(a, b, c, d) = (0, 0, 1, 1)$ and $S = 2$ for $(a, b, c, d) = (0, 1, 0, 1)$, due to continuity all the values from $(1, 2)$ are obtained, for example, for $(a, b, c, d) = (x(1 - x), x, 1 - x, 1)$, where x goes through $(0, 1)$.

Second solution. Set

$$S_1 = \frac{a}{a+b+d} + \frac{c}{b+c+d} \qquad \text{and} \qquad S_2 = \frac{b}{a+b+c} + \frac{d}{a+c+d}.$$

We may assume without loss of generality that $a + b + c + d = 1$. Putting $a + c = x$ and $b + d = y$ (then $x + y = 1$), we obtain that the set of values of

$$S_1 = \frac{a}{1-c} + \frac{c}{1-a} = \frac{2ac+x-x^2}{ac+1-x}$$

is $\left(x, \frac{2x}{2-x}\right]$. Having the analogous result for S_2 in mind, we conclude that the values that $S = S_1 + S_2$ can take are $\left(x+y, \frac{2x}{2-x} + \frac{2y}{2-y}\right]$. Since $x+y=1$ and

$$\frac{2x}{2-x} + \frac{2y}{2-y} = \frac{4-4xy}{2+xy} \leq 2$$

with equality for $xy = 0$, the desired set of values for S is $(1,2)$.

9. There exist real numbers a,b,c with $\tan a = x$, $\tan b = y$, $\tan c = z$. Then using the additive formula for tangents we obtain

$$\tan(a+b+c) = \frac{x+y+z-xyz}{1-xy-xz-yz}.$$

We are given that $xyz = x+y+z$. In this case $xy+yz+zx = 1$ is impossible; otherwise, x,y,z would be the zeros of a cubic polynomial $t^3 - \lambda t^2 + t - \lambda = (t^2+1)(t-\lambda)$ (where $\lambda = xyz$), which has only one real root. It follows that

$$x+y+z = xyz \iff \tan(a+b+c) = 0. \tag{1}$$

Hence $a+b+c = k\pi$ for some $k \in \mathbb{Z}$. We note that $\frac{3x-x^3}{1-3x^2}$ actually expresses $\tan 3a$. Since $3a + 3b + 3c = 3k\pi$, the result follows from (1) for the numbers $\frac{3x-x^3}{1-3x^2}, \frac{3y-y^3}{1-3y^2}, \frac{3z-z^3}{1-3z^2}$.

10. If we set $\angle ACD = \gamma_1$ and $\angle BCD = \gamma_2$ for a point D on the segment AB, then by the sine theorem,

$$f(D) = \frac{CD^2}{AD \cdot BD} = \frac{CD}{AD} \cdot \frac{CD}{BD} = \frac{\sin\alpha \sin\beta}{\sin\gamma_1 \sin\gamma_2}.$$

The denominator of the last fraction is

$$\sin\gamma_1 \sin\gamma_2 = \frac{1}{2}(\cos(\gamma_1 - \gamma_2) - \cos(\gamma_1 + \gamma_2))$$
$$= \frac{1}{2}(\cos(\gamma_1 - \gamma_2) - \cos\gamma) \leq \frac{1 - \cos\gamma}{2} = \sin^2\frac{\gamma}{2}.$$

Now we deduce that the set of values of $f(D)$ is the interval $\left[\frac{\sin\alpha\sin\beta}{\sin^2\frac{\gamma}{2}}, +\infty\right)$. Hence $f(D) = 1$ (equivalently, $CD^2 = AD \cdot BD$) is possible if and only if $\sin\alpha\sin\beta \leq \sin^2\frac{\gamma}{2}$, i.e.,

$$\sqrt{\sin\alpha\sin\beta} \leq \sin\frac{\gamma}{2}.$$

Second solution. Let E be the second point of intersection of the line CD with the circumcircle k of ABC. Since $AD \cdot BD = CD \cdot ED$ (power of D with respect to k),

$CD^2 = AD \cdot BD$ ie equivalent to $ED = CD$. Clearly the ratio $\frac{ED}{CD}$ $(D \in AB)$ takes a maximal value when E is the midpoint of the arc AB not containing C. (This follows from $ED : CD = E'D : C'D$ when C' and E' are respectively projections from C and E onto AB.) On the other hand, it is directly shown that in this case

$$\frac{ED}{CD} = \frac{\sin^2 \frac{\gamma}{2}}{\sin \alpha \sin \beta},$$

and the assertion follows.

11. First, we notice that $a_1 + a_2 + \cdots + a_p = 32$. The numbers a_i are distinct, and consequently $a_i \geq i$ and $a_1 + \cdots + a_p \geq p(p+1)/2$. Therefore $p \leq 7$.

The number 32 can be represented as a sum of 7 mutually distinct positive integers in the following ways:

(1) $32 = 1+2+3+4+5+6+11;$
(2) $32 = 1+2+3+4+5+7+10;$
(3) $32 = 1+2+3+4+5+8+9;$
(4) $32 = 1+2+3+4+6+7+9;$
(5) $32 = 1+2+3+5+6+7+8.$

The case (1) is eliminated because there is no rectangle with 22 cells on an 8×8 chessboard. In the other cases the partitions are realized as below.

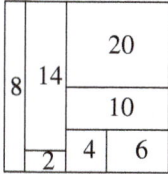   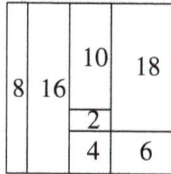      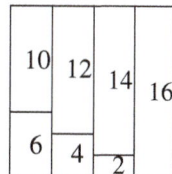

Case (2)   Case (3)   Case (4)   Case (5)

12. We say that a word is *good* if it doesn't contain any nonallowed word. Let a_n be the number of good words of length n. If we prolong any good word of length n by adding one letter to its end (there are $3a_n$ words that can be so obtained), we get either

(i) a good word of length $n+1$, or
(ii) an $(n+1)$-letter word of the form XY, where X is a good word and Y a nonallowed word.

The number of words of type (ii) with word Y of length k is exactly a_{n+1-k}; hence the total number of words of kind (ii) doesn't exceed $a_{n-1} + \cdots + a_1 + a_0$ (where $a_0 = 1$). Hence

$$a_{n+1} \geq 3a_n - (a_{n-1} + \cdots + a_1 + a_0), \qquad a_0 = 1, \ a_1 = 3. \qquad (1)$$

We prove by induction that $a_{n+1} > 2a_n$ for all n. For $n = 1$ the claim is trivial. If it holds for $i \leq n$, then $a_i \leq 2^{i-n} a_n$; thus we obtain from (1)

$$a_{n+1} > a_n \left(3 - \frac{1}{2} - \frac{1}{2^2} - \cdots - \frac{1}{2^n} \right) > 2a_n.$$

Therefore $a_n \geq 2^n$ for all n (moreover, one can show from (1) that $a_n \geq (n+2)2^{n-1}$); hence there exist good words of length n.

Remark. If there are two nonallowed words (instead of one) of each length greater than 1, the statement of the problem need not remain true.

4.17 Solutions to the Shortlisted Problems of IMO 1975

1. First, we observe that there cannot exist three routes of the form (A,B,C), (A,B,D), (A,C,D), for if E,F are the remaining two ports, there can be only one route covering A,E, namely, (A,E,F). Thus if (A,B,C), (A,B,D) are two routes, the one covering A,C must be w.l.o.g. (A,C,E). The other roots are uniquely determined: These are (A,D,F), (A,E,F), (B,D,E), (B,E,F), (B,C,F), (C,D,E), (C,D,F).

2. Since there are finitely many arrangements of the z_i's, assume that $z_1, \ldots, z_n$ is the one for which $\sum_{i=1}^n (x_i - z_i)^2$ is minimal. We claim that in this case if $i < j$ and $x_i \neq x_j$ then $z_i \geq z_j$, from which the claim of the problem directly follows. Indeed, otherwise we would have

$$
\begin{aligned}
(x_i - z_j)^2 + (x_j - z_i)^2 &= (x_i - z_i)^2 + (x_j - z_j)^2 \\
&\quad + 2(x_i z_i + x_j z_j - x_i z_j - x_j z_i) \\
&= (x_i - z_i)^2 + (x_j - z_j)^2 + 2(x_i - x_j)(z_i - z_j) \\
&\leq (x_i - z_i)^2 + (x_j - z_j)^2,
\end{aligned}
$$

contradicting the assumption.

3. From $\left((k+1)^{2/3} + (k+1)^{1/3}k^{1/3} + k^{2/3}\right)\left((k+1)^{1/3} - k^{1/3}\right) = 1$ and $3k^{2/3} < (k+1)^{2/3} + (k+1)^{1/3}k^{1/3} + k^{2/3} < 3(k+1)^{2/3}$ we obtain

$$
3\left((k+1)^{1/3} - k^{1/3}\right) < k^{-2/3} < 3\left(k^{1/3} - (k-1)^{1/3}\right).
$$

Summing from 1 to n we get

$$
1 + 3\left((n+1)^{1/3} - 2^{1/3}\right) < \sum_{k=1}^n k^{-2/3} < 1 + 3(n^{1/3} - 1).
$$

In particular, for $n = 10^9$ this inequality gives

$$
2997 < 1 + 3\left((10^9 + 1)^{1/3} - 2^{1/3}\right) < \sum_{k=1}^{10^9} k^{-2/3} < 2998.
$$

Therefore $\left[\sum_{k=1}^{10^9} k^{-2/3}\right] = 2997$.

4. Put $\Delta a_n = a_n - a_{n+1}$. By the imposed condition, $\Delta a_n > \Delta a_{n+1}$. Suppose that for some n, $\Delta a_n < 0$: Then for each $k \geq n$, $\Delta a_k < \Delta a_n$; hence $a_n - a_{n+m} = \Delta a_n + \cdots + \Delta a_{n+m-1} < m\Delta a_n$. Thus for sufficiently large m it holds that $a_n - a_{n+m} < -1$, which is impossible. This proves the first part of the inequality. Next one observes that

$$
n \geq \sum_{k=1}^n a_k = na_{n+1} + \sum_{k=1}^n k\Delta a_k \geq (1 + 2 + \cdots + n)\Delta a_n = \frac{n(n+1)}{2}\Delta a_n.
$$

Hence $(n+1)\Delta a_n \leq 2$.

5. There are exactly $8 \cdot 9^{k-1}$ k-digit numbers in M (the first digit can be chosen in 8 ways, while any other position admits 9 possibilities). The least of them is 10^k, and hence

$$\sum_{x_j<10^k} \frac{1}{x_j} = \sum_{i=1}^{k} \sum_{10^{i-1}\leq x_j<10^i} \frac{1}{x_j} < \sum_{i=1}^{k} \sum_{10^{i-1}\leq x_j<10^i} \frac{1}{10^{i-1}}$$

$$= \sum_{i=1}^{k} \frac{8 \cdot 9^{i-1}}{10^{i-1}} = 80\left(1 - \frac{9^k}{10^k}\right) < 80.$$

6. Let us denote by C the sum of digits of B. We know that $16^{16} \equiv A \equiv B \equiv C$ (mod 9). Since $16^{16} = 2^{64} = 2^{6 \cdot 10 + 4} \equiv 2^4 \equiv 7$ (mod 9), we get $C \equiv 7$ (mod 9). Moreover, $16^{16} < 100^{16} = 10^{32}$, hence A cannot exceed $9 \cdot 32 = 288$; consequently, B cannot exceed 19 and C is at most 10. Therefore $C = 7$.

7. We use induction on m. Denote by S_m the left-hand side of the equality to be proved. First $S_0 = (1-y)(1+y+\cdots+y^n)+y^{n+1} = 1$, since $x = 1-y$. Furthermore,

$$S_{m+1} - S_m$$
$$= \binom{m+n+1}{m+1}x^{m+1}y^{n+1} + x^{m+1}\sum_{j=0}^{n}\left(\binom{m+1+j}{j}xy^j - \binom{m+j}{j}y^j\right)$$
$$= \binom{m+n+1}{m+1}x^{m+1}y^{n+1}$$
$$\quad + x^{m+1}\sum_{j=0}^{n}\left(\binom{m+1+j}{j}y^j - \binom{m+j}{j}y^j - \binom{m+1+j}{j}y^{j+1}\right)$$
$$= x^{m+1}\left[\binom{m+n+1}{n}y^{n+1} + \sum_{j=0}^{n}\left(\binom{m+j}{j-1}y^j - \binom{m+j+1}{j}y^{j+1}\right)\right]$$
$$= 0;$$

i.e., $S_{m+1} = S_m = 1$ for every m.

Second solution. Let us be given an unfair coin that, when tossed, shows heads with probability x and tails with probability y. Note that $x^{m+1}\binom{m+j}{j}y^j$ is the probability that until the moment when the $(m+1)$th head appears, exactly j tails ($j < n+1$) have appeared. Similarly, $y^{n+1}\binom{n+i}{i}x^i$ is the probability that exactly i heads will appear before the $(n+1)$th tail occurs. Therefore, the above sum is the probability that either $m+1$ heads will appear before $n+1$ tails, or vice versa, and this probability is clearly 1.

8. Denote by K and L the feet of perpendiculars from the points P and Q to the lines BC and AC respectively.

Let M and N be the points on AB (ordered $A-N-M-B$) such that the triangle RMN is isosceles with $\angle R = 90°$. By sine theorem we have $\frac{BM}{BA} = \frac{BM}{BR}$. $\frac{BR}{BA} = \frac{\sin 15°}{\sin 45°}$. Since $\frac{BK}{BC} = \frac{\sin 45° \sin 30°}{\cos 15°} = \frac{\sin 15°}{\sin 45°}$, we deduce that $MK \parallel AC$ and $MK = AL$. Similarly, $NL \parallel BC$

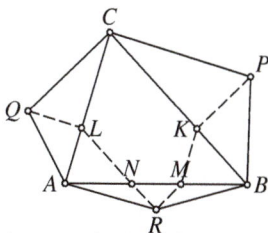

and $NL = BK$. It follows that the vectors $\overrightarrow{RN}$, $\overrightarrow{NL}$, and $\overrightarrow{LQ}$ are the images of $\overrightarrow{RM}$, $\overrightarrow{KP}$, and $\overrightarrow{MK}$ respectively under a rotation of $90°$, and consequently the same holds for their sums $\overrightarrow{RQ}$ and $\overrightarrow{RP}$. Therefore, $QR = RP$ and $\angle QRP = 90°$.

Second solution. Let ABS be the equilateral triangle constructed in the exterior of $\triangle ABC$. Obviously, the triangles BPC, BRS, ARS, AQC are similar. Let f be the rotational homothety centered at B that maps P onto C, and let g be the rotational homothety about A that maps C onto Q. The composition $h = g \circ f$ is also a rotational homothety; its angle is $\angle PBC + \angle CAQ = 90°$, and the coefficient is $\frac{BC}{BP} \cdot \frac{AQ}{AC} = 1$. Moreover, R is a fixed point of h because $f(R) = S$ and $g(S) = R$. Hence R is the center of h, and the statement follows from $h(P) = Q$.

Remark. There are two more possible approaches: One includes using complex numbers and the other one is mere calculating of RP, RQ, PQ by the cosine theorem.

Second remark. The problem allows a generalization: Given that $\angle CBP = \angle CAQ = \alpha$, $\angle BCP = \angle ACQ = \beta$, and $\angle RAB = \angle RBA = 90° - \alpha - \beta$, show that $RP = RQ$ and $\angle PRQ = 2\alpha$.

9. Suppose n is the natural number with $na \le 1 < (n+1)a$. Let $f_a(x) = f(x-a)$. If a function f with the desired properties exists, then $f_a(a) = 0$ and let w.l.o.g. $f(a) > 0$, or equivalently, let the graph of f_a lie below the graph of f. In this case also $f(2a) > f(a)$, since otherwise, the graphs of f and f_a would intersect between a and $2a$. Continuing in this way we are led to $0 = f(0) < f(a) < f(2a) < \cdots < f(na)$. Thus if $na = 1$, i.e., $a = 1/n$, such an f does not exist. On the other hand, if $a \ne 1/n$, then we similarly obtain $f(1) > f(1-a) > f(1-2a) > \cdots > f(1-na)$. Choosing values of f at $ia, 1-ia$, $i = 1,\ldots,n$, so that they satisfy $f(1-na) < \cdots < f(1-a) < 0 < f(a) < \cdots < f(na)$, we can extend f to other values of $[0,1]$ by linear interpolation. A function obtained this way has the desired property.

10. We shall prove that for all x, y with $x + y = 1$ it holds that $f(x,y) = x - 2y$. In this case $f(x,y) = f(x, 1-x)$ can be regarded as a polynomial in $z = x - 2y = 3x - 2$, say $f(x, 1-x) = F(z)$. Putting in the given relation $a = b = x/2$, $c = 1-x$, we obtain $f(x, 1-x) + 2f(1-x/2, x/2) = 0$; hence $F(z) + 2F(-z/2) = 0$. Now $F(1) = 1$, and we get that for all k, $F((-2)^k) = (-2)^k$. Thus $F(z) = z$ for infinitely many values of z; hence $F(z) \equiv z$. Consequently $f(x,y) = x - 2y$ if $x + y = 1$.

For general x, y with $x + y \neq 0$, since f is homogeneous, we have $f(x,y) = (x+y)^n f\left(\frac{x}{x+y}, \frac{y}{x+y}\right) = (x+y)^n \left(\frac{x}{x+y} - 2\frac{y}{x+y}\right) = (x+y)^{n-1}(x - 2y)$. The same is true for $x + y = 0$, because f is a polynomial.

11. Let (a_{k_i}) be the subsequence of (a_k) consisting of all a_k's that give remainder r upon division by a_1. For every $i > 1$, $a_{k_i} \equiv a_{k_1} \pmod{a_1}$; hence $a_{k_i} = a_{k_1} + y a_1$ for some integer $y > 0$. It follows that for every $r = 0, 1, \ldots, a_1 - 1$ there is exactly one member of the corresponding $(a_{k_i})_{i \geq 1}$ that cannot be represented as $x a_l + y a_m$, and hence at most $a_1 + 1$ members of (a_k) in total are not representable in the given form.

12. Since $\sin 2x_i = 2 \sin x_i \cos x_i$ and $\sin(x_i + x_{i+1}) + \sin(x_i - x_{i+1}) = 2 \sin x_i \cos x_{i+1}$, the inequality from the problem is equivalent to

$$(\cos x_1 - \cos x_2) \sin x_1 + (\cos x_2 - \cos x_3) \sin x_2 + \cdots$$

$$\cdots + (\cos x_{v-1} - \cos x_v) \sin x_{v-1} < \frac{\pi}{4}. \tag{1}$$

Consider the unit circle with center at $O(0,0)$ and points $M_i(\cos x_i, \sin x_i)$ on it. Also, choose the points $N_i(\cos x_i, 0)$ and $M_i'(\cos x_{i+1}, \sin x_i)$. It is clear that $(\cos x_i - \cos x_{i+1}) \sin x_i$ is equal to the area of the rectangle $M_i N_i N_{i+1} M_i'$. Since all these rectangles are disjoint and lie inside the quarter circle in the first quadrant whose area is $\frac{\pi}{4}$, inequality (1) follows.

13. Suppose that $A_k A_{k+1} \cap A_m A_{m+1} \neq \emptyset$ for some $k, m > k + 1$. Without loss of generality we may suppose that $k = 0$, $m = n - 1$ and that no two segments $A_k A_{k+1}$ and $A_m A_{m+1}$ intersect for $0 \leq k < m - 1 < n - 1$ except for $k = 0$, $m = n - 1$. Also, shortening $A_0 A_1$, we may suppose that $A_0 \in A_{n-1} A_n$. Finally, we may reduce the problem to the case that $A_0 \ldots A_{n-1}$ is convex: Otherwise, the segment $A_{n-1} A_n$ can be prolonged so that it intersects some $A_k A_{k+1}$, $0 < k < n - 2$.
If $n = 3$, then $A_1 A_2 \geq 2 A_0 A_1$ implies $A_0 A_2 > A_0 A_1$, hence $\angle A_0 A_1 A_2 > \angle A_1 A_2 A_3$, a contradiction.
Let $n = 4$. From $A_3 A_2 > A_1 A_2$ we conclude that $\angle A_3 A_1 A_2 > \angle A_1 A_3 A_2$. Using the inequality $\angle A_0 A_3 A_2 > \angle A_0 A_1 A_2$ we obtain that $\angle A_0 A_3 A_1 > \angle A_0 A_1 A_3$ implying $A_0 A_1 > A_0 A_3$. Now we have $A_2 A_3 < A_3 A_0 + A_0 A_1 + A_1 A_2 < 2 A_0 A_1 + A_1 A_2 \leq 2 A_1 A_2 \leq A_2 A_3$, which is not possible.
Now suppose $n \geq 5$. If α_i is the exterior angle at A_i, then $\alpha_1 > \cdots > \alpha_{n-1}$; hence $\alpha_{n-1} < \frac{360°}{n-1} \leq 90°$. Consequently $\angle A_{n-2} A_{n-1} A_0 \geq 90°$ and $A_0 A_{n-2} > A_{n-1} A_{n-2}$. On the other hand, $A_0 A_{n-2} < A_0 A_1 + A_1 A_2 + \cdots + A_{n-3} A_{n-2} < \left(\frac{1}{2^{n-2}} + \frac{1}{2^{n-3}} + \cdots + \frac{1}{2}\right) A_{n-1} A_{n-2} < A_{n-1} A_{n-2}$, which contradicts the previous relation.

14. We shall prove that for every $n \in \mathbb{N}$, $\sqrt{2n + 25} \leq x_n \leq \sqrt{2n + 25} + 0.1$. Note that for $n = 1000$ this gives us exactly the desired inequalities.
First, notice that the recurrent relation is equivalent to

$$2 x_k (x_{k+1} - x_k) = 2. \tag{1}$$

Since $x_0 < x_1 < \cdots < x_k < \cdots$, from (1) we get $x_{k+1}^2 - x_k^2 = (x_{k+1} + x_k)(x_{k+1} - x_k) > 2$. Adding these up we obtain $x_n^2 \geq x_0^2 + 2n$, which proves the first inequality.

On the other hand, $x_{k+1} = x_k + \frac{1}{x_k} \leq x_k + 0.2$ (for $x_k \geq 5$), and one also deduces from (1) that $x_{k+1}^2 - x_k^2 - 0.2(x_{k+1} - x_k) = (x_{k+1} + x_k - 0.2)(x_{k+1} - x_k) \leq 2$. Again, adding these inequalities up, $(k = 0, \ldots, n-1)$ yields

$$x_n^2 \leq 2n + x_0^2 + 0.2(x_n - x_0) = 2n + 24 + 0.2x_n.$$

Solving the corresponding quadratic equation, we obtain

$$x_n < 0.1 + \sqrt{2n + 24.01} < 0.1 + \sqrt{2n + 25}.$$

15. Assume that the center of the circle is at the origin $O(0,0)$, and that the points $A_1, A_2, \ldots, A_{1975}$ are arranged on the upper half-circle so that $\angle A_i O A_1 = \alpha_i$ ($\alpha_1 = 0$). The distance $A_i A_j$ equals $2\sin\frac{\alpha_j - \alpha_i}{2} = 2\sin\frac{\alpha_j}{2}\cos\frac{\alpha_i}{2} - \cos\frac{\alpha_j}{2}\sin\frac{\alpha_i}{2}$, and it will be rational if all $\sin\frac{\alpha_k}{2}, \cos\frac{\alpha_k}{2}$ are rational.

Finally, observe that there exist infinitely many angles α such that both $\sin\alpha$, $\cos\alpha$ are rational, and that such α can be arbitrarily small. For example, take α so that $\sin\alpha = \frac{2t}{t^2+1}$ and $\cos\alpha = \frac{t^2-1}{t^2+1}$ for any $t \in \mathbb{Q}$.

4.18 Solutions to the Shortlisted Problems of IMO 1976

1. Let r denote the common inradius. Some two of the four triangles with the inradii ρ have cross angles at M: Suppose these are $\triangle AMB_1$ and $\triangle BMA_1$. We shall show that $\triangle AMB_1 \cong \triangle BMA_1$. Indeed, the altitudes of these two triangles are both equal to r, the inradius of $\triangle ABC$, and their interior angles at M are equal to some angle φ. If P is the point of tangency of the incircle of $\triangle A_1MB$ with MB, then $\frac{r}{\rho} = \frac{A_1M+BM+A_1B}{A_1B}$, which also implies $\frac{r-2\rho}{\rho} = \frac{A_1M+BM-A_1B}{A_1B} = \frac{2MP}{A_1B} = \frac{2r\cot(\varphi/2)}{A_1B}$. Since similarly $\frac{r-2\rho}{\rho} = \frac{2r\cot(\varphi/2)}{B_1A}$, we obtain $A_1B = B_1A$ and consequently $\triangle AMB_1 \cong \triangle BMA_1$. Thus $\angle BAC = \angle ABC$ and $CC_1 \perp AB$. There are two alternatives for the other two incircles:

   (i) If the inradii of AMC_1 and AMB_1 are equal to r, it is easy to obtain that $\triangle AMC_1 \cong \triangle AMB_1$. Hence $\angle AB_1M = \angle AC_1M = 90°$, and $\triangle ABC$ is equilateral.

   (ii) The inradii of AMB_1 and CMB_1 are equal to r. Put $x = \angle MAC_1 = \angle MBC_1$. In this case $\varphi = 2x$ and $\angle B_1MC = 90° - x$. Now we have $\frac{AB_1}{CB_1} = \frac{S_{AMB_1}}{S_{CMB_1}} = \frac{AM+MB_1+AB_1}{CM+MB_1+CB_1} = \frac{AM+MB_1-AB_1}{CM+MB_1-CB_1} = \frac{\cot x}{\cot(45°-x/2)}$. On the other hand, we have $\frac{AB_1}{CB_1} = \frac{AB}{BC} = 2\cos 2x$. Thus we have an equation for x: $\tan(45° - x/2) = 2\cos 2x \tan x$, or equivalently

   $$2\tan\left(45° - \frac{x}{2}\right) \sin\left(45° - \frac{x}{2}\right) \cos\left(45° - \frac{x}{2}\right) = 2\cos 2x \sin x.$$

   Hence $\sin 3x - \sin x = 2\sin^2\left(45° - \frac{x}{2}\right) = 1 - \sin x$, implying $\sin 3x = 1$, i.e., $x = 30°$. Therefore $\triangle ABC$ is equilateral.

2. Let us put $b_i = i(n+1-i)/2$, and let $c_i = a_i - b_i$, $i = 0, 1, \ldots, n+1$. It is easy to verify that $b_0 = b_{n+1} = 0$ and $b_{i-1} - 2b_i + b_{i+1} = -1$. Subtracting this inequality from $a_{i-1} - 2a_i + a_{i+1} \geq -1$, we obtain $c_{i-1} - 2c_i + c_{i+1} \geq 0$, i.e., $2c_i \leq c_{i-1} + c_{i+1}$. We also have $c_0 = c_{n+1} = 0$.
   Suppose that there exists $i \in \{1, \ldots, n\}$ for which $c_i > 0$, and let c_k be the maximal such c_i. Assuming w.l.o.g. that $c_{k-1} < c_k$, we obtain $c_{k-1} + c_{k+1} < 2c_k$, which is a contradiction. Hence $c_i \leq 0$ for all i; i.e., $a_i \leq b_i$.
   Similarly, considering the sequence $c_i' = a_i + b_i$ one can show that $c_i' \geq 0$, i.e., $a_i \geq -b_i$ for all i. This completes the proof.

3. (a) Let $ABCD$ be a quadrangle with $16 = d = AB + CD + AC$, and let S be its area. Then $S \leq (AC \cdot AB + AC \cdot CD)/2 = AC(d - AC)/2 \leq d^2/8 = 32$, where equality occurs if and only if $AB \perp AC \perp CD$ and $AC = AB + CD = 8$. In this case $BD = 8\sqrt{2}$.

   (b) Let A' be the point with $\overrightarrow{DA'} = \overrightarrow{AC}$. The triangular inequality implies $AD + BC \geq AA' = 8\sqrt{5}$. Thus the perimeter attains its minimum for $AB = CD = 4$.

   (c) Let us assume w.l.o.g. that $CD \leq AB$. Then C lies inside $\triangle BDA'$ and hence $BC + AD = BC + CA' < BD + DA'$. The maximal value $BD + DA'$ of $BC + AD$ is attained when C approaches D, making a degenerate quadrangle.

4. The first few values are easily verified to be $2^{r_n} + 2^{-r_n}$, where $r_0 = 0$, $r_1 = r_2 = 1$, $r_3 = 3$, $r_4 = 5$, $r_5 = 11, \ldots$. Let us put $u_n = 2^{r_n} + 2^{-r_n}$ (we will show that r_n exists and is integer for each n). A simple calculation gives us $u_n(u_{n-1}^2 - 2) = 2^{r_n + 2r_{n-1}} + 2^{-r_n - 2r_{n-1}} + 2^{r_n - 2r_{n-1}} + 2^{-r_n + 2r_{n-1}}$. If an array q_n, with $q_0 = 0$ and $q_1 = 1$, is set so as to satisfy the linear recurrence $q_{n+1} = q_n + 2q_{n-1}$, then it also satisfies $q_n - 2q_{n-1} = -(q_{n-1} - 2q_{n-2}) = \cdots = (-1)^{n-1}(q_1 - 2q_0) = (-1)^{n-1}$. Assuming inductively up to n $r_i = q_i$, the expression for $u_n(u_{n-1}^2 - 2) = u_{n+1} + u_1$ reduces to $2^{q_{n+1}} + 2^{-q_{n+1}} + u_1$. Therefore, $r_{n+1} = q_{n+1}$. The solution to this linear recurrence with $r_0 = 0$, $r_1 = 1$ is $r_n = q_n = \frac{2^n - (-1)^n}{3}$, and since $\lfloor u_n \rfloor = 2^{r_n}$ for $n \geq 0$, the result follows.

Remark. One could simply guess that $u_n = 2^{r_n} + 2^{-r_n}$ for $r_n = \frac{2^n - (-1)^n}{3}$, and then prove this result by induction.

5. If one substitutes an integer q-tuple $(x_1, \ldots, x_q)$ satisfying $|x_i| \leq p$ for all i in an equation of the given system, the absolute value of the right-hand member never exceeds pq. So for the right-hand member of the system there are $(2pq + 1)^p$ possibilities There are $(2p+1)^q$ possible q-tuples $(x_1, \ldots, x_q)$. Since $(2p+1)^q > (2pq + 1)^p$, there are at least two q-tuples $(y_1, \ldots, y_q)$ and $(z_1, \ldots, z_q)$ giving the same right-hand members in the given system. The difference $(x_1, \ldots, x_q) = (y_1 - z_1, \ldots, y_q - z_q)$ thus satisfies all the requirements of the problem.

6. Suppose $a_1 \leq a_2 \leq a_3$ are the dimensions of the box. If we set $b_i = [a_i/\sqrt[3]{2}]$, the condition of the problem is equivalent to $\frac{a_1}{b_1} \cdot \frac{a_2}{b_2} \cdot \frac{a_3}{b_3} = 5$. We list some values of $a, b = [a/\sqrt[3]{2}]$ and a/b:

a	2	3	4	5	6	7	8	9	10
b	1	2	3	3	4	5	6	7	7
a/b	2	1.5	1.33	1.67	1.5	1.4	1.33	1.29	1.43

We note that if $a > 2$, then $a/b \leq 5/3$, and if $a > 5$, then $a/b \leq 3/2$. If $a_1 > 2$, then $\frac{a_1}{b_1} \cdot \frac{a_2}{b_2} \cdot \frac{a_3}{b_3} < (5/3)^3 < 5$, a contradiction. Hence $a_1 = 2$. If also $a_2 = 2$, then $a_3/b_3 = 5/4 < \sqrt[3]{2}$, which is impossible. Also, if $a_2 \geq 6$, then $\frac{a_2}{b_2} \cdot \frac{a_3}{b_3} \leq (1.5)^2 < 2.5$, again a contradiction. We thus have the following cases:
  (i) $a_1 = 2, a_2 = 3$, then $a_3/b_3 = 5/3$, which holds only if $a_3 = 5$;
  (ii) $a_1 = 2, a_2 = 4$, then $a_3/b_3 = 15/8$, which is impossible;
  (iii) $a_1 = 2, a_2 = 5$, then $a_3/b_3 = 3/2$, which holds only if $a_3 = 6$.
The only possible sizes of the box are therefore $(2, 3, 5)$ and $(2, 5, 6)$.

7. The map T transforms the interval $(0, a]$ onto $(1 - a, 1]$ and the interval $(a, 1]$ onto $(0, 1 - a]$. Clearly T preserves the measure. Since the measure of the interval $[0, 1]$ is finite, there exist two positive integers k, $l > k$ such that $T^k(J)$ and $T^l(J)$ are not disjoint. But the map T is bijective; hence $T^{l-k}(J)$ and J are not disjoint.

8. Every polynomial with real coefficients can be factored as a product of linear and quadratic polynomials with real coefficients. Thus it suffices to prove the result only for a quadratic polynomial $P(x) = x^2 - 2ax + b^2$, with $a > 0$ and $b^2 > a^2$. Using the identity

$$(x^2+b^2)^{2n} - (2ax)^{2n} = (x^2 - 2ax + b^2) \sum_{k=0}^{2n-1} (x^2+b^2)^k (2ax)^{2n-k-1}$$

we have solved the problem if we can choose n such that $b^{2n}\binom{2n}{n} > 2^{2n}a^{2n}$. However, it is is easy to show that $2n\binom{2n}{n} < 2^{2n}$; hence it is enough to take n such that $(b/a)^{2n} > 2n$. Since $\lim_{n\to\infty}(2n)^{1/(2n)} = 1 < b/a$, such an n always exists.

9. The equation $P_n(x) = x$ is of degree 2^n, and has at most 2^n distinct roots. If $x > 2$, then by simple induction $P_n(x) > x$ for all n. Similarly, if $x < -1$, then $P_1(x) > 2$, which implies $P_n(x) > 2$ for all n. It follows that all real roots of the equation $P_n(x) = x$ lie in the interval $[-2, 2]$, and thus have the form $x = 2\cos t$. Observe that $P_1(2\cos t) = 4\cos^2 t - 2 = 2\cos 2t$, and in general $P_n(2\cos t) = 2\cos 2^n t$. Our equation becomes

$$\cos 2^n t = \cos t,$$

which indeed has 2^n different solutions $t = \frac{2\pi m}{2^n-1}$ ($m = 0,1,\ldots,2^{n-1}-1$) and $t = \frac{2\pi m}{2^n+1}$ ($m = 1,2,\ldots,2^{n-1}$).

10. Let $a_1 \le a_2 \le \cdots \le a_n$ be positive integers whose sum is 1976. Let M denote the maximal value of $a_1 a_2 \cdots a_n$. We make the following observations:
    (1) $a_1 = 1$ does not yield the maximum, since replacing $1, a_2$ by $1 + a_2$ increases the product.
    (2) $a_j - a_i \ge 2$ does not yield the maximal value, since replacing a_i, a_j by $a_i + 1, a_j - 1$ increases the product.
    (3) $a_i \ge 5$ does not yield the maximal value, since $2(a_i - 2) = 2a_i - 4 > a_i$.
    Since $4 = 2^2$, we may assume that all a_i are either 2 or 3, and $M = 2^k 3^l$, where $2k + 3l = 1976$.
    (4) $k \ge 3$ does not yield the maximal value, since $2 \cdot 2 \cdot 2 < 3 \cdot 3$.
    Hence $k \le 2$ and $2k \equiv 1976 \pmod 3$ gives us $k = 1$, $l = 658$ and $M = 2 \cdot 3^{658}$.

11. We shall show by induction that $5^{2^k} - 1 = 2^{k+2}q_k$ for each $k = 0,1,\ldots$, where $q_k \in \mathbb{N}$. Indeed, the statement is true for $k = 0$, and if it holds for some k then $5^{2^{k+1}} - 1 = \left(5^{2^k}+1\right)\left(5^{2^k}-1\right) = 2^{k+3}d_{k+1}$ where $d_{k+1} = \left(5^{2^k}+1\right)d_k/2$ is an integer by the inductive hypothesis.
    Let us now choose $n = 2^k + k + 2$. We have $5^n = 10^{k+2}q_k + 5^{k+2}$. It follows from $5^4 < 10^3$ that 5^{k+2} has at most $[3(k+2)/4] + 2$ nonzero digits, while $10^{k+2}q_k$ ends in $k+2$ zeros. Hence the decimal representation of 5^n contains at least $[(k+2)/4] - 2$ consecutive zeros. Now it suffices to take $k > 4 \cdot 1978$.

12. Suppose the decomposition into k polynomials is possible. The sum of coefficients of each polynomial $a_1 x + a_2 x^2 + \cdots + a_n x^n$ equals $1 + \cdots + n = n(n+1)/2$ while the sum of coefficients of $1976(x + x^2 + \cdots + x^n)$ is $1976n$. Hence we must have $1976n = kn(n+1)/2$, which reduces to $(n+1) \mid 3952 = 2^4 \cdot 13 \cdot 19$. In other words, n is of the form $n = 2^\alpha 13^\beta 19^\gamma - 1$, with $0 \le \alpha \le 4, 0 \le \beta \le 1, 0 \le \gamma \le 1$.

We can immediately eliminate the values $n = 0$ and $n = 3951$ that correspond to $\alpha = \beta = \gamma = 0$ and $\alpha = 4$, $\beta = \gamma = 1$.

We claim that all other values n are permitted. There are two cases.

$\alpha \leq 3$. In this case $k = 3952/(n+1)$ is even. The simple choice of the polynomials $P = x + 2x^2 + \cdots + nx^n$ and $P' = nx + (n-1)x^2 + \cdots + x^n$ suffices, since $k(P+P')/2 = 1976(x+x^2+\cdots+x^n)$.

$\alpha = 4$. Then k is odd. Consider $(k-3)/2$ pairs (P,P') of the former case and

$$P_1 = \left[nx + (n-1)x^3 + \cdots + \tfrac{n+1}{2}x^n\right]$$
$$\quad + \left[\tfrac{n-1}{2}x^2 + \tfrac{n-3}{2}x^4 + \cdots + x^{n-1}\right];$$
$$P_2 = \left[\tfrac{n+1}{2}x + \tfrac{n-1}{2}x^3 + \cdots + x^n\right]$$
$$\quad + \left[nx^2 + (n-1)x^4 + \cdots + \tfrac{n+3}{2}x^{n-1}\right].$$

Then $P + P_1 + P_2 = 3(n+1)(x+x^2+\cdots+x^n)/2$ and therefore $(k-3)(P+P')/2 + (P+P_1+P_2) = 1976(x+x^2+\cdots+x^n)$.

It follows that the desired decomposition is possible if and only if $1 < n < 3951$ and $n+1 \mid 2 \cdot 1976$.

4.19 Solutions to the Longlisted Problems of IMO 1977

1. Let P be the projection of S onto the plane $ABCDE$. Obviously $BS > CS$ is equivalent to $BP > CP$. The conditions of the problem imply that $PA > PB$ and $PA > PE$. The locus of such points P is the region of the plane that is determined by the perpendicular bisectors of segments AB and AE and that contains the point diametrically opposite A. But since $AB < DE$, the whole of this region lies on one side of the perpendicular bisector of BC. The result follows immediately.

   Remark. The assumption $BC < CD$ is redundant.

2. We shall prove by induction on n that $f(x) > f(n)$ whenever $x > n$. The case $n = 0$ is trivial. Suppose that $n \geq 1$ and that $x > k$ implies $f(x) > f(k)$ for all $k < n$. It follows that $f(x) \geq n$ holds for all $x \geq n$. Let $f(m) = \min_{x \geq n} f(x)$. If we suppose that $m > n$, then $m - 1 \geq n$ and consequently $f(m - 1) \geq n$. But in this case the inequality $f(m) > f(f(m-1))$ contradicts the minimality property of m. The inductive proof is thus completed.

   It follows that f is strictly increasing, so $f(n+1) > f(f(n))$ implies that $n+1 > f(n)$. But since $f(n) \geq n$ we must have $f(n) = n$.

3. Let $v_1, v_2, \ldots, v_k$ be k persons who are not acquainted with each other. Let us denote by m the number of acquainted couples and by d_j the number of acquaintances of person v_j. Then

$$m \leq d_{k+1} + d_{k+2} + \cdots + d_n \leq d(n - k) \leq k(n - k) \leq \left(\frac{k + (n - k)}{2} \right)^2 = \frac{n^2}{4}.$$

4. Consider any vertex v_n from which the maximal number d of segments start, and suppose it is not a vertex of a triangle. Let $\mathscr{A} = \{v_1, v_2, \ldots, v_d\}$ be the set of points that are connected to v_n, and let $\mathscr{B} = \{v_{d+1}, v_{d+2}, \ldots, v_n\}$ be the set of the other points. Since v_n is not a vertex of a triangle, there is no segment both of whose vertices lie in $\mathscr{A}$; i.e., each segment has an end in $\mathscr{B}$. Thus, if d_j denotes the number of segments at v_j and m denotes the total number of segments, we have $m \leq d_{d+1} + d_{d+2} + \cdots + d_n \leq d(n - d) \leq \lceil n^2/4 \rceil = m$. This means that each inequality must be equality, implying that each point in $\mathscr{B}$ is a vertex of d segments, and each of these segments has the other end in $\mathscr{A}$. Then there is no triangle at all, which is a contradiction.

5. Let us denote by I and E the sets of interior boundary points and exterior boundary points. Let $ABCD$ be the square inscribed in the circle k with sides parallel to the coordinate axes. Lines AB, BC, CD, DA divide the plane into 9 regions: $\mathscr{R}, \mathscr{R}_A, \mathscr{R}_B, \mathscr{R}_C, \mathscr{R}_D, \mathscr{R}_{AB}, \mathscr{R}_{BC}, \mathscr{R}_{CD}, \mathscr{R}_{DA}$. There is a unique pair of lattice points $A_I \in \mathscr{R}$, $A_E \in \mathscr{R}_A$ that are opposite vertices of a

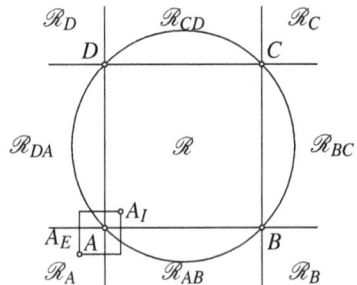

unit square. We similarly define $B_I, C_I, D_I, B_E, C_E, D_E$. Let us form a graph G by connecting each point from E lying in $\mathcal{R}_{AB}$ (respectively $\mathcal{R}_{BC}, \mathcal{R}_{CD}, \mathcal{R}_{DA}$) to its upper (respectively left, lower, right) neighbor point (which clearly belongs to I). It is easy to see that:

(i) All vertices from I other than A_I, B_I, C_I, D_I have degree 1.

(ii) A_E is not in E if and only if $A_I \in I$ and $\deg A_I = 2$.

(iii) No other lattice points inside $\mathcal{R}_A$ belong to E.

Thus if m is the number of edges of the graph G and s is the number of points among A_E, B_E, C_E, and D_E that are in E, using (i)–(iii) we easily obtain $|E| = m + s$ and $|I| = m - (4 - s) = |E| + 4$.

6. Let $\langle y \rangle$ denote the distance from $y \in \mathbb{R}$ to the closest even integer. We claim that

$$\langle 1 + \cos x \rangle \leq \sin x \qquad \text{for all } x \in [0, \pi].$$

Indeed, if $\cos x \geq 0$, then $\langle 1 + \cos x \rangle = 1 - \cos x \leq 1 - \cos^2 x = \sin^2 x \leq \sin x$; the proof is similar if $\cos x < 0$.

We note that $\langle x + y \rangle \leq \langle x \rangle + \langle y \rangle$ holds for all $x, y \in \mathbb{R}$. Therefore

$$\sum_{j=1}^{n} \sin x_j \geq \sum_{j=1}^{n} \langle 1 + \cos x_j \rangle \geq \left\langle \sum_{j=1}^{n} (1 + \cos x_j) \right\rangle = 1.$$

7. Let us suppose that $c_1 \leq c_2 \leq \cdots \leq c_n$ and that $c_1 < 0 < c_n$. There exists k, $1 \leq k < n$, such that $c_k \leq 0 < c_{k+1}$. Then we have

$$
\begin{aligned}
(n-1)(c_1^2 + c_2^2 + \cdots + c_n^2) &\geq k(c_1^2 + \cdots + c_k^2) + (n-k)(c_{k+1}^2 + \cdots + c_n^2) \\
&\geq (c_1 + \cdots + c_k)^2 + (c_{k+1} + \cdots + c_n)^2 \\
&= (c_1 + \cdots + c_n)^2 \\
&\quad -2(c_1 + \cdots + c_k)(c_{k+1} + \cdots + c_n),
\end{aligned}
$$

from which we obtain $(c_1 + \cdots + c_k)(c_{k+1} + \cdots + c_n) \geq 0$, a contradiction.

Second solution. By the given condition and the inequality between arithmetic and quadratic mean we have

$$
\begin{aligned}
(c_1 + \cdots + c_n)^2 &= (n-1)(c_1^2 + \cdots + c_{n-1}^2) + (n-1)c_n^2 \\
&\geq (c_1 + \cdots + c_{n-1})^2 + (n-1)c_n^2,
\end{aligned}
$$

which is equivalent to $2(c_1 + c_2 + \cdots + c_n)c_n \geq nc_n^2$. Similarly, $2(c_1 + c_2 + \cdots + c_n)c_i \geq nc_i^2$ for all $i = 1, \ldots, n$. Hence all c_i are of the same sign.

8. There is exactly one point satisfying the given condition on each face of the hexahedron. Namely, on the face ABD it is the point that divides the median from D in the ratio $32 : 3$.

9. A necessary and sufficient condition for $\mathcal{M}$ to be nonempty is that $1/\sqrt{10} \leq t \leq 1$.

10. Integers a, b, q, r satisfy

$$a^2 + b^2 = (a+b)q + r, \quad 0 \le r < a+b, \quad q^2 + r = 1977.$$

From $q^2 \le 1977$ it follows that $q \le 44$, and consequently $a^2 + b^2 < 45(a+b)$. Having in mind the inequality $(a+b)^2 \le 2(a^2+b^2)$, we get $(a+b)^2 < 90(a+b)$, i.e., $a+b < 90$ and consequently $r < 90$. Now from $q^2 = 1977 - r > 1977 - 90 = 1887$ it follows that $q > 43$; hence $q = 44$ and $r = 41$. It remains to find positive integers a and b satisfying $a^2 + b^2 = 44(a+b) + 41$, or equivalently

$$(a-22)^2 + (b-22)^2 = 1009.$$

The Diophantine equation $A^2 + B^2 = 1009$ has only two pairs of positive solutions: $(15, 28)$ and $(28, 15)$. Hence $(|a-22|, |b-22|) \in \{(15, 28), (28, 15)\}$, which implies $(a, b) \in \{(7, 50), (37, 50), (50, 7), (50, 37)\}$.

11. (a) Suppose to the contrary that none of the numbers $z_0, z_1, \ldots, z_{n-1}$ is divisible by n. Then two of these numbers, say z_k and z_l $(0 \le k < l \le n-1)$, are congruent modulo n, and thus $n \mid z_l - z_k = z^{k+1} z_{l-k-1}$. But since $(n, z) = 1$, this implies $n \mid z_{l-k-1}$, which is a contradiction.

(b) Again suppose the contrary, that none of $z_0, z_1, \ldots, z_{n-2}$ is divisible by n. Since $(z-1, n) = 1$, this is equivalent to $n \nmid (z-1)z_j$, i.e., $z^k \not\equiv 1 \pmod{n}$ for all $k = 1, 2, \ldots, n-1$. But since $(z, n) = 1$, we also have that $z^k \not\equiv 0 \pmod{n}$. It follows that there exist k, l, $1 \le k < l \le n-1$ such that $z^k \equiv z^l$, i.e., $z^{l-k} \equiv 1 \pmod{n}$, which is a contradiction.

12. According to part (a) of the previous problem we can conclude that $T = \{n \in \mathbb{N} \mid (n, z) = 1\}$.

13. The figure Φ contains two points A and B having maximum distance. Let h be the semicircle with diameter AB that lies in Φ, and let k be the circle containing h. Consider any point M inside k. The line passing through M that is orthogonal to AM meets h in some point P (because $\angle AMB > 90°$). Let h' and $\overline{h}'$ be the two semicircles with diameter AP, where $M \in h'$. Since $\overline{h}'$ contains a point C such that $BC > AB$, it cannot be contained in Φ, implying that $h' \subset \Phi$. Hence M belongs to Φ. Since Φ contains no points outside the circle k, it must coincide with the disk determined by k. On the other hand, any disk has the required property.

14. We prove by induction on n that independently of the word w_0, the given algorithm generates all words of length n. This is clear for $n = 1$. Suppose now the statement is true for $n - 1$, and that we are given a word $w_0 = c_1 c_2 \ldots c_n$ of length n. Obviously, the words $w_0, w_1, \ldots, w_{2^{n-1}-1}$ all have the nth digit c_n, and by the inductive hypothesis these are all words whose nth digit is c_n. Similarly, by the inductive hypothesis $w_{2^{n-1}}, \ldots, w_{2^n-1}$ are all words whose nth digit is $1 - c_n$, and the induction is complete.

15. Each segment is an edge of at most two squares and a diagonal of at most one square. Therefore $p_k = 0$ for $k > 3$, and we have to prove that

$$p_0 = p_2 + 2p_3. \tag{1}$$

Let us calculate the number $q(n)$ of considered squares. Each of these squares is inscribed in a square with integer vertices and sides parallel to the coordinate axes. There are $(n-s)^2$ squares of side s with integer vertices and sides parallel to the coordinate axes, and each of them circumscribes exactly s of the considered squares. It follows that $q(n) = \sum_{s=1}^{n-1}(n-s)^2 s = n^2(n^2-1)/12$. Computing the number of edges and diagonals of the considered squares in two ways, we obtain that

$$p_1 + 2p_2 + 3p_3 = 6q(n). \tag{2}$$

On the other hand, the total number of segments with endpoints in the considered integer points is given by

$$p_0 + p_1 + p_2 + p_3 = \binom{n^2}{2} = \frac{n^2(n^2-1)}{2} = 6q(n). \tag{3}$$

Now (1) follows immediately from (2) and (3).

16. For $i=k$ and $j=l$ the system is reduced to $1 \le i,j \le n$, and has exactly n^2 solutions. Let us assume that $i \ne k$ or $j \ne l$. The points $A(i,j)$, $B(k,l)$, $C(-j+k+l, i-k+l)$, $D(i-j+l, i+j-k)$ are vertices of a negatively oriented square with integer vertices lying inside the square $[1,n] \times [1,n]$, and each of these squares corresponds to exactly 4 solutions to the system. By the previous problem there are exactly $q(n) = n^2(n^2-1)/12$ such squares. Hence the number of solutions is equal to $n^2 + 4q(n) = n^2(n^2+2)/3$.

17. Centers of the balls that are tangent to K are vertices of a regular polyhedron with triangular faces, with edge length $2R$ and radius of circumscribed sphere $r+R$. Therefore the number n of these balls is 4, 6, or 20. It is straightforward to obtain that:
   (i) If $n=4$, then $r+R = 2R(\sqrt{6}/4)$, whence $R = r(2+\sqrt{6})$.
   (ii) If $n=6$, then $r+R = 2R(\sqrt{2}/2)$, whence $R = r(1+\sqrt{2})$.
   (iii) If $n=20$, then $r+R = 2R\sqrt{5+\sqrt{5}}/8$. In this case we can conclude that
   $$R = r\left[\sqrt{5-2\sqrt{5}} + (3-\sqrt{5})/2\right].$$

18. Let U be the midpoint of the segment AB. The point M belongs to CU and $CM = (\sqrt{5}-1)CU/2$, $r = CU\sqrt{\sqrt{5}-2}$.

19. We shall prove the statement by induction on m. For $m=2$ it is trivial, since each power of 5 greater than 5 ends in 25. Suppose that the statement is true for some $m \ge 2$, and that the last m digits of 5^n alternate in parity. It can be shown by induction that the maximum power of 2 that divides $5^{2^{m-2}} - 1$ is 2^m, and consequently the difference $5^{n+2^{m-2}} - 5^n$ is divisible by 10^m but not by $2 \cdot 10^m$. It follows that the last m digits of the numbers $5^{n+2^{m-2}}$ and 5^n coincide, but the digits at the position $m+1$ have opposite parity. Hence the last $m+1$ digits of one of these two powers of 5 alternate in parity. The inductive proof is completed.

20. There exist u, v such that $a\cos x + b\sin x = r\cos(x-u)$ and $A\cos 2x + B\sin 2x = R\cos 2(x-v)$, where $r = \sqrt{a^2+b^2}$ and $R = \sqrt{A^2+B^2}$. Then $1 - f(x) = r\cos(x-u) + R\cos 2(x-v) \le 1$ holds for all $x \in \mathbb{R}$.

There exists $x \in \mathbb{R}$ such that $\cos(x-u) \ge 0$ and $\cos 2(x-v) = 1$ (indeed, either $x = v$ or $x = v+\pi$ works). It follows that $R \le 1$. Similarly, there exists $x \in \mathbb{R}$ such that $\cos(x-u) = 1/\sqrt{2}$ and $\cos 2(x-v) \ge 0$ (either $x = u - \pi/4$ or $x = u + \pi/4$ works). It follows that $r \le \sqrt{2}$.

Remark. The proposition of this problem contained as an addendum the following, more difficult, inequality:

$$\sqrt{a^2+b^2} + \sqrt{A^2+B^2} \le 2.$$

The proof follows from the existence of $x \in \mathbb{R}$ such that $\cos(x-u) \ge 1/2$ and $\cos 2(x-v) \ge 1/2$.

21. Let us consider the vectors $v_1 = (x_1, x_2, x_3)$, $v_2 = (y_1, y_2, y_3)$, $v_3 = (1,1,1)$ in space. The given equalities express the condition that these three vectors are mutually perpendicular. Also, $\frac{x_1^2}{x_1^2+x_2^2+x_3^2}$, $\frac{y_1^2}{y_1^2+y_2^2+y_3^2}$, and $1/3$ are the squares of the projections of the vector $(1,0,0)$ onto the directions of v_1, v_2, v_3, respectively. The result follows from the fact that the sum of squares of projections of a unit vector on three mutually perpendicular directions is 1.

22. Since the quadrilateral OA_1BB_1 is cyclic, $\angle OA_1B_1 = \angle OBC$. By using the analogous equalities we obtain $\angle OA_4B_4 = \angle OB_3C_3 = \angle OC_2D_2 = \angle OD_1A_1 = \angle OAB$, and similarly $\angle OB_4A_4 = \angle OBA$. Hence $\triangle OA_4B_4 \sim \triangle OAB$. Analogously, we have for the other three pairs of triangles $\triangle OB_4C_4 \sim \triangle OBC$, $\triangle OC_4D_4 \sim \triangle OCD$, $\triangle OD_4A_4 \sim \triangle ODA$, and consequently $ABCD \sim A_4B_4C_4D_4$.

23. Every polynomial $q(x_1, \ldots, x_n)$ with integer coefficients can be expressed in the form $q = r_1 + x_1 r_2$, where r_1, r_2 are polynomials in $x_1, \ldots, x_n$ with integer coefficients in which the variable x_1 occurs only with even exponents. Thus if $q_1 = r_1 - x_1 r_2$, the polynomial $qq_1 = r_1^2 - x_1^2 r_2^2$ contains x_1 only with even exponents. We can continue inductively constructing polynomials q_j, $j = 2, 3, \ldots, n$, such that $qq_1q_2 \cdots q_j$ contains each of the variables $x_1, x_2, \ldots, x_j$ only with even exponents. Thus the polynomial $qq_1 \cdots q_n$ is a polynomial in $x_1^2, \ldots, x_n^2$.

The polynomials f and g exist for every $n \in \mathbb{N}$. In fact, it suffices to construct $q_1, \ldots, q_n$ for the polynomial $q = x_1 + \cdots + x_n$ and take $f = q_1 q_2 \cdots q_n$.

24. Setting $x = y = 0$ gives us $f(0) = 0$. Let us put $g(x) = \arctan f(x)$. The given functional equation becomes $\tan g(x+y) = \tan(g(x) + g(y))$; hence

$$g(x+y) = g(x) + g(y) + k(x,y)\pi,$$

where $k(x,y)$ is an integer function. But $k(x,y)$ is continuous and $k(0,0) = 0$, therefore $k(x,y) = 0$. Thus we obtain the classical Cauchy's functional equation $g(x+y) = g(x) + g(y)$ on the interval $(-1, 1)$, all of whose continuous solutions are of the form $g(x) = ax$ for some real a. Moreover, $g(x) \in (-\pi, \pi)$ implies $|a| \le \pi/2$.

Therefore $f(x) = \tan ax$ for some $|a| \le \pi/2$, and this is indeed a solution to the given equation.

25. Let

$$f_n(z) = z^n + a \sum_{k=1}^{n} \binom{n}{k} (a - kb)^{k-1}(z + kb)^{n-k}.$$

We shall prove by induction on n that $f_n(z) = (z + a)^n$. This is trivial for $n = 1$. Suppose that the statement is true for some positive integer $n - 1$. Then

$$f_n'(z) = nz^{n-1} + a \sum_{k=1}^{n-1} \binom{n}{k}(n-k)(a-kb)^{k-1}(z+kb)^{n-k-1}$$

$$= nz^{n-1} + na \sum_{k=1}^{n-1} \binom{n-1}{k}(a-kb)^{k-1}(z+kb)^{n-k-1}$$

$$= nf_{n-1}(z) = n(z+a)^{n-1}.$$

It remains to prove that $f_n(-a) = 0$. For $z = -a$ we have by the lemma of (SL81-13),

$$f_n(-a) = (-a)^n + a \sum_{k=1}^{n} \binom{n}{k}(-1)^{n-k}(a-kb)^{n-1}$$

$$= a \sum_{k=0}^{n} \binom{n}{k}(-1)^{n-k}(a-kb)^{n-1} = 0.$$

26. The result is an immediate consequence (for $G = \{-1,1\}$) of the following generalization.

(1) Let G be a proper subgroup of $\mathbb{Z}_n^*$ (the multiplicative group of residue classes modulo n coprime to n), and let V be the union of elements of G. A number $m \in V$ is called indecomposable in V if there do not exist numbers $p, q \in V$, $p, q \notin \{-1, 1\}$, such that $pq = m$. There exists a number $r \in V$ that can be expressed as a product of elements indecomposable in V in more than one way.

First proof. We shall start by proving the following lemma.

Lemma. There are infinitely many primes not in V that do not divide n.

Proof. There is at least one such prime: In fact, any number other than ± 1 not in V must have a prime factor not in V, since V is closed under multiplication. If there were a finite number of such primes, say $p_1, p_2, \ldots, p_k$, then one of the numbers $p_1 p_2 \cdots p_k + n$, $p_1^2 p_2 \cdots p_k + n$ is not in V and is coprime to n and $p_1, \ldots, p_k$, which is a contradiction.

[This lemma is actually a direct consequence of Dirichlet's theorem.]

Let us consider two such primes p, q that are congruent modulo n. Let p^k be the least power of p that is in V. Then p^k, q^k, $p^{k-1}q$, pq^{k-1} belong to V and are indecomposable in V. It follows that

$$r = p^k \cdot q^k = p^{k-1}q \cdot pq^{k-1}$$

has the desired property.

Second proof. Let p be any prime not in V that does not divide n, and let p^k be the least power of p that is in V. Obviously p^k is indecomposable in V. Then the number

$$r = p^k \cdot (p^{k-1} + n)(p+n) = p(p^{k-1} + n) \cdot p^{k-1}(p+n)$$

has at least two different factorizations into indecomposable factors.

27. The result is a consequence of the generalization from the previous problem for $G = \{1\}$.

    Remark. There is an explicit example: $r = (n-1)^2 \cdot (2n-1)^2 = [(n-1)(2n-1)]^2$.

28. The recurrent relations give us that

$$x_{i+1} = \left[\frac{x_i + [n/x_i]}{2}\right] = \left[\frac{x_i + n/x_i}{2}\right] \geq [\sqrt{n}].$$

    On the other hand, if $x_i > [\sqrt{n}]$ for some i, then we have $x_{i+1} < x_i$. This follows from the fact that $x_{i+1} < x_i$ is equivalent to $x_i > (x_i + n/x_i)/2$, i.e., to $x_i^2 > n$. Therefore $x_i = [\sqrt{n}]$ holds for at least one $i \leq n - [\sqrt{n}] + 1$.

    Remark. If $n+1$ is a perfect square, then $x_i = [\sqrt{n}]$ implies $x_{i+1} = [\sqrt{n}] + 1$. Otherwise, $x_i = [\sqrt{n}]$ implies $x_{i+1} = [\sqrt{n}]$.

29. Let us denote the midpoints of segments $LM, AN, BL, MN, BK, CM, NK, CL, DN, KL, DM, AK$ by $P_1, P_2, P_3, P_4, P_5, P_6, P_7, P_8, P_9, P_{10}, P_{11}, P_{12}$, respectively. We shall prove that the dodecagon $P_1 P_2 P_3 \ldots P_{11} P_{12}$ is regular. From $BL = BA$ and $\angle ABL = 30°$ it follows that $\angle BAL = 75°$. Similarly $\angle DAM = 75°$, and therefore $\angle LAM = 60°$, which together with the fact $AL = AM$ implies that $\triangle ALM$ is equilateral. Now, from the triangles OLM and ALN, we deduce $OP_1 = LM/2$, $OP_2 = AL/2$ and $OP_2 \parallel AL$.

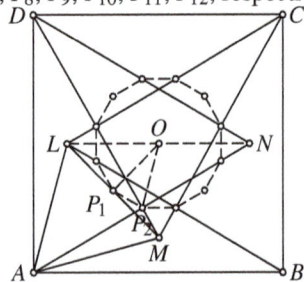

    Hence $OP_1 = OP_2$, $\angle P_1 OP_2 = \angle P_1 AL = 30°$ and $\angle P_2 OM = \angle LAD = 15°$. The desired result follows from symmetry.

30. Suppose $\angle SBA = x$. By the trigonometric form of Ceva's theorem we have

$$\frac{\sin(96° - x)}{\sin x} \frac{\sin 18°}{\sin 12°} \frac{\sin 6°}{\sin 48°} = 1. \tag{1}$$

    We claim that $x = 12°$ is a solution of this equation. To prove this, it is enough to show that $\sin 84° \sin 6° \sin 18° = \sin 48° \sin 12° \sin 12°$, which is equivalent to $\sin 18° = 2\sin 48° \sin 12° = \cos 36° - \cos 60°$. The last equality can be checked directly.

Since the equation is equivalent to

$$(\sin 96^\circ \cot x - \cos 96^\circ) \sin 6^\circ \sin 18^\circ = \sin 48^\circ \sin 12^\circ,$$

the solution $x \in [0, \pi)$ is unique. Hence $x = 12^\circ$.

Second solution. We know that if a, b, c, a', b', c' are points on the unit circle in the complex plane, the lines aa', bb', cc' are concurrent if and only if

$$(a - b')(b - c')(c - a') = (a - c')(b - a')(c - b'). \tag{1}$$

We shall prove that $x = 12^\circ$. We may suppose that ABC is the triangle in the complex plane with vertices $a = 1$, $b = \varepsilon^9$, $c = \varepsilon^{14}$, where $\varepsilon = \cos \frac{\pi}{15} + i \sin \frac{\pi}{15}$. If $a' = \varepsilon^{12}$, $b' = \varepsilon^{28}$, $c' = \varepsilon$, our task is the same as proving that lines aa', bb', cc' are concurrent, or by (1) that

$$(1 - \varepsilon^{28})(\varepsilon^9 - \varepsilon)(\varepsilon^{14} - \varepsilon^{12}) - (1 - \varepsilon)(\varepsilon^9 - \varepsilon^{12})(\varepsilon^{14} - \varepsilon^{28}) = 0.$$

The last equality holds, since the left-hand side is divisible by the minimum polynomial of ε: $z^8 + z^7 - z^5 - z^4 - z^3 + z + 1$.

31. We obtain from (1) that $f(1, c) = f(1, c)f(1, c)$; hence $f(1, c) = 1$ and consequently $f(-1, c)f(-1, c) = f(1, c) = 1$, i.e. $f(-1, c) = 1$. Analogously, $f(c, 1) = f(c, -1) = 1$.
Clearly $f(1, 1) = f(-1, 1) = f(1, -1) = 1$. Now let us assume that $a \neq 1$. Observe that $f(x^{-1}, y) = f(x, y^{-1}) = f(x, y)^{-1}$. Thus by (1) and (2) we get

$$1 = f(a, 1 - a)f(1/a, 1 - 1/a)$$
$$= f(a, 1 - a)f\left(a, \frac{1}{1 - 1/a}\right) = f\left(a, \frac{1 - a}{1 - 1/a}\right) = f(a, -a).$$

We now have $f(a, a) = f(a, -1)f(a, -a) = 1 \cdot 1 = 1$ and $1 = f(ab, ab) = f(a, ab)f(b, ab) = f(a, a)f(a, b)f(b, a)f(b, b) = f(a, b)f(b, a)$.

32. It is a known result that among six persons there are 3 mutually acquainted or 3 mutually unacquainted. By the condition of the problem the last case is excluded. If there is a man in the room who is not acquainted with four of the others, then these four men are mutually acquainted. Otherwise, each man is acquainted with at least five others, and since the sum of numbers of acquaintances of all men in the room is even, one of the men is acquainted with at least six men. Among these six there are three mutually acquainted, and they together with the first one make a group of four mutually acquainted men.

33. Let r be the radius of K and $s > \sqrt{2}/r$ an integer. Consider the points $A_k(ka_1 - [ka_1], ka_2 - [ka_2])$, where $k = 0, 1, 2, \ldots, s^2$. Since all these points are in the unit square, two of them, say A_p, A_q, $q > p$, are in a small square with side $1/s$, and consequently $A_p A_q \leq \sqrt{2}/s < r$. Therefore, for $n = q - p$, $m_1 = [qa_1] - [pa_1]$ and $m_2 = [qa_2] - [pa_2]$ the distance between the points $n(a_1, a_2)$ and (m_1, m_2) is less then r, i.e., the point (m_1, m_2) is in the circle $K + n(a_1, a_2)$.

34. Let A be the set of the 2^n sequences of n terms equal to ± 1. Since there are k^2 products ab with $a, b \in B$, by the pigeonhole principle there exists $c \in A$ such that $ab = c$ holds for at most $k^2/2^n$ pairs $(a, b) \in B \times B$. Then $cb \in B$ holds for at most $k^2/2^n$ values $b \in B$, which means that $|B \cap cB| \leq k^2/2^n$.

35. The solutions are 0 and $N_k = 10\underbrace{99\ldots9}_{k}89$, where $k = 0, 1, 2, \ldots$.

    Remark. If we omit the condition that at most one of the digits is zero, the solutions are numbers of the form $N_{k_1} N_{k_2} \ldots N_{k_r}$, where $k_1 = k_r$, $k_2 = k_{r-1}$ etc. The more general problem $k \cdot \overline{a_1 a_2 \ldots a_n} = \overline{a_n \ldots a_2 a_1}$ has solutions only for $k = 9$ and for $k = 4$ (namely 0, 2199...978 and combinations as above).

36. It can be shown by simple induction that $S^m(a_1, \ldots, a_{2^n}) = (b_1, \ldots, b_{2^n})$, where

$$b_k = \prod_{i=0}^{m} a_{k+i}^{\binom{m}{i}} \quad \text{(assuming that } a_{k+2^n} = a_k\text{).}$$

    If we take $m = 2^n$ all the binomial coefficients $\binom{m}{i}$ apart from $i = 0$ and $i = m$ will be even, and thus $b_k = a_k a_{k+m} = 1$ for all k.

37. We look for a solution with $x_1^{A_1} = \cdots = x_n^{A_n} = n^{A_1 A_2 \cdots A_n x}$ and $x_{n+1} = n^y$. In order for this to be a solution we must have $A_1 A_2 \cdots A_n x + 1 = A_{n+1} y$. This equation has infinitely many solutions (x, y) in $\mathbb{N}$, since $A_1 A_2 \cdots A_n$ and A_{n+1} are coprime.

38. The condition says that the quadratic equation $f(x) = 0$ has distinct real solutions, where

$$f(x) = 3x^2 \sum_{j=1}^{n} m_j - 2x \sum_{j=1}^{n} m_j(a_j + b_j + c_j) + \sum_{j=1}^{n} m_j(a_j b_j + b_j c_j + c_j a_j).$$

    It is easy to verify that the function f is the derivative of

$$F(x) = \sum_{j=1}^{n} m_j(x - a_j)(x - b_j)(x - c_j).$$

    Since $F(a_1) \leq 0 \leq F(a_n)$, $F(b_1) \leq 0 \leq F(b_n)$ and $F(c_1) \leq 0 \leq F(c_n)$, $F(x)$ has three distinct real roots, and hence by Rolle's theorem its derivative $f(x)$ has two distinct real roots.

39. By the pigeonhole principle, we can find 5 distinct points among the given 37 such that their x-coordinates are congruent and their y-coordinates are congruent modulo 3. Now among these 5 points either there exist three with z-coordinates congruent modulo 3, or there exist three whose z-coordinates are congruent to 0, 1, 2 modulo 3. These three points are the desired ones.

    Remark. The minimum number n such that among any n integer points in space one can find three points whose barycenter is an integer point is $n = 19$. Each proof of this result seems to consist in studying a great number of cases.

40. Let us divide the chessboard into 16 squares $Q_1, Q_2, \ldots, Q_{16}$ of size 2×2. Let s_k be the sum of numbers in Q_k, and let us assume that $s_1 \geq s_2 \geq \cdots \geq s_{16}$. Since $s_4 + s_5 + \cdots + s_{16} \geq 1 + 2 + \cdots + 52 = 1378$, we must have $s_4 \geq 100$ and hence $s_1, s_2, s_3 \geq 100$ as well.

41. The considered sums are congruent modulo N to $S_k = \sum_{i=1}^{N}(i+k)a_i$, $k = 0, 1, \ldots, N-1$. Since $S_k = S_0 + k(a_1 + \cdots + a_n) = S_0 + k$, all these sums give distinct residues modulo N and therefore are distinct.

42. It can be proved by induction on n that
$$\{a_{n,k} \mid 1 \leq k \leq 2^n\} = \{2^m \mid m = 3^n + 3^{n-1}s_1 + \cdots + 3^1 s_{n-1} + s_n \ (s_i = \pm 1)\}.$$

Thus the result is an immediate consequence of the following lemma.

Lemma. Each positive integer s can be uniquely represented in the form
$$s = 3^n + 3^{n-1}s_1 + \cdots + 3^1 s_{n-1} + s_n, \quad \text{where } s_i \in \{-1, 0, 1\}. \quad (1)$$

Proof. Both the existence and the uniqueness can be shown by simple induction on s. The statement is trivial for $s = 1$, while for $s > 1$ there exist $q \in \mathbb{N}$, $r \in \{-1, 0, 1\}$ such that $s = 3q + r$, and q has a unique representation of the form (1).

43. Since $k(k+1)\cdots(k+p) = (p+1)!\binom{k+p}{p+1} = (p+1)!\left[\binom{k+p+1}{p+2} - \binom{k+p}{p+2}\right]$, it follows that
$$\sum_{k=1}^{n} k(k+1)\cdots(k+p) = (p+1)!\binom{n+p+1}{p+2} = \frac{n(n+1)\cdots(n+p+1)}{p+2}.$$

44. Let $d(X, \sigma)$ denote the distance from a point X to a plane σ. Let us consider the pair (A, π) where $A \in E$ and π is a plane containing some three points $B, C, D \in E$ such that $d(A, \pi)$ is the smallest possible. We may suppose that B, C, D are selected such that $\triangle BCD$ contains no other points of E. Let A' be the projection of A on π, and let l_b, l_c, l_d be lines through B, C, D parallel to CD, DB, BC respectively. If A' is in the half-plane determined by l_d not containing BC, then $d(D, ABC) \leq d(A', ABC) < d(A, BCD)$, which is impossible. Similarly, A' lies in the half-planes determined by l_b, l_c that contain D, and hence A' is inside the triangle bordered by l_b, l_c, l_d. The minimality property of (A, π) and the way in which BCD was selected guarantee that $E \cap T = \{A, B, C, D\}$.

45. As in the previous problem, let us choose the pair (A, π) such that $d(A, \pi)$ is minimal. If π contains only three points of E, we are done. If not, there are four points in $E \cap P$, say A_1, A_2, A_3, A_4, such that the quadrilateral $Q = A_1 A_2 A_3 A_4$ contains no other points of E. Suppose Q is not convex, and that w.l.o.g. A_1 is inside the triangle $A_2 A_3 A_4$. If A_0 is the projection of A on P, the point A_1 belongs to one of the triangles $A_0 A_2 A_3$, $A_0 A_3 A_4$, $A_0 A_4 A_2$, say $A_0 A_2 A_3$. Then $d(A_1, AA_2 A_3) \leq d(A_0, AA_2 A_3) < AA_0$, which is impossible. Hence Q is convex. Also, by the minimality property of (A, π) the pyramid $AA_1 A_2 A_3 A_4$ contains no other points of E.

46. We need to consider only the case $t > |x|$. There is no loss of generality in assuming $x > 0$.

To obtain the estimate from below, set

$$a_1 = f\left(-\frac{x+t}{2}\right) - f(-(x+t)), \qquad a_2 = f(0) - f\left(-\frac{x+t}{2}\right),$$
$$a_3 = f\left(\frac{x+t}{2}\right) - f(0), \qquad a_4 = f(x+t) - f\left(\frac{x+t}{2}\right).$$

Since $-(x+t) < x-t$ and $x < (x+t)/2$, we have $f(x) - f(x-t) \le a_1 + a_2 + a_3$. Since $2^{-1} < a_{j+1}/a_j < 2$, it follows that

$$g(x,t) > \frac{a_4}{a_1 + a_2 + a_3} > \frac{a_3/2}{4a_3 + 2a_3 + a_3} = 14^{-1}.$$

To obtain the estimate from above, set

$$b_1 = f(0) - f\left(-\frac{x+t}{3}\right), \qquad b_2 = f\left(\frac{x+t}{3}\right) - f(0),$$
$$b_3 = f\left(\frac{2(x+t)}{3}\right) - f\left(\frac{x+t}{3}\right), \qquad b_4 = f(x+t) - f\left(\frac{2(x+t)}{3}\right).$$

If $t < 2x$, then $x - t < -(x+t)/3$ and therefore $f(x) - f(x-t) \ge b_1$. If $t \ge 2x$, then $(x+t)/3 \le x$ and therefore $f(x) - f(x-t) \ge b_2$. Since $2^{-1} < b_{j+1}/b_j < 2$, we get

$$g(x,t) < \frac{b_2 + b_3 + b_4}{\min\{b_1, b_2\}} < \frac{b_2 + 2b_2 + 4b_2}{b_2/2} = 14.$$

47. M lies on AB and N lies on BC. If $CQ \le 2CD/3$, then $BM = CQ/2$. If $CQ > 2CD/3$, then N coincides with C.

48. Let a plane cut the edges AB, BC, CD, DA at points K, L, M, N respectively.

Let D', A', B' be distinct points in the plane ABC such that the triangles BCD', $CD'A'$, $D'A'B'$ are equilateral, and $M' \in [CD']$, $N' \in [D'A']$, and $K' \in [A'B']$ such that $CM' = CM$, $A'N' = AN$, and $A'K' = AK$. The perimeter P of the quadrilateral $KLMN$ is equal to the length of the polygonal line $KLM'N'K'$, which is not less than KK'.

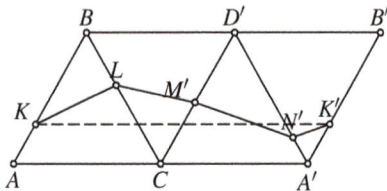

It follows that $P \ge 2a$.

Let us consider all quadrilaterals $KLMN$ that are obtained by intersecting the tetrahedron by a plane parallel to a fixed plane α. The lengths of the segments KL, LM, MN, NK are linear functions in AK, and so is P. Thus P takes its maximum at an endpoint of the interval, i.e., when the plane $KLMN$ passes through one of the vertices A, B, C, D, and it is easy to see that in this case $P \le 3a$.

49. If one of p,q, say p, is zero, then $-q$ is a perfect square. Conversely, $(p,q) = (0,-t^2)$ and $(p,q) = (-t^2,0)$ satisfy the conditions for $t \in \mathbb{Z}$.

We now assume that p,q are nonzero. If the trinomial $x^2 + px + q$ has two integer roots x_1, x_2, then $|q| = |x_1 x_2| \geq |x_1| + |x_2| - 1 \geq |p| - 1$. Similarly, if $x^2 + qx + p$ has integer roots, then $|p| \geq |q| - 1$ and $q^2 - 4p$ is a square. Thus we have two cases to investigate:

   (i) $|p| = |q|$. Then $p^2 - 4q = p^2 \pm 4p$ is a square, so $(p,q) = (4,4)$.
   (ii) $|p| = |q| \pm 1$. The solutions for (p,q) are $(t, -1-t)$ for $t \in \mathbb{Z}$ and $(5,6)$, $(6,5)$.

50. Suppose that $P_n(x) = n$ for $x \in \{x_1, x_2, \ldots, x_n\}$. Then

$$P_n(x) = (x - x_1)(x - x_2) \cdots (x - x_n) + n.$$

From $P_n(0) = 0$ we obtain $n = |x_1 x_2 \cdots x_n| \geq 2^{n-2}$ (because at least $n-2$ factors are different from ± 1) and therefore $n \geq 2^{n-2}$. It follows that $n \leq 4$.

For each positive integer $n \leq 4$ there exists a polynomial P_n. Here is the list of such polynomials:

$$n = 1: \pm x, \qquad\qquad n = 2: 2x^2, \ x^2 \pm x, \ -x^2 \pm 3x,$$
$$n = 3: \pm(x^3 - x) + 3x^2, \qquad n = 4: -x^4 + 5x^2.$$

51. We shall use the following algorithm:

> Choose a segment of maximum length ("basic" segment) and put on it unused segments of the opposite color without overlapping, each time of the maximum possible length, as long as it is possible. Repeat the procedure with remaining segments until all the segments are used.

Let us suppose that the last basic segment is black. Then the length of the used part of any white basic segment is greater than the free part, and consequently at least one-half of the length of the white segments has been used more than once. Therefore all basic segments have total length at most 1.5 and can be distributed on a segment of length 1.51.

On the other hand, if we are given two white segments of lengths 0.5 and two black segments of lengths 0.999 and 0.001, we cannot distribute them on a segment of length less than 1.499.

52. The maximum and minimum are $2R\sqrt{4 - 2k^2}$ and $2R\left(1 + \sqrt{1 - k^2}\right)$ respectively.

53. The discriminant of the given equation considered as a quadratic equation in b is $196 - 75a^2$. Thus $75a^2 \leq 196$ and hence $-1 \leq a \leq 1$. Now the integer solutions of the given equation are easily found: $(-1,3)$, $(0,0)$, $(1,2)$.

54. We shall use the following lemma.

Lemma. If a real function f is convex on the interval I and $x,y,z \in I$, $x \leq y \leq z$, then

$$(y - z)f(x) + (z - x)f(y) + (x - y)f(z) \leq 0.$$

Proof. The inequality is obvious for $x=y=z$. If $x<z$, then there exist p,r such that $p+r=1$ and $y=px+rz$. Then by Jensen's inequality $f(px+rz) \leq pf(x)+rf(z)$, which is equivalent to the statement of the lemma.

By applying the lemma to the convex function $-\ln x$ we obtain $x^y y^z z^x \geq y^x z^y x^z$ for any $0<x\leq y\leq z$. Multiplying the inequalities $a^b b^c c^a \geq b^a c^b a^c$ and $a^c c^d d^a \geq c^a d^c a^d$ we get the desired inequality.

Remark. Similarly, for $0 < a_1 \leq a_2 \leq \cdots \leq a_n$ it holds that $a_1^{a_2} a_2^{a_3} \cdots a_n^{a_1} \geq a_2^{a_1} a_3^{a_2} \cdots a_1^{a_n}$.

55. The statement is true without the assumption that $O \in BD$. Let $BP \cap DN = \{K\}$. If we denote $\overrightarrow{AB} = a$, $\overrightarrow{AD} = b$ and $\overrightarrow{AO} = \alpha a + \beta b$ for some $\alpha, \beta \in \mathbb{R}$, $1/\alpha + 1/\beta \neq 1$, by straightforward calculation we obtain that

$$\overrightarrow{AK} = \frac{\alpha}{\alpha+\beta-\alpha\beta} a + \frac{\beta}{\alpha+\beta-\alpha\beta} b = \frac{1}{\alpha+\beta-\alpha\beta} \overrightarrow{AO}.$$

Hence A, K, O are collinear.

56. See the solution to (LL67-36).

57. Suppose that there exists a sequence of 17 terms $a_1, a_2, \ldots, a_{17}$ satisfying the required conditions. Then the sum of terms in each row of the rectangular array below is positive, while the sum of terms in each column is negative, which is a contradiction.

$$a_1\ a_2\ \ldots\ a_{11}$$
$$a_2\ a_3\ \ldots\ a_{12}$$
$$\vdots\ \vdots\qquad \vdots$$
$$a_7\ a_8\ \ldots\ a_{17}$$

On the other hand, there exist 16-term sequences with the required property. An example is $5,5,-13,5,5,5,-13,5,5,-13,5,5,5,-13,5,5$ which can be obtained by solving the system of equations $\sum_{i=k}^{k+10} a_i = 1$ $(k = 1,2,\ldots,6)$ and $\sum_{i=l}^{l+6} a_i = -1$ $(l = 1,2,\ldots,10)$.

Second solution. We shall prove a stronger statement: If 7 and 11 in the question are replaced by any positive integers m,n, then the maximum number of terms is $m+n-(m,n)-1$.

Let $a_1, a_2, \ldots, a_l$ be a sequence of real numbers, and let us define $s_0 = 0$ and $s_k = a_1 + \cdots + a_k$ $(k=1,\ldots,l)$. The given conditions are equivalent to $s_k > s_{k+m}$ for $0 \leq k \leq l-m$ and $s_k < s_{k+n}$ for $0 \leq k \leq l-n$.

Let $d = (m,n)$ and $m = m'd$, $n = n'd$. Suppose that there exists a sequence (a_k) of length greater than or equal to $l = m+n-d$ satisfying the required conditions. Then the $m'+n'$ numbers $s_0, s_d, \ldots, s_{(m'+n'-1)d}$ satisfy n' inequalities $s_{k+m} < s_k$ and m' inequalities $s_k < s_{k+n}$. Moreover, each term s_{kd} appears twice in these inequalities: once on the left-hand and once on the right-hand side. It follows that there exists a ring of inequalities $s_{i_1} < s_{i_2} < \cdots < s_{i_k} < s_{i_1}$, giving a contradiction. On the other hand, suppose that such a ring of inequalities can be made also for $l = m+n-d-1$, say $s_{i_1} < s_{i_2} < \cdots < s_{i_k} < s_{i_1}$. If there are p inequalities of the form $a_{k+m} < a_k$ and q inequalities of the form $a_{k+n} > a_k$ in the ring, then $qn = rm$, which implies $m' \mid q$, $n' \mid p$ and thus $k = p + q \geq m' + n'$. But since all $i_1, i_2, \ldots, i_k$ are congruent modulo d, we have $k \leq m' + n' - 1$, a contradiction. Hence there exists a sequence of length $m + n - d - 1$ with the required property.

58. The following inequality (Finsler and Hadwiger, 1938) is sharper than the one we have to prove:

$$2ab + 2bc + 2ca - a^2 - b^2 - c^2 \geq 4S\sqrt{3}. \tag{1}$$

First proof. Let us set $2x = b + c - a$, $2y = c + a - b$, $2z = a + b - c$. Then $x, y, z > 0$ and the inequality (1) becomes

$$y^2 z^2 + z^2 x^2 + x^2 y^2 \geq xyz(x + y + z),$$

which is equivalent to the obvious inequality $(xy - yz)^2 + (yz - zx)^2 + (zx - xy)^2 \geq 0$.

Second proof. Using the known relations for a triangle

$$a^2 + b^2 + c^2 = 2s^2 - 2r^2 - 8rR,$$
$$ab + bc + ca = s^2 + r^2 + 4rR,$$
$$S = rs,$$

where r and R are the radii of the incircle and the circumcircle, s the semiperimeter and S the area, we can transform (1) into

$$s\sqrt{3} \leq 4R + r.$$

The last inequality is a consequence of the inequalities $2r \leq R$ and $s^2 \leq 4R^2 + 4Rr + 3r^2$, where the last one follows from the equality $HI^2 = 4R^2 + 4Rr + 3r^2 - s^2$ (H and I being the orthocenter and the incenter of the triangle).

59. Let us consider the set R of pairs of coordinates of the points from E reduced modulo 3. If some element of R occurs thrice, then the corresponding points are vertices of a triangle with integer barycenter. Also, no three elements from E can have distinct x-coordinates and distinct y-coordinates. By an easy discussion we can conclude that the set R contains at most four elements. Hence $|E| \leq 8$.
An example of a set E consisting of 8 points that satisfies the required condition is

$$E = \{(0,0), (1,0), (0,1), (1,1), (3,6), (4,6), (3,7), (4,7)\}.$$

60. By Lagrange's interpolation formula we have

$$F(x) = \sum_{j=0}^{n} F(x_j) \frac{\prod_{i \neq j}(x - x_j)}{\prod_{i \neq j}(x_i - x_j)}.$$

Since the leading coefficient in $F(x)$ is 1, it follows that

$$1 = \sum_{j=0}^{n} \frac{F(x_j)}{\prod_{i \neq j}(x_i - x_j)}.$$

Since

$$\left| \prod_{i \neq j}(x_i - x_j) \right| = \prod_{i=0}^{j-1} |x_i - x_j| \prod_{i=j+1}^{n} |x_i - x_j| \geq j!(n-j)!,$$

we have

$$1 \leq \sum_{j=0}^{n} \frac{|F(x_j)|}{\left| \prod_{i \neq j}(x_i - x_j) \right|} \leq \frac{1}{n!} \sum_{j=0}^{n} \binom{n}{j} |F(x_j)| \leq \frac{2^n}{n!} \max |F(x_j)|.$$

Now the required inequality follows immediately.

4.20 Solutions to the Shortlisted Problems of IMO 1978

1. There exists an M_s that contains at least $2n/k = 2(k^2 + 1)$ elements. It follows that M_s contains either at least $k^2 + 1$ even numbers or at least $k^2 + 1$ odd numbers. In the former case, consider the predecessors of those $k^2 + 1$ numbers: among them, at least $\frac{k^2+1}{k+1} > k$, i.e., at least $k + 1$, belong to the same subset, say M_t. Then we choose s, t. The latter case is similar.

   Second solution. For all $i, j \in \{1, 2, \ldots, k\}$, consider the set $N_{ij} = \{r \mid 2r \in M_i, 2r - 1 \in M_j\}$. Then $\{N_{ij} \mid i, j\}$ is a partition of $\{1, 2, \ldots, n\}$ into k^2 subsets. For $n \geq k^3 + 1$ one of these subsets contains at least $k + 1$ elements, and the statement follows.

   Remark. The statement is not necessarily true when $n = k^3$.

2. Consider the transformation ϕ of the plane defined as the homothety $\mathscr{H}$ with center B and coefficient 2 followed by the rotation $\mathscr{R}$ about the center O through an angle of $60°$. Being direct, this mapping must be a rotational homothety. We also see that $\mathscr{H}$ maps S into the point symmetric to S with respect to OA, and $\mathscr{R}$ takes it back to S. Hence S is a fixed point, and is consequently also the center of ϕ. Therefore ϕ is the rotational homothety about S with the angle $60°$ and coefficient 2. (In fact, this could also be seen from the fact that ϕ preserves angles of triangles and maps the segment SR onto SB, where R is the midpoint of AB.)

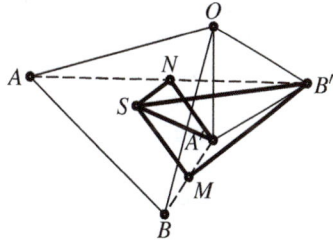

   Since $\phi(M) = B'$, we conclude that $\angle MSB' = 60°$ and $SB'/SM = 2$. Similarly, $\angle NSA' = 60°$ and $SA'/SN = 2$, so triangles MSB' and NSA' are indeed similar.

   Second solution. Probably the simplest way here is using complex numbers. Put the origin at O and complex numbers a, a' at points A, A', and denote the primitive sixth root of 1 by ω. Then the numbers at B, B', S and N are ωa, $\omega a'$, $(a + \omega a)/3$, and $(a + \omega a')/2$ respectively. Now it is easy to verify that $(n - s) = \omega(a' - s)/2$, i.e., that $\angle NSA' = 60°$ and $SA'/SN = 2$.

3. What we need are m, n for which $1978^m(1978^{n-m} - 1)$ is divisible by $1000 = 8 \cdot 125$. Since $1978^{n-m} - 1$ is odd, it follows that 1978^m is divisible by 8, so $m \geq 3$.

   Also, $1978^{n-m} - 1$ is divisible by 125, i.e., $1978^{n-m} \equiv 1 \pmod{125}$. Note that $1978 \equiv -2 \pmod 5$, and consequently also $-2^{n-m} \equiv 1$. Hence $4 \mid n - m = 4k, k \geq 1$. It remains to find the least k such that $1978^{4k} \equiv 1 \pmod{125}$. Since $1978^4 \equiv (-22)^4 = 484^2 \equiv (-16)^2 = 256 \equiv 6$, we reduce it to $6^k \equiv 1$. Now $6^k = (1+5)^k \equiv 1 + 5k + 25\binom{k}{2} \pmod{125}$, which reduces to $125 \mid 5k(5k - 3)$. But $5k - 3$ is not divisible by 5, and so $25 \mid k$. Therefore $100 \mid n - m$, and the desired values are $m = 3, n = 103$.

4. Let γ, φ be the angles of T_1 and T_2 opposite to c and w respectively. By the cosine theorem, the inequality is transformed into

$$a^2(2v^2 - 2uv\cos\varphi) + b^2(2u^2 - 2uv\cos\varphi)$$
$$+ 2(a^2 + b^2 - 2ab\cos\gamma)uv\cos\varphi \geq 4abuv\sin\gamma\sin\varphi.$$

This is equivalent to $2(a^2v^2 + b^2u^2) - 4abuv(\cos\gamma\cos\varphi + \sin\gamma\sin\varphi) \geq 0$, i.e., to

$$2(av - bu)^2 + 4abuv(1 - \cos(\gamma - \varphi)) \geq 0,$$

which is clearly satisfied. Equality holds if and only if $\gamma = \varphi$ and $a/b = u/v$, i.e., when the triangles are similar, a corresponding to u and b to v.

5. We first explicitly describe the elements of the sets M_1, M_2.

   $x \notin M_1$ is equivalent to $x = a + (a+1) + \cdots + (a+n-1) = n(2a+n-1)/2$ for some natural numbers n, a, $n \geq 2$. Among n and $2a + n - 1$, one is odd and the other even, and both are greater than 1; so x has an odd factor ≥ 3. On the other hand, for every x with an odd divisor $p > 3$ it is easy to see that there exist corresponding a, n. Therefore $M_1 = \{2^k \mid k = 0, 1, 2, \dots\}$.

   $x \notin M_2$ is equivalent to $x = a + (a+2) + \cdots + (a+2(n-1)) = n(a+n-1)$, where $n \geq 2$, i.e. to x being composite. Therefore $M_2 = \{1\} \cup \{p \mid p = \text{prime}\}$.

   $x \notin M_3$ is equivalent to $x = a + (a+3) + \cdots + (a+3(n-1)) = n(2a+3(n-1))/2$.

   It remains to show that every $c \in M_3$ can be written as $c = 2^k p$ with p prime. Suppose the opposite, that $c = 2^k pq$, where p, q are odd and $q \geq p \geq 3$. Then there exist positive integers a, n ($n \geq 2$) such that $c = n(2a + 3(n-1))/2$ and hence $c \notin M_3$. Indeed, if $k = 0$, then $n = 2$ and $2a + 3 = pq$ work; otherwise, setting $n = p$ one obtains $a = 2^k q - 3(p-1)/2 \geq 2q - 3(p-1)/2 \geq (p+3)/2 > 1$.

6. For fixed n and the set $\{\varphi(1), \dots, \varphi(n)\}$, there are finitely many possibilities for a mapping φ on $\{1, \dots, n\}$. Suppose φ is the one among these for which $\sum_{k=1}^{n} \varphi(k)/k^2$ is minimal. If $i < j$ and $\varphi(i) > \varphi(j)$ for some $i, j \in \{1, \dots, n\}$, define ψ as $\psi(i) = \varphi(j)$, $\psi(j) = \varphi(i)$, and $\psi(k) = \varphi(k)$ for all other k. Then

$$\sum \frac{\varphi(k)}{k^2} - \sum \frac{\psi(k)}{k^2} = \left(\frac{\varphi(i)}{i^2} + \frac{\varphi(j)}{j^2} \right) - \left(\frac{\varphi(i)}{j^2} + \frac{\varphi(j)}{i^2} \right)$$
$$= (i - j)(\varphi(j) - \varphi(i)) \frac{i+j}{i^2 j^2} > 0,$$

which contradicts the assumption. This shows that $\varphi(1) < \cdots < \varphi(n)$, and consequently $\varphi(k) \geq k$ for all k. Hence

$$\sum_{k=1}^{n} \frac{\varphi(k)}{k^2} \geq \sum_{k=1}^{n} \frac{k}{k^2} = \sum_{k=1}^{n} \frac{1}{k}.$$

7. Let $x = OA$, $y = OB$, $z = OC$, $\alpha = \angle BOC$, $\beta = \angle COA$, $\gamma = \angle AOB$. The conditions yield the equation $x + y + \sqrt{x^2 + y^2 - 2xy\cos\gamma} = 2p$, which transforms to $(2p - x - y)^2 = x^2 + y^2 - 2xy\cos\gamma$, i.e. $(p - x)(p - y) = xy(1 - \cos\gamma)$. Thus

$$\frac{p-x}{x} \cdot \frac{p-y}{y} = 1 - \cos\gamma,$$

and analogously $\frac{p-y}{y} \cdot \frac{p-z}{z} = 1 - \cos\alpha$, $\frac{p-z}{z} \cdot \frac{p-x}{x} = 1 - \cos\beta$. Setting $u = \frac{p-x}{x}$, $v = \frac{p-y}{y}$, $w = \frac{p-z}{z}$, the above system becomes

$$uv = 1 - \cos\gamma, \qquad vw = 1 - \cos\alpha, \qquad wu = 1 - \cos\beta.$$

This system has a unique solution in positive real numbers u, v, w:

$$u = \sqrt{\frac{(1 - \cos\beta)(1 - \cos\gamma)}{1 - \cos\alpha}}, \quad v = \sqrt{\frac{(1 - \cos\gamma)(1 - \cos\alpha)}{1 - \cos\beta}},$$

$$w = \sqrt{\frac{(1 - \cos\alpha)(1 - \cos\beta)}{1 - \cos\gamma}}.$$

Finally, the values of x, y, z are uniquely determined from u, v, w.

Remark. It is not necessary that the three lines be in the same plane. Also, there could be any odd number of lines instead of three.

8. Take the subset $\{a_i\} = \{1, 7, 11, 13, 17, 19, 23, 29, \ldots, 30m - 1\}$ of S containing all the elements of S that are not multiples of 3. There are $8m$ such elements. Every element in S can be uniquely expressed as $3^t a_i$ for some i and $t \geq 0$. In a subset of S with $8m + 1$ elements, two of them will have the same a_i, hance one will divide the other.

On the other hand, for each $i = 1, 2, \ldots, 8m$ choose $t \geq 0$ such that $10m < b_i = 3^t a_i < 30m$. Then there are $8m$ b_i's in the interval $(10m, 30m)$, and the quotient of any two of them is less than 3, so none of them can divide any other. Thus the answer is $8m$.

9. Since the nth missing number (gap) is $f(f(n)) + 1$ and $f(f(n))$ is a member of the sequence, there are exactly $n - 1$ gaps less than $f(f(n))$. This leads to

$$f(f(n)) = f(n) + n - 1. \tag{1}$$

Since 1 is not a gap, we have $f(1) = 1$. The first gap is $f(f(1)) + 1 = 2$. Two consecutive integers cannot both be gaps (the predecessor of a gap is of the form $f(f(m))$). Now we deduce $f(2) = 3$; a repeated application of the formula above gives $f(3) = 3 + 1 = 4$, $f(4) = 4 + 2 = 6$, $f(6) = 9$, $f(9) = 14$, $f(14) = 22$, $f(22) = 35$, $f(35) = 56$, $f(56) = 90$, $f(90) = 145$, $f(145) = 234$, $f(234) = 378$. Also, $f(f(35)) + 1 = 91$ is a gap, so $f(57) = 92$. Then by (1), $f(92) = 148$, $f(148) = 239$, $f(239) = 386$. Finally, here $f(f(148)) + 1 = 387$ is a gap, so $f(240) = 388$.

Second solution. As above, we arrive at formula (1). Then by simple induction it follows that $f(F_n + 1) = F_{n+1} + 1$, where F_k is the Fibonacci sequence ($F_1 = F_2 = 1$).

We now prove by induction (on n) that $f(F_n + x) = F_{n+1} + f(x)$ for all x with $1 \le x \le F_{n-1}$. This is trivially true for $n = 0, 1$. Supposing that it holds for $n - 1$, we shall prove it for n:

(i) If $x = f(y)$ for some y, then by the inductive assumption and (1)

$$f(F_n + x) = f(F_n + f(y)) = f(f(F_{n-1} + y))$$
$$= F_n + f(y) + F_{n-1} + y - 1 = F_{n+1} + f(x).$$

(ii) If $x = f(f(y)) + 1$ is a gap, then $f(F_n + x - 1) + 1 = F_{n+1} + f(x-1) + 1$ is a gap also:

$$F_{n+1} + f(x) + 1 = F_{n+1} + f(f(f(y))) + 1$$
$$= f(F_n + f(f(y))) + 1 = f(f(F_{n-1} + f(y))) + 1.$$

It follows that $f(F_n + x) = F_{n+1} + f(x-1) + 2 = F_{n+1} + f(x)$.

Now, since we know that each positive integer x is expressible as $x = F_{k_1} + F_{k_2} + \cdots + F_{k_r}$, where $0 < k_r \ne 2$, $k_i \ge k_{i+1} + 2$, we obtain $f(x) = F_{k_1+1} + F_{k_2+1} + \cdots + F_{k_r+1}$. Particularly, $240 = 233 + 5 + 2$, so $f(240) = 377 + 8 + 3 = 388$.

Remark. It can be shown that $f(x) = [\alpha x]$, where $\alpha = (1 + \sqrt{5})/2$.

10. Assume the opposite. One of the countries, say A, contains at least 330 members $a_1, a_2, \ldots, a_{330}$ of the society ($6 \cdot 329 = 1974$). Consider the differences $a_{330} - a_i$, $i = 1, 2, \ldots, 329$: the members with these numbers are not in A, so at least 66 of them, $a_{330} - a_{i_1}, \ldots, a_{330} - a_{i_{66}}$, belong to the same country, say B. Then the differences $(a_{i_{66}} - a_{330}) - (a_{i_j} - a_{330}) = a_{i_{66}} - a_{i_j}$, $j = 1, 2, \ldots, 65$, are neither in A nor in B. Continuing this procedure, we find that 17 of these differences are in the same country, say C, then 6 among 16 differences of themselves in a country D, and 3 among 5 differences of themselves in E; finally, two differences of these 3 differences belong to country F, so that the difference of themselves cannot be in any country. This is a contradiction.

Remark. The following stronger ($[6!e] = 1957$) statement can be proved in the same way.

Schur's lemma. If n is a natural number and e the logarithm base, then for every partition of the set $\{1, 2, \ldots, [en!]\}$ into n subsets one of these subsets contains some two elements and their difference.

11. Set $F(x) = f_1(x)f_2(x) \cdots f_n(x)$: we must prove concavity of $F^{1/n}$. By the assumption,

$$F(\theta x + (1 - \theta)y) \ge \prod_{i=1}^{n} [\theta f_i(x) + (1 - \theta)f(y)]$$

$$= \sum_{k=0}^{n} \theta^k (1 - \theta)^{n-k} \sum f_{i_1}(x) \ldots f_{i_k}(x) f_{i_{k+1}}(y) f_{i_n}(y),$$

where the second sum goes through all $\binom{n}{k}$ k-subsets $\{i_1,\ldots,i_k\}$ of $\{1,\ldots,n\}$. The inequality between the arithmetic and geometric means now gives us

$$\sum f_{i_1}(x)f_{i_2}(x)\cdots f_{i_k}(x)f_{i_{k+1}}(y)f_{i_n}(y) \geq \binom{n}{k}F(x)^{k/n}F(y)^{(n-k)/n}.$$

Inserting this in the above inequality and using the binomial formula, we finally obtain

$$F(\theta x+(1-\theta)y) \geq \sum_{k=0}^{n} \theta^k(1-\theta)^{n-k}\binom{n}{k}F(x)^{k/n}F(y)^{(n-k)/n}$$
$$= \left(\theta F(x)^{1/n}+(1-\theta)F(y)^{1/n}\right)^n,$$

which proves the assertion.

12. Let O be the center of the smaller circle, T its contact point with the circumcircle of ABC, and J the midpoint of segment BC. The figure is symmetric with respect to the line through A,O,J,T.

A homothety centered at A taking T into J will take the smaller circle into the incircle of ABC, hence will take O into the incenter I. On the other hand, $\angle ABT = \angle ACT = 90°$ implies that the quadrilaterals $ABTC$ and $APOQ$ are similar. Hence the above homothety also maps O to the midpoint of PQ. This finishes the proof.

Remark. The assertion is true for a nonisosceles triangle ABC as well, and this (more difficult) case is a matter of SL93-3.

13. *Lemma.* If $MNPQ$ is a rectangle and O any point in space, then $OM^2 + OP^2 = ON^2 + OQ^2$.

Proof. Let O_1 be the projection of O onto $MNPQ$, and m,n,p,q denote the distances of O_1 from MN,NP,PQ,QM, respectively. Then $OM^2 = OO_1^2 + q^2 + m^2$, $ON^2 = OO_1^2 + m^2 + n^2$, $OP^2 = OO_1^2 + n^2 + p^2$, $OQ^2 = OO_1^2 + p^2 + q^2$, and the lemma follows immediately.

Now we return to the problem. Let O be the center of the given sphere S, and X the point opposite P in the face of the parallelepiped through P,A,B. By the lemma, we have $OP^2 + OQ^2 = OC^2 + OX^2$ and $OP^2 + OX^2 = OA^2 + OB^2$. Hence $2OP^2 + OQ^2 = OA^2 + OB^2 + OC^2 = 3R^2$, i.e. $OQ = \sqrt{3R^2 - OP^2} > R$.

We claim that the locus of Q is the whole sphere $(O, \sqrt{3R^2 - OP^2})$. Choose any point Q on this sphere. Since $OQ > R > OP$, the sphere with diameter PQ intersects S on a circle. Let C be an arbitrary point on this circle, and X the point opposite C in the rectangle $PCQX$. By the lemma, $OP^2 + OQ^2 = OC^2 + OX^2$, hence $OX^2 = 2R^2 - OP^2 > R^2$. The plane passing through P and perpendicular to PC intersects S in a circle γ; both P,X belong to this plane, P being inside and X outside the circle, so that the circle with diameter PX intersects γ at some point B. Finally, we choose A to be the point opposite B in the rectangle $PBXA$: we deduce that $OA^2 + OB^2 = OP^2 + OX^2$, and consequently $A \in S$. By the construction, there is a rectangular parallelepiped through P,A,B,C,X,Q.

14. We label the cells of the cube by (a_1, a_2, a_3), $a_i \in \{1, 2, \ldots, 2n+1\}$, in a natural way: for example, as Cartesian coordinates of centers of the cells $((1,1,1)$ is one corner, etc.). Notice that there should be $(2n+1)^3 - 2n(2n+1) \cdot 2(n+1) = 2n+1$ void cells, i.e., those not covered by any piece of soap.

$n = 1$. In this case, six pieces of soap $1 \times 2 \times 2$ can be placed on the following positions: $[(1,1,1),(2,2,1)]$, $[(3,1,1),(3,2,2)]$, $[(2,3,1),(3,3,2)]$ and the symmetric ones with respect to the center of the box. (Here $[A, B]$ denotes the rectangle with opposite corners at A, B.)

n is even. Each of the $2n+1$ planes $P_k = \{(a_1, a_2, k) \mid a_i = 1, \ldots, 2n+1\}$ can receive $2n$ pieces of soap: In fact, P_k can be partitioned into four $n \times (n+1)$ rectangles at the corners and the central cell, while an $n \times (n+1)$ rectangle can receive $n/2$ pieces of soap.

n is odd, $n > 1$. Let us color a cell (a_1, a_2, a_3) blue, red, or yellow if exactly three, two or one a_i respectively is equal to $n+1$. Thus there are 1 blue, $6n$ red, and $12n^2$ yellow cells. We notice that each piece of soap must contain at least one colored cell (because $2(n+1) > 2n+1$). Also, every piece of soap contains an even number (actually, $1 \cdot 2$, $1(n+1)$, or $2(n+1)$) of cells in P_k. On the other hand, $2n+1$ cells are void, i.e., one in each plane. There are several cases for a piece of soap S:

  (i) S consists of 1 blue, $n+1$ red and n yellow cells;
 (ii) S consists of 2 red and $2n$ yellow cells (and no blue cells);
(iii) S contains 1 red cell, $n+1$ yellow cells, and the rest are uncolored;
 (iv) S contains 2 yellow cells and no blue or red ones.

From the descriptions of the last three cases, we can deduce that if S contains r red cells and no blue, then it contains exactly $2 + (n-1)r$ red ones. $(*)$

Now, let $B_1, \ldots, B_k$ be all boxes put in the cube, with a possible exception for the one covering the blue cell: thus $k = 2n(2n+1)$ if the blue cell is void, or $k = 2n(2n+1) - 1$ otherwise. Let r_i and y_i respectively be the numbers of red and yellow cells inside B_i. By $(*)$ we have $y_1 + \cdots + y_k = 2k + (n-1)(r_1 + \cdots + r_k)$. If the blue cell is void, then $r_1 + \cdots + r_k = 6n$ and consequently $y_1 + \cdots + y_k = 4n(2n+1) + 6n(n-1) = 14n^2 - 2n$, which is impossible because there are only $12n^2 < 14n^2 - 2n$ yellow cells. Otherwise, $r_1 + \cdots + r_k \geq 5n - 2$ (because $n+1$ red cells are covered by the box containing the blue cell, and one can be void) and consequently $y_1 + \cdots + y_k \geq 4n(2n+1) - 2 + (n-1)(5n-2) = 13n^2 - 3n$; since there are n more yellow cells in the box containing the blue one, this counts for $13n^2 - 2n > 12n^2$ ($n \geq 3$), again impossible.

Remark. The following solution of the case n odd is simpler, but does not work for $n = 3$. For $k = 1, 2, 3$, let m_k be the number of pieces whose long sides are perpendicular to the plane $\pi_k(a_k = n+1)$. Each of these m_k pieces covers exactly 2 cells of π_k, while any other piece covers $n+1$, $2(n+1)$, or none. It follows that $4n^2 + 4n - 2m_k$ is divisible by $n+1$, and so is $2m_k$. This further implies that

$2m_1 + 2m_2 + 2m_3 = 4n(2n+1)$ is a multiple of $n+1$, which is impossible for each odd n except $n = 1$ and $n = 3$.

15. Let $C_n = \{a_1,\ldots,a_n\}$ $(C_0 = \emptyset)$ and $P_n = \{f(B) \mid B \subseteq C_n\}$. We claim that P_n contains at least $n+1$ distinct elements. First note that $P_0 = \{0\}$ contains one element. Suppose that $P_{n+1} = P_n$ for some n. Since $P_{n+1} \supseteq \{a_{n+1} + r \mid r \in P_n\}$, it follows that for each $r \in P_n$, also $r + a_{n+1} \in P_n$. Then obviously $0 \in P_n$ implies $ka_{n+1} \in P_n$ for all k; therefore $P_n = P$ has at least $p \geq n+1$ elements. Otherwise, if $P_{n+1} \supset P_n$ for all n, then $|P_{n+1}| \geq |P_n| + 1$ and hence $|P_n| \geq n+1$, as claimed. Consequently, $|P_{p-1}| \geq p$. (All the operations here are performed modulo p.)

16. Clearly $|x| \leq 1$. As x runs over $[-1, 1]$, the vector $u = (ax, a\sqrt{1-x^2})$ runs over all vectors of length a in the plane having a nonnegative vertical component. Putting $v = (by, b\sqrt{1-y^2})$, $w = (cz, c\sqrt{1-z^2})$, the system becomes $u + v = w$, with vectors u, v, w of lengths a, b, c respectively in the upper half-plane. Then a, b, c are sides of a (possibly degenerate) triangle; i.e, $|a - b| \leq c \leq a + b$ is a necessary condition.

Conversely, if a, b, c satisfy this condition, one constructs a triangle OMN with $OM = a$, $ON = b$, $MN = c$. If the vectors $\overrightarrow{OM}, \overrightarrow{ON}$ have a positive nonnegative component, then so does their sum. For every such triangle, putting $u = \overrightarrow{OM}$, $v = \overrightarrow{ON}$, and $w = \overrightarrow{OM} + \overrightarrow{ON}$ gives a solution, and every solution is given by one such triangle. This triangle is uniquely determined up to congruence: $\alpha = \angle MON = \angle(u,v)$ and $\beta = \angle(u,w)$.

Therefore, all solutions of the system are

$$x = \cos t, \quad y = \cos(t+\alpha), \quad z = y = \cos(t+\beta), \quad t \in [0, \pi - \alpha] \text{ or}$$
$$x = \cos t, \quad y = \cos(t-\alpha), \quad z = y = \cos(t-\beta), \quad t \in [\alpha, \pi].$$

17. Let $z_0 \geq 1$ be a positive integer. Supposing that the statement is true for all triples (x, y, z) with $z < z_0$, we shall prove that it is true for $z = z_0$ too.

If $z_0 = 1$, verification is trivial, while $x_0 = y_0$ is obviously impossible. So let there be given a triple (x_0, y_0, z_0) with $z_0 > 1$ and $x_0 < y_0$, and define another triple (x, y, z) by

$$x = z_0, \quad y = x_0 + y_0 - 2z_0, \quad \text{and} \quad z = z_0 - x_0.$$

Then x, y, z are positive integers. This is clear for x, z, while $y = x_0 + y_0 - 2z_0 \geq 2(\sqrt{x_0 y_0} - z_0) > 2(z_0 - z_0) = 0$. Moreover, $xy - z^2 = x_0(x_0 + y_0 - 2z_0) - (z_0 - x_0)^2 = x_0 y_0 - z_0^2 = 1$ and $z < z_0$, so that by the assumption, the statement holds for x, y, z. Thus for some nonnegative integers a, b, c, d we have

$$x = a^2 + b^2, \quad y = c^2 + d^2, \quad z = ac + bd.$$

But then we obtain representations of this sort for x_0, y_0, z_0 too:

$$x_0 = a^2 + b^2, \quad y_0 = (a+c)^2 + (b+d)^2, \quad z_0 = a(a+c) + b(b+d).$$

For the second part of the problem, we note that for $z = (2q)!$,

$$z^2 = (2q)!(2q)(2q-1)\cdots 1 \equiv (2q)!\cdot(-(2q+1))(-(2q+2))\cdots(-4q)$$
$$= (-1)^{2q}(4q)! \equiv -1 \pmod{p},$$

by Wilson's theorem. Hence $p \mid z^2+1 = py$ for some positive integer $y > 0$. Now it follows from the first part that there exist integers a,b such that $x = p = a^2+b^2$.

Second solution. Another possibility is using arithmetic of Gaussian integers.

Lemma. Suppose m,n,p,q are elements of $\mathbb{Z}$ or any other unique factorization domain, with $mn = pq$. then there exist elements a,b,c,d such that $m = ab$, $n = cd$, $p = ac$, $q = bd$.

Proof is direct, for example using factorization of a,b,c,d into primes.

We now apply this lemma to the Gaussian integers in our case (because $\mathbb{Z}[i]$ has the unique factorization property), having in mind that $xy = z^2+1 = (z+i)(z-i)$. We obtain

$$(1)\ \ x = ab, \ \ (2)\ \ y = cd, \ \ (3)\ \ z+i = ac, \ \ (4)\ \ z-i = bd$$

for some $a,b,c,d \in \mathbb{Z}[i]$. Let $a = a_1+a_2i$, etc. By (3) and (4), $\gcd(a_1,a_2) = \cdots = \gcd(d_1,d_2)$. Then (1) and (2) give us $b = \bar{a}$, $c = \bar{d}$. The statement follows at once: $x = ab = a\bar{a} = a_1^2 + a_2^2$, $y = d\bar{d} = d_1^2 + d_2^2$ and $z+i = (a_1d_1 + a_2d_2) + \imath(a_2d_1 - a_1d_2) \Rightarrow z = a_1d_1 + a_2d_2$.

4.21 Solutions to the Shortlisted Problems of IMO 1979

1. We prove more generally, by induction on n, that any $2n$-gon with equal edges and opposite edges parallel to each other can be dissected. For $n = 2$ the only possible such $2n$-gon is a single lozenge, so our theorem holds in this case. We will now show that it holds for general n. Assume by induction that it holds for $n - 1$. Let $A_1A_2 \ldots A_{2n}$ be an arbitrary $2n$-gon with equal edges and opposite edges parallel to each other. Then we can construct points B_i for $i = 3, 4, \ldots, n$ such that $\overrightarrow{A_iB_i} = \overrightarrow{A_2A_1} = \overrightarrow{A_{n+1}A_{n+2}}$. We set $B_2 = A_{2n+1} = A_1$ and $B_{n+1} = A_{n+2}$. It follows that $A_iB_iB_{i+1}A_{i+1}$ for $i = 2, 3, 4, \ldots, n$ are all lozenges. It also follows that B_iB_{i+1} for $i = 2, 3, 4, \ldots, n$ are equal to the edges of $A_1A_2 \ldots A_{2n}$ and parallel to A_iA_{i+1} and hence to $A_{n+i}A_{n+i+1}$. Thus $B_2 \ldots B_{n+1}A_{n+3} \ldots A_{2n}$ is a $2(n-1)$-gon with equal edges and opposite sides parallel and hence, by the induction hypothesis, can be dissected into lozenges. We have thus provided a dissection for $A_1A_2 \ldots A_{2n}$. This completes the proof.

2. The only way to arrive at the latter alternative is to draw four different socks in the first drawing or to draw only one pair in the first drawing and then draw two different socks in the last drawing. We will call these probabilities respectively p_1, p_2, p_3. We calculate them as follows:

$$p_1 = \frac{\binom{5}{4}2^4}{\binom{10}{4}} = \frac{8}{21}, \quad p_2 = \frac{5\binom{4}{2}2^2}{\binom{10}{4}} = \frac{4}{7}, \quad p_3 = \frac{4}{\binom{6}{2}} = \frac{4}{15}.$$

We finally calculate the desired probability: $P = p_1 + p_2p_3 = \frac{8}{15}$.

3. An obvious solution is $f(x) = 0$. We now look for nonzero solutions. We note that plugging in $x = 0$ we get $f(0)^2 = f(0)$; hence $f(0) = 0$ or $f(0) = 1$. If $f(0) = 0$, then f is of the form $f(x) = x^k g(x)$, where $g(0) \neq 0$. Plugging this formula into $f(x)f(2x^2) = f(2x^3 + x)$ we get

$$2^k x^{2k} g(x)g(2x^2) = (2x^2 + 1)^k g(2x^3 + x).$$

Plugging in $x = 0$ gives us $g(0) = 0$, which is a contradiction. Hence $f(0) = 1$. For an arbitrary root α of the polynomial f, $2\alpha^3 + \alpha$ must also be a root. Let α be a root of the largest modulus. If $|\alpha| > 1$ then $|2\alpha^3 + \alpha| > 2|\alpha|^3 - |\alpha| > |\alpha|$, which is impossible. It follows that $|\alpha| \leq 1$ and hence all roots of f have modules less than or equal to 1. But the product of all roots of f is $|f(0)| = 1$, which implies that all the roots have modulus 1. Consequently, for a root α it holds that $|\alpha| = |2\alpha^3 - \alpha| = 1$. This is possible only if $\alpha = \pm \imath$. Since the coefficients of f are real it follows that f must be of the form $f(x) = (x^2 + 1)^k$ where $k \in \mathbb{N}_0$. These polynomials satisfy the original formula. Hence, the solutions for f are $f(x) = 0$ and $f(x) = (x^2 + 1)^k$, $k \in \mathbb{N}_0$.

4. Let us prove first that the edges $A_1A_2, A_2A_3, \ldots, A_5A_1$ are of the same color. Assume the contrary, and let w.l.o.g. A_1A_2 be red and A_2A_3 be green. Three of the segments A_2B_l ($l = 1, 2, 3, 4, 5$), say A_2B_i, A_2B_j, A_2B_k, have to be of the same

color, let it w.l.o.g. be red. Then A_1B_i, A_1B_j, A_1B_k must be green. At least one of the sides of triangle $B_iB_jB_k$, say B_iB_j, must be an edge of the prism. Then looking at the triangles $A_1B_iB_j$ and $A_2B_iB_j$ we deduce that B_iB_j can be neither green nor red, which is a contradiction. Hence all five edges of the pentagon $A_1A_2A_3A_4A_5$ have the same color. Similarly, all five edges of $B_1B_2B_3B_4B_5$ have the same color.

We now show that the two colors are the same. Assume otherwise, i.e., that w.l.o.g. the A edges are painted red and the B edges green. Let us call segments of the form A_iB_j diagonal (i and j may be equal). We now count the diagonal segments by grouping the red segments based on their A point, and the green segments based on their B point. As above, the assumption that three of A_iB_j for fixed i are red leads to a contradiction. Hence at most two diagonal segments out of each A_i may be red, which counts up to at most 10 red segments. Similarly, at most 10 diagonal segments can be green. But then we can paint at most 20 diagonal segments out of 25, which is a contradiction. Hence all edges in the pentagons $A_1A_2A_3A_4A_5$ and $B_1B_2B_3B_4B_5$ have the same color.

5. Let $A = \{x \mid (x,y) \in M\}$ and $B = \{y \mid (x,y) \in M\}$. Then A and B are disjoint and hence

$$|M| \le |A| \cdot |B| \le \frac{(|A|+|B|)^2}{4} \le \left[\frac{n^2}{4}\right].$$

These cardinalities can be achieved for $M = \{(a,b) \mid a = 1,2,\ldots,[n/2], b = [n/2]+1,\ldots,n\}$.

6. Setting $q = x^2 + x - p$, the given equation becomes

$$\sqrt{(x+1)^2 - 2q} + \sqrt{(x+2)^2 - q} = \sqrt{(2x+3)^2 - 3q}. \tag{1}$$

Taking squares of both sides we get $2\sqrt{((x+1)^2 - 2q)((x+2)^2 - q)} = 2(x+1)(x+2)$. Taking squares again we get

$$q\left(2q - 2(x+2)^2 - (x+1)^2\right) = 0.$$

If $2q = 2(x+2)^2 + (x+1)^2$, at least one of the expressions under the three square roots in (1) is negative, and in that case the square root is not well-defined. Thus, we must have $q = 0$.

Now (1) is equivalent to $|x+1| + |x+2| = |2x+3|$, which holds if and only if $x \notin (-2,-1)$. The number of real solutions x of $q = x^2 + x - p = 0$ which are not in the interval $(-2,-1)$ is zero if $p < -1/4$, one if $p = -1/4$ or $0 < p < 2$, and two otherwise.

Hence, the answer is $-1/4 < p \le 0$ or $p \ge 2$.

7. We denote the sum mentioned above by S. We have the following equalities:

$$S = 1 - \frac{1}{2} + \frac{1}{3} - \frac{1}{4} + \cdots - \frac{1}{1318} + \frac{1}{1319}$$

$$= 1 + \frac{1}{2} + \cdots + \frac{1}{1319} - 2\left(\frac{1}{2} + \frac{1}{4} + \cdots + \frac{1}{1318}\right)$$

$$= 1 + \frac{1}{2} + \cdots + \frac{1}{1319} - \left(1 + \frac{1}{2} + \cdots + \frac{1}{659}\right)$$

$$= \frac{1}{660} + \frac{1}{661} + \cdots + \frac{1}{1319}$$

$$= \sum_{i=660}^{989} \frac{1}{i} + \frac{1}{1979 - i} = \sum_{i=660}^{989} \frac{1979}{i \cdot (1979 - i)}$$

Since no term in the sum contains a denominator divisible by 1979 (1979 is a prime number), it follows that when S is represented as p/q the numerator p will have to be divisible by 1979.

8. By the definition of f, it holds that $f(0.b_1 b_2 \ldots) = 3b_1/4 + f(0.b_2 b_3 \ldots)/4 = 0.b_1 b_1 + f(0.b_2 b_3 \ldots)/4$. Continuing this argument we obtain

$$f(0.b_1 b_2 b_3 \ldots) = 0.b_1 b_1 \ldots b_n b_n + \frac{1}{2^{2n}} f(0.b_{n+1} b_{n+2} \ldots). \tag{1}$$

The binary representation of every rational number is eventually periodic. Let us first determine $f(x)$ for a rational x with the periodic representation $x = 0.\overline{b_1 b_2 \ldots b_n}$. Using (1) we obtain $f(x) = 0.b_1 b_1 \ldots b_n b_n + f(x)/2^{2n}$, and hence $f(x) = \frac{2^n}{2^n - 1} 0.b_1 b_1 \ldots b_n b_n = 0.\overline{b_1 b_1 \ldots b_n b_n}$.

Now let $x = 0.a_1 a_2 \ldots a_k \overline{b_1 b_2 \ldots b_n}$ be an arbitrary rational number. Then it follows from (1) that

$$f(x) = 0.a_1 a_1 \ldots a_k a_k + \frac{1}{2^{2n}} f(0.\overline{b_1 b_2 \ldots b_n}) = 0.a_1 a_1 \ldots a_k a_k \overline{b_1 b_1 \ldots b_n b_n}.$$

Hence $f(0.b_1 b_2 \ldots) = 0.b_1 b_1 b_2 b_2 \ldots$ for every rational number $0.b_1 b_2 \ldots$.

9. Let us number the vertices, starting from S and moving clockwise. In that case $S = 1$ and $F = 5$. After an odd number of moves to a neighboring point we can be only on an even point, and hence it follows that $a_{2n-1} = 0$ for all $n \in \mathbb{N}$. Let us define respectively z_n and w_n as the number of paths from S to S in $2n$ moves and the number of paths from S to points 3 and 7 in $2n$ moves. We easily derive the following recurrence relations:

$$a_{2n+2} = w_n, \quad w_{n+1} = 2w_n + 2z_n, \quad z_{n+1} = 2z_n + w_n, \quad n = 0, 1, 2, \ldots .$$

By subtracting the second equation from the third we get $z_{n+1} = w_{n+1} - w_n$. By plugging this equation into the formula for w_{n+2} we get $w_{n+2} - 4w_{n+1} + 2w_n = 0$. The roots of the characteristic equation $r^2 - 4r + 2 = 0$ are $x = 2 + \sqrt{2}$ and $y = 2 - \sqrt{2}$. From the conditions $w_0 = 0$ and $w_1 = 2$ we easily obtain $a_{2n} = w_{n-1} = (x^{n-1} - y^{n-1})/\sqrt{2}$.

10. In the cases $a = \overrightarrow{0}$, $b = \overrightarrow{0}$, and $a \parallel b$ the inequality is trivial. Otherwise, let us consider a triangle ABC such that $\overrightarrow{CB} = a$ and $\overrightarrow{CA} = b$. From this point on we shall refer to α, β, γ as angles of ABC. Since $|a \times b| = |a||b| \sin \gamma$, our inequality reduces to $|a||b| \sin^3 \gamma \le 3\sqrt{3}|c|^2/8$, which is further reduced to

$$\sin \alpha \sin \beta \sin \gamma \le \frac{3\sqrt{3}}{8}$$

using the sine law. The last inequality follows immediately from Jensen's inequality applied to the function $f(x) = \ln \sin x$, which is concave for $0 < x < \pi$ because $f'(x) = \cot x$ is strictly decreasing.

11. Let us define $y_i = x_i^2$. We thus have $y_1 + y_2 + \cdots + y_n = 1$, $y_i \ge 1/n^2$, and $P = \sqrt{y_1 y_2 \cdots y_n}$.
The upper bound is obtained immediately from the AM–GM inequality: $P \le 1/n^{n/2}$, where equality holds when $x_i = \sqrt{y_i} = 1/\sqrt{n}$.
For the lower bound, let us assume w.l.o.g. that $y_1 \ge y_2 \ge \cdots \ge y_n$. We note that if $a \ge b \ge 1/n^2$ and $s = a + b > 2/n^2$ is fixed, then $ab = (s^2 - (a-b)^2)/4$ is minimized when $|a - b|$ is maximized, i.e., when $b = 1/n^2$. Hence $y_1 y_2 \cdots y_n$ is minimal when $y_2 = y_3 = \cdots = y_n = 1/n^2$. Then $y_1 = (n^2 - n + 1)/n^2$ and therefore $P_{\min} = \sqrt{n^2 - n + 1}/n^n$.

12. The first criterion ensures that all sets in an S-family are distinct. Since the number of different families of subsets is finite, h has to exist. In fact, we will show that $h = 11$. First of all, if there exists $X \in F$ such that $|X| \ge 5$, then by (3) there exists $Y \in F$ such that $X \cup Y = R$. In this case $|F|$ is at most 2. Similarly, for $|X| = 4$, for the remaining two elements either there exists a subset in F that contains both, in which case we obtain the previous case, or there exist different Y and Z containing them, in which case $X \cup Y \cup Z = R$, which must not happen. Hence we can assume $|X| \le 4$ for all $X \in F$.
Assume $|X| = 1$ for some X. In that case other sets must not contain that subset and hence must be contained in the remaining 5-element subset. These elements must not be subsets of each other. From elementary combinatorics, the largest number of subsets of a 5-element set of which none is subset of another is $\binom{5}{2} = 10$. This occurs when we take all 2-element subsets. These subsets also satisfy (2). Hence $|F|_{\max} = 11$ in this case.
Otherwise, let us assume $|X| = 3$ for some X. Let us define the following families of subsets: $G = \{Z = Y \setminus X \mid Y \in F\}$ and $H = \{Z = Y \cap X \mid Y \in F\}$. Then no two sets in G must complement each other in $R \setminus X$, and G must cover this set. Hence G contains exactly the sets of each of the remaining 3 elements. For each element of G no two sets in H of which one is a subset of another may be paired with it. There can be only 3 such subsets selected within a 3-element set X. Hence the number of remaining sets is smaller than $3 \cdot 3 = 9$. Hence in this case $|F|_{\max} = 10$.
In the remaining case all subsets have two elements. There are $\binom{6}{2} = 15$ of them. But for every three that complement each other one must be discarded; hence the maximal number for F in this case is $2 \cdot 15/3 = 10$.
It follows that $h = 11$.

13. From elementary trigonometry we have $\sin 3t = 3\sin t - 4\sin^3 t$. Hence, if we denote $y = \sin 20°$, we have $\sqrt{3}/2 = \sin 60° = 3y - 4y^3$. Obviously $0 < y < 1/2 = \sin 30°$. The function $f(x) = 3x - 4x^3$ is strictly increasing on $[0, 1/2)$ because $f'(x) = 3 - 12x^2 > 0$ for $0 \leq x < 1/2$. Now the desired inequality $\frac{20}{60} = \frac{1}{3} < \sin 20° < \frac{21}{60} = \frac{7}{20}$ follows from

$$f\left(\frac{1}{3}\right) < \frac{\sqrt{3}}{2} < f\left(\frac{7}{20}\right),$$

which is directly verified.

14. Let us assume that $a \in \mathbb{R} \setminus \{1\}$ is such that there exist a and x such that $x = \log_a x$, or equivalently $f(x) := \ln x/x = \ln a$. Then a is a value of the function $f(x)$ for $x \in \mathbb{R}^+ \setminus \{1\}$, and the converse also holds.

First we observe that $f(x)$ tends to $-\infty$ as $x \to 0$ and $f(x)$ tends to 0 as $x \to 1$. Since $f(x) > 0$ for $x > 1$, the function $f(x)$ takes its maximum at a point x for which $f'(x) = (1 - \ln x)/x^2 = 0$. Hence

$$\max f(x) = f(e) = e^{1/e}.$$

It follows that the set of values of $f(x)$ for $x \in \mathbb{R}^+$ is the interval $(-\infty, e^{1/e})$, and consequently the desired set of bases a of logarithms is $(0, 1) \cup (1, e^{1/e}]$.

15. We note that

$$\sum_{i=1}^{5} i(a - i^2)^2 x_i = a^2 \sum_{i=1}^{5} i x_i - 2a \sum_{i=1}^{5} i^3 x_i + \sum_{i=1}^{5} i^5 x_i = a^2 \cdot a - 2a \cdot a^2 + a^3 = 0.$$

Since the terms in the sum on the left are all nonnegative, it follows that all the terms have to be 0. Thus, either $x_i = 0$ for all i, in which case $a = 0$, or $a = j^2$ for some j and $x_i = 0$ for $i \neq j$. In this case, $x_j = a/j = j$. Hence, the only possible values of a are $\{0, 1, 4, 9, 16, 25\}$.

16. Obviously, no two elements of F can be complements of each other. If one of the sets has one element, then the conclusion is trivial. If there exist two different 2-element sets, then they must contain a common element, which in turn must then be contained in all other sets. Thus we can assume that there exists at most one 2-element subset of K in F. Since there can be at most 6 subsets of more than 3 elements of a 5-element set, it follows that at least 9 out of 10 possible 3-element subsets of K belong to F. Let us assume, without loss of generality, that all sets but $\{c,d,e\}$ belong to F. Then sets $\{a,b,c\}$, $\{a,d,e\}$, and $\{b,c,d\}$ have no common element, which is a contradiction. Hence it follows that all sets have a common element.

17. Let K, L, and M be intersections of CQ and BR, AR and CP, and AQ and BP, respectively. Let $\angle X$ denote the angle of the hexagon $KQMPLR$ at the vertex X, where X is one of the six points. By an elementary calculation of angles we get

$$\angle K = 140°, \angle L = 130°, \angle M = 150°, \angle P = 100°, \angle Q = 95°, \angle R = 105°.$$

Since $\angle KBC = \angle KCB$, it follows that K is on the symmetry line of ABC through A. Analogous statements hold for L and M. Let K_R and K_Q be points symmetric to K with respect to AR and AQ, respectively. Since $\angle AK_QQ = \angle AK_QK_R = 70°$ and $\angle AK_RR = \angle AK_RK_Q = 70°$, it follows that the points K_R, R, Q, and K_Q are collinear. Hence $\angle QRK = 2\angle R - 180°$ and $\angle RQK = 2\angle Q - 180°$. In the same way we conclude that $\angle PRL = 2\angle R - 180°$, $\angle RPL = 2\angle P - 180°$, $\angle QPM = 2\angle P - 180°$ and $\angle PQM = 2\angle Q - 180°$. From these formulas we easily get $\angle RPQ = 60°$, $\angle RQP = 75°$, and $\angle QRP = 45°$.

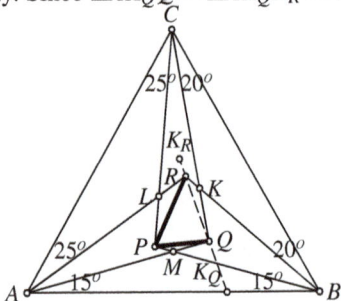

18. Let us write all a_i in binary representation. For $S \subseteq \{1,2,\ldots,m\}$ let us define $b(S)$ as the number in whose binary representation ones appear in exactly the slots where ones appear in all a_i where $i \subseteq S$ and don't appear in any other a_i. Some $b(S)$, including $b(\emptyset)$, will equal 0, and hence there are fewer than 2^m different positive $b(S)$. We note that no two positive $b(S_1)$ and $b(S_2)$ $(S_1 \neq S_2)$ have ones in the same decimal places. Hence sums of distinct $b(S)$'s are distinct. Moreover

$$a_i = \sum_{i \in S} b(S)$$

and hence the positive $b(S)$ are indeed the numbers $b_1,\ldots,b_n$ whose existence we had to prove.

19. Let us define i_j for two positive integers i and j in the following way: $i_1 = i$ and $i_{j+1} = i^{i_j}$ for all positive integers j. Thus we must find the smallest m such that $100_m > 3_{100}$. Since $100_1 = 100 > 27 = 3_2$, we inductively have $100_j = 10^{100_{j-1}} > 3^{100_{j-1}} > 3^{3_j} = 3_{j+1}$ and hence $m \leq 99$. We now prove that $m = 99$ by proving $100_{98} < 3_{100}$. We note that $(100_1)^2 = 10^4 < 27^4 = 3^{12} < 3^{27} = 3_3$. We also note for $d > 12$ (which trivially holds for all $d = 100_i$) that if $c > d^2$, then we have

$$3^c > 3^{d^2} > 3^{12d} = (3^{12})^d > 10000^d = (100^d)^2.$$

Hence from $3_3 > (100_1)^2$ it inductively follows that $3_j > (100_{j-2})^2 > 100_{j-2}$ and hence that $100_{99} > 3_{100} > 100_{98}$. Hence $m = 99$.

20. Let $x_k = \max\{x_1,x_2,\ldots,x_n\}$. Then $x_ix_{i+1} \leq x_ix_k$ for $i = 1,2,\ldots,k-1$ and $x_ix_{i+1} \leq x_kx_{i+1}$ for $i = k,\ldots,n-1$. Summing up these inequalities for $i = 1,2,\ldots,n-1$ we obtain

$$\sum_{i=1}^{n-1} \leq x_k(x_1 + \cdots + x_{k-1} + x_{k+1} + \cdots + x_n) = x_k(a - x_k) \leq \frac{a^2}{4}.$$

We note that the value $a^2/4$ is attained for $x_1 = x_2 = a/2$ and $x_3 = \cdots = x_n = 0$. Hence $a^2/4$ is the required maximum.

21. Denote $m = 10^6$ and let $f(n)$ be the number of different ways $n \in \mathbb{N}$ can be expressed as $x^2 + y^3$ with $x, y \in \{0, 1, \ldots, m\}$. Clearly $f(n) = 0$ for $n < 0$ or $n > m^2 + m^3$. The first equation can be written as $x^2 + t^3 = y^2 + z^3 = n$, whereas the second equation can be written as $x^2 + t^3 = n+1$, $y^2 + z^3 = n$. Hence we obtain the following formulas for M and N:

$$M = \sum_{i=0}^{m} f(i)^2, \quad N = \sum_{i=0}^{m-1} f(i) f(i+1).$$

Using the AM–GM inequality we get

$$N = \sum_{i=0}^{m-1} f(i) f(i+1)$$

$$\leq \sum_{i=0}^{m-1} \frac{f(i)^2 + f(i+1)^2}{2} = \frac{f(0)^2}{2} + \sum_{i=1}^{m-1} f(i)^2 + \frac{f(m)^2}{2} < M.$$

The last inequality is strict, since $f(0) = 1 > 0$. This completes our proof.

22. Let the centers of the two circles be denoted by O and O_1 and their respective radii by r and r_1, and let the positions of the points on the circles at time t be denoted by $M(t)$ and $N(t)$. Let Q be the point such that OAO_1Q is a parallelogram. We will show that Q is the point P we are looking for, i.e., that $QM(t) = QN(t)$ for all t. We note that $OQ = O_1A = r_1$, $O_1Q = OA = r$ and $\angle QOA = \angle QO_1A = \phi$. Since the two points return to A at the same time, it follows that $\angle M(t)OA = \angle N(t)O_1A = \omega t$. Therefore $\angle QOM(t) = \angle QO_1N(t) = \phi + \omega t$, from which it follows that $\triangle QOM(t) \cong \triangle QO_1N(t)$. Hence $QM(t) = QN(t)$, as we claimed.

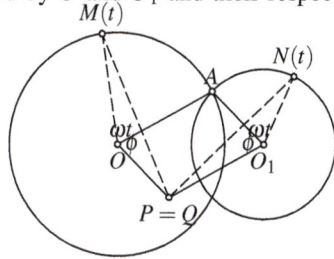

23. It is easily verified that no solutions exist for $n \leq 8$. Let us now assume that $n > 8$. We note that $2^8 + 2^{11} + 2^n = 2^8 \cdot (9 + 2^{n-8})$. Hence $9 + 2^{n-8}$ must also be a square, say $9 + 2^{n-8} = x^2$, $x \in \mathbb{N}$, i.e., $2^{n-8} = x^2 - 9 = (x-3)(x+3)$. Thus $x - 3$ and $x + 3$ are both powers of 2, which is possible only for $x = 5$ and $n = 12$. Hence, $n = 12$ is the only solution.

24. Clearly O is the midpoint of BC. Let M and N be the points of tangency of the circle with AB and AC, respectively, and let $\angle BAC = 2\phi$. Then $\angle BOM = \angle CON = \phi$.

Let us assume that PQ touches the circle in X. If we set $\angle POM = \angle POX = x$ and $\angle QON = \angle QOX = y$, then $2x + 2y = \angle MON = 180° - 2\phi$, i.e., $y = 90° - \phi - x$. It follows that $\angle OQC = 180° - \angle QOC - \angle OCQ = 180° - (\phi + y) - (90° - \phi) = 90° - y = x + \phi = \angle BOP$. Hence the triangles BOP and CQO are similar, and consequently $BP \cdot CQ = BO \cdot CO = (BC/2)^2$.

Conversely, let $BP \cdot CQ = (BC/2)^2$ and let Q' be the point on (AC) such that PQ' is tangent to the circle. Then $BP \cdot CQ' = (BC/2)^2$, which implies $Q \equiv Q'$.

25. Let us first look for such a point R on a ray l in π going through P. Let $\angle QPR = 2\theta$. Consider a point Q' on the extension of l beyond P such that $Q'P = QP$. Then we have

$$\frac{QP + PR}{QR} = \frac{RQ'}{QR} = \frac{\sin \angle Q'QR}{\sin \angle QQ'R}.$$

Since $\angle QQ'R$ is fixed, the maximum of the expression occurs when $\angle Q'QR = 90°$, i.e., when $PR = PQ$. In this case, $(QP + PR)/QR = 1/\sin \theta$. Looking at all possible rays l, we see that θ is minimal when l contains the projection of PQ onto π. Hence, if $PQ \not\perp \pi$, the desired point R is the point on the projection of ray PQ onto π such that $PR = PQ$; otherwise, R is any point of the circle $k(P, PQ)$.

26. Let us assume that $f(x+y) = f(x) + f(y)$ for all reals. In this case we trivially apply the equation to get $f(x+y+xy) = f(x+y) + f(xy) = f(x) + f(y) + f(xy)$. Hence the equivalence is proved in the first direction.

Now let us assume that $f(x+y+xy) = f(x) + f(y) + f(xy)$ for all reals. Plugging in $x = y = 0$ we get $f(0) = 0$. Plugging in $y = -1$ we get $f(x) = -f(-x)$. Plugging in $y = 1$ we get $f(2x+1) = 2f(x) + f(1)$ and hence $f(2(u+v+uv) + 1) = 2f(u+v+uv) + f(1) = 2f(uv) + 2f(u) + 2f(v) + f(1)$ for all real u and v. On the other hand, plugging in $x = u$ and $y = 2v+1$ we get $f(2(u+v+uv) + 1) = f(u + (2v+1) + u(2v+1)) = f(u) + 2f(v) + f(1) + f(2uv+u)$. Hence it follows that $2f(uv) + 2f(u) + 2f(v) + f(1) = f(u) + 2f(v) + f(1) + f(2uv+u)$, i.e.,

$$f(2uv + u) = 2f(uv) + f(u). \tag{1}$$

Plugging in $v = -1/2$ we get $0 = 2f(-u/2) + f(u) = -2f(u/2) + f(u)$. Hence, $f(u) = 2f(u/2)$ and consequently $f(2x) = 2f(x)$ for all reals. Now (1) reduces to $f(2uv + u) = f(2uv) + f(u)$. Plugging in $u = y$ and $x = 2uv$, we obtain $f(x) + f(y) = f(x+y)$ for all nonzero reals x and y. Since $f(0) = 0$, it trivially holds that $f(x+y) = f(x) + f(y)$ when one of x and y is 0.

Second solution. Assume that $f(x+y+xy) = f(x) + f(y) + f(xy)$ for all x, y. Substituting $(x, -y)$ in the functional equation and adding it to the original equation yields

$$f(x-t) + f(x+t) = 2f(x), \quad \text{where } t = y(x+1). \tag{2}$$

Thus (2) holds whenever $x \neq -1$. Similarly, for $t \neq -1$ we have $f(t-x) + f(x+t) = 2f(t)$. Summing these two equalities and using $f(y) = -f(-y)$ as shown above we obtain $f(x) + f(t) = f(x+t)$ for all $x, t \neq -1$. The case $x = -1$ or $t = -1$ is easy to handle with, as $f(-1) = 2f(-\frac{1}{2})$.

4.22 Solutions to the Shortlisted Problems of IMO 1981

1. Assume that the set $\{a-n+1, a-n+2,\ldots,a\}$ of n consecutive numbers satisfies the condition $a \mid \operatorname{lcm}[a-n+1,\ldots,a-1]$. Let $a = p_1^{\alpha_1} p_2^{\alpha_2}\ldots p_r^{\alpha_r}$ be the canonic representation of a, where $p_1 < p_2 < \cdots < p_r$ are primes and $\alpha_1,\cdots,\alpha_r > 0$. Then for each $j = 1,2,\ldots,r$, there exists m, $m = 1,2,\ldots,n-1$, such that $p_j^{\alpha_j} \mid a-m$, i.e., such that $p_j^{\alpha_j} \mid m$. Thus $p_j^{\alpha_j} \leq n-1$. If $r = 1$, then $a = p_1^{\alpha_1} \leq n-1$, which is impossible. Therefore $r \geq 2$. But then there must exist two distinct prime numbers less than n; hence $n \geq 4$.

   For $n = 4$, we must have $p_1^{\alpha_1}, p_2^{\alpha_2} \leq 3$, which leads to $p_1 = 2$, $p_2 = 3$, $\alpha_1 = \alpha_2 = 1$. Therefore $a = 6$, and $\{3,4,5,6\}$ is a unique set satisfying the condition of the problem.

   For every $n \geq 5$ there exist at least two such sets. In fact, for $n = 5$ we easily find two sets: $\{2,3,4,5,6\}$ and $\{8,9,10,11,12\}$. Suppose that $n \geq 6$. Let r,s,t be natural numbers such that $2^r \leq n-1 < 2^{r+1}$, $3^s \leq n-1 < 3^{s+1}$, $5^t \leq n-1 < 5^{t+1}$. Taking $a = 2^r \cdot 3^s$ and $a = 2^r \cdot 5^t$ we obtain two distinct sets with the required property. Thus the answers are (a) $n \geq 4$ and (b) $n = 4$.

2. *Lemma.* Let E, F, G, H, I, and K be points on edges AB, BC, CD, DA, AC, and BD of a tetrahedron. Then there is a sphere that touches the edges at these points if and only if

$$AE = AH = AI, \quad BE = BF = BK,$$
$$CF = CG = CI, \quad DG = DH = DK. \qquad (*)$$

   Proof. The "only if" side of the equivalence is obvious.

   We now assume $(*)$. Denote by $\varepsilon, \phi, \gamma, \eta, \iota$, and κ planes through E, F, G, H, I, K perpendicular to AB, BC, CD, DA, AC and BD respectively. Since the three planes ε, η, and ι are not mutually parallel, they intersect in a common point O. Clearly, $\triangle AEO \cong \triangle AHO \cong \triangle AIO$; hence $OE = OH = OI = r$, and the sphere $\sigma(O,r)$ is tangent to AB, AD, AC. To prove that σ is also tangent to BC, CD, BD it suffices to show that planes ϕ, γ, and κ also pass through O. Without loss of generality we can prove this for just ϕ. By the conditions for E,F,I, these are exactly the points of tangency of the incircle of $\triangle ABC$ and its sides, and if S is the incenter, then $SE \perp AB$, $SF \perp BC$, $SI \perp AC$. Hence ε, ι, and ϕ all pass through S and are perpendicular to the plane ABC, and consequently all share the line l through S perpendicular to ABC. Since $l = \varepsilon \cap \iota$, the point O will be situated on l, and hence ϕ will also contain O. This completes our proof of the lemma.

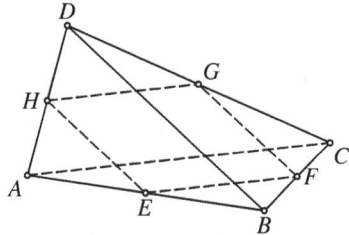

   Let $AH = AE = x$, $BE = BF = y$, $CF = CG = z$, and $DG = DH = w$. If the sphere is also tangent to AC at some point I, then $AI = x$ and $IC = z$. Using the stated lemma it suffices to prove that if $AC = x+z$, then $BD = y+w$.

Let $EF = FG = GH = HI = t$, $\angle BAD = \alpha$, $\angle ABC = \beta$, $\angle BCD = \gamma$, and $\angle ADC = \delta$. We get

$$t^2 = EH^2 = AE^2 + AH^2 - 2 \cdot AE \cdot AH \cos \alpha = 2x^2(1 - \cos \alpha).$$

We similarly conclude that $t^2 = 2y^2(1 - \cos \beta) = 2z^2(1 - \cos \gamma) = 2w^2(1 - \cos \delta)$. Further, using that $AB = x + y$, $BC = y + z$, $\cos \beta = 1 - t^2/2y^2$, we obtain

$$AC^2 = AB^2 + BC^2 - 2AB \cdot BC \cos \beta = (x - z)^2 + t^2 \left(\frac{x}{y} + 1 \right) \left(\frac{z}{y} + 1 \right).$$

Analogously, from the triangle ADC we get $AC^2 = (x - z)^2 + t^2(x/w + 1)(z/w + 1)$, which gives $(x/y + 1)(z/y + 1) = (x/w + 1)(z/w + 1)$. Since $f(s) = (x/s + 1)(z/s + 1)$ is a decreasing function in s, it follows that $y = w$; similarly $x = z$. Hence $CF = CG = x$ and $DG = DH = y$. Hence $AC \parallel EF$ and $AC : t = AC : EF = AB : EB = (x + y) : y$; i.e., $AC = t(x + y)/y$. Similarly, from the triangle ABD, we get that $BD = t(x + y)/x$. Hence if $AC = x + z = 2x$, it follows that $2x = t(x + y)/y \Rightarrow 2xy = t(x + y) \Rightarrow BD = t(x + y)/x = 2y = y + w$. This completes the proof.

Second solution. Without loss of generality, assume that $EF = 2$. Consider the Cartesian system in which points O, E, F, G, H respectively have coordinates $(0, 0, 0)$, $(-1, -1, a)$, $(1, -1, a)$, $(1, 1, a)$, $(-1, 1, a)$. Line AH is perpendicular to OH and AE is perpendicular to OE; hence from Pythagoras's theorem $AO^2 = AH^2 + HO^2 = AE^2 + EO^2 = AE^2 + HO^2$, which implies $AH = AE$. Therefore the y-coordinate of A is zero; analogously the x-coordinates of B and D and the y-coordinate of C are 0. Let A have coordinates $(x_0, 0, z_1)$: then $\overrightarrow{EA}(x_0 + 1, 1, z_1 - a) \perp \overrightarrow{EO}(1, 1, -a)$, i.e., $\overrightarrow{EA} \cdot \overrightarrow{EO} = x_0 + 2 + a(a - z_1) = 0$. Similarly, for $B(0, y_0, z_2)$ we have $y_0 + 2 + a(a - z_2) = 0$. This gives us

$$z_1 = \frac{x_0 + a^2 + 2}{a}, \qquad\qquad z_2 = \frac{y_0 + a^2 + 2}{a}. \qquad (1)$$

We haven't used yet that $A(x_0, 0, z_1)$, $E(-1, -1, a)$ and $B(0, y_0, z_2)$ are collinear, so let A', B', E' be the feet of perpendiculars from A, B, E to the plane xy. The line $A'B'$, given by $y_0 x + x_0 y = x_0 y_0$, $z = 0$, contains the point $E'(-1, -1, 0)$, from which we obtain

$$(x_0 + 1)(y_0 + 1) = 1. \qquad (2)$$

In the same way, from the points B and C we get relations similar to (1) and (2) and conclude that C has the coordinates $C(-x_0, 0, z_1)$. Similarly we get $D(0, -y_0, z_2)$. The condition that AC is tangent to the sphere $\sigma(O, OE)$ is equivalent to $z_1 = \sqrt{a^2 + 2}$, i.e., to $x_0 = a\sqrt{a^2 + 2} - (a^2 + 2)$. But then (2) implies that $y_0 = -a\sqrt{a^2 + 2} - (a^2 + 2)$ and $z_2 = -\sqrt{a^2 + 2}$, which means that the sphere σ is tangent to BD as well. This finishes the proof.

3. Denote $\max(a + b + c, b + c + d, c + d + e, d + e + f, e + f + g)$ by p. We have

$$(a + b + c) + (c + d + e) + (e + f + g) = 1 + c + e \le 3p,$$

which implies that $p \geq 1/3$. However, $p = 1/3$ is achieved by taking $(a, b, c, d, e, f, g) = (1/3, 0, 0, 1/3, 0, 0, 1/3)$. Therefore the answer is $1/3$.

Remark. In fact, one can prove a more general statement in the same way. Given positive integers n, k, $n \geq k$, if $a_1, a_2, \ldots, a_n$ are nonnegative real numbers with sum 1, then the minimum value of $\max_{i=1,\ldots,n-k+1}\{a_i + a_{i+1} + \cdots + a_{i+k-1}\}$ is $1/r$, where r is the integer with $k(r-1) < n \leq kr$.

4. We shall use the known formula for the Fibonacci sequence

$$f_n = \frac{1}{\sqrt{5}}(\alpha^n - (-1)^n\alpha^{-n}), \qquad \text{where } \alpha = \frac{1+\sqrt{5}}{2}. \tag{1}$$

(a) Suppose that $af_n + bf_{n+1} = f_{k_n}$ for all n, where $k_n > 0$ is an integer depending on n. By (1), this is equivalent to $a(\alpha^n - (-1)^n\alpha^{-n}) + b(\alpha^{n+1} + (-1)^n\alpha^{-n-1}) = \alpha^{k_n} - (-1)^{k_n}\alpha^{-k_n}$, i.e.,

$$\alpha^{k_n-n} = a + b\alpha - \alpha^{-2n}(-1)^n(a - b\alpha^{-1} - (-\alpha)^{n-k_n}) \to a + b\alpha \tag{2}$$

as $n \to \infty$. Hence, since k_n is an integer, $k_n - n$ must be constant from some point on: $k_n = n + k$ and $\alpha^k = a + b\alpha$. Then it follows from (2) that $\alpha^{-k} = a - b\alpha^{-1}$, and from (1) we conclude that $af_n + bf_{n+1} = f_{k+n}$ holds for every n. Putting $n = 1$ and $n = 2$ in the previous relation and solving the obtained system of equations we get $a = f_{k-1}$, $b = f_k$. It is easy to verify that such a and b satisfy the conditions.

(b) As in (a), suppose that $uf_n^2 + vf_{n+1}^2 = f_{l_n}$ for all n. This leads to

$$u + v\alpha^2 - \sqrt{5}\alpha^{l_n-2n} = 2(u-v)(-1)^n\alpha^{-2n}$$
$$-(u\alpha^{-4n} + v\alpha^{-4n-2} + (-1)^{l_n}\sqrt{5}\alpha^{-l_n-2n})$$
$$\to 0,$$

as $n \to \infty$. Thus $u + v\alpha^2 = \sqrt{5}\alpha^{l_n-2n}$, and $l_n - 2n = k$ is equal to a constant. Putting this into the above equation and multiplying by α^{2n} we get $u - v \to 0$ as $n \to \infty$, i.e., $u = v$. Finally, substituting $n = 1$ and $n = 2$ in $uf_n^2 + uf_{n+1}^2 = f_{l_n}$ we easily get that the only possibility is $u = v = 1$ and $k = 1$. It is easy to verify that such u and v satisfy the conditions.

5. There are four types of small cubes upon disassembling:
   (1) 8 cubes with three faces, painted black, at one corner;
   (2) 12 cubes with two black faces, both at one edge;
   (3) 6 cubes with one black face;
   (4) 1 completely white cube.

All cubes of type (1) must go to corners, and be placed in a correct way (one of three): for this step we have $3^8 \cdot 8!$ possibilities. Further, all cubes of type (2) must go in a correct way (one of two) to edges, admitting $2^{12} \cdot 12!$ possibilities; similarly, there are $4^6 \cdot 6!$ ways for cubes of type (3), and 24 ways for the cube of type (4). Thus the total number of good reassemblings is $3^8 8! \cdot 2^{12} 12! \cdot 4^6 6! \cdot$

24, while the number of all possible reassemblings is $24^{27} \cdot 27!$. The desired probability is $\frac{3^8 8! \cdot 2^{12} 12! \cdot 4^6 6! \cdot 24}{24^{27} \cdot 27!}$. It is not necessary to calculate these numbers to find out that the blind man practically has no chance to reassemble the cube in a right way: in fact, the probability is of order $1.8 \cdot 10^{-37}$.

6. Assume w.l.o.g. that $n = \deg P \geq \deg Q$, and let $P_0 = \{z_1, z_2, \ldots, z_k\}$, $P_1 = \{z_{k+1}, z_{k+2}, \ldots z_{k+m}\}$. The polynomials P and Q match at $k + m$ points z_1, z_2, $\ldots$, z_{k+m}; hence if we prove that $k + m > n$, the result will follow.
By the assumption,

$$P(x) = (x - z_1)^{\alpha_1} \cdots (x - z_k)^{\alpha_k} = (x - z_{k+1})^{\alpha_{k+1}} \cdots (x - z_{k+m})^{\alpha_{k+m}} + 1$$

for some positive integers $\alpha_1, \ldots, \alpha_{k+m}$. Let us consider $P'(x)$. As we know, it is divisible by $(x - z_i)^{\alpha_i - 1}$ for $i = 1, 2, \ldots, k + m$; i.e.,

$$\prod_{i=1}^{k+m} (x - z_i)^{\alpha_i - 1} \mid P'(x).$$

Therefore $2n - k - m = \deg \prod_{i=1}^{k+m} (x - z_i)^{\alpha_i - 1} \leq \deg P' = n - 1$, i.e., $k + m \geq n + 1$, as we claimed.

7. We immediately find that $f(1, 0) = f(0, 1) = 2$. Then $f(1, y + 1) = f(0, f(1, y)) = f(1, y) + 1$; hence $f(1, y) = y + 2$ for $y \geq 0$. Next we find that $f(2, 0) = f(1, 1) = 3$ and $f(2, y + 1) = f(1, f(2, y)) = f(2, y) + 2$, from which $f(2, y) = 2y + 3$. Particularly, $f(2, 2) = 7$. Further, $f(3, 0) = f(2, 1) = 5$ and $f(3, y + 1) = f(2, f(3, y)) = 2f(3, y) + 3$. This gives by induction $f(3, y) = 2^{y+3} - 3$. For $y = 3$, $f(3, 3) = 61$. Finally, from $f(4, 0) = f(3, 1) = 13$ and $f(4, y + 1) = f(3, f(4, y)) = 2^{f(4,y)+3} - 3$, we conclude that

$$f(4, y) = 2^{2^{\cdot^{\cdot^2}}} - 3 \quad (y + 3 \text{ twos}).$$

8. Since the number k, $k = 1, 2, \ldots, n - r + 1$, is the minimum in exactly $\binom{n-k}{r-1}$ r-element subsets of $\{1, 2, \ldots, n\}$, it follows that

$$f(n, r) = \frac{1}{\binom{n}{r}} \sum_{k=1}^{n-r+1} k \binom{n-k}{r-1}.$$

Using the equality $\binom{r+j}{j} = \sum_{i=0}^{j} \binom{r+i-1}{r-1}$, we get

$$\sum_{k=1}^{n-r+1} k \binom{n-k}{r-1} = \sum_{j=0}^{n-r} \left(\sum_{i=0}^{j} \binom{r+i-1}{r-1} \right)$$

$$= \sum_{j=0}^{n-r} \binom{r+j}{r} = \binom{n+1}{r+1} = \frac{n+1}{r+1} \binom{n}{r}.$$

Therefore $f(n, r) = (n + 1)/(r + 1)$.

9. If we put $1 + 24a_n = b_n^2$, the given recurrent relation becomes

$$\frac{2}{3}b_{n+1}^2 = \frac{3}{2} + \frac{b_n^2}{6} + b_n = \frac{2}{3}\left(\frac{3}{2} + \frac{b_n}{2}\right)^2, \quad \text{i.e.,} \quad b_{n+1} = \frac{3 + b_n}{2}, \quad (1)$$

where $b_1 = 5$. To solve this recurrent equation, we set $c_n = 2^{n-1}b_n$. From (1) we obtain

$$c_{n+1} = c_n + 3 \cdot 2^{n-1} = \cdots = c_1 + 3(1 + 2 + 2^2 + \cdots + 2^{n-1})$$
$$= 5 + 3(2^n - 1) = 3 \cdot 2^n + 2.$$

Therefore $b_n = 3 + 2^{-n+2}$ and consequently

$$a_n = \frac{b_n^2 - 1}{24} = \frac{1}{3}\left(1 + \frac{3}{2^n} + \frac{1}{2^{2n-1}}\right) = \frac{1}{3}\left(1 + \frac{1}{2^{n-1}}\right)\left(1 + \frac{1}{2^n}\right).$$

10. It is easy to see that partitioning into $p = 2k$ squares is possible for $k \geq 2$ (Fig. 1). Furthermore, whenever it is possible to partition the square into p squares, there is a partition of the square into $p + 3$ squares: namely, in the partition into p squares, divide one of them into four new squares.

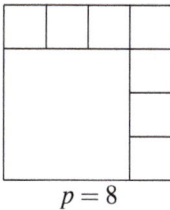

p = 8

Fig. 1

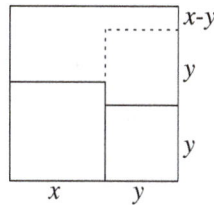

Fig. 2

This implies that both $p = 2k$ and $p = 2k + 3$ are possible if $k \geq 2$, and therefore all $p \geq 6$ are possible.

On the other hand, partitioning the square into 5 squares is not possible. Assuming it is possible, one of its sides would be covered by exactly two squares, which cannot be of the same size (Fig. 2). The rest of the big square cannot be partitioned into three squares. Hence, the answer is $n = 6$.

11. Let us denote the center of the semicircle by O, and $\angle AOB = 2\alpha$, $\angle BOC = 2\beta$, $AC = m$, $CE = n$.
We claim that $a^2 + b^2 + n^2 + abn = 4$. Indeed, since $a = 2\sin\alpha$, $b = 2\sin\beta$, $n = 2\cos(\alpha + \beta)$, we have

$$a^2 + b^2 + n^2 + abn$$
$$= 4(\sin^2\alpha + \sin^2\beta + \cos^2(\alpha + \beta) + 2\sin\alpha\sin\beta\cos(\alpha + \beta))$$
$$= 4 + 4\left(-\frac{\cos 2\alpha}{2} - \frac{\cos 2\beta}{2} + \cos(\alpha + \beta)\cos(\alpha - \beta)\right)$$
$$= 4 + 4(\cos(\alpha + \beta)\cos(\alpha - \beta) - \cos(\alpha + \beta)\cos(\alpha - \beta)) = 4.$$

Analogously, $c^2 + d^2 + m^2 + cdm = 4$. By adding both equalities and subtracting $m^2 + n^2 = 4$ we obtain

$$a^2 + b^2 + c^2 + d^2 + abn + cdm = 4.$$

Since $n > c$ and $m > b$, the desired inequality follows.

12. We will solve the contest problem (in which $m, n \in \{1, 2, \dots, 1981\}$). For $m = 1$, n can be either 1 or 2. If $m > 1$, then $n(n - m) = m^2 \pm 1 > 0$; hence $n - m > 0$. Set $p = n - m$. Since $m^2 - mp - p^2 = m^2 - p(m + p) = -(n^2 - nm - m^2)$, we see that (m, n) is a solution of the equation if and only if (p, m) is a solution too. Therefore, all the solutions of the equation are given as two consecutive members of the Fibonacci sequence

$$1, 1, 2, 3, 5, 8, 13, 21, 34, 55, 89, 144, 233, 377, 610, 987, 1597, 2584, \dots .$$

So the required maximum is $987^2 + 1597^2$.

13. *Lemma.* For any polynomial P of degree at most n,

$$\sum_{i=0}^{n+1} (-1)^i \binom{n+1}{i} P(i) = 0. \tag{1}$$

Proof. We shall use induction on n. For $n = 0$ it is trivial. Assume that it is true for $n = k$ and suppose that $P(x)$ is a polynomial of degree $k + 1$. Then $P(x) - P(x + 1)$ clearly has degree at most k; hence (1) gives

$$0 = \sum_{i=0}^{k+1} (-1)^i \binom{k+1}{i} (P(i) - P(i+1))$$

$$= \sum_{i=0}^{k+1} (-1)^i \binom{k+1}{i} P(i) + \sum_{i=1}^{k+2} (-1)^i \binom{k+1}{i-1} P(i)$$

$$= \sum_{i=0}^{k+2} (-1)^i \binom{k+2}{i} P(i).$$

This completes the proof of the lemma.

Now we apply the lemma to obtain the value of $P(n + 1)$. Since $P(i) = \binom{n+1}{i}^{-1}$ for $i = 0, 1, \dots, n$, we have

$$0 = \sum_{i=0}^{n+1} (-1)^i \binom{n+1}{i} P(i) = (-1)^{n+1} P(n+1) + \begin{cases} 1, & 2 \mid n; \\ 0, & 2 \nmid n. \end{cases}$$

It follows that $P(n+1) = \begin{cases} 1, & 2 \mid n; \\ 0, & 2 \nmid n. \end{cases}$

14. We need the following lemma.

Lemma. If a convex quadrilateral $PQRS$ satisfies $PS = QR$ and $\angle SPQ \geq \angle RQP$, then $\angle QRS \geq \angle PSR$.

Proof. If the lines PS and QR are parallel, then this quadrilateral is a parallelogram, and the statement is trivial. Otherwise, let X be the point of intersection of lines PS and QR.

Assume that $\angle SPQ + \angle RQP > 180°$. Then $\angle XPQ \leq \angle XQP$ implies that $XP \geq XQ$, and consequently $XS \geq XR$. Hence, $\angle QRS = \angle XRS \geq \angle XSR = \angle PSR$.

Similarly, if $\angle SPQ + \angle RQP < 180°$, then $\angle XPQ \geq \angle XQP$, from which it follows that $XP \leq XQ$, and thus $XS \leq XR$; hence $\angle QRS = 180° - \angle XRS \geq 180° - \angle XSR = \angle PSR$.

Now we apply the lemma to the quadrilateral $ABCD$. Since $\angle B \geq \angle C$ and $AB = CD$, it follows that $\angle CDA \geq \angle BAD$, which together with $\angle EDA = \angle EAD$ gives $\angle D \geq \angle A$. Thus $\angle A = \angle B = \angle C = \angle D$. Analogously, by applying the lemma to $BCDE$ we obtain $\angle E \geq \angle B$, and hence $\angle B = \angle C = \angle D = \angle E$.

15. Set $BC = a$, $CA = b$, $AB = c$, and denote the area of $\triangle ABC$ by P, and $a/PD + b/PE + c/PF$ by S. Since $a \cdot PD + b \cdot PE + c \cdot PF = 2P$, by the Cauchy–Schwarz inequality we have

$$2PS = (a \cdot PD + b \cdot PE + c \cdot PF)\left(\frac{a}{PD} + \frac{b}{PE} + \frac{c}{PF}\right) \geq (a+b+c)^2,$$

with equality if and only if $PD = PE = PF$, i.e., P is the incenter of $\triangle ABC$. In that case, S attains its minimum:

$$S_{\min} = \frac{(a+b+c)^2}{2P}.$$

16. The sequence $\{u_n\}$ is bounded, whatever u_1 is. Indeed, assume the opposite, and let u_m be the first member of the sequence such that $|u_m| > \max\{2, |u_1|\}$. Then $|u_{m-1}| = |u_m^3 - 15/64| > |u_m|$, which is impossible.

Next, let us see for what values of u_m, u_{m+1} is greater, equal, or smaller, respectively.

If $u_{m+1} = u_m$, then $u_m = u_{m+1}^3 - 15/64 = u_m^3 - 15/64$; i.e., u_m is a root of $x^3 - x - 15/64 = 0$. This equation factors as $(x + 1/4)(x^2 - x/4 - 15/16) = 0$, and hence u_m is equal to $x_1 = (1 - \sqrt{61})/8$, $x_2 = -1/4$, or $x_3 = (1 + \sqrt{61})/8$, and these are the only possible limits of the sequence.

Each of $u_{m+1} > u_m$, $u_{m+1} < u_m$ is equivalent to $u_m^3 - u_m - 15/64 < 0$ and $u_m^3 - u_m - 15/64 > 0$ respectively. Thus the former is satisfied for u_m in the interval $I_1 = (-\infty, x_1)$ or $I_3 = (x_2, x_3)$, while the latter is satisfied for u_m in $I_2 = (x_1, x_2)$ or $I_4 = (x_3, \infty)$. Moreover, since the function $f(x) = \sqrt[3]{x + 15/64}$ is strictly increasing with fixed points x_1, x_2, x_3, it follows that u_m will never escape from the interval I_1, I_2, I_3, or I_4 to which it belongs initially. Therefore:

(1) if u_1 is one of x_1, x_2, x_3, the sequence $\{u_m\}$ is constant;
(2) if $u_1 \in I_1$, then the sequence is strictly increasing and tends to x_1;
(3) if $u_1 \in I_2$, then the sequence is strictly decreasing and tends to x_1;
(4) if $u_1 \in I_3$, then the sequence is strictly increasing and tends to x_3;
(5) if $u_1 \in I_4$, then the sequence is strictly decreasing and tends to x_3.

17. Let us denote by S_A, S_B, S_C the centers of the given circles, where S_A lies on the bisector of $\angle A$, etc. Then $S_A S_B \parallel AB$, $S_B S_C \parallel BC$, $S_C S_A \parallel CA$, so that the inner bisectors of the angles of triangle ABC are also inner bisectors of the angles of $\triangle S_A S_B S_C$. These two triangles thus have a common incenter S, which is also the center of the homothety χ mapping $\triangle S_A S_B S_C$ onto $\triangle ABC$.

The point O is the circumcenter of triangle $S_A S_B S_C$, and so is mapped by χ onto the circumcenter P of ABC. This means that O, P, and the center S of χ are collinear.

18. Let C be the convex hull of the set of the planets: its border consists of parts of planes, parts of cylinders, and parts of the surfaces of some planets. These parts of planets consist exactly of all the invisible points; any point on a planet that is inside C is visible. Thus it remains to show that the areas of all the parts of planets lying on the border of C add up to the area of one planet.

As we have seen, an invisible part of a planet is bordered by some main spherical arcs, parallel two by two. Now fix any planet P, and translate these arcs onto arcs on the surface of P. All these arcs partition the surface of P into several parts, each of which corresponds to the invisible part of one of the planets. This correspondence is bijective, and therefore the statement follows.

19. Consider the partition of plane π into regular hexagons, each having inradius 2. Fix one of these hexagons, denoted by γ. For any other hexagon x in the partition, there exists a unique translation τ_x taking it onto γ. Define the mapping $\varphi : \pi \to \gamma$ as follows: If A belongs to the interior of a hexagon x, then $\varphi(A) = \tau_x(A)$ (if A is on the border of some hexagon, it does not actually matter where its image is).

The total area of the images of the union of the given circles equals S, while the area of the hexagon γ is $8\sqrt{3}$. Thus there exists a point B of γ that is covered at least $\frac{S}{8\sqrt{3}}$ times, i.e., such that $\varphi^{-1}(B)$ consists of at least $\frac{S}{8\sqrt{3}}$ distinct points of the plane that belong to some of the circles. For any of these points, take a circle that contains it. All these circles are disjoint, with total area not less than $\frac{\pi}{8\sqrt{3}} S \geq 2S/9$.

Remark. The statement becomes false if the constant $2/9$ is replaced by any number greater than $1/4$. In that case a counterexample is, for example, a set of unit circles inside a circle of radius 2 covering a sufficiently large part of its area.

4.23 Solutions to the Shortlisted Problems of IMO 1982

1. From $f(1) + f(1) \leq f(2) = 0$ we obtain $f(1) = 0$. Since $0 < f(3) \leq f(1) + f(2) + 1$, it follows that $f(3) = 1$. Note that if $f(3n) \geq n$, then $f(3n+3) \geq f(3n) + f(3) \geq n + 1$. Hence by induction $f(3n) \geq n$ holds for all $n \in \mathbb{N}$. Moreover, if the inequality is strict for some n, then it is so for all integers greater than n as well. Since $f(9999) = 3333$, we deduce that $f(3n) = n$ for all $n \leq 3333$. By the given condition, we have $3f(n) \leq f(3n) \leq 3f(n) + 2$. Therefore $f(n) = [f(3n)/3] = [n/3]$ for $n \leq 3333$. In particular, $f(1982) = [1982/3] = 660$.

2. Since K does not contain a lattice point other than $O(0,0)$, it is bounded by four lines u, v, w, x that pass through the points $U(1,0)$, $V(0,1)$, $W(-1,0)$, $X(0,-1)$ respectively. Let $PQRS$ be the quadrilateral formed by these lines, where $U \in SP$, $V \in PQ$, $W \in QR$, $X \in RS$.

   If one of the quadrants, say Q_1, contains no vertices of $PQRS$, then $K \cap Q_1$ is contained in $\triangle OUV$ and hence has area less than $1/2$. Consequently the area of K is less than 2.

   Let us now suppose that P, Q, R, S lie in different quadrants. One of the angles of $PQRS$ is at least $90°$: let it be $\angle P$. Then $S_{UPV} \leq PU \cdot PV/2 \leq (PU^2 + PV^2)/4 \leq UV^2/4 = 1/2$, which implies that $S_{K \cap Q_1} < S_{OUPV} \leq 1$. Hence the area of K is less than 4.

3. (a) By the Cauchy–Schwarz inequality we have $(x_0^2/x_1 + \cdots + x_{n-1}^2/x_n) \cdot (x_1 + \cdots + x_n) \geq (x_0 + \cdots + x_{n-1})^2$. Let us set $X_{n-1} = x_1 + x_2 + \cdots + x_{n-1}$. Using $x_0 = 1$, the last inequality can be rewritten as

   $$\frac{x_0^2}{x_1} + \cdots + \frac{x_{n-1}^2}{x_n} \geq \frac{(1 + X_{n-1})^2}{X_{n-1} + x_n} \geq \frac{4X_{n-1}}{X_{n-1} + x_n} = \frac{4}{1 + x_n/X_{n-1}}. \quad (1)$$

   Since $x_n \leq x_{n-1} \leq \cdots \leq x_1$, it follows that $X_{n-1} \geq (n-1)x_n$. Now (1) yields $x_0^2/x_1 + \cdots + x_{n-1}^2/x_n \geq 4(n-1)/n$, which exceeds 3.999 for $n > 4000$.

   (b) The sequence $x_n = 1/2^n$ obviously satisfies the required condition.

   Second solution to part (a). For each $n \in \mathbb{N}$, let us find a constant c_n such that the inequality $x_0^2/x_1 + \cdots + x_{n-1}^2/x_n \geq c_n x_0$ holds for any sequence $x_0 \geq x_1 \geq \cdots \geq x_n > 0$.

   For $n = 1$ we can take $c_1 = 1$. Assuming that c_n exists, we have

   $$\frac{x_0^2}{x_1} + \left(\frac{x_1^2}{x_2} + \cdots + \frac{x_n^2}{x_{n+1}}\right) \geq \frac{x_0^2}{x_1} + c_n x_1 \geq 2\sqrt{x_0^2 c_n} = x_0 \cdot 2\sqrt{c_n}.$$

   Thus we can take $c_{n+1} = 2\sqrt{c_n}$. Then inductively $c_n = 2^{2 - 1/2^{n-2}}$, and since $c_n \to 4$ as $n \to \infty$, the result follows.

   Third solution. Since $\{x_n\}$ is decreasing, there exists $\lim_{n \to \infty} x_n = x \geq 0$. If $x > 0$, then $x_{n-1}^2/x_n \geq x_n \geq x$ holds for each n, and the result is trivial. If otherwise $x = 0$, then we note that $x_{n-1}^2/x_n \geq 4(x_{n-1} - x_n)$ for each n, with equality if and only if $x_{n-1} = 2x_n$. Hence

$$\lim_{n\to\infty}\sum_{k=1}^{n}\frac{x_{k-1}^2}{x_k}\geq\lim_{n\to\infty}\sum_{k=1}^{n}4(x_{k-1}-x_k)=4x_0=4.$$

Equality holds if and only if $x_{n-1}=2x_n$ for all n, and consequently $x_n=1/2^n$.

4. Suppose that a satisfies the requirements of the problem and that x, qx, q^2x, q^3x are the roots of the given equation. Then $x\neq 0$ and we may assume that $|q|>1$, so that $|x|<|qx|<|q^2x|<|q^3x|$. Since the equation is symmetric, $1/x$ is also a root and therefore $1/x=q^3x$, i.e., $q=x^{-2/3}$. It follows that the roots are $x,x^{1/3},x^{-1/3},x^{-1}$. Now by Viète's formula we have $x+x^{1/3}+x^{-1/3}+x^{-1}=a/16$ and $x^{4/3}+x^{2/3}+2+x^{-2/3}+x^{-4/3}=(2a+17)/16$. On setting $z=x^{1/3}+x^{-1/3}$ these equations become

$$z^3-2z=a/16,$$
$$(z^2-2)^2+z^2-2=(2a+17)/16.$$

Substituting $a=16(z^3-2z)$ in the second equation leads to $z^4-2z^3-3z^2+4z+15/16=0$. We observe that this polynomial factors as $(z+3/2)(z-5/2)(z^2-z-1/4)$. Since $|z|=|x^{1/3}+x^{-1/3}|\geq 2$, the only viable value is $z=5/2$. Consequently $a=170$ and the roots are $1/8,1/2,2,8$.

5. Notice that $\triangle A_5B_4A_4\cong\triangle A_3B_2A_2$. We know that $\angle A_5A_3A_2=90°$ and that $\angle A_2B_4A_4$ is equal to the sum of the angles $\angle A_2B_4A_3$ and $\angle A_3B_4A_4$. Clearly, $\angle A_2B_4A_3=90°-\angle B_2A_2A_3$ and $\angle A_3B_4A_4=\angle B_4A_5A_4+\angle A_5A_4B_4$. Hence we conclude that $\angle A_2B_4A_4=90°+\angle B_4A_5A_4=120°$. Hence B_4 belongs to the circle with center A_3 and radius A_3A_4, so $A_3A_4=A_3B_4$. Thus $\lambda=A_3B_4/A_3A_5=A_3A_4/A_3A_5=1/\sqrt{3}$.

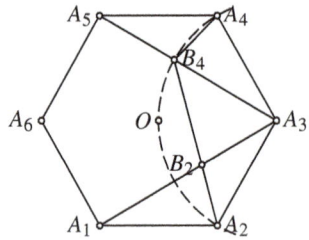

6. Denote by $d(U,V)$ the distance between points or sets of points U and V. For $P,Q\in L$ we shall denote by L_{PQ} the part of L between points P and Q and by l_{PQ} the length of this part. Let us denote by S_i $(i=1,2,3,4)$ the vertices of S and by T_i points of L such that $S_iT_i\leq 1/2$ in such a way that $l_{A_0T_1}$ is the least of the $l_{A_0T_i}$'s, S_2 and S_4 are neighbors of S_1, and $l_{A_0T_2}<l_{A_0T_4}$.

Now we shall consider the points of the segment S_1S_4. Let D and E be the sets of points defined as follows: $D=\{X\in[S_1S_4]\mid d(X,L_{A_0T_2})\leq 1/2\}$ and $E=\{X\in[S_1S_4]\mid d(X,L_{T_2A_n})\leq 1/2\}$. Clearly D and E are closed, nonempty (indeed, $S_1\in D$ and $S_4\in E$) subsets of $[S_1S_4]$. Since their union is a connected set S_1S_4, it follows that they must have a nonempty intersection. Let $P\in D\cap E$. Then there exist points $X\in L_{A_0T_2}$ and $Y\in L_{T_2A_n}$ such that $d(P,X)\leq 1/2$, $d(P,Y)\leq 1/2$, and consequently $d(X,Y)\leq 1$. On the other hand, T_2 lies between X and Y on L, and thus $L_{XY}=L_{XT_2}+L_{T_2Y}\geq XT_2+T_2Y\geq(PS_2-XP-S_2T_2)+(PS_2-YP-S_2T_2)\geq 99+99=198$.

7. Let a, b, ab be the roots of the cubic polynomial $p(x) = (x-a)(x-b)(x-ab)$. Observe that

$$2p(-1) = -2(1+a)(1+b)(1+ab);$$
$$p(1) + p(-1) - 2(1+p(0)) = -2(1+a)(1+b).$$

The statement of the problem is trivial if both the expressions are equal to zero. Otherwise, the quotient $\frac{2p(-1)}{p(1)+p(-1)-2(1+p(0))} = 1 + ab$ is rational and consequently ab is rational. But since $(ab)^2 = -p(0)$ is an integer, it follows that ab is also an integer. This completes the proof.

8. Let $\mathscr{F}$ be the given figure. Consider any chord AB of the circumcircle γ that supports $\mathscr{F}$. The other supporting lines to $\mathscr{F}$ from A and B intersect γ again at D and C respectively so that $\angle DAB = \angle ABC = 90°$. Then $ABCD$ is a rectangle, and hence CD must support $\mathscr{F}$ as well, from which it follows that $\mathscr{F}$ is inscribed in the rectangle $ABCD$ touching each of its sides. We easily conclude that $\mathscr{F}$ is the intersection of all such rectangles. Now, since the center O of γ is the center of symmetry of all these rectangles, it must be so for their intersection $\mathscr{F}$ as well.

9. Let X and Y be the midpoints of the segments AP and BP. Then $DYPX$ is a parallelogram. Since X and Y are the circumcenters of the triangles APM and BPL, we conclude that $XM = XP = DY$ and $YL = YP = DX$. Furthermore, we have $\angle DXM = \angle DXP + \angle PXM = \angle DXP + 2\angle PAM$. Similarly, $\angle DYL = \angle DYP + 2\angle PBL$ hence $\angle DXM = \angle DYL$. Therefore, the triangles DXM and LYD are congruent, implying $DM = DL$.

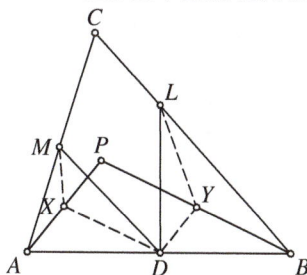

10. If the two balls taken from the box are both white, then the number of white balls decreases by two; otherwise, it remains unchanged. Hence the parity of the number of white balls does not change during the procedure. Therefore if p is even, the last ball cannot be white; the probability is 0. If p is odd, the last ball has to be white; the probability is 1.

11. (a) Suppose $\{a_1, a_2, \ldots, a_n\}$ is the arrangement that yields the maximal value Q_{max} of Q. Note that the value of Q for the rearrangement $\{a_1, \ldots, a_{i-1}, a_j, a_{j-1}, \ldots, a_i, a_{j+1}, \ldots, a_n\}$ equals $Q_{max} - (a_i - a_j)(a_{i-1} - a_{j+1})$, where $1 < i < j < n$. Hence $(a_i - a_j)(a_{i-1} - a_{j+1}) \geq 0$ for all $1 < i < j < n$. We may suppose w.l.o.g. that $a_1 = 1$. Let $a_i = 2$. If $2 < i < n$, then $(a_2 - a_i)(a_1 - a_{i+1}) < 0$, which is impossible. Therefore i is either 2 or n; let w.l.o.g. $a_n = 2$. Further, if $a_j = 3$ for $2 < j < n$, then $(a_1 - a_{j+1})(a_2 - a_j) < 0$, which is impossible; therefore $a_2 = 3$. Continuing this argument we obtain that $A = \{1, 3, 5, \ldots, 2[(n-1)/2] + 1, 2[n/2], \ldots, 4, 2\}$.

   (b) A similar argument leads to the minimizing rearrangement $\{1, n, 2, n-1, \ldots, [n/2]+1\}$.

12. Let y be the line perpendicular to L passing through the center of C. It can be shown by a continuity argument that there exists a point $Y \in y$ such that an inversion Ψ centered at Y maps C and L onto two concentric circles $\widehat{C}$ and $\widehat{L}$. Let $\widehat{X}$ denote the image of an object X under Ψ. Then the circles $\widehat{C_i}$ touch $\widehat{C}$ externally and $\widehat{L}$ internally, and all have the same radius. Let us now rotate the picture around the common center Z of $\widehat{C}$ and $\widehat{L}$ so that $\widehat{C_3}$ passes through Y. Applying the inversion Ψ again on the picture thus obtained, $\widehat{C}$ and $\widehat{L}$ go back to C and L, but $\widehat{C_3}$ goes to a line C_3' parallel to L, while the images of $\widehat{C_1}$ and $\widehat{C_2}$ go to two equal circles C_1' and C_2' touching L, C_3', and C. This way we have achieved that C_3 becomes a line.

Denote by O_1, O_2, O respectively the centers of the circles C_1', C_2', C and by T the point of tangency of the circles C_1' and C_2'. If x is the common radius of the circles C_1' and C_2', then from $\triangle O_1 T O$ we obtain that $(x-1)^2 + x^2 = (x+1)^2$,

and thus $x = 4$. Hence the distance of O from L equals $2x - 1 = 7$.

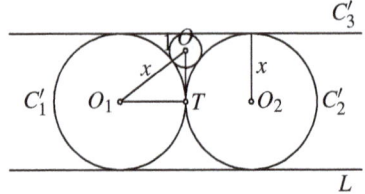

13. The points S_1, S_2, S_3 clearly lie on the inscribed circle. Let $\widehat{XY}$ denote the oriented arc XY. The arcs $\widehat{T_2 S_1}$ and $\widehat{T_1 T_3}$ are equal, since they are symmetric with respect to the bisector of $\angle A_1$. Similarly, $\widehat{T_3 T_2} = \widehat{S_2 T_1}$. Hence $\widehat{T_3 S_1} = \widehat{T_3 T_2} + \widehat{T_2 S_1} = \widehat{S_2 T_1} + \widehat{T_1 T_3} = \widehat{S_2 T_3}$. It follows that $S_1 S_2$ is parallel to $A_1 A_2$, and con-

sequently $S_1 S_2 \parallel M_1 M_2$. Analogously $S_1 S_3 \parallel M_1 M_3$ and $S_2 S_3 \parallel M_2 M_3$.
Since the circumcircles of $\triangle M_1 M_2 M_3$ and $\triangle S_1 S_2 S_3$ are not equal, these triangles are not congruent and hence they must be homothetic. Then all the lines $M_i S_i$ pass through the center of homothety.

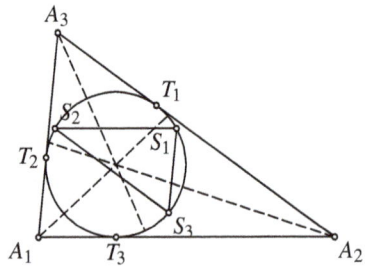

Second solution. Set the complex plane so that the incenter of $\triangle A_1 A_2 A_3$ is the unit circle centered at the origin. Let t_i, s_i respectively denote the complex numbers of modulus 1 corresponding to T_i, S_i. Clearly $t_1 \bar{t_1} = t_2 \bar{t_2} = t_3 \bar{t_3} = 1$. Since $T_2 T_3$ and $T_1 S_1$ are parallel, we obtain $t_2 t_3 = t_1 s_1$, or $s_1 = t_2 t_3 \bar{t_1}$. Similarly $s_2 = t_1 t_3 \bar{t_2}$, $s_3 = t_1 t_2 \bar{t_3}$, from which it follows that $s_2 - s_3 = t_1 (t_3 \bar{t_2} - t_2 \bar{t_3})$. Since the number in parentheses is strictly imaginary, we conclude that $OT_1 \perp S_2 S_3$ and consequently $S_2 S_3 \parallel A_2 A_3$. We proceed as in the first solution.

14. (a) If any two of A_1, B_1, C_1, D_1 coincide, say $A_1 \equiv B_1$, then $ABCD$ is inscribed in a circle centered at A_1 and hence all A_1, B_1, C_1, D_1 coincide.
    Assume now the opposite, and let w.l.o.g. $\angle DAB + \angle DCB < 180°$. Then A is outside the circumcircle of $\triangle BCD$, so $A_1 A > A_1 C$. Similarly, $C_1 C > C_1 A$. Hence the perpendicular bisector l_{AC} of AC separates points A_1 and C_1. Since B_1, D_1 lie on l_{AC}, this means that A_1 and C_1 are on opposite sides

B_1D_1. Similarly one can show that B_1 and D_1 are on opposite sides of A_1C_1.

(b) Since $A_2B_2 \perp C_1D_1$ and $C_1D_1 \perp AB$, it follows that $A_2B_2 \parallel AB$. Similarly $A_2C_2 \parallel AC$, $A_2D_2 \parallel AD$, $B_2C_2 \parallel BC$, $B_2D_2 \parallel BD$, and $C_2D_2 \parallel CD$. Hence $\triangle A_2B_2C_2 \sim \triangle ABC$ and $\triangle A_2D_2C_2 \sim \triangle ADC$, and the result follows.

15. Let $a = k/n$, where $n, k \in \mathbb{N}$, $n \geq k$. Putting $t^n = s$, the given inequality becomes $\frac{1-t^k}{1-t^n} \leq (1+t^n)^{k/n-1}$, or equivalently

$$(1+t+\cdots+t^{k-1})^n(1+t^n)^{n-k} \leq (1+t+\cdots+t^{n-1})^n.$$

This is clearly true for $k = n$. Therefore it is enough to prove that the left-hand side of the above inequality is an increasing function of k. We are led to show that $(1+t+\cdots+t^{k-1})^n(1+t^n)^{n-k} \leq (1+t+\cdots+t^k)^n(1+t^n)^{n-k-1}$. This is equivalent to $1+t^n \leq A^n$, where $A = \frac{1+t+\cdots+t^k}{1+t+\cdots+t^{k-1}}$. But this easily follows, since

$$A^n - t^n = (A-t)(A^{n-1} + A^{n-2}t + \cdots + t^{n-1})$$
$$\geq (A-t)(1+t+\cdots+t^{n-1}) = \frac{1+t+\cdots+t^{n-1}}{1+t+\cdots+t^{k-1}} \geq 1.$$

Remark. The original problem asked to prove the inequality for real a.

16. It is easy to verify that whenever (x,y) is a solution of the equation $x^3 - 3xy^2 + y^3 = n$, so are the pairs $(y-x,-x)$ and $(-y,x-y)$. No two of these three solutions are equal unless $x = y = n = 0$.
Observe that $2891 \equiv 2 \pmod 9$. Since $x^3, y^3 \equiv 0, \pm 1 \pmod 9$, $x^3 - 3xy^2 + y^3$ cannot give the remainder 2 when divided by 9. Hence the above equation for $n = 2891$ has no integer solutions.

17. Let A be the origin of the Cartesian plane. Suppose that $BC : AC = k$ and that (a,b) and (a_1,b_1) are coordinates of the points C and C_1, respectively. Then the coordinates of the point B are $(a,b) + k(-b,a) = (a-kb, b+ka)$, while the coordinates of B_1 are $(a_1,b_1) + k(b_1,-a_1) = (a+kb_1, b_1 - ka_1)$. Thus the lines BC_1 and CB_1 are given by the equations $\frac{x-a_1}{y-b_1} = \frac{x-(a-kb)}{y-(b+ka)}$ and $\frac{x-a}{y-b} = \frac{x-(a_1+kb_1)}{y-(b_1-ka_1)}$ respectively. After multiplying, these equations transform into the forms

$$BC_1 : \quad kax + kby = kaa_1 + kbb_1 + ba_1 - ab_1 - (b-b_1)x + (a-a_1)y$$
$$CB_1 : \quad ka_1x + kb_1y = kaa_1 + kbb_1 + ba_1 - ab_1 - (b-b_1)x + (a-a_1)y.$$

The coordinates (x_0,y_0) of the point M satisfy these equations, from which we deduce that $kax_0 + kby_0 = ka_1x_0 + kb_1y_0$. This yields $\frac{x_0}{y_0} = -\frac{b_1-b}{a_1-a}$, implying that the lines CC_1 and AM are perpendicular.

18. Set the coordinate system with the axes x, y, z along the lines l_1, l_2, l_3 respectively. The coordinates (a,b,c) of M satisfy $a^2 + b^2 + c^2 = R^2$, and so S_M is given by the equation $(x-a)^2 + (y-b)^2 + (z-c)^2 = R^2$. Hence the coordinates of P_1 are $(x,0,0)$ with $(x-a)^2 + b^2 + c^2 = R^2$, implying that either $x = 2a$ or $x = 0$.

Thus by the definition we obtain $x = 2a$. Similarly, the coordinates of P_2 and P_3 are $(0,2b,0)$ and $(0,0,2c)$ respectively. Now, the centroid of $\triangle P_1P_2P_3$ has the coordinates $(2a/3, 2b/3, 2c/3)$. Therefore the required locus of points is the sphere with center O and radius $2R/3$.

19. Let us set $x = m/n$. Since $f(x) = (m+n)/\sqrt{m^2+n^2} = (x+1)/\sqrt{1+x^2}$ is a continuous function of x, $f(x)$ takes all values between any two values of f; moreover, the corresponding x can be rational. This completes the proof.

    Remark. Since f is increasing for $x \geq 1$, $1 \leq x < z < y$ implies $f(x) < f(z) < f(y)$.

20. Since MN is the image of AC under rotation about B for $60°$, we have $MN = AC$. Similarly, PQ is the image of AC under rotation about D through $60°$, from which it follows that $PQ \parallel MN$. Hence either M,N,P,Q are collinear or $MNPQ$ is a parallelogram.

4.24 Solutions to the Shortlisted Problems of IMO 1983

1. Suppose that there are n airlines $A_1, \dots, A_n$ and $N > 2^n$ cities. We shall prove that there is a round trip by at least one A_i containing an odd number of stops.

   For $n = 1$ the statement is trivial, since one airline serves at least 3 cities and hence $P_1 P_2 P_3 P_1$ is a round trip with 3 landings. We use induction on n, and assume that $n > 1$. Suppose the contrary, that all round trips by A_n consist of an even number of stops. Then we can separate the cities into two nonempty classes $Q = \{Q_1, \dots, Q_r\}$ and $R = \{R_1, \dots, R_s\}$ (where $r + s = N$), so that each flight by A_n runs between a Q-city and an R-city. (Indeed, take any city Q_1 served by A_n; include each city linked to Q_1 by A_n in R, then include in Q each city linked by A_n to any R-city, etc. Since all round trips are even, no contradiction can arise.) At least one of r, s is larger than 2^{n-1}, say $r > 2^{n-1}$. But, only $A_1, \dots, A_{n-1}$ run between cities in $\{Q_1, \dots, Q_r\}$; hence by the induction hypothesis at least one of them flies a round trip with an odd number of landings, a contradiction. It only remains to notice that for $n = 10$, $2^n = 1024 < 1983$.

   Remark. If there are $N = 2^n$ cities, there is a schedule with n airlines that contain no odd round trip by any of the airlines. Let the cities be P_k, $k = 0, \dots, 2^n - 1$, and write k in the binary system as an n-digit number $\overline{a_1 \dots a_n}$ (e.g., $1 = (0 \dots 001)_2$). Link P_k and P_l by A_i if the ith digits k and l are distinct but the first $i - 1$ digits are the same. All round trips under A_i are even, since the ith digit alternates.

2. By definition, $\sigma(n) = \sum_{d \mid n} d = \sum_{d \mid n} n/d = n \sum_{d \mid n} 1/d$, hence $\sigma(n)/n = \sum_{d \mid n} 1/d$. In particular, $\sigma(n!)/n! = \sum_{d \mid n!} 1/d \geq \sum_{k=1}^{n} 1/k$. It follows that the sequence $\sigma(n)/n$ is unbounded, and consequently there exist an infinite number of integers n such that $\sigma(n)/n$ is strictly greater than $\sigma(k)/k$ for $k < n$.

3. (a) A circle is not Pythagorean. Indeed, consider the partition into two semicircles each closed at one and open at the other end.

   (b) An equilateral triangle, call it PQR, is Pythagorean. Let P', Q', and R' be the points on QR, RP, and PQ such that $PR' : R'Q = QP' : P'R = RQ' : Q'P = 1 : 2$. Then $Q'R' \perp PQ$, etc. Suppose that PQR is not Pythagorean, and consider a partition into A, B, neither of which contains the vertices of a right-angled triangle. At least two of P', Q', and R' belong to the same class, say $P', Q' \in A$. Then $[PR] \setminus \{Q'\} \subset B$ and hence $R' \in A$ (otherwise, if R'' is the foot of the perpendicular from R' to PR, $\triangle RR'R''$ is right-angled with all vertices in B). But this implies again that $[PQ] \setminus \{R'\} \subset B$, and thus B contains vertices of a rectangular triangle. This is a contradiction.

4. The rotational homothety centered at C that sends B to R also sends A to Q; hence the triangles ABC and QRC are similar. For the same reason, $\triangle ABC$ and $\triangle PBR$ are similar. Moreover, $BR = CR$; hence $\triangle CRQ \cong \triangle RBP$. Thus $PR = QC = AQ$ and $QR = PB = PA$, so $APQR$ is a parallelogram.

5. Each natural number p can be written uniquely in the form $p = 2^q(2r - 1)$. We call $2r - 1$ the odd part of p. Let $A_n = (a_1, a_2, \dots, a_n)$ be the first sequence. Clearly the terms of A_n must have different odd parts, so those parts must be at

least $1,3,\ldots,2n-1$. Being the first sequence, A_n must have the numbers $2n-1,2n-3,\ldots,2k+1$ as terms, where $k=[n+1/3]$ (then $3(2k-1)<2n-1<3(2k+1)$). Smaller odd numbers $2s+1$ (with $s<k$) obviously cannot be terms of A_n. In this way we have obtained the $n-k$ odd numbers of A_n. The other k terms must be even, and by the same reasoning as above they must be precisely the terms of $2A_k$ (twice the terms of A_k). Therefore A_n is defined recursively as

$$A_0 = \emptyset, \quad A_1 = \{1\}, \quad A_2 = \{3,2\};$$
$$A_n = \{2n-1,2n-3,\ldots,2k+1\} \cup 2A_k.$$

6. The existence of r: Let $S = \{x_1 + x_2 + \cdots + x_i - 2i \mid i = 1,2,\ldots,n\}$. Let $\max S$ be attained for the first time at r'.

If $r' = n$, then $x_1 + x_2 + \cdots + x_i - 2i < 2$ for $1 \le i \le n-1$, so one can take $r = r'$. Suppose that $r' < n$. Then for $l < n - r'$ we have $x_{r'+1} + x_{r'+2} + \cdots + x_{r'+l} = (x_1 + \cdots + x_{r'+l} - 2(r'+l)) - (x_1 + \cdots + x_{r'} - 2r') + 2l \le 2l$; also, for $i < r'$ we have $(x_{r'+1} + \cdots + x_n) + (x_1 + \cdots + x_i - 2i) < (x_{r'+1} + \cdots + x_n) + (x_1 + \cdots + x_{r'} - 2r') = (x_1 + \cdots + x_n) - 2r' = 2(n-r') + 2 \Rightarrow x_{r'+1} + \cdots + x_n + x_1 + \cdots + x_i \le 2(n+i-r') + 1$, so we can again take $r = r'$.

For the second part of the problem, we relabel the sequence so that $r = 0$ works. Suppose that the inequalities are strict. We have $x_1 + x_2 + \cdots + x_k \le 2k$, $k = 1,\ldots,n-1$. Now, $2n+2 = (x_1 + \cdots + x_k) + (x_{k+1} + \cdots + x_n) \le 2k + x_{k+1} + \cdots + x_n \Rightarrow x_{k+1} + \cdots + x_n \ge 2(n-k) + 2 > 2(n-k) + 1$. So we cannot begin with x_{k+1} for any $k > 0$.

Now assume that there is an equality for some k. There are two cases:

(i) Suppose $x_1 + x_2 + \cdots + x_i \le 2i$ ($i = 1,\ldots,k$) and $x_1 + \cdots + x_k = 2k + 1$, $x_1 + \cdots + x_{k+l} \le 2(k+l) + 1$ ($1 \le l \le n-1-k$). For $i \le k-1$ we have $x_{i+1} + \cdots + x_n = 2(n+1) - (x_1 + \cdots + x_i) > 2(n-i) + 1$, so we cannot take $r = i$. If there is a $j \ge 1$ such that $x_1 + x_2 + \cdots + x_{k+j} \le 2(k+j)$, then also $x_{k+j+1} + \cdots + x_n > 2(n-k-j) + 1$. If $(\forall j \ge 1)\, x_1 + \cdots + x_{k+j} = 2(k+j) + 1$, then $x_n = 3$ and $x_{k+1} = \cdots = x_{n-1} = 2$. In this case we directly verify that we cannot take $r = k + j$. However, we can also take $r = k$: for $k + l \le n - 1$, $x_{k+1} + \cdots + x_{k+l} \le 2(k+l) + 1 - (2k+1) = 2l$, also $x_{k+1} + \cdots + x_n = 2(n-k) + 1$, and moreover $x_1 \le 2$, $x_1 + x_2 \le 4, \ldots$.

(ii) Suppose $x_1 + \cdots + x_i \le 2i$ ($1 \le i \le n-2$) and $x_1 + \cdots + x_{n-1} = 2n - 1$. Then we can obviously take $r = n - 1$. On the other hand, for any $1 \le i \le n-2$, $x_{i+1} + \cdots + x_{n-1} + x_n = (x_1 + \cdots + x_{n-1}) - (x_1 + \cdots + x_i) + 3 > 2(n-i) + 1$, so we cannot take another $r \ne 0$.

7. Clearly, each a_n is positive and $\sqrt{a_{n+1}} = \sqrt{a_n}\sqrt{a+1} + \sqrt{a_n+1}\sqrt{a}$. Notice that $\sqrt{a_{n+1}+1} = \sqrt{a+1}\sqrt{a_n+1} + \sqrt{a}\sqrt{a_n}$. Therefore

$$(\sqrt{a+1} - \sqrt{a})(\sqrt{a_n+1} - \sqrt{a_n})$$
$$= (\sqrt{a+1}\sqrt{a_n+1} + \sqrt{a}\sqrt{a_n}) - (\sqrt{a_n}\sqrt{a+1} + \sqrt{a_n+1}\sqrt{a})$$
$$= \sqrt{a_{n+1}+1} - \sqrt{a_{n+1}}.$$

By induction, $\sqrt{a_{n+1}} - \sqrt{a_n} = \left(\sqrt{a+1} - \sqrt{a}\right)^n$. Similarly, $\sqrt{a_{n+1}} + \sqrt{a_n} = \left(\sqrt{a+1} + \sqrt{a}\right)^n$. Hence,

$$\sqrt{a_n} = \frac{1}{2}\left[\left(\sqrt{a+1} + \sqrt{a}\right)^n - \left(\sqrt{a+1} - \sqrt{a}\right)^n\right],$$

from which the result follows.

8. Situations in which the condition of the statement is fulfilled are the following:

S_1: $N_1(t) = N_2(t) = N_3(t)$

S_2: $N_i(t) = N_j(t) = h$, $N_k(t) = h+1$, where (i, j, k) is a permutation of the set $\{1,2,3\}$. In this case the first student to leave must be from row k. This leads to the situation S_1.

S_3: $N_i(t) = h, N_j(t) = N_k(t) = h+1$, $((i, j, k)$ is a permutation of the set $\{1,2,3\})$. In this situation the first student leaving the room belongs to row j (or k) and the second to row k (or j). After this we arrive at the situation S_1.

Hence, the initial situation is S_1 and after each triple of students leaving the room the situation S_1 must recur. We shall compute the probability P_h that from a situation S_1 with $3h$ students in the room ($h \le n$) one arrives at a situation S_1 with $3(h-1)$ students in the room:

$$P_h = \frac{(3h) \cdot (2h) \cdot h}{(3h) \cdot (3h-1) \cdot (3h-2)} = \frac{3!h^3}{3h(3h-1)(3h-2)}.$$

Since the room becomes empty after the repetition of n such processes, which are independent, we obtain for the probability sought

$$P = \prod_{h=1}^{n} P_h = \frac{(3!)^n (n!)^3}{(3n)!}.$$

9. For any triangle of sides a, b, c there exist 3 nonnegative numbers x, y, z such that $a = y+z$, $b = z+x$, $c = x+y$ (these numbers correspond to the division of the sides of a triangle by the point of contact of the incircle). The inequality becomes

$$(y+z)^2(z+x)(y-x) + (z+x)^2(x+y)(z-y) + (x+y)^2(y+z)(x-z) \ge 0.$$

Expanding, we get $xy^3 + yz^3 + zx^3 \ge xyz(x+y+z)$. This follows from Cauchy's inequality $(xy^3 + yz^3 + zx^3)(z+x+y) \ge \left(\sqrt{xyz}(x+y+z)\right)^2$ with equality if and only if $xy^3/z = yz^3/x = zx^3/y$, or equivalently $x = y = z$, i.e., $a = b = c$.

10. Choose $P(x) = \frac{p}{q}\left((qx-1)^{2n+1} + 1\right)$, $I = [1/2q, 3/2q]$. Then all the coefficients of P are integers, and

$$\left|P(x) - \frac{p}{q}\right| = \left|\frac{p}{q}(qx-1)^{2n+1}\right| \le \left|\frac{p}{q}\right|\frac{1}{2^{2n+1}},$$

for $x \in I$. The desired inequality follows if n is chosen large enough.

11. First suppose that the binary representation of x is finite: $x = 0, a_1 a_2 \ldots a_n = \sum_{j=1}^{n} a_j 2^{-j}$, $a_i \in \{0, 1\}$. We shall prove by induction on n that

$$f(x) = \sum_{j=1}^{n} b_0 \ldots b_{j-1} a_j, \quad \text{where } b_k = \begin{cases} -b & \text{if } a_k = 0, \\ 1 - b & \text{if } a_k = 1. \end{cases}$$

(Here $a_0 = 0$.) Indeed, by the recursion formula,

$$a_1 = 0 \Rightarrow f(x) = bf\left(\sum_{j=1}^{n-1} a_{j+1} 2^{-j}\right) = b \sum_{j=1}^{n-1} b_1 \ldots b_j a_{j+1}$$

$$\text{hence } f(x) = \sum_{j=0}^{n-1} b_0 \ldots b_j a_{j+1} \text{ as } b_0 = b_1 = b;$$

$$a_1 = 1 \Rightarrow f(x) = b + (1-b)f\left(\sum_{j=1}^{n-1} a_{j+1} 2^{-j}\right) = \sum_{j=0}^{n-1} b_0 \ldots b_j a_{j+1},$$

$$\text{as } b_0 = b, b_1 = 1 - b.$$

Clearly, $f(0) = 0$, $f(1) = 1$, $f(1/2) = b > 1/2$. Assume $x = \sum_{j=0}^{n} a_j 2^{-j}$, and for $k \geq 2$, $v = x + 2^{-n-k+1}$, $u = x + 2^{-n-k} = (v+x)/2$. Then $f(v) = f(x) + b_0 \ldots b_n b^{k-2}$ and $f(u) = f(x) + b_0 \ldots b_n b^{k-1} > (f(v) + f(x))/2$. This means that the point $(u, f(u))$ lies above the line joining $(x, f(x))$ and $(v, f(v))$. By induction, every $(x, f(x))$, where x has a finite binary expansion, lies above the line joining $(0,0)$ and $(1/2, b)$ if $0 < x < 1/2$, or above the line joining $(1/2, b)$ and $(1,1)$ if $1/2 < x < 1$. It follows immediately that $f(x) > x$. For the second inequality, observe that

$$f(x) - x = \sum_{j=1}^{\infty} (b_0 \ldots b_{j-1} - 2^{-j}) a_j$$

$$< \sum_{j=1}^{\infty} (b^j - 2^{-j}) a_j < \sum_{j=1}^{\infty} (b^j - 2^{-j}) = \frac{b}{1-b} - 1 = c.$$

By continuity, these inequalities also hold for x with infinite binary representations.

12. Putting $y = x$ in (i) we see that there exist positive real numbers z such that $f(z) = z$ (this is true for every $z = xf(x)$). Let a be any of them. Then $f(a^2) = f(af(a)) = af(a) = a^2$, and by induction, $f(a^n) = a^n$. If $a > 1$, then $a^n \to +\infty$ as $n \to \infty$, and we have a contradiction with (ii). Again, $a = f(a) = f(1 \cdot a) = af(1)$, so $f(1) = 1$. Then, $af(a^{-1}) = f(a^{-1}f(a)) = f(1) = 1$, and by induction, $f(a^{-n}) = a^{-n}$. This shows that $a \not< 1$. Hence, $a = 1$. It follows that $xf(x) = 1$, i.e., $f(x) = 1/x$ for all x. This function satisfies (i) and (ii), so $f(x) = 1/x$ is the unique solution.

13. Given any coloring of the $3 \times 1983 - 2$ points of the axes, we prove that there is a unique coloring of E having the given property and extending this coloring.

The first thing to notice is that given any rectangle R_1 parallel to a coordinate plane and whose edges are parallel to the axes, there is an even number r_1 of red vertices on R_1. Indeed, let R_2 and R_3 be two other rectangles that are translated from R_1 orthogonally to R_1 and let r_2, r_3 be the numbers of red vertices on R_2 and R_3 respectively. Then $r_1 + r_2$, $r_1 + r_3$, and $r_2 + r_3$ are multiples of 4, so $r_1 = (r_1 + r_2 + r_1 + r_3 - r_2 - r_3)/2$ is even.

Since any point of a coordinate plane is a vertex of a rectangle whose remaining three vertices lie on the corresponding axes, this determines uniquely the coloring of the coordinate planes. Similarly, the coloring of the inner points of the parallelepiped is completely determined. The solution is hence $2^{3 \times 1983 - 2} = 2^{5947}$.

14. Let T_n be the set of all nonnegative integers whose ternary representations consist of at most n digits and do not contain a digit 2. The cardinality of T_n is 2^n, and the greatest integer in T_n is $11\ldots1 = 3^0 + 3^1 + \cdots + 3^{n-1} = (3^n - 1)/2$. We claim that there is no arithmetic triple in T_n. To see this, suppose $x, y, z \in T_n$ and $2y = x + z$. Then $2y$ has only 0's and 2's in its ternary representation, and a number of this form can be the sum of two integers $x, z \in T_n$ in only one way, namely $x = z = y$. But $|T_{10}| = 2^{10} = 1024$ and $\max T_{10} = (3^{10} - 1)/2 = 29524 < 30000$. Thus the answer is yes.

15. There is no such set. Suppose that M satisfies the conditions (i) and (ii) and let $q_n = |\{a \in M : a \le n\}|$. Consider the differences $b - a$, where $a, b \in M$ and $10 < a < b \le k$. They are all positive and less than k, and (ii) implies that they are $\binom{q_k - q_{10}}{2}$ different integers. Hence $\binom{q_k - q_{10}}{2} < k$, so $q_k \le \sqrt{2k} + 10$. It follows from (i) that among the numbers of the form $a + b$, where $a, b \in M$, $a \le b \le n$, or $a \le n < b \le 2n$, there are all integers from the interval $[2, 2n + 1]$. Thus $\binom{q_n + 1}{2} + q_n(q_{2n} - q_n) \ge 2n$ for every $n \in \mathbb{N}$. Set $Q_k = \sqrt{2k} + 10$. We have

$$\binom{q_n + 1}{2} + q_n(q_{2n} - q_n) = \frac{1}{2}q_n + \frac{1}{2}q_n(2q_{2n} - q_n)$$

$$\le \frac{1}{2}q_n + \frac{1}{2}q_n(2Q_{2n} - q_n)$$

$$\le \frac{1}{2}Q_n + \frac{1}{2}Q_n(2Q_{2n} - Q_n)$$

$$\le 2(\sqrt{2} - 1)n + (20 + \sqrt{2}/2)\sqrt{n} + 55,$$

which is less than n for n large enough, a contradiction.

16. Set $h_{n,i}(x) = x^i + \cdots + x^{n-i}$, $2i \le n$. The set $F(n)$ is the set of linear combinations with nonnegative coefficients of the $h_{n,i}$'s. This is a convex cone. Hence, it suffices to prove that $h_{n,i}h_{m,j} \in F(m+n)$. Indeed, setting $p = n - 2i$ and $q = m - 2j$ and assuming $p \le q$ we obtain

$$h_{n,i}(x)h_{m,j}(x) = (x^i + \cdots + x^{i+p})(x^j + \cdots + x^{j+q}) = \sum_{k=i+j}^{n-i+j} h_{m+n,k},$$

which proves the claim.

17. Set $a = \min P_i P_j$, $b = \max P_i P_j$. We use the following lemma.

Lemma. There exists a disk of radius less than or equal to $b/\sqrt{3}$ containing all the P_i's.

Assuming that this is proved, the disks with center P_i and radius $a/2$ are disjoint and included in a disk of radius $b/\sqrt{3}+a/2$; hence comparing areas,

$$n\pi \cdot \frac{a^2}{4} < \pi \cdot \left(\frac{b}{\sqrt{3}}+a/2\right)^2 \quad \text{and} \quad b > \sqrt{3}/2 \cdot (\sqrt{n}-1)a.$$

Proof of the lemma. If a nonobtuse triangle with sides $a \ge b \ge c$ has a circumscribed circle of radius R, we have $R = a/(2\sin\alpha) \le a/\sqrt{3}$. Now we show that there exists a disk D of radius R containing $A = \{P_1,\ldots,P_n\}$ whose border C is such that $C \cap A$ is not included in an open semicircle, and hence contains either two diametrically opposite points and $R \le b/2$, or an acute-angled triangle and $R \le b/\sqrt{3}$.

Among all disks whose borders pass through three points of A and that contain all of A, let D be the one of least radius. Suppose that $C \cap A$ is contained in an arc of central angle less than $180°$, and that P_i, P_j are its endpoints. Then there exists a circle through P_i, P_j of smaller radius that contains A, a contradiction. Thus D has the required property, and the assertion follows.

18. Let (x_0, y_0, z_0) be one solution of $bcx + cay + abz = n$ (not necessarily nonnegative). By subtracting $bcx_0 + cay_0 + abz_0 = n$ we get

$$bc(x-x_0) + ca(y-y_0) + ab(z-z_0) = 0.$$

Since $(a,b) = (a,c) = 1$, we must have $a|x-x_0$ or $x - x_0 = as$. Substituting this in the last equation gives

$$bcs + c(y-y_0) + b(z-z_0) = 0.$$

Since $(b,c) = 1$, we have $b|y-y_0$ or $y - y_0 = bt$. If we substitute this in the last equation we get $bcs + bct + b(z-z_0) = 0$, or $cs + ct + z - z_0 = 0$, or $z - z_0 = -c(s+t)$. In $x = x_0 + as$ and $y = y_0 + bt$, we can choose s and t such that $0 \le x \le a-1$ and $0 \le y \le b-1$. If $n > 2abc - bc - ca - ab$, then $abz = n - bcx - acy > 2abc - ab - bc - ca - bc(a-1) - ca(b-1) = -ab$ or $z > -1$, i.e., $z \ge 0$. Hence, it is representable as $bcx + cay + abz$ with $x, y, z \ge 0$.

Now we prove that $2abc - bc - ca - ab$ is not representable as $bcx + cay + abz$ with $x, y, z \ge 0$. Suppose that $bcx + cay + abz = 2abc - ab - bc - ca$ with $x, y, z \ge 0$. Then

$$bc(x+1) + ca(y+1) + ab(z+1) = 2abc$$

with $x+1, y+1, z+1 \ge 1$. Since $(a,b) = (a,c) = 1$, we have $a|x+1$ and thus $a \le x+1$. Similarly $b \le y+1$ and $c \le z+1$. Thus $bca + cab + abc \le 2abc$, a contradiction.

19. For all n, there exists a unique polynomial P_n of degree n such that $P_n(k) = F_k$ for $n+2 \le k \le 2n+2$ and $P_n(2n+3) = F_{2n+3} - 1$. For $n = 0$, we have $F_1 = $

$F_2 = 1$, $F_3 = 2$, $P_0 = 1$. Now suppose that P_{n-1} has been constructed and let P_n be the polynomial of degree n satisfying $P_n(X+2) - P_n(X+1) = P_{n-1}(X)$ and $P_n(n+2) = F_{n+2}$. (The mapping $\mathbb{R}_n[X] \to \mathbb{R}_{n-1}[X] \times \mathbb{R}$, $P \mapsto (Q, P(n+2))$, where $Q(X) = P(X+2) - P(X+1)$, is bijective, since it is injective and those two spaces have the same dimension; clearly $\deg Q = \deg P - 1$.) Thus for $n+2 \le k \le 2n+2$ we have $P_n(k+1) = P_n(k) + F_{k-1}$ and $P_n(n+2) = F_{n+2}$; hence by induction on k, $P_n(k) = F_k$ for $n+2 \le k \le 2n+2$ and

$$P_n(2n+3) = F_{2n+2} + P_{n-1}(2n+1) = F_{2n+3} - 1.$$

Finally, P_{990} is exactly the polynomial P of the terms of the problem, for $P_{990} - P$ has degree less than or equal to 990 and vanishes at the 991 points $k = 992, \ldots, 1982$.

20. If $(x_1, x_2, \ldots, x_n)$ satisfies the system with parameter a, then $(-x_1, -x_2, \ldots, -x_n)$ satisfies the system with parameter $-a$. Hence it is sufficient to consider only $a \ge 0$.

    Let $(x_1, \ldots, x_n)$ be a solution. Suppose $x_1 \le a$, $x_2 \le a, \ldots, x_n \le a$. Summing the equations we get

    $$(x_1 - a)^2 + \cdots + (x_n - a)^2 = 0$$

    and see that $(a, a, \ldots, a)$ is the only such solution. Now suppose that $x_k \ge a$ for some k. According to the kth equation,

    $$x_{k+1}|x_{k+1}| = x_k^2 - (x_k - a)^2 = a(2x_k - a) \ge a^2,$$

    which implies that $x_{k+1} \ge a$ as well (here $x_{n+1} = x_1$). Consequently, all $x_1, x_2, \ldots, x_n$ are greater than or equal to a, and as above $(a, a, \ldots, a)$ is the only solution.

21. Using the identity

    $$a^n - b^n = (a - b) \sum_{m=0}^{n-1} a^{n-m-1} b^m$$

    with $a = k^{1/n}$ and $b = (k-1)^{1/n}$ one obtains

    $$1 < \left(k^{1/n} - (k-1)^{1/n}\right) nk^{1-1/n} \text{ for all integers } n > 1 \text{ and } k \ge 1.$$

    This gives us the inequality $k^{1/n-1} < n\left(k^{1/n} - (k-1)^{1/n}\right)$ if $n > 1$ and $k \ge 1$. In a similar way one proves that $n\left((k+1)^{1/n} - k^{1/n}\right) < k^{1/n-1}$ if $n > 1$ and $k \ge 1$. Hence for $n > 1$ and $m > 1$ it holds that

    $$n \sum_{k=1}^{m} \left((k+1)^{1/n} - k^{1/n}\right) < \sum_{k=1}^{m} k^{1/n-1}$$
    $$< n \sum_{k=2}^{m} \left(k^{1/n} - (k-1)^{1/n}\right) + 1,$$

    or equivalently,

$$n\left((m+1)^{1/n}-1\right) < \sum_{k=1}^{m} k^{1/n-1} < n\left(m^{1/n}-1\right)+1.$$

The choice $n = 1983$ and $m = 2^{1983}$ then gives

$$1983 < \sum_{k=1}^{2^{1983}} k^{1/1983-1} < 1984.$$

Therefore the greatest integer less than or equal to the given sum is 1983.

22. Decompose n into $n = st$, where the greatest common divisor of s and t is 1 and where $s > 1$ and $t > 1$. For $1 \le k \le n$ put $k = vs + u$, where $0 \le v \le t-1$ and $1 \le u \le s$, and let $a_k = a_{vs+u}$ be the unique integer in the set $\{1,2,3,\ldots,n\}$ such that $vs + ut - a_{vs+u}$ is a multiple of n. To prove that this construction gives a permutation, assume that $a_{k_1} = a_{k_2}$, where $k_i = v_i s + u_i$, $i = 1, 2$. Then $(v_1 - v_2)s + (u_1 - u_2)t$ is a multiple of $n = st$. It follows that t divides $(v_1 - v_2)$, while $|v_1 - v_2| \le t-1$, and that s divides $(u_1 - u_2)$, while $|u_1 - u_2| \le s-1$. Hence, $v_1 = v_2$, $u_1 = u_2$, and $k_1 = k_2$. We have proved that $a_1, \ldots, a_n$ is a permutation of $\{1, 2, \ldots, n\}$ and hence

$$\sum_{k=1}^{n} k \cos\frac{2\pi a_k}{n} = \sum_{v=0}^{t-1}\left(\sum_{u=1}^{s}(vs+u)\cos\left(\frac{2\pi v}{t}+\frac{2\pi u}{s}\right)\right).$$

Using $\sum_{u=1}^{s}\cos(2\pi u/s) = \sum_{u=1}^{s}\sin(2\pi u/s) = 0$ and the additive formulas for cosine, one finds that

$$\sum_{k=1}^{n} k\cos\frac{2\pi a_k}{n} = \sum_{v=0}^{t-1}\left(\cos\frac{2\pi v}{t}\sum_{u=1}^{s} u\cos\frac{2\pi u}{s} - \sin\frac{2\pi v}{t}\sum_{u=1}^{s} u\sin\frac{2\pi u}{s}\right)$$

$$= \left(\sum_{u=1}^{s} u\cos\frac{2\pi u}{s}\right)\left(\sum_{v=0}^{t-1}\cos\frac{2\pi v}{t}\right)$$

$$- \left(\sum_{u=1}^{s} u\sin\frac{2\pi u}{s}\right)\left(\sum_{v=0}^{t-1}\sin\frac{2\pi v}{t}\right) = 0.$$

23. We note that $\angle O_1 K O_2 = \angle M_1 K M_2$ is equivalent to $\angle O_1 K M_1 = \angle O_2 K M_2$. Let S be the intersection point of the common tangents, and let L be the second point of intersection of SK and W_1. Since $\triangle SO_1 P_1 \sim \triangle SP_1 M_1$, we have $SK \cdot SL = SP_1^2 = SO_1 \cdot SM_1$ which implies that points O_1, L, K, M_1 lie on a circle. Hence $\angle O_1 K M_1 = \angle O_1 L M_1 = \angle O_2 K M_2$.

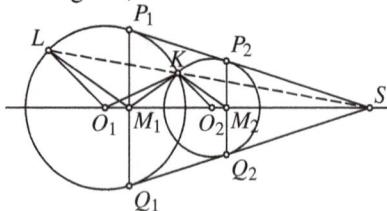

24. See the solution of (SL91-15).

25. Suppose the contrary, that $\mathbb{R}^3 = P_1 \cup P_2 \cup P_3$ is a partition such that $a_1 \in \mathbb{R}^+$ is not realized by P_1, $a_2 \in \mathbb{R}^+$ is not realized by P_2 and $a_3 \in \mathbb{R}^+$ not realized by P_3, where w.l.o.g. $a_1 \geq a_2 \geq a_3$.

If $P_1 = \emptyset = P_2$, then $P_3 = \mathbb{R}^3$, which is impossible.

If $P_1 = \emptyset$, and $X \in P_2$, the sphere centered at X with radius a_2 is included in P_3 and $a_3 \leq a_2$ is realized, which is impossible.

If $P_1 \neq \emptyset$, let $X_1 \in P_1$. The sphere S centered in X_1, of radius a_1 is included in $P_2 \cap P_3$. Since $a_1 \geq a_3$, $S \not\subset P_3$. Let $X_2 \in P_2 \cap S$. The circle $\{Y \in S \mid d(X_2, Y) = a_2\}$ is included in P_3, but $a_2 \leq a_1$; hence it has radius $r = a_2\sqrt{1 - a_2^2/(4a_1^2)} \geq a_2\sqrt{3}/2$ and $a_3 \leq a_2 \leq a_2\sqrt{3} < 2r$; hence a_3 is realized by P_3.

4.25 Solutions to the Shortlisted Problems of IMO 1984

1. This is the same problem as (SL83-20).

2. (a) For $m = t(t-1)/2$ and $n = t(t+1)/2$ we have $4mn - m - n = (t^2-1)^2 - 1$.

   (b) Suppose that $4mn - m - n = p^2$, or equivalently, $(4m-1)(4n-1) = 4p^2 + 1$. The number $4m - 1$ has at least one prime divisor, say q, that is of the form $4k + 3$. Then $4p^2 \equiv -1 \pmod q$. However, by Fermat's theorem we have

$$1 \equiv (2p)^{q-1} = \left(4p^2\right)^{\frac{q-1}{2}} \equiv (-1)^{\frac{q-1}{2}} \pmod q,$$

   which is impossible since $(q-1)/2 = 2k + 1$ is odd.

3. From the equality $n = d_6^2 + d_7^2 - 1$ we see that d_6 and d_7 are relatively prime and $d_7 \mid d_6^2 - 1 = (d_6 - 1)(d_6 + 1)$, $d_6 \mid d_7^2 - 1 = (d_7 - 1)(d_7 + 1)$.

   Suppose that $d_6 = ab$, $d_7 = cd$ with $1 < a < b$, $1 < c < d$. Then n has 7 divisors smaller than d_7, namely $1, a, b, c, d, ab, ac$, which is impossible. Hence, one of the two numbers d_6 and d_7 is either a prime p or the square p^2 of a prime $p \neq 2$. Let it be d_i, $\{i, j\} = \{6, 7\}$; then $d_i \mid (d_j - 1)(d_j + 1)$ implies that $d_j \equiv \pm 1 \pmod{d_i}$, and consequently $(d_i^2 - 1)/d_j \equiv \pm 1$ as well. But either d_j or $(d_i^2 - 1)/d_j$ is less than d_i, and therefore equals $d_i - 1$ or equals 1. The only nontrivial possibilities are $(d_i^2 - 1)/d_j = 1$ and $d_j = d_i \pm 1$. In the first case we get $d_i < d_j$; hence $d_7 = d_6^2 - 1 = (d_6 - 1)(d_6 + 1)$; hence $d_6 + 1$ is a divisor of n that is between d_6 and d_7. This is impossible. We thus conclude that $d_7 = d_6 + 1$. Setting $d_6 = x$, $d_7 = x + 1$ we obtain that $n = x^2 + (x+1)^2 - 1 = 2x(x+1)$ is even.

   (i) Assume that one of $x, x+1$ is a prime p. The other one has at most 6 divisors and hence must be of the form $2^3, 2^4, 2^5, 2q, 2q^2, 4q$, where q is an odd prime. The numbers 2^3 and 2^4 are easily eliminated, while 2^5 yields the solution $x = 31$, $x + 1 = 32$, $n = 1984$. Also, $2q$ is eliminated because $n = 4pq$ then has only 4 divisors less than x; $2q^2$ is eliminated because $n = 4pq^2$ has at least 6 divisors less than x; $4q$ is also eliminated because $n = 8pq$ has 6 divisors less than x.

   (ii) Assume that one of $x, x+1$ is p^2. The other one has at most 5 divisors (p excluded), and hence is of the form $2^3, 2^4, 2q$, where q is an odd prime. The number 2^3 yields the solution $x = 8$, $x + 1 = 9$, $n = 144$, while 2^4 is easily eliminated. Also, the number $2q$ is eliminated because $n = 4p^2q$ has 6 divisors less than x.

   Thus there are two solutions in total: 144 and 1984.

4. Consider the convex n-gon $A_1 A_2 \ldots A_n$ (the indices are considered modulo n). For any diagonal $A_i A_j$ we have $A_i A_j + A_{i+1} A_{j+1} > A_i A_{i+1} + A_j A_{j+1}$. Summing all such $n(n-3)/2$ inequalities, we obtain $2d > (n-3)p$, proving the first inequality.

   Let us now prove the second inequality. We notice that for each diagonal $A_i A_{i+j}$ (we may assume w.l.o.g. that $j \leq \lfloor n/2 \rfloor$) the following relation holds:

$$A_i A_{i+j} < A_i A_{i+1} + \cdots + A_{i+j-1} A_{i+j}. \tag{1}$$

If $n = 2k+1$, then summing the inequalities (1) for $j = 2,3,\ldots,k$ and $i = 1,2,\ldots,n$ yields $d < (2+3+\cdots+k)p = ([n/2][n+1/2]-2)p/2$.

If $n = 2k$, then summing the inequalities (1) for $j = 2,3,\ldots,k-1, i = 1,2,\ldots,n$ and for $j = k, i = 1,2,\ldots,k$ again yields $d < (2+3+\cdots+(k-1)+k/2)p = \frac{1}{2}([n/2][n+1/2]-2)p$.

5. Let $f(x,y,z) = xy+yz+zx-2xyz$. The first inequality follows immediately by adding $xy \geq xyz$, $yz \geq xyz$, and $zx \geq xyz$ (in fact, a stronger inequality $xy+yz+zx-9xyz \geq 0$ holds).

Assume w.l.o.g. that z is the smallest of x,y,z. Since $xy \leq (x+y)^2/4 = (1-z)^2/4$ and $z \leq 1/2$, we have

$$xy+yz+zx-2xyz = (x+y)z+xy(1-2z)$$
$$\leq (1-z)z+\frac{(1-z)^2(1-2z)}{4}$$
$$= \frac{7}{27}-\frac{(1-2z)(1-3z)^2}{108} \leq \frac{7}{27}.$$

6. From the given recurrence we infer $f_{n+1} - f_n = f_n - f_{n-1} + 2$. Consequently, $f_{n+1} - f_n = (f_2 - f_1) + 2(n-1) = c - 1 + 2(n-1)$. Summing up for $n = 1,2,\ldots,k-1$ yields the explicit formula

$$f_k = f_1 + (k-1)(c-1) + (k-1)(k-2) = k^2 + bk - b,$$

where $b = c - 4$. Now we easily obtain $f_k f_{k+1} = k^4 + 2(b+1)k^3 + (b^2+b+1)k^2 - (b^2+b)k - b$. We are looking for an r for which the last expression equals f_r. Setting $r = k^2 + pk + q$ we get by a straightforward calculation that $p = b+1$, $q = -b$, and $r = k^2 + (b+1)k - b = f_k + k$. Hence $f_k f_{k+1} = f_{f_k+k}$ for all k.

7. It clearly suffices to solve the problem for the remainders modulo 4 (16 of each kind).

   (a) The remainders can be placed as shown in Figure 1, so that they satisfy the conditions.

   ```
   1 0 3 2 1 0 3 2
   2 3 0 1 2 3 0 1
   3 2 1 0 3 2 1 0
   0 1 2 3 0 1 2 3
   1 0 3 2 1 0 3 2
   2 3 0 1 2 3 0 1
   3 2 1 0 3 2 1 0
   0 1 2 3 0 1 2 3
   ```

   ```
       p
     q r s
       t
   ```

   Fig. 1   Fig. 2

   (b) Suppose that the required numbering exists. Consider a part of the chessboard as in Figure 2. By the stated condition, all the numbers $p+q+r+s, q+r+s+t, p+q+r+t, p+r+s+t$ give the same remainder modulo 4, and so do p,q,r,s. We deduce that all numbers on black cells of the board, except possibly the two corner cells, give the same remainder, which is impossible.

8. Suppose that the statement of the problem is false. Consider two arbitrary circles $R = (O, r)$ and $S = (O, s)$ with $0 < r < s < 1$. The point $X \in R$ with $\alpha(X) = r(s-r) < 2\pi$ satisfies that $C(X) = S$. It follows that the color of the point X does not appear on S. Consequently, the set of colors that appear on R is not the same as the set of colors that appear on S. Hence any two distinct circles with center at O and radii less than 1 have distinct sets of colors. This is a contradiction, since there are infinitely many such circles but only finitely many possible sets of colors.

9. Let us show first that the system has at most one solution. Suppose that (x, y, z) and (x', y', z') are two distinct solutions and that w.l.o.g. $x < x'$. Then the second and third equation imply that $y > y'$ and $z > z'$, but then $\sqrt{y-a} + \sqrt{z-a} > \sqrt{y'-a} + \sqrt{z'-a}$, which is a contradiction.
   We shall now prove the existence of at least one solution. Let P be an arbitrary point in the plane and K, L, M points such that $PK = \sqrt{a}$, $PL = \sqrt{b}$, $PM = \sqrt{c}$, and $\angle KPL = \angle LPM = \angle MPK = 120°$. The lines through K, L, M perpendicular respectively to PK, PL, PM form an equilateral triangle ABC, where $K \in BC$, $L \in AC$, and $M \in AB$. Since its area equals $AB^2 \sqrt{3}/4 = S_{\triangle BPC} + S_{\triangle APC} + S_{\triangle APB} = AB\left(\sqrt{a} + \sqrt{b} + \sqrt{c}\right)/2$, it follows that $AB = 1$. Therefore $x = PA^2$, $y = PB^2$, and $z = PC^2$ is a solution of the system (indeed, $\sqrt{y-a} + \sqrt{z-a} = \sqrt{PB^2 - PK^2} + \sqrt{PC^2 - PK^2} = BK + CK = 1$, etc.).

10. Suppose that the product of some five consecutive numbers is a square. It is easily seen that among them at least one, say n, is divisible neither by 2 nor 3. Since n is coprime to the remaining four numbers, it is itself a square of a number m of the form $6k \pm 1$. Thus $n = (6k \pm 1)^2 = 24r + 1$, where $r = k(3k \pm 1)/2$. Note that neither of the numbers $24r - 1$, $24r + 5$ is one of our five consecutive numbers because it is not a square. Hence the five numbers must be $24r, 24r + 1, \ldots, 24r + 4$. However, the number $24r + 4 = (6k \pm 1)^2 + 3$ is divisible by $6r + 1$, which implies that it is a square as well. It follows that these two squares are 1 and 4, which is impossible.

11. Suppose that an integer x satisfies the equation. Then the numbers $x - a_1, x - a_2, \ldots, x - a_{2n}$ are $2n$ distinct integers whose product is $1 \cdot (-1) \cdot 2 \cdot (-2) \cdots n \cdot (-n)$.
   From here it is obvious that the numbers $x - a_1, x - a_2, \ldots, x - a_{2n}$ are some reordering of the numbers $-n, -n+1, \ldots, -1, 1, \ldots, n-1, n$. It follows that their sum is 0, and therefore $x = (a_1 + a_2 + \cdots + a_{2n})/2n$. This is the only solution if $\{a_1, a_2, \ldots, a_{2n}\} = \{x - n, \ldots, x - 1, x + 1, \ldots, x + n\}$ for some $x \in \mathbb{N}$. Otherwise there is no solution.

12. By the binomial formula we have
$$(a+b)^7 - a^7 - b^7 = 7ab[(a^5 + b^5) + 3ab(a^3 + b^3) + 5a^2b^2(a+b)]$$
$$= 7ab(a+b)(a^2 + ab + b^2)^2.$$

Thus it will be enough to find a and b such that $7 \nmid a,b$ and $7^3 \mid a^2 + ab + b^2$. Such numbers must satisfy $(a+b)^2 > a^2 + ab + b^2 \geq 7^3 = 343$, implying $a + b \geq 19$. Trying $a = 1$ we easily find the example $(a,b) = (1,18)$.

13. Let Z be the given cylinder of radius r, altitude h, and volume $\pi r^2 h = 1$, k_1 and k_2 the circles surrounding its bases, and V the volume of an inscribed tetrahedron $ABCD$.

    We claim that there is no loss of generality in assuming that A,B,C,D all lie on $k_1 \cup k_2$. Indeed, if the vertices A,B,C are fixed and D moves along a segment EF parallel to the axis of the cylinder ($E \in k_1$, $F \in k_2$), the maximum distance of D from the plane ABC (and consequently the maximum value of V) is achieved either at E or at F. Hence we shall consider only the following two cases:

    (i) $A,B \in k_1$ and $C,D \in k_2$. Let P,Q be the projections of A,B on the plane of k_2, and R,S the projections of C,D on the plane of k_1, respectively. Then V is one-third of the volume V' of the prism $ARBSCPDQ$ with bases $ARBS$ and $CPDQ$. The area of the quadrilateral $ARBS$ inscribed in k_1 does not exceed the area of the square inscribed therein, which is $2r^2$. Hence $3V = V' \leq 2r^2 h = 2/\pi$.

    (ii) $A,B,C \in k_1$ and $D \in k_2$. The area of the triangle ABC does not exceed the area of an equilateral triangle inscribed in k_1, which is $3\sqrt{3}r^2/4$. Consequently, $V \leq \frac{\sqrt{3}}{4} r^2 h = \frac{\sqrt{3}}{4\pi} < \frac{2}{3\pi}$.

14. Let M and N be the midpoints of AB and CD, and let M',N' be their projections on CD and AB, respectively. We know that $MM' = AB$, and hence

$$S_{ABCD} = S_{AMD} + S_{BMC} + S_{CMD} = \frac{1}{2}(S_{ABD} + S_{ABC}) + \frac{1}{4}AB \cdot CD. \qquad (1)$$

The line AB is tangent to the circle with diameter CD if and only if $NN' = CD/2$, or equivalently,

$$S_{ABCD} = S_{AND} + S_{BNC} + S_{ANB} = \frac{1}{2}(S_{BCD} + S_{ACD}) + \frac{1}{4}AB \cdot CD.$$

By (1), this is further equivalent to $S_{ABC} + S_{ABD} = S_{BCD} + S_{ACD}$. But since $S_{ABC} + S_{ACD} = S_{ABD} + S_{BCD} = S_{ABCD}$, this reduces to $S_{ABC} = S_{BCD}$, i.e., to $BC \parallel AD$.

15. (a) Since rotation by $60°$ around A transforms the triangle CAF into $\triangle EAB$, it follows that $\angle(CF,EB) = 60°$. We similarly deduce that $\angle(EB,AD) = \angle(AD,FC) = 60°$. Let S be the intersection point of BE and AD. Since $\angle CSE = \angle CAE = 60°$, we have that $EASC$ is cyclic. Therefore $\angle(AS,SC) = 60° = \angle(AD,FC)$, which implies that S lies on CF as well.

    (b) A rotation of $EASC$ around E by $60°$ transforms A into C and S into a point T for which $SE = ST = SC + CT = SC + SA$. Summing the equality $SE = SC + SA$ and the analogous equalities $SD = SB + SC$ and $SF = SA + SB$ yields the result.

16. From the first two conditions we can easily conclude that $a + d > b + c$ (indeed, $(d+a)^2 - (d-a)^2 = (c+b)^2 - (c-b)^2 = 4ad = 4bc$ and $d - a > c - b > 0$). Thus $k > m$.

From $d = 2^k - a$ and $c = 2^m - b$ we get $a(2^k - a) = b(2^m - b)$, or equivalently,

$$(b+a)(b-a) = 2^m(b - 2^{k-m}a). \tag{1}$$

Since $2^{k-m}a$ is even and b is odd, the highest power of 2 that divides the right-hand side of (1) is m. Hence $(b+a)(b-a)$ is divisible by 2^m but not by 2^{m+1}, which implies $b+a = 2^{m_1}p$ and $b-a = 2^{m_2}q$, where $m_1, m_2 \geq 1$, $m_1 + m_2 = m$, and p, q are odd.

Furthermore, $b = (2^{m_1}p + 2^{m_2}q)/2$ and $a = (2^{m_1}p - 2^{m_2}q)/2$ are odd, so either $m_1 = 1$ or $m_2 = 1$. Note that $m_1 = 1$ is not possible, since it would imply that $b - a = 2^{m-1}q \geq 2^{m-1}$, although $b + c = 2^m$ and $b < c$ imply that $b < 2^{m-1}$. Hence $m_2 = 1$ and $m_1 = m - 1$. Now since $a + b < b + c = 2^m$, we obtain $a + b = 2^{m-1}$ and $b - a = 2q$, where q is an odd integer. Substituting these into (1) and dividing both sides by 2^m we get

$$q = 2^{m-2} + q - 2^{k-m}a \quad \Longrightarrow \quad 2^{k-m}a = 2^{m-2}.$$

Since a is odd and $k > m$, it follows that $a = 1$.

Remark. Now it is not difficult to prove that all quadruples (a, b, c, d) that satisfy the given conditions are of the form $(1, 2^{m-1} - 1, 2^{m-1} + 1, 2^{2m-2} - 1)$, where $m \in \mathbb{N}$, $m \geq 3$.

17. For any $m = 0, 1, \ldots, n-1$, we shall find the number of permutations $(x_1, x_2, \ldots, x_n)$ with exactly k discordant pairs such that $x_n = n - m$. This x_n is a member of exactly m discordant pairs, and hence the permutation $(x_1, \ldots, x_{n-1})$ of the set $\{1, 2, \ldots, n\} \setminus \{m\}$ must have exactly $k - m$ discordant pairs: there are $d(n-1, k-m)$ such permutations. Therefore

$$d(n,k) = d(n-1,k) + d(n-1,k-1)\cdots + d(n-1,k-n+1)$$
$$= d(n-1,k) + d(n,k-1)$$

(note that $d(n,k)$ is 0 if $k < 0$ or $k > \binom{n}{2}$).

We now proceed to calculate $d(n,2)$ and $d(n,3)$. Trivially, $d(n,0) = 1$. It follows that $d(n,1) = d(n-1,1) + d(n,0) = d(n-1,1) + 1$, which yields $d(n,1) = d(1,1) + n - 1 = n - 1$.

Further, $d(n,2) = d(n-1,2) + d(n,1) = d(n-1,2) + n - 1 = d(2,2) + 2 + 3 + \cdots + n - 1 = (n^2 - n - 2)/2$.

Finally, using the known formula $1^2 + 2^2 + \cdots + k^2 = k(k+1)(2k+1)/6$, we have $d(n,3) = d(n-1,3) + d(n,2) = d(n-1,3) + (n^2 - n - 2)/2 = d(2,3) + \sum_{i=3}^{n} (n^2 - n - 2)/2 = (n^3 - 7n + 6)/6$.

18. Suppose that circles $k_1(O_1, r_1)$, $k_2(O_2, r_2)$, and $k_3(O_3, r_3)$ touch the edges of the angles $\angle BAC$, $\angle ABC$, and $\angle ACB$, respectively. Denote also by O and r the center and radius of the incircle. Let P be the point of tangency of the incircle with AB and let F be the foot of the perpendicular from O_1 to OP. From $\triangle O_1 FO$ we obtain $\cot(\alpha/2) = 2\sqrt{rr_1}/(r - r_1)$ and analogously $\cot(\beta/2) = 2\sqrt{rr_2}/(r - r_2)$, $\cot(\gamma/2) = 2\sqrt{rr_3}/(r - r_3)$. We will now use a well-known trigonometric identity for the angles of a triangle:

$$\cot\frac{\alpha}{2} + \cot\frac{\beta}{2} + \cot\frac{\gamma}{2} = \cot\frac{\alpha}{2}\cdot\cot\frac{\beta}{2}\cdot\cot\frac{\gamma}{2}.$$

(This identity follows from $\tan(\gamma/2) = \cot(\alpha/2 + \beta/2)$ and the formula for the cotangent of a sum.)

Plugging in the obtained cotangents, we get

$$\frac{2\sqrt{rr_1}}{r-r_1} + \frac{2\sqrt{rr_2}}{r-r_2} + \frac{2\sqrt{rr_3}}{r-r_3} = \frac{2\sqrt{rr_1}}{r-r_1}\cdot\frac{2\sqrt{rr_2}}{r-r_2}\cdot\frac{2\sqrt{rr_3}}{r-r_3} \Rightarrow$$
$$\sqrt{r_1}(r-r_2)(r-r_3) + \sqrt{r_2}(r-r_1)(r-r_3)$$
$$+ \sqrt{r_3}(r-r_1)(r-r_2) = 4r\sqrt{r_1 r_2 r_3}.$$

For $r_1 = 1$, $r_2 = 4$, and $r_3 = 9$ we get

$$(r-4)(r-9) + 2(r-1)(r-9) + 3(r-1)(r-4) = 24r \Rightarrow 6(r-1)(r-11) = 0.$$

Clearly, $r = 11$ is the only viable value for r.

19. First, we shall prove that the numbers in the nth row are exactly the numbers

$$\frac{1}{n\binom{n-1}{0}}, \ \frac{1}{n\binom{n-1}{1}}, \ \frac{1}{n\binom{n-1}{2}}, \ \dots, \ \frac{1}{n\binom{n-1}{n-1}}. \tag{1}$$

The proof of this fact can be done by induction. For small n, the statement can be easily verified. Assuming that the statement is true for some n, we have that the kth element in the $(n+1)$st row is, as is directly verified,

$$\frac{1}{n\binom{n-1}{k-1}} - \frac{1}{(n+1)\binom{n}{k-1}} = \frac{1}{(n+1)\binom{n}{k}}.$$

Thus (1) is proved. Now the geometric mean of the elements of the nth row becomes:

$$\frac{1}{n\sqrt[n]{\binom{n-1}{0}\cdot\binom{n-1}{1}\cdots\binom{n-1}{n-1}}} \geq \frac{1}{n\left(\frac{\binom{n-1}{0}+\binom{n-1}{1}+\cdots+\binom{n-1}{n-1}}{n}\right)} = \frac{1}{2^{n-1}}.$$

The desired result follows directly from substituting $n = 1984$.

20. Define the set $S = \mathbb{R}^+ \setminus \{1\}$. The given inequality is equivalent to $\ln b/\ln a < \ln(b+1)/\ln(a+1)$.

If $b = 1$, it is obvious that each $a \in S$ satisfies this inequality. Suppose now that b is also in S.

Let us define on S a function $f(x) = \ln(x+1)/\ln x$. Since $\ln(x+1) > \ln x$ and $1/x > 1/x+1 > 0$, we have

$$f'(x) = \frac{\frac{\ln x}{x+1} - \frac{\ln(x+1)}{x}}{\ln^2 x} < 0 \quad \text{for all } x.$$

Hence f is always decreasing. We also note that $f(x) < 0$ for $x < 1$ and that $f(x) > 0$ for $x > 1$ (at $x = 1$ there is a discontinuity).

Let us assume $b > 1$. From $\ln b / \ln a < \ln(b+1) / \ln(a+1)$ we get $f(b) > f(a)$. This holds for $b > a$ or for $a < 1$.

Now let us assume $b < 1$. This time we get $f(b) < f(a)$. This holds for $a < b$ or for $a > 1$.

Hence all the solutions to $\log_a b < \log_{a+1}(b+1)$ are $\{b = 1, a \in S\}$, $\{a > b > 1\}$, $\{b > 1 > a\}$, $\{a < b < 1\}$, and $\{b < 1 < a\}$.

4.26 Solutions to the Shortlisted Problems of IMO 1985

1. Since there are 9 primes ($p_1 = 2 < p_2 = 3 < \cdots < p_9 = 23$) less than 26, each number $x_j \in M$ is of the form $\prod_{i=1}^{9} p_i^{a_{ij}}$, where $0 \leq a_{ij}$. Now, $x_j x_k$ is a square if $a_{ij} + a_{ik} \equiv 0 \pmod 2$ for $i = 1, \ldots, 9$. Since the number of distinct ninetuples modulo 2 is 2^9, any subset of M with at least 513 elements contains two elements with square product. Starting from M and eliminating such pairs, one obtains $(1985 - 513)/2 = 736 > 513$ distinct two-element subsets of M each having a square as the product of elements. Reasoning as above, we find at least one (in fact many) pair of such squares whose product is a fourth power.

2. The polyhedron has $3 \cdot 12/2 = 18$ edges, and by Euler's formula, 8 vertices. Let v_1 and v_2 be the numbers of vertices at which respectively 3 and 6 edges meet. Then $v_1 + v_2 = 8$ and $3v_1 + 6v_2 = 2 \cdot 18$, implying that $v_1 = 4$. Let A, B, C, D be the vertices at which three edges meet. Since the dihedral angles are equal, all the edges meeting at A, say AE, AF, AG, must have equal length, say x. (If $x = AE = AF \neq AG = y$, and AEF, AFG, and AGE are isosceles, $\angle EAF \neq \angle FAG$, in contradiction to the equality of the dihedral angles.) It is easy to see that at E, F, and G six edges meet. One proceeds to conclude that if H is the fourth vertex of this kind, $EFGH$ must be a regular tetrahedron of edge length y, and the other vertices A, B, C, and D are tops of isosceles pyramids based on EFG, EFH, FGH, and GEH. Let the plane through A, B, C meet EF, HF, and GF, at E', H', and G'. Then $AE'BH'CG'$ is a regular hexagon, and since $x = FA = FE'$, we have $E'G' = x$ and $AE' = x/\sqrt{3}$. From the isosceles triangles AEF and FAE' we obtain finally, with $\angle EFA = \alpha$,

$$\frac{y}{2x} = \cos\alpha = 1 - 2\sin^2(\alpha/2), \quad x/(2x\sqrt{3}) = \sin(\alpha/2),$$

and $y/x = 5/3$.

3. We shall write $P \equiv Q$ for two polynomials P and Q if $P(x) - Q(x)$ has even coefficients.
   We observe that $(1+x)^{2^m} \equiv 1 + x^{2^m}$ for every $m \in \mathbb{N}$. Consequently, for every polynomial p with degree less than $k = 2^m$, $w(p \cdot q_k) = 2w(p)$.
   Now we prove the inequality from the problem by induction on i_n. If $i_n \leq 1$, the inequality is trivial. Assume it is true for any sequence with $i_1 < \cdots < i_n < 2^m$ ($m \geq 1$), and let there be given a sequence with $k = 2^m \leq i_n < 2^{m+1}$. Consider two cases.
   (i) $i_1 \geq k$. Then $w(q_{i_1} + \cdots + q_{i_n}) = 2w(q_{i_1-k} + \cdots + q_{i_n-k}) \geq 2w(q_{i_1-k}) = w(q_{i_1})$.
   (ii) $i_1 < k$. Then the polynomial $p = q_{i_1} + \cdots + q_{i_n}$ has the form

$$p = \sum_{i=0}^{k-1} a_i x^i + (1+x)^k \sum_{i=0}^{k-1} b_i x^i \equiv \sum_{i=0}^{k-1} \left[(a_i + b_i)x^i + b_i x^{i+k} \right].$$

   Whenever some a_i is odd, either $a_i + b_i$ or b_i in the above sum will be odd. It follows that $w(p) \geq w(q_{i_1})$, as claimed.

The proof is complete.

4. Let $\langle x \rangle$ denote the residue of an integer x modulo n. Also, we write $a \sim b$ if a and b receive the same color. We claim that all the numbers $\langle ij \rangle$, $i = 1, 2, \ldots, n-1$, are of the same color. Since j and n are coprime, this will imply the desired result.

We use induction on i. For $i = 1$ the statement is trivial. Assume now that the statement is true for $i = 1, \ldots, k-1$. For $1 < k < n$ we have $\langle kj \rangle \neq j$. If $\langle kj \rangle > j$, then by (ii), $\langle kj \rangle \sim \langle kj \rangle - j = \langle (k-1)j \rangle$. If otherwise $\langle kj \rangle < j$, then by (ii) and (i), $\langle kj \rangle \sim j - \langle kj \rangle \sim n - j + \langle kj \rangle = \langle (k-1)j \rangle$. This completes the induction.

5. Let w.l.o.g. circle C have unit radius. For each $m \in \mathbb{R}$, the locus of points M such that $f(M) = m$ is the circle C_m with radius $r_m = m/(m+1)$, that is tangent to C at A. Let O_m be the center of C_m. We have to show that if $M \in C_m$ and $N \in C_n$, where $m, n > 0$, then the midpoint P of MN lies inside the circle $C_{(m+n)/2}$. This is trivial if $m = n$, so let $m \neq n$.

For fixed M, P is in the image C'_n of C_n under the homothety with center M and coefficient $1/2$. The center of the circle C'_n is at the midpoint of $O_n M$. If we let both M and N vary, P will be on the union of circles with radius $r_n/2$ and centers in the image of C_m under the homothety with center O_n and coefficient $1/2$. Hence P is not outside the circle centered at the midpoint $O_m O_n$ and with radius $(r_m + r_n)/2$. It remains to show that $r_{(m+n)/2} > (r_m + r_n)/2$. But this inequality is easily reduced to $(m-n)^2 > 0$, which is true.

6. Let us set

$$x_{n,i} = \sqrt[i]{i + \sqrt[i+1]{i+1 + \cdots + \sqrt[n]{n}}},$$

$$y_{n,i} = x_{n+1,i}^{i-1} + x_{n+1,i}^{i-2} x_{n,i} + \cdots + x_{n,i}^{i-1}.$$

In particular, $x_{n,2} = x_n$ and $x_{n,i} = 0$ for $i > n$. We observe that for $n \geq i \geq 2$,

$$x_{n+1,i} - x_{n,i} = \frac{x_{n+1,i}^i - x_{n,i}^i}{y_{n,i}} = \frac{x_{n+1,i+1} - x_{n,i+1}}{y_{n,i}}.$$

Since $y_{n,i} > i x_{n,i}^{i-1} \geq i^{1+(i-1)/i} \geq i^{3/2}$ and $x_{n+1,n+1} - x_{n,n+1} = \sqrt[n+1]{n+1}$, simple induction gives

$$x_{n+1} - x_n \leq \frac{\sqrt[n+1]{n+1}}{(n!)^{3/2}} < \frac{1}{n!} \quad \text{for } n > 2.$$

The inequality for $n = 2$ is directly verified.

7. Let $k_i \geq 0$ be the largest integer such that $p^{k_i} \mid x_i$, $i = 1, \ldots, n$, and $y_i = x_i/p^{k_i}$. We may assume that $k = k_1 + \cdots + k_n$. All the y_i must be distinct. Indeed, if $y_i = y_j$ and $k_i > k_j$, then $x_i \geq p x_j \geq 2 x_j \geq 2 x_1$, which is impossible. Thus $y_1 y_2 \ldots y_n = P/p^k \geq n!$.

If equality holds, we must have $y_i = 1$, $y_j = 2$ and $y_k = 3$ for some i, j, k. Thus $p \geq 5$, which implies that either $y_i/y_j \leq 1/2$ or $y_i/y_j \geq 5/2$, which is impossible. Hence the inequality is strict.

8. Among ten consecutive integers that divide n, there must exist numbers divisible by 2^3, 3^2, 5, and 7. Thus the desired number has the form

$$n = 2^{\alpha_1} 3^{\alpha_2} 5^{\alpha_3} 7^{\alpha_4} 11^{\alpha_5} \cdots, \quad \text{where } \alpha_1 \geq 3, \ \alpha_2 \geq 2, \ \alpha_3 \geq 1, \ \alpha_4 \geq 1.$$

Since n has $(\alpha_1 + 1)(\alpha_2 + 1)(\alpha_3 + 1) \cdots$ distinct factors, and $(\alpha_1 + 1)(\alpha_2 + 1)(\alpha_3 + 1)(\alpha_4 + 1) \geq 48$, we must have $(\alpha_5 + 1) \cdots \leq 3$. Hence at most one α_j, $j > 4$, is positive, and in the minimal n this must be α_5. Checking through the possible combinations satisfying $(\alpha_1 + 1)(\alpha_2 + 1) \cdots (\alpha_5 + 1) = 144$ one finds that the minimal n is $2^5 \cdot 3^2 \cdot 5 \cdot 7 \cdot 11 = 110880$.

9. Let $\vec{a}, \vec{b}, \vec{c}, \vec{d}$ denote the vectors $\overrightarrow{OA}, \overrightarrow{OB}, \overrightarrow{OC}, \overrightarrow{OD}$ respectively. Then $|\vec{a}| = |\vec{b}| = |\vec{c}| = |\vec{d}| = 1$. The centroids of the faces are $(\vec{b} + \vec{c} + \vec{d})/3$, $(\vec{a} + \vec{c} + \vec{d})/3$, etc., and each of these is at distance $1/3$ from $P = (\vec{a} + \vec{b} + \vec{c} + \vec{d})/3$; hence the required radius is $1/3$. To compute $|P|$ as a function of the edges of $ABCD$, observe that $AB^2 = (\vec{b} - \vec{a})^2 = 2 - 2\vec{a} \cdot \vec{b}$ etc. Now

$$P^2 = \frac{|\vec{a} + \vec{b} + \vec{c} + \vec{d}|^2}{9}$$
$$= \frac{16 - 2(AB^2 + BC^2 + AC^2 + AD^2 + BD^2 + CD^2)}{9}.$$

10. If M is at a vertex of the regular tetrahedron $ABCD$ ($AB = 1$), then one can take M' at the center of the opposite face of the tetrahedron.

Let M be on the face (ABC) of the tetrahedron, excluding the vertices. Consider a continuous map f of $\mathbb{C}$ onto the surface S of $ABCD$ that maps $m + n e^{i\pi/3}$ for $m, n \in \mathbb{Z}$ onto A, B, C, D if $(m, n) \equiv (1, 1)$, $(1, 0)$, $(0, 1)$, $(0, 0)$ (mod 2) respectively, and maps each unit equilateral triangle with vertices of the form $m + n e^{i\pi/3}$ isometrically onto the corresponding face of $ABCD$. The point M then has one preimage M_j, $j = 1, 2, \ldots, 6$, in each of the six preimages of $\triangle ABC$ having two vertices on the unit circle. The M_j's form a convex centrally symmetric (possibly degenerate) hexagon. Of the triangles formed by two adjacent sides of this hexagon consider the one, say $M_1 M_2 M_3$, with the smallest radius of circumcircle and denote by $\widehat{M'}$ its circumcenter. Then we can choose $M' = f(\widehat{M'})$. Indeed, the images of the segments $M_1 \widehat{M'}$, $M_2 \widehat{M'}$, $M_3 \widehat{M'}$ are three different shortest paths on S from M to M'.

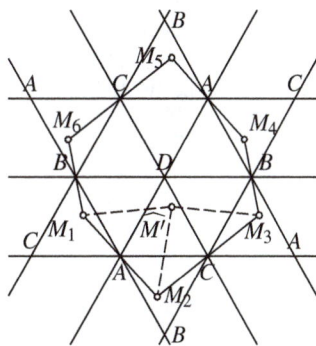

11. Let $-x_1,\ldots,-x_6$ be the roots of the polynomial. Let $s_{k,i}$ ($k \le i \le 6$) denote the sum of all products of k of the numbers $x_1,\ldots,x_i$. By Vieta's formula we have $a_k = s_{k,6}$ for $k = 1,\ldots,6$. Since $s_{k,i} = s_{k-1,i-1}x_i + s_{k,i-1}$, one can compute the a_k by the following scheme (the horizontal and vertical arrows denote multiplications and additions respectively):

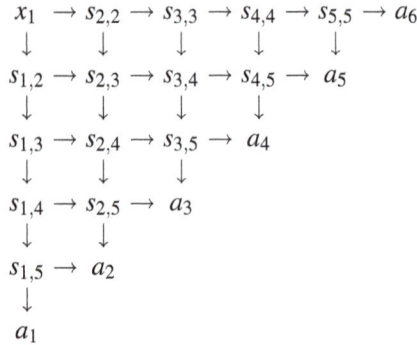

$$
\begin{array}{ccccccccccc}
x_1 & \to & s_{2,2} & \to & s_{3,3} & \to & s_{4,4} & \to & s_{5,5} & \to & a_6 \\
\downarrow & & \downarrow & & \downarrow & & \downarrow & & \downarrow & & \\
s_{1,2} & \to & s_{2,3} & \to & s_{3,4} & \to & s_{4,5} & \to & a_5 & & \\
\downarrow & & \downarrow & & \downarrow & & \downarrow & & & & \\
s_{1,3} & \to & s_{2,4} & \to & s_{3,5} & \to & a_4 & & & & \\
\downarrow & & \downarrow & & \downarrow & & & & & & \\
s_{1,4} & \to & s_{2,5} & \to & a_3 & & & & & & \\
\downarrow & & \downarrow & & & & & & & & \\
s_{1,5} & \to & a_2 & & & & & & & & \\
\downarrow & & & & & & & & & & \\
a_1 & & & & & & & & & &
\end{array}
$$

12. We shall prove by induction on m that $P_m(x,y,z)$ is symmetric and that

$$(x+y)P_m(x,z,y+1) - (x+z)P_m(x,y,z+1) = (y-z)P_m(x,y,z) \qquad (1)$$

holds for all x,y,z. This is trivial for $m = 0$. Assume now that it holds for $m = n-1$.

Since obviously $P_n(x,y,z) = P_n(y,x,z)$, the symmetry of P_n will follow if we prove that $P_n(x,y,z) = P_n(x,z,y)$. Using (1) we have $P_n(x,z,y) - P_n(x,y,z) = (y+z)[(x+y)P_{n-1}(x,z,y+1) - (x+z)P_{n-1}(x,y,z+1)] - (y^2 - z^2)P_{n-1}(x,y,z) = (y+z)(y-z)P_{n-1}(x,y,z) - (y^2 - z^2)P_{n-1}(x,y,z) = 0$. It remains to prove (1) for $m = n$. Using the already established symmetry we have

$$
\begin{aligned}
&(x+y)P_n(x,z,y+1) - (x+z)P_n(x,y,z+1) \\
&= (x+y)P_n(y+1,z,x) - (x+z)P_n(z+1,y,x) \\
&= (x+y)[(y+x+1)(z+x)P_{n-1}(y+1,z,x+1) - x^2 P_{n-1}(y+1,z,x)] \\
&\quad -(x+z)[(z+x+1)(y+x)P_{n-1}(z+1,y,x+1) - x^2 P_{n-1}(z+1,y,x)] \\
&= (x+y)(x+z)(y-z)P_{n-1}(x+1,y,z) - x^2(y-z)P_{n-1}(x,y,z) \\
&= (y-z)P_n(z,y,x) = (y-z)P_n(x,y,z),
\end{aligned}
$$

as claimed.

13. If m and n are relatively prime, there exist positive integers p,q such that $pm = qn+1$. Thus by putting m balls in some boxes p times we can achieve that one box receives $q+1$ balls while all others receive q balls. Repeating this process sufficiently many times, we can obtain an equal distribution of the balls.

Now assume $\gcd(m,n) > 1$. If initially there is only one ball in the boxes, then after k operations the number of balls will be $1 + km$, which is never divisible by n. Hence the task cannot be done.

14. It suffices to prove the existence of a good point in the case of exactly 661 −1's.
We prove by induction on k that in any arrangement with $3k + 2$ points k of
which are −1's a good point exists. For $k = 1$ this is clear by inspection. Assume
that the assertion holds for all arrangements of $3n + 2$ points and consider an
arrangement of $3(n + 1) + 2$ points. Now there exists a sequence of consecutive
−1's surrounded by two +1's. There is a point P which is good for the arrangement obtained by removing the two +1's bordering the sequence of −1's and
one of these −1's. Since P is out of this sequence, clearly the removal either
leaves a partial sum as it was or diminishes it by 1, so P is good for the original
arrangement.

Second solution. Denote the number on an arbitrary point by a_1, and the numbers on successive points going in the positive direction by $a_2, a_3, \ldots$ (in particular, $a_{k+1985} = a_k$). We define the partial sums $s_0 = 0$, $s_n = a_1 + a_2 + \cdots + a_n$ for
all positive integers n; then $s_{k+1985} = s_k + s_{1985}$ and $s_{1985} \geq 663$. Since $s_{1985m} \geq$
$663m$ and $3 \cdot 663m > 1985(m + 2) + 1$ for large m, not all values $0, 1, 2, \ldots 663m$
can appear thrice among the $1985(m + 2) + 1$ sums $s_{-1985}, s_{-1984}, \ldots, s_{1985(m+1)}$
(and none of them appears out of this set). Thus there is an integral value $s > 0$
that appears at most twice as a partial sum, say $s_k = s_l = s$, $k < l$. Then either a_k
or a_l is a good point. Actually, $s_i > s$ must hold for all $i > l$, and $s_i < s$ for all
$i < k$ (otherwise, the sum s would appear more than twice). Also, for the same
reason there cannot exist indices p, q between k and l such that $s_p > s$ and $s_q < s$;
i.e., for $k < p < l$, s_p's are either all greater than or equal to s, or smaller than or
equal to s. In the former case a_k is good, while in the latter a_l is good.

15. There is no loss of generality if we assume $K = ABCD$, $K' = AB'C'D'$, and
that K' is obtained from K by a clockwise rotation around A by ϕ, $0 \leq \phi \leq$
$\pi/4$. Let $C'D'$, $B'C'$, and the parallel
to AB through D' meet the line BC at
E, F, and G respectively. Let us now
choose points $E' \in AB'$, $G' \in AB$, $C'' \in$
AD', and $E'' \in AD$ such that the triangles $AE'G'$ and $AC''E''$ are translates

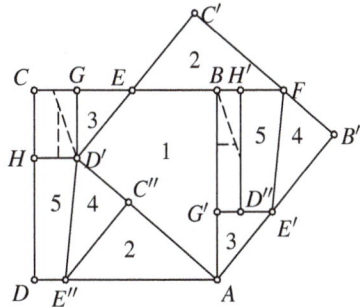

of the triangles $D'EG$ and $FC'E$ respectively. Since $AE' = D'E$ and $AC'' = FC'$,
we have $C''E'' = C'E = B'E'$ and $C''D' = B'F$, which imply that $\triangle E''C''D'$ is a
translate of $\triangle E'B'F$, and consequently $E''D' = E'F$ and $E''D' \parallel E'F$. It follows
that there exist points $H \in CD$, $H' \in BF$, and $D'' \in E'G'$ such that $E''D'HD$
is a translate of $E'FH'D''$. The remaining parts of K and K' are the rectangles
$D'GCH$ and $D''H'BG'$ of equal area.

We shall now show that two rectangles with parallel sides and equal areas can be
decomposed into translation invariant parts. Let the sides of the rectangles $XYZT$
and $X'Y'Z'T'$ $(XY \parallel X'Y')$ satisfy $X'Y' < XY$, $Y'Z' > YZ$, and $X'Y' \cdot Y'Z' = XY \cdot$
YZ. Suppose that $2X'Y' > XY$ (otherwise, we may cut off congruent rectangles
from both the original ones until we reduce them to the case of $2X'Y' > XY$).

Let $U \in XY$ and $V \in ZT$ be points such that $YU = TV = X'Y'$ and $W \in XV$ be a point such that $UW \parallel XT$. Then translating $\triangle XUW$ to a triangle VZR and $\triangle XVT$ to a triangle WRS results in a rectangle $UYRS$ congruent to $X'Y'Z'T'$. Thus we have partitioned K and K' into translation-invariant parts. Although not all the parts are triangles, we may simply triangulate them.

16. Let the three circles be $\alpha(A,a)$, $\beta(B,b)$, and $\gamma(C,c)$, and assume $c \leq a,b$. We denote by $\mathscr{R}_{X,\varphi}$ the rotation around X through an angle φ. Let PQR be an equilateral triangle, say of positive orientation (the case of negatively oriented $\triangle PQR$ is analogous), with $P \in \alpha$, $Q \in \beta$, and $R \in \gamma$. Then $Q = \mathscr{R}_{P,-60°}(R) \in \mathscr{R}_{P,-60°}(\gamma) \cap \beta$.
Since the center of $\mathscr{R}_{P,-60°}(\gamma)$ is $\mathscr{R}_{P,-60°}(C) = \mathscr{R}_{C,60°}(P)$ and it belongs to $\mathscr{R}_{C,60°}(\alpha)$, the union of circles $\mathscr{R}_{P,-60°}(\gamma)$ as P varies on α is the annulus $\mathscr{U}$ with center $A' = \mathscr{R}_{C,60°}(A)$ and radii $a - c$ and $a + c$. Hence there is a solution if and only if $\mathscr{U} \cap \beta$ is nonempty.

17. The statement of the problem is equivalent to the statement that there is one and only one a such that $1 - 1/n < f_n(a) < 1$ for all n. We note that each f_n is a polynomial with positive coefficients, and therefore increasing and convex in $\mathbb{R}^+$.
Define x_n and y_n by $f_n(x_n) = 1 - 1/n$ and $f_n(y_n) = 1$. Since

$$f_{n+1}(x_n) = \left(1 - \frac{1}{n}\right)^2 + \left(1 - \frac{1}{n}\right)\frac{1}{n} = 1 - \frac{1}{n}$$

and $f_{n+1}(y_n) = 1 + 1/n$, it follows that $x_n < x_{n+1} < y_{n+1} < y_n$. Moreover, the convexity of f_n together with the fact that $f_n(x) > x$ for all $x > 0$ implies that $y_n - x_n < f_n(y_n) - f_n(x_n) = 1/n$. Therefore the sequences have a common limit a, which is the only number lying between x_n and y_n for all n. By the definition of x_n and y_n, the statement immediately follows.

18. Set $y_i = \frac{x_i^2}{x_{i+1}x_{i+2}}$, where $x_{n+i} = x_i$. Then $\prod_{i=1}^{n} y_i = 1$ and the inequality to be proved becomes $\sum_{i=1}^{n} \frac{y_i}{1+y_i} \leq n - 1$, or equivalently

$$\sum_{i=1}^{n} \frac{1}{1+y_i} \geq 1.$$

We prove this inequality by induction on n.
Since $\frac{1}{1+y} + \frac{1}{1+y^{-1}} = 1$, the inequality is true for $n = 2$. Assume that it is true for $n - 1$, and let there be given $y_1, \ldots, y_n > 0$ with $\prod_{i=1}^{n} y_i = 1$. Then $\frac{1}{1+y_{n-1}} + \frac{1}{1+y_n} > \frac{1}{1+y_{n-1}y_n}$, which is equivalent to $1 + y_n y_{n-1}(1 + y_n + y_{n-1}) > 0$. Hence by the inductive hypothesis

$$\sum_{i=1}^{n} \frac{1}{1+y_i} \geq \sum_{i=1}^{n-2} \frac{1}{1+y_i} + \frac{1}{1+y_{n-1}y_n} \geq 1.$$

Remark. The constant $n-1$ is best possible (take for example $x_i = a^i$ with a arbitrarily large).

19. Suppose that for some $n > 6$ there is a regular n-gon with vertices having integer coordinates, and that $A_1 A_2 \ldots A_n$ is the smallest such n-gon, of side length a. If O is the origin and B_i the point such that $\overrightarrow{OB_i} = \overrightarrow{A_{i-1}A_i}$, $i = 1, 2, \ldots, n$ (where $A_0 = A_n$), then B_i has integer coordinates and $B_1 B_2 \ldots B_n$ is a regular polygon of side length $2a\sin(\pi/n) < a$, which is impossible.
It remains to analyze the cases $n \le 6$. If $\mathscr{P}$ is a regular n-gon with $n = 3, 5, 6$, then its center C has rational coordinates. We may suppose that C also has integer coordinates and then rotate $\mathscr{P}$ around C thrice through $90°$, thus obtaining a regular 12-gon or 20-gon, which is impossible. Hence we must have $n = 4$ which is indeed a solution.

20. Let O be the center of the circle touching the three sides of $BCDE$ and let $F \in (ED)$ be the point such that $EF = EB$. Then $\angle EFB = 90° - \angle E/2 = \angle C/2 = \angle OCB$, which implies that B, C, F, O lie on a circle. It follows that $\angle DFC = \angle OBC = \angle B/2 = 90° - \angle D/2$ and consequently $\angle DCF = \angle DFC$. Hence $ED = EF + FD = EB + CD$.

Second solution. Let r be the radius of the small circle and let M, N be the points of tangency of the circle with BE and CD respectively. Then

$$EM = r\cot E, \, DN = r\cot D, \, MB = r\cot\frac{\angle B}{2} = r\tan\frac{\angle D}{2}, \, NC = r\tan\frac{\angle E}{2},$$
$$\text{and } ED = EO + OD = \frac{r}{\sin D} + \frac{r}{\sin E}.$$

The statement follows from the identity $\cot x + \tan(x/2) = 1/\sin x$.

21. Let B_1 and C_1 be the points on the rays AC and AB respectively such that $XB_1 = XC = XB = XC_1$. Then $\angle XB_1C = \angle XCB_1 = \angle ABC$ and $\angle XC_1B = \angle XBC_1 = \angle ACB$, which imply that B_1, X, C_1 are collinear and $\triangle AB_1C_1 \sim \triangle ABC$. Moreover, X is the midpoint of B_1C_1 because $XB_1 = XC_1$, from which we conclude that $\triangle AXB_1 \sim \triangle AMB$. Therefore $\angle CAX = \angle BAM$ and

$$\frac{AM}{AX} = \frac{BM}{XB_1} = \frac{BM}{BC} = \cos\alpha.$$

22. Assume that $\triangle ABC$ is acute (the case of an obtuse $\triangle ABC$ is similar). Let S and R be the centers of the circumcircles of $\triangle ABC$ and $\triangle KBN$, respectively. Since $\angle BNK = \angle BAC$, the triangles BNK and BAC are similar. Now we have $\angle CBR = \angle ABS = 90° - \angle ACB$, which gives us $BR \perp AC$ and consequently $BR \parallel OS$. Similarly $BS \perp KN$ implies that $BS \parallel OR$. Hence $BROS$ is a parallelogram. Let L be the point symmetric to B with respect to R. Then $RLOS$ is also a parallelogram, and since $SR \perp BM$, we obtain $OL \perp BM$. However, we also have $LM \perp BM$, from which we conclude that O, L, M are collinear and $OM \perp BM$.

Second solution. The lines BM, NK, and CA are the radical axes of pairs of the three circles, and hence they intersect at a single point P. Also, the quadrilateral $MNCP$ is cyclic. Let $OA = OC = OK = ON = r$. We then have $BM \cdot BP = BN \cdot BC = OB^2 - r^2$, $PM \cdot PB = PN \cdot PK = OP^2 - r^2$. It follows that $OB^2 - OP^2 = BP(BM - PM) = BM^2 - PM^2$, which implies that $OM \perp MB$.

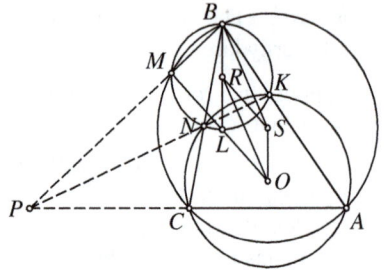

4.27 Solutions to the Shortlisted Problems of IMO 1986

1. If $w > 2$, then setting in (i) $x = w - 2, y = 2$, we get $f(w) = f((w-2)f(w))f(2) = 0$. Thus

$$f(x) = 0 \quad \text{if and only if} \quad x \geq 2.$$

   Now let $0 \leq y < 2$ and $x \geq 0$. The LHS in (i) is zero if and only if $xf(y) \geq 2$, while the RHS is zero if and only if $x + y \geq 2$. It follows that $x \geq 2/f(y)$ if and only if $x \geq 2 - y$. Therefore

$$f(y) = \begin{cases} \frac{2}{2-y} & \text{for } 0 \leq y < 2; \\ 0 & \text{for } y \geq 2. \end{cases}$$

   The confirmation that f satisfies the given conditions is straightforward.

2. No. If a were rational, its decimal expansion would be periodic from some point. Let p be the number of decimals in the period. Since $f(10^{2p})$ has $2np$ zeros, it contains a full periodic part; hence the period would consist only of zeros, which is impossible.

3. Let E be the point where the boy turned westward, reaching the shore at D. Let the ray DE cut AC at F and the shore again at G. Then $EF = AE = x$ (because AEF is an equilateral triangle) and $FG = DE = y$. From $AE \cdot EB = DE \cdot EG$ we obtain $x(86 - x) = y(x + y)$. If x is odd, then $x(86 - x)$ is odd, while $y(x + y)$ is even. Hence x is even, and so y must also be even. Let $y = 2y_1$. The above equation can be rewritten as

$$(x + y_1 - 43)^2 + (2y_1)^2 = (43 - y_1)^2.$$

   Since $y_1 < 43$, we have $(2y_1, 43 - y_1) = 1$, and thus $(|x + y_1 - 43|, 2y_1, 43 - y_1)$ is a primitive Pythagorean triple. Consequently there exist integers $a > b > 0$ such that $y_1 = ab$ and $43 - y_1 = a^2 + b^2$. We obtain that $a^2 + b^2 + ab = 43$, which has the unique solution $a = 6, b = 1$. Hence $y = 12$ and $x = 2$ or $x = 72$.

   Remark. The Diophantine equation $x(86 - x) = y(x + y)$ can be also solved directly. Namely, we have that $x(344 - 3x) = (2y + x)^2$ is a square, and since x is even, we have $(x, 344 - 3x) = 2$ or 4. Consequently $x, 344 - 3x$ are either both squares or both two times squares. The rest is easy.

4. Let $x = p^\alpha x', y = p^\beta y', z = p^\gamma z'$ with $p \nmid x'y'z'$ and $\alpha \geq \beta \geq \gamma$. From the given equation it follows that $p^n(x + y) = z(xy - p^n)$ and consequently $z' \mid x + y$. Since also $p^\gamma \mid x + y$, we have $z \mid x + y$, i.e., $x + y = qz$. The given equation together with the last condition gives us

$$xy = p^n(q + 1) \quad \text{and} \quad x + y = qz. \tag{1}$$

   Conversely, every solution of (1) gives a solution of the given equation.
   For $q = 1$ and $q = 2$ we obtain the following classes of $n + 1$ solutions each:

$$q = 1: (x,y,z) = (2p^i, p^{n-i}, 2p^i + p^{n-i}) \qquad \text{for } i = 0,1,2,\ldots,n;$$
$$q = 2: (x,y,z) = \left(3p^j, p^{n-j}, \tfrac{3p^j + p^{n-j}}{2}\right) \qquad \text{for } j = 0,1,2,\ldots,n.$$

For $n = 2k$ these two classes have a common solution $(2p^k, p^k, 3p^k)$; otherwise, all these solutions are distinct. One further solution is given by $(x,y,z) = (1, p^n(p^n + 3)/2, p^2 + 2)$, not included in the above classes for $p > 3$. Thus we have found $2(n+1)$ solutions.

Another type of solution is obtained if we put $q = p^k + p^{n-k}$. This yields the solutions

$$(x,y,z) = (p^k, p^n + p^{n-k} + p^{2n-2k}, p^{n-k} + 1) \qquad \text{for } k = 0,1,\ldots,n.$$

For $k < n$ these are indeed new solutions. So far, we have found $3(n+1) - 1$ or $3(n+1)$ solutions. One more solution is given by $(x,y,z) = (p, p^n + p^{n-1}, p^{n-1} + p^{n-2} + 1)$.

5. Suppose that for every $a,b \in \{2,5,13,d\}$, $a \neq b$, the number $ab - 1$ is a perfect square. In particular, for some integers x,y,z we have

$$2d - 1 = x^2, \qquad 5d - 1 = y^2, \qquad 13d - 1 = z^2.$$

Since x is clearly odd, $d = (x^2 + 1)/2$ is also odd because $4 \nmid x^2 + 1$. It follows that y and z are even, say $y = 2y_1$ and $z = 2z_1$. Hence $(z_1 - y_1)(z_1 + y_1) = (z^2 - y^2)/4 = 2d$. But in this case one of the factors $z_1 - y_1$, $z_1 + y_1$ is odd and the other one is even, which is impossible.

6. There are five such numbers:

$$69300 = 2^2 \cdot 3^2 \cdot 5^2 \cdot 7 \cdot 11 : \quad 3 \cdot 3 \cdot 3 \cdot 2 \cdot 2 = 108 \text{ divisors};$$
$$50400 = 2^5 \cdot 3^2 \cdot 5^2 \cdot 7 : \quad 6 \cdot 3 \cdot 3 \cdot 2 = 108 \text{ divisors};$$
$$60480 = 2^6 \cdot 3^3 \cdot 5 \cdot 7 : \quad 7 \cdot 4 \cdot 2 \cdot 2 = 112 \text{ divisors};$$
$$55440 = 2^4 \cdot 3^2 \cdot 5 \cdot 7 \cdot 11 : \quad 5 \cdot 3 \cdot 2 \cdot 2 \cdot 2 = 120 \text{ divisors};$$
$$65520 = 2^4 \cdot 3^2 \cdot 5 \cdot 7 \cdot 13 : \quad 5 \cdot 3 \cdot 2 \cdot 2 \cdot 2 = 120 \text{ divisors}.$$

7. Let $P(x) = (x - x_0)(x - x_1)\cdots(x - x_n)(x - x_{n+1})$. Then

$$P'(x) = \sum_{j=0}^{n+1} \frac{P(x)}{x - x_j} \qquad \text{and} \qquad P''(x) = \sum_{j=0}^{n+1}\sum_{k \neq j} \frac{P(x)}{(x - x_j)(x - x_k)}.$$

Therefore

$$P''(x_i) = 2P'(x_i) \sum_{j \neq i} \frac{1}{(x_i - x_j)}$$

for $i = 0,1,\ldots,n+1$, and the given condition implies $P''(x_i) = 0$ for $i = 1,2,\ldots,n$. Consequently,

$$x(x-1)P''(x) = (n+2)(n+1)P(x). \qquad (1)$$

It is easy to observe that there is a unique monic polynomial of degree $n+2$ satisfying differential equation (1). On the other hand, the polynomial $Q(x) = (-1)^n P(1-x)$ also satisfies this equation, is monic, and $\deg Q = n+2$. Therefore $(-1)^n P(1-x) = P(x)$, and the result follows.

8. We shall solve the problem in the alternative formulation. Let $L_G(v)$ denote the length of the longest directed chain of edges in the given graph G that begins in a vertex v and is arranged decreasingly relative to the numbering. By the pigeonhole principle it suffices to show that $\sum_v L(v) \geq 2q$ in every such graph. We do this by induction on q.

For $q = 1$ the claim is obvious. We assume that it is true for $q-1$ and consider a graph G with q edges numbered $1, \ldots, q$. Let the edge number q connect vertices u and w. Removing this edge, we get a graph G' with $q-1$ edges. We then have

$$L_G(u) \geq L_{G'}(w) + 1, \ L_G(w) \geq L_{G'}(u) + 1, \ L_G(v) \geq L_{G'}(v) \text{ for other } v.$$

Since $\sum L_{G'}(v) \geq 2(q-1)$ by inductive assumption, it follows that $\sum L_G(v) \geq 2(q-1) + 2 = 2q$ as desired.

Second solution. Let us place a spider at each vertex of the graph. Let us now interchange the positions of the two spiders at the endpoints of each edge, listing the edges increasingly with respect to the numbering. This way we will move spiders exactly $2q$ times (two for each edge). Hence there is a spider that will be moved at least $2q/n$ times. All that remains is to notice that the path of each spider consists of edges numbered in increasing order.

Remark. A chain of the stated length having all vertices distinct does not necessarily exist. An example is $n = 4$, $q = 6$ with the numbering following the order ab, cd, ac, bd, ad, bc.

9. We shall use induction on the number n of points. The case $n = 1$ is trivial. Let us suppose that the statement is true for all $1, 2, \ldots, n-1$, and that we are given a set T of n points.

If there exists a point $P \in T$ and a line l that is parallel to an axis and contains P and no other points of T, then by the inductive hypothesis we can color the set $T \setminus \{P\}$ and then use a suitable color for P. Let us now suppose that whenever a line parallel to an axis contains a point of T, it contains another point of T. It follows that for an arbitrary point $P_0 \in T$ we can choose points $P_1, P_2, \ldots$ such that $P_k P_{k+1}$ is parallel to the x-axis for k even, and to the y-axis for k odd. We eventually come to a pair of integers (r, s) of the same parity, $0 \leq r < s$, such that lines $P_r P_{r+1}$ and $P_s P_{s+1}$ coincide. Hence the closed polygonal line $P_{r+1} P_{r+2} \ldots P_s P_{r+1}$ is of even length. Thus we may color the points of this polygonal line alternately and then apply the inductive assumption for the rest of the set T. The induction is complete.

Second solution. Let $P_1, P_2, \ldots, P_k$ be the points lying on a line l parallel to an axis, going from left to right or from up to down. We draw segments joining P_1 with P_2, P_3 with P_4, and generally P_{2i-1} with P_{2i}. Having this done for every such line l, we obtain a set of segments forming certain polygonal lines. If one of these polygonal lines is closed, then it must have an even number of vertices. Thus, we can color the vertices on each of the polygonal lines alternately (a point not lying on any of the polygonal lines may be colored arbitrarily). The obtained coloring satisfies the conditions.

10. The set $X = \{1, \ldots, 1986\}$ splits into triads $T_1, \ldots, T_{662}$, where $T_j = \{3j-2, 3j-1, 3j\}$.

Let $\mathscr{F}$ be the family of all k-element subsets P such that $|P \cap T_j| = 1$ or 2 for some index j. If j_0 is the smallest such j_0, we define P' to be the k-element set obtained from P by replacing the elements of $P \cap T_{j_0}$ by the ones following cyclically inside T_{j_0}. Let $s(P)$ denote the remainder modulo 3 of the sum of elements of P. Then $s(P), s(P'), s(P'')$ are distinct, and $P''' = P$. Thus the operator $'$ gives us a bijective correspondence between the sets $X \in \mathscr{F}$ with $s(P) = 0$, those with $s(P) = 1$, and those with $s(P) = 2$.

If $3 \nmid k$ is not divisible by 3, then each k-element subset of X belongs to $\mathscr{F}$, and the game is fair. If $3 \mid k$, then k-element subsets not belonging to $\mathscr{F}$ are those that are unions of several triads. Since every such subset has the sum of elements divisible by 3, it follows that player A has the advantage.

11. Let X be a finite set in the plane and l_k a line containing exactly k points of X ($k = 1, \ldots, n$). Then l_n contains n points, l_{n-1} contains at least $n-2$ points not lying on l_n, l_{n-2} contains at least $n-4$ points not lying on l_n or l_{n-1}, etc. It follows that

$$|X| \geq g(n) = n + (n-2) + (n-4) + \cdots + \left(n - 2\left[\tfrac{n}{2}\right]\right).$$

Hence $f(n) \geq g(n) = \left[\tfrac{n+1}{2}\right]\left[\tfrac{n+2}{2}\right]$, where the last equality is easily proved by induction.

We claim that $f(n) = g(n)$. To prove this, we shall inductively construct a set X_n of cardinality $g(n)$ with the required property. For $n \leq 2$ a one-point and two-point set satisfy the requirements. Assume that X_n is a set of $g(n)$ points and that l_k is a line containing exactly k points of X_n, $k = 1, \ldots, n$. Consider any line l not parallel to any of the l_k's and not containing any point of X_n or any intersection point of the l_k. Let l intersect l_k in a point P_k, $k = 1, \ldots, n$, and let P_{n+1}, P_{n+2} be two points on l other than $P_1, \ldots, P_n$. We define $X_{n+2} = X_n \cup \{P_1, \ldots, P_{n+2}\}$. The set X_{n+2} consists of $g(n) + (n+2) = g(n+2)$ points. Since the lines $l, l_n, \ldots, l_2, l_1$ meet X_n in $n+2, n+1, \ldots, 3, 2$ points respectively (and there clearly exists a line containing only one point of X_{n+2}), this set also meets the demands.

12. We define $f(x_1, \ldots, x_5) = \sum_{i=1}^{5}(x_{i+1} - x_{i-1})^2$ ($x_0 = x_5$, $x_6 = x_1$). Assuming that $x_3 < 0$, according to the rules the lattice vector $X = (x_1, x_2, x_3, x_4, x_5)$ changes into $Y = (x_1, x_2 + x_3, -x_3, x_4 + x_3, x_5)$. Then

$$\begin{aligned}
f(Y) - f(X) &= (x_2 + x_3 - x_5)^2 + (x_1 + x_3)^2 + (x_2 - x_4)^2 \\
&\quad + (x_3 + x_5)^2 + (x_1 - x_3 - x_4)^2 - (x_2 - x_5)^2 \\
&\quad - (x_3 - x_1)^2 - (x_4 - x_2)^2 - (x_5 - x_3)^2 - (x_1 - x_4)^2 \\
&= 2x_3(x_1 + x_2 + x_3 + x_4 + x_5) = 2x_3 S < 0.
\end{aligned}$$

Thus f strictly decreases after each step, and since it takes only nonnegative integer values, the number of steps must be finite.

Remark. One could inspect the behavior of $g(X) = \sum_{i=1}^{5}\sum_{j=1}^{5}|x_i+x_{i+1}+\cdots+x_{j-1}|$ instead. Then $g(Y)-g(X) = |S+x_3|-|S-x_3| > 0$.

13. Let us consider the infinite integer lattice and assume that having reached a point (x,n) or (n,y), the particle continues moving east and north following the rules of the game. The required probability p_k is equal to the probability of getting to one of the points $E_1(n,n+k)$, $E_2(n+k,n)$, but without passing through $(n,n+k-1)$ or $(n+k-1,n)$. Thus p is equal to the probability p_1 of getting to $E_1(n,n+k)$ via $D_1(n-1,n+k)$ plus the probability p_2 of getting to $E_2(n+k,n)$ via $D_2(n+k,n-1)$. Both p_1 and p_2 are easily seen to be equal to $\binom{2n+k-1}{n-1}2^{-2n-k}$, and therefore $p = \binom{2n+k-1}{n-1}2^{-2n-k+1}$.

14. We shall use the following simple fact.

   Lemma. If $\hat{k}$ is the image of a circle k under an inversion centered at a point Z, and O_1, O_2 are centers of k and $\hat{k}$, then O_1, O_2, and Z are collinear.

   Proof. The result follows immediately from the symmetry with respect to the line ZO_1.

   Let I be the center of the inscribed circle i. Since $IX \cdot IA = IE^2$, the inversion with respect to i takes points A into X, and analogously B,C into Y,Z respectively. It follows from the lemma that the center of circle ABC, the center of circle XYZ, and point I are collinear.

15. (a) This is the same problem as *SL82-14*.

   (b) If S is the midpoint of AC, we have $B'S = AC\frac{\cos\angle D}{2\sin\angle D}$, $D'S = AC\frac{\cos\angle B}{2\sin\angle B}$, $B'D' = AC\left|\frac{\sin(\angle B+\angle D)}{2\sin\angle B\sin\angle D}\right|$. These formulas are true also if $\angle B > 90°$ or $\angle D > 90°$. We similarly obtain that $A''C'' = B'D'\left|\frac{\sin(\angle A'+\angle C')}{2\sin\angle A'\sin\angle C'}\right|$. Therefore

   $$A''C'' = AC\frac{\sin^2(\angle A+\angle C)}{4\sin\angle A\sin\angle B\sin\angle C\sin\angle D}.$$

16. Let Z be the center of the polygon. Suppose that at some moment we have $A \in P_{i-1}P_i$ and $B \in P_iP_{i+1}$, where P_{i-1}, P_i, P_{i+1} are adjacent vertices of the polygon. Since $\angle AOB = 180° - \angle P_{i-1}P_iP_{i+1}$, the quadrilateral AP_iBO is cyclic. Hence $\angle AP_iO = \angle ABO = \angle AP_iZ$, which means that $O \in P_iZ$. More-

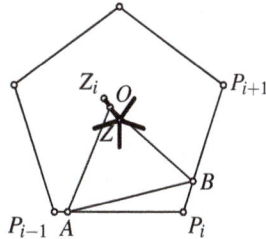

over, from $OP_i = 2r\sin\angle P_iAO$, where r is the radius of circle AP_iBO, we obtain that $ZP_i \le OP_i \le ZP_i/\cos(\pi/n)$. Thus O traces a segment ZZ_i as A and B move along $P_{i-1}P_i$ and P_iP_{i+1} respectively, where Z_i is a point on the ray P_iZ with $P_iZ_i\cos(\pi/n) = P_iZ$. When A,B move along the whole circumference of the polygon, O traces an asterisk consisting of n segments of equal length emanating from Z and pointing away from the vertices.

17. We use complex numbers to represent the position of a point in the plane. For convenience, let $A_1, A_2, A_3, A_4, A_5, \ldots$ be $A, B, C, A, B, \ldots$ respectively, and let P_0 be the origin. After the kth step, the position of P_k will be $P_k = A_k + (P_{k-1} - A_k)u$, $k = 1, 2, 3, \ldots$, where $u = e^{4\pi i/3}$. We easily obtain

$$P_k = (1-u)(A_k + uA_{k-1} + u^2 A_{k-2} + \cdots + u^{k-1}A_1).$$

The condition $P_0 \equiv P_{1986}$ is equivalent to $A_{1986} + uA_{1985} + \cdots + u^{1984}A_2 + u^{1985}A_1 = 0$, which, having in mind that $A_1 = A_4 = A_7 = \cdots$, $A_2 = A_5 = A_8 = \cdots$, $A_3 = A_6 = A_9 = \cdots$, reduces to

$$662(A_3 + uA_2 + u^2 A_1) = (1 + u^3 + \cdots + u^{1983})(A_3 + uA_2 + u^2 A_1) = 0.$$

It follows that $A_3 - A_1 = u(A_1 - A_2)$, and the assertion follows.

Second solution. Let f_P denote the rotation with center P through $120°$ clockwise. Let $f_1 = f_A$. Then $f_1(P_0) = P_1$. Let $B' = f_1(B)$, $C' = f_1(C)$, and $f_2 = f_{B'}$. Then $f_2(P_1) = P_2$ and $f_2(AB'C') = A'B'C''$. Finally, let $f_3 = f_{C''}$ and $f_3(A'B'C'') = A''B''C''$. Then $g = f_3 f_2 f_1$ is a translation sending P_0 to P_3 and C to C''. Now $P_{1986} = P_0$ implies that g^{662} is the identity, and thus $C = C''$.

Let K be such that ABK is equilateral and positively oriented. We observe that $f_2 f_1(K) = K$; therefore the rotation $f_2 f_1$ satisfies $f_2 f_1(P) \neq P$ for $P \neq K$. Hence $f_2 f_1(C) = C'' = C$ implies $K = C$.

18. We shall use the following criterion for a quadrangle to be circumscribable.

   Lemma. The quadrangle $AYDZ$ is circumscribable if and only if $DB - DC = AB - AC$.

   Proof. Suppose that $AYDZ$ is circumscribable and that the incircle is tangent to AZ, ZD, DY, YA at M, N, P, Q respectively. Then $DB - DC = PB - NC = MB - QC = AB - AC$. Conversely, assume that $DB - DC = AB - AC$ and let
   a tangent from D to the incircle of the triangle ACZ meet CZ and CA at $D' \neq Z$ and $Y' \neq A$ respectively. According to the first part we have $D'B - D'C = AB - AC$. It follows that $|D'B - DB| = |D'C - DC| = DD'$, implying that $D' \equiv D$.

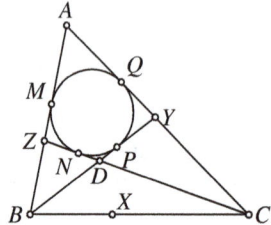

   Let us assume that $DZBX$ and $DXCY$ are circumscribable. Using the lemma we obtain $DC - DA = BC - BA$ and $DA - DB = CA - CB$. Adding these two inequalities yields $DC - DB = AC - AB$, and the statement follows from the lemma.

19. Let M and N be the midpoints of segments AB and CD, respectively. The given conditions imply that $\triangle ABD \cong \triangle BAC$ and $\triangle CDA \cong \triangle DCB$; hence $MC = MD$ and $NA = NB$. It follows that M and N both lie on the perpendicular bisectors of AB and CD, and consequently MN is the common perpendicular bisector of AB and CD. Points B and C are symmetric to A and D with respect to MN. Now if P

is a point in space and P' the point symmetric to P with respect to MN, we have $BP = AP'$, $CP = DP'$, and thus $f(P) = AP + AP' + DP + DP'$. Let PP' intersect MN in Q. Then $AP + AP' \geq 2AQ$ and $DP + DP' \geq 2DQ$, from which it follows that $f(P) \geq 2(AQ + DQ) = f(Q)$. It remains to minimize $f(Q)$ with Q moving along the line MN.

Let us rotate point D around MN to a point D' that belongs to the plane AMN, on the side of MN opposite to A. Then $f(Q) = 2(AQ + D'Q) \geq AD'$, and equality occurs when Q is the intersection of AD' and MN. Thus $\min f(Q) = AD'$. We note that $4MD^2 = 2AD^2 + 2BD^2 - AB^2 = 2a^2 + 2b^2 - AB^2$ and $4MN^2 = 4MD^2 - CD^2 = 2a^2 + 2b^2 - AB^2 - CD^2$. Now, $AD'^2 = (AM + D'N)^2 + MN^2$, which together with $AM + D'N = (a+b)/2$ gives us

$$AD'^2 = \frac{a^2 + b^2 + AB \cdot CD}{2} = \frac{a^2 + b^2 + c^2}{2}.$$

We conclude that $\min f(Q) = \sqrt{(a^2 + b^2 + c^2)/2}$.

20. If the faces of the tetrahedron $ABCD$ are congruent triangles, we must have $AB = CD$, $AC = BD$, and $AD = BC$. Then the sum of angles at A is $\angle BAC + \angle CAD + \angle DAB = \angle BDC + \angle CBD + \angle DCB = 180°$.

We now assume that the sum of angles at each vertex is $180°$. Let us construct triangles BCD', CAD'', ABD''' in the plane ABC, exterior to $\triangle ABC$, such that $\triangle BCD' \cong \triangle BCD$, $\triangle CAD'' \cong \triangle CAD$, and $\triangle ABD''' \cong \triangle ABD$. Then by the assumption, $A \in D''D'''$, $B \in D'''D'$, and $C \in D'D''$. Since also $D''A = D'''A = DA$, etc., A, B, C are the midpoints of segments $D''D''', D'''D', D'D''$ respectively. Thus the triangles ABC, BCD', CAD'', ABD''' are congruent, and the statement follows.

21. Since the sum of all edges of $ABCD$ is 3, the statement of the problem is an immediate consequence of the following statement:

Lemma. Let r be the inradius of a triangle with sides a, b, c. Then $a + b + c \geq 6\sqrt{3} \cdot r$, with equality if and only if the triangle is equilateral.

Proof. If S and p denotes the area and semiperimeter of the triangle, by Heron's formula and the AM–GM inequality we have

$$pr = S = \sqrt{p(p-a)(p-b)(p-c)}$$
$$\leq \sqrt{p \left(\frac{(p-a) + (p-b) + (p-c)}{3} \right)^3} = \sqrt{\frac{p^4}{27}} = \frac{p^2}{3\sqrt{3}},$$

i.e., $p \geq 3\sqrt{3} \cdot r$, which is equivalent to the claim.

4.28 Solutions to the Shortlisted Problems of IMO 1987

1. By (ii), $f(x) = 0$ has at least one solution, and there is the greatest among them, say x_0. Then by (v), for any x,

$$0 = f(x)f(x_0) = f(xf(x_0) + x_0 f(x) - x_0 x) = f(x_0(f(x) - x)). \qquad (1)$$

It follows that $x_0 \geq x_0(f(x) - x)$.
Suppose $x_0 > 0$. By (i) and (iii), since $f(x_0) - x_0 < 0 < f(0) - 0$, there is a number z between 0 and x_0 such that $f(z) = z$. By (1), $0 = f(x_0(f(z) - z)) = f(0) = 1$, a contradiction. Hence, $x_0 < 0$. Now the inequality $x_0 \geq x_0(f(x) - x)$ gives $f(x) - x \geq 1$ for all x; so, $f(1987) \geq 1988$. Therefore $f(1987) = 1988$.

2. Let d_i denote the number of cliques of which person i is a member. Clearly $d_i \geq 2$. We now distinguish two cases:
   (i) For some i, $d_i = 2$. Suppose that i is a member of two cliques, C_p and C_q. Then $|C_p| = |C_q| = n$, since for each couple other than i and his/her spouse, one member is in C_p and one in C_q. There are thus $(n-1)(n-2)$ pairs (r, s) of nonspouse persons distinct from i, where $r \in C_p$, $s \in C_q$. We observe that each such pair accounts for a different clique. Otherwise, we find two members of C_p or C_q who belong to one other clique. It follows that $k \geq 2 + (n-1)(n-2) \geq 2n$ for $n \geq 4$.
   (ii) For every i, $d_i \geq 3$. Suppose that $k < 2n$. For $i = 1, 2, \ldots, 2n$ assign to person i an indeterminant x_i, and for $j = 1, 2, \ldots, k$ set $y = \sum_{i \in C_j} x_i$. From linear algebra, we know that if $k < 2n$, then there exist $x_1, x_2, \ldots, x_{2n}$, not all zero, such that $y_1 = y_2 = \cdots = y_k = 0$.
   On the other hand, suppose that $y_1 = y_2 = \cdots = y_k = 0$. Let M be the set of the couples and M' the set of all other pairs of persons. Then

   $$0 = \sum_{j=1}^{k} y_j^2 = \sum_{i=1}^{2n} d_i x_i^2 + 2 \sum_{(i,j) \in M'} x_i x_j$$

   $$= \sum_{i=1}^{2n} (d_i - 2) x_i^2 + (x_1 + x_2 + \cdots + x_{2n})^2 + \sum_{(i,j) \in M} (x_i - x_j)^2$$

   $$\geq \sum_{i=1}^{2n} x_i^2 > 0,$$

   if not all $x_1, x_2, \ldots, x_{2n}$ are zero, which is a contradiction. Hence $k \geq 2n$.

   Remark. The condition $n \geq 4$ is essential. For a party attended by 3 couples $\{(1,4), (2,5), (3,6)\}$, there is a collection of 4 cliques satisfying the conditions: $\{(1,2,3), (3,4,5), (5,6,1), (2,4,6)\}$.

3. The answer: yes. Set

$$p(k,m) = k + [1 + 2 + \cdots + (k+m)] = \frac{(k+m)^2 + 3k + m}{2}.$$

It is obviously of the desired type.

4. Setting $x_1 = \overrightarrow{AB}$, $x_2 = \overrightarrow{AD}$, $x_3 = \overrightarrow{AE}$, we have to prove that

$$|x_1 + x_2| + |x_2 + x_3| + |x_3 + x_1| \le |x_1| + |x_2| + |x_3| + |x_1 + x_2 + x_3|.$$

We have

$$(|x_1| + |x_2| + |x_3|)^2 - |x_1 + x_2 + x_3|^2$$
$$= 2\sum_{1 \le i < j \le 3} (|x_i||x_j| - \langle x_i, x_j \rangle) = \sum_{1 \le i < j \le 3} \left[(|x_i| + |x_j|)^2 - |x_i + x_j|^2 \right]$$
$$= \sum_{1 \le i < j \le 3} (|x_i| + |x_j| + |x_i + x_j|)(|x_i| + |x_j| - |x_i + x_j|).$$

The following two inequalities are obvious:

$$|x_i| + |x_j| - |x_i + x_j| \ge 0, \tag{1}$$

$$|x_i| + |x_j| + |x_i + x_j| \le |x_1| + |x_2| + |x_3| + |x_1 + x_2 + x_3|. \tag{2}$$

It follows that

$$(|x_1| + |x_2| + |x_3|)^2 - |x_1 + x_2 + x_3|^2$$
$$\le \left(\sum_{i=1}^{3} |x_i| + \left| \sum_{i=1}^{3} x_i \right| \right) \left(2\sum_{i=1}^{3} |x_i| - \sum_{1 \le i < j \le 3} |x_i + x_j| \right),$$

and dividing by the positive number $\sum_{i=1}^{3} |x_i| + \left| \sum_{i=1}^{3} x_i \right|$ we obtain

$$\sum_{i=1}^{3} |x_i| - \left| \sum_{i=1}^{3} x_i \right| \le 2\sum_{i=1}^{3} |x_i| - \sum_{1 \le i < j \le 3} |x_i + x_j|.$$

The inequality is proven. Let us analyze the cases of equality. If one of the vectors is null, then equality obviously holds. Suppose that $x_i \ne 0$, $i = 1, 2, 3$. For every i, j, at least one of (1) and (2) is equality. Equality in (1) holds if and only if x_i and x_j are collinear with the same direction, while in (2) it holds if and only if $-x_k$ and $x_1 + x_2 + x_3$ are collinear with the same direction. If not all the vectors are collinear, then there are at least two distinct pairs x_i, x_j, $i < j$, for which (2) is an equality, so at least two of x_i are collinear with $x_1 + x_2 + x_3$, but then so is the third; hence, the sum $x_1 + x_2 + x_3$ must be 0. Thus the cases of equality are (a) the vectors are collinear with the same direction; (b) the vectors are collinear, two of them have the same direction, say x_i, x_j, and $|x_k| \ge |x_i| + |x_j|$; (c) one of the vectors is 0; (d) their sum is 0.

Second solution. The following technique, although not quite elementary, is often used to effectively reduce geometric inequalities of first degree, like this one, to the one-dimensional case.

Let σ be a fixed sphere with center O. For an arbitrary segment d in space, and any line l, we denote by $\pi_l(d)$ the length of the projection of d onto l. Consider the integral of lengths of these projections on all possible directions of OP, with

P moving on the sphere: $\int_\sigma \pi_{OP}(d)\,d\sigma$. It is clear that this value depends only on the length of d (because of symmetry); hence

$$\int_\sigma \pi_{OP}\,d\sigma = c \cdot |d| \qquad \text{for some constant } c \neq 0. \tag{1}$$

Notice that by the one-dimensional case, for any point $P \in \sigma$,

$$\pi_{OP}(x_1) + \pi_{OP}(x_2) + \pi_{OP}(x_3) + \pi_{OP}(x_1 + x_2 + x_3)$$

$$\geq \pi_{OP}(x_1 + x_2) + \pi_{OP}(x_1 + x_3) + \pi_{OP}(x_2 + x_3).$$

By integration on σ, using (1), we obtain

$$c(|x_1| + |x_2| + |x_3| + |x_1 + x_2 + x_3|) \geq c(|x_1 + x_2| + |x_1 + x_3| + |x_2 + x_3|).$$

5. Assuming the notation $a = \overline{BC}$, $b = \overline{AC}$, $c = \overline{AB}$; $x = \overline{BL}$, $y = \overline{CM}$, $z = \overline{AN}$, from the Pythagorean theorem we obtain

$$(a-x)^2 + (b-y)^2 + (c-z)^2 = x^2 + y^2 + z^2$$

$$= \frac{x^2 + (a-x)^2 + y^2 + (b-y)^2 + z^2 + (c-z)^2}{2}.$$

Since $x^2 + (a-x)^2 = a^2/2 + (a-2x)^2/2 \geq a^2/2$ and similarly $y^2 + (b-y)^2 \geq b^2/2$ and $z^2 + (c-z)^2 \geq c^2/2$, we get

$$x^2 + y^2 + z^2 \geq \frac{a^2 + b^2 + c^2}{4}.$$

Equality holds if and only if P is the circumcenter of the triangle ABC, i.e., when $x = a/2$, $y = b/2$, $z = c/2$.

6. Suppose w.l.o.g. that $a \geq b \geq c$. Then $1/(b+c) \geq 1/(a+c) \geq 1/(a+b)$. Chebyshev's inequality yields

$$\frac{a^n}{b+c} + \frac{b^n}{a+c} + \frac{c^n}{a+b} \geq \frac{1}{3}(a^n + b^n + c^n)\left(\frac{1}{b+c} + \frac{1}{a+c} + \frac{1}{a+b}\right). \tag{1}$$

By the Cauchy-Schwarz inequality we have

$$2(a+b+c)\left(\frac{1}{b+c} + \frac{1}{a+c} + \frac{1}{a+b}\right) \geq 9,$$

and the mean inequality yields $(a^n + b^n + c^n)/3 \geq [(a+b+c)/3]^n$. We obtain from (1) that

$$\frac{a^n}{b+c} + \frac{b^n}{a+c} + \frac{c^n}{a+b} \geq \left(\frac{a+b+c}{3}\right)^n\left(\frac{1}{b+c} + \frac{1}{a+c} + \frac{1}{a+b}\right)$$

$$\geq \frac{3}{2}\left(\frac{a+b+c}{3}\right)^{n-1} = \left(\frac{2}{3}\right)^{n-2} S^{n-1}.$$

7. For all real numbers v the following inequality holds:

$$\sum_{0\le i<j\le 4}(v_i-v_j)^2 \le 5\sum_{i=0}^{4}(v_i-v)^2. \tag{1}$$

Indeed,

$$\sum_{0\le i<j\le 4}(v_i-v_j)^2 = \sum_{0\le i<j\le 4}[(v_i-v)-(v_j-v)]^2$$

$$= 5\sum_{i=0}^{4}(v_i-v)^2 - \left(\sum_{i=0}^{4}(v_i-v)\right)^2 \le 5\sum_{i=0}^{4}(v_i-v)^2.$$

Let us first take v_i's, satisfying condition (1), so that w.l.o.g. $v_0 \le v_1 \le v_2 \le v_3 \le v_4 \le 1+v_0$. Defining $v_5 = 1+v_0$, we see that one of the differences $v_{j+1}-v_j$, $j=0,\ldots,4$, is at most $1/5$. Take $v=(v_{j+1}+v_j)/2$, and then place the other three v_j's in the segment $[v-1/2, v+1/2]$. Now we have $|v-v_j| \le 1/10$, $|v-v_{j+1}| \le 1/10$, and $|v-v_k| \le 1/2$, for any k different from $j, j+1$. The v_i's thus obtained have the required property. In fact, using the inequality (1), we obtain

$$\sum_{0\le i<j\le 4}(v_i-v_j)^2 \le 5\left(2\left(\frac{1}{10}\right)^2+3\left(\frac{1}{2}\right)^2\right)=3.85<4.$$

Remark. The best possible estimate for the right-hand side is 2.

8. (a) Consider

$$a_i = ik+1, \quad i=1,2,\ldots,m; \qquad b_j = jm+1, \quad j=1,2,\ldots,k.$$

Assume that $mk \mid a_ib_j - a_sb_t = (ik+1)(jm+1) - (sk+1)(tm+1) = km(ij-st) + m(j-t) + k(i-s)$. Since m divides this sum, we get that $m \mid k(i-s)$, or, together with $\gcd(k,m)=1$, that $i=s$. Similarly $j=t$, which proves part (a).

(b) Suppose the opposite, i.e., that all the residues are distinct. Then the residue 0 must also occur, say at a_1b_1: $mk \mid a_1b_1$; so, for some a' and b', $a' \mid a_1$, $b' \mid b_1$, and $a'b' = mk$. Assuming that for some $i,s \ne i$, $a' \mid a_i - a_s$, we obtain $mk = a'b' \mid a_ib_1 - a_sb_1$, a contradiction. This shows that $a' \ge m$ and similarly $b' \ge k$, and thus from $a'b' = mk$ we have $a' = m$, $b' = k$. We also get $(*)$: all a_i's give distinct residues modulo $m = a'$, and all b_j's give distinct residues modulo $k = b'$.

Now let p be a common prime divisor of m and k. By $(*)$, exactly $\frac{p-1}{p}m$ of a_i's and exactly $\frac{p-1}{p}k$ of b_j's are not divisible by p. Therefore there are precisely $\frac{(p-1)^2}{p^2}mk$ products a_ib_j that are not divisible by p, although from the assumption that they all give distinct residues it follows that the number of such products is $\frac{p-1}{p}mk \ne \frac{(p-1)^2}{p^2}mk$. We have arrived at a contradiction, thus proving (b).

9. The answer is yes. Consider the curve

$$C = \{(x,y,z) \mid x = t, y = t^3, z = t^5, \ t \in \mathbb{R}\}.$$

Any plane defined by an equation of the form $ax + by + cz + d = 0$ intersects the curve C at points (t, t^3, t^5) with t satisfying $ct^5 + bt^3 + at + d = 0$. This last equation has at least one but only finitely many solutions.

10. Denote by r, R (take w.l.o.g. $r < R$) the radii and by A, B the centers of the spheres S_1, S_2 respectively. Let s be the common radius of the spheres in the ring, C the center of one of them, say S, and D the foot of the perpendicular from C to AB. The centers of the spheres in the ring form a regular n-gon with center D, and thus $\sin(\pi/n) = s/CD$. Using the Heron's formula on the triangle ABC, we obtain $(r+R)^2 CD^2 = 4rRs(r+R+s)$, hence

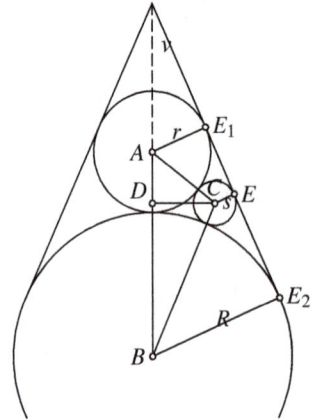

$$\sin^2 \frac{\pi}{n} = \frac{s^2}{CD^2} = \frac{(r+R)^2 s}{4(r+R+s)rR}. \tag{1}$$

Choosing the unit of length so that $r + R = 2$, for simplicity of writing, we write (1) as $1/\sin^2(\pi/n) = rR(1 + 2/s)$. Let now v be half the angle at the top of the cone. Then clearly $R - r = (R+r)\sin v = 2\sin v$, giving us $R = 1 + \sin v, r = 1 - \sin v$. It follows that

$$\frac{1}{\sin^2 \frac{\pi}{n}} = \left(1 + \frac{2}{s}\right) \cos^2 v. \tag{2}$$

We need to express s as a function of R and r. Let E_1, E_2, E be collinear points of tangency of S_1, S_2, and S with the cone. Obviously, $E_1 E_2 = E_1 E + E_2 E$, i.e., $2\sqrt{rs} + 2\sqrt{Rs} = 2\sqrt{Rr} = (R+r)\cos v = 2\cos v$. Hence,

$$\cos^2 v = s(\sqrt{R} + \sqrt{r})^2 = s(R + r + 2\sqrt{Rr}) = s(2 + 2\cos v).$$

Substituting this into (2), we obtain $2 + \cos v = 1/\sin(\pi/n)$. Therefore $1/3 < \sin(\pi/n) < 1/2$, and we conclude that the possible values for n are 7, 8, and 9.

11. Let A_1 be the set that contains 1, and let the minimal element of A_2 be less than that of A_3. We shall construct the partitions with required properties by allocating successively numbers to the subsets that always obey the rules. The number 1 must go to A_1; we show that for every subsequent number we have exactly two possibilities. Actually, while A_2 and A_3 are both empty, every successive number can enter either A_1 or A_2. Further, when A_2 is no longer empty, we use induction

on the number to be placed, denote it by m: if m can enter A_i or A_j but not A_k, and it enters A_i, then $m+1$ can be placed in A_i or A_k, but not in A_j. The induction step is finished. This immediately gives us that the final answer is 2^{n-1}.

12. Here all angles will be oriented and measured counterclockwise. Note that $\angle CA'B = \angle AB'C = \angle BC'A = \pi/3$. Let a',b',c' denote respectively the inner bisectors of angles A',B',C' in triangle $A'B'C'$. The lines a', b', c' meet at the centroid X of $A'B'C'$, and $\angle(a',b') = \angle(b',c') = \angle(c',a') = 2\pi/3$. Now let K,L,M be the points such that $KB = KC$, $LC = LA$, $MA = MB$, and $\angle BKC = \angle CLA = \angle AMB = 2\pi/3$, and let C_1, C_2, C_3 be the circles circumscribed about triangles BKC, CLA, and AMB respectively. These circles are characterized by $C_1 = \{Z \mid \angle BZC = 2\pi/3\}$, etc.; hence we deduce that they meet at a point P such that $\angle BPC = \angle CPA = \angle APB = 2\pi/3$ (Torricelli's point). Points A',B',C' run over $C_1 \smallsetminus \{P\}, C_2 \smallsetminus \{P\}, C_3 \smallsetminus \{P\}$ respectively. As for a',b',c', we see that $K \in a'$, $L \in b'$, $M \in c'$, and also that they can take all possible directions except KP, LP, MP respectively (if $K = P$, KP is assumed to be the corresponding tangent at K). Then, since $\angle KXL = 2\pi/3$, X runs over the circle defined by $\{Z \mid \angle KZL = 2\pi/3\}$, without P. But analogously, X runs over the circle $\{Z \mid \angle LZM = 2\pi/3\}$, from which we can conclude that these two circles are the same, both equal to the circumcircle of KLM, and consequently also that triangle KLM is equilateral (which is, anyway, a well-known fact). Therefore, the locus of the points X is the circumcircle of KLM minus point P.

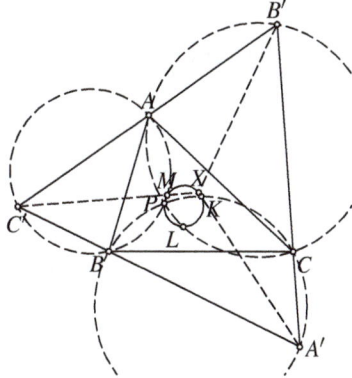

13. We claim that the points $P_i(i, i^2)$, $i = 1, 2, \ldots, 1987$, satisfy the conditions. In fact:
   (i) $\overline{P_iP_j} = \sqrt{(i-j)^2 + (i^2 - j^2)^2} = |i - j|\sqrt{1 + (i+j)^2}$.
   It is known that for each positive integer n, $\sqrt{n}$ is either an integer or an irrational number. Since $i + j < \sqrt{1 + (i+j)^2} < i + j + 1$, $\sqrt{1 + (i+j)^2}$ is not an integer, it is irrational, and so is $\overline{P_iP_j}$.
   (ii) The area A of the triangle $P_iP_jP_k$, for distinct i, j, k, is given by

$$A = \left| \frac{i^2 + j^2}{2}(i - j) + \frac{j^2 + k^2}{2}(j - k) + \frac{k^2 + i^2}{2}(k - i) \right|$$
$$= \left| \frac{(i-j)(j-k)(k-i)}{2} \right| \in \mathbb{Q} \smallsetminus \{0\},$$

also showing that this triangle is nondegenerate.

14. Let x_n be the total number of counted words of length n, and y_n, z_n, u_n, z_n, y_n the numbers of counted words of length n starting with $0, 1, 2, 3, 4$, respectively (indeed, by symmetry, words starting with 0 are equally numbered as those starting

with 4, etc.). We have the clear relations

$$(1)\ y_n = z_{n-1}; \qquad (2)\ z_n = y_{n-1} + u_{n-1};$$
$$(3)\ u_n = 2z_{n-1}; \qquad (4)\ x_n = 2y_n + 2z_n + u_n.$$

From (1), (2), and (3) we get $z_n = z_{n-2} + 2z_{n-2} = 3z_{n-2}$, with $z_1 = 1$, $z_2 = 2$, which gives

$$z_{2n} = 2 \cdot 3^{n-1}, \qquad z_{2n+1} = 3^n.$$

Then (1), (3), and (4) obviously imply

$$y_{2n} = 3^{n-1}, \qquad y_{2n+1} = 2 \cdot 3^{n-1};$$
$$u_{2n} = 2 \cdot 3^{n-1}, \qquad u_{2n+1} = 4 \cdot 3^{n-1};$$
$$x_{2n} = 8 \cdot 3^{n-1}, \qquad x_{2n+1} = 14 \cdot 3^{n-1};$$

with the initial number $x_1 = 5$.

15. Since $x_1^2 + x_2^2 + \cdots + x_n^2 = 1$, we get by the Cauchy-Schwarz inequality

$$|x_1| + |x_2| + \cdots + |x_n| \le \sqrt{n(x_1^2 + x_2^2 + \cdots + x_n^2)} = \sqrt{n}.$$

Hence all k^n sums of the form $e_1 x_1 + e_2 x_2 + \cdots + e_n x_n$, with $e_i \in \{0, 1, 2, \ldots, k-1\}$, must lie in some closed interval $\Im$ of length $(k-1)\sqrt{n}$. This interval can be covered with $k^n - 1$ closed subintervals of length $\frac{k-1}{k^n-1}\sqrt{n}$. By the pigeonhole principle there must be two of these sums lying in the same subinterval. Their difference, which is of the form $e_1 x_1 + e_2 x_2 + \cdots + e_n x_n$ where $e_i \in \{0, \pm 1, \ldots, \pm(k-1)\}$, satisfies

$$|e_1 x_1 + e_2 x_2 + \cdots + e_n x_n| \le \frac{(k-1)\sqrt{n}}{k^n - 1}.$$

16. We assume that $S = \{1, 2, \ldots, n\}$, and use the obvious fact

$$\sum_{k=0}^{n} p_n(k) = n! \tag{0}$$

(a) To each permutation π of S we assign an n-vector $(e_1, e_2, \ldots, e_n)$, where e_i is 1 if i is a fixed point of π, and 0 otherwise. Since exactly $p_n(k)$ of the assigned vectors contain exactly k "1"s, the considered sum $\sum_{k=0}^{n} k p_n(k)$ counts all the "1"s occurring in all the $n!$ assigned vectors. But for each i, $1 \le i \le n$, there are exactly $(n-1)!$ permutations that fix i; i.e., exactly $(n-1)!$ of the vectors have $e_i = 1$. Therefore the total number of "1"s is $n \cdot (n-1)! = n!$, implying

$$\sum_{k=0}^{n} k p_n(k) = n!. \tag{1}$$

(b) In this case, to each permutation π of S we assign a vector $(d_1,\ldots,d_n)$ instead, with $d_i = k$ if i is a fixed point of π, and $d_i = 0$ otherwise, where k is the number of fixed points of π.

Let us count the sum Z of all components d_i for all the $n!$ permutations. There are $p_n(k)$ such vectors with exactly k components equal to k, and sums of components equal to k^2. Thus, $Z = \sum_{k=0}^n k^2 p_n(k)$.

On the other hand, we may first calculate the sum of all components d_i for fixed i. In fact, the value $d_i = k > 0$ will occur exactly $p_{n-1}(k-1)$ times, so that the sum of the d_i's is $\sum_{k=1}^n k p_{n-1}(k-1) = \sum_{k=0}^{n-1}(k+1)p_{n-1}(k) = 2(n-1)!$. Summation over i yields

$$Z = \sum_{k=0}^n k^2 p_n(k) = 2n!. \qquad (2)$$

From (0), (1), and (2), we conclude that

$$\sum_{k=0}^n (k-1)^2 p_n(k) = \sum_{k=0}^n k^2 p_n(k) - 2\sum_{k=0}^n k p_n(k) + \sum_{k=0}^n p_n(k) = n!.$$

Remark. Only the first part of this problem was given on the IMO.

17. The number of 4-colorings of the set M is equal to 4^{1987}. Let A be the number of arithmetic progressions in M with 10 terms. The number of colorings containing a monochromatic arithmetic progression with 10 terms is less than $4A \cdot 4^{1977}$. So, if $A < 4^9$, then there exist 4-colorings with the required property.

Now we estimate the value of A. If the first term of a 10-term progression is k and the difference is d, then $1 \leq k \leq 1978$ and $d \leq \left[\frac{1987-k}{9}\right]$; hence

$$A = \sum_{k=1}^{1978}\left[\frac{1987-k}{9}\right] < \frac{1986+1985+\cdots+9}{9} = \frac{1995 \cdot 1978}{18} < 4^9.$$

18. Note first that the statement that some $a+x, a+y, a+x+y$ belong to a class C is equivalent to the following statement:

(1) There are positive integers $p, q \in C$ such that $p < q \leq 2p$.

Indeed, given p, q, take simply $x = y = q - p$, $a = 2p - q$; conversely, if a, x, y $(x \leq y)$ exist such that $a+x, a+y, a+x+y \in C$, take $p = a+y$, $q = a+x+y$: clearly, $p < q \leq 2p$.

We will show that $h(r) = 2r$. Let $\{1, 2, \ldots, 2r\} = C_1 \cup C_2 \cup \cdots \cup C_r$ be an arbitrary partition into r classes. By the pigeonhole principle, two among the $r+1$ numbers $r, r+1, \ldots, 2r$ belong to the same class, say $i, j \in C_k$. If w.l.o.g. $i < j$, then obviously $i < j \leq 2i$, and so by (1) this C_k has the required property. On the other hand, we consider the partition

$$\{1, 2, \ldots, 2r - t\} = \bigcup_{k=1}^{r-t}\{k, k+r\} \cup \{r-t+1\} \cup \cdots \cup \{r\}$$

and prove that (1), and thus also the required property, does not hold. In fact, none of the classes in the partition contains p and q with $p < q \leq 2p$, because $k + r > 2k$.

19. The facts given in the problem allow us to draw a triangular pyramid with angles $2\alpha, 2\beta, 2\gamma$ at the top and lateral edges of length $1/2$. At the base there is a triangle whose side lengths are exactly $\sin\alpha, \sin\beta, \sin\gamma$. The area of this triangle does not exceed the sum of areas of the lateral sides, which equals $(\sin 2\alpha + \sin 2\beta + \sin 2\gamma)/8$.

20. Let y be the smallest nonnegative integer with $y \leq p - 2$ for which $f(y)$ is a composite number. Denote by q the smallest prime divisor of $f(y)$. We claim that $y < q$.

    Suppose the contrary, that $y \geq q$. Let r be a positive integer such that $y \equiv r \pmod{q}$. Then $f(y) \equiv f(r) \equiv 0 \pmod{q}$, and since $q \leq y \leq p - 2 \leq f(r)$, we conclude that $q \mid f(r)$, which is a contradiction to the minimality of y.

    Now, we will prove that $q > 2y$. Suppose the contrary, that $q \leq 2y$. Since

    $$f(y) - f(x) = (y - x)(y + x + 1),$$

    we observe that $f(y) - f(q - 1 - y) = (2y - q + 1)q$, from which it follows that $f(q - 1 - y)$ is divisible by q. But by the assumptions, $q - 1 - y < y$, implying that $f(q - 1 - y)$ is prime and therefore equal to q. This is impossible, because

    $$f(q - 1 - y) = (q - 1 - y)^2 + (q - 1 - y) + p > q + p - y - 1 \geq q.$$

    Therefore $q \geq 2y + 1$. Now, since $f(y)$, being composite, cannot be equal to q, and q is its smallest prime divisor, we obtain that $f(y) \geq q^2$. Consequently,

    $$y^2 + y + p \geq q^2 \geq (2y + 1)^2 = 4y^2 + 4y + 1 \Rightarrow 3(y^2 + y) \leq p - 1,$$

    and from this we easily conclude that $y < \sqrt{p/3}$, which contradicts the condition of the problem. In this way, all the numbers

    $$f(0), f(1), \ldots, f(p - 2)$$

    must be prime.

21. Let P be the second point of intersection of segment BC and the circle circumscribed about quadrilateral $AKLM$. Denote by E the intersection point of the lines KN and BC and by F the intersection point of the lines MN and BC. Then $\angle BCN = \angle BAN$ and $\angle MAL = \angle MPL$, as angles on the same arc. Since AL is a bisector, $\angle BCN = \angle BAL = \angle MAL = \angle MPL$, and consequently $PM \parallel NC$. Similarly we prove $KP \parallel BN$. Then the quadrilaterals $BKPN$ and $NPMC$ are trapezoids; hence

$$S_{BKE} = S_{NPE} \quad \text{and} \quad S_{NPF} = S_{CMF}.$$

Therefore $S_{ABC} = S_{AKNM}$.

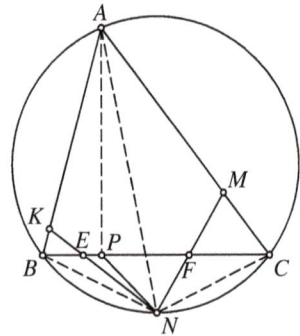

22. Suppose that there exists such function f. Then we obtain

$$f(n+1987) = f(f(f(n))) = f(n) + 1987 \quad \text{for all } n \in \mathbb{N},$$

and from here, by induction, $f(n+1987t) = f(n) + 1987t$ for all $n,t \in \mathbb{N}$.
Further, for any $r \in \{0, 1, \ldots, 1986\}$, let $f(r) = 1987k + l$, $k, l \in \mathbb{N}$, $l \leq 1986$. We have

$$r + 1987 = f(f(r)) = f(l + 1987k) = f(l) + 1987k,$$

and consequently there are two possibilities:
  (i) $k = 1 \Rightarrow f(r) = l + 1987$ and $f(l) = r$;
  (ii) $k = 0 \Rightarrow f(r) = l$ and $f(l) = r + 1987$;
in both cases, $r \neq l$. In this way, the set $\{0, 1, \ldots, 1986\}$ decomposes into pairs $\{a, b\}$ such that

$$f(a) = b \text{ and } f(b) = a + 1987, \quad \text{or} \quad f(b) = a \text{ and } f(a) = b + 1987.$$

But the set $\{0, 1, \ldots, 1986\}$ has an odd number of elements, and cannot be decomposed into pairs. Contradiction.

23. If we prove the existence of $p, q \in \mathbb{N}$ such that the roots r, s of

$$f(x) = x^2 - kp \cdot x + kq = 0$$

are irrational real numbers with $0 < s < 1$ (and consequently $r > 1$), then we are done, because from $r + s, rs \equiv 0 \pmod{k}$ we get $r^m + s^m \equiv 0 \pmod{k}$, and $0 < s^m < 1$ yields the assertion.
To prove the existence of such natural numbers p and q, we can take them such that $f(0) > 0 > f(1)$, i.e.,

$$kq > 0 > k(q-p) + 1 \quad \Rightarrow \quad p > q > 0.$$

The irrationality of r can be obtained by taking $q = p - 1$, because the discriminant $D = (kp)^2 - 4kp + 4k$, for $(kp-2)^2 < D < (kp-1)^2$, is not a perfect square for $p \geq 2$.

4.29 Solutions to the Shortlisted Problems of IMO 1988

1. Assume that p and q are real and $b_0, b_1, b_2, \dots$ is a sequence such that $b_n = pb_{n-1} + qb_{n-2}$ for all $n > 1$. From the equalities $b_n = pb_{n-1} + qb_{n-2}$, $b_{n+1} = pb_n + qb_{n-1}$, $b_{n+2} = pb_{n+1} + qb_n$, eliminating b_{n+1} and b_{n-1} we obtain that $b_{n+2} = (p^2 + 2q)b_n - q^2b_{n-2}$. So the sequence $b_0, b_2, b_4, \dots$ has the property

$$b_{2n} = Pb_{2n-2} + Qb_{2n-4}, \qquad P = p^2 + 2q, \quad Q = -q^2. \tag{1}$$

We shall solve the problem by induction. The sequence a_n has $p = 2$, $q = 1$, and hence $P = 6$, $Q = -1$.

Let $k = 1$. Then $a_0 = 0$, $a_1 = 1$, and a_n is of the same parity as a_{n-2}; i.e., it is even if and only if n is even.

Let $k \geq 1$. We assume that for $n = 2^k m$, the numbers a_n are divisible by 2^k, but divisible by 2^{k+1} if and only if m is even. We assume also that the sequence $c_0, c_1, \dots$, with $c_m = a_{m \cdot 2^k}$, satisfies the condition $c_n = pc_{n-1} - c_{n-2}$, where $p \equiv 2 \pmod 4$ (for $k = 1$ it is true). We shall prove the same statement for $k + 1$. According to (1), $c_{2n} = Pc_{2n-2} - c_{2n-4}$, where $P = p^2 - 2$. Obviously $P \equiv 2 \pmod 4$. Since $P = 4s + 2$ for some integer s, and $c_{2n} = 2^{k+1}d_{2n}$, $c_0 = 0$, $c_1 \equiv 2^k \pmod{2^{k+1}}$, and $c_2 = pc_1 \equiv 2^{k+1} \pmod{2^{k+2}}$, we have

$$c_{2n} = (4s+2)2^{k+1}d_{2n-2} - c_{2n-4} \equiv c_{2n-4} \pmod{2^{k+2}},$$

i.e., $0 \equiv c_0 \equiv c_4 \equiv c_8 \equiv \cdots$ and $2^{k+1} \equiv c_2 \equiv c_6 \equiv \cdots \pmod{2^{k+2}}$, which proves the statement.

Second solution. The recursion is solved by

$$a_n = \frac{1}{2\sqrt 2}\left((1+\sqrt 2)^n - (1-\sqrt 2)^n\right) = \binom{n}{1} + 2\binom{n}{3} + 2^2\binom{n}{5} + \cdots.$$

Let $n = 2^k m$ with m odd; then for $p > 0$ the summand

$$2^p \binom{n}{2p+1} = 2^{k+p} m \frac{(n-1)\dots(n-2p)}{(2p+1)!} = 2^{k+p}\frac{m}{2p+1}\binom{n-1}{2p}$$

is divisible by 2^{k+p}, because the denominator $2p+1$ is odd. Hence

$$a_n = n + \sum_{p>0} 2^p \binom{n}{2p+1} = 2^k m + 2^{k+1} N$$

for some integer N, so that a_n is exactly divisible by 2^k.

Third solution. It can be proven by induction that $a_{2n} = 2a_n(a_n + a_{n+1})$. The required result follows easily, again by induction on k.

2. For polynomials $f(x), g(x)$ with integer coefficients, we use the notation $f(x) \sim g(x)$ if all the coefficients of $f - g$ are even. Let $n = 2^s$. It is immediately shown by induction that $(x^2 + x + 1)^{2^s} \sim x^{2^{s+1}} + x^{2^s} + 1$, and the required number for $n = 2^s$ is 3.

Let $n = 2^s - 1$. If s is odd, then $n \equiv 1 \pmod 3$, while for s even, $n \equiv 0 \pmod 3$. Consider the polynomial

$$R_s(x) = \begin{cases} (x+1)(x^{2n-1}+x^{2n-4}+\cdots+x^{n+3})+x^{n+1} \\ +x^n+x^{n-1}+(x+1)(x^{n-4}+x^{n-7}+\cdots+1), & 2\nmid s; \\ \\ (x+1)(x^{2n-1}+x^{2n-4}+\cdots+x^{n+2})+x^n \\ +(x+1)(x^{n-3}+x^{n-6}+\cdots+1), & 2\mid s. \end{cases}$$

It is easily checked that $(x^2+x+1)R_s(x) \sim x^{2^{s+1}}+x^{2^s}+1 \sim (x^2+x+1)^{2^s}$, so that $R_s(x) \sim (x^2+x+1)^{2^s-1}$. In this case, the number of odd coefficients is $(2^{s+2}-(-1)^s)/3$.

Now we pass to the general case. Let the number n be represented in the binary system as

$$n = \underbrace{11\ldots1}_{a_k}\underbrace{00\ldots0}_{b_k}\underbrace{11\ldots1}_{a_{k-1}}\underbrace{00\ldots0}_{b_{k-1}}\ldots\underbrace{11\ldots1}_{a_1}\underbrace{00\ldots0}_{b_1},$$

$b_i > 0 \ (i > 1)$, $b_1 \geq 0$, and $a_i > 0$. Then $n = \sum_{i=1}^k 2^{s_i}(2^{a_i}-1)$, where $s_i = b_1 + a_1 + b_2 + a_2 + \cdots + b_i$, and hence

$$u_n(x) = (x^2+x+1)^n = \prod_{i=1}^k (x^2+x+1)^{2^{s_i}(2^{a_i}-1)} \sim \prod_{i=1}^k R_{a_i}(x^{2^{s_i}}).$$

Let $R_{a_i}(x^{2^{s_i}}) \sim x^{r_{i,1}}+\cdots+x^{r_{i,d_i}}$; clearly $r_{i,j}$ is divisible by 2^{s_i} and $r_{i,j} \leq 2^{s_i+1}(2^{a_i}-1) < 2^{s_i+1}$, so that for any j, $r_{i,j}$ can have nonzero binary digits only in some position t, $s_i \leq t \leq s_{i+1}-1$. Therefore, in

$$\prod_{i=1}^k R_{a_i}(x^{2^{s_i}}) \sim \prod_{i=1}^k (x^{r_{i,1}}+\cdots+x^{r_{i,d_i}}) = \sum_{i=1}^k \sum_{p_i=1}^{d_i} x^{r_{1,p_1}+r_{2,p_2}+\cdots+r_{k,p_k}}$$

all the exponents $r_{1,p_1}+r_{2,p_2}+\cdots+r_{k,p_k}$ are different, so that the number of odd coefficients in $u_n(x)$ is

$$\prod_{i=1}^k d_i = \prod_{i=1}^k \frac{2^{a_i+2}-(-1)^{a_i}}{3}.$$

3. Let R be the circumradius, r the inradius, s the semiperimeter, Δ the area of ABC and Δ' the area of $A'B'C'$. The angles of triangle $A'B'C'$ are $A' = 90° - A/2$, $B' = 90° - B/2$, and $C' = 90° - C/2$, and hence

$$\Delta = 2R^2 \sin A \sin B \sin C$$

and $$\Delta' = 2R^2 \sin A' \sin B' \sin C' = 2R^2 \cos\frac{A}{2}\cos\frac{B}{2}\cos\frac{C}{2}.$$

Hence,

$$\frac{\Delta}{\Delta'} = \frac{\sin A \sin B \sin C}{\cos\frac{A}{2}\cos\frac{B}{2}\cos\frac{C}{2}} = 8\sin\frac{A}{2}\sin\frac{B}{2}\sin\frac{C}{2} = \frac{2r}{R}.$$

Here we used that $r = AI\sin(A/2) = \cdots = 4R\sin(A/2)\cdot\sin(B/2)\cdot\sin(C/2)$. Euler's inequality $2r \leq R$ shows that $\Delta \leq \Delta'$.

Second solution. Let H be orthocenter of triangle ABC, and H_a, H_b, H_c points symmetric to H with respect to BC, CA, AB, respectively. Since $\angle BH_aC =$

$\angle BHC = 180° - \angle A$, points H_a, H_b, H_c lie on the circumcircle of ABC, and the area of the hexagon $AH_cBH_aCH_b$ is double the area of ABC. (1)

Let us apply the analogous result for the triangle $A'B'C'$. Since its orthocenter is the incenter I of ABC, and the point symmetric to I with respect to $B'C'$ is the point A, we find by (1) that the area of the hexagon $AC'BA'CB'$ is double the area of $A'B'C'$.

But it is clear that the area of $\triangle CH_aB$ is less than or equal to the area of $\triangle CA'B$ etc.; hence, the area of $AH_cBH_aCH_b$ does not exceed the area of $AC'BA'CB'$. The statement follows immediately.

4. Suppose that the numbers of any two neighboring squares differ by at most $n-1$. For $k = 1, 2, \ldots, n^2 - n$, let A_k, B_k, and C_k denote, respectively, the sets of squares numbered by $1, 2, \ldots, k$; of squares numbered by $k+n, k+n+1, \ldots, n^2$; and of squares numbered by $k+1, \ldots, k+n-1$. By the assumption, the squares from A_k and B_k have no edge in common; C_k has $n-1$ elements only. Consequently, for each k there exists a row and a column all belonging either to A_k, or to B_k. For $k = 1$, it must belong to B_k, while for $k = n^2 - n$ it belongs to A_k. Let k be the smallest index such that A_k contains a whole row and a whole column. Since B_{k-1} has that property too, it must have at least two squares in common with A_k, which is impossible.

5. Let $n = 2k$ and let $A = \{A_1, \ldots, A_{2k+1}\}$ denote the family of sets with the desired properties. Since every element of their union B belongs to at least two sets of A, it follows that $A_j = \bigcup_{i \neq j} A_i \cap A_j$ holds for every $1 \leq j \leq 2k+1$. Since each intersection in the sum has at most one element and A_j has $2k$ elements, it follows that every element of A_j, i.e., in general of B, is a member of exactly two sets.

We now prove that k is even, assuming that the marking described in the problem exists. We have already shown that for every two indices $1 \leq j \leq 2k+1$ and $i \neq j$ there exists a unique element contained in both A_i and A_j. On a $2k \times 2k$ matrix let us mark in the ith column and jth row for $i \neq j$ the number that was joined to the element of B in $A_i \cap A_j$. In the ith row and column let us mark the number of the element of B in $A_i \cap A_{2k+1}$. In each row from the conditions of the marking there must be an even number of zeros. Hence, the total number of zeros in the matrix is even. The matrix is symmetric with respect to its main diagonal; hence it has an even number of zeros outside its main diagonal. Hence, the number of zeros on the main diagonal must also be even and this number equals the number of elements in A_{2k+1} that are marked with 0, which is k. Hence k must be even.

For even k we note that the dimensions of a $2k \times 2k$ matrix are divisible by 4. Tiling the entire matrix with the 4×4 submatrix

$$Q = \begin{bmatrix} 0 & 1 & 0 & 1 \\ 1 & 0 & 1 & 0 \\ 0 & 1 & 1 & 0 \\ 1 & 0 & 0 & 1 \end{bmatrix},$$

we obtain a marking that indeed satisfies all the conditions of the problem; hence we have shown that the marking is possible if and only if k is even.

6. Let ω be the plane through AB, parallel to CD. Define the point transformation $f : X \mapsto X'$ in space as follows. If $X \in KL$, then $X' = X$; otherwise, let ω_X be the plane through X parallel to ω: then X' is the point symmetric to X with respect to the intersection point of KL with ω_X. Clearly, $f(A) = B$, $f(B) = A$, $f(C) = D$, $f(D) = C$; hence f maps the tetrahedron onto itself.

We shall show that f preserves volumes. Let $s : X \mapsto X''$ denote the symmetry with respect to KL, and g the transformation mapping X'' into X'; then $f = g \circ s$. If points $X_1'' = s(X_1)$ and $X_2'' = s(X_2)$ have the property that $X_1'' X_2''$ is parallel to KL, then the segments $X_1'' X_2''$ and $X_1' X_2'$ have the same length and lie on the same line. Then by Cavalieri's principle g preserves volume, and so does f.

Now, if α is any plane containing the line KL, the two parts of the tetrahedron on which it is partitioned by α are transformed into each other by f, and therefore have the same volumes.

Second solution. Suppose w.l.o.g. that the plane α through KL meets the interiors of edges AC and BD at X and Y. Let $\overrightarrow{AX} = \lambda \overrightarrow{AC}$ and $\overrightarrow{BY} = \mu \overrightarrow{BD}$, for $0 \le \lambda, \mu \le 1$. Then the vectors $\overrightarrow{KX} = \lambda \overrightarrow{AC} - \overrightarrow{AB}/2$, $\overrightarrow{KY} = \mu \overrightarrow{BD} + \overrightarrow{AB}/2$, $\overrightarrow{KL} = \overrightarrow{AC}/2 + \overrightarrow{BD}/2$ are coplanar; i.e., there exist real numbers a, b, c, not all zero, such that

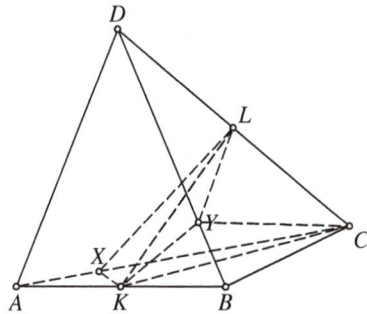

$$\overrightarrow{0} = a\overrightarrow{KX} + b\overrightarrow{KY} + c\overrightarrow{KL} = (\lambda a + c/2)\overrightarrow{AC} + (\mu b + c/2)\overrightarrow{BD} + \frac{b-a}{2}\overrightarrow{AB}.$$

Since $\overrightarrow{AC}, \overrightarrow{BD}, \overrightarrow{AB}$ are linearly independent, we must have $a = b$ and $\lambda = \mu$. We need to prove that the volume of the polyhedron $KXLYBC$, which is one of the parts of the tetrahedron $ABCD$ partitioned by α, equals half of the volume V of $ABCD$. Indeed, we obtain

$$V_{KXLYBC} = V_{KXLC} + V_{KBYLC} = \frac{1}{4}(1-\lambda)V + \frac{1}{4}(1+\mu)V = \frac{1}{2}V.$$

7. The algebraic equation $x^3 - 3x^2 + 1 = 0$ admits three real roots β, γ, a, with

$$-0.6 < \beta < -0.5, \quad 0.6 < \gamma < 0.7, \quad \sqrt{8} < a < 3.$$

Define, for all integers n,

$$u_n = \beta^n + \gamma^n + a^n.$$

It holds that $u_{n+3} = 3u_{n+2} - u_n$.

Obviously, $0 < \beta^n + \gamma^n < 1$ for all $n \ge 2$, and we see that $u_n - 1 = [a^n]$ for $n \ge 2$. It is now a question whether $u_{1788} - 1$ and $u_{1988} - 1$ are divisible by 17. Working modulo 17, we get $u_0 \equiv 3$, $u_1 \equiv 3$, $u_2 \equiv 9$, $u_3 \equiv 7$, $u_4 \equiv 1, \ldots, u_{16} \equiv 3$, $u_{17} \equiv 3$, $u_{18} \equiv 9$. Thus, u_n is periodic modulo 17, with period 16. Since $1788 =$

$16 \cdot 111 + 12$, $1988 = 16 \cdot 124 + 4$, it follows that $u_{1788} \equiv u_{12} \equiv 1$ and $u_{1988} \equiv u_4 = 1$. So, $[a^{1788}]$ and $[a^{1988}]$ are divisible by 17.

Second solution. The polynomial $x^3 - 3x^2 + 1$ allows the factorization modulo 17 as $(x-4)(x-5)(x+6)$. Hence it is easily seen that $u_n \equiv 4^n + 5^n + (-6)^n$. Fermat's theorem gives us $4^n \equiv 5^n \equiv (-6)^n \equiv 1$ for $16 \mid n$, and the rest follows easily.

Remark. In fact, the roots of $x^3 - 3x^2 + 1 = 0$ are $\frac{1}{2\sin 10°}$, $\frac{1}{2\sin 50°}$, and $-\frac{1}{2\sin 70°}$.

8. Consider first the case that the vectors are on the same line. Then if e is a unit vector, we can write $u_1 = x_1 e, \ldots, u_n = x_n e$ for scalars x_i, $|x_i| \leq 1$, with zero sum. It is now easy to permute $x_1, x_2, \ldots, x_n$ into $z_1, z_2, \ldots z_n$ so that $|z_1| \leq 1, |z_1 + z_2| \leq 1, \ldots, |z_1 + z_2 + \cdots + z_{n-1}| \leq 1$. Indeed, suppose w.l.o.g. that $z_1 = x_1 \geq 0$; then we choose $z_2, \ldots, z_r$ from the x_i's to be negative, until we get to the first r with $x_1 + x_2 + \cdots + x_r \leq 0$; we continue successively choosing positive z_j's from the remaining x_i's until we get the first partial sum that is positive, and so on. It is easy to verify that $|z_1 + z_2 + \cdots + z_j| \leq 1$ for all $j = 1, 2, \ldots, n$.

Now we pass to the general case. Let s be the longest vector that can be obtained by summing a subset of $u_1, \ldots, u_m$, and assume w.l.o.g. that $s = u_1 + \cdots + u_p$. Further, let δ and δ' respectively be the lines through the origin O in the direction of s and perpendicular to s, and e, e' respectively the unit vectors on δ and δ'. Put $u_i = x_i e + y_i e'$, $i = 1, 2, \ldots, m$. By the definition of δ and δ', we have $|x_i|, |y_i| \leq 1$; $x_1 + \cdots + x_m = y_1 + \cdots + y_m = 0$; $y_1 + \cdots + y_p = y_{p+1} + \cdots + y_m = 0$; we also have $x_{p+1}, \ldots, x_m \leq 0$ (otherwise, if $x_i > 0$ for some i, then $|s + v_i| > |s|$), and similarly $x_1, \ldots, x_p \geq 0$. Finally, suppose by the one-dimensional case that $y_1, \ldots, y_p$ and $y_{p+1}, \ldots, y_m$ are permuted in such a way that all the sums $y_1 + \cdots + y_i$ and $y_{p+1} + \cdots + y_{p+i}$ are ≤ 1 in absolute value.

We apply the construction of the one-dimensional case to $x_1, \ldots, x_m$ taking, as described above, positive z_i's from $x_1, x_2, \ldots, x_p$ and negative ones from $x_{p+1}, \ldots, x_m$, but so that the order is preserved; this way we get a permutation $x_{\sigma_1}, x_{\sigma_2}, \ldots, x_{\sigma_m}$. It is then clear that each sum $y_{\sigma_1} + y_{\sigma_2} + \cdots + y_{\sigma_k}$ decomposes into the sum $(y_1 + y_2 + \cdots + y_l) + (y_{p+1} + \cdots + y_{p+n})$ (because of the preservation of order), and that each of these sums is less than or equal to 1 in absolute value. Thus each sum $u_{\sigma_1} + \cdots + u_{\sigma_k}$ is composed of a vector of length at most 2 and an orthogonal vector of length at most 1, and so is itself of length at most $\sqrt{5}$.

9. Let us assume $\frac{a^2 + b^2}{ab + 1} = k \in \mathbb{N}$. We then have $a^2 - kab + b^2 = k$. Let us assume that k is not an integer square, which implies $k \geq 2$. Now we observe the minimal pair (a, b) such that $a^2 - kab + b^2 = k$ holds. We may assume w.l.o.g. that $a \geq b$. For $a = b$ we get $k = (2-k)a^2 \leq 0$; hence we must have $a > b$.

Let us observe the quadratic equation $x^2 - kbx + b^2 - k = 0$, which has solutions a and a_1. Since $a + a_1 = kb$, it follows that $a_1 \in \mathbb{Z}$. Since $a > kb$ implies $k > a + b^2 > kb$ and $a = kb$ implies $k = b^2$, it follows that $a < kb$ and thus $b^2 > k$. Since $aa_1 = b^2 - k > 0$ and $a > 0$, it follows that $a_1 \in \mathbb{N}$ and $a_1 = \frac{b^2 - k}{a} < \frac{a^2 - 1}{a} < a$. We have thus found an integer pair (a_1, b) with $0 < a_1 < a$ that satisfies the original equation. This is a contradiction of the initial assumption that (a,b) is minimal. Hence k must be an integer square.

10. We claim that if the family $\{A_1,\ldots,A_t\}$ separates the n-set N, then $2^t \geq n$. The proof goes by induction. The case $t = 1$ is clear, so suppose that the claim holds for $t - 1$. Since A_t does not separate elements of its own or its complement, it follows that $\{A_1,\ldots,A_{t-1}\}$ is separating for both A_t and $N \smallsetminus A_t$, so that $|A_t|$, $|N \smallsetminus A_t| \leq 2^{t-1}$. Then $|N| \leq 2 \cdot 2^{t-1} = 2^t$, as claimed.
Also, if the set N with $N = 2^t$ is separated by $\{A_1,\ldots,A_t\}$, then (precisely) one element of N is not covered. To show this, we again use induction. This is trivial for $t = 1$, so let $t \geq 1$. Since $A_1,\ldots,A_{t-1}$ separate both A_t and $N \smallsetminus A_t$, $N \smallsetminus A_t$ must have exactly 2^{t-1} elements, and thus one of its elements is not covered by $A_1,\ldots,A_{t-1}$, and neither is covered by A_t. We conclude that a separating and covering family of t subsets can exist only if $n \leq 2^t - 1$.
We now construct such subsets for the set N if $2^{t-1} \leq n \leq 2^t - 1, t \geq 1$. For $t = 1$, put $A_1 = \{1\}$. In the step from t to $t+1$, let $N = N' \cup N'' \cup \{y\}$, where $|N'|, |N''| \leq 2^{t-1}$; let $A_1',\ldots,A_t'$ be subsets covering and separating N' and $A_1'',\ldots,A_t''$ such subsets for N''. Then the subsets $A_i = A_i' \cup A_i''$ $(i = 1,\ldots,t)$ and $A_{t+1} = N'' \cup \{y\}$ obviously separate and cover N.
The answer: $t = [\log_2 n] + 1$.

Second solution. Suppose that the sets $A_1,\ldots,A_t$ cover and separate N. Label each element $x \in N$ with a string $(x_1 x_2 \ldots x_t)$ of 0's and 1's, where x_i is 1 when $x \in A_i$, 0 otherwise. Since the A_i's separate, these strings are distinct; since they cover, the string $(00\ldots0)$ does not occur. Hence $n \leq 2^t - 1$. Conversely, for $2^{t-1} \leq n < 2^t$, represent the elements of N in base 2 as strings of 0's and 1's of length t. For $1 \leq i \leq t$, take A_i to be the set of numbers in N whose binary string has a 1 in the ith place. These sets clearly cover and separate.

11. The answer is 32. Write the combinations as triples $k = (x,y,z)$, $0 \leq x,y,z \leq 7$. Define the sets $K_1 = \{(1,0,0), (0,1,0), (0,0,1), (1,1,1)\}$, $K_2 = \{(2,0,0), (0,2,0), (0,0,2), (2,2,2)\}$, $K_3 = \{(0,0,0),(4,4,4)\}$, and $K = \{k = k_1 + k_2 + k_3 \mid k_i \in K_i, i = 1,2,3\}$. There are 32 combinations in K. We shall prove that these combinations will open the safe in every case.
Let $t = (a,b,c)$ be the right combination. Set $k_3 = (0,0,0)$ if at least two of a,b,c are less than 4, and $k_3 = (4,4,4)$ otherwise. In either case, the difference $t - k_3$ contains two nonnegative elements not greater than 3. Choosing a suitable k_2 we can achieve that $t - k_3 - k_2$ contains two elements that are 0, 1. So, there exists k_1 such that $t - k_3 - k_2 - k_1 = t - k$ contains two zeros, for $k \in K$. This proves that 32 is sufficient.
Suppose that K is a set of at most 31 combinations. We say that $k \in K$ *covers* the combination k_1 if k and k_1 differ in at most one position. One of the eight sets $M_i = \{(i,y,z) \mid 0 \leq y,z \leq 7\}, i = 0,1,\ldots,7$, contains at most three elements of K. Suppose w.l.o.g. that this is M_0. Further, among the eight sets $N_j = \{(0,j,z) \mid 0 \leq z \leq 7\}, j = 0,\ldots,7$, there are at least five, say w.l.o.g. $N_0,\ldots,N_4$, not containing any of the combinations from K.

Of the 40 elements of the set $N = \{(0,y,z) \mid 0 \le y \le 4,\ 0 \le z \le 7\}$, at most $5 \cdot 3 = 15$ are covered by $K \cap M_0$, and at least 25 aren't. Consequently, the intersection of K with $L = \{(x,y,z) \mid 1 \le x \le 7, 0 \le y \le 4, 0 \le z \le 7\}$ contains at least 25 elements. So K has at most $31 - 25 = 6$ elements in the set $P = \{(x,y,z) \mid 0 \le x \le 7, 5 \le y \le 7, 0 \le z \le 7\}$. This implies that for some $j \in \{5,6,7\}$, say w.l.o.g. $j = 7$, K contains at most two elements in $Q_j = \{(x,y,z) \mid 0 \le x,z \le 7,\ y = j\}$; denote them by l_1, l_2. Of the 64 elements of Q_7, at most 30 are covered by l_1 and l_2. But then there remain 34 uncovered elements, which must be covered by different elements of $K \setminus Q_7$, having itself at most 29 elements. Contradiction.

12. Let $E(XYZ)$ stand for the area of a triangle XYZ. We have

$$\frac{E_1}{E} = \frac{E(AMR)}{E(AMK)} \cdot \frac{E(AMK)}{E(ABK)} \cdot \frac{E(ABK)}{E(ABC)} = \frac{MR}{MK} \cdot \frac{AM}{AB} \cdot \frac{BK}{BC} \quad \Rightarrow$$

$$\left(\frac{E_1}{E}\right)^{1/3} \le \frac{1}{3}\left(\frac{MR}{MK} + \frac{AM}{AB} + \frac{BK}{BC}\right).$$

We similarly obtain

$$\left(\frac{E_2}{E}\right)^{1/3} \le \frac{1}{3}\left(\frac{KR}{MK} + \frac{BM}{AB} + \frac{CK}{BC}\right).$$

Therefore $(E_1/E)^{1/3} + (E_2/E)^{1/3} \le 1$, i.e., $\sqrt[3]{E_1} + \sqrt[3]{E_2} \le \sqrt[3]{E}$. Analogously, $\sqrt[3]{E_3} + \sqrt[3]{E_4} \le \sqrt[3]{E}$ and $\sqrt[3]{E_5} + \sqrt[3]{E_6} \le \sqrt[3]{E}$; hence

$$8\sqrt[6]{E_1 E_2 E_3 E_4 E_5 E_6}$$
$$= 2(\sqrt[3]{E_1}\sqrt[3]{E_2})^{1/2} \cdot 2(\sqrt[3]{E_3}\sqrt[3]{E_4})^{1/2} \cdot 2(\sqrt[3]{E_5}\sqrt[3]{E_6})^{1/2}$$
$$\le (\sqrt[3]{E_1} + \sqrt[3]{E_2}) \cdot (\sqrt[3]{E_3} + \sqrt[3]{E_4}) \cdot (\sqrt[3]{E_5} + \sqrt[3]{E_6}) \le E.$$

13. Let $AB = c$, $AC = b$, $\angle CBA = \beta$, $BC = a$, and $AD = h$.
Let r_1 and r_2 be the inradii of ABD and ADC respectively and O_1 and O_2 the centers of the respective incircles. It is obvious that $r_1/r_2 = c/b$. We also have $DO_1 = \sqrt{2}r_1$, $DO_2 = \sqrt{2}r_2$, and $\angle O_1DA = \angle O_2DA = 45°$. Hence $\angle O_1DO_2 = 90°$ and $DO_1/DO_2 = c/b$ from which it follows that $\triangle O_1DO_2 \sim \triangle BAC$. We now define P as the intersection of the circumcircle of $\triangle O_1DO_2$ with DA. From the above similarity we have $\angle DPO_2 = \angle DO_1O_2 = \beta = \angle DAC$. It follows that $PO_2 \parallel AC$ and from $\angle O_1PO_2 = 90°$ it also follows that $PO_1 \parallel AB$. We also have $\angle PO_1O_2 = \angle PO_2O_1 = 45°$; hence $\angle LKA = \angle KLA = 45°$, and thus $AK = AL$. From $\angle O_1KA = \angle O_1DA = 45°$, $O_1A = O_1A$, and $\angle O_1KA = \angle O_1DA$ we have $\triangle O_1KA \cong \triangle O_1DA$ and hence $AL = AK = AD = h$. Thus

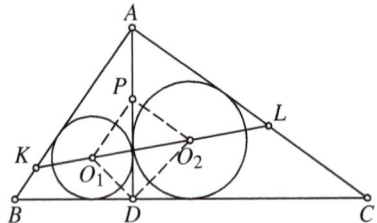

$$\frac{E}{E_1} = \frac{ah/2}{h^2/2} = \frac{a}{h} = \frac{a^2}{ah} = \frac{b^2+c^2}{bc} \geq 2 \,.$$

Remark. It holds that for an arbitrary triangle ABC, $AK = AL$ if and only if $AB = AC$ or $\angle BAC = 90°$.

14. Consider an array $[a_{ij}]$ of the given property and denote the sums of the rows and the columns by r_i and c_j respectively. Among the r_i's and c_j's, one element of $[-n, n]$ is missing, so that there are at least n nonnegative and n nonpositive sums. By permuting rows and columns we can obtain an array in which $r_1, \ldots, r_k$ and $c_1, \ldots, c_{n-k}$ are nonnegative. Clearly

$$\sum_{i=1}^{n} |r_i| + \sum_{j=1}^{n} |c_j| \geq \sum_{r=-n}^{n} |r| - n = n^2.$$

But on the other hand,

$$\sum_{i=1}^{n} |r_i| + \sum_{j=1}^{n} |c_j| = \sum_{i=1}^{k} r_i - \sum_{i=k+1}^{n} r_i + \sum_{j=1}^{n-k} c_j - \sum_{j=n-k+1}^{n} c_j =$$
$$= \sum_{i \leq k} a_{ij} - \sum_{i > k} a_{ij} + \sum_{j \leq n-k} a_{ij} - \sum_{j > n-k} a_{ij} =$$
$$= 2 \sum_{i=1}^{k} \sum_{j=1}^{n-k} a_{ij} - 2 \sum_{i=k+1}^{n} \sum_{j=n-k+1}^{n} a_{ij} \leq 4k(n-k).$$

This yields $n^2 \leq 4k(n-k)$, i.e., $(n-2k)^2 \leq 0$, and thus n must be even.
We proceed to show by induction that for all even n an array of the given type exists. For $n = 2$ the array in Fig. 1 is good. Let such an $n \times n$ array be given for some even $n \geq 2$, with $c_1 = n$, $c_2 = -n+1$, $c_3 = n-2, \ldots, c_{n-1} = 2$, $c_n = -1$ and $r_1 = n-1$, $r_2 = -n+2, \ldots, r_{n-1} = 1$, $r_n = 0$. Upon enlarging this array as indicated in Fig. 2, the positive sums are increased by 2, the nonpositive sums are decreased by 2, and the missing sums $-1, 0, 1, 2$ occur in the new rows and columns, so that the obtained array $(n+2) \times (n+2)$ is of the same type.

Fig. 1

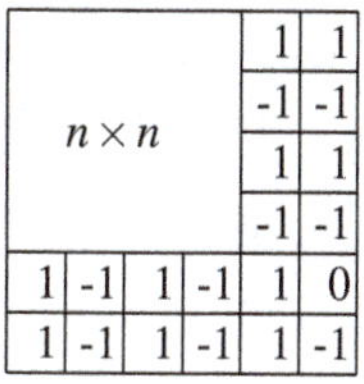

Fig. 2

15. Referring to the description of L_A, we have $\angle AMN = \angle AHN = 90° - \angle HAC = \angle C$, and similarly $\angle ANM = \angle B$. Since the triangle ABC is acute-angled, the line L_A lies inside the angle A. Hence if $P = L_A \cap BC$ and $Q = L_B \cap AC$, we get $\angle BAP = 90° - \angle C$; hence AP passes through the circumcenter O of $\triangle ABC$.

Similarly we prove that L_B and L_C contains the circumcenter O also. It follows that L_A, L_B and L_C intersect at the point O.

Remark. Without identifying the point of intersection, one can prove the concurrence of the three lines using Ceva's theorem, in usual or trigonometric form.

16. Let $f(x) = \sum_{k=1}^{70} \frac{k}{x-k}$. For all integers $i = 1, \ldots, 70$ we have that $f(x)$ tends to plus infinity as x tends downward to i, and $f(x)$ tends to minus infinity as x tends upward to i. As x tends to infinity, $f(x)$ tends to 0. Hence it follows that there exist $x_1, x_2, \ldots, x_{70}$ such that $1 < x_1 < 2 < x_2 < 3 < \cdots < x_{69} < 70 < x_{70}$ and $f(x_i) = \frac{5}{4}$ for all $i = 1, \ldots, 70$. Then the solution to the inequality is given by $S = \bigcup_{i=1}^{70} (i, x_i]$.

For numbers x for which $f(x)$ is well-defined, the equality $f(x) = \frac{5}{4}$ is equivalent to

$$p(x) = \prod_{j=1}^{70}(x-j) - \frac{4}{5}\sum_{k=1}^{70} k \prod_{\substack{j=1 \\ j \neq k}}^{70}(x-j) = 0.$$

The numbers $x_1, x_2, \ldots, x_{70}$ are then the zeros of this polynomial. The sum $\sum_{i=1}^{70} x_i$ is then equal to minus the coefficient of x^{69} in p, which is $\sum_{i=1}^{70} \left(i + \frac{4}{5}i\right)$. Finally,

$$|S| = \sum_{i=1}^{70}(x_i - i) = \frac{4}{5} \cdot \sum_{i=1}^{70} i = \frac{4}{5} \cdot \frac{70 \cdot 71}{2} = 1988.$$

17. Let AC and AD meet BE in R, S, respectively. Then by the conditions of the problem,

$$\angle AEB = \angle EBD = \angle BDC = \angle DBC = \angle ADB = \angle EAD = \alpha,$$
$$\angle ABE = \angle BEC = \angle ECD = \angle CED = \angle ACE = \angle BAC = \beta,$$
$$\angle BCA = \angle CAD = \angle ADE = \gamma.$$

Since $\angle SAE = \angle SEA$, it follows that $AS = SE$, and analogously $BR = RA$. But $BSDC$ and $REDC$ are parallelograms; hence $BS = CD = RE$, giving us $BR = SE$ and $AR = AS$. Then also $AC = AD$, because $RS \parallel CD$. We deduce that $2\beta = \angle ACD = \angle ADC = 2\alpha$, i.e., $\alpha = \beta$.

It will be sufficient to show that $\alpha = \gamma$, since that will imply $\alpha = \beta = \gamma = 36°$. We have that the sum of the interior angles of ACD is $4\alpha + \gamma = 180°$. We have

$$\frac{\sin \gamma}{\sin \alpha} = \frac{AE}{DE} = \frac{AE}{CD} = \frac{AE}{RE} = \frac{\sin(2\alpha + \gamma)}{\sin(\alpha + \gamma)},$$

i.e., $\cos\alpha - \cos(\alpha + 2\gamma) = 2\sin\gamma\sin(\alpha + \gamma) = 2\sin\alpha\sin(2\alpha + \gamma) = \cos(\alpha + \gamma) - \cos(3\alpha + \gamma)$. From $4\alpha + \gamma = 180°$ we obtain $-\cos(3\alpha + \gamma) = \cos\alpha$. Hence

$$\cos(\alpha + \gamma) + \cos(\alpha + 2\gamma) = 2\cos\frac{\gamma}{2}\cos\frac{2\alpha + 3\gamma}{2} = 0,$$

so that $2\alpha + 3\gamma = 180°$. It follows that $\alpha = \gamma$.

Second solution. We have $\angle BEC = \angle ECD = \angle DEC = \angle ECA = \angle CAB$, and hence the trapezoid $BAEC$ is cyclic; consequently, $AE = BC$. Similarly $AB = ED$, and $ABCD$ is cyclic as well. Thus $ABCDE$ is cyclic and has all sides equal; i.e., it is regular.

18. (a) Define $\angle APO = \phi$ and $S = AB^2 + AC^2 + BC^2$. We calculate $PA = 2r\cos\phi$ and $PB, PC = \sqrt{R^2 - r^2\cos^2\phi} \pm r\sin\phi$. We also have $AB^2 = PA^2 + PB^2$, $AC^2 = PA^2 + PC^2$ and $BC = BP + PC$. Combining all these we obtain

$$\begin{aligned} S &= AB^2 + AC^2 + BC^2 = 2(PA^2 + PB^2 + PC^2 + PB \cdot PC) \\ &= 2(4r^2\cos^2\phi + 2(R^2 - r^2\cos^2\phi + r^2\sin^2\phi) + R^2 - r^2) \\ &= 6R^2 + 2r^2. \end{aligned}$$

Hence it follows that S is constant; i.e., it does not depend on ϕ.

   (b) Let B_1 and C_1 respectively be points such that $APBB_1$ and $APCC_1$ are rectangles. It is evident that B_1 and C_1 lie on the larger circle and that $\overrightarrow{PU} = \frac{1}{2}\overrightarrow{PB_1}$ and $\overrightarrow{PV} = \frac{1}{2}\overrightarrow{PC_1}$. It is evident that we can arrange for an arbitrary point on the larger circle to be B_1 or C_1. Hence, the locus of U and V is equal to the circle obtained when the larger circle is shrunk by a factor of $1/2$ with respect to point P.

19. We will show that $f(n) = n$ for every n (thus also $f(1988) = 1988$).

   Let $f(1) = r$ and $f(2) = s$. We obtain respectively the following equalities: $f(2r) = f(r + r) = 2$; $f(2s) = f(s + s) = 4$; $f(4) = f(2 + 2) = 4r$; $f(8) = f(4 + 4) = 4s$; $f(5r) = f(4r + r) = 5$; $f(r + s) = 3$; $f(8) = f(5 + 3) = 6r + s$. Then $4s = 6r + s$, which means that $s = 2r$.

   Now we prove by induction that $f(nr) = n$ and $f(n) = nr$ for every $n \geq 4$. First we have that $f(5) = f(2 + 3) = 3r + s = 5r$, so that the statement is true for $n = 4$ and $n = 5$. Suppose that it holds for $n - 1$ and n. Then $f(n + 1) = f(n - 1 + 2) = (n - 1)r + 2r = (n + 1)r$, and $f((n + 1)r) = f((n - 1)r + 2r) = (n - 1) + 2 = n + 1$. This completes the induction.

   Since $4r \geq 4$, we have that $f(4r) = 4r^2$, and also $f(4r) = 4$. Then $r = 1$, and consequently $f(n) = n$ for every natural number n.

   Second solution. $f(f(1) + n + m) = f(f(1) + f(f(n) + f(m))) = 1 + f(n) + f(m)$, so $f(n) + f(m)$ is a function of $n + m$. Hence $f(n + 1) + f(1) = f(n) + f(2)$ and $f(n + 1) - f(n) = f(2) - f(1)$, implying that $f(n) = An + B$ for some constants A, B. It is easy to check that $A = 1$, $B = 0$ is the only possibility.

20. Suppose that $A_n = \{1, 2, \ldots, n\}$ is partitioned into B_n and C_n, and that neither B_n nor C_n contains 3 distinct numbers one of which is equal to the product of the other two. If $n \geq 96$, then the divisors of 96 must be split up. Let w.l.o.g. $2 \in B_n$. There are four cases.

   (i) $3 \in B_n, 4 \in B_n$. Then $6, 8, 12 \in C_n \Rightarrow 48, 96 \in B_n$. A contradiction for $96 = 2 \cdot 48$.

   (ii) $3 \in B_n, 4 \in C_n$. Then $6 \in C_n$, $24 \in B_n$, $8, 12, 48 \in C_n$. A contradiction for $48 = 6 \cdot 8$.

(iii) $3 \in C_n, 4 \in B_n$. Then $8 \in C_n$, $24 \in B_n$, $6, 48 \in C_n$. A contradiction for $48 = 6 \cdot 8$.

(iv) $3 \in C_n, 4 \in C_n$. Then $12 \in B_n$, $6, 24 \in C_n$. A contradiction for $24 = 4 \cdot 6$.

If $n = 95$, there is a very large number of ways of partitioning A_n. For example, $B_n = \{1, p, p^2, p^3 q^2, p^4 q, p^2 qr \mid p, q, r$ are distinct primes$\}$, $C_n = \{p^3, p^4, p^5, p^6, pq, p^2 q, p^3 q, p^2 q^2, pqr \mid p, q, r$ are distinct primes$\}$. Then $B_{95} = \{1, 2, 3, 4, 5, 7, 9, 11, 13, 17, 19, 23, 25, 29, 31, 37, 41, 43, 47, 48, 49, 53, 59, 60, 61, 67, 71, 72, 73, 79, 80, 83, 84, 89, 90\}$.

21. Let X be the set of all ordered triples $a = (a_1, a_2, a_3)$ for $a_i \in \{0, 1, \ldots, 7\}$. Write $a \prec b$ if $a_i \le b_i$ for $i = 1, 2, 3$ and $a \ne b$. Call a subset $Y \subset X$ *independent* if there are no $a, b \in Y$ with $a \prec b$. We shall prove that an independent set contains at most 48 elements.

For $j = 0, 1, \ldots, 21$ let $X_j = \{(a_1, a_2, a_3) \in X \mid a_1 + a_2 + a_3 = j\}$. If $x \prec y$ and $x \in X_j$, $y \in X_{j+1}$ for some j, then we say that y is a *successor* of x, and x a *predecessor* of y.

    Lemma. If A is an m-element subset of X_j and $j \le 10$, then there are at least m distinct successors of the elements of A.

    Proof. For $k = 0, 1, 2, 3$ let $X_{j,k} = \{(a_1, a_2, a_3) \in X_j \mid \min(a_1, a_2, a_3, 7 - a_1, 7 - a_2, 7 - a_3) = k\}$. It is easy to verify that every element of $X_{j,k}$ has at least two successors in $X_{j+1,k}$ and every element of $X_{j+1,k}$ has at most two predecessors in $X_{j,k}$. Therefore the number of elements of $A \cap X_{j,k}$ is not greater than the number of their successors. Since X_j is a disjoint union of $X_{j,k}$, $k = 0, 1, 2, 3$, the lemma follows.

Similarly, elements of an m-element subset of X_j, $j \ge 11$, have at least m predecessors.

Let Y be an independent set, and let p, q be integers such that $p < 10 < q$. We can transform Y by replacing all the elements of $Y \cap X_p$ with their successors, and all the elements of $Y \cap X_q$ with their predecessors. After this transformation Y will still be independent, and by the lemma its size will not be reduced. Every independent set can be eventually transformed in this way into a subset of X_{10}, and X_{10} has exactly 48 elements.

22. Set $X = \sum_{i=1}^{p} x_i$ and w.l.o.g. assume that $X \ge 0$ (if $(x_1, \ldots, x_p)$ is a solution, then $(-x_1, \ldots, -x_p)$ is a solution too). Since $x^2 \ge x$ for all integers x, it follows that $\sum_{i=1}^{p} x_i^2 \ge X$.

If the last inequality is an equality, then all x_i's are 0 or 1; then, taking that there are a 1's, the equation becomes $4p + 1 = 4(a + 1) + \frac{4}{a-1}$, which forces $p = 6$ and $a = 5$.

Otherwise, we have $X + 1 \le \sum_{i=1}^{p} x_i^2 = \frac{4}{4p+1} X^2 + 1$, so $X \ge p + 1$. Also, by the Cauchy–Schwarz inequality, $X^2 \le p \sum_{i=1}^{p} x_i^2 = \frac{4p}{4p+1} X^2 + p$, so $X^2 \le 4p^2 + p$ and $X \le 2p$. Thus $1 \le X/p \le 2$. However,

$$\sum_{i=1}^{p}\left(x_i-\frac{X}{p}\right)^2 = \sum x_i^2 - \frac{2X}{p}\sum x_i + \frac{X^2}{p}$$

$$= \sum x_i^2 - p\frac{X^2}{p^2} = 1 - \frac{X^2}{p(4p+1)} < 1,$$

and we deduce that $-1 < x_i - X/p < 1$ for all i. This finally gives $x_i \in \{1,2\}$. Suppose there are b 2's. Then $3b + p = 4(b+p)^2/(4p+1) + 1$, so $p = b + 1/(4b-3)$, which leads to $p = 2, b = 1$.

Thus there are no solutions for any $p \notin \{2,6\}$.

Remark. The condition $p = n(n+1)$, $n \geq 3$, was unnecessary in the official solution, too (its only role was to simplify showing that $X \neq p-1$).

23. Denote by R the intersection point of lines AQ and BC. We know that $BR : RC = c : b$ and $AQ : QR = (b+c) : a$. By applying Stewart's theorem to $\triangle PBC$ and $\triangle PAR$ we obtain

$$a \cdot AP^2 + b \cdot BP^2 + c \cdot CP^2 = aPA^2 + (b+c)PR^2 + (b+c)RB \cdot RC$$
$$= (a+b+c)QP^2 + (b+c)RB \cdot RC + (a+b+c)QA \cdot QR. \tag{1}$$

On the other hand, putting $P = Q$ into (1), we get that

$$a \cdot AQ^2 + b \cdot BQ^2 + c \cdot CQ^2 = (b+c)RB \cdot RC + (a+b+c)QA \cdot QR,$$

and the required statement follows.

Second solution. At vertices A, B, C place weights equal to a, b, c in some units respectively, so that Q is the center of gravity of the system. The left side of the equality to be proved is in fact the moment of inertia of the system about the axis through P and perpendicular to the plane ABC. On the other side, the right side expresses the same, due to the parallel axes theorem.

Alternative approach. Analytical geometry. The fact that all the variable segments appear squared usually implies that this is a good approach. Assign coordinates $A(x_a, y_a)$, $B(x_b, y_b)$, $C(x_c, y_c)$, and $P(x, y)$, use that $(a+b+c)\mathbf{Q} = a\mathbf{A} + b\mathbf{B} + c\mathbf{C}$, and calculate. Alternatively, differentiate $f(x,y) = a \cdot AP^2 + b \cdot BP^2 + c \cdot CP^2 - (a+b+c)QP^2$ and show that it is constant.

24. The first condition means in fact that $a_k - a_{k+1}$ is decreasing. In particular, if $a_k - a_{k+1} = -\delta < 0$, then $a_k - a_{k+m} = (a_k - a_{k+1}) + \cdots + (a_{k+m-1} - a_{k+m}) < -m\delta$, which implies that $a_{k+m} > a_k + m\delta$, and consequently $a_{k+m} > 1$ for large enough m, a contradiction. Thus $a_k - a_{k+1} \geq 0$ for all k.

Suppose that $a_k - a_{k+1} > 2/k^2$. Then for all $i < k$, $a_i - a_{i+1} > 2/k^2$, so that $a_i - a_{k+1} > 2(k+1-i)/k^2$, i.e., $a_i > 2(k+1-i)/k^2$, $i = 1, 2, \ldots, k$. But this implies $a_1 + a_2 + \cdots + a_k > 2/k^2 + 4/k^2 + \cdots + 2k/k^2 = k(k+1)/k^2$, which is impossible. Therefore $a_k - a_{k+1} \leq 2/k^2$ for all k.

25. Observe that $1001 = 7 \cdot 143$, i.e., $10^3 = -1 + 7a$, $a = 143$. Then by the binomial theorem, $10^{21} = (-1+7a)^7 = -1 + 7^2b$ for some integer b, so that we also have

$10^{21n} \equiv -1 \pmod{49}$ for any odd integer $n > 0$. Hence $N = \frac{9}{49}(10^{21n}+1)$ is an integer of $21n$ digits, and $N(10^{21n}+1) = \left(\frac{3}{7}(10^{21n}+1)\right)^2$ is a double number that is a perfect square.

26. In the sequel, $\overline{a_1 a_2 \ldots a_\alpha}$ will be used to representation a number whose binary digits are $a_1, \ldots, a_\alpha$. We will show by induction that if $n = \overline{c_k c_{k-1} \ldots c_0} = \sum_{i=0}^{k} c_i 2^i$ is the binary representation of n ($c_i \in \{0,1\}$), then $f(n) = \overline{c_0 c_1 \ldots c_k} = \sum_{i=0}^{k} c_i 2^{k-i}$ is the number whose binary representation is the palindrome of the binary representation of n. This evidently holds for $n \in \{1,2,3\}$. Let us assume that the claim holds for all numbers up to $n-1$ and show it holds for $n = \overline{c_k c_{k-1} \ldots c_0}$. We observe three cases:
   (i) $c_0 = 0 \Rightarrow n = 2m \Rightarrow f(n) = f(m) = \overline{0 c_1 \ldots c_k} = \overline{c_0 c_1 \ldots c_k}$.
   (ii) $c_0 = 1, c_1 = 0 \Rightarrow n = 4m+1 \Rightarrow f(n) = 2f(2m+1) - f(m) = 2 \cdot \overline{1 c_2 \ldots c_k} - \overline{c_2 \ldots c_k} = 2^k + 2 \cdot \overline{c_2 \ldots c_k} - \overline{c_2 \ldots c_k} = \overline{10 c_2 \ldots c_k} = \overline{c_0 c_1 \ldots c_k}$.
   (iii) $c_0 = 1, c_1 = 1 \Rightarrow n = 4m+3 \Rightarrow f(n) = 3f(2m+1) - 2f(m) = 3 \cdot \overline{1 c_2 \ldots c_k} - 2 \cdot \overline{c_2 \ldots c_k} = 2^k + 2^{k-1} + 3 \cdot \overline{c_2 \ldots c_k} - 2 \cdot \overline{c_2 \ldots c_k} = \overline{11 c_2 \ldots c_k} = \overline{c_0 c_1 \ldots c_k}$.
We thus have to find the number of palindromes in binary representation smaller than $1988 = \overline{11111000100}$. We note that for all $m \in \mathbb{N}$ the numbers of $2m$- and $(2m-1)$-digit binary palindromes are both equal to 2^{m-1}. We also note that $\overline{11111011111}$ and $\overline{11111111111}$ are the only 11-digit palindromes larger than 1988. Hence we count all palindromes of up to 11 digits and exclude the largest two. The number of $n \leq 1988$ such that $f(n) = n$ is thus equal to $1+1+2+2+4+4+8+8+16+16+32-2 = 92$.

27. Consider a Cartesian system with the x-axis on the line BC and origin at the foot of the perpendicular from A to BC, so that A lies on the y-axis. Let A be $(0,\alpha)$, $B(-\beta,0)$, $C(\gamma,0)$, where $\alpha, \beta, \gamma > 0$ (because ABC is acute-angled). Then

$$\tan B = \frac{\alpha}{\beta}, \quad \tan C = \frac{\alpha}{\gamma} \quad \text{and} \quad \tan A = -\tan(B+C) = \frac{\alpha(\beta+\gamma)}{\alpha^2 - \beta\gamma};$$

here $\tan A > 0$, so $\alpha^2 > \beta\gamma$. Let L have equation $x\cos\theta + y\sin\theta + p = 0$. Then

$$u^2 \tan A + v^2 \tan B + w^2 \tan C$$
$$= \frac{\alpha(\beta+\gamma)}{\alpha^2 - \beta\gamma}(\alpha\sin\theta + p)^2 + \frac{\alpha}{\beta}(-\beta\cos\theta + p)^2 + \frac{\alpha}{\gamma}(\gamma\cos\theta + p)^2$$
$$= (\alpha^2\sin^2\theta + 2\alpha p\sin\theta + p^2)\frac{\alpha(\beta+\gamma)}{\alpha^2 - \beta\gamma} + \alpha(\beta+\gamma)\cos^2\theta + \frac{\alpha(\beta+\gamma)}{\beta\gamma}p^2$$
$$= \frac{\alpha(\beta+\gamma)}{\beta\gamma(\alpha^2 - \beta\gamma)}(\alpha^2 p^2 + 2\alpha p\beta\gamma\sin\theta + \alpha^2\beta\gamma\sin^2\theta + \beta\gamma(\alpha^2 - \beta\gamma)\cos^2\theta)$$
$$= \frac{\alpha(\beta+\gamma)}{\beta\gamma(\alpha^2 - \beta\gamma)}\left[(\alpha p + \beta\gamma\sin\theta)^2 + \beta\gamma(\alpha^2 - \beta\gamma)\right] \geq \alpha(\beta+\gamma) = 2\Delta,$$

with equality when $\alpha p + \beta\gamma\sin\theta = 0$, i.e., if and only if L passes through $(0, \beta\gamma/\alpha)$, which is the orthocenter of the triangle.

28. The sequence is uniquely determined by the conditions, and $a_1 = 2$, $a_2 = 7$, $a_3 = 25$, $a_4 = 89$, $a_5 = 317,\ldots$; it satisfies $a_n = 3a_{n-1} + 2a_{n-2}$ for $n = 3, 4, 5$. We show that the sequence b_n given by $b_1 = 2$, $b_2 = 7$, $b_n = 3b_{n-1} + 2b_{n-2}$ has the same inequality property, i.e., that $b_n = a_n$:

$$b_{n+1}b_{n-1} - b_n^2 = (3b_n + 2b_{n-1})b_{n-1} - b_n(3b_{n-1} + 2b_{n-2}) = -2(b_nb_{n-2} - b_{n-1}^2)$$

for $n > 2$ gives that $b_{n+1}b_{n-1} - b_n^2 = (-2)^{n-2}$ for all $n \geq 2$. But then

$$\left| b_{n+1} - \frac{b_n^2}{b_{n-1}} \right| = \frac{2^{n-2}}{b_{n-1}} < \frac{1}{2},$$

since it is easily shown that $b_{n-1} > 2^{n-1}$ for all n. It is obvious that $a_n = b_n$ are odd for $n > 1$.

29. Let the first train start from Signal 1 at time 0, and let t_j be the time it takes for the jth train in the series to travel from one signal to the next. By induction on k, we show that Train k arrives at signal n at time $s_k + (n-2)m_k$, where $s_k = t_1 + \cdots + t_k$ and $m_k = \max_{j=1,\ldots,k} t_j$.

For $k = 1$ the statement is clear. We now suppose that it is true for k trains and for every n, and add a $(k+1)$th train behind the others at Signal 1. There are two cases to consider:

(i) $t_{k+1} \geq m_k$, i.e., $m_{k+1} = t_{k+1}$. Then Train $k+1$ leaves Signal 1 when all the others reach Signal 2, which by the induction happens at time s_k. Since by the induction hypothesis Train k arrives at Signal $i+1$ at time $s_k + (i-1)m_k \leq s_k + (i-1)t_{k+1}$, Train $k+1$ is never forced to stop. The journey finishes at time $s_k + (n-1)t_{k+1} = s_{k+1} + (n-2)m_{k+1}$.

(ii) $t_{k+1} < m_k$, i.e., $m_{k+1} = m_k$. Train $k+1$ leaves Signal 1 at time s_k, and reaches Signal 2 at time $s_k + t_{k+1}$, but must wait there until all the other trains get to Signal 3, i.e., until time $s_k + m_k$ (by the induction hypothesis). So it reaches Signal 3 only at time $s_k + m_k + t_{k+1}$. Similarly, it gets to Signal 4 at time $s_k + 2m_k + t_{k+1}$, etc. Thus the entire schedule finishes at time $s_k + (n-2)m_k + t_{k+1} = s_{k+1} + (n-2)m_{k+1}$.

30. Let Δ_1, s_1, r' denote the area, semiperimeter, and inradius of triangle ABM, Δ_2, s_2, r' the same quantities for triangle MBC, and Δ, s, r those for $\triangle ABC$. Also, let P' and Q' be the points of tangency of the incircle of $\triangle ABM$ with the side AB and of the incircle of $\triangle MBC$ with the side BC, respectively, and let P, Q be the points of tangency of the incircle of $\triangle ABC$ with the sides AB, BC. We have $\Delta_1 = s_1 r'$, $\Delta_2 = s_2 r'$, $\Delta = sr$, so that $sr = (s_1 + s_2)r'$. Then

$$s_1 + s_2 = s + BM \quad \Rightarrow \quad \frac{r'}{r} = \frac{s}{s + BM}. \qquad (1)$$

On the other hand, from the similarity of the triangles it follows that $AP'/AP = CQ'/CQ = r'/r$. By a well-known formula we find that $AP = s - BC$, $CQ = s - AB$, $AP' = s_1 - BM$, $CQ' = s_2 - BM$, and therefore deduce that

$$\frac{r'}{r} = \frac{s_1 - BM}{s - BC} = \frac{s_2 - BM}{s - AB} \Rightarrow \frac{r'}{r} = \frac{s_1 + s_2 - 2BM}{2s - AB - BC} = \frac{s - BM}{AC}. \tag{2}$$

It follows from (1) and (2) that $(s - BM)/AC = s/(s + BM)$, giving us $s^2 - BM^2 = s \cdot AC$. Finally,

$$BM^2 = s(s - AC) = s \cdot BP = s \cdot r\cot\frac{B}{2} = \Delta\cot\frac{B}{2}.$$

31. Denote the number of participants by $2n$, and assign to each seat one of the numbers $1, 2, \ldots, 2n$. Let the participant who was sitting at the seat k before the break move to seat $\pi(k)$. It suffices to prove that for every permutation π of the set $\{1, 2, \ldots, 2n\}$, there exist distinct i, j such that $\pi(i) - \pi(j) = \pm(i - j)$, the differences being calculated modulo $2n$.

If there are distinct i and j such that $\pi(i) - i = \pi(j) - j$ modulo $2n$, then we are done.

Suppose that all the differences $\pi(i) - i$ are distinct modulo $2n$. Then they take values $0, 1, \ldots, 2n - 1$ in some order, and consequently

$$\sum_{i=1}^{2n}(\pi(i) - i) = 0 + 1 + \cdots + (2n - 1) \equiv n(2n - 1) \pmod{2n}.$$

On the other hand, $\sum_{i=1}^{2n}(\pi(i) - i) = \sum\pi(i) - \sum i = 0$, which is a contradiction because $n(2n - 1)$ is not divisible by $2n$.

Remark. For an odd number of participants, the statement is false. For example, the permutation $(a, 2a, \ldots, (2n + 1)a)$ of $(1, 2, \ldots, 2n + 1)$ modulo $2n + 1$ does not satisfy the statement when $\gcd(a^2 - 1, 2n + 1) = 1$. Check that such an a always exists.

4.30 Solutions to the Shortlisted Problems of IMO 1989

1. Let I denote the intersection of the three internal bisectors. Then $IA_1 = A_1A^0$.
   One way to prove this is to real-
   ize that the circumcircle of $\triangle ABC$
   is the nine-point circle of $\triangle A^0B^0C^0$,
   so it bisects IA^0, since I is the orthocenter of $A^0B^0C^0$. Another way
   is through noting that $IA_1 = A_1B$.
   This is a consequence of $\angle A_1IB = \angle IBA_1 = (\angle A + \angle B)/2$, and $A_1B = A_1A^0$ which follows from $\angle A_1A^0B = \angle A_1BA^0 = 90° - \angle IBA_1$. Hence, we
   obtain $S_{IA_1B} = S_{A^0A_1B}$. Repeating this
   argument for the six triangles that have a vertex at I and adding them up gives
   us $S_{A^0B^0C^0} = 2S_{AC_1BA_1CB_1}$. To prove $S_{AC_1BA_1CB_1} \geq 2S_{ABC}$, draw the three altitudes
   in triangle ABC intersecting in H. Let X, Y, and Z be the symmetric points of H
   with respect the sides BC, CA, and AB, respectively. Then X, Y, Z are points on
   the circumcircle of $\triangle ABC$ (because $\angle BXC = \angle BHC = 180° - \angle A$). Since A_1 is
   the midpoint of the arc BC, we have $S_{BA_1C} \geq S_{BXC}$. Hence

$$S_{AC_1BA_1CB_1} \geq S_{AZBXCY} = 2(S_{BHC} + S_{CHA} + S_{AHB}) = 2S_{ABC}.$$

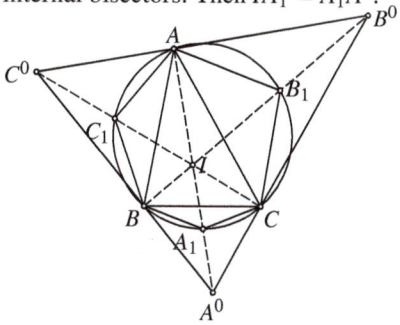

2. Let the carpet have width x, length y. Suppose that the carpet $EFGH$ lies in a
   room $ABCD$, E being on AB, F on BC, G on CD, and H on DA. Then $\triangle AEH \equiv \triangle CGF \sim \triangle BFE \equiv \triangle DHG$. Let $\frac{y}{x} = k$, $AE = a$ and $AH = b$. In that case $BE = kb$
   and $DH = ka$.
   Thus $a + kb = 50$, $ka + b = 55$, whence $a = \frac{55k - 50}{k^2 - 1}$ and $b = \frac{50k - 55}{k^2 - 1}$. Hence $x^2 = a^2 + b^2 = \frac{5525k^2 - 11000k + 5525}{(k^2 - 1)^2}$, i.e.,

$$x^2(k^2 - 1)^2 = 5525k^2 - 11000k + 5525.$$

Similarly, from the equations for the second storeroom, we get

$$x^2(k^2 - 1)^2 = 4469k^2 - 8360k + 4469.$$

Combining the two equations, we get $5525k^2 - 11000k + 5525 = 4469k^2 - 8360k + 4469$, which implies $k = 2$ or $1/2$. Without loss of generality we have
$y = 2x$ and $a + 2b = 50$, $2a + b = 55$; hence $a = 20$, $b = 15$, $x = \sqrt{15^2 + 20^2} = 25$,
and $y = 50$. We have thus shown that the carpet is 25 feet by 50 feet.

3. Let the carpet have width x, length y. Let the length of the storerooms be q.
   Let $y/x = k$. Then, as in the previous problem, $(kq - 50)^2 + (50k - q)^2 = (kq - 38)^2 + (38k - q)^2$, i.e.,

$$kq = 22(k^2 + 1). \tag{1}$$

Also, as before, $x^2 = \left(\frac{kq - 50}{k^2 - 1}\right)^2 + \left(\frac{50k - q}{k^2 - 1}\right)^2$, i.e.,

$$x^2(q^2 - 1)^2 = (k^2 + 1)(q^2 - 1900), \tag{2}$$

which, together with (1), yields

$$x^2 k^2 (k^2 - 1)^2 = (k^2 + 1)(484k^4 - 932k^2 + 484).$$

Since k is rational, let $k = c/d$, where c and d are integers with $\gcd(c,d) = 1$. Then we obtain

$$x^2 c^2 (c^2 - d^2)^2 = c^2 (484c^4 - 448c^2 d^2 - 448d^4) + 484d^6.$$

We thus have $c^2 \mid 484d^6$, but since $(c,d) = 1$, we have $c^2 \mid 484 \Rightarrow c \mid 22$. Analogously, $d \mid 22$; thus $k = 1, 2, 11, 22, \frac{1}{2}, \frac{1}{11}, \frac{1}{22}, \frac{2}{11}, \frac{11}{2}$. Since reciprocals lead to the same solution, we need only consider $k \in \{1, 2, 11, 22, \frac{11}{2}\}$, yielding $q = 44, 55, 244, 485, 125$, respectively. We can test these values by substituting them into (2). Only $k = 2$ gives us an integer solution, namely $x = 25, y = 50$.

4. First we note that for every integer $k > 0$ and prime number p, p^k doesn't divide $k!$. This follows from the fact that the highest exponent r of p for which $p^r \mid k!$ is

$$r = \left[\frac{k}{p}\right] + \left[\frac{k}{p^2}\right] + \cdots < \frac{k}{p} + \frac{k}{p^2} + \cdots = \frac{k}{p-1} < k.$$

Now suppose that α is a rational root of the given equation. Then

$$\alpha^n + \frac{n!}{(n-1)!}\alpha^{n-1} + \cdots + \frac{n!}{2!}\alpha^2 + \frac{n!}{1!}\alpha + n! = 0, \tag{1}$$

from which we can conclude that α must be an integer, not equal to ± 1. Let p be a prime divisor of n and let r be the highest exponent of p for which $p^r \mid n!$. Then $p \mid \alpha$. Since $p^k \mid \alpha^k$ and $p^k \nmid k!$, we obtain that $p^{r+1} \mid n!\alpha^k/k!$ for $k = 1, 2, \ldots, n$. But then it follows from (1) that $p^{r+1} \mid n!$, a contradiction.

5. According to the Cauchy–Schwarz inequality,

$$\left(\sum_{i=1}^n a_i\right)^2 \leq \left(\sum_{i=1}^n a_i^2\right)\left(\sum_{i=1}^n 1^2\right) = n\left(\sum_{i=1}^n a_i^2\right).$$

Since $r_1 + \cdots + r_n = -n$, applying this inequality we obtain $r_1^2 + \ldots + r_n^2 \geq n$, and applying it three more times, we obtain

$$r_1^{16} + \cdots + r_n^{16} \geq n,$$

with equality if and only if $r_1 = r_2 = \ldots = r_n = -1$ and $p(x) = (x+1)^n$.

6. Let us denote the measures of the inner angles of the triangle ABC by α, β, γ. Then $P = r^2(\sin 2\alpha + \sin 2\beta + \sin 2\gamma)/2$. Since the inner angles of the triangle $A'B'C'$ are $(\beta + \gamma)/2, (\gamma + \alpha)/2, (\alpha + \beta)/2$, we also have $Q = r^2[\sin(\beta + \gamma) +$

$\sin(\gamma+\alpha)+\sin(\alpha+\beta)]/2$. Applying the AM–GM mean inequality, we now obtain

$$16Q^3 = \frac{16}{8}r^6(\sin(\beta+\gamma)+\sin(\gamma+\alpha)+\sin(\alpha+\beta))^3$$
$$\geq 54r^6\sin(\beta+\gamma)\sin(\gamma+\alpha)\sin(\alpha+\beta)$$
$$= 27r^6[\cos(\alpha-\beta)-\cos(\alpha+\beta+2\gamma)]\sin(\alpha+\beta)$$
$$= 27r^6[\cos(\alpha-\beta)+\cos\gamma]\sin(\alpha+\beta)$$
$$= \frac{27}{2}r^6[\sin(\alpha+\beta+\gamma)+\sin(\alpha+\beta-\gamma)+\sin2\alpha+\sin2\beta]$$
$$= \frac{27}{2}r^6[\sin(2\gamma)+\sin2\alpha+\sin2\beta] = 27r^4P.$$

This completes the proof.

7. Assume that P_1 and P_2 are points inside E, and that the line P_1P_2 intersects the perimeter of E at Q_1 and Q_2. If we prove the statement for Q_1 and Q_2, we are done, since these arcs can be mapped homothetically to join P_1 and P_2.
   Let V_1,V_2 be two vertices of E. Then applying two homotheties to the inscribed circle of E one can find two arcs (one of them may be a side of E) joining these two points, both tangent to the sides of E that meet at V_1 and V_2. If A is any point of the side V_2V_3, two homotheties with center V_1 take the arcs joining V_1 to V_2 and V_3 into arcs joining V_1 to A; their angle of incidence at A remains $(1-2/n)\pi$.
   Next, for two arbitrary points Q_1 and Q_2 on two different sides V_1V_2 and V_3V_4, we join V_1 and V_2 to Q_2 with pairs of arcs that meet at Q_2 and have an angle of incidence $(1-2/n)\pi$. The two arcs that meet the line Q_1Q_2 again outside E meet at Q_2 at an angle greater than or equal to $(1-2/n)\pi$. Two homotheties with center Q_2 carry these arcs to ones meeting also at Q_1 with the same angle of incidence.

8. Let A,B,C,D denote the vertices of R. We consider the set $\mathscr{S}$ of all points E of the plane that are vertices of at least one rectangle, and its subset $\mathscr{S}'$ consisting of those points in $\mathscr{S}$ that have both coordinates integral in the orthonormal coordinate system with point A as the origin and lines AB,AD as axes.
   First, to each $E \in \mathscr{S}$ we can assign an integer n_E as the number of rectangles R_i with one vertex at E. It is easy to check that $n_E = 1$ if E is one of the vertices A,B,C,D; in all other cases n_E is either 2 or 4.
   Furthermore, for each rectangle R_i we define $f(R_i)$ as the number of vertices of R_i that belong to $\mathscr{S}'$. Since every R_i has at least one side of integer length, $f(R_i)$ can take only values 0, 2, or 4. Therefore we have

$$\sum_{i=1}^n f(R_i) \equiv 0 \pmod 2.$$

On the other hand, $\sum_{i=1}^n f(R_i)$ is equal to $\sum_{E\in\mathscr{S}'} n_E$, implying that

$$\sum_{E\in\mathscr{S}'} n_E \equiv 0 \pmod 2.$$

However, since $n_A = 1$, at least one other n_E, where $E \in \mathscr{S}'$, must be odd, and that can happen only for E being $B, C,$ or D. We conclude that at least one of the sides of R has integral length.

Second solution. Consider the coordinate system introduced above. If D is a rectangle whose sides are parallel to the axes of the system, it is easy to prove that

$$\int_D \sin(2\pi(x+\alpha))\sin(2\pi(y+\beta))\,dxdy = 0, \quad \text{for all } \alpha, \beta \in \mathbb{R}$$

if and only if at least one side of D has integral length. This holds for all R_i's therefore, adding up these equalities for each α and β we get

$$\int_R \sin(2\pi(x+\alpha))\sin(2\pi(y+\beta))\,dxdy = 0.$$

Thus, R also has a side of integral length.

9. From $a_{n+1} + b_{n+1}\sqrt[3]{2} + c_{n+1}\sqrt[3]{4} = (a_n + b_n\sqrt[3]{2} + c_n\sqrt[3]{4})(1 + 4\sqrt[3]{2} - 4\sqrt[3]{4})$ we obtain $a_{n+1} = a_n - 8b_n + 8c_n$. Since $a_0 = 1$, a_n is odd for all n.

   For an integer $k > 0$, we can write $k = 2^l k'$, k' being odd and l a nonnegative integer. Let us set $v(k) = l$, and define $\beta_n = v(b_n), \gamma_n = v(c_n)$. We prove the following lemmas:

   Lemma 1. For every integer $p \geq 0$, b_{2^p} and c_{2^p} are nonzero, and $\beta_{2^p} = \gamma_{2^p} = p + 2$.

   Proof. By induction on p. For $p = 0$, $b_1 = 4$ and $c_1 = -4$, so the assertion is true. Suppose that it holds for p. Then

   $$(1 + 4\sqrt[3]{2} - 4\sqrt[3]{4})^{2^{p+1}} = (a + 2^{p+2}(b'\sqrt[3]{2} + c'\sqrt[3]{4}))^2 \text{ with } a, b', \text{ and } c' \text{ odd.}$$

   Then we easily obtain that $(1 + 4\sqrt[3]{2} - 4\sqrt[3]{4})^{2^{p+1}} = A + 2^{p+3}(B\sqrt[3]{2} + C\sqrt[3]{4})$, where $A, B = ab' + 2^{p+1}E, C = ac' + 2^{p+1}F$ are odd. Therefore Lemma 1 holds for $p + 1$.

   Lemma 2. Suppose that for integers $n, m \geq 0$, $\beta_n = \gamma_n = \lambda > \beta_m = \gamma_m = \mu$. Then b_{n+m}, c_{n+m} are nonzero and $\beta_{n+m} = \gamma_{n+m} = \mu$.

   Proof. Calculating $(a' + 2^\lambda(b'\sqrt[3]{2} + c'\sqrt[3]{4}))(a'' + 2^\mu(b''\sqrt[3]{2} + c''\sqrt[3]{4}))$, with a', b', c', a'', b'', c'' odd, we easily obtain the product $A + 2^\mu(B\sqrt[3]{2} + C\sqrt[3]{4})$, where $A, B = a'b'' + 2^{\lambda-\mu}E$, and $C = a'c'' + 2^{\lambda-\mu}F$ are odd, which proves Lemma 2.

   Since every integer $n > 0$ can be written as $n = 2^{p_r} + \cdots + 2^{p_1}$, with $0 \leq p_1 < \cdots < p_r$, from Lemmas 1 and 2 it follows that c_n is nonzero, and that $\gamma_n = p_1 + 2$.

   Remark. b_{1989} and c_{1989} are divisible by 4, but not by 8.

10. Plugging in $wz + a$ instead of z into the functional equation, we obtain

$$f(wz + a) + f(w^2z + wa + a) = g(wz + a). \tag{1}$$

By repeating this process, this time in (1), we get

$$f(w^2z + wa + a) + f(z) = g(w^2z + wa + a). \tag{2}$$

Solving the system of linear equations (1), (2) and the original functional equation, we easily get

$$f(z) = \frac{g(z) + g(w^2 z + wa + a) - g(wz + a)}{2}.$$

This function thus uniquely satisfies the original functional equation.

11. Call a binary sequence S of length n *repeating* if for some $d \mid n$, $d > 1$, S can be split into d identical blocks. Let x_n be the number of nonrepeating binary sequences of length n. The total number of binary sequences of length n is obviously 2^n. Any sequence of length n can be produced by repeating its unique longest nonrepeating initial block according to need. Hence, we obtain the recursion relation $\sum_{d \mid n} x_d = 2^n$. This, along with $x_1 = 2$, gives us $a_n = x_n$ for all n.

We now have that the sequences counted by x_n can be grouped into groups of n, the sequences in the same group being cyclic shifts of each other. Hence, $n \mid x_n = a_n$.

12. Assume that each car starts with a unique ranking number. Suppose that while turning back at a meeting point two cars always exchanged their ranking numbers. We can observe that ranking numbers move at a constant speed and direction. One hour later, after several exchanges, each starting point will be occupied by a car of the same ranking number and proceeding in the same direction as the one that started from there one hour ago.

We now give the cars back their original ranking numbers. Since the sequence of the cars along the track cannot be changed, the only possibility is that the original situation has been rotated, maybe onto itself. Hence for some $d \mid n$, after d hours each car will be at its starting position and orientation.

13. Let us construct the circles σ_1 with center A and radius $R_1 = AD$, σ_2 with center B and radius $R_2 = BC$, and σ_3 with center P and radius x. The points C and D lie on σ_2 and σ_1 respectively, and CD is tangent to σ_3. From this it is plain that the greatest value of x occurs when CD is also tangent to σ_1 and σ_2. We shall show that in this case the required inequality is really an equality, i.e., that $\frac{1}{\sqrt{x}} = \frac{1}{\sqrt{AD}} + \frac{1}{\sqrt{BC}}$. Then the inequality will immediately follow.
Denote the point of tangency of CD with σ_3 by M. By the Pythagorean theorem we have $CD = \sqrt{(R_1 + R_2)^2 - (R_1 - R_2)^2} = 2\sqrt{R_1 R_2}$. On the other hand, $CD = CM + MD = 2\sqrt{R_2 x} + 2\sqrt{R_1 x}$. Hence, we obtain $\frac{1}{\sqrt{x}} = \frac{1}{\sqrt{R_1}} + \frac{1}{\sqrt{R_2}}$.

14. *Lemma 1.* In a quadrilateral $ABCD$ circumscribed about a circle, with points of tangency P, Q, R, S on DA, AB, BC, CD respectively, the lines AC, BD, PR, QS concur.
Proof. Follows immediately, for example, from Brianchon's theorem.

Lemma 2. Let a variable chord XY of a circle $C(I, r)$ subtend a right angle at a fixed point Z within the circle. Then the locus of the midpoint P of XY is a circle whose center is at the midpoint M of IZ and whose radius is $\sqrt{r^2/2 - IZ^2/4}$.

Proof. From $\angle XZY = 90°$ follows $\overrightarrow{ZX} \cdot \overrightarrow{ZY} = (\overrightarrow{IX} - \overrightarrow{IZ}) \cdot (\overrightarrow{IY} - \overrightarrow{IZ}) = 0$. Therefore,

$$\overrightarrow{MP}^2 = (\overrightarrow{MI} + \overrightarrow{IP})^2 = \frac{1}{4}(-\overrightarrow{IZ} + \overrightarrow{IX} + \overrightarrow{IY})^2$$

$$= \frac{1}{4}(IX^2 + IY^2 - IZ^2 + 2(\overrightarrow{IX} - \overrightarrow{IZ}) \cdot (\overrightarrow{IY} - \overrightarrow{IZ}))$$

$$= \frac{1}{2}r^2 - \frac{1}{4}IZ^2.$$

Lemma 3. Using notation as in Lemma 1, if $ABCD$ is cyclic, PR is perpendicular to QS.

Proof. Consider the inversion in $C(I,r)$, mapping A to A' etc. (P,Q,R,S are fixed). As is easily seen, A', B', C', D' will lie at the midpoints of PQ, QR, RS, SP, respectively. $A'B'C'D'$ is a parallelogram, but also cyclic, since inversion preserves circles; thus it must be a rectangle, and so $PR \perp QS$.

Now we return to the main result. Let I and O be the incenter and circumcenter, Z the intersection of the diagonals, and P,Q,R,S,A',B',C',D' points as defined in Lemmas 1 and 3. From Lemma 3, the chords PQ,QR,RS,SP subtend $90°$ at Z. Therefore by Lemma 2 the points A',B',C',D' lie on a circle whose center is the midpoint Y of IZ. Since this circle is the image of the circle $ABCD$ under the considered inversion (centered at I), it follows that I,O,Y are collinear, and hence so are I,O,Z.

Remark. This is the famous Newton's theorem for bicentric quadrilaterals.

15. By Cauchy's inequality, $44 < \sqrt{1989} < a+b+c+d \le \sqrt{2 \cdot 1989} < 90$. Since $m^2 = a+b+c+d$ is of the same parity as $a^2+b^2+c^2+d^2 = 1989$, m^2 is either 49 or 81. Let $d = \max\{a,b,c,d\}$.
Suppose that $m^2 = 49$. Then $(49-d)^2 = (a+b+c)^2 > a^2+b^2+c^2 = 1989-d^2$, and so $d^2 - 49d + 206 > 0$. This inequality does not hold for $5 \le d \le 44$. Since $d \ge \sqrt{1989/4} > 22$, d must be at least 45, which is impossible because $45^2 > 1989$. Thus we must have $m^2 = 81$ and $m = 9$. Now, $4d > 81$ implies $d \ge 21$. On the other hand, $d < \sqrt{1989}$, and hence $d = 25$ or $d = 36$. Suppose that $d = 25$ and put $a = 25 - p, b = 25 - q, c = 25 - r$ with $p,q,r \ge 0$. From $a+b+c = 56$ it follows that $p+q+r = 19$, which, together with $(25-p)^2 + (25-q)^2 + (25-r)^2 = 1364$, gives us $p^2 + q^2 + r^2 = 439 > 361 = (p+q+r)^2$, a contradiction. Therefore $d = 36$ and $n = 6$.

Remark. A little more calculation yields the unique solution $a = 12$, $b = 15$, $c = 18$, $d = 36$.

16. Define $S_k = \sum_{i=0}^{k} a_i$ $(k = 0, 1, \ldots, n)$ and $S_{-1} = 0$. We note that $S_{n-1} = S_n$. Hence

$$S_n = \sum_{k=0}^{n-1} a_k = nc + \sum_{k=0}^{n-1}\sum_{i=k}^{n-1} a_{i-k}(a_i+a_{i+1})$$

$$= nc + \sum_{i=0}^{n-1}\sum_{k=0}^{i} a_{i-k}(a_i+a_{i+1}) = nc + \sum_{i=0}^{n-1}(a_i+a_{i+1})\sum_{k=0}^{i} a_{i-k}$$

$$= nc + \sum_{i=0}^{n-1}(S_{i+1}-S_{i-1})S_i = nc + S_n^2,$$

i.e., $S_n^2 - S_n + nc = 0$. Since S_n is real, the discriminant of the quadratic equation must be positive, and hence $c \le \frac{1}{4n}$.

17. A figure consisting of 9 lines is shown below.

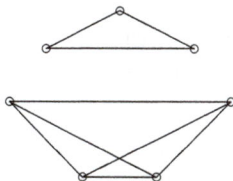

Now we show that 8 lines are not sufficient. Assume the opposite. By the pigeon-hole principle, there is a vertex, say A, that is joined to at most 2 other vertices. Let B,C,D,E denote the vertices to which A is not joined, and F,G the other two vertices. Then any two vertices of B,C,D,E must be mutually joined for an edge to exist within the triangle these two points form with A. This accounts for 6 segments. Since only two segments remain, among A, F, and G at least two are not joined. Taking these two and one of B,C,D,E that is not joined to any of them (it obviously exists), we get a triple of points, no two of which are joined; a contradiction.

Second solution. Since (i) is equivalent to the fact that no three points make a "blank triangle," by Turan's theorem the number of "blank edges" cannot exceed $[7^2/4] = 12$, leaving at least $7 \cdot 6/2 - 12 = 9$ segments. For general n, the answer is $[(n-1)/2]^2$.

18. Consider the triangle MA_iM_i. Obviously, the point M_i is the image of A_i under the composition C of rotation $R_M^{\alpha/2-90°}$ and homothety $H_M^{2\sin(\alpha/2)}$. Therefore, the polygon $M_1M_2\ldots M_n$ is obtained as the image of $A_1A_2\ldots A_n$ under the rotational homothety C with coefficient $2\sin(\alpha/2)$. Therefore $S_{M_1M_2\ldots M_n} = 4\sin^2(\alpha/2)\cdot S$.

19. Let us color the board in a chessboard fashion. Denote by S_b and S_w respectively the sum of numbers in the black and in the white squares. It is clear that every allowed move leaves the difference $S_b - S_w$ unchanged. Therefore a necessary condition for annulling all the numbers is $S_b = S_w$.
We now show it is sufficient. Assuming $S_b = S_w$ let us observe a triple of (different) cells a, b, c with respective values x_a, x_b, x_c where a and c are both adjacent to b. We first prove that we can reduce x_a to be 0 if $x_a > 0$. If $x_a \le x_b$, we subtract x_a from both a and b. If $x_a > x_b$, we add $x_a - x_b$ to b and c and proceed as in the previous case. Applying the reduction in sequence, along the entire board, we

reduce all cells except two neighboring cells to be 0. Since $S_b = S_w$ is invariant, the two cells must have equal values and we can thus reduce them both to 0.

20. Suppose $k \geq 1/2 + \sqrt{2n}$. Consider a point P in S. There are at least k points in S having all the same distance to P, so there are at least $\binom{k}{2}$ pairs of points A, B with $AP = BP$. Since this is true for every point $P \in S$, there are at least $n\binom{k}{2}$ triples of points (A, B, P) for which $AP = BP$ holds. However,

$$n\binom{k}{2} = n\frac{k(k-1)}{2} \geq \frac{n}{2}\left(\sqrt{2n} + \frac{1}{2}\right)\left(\sqrt{2n} - \frac{1}{2}\right)$$
$$= \frac{n}{2}\left(2n - \frac{1}{4}\right) > n(n-1) = 2\binom{n}{2}.$$

Since $\binom{n}{2}$ is the number of all possible pairs (A, B) with $A, B \in S$, there must exist a pair of points A, B with more than two points P_i such that $AP_i = BP_i$. These points P_i are collinear (they lie on the perpendicular bisector of AB), contradicting condition (i).

21. In order to obtain a triangle as the intersection we must have three points P, Q, R on three sides of the tetrahedron passing through one vertex, say T. It is clear that we may suppose w.l.o.g. that P is a vertex, and Q and R lie on the edges TP_1 and TP_2 (P_1, P_2 are vertices) or on their extensions respectively. Suppose that $\overrightarrow{TQ} = \lambda \overrightarrow{TP_1}$ and $\overrightarrow{TR} = \mu \overrightarrow{TP_2}$, where $\lambda, \mu > 0$. Then

$$\cos \angle QPR = \frac{\overrightarrow{PQ} \cdot \overrightarrow{PR}}{PQ \cdot PR} = \frac{(\lambda-1)(\mu-1)+1}{2\sqrt{\lambda^2 - \lambda + 1}\sqrt{\mu^2 - \mu + 1}}.$$

In order to obtain an obtuse angle (with $\cos < 0$) we must choose $\mu < 1$ and $\lambda > \frac{2-\mu}{1-\mu} > 1$. Since $\sqrt{\lambda^2 - \lambda + 1} > \lambda - 1$ and $\sqrt{\mu^2 - \mu + 1} > 1 - \mu$, we get that for $(\lambda - 1)(\mu - 1) + 1 < 0$,

$$\cos \angle QPR > \frac{1 - (1-\mu)(\lambda-1)}{2(1-\mu)(\lambda-1)} > -\frac{1}{2}; \qquad \text{hence } \angle QPR < 120°.$$

Remark. After obtaining the formula for $\cos \angle QPR$, the official solution was as follows: For fixed $\mu_0 < 1$ and $\lambda > 1$, $\cos \angle QPR$ is a decreasing function of λ: indeed,

$$\frac{\partial \cos \angle QPR}{\partial \lambda} = \frac{\mu - (3-\mu)\lambda}{4(\lambda^2 - \lambda + 1)^{3/2}(\mu^2 - \mu + 1)^{1/2}} < 0.$$

Similarly, for a fixed, sufficiently large λ_0, $\cos \angle QPR$ is decreasing for μ decreasing to 0. Since $\lim_{\lambda \to 0, \mu \to 0+} \cos \angle QPR = -1/2$, we conclude that $\angle QPR < 120°$.

22. The statement remains valid if 17 is replaced by any divisor k of $1989 = 3^2 \cdot 13 \cdot 17$, $1 < k < 1989$, so let k be one such divisor. The set $\{1, 2, \ldots, 1989\}$ can be partitioned as $\{1, 2, \ldots, 3k\} \cup \bigcup_{j=1}^{L}\{(2j+1)k+1, (2j+1)k+2, \ldots, (2j+1)k+2k\} = X \cup Y_1 \cup \cdots \cup Y_L$, where $L = (1989 - 3k)/2k$. The required statement will be an obvious consequence of the following two claims.

Claim 1. The set $X = \{1, 2, \ldots, 3k\}$ can be partitioned into k disjoint subsets, each having 3 elements and the same sum.

Proof. Since k is odd, let $t = k - 1/2$ and $X = \{1, 2, \ldots, 6t + 3\}$. For $l = 1, 2, \ldots, t$, define

$$X_{2l-1} = \{l, 3t + 1 + l, 6t + 5 - 2l\},$$
$$X_{2l} = \{t + 1 + l, 2t + 1 + l, 6t + 4 - 2l\}$$
$$X_{2t+1} = X_k = \{t + 1, 4t + 2, 4t + 3\}.$$

It is easily seen that these three subsets are disjoint and that the sum of elements in each set is $9t + 6$.

Claim 2. Each $Y_j = \{(2j + 1)k + 1, \ldots, (2j + 1)k + 2k\}$ can be partitioned into k disjoint subsets, each having 2 elements and the same sum.

Proof. The obvious partitioning works: $Y_j = \{(2j + 1)k + 1, (2j + 1)k + 2k\} \cup \cdots \cup \{(2j + 1)k + k, (2j + 1)k + (k + 1)\}$.

23. Two numbers $x, y \in \{1, \ldots, 2n\}$ will be called *twins* if $|x - y| = n$. Then the set $\{1, \ldots, 2n\}$ splits into n pairs of twins. A permutation $(x_1, \ldots, x_{2n})$ of this set is said to be of type T_k if $|x_i - x_{i+1}| = n$ holds for exactly k indices i (thus a permutation of type T_0 contains no pairs of neighboring twins). Denote by $F_k(n)$ the number of T_k-type permutations of $\{1, \ldots, 2n\}$.

Let $(x_1, \ldots, x_{2n})$ be a permutation of type T_0. Removing x_{2n} and its twin, we obtain a permutation of $2n - 2$ elements consisting of $n - 1$ pairs of twins. This new permutation is of one of the following types:

(i) type T_0: x_{2n} can take $2n$ values, and its twin can take any of $2n - 2$ positions;

(ii) type T_1: x_{2n} can take any one of $2n$ values, but its twin must be placed to separate the unique pair of neighboring twins in the new permutation.

The recurrence formula follows:

$$F_0(n) = 2n[(2n - 2)F_0(n - 1) + F_1(n - 1)]. \tag{1}$$

Now let $(x_1, \ldots, x_{2n})$ be a permutation of type T_1, and let (x_j, x_{j+1}) be the unique neighboring twin pair. Similarly, on removing this pair we get a permutation of $2n - 2$ elements, either of type T_0 or of type T_1. The pair (x_j, x_{j+1}) is chosen out of n twin pairs and can be arranged in two ways. Also, in the first case it can be placed anywhere ($2n - 1$ possible positions), but in the second case it must be placed to separate the unique pair of neighboring twins. Hence,

$$F_1(n) = 2n[(2n - 1)F_0(n - 1) + F_1(n - 1)] = F_0(n) + 2nF_0(n - 1). \tag{2}$$

This implies that $F_0(n) < F_1(n)$. Therefore the permutations with at least one neighboring twin pair are more numerous than those with no such pairs.

Remark 1. As in the official solution, formulas (1) and (2) together give for F_0 the recurrence

$$F_0(n) = 2n[(2n - 1)F_0(n - 1) + (2n - 2)F_0(n - 2)].$$

For the ratio $p_n = F_0(n)/(2n)!$, simple algebraic manipulation yields $p_n = p_{n-1} + \frac{p_{n-2}}{(2n-3)(2n-1)}$. Since $p_1 = 0$, we get

$$p_n < p_{n-1} + \frac{1}{(2n-3)(2n-1)} = p_{n-1} + \frac{1}{2(2n-3)} - \frac{1}{2(2n-1)} < \cdots < \frac{1}{2}.$$

Remark 2. Using the inclusion–exclusion principle, the following formula can be obtained:

$$F_0(n) = 2^0 \binom{n}{0}(2n)! - 2^1 \binom{n}{1}(2n-1)! + 2^2 \binom{n}{2}(2n-2)! - \cdots$$
$$\cdots + (-1)^{n-1} 2^n \binom{n}{n} n!.$$

One consequence is that in fact, $\lim_{n \to \infty} p_n = 1/e$.

Second solution. Let $f : T_0 \to T_1$ be the mapping defined as follows: if $(x_1, x_2, \ldots, x_{2n}) \in T_0$ and x_k, $k > 2$, is the twin of x_1, then

$$f(x_1, x_2, \ldots, x_{2n}) = (x_2, \ldots, x_{k-1}, x_1, x_k, \ldots, x_{2n}).$$

The mapping f is injective, but not surjective. Thus $F_0(n) < F_1(n)$.

24. Instead of Euclidean distance, we will use the angles $\angle A_i O A_j$, O denoting the center of the sphere. Let $\{A_1, \ldots, A_5\}$ be any set for which $\min_{i \neq j} \angle A_i O A_j \geq \pi/2$ (such a set exists: take for example five vertices of an octagon). We claim that two of the A_i's must be antipodes, thus implying that $\min_{i \neq j} \angle A_i O A_j$ is exactly equal to $\pi/2$, and consequently that $\min_{i \neq j} A_i A_j = \sqrt{2}$.
Suppose no two of the five points are antipodes. Visualize A_5 as the south pole. Then $A_1, \ldots, A_4$ lie in the northern hemisphere, including the equator (but excluding the north pole). No two of $A_1, \ldots, A_4$ can lie in the interior of a quarter of this hemisphere, which means that any two of them differ in longitude by at least $\pi/2$. Hence, they are situated on four meridians that partition the sphere into quarters. Finally, if one of them does not lie on the equator, its two neighbors must. Hence, in any case there will exist an antipodal pair, giving us a contradiction.

25. We may assume w.l.o.g. that $a > 0$ (because $a, b < 0$ is impossible, and $a, b \neq 0$ from the condition of the problem). Let $(x_0, y_0, z_0, w_0) \neq (0, 0, 0, 0)$ be a solution of $x^2 - ay^2 - bz^2 + abw^2$. Then

$$x_0^2 - ay_0^2 = b(z_0^2 - aw_0^2).$$

Multiplying both sides by $(z_0^2 - aw_0^2)$, we get

$$(x_0^2 - ay_0^2)(z_0^2 - aw_0^2) - b(z_0^2 - aw_0^2)^2 = 0$$
$$\Leftrightarrow (x_0 z_0 - ay_0 w_0)^2 - a(y_0 z_0 - x_0 w_0)^2 - b(z_0^2 - aw_0^2)^2 = 0.$$

Hence, for $x_1 = x_0 z_0 - ay_0 w_0$, $y_1 = y_0 z_0 - x_0 w_0$, $z_1 = z_0^2 - aw_0^2$, we have

$$x_1^2 - ay_1^2 - bz_1^2 = 0.$$

If (x_1, y_1, z_1) is the trivial solution, then $z_1 = 0$ implies $z_0 = w_0 = 0$ and similarly $x_0 = y_0 = 0$ because a is not a perfect square. This contradicts the initial assumption.

26. By the Cauchy–Schwarz inequality,

$$\left(\sum_{i=1}^n x_i\right)^2 \le n \sum_{i=1}^n x_i^2.$$

Since $\sum_{i=1}^n x_i = a - x_0$ and $\sum_{i=1}^n x_i^2 = b - x_0^2$, we have $(a - x_0)^2 \le n(b - x_0^2)$, i.e.,

$$(n+1)x_0^2 - 2ax_0 + (a^2 - nb) \le 0.$$

The discriminant of this quadratic is $D = 4n(n+1)\left[b - a^2/(n+1)\right]$, so we conclude that
   (i) if $a^2 > (n+1)b$, then such an x_0 does not exist;
   (ii) if $a^2 = (n+1)b$, then $x_0 = a/n+1$;   and
   (iii) if $a^2 < (n+1)b$, then $\frac{a-\sqrt{D}/2}{n+1} \le x_0 \le \frac{a+\sqrt{D}/2}{n+1}$.
It is easy to see that these conditions for x_0 are also sufficient.

27. Let n be the required exponent, and suppose $n = 2^k q$, where q is an odd integer. Then we have

$$m^n - 1 = (m^{2^k} - 1)[(m^{2^k(q-1)} + \cdots + m^{2^k} + 1] = (m^{2^k} - 1)A,$$

where A is odd. Therefore $m^n - 1$ and $m^{2^k} - 1$ are divisible by the same power of 2, and so $n = 2^k$.
Next, we observe that

$$m^{2^k} - 1 = (m^{2^{k-1}} - 1)(m^{2^{k-1}} + 1) = \cdots$$
$$= (m^2 - 1)(m^2 + 1)(m^4 + 1)\cdots(m^{2^{k-1}} + 1).$$

Let s be the maximal positive integer for which $m \equiv \pm 1 \pmod{2^s}$. Then $m^2 - 1$ is divisible by 2^{s+1} and not divisible by 2^{s+2}. All the numbers $m^2 + 1, m^4 + 1, \ldots, m^{2^{k-1}} + 1$ are divisible by 2 and not by 4. Hence $m^{2^k} - 1$ is divisible by 2^{s+k} and not by 2^{s+k+1}.
It follows from the above consideration that the smallest exponent n equals 2^{1989-s} if $s \le 1989$, and $n = 1$ if $s > 1989$.

28. Assume w.l.o.g. that the rays OA_1, OA_2, OA_3, OA_4 are arranged clockwise. Setting $OA_1 = a$, $OA_2 = b$, $OA_3 = c$, $OA_4 = d$, and $\angle A_1OA_2 = x$, $\angle A_2OA_3 = y$, $\angle A_3OA_4 = z$, we have

$$S_1 = \sigma(OA_1A_2) = \frac{1}{2}ab|\sin x|, \ S_2 = \sigma(OA_1A_3) = \frac{1}{2}ac|\sin(x+y)|,$$

$$S_3 = \sigma(OA_1A_4) = \frac{1}{2}ad|\sin(x+y+z)|, \ S_4 = \sigma(OA_2A_3) = \frac{1}{2}bc|\sin y|,$$

$$S_5 = \sigma(OA_2A_4) = \frac{1}{2}bd|\sin(y+z)|, \ S_6 = \sigma(OA_3A_4) = \frac{1}{2}cd|\sin z|.$$

Since $\sin(x+y+z)\sin y + \sin x\sin z = \sin(x+y)\sin(y+z)$, it follows that there exists a choice of $k,l \in \{0,1\}$ such that

$$S_1S_6 + (-1)^kS_2S_5 + (-1)^lS_3S_4 = 0.$$

For example (w.l.o.g.), if $S_3S_4 = S_1S_6 + S_2S_5$, we have

$$\left(\max_{1\le i\le 6} S_i\right)^2 \ge S_3S_4 = S_1S_6 + S_2S_5 \ge 1+1 = 2,$$

i.e., $\max_{1\le i\le 6} S_i \ge \sqrt{2}$ as claimed.

29. Let P_i, sitting at the place A, and P_j sitting at B, be two birds that can see each other. Let k and l respectively be the number of birds visible from B but not from A, and the number of those visible from A but not from B. Assume that $k \ge l$. Then if all birds from B fly to A, each of them will see l new birds, but won't see k birds anymore. Hence the total number of mutually visible pairs does not increase, while the number of distinct positions occupied by at least one bird decreases by one. Repeating this operation as many times as possible one can arrive at a situation in which two birds see each other if and only if they are in the same position. The number of such distinct positions is at most 35, while the total number of mutually visible pairs is not greater than at the beginning. Thus the problem is equivalent to the following one:

(1) If $x_i \ge 0$ are integers with $\sum_{j=1}^{35} x_j = 155$, find the least possible value of $\sum_{j=1}^{35} (x_j^2 - x_j)/2$.

If $x_j \ge x_i + 2$ for some i, j, then the sum of $(x_j^2 - x_j)/2$ decreases (for $x_j - x_i - 2$) if x_i, x_j are replaced with $x_i + 1$, $x_j - 1$. Consequently, our sum attains its minimum when the x_i's differ from each other by at most 1. In this case, all the x_i's are equal to either $[155/35] = 4$ or $[155/35] + 1 = 5$, where $155 = 20\cdot 4 + 15\cdot 5$. It follows that the (minimum possible) number of mutually visible pairs is $20\cdot \frac{4\cdot 3}{2} + 15\cdot \frac{5\cdot 4}{2} = 270$.

Second solution for (1). Considering the graph consisting of birds as vertices and pairs of mutually nonvisible birds as edges, we see that there is no complete 36-subgraph. Turán's theorem gives the answer immediately. (See problem (SL89-17).)

30. For all n such N exists. For a given n choose $N = (n+1)!^2 + 1$. Then $1 + j$ is a proper factor of $N + j$ for $1 \le j \le n$. So if $N + j = p^m$ is a power of a prime p, then $1 + j = p^r$ for some integer r, $1 \le r < m$. But then p^{r+1} divides both $(n+1)!^2 = N - 1$ and $p^m = N + j$, implying that $p^{r+1} \mid 1 + j$, which is impossible. Thus none of $N+1, N+2, \ldots, N+n$ is a power of a prime.

Second solution. Let $p_1, p_2, \ldots, p_{2n}$ be distinct primes. By the Chinese remainder theorem, there exists a natural number N such that $p_1p_2 \mid N+1$, $p_3p_4 \mid N+2$, $\ldots$, $p_{2n-1}p_{2n} \mid N+n$, and then obviously none of the numbers $N+1, \ldots, N+n$ can be a power of a prime.

31. Let us denote by N_{pqr} the number of solutions for which $a_p/x_p \geq a_q/x_q \geq a_r/x_r$, where (p,q,r) is one of six permutations of $(1,2,3)$. It is clearly enough to prove that $N_{pqr} + N_{qpr} \leq 2a_1a_2(3 + \ln(2a_1))$.

First, from

$$\frac{3a_p}{x_p} \geq \frac{a_p}{x_p} + \frac{a_q}{x_q} + \frac{a_r}{x_r} = 1 \quad \text{and} \quad \frac{a_p}{x_p} < 1$$

we get $a_p + 1 \leq x_p \leq 3a_p$. Similarly, for fixed x_p we have

$$\frac{2a_q}{x_q} \geq \frac{a_q}{x_q} + \frac{a_r}{x_r} = 1 - \frac{a_p}{x_p} \quad \text{and} \quad \frac{a_q}{x_q} \leq \min\left(\frac{a_p}{x_p}, 1 - \frac{a_p}{x_p}\right),$$

which gives $\max\{a_q \cdot x_p/a_p, a_q \cdot x_p/(x_p - a_p)\} \leq x_q \leq 2a_q \cdot x_p/(x_p - a_p)$, i.e., if $a_p + 1 \leq x_p \leq 2a_p$ there are at most $a_q \cdot x_p/(x_p - a_p) + 1/2$ possible values for x_q (because there are $[2x] - [x] = [x + 1/2]$ integers between x and $2x$), and if $2a_p + 1 \leq x_p \leq 3a_p$, at most $2a_q \cdot x_p/(x_p - a_p) - a_q \cdot x_p/a_p + 1$ possible values. Given x_p and x_q, x_r is uniquely determined. Hence

$$N_{pqr} \leq \sum_{x_p=a_p+1}^{2a_p} \left(\frac{a_q \cdot x_p}{x_p - a_p} + \frac{1}{2}\right) + \sum_{x_p=2a_p+1}^{3a_p} \left(\frac{2a_q \cdot x_p}{x_p - a_p} - \frac{a_q \cdot x_p}{a_p} + 1\right)$$

$$= \frac{3a_p}{2} + a_q \sum_{k=1}^{a_p} \left[\frac{k + a_p}{k} + \left(\frac{2(k + 2a_p)}{k + a_p} - \frac{k + 2a_p}{a_p}\right)\right]$$

$$= \frac{3a_p}{2} + a_q \sum_{k=1}^{a_p} \left[1 - \frac{k}{a_p} + a_p\left(\frac{1}{k} + \frac{2}{k + a_p}\right)\right]$$

$$= \frac{3a_p}{2} - \frac{a_q}{2} + a_p a_q \left(\frac{1}{2} + \sum_{k=1}^{a_p}\left(\frac{1}{k} + \frac{2}{k + a_p}\right)\right)$$

$$\leq a_p a_q \left(\frac{3}{2a_q} - \frac{1}{2a_p} + \ln(2a_p) + \frac{5}{2} - \ln 2\right).$$

Here we have used $\sum_{k=1}^{n}(1/k + 2/(k+n)) \leq \ln(2n) + 2 - \ln 2$ (this can be proved by induction). Hence,

$$N_{pqr} + N_{qpr} \leq 2a_p a_q(1 + 0.5 + \ln(2a_p) + 2 - \ln 2) < 2a_1a_2(2.81 + \ln(2a_1)).$$

Remark. The official solution was somewhat simpler, but used that the interval $(x, 2x]$, for real x, cannot contain more than x integers, which is false in general. Thus it could give only a weaker estimate $N \leq 6a_1a_2(9/2 - \ln 2 + \ln(2a_1))$.

32. Let CC' be an altitude, and R the circumradius. Then, since $AH = R$, we have $AC' = |R\sin B|$ and hence (1) $CC' = |R\sin B \tan A|$. On the other hand, $CC' = |BC\sin B| = 2|R\sin A \sin B|$, which together with (1) yields $2|\sin A| = |\tan A| \Rightarrow |\cos A| = 1/2$. Hence, $\angle A$ is $60°$. (Without the condition that the triangle is acute, $\angle A$ could also be $120°$.)

Second Solution. For a point X, let $\overline{X}$ denote the vector OX. Then $|\overline{A}| = |\overline{B}| = |\overline{C}| = R$ and $\overline{H} = \overline{A} + \overline{B} + \overline{C}$, and moreover,

$$R^2 = (\overline{H} - \overline{A})^2 = (\overline{B} + \overline{C})^2 = 2\overline{B}^2 + 2\overline{C}^2 - (\overline{B} - \overline{C})^2 = 4R^2 - BC^2.$$

It follows that $\sin A = \frac{BC}{2R} = \sqrt{3}/2$, i.e., that $\angle A = 60°$.

Third Solution. Let A_1 be the midpoint of BC. It is well known that $AH = 2OA_1$, and since $AH = AO = BO$, it means that in the right-angled triangle BOA_1 the relation $BO = 2OA_1$ holds. Thus $\angle BOA_1 = \angle A = 60°$.

4.31 Solutions to the Shortlisted Problems of IMO 1990

1. Let N be a number that can be written as a sum of 1990 consecutive integers and as a sum of consecutive positive integers in exactly 1990 ways. The former requirement gives us $N = m + (m+1) + \cdots + (m+1989) = 995(2m+1989)$ for some m. Thus $2 \nmid N$, $5 \mid N$, and $199 \mid N$. The latter requirement tells us that there are exactly 1990 ways to express N as $n + (n+1) + \cdots + (n+k)$, or equivalently, express $2N$ as $(k+1)(2n+k)$. Since N is odd, it follows that one of the factors $k+1$ and $2n+k$ is odd and the other is divisible by 2, but not by 4. Evidently $k+1 < 2n+k$. On the other hand, every factorization $2N = ab$, $1 < a < b$, corresponds to a single pair (n,k), where $n = \frac{b-a+1}{2}$ (which is an integer) and $k = a - 1$. The number of such factorizations is equal to $d(2N)/2 - 1$ because $a = b$ is impossible (here $d(x)$ denotes the number of positive divisors of an $x \in \mathbb{N}$). Hence we must have $d(2N) = 2 \cdot 1991 = 3982$. Now let $2N = 2 \cdot 5^{e_1} \cdot 199^{e_2} \cdot p_3^{e_3} \cdots p_r^{e_r}$ be a factorization of $2N$ into prime numbers, where $p_3, \ldots, p_r$ are distinct primes other than 2, 5, and 199 and $e_1, \cdots, e_r$ are positive integers. Then $d(2N) = 2(e_1+1)(e_2+1) \cdots (e_r+1)$, from which we deduce $(e_1+1)(e_2+1) \cdots (e_r+1) = 1991 = 11 \cdot 181$. We thus get $\{e_1, e_2\} = \{10, 180\}$ and $e_3 = \cdots = e_r = 0$. Hence $N = 5^{10} \cdot 199^{180}$ and $N = 5^{180} \cdot 199^{10}$ are the only possible solutions. These numbers indeed satisfy the desired properties.

2. We will call a cycle with m committees and n countries an (m,n) cycle. We will number the delegates from each country with numbers $1, 2, 3$ and denote committees by arrays of these integers (of length n) defining which of the delegates from each country is in the committee. We will first devise methods of constructing larger cycles out of smaller cycles.

   Let $A_1, \ldots, A_m$ be an (m,n) cycle, where m is odd. Then the following is a $(2m, n+1)$ cycle:

   $$(A_1, 1), (A_2, 2), \ldots, (A_m, 1), (A_1, 2), (A_2, 1), \ldots, (A_m, 2).$$

   Also, let $A_1, \ldots, A_m$ be an (m,n) cycle and $k \leq m$ an even integer. Then the cycle

   $$(A_1, 3), (A_2, 1), (A_3, 2), \ldots, (A_{k-2}, 1), (A_{k-1}, 2),$$
   $$(A_k, 3), (A_{k-1}, 1), (A_{k-2}, 2), \ldots, (A_2, 2)$$

   is a $(2(k-1), n+1)$ cycle.

   Starting from the $((1),(2),(3))$ cycle with parameters $(3,1)$ we can sequentially construct larger cycles using the shown methods. The obtained cycles have parameters as follows:

   $$(6,2), (10,3), \ldots, (2^k + 2, k), \ldots, (1026, 10), (1990, 11).$$

   Thus there exists a cycle of 1990 committees with 11 countries.

3. A segment connecting two points which divides the given circle into two arcs one of which contains exactly n points in its interior we will call a *good* segment. Good segments determine one or more closed polygonal lines that we will call

stars. Let us compute the number of stars. Note first that $\gcd(n+1, 2n-1) = \gcd(n+1, 3)$.

(i) Suppose that $3 \nmid n+1$. Then the good segments form a single star. Among any n points, two will be adjacent vertices of the star. On the other hand, we can select $n-1$ alternate points going along the star, and in this case no two points lie on a good segment. Hence $N = n$.

(ii) If $3 \mid n+1$, we obtain three stars of $\left[\frac{2n-1}{3}\right]$ vertices. If more than $\left[\frac{2n-1}{6}\right] = \frac{n-2}{3}$ points are chosen on any of the stars, then two of them will be connected with a good segment. On the other hand, we can select $\frac{n-2}{3}$ alternate points on each star, which adds up to $n-2$ points in total, no two of which lie on a good segment. Hence $N = n-1$.

To sum up, $N = n$ for $3 \nmid 2n-1$ and $N = n-1$ for $3 \mid 2n-1$.

4. Assuming that A_1 is not such a set A_i, it follows that for every m there exist m consecutive numbers not in A_1. It follows that $A_2 \cup A_3 \cup \cdots \cup A_r$ contains arbitrarily long sequences of numbers. Inductively, let us assume that $A_j \cup A_{j+1} \cup \cdots \cup A_r$ contains arbitrarily long sequences of consecutive numbers and none of $A_1, A_2, \ldots, A_{j-1}$ is the desired set A_i. Let us assume that A_j is also not A_i. Hence for each m there exists $k(m)$ such that among $k(m)$ elements of A_j there exist two consecutive elements that differ by at least m. Let us consider $m \cdot k(m)$ consecutive numbers in $A_j \cup \cdots \cup A_r$, which exist by the induction hypothesis. Then either A_j contains fewer than $k(m)$ of these integers, in which case $A_{j+1} \cup \cdots \cup A_r$ contains m consecutive integers by the pigeonhole principle or A_j contains $k(m)$ integers among which there exists a gap of length m of consecutive integers that belong to $A_{j+1} \cup \cdots \cup A_r$. Hence we have proven that $A_{j+1} \cup \cdots \cup A_r$ contains sequences of integers of arbitrary length. By induction, assuming that $A_1, A_2, \ldots, A_{r-1}$ do not satisfy the conditions to be the set A_i, it follows that A_r contains sequences of consecutive integers of arbitrary length and hence satisfies the conditions necessary for it to be the set A_i.

5. Let O be the circumcenter of ABC, E the midpoint of OH, and R and r the radii of the circumcircle and incircle respectively. We use the following facts from elementary geometry: $\overrightarrow{OH} = 3\overrightarrow{OG}$, $OK^2 = R^2 - 2Rr$, and $KE = \frac{R}{2} - r$. Hence $\overrightarrow{KH} = 2\overrightarrow{KE} - \overrightarrow{KO}$ and $\overrightarrow{KG} = \frac{2\overrightarrow{KE} + \overrightarrow{KO}}{3}$. We then obtain

$$\overrightarrow{KH} \cdot \overrightarrow{KG} = \frac{1}{3}(4KE^2 - KO^2) = -\frac{2}{3}r(R - 2r) < 0.$$

Hence $\cos \angle GKH < 0 \Rightarrow \angle GKH > 90°$.

6. Let W denote the set of all n_0 for which player A has a winning strategy, L the set of all n_0 for which player B has a winning strategy, and T the set of all n_0 for which a tie is ensured.

Lemma. Assume $\{m, m+1, \ldots 1990\} \subseteq W$ and that there exists $s \leq 1990$ such that $s/p^r \geq m$, where p^r is the largest degree of a prime that divides s. Then all integers x such that $\sqrt{s} \leq x < m$ also belong in W.

Proof. Starting from x, player A can choose s, and by definition of s, player B cannot choose a number smaller than m. This ensures player A the victory. We now have trivially that since $45^2 = 2025 > 1990$, it follows that for $n_0 \in \{45, \ldots, 1990\}$ the player A can choose 1990 in the first move. Therefore $\{45, \ldots, 1990\} \subseteq W$. Using $m = 45$ and selecting $s = 420 = 2^2 \cdot 3 \cdot 5 \cdot 7$ we apply the lemma to get that all integers x such that $\sqrt{420} < 21 \leq x \leq 1990$ are in W. Again, using $m = 21$ and selecting $s = 168 = 2^3 \cdot 3 \cdot 7$ we apply the lemma to get that all integers x such that $\sqrt{168} < 13 \leq x \leq 1990$ are in W. Selecting $s = 105$ we obtain the new value for m at $m = 11$. Selecting $s = 60$ we obtain $m = 8$. Thus $\{8, \ldots, 1990\} \subseteq W$.

For $n_0 > 1990$ there exists $r \in N$ such that $2^r \cdot 3^2 < n_0 \leq 2^{r+1} \cdot 3^2 < n_0^2$. Player A can take $n_1 = 2^{r+1} \cdot 3^2$. The number player B selects has to satisfy $8 \leq n_2 < n_0$. After finitely many steps he will select $8 \leq n_{2r} \leq 1990$, and A will have a winning strategy. Hence all $m \geq 8$ belong to W.

Now let us consider the case $n_0 \leq 5$. Since the smallest number divisible by three different primes is 30 and $n_0^2 \leq 5^2 = 25 < 30$, it follows that n_1 is of the form $n_1 = p^r$ or $n_1 = p^r \cdot q^s$, where p and q are two different primes. In the first case player B can choose 1 and win, while in the second case he can select the smaller of p^r, q^s, which is also smaller than $\sqrt{n_1} \leq n_0$. Thus player B can eventually reach $n_{2k} = 1$. Thus $\{2, 3, 4, 5\} \subseteq L$.

Finally, for $n_0 = 6$ or $n_0 = 7$ player A must select a number divisible by at least three primes, which must be $30 = 2 \cdot 3 \cdot 5$ or $42 = 2 \cdot 3 \cdot 7$; otherwise, B can select a degree of a prime smaller than n_0, yielding $n_2 < 6$ and victory for B. Player B must select a number smaller than 8. Hence, he has to select 6 in both cases. Afterwards, to avoid losing the game, player A will always choose 30 and player B always 6. In this case we would have a tie. Hence $T \subseteq \{6, 7\}$.

Considering that we have accounted for all integers $n_0 > 1$, the final solution is $L = \{2, 3, 4, 5\}$, $T = \{6, 7\}$, and $W = \{x \in \mathbb{N} \mid x \geq 8\}$.

7. Let $f(n) = g(n)2^{n^2}$ for all n. The recursion then transforms into $g(n+2) - 2g(n+1) + g(n) = n \cdot 16^{-n-1}$ for $n \in \mathbb{N}_0$. By summing this equation from 0 to $n-1$, we get

$$g(n+1) - g(n) = \frac{1}{15^2} \cdot (1 - (15n+1)16^{-n}).$$

By summing up again from 0 to $n-1$ we get $g(n) = \frac{1}{15^3} \cdot (15n - 32 + (15n + 2)16^{-n+1})$. Hence

$$f(n) = \frac{1}{15^3} \cdot (15n + 2 + (15n - 32)16^{n-1}) \cdot 2^{(n-2)^2}.$$

Now let us look at the values of $f(n)$ modulo 13:

$$f(n) \equiv 15n + 2 + (15n - 32)16^{n-1} \equiv 2n + 2 + (2n - 6)3^{n-1}.$$

We have $3^3 \equiv 1 \pmod{13}$. Plugging in $n \equiv 1 \pmod{13}$ and $n \equiv 1 \pmod 3$ for $n = 1990$ gives us $f(1990) \equiv 0 \pmod{13}$. We similarly calculate $f(1989) \equiv 0$ and $f(1991) \equiv 0 \pmod{13}$.

8. Since $2^{1990} < 8^{700} < 10^{700}$, we have $f_1(2^{1990}) < (9 \cdot 700)^2 < 4 \cdot 10^7$. We then have $f_2(2^{1990}) < (3+9 \cdot 7)^2 < 4900$ and finally $f_3(2^{1990}) < (3+9 \cdot 3)^2 = 30^2$. It is easily shown that $f_k(n) \equiv f_{k-1}(n)^2 \pmod 9$. Since $2^6 \equiv 1 \pmod 9$, we have $2^{1990} \equiv 2^4 \equiv 7$ (all congruences in this problem will be mod 9). It follows that $f_1(2^{1990}) \equiv 7^2 \equiv 4$ and $f_2(2^{1990}) \equiv 4^2 \equiv 7$. Indeed, it follows that $f_{2k}(2^{1990}) \equiv 7$ and $f_{2k+1}(2^{1990}) \equiv 4$ for all integer $k > 0$. Thus $f_3(2^{1990}) = r^2$ where $r < 30$ is an integer and $r \equiv f_2(2^{1990}) \equiv 7$. It follows that $r \in \{7, 16, 25\}$ and hence $f_3(2^{1990}) \in \{49, 256, 625\}$. It follows that $f_4(2^{1990}) = 169$, $f_5(2^{1990}) = 256$, and inductively $f_{2k}(2^{1990}) = 169$ and $f_{2k+1}(2^{1990}) = 256$ for all integer $k > 1$. Hence $f_{1991}(2^{1990}) = 256$.

9. Let a, b, c be the lengths of the sides of $\triangle ABC$, $s = \frac{a+b+c}{2}$, r the inradius of the triangle, and c_1 and b_1 the lengths of AB_2 and AC_2 respectively. As usual we will denote by $S(XYZ)$ the area of $\triangle XYZ$. We have

$$S(AC_1B_2) = \frac{AC_1 \cdot AB_2}{AC \cdot AB} S(ABC) = \frac{c_1 rs}{2b},$$

$$S(AKB_2) = \frac{c_1 r}{2}, \quad S(AC_1K) = \frac{cr}{4}.$$

From $S(AC_1B_2) = S(AKB_2) + S(AC_1K)$ we get $\frac{c_1 rs}{2b} = \frac{c_1 r}{2} + \frac{cr}{4}$; therefore $(a-b+c)c_1 = bc$. By looking at the area of $\triangle AB_1C_2$ we similarly obtain $(a+b-c)b_1 = bc$. From these two equations and from $S(ABC) = S(AB_2C_2)$, from which we have $b_1c_1 = bc$, we obtain

$$a^2 - (b-c)^2 = bc \Rightarrow \frac{b^2 + c^2 - a^2}{2bc} = \cos(\angle BAC) = \frac{1}{2} \Rightarrow \angle BAC = 60°.$$

10. Let r be the radius of the base and h the height of the cone. We may assume w.l.o.g. that $r = 1$. Let A be the top of the cone, BC the diameter of the circumference of the base such that the plane touches the circumference at B, O the center of the base, and H the midpoint of OA (also belonging to the plane). Let BH cut the sheet of the cone at D. By applying Menelaus's theorem to $\triangle AOC$ and $\triangle BHO$, we conclude that $\frac{AD}{DC} = \frac{CB}{BO} \cdot \frac{OH}{HA} = \frac{1}{2}$ and $\frac{HD}{DB} = \frac{HA}{AO} \cdot \frac{OC}{CB} = \frac{1}{4}$. The plane cuts the cone in an ellipse whose major axis is BD. Let E be the center of this ellipse and FG its minor axis. We have $\frac{BE}{ED} = \frac{1}{2}$. Let E', F', G' be radial projections of E, F, G from A onto the base of the cone. Then E sits on BC. Let $h(X)$ denote the height of a point X with respect to the base of the cone. We have $h(E) = h(D)/2 = h/3$. Hence $EF = 2E'F'/3$. Applying Menelaus's theorem to $\triangle BHO$ we get $\frac{OE'}{E'B} = \frac{BE}{EH} \cdot \frac{HA}{AO} = 1$. Hence $EF = \frac{2}{3}\frac{\sqrt{3}}{2} = \frac{1}{\sqrt{3}}$.
Let d denote the distance from A to the plane. Let V_1 and V denote the volume of the cone above the plane (on the same side of the plane as A) and the total volume of the cone. We have

$$\frac{V_1}{V} = \frac{BE \cdot EF \cdot d}{h} = \frac{(2BH/3)(1/\sqrt{3})(2S_{AHB}/BH)}{h}$$

$$= \frac{(2/3)(1/\sqrt{3})(h/2)}{h} = \frac{1}{3\sqrt{3}}.$$

Since this ratio is smaller than $1/2$, we have indeed selected the correct volume for our ratio.

11. Assume $\mathscr{B}(A,E,M,B)$. Since A,B,C,D lie on a circle, we have $\angle GCE = \angle MBD$ and $\angle MAD = \angle FCE$. Since FD is tangent to the circle around $\triangle EMD$ at E, we have $\angle MDE = \angle FEB = \angle AEG$. Consequently, $\angle CEF = 180° - \angle CEA - \angle FEB = 180° - \angle MED - \angle MDE = \angle EMD$ and $\angle CEG = 180° - \angle CEF = 180° - \angle EMD = \angle DMB$. Therefore $\triangle CEF \sim \triangle AMD$ and $\triangle CEG \sim \triangle BMD$. From the first similarity we obtain $CE \cdot MD = AM \cdot EF$, and from the second we obtain $CE \cdot MD = BM \cdot EG$. Hence

$$AM \cdot EF = BM \cdot EG \Rightarrow$$

$$\frac{GE}{EF} = \frac{AM}{BM} = \frac{\lambda}{1-\lambda}.$$

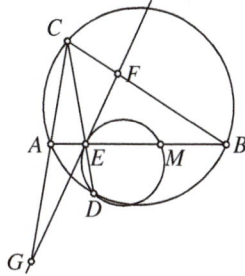

If $\mathscr{B}(A,M,E,B)$, interchanging the roles of A and B we similarly obtain $\frac{GE}{EF} = \frac{\lambda}{1-\lambda}$.

12. Let $d(X,l)$ denote the distance of a point X from a line l. Using the elementary facts that $AF : FC = c : a$ and $BD : DC = c : b$, we obtain $d(F,L) = \frac{a}{a+c}h_c$ and $d(D,L) = \frac{b}{b+c}h_c$, where h_a is the altitude of $\triangle ABC$ from A. We also have $\angle FGC = \beta/2$, $\angle DEC = \alpha/2$. It follows that

$$DE = \frac{d(D,L)}{\sin(\alpha/2)} \quad \text{and} \quad FG = \frac{d(F,L)}{\sin(\beta/2)}. \tag{1}$$

Now suppose that $a > b$. Since the function $f(x) = \frac{x}{x+c}$ is strictly increasing, we deduce $d(F,L) > d(D,L)$. Furthermore, $\sin(\alpha/2) > \sin(\beta/2)$, so we get from (1) that $FG > DE$.

Similarly, $a < b$ implies $FG < DE$. Hence we must have $a = b$, i.e., $AC = BC$.

13. We will call the ground the "zeroth" rung. We will prove that the minimum n is $n = a + b - (a,b)$. It is plain that if $(a,b) = k > 1$, the scientist can climb only onto the rungs divisible by k and we can just observe these rungs to obtain the situation equivalent to $a' = a/k, b' = b/k$, and $n' = a' + b' - 1$. Thus let us assume that $(a,b) = 1$ and show that $n = a + b - 1$.

We obviously have $n > a$. Consider $n = a + b - k$, $k \geq 1$, and let us assume without loss of generality that $a > b$ (otherwise, we can reverse the problem starting from the top rung in our round trip). Then we can uniquely define the numbers r_i, $0 \leq r_i < b$, by $r_i \equiv ia \pmod{b}$. We now describe the only possible sequence of moves. From a position $0 \leq p \leq b - k$ we can move only a rungs upward and for $p > b - 1$ we can move only b rungs downward. If we end up at $b - k < p \leq b - 1$, we are stuck. Hence, given that we are at r_i, if $r_i \leq b - k$, we can move to $a + r_i$, and when we descend as far as we can go we will end up at $r_{i+1} \equiv a + r_i \pmod{b}$.

If the mathematician climbs to the highest rung and then comes back to $r_i = 0$, then we deduce $b \mid ia$, so $i \geq b$. But since $(a,b) = 1$, there exists $0 < j < b$ such that $r_j \equiv ja \equiv b - 1 \pmod{b}$. Thus the mathematician has visited the position $b - 1$. For him not to get stuck we must have $k \leq 1$ and $n \geq a + b - 1$. For $n = a + b - 1$ by induction he can come to any position r_i, $i \geq 0$, so he eventually comes to $r_j = b - 1$, climbs to the highest rung, and then continues until he gets to $r_b = 0$. Hence the answer to the problem is $n = a + b - 1$.

14. Let V be the set of all midpoints of bad sides, and E the set of segments connecting two points in V that belong to the same triangle. Each edge in E is parallel to exactly one good side and thus is parallel to the coordinate grid and has half-integer coordinates. Thus, the edges of E are a subset of the grid formed by joining the centers of the squares in the original grid to each other. Let G be a graph whose set of vertices is V and set of edges is E. The degree of each vertex X, denoted by $d(X)$, is 0, 1, or 2. We observe the following cases:

(i) $d(X) = 0$ for some X. Then both triangles containing X have two good sides.

(ii) $d(X) = 1$ for some X. Since $\sum_{X \in V} d(X) = 2|E|$ is even, it follows that at least another vertex Y has the degree 1. Hence both X and Y belong to triangles having two good sides.

(iii) $d(X) = 2$ for all $X \in V$. We will show that this case cannot occur. We prove first that centers of all the squares of the $m \times n$ board belong to $V \cup E$. A bad side contains no points with half-integer coordinates in its interior other than its midpoint. Therefore either a point X is in V, or it lies on the segment connecting the midpoints of the two bad sides. Evidently, the graph G can be partitioned into disjoint cycles. Each center of a square is passed exactly once in exactly one cycle. Let us color the board black and white in a standard chessboard fashion. Each cycle passes through centers that must alternate in color, and hence it contains an equal number of black and white centers. Consequently, the numbers of black and white squares on the entire board must be equal, contradicting the condition that m and n are odd.

Our proof is thus completed.

15. Let $S(Z)$ denote the sum of all the elements of a set Z. We have $S(X) = (k+1) \cdot 1990 + \frac{k(k+1)}{2}$. To partition the set into two parts with equal sums, $S(X)$ must be even and hence $\frac{k(k+1)}{2}$ must be even. Hence k is of the form $4r$ or $4r + 3$, where r is an integer.

For $k = 4r + 3$ we can partition X into consecutive fourtuplets $\{1990 + 4l, 1990 + 4l + 1, 1990 + 4l + 2, 1990 + 4l + 3\}$ for $0 \leq l \leq r$ and put $1990 + 4l, 1990 + 4l + 3 \in A$ and $1990 + 4l + 1, 1990 + 4l + 2 \in B$ for all l. This would give us $S(A) = S(B) = (3980 + 4r + 3)(r + 1)$.

For $k = 4r$ the numbers of elements in A and B must differ. Let us assume without loss of generality $|A| < |B|$. Then $S(A) \leq (1990 + 2r + 1) + (1990 + 2r + 2) + \cdots + (1990 + 4r)$ and $S(B) \geq 1990 + 1991 + \cdots + (1990 + 2r)$. Plugging these inequalities into the condition $S(A) = S(B)$ gives us $r \geq 23$ and consequently $k \geq 92$. We note that $B = \{1990, 1991, \ldots, 2034, 2052, 2082\}$ and $A = \{2035, 2036, \ldots, 2051, 2053, \ldots, 2081\}$ is a partition for $k = 92$ that satisfies $S(A) = S(B)$. To construct a partition out of higher $k = 4r$ we use the $k = 92$ partition for the first 93 elements and construct for the remaining elements as was done for $k = 4r + 3$.

Hence we can construct a partition exactly for the integers k of the form $k = 4r + 3$, $r \geq 0$, and $k = 4r$, $r \geq 23$.

16. Let $A_0 A_1 \ldots A_{1989}$ be the desired 1990-gon. We also define $A_{1990} = A_0$. Let O be an arbitrary point. For $1 \leq i \leq 1990$ let B_i be a point such that $\overrightarrow{OB_i} = \overrightarrow{A_{i-1}A_i}$. We define $B_0 = B_{1990}$. The points B_i must satisfy the following properties: $\angle B_i O B_{i+1} = \frac{2\pi}{1990}$, $0 \leq i \leq 1989$, lengths of OB_i are a permutation of $1^2, 2^2, \ldots, 1989^2, 1990^2$, and $\sum_{i=0}^{1989} \overrightarrow{OB_i} = \overrightarrow{0}$. Conversely, any such set of points B_i corresponds to a desired 1990-gon. Hence, our goal is to construct vectors $\overrightarrow{OB_i}$ satisfying all the stated properties.

Let us group vectors of lengths $(2n-1)^2$ and $(2n)^2$ into pairs and put them diametrically opposite each other. The length of the resulting vectors is $4n-1$. The problem thus reduces to arranging vectors of lengths $3, 7, 11, \ldots, 3979$ at mutual angles of $\frac{2\pi}{995}$ such that their sum is $\overrightarrow{0}$. We partition the 995 directions into 199 sets of five directions at mutual angles $\frac{2\pi}{5}$. The directions when intersected with a unit circle form a regular pentagon. We group the set of lengths of vectors $3, 7, \ldots, 3979$ into 199 sets of five consecutive elements of the set. We place each group of lengths on directions belonging to the same group of directions, thus constructing five vectors. We use that $\overrightarrow{OC_1} + \cdots + \overrightarrow{OC_n} = 0$ where O is the center of a regular n-gon $C_1 \ldots C_n$. In other words, vectors of equal lengths along directions that form a regular n-gon cancel each other out. Such are the groups of five directions. Hence, we can assume for each group of five lengths for its lengths to be $\{0, 4, 8, 12, 16\}$. We place these five lengths in a random fashion on a single group of directions. We then rotate the configuration clockwise by $\frac{2\pi}{199}$ to cover other groups of directions and repeat until all groups of directions are exhausted. It follows that all vectors of each of the lengths $\{0, 4, 8, 12, 16\}$ will form a regular 199-gon and will thus cancel each other out.

We have thus constructed a way of obtaining points B_i and have hence shown the existence of the 1990-gon satisfying (i) and (ii).

17. Let us set a coordinate system denoting the vertices of the block. The vertices of the unit cubes of the block can be described as $\{(x, y, z) \mid 0 \leq x \leq p, 0 \leq y \leq q, 0 \leq z \leq r\}$, and we restrict our attention to only these points. Suppose the point A is fixed at (a, b, c). Then for every other necklace point (x, y, z) numbers $x - a$, $y - b$, and $z - c$ must be of equal parity. Conversely, every point (x, y, z) such that $x - a$, $y - b$, and $z - c$ are of the same parity has to be a necklace point. Consider the graph G whose vertices are all such points and edges are all diagonals of the unit cubes through these points. In part (a) we are looking for an open or closed Euler path, while in part (b) we are looking for a closed Euler path.

Necklace points in the interior of the (p,q,r) box have degree 8, points on the surface have degree 4, points on the edge have degree 2, and points on the corner have degree 1. A closed Euler path can be formed if and only if all vertices are of an even degree, while an open Euler path can be formed if and only if exactly two vertices have an odd degree. Hence the problem in part (a) amounts to being able to choose a point A such that 0 or 2 corner vertices are necklace vertices, whereas in part (b) no corner points can be necklace vertices. We distinguish two cases.

(i) At least two of p,q,r, say p,q, are even. We can choose $a = 1$, $b = c = 0$. In this case none of the corners is a necklace point. Hence a closed Euler path exists.

(ii) At most one of p,q,r is even. However one chooses A, exactly two necklace points are at the corners. Hence, an open Euler path exists, but it is impossible to form a closed path.

Hence, in part (a), a box can be made of all (p,q,r) and in part (b) only those (p,q,r) where at least two of the numbers are even.

18. Clearly, it suffices to consider the case $(a,b) = 1$. Let S be the set of integers such that $M - b \leq x \leq M + a - 1$. Then $f(S) \subseteq S$ and $0 \in S$. Consequently, $f^k(0) \in S$. Let us assume for $k > 0$ that $f^k(0) = 0$. Since $f(m) = m + a$ or $f(m) = m - b$, it follows that k can be written as $k = r + s$, where $ra - sb = 0$. Since a and b are relatively prime, it follows that $k \geq a + b$.

Let us now prove that $f^{a+b}(0) = 0$. In this case $a + b = r + s$ and hence $f^{a+b}(0) = (a + b - s)a - sb = (a + b)(a - s)$. Since $a + b \mid f^{a+b}(0)$ and $f^{a+b}(0) \in S$, it follows that $f^{a+b}(0) = 0$. Thus for $(a,b) = 1$ it follows that $k = a + b$. For other a and b we have $k = \frac{a+b}{(a,b)}$.

19. Let d_1, d_2, d_3, d_4 be the distances of the point P to the tetrahedron. Let d be the height of the regular tetrahedron. Let $x_i = d_i/d$. Clearly, $x_1 + x_2 + x_3 + x_4 = 1$, and given this condition, the parameters vary freely as we vary P within the tetrahedron. The four tetrahedra have volumes x_1^3, x_2^3, x_3^3, and x_4^3, and the four parallelepipeds have volumes of $6x_2x_3x_4$, $6x_1x_3x_4$, $6x_1x_2x_4$, and $6x_1x_2x_3$. Hence, using $x_1 + x_2 + x_3 + x_4 = 1$ and setting $g(x) = x^2(1 - x)$, we directly verify that

$$f(P) = f(x_1, x_2, x_3, x_4) = 1 - \sum_{i=1}^{4} x_i^3 - 6 \sum_{1 \leq i < j < k \leq 4} x_i x_j x_k$$
$$= 3(g(x_1) + g(x_2) + g(x_3) + g(x_4)) .$$

We note that $g(0) = 0$ and $g(1) = 0$. Hence, as x_1 tends to 1 and other variables tend to 0, $f(x_1, x_2, x_3, x_4) = 0$. Thus $f(P)$ is sharply bounded downwards at 0. We now find an upper bound. We note that

$$g(x_i + x_j) = (x_i + x_j)^2(1 - x_1 - x_2)$$
$$= g(x_i) + g(x_j) + 2x_i x_j \left(1 - \frac{3}{2}(x_i + x_j)\right);$$

thus for $x_i + x_j \leq 2/3$ and $x_i, x_j > 0$ we have $g(x_i + x_j) + g(0) \geq g(x_i) + g(x_j)$. Equality holds only when $x_i + x_j = 2/3$.

Assuming without loss of generality $x_1 \geq x_2 \geq x_3 \geq x_4$, we have $g(x_1) + g(x_2) + g(x_3) + g(x_4) < g(x_1) + g(x_2) + g(x_3 + x_4)$. Assuming $y_1 + y_2 + y_3 = 1$ and $y_1 \geq y_2 \geq y_3$, we have $g(y_1) + g(y_2) + g(y_3) \leq g(y_1) + g(y_2 + y_3)$. Hence $g(x_1) + g(x_2) + g(x_3) + g(x_4) < g(x) + g(1 - x)$ for some x. We also have $g(x) + g(1 - x) = x(1 - x) \leq 1/4$. Hence $f(P) \leq 3/4$. Equality holds for $x_1 = x_2 = 1/2$, $x_3 = x_4 = 0$ (corresponding to the midpoint of an edge), and as the variables converge to these values, $f(P)$ converges to $3/4$. Hence the bounds for $f(P)$ are

$$0 < f(P) < \frac{3}{4}.$$

20. Let n be the unique integer such that $2^{n-1} \leq k < 2^n$. Let $S(n)$ be the set of numbers less than 10^n that are written with only the digits $\{0, 1\}$ in the decimal system. Evidently $|S(n)| = 2^n > k$ and hence there exist two numbers $x, y \in S(n)$ such that $k \mid x - y$.

Let us show that $w = |x - y|$ is the desired number. By definition $k \mid w$. We also have

$$w < 1.2 \cdot 10^{n-1} \leq 1.2 \cdot (2^3 \sqrt{2})^{n-1} \leq 1.2 \cdot k^3 \sqrt{k} \leq k^4.$$

Finally, since $x, y \in S(n)$, it follows that $w = |x - y|$ can be written using only the digits $\{0, 1, 8, 9\}$. This completes the proof.

21. We must solve the congruence $(1 + 2^p + 2^{n-p})N \equiv 1 \pmod{2^n}$. Since $(1 + 2^p + 2^{n-p})$ and 2^n are coprime, there clearly exists a unique N satisfying this equation and $0 < N < 2^n$.

Let us assume $n = mp$. Then we have $(1 + 2^p)\left(\sum_{j=0}^{m-1} (-1)^j 2^{jp}\right) \equiv 1 \pmod{2^n}$ and $(1 + 2^{n-p})(1 - 2^{n-p}) \equiv 1 \pmod{2^n}$. By multiplying the two congruences we obtain

$$(1 + 2^p)(1 + 2^{n-p})(1 - 2^{n-p})\left(\sum_{j=0}^{m-1} (-1)^j 2^{jp}\right) \equiv 1 \pmod{2^n}.$$

Since $(1 + 2^p)(1 + 2^{n-p}) \equiv (1 + 2^p + 2^{n-p}) \pmod{2^n}$, it follows that $N \equiv (1 - 2^{n-p})\left(\sum_{j=0}^{m-1} (-1)^j 2^{jp}\right) \pmod{2^n}$. The integer $N = \sum_{j=0}^{m-1} (-1)^j 2^{jp} - 2^{n-p} + 2^n$ satisfies the congruence and $0 < N \leq 2^n$. Using that for $a > b$ we have in binary representation

$$2^a - 2^b = \underbrace{11\ldots11}_{a-b \text{ times}}\underbrace{00\ldots00}_{b \text{ times}},$$

the binary representation of N is calculated as follows:

$$N = \begin{cases} \underbrace{11\ldots11}_{p \text{ times}}\underbrace{11\ldots11}_{p \text{ times}}\underbrace{00\ldots00}_{p \text{ times}}\ldots\underbrace{11\ldots11}_{p \text{ times}}\underbrace{00\ldots00}_{p-1 \text{ times}}1, & 2 \nmid \frac{n}{p}, \\[2em] \underbrace{11\ldots11}_{p-1 \text{ times}}\underbrace{00\ldots00}_{p+1 \text{ times}}\underbrace{11\ldots11}_{p \text{ times}}\underbrace{00\ldots00}_{p \text{ times}}\ldots\underbrace{11\ldots11}_{p \text{ times}}\underbrace{00\ldots00}_{p-1 \text{ times}}1, & 2 \mid \frac{n}{p}. \end{cases}$$

22. We can assume without loss of generality that each connection is serviced by only one airline and the problem reduces to finding two disjoint monochromatic cycles of the same color and of odd length on a complete graph of 10 points colored by two colors. We use the following two standard lemmas:

Lemma 1. Given a complete graph on six points whose edges are colored with two colors there exists a monochromatic triangle.

Proof. Let us denote the vertices by $c_1, c_2, c_3, c_4, c_5, c_6$. By the pigeonhole principle at least three vertices out of c_1, say c_2, c_3, c_4, are of the same color, let us call it red. Assuming that at least one of the edges connecting points c_2, c_3, c_4 is red, the connected points along with c_1 form a red triangle. Otherwise, edges connecting c_2, c_3, c_4 are all of the opposite color, let us call it blue, and hence in all cases we have a monochromatic triangle.

Lemma 2. Given a complete graph on five points whose edges are colored with two colors there exists a monochromatic triangle or a monochromatic cycle of length five.

Proof. Let us denote the vertices by c_1, c_2, c_3, c_4, c_5. Assume that out of a point c_i three vertices are of the same color. We can then proceed as in Lemma 1 to obtain a monochromatic triangle. Otherwise, each point is connected to other points with exactly two red and two blue vertices. Hence, we obtain monochromatic cycles starting from a single point and moving along the edges of the same color. Since each cycle must be of length at least three (i.e., we cannot have more than one cycle of one color), it follows that for both red and blue we must have one cycle of length five of that color.

We now apply the lemmas. Let us denote the vertices by $c_1, c_2, \ldots, c_{10}$. We apply Lemma 1 to vertices $c_1, \ldots, c_6$ to obtain a monochromatic triangle. Out of the seven remaining vertices we select 6 and again apply Lemma 1 to obtain another monochromatic triangle. If they are of the same color, we are done. Otherwise, out of the nine edges connecting the two triangles of opposite color at least 5 are of the same color, we can assume blue w.l.o.g., and hence a vertex of a red triangle must contain at least two blue edges whose endpoints are connected with a blue edge. Hence there exist two triangles of different colors joined at a vertex. These take up five points. Applying Lemma 2 on the five remaining points, we obtain a monochromatic cycle of odd length that is of the same color as one of the two joined triangles and disjoint from both of them.

23. Let us assume $n > 1$. Obviously n is odd. Let $p \geq 3$ be the smallest prime divisor of n. In this case $(p-1, n) = 1$. Since $2^n + 1 \mid 2^{2n} - 1$, we have that $p \mid 2^{2n} - 1$. Thus it follows from Fermat's little theorem and elementary number theory that $p \mid (2^{2n} - 1, 2^{p-1} - 1) = 2^{(2n, p-1)} - 1$. Since $(2n, p-1) \leq 2$, it follows that $p \mid 3$ and hence $p = 3$.

Let us assume now that n is of the form $n = 3^k d$, where $2, 3 \nmid d$. We first prove that $k = 1$.

Lemma. If $2^m - 1$ is divisible by 3^r, then m is divisible by 3^{r-1}.

Proof. This is the lemma from (SL97-14) with $p = 3$, $a = 2^2$, $k = m$, $\alpha = 1$, and $\beta = r$.

Since 3^{2k} divides $n^2 \mid 2^{2n} - 1$, we can apply the lemma to $m = 2n$ and $r = 2k$ to conclude that $3^{2k-1} \mid n = 3^k d$. Hence $k = 1$.

Finally, let us assume $d > 1$ and let q be the smallest prime factor of d. Obviously q is odd, $q \geq 5$, and $(n, q-1) \in \{1, 3\}$. We then have $q \mid 2^{2n} - 1$ and $q \mid 2^{q-1} - 1$. Consequently, $q \mid 2^{(2n,q-1)} - 1 = 2^{2(n,q-1)} - 1$, which divides $2^6 - 1 = 63 = 3^2 \cdot 7$, so we must have $q = 7$. However, in that case we obtain $7 \mid n \mid 2^n + 1$, which is a contradiction, since powers of two can only be congruent to $1, 2$ and 4 modulo 7. It thus follows that $d = 1$ and $n = 3$. Hence $n > 1 \Rightarrow n = 3$.

It is easily verified that $n = 1$ and $n = 3$ are indeed solutions. Hence these are the only solutions.

24. Let us denote $A = b+c+d$, $B = a+c+d$, $C = a+b+d$, $D = a+b+c$. Since $ab + bc + cd + da = 1$ the numbers A, B, C, D are all positive. By trivially applying the AM-GM inequality we have:

$$a^2 + b^2 + c^2 + d^2 \geq ab + bc + cd + da = 1 .$$

We will prove the inequality assuming only that A, B, C, D are positive and $a^2 + b^2 + c^2 + d^2 \geq 1$. In this case we may assume without loss of generality that $a \geq b \geq c \geq d \geq 0$. Hence $a^3 \geq b^3 \geq c^3 \geq d^3 \geq 0$ and $\frac{1}{A} \geq \frac{1}{B} \geq \frac{1}{C} \geq \frac{1}{D} > 0$. Using the Chebyshev and Cauchy inequalities we obtain:

$$\frac{a^3}{A} + \frac{b^3}{B} + \frac{c^3}{C} + \frac{d^3}{D}$$
$$\geq \frac{1}{4}(a^3 + b^3 + c^3 + d^3)\left(\frac{1}{A} + \frac{1}{B} + \frac{1}{C} + \frac{1}{D}\right)$$
$$\geq \frac{1}{16}(a^2 + b^2 + c^2 + d^2)(a+b+c+d)\left(\frac{1}{A} + \frac{1}{B} + \frac{1}{C} + \frac{1}{D}\right)$$
$$= \frac{1}{48}(a^2 + b^2 + c^2 + d^2)(A+B+C+D)\left(\frac{1}{A} + \frac{1}{B} + \frac{1}{C} + \frac{1}{D}\right) \geq \frac{1}{3} .$$

This completes the proof.

25. Plugging in $x = 1$ we get $f(f(y)) = f(1)/y$ and hence $f(y_1) = f(y_2)$ implies $y_1 = y_2$ i.e. that the function is bijective. Plugging in $y = 1$ gives us $f(xf(1)) = f(x) \Rightarrow xf(1) = x \Rightarrow f(1) = 1$. Hence $f(f(y)) = 1/y$. Plugging in $y = f(z)$ implies $1/f(z) = f(1/z)$. Finally setting $y = f(1/t)$ into the original equation gives us $f(xt) = f(x)/f(1/t) = f(x)f(t)$.

Conversely, any functional equation on $\mathbb{Q}^+$ satisfying (i) $f(xt) = f(x)f(t)$ and (ii) $f(f(x)) = \frac{1}{x}$ for all $x, t \in \mathbb{Q}^+$ also satisfies the original functional equation: $f(xf(y)) = f(x)f(f(y)) = \frac{f(x)}{y}$. Hence it suffices to find a function satisfying (i) and (ii).

We note that all elements $q \in \mathbb{Q}^+$ are of the form $q = \prod_{i=1}^n p_i^{a_i}$ where p_i are prime and $a_i \in \mathbb{Z}$. The criterion (i) implies $f(q) = f(\prod_{i=1}^n p_i^{a_i}) = \prod_{i=1}^n f(p_i)^{a_i}$. Thus it is sufficient to define the function on all primes. For the function to satisfy (ii) it is necessary and sufficient for it to satisfy $f(f(p)) = \frac{1}{p}$ for all primes p. Let q_i denote the i-th smallest prime. We define our function f as follows:

$$f(q_{2k-1}) = q_{2k}, \quad f(q_{2k}) = \frac{1}{q_{2k-1}}, \quad k \in \mathbb{N}.$$

Such a function clearly satisfies (ii) and along with the additional condition $f(xt) = f(x)f(t)$ it is well defined for all elements of $\mathbb{Q}^+$ and it satisfies the original functional equation.

26. We note that $|P(x)/x| \to \infty$. Hence, there exists an integer number M such that $M > |q_1|$ and $|P(x)| \le |x| \Rightarrow |x| < M$. It follows that $|q_i| < M$ for all $i \in \mathbb{N}$ because assuming $|q_i| \ge M$ for some i we get $|q_{i-1}| = |P(q_i)| > |q_i| \ge M$ and this ultimately contradicts $|q_1| < M$.

Let us define $q_1 = \frac{r}{s}$ and $P(x) = \frac{ax^3 + bx^2 + cx + d}{e}$ where r, s, a, b, c, d, e are all integers. For $N = sa$ we shall prove by induction that Nq_i is an integer for all $i \in \mathbb{N}$. By definition $N \ne 0$.

For $i = 1$ this obviously holds. Assume it holds for some $i \in \mathbb{N}$. Then using $q_i = P(q_{i+1})$ we have that Nq_{i+1} is a zero of the polynomial

$$Q(x) = \frac{e}{a} N^3 \left(P\left(\frac{x}{N}\right) - q_i \right)$$
$$= x^3 + (sb)x^2 + (s^2ac)x + (s^3a^2d - s^2ae(Nq_i)).$$

Since $Q(x)$ is a monic polynomial with integer coefficients (a conclusion for which we must assume the induction hypothesis) and Nq_{i+1} is rational it follows by the rational root theorem that Nq_{i+1} is an integer.

It follows that all q_i are multiples of $1/N$. Since $-M < q_i < M$ we conclude that q_i can take less than $T = 2M|N|$ distinct values. Therefore for each j there are m_j and $m_j + k_j$ ($k_j > 0$) both belonging to the set $\{jT + 1, jT + 2, \ldots, jT + T\}$ such that $q_{m_j} = q_{m_j + k_j}$. Since $k_j < T$ for all k_j it follows that there exists a positive integer k which appears an infinite number of times in the sequence k_j, i.e. there exist infinitely many integers m such that $q_m = q_{m+k}$. Moreover, $q_m = q_{m+k}$ clearly implies $q_n = q_{n+k}$ for all $n \le m$. Hence $q_n = q_{n+k}$ holds for all n.

27. Let us denote by $A_n(k)$ the n-digit number which consists of $n - 1$ ones and one digit seven in the $k + 1$-th rightmost position ($0 \le k < n$). Then $A_n(k) = (10^n + 54 \cdot 10^k - 1)/9$.

We note that if $3 \mid n$ we have that $3 \mid A_n(k)$ for all k. Hence n cannot be divisible by 3.

Now let $3 \nmid n$. We claim that for each such $n \ge 5$, there exists $k < n$ for which $7 \mid A_n(k)$. We see that $A_n(k)$ is divisible by 7 if and only if $10^n - 1 \equiv 2 \cdot 10^k \pmod{7}$. There are several cases.

$n \equiv 1 \pmod 6$. Then $10^n - 1 \equiv 2 \equiv 2 \cdot 10^0$, so $7 \mid A_n(0)$.
$n \equiv 2 \pmod 6$. Then $10^n - 1 \equiv 1 \equiv 2 \cdot 10^4$, so $7 \mid A_n(4)$.
$n \equiv 4 \pmod 6$. Then $10^n - 1 \equiv 3 \equiv 2 \cdot 10^5$, so $7 \mid A_n(5)$.
$n \equiv 5 \pmod 6$. Then $10^n - 1 \equiv 4 \equiv 2 \cdot 10^2$, so $7 \mid A_n(2)$.

The remaining cases are $n = 1, 2, 4$. For $n = 4$ the number $1711 = 29 \cdot 59$ is composite, while it is easily checked that $n = 1$ and $n = 2$ are solutions. Hence the answer is $n = 1, 2$.

28. Let us first prove the following lemma.

   Lemma. Let $(b'/a', d'/c')$ and $(b''/a'', d''/c'')$ be two points with rational coordinates where the fractions given are irreducible. If both a' and c' are odd and the distance between the two points is 1 then it follows that a'' and c'' are odd, and that $b' + d'$ and $b'' + d''$ are of a different parity.

   Proof. Let b/a and d/c be irreducible fractions such that $b'/a' - b''/a'' = b/a$ and $d'/c' - d''/c'' = d/c$. Then it follows that $b^2/a^2 + d^2/c^2 = 1 \Rightarrow b^2c^2 + a^2d^2 = a^2c^2$. Since $(a,b) = 1$ and $(c,d) = 1$ it follows that $a \mid c$, $c \mid a$ and hence $a = c$. Consequently $b^2 + d^2 = a^2$. Since a is mutually co-prime to b and d it follows that a and $b+d$ are odd. From $b''/a'' = b/a + b'/a'$ we get that $a'' \mid aa'$, so a'' is odd. Similarly, c'' is odd as well. Now it follows that $b'' \equiv b + b'$ and similarly $d'' \equiv d + d'$ (mod 2). Hence $b'' + d'' \equiv b' + d' + b + d \equiv b' + d' + 1$ (mod 2), from which it follows that $b' + d'$ and $b'' + d''$ are of a different parity.

   Without loss of generality we start from the origin of the coordinate system $(0/1, 0/1)$. Initially $b + d = 0$ and after moving to each subsequent point along the broken line $b + d$ changes parity by the lemma. Hence it will not be possible to return to the origin after an odd number of steps since $b + d$ will be odd.

4.32 Solutions to the Shortlisted Problems of IMO 1991

1. All the angles $\angle PP_1C$, $\angle PP_2C$, $\angle PQ_1C$, $\angle PQ_2C$ are right, hence P_1, P_2, Q_1, Q_2 lie on the circle with diameter PC. The result now follows immediately from Pascal's theorem applied to the hexagon $P_1PP_2Q_1CQ_2$. It tells us that the points of intersection of the three pairs of lines P_1C, PQ_1 (intersection A), P_1Q_2, P_2Q_1 (intersection X) and PQ_2, P_2C (intersection B) are collinear.

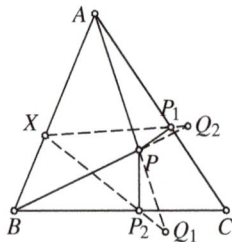

2. Let HQ meet PB at Q' and HR meet PC at R'. From $MP = MB = MC$ we have $\angle BPC = 90^\circ$. So $PR'HQ'$ is a rectangle. Since PH is perpendicular to BC, it follows that the circle with diameter PH, through P, R', H, Q', is tangent to BC. It is now sufficient to show that QR is parallel to $Q'R'$. Let CP meet AB at X, and BP meet AC at Y. Since P is on the median, it follows (for example, by Ceva's theorem) that $AX/XB = AY/YC$, i.e. that XY is parallel to BC.

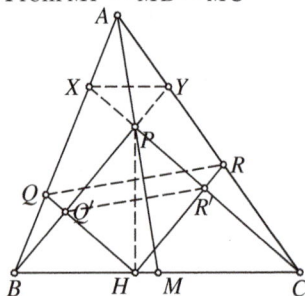

Therefore, $PY/BP = PX/CP$. Since HQ is parallel to CX, we have $QQ'/HQ' = PX/CP$ and similarly $RR'/HR' = PY/BP$. It follows that $QQ'/HQ' = RR'/HR'$, hence QR is parallel to $Q'R'$ as required.

Second solution. It suffices to show that $\angle RHC = \angle RQH$, or equivalently $RH : QH = PC : PB$. We assume $PC : PB = 1 : x$. Let $X \in AB$ and $Y \in AC$ be points such that $MX \perp PB$ and $MY \perp PC$. Since MX bisects $\angle AMB$ and MY bisects AMC, we deduce

$$AX : XB = AM : MB = AY : YC \Rightarrow XY \parallel BC \Rightarrow$$
$$\Rightarrow \triangle XYM \sim \triangle CBP \Rightarrow XM : MY = 1 : x.$$

Now from $CH : HB = 1 : x^2$ we obtain $RH : MY = CH : CM = 1 : \frac{1+x^2}{2}$ and $QH : MX = BH : BM = x^2 : \frac{1+x^2}{2}$. Therefore

$$RH : QH = \frac{2}{1+x^2}MY : \frac{2x^2}{1+x^2}MX = 1 : x.$$

3. Consider the problem with the unit circle on the complex plane. For convenience, we use the same letter for a point in the plane and its corresponding complex number.

Lemma 1. Line $l(S, PQR)$ contains the point $Z = \frac{P+Q+R+S}{2}$.

Proof. Suppose P', Q', R' are the feet of perpendiculars from S to QR, RP, PQ respectively. It suffices to show that P', Q', R', Z are on the same line. Let us first represent P' by Q, R, S. Since $P' \in QR$, we have $\frac{P'-Q}{R-Q} = \overline{\left(\frac{P'-Q}{R-Q}\right)}$, that is,

$$(P' - Q)(\overline{R} - \overline{Q}) = (\overline{P'} - \overline{Q})(R - Q). \tag{1}$$

On the other hand, since $SP' \perp QR$, the ratio $\frac{P'-S}{R-Q}$ is purely imaginary. Thus

$$(P' - S)(\overline{R} - \overline{Q}) = -(\overline{P'} - \overline{S})(R - Q). \tag{2}$$

Eliminating $\overline{P'}$ from (1) and (2) and using the fact that $\overline{X} = X^{-1}$ for X on the unit circle, we obtain $P' = (Q + R + S - QR/S)/2$ and analogously $Q' = (P + R + S - PR/S)/2$ and $R' = (P + Q + S - PQ/S)/2$. Hence $Z - P' = (P + QR/S)/2, Z - Q' = (Q + PR/S)/2$ and $Z - R' = (R + PQ/S)/2$. Setting $P = p^2$, $Q = q^2$, $R = r^2$, $S = s^2$ we obtain $Z - P' = \frac{pqr}{2s}\left(\frac{ps}{qr} + \frac{qr}{ps}\right)$, $Z - Q' = \frac{pqr}{2s}\left(\frac{qs}{pr} + \frac{pr}{qs}\right)$ and $Z - P' = \frac{pqr}{2s}\left(\frac{rs}{pq} + \frac{pq}{rs}\right)$.

Since $x + x^{-1} = 2\mathrm{Re}\,x$ is real for all x on the unit circle, it follows that the ratio of every pair of these differences is real, which means that Z, P', Q', R' belong to the same line.

Lemma 2. If P, Q, R, S are four different points on a circle, then the lines $l(P, QRS)$, $l(Q, RSP)$, $l(R, SPQ)$, $l(S, PQR)$ intersect at one point.

Proof. By Lemma 1, they all pass through $\frac{P+Q+R+S}{2}$.

Now we can find the needed conditions for A, B, ..., F. In fact, the lines $l(A, BDF)$, $l(D, ABF)$ meet at $Z_1 = \frac{A+B+D+F}{2}$, and $l(B, ACE)$, $l(E, ABC)$ meet at $Z_2 = \frac{A+B+C+E}{2}$. Hence, $Z_1 \equiv Z_2$ if and only if $D - C = E - F \Leftrightarrow CDEF$ is a rectangle.

Remark. The line $l(S, PQR)$ is widely known as Simson's line; the proof that the feet of perpendiculars are collinear is straightforward. The key claim, Lemma 1, is a known property of Simson's lines, and can be shown elementarily:

* $l(S, PQR)$ passes through the midpoint X of HS, where H is the orthocenter of PQR.

4. Assume the contrary, that $\angle MAB$, $\angle MBC$, $\angle MCA$ are all greater than $30°$. By the sine Ceva theorem, it holds that

$$\sin \angle MAC \sin \angle MBA \sin \angle MCB$$
$$= \sin \angle MAB \sin \angle MBC \sin \angle MCA > \sin^3 30° = \frac{1}{8}. \tag{$*$}$$

On the other hand, since $\angle MAC + \angle MBA + \angle MCB < 180° - 3 \cdot 30° = 90°$, Jensen's inequality applied on the concave function $\ln \sin x$ ($x \in [0, \pi]$) gives us $\sin \angle MAC \sin \angle MBA \sin \angle MCB < \sin^3 30°$, contradicting $(*)$.

Second solution. Denote the intersections of PA, PB, PC with BC, CA, AB by A_1, B_1, C_1, respectively. Suppose that each of the angles $\angle PAB$, $\angle PBC$, $\angle PCA$ is greater than $30°$ and denote $PA = 2x$, $PB = 2y$, $PC = 2z$. Then $PC_1 > x$, $PA_1 > y$, $PB_1 > z$. On the other hand, we know that

$$\frac{PC_1}{PC+PC_1}+\frac{PA_1}{PA+PA_1}+\frac{PB_1}{PB+PB_1}=\frac{S_{ABP}}{S_{ABC}}+\frac{S_{PBC}}{S_{ABC}}+\frac{S_{APC}}{S_{ABC}}=1.$$

Since the function $\frac{t}{p+t}$ is increasing, we obtain $\frac{x}{2z+x}+\frac{y}{2x+y}+\frac{z}{2y+z}<1$. But on the contrary, Cauchy-Schwartz inequality (or alternatively Jensen's inequality) yields

$$\frac{x}{2z+x}+\frac{y}{2x+y}+\frac{z}{2y+z}\geq\frac{(x+y+z)^2}{x(2z+x)+y(2x+y)+z(2y+z)}=1.$$

5. Let P_1 be the point on the side BC such that $\angle BFP_1=\beta/2$. Then $\angle BP_1F=180^o-3\beta/2$, and the sine law gives us $\frac{BF}{BP_1}=\frac{\sin(3\beta/2)}{\sin(\beta/2)}=3-4\sin^2(\beta/2)=1+2\cos\beta$.
   Now we calculate $\frac{BF}{BP}$. We have $\angle BIF=120^o-\beta/2$, $\angle BFI=60^o$ and $\angle BIC=120^o$, $\angle BCI=\gamma/2=60^o-\beta/2$. By the sine law,

   $$BF=BI\frac{\sin(120^o-\beta/2)}{\sin 60^o},\qquad BP=\frac{1}{3}BC=BI\frac{\sin 120^o}{3\sin(60^o-\beta/2)}.$$

   This implies $\frac{BF}{BP}=\frac{3\sin(60^o-\beta/2)\sin(60^o+\beta/2)}{\sin^2 60^o}=4\sin(60^o-\beta/2)\sin(60^o+\beta/2)=2(\cos\beta-\cos 120^o)=2\cos\beta+1=\frac{BF}{BP_1}$. Therefore $P\equiv P_1$.

6. Let a,b,c be sides of the triangle. Let A_1 be the intersection of line AI with BC. By the known fact, $BA_1:A_1C=c:b$ and $AI:IA_1=AB:BA_1$, hence $BA_1=\frac{ac}{b+c}$ and $\frac{AI}{IA_1}=\frac{AB}{BA_1}=\frac{b+c}{a}$. Consequently $\frac{AI}{IA}=\frac{b+c}{a+b+c}$.
   Put $a=n+p$, $b=p+m$, $c=m+n$: it is obvious that m,n,p are positive. Our inequality becomes

   $$2<\frac{(2m+n+p)(m+2n+p)(m+n+2p)}{(m+n+p)^3}\leq\frac{64}{27}.$$

   The right side inequality immediately follows from the inequality between arithmetic and geometric means applied on $2m+n+p$, $m+2n+p$ and $m+n+2p$. For the left side inequality, denote by $T=m+n+p$. Then we can write $(2m+n+p)(m+2n+p)(m+n+2p)=(T+m)(T+n)(T+p)$ and

   $$(T+m)(T+n)(T+p)=T^3+(m+n+p)T^2+(mn+np+pn)T+mnp>2T^3.$$

   Remark. The inequalities cannot be improved. In fact, $\frac{AI\cdot BI\cdot CI}{I_AI_BI_C}$ is equal to $8/27$ for $a=b=c$, while it can be arbitrarily close to $1/4$ if $a=b$ and c is sufficiently small.

7. The given equations imply $AB=CD$, $AC=BD$, $AD=BC$. Let L_1,M_1,N_1 be the midpoints of AD,BD,CD respectively. Then the above equalities yield

   $$L_1M_1=AB/2=LM,\ L_1M_1\parallel AB\parallel LM;$$
   $$L_1M=CD/2=LM_1,\ L_1M\parallel CD\parallel LM_1.$$

Thus L, M, L_1, M_1 are coplanar and LML_1M_1 is a rhombus as well as MNM_1N_1 and LNL_1N_1. Then the segments LL_1, MM_1, NN_1 have the common midpoint Q and $QL \perp QM$, $QL \perp QN$, $QM \perp QN$. We also infer that the line NN_1 is perpendicular to the plane LML_1M_1 and hence to the line AB. Thus $QA = QB$, and similarly, $QB = QC = QD$, hence Q is just the center O, and $\angle LOM = \angle MON = \angle NOL = 90°$.

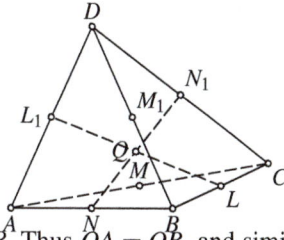

8. Let $P_1(x_1, y_1), P_2(x_2, y_2), \ldots, P_n(x_n, y_n)$ be the n points of S in the coordinate plane. We may assume $x_1 < x_2 < \cdots < x_n$ (choosing adequate axes and renumbering the points if necessary). Define d to be half the minimum distance of P_i from the line $P_j P_k$, where i, j, k go through all possible combinations of mutually distinct indices.

   First we define a set T containing $2n - 4$ points:

   $$T = \{(x_i, y_i - d), (x_i, y_i + d) \mid i = 2, 3, \ldots, n - 1\}.$$

   Consider any triangle $P_k P_l P_m$, where $k < l < m$. Its interior contains at least one of the two points $(x_l, y_l \pm d)$, so T is a set of $2n - 4$ points with the required property. However, at least one of the points of T is useless. The convex hull of S is a polygon with at least three points in S as vertices. Let P_j be a vertex of that hull distinct from P_1 and P_n. Clearly one of the points $(x_j, y_j \pm d)$ lies outside the convex hull, and thus can be left out. The remaining set of $2n - 5$ points satisfies the conditions.

9. Let A_1, A_2 be two points of E which are joined. In $E \setminus \{A_1, A_2\}$, there are at most 397 points to which A_1 is not joined, and at most as much to which A_2 is not joined. Consequently, there exists a point A_3 which is joined to both A_1 and A_2. There are at most $3 \cdot 397 = 1191$ points of $E \setminus \{A_1, A_2, A_3\}$ to which at least one of A_1, A_2, A_3 is not joined, hence it is possible to choose a point A_4 joined to A_1, A_2, A_3. Similarly, there exists a point A_5 which is joined to all A_1, A_2, A_3, A_4. Finally, among the remaining 1986 points, there are at most $5 \cdot 397 = 1985$ which are not joined to one of the points $A_1, \ldots, A_5$. Thus there is at least one point A_6 joined to all $A_1, \ldots, A_5$. It is clear that $A_1, \ldots, A_6$ are pairwise joined.

   Solution of the alternative version. Let be given 1991 points instead. Number the points from 1 to 1991, and join i and j if and only if $i - j$ is not a multiple of 5. Then each i is joined to 1592 or 1593 other points, and obviously among any six points there are two which are not joined.

10. We start at some vertex v_0 and walk along distinct edges of the graph, numbering them $1, 2, \ldots$ in the order of appearance, until this is no longer possible without reusing an edge. If there are still edges which are not numbered, one of them has a vertex which has already been visited (else G would not be connected). Starting from this vertex, we continue to walk along unused edges resuming the

numbering, until we eventually get stuck. Repeating this procedure as long as possible, we shall number all the edges.

Let v be a vertex which is incident with $e \geq 2$ edges. If $v = v_0$, then it is on the edge 1, so the gcd at v is 1. If $v \neq v_0$, suppose that it was reached for the first time by the edge r. At that time there was at least one unused edge incident with v (as $e \geq 2$), hence one of them was labelled by $r + 1$. The gcd at v again equals $\gcd(r, r+1) = 1$.

11. To start with, observe that $\frac{1}{n-m}\binom{n-m}{m} = \frac{1}{n}\left[\binom{n-m}{m} + \binom{n-m-1}{m-1}\right]$.

For $n = 1, 2, \ldots$ set $S_n = \sum_{m=0}^{[n/2]} (-1)^m \binom{n-m}{m}$. Using the identity $\binom{m}{k} = \binom{m-1}{k} + \binom{m-1}{k-1}$ we obtain the following relation for S_n:

$$S_{n+1} = \sum_m (-1)^m \binom{n-m+1}{m}$$
$$= \sum_m (-1)^m \binom{n-m}{m} + \sum_m (-1)^m \binom{n-m}{m-1} = S_n - S_{n-1}.$$

Since the initial members of the sequence S_n are $1, 1, 0, -1, -1, 0, 1, 1, \ldots$, we thus find that S_n is periodic with period 6.

Now the sum from the problem reduces to

$$\frac{1}{1991}\binom{1991}{0} - \frac{1}{1991}\left[\binom{1990}{1} + \binom{1989}{0}\right] + \cdots - \frac{1}{1991}\left[\binom{996}{995} + \binom{995}{994}\right]$$

$$= \frac{1}{1991}(S_{1991} - S_{1989}) = \frac{1}{1991}(0 - (-1)) = \frac{1}{1991}.$$

12. Let A_m be the set of those elements of S which are divisible by m. By the inclusion-exclusion principle, the number of elements divisible by 2, 3, 5 or 7 equals

$$|A_2 \cup A_3 \cup A_5 \cup A_7|$$
$$= |A_2| + |A_3| + |A_5| + |A_7| - |A_6| - |A_{10}| - |A_{14}| - |A_{15}|$$
$$\quad - |A_{21}| - |A_{35}| + |A_{30}| + |A_{42}| + |A_{70}| + |A_{105}| - |A_{210}|$$
$$= 140 + 93 + 56 + 40 - 46 - 28 - 20 - 18$$
$$\quad - 13 - 8 + 9 + 6 + 4 + 2 - 1 = 216.$$

Among any five elements of the set $A_2 \cup A_3 \cup A_5 \cup A_7$, one of the sets A_2, A_3, A_5, A_7 contains at least two, and those two are not relatively prime. Therefore $n > 216$. We claim that the answer is $n = 217$. First notice that the set $A_2 \cup A_3 \cup A_5 \cup A_7$ consists of four prime $(2, 3, 5, 7)$ and 212 composite numbers. The set $S \setminus A$ contains exactly 8 composite numbers: namely, $11^2, 11 \cdot 13, 11 \cdot 17, 11 \cdot 19, 11 \cdot 23, 13^2, 13 \cdot 17, 13 \cdot 19$. Thus S consists of the unity, 220 composite numbers and 59 primes.

Let A be a 217-element subset of S, and suppose that there are no five pairwise relatively prime numbers in A. Then A can contain at most 4 primes (or unity and three primes) and at least 213 composite numbers. Hence the set $S \setminus A$ contains at most 7 composite numbers. Consequently, at least one of the following 8 five-element sets is disjoint with $S \setminus A$, and is thus entirely contained in A:

$$\{2 \cdot 23, 3 \cdot 19, 5 \cdot 17, 7 \cdot 13, 11 \cdot 11\}, \quad \{2 \cdot 29, 3 \cdot 23, 5 \cdot 19, 7 \cdot 17, 11 \cdot 13\},$$
$$\{2 \cdot 31, 3 \cdot 29, 5 \cdot 23, 7 \cdot 19, 11 \cdot 17\}, \quad \{2 \cdot 37, 3 \cdot 31, 5 \cdot 29, 7 \cdot 23, 11 \cdot 19\},$$
$$\{2 \cdot 41, 3 \cdot 37, 5 \cdot 31, 7 \cdot 29, 11 \cdot 23\}, \quad \{2 \cdot 43, 3 \cdot 41, 5 \cdot 37, 7 \cdot 31, 13 \cdot 17\},$$
$$\{2 \cdot 47, 3 \cdot 43, 5 \cdot 41, 7 \cdot 37, 13 \cdot 19\}, \quad \{2 \cdot 2, 3 \cdot 3, 5 \cdot 5, 7 \cdot 7, 13 \cdot 13\}.$$

As each of these sets consists of five numbers relatively prime in pairs, the claim is proved.

13. Call a sequence $e_1, \ldots, e_n$ *good* if $e_1 a_1 + \cdots + e_n a_n$ is divisible by n. Among the sums $s_0 = 0$, $s_1 = a_1$, $s_2 = a_1 + a_2$, $\ldots$, $s_n = a_1 + \cdots + a_n$, two give the same remainder modulo n, and their difference corresponds to a good sequence. To show that, permuting the a_i's, we can find $n - 1$ different sequences, we use the following

Lemma. Let A be a $k \times n$ ($k \le n - 2$) matrix of zeros and ones, whose every row contains at least one 0 and at least two 1's. Then it is possible to permute columns of A is such a way that in any row 1's do not form a block.

Proof. We will use the induction on k. The case $k = 1$ and arbitrary $n \ge 3$ is trivial. Suppose that $k \ge 2$ and that for $k - 1$ and any $n \ge k + 1$ the lemma is true. Consider a $k \times n$ matrix A, $n \ge k + 2$. We mark an element a_{ij} if either it is the only zero in the i-th row, or one of the 1's in the row if it contains exactly two 1's. Since $n \ge 4$, every row contains at most two marked elements, which adds up to at most $2k < 2n$ marked elements in total. It follows that there is a column with at most one marked element. Assume w.l.o.g. that it is the first column and that a_{1j} isn't marked for $j > 1$. The matrix B, obtained by omitting the first row and first column from A, satisfies the conditions of the lemma. Therefore, we can permute columns of B and get the required form. Considered as a permutation of column of A, this permutation may leave a block of 1's only in the first row of A. In the case that it is so, if $a_{11} = 1$ we put the first column in the last place, otherwise we put it between any two columns having 1's in the first row. The obtained matrix has the required property.

Suppose now that we have got k different nontrivial good sequences $e_1^i, \ldots, e_n^i$, $i = 1, \ldots, k$, and that $k \le n - 2$. The matrix $A = (e_j^i)$ fulfills the conditions of Lemma, hence there is a permutation σ from Lemma. Now among the sums $s_0 = 0$, $s_1 = a_{\sigma(1)}$, $s_2 = a_{\sigma(1)} + a_{\sigma(2)}$, $\ldots$, $s_n = a_{\sigma(1)} + \cdots + a_{\sigma(n)}$, two give the same remainder modulo n. Let $s_p \equiv s_q \pmod{n}$, $p < q$. Then $n \mid s_q - s_p = a_{\sigma(p+1)} + \cdots + a_{\sigma(q)}$, and this yields a good sequence $e_1, \ldots, e_n$ with $e_{\sigma(p+1)} = \cdots = e_{\sigma(q)} = 1$ and other e's equal to zero. Since from the construction we see that none of the sequences $e_{\sigma(j)^i}$ has all 1's in a block, in this way we have got a new nontrivial good sequence, and we can continue this procedure until there are $n - 1$ sequences. Together with the trivial $0, \ldots, 0$ sequence, we have found n good sequences.

14. Suppose that $f(x_0), f(x_0+1), \ldots, f(x_0+2p-2)$ are squares. If $p \mid a$ and $p \nmid b$, then $f(x) \equiv bx + c \pmod{p}$ for $x = x_0, \ldots, x_0 + p - 1$ form a complete system of residues modulo p. However, a square is always congruent to exactly one of the $\frac{p+1}{2}$ numbers $0, 1^2, 2^2, \ldots, (\frac{p-1}{2})^2$ and thus cannot give every residue modulo p. Also, if $p \mid a$ and $p \mid b$, then $p \mid b^2 - 4ac$.

We now assume $p \nmid a$. The following identities hold for any quadric polynomial:

$$4a \cdot f(x) = (2ax + b)^2 - (b^2 - 4ac) \tag{1}$$

and

$$f(x+p) - f(x) = p(2ax + b) + p^2 a. \tag{2}$$

Suppose that there is an y, $x_0 \le y \le x_0 + p - 2$, for which $f(y)$ is divisible by p. Then both $f(y)$ and $f(y+p)$ are squares divisible by p, and therefore both are divisible by p^2. But relation (2) implies that $p \mid 2ay + b$, and hence by (1) $b^2 - 4ac$ is divisible by p as well.

Therefore it suffices to show that such an y exists, and for that aim we prove that there are two such y in $[x_0, x_0 + p - 1]$. Assume the opposite. Since for $x = x_0, x_0 + 1, \ldots, x_0 + p - 1$ $f(x)$ is congruent modulo p to one of the $\frac{p-1}{2}$ numbers $1^2, 2^2, \ldots, (\frac{p-1}{2})^2$, it follows by the pigeon-hole principle that for some mutually distinct $u, v, w \in \{x_0, \ldots, x_0 + p - 1\}$ we have $f(u) \equiv f(v) \equiv f(w) \pmod{p}$. Consequently the difference $f(u) - f(v) = (u - v)(a(u + v) + b)$ is divisible by p, but it is clear that $p \nmid u - v$, hence $a(u + v) \equiv -b \pmod{p}$. Similarly $a(u + w) \equiv -b \pmod{p}$, which together with the previous congruence yields $p \mid a(v - w) \Rightarrow p \mid v - w$ which is clearly impossible. It follows that $p \mid f(y_1)$ for at least one y_1, $x_0 \le y_1 < x_0 + p$.

If y_2, $x_0 \le y_2 < x_0 + p$ is such that $a(y_1 + y_2) + b \equiv 0 \pmod{p}$, we have $p \mid f(y_1) - f(y_2) \Rightarrow p \mid f(y_2)$. If $y_1 = y_2$, then by (1) $p \mid b^2 - 4ac$. Otherwise, among y_1, y_2 one belongs to $[x_0, x_0 + p - 2]$ as required.

Second solution. Using Legendre's symbols $\left(\frac{a}{p}\right)$ for quadratic residues we can prove a stronger statement for $p \ge 5$. It can be shown that

$$\sum_{x=0}^{p-1} \left(\frac{ax^2 + bx + c}{p}\right) = -\left(\frac{a}{p}\right) \quad \text{if} \quad p \nmid b^2 - 4ac,$$

hence for at most $\frac{p+3}{2}$ values of x between x_0 and $x_0 + p - 1$ inclusive, $ax^2 + bx + c$ is a quadratic residue or 0 modulo p. Therefore, if $p \ge 5$ and $f(x)$ is a square for $\frac{p+5}{2}$ consecutive values, then $p \mid b^2 - 4ac$.

15. Assume that the sequence has the period T. We can find integers $k > m > 0$, as large as we like, such that $10^k \equiv 10^m \pmod{T}$, using for example Euler's theorem. It is obvious that $a_{10^k - 1} = a_{10^k}$ and hence, taking k sufficiently large and using the periodicity, we see that

$$a_{2 \cdot 10^k - 10^m - 1} = a_{10^k - 1} = a_{10^k} = a_{2 \cdot 10^k - 10^m}.$$

Since $(2 \cdot 10^k - 10^m)! = (2 \cdot 10^k - 10^m)(2 \cdot 10^k - 10^m - 1)!$ and the last nonzero digit of $2 \cdot 10^k - 10^m$ is nine, we must have $a_{2 \cdot 10^k - 10^m - 1} = 5$ (if s is a digit, the last digit of $9s$ is s only if $s = 5$). But this means that 5 divides $n!$ with a greater power than 2 does, which is impossible. Indeed, if the exponents of these powers are α_2, α_5 respectively, then $\alpha_5 = [n/5] + [n/5^2] + \cdots \le \alpha_2 = [n/2] + [n/2^2] + \cdots$.

16. Let p be the least prime number that does not divide n: thus $a_1 = 1$ and $a_2 = p$. Since $a_2 - a_1 = a_3 - a_2 = \cdots = r$, the a_i's are $1, p, 2p - 1, 3p - 2, \ldots$. We have the following cases:

   $p = 2$. Then $r = 1$ and the numbers $1, 2, 3, \ldots, n - 1$ are relatively prime to n, hence n is a prime.

   $p = 3$. Then $r = 2$, so every odd number less than n is relatively prime to n, from which we deduce that n has no odd divisors. Therefore $n = 2^k$ for some $k \in \mathbb{N}$.

   $p > 3$. Then $r = p - 1$ and $a_{k+1} = a_1 + k(p-1) = 1 + k(p-1)$. Since $n-1$ also must belong to the progression, we have $p - 1 \mid n - 2$. Let q be any prime divisor of $p - 1$. Then also $q \mid n - 2$. On the other hand, since $q < p$, it must divide n too, therefore $q \mid 2$, i.e. $q = 2$. This means that $p - 1$ has no prime divisors other than 2 and thus $p = 2^l + 1$ for some $l \ge 2$. But in order for p to be prime, l must be even (because $3 \mid 2^l + 1$ for l odd). Now we recall that $2p - 1$ is also relatively prime to n; but $2p - 1 = 2^{l+1} + 1$ is divisible by 3, which is a contradiction because $3 \mid n$.

17. Taking the equation $3^x + 4^y = 5^z$ $(x, y, z > 0)$ modulo 3, we get that $5^z \equiv 1 \pmod{3}$, hence z is even, say $z = 2z_1$. The equation then becomes $3^x = 5^{2z_1} - 4^y = (5^{z_1} - 2^y)(5^{z_1} + 2^y)$. Each factor $5^{z_1} - 2^y$ and $5^{z_1} + 2^y$ is a power of 3, for which the only possibility is $5^{z_1} + 2^y = 3^x$ and $5^{z_1} - 2^y = 1$. Again modulo 3 these equations reduce to $(-1)^{z_1} + (-1)^y = 0$ and $(-1)^{z_1} - (-1)^y = 1$, implying that z_1 is odd and y is even. Particularly, $y \ge 2$. Reducing the equation $5^{z_1} + 2^y = 3^x$ modulo 4 we get that $3^x \equiv 1$, hence x is even. Now if $y > 2$, modulo 8 this equation yields $5 \equiv 5^{z_1} \equiv 3^x \equiv 1$, a contradiction. Hence $y = 2$, $z_1 = 1$. The only solution of the original equation is $x = y = z = 2$.

18. For integers $a > 0$, $n > 0$ and $\alpha \ge 0$, we shall write $a^\alpha \parallel n$ when $a^\alpha \mid n$ and $a^{\alpha+1} \nmid n$.

   Lemma. For every odd number $a \ge 3$ and an integer $n \ge 0$ it holds that

   $$a^{n+1} \parallel (a+1)^{a^n} - 1 \quad \text{and} \quad a^{n+1} \parallel (a-1)^{a^n} + 1.$$

   Proof. We shall prove the first relation by induction (the second is analogous). For $n = 0$ the statement is obvious. Suppose that it holds for some n, i.e. that $(1+a)^{a^n} = 1 + Na^{n+1}$, $a \nmid N$. Then

   $$(1+a)^{a^{n+1}} = (1 + Na^{n+1})^a = 1 + a \cdot Na^{n+1} + \binom{a}{2}N^2 a^{2n+2} + Ma^{3n+3}$$

for some integer M. Since $\binom{a}{2}$ is divisible by a for a odd, we deduce that the part of the above sum behind $1 + a \cdot N a^{n+1}$ is divisible by a^{n+3}. Hence $(1+a)^{a^{n+1}} = 1 + N' a^{n+2}$, where $a \nmid N'$.

It follows immediately from Lemma that

$$1991^{1993} \parallel 1990^{1991^{1992}} + 1 \quad \text{and} \quad 1991^{1991} \parallel 1992^{1991^{1990}} - 1.$$

Adding these two relations we obtain immediately that $k = 1991$ is the desired value.

19. Set $x = \cos(\pi a)$. The given equation is equivalent to $4x^3 + 4x^2 - 3x - 2 = 0$, which factorizes as $(2x + 1)(2x^2 + x - 2) = 0$.

The case $2x + 1 = 0$ yields $\cos(\pi a) = -1/2$ and $a = 2/3$. It remains to show that if x satisfies $2x^2 + x - 2 = 0$ then a is not rational. The polynomial equation $2x^2 + x - 2 = 0$ has two real roots, $x_{1,2} = \frac{-1 \pm \sqrt{17}}{4}$, and since $|x| \leq 1$ we must have $x = \cos \pi a = \frac{-1 + \sqrt{17}}{4}$.

We now prove by induction that, for every integer $n \geq 0$, $\cos(2^n \pi a) = \frac{a_n + b_n \sqrt{17}}{4}$ for some odd integers a_n, b_n. The case $n = 0$ is trivial. Also, if $\cos(2^n \pi a) = \frac{a_n + b_n \sqrt{17}}{4}$, then

$$\cos(2^{n+1} \pi a) = 2 \cos^2(2^n \pi a) - 1$$
$$= \frac{1}{4} \left(\frac{a_n^2 + 17 b_n^2 - 8}{2} + a_n b_n \sqrt{17} \right) = \frac{a_{n+1} + b_{n+1} \sqrt{17}}{4}.$$

By the inductive step that a_n, b_n are odd, it is obvious that a_{n+1}, b_{n+1} are also odd. This proves the claim.

Note also that, since $a_{n+1} = \frac{1}{2}(a_n^2 + 17 b_n^2 - 8) > a_n$, the sequence $\{a_n\}$ is strictly increasing. Hence the set of values of $\cos(2^n \pi a)$, $n = 0, 1, 2, \ldots$, is infinite (because $\sqrt{17}$ is irrational). However, if a were rational, then the set of values of $\cos m\pi a$, $m = 1, 2, \ldots$, would be finite, a contradiction. Therefore the only possible value for a is $2/3$.

20. We prove the result with 1991 replaced by any positive integer k. For natural numbers p, q, let $\varepsilon = (\alpha p - [\alpha p])(\alpha q - [\alpha q])$. Then $0 < \varepsilon < 1$ and

$$\varepsilon = \alpha^2 pq - \alpha(p[\alpha q] + q[\alpha p]) + [\alpha p][\alpha q].$$

Multiplying this equality by $\alpha - k$ and using $\alpha^2 = k\alpha + 1$, i.e. $\alpha(\alpha - k) = 1$, we get

$$(\alpha - k)\varepsilon = \alpha(pq + [\alpha p][\alpha q]) - (p[\alpha q] + q[\alpha p] + k[\alpha p][\alpha q]).$$

Since $0 < (\alpha - k)\varepsilon < 1$, we have $[\alpha(p * q)] = p[\alpha q] + q[\alpha p] + k[\alpha p][\alpha q]$. Now

$$(p * q) * r = (p * q)r + [\alpha(p * q)][\alpha r] =$$
$$= pqr + [\alpha p][\alpha q]r + [\alpha q][\alpha r]p + [\alpha r][\alpha p]q + k[\alpha p][\alpha q][\alpha r].$$

Since the last expression is symmetric, the same formula is obtained for $p * (q * r)$.

21. The polynomial $g(x)$ factorizes as $g(x) = f(x)^2 - 9 = (f(x) - 3)(f(x) + 3)$. If one of the equations $f(x) + 3 = 0$ and $f(x) - 3 = 0$ has no integer solutions, then the number of integer solutions of $g(x) = 0$ clearly does not exceed 1991.

Suppose now that both $f(x) + 3 = 0$ and $f(x) - 3 = 0$ have integer solutions. Let $x_1, \ldots, x_k$ be distinct integer solutions of the former, and $x_{k+1}, \ldots, x_{k+l}$ be distinct integer solutions of the latter equation. There exist monic polynomials $p(x), q(x)$ with integer coefficients such that $f(x) + 3 = (x - x_1)(x - x_2) \ldots (x - x_k)p(x)$ and $f(x) - 3 = (x - x_{k+1})(x - x_{k+2}) \ldots (x - x_{k+l})q(x)$. Thus we obtain

$$(x - x_1)(x - x_2) \ldots (x - x_k)p(x) - (x - x_{k+1})(x - x_{k+2}) \ldots (x - x_{k+l})q(x) = 6.$$

Putting $x = x_{k+1}$ we get $(x_{k+1} - x_1)(x_{k+1} - x_2) \cdots (x_{k+1} - x_k) \mid 6$, and since the product of more than four distinct integers cannot divide 6, this implies $k \le 4$. Similarly $l \le 4$; hence $g(x) = 0$ has at most 8 distinct integer solutions.

Remark. The proposer provided a solution for the upper bound of 1995 roots which was essentially the same as that of (IMO74-6).

22. Suppose w.l.o.g. that the center of the square is at the origin $O(0,0)$. We denote the curve $y = f(x) = x^3 + ax^2 + bx + c$ by γ and the vertices of the square by A, B, C, D in this order.

At first, the symmetry with respect to the point O maps γ into the curve $\bar{\gamma}$ ($y = f(-x) = x^3 - ax^2 + bx - c$). Obviously $\bar{\gamma}$ also passes through A, B, C, D, and thus has four different intersection points with γ. Then $2ax^2 + 2c$ has at least four distinct solution, which implies $a = c = 0$. Particularly, γ passes through O and intersects all quadrants, and hence $b < 0$.

Further, the curve γ', obtained by rotation of γ around O for $90°$, has an equation $-x = f(y)$ and also contains the points A, B, C, D and O. The intersection points (x, y) of $\gamma \cap \gamma'$ are determined by $-x = f(f(x))$, and hence they are roots of a polynomial $p(x) = f(f(x)) + x$ of 9-th degree. But the number of times that one cubic actually crosses the other in each quadrant is in the general case even (draw the picture!), and since $ABCD$ is the only square lying on $\gamma \cap \gamma'$, the intersection points A, B, C, D must be double. It follows that

$$p(x) = x[(x - r)(x + r)(x - s)(x + s)]^2, \tag{1}$$

where r, s are the x-coordinates of A and B. On the other hand, $p(x)$ is defined by $(x^3 + bx)^3 + b(x^3 + bx) + x$, and therefore equating of coefficients with (1) yields

$$3b = -2(r^2 + s^2), \qquad 3b^2 = (r^2 + s^2)^2 + 2r^2s^2,$$
$$b(b^2 + 1) = -2r^2s^2(r^2 + s^2), \qquad b^2 + 1 = r^4s^4.$$

Straightforward solving this system of equations gives $b = -\sqrt{8}$ and $r^2 + s^2 = \sqrt{18}$.

The line segment from O to (r, s) is half a diagonal of the square, and thus a side of the square has length $a = \sqrt{2(r^2 + s^2)} = \sqrt[4]{72}$.

23. From (i), replacing m by $f(f(m))$, we get

$$f(f(f(m)) + f(f(n))) = -f(f(f(f(m)) + 1)) - n;$$
analogously $$f(f(f(n)) + f(f(m))) = -f(f(f(f(n)) + 1)) - m.$$

From these relations we get $f(f(f(f(m)) + 1)) - f(f(f(f(n)) + 1)) = m - n$. Again from (i),

$$f(f(f(f(m)) + 1)) = f(-m - f(f(2)))$$
and $$f(f(f(f(n)) + 1)) = f(-n - f(f(2))).$$

Setting $f(f(2)) = k$ we obtain $f(-m - k) - f(-n - k) = m - n$ for all integers m, n. This implies $f(m) = f(0) - m$. Then also $f(f(m)) = m$, and using this in (i) we finally get

$$f(n) = -n - 1 \quad \text{for all integers } n.$$

Particularly $f(1991) = -1992$.
From (ii) we obtain $g(n) = g(-n - 1)$ for all integers n. Since g is a polynomial, it must also satisfy $g(x) = g(-x - 1)$ for all real x. Let us now express g as a polynomial on $x + 1/2$: $g(x) = h(x + 1/2)$. Then h satisfies $h(x + 1/2) = h(-x - 1/2)$, i.e. $h(y) = h(-y)$, hence it is a polynomial in y^2; thus g is a polynomial in $(x + 1/2)^2 = x^2 + x + 1/4$. Hence $g(n) = p(n^2 + n)$ (for some polynomial p) is the most general form of g.

24. Let $y_k = a_k - a_{k+1} + a_{k+2} - \cdots + a_{k+n-1}$ for $k = 1, 2, \ldots, n$, where we define $x_{i+n} = x_i$ for $1 \le i \le n$. We then have $y_1 + y_2 = 2a_1, y_2 + y_3 = 2a_2, \ldots, y_n + y_1 = 2a_n$.

   (i) Let $n = 4k - 1$ for some integer $k > 0$. Then for each $i = 1, 2, \ldots, n$ we have that $y_i = (a_i + a_{i+1} + \cdots + a_{i-1}) - 2(a_{i+1} + a_{i+3} + \cdots + a_{i-2}) = 1 + 2 + \cdots + (4k - 1) - 2(a_{i+1} + a_{i+3} + \cdots + a_{i-2})$ is even. Suppose now that $a_1, \ldots, a_n$ is a good permutation. Then each y_i is positive and even, so $y_i \ge 2$. But for some $t \in \{1, \ldots, n\}$ we must have $a_t = 1$, and thus $y_t + y_{t+1} = 2a_t = 2$ which is impossible. Hence the numbers $n = 4k - 1$ are not good.

   (ii) Let $n = 4k + 1$ for some integer $k > 0$. Then $2, 4, \ldots, 4k, 4k + 1, 4k - 1, \ldots, 3, 1$ is a permutation with the desired property. Indeed, in this case $y_1 = y_{4k+1} = 1, y_2 = y_{4k} = 3, \ldots, y_{2k} = y_{2k+2} = 4k - 1, y_{2k+1} = 4k + 1$.
   Therefore all nice numbers are given by $4k + 1, k \in \mathbb{N}$.

25. Since replacing x_1 by 1 can only reduce the set of indices i for which the desired inequality holds, we may assume $x_1 = 1$. Similarly we may assume $x_n = 0$. Now we can let i be the largest index such that $x_i > 1/2$. Then $x_{i+1} \le 1/2$, hence

$$x_i(1 - x_{i+1}) \ge \frac{1}{4} = \frac{1}{4}x_1(1 - x_n).$$

26. Without loss of generality we can assume $b_1 \ge b_2 \ge \cdots \ge b_n$. We denote by A_i the product $a_1 a_2 \ldots a_{i-1} a_{i+1} \ldots a_n$. If for some $i < j$ holds $A_i < A_j$, then $b_i A_i + b_j A_j \le b_i A_j + b_j A_i$ (or equivalently $(b_i - b_j)(A_i - A_j) \le 0$). Therefore the sum $\sum_{i=1}^{n} b_i A_i$ does not decrease when we rearrange the numbers $a_1, \ldots, a_n$ so that $A_1 \ge \cdots \ge A_n$, and consequently $a_1 \le \cdots \le a_n$. Further, for fixed a_i's and $\sum b_i =$

1, the sum $\sum_{i=1}^{n} b_i A_i$ is maximal when b_1 takes the largest possible value, i.e. $b_1 = p$, b_2 takes the remaining largest possible value $b_2 = 1 - p$, whereas $b_3 = \cdots = b_n = 0$. In this case

$$\sum_{i=1}^{n} b_i A_i = pA_1 + (1-p)A_2 = a_3 \ldots a_n(pa_2 + (1-p)a_1)$$

$$\leq p(a_1 + a_2)a_3 \ldots a_n \leq \frac{p}{(n-1)^{n-1}},$$

using the inequality between the geometric and arithmetic means for $a_3, \ldots, a_n$, $a_1 + a_2$.

27. Write $F(x_1, \ldots, x_n) = \sum_{i<j} x_i x_j (x_i + x_j)$. Choose an n-tuple $(x_1, \ldots, x_n)$, $\sum_{i=1}^{n} x_i = 1$, $x_i \geq 0$ with at least three nonzero components, and assume w.l.o.g. that $x_1 \geq \cdots \geq x_{k-1} \geq x_k \geq x_{k+1} = \cdots = x_n = 0$. We claim that replacing x_{k-1}, x_k with $x_{k-1} + x_k, 0$ the value of F increases. Write for brevity $x_{k-1} = a$, $x_k = b$. Then

$$F(\ldots, a+b, 0, 0, \ldots) - F(\ldots, a, b, 0, \ldots)$$

$$= \sum_{i=1}^{k-2} x_i(a+b)(x_i + a + b) - \sum_{i=1}^{k-2} [x_i a(x_i + a) + x_i b(x_i + b)] - ab(a+b)$$

$$= ab\left(2\sum_{i=1}^{k-2} x_i - a - b\right) = ab(2 - 3(a+b)) > 0,$$

because $x_{k-1} + x_k \leq \frac{2}{3}(x_1 + x_{k-1} + x_{k-2}) \leq \frac{2}{3}$. Repeating this procedure we can reduce the number of nonzero x_i's to two, increasing the value of F in each step. It remains to maximize F over n-tuples $(x_1, x_2, 0, \ldots, 0)$ with $x_1, x_2 \geq 0$, $x_1 + x_2 = 1$: in this case F equals $x_1 x_2$ and attains its maximum value $\frac{1}{4}$ when $x_1 = x_2 = \frac{1}{2}$, $x_3 = \ldots, x_n = 0$.

28. Let $x_n = c(n\sqrt{2} - [n\sqrt{2}])$ for some constant $c > 0$. For $i > j$, putting $p = [i\sqrt{2}] - [j\sqrt{2}]$, we have

$$|x_i - x_j| = c|(i-j)\sqrt{2} - p| = \frac{|2(i-j)^2 - p^2|c}{(i-j)\sqrt{2} + p} \geq \frac{c}{(i-j)\sqrt{2} + p} \geq \frac{c}{4(i-j)},$$

because $p < (i-j)\sqrt{2} + 1$. Taking $c = 4$, we obtain that for any $i > j$, $(i-j)|x_i - x_j| \geq 1$. Of course, this implies $(i-j)^a |x_i - x_j| \geq 1$ for any $a > 1$.

Remark. The constant 4 can be replaced with $3/2 + \sqrt{2}$.

Second solution. Another example of a sequence $\{x_n\}$ is constructed in the following way: $x_1 = 0$, $x_2 = 1$, $x_3 = 2$ and $x_{3^k i + m} = x_m + \frac{i}{3^k}$ for $i = 1, 2$ and $1 \leq m \leq 3^k$. It is easily shown that $|i - j| \cdot |x_i - x_j| \geq 1/3$ for any $i \neq j$.

Third solution. If $n = b_0 + 2b_1 + \cdots + 2^k b_k$, $b_i \in \{0, 1\}$, then one can set $x_n = b_0 + 2^{-a}b_1 + \cdots + 2^{-ka}b_k$. In this case it holds that $|i - j|^a |x_i - x_j| \geq \frac{2^a - 2}{2^a - 1}$.

29. One easily observes that the following sets are super-invariant: one-point set, its complement, closed and open half-lines or their complements, and the whole real line. To show that these are the only possibilities, we first observe that S is super-invariant if and only if for each $a > 0$ there is a b such that $x \in S \Leftrightarrow ax + b \in S$.

   (i) Suppose that for some a there are two such b's: b_1 and b_2. Then $x \in S \Leftrightarrow ax + b_1 \in S$ and $x \in S \Leftrightarrow ax + b_2 \in S$, which implies that S is periodic: $y \in S \Leftrightarrow y + \frac{b_1 - b_2}{a} \in S$. Since S is identical to a translate of any stretching of S, all positive numbers are periods of S. Therefore $S \equiv \mathbb{R}$.

   (ii) Assume that, for each a, $b = f(a)$ is unique. Then for any a_1 and a_2,

$$x \in S \Leftrightarrow a_1 x + f(a_1) \in S \Leftrightarrow a_1 a_2 x + a_2 f(a_1) + f(a_2) \in S$$
$$\Leftrightarrow a_2 x + f(a_2) \in S \Leftrightarrow a_1 a_2 x + a_1 f(a_2) + f(a_1) \in S.$$

   As above it follows that $a_1 f(a_2) + f(a_1) = a_2 f(a_1) + f(a_2)$, or equivalently $f(a_1)(a_2 - 1) = f(a_2)(a_1 - 1)$. Hence (for some c), $f(a) = c(a - 1)$ for all a. Now $x \in S \Leftrightarrow ax + c(a - 1) \in S$ actually means that $y - c \in S \Leftrightarrow ay - c \in S$ for all a. Then it is easy to conclude that $\{y - c \mid y \in S\}$ is either a half-line or the whole line, and so is S.

30. Let a and b be the integers written by A and B respectively, and let $x < y$ be the two integers written by the referee. Suppose that none of A and B ever answers "yes".

   Initially, regardless of a, A knows that $0 \le b \le y$ and answers "no". In the second step, B knows that A obtained $0 \le b \le y$, but if a were greater than x, A would know that $a + b = y$ and would thus answer "yes". So B concludes $0 \le a \le x$ but answers "no". The process continues.

   Suppose that, in the n-th step, A knows that B obtained $r_{n-1} \le a \le s_{n-1}$. If $b > x - r_{n-1}$, B would know that $a + b > x$ and hence $a + b = y$, while if $b < y - s_{n-1}$, B would know that $a + b < y$, i.e. $a + b = x$: in both cases he would be able to guess a. However, B answered "no", from which A concludes $y - s_{n-1} \le b \le x - r_{n-1}$. Put $r_n = y - s_{n-1}$ and $s_n = x - r_{n-1}$.

   Similarly, in the next step B knows that A obtained $r_n \le b \le s_n$ and, since A answered "no", concludes $y - s_n \le a \le x - r_n$. Put $r_{n+1} = y - s_n$ and $s_{n+1} = x - r_n$. Notice that in both cases $s_{i+1} - r_{i+1} = s_i - r_i - (y - x)$. Since $y - x > 0$, there exists an m for which $s_m - r_m < 0$, a contradiction.

4.33 Solutions to the Shortlisted Problems of IMO 1992

1. Assume that a pair (x, y) with $x < y$ satisfies the required conditions. We claim that the pair (y, x_1) also satisfies the conditions, where $x_1 = \frac{y^2 + m}{x}$ (note that $x_1 > y$ is a positive integer). This will imply the desired result, since starting from the pair $(1, 1)$ we can obtain arbitrarily many solutions.
First, we show that $\gcd(x_1, y) = 1$. Suppose to the contrary that $\gcd(x_1, y) = d > 1$. Then $d \mid x_1 \mid y^2 + m \Rightarrow d \mid m$, which implies $d \mid y \mid x^2 + m \Rightarrow d \mid x$. But this last is impossible, since $\gcd(x, y) = 1$. Thus it remains to show that $x_1 \mid y^2 + m$ and $y \mid x_1^2 + m$. The former relation is obvious. Since $\gcd(x, y) = 1$, the latter is equivalent to $y \mid (xx_1)^2 + mx^2 = y^4 + 2my^2 + m^2 + mx^2$, which is true because $y \mid m(m + x^2)$ by the assumption. Hence (y, x_1) indeed satisfies all the required conditions.

Remark. The original problem asked to prove the existence of a pair (x, y) of positive integers satisfying the given conditions such that $x + y \leq m + 1$. The problem in this formulation is trivial, since the pair $x = y = 1$ satisfies the conditions. Moreover, this is sometimes the only solution with $x + y \leq m + 1$. For example, for $m = 3$ the least nontrivial solution is $(x_0, y_0) = (1, 4)$.

2. Let us define x_n inductively as $x_n = f(x_{n-1})$, where $x_0 \geq 0$ is a fixed real number. It follows from the given equation in f that $x_{n+2} = -ax_{n+1} + b(a+b)x_n$. The general solution to this equation is of the form

$$x_n = \lambda_1 b^n + \lambda_2 (-a - b)^n,$$

where $\lambda_1, \lambda_2 \in \mathbb{R}$ satisfy $x_0 = \lambda_1 + \lambda_2$ and $x_1 = \lambda_1 b - \lambda_2(a+b)$. In order to have $x_n \geq 0$ for all n we must have $\lambda_2 = 0$. Hence $x_0 = \lambda_1$ and $f(x_0) = x_1 = \lambda_1 b = bx_0$. Since x_0 was arbitrary, we conclude that $f(x) = bx$ is the only possible solution of the functional equation. It is easily verified that this is indeed a solution.

3. Consider two squares $AB'CD'$ and $A'BC'D$. Since $AC \perp BD$, these two squares are homothetic, which implies that the lines AA', BB', CC', DD' are concurrent at a certain point O.
Since the rotation about A by $90°$ takes $\triangle ABK$ into $\triangle AFD$, it follows that $BK \perp DF$. Denote by T the intersection of BK and DF. The rotation about some point X by $90°$ maps BK into DF if and only if TX bisects an angle between BK and DF. Therefore $\angle FTA = \angle ATK = 45°$. Moreover, the quadrilateral $BA'DT$ is cyclic. Therefore $\angle BTA' = BDA' = 45°$ and consequently that the points A, T, A' are collinear. It follows that the point O lies on a bisector of $\angle BTD$ and therefore the rotation $\mathscr{R}$ about O by $90°$ takes BK into DF. Analogously, $\mathscr{R}$ maps the lines CE, DG, AI into AH, BJ, CL. Hence the quadrilateral $P_1Q_1R_1S_1$ is the image of the quadrilateral $P_2Q_2R_2S_2$, and the result follows.

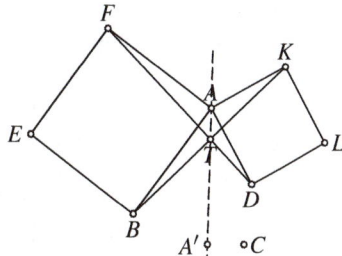

4. There are 36 possible edges in total. If not more than 3 edges are left undrawn, then we can choose 6 of the given 9 points no two of which are connected by an undrawn edge. These 6 points together with the edges between them form a two-colored complete graph, and thus by a well-known result there exists at least one monochromatic triangle. It follows that $n \leq 33$.

In order to show that $n = 33$, we shall give an example of a graph with 32 edges that does not contain a monochromatic triangle. Let us start with a complete graph C_5 with 5 vertices. Its edges can be colored in two colors so that there is no monochromatic triangle (Fig. 1). Furthermore, given a graph $\mathcal{H}$ with k vertices without monochromatic triangles, we can add to it a new vertex, join it to all vertices of $\mathcal{H}$ except A, and color each edge BX in the same way as AX. The obtained graph obviously contains no monochromatic triangles. Applying this construction four times to the graph C_5 we arrive to an example like that shown on Fig. 2.

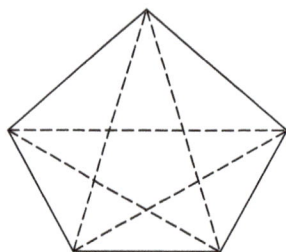

Fig. 1

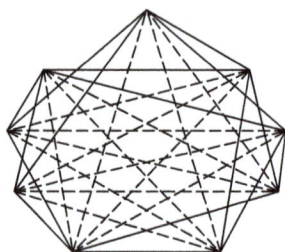

Fig. 2

Second solution. For simplicity, we call the colors red and blue.

Let $r(k,l)$ be the least positive integer r such that each complete r-graph whose edges are colored in red and blue contains either a complete red k-graph or a complete blue l-graph. Also, let $t(n,k)$ be the greatest possible number of edges in a graph with n vertices that does not contain a complete k-graph. These numbers exist by the theorems of Ramsey and Turán.

Let us assume that $r(k,l) < n$. Every graph with n vertices and $t(n,r(k,l)) + 1$ edges contains a complete subgraph with $r(k,l)$ vertices, and this subgraph contains either a red complete k-graph or a blue complete l-graph.

We claim that $t(n,r(k,l)) + 1$ is the smallest number of edges with the above property. By the definition of $r(k,l)$ there exists a coloring of the complete graph H with $r(k,l) - 1$ vertices in two colors such that no red complete k-graph or blue complete l-graph exists. Let c_{ij} be the color in which the edge (i,j) of H is colored, $1 \leq i < j \leq r(k,l) - 1$. Consider a complete $r(k,l) - 1$-partite graph G with n vertices and exactly $t(n,r(k,l))$ edges and denote its partitions by P_i, $i = 1,\ldots,r(k,l) - 1$. If we color each edge of H between P_i and P_j ($j < i$) in the color c_{ij}, we obviously obtain a graph with n vertices and $t(n,r(k,l))$ edges in two colors that contains neither a red complete k-graph nor a blue complete l-graph.

Therefore the answer to our problem is $t(9,r(3,3)) + 1 = t(9,6) + 1 = 33$.

5. Denote by K, L, M, and N the midpoints of the sides AB, BC, CD, and DA, respectively. The quadrilateral $KLMN$ is a rhombus. We shall prove that $O_1 O_3 \parallel KM$. Similarly, $O_2 O_4 \parallel LN$, and the desired result follows immediately.

We have $\overrightarrow{O_1 O_3} = \overrightarrow{KM} + \left(\overrightarrow{O_1 K} + \overrightarrow{MO_3}\right)$. Assume that $ABCD$ is positively oriented. A rotational homothety $\mathcal{R}$ with angle $-90°$ and coefficient $1/\sqrt{3}$ takes the vectors $\overrightarrow{BK}$ and $\overrightarrow{CM}$ into $\overrightarrow{O_1 K}$ and $\overrightarrow{MO_3}$ respectively. Therefore

$$\overrightarrow{O_1 O_3} = \overrightarrow{KM} + (\overrightarrow{O_1 K} + \overrightarrow{MO_3}) = \overrightarrow{KM} + \mathcal{R}(\overrightarrow{BK} + \overrightarrow{CM})$$
$$= \overrightarrow{KM} + \frac{1}{2}\mathcal{R}(\overrightarrow{BA} + \overrightarrow{CD}) = \overrightarrow{KM} + \mathcal{R}(\overrightarrow{LN}).$$

Since $LN \perp KM$, it follows that $\mathcal{R}(LN)$ is parallel to KM and so is $O_1 O_3$.

6. It is easy to see that f is injective and surjective. From $f(x^2 + f(y)) = f((-x)^2 + f(y))$ it follows that $f(x)^2 = (f(-x))^2$, which implies $f(-x) = -f(x)$ because f is injective. Furthermore, there exists $z \in \mathbb{R}$ such that $f(z) = 0$. From $f(-z) = -f(z) = 0$ we deduce that $z = 0$. Now we have $f(x^2) = f(x^2 + f(0)) = 0 + (f(x))^2 = f(x)^2$, and consequently $f(x) = f(\sqrt{x})^2 > 0$ for all $x > 0$. It also follows that $f(x) < 0$ for $x < 0$. In other words, f preserves sign. Now setting $x > 0$ and $y = -f(x)$ in the given functional equation we obtain

$$f(x - f(x)) = f(\sqrt{x}^2 + f(-x)) = -x + f(\sqrt{x})^2 = -(x - f(x)).$$

But since f preserves sign, this implies that $f(x) = x$ for $x > 0$. Moreover, since $f(-x) = -f(x)$, it follows that $f(x) = x$ for all x. It is easily verified that this is indeed a solution.

7. Let G_1, G_2 touch the chord BC at P, Q and touch the circle G at R, S respectively. Let D be the midpoint of the complementary arc BC of G. The homothety centered at R mapping G_1 onto G also maps the line BC onto a tangent of G parallel to BC. It follows that this line touches G at point D, which is therefore the image of P under the homothety. Hence R, P, and D are collinear. Since $\angle DBP = \angle DCB = \angle DRB$, it follows that $\triangle DBP \sim \triangle DRB$ and consequently that $DP \cdot DR = DB^2$. Similarly, points S, Q, D are collinear and satisfy $DQ \cdot DS = DB^2 = DP \cdot DR$. Hence D lies on the radical axis of the circles G_1 and G_2, i.e., on their common tangent AW, which also implies that AW bisects the angle BAD. Furthermore, since $DB = DC = DW = \sqrt{DP \cdot DR}$, it follows from the lemma of (SL99-14) that W is the incenter of $\triangle ABC$.

Remark. According to the third solution of (SL93-3), both PW and QW contain the incenter of $\triangle ABC$, and the result is immediate. The problem can also be solved by inversion centered at W.

8. For simplicity, we shall write n instead of 1992.

Lemma. There exists a tangent n-gon $A_1 A_2 \dots A_n$ with sides $A_1 A_2 = a_1$, $A_2 A_3 = a_2, \dots, A_n A_1 = a_n$ if and only if the system

$$x_1 + x_2 = a_1, \ x_2 + x_3 = a_2, \ , \dots, \ x_n + x_1 = a_n \tag{1}$$

has a solution $(x_1,\ldots,x_n)$ in positive reals.

Proof. Suppose that such an n-gon $A_1A_2\ldots A_n$ exists. Let the side A_iA_{i+1} touch the inscribed circle at point P_i (where $A_{n+1}=A_1$). Then $x_1=A_1P_n=A_1P_1$, $x_2=A_2P_1=A_2P_2,\ \ldots,\ x_n=A_nP_{n-1}=A_nP_n$ is clearly a positive solution of the system (1).

Now suppose that the system (1) has a positive real solution $(x_1,\ldots,\ x_n)$. Let us draw a polygonal line $A_1\ A_2\ \ldots\ A_{n+1}$ touching a circle of radius r at points $P_1,P_2,\ldots,P_n$ respectively such that $A_1P_1=A_{n+1}P_n=x_1$ and $A_iP_i=A_iP_{i-1}=x_i$ for $i=2,\ldots,n$. Observe that

$$OA_1 = OA_{n+1} = \sqrt{x_1^2 + r^2}$$

and the function $f(r) = \angle A_1OA_2 + \angle A_2OA_3 + \cdots + \angle A_nOA_{n+1} = 2\cdot(\arctan\frac{x_1}{r}+\cdots+\arctan\frac{x_n}{r})$ is continuous. Thus $A_1A_2\ldots A_{n+1}$ is a closed simple polygonal line if and only if $f(r) = 360°$. But such an r exists, since $f(r)\to 0$ when $r\to\infty$, and $f(r)\to\infty$ when $r\to 0$. This proves the second direction of the lemma.

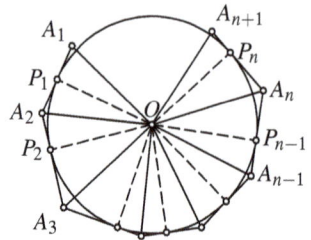

For $n=4k$, the system (1) is solvable in positive reals if $a_i=i$ for $i\equiv 1,2\ (\mathrm{mod}\ 4)$, $a_i=i+1$ for $i\equiv 3$ and $a_i=i-1$ for $i\equiv 0\ (\mathrm{mod}\ 4)$. Indeed, one solution is given by $x_i=1/2$ for $i\equiv 1$, $x_i=3/2$ for $i\equiv 3$ and $x_i=i-3/2$ for $i\equiv 0,2\ (\mathrm{mod}\ 4)$.

Remark. For $n=4k+2$ there is no such n-gon. In fact, solvability of the system (1) implies $a_1+a_3+\cdots=a_2+a_4+\cdots$, while in the case $n=4k+2$ the sum $a_1+a_2+\cdots+a_n$ is odd.

9. Since the equation $x^3-x-c=0$ has only one real root for every $c>2/(3\sqrt{3})$, α is the unique real root of $x^3-x-33^{1992}=0$. Hence $f^n(\alpha)=f(\alpha)=\alpha$.

Remark. Consider any irreducible polynomial $g(x)$ in the place of x^3-x-33^{1992}. The problem amounts to proving that if α and $f(\alpha)$ are roots of g, then any $f^{(n)}(\alpha)$ is also a root of g. In fact, since $g(f(x))$ vanishes at $x=\alpha$, it must be divisible by the minimal polynomial of α, that is, $g(x)$. It follows by induction that $g(f^{(n)}(x))$ is divisible by $g(x)$ for all $n\in\mathbb{N}$, and hence $g(f^{(n)}(\alpha))=0$.

10. Let us set $S(x) = \{(y,z)\mid (x,y,z)\in V\}$, $S_y(x)=\{z\mid (x,z)\in S_y\}$ and $S_z(x)=\{y\mid (x,y)\in S_z\}$. Clearly $S(x)\subset S_x$ and $S(x)\subset S_y(x)\times S_z(x)$. It follows that

$$|V| = \sum_x |S(x)| \le \sum_x \sqrt{|S_x||S_y(x)||S_z(x)|}$$
$$= \sqrt{|S_x|}\sum_x \sqrt{|S_y(x)||S_z(x)|}. \tag{1}$$

Using the Cauchy–Schwarz inequality we also get

$$\sum_x \sqrt{|S_y(x)||S_z(x)|} \le \sqrt{\sum_x |S_y(x)|}\sqrt{\sum_x |S_z(x)|} = \sqrt{|S_y||S_z|}. \qquad (2)$$

Now (1) and (2) together yield $|V| \le \sqrt{|S_x||S_y||S_z|}$.

11. Let I be the incenter of $\triangle ABC$. Since $90° + \alpha/2 = \angle BIC = \angle DIE = 138°$, we obtain that $\angle A = 96°$.

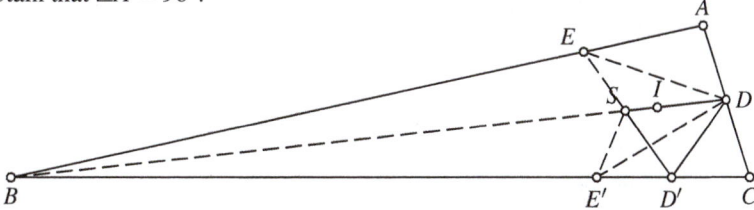

Let D' and E' be the points symmetric to D and E with respect to CE and BD respectively, and let S be the intersection point of ED' and BD. Then $\angle BDE' = 24°$ and $\angle D'DE' = \angle D'DE - \angle E'DE = 24°$,which means that DE' bisects the angle SDD'. Moreover, $\angle E'SB = \angle ESB = \angle EDS + \angle DES = 60°$ and hence SE' bisects the angle $D'SB$. It follows that E' is the excenter of $\triangle D'DS$ and consequently $\angle D'DC = \angle DD'C = \angle SD'E' = (180° - 72°)/2 = 54°$. Finally, $\angle C = 180° - 2 \cdot 54° = 72°$ and $\angle B = 12°$.

12. Let us set $\deg f = n$ and $\deg g = m$. We shall prove the result by induction on n. If $n < m$, then $\deg_x[f(x) - f(y)] < \deg_x[g(x) - g(y)]$, which implies that $f(x) - f(y) = 0$, i.e., that f is constant. The statement trivially holds.
Assume now that $n \ge m$. Transition to $f_1(x) = f(x) - f(0)$ and $g_1(x) = g(x) - g(0)$ allows us to suppose that $f(0) = g(0) = 0$. Then the given condition for $y = 0$ gives us $f(x) = f_1(x)g(x)$, where $f_1(x) = a(x,0)$ and $\deg f_1 = n - m$. We now have

$$a(x,y)(g(x) - g(y)) = f(x) - f(y) = f_1(x)g(x) - f_1(y)g(y)$$
$$= [f_1(x) - f_1(y)]g(x) + f_1(y)[g(x) - g(y)].$$

Since $g(x)$ is relatively prime to $g(x) - g(y)$, it follows that $f_1(x) - f_1(y) = b(x,y)(g(x) - g(y))$ for some polynomial $b(x,y)$. By the induction hypothesis there exists a polynomial h_1 such that $f_1(x) = h_1(g(x))$ and consequently $f(x) = g(x) \cdot h_1(g(x)) = h(g(x))$ for $h(t) = th_1(t)$. Thus the induction is complete.

13. Let us define

$$F(p,q,r) = \frac{(pqr - 1)}{(p-1)(q-1)(r-1)}$$
$$= 1 + \frac{1}{p-1} + \frac{1}{q-1} + \frac{1}{r-1}$$
$$+ \frac{1}{(p-1)(q-1)} + \frac{1}{(q-1)(r-1)} + \frac{1}{(r-1)(p-1)}.$$

Obviously F is a decreasing function of p, q, r. Suppose that $1 < p < q < r$ are integers for which $F(p, q, r)$ is an integer. Observe that p, q, r are either all even or all odd. Indeed, if for example p is odd and q is even, then $pqr - 1$ is odd while $(p-1)(q-1)(r-1)$ is even, which is impossible. Also, if p, q, r are even then $F(p, q, r)$ is odd.

If $p \geq 4$, then $1 < F(p, q, r) \leq F(4, 6, 8) = 191/105 < 2$, which is impossible. Hence $p \leq 3$.

Let $p = 2$. Then q, r are even and $1 < F(2, q, r) \leq F(2, 4, 6) = 47/15 < 4$. Therefore $F(2, q, r) = 3$. This equality reduces to $(q-3)(r-3) = 5$, with the unique solution $q = 4$, $r = 8$.

Let $p = 3$. Then q, r are odd and $1 < F(3, q, r) \leq F(3, 5, 7) = 104/48 < 3$. Therefore $F(3, q, r) = 2$. This equality reduces to $(q-4)(r-4) = 11$, which leads to $q = 5$, $r = 15$.

Hence the only solutions (p, q, r) of the problem are $(2, 4, 8)$ and $(3, 5, 15)$.

14. We see that $x_1 = 2^0$. Suppose that for some $m, r \in \mathbb{N}$ we have $x_m = 2^r$. Then inductively $x_{m+i} = 2^{r-i}(2i+1)$ for $i = 1, 2, \ldots, r$ and $x_{m+r+1} = 2^{r+1}$. Since every natural number can be uniquely represented as the product of an odd number and a power of two, we conclude that every natural number occurs in our sequence exactly once.

Moreover, it follows that $2k - 1 = x_{k(k+1)/2}$. Thus $x_n = 1992 = 2^3 \cdot 249$ implies that $x_{n+3} = 255 = 2 \cdot 128 - 1 = x_{128 \cdot 129/2} = x_{8256}$. Hence $n = 8253$.

15. The result follows from the following lemma by taking $n = \frac{1992 \cdot 1993}{2}$ and $M = \{d, 2d, \ldots, 1992d\}$.

    Lemma. For every $n \in \mathbb{N}$ there exists a natural number d such that all the numbers $d, 2d, \ldots, nd$ are of the form m^k $(m, k \in \mathbb{N}, k \geq 2)$.

    Proof. Let $p_1, p_2, \ldots, p_n$ be distinct prime numbers. We shall find d in the form $d = 2^{\alpha_2} 3^{\alpha_3} \cdots n^{\alpha_n}$, where $\alpha_i \geq 0$ are integers such that kd is a perfect p_kth power. It is sufficient to find α_i, $i = 2, 3, \ldots, n$, such that $\alpha_i \equiv 0 \pmod{p_j}$ if $i \neq j$ and $\alpha_i \equiv -1 \pmod{p_j}$ if $i = j$. But the existence of such α_i's is an immediate consequence of the Chinese remainder theorem.

16. Observe that $x^4 + x^3 + x^2 + x + 1 = (x^2 + 3x + 1)^2 - 5x(x+1)^2$. Thus for $x = 5^{25}$ we have

$$N = x^4 + x^3 + x^2 + x + 1$$
$$= (x^2 + 3x + 1 - 5^{13}(x+1))(x^2 + 3x + 1 + 5^{13}(x+1)) = A \cdot B.$$

Clearly, both A and B are positive integers greater than 1.

17. (a) Let $n = \sum_{i=1}^{k} 2^{a_i}$, so that $\alpha(n) = k$. Then

$$n^2 = \sum_i 2^{2a_i} + \sum_{i<j} 2^{a_i + a_j + 1}$$

has at most $k + \binom{k}{2} = \frac{k(k+1)}{2}$ binary ones.

(b) The above inequality is an equality for all numbers $n_k = 2^k$.

(c) Put $n_m = 2^{2^m-1} - \sum_{j=1}^m 2^{2^m-2^j}$, where $m > 1$. It is easy to see that $\alpha(n_m) = 2^m - m$. On the other hand, squaring and simplifying yields $n_m^2 = 1 + \sum_{i<j} 2^{2^{m+1}+1-2^i-2^j}$. Hence $\alpha(n_m^2) = 1 + \frac{m(m+1)}{2}$ and thus

$$\frac{\alpha(n_m^2)}{\alpha(n_m)} = \frac{2+m(m+1)}{2(2^m-m)} \to 0 \quad \text{as } m \to \infty.$$

Solution to the alternative parts.

(d) Let $n = \sum_{i=1}^n 2^{2^i}$. Then $n^2 = \sum_{i=1}^n 2^{2^{i+1}} + \sum_{i<j} 2^{2^i+2^j+1}$ has exactly $\frac{k(k+1)}{2}$ binary ones, and therefore $\frac{\alpha(n^2)}{\alpha(n)} = \frac{2k}{k(k+1)} \to \infty$.

(e) Consider the sequence n_i constructed in part (c). Let $\theta > 1$ be a constant to be chosen later, and let $N_i = 2^{m_i} n_i - 1$ where $m_i > \alpha(n_i)$ is such that $m_i/\alpha(n_i) \to \theta$ as $i \to \infty$. Then $\alpha(N_i) = \alpha(n_i) + m_i - 1$, whereas $N_i^2 = 2^{2m_i} n_i^2 - 2^{m_i+1} n_i + 1$ and $\alpha(N_i^2) = \alpha(n_i^2) - \alpha(n_i) + m_i$. It follows that

$$\lim_{i\to\infty} \frac{\alpha(N_i^2)}{\alpha(N_i)} = \lim_{i\to\infty} \frac{\alpha(n_i^2) + (\theta-1)\alpha(n_i)}{(1+\theta)\alpha(n_i)} = \frac{\theta-1}{\theta+1},$$

which is equal to $\gamma \in [0,1]$ for $\theta = \frac{1+\gamma}{1-\gamma}$ (for $\gamma = 1$ we set $m_i/\alpha(n_i) \to \infty$).

(f) Let be given a sequence $(n_i)_{i=1}^\infty$ with $\alpha(n_i^2)/\alpha(n_i) \to \gamma$. Taking $m_i > \alpha(n_i)$ and $N_i = 2^{m_i} n_i + 1$ we easily find that $\alpha(N_i) = \alpha(n_i) + 1$ and $\alpha(N_i^2) = \alpha(n_i^2) + \alpha(n_i) + 1$. Hence $\alpha(N_i^2)/\alpha(N_i) = \gamma + 1$. Continuing this procedure we can construct a sequence t_i such that $\alpha(t_i^2)/\alpha(t_i) = \gamma + k$ for an arbitrary $k \in \mathbb{N}$.

18. Let us define inductively $f^1(x) = f(x) = \frac{1}{x+1}$ and $f^n(x) = f(f^{n-1}(x))$, and let $g_n(x) = x + f(x) + f^2(x) + \cdots + f^n(x)$. We shall prove first the following statement.

Lemma. The function $g_n(x)$ is strictly increasing on $[0,1]$, and $g_{n-1}(1) = F_1/F_2 + F_2/F_3 + \cdots + F_n/F_{n+1}$.

Proof. Since $f(x) - f(y) = \frac{y-x}{(1+x)(1+y)}$ is smaller in absolute value than $x - y$, it follows that $x > y$ implies $f^{2k}(x) > f^{2k}(y)$ and $f^{2k+1}(x) < f^{2k+1}(y)$, and moreover that for every integer $k \geq 0$,

$$[f^{2k}(x) - f^{2k}(y)] + [f^{2k+1}(x) - f^{2k+1}(y)] > 0.$$

Hence if $x > y$, we have $g_n(x) - g_n(y) = (x - y) + [f(x) - f(y)] + \cdots + [f^n(x) - f^n(y)] > 0$, which yields the first part of the lemma.

The second part follows by simple induction, since $f^k(1) = F_{k+1}/F_{k+2}$.

If some $x_i = 0$ and consequently $x_j = 0$ for all $j \geq i$, then the problem reduces to the problem with $i - 1$ instead of n. Thus we may assume that all $x_1, \ldots, x_n$ are different from 0. If we write $a_i = [1/x_i]$, then $x_i = \frac{1}{a_i + x_{i+1}}$. Thus we can regard x_i as functions of x_n depending on $a_1, \ldots, a_{n-1}$.

Suppose that $x_n, a_{n-1}, \ldots, a_3, a_2$ are fixed. Then $x_2, x_3, \ldots, x_n$ are all fixed, and $x_1 = \frac{1}{a_1 + x_2}$ is maximal when $a_1 = 1$. Hence the sum $S = x_1 + x_2 + \cdots + x_n$ is maximized for $a_1 = 1$.

We shall show by induction on i that S is maximized for $a_1 = a_2 = \cdots = a_i = 1$. In fact, assuming that the statement holds for $i-1$ and thus $a_1 = \cdots = a_{i-1} = 1$, having $x_n, a_{n-1}, \ldots, a_{i+1}$ fixed we have that $x_n, \ldots, x_{i+1}$ are also fixed, and that $x_{i-1} = f(x_i), \ldots, x_1 = f^{i-1}(x_i)$. Hence by the lemma, $S = g_{i-1}(x_i) + x_{i+1} + \cdots + x_n$ is maximal when $x_i = \frac{1}{a_i + x_{i+1}}$ is maximal, that is, for $a_i = 1$. Thus the induction is complete.

It follows that $x_1 + \cdots + x_n$ is maximal when $a_1 = \cdots = a_{n-1} = 1$, so that $x_1 + \cdots + x_n = g_{n-1}(x_1)$. By the lemma, the latter does not exceed $g_{n-1}(1)$. This completes the proof.

Remark. The upper bound is the best possible, because it is approached by taking x_n close to 1 and inductively (in reverse) defining $x_{i-1} = \frac{1}{1+x_i} = \frac{1}{a_i + x_i}$.

19. Observe that $f(x) = (x^4 + 2x^2 + 3)^2 - 8(x^2 - 1)^2 = [x^4 + 2(1 - \sqrt{2})x^2 + 3 + 2\sqrt{2}][x^4 + 2(1 + \sqrt{2})x^2 + 3 - 2\sqrt{2}]$. Now it is easy to find that the roots of f are
$$x_{1,2,3,4} = \pm i\left(i\sqrt[4]{2} \pm 1\right) \quad \text{and} \quad x_{5,6,7,8} = \pm i\left(\sqrt[4]{2} \pm 1\right).$$

In other words, $x_k = \alpha_i + \beta_j$, where $\alpha_i^2 = -1$ and $\beta_j^4 = 2$.
We claim that any root of f can be obtained from any other using rational functions. In fact, we have
$$x^3 = -\alpha_i - 3\beta_j + 3\alpha_i\beta_j^2 + \beta_j^3,$$
$$x^5 = 11\alpha_i + 7\beta_j - 10\alpha_i\beta_j^2 - 10\beta_j^3$$
$$x^7 = -71\alpha_i - 49\beta_j + 35\alpha_i\beta_j^2 + 37\beta_j^3,$$

from which we easily obtain that
$$\alpha_i = 24^{-1}(127x + 5x^3 + 19x^5 + 5x^7), \quad \beta_j = 24^{-1}(151x + 5x^3 + 19x^5 + 5x^7).$$

Since all other values of α and β can be obtained as rational functions of α_i and β_j, it follows that all the roots x_l are rational functions of a particular root x_k. We now note that if x_1 is an integer such that $f(x_1)$ is divisible by p, then $p > 3$ and $x_1 \in \mathbb{Z}_p$ is a root of the polynomial f. By the previous consideration, all remaining roots $x_2, \ldots, x_8$ of f over the field $\mathbb{Z}_p$ are rational functions of x_1, since 24 is invertible in $\mathbb{Z}_p$. Then $f(x)$ factors as
$$f(x) = (x - x_1)(x - x_2) \cdots (x - x_8),$$

and the result follows.

20. Denote by U the point of tangency of the circle C and the line l. Let X and U' be the points symmetric to U with respect to S and M respectively; these points do not depend on the choice of P. Also, let C' be the excircle

of $\triangle PQR$ corresponding to P, S' the center of C', and W, W' the points of tangency of C and C' with the line PQ respectively. Obviously, $\triangle WSP \sim \triangle W'S'P$. Since $SX \parallel S'U'$ and $SX : S'U' = SW : S'W' = SP : S'P$, we deduce that $\triangle SXP \sim \triangle S'U'P$, and consequently that P lies on the line XU'. On the other hand, it is easy to show that each point P of the ray $U'X$ over X satisfies the required condition. Thus the desired locus is the extension of $U'X$ over X.

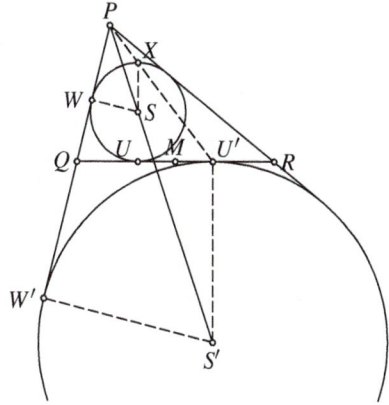

21. (a) Representing n^2 as a sum of $n^2 - 13$ squares is equivalent to representing 13 as a sum of numbers of the form $x^2 - 1$, $x \in \mathbb{N}$, such as $0, 3, 8, 15, \ldots$. But it is easy to check that this is impossible, and hence $s(n) \le n^2 - 14$.
    (b) Let us prove that $s(13) = 13^2 - 14 = 155$. Observe that

$$13^2 = 8^2 + 8^2 + 4^2 + 4^2 + 3^2$$
$$= 8^2 + 8^2 + 4^2 + 4^2 + 2^2 + 2^2 + 1^2$$
$$= 8^2 + 8^2 + 4^2 + 3^2 + 3^2 + 2^2 + 1^2 + 1^2 + 1^2.$$

Given any representation of n^2 as a sum of m squares one of which is even, we can construct a representation as a sum of $m + 3$ squares by dividing the even square into four equal squares. Thus the first equality enables us to construct representations with $5, 8, 11, \ldots, 155$ squares, the second to construct ones with $7, 10, 13, \ldots, 154$ squares, and the third with $9, 12, \ldots, 153$ squares. It remains only to represent 13^2 as a sum of $k = 2, 3, 4, 6$ squares. This can be done as follows:

$$13^2 = 12^2 + 5^2 = 12^2 + 4^2 + 3^2$$
$$= 11^2 + 4^2 + 4^2 + 4^2 = 12^2 + 3^2 + 2^2 + 2^2 + 2^2 + 2^2.$$

(c) We shall prove that whenever $s(n) = n^2 - 14$ for some $n \ge 13$, it also holds that $s(2n) = (2n)^2 - 14$. This will imply that $s(n) = n^2 - 14$ for any $n = 2^t \cdot 13$.
If $n^2 = x_1^2 + \cdots + x_r^2$, then we have

$$(2n)^2 = (2x_1)^2 + \cdots + (2x_r)^2.$$

Replacing $(2x_i)^2$ with $x_i^2 + x_i^2 + x_i^2 + x_i^2$ as long as it is possible we can obtain representations of $(2n)^2$ consisting of $r, r + 3, \ldots, 4r$ squares. This gives representations of $(2n)^2$ into k squares for any $k \le 4n^2 - 62$. Further, we observe that each number $m \ge 14$ can be written as a sum of $k \ge m$ numbers of the form $x^2 - 1$, $x \in \mathbb{N}$, which is easy to verify. Therefore if

$k \leq 4n^2 - 14$, it follows that $4n^2 - k$ is a sum of k numbers of the form $x^2 - 1$ (since $k \geq 4n^2 - k \geq 14$), and consequently $4n^2$ is a sum of k squares.

Remark. One can find exactly the value of $s(n)$ for each n:

$$s(n) = \begin{cases} 1, & \text{if } n \text{ has no prime divisor congruent to } 1 \text{ mod } 4; \\ 2, & \text{if } n \text{ is of the form } 5 \cdot 2^k, k \text{ a positive integer;} \\ n^2 - 14, & \text{otherwise.} \end{cases}$$

4.34 Solutions to the Shortlisted Problems of IMO 1993

1. First we notice that for a rational point O (i.e., with rational coordinates), there exist 1993 rational points in each quadrant of the unit circle centered at O. In fact, it suffices to take

$$X = \left\{ O + \left(\pm \frac{t^2 - 1}{t^2 + 1}, \pm \frac{2t}{t^2 + 1} \right) \Big| \, t = 1, 2, \ldots, 1993 \right\}.$$

Now consider the set $A = \{(i/q, j/q) \mid i, j = 0, 1, \ldots, 2q\}$, where $q = \prod_{t=1}^{1993}(t^2 + 1)$. We claim that A gives a solution for the problem. Indeed, for any $P \in A$ there is a quarter of the unit circle centered at P that is contained in the square $[0, 2] \times [0, 2]$. As explained above, there are 1993 rational points on this quarter circle, and by definition of q they all belong to A.

Remark. Substantially the same problem was proposed by Bulgaria for IMO 71: see (SL71-2), where we give another possible construction of a set A.

2. It is well known that $r \leq \frac{1}{2}R$. Therefore $\frac{1}{3}(1 + r)^2 \leq \frac{1}{3}\left(1 + \frac{1}{2}\right)^2 = \frac{3}{4}$.
   It remains only to show that $p \leq \frac{1}{4}$. We note that p does not exceed one half of the circumradius of $\triangle A'B'C'$. However, by the theorem on the nine-point circle, this circumradius is equal to $\frac{1}{2}R$, and the conclusion follows.

   Second solution. By a well-known relation we have $\cos A + \cos B + \cos C = 1 + \frac{r}{R}$ $(= 1 + r$ when $R = 1)$. Next, recalling that the incenter of $\triangle A'B'C'$ is at the orthocenter of $\triangle ABC$, we easily obtain $p = 2\cos A \cos B \cos C$. Cosines of angles of a triangle satisfy the identity $\cos^2 A + \cos^2 B + \cos^2 C + 2\cos A \cos B \cos C = 1$ (the proof is straightforward: see (SL81-11)). Thus

$$p + \frac{1}{3}(1 + r)^2 = 2\cos A \cos B \cos C + \frac{1}{3}(\cos A + \cos B + \cos C)^2$$
$$\leq 2\cos A \cos B \cos C + \cos^2 A + \cos^2 B + \cos^2 C = 1.$$

3. Let O_1 and ρ be the center and radius of k_c. It is clear that C, I, O_1 are collinear and $CI/CO_1 = r/\rho$. By Stewart's theorem applied to $\triangle OCO_1$,

$$OI^2 = \frac{r}{\rho}OO_1^2 + \left(1 - \frac{r}{\rho}\right)OC^2 - CI \cdot IO_1. \tag{1}$$

Since $OO_1 = R - \rho$, $OC = R$ and by Euler's formula $OI^2 = R^2 - 2Rr$, substituting these values in (1) gives $CI \cdot IO_1 = r\rho$, or equivalently $CO_1 \cdot IO_1 = \rho^2 = DO_1^2$. Hence the triangles CO_1D and DO_1I are similar, implying $\angle DIO_1 = 90°$. Since $CD = CE$ and the line CO_1 bisects the segment DE, it follows that I is the midpoint of DE.

Second solution. Under the inversion with center C and power ab, k_c is transformed into the excircle of $\widehat{A}\widehat{B}C$ corresponding to C. Thus $CD = \frac{ab}{s}$, where s is the common semiperimeter of $\triangle ABC$ and $\triangle \widehat{A}\widehat{B}C$, and consequently the distance from D to BC is $\frac{ab}{s}\sin C = \frac{2S_{ABC}}{s} = 2r$. The statement follows immediately.

Third solution. We shall prove a stronger statement: Let $ABCD$ be a convex quadrilateral inscribed in a circle k, and k' the circle that is tangent to segments BO, AO at K, L respectively (where $O = BD \cap AC$), and internally to k at M. Then KL contains the incenters I, J of $\triangle ABC$ and $\triangle ABD$.

Let K', K'', L', L'', N denote the midpoints of arcs BC, BD, AC, AD, AB that don't contain M; X', X'' the points on k defined by $X'N = NX'' = K'K'' = L'L''$ (as oriented arcs); and set $S = AK' \cap BL''$, $\overline{M} = NS \cap k$, $\overline{K} = K''M \cap BO$, $\overline{L} = L'M \cap AO$.

It is clear that $I = AK' \cap BL'$, $J = AK'' \cap BL''$. Furthermore, $X'\overline{M}$ contains I (to see this, use the fact that for A, B, C, D, E, F on k, lines AD, BE, CF are concurrent if and only if $AB \cdot CD \cdot EF = BC \cdot DE \cdot FA$, and then express $A\overline{M}/\overline{M}B$ by applying this rule to $AMBK'NL''$ and show that $AK', \overline{M}X', BL'$ are concurrent). Similarly,

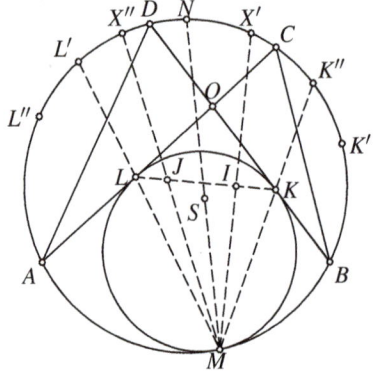

$X''\overline{M}$ contains J. Now the points $B, \overline{K}, I, S, \overline{M}$ lie on a circle ($\angle B\overline{K}M = \angle BIM = \angle BSM$), and points A, $\overline{L}$, J, S, $\overline{M}$ do so as well. Lines $I\overline{K}, J\overline{L}$ are parallel to $K''L'$ (because $\angle \overline{M}KI = \angle \overline{M}BI = \angle \overline{M}K''L'$). On the other hand, the quadrilateral $ABIJ$ is cyclic, and simple calculation with angles shows that IJ is also parallel to $K''L'$. Hence $\overline{K}, I, J, \overline{L}$ are collinear. Finally, $\overline{K} \equiv K$, $\overline{L} \equiv L$, and $\overline{M} \equiv M$ because the homothety centered at M that maps k' to k sends K to K'' and L to L' (thus M, K, K'', as well as M, L, L', must be collinear). As is seen now, the deciphered picture yields many other interesting properties. Thus, for example, N, S, M are collinear, i.e., $\angle AMS = \angle BMS$.

Fourth solution. We give an alternative proof of the more general statement in the third solution. Let W be the foot of the perpendicular from B to AC. We define $q = CW$, $h = BW$, $t = OL = OK$, $x = AL$, $\theta = \angle WBO$ (θ is negative if $\mathcal{B}(O, W, A)$, $\theta = 0$ if $W = O$), and as usual, $a = BC$, $b = AC$, $c = AB$. Let $\alpha = \angle KLC$ and $\beta = \angle ILC$ (both angles must be acute). Our goal is to prove $\alpha = \beta$. We note that $90° - \theta = 2\alpha$. One easily gets

$$\tan \alpha = \frac{\cos \theta}{1 + \sin \theta}, \quad \tan \beta = \frac{\frac{2S_{ABC}}{a+b+c}}{\frac{b+c-a}{2} - x}. \tag{1}$$

Applying Casey's theorem to A, B, C, k', we get $AC \cdot BK + AL \cdot BC = AB \cdot CL$, i.e., $b\left(\frac{h}{\cos \theta} - t\right) + xa = c(b - x)$. Using that $t = b - x - q - h\tan\theta$ we get

$$x = \frac{b(b+c-q) - bh\left(\frac{1}{\cos \theta} + \tan \theta\right)}{a+b+c}. \tag{2}$$

Plugging (2) into the second equation of (1) and using $bh = 2S_{ABC}$ and $c^2 = b^2 + a^2 - 2bq$, we obtain $\tan \alpha = \tan \beta$, i.e., $\alpha = \beta$, which completes our proof.

4. Let h be the altitude from A and $\varphi = \angle BAD$. We have $BM = \frac{1}{2}(BD + AB - AD)$ and $MD = \frac{1}{2}(BD - AB + AD)$, so

$$\frac{1}{MB} + \frac{1}{MD} = \frac{BD}{MB \cdot MD} = \frac{4BD}{BD^2 - AB^2 - AD^2 + 2AB \cdot AD}$$
$$= \frac{4BD}{2AB \cdot AD(1 - \cos\varphi)} = \frac{2BD\sin\varphi}{2S_{ABD}(1 - \cos\varphi)}$$
$$= \frac{2BD\sin\varphi}{BD \cdot h(1 - \cos\varphi)} = \frac{2}{h\tan\frac{\varphi}{2}}.$$

It follows that $\frac{1}{MB} + \frac{1}{MD}$ depends only on h and φ. Specially, $\frac{1}{NC} + \frac{1}{NE} = \frac{2}{h\tan(\varphi/2)}$ as well.

5. For $n = 1$ the game is trivially over. If $n = 2$, it can end, for example, in the following way:

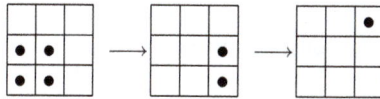

Fig. 1

The sequence of moves shown in Fig. 2 enables us to remove three pieces placed in a 1×3 rectangle, using one more piece and one more free cell. In that way, for any $n \geq 4$ we can reduce an $(n+3) \times (n+3)$ square to an $n \times n$ square (Fig. 3). Therefore the game can end for every n that is not divisible by 3.

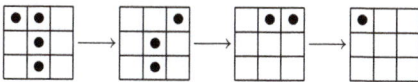

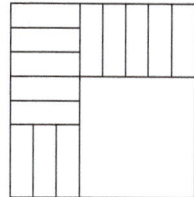

Fig. 2   Fig. 3

Suppose now that one can play the game on a $3k \times 3k$ square so that at the end only one piece remains. Denote the cells by (i, j), $i, j \in \{1, \ldots, 3k\}$, and let S_0, S_1, S_2 denote the numbers of pieces on those squares (i, j) for which $i + j$ gives remainder 0, 1, 2 respectively upon division by 3. Initially $S_0 = S_1 = S_2 = 3k^2$. After each move, two of S_0, S_1, S_2 diminish and one increases by one. Thus each move reverses the parity of the S_i's, so that S_0, S_1, S_2 are always of the same parity. But in the final position one of the S_i's must be equal to 1 and the other two must be 0, which is impossible.

6. Notice that for $\alpha = \frac{1+\sqrt{5}}{2}$, $\alpha^2 n = \alpha n + n$ for all $n \in \mathbb{N}$. We shall show that $f(n) = \left[\alpha n + \frac{1}{2}\right]$ (the closest integer to αn) satisfies the requirements. Observe that f

is strictly increasing and $f(1) = 2$. By the definition of f, $|f(n) - \alpha n| \leq \frac{1}{2}$ and $f(f(n)) - f(n) - n$ is an integer. On the other hand,

$$
\begin{aligned}
|f(f(n)) - f(n) - n| &= |f(f(n)) - f(n) - \alpha^2 n + \alpha n| \\
&= |f(f(n)) - \alpha f(n) + \alpha f(n) - \alpha^2 n - f(n) + \alpha n| \\
&= |(\alpha - 1)(f(n) - \alpha n) + (f(f(n)) - \alpha f(n))| \\
&\leq (\alpha - 1)|f(n) - \alpha n| + |f(f(n)) - \alpha f(n)| \\
&\leq \frac{1}{2}(\alpha - 1) + \frac{1}{2} = \frac{1}{2}\alpha < 1,
\end{aligned}
$$

which implies that $f(f(n)) - f(n) - n = 0$.

7. Multiplying by a and c the equation

$$ax^2 + 2bxy + cy^2 = P^k n, \tag{1}$$

gives $(ax + by)^2 + Py^2 = aP^k n$ and $(bx + cy)^2 + Px^2 = cP^k n$.
It follows immediately that $M(n)$ is finite; moreover, $(ax + by)^2$ and $(bx + cy)^2$ are divisible by P, and consequently $ax + by$, $bx + cy$ are divisible by P because P is not divisible by a square greater than 1. Thus there exist integers X, Y such that $bx + cy = PX$, $ax + by = -PY$. Then $x = -bX - cY$ and $y = aX + bY$. Introducing these values into (1) and simplifying the expression obtained we get

$$aX^2 + 2bXY + cY^2 = P^{k-1}n. \tag{2}$$

Hence $(x, y) \mapsto (X, Y)$ is a bijective correspondence between integral solutions of (1) and (2), so that $M(P^k n) = M(P^{k-1}n) = \cdots = M(n)$.

8. Suppose that $f(n) = 1$ for some $n > 0$. Then $f(n+1) = n+2$, $f(n+2) = 2n+4$, $f(n+3) = n+1$, $f(n+4) = 2n+5$, $f(n+5) = n$, and so by induction $f(n + 2k) = 2n + 3 + k$, $f(n + 2k - 1) = n + 3 - k$ for $k = 1, 2, \ldots, n+2$. Particularly, $n' = 3n + 3$ is the smallest value greater than n for which $f(n') = 1$. It follows that all numbers n with $f(n) = 1$ are given by $n = b_i$, where $b_0 = 1$, $b_n = 3b_{n-1} + 3$. Furthermore, $b_n = 3 + 3b_{n-1} = 3 + 3^2 + 3^2 b_{n-2} = \cdots = 3 + 3^2 + \cdots + 3^n + 3^n = \frac{1}{2}(5 \cdot 3^n - 3)$.
It is seen from above that if $n \leq b_i$, then $f(n) \leq f(b_i - 1) = b_i + 1$. Hence if $f(n) = 1993$, then $n \geq b_i \geq 1992$ for some i. The smallest such b_i is $b_7 = 5466$, and $f(b_i + 2k - 1) = b_i + 3 - k = 1993$ implies $k = 3476$. Thus the least integer in S is $n_1 = 5466 + 2 \cdot 3476 - 1 = 12417$.
All the elements of S are given by $n_i = b_{i+6} + 2k - 1$, where $b_{i+6} + 3 - k = 1993$, i.e., $k = b_{i+6} - 1990$. Therefore $n_i = 3b_{i+6} - 3981 = \frac{1}{2}(5 \cdot 3^{i+7} - 7971)$. Clearly S is infinite and $\lim_{i \to \infty} \frac{n_{i+1}}{n_i} = 3$.

9. We shall first complete the "multiplication table" for the sets A, B, C. It is clear that this multiplication is commutative and associative, so that we have the following relations:

$$
\begin{aligned}
AC &= (AB)B = BB = C; \\
A^2 &= AA = (AB)C = BC = A; \\
C^2 &= CC = B(BC) = BA = B.
\end{aligned}
$$

(a) Now put 1 in A and distribute the primes arbitrarily in A, B, C. This distribution uniquely determines the partition of $\mathbb{Q}^+$ with the stated property. Indeed, if an arbitrary rational number

$$x = p_1^{\alpha_1} \cdots p_k^{\alpha_k} q_1^{\beta_1} \cdots q_l^{\beta_l} r_1^{\gamma_1} \cdots r_m^{\gamma_m}$$

is given, where $p_i \in A$, $q_i \in B$, $r_i \in C$ are primes, it is easy to see that x belongs to A, B, or C according as $\beta_1 + \cdots + \beta_l + 2\gamma_1 + \cdots + 2\gamma_m$ is congruent to 0, 1, or 2 (mod 3).

(b) In every such partition, cubes all belong to A. In fact, $A^3 = A^2 A = AA = A$, $B^3 = B^2 B = CB = A$, $C^3 = C^2 C = BC = A$.

(c) By (b) we have $1, 8, 27 \in A$. Then $2 \notin A$, and since the problem is symmetric with respect to B, C, we can assume $2 \in B$ and consequently $4 \in C$. Also $7 \notin A$, and also $7 \notin B$ (otherwise, $28 = 4 \cdot 7 \in A$ and $27 \in A$), so $7 \in C$, $14 \in A$, $28 \in B$. Further, we see that $3 \notin A$ (since otherwise $9 \in A$ and $8 \in A$). Put 3 in C. Then $5 \notin B$ (otherwise $15 \in A$ and $14 \in A$), so let $5 \in C$ too. Consequently $6, 10 \in A$. Also $13 \notin A$, and $13 \notin C$ because $26 \notin A$, so $13 \in B$. Now it is easy to distribute the remaining primes $11, 17, 19, 23, 29, 31$: one possibility is

$$A = \{1, 6, 8, 10, 14, 19, 23, 27, 29, 31, 33, \dots\},$$
$$C = \{3, 4, 5, 7, 18, 22, 24, 26, 30, 32, 34, \dots\},$$
$$B = \{2, 9, 11, 12, 13, 15, 16, 17, 20, 21, 25, 28, 35, \dots\}.$$

Remark. It can be proved that $\min\{n \in \mathbb{N} \mid n \in A, n+1 \in A\} \leq 77$.

10. (a) Let $n = p$ be a prime and let $p \mid a^p - 1$. By Fermat's theorem $p \mid a^{p-1} - 1$, so that $p \mid a^{\gcd(p, p-1)} - 1 = a - 1$, i.e., $a \equiv 1 \pmod{p}$. Since then $a^i \equiv 1 \pmod{p}$, we obtain $p \mid a^{p-1} + \cdots + a + 1$ and hence $p^2 \mid a^p - 1 = (a-1)(a^{p-1} + \cdots + a + 1)$.

(b) Let $n = p_1 \cdots p_k$ be a product of distinct primes and let $n \mid a^n - 1$. Then from $p_i \mid a^n - 1 = (a^{(n/p_i)})^{p_i} - 1$ and part (a) we conclude that $p_i^2 \mid a^n - 1$. Since this is true for all indices i, we also have $n^2 \mid a^n - 1$; hence n has the property P.

11. Due to the extended Eisenstein criterion, f must have an irreducible factor of degree not less than $n - 1$. Since f has no integral zeros, it must be irreducible.

Second solution. The proposer's solution was as follows. Suppose that $f(x) = g(x)h(x)$, where g, h are nonconstant polynomials with integer coefficients. Since $|f(0)| = 3$, either $|g(0)| = 1$ or $|h(0)| = 1$. We may assume $|g(0)| = 1$ and that $g(x) = (x - \alpha_1) \cdots (x - \alpha_k)$. Then $|\alpha_1 \cdots \alpha_k| = 1$. Since $\alpha_i^{n-1}(\alpha_i + 5) = -3$, taking the product over $i = 1, 2, \dots, k$ yields $|(\alpha_1 + 5) \cdots (\alpha_k + 5)| = |g(-5)| = 3^k$. But $f(-5) = g(-5)h(-5) = 3$, so the only possibility is $\deg g = k = 1$. This is impossible, because f has no integral zeros.

Remark. Generalizing this solution, it can be shown that if a, m, n are positive integers and $p < a - 1$ is a prime, then $F(x) = x^m(x+a)^n + p$ is irreducible. The details are left to the reader.

12. Let $x_1 < x_2 < \cdots < x_n$ be the elements of S. We use induction on n. The result is trivial for $k = 1$ or $n = k$, so assume that it is true for $n - 1$ numbers. Then there exist $m = (k-1)(n-k)+1$ distinct sums of $k - 1$ numbers among $x_2, \ldots, x_n$; call these sums S_i, $S_1 < S_2 < \cdots < S_m$. Then $x_1 + S_1, x_1 + S_2, \ldots, x_1 + S_m$ are distinct sums of k of the numbers $x_1, x_2, \ldots, x_n$. However, the biggest of these sums is

$$x_1 + S_m \le x_1 + x_{n-k+2} + x_{n-k+3} + \cdots + x_n;$$

hence we can find $n - k$ sums that are greater and thus not included here: $x_2 + x_{n-k+2} + \cdots + x_n, x_3 + x_{n-k+2} + \cdots + x_n, \ldots, x_{n-k+1} + x_{n-k+2} + \cdots + x_n$. This counts for $k(n - k) + 1$ sums in total.

Remark. Equality occurs if S is an arithmetic progression.

13. For an odd integer $N > 1$, let $S_N = \{(m,n) \in S \mid m+n = N\}$. If $f(m,n) = (m_1, n_1)$, then $m_1 + n_1 = m + n$ with m_1 odd and $m_1 \le \frac{n}{2} < \frac{N}{2} < n_1$, so f maps S_N to S_N. Also f is bijective, since if $f(m,n) = (m_1, n_1)$, then n is uniquely determined as the even number of the form $2^k m_1$ that belongs to the interval $[\frac{N+1}{2}, N]$, and this also determines m.

Note that S_N has at most $[\frac{N+1}{4}]$ elements, with equality if and only if N is prime. Thus if $(m,n) \in S_N$, there exist s, r with $1 \le s < r \le [\frac{N+5}{4}]$ such that $f^s(m,n) = f^r(m,n)$. Consequently $f^t(m,n) = (m,n)$, where $t = r - s$, $0 < t \le [\frac{N+1}{4}] = [\frac{m+n+1}{4}]$.

Suppose that $(m,n) \in S_N$ and t is the least positive integer with $f^t(m,n) = (m,n)$. We write $(m,n) = (m_0, n_0)$ and $f^i(m,n) = (m_i, n_i)$ for $i = 1, \ldots, t$. Then there exist positive integers a_i such that $2^{a_i} m_i = n_{i-1}$, $i = 1, \ldots, t$. Since $m_t = m_0$, multiplying these equalities gives

$$2^{a_1 + a_2 + \cdots + a_t} m_0 m_1 \cdots m_{t-1} = n_0 n_1 \cdots n_{t-1}$$
$$\equiv (-1)^t m_0 m_1 \cdots m_{t-1} \pmod{N}. \tag{1}$$

It follows that $N \mid 2^k \pm 1$ and consequently $N \mid 2^{2k} - 1$, where $k = a_1 + \cdots + a_t$. On the other hand, it also follows that $2^k \mid n_0 n_1 \cdots n_{t-1} \mid (N-1)(N-3) \cdots (N - 2[N/4])$. But since

$$\frac{(N-1)(N-3) \cdots (N - 2[\frac{N}{4}])}{1 \cdot 3 \cdots (2[\frac{N-2}{4}] + 1)} = \frac{2 \cdot 4 \cdots (N-1)}{1 \cdot 2 \cdots \frac{N-1}{2}} = 2^{\frac{N-1}{2}},$$

we conclude that $0 < k \le \frac{N-1}{2}$, where equality holds if and only if $\{n_1, \ldots, n_t\}$ is the set of all even integers from $\frac{N+1}{2}$ to $N - 1$, and consequently $t = \frac{N+1}{4}$. Now if $N \nmid 2^h - 1$ for $1 \le h < N - 1$, we must have $2k = N - 1$. Therefore $t = \frac{N+1}{4}$.

14. We first assume that all angles of triangle ABC are less than $120°$. Consider the Torricelli point T of the triangle. It holds that $\angle(AT, EF) = \angle(BT, FD) = \angle(CT, DE) = \theta$ for some angle θ. Therefore

$$2S = 2(S_{AETF} + S_{BFTD} + S_{CDTE})$$
$$= (AT \cdot EF + BT \cdot FD + CT \cdot DE) \sin\theta \tag{1}$$
$$= (AT + BT + CT)DE \sin\theta \le (AT + BT + CT)DE.$$

On the other hand, by the cosine theorem we get

$$AT^2 + AT \cdot BT + BT^2 = c^2,$$
$$BT^2 + BT \cdot CT + CT^2 = a^2,$$
$$CT^2 + CT \cdot AT + AT^2 = b^2,$$
$$3(AT \cdot BT + BT \cdot CT + CT \cdot AT) = 4\sqrt{3}(S_{ATB} + S_{BTC} + S_{CTA})$$
$$= 4\sqrt{3}S.$$

Adding these four equalities, we obtain $2(AT + BT + CT)^2 = a^2 + b^2 + c^2 + 4\sqrt{3}S$, which together with (1) implies the desired inequality.

Assume now that $\angle C \geq 120°$ and take T to be the point lying on the same side of AB as C such that $\angle BTC = \angle CTA = 60°$ (if $\angle C = 120°$, take $T \equiv C$). In this case it is shown as above that

$$2S \leq (AT + BT - CT)DE$$
$$\text{and} \quad 2(AT + BT - CT)^2 = a^2 + b^2 + c^2 + 4\sqrt{3}S,$$

and the inequality follows as before.

15. Denote by $d(PQR)$ the diameter of a triangle PQR. It is clear that $d(PQR) \cdot m(PQR) = 2S_{PQR}$. So if the point X lies inside the triangle ABC or on its boundary, we have $d(ABX), d(BCX), d(CAX) \leq d(ABC)$, which implies

$$m(ABX) + m(BCX) + m(CAX) = \frac{2S_{ABX}}{d(ABX)} + \frac{2S_{BCX}}{d(BCX)} + \frac{2S_{CAX}}{d(CAX)}$$
$$\geq \frac{2S_{ABX} + 2S_{BCX} + 2S_{CAX}}{d(ABC)}$$
$$= \frac{2S_{ABC}}{d(ABC)} = m(ABC).$$

If X is outside $\triangle ABC$ but inside the angle BAC, consider the point Y of intersection of AX and BC. Then $m(ABX) + m(BCX) + m(CAX) \geq m(ABY) + m(BCY) + m(CAY) \geq m(ABC)$. Also, if X is inside the opposite angle of $\angle BAC$ (i.e., $\angle DAE$, where $\mathcal{B}(D,A,B)$ and $\mathcal{B}(E,A,C)$), then $m(ABX) + m(BCX) + m(CAX) \geq m(BCX) \geq m(ABC)$. Since these are essentially all possible different positions of point X, we have finished the proof.

16. Let $S_n = \{A = (a_1,\ldots,a_n) \mid 0 \leq a_i < i\}$. For each $A = (a_1,\ldots,a_n)$, denote $A' = (a_1,\ldots,a_{n-1})$, so we can write $A = (A', a_n)$. The proof of the statement from the problem will be given by induction on n. For $n = 2$ there are two possibilities for A_0, so one directly checks that $A_2 = A_0$. Now assume that $n \geq 3$ and that $A_0 = (A_0', a_{0n}) \in S_n$. It is clear that then any A_i is in S_n too. By the induction hypothesis there exists $k \in \mathbb{N}$ such that $A_k' = A_{k+2}' = A_{k+4}' = \cdots$ and $A_{k+1}' = A_{k+3}' = \cdots$. Observe that if we increase (decrease) a_{kn}, $a_{k+1,n}$ will decrease (respectively increase), and this will also increase (respectively decrease) $a_{k+2,n}$. Hence $a_{kn}, a_{k+2,n}, a_{k+4,n}, \ldots$ is monotonically increasing or decreasing,

and since it is bounded (by 0 and $n-1$), it follows that we will eventually have $a_{k+2i,n} = a_{k+2i+2,n} = \cdots$. Consequently $A_{k+2i} = A_{k+2i+2}$.

17. We introduce the rotation operation Rot to the left by one, so that $\text{Step}_j = \text{Rot}^{-j} \circ \text{Step}_0 \circ \text{Rot}^j$. Now writing $\text{Step}^* = \text{Rot} \circ \text{Step}_0$, the problem is transformed into the question whether there is an $M(n)$ such that all lamps are on again after $M(n)$ successive applications of Step^*.

    We operate in the field $\mathbb{Z}_2$, representing off by 0 and on by 1. So if the status of L_j at some moment is given by $v_j \in \mathbb{Z}_2$, the effect of Step_j is that v_j is replaced by $v_j + v_{j-1}$. With the n-tuple $v_0, \ldots, v_{n-1}$ we associate the polynomial

    $$P(x) = v_{n-1}x^{n-1} + v_0 x^{n-2} + v_1 x^{n-3} + \cdots + v_{n-2}.$$

    By means of Step^*, this polynomial is transformed into the polynomial $Q(x)$ over $\mathbb{Z}$ of degree less than n that satisfies $Q(x) \equiv xP(x) \pmod{x^n + x^{n-1} + 1}$. From now on, the sign $\equiv$ always stands for congruence with this modulus.

    (a) It suffices to show the existence of $M(n)$ with $x^{M(n)} \equiv 1$. Because the number of residue classes is finite, there are r, q, with $r < q$, such that $x^q \equiv x^r$, i.e., $x^r(x^{q-r} - 1) = 0$. One can take $M(n) = q - r$. (Or simply note that there are only finitely many possible configurations; since each operation is bijective, the configuration that reappears first must be on, on, $\ldots$, on.)

    (b) We shall prove that if $n = 2^k$, then $x^{n^2-1} \equiv 1$. We have $x^{n^2} \equiv (x^{n-1}+1)^n \equiv x^{n^2-n} + 1$, because all binomial coefficients of order $n = 2^k$ are even, apart from the first one and the last one. Since also $x^{n^2} \equiv x^{n^2-1} + x^{n^2-n}$, this is what we wanted.

    (c) Now if $n = 2^k + 1$, we prove that $x^{n^2-n+1} \equiv 1$. We have $x^{n^2-1} \equiv (x^{n+1})^{n-1} \equiv (x + x^n)^{n-1} \equiv x^{n-1} + x^{n^2-n}$ (again by evenness of binomial coefficients of order $n - 1 = 2^k$). Together with $x^{n^2} \equiv x^{n^2-1} + x^{n^2-n}$, this leads to $x^{n^2} \equiv x^{n-1}$.

18. Let B_n be the set of sequences with the stated property ($S_n = |B_n|$). We shall prove by induction on n that $S_n \geq \frac{3}{2}S_{n-1}$ for every n.

    Suppose that for every $i \leq n$, $S_i \geq \frac{3}{2}S_{i-1}$, and consequently $S_i \leq \left(\frac{2}{3}\right)^{n-i} S_n$. Let us consider the $2S_n$ sequences obtained by putting 0 or 1 at the end of any sequence from B_n. If some sequence among them does not belong to B_{n+1}, then for some $k \geq 1$ it can be obtained by extending some sequence from B_{n+1-6k} by a sequence of k terms repeated six times. The number of such sequences is $2^k S_{n+1-6k}$. Hence the number of sequences not satisfying our condition is not greater than

    $$\sum_{k\geq 1} 2^k S_{n+1-6k} \leq \sum_{k\geq 1} 2^k \left(\frac{2}{3}\right)^{6k-1} S_n = \frac{3}{2}S_n \frac{2(2/3)^6}{1 - 2(2/3)^6} = \frac{192}{601}S_n < \frac{1}{2}S_n.$$

    Therefore S_{n+1} is not smaller than $2S_n - \frac{1}{2}S_n = \frac{3}{2}S_n$. Thus we have $S_n \geq \left(\frac{3}{2}\right)^n$.

19. Let s be the minimum number of nonzero digits that can appear in the b-adic representation of any number divisible by $b^n - 1$. Among all numbers divisible by $b^n - 1$ and having s nonzero digits in base b, we choose the number A with

the minimum sum of digits. Let $A = a_1 b^{n_1} + \cdots + a_s b^{n_s}$, where $0 < a_i \leq b - 1$ and $n_1 > n_2 > \cdots > n_s$.

First, suppose that $n_i \equiv n_j \pmod{n}$, $i \neq j$. Consider the number

$$B = A - a_i b^{n_i} - a_j b^{n_j} + (a_i + a_j) b^{n_j + kn},$$

with k chosen large enough so that $n_j + kn > n_1$: this number is divisible by $b^n - 1$ as well. But if $a_i + a_j < b$, then B has $s - 1$ digits in base b, which is impossible; on the other hand, $a_i + a_j \geq b$ is also impossible, for otherwise B would have sum of digits less for $b - 1$ than that of A (because B would have digits 1 and $a_i + a_j - b$ in the positions $n_j + kn + 1$, $n_j + kn$). Therefore $n_i \not\equiv n_j$ if $i \neq j$.

Let $n_i \equiv r_i$, where $r_i \in \{0, 1, \ldots, n - 1\}$ are distinct. The number $C = a_1 b^{r_1} + \cdots + a_s b^{r_s}$ also has s digits and is divisible by $b^n - 1$. But since $C < b^n$, the only possibility is $C = b^n - 1$ which has exactly n digits in base b. It follows that $s = n$.

20. For every real x we shall denote by $\lfloor x \rfloor$ and $\lceil x \rceil$ the greatest integer less than or equal to x and the smallest integer greater than or equal to x respectively. The condition $c_i + nk_i \in [1 - n, n]$ is equivalent to $k_i \in I_i = \left[\frac{1 - c_i}{n} - 1, 1 - \frac{c_i}{n}\right]$. For every c_i, this interval contains two integers (not necessarily distinct), namely $p_i = \left\lceil \frac{1 - c_i}{n} - 1 \right\rceil \leq q_i = \left\lfloor 1 - \frac{c_i}{n} \right\rfloor$. In order to show that there exist integers $k_i \in I_i$ with $\sum_{i=1}^{n} k_i = 0$, it is sufficient to show that $\sum_{i=1}^{n} p_i \leq 0 \leq \sum_{i=1}^{n} q_i$. Since $p_i < \frac{1 - c_i}{n}$, we have

$$\sum_{i=1}^{n} p_i < 1 - \sum_{i=1}^{n} \frac{c_i}{n} \leq 1,$$

and consequently $\sum_{i=1}^{n} p_i \leq 0$ because the p_i's are integers. On the other hand, $q_i > -\frac{c_i}{n}$ implies

$$\sum_{i=1}^{n} q_i > -\sum_{i=1}^{n} \frac{c_i}{n} \geq -1,$$

which leads to $\sum_{i=1}^{n} q_i \geq 0$. The proof is complete.

21. Assume that S is a circle with center O that cuts S_i diametrically in points P_i, Q_i, $i \in \{A, B, C\}$, and denote by r_i, r the radii of S_i and S respectively. Since OA is perpendicular to $P_A Q_A$, it follows by Pythagoras's theorem that $OA^2 + AP_A^2 = OP_A^2$, i.e., $r_A^2 + OA^2 = r^2$. Analogously $r_B^2 + OB^2 = r^2$ and $r_C^2 + OC^2 = r^2$. Thus if O_A, O_B, O_C are the feet of perpendiculars from O to BC, CA, AB respectively, then $O_C A^2 - O_C B^2 = r_B^2 - r_A^2$. Since the left-hand side is a monotonic function of $O_C \in AB$, the point O_C is uniquely determined by the imposed conditions. The same holds for O_A and O_B. If A, B, C are not collinear, then the positions of O_A, O_B, O_C uniquely determine the point O, and therefore the circle S also. On the other hand, if A, B, C are collinear, all one can deduce is that O lies on the lines l_A, l_B, l_C through O_A, O_B, O_C, perpendicular to BC, CA, AB respectively.

Hence, l_A, l_B, l_C are parallel, so O can be either anywhere on the line if these lines coincide, or nowhere if they don't coincide. So if there exists more than one circle S, A, B, C lie on a line and the foot O' of the perpendicular from O to the line ABC is fixed. If X, Y are the intersection points of S and the line ABC, then $r^2 = OX^2 = OA^2 + r_A^2$ and consequently $O'X^2 = O'A^2 + r_A^2$, which implies that X, Y are fixed.

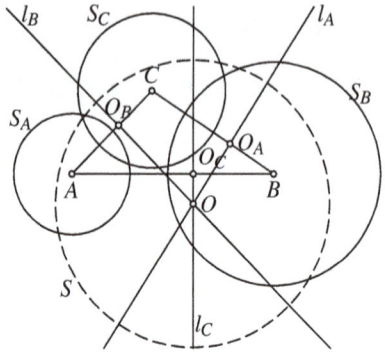

22. Let M be the point inside $\angle ADB$ that satisfies $DM = DB$ and $DM \perp DB$. Then $\angle ADM = \angle ACB$ and $AD/DM = AC/CB$. It follows that the triangles ADM, ACB are similar; hence $\angle CAD = \angle BAM$ (because $\angle CAB = \angle DAM$) and $AB/AM = AC/AD$. Consequently the triangles CAD, BAM are similar and therefore $\frac{AC}{AB} = \frac{CD}{BM} = \frac{CD}{\sqrt{2BD}}$. Hence $\frac{AB \cdot CD}{AC \cdot BD} = \sqrt{2}$. Let CT, CU be the tangents at C to the circles ACD, BCD respectively. Then (in oriented angles) $\angle TCU = \angle TCD + \angle DCU = \angle CAD + \angle CBD = 90°$, as required.

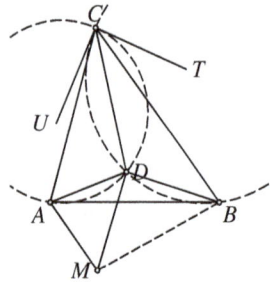

Second solution to the first part. Denote by E, F, G the feet of the perpendiculars from D to BC, CA, AB. Consider the pedal triangle EFG. Since $FG = AD \sin \angle A$, from the sine theorem we have $FG : GE : EF = (CD \cdot AB) : (BD \cdot AC) : (AD \cdot BC)$. Thus $EG = FG$. On the other hand, $\angle EGF = \angle EGD + \angle DGF = \angle CBD + \angle CAD = 90°$ implies that $EF : EG = \sqrt{2} : 1$; hence the required ratio is $\sqrt{2}$.

Third solution to the first part. Under inversion centered at C and with power $r^2 = CA \cdot CB$, the triangle DAB maps into a right-angled isosceles triangle $D^*A^*B^*$, where

$$D^*A^* = \frac{AD \cdot BC}{CD}, \quad D^*B^* = \frac{AC \cdot BD}{CD}, \quad A^*B^* = \frac{AB \cdot CD}{CD}.$$

Thus $D^*B^* : A^*B^* = \sqrt{2}$, and this is the required ratio.

23. Let the given numbers be $a_1, \ldots, a_n$. Put $s = a_1 + \cdots + a_n$ and $m = \text{lcm}(a_1, \ldots, a_n)$ and write $m = 2^k r$ with $k \geq 0$ and r odd. Let the binary expansion of r be $r = 2^{k_0} + 2^{k_1} + \cdots + 2^{k_t}$, with $0 = k_0 < \cdots < k_t$. Adjoin to the set $\{a_1, \ldots, a_n\}$ the numbers $2^{k_i} s$, $i = 1, 2, \ldots, t$. The sum of the enlarged set is rs. Finally, adjoin $rs, 2rs, 2^2 rs, \ldots, 2^{l-1} rs$ for $l = \max\{k, k_t\}$. The resulting set has sum $2^l rs$, which is divisible by m and so by each of a_j, and also by the $2^i s$ above and by $rs, 2rs, \ldots, 2^{l-1} rs$. Therefore this is a DS-set.

Second solution. We show by induction that there is a *DS*-set containing 1 and n. For $n = 2,3$, take $\{1,2,3\}$. Assume that $\{1,n,b_1,\ldots,b_k\}$ is a *DS*-set. Then $\{1,n+1,n,2(n+1)n,2(n+1)b_1,\ldots,2(n+1)b_k\}$ is a *DS*-set too.

For given $a_1,\ldots,a_n$ let m be a sufficiently large common multiple of the a_i's such that $u = m - (a_1 + \cdots + a_n) \neq a_i$ for all i. There exist $b_1,\ldots,b_k$ such that $\{1,u,b_1,\ldots,b_k\}$ is a *DS*-set. It is clear that $\{a_1,\ldots,a_n,u,mu,mb_1,\ldots,mb_k\}$ is a *DS*-set containing $a_1,\ldots,a_n$.

24. By the Cauchy–Schwarz inequality, if $x_1,x_2,\ldots,x_n$ and $y_1,y_2,\ldots,y_n$ are positive numbers, then

$$\left(\sum_{i=1}^{n}\frac{x_i}{y_i}\right)\left(\sum_{i=1}^{n}x_iy_i\right) \geq \left(\sum_{i=1}^{n}x_i\right)^2.$$

Applying this to the numbers a,b,c,d and $b+2c+3d, c+2d+3a, d+2a+3b, a+2b+3c$ (here $n=4$), we obtain

$$\frac{a}{b+2c+3d} + \frac{b}{c+2d+3a} + \frac{c}{d+2a+3b} + \frac{d}{a+2b+3c}$$

$$\geq \frac{(a+b+c+d)^2}{4(ab+ac+ad+bc+bd+cd)} \geq \frac{2}{3}.$$

The last inequality follows, for example, from $(a-b)^2 + (a-c)^2 + \cdots + (c-d)^2 \geq 0$. Equality holds if and only if $a = b = c = d$.

Second solution. Putting $A = b+2c+3d$, $B = c+2d+3a$, $C = d+2a+3b$, $D = a+2b+3c$, our inequality transforms into

$$\frac{-5A+7B+C+D}{24A} + \frac{-5B+7C+D+A}{24B}$$

$$+ \frac{-5C+7D+A+B}{24C} + \frac{-5D+7A+B+C}{24D} \geq \frac{2}{3}.$$

This follows from the arithmetic-geometric mean inequality, since $\frac{B}{A}+\frac{C}{B}+\frac{D}{C}+\frac{A}{D} \geq 4$, etc.

25. We need only consider the case $a > 1$ (since the case $a < -1$ is reduced to $a > 1$ by taking $a' = -a$, $x_i' = -x_i$). Since the left sides of the equations are nonnegative, we have $x_i \geq -\frac{1}{a} > -1$, $i = 1,\ldots,1000$. Suppose w.l.o.g. that $x_1 = \max\{x_i\}$. In particular, $x_1 \geq x_2,x_3$. If $x_1 \geq 0$, then we deduce that $x_{1000}^2 \geq 1 \Rightarrow x_{1000} \geq 1$; further, from this we deduce that $x_{999} > 1$ etc., so either $x_i > 1$ for all i or $x_i < 0$ for all i.

   (i) $x_i > 1$ for every i. Then $x_1 \geq x_2$ implies $x_1^2 \geq x_2^2$, so $x_2 \geq x_3$. Thus $x_1 \geq x_2 \geq \cdots \geq x_{1000} \geq x_1$, and consequently $x_1 = \cdots = x_{1000}$. In this case the only solution is $x_i = \frac{1}{2}(a + \sqrt{a^2+4})$ for all i.

   (ii) $x_i < 0$ for every i. Then $x_1 \geq x_3$ implies $x_1^2 \leq x_3^2 \Rightarrow x_2 \leq x_4$. Similarly, this leads to $x_3 \geq x_5$, etc. Hence $x_1 \geq x_3 \geq x_5 \geq \cdots \geq x_{999} \geq x_1$ and $x_2 \leq x_4 \leq \cdots \leq x_2$, so we deduce that $x_1 = x_3 = \cdots$ and $x_2 = x_4 = \cdots$. Therefore the system is reduced to $x_1^2 = ax_2 + 1$, $x_2^2 = ax_1 + 1$. Subtracting these equations, one obtains $(x_1 - x_2)(x_1 + x_2 + a) = 0$. There are two possibilities:

(1) If $x_1 = x_2$, then $x_1 = x_2 = \cdots = \frac{1}{2}(a - \sqrt{a^2 + 4})$.

(2) $x_1 + x_2 + a = 0$ is equivalent to $x_1^2 + ax_1 + (a^2 - 1) = 0$. The discriminant of the last equation is $4 - 3a^2$. Therefore if $a > \frac{2}{\sqrt{3}}$, this case yields no solutions, while if $a \leq \frac{2}{\sqrt{3}}$, we obtain $x_1 = \frac{1}{2}(-a - \sqrt{4 - 3a^2})$, $x_2 = \frac{1}{2}(-a + \sqrt{4 - 3a^2})$, or vice versa.

26. Set

$$f(a,b,c,d) = abc + bcd + cda + dab - \frac{176}{27}abcd$$

$$= ab(c+d) + cd\left(a + b - \frac{176}{27}ab\right).$$

If $a + b - \frac{176}{a}b \leq 0$, by the arithmetic-geometric inequality we have $f(a,b,c,d) \leq ab(c+d) \leq \frac{1}{27}$.

On the other hand, if $a + b - \frac{176}{27}ab > 0$, the value of f increases if c,d are replaced by $\frac{c+d}{2}, \frac{c+d}{2}$. Consider now the following fourtuplets:

$$P_0(a,b,c,d),\ P_1\left(a,b,\frac{c+d}{2},\frac{c+d}{2}\right),\ P_2\left(\frac{a+b}{2},\frac{a+b}{2},\frac{c+d}{2},\frac{c+d}{2}\right),$$

$$P_3\left(\frac{1}{4},\frac{a+b}{2},\frac{c+d}{2},\frac{1}{4}\right),\ P_4\left(\frac{1}{4},\frac{1}{4},\frac{1}{4},\frac{1}{4}\right)$$

From the above considerations we deduce that for $i = 0,1,2,3$ either $f(P_i) \leq f(P_{i+1})$, or directly $f(P_i) \leq 1/27$. Since $f(P_4) = 1/27$, in every case we are led to

$$f(a,b,c,d) = f(P_0) \leq \frac{1}{27}.$$

Equality occurs only in the cases $(0,1/3,1/3,1/3)$ (with permutations) and $(1/4,1/4,1/4,1/4)$.

Remark. Lagrange multipliers also work. On the boundary of the set one of the numbers a,b,c,d is 0, and the inequality immediately follows, while for an extremum point in the interior, among a,b,c,d there are at most two distinct values, in which case one easily verifies the inequality.

4.35 Solutions to the Shortlisted Problems of IMO 1994

1. Obviously $a_0 > a_1 > a_2 > \cdots$. Since $a_k - a_{k+1} = 1 - \frac{1}{a_k+1}$, we have $a_n = a_0 + (a_1 - a_0) + \cdots + (a_n - a_{n-1}) = 1994 - n + \frac{1}{a_0+1} + \cdots + \frac{1}{a_{n-1}+1} > 1994 - n$. Also, for $1 \le n \le 998$,

$$\frac{1}{a_0+1} + \cdots + \frac{1}{a_{n-1}+1} < \frac{n}{a_{n-1}+1} < \frac{998}{a_{997}+1} < 1$$

because as above, $a_{997} > 997$. Hence $\lfloor a_n \rfloor = 1994 - n$.

2. We may assume that $a_1 > a_2 > \cdots > a_m$. We claim that for $i = 1, \ldots, m$, $a_i + a_{m+1-i} \ge n + 1$. Indeed, otherwise $a_i + a_{m+1-i}, \ldots, a_i + a_{m-1}, a_i + a_m$ are i different elements of A greater than a_i, which is impossible. Now by adding for $i = 1, \ldots, m$ we obtain $2(a_1 + \cdots + a_m) \ge m(n+1)$, and the result follows.

3. The last condition implies that $f(x) = x$ has at most one solution in $(-1, 0)$ and at most one solution in $(0, \infty)$. Suppose that for $u \in (-1, 0)$, $f(u) = u$. Then putting $x = y = u$ in the given functional equation yields $f(u^2 + 2u) = u^2 + 2u$. Since $u \in (-1, 0) \Rightarrow u^2 + 2u \in (-1, 0)$, we deduce that $u^2 + 2u = u$, i.e., $u = -1$ or $u = 0$, which is impossible. Similarly, if $f(v) = v$ for $v \in (0, \infty)$, we are led to the same contradiction.

   However, for all $x \in S$ we have

   $$f(x + (1+x)f(x)) = x + (1+x)f(x),$$

   so we must have $x + (1+x)f(x) = 0$. Therefore $f(x) = -\frac{x}{1+x}$ for all $x \in S$. It is directly verified that this function satisfies all the conditions.

4. Suppose that $\alpha = \beta$. The given functional equation for $x = y$ yields $f(x/2) = x^{-\alpha} f(x)^2/2$; hence the functional equation can be written as

   $$f(x)f(y) = \frac{1}{2} x^\alpha y^{-\alpha} f(y)^2 + \frac{1}{2} y^\alpha x^{-\alpha} f(x)^2,$$

   i.e.,

   $$\left((x/y)^{\alpha/2} f(y) - (y/x)^{\alpha/2} f(x) \right)^2 = 0.$$

   Hence $f(x)/x^\alpha = f(y)/y^\alpha$ for all $x, y \in \mathbb{R}^+$, so $f(x) = \lambda x^\alpha$ for some λ. Substituting into the functional equation we obtain that $\lambda = 2^{1-\alpha}$ or $\lambda = 0$. Thus either $f(x) \equiv 2^{1-\alpha} x^\alpha$ or $f(x) \equiv 0$.
   Now let $\alpha \ne \beta$. Interchanging x with y in the given equation and subtracting these equalities from each other, we get $(x^\alpha - x^\beta)f(y/2) = (y^\alpha - y^\beta)f(x/2)$, so for some constant $\lambda \ge 0$ and all $x \ne 1$, $f(x/2) = \lambda(x^\alpha - x^\beta)$. Substituting this into the given equation, we obtain that only $\lambda = 0$ is possible, i.e., $f(x) \equiv 0$.

5. If $f^{(n)}(x) = \frac{p_n(x)}{q_n(x)}$ for some positive integer n and polynomials p_n, q_n, then

   $$f^{(n+1)}(x) = f\left(\frac{p_n(x)}{q_n(x)} \right) = \frac{p_n(x)^2 + q_n(x)^2}{2p_n(x)q_n(x)}.$$

Note that $f^{(0)}(x) = x/1$. Thus $f^{(n)}(x) = \frac{p_n(x)}{q_n(x)}$, where the sequence of polynomials p_n, q_n is defined recursively by

$$p_0(x) = x, \quad q_0(x) = 1, \quad \text{and}$$

$$p_{n+1}(x) = p_n(x)^2 + q_n(x)^2, \quad q_{n+1}(x) = 2p_n(x)q_n(x).$$

Furthermore, $p_0(x) \pm q_0(x) = x \pm 1$ and $p_{n+1}(x) \pm q_{n+1}(x) = p_n(x)^2 + q_n(x)^2 \pm 2p_n(x)q_n(x) = (p_n(x) \pm q_n(x))^2$, so $p_n(x) \pm q_n(x) = (x \pm 1)^{2^n}$ for all n. Hence

$$p_n(x) = \frac{(x+1)^{2^n} + (x-1)^{2^n}}{2} \quad \text{and} \quad q_n(x) = \frac{(x+1)^{2^n} - (x-1)^{2^n}}{2}.$$

Finally,

$$\frac{f^{(n)}(x)}{f^{(n+1)}(x)} = \frac{p_n(x)q_{n+1}(x)}{q_n(x)p_{n+1}(x)} = \frac{2p_n(x)^2}{p_{n+1}(x)} = \frac{((x+1)^{2^n} + (x-1)^{2^n})^2}{(x+1)^{2^{n+1}} + (x-1)^{2^{n+1}}}$$

$$= 1 + \frac{2\left(\frac{x+1}{x-1}\right)^{2^n}}{1 + \left(\frac{x+1}{x-1}\right)^{2^{n+1}}} = 1 + \frac{1}{f\left(\left(\frac{x+1}{x-1}\right)^{2^n}\right)}.$$

6. Call the first and second player M and N respectively. N can keep $A \leq 6$. Indeed, let 10 dominoes be placed as shown in the picture, and whenever M marks a 1 in a cell of some domino, let N mark 0 in the other cell of that domino if it is still empty. Since any 3×3 square contains at least three complete dominoes, there are at least three 0's inside. Hence $A \leq 6$.

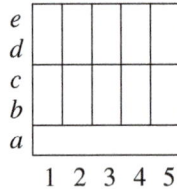

We now show that M can make $A = 6$. Let him start by marking 1 in $c3$. By symmetry, we may assume that N's response is made in row 4 or 5. Then M marks 1 in $c2$. If N puts 0 in $c1$, then M can always mark two 1's in $b \times \{1, 2, 3\}$ as well as three 1's in $\{a, d\} \times \{1, 2, 3\}$. Thus either $\{a, b, c\} \times \{1, 2, 3\}$ or $\{b, c, d\} \times \{1, 2, 3\}$ will contain six 1's. However, if N does not play his second move in $c1$, then M plays there, and thus he can easily achieve to have six 1's either in $\{a, b, c\} \times \{1, 2, 3\}$ or $\{c, d, e\} \times \{1, 2, 3\}$.

7. Let $a_1, a_2, \ldots, a_m$ be the ages of the male citizens ($m \geq 1$). We claim that the age of each female citizen can be expressed in the form $c_1 a_1 + \cdots + c_m a_m$ for some constants $c_i \geq 0$, and we will prove this by induction on the number n of female citizens.

The claim is clear if $n = 1$. Suppose it holds for n and consider the case of $n + 1$ female citizens. Choose any of them, say A of age x who knows k citizens (at least one male). By the induction hypothesis, the age of each of the other n females is expressible as $c_1 a_1 + \cdots + c_m a_m + c_0 x$, where $c_i \geq 0$ and $c_0 + c_1 + \cdots + c_m = 1$. Consequently, the sum of ages of the k citizens who know A is

$kx = b_1a_1 + \cdots + b_ma_m + b_0x$ for some constants $b_i \geq 0$ with sum k. But A knows at least one male citizen (who does not contribute to the coefficient of x), so $b_0 \leq k - 1$. Hence $x = \frac{b_1a_1 + \cdots + b_ma_m}{k - b_0}$, and the claim follows.

8. (a) Let a, b, c, $a \leq b \leq c$ be the amounts of money in dollars in Peter's first, second, and third account, respectively. If $a = 0$, then we are done, so suppose that $a > 0$. Let Peter make transfers of money into the first account as follows. Write $b = aq + r$ with $0 \leq r < a$ and let $q = m_0 + 2m_1 + \cdots + 2^k m_k$ be the binary representation of q ($m_i \in \{0, 1\}$, $m_k = 1$). In the ith transfer, $i = 1, 2, \ldots, k+1$, if $m_i = 1$ he transfers money from the second account, while if $m_i = 0$ he does so from the third. In this way he has transferred exactly $(m_0 + 2m_1 + \cdots + 2^k m_k)a$ dollars from the second account, thus leaving r dollars in it, $r < a$. Repeating this procedure, Peter can diminish the amount of money in the smallest account to zero, as required.

   (b) If Peter has an odd number of dollars, he clearly cannot transfer his money into one account.

9. (a) For $i = 1, \ldots, n$, let d_i be 0 if the card i is in the ith position, and 1 otherwise. Define $b = d_1 + 2d_2 + 2^2 d_3 + \cdots + 2^{n-1} d_n$, so that $0 \leq b \leq 2^n - 1$, and $b = 0$ if and only if the game is over. After each move some digit d_l changes from 1 to 0 while $d_{l+1}, d_{l+2}, \ldots$ remain unchanged. Hence b decreases after each move, and consequently the game ends after at most $2^n - 1$ moves.

   (b) Suppose the game lasts exactly $2^n - 1$ moves. Then each move decreases b for exactly one, so playing the game in reverse (starting from the final configuration), every move is uniquely determined. It follows that if the configuration that allows a game lasting $2^n - 1$ moves exists, it must be unique.

   Consider the initial configuration $0, n, n - 1, \ldots, 2, 1$. We prove by induction that the game will last exactly $2^n - 1$ moves, and that the card 0 will get to the 0th position only in the last move. This is trivial for $n = 1$, so suppose that the claim is true for some $n = m - 1 \geq 1$ and consider the case $n = m$. Obviously the card 0 does not move until the card m gets to the 0-th position. But if we ignore the card 0 and consider the card m to be the card 0, the induction hypothesis gives that the card m will move to the 0th position only after $2^{m-1} - 1$ moves. After these $2^{m-1} - 1$ moves, we come to the configuration $0, m - 1, \ldots, 2, 1, m$. The next move yields $m, 0, m - 1, \ldots, 2, 1$, so by the induction hypothesis again we need $2^{m-1} - 1$ moves more to finish the game.

10. (a) The case $n > 1994$ is trivial. Suppose that $n = 1994$. Label the girls G_1 to G_{1994}, and let G_1 initially hold all the cards. At any moment give to each card the value i, $i = 1, \ldots, 1994$, if G_i holds it. Define the characteristic C of a position as the sum of all these values. Initially $C = 1994$. In each move, if G_i passes cards to G_{i-1} and G_{i+1} (where $G_0 = G_{1994}$ and $G_{1995} = G_1$), C changes for ± 1994 or does not change, so that it remains divisible by

1994. But if the game ends, the characteristic of the final position will be $C = 1 + 2 + \cdots + 1994 = 997 \cdot 1995$, which is not divisible by 1994.

(b) Whenever a card is passed from one girl to another for the first time, let the girls sign their names on it. Thereafter, if one of them passes a card to her neighbor, we shall assume that the passed card is exactly the one signed by both of them. Thus each signed card is stuck between two neighboring girls, so if $n < 1994$, there are two neighbors who never exchange cards. Consequently, there is a girl G who played only a finite number of times. If her neighbor plays infinitely often, then after her last move, G will continue to accumulate cards indefinitely, which is impossible. Hence every girl plays finitely many times.

11. Tile the table with dominoes and numbers as shown in the picture. The second player will not lose if whenever the first player plays in a cell of a domino, he plays in the other cell of the same domino. However, if the first player plays in a cell with a number, the second plays in the cell with same number that is diagonally adjacent.

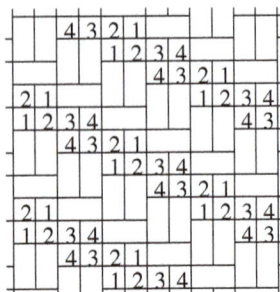

12. Define S_n recursively as follows: Let $S_2 = \{(0,0),(1,1)\}$ and $S_{n+1} = S_n \cup T_n$, where $T_n = \{(x + 2^{n-1}, y + M_n) \mid (x,y) \in S_n\}$, with M_n chosen large enough so that the entire set T_n above lies above every line passing through two points of S_n. By definition, S_n has exactly 2^{n-1} points and contains no three collinear points. We claim that no $2n$ points of this set are the vertices of a convex $2n$-gon.

Consider an arbitrary convex polygon $\mathcal{P}$ with vertices in S_n. Join by a diagonal d the two vertices of $\mathcal{P}$ having the smallest and greatest x-coordinates. This diagonal divides $\mathcal{P}$ into two convex polygons $\mathcal{P}_1, \mathcal{P}_2$, the former lying above d. We shall show by induction that both $\mathcal{P}_1, \mathcal{P}_2$ have at most n vertices. Assume to the contrary that $\mathcal{P}_1$ has at least $n + 1$ vertices $A_1(x_1, y_1), \ldots, A_{n+1}(x_{n+1}, y_{n+1})$ in S_n, with $x_1 < \cdots < x_{n+1}$. It follows that

$$\frac{y_2 - y_1}{x_2 - x_1} > \cdots > \frac{y_{n+1} - y_n}{x_{n+1} - x_n}.$$

By the induction hypothesis, not more than $n - 1$ of these vertices belong to S_{n-1} or T_{n-1}, so let $A_{k-1}, A_k \in S_{n-1}$, $A_{k+1} \in T_{n-1}$. But by the construction of T_{n-1}, $\frac{y_{k+1} - y_k}{x_{k+1} - x_k} > \frac{y_k - y_{k-1}}{x_k - x_{k-1}}$, which gives a contradiction. Similarly, $\mathcal{P}_2$ has no more than n vertices, and therefore $\mathcal{P}$ itself has at most $2n - 2$ vertices.

13. Extend AD and BC to meet at P, and let Q be the foot of the perpendicular from P to AB. Denote by O the center of Γ. Since $\triangle PAQ \sim \triangle OAD$ and $\triangle PBQ \sim \triangle OBC$, we obtain $\frac{AQ}{AD} = \frac{PQ}{OD} = \frac{PQ}{OC} = \frac{BQ}{BC}$. Therefore $\frac{AQ}{QB} \cdot \frac{BC}{CP} \cdot \frac{PD}{DA} = 1$, so by the converse Ceva theorem, AC, BD, and PQ are concurrent. It follows that $Q \equiv F$.

Finally, since the points O, C, P, D, F are concyclic, we have $\angle DFP = \angle DOP = \angle POC = \angle PFC$.

14. Although it does not seem to have been noticed at the jury, the statement of the problem is *false*. For $A(0,0), B(0,4), C(1,4), D(7,0)$, we have $M(4,2)$, $P(2,1)$, $Q(2,3)$ and $N(9/2, 1/2) \notin \triangle ABM$.

   The official solution, if it can be called so, actually shows that N lies inside $ABCD$ and goes as follows: The case $AD = BC$ is trivial, so let $AD > BC$. Let L be the midpoint of AB. Complete the parallelograms $ADMX$ and $BCMY$. Now $N = DX \cap CY$, so let CY and DX intersect AB at K and H respectively. From $LX = LY$ and

$$\frac{HL}{LX} = \frac{HA}{AD} < \frac{LA}{AD} < \frac{KB}{AD} < \frac{KB}{BC} = \frac{KL}{LY}$$

   we get $HL < KL$, and the statement follows.

15. We shall prove that AD is a common tangent of ω and ω_2. Denote by K, L the points of tangency of ω with l_1 and l_2 respectively. Let r, r_1, r_2 be the radii of $\omega, \omega_1, \omega_2$ respectively, and set $KA = x$, $LB = y$. It will be enough if we show that $xy = 2r^2$, since this will imply that $\triangle KLB$ and $\triangle AKO$ are similar, where O is the center of ω, and consequently that $OA \perp KD$ (because $D \in KB$). Now if O_1 is the center of ω_1, we have $x^2 = KA^2 = OO_1^2 - (KO - AO_1)^2 = (r+r_1)^2 - (r - r_1)^2 = 4rr_1$ and analogously $y^2 = 4rr_2$. But we also have $(r_1 + r_2)^2 = O_1O_2^2 = (x-y)^2 + (2r - r_1 - r_2)^2$, so $x^2 - 2xy + y^2 = 4r(r_1 + r_2 - r)$, from which we obtain $xy = 2r^2$ as claimed. Hence AD is tangent to both ω, ω_2, and similarly BC is tangent to ω, ω_1.

   It follows that Q lies on the radical axes of pairs of circles (ω, ω_1) and (ω, ω_2). Therefore Q also lies on the radical axis of (ω_1, ω_2), i.e., on the common tangent at E of ω_1 and ω_2. Hence $QC = QD = QE$.

   Second solution. An inversion with center at D maps ω and ω_2 to parallel lines, ω_1 and l_2 to disjoint equal circles touching ω, ω_2, and l_1 to a circle externally tangent to ω_1, l_2, and to ω. It is easy to see that the obtained picture is symmetric (with respect to a diameter of l_1), and that line AD is parallel to the lines ω and ω_2. Going back to the initial picture, this means that AD is a common tangent of ω and ω_2. The end is like that in the first solution.

16. First, assume that $\angle OQE = 90°$. Extend PN to meet AC at R. Then $OEPQ$ and $ORFQ$ are cyclic quadrilaterals; hence we have $\angle OEQ = \angle OPQ = \angle ORQ = \angle OFQ$. It follows that $\triangle OEQ \cong \triangle OFQ$ and $QE = QF$.

   Now suppose $QE = QF$. Let S be the point symmetric to A with respect to Q, so that the quadrilateral $AESF$ is a parallelogram. Draw the line $E'F'$ through Q so that $\angle OQE' = 90°$ and $E' \in AB$, $F' \in AC$. By the first part $QE' = QF'$; hence $AE'SF'$ is also a parallelogram. It follows that $E \equiv E'$, $F \equiv F'$, and $\angle OQE = 90°$.

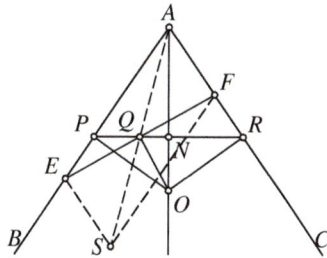

17. We first prove that AB cuts OE in a fixed point H. Note that $\angle OAH = \angle OMA = \angle OEA$ (because O,A,E,M lie on a circle); hence $\triangle OAH \sim \triangle OEA$. This implies $OH \cdot OE = OA^2$, i.e., H is fixed. Let the lines AB and CD meet at K. Since $EAOBM$ and $ECDM$ are cyclic, we have $\angle EAK = \angle EMB = \angle ECK$, so $ECAK$ is cyclic.

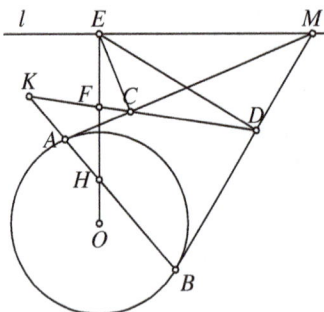

Therefore $\angle EKA = 90°$, hence $EKBD$ is also cyclic and $EK \parallel OM$. Then $\angle EKF = \angle EBD = \angle EOM = \angle OEK$, from which we deduce that $KF = FE$. However, since $\angle EKH = 90°$, the point F is the midpoint of EH; hence it is fixed.

18. Since for each of the subsets $\{1,4,9\}$, $\{2,6,12\}$, $\{3,5,15\}$ and $\{7,8,14\}$ the product of its elements is a square and these subsets are disjoint, we have $|M| \leq 11$. Suppose that $|M| = 11$. Then $10 \in M$ and none of the disjoint subsets $\{1,4,9\},\{2,5\},\{6,15\},\{7,8,14\}$ is a subset of M. Consequently $\{3,12\} \subset M$, so none of $\{1\},\{4\},\{9\},\{2,6\},\{5,15\}$, and $\{7,8,14\}$ is a subset of M: thus $|M| \leq 9$, a contradiction. It follows that $|M| \leq 10$, and this number is attained in the case $M = \{1,4,5,6,7,10,11,12,13,14\}$.

19. Since $mn - 1$ and m^3 are relatively prime, $mn - 1$ divides $n^3 + 1$ if and only if it divides $m^3(n^3 + 1) = (m^3n^3 - 1) + m^3 + 1$. Thus

$$\frac{n^3+1}{mn-1} \in \mathbb{Z} \Leftrightarrow \frac{m^3+1}{mn-1} \in \mathbb{Z};$$

hence we may assume that $m \geq n$. If $m = n$, then $\frac{n^3+1}{n^2-1} = n + \frac{1}{n-1}$ is an integer, so $m = n = 2$. If $n = 1$, then $\frac{2}{m-1} \in \mathbb{Z}$, which happens only when $m = 2$ or $m = 3$. Now suppose $m > n \geq 2$. Since $m^3 + 1 \equiv 1$ and $mn - 1 \equiv -1 \pmod{n}$, we deduce $\frac{n^3+1}{mn-1} = kn - 1$ for some integer $k > 0$. On the other hand, $kn - 1 < \frac{n^3+1}{n^2-1} = n + \frac{1}{n-1} \leq 2n - 1$ gives that $k = 1$, and therefore $n^3 + 1 = (mn-1)(n-1)$. This yields $m = \frac{n^2+1}{n-1} = n + 1 + \frac{2}{n-1} \in \mathbb{N}$, so $n \in \{2,3\}$ and $m = 5$. The solutions with $m < n$ are obtained by symmetry.
There are 9 solutions in total: $(1,2)$, $(1,3)$, $(2,1)$, $(3,1)$, $(2,2)$, $(2,5)$, $(3,5)$, $(5,2)$, $(5,3)$.

20. Let A be the set of all numbers of the form $p_1 p_2 \ldots p_{p_1}$, where $p_1 < p_2 < \cdots < p_{p_1}$ are primes. In other words, $A = \{2 \cdot 3, 2 \cdot 5, \ldots\} \cup \{3 \cdot 5 \cdot 7, 3 \cdot 5 \cdot 11, \ldots\} \cup \{5 \cdot 7 \cdot 11 \cdot 13 \cdot 17, \ldots\} \cup \cdots$.
This set satisfies the requirements of the problem. Indeed, for any infinite set of primes $P = \{q_1, q_2, \ldots\}$ (where $q_1 < q_2 < \cdots$) we have

$$m = q_1 q_2 \cdots q_{q_1} \in A \quad \text{and} \quad n = q_2 q_3 \cdots q_{q_1+1} \notin A.$$

21. Note first that $y_n = 2^k$ ($k \geq 2$) and $z_k \equiv 1 \pmod 4$ for all n, so if x_n is odd, x_{n+1} will be even. Further, it is shown by induction on n that $y_n > z_n$ when x_{n-1} is even and $2y_n > z_n > y_n$ when x_{n-1} is odd. In fact, $n = 1$ is the trivial case, while if it holds for $n \geq 1$, then $y_{n+1} = 2y_n > z_n = z_{n+1}$ if x_n is even, and $2y_{n+1} = 2y_n > y_n + z_n = z_{n+1}$ if x_n is odd (since then x_{n-1} is even).
If $x_1 = 0$, then $x_0 = 3$ is good. Suppose $x_n = 0$ for some $n \geq 2$. Then x_{n-1} is odd and x_{n-2} is even, so that $y_{n-1} > z_{n-1}$. We claim that a pair (y_{n-1}, z_{n-1}), where $2^k = y_{n-1} > z_{n-1} > 0$ and $z_{n-1} \equiv 1 \pmod 4$, uniquely determines $x_0 = f(y_{n-1}, z_{n-1})$. We see that $x_{n-1} = \frac{1}{2}y_{n-1} + z_{n-1}$, and define (x_k, y_k, z_k) backwards as follows, until we get $(y_k, z_k) = (4, 1)$. If $y_k > z_k$, then x_{k-1} must have been even, so we define $(x_{k-1}, y_{k-1}, z_{k-1}) = (2x_k, y_k/2, z_k)$; otherwise x_{k-1} must have been odd, so we put $(x_{k-1}, y_{k-1}, z_{k-1}) = (x_k - y_k/2 + z_k, y_k, z_k - y_k)$. We eventually arrive at $(y_0, z_0) = (4, 1)$ and a good integer $x_0 = f(y_{n-1}, z_{n-1})$, as claimed. Thus for example $(y_{n-1}, z_{n-1}) = (64, 61)$ implies $x_{n-1} = 93$, $(x_{n-2}, y_{n-2}, z_{n-2}) = (186, 32, 61)$ etc., and $x_0 = 1953$, while in the case of $(y_{n-1}, z_{n-1}) = (128, 1)$ we get $x_0 = 2080$.
Note that $y' > y \Rightarrow f(y', z') > f(y, z)$ and $z' > z \Rightarrow f(y, z') > f(y, z)$. Therefore there are no y, z for which $1953 < f(y, z) < 2080$. Hence all good integers less than or equal to 1994 are given as $f(y, z)$, $y = 2^k \leq 64$ and $0 < z \equiv 1 \pmod 4$, and the number of such (y, z) equals $1 + 2 + 4 + 8 + 16 = 31$. So the answer is 31.

22. (a) Denote by $b(n)$ the number of 1's in the binary representation of n. Since $b(2k + 2) = b(k + 1)$ and $b(2k + 1) = b(k) + 1$, we deduce that

$$f(k+1) = \begin{cases} f(k) + 1, & \text{if } b(k) = 2; \\ f(k), & \text{otherwise.} \end{cases} \tag{1}$$

The set of k's with $b(k) = 2$ is infinite, so it follows that $f(k)$ is unbounded. Hence f takes all natural values.

(b) Since f is increasing, k is a unique solution of $f(k) = m$ if and only if $f(k-1) < f(k) < f(k+1)$. By (1), this inequality is equivalent to $b(k-1) = b(k) = 2$. It is easy to see that then $k - 1$ must be of the form $2^t + 1$ for some t. In this case, $\{k+1, \ldots, 2k\}$ contains the number $2^{t+1} + 3 = 10 \ldots 011_2$ and $\frac{t(t-1)}{2}$ binary $(t+1)$-digit numbers with three 1's, so $m = f(k) = \frac{t(t-1)}{2} + 1$.

23. (a) Let p be a prime divisor of x_i, $i > 1$, and let $x_j \equiv u_j \pmod p$ where $0 \leq u_j \leq p - 1$ (particularly $u_i \equiv 0$). Then $u_{j+1} \equiv u_j u_{j-1} + 1 \pmod p$. The number of possible pairs (u_j, u_{j+1}) is finite, so u_j is eventually periodic. We claim that for some $d_p > 0$, $u_{i+d_p} = 0$. Indeed, suppose the contrary and let $(u_m, u_{m+1}, \ldots, u_{m+d-1})$ be the first period for $m \geq i$. Then $m \neq i$. By the assumption $u_{m-1} \not\equiv u_{m+d-1}$, but $u_{m-1} u_m \equiv u_{m+1} - 1 \equiv u_{m+d+1} - 1 \equiv u_{m+d-1} u_{m+d} \equiv u_{m+d-1} u_m \pmod p$, which is impossible if $p \nmid u_m$. Hence there is a d_p with $u_i = u_{i+d_p} = 0$ and moreover $u_{i+1} = u_{i+d_p+1} = 1$, so the sequence u_j is periodic with period d_p starting from u_i. Let m be the least

common multiple of all d_p's, where p goes through all prime divisors of x_i. Then the same primes divide every x_{i+km}, $k = 1, 2, \ldots$, so for large enough k and $j = i + km$, $x_i^i \mid x_j^j$.

(b) If $i = 1$, we cannot deduce that $x_{i+1} \equiv 1 \pmod{p}$. The following example shows that the statement from (a) need not be true in this case. Take $x_1 = 22$ and $x_2 = 9$. Then x_n is even if and only if $n \equiv 1 \pmod{3}$, but modulo 11 the sequence $\{x_n\}$ is $0, 9, 1, 10, 0, 1, 1, 2, 3, 7, 0, \ldots$, so $11 \mid x_n$ $(n > 1)$ if and only if $n \equiv 5 \pmod{6}$. Thus for no $n > 1$ can we have $22 \mid x_n$.

24. A multiple of 10 does not divide any wobbly number. Also, if $25 \mid n$, then every multiple of n ends with 25, 50, 75, or 00; hence it is not wobbly. We now show that every other number n divides some wobbly number.

(i) Let n be odd and not divisible by 5. For any $k \geq 1$ there exists l such that $(10^k - 1)n$ divides $10^l - 1$, and thus also divides $10^{kl} - 1$. Consequently, $v_k = \frac{10^{kl}-1}{10^k-1}$ is divisible by n, and it is wobbly when $k = 2$ (indeed, $v_2 = 101\ldots01$).

If n is divisible by 5, one can simply take $5v_2$ instead.

(ii) Let n be a power of 2. We prove by induction on m that 2^{2m+1} has a wobbly multiple w_m with exactly m nonzero digits. For $m = 1$, take $w_1 = 8$. Suppose that for some $m \geq 1$ there is a wobbly $w_m = 2^{2m+1}d_m$. Then the numbers $a \cdot 10^{2m} + w_m$ are wobbly and divisible by 2^{2m+1} when $a \in \{2, 4, 6, 8\}$. Moreover, one of these numbers is divisible by 2^{2m+3}. Indeed, it suffices to choose a such that $\frac{a}{2} + d_m$ is divisible by 4. This proves the induction step.

(iii) Let $n = 2^m r$, where $m \geq 1$ and r is odd, $5 \nmid r$. Then $v_{2m}w_m$ is wobbly and divisible by both 2^m and r (using notation from (i), $r \mid v_{2m}$).

4.36 Solutions to the Shortlisted Problems of IMO 1995

1. Let $x = \frac{1}{a}, y = \frac{1}{b}, z = \frac{1}{c}$. Then $xyz = 1$ and

$$S = \frac{1}{a^3(b+c)} + \frac{1}{b^3(c+a)} + \frac{1}{c^3(a+b)} = \frac{x^2}{y+z} + \frac{y^2}{z+x} + \frac{z^2}{x+y}.$$

We must prove that $S \geq \frac{3}{2}$. From the Cauchy–Schwarz inequality,

$$[(y+z)+(z+x)+(x+y)] \cdot S \geq (x+y+z)^2 \quad \Rightarrow \quad S \geq \frac{x+y+z}{2}.$$

It follows from the A-G mean inequality that $\frac{x+y+z}{2} \geq \frac{3}{2}\sqrt[3]{xyz} = \frac{3}{2}$; hence the proof is complete. Equality holds if and only if $x = y = z = 1$, i.e., $a = b = c = 1$.

Remark. After reducing the problem to $\frac{x^2}{y+z} + \frac{y^2}{z+x} + \frac{z^2}{x+y} \geq \frac{3}{2}$, we can solve the problem using Jensen's inequality applied to the function $g(u,v) = u^2/v$. The problem can also be solved using Muirhead's inequality.

2. We may assume $c \geq 0$ (otherwise, we may simply put $-y_i$ in the place of y_i). Also, we may assume $a \geq b$. If $b \geq c$, it is enough to take $n = a+b-c$, $x_1 = \cdots = x_a = 1$, $y_1 = \cdots = y_c = y_{a+1} = \cdots = y_{a+b-c} = 1$, and the other x_i's and y_i's equal to 0, so we need only consider the case $a > c > b$.
   We proceed to prove the statement of the problem by induction on $a+b$. The case $a+b = 1$ is trivial. Assume that the statement is true when $a+b \leq N$, and let $a+b = N+1$. The triple $(a+b-2c, b, c-b)$ satisfies the condition (since $(a+b-2c)b - (c-b)^2 = ab - c^2$), so by the induction hypothesis there are n-tuples $(x_i)_{i=1}^n$ and $(y_i)_{i=1}^n$ with the wanted property. It is easy to verify that $(x_i + y_i)_{i=1}^n$ and $(y_i)_{i=1}^n$ give a solution for (a,b,c).

3. Write $A_i = \frac{a_i^2 + a_{i+1}^2 - a_{i+2}^2}{a_i + a_{i+1} - a_{i+2}} = a_i + a_{i+1} + a_{i+2} - \frac{2a_i a_{i+1}}{a_i + a_{i+1} - a_{i+2}}$. Since $2a_i a_{i+1} \geq 4(a_i + a_{i+1} - 2)$ (which is equivalent to $(a_i - 2)(a_{i+1} - 2) \geq 0$), it follows that $A_i \leq a_i + a_{i+1} + a_{i+2} - 4\left(1 + \frac{a_{i+2}-2}{a_i + a_{i+1} - a_{i+2}}\right) \leq a_i + a_{i+1} + a_{i+2} - 4\left(1 + \frac{a_{i+2}-2}{4}\right)$, because $1 \leq a_i + a_{i+1} - a_{i+2} \leq 4$. Therefore $A_i \leq a_i + a_{i+1} - 2$, so $\sum_{i=1}^n A_i \leq 2s - 2n$ as required.

4. The second equation is equivalent to $\frac{a^2}{yz} + \frac{b^2}{zx} + \frac{c^2}{xy} + \frac{abc}{xyz} = 4$. Let $x_1 = \frac{a}{\sqrt{yz}}$, $y_1 = \frac{b}{\sqrt{zx}}$, $z_1 = \frac{c}{\sqrt{xy}}$. Then $x_1^2 + y_1^2 + z_1^2 + x_1 y_1 z_1 = 4$, where $0 < x_1, y_1, z_1 < 2$. Regarding this as a quadratic equation in z_1, the discriminant $(4 - x_1^2)(4 - y_1^2)$ suggests that we let $x_1 = 2\sin u$, $y_1 = 2\sin v$, $0 < u, v < \pi/2$. Then it is directly shown that z_1 will be exactly $2\cos(u+v)$ as the only positive solution of the quadratic equation.
   Thus $a = 2\sqrt{yz}\sin u$, $b = 2\sqrt{xz}\sin v$, $c = 2\sqrt{xy}(\cos u\cos v - \sin u\sin v)$, so from $x+y+z-a-b-c = 0$ we obtain

$$(\sqrt{x}\cos v - \sqrt{y}\cos u)^2 + (\sqrt{x}\sin v + \sqrt{y}\sin u - \sqrt{z})^2 = 0,$$

which implies

$$\sqrt{z} = \sqrt{x}\sin v + \sqrt{y}\sin u = \frac{1}{2}(y_1\sqrt{x}+x_1\sqrt{y}) = \frac{1}{2}\left(\frac{b}{\sqrt{zx}}\sqrt{x}+\frac{a}{\sqrt{yz}}\sqrt{y}\right).$$

Therefore $z = \frac{a+b}{2}$. Similarly, $x = \frac{b+c}{2}$ and $y = \frac{c+a}{2}$. It is clear that the triple $(x,y,z) = \left(\frac{b+c}{2},\frac{c+a}{2},\frac{a+b}{2}\right)$ is indeed a (unique) solution of the given system of equations.

Second solution. Put $x = \frac{b+c}{2} - u$, $y = \frac{c+a}{2} - v$, $z = \frac{a+b}{2} - w$, where $u \le \frac{b+c}{2}$, $v \le \frac{c+a}{2}$, $w \le \frac{a+b}{2}$ and $u+v+w=0$. The equality $abc+a^2x+b^2y+c^2z=4xyz$ becomes $2(au^2+bv^2+cw^2+2uvw)=0$. Now $uvw > 0$ is clearly impossible. On the other hand, if $uvw \le 0$, then two of u,v,w are nonnegative, say $u,v \ge 0$. Taking into account $w = -u-v$, the above equality reduces to $2[(a+c-2v)u^2 + (b+c-2u)v^2+2cuv] = 0$, so $u = v = 0$.

Third solution. The fact that we are given two equations and three variables suggests that this is essentially a problem on inequalities. Setting $f(x,y,z) = 4xyz - a^2x - b^2y - c^2z$, we should show that $\max f(x,y,z) = abc$, for $0 < x,y,z$, $x+y+z = a+b+c$, and find when this value is attained. Thus we apply Lagrange multipliers to $F(x,y,z) = f(x,y,z) - \lambda(x+y+z-a-b-c)$, and obtain that f takes a maximum at (x,y,z) such that $4yz - a^2 = 4zx - b^2 = 4xy - c^2 = \lambda$ and $x+y+z = a+b+c$. The only solution of this system is $(x,y,z) = \left(\frac{b+c}{2},\frac{c+a}{2},\frac{a+b}{2}\right)$.

5. Suppose that a function f satisfies the condition, and let c be the least upper bound of $\{f(x) \mid x \in \mathbb{R}\}$. We have $c \ge 2$, since $f(2) = f(1+1/1^2) = f(1) + f(1)^2 = 2$. Also, since c is the least upper bound, for each $k = 1,2,\dots$ there is an $x_k \in \mathbb{R}$ such that $f(x_k) \ge c - 1/k$. Then

$$c \ge f\left(x_k + \frac{1}{x_k^2}\right) \ge c - \frac{1}{k} + f\left(\frac{1}{x_k}\right)^2 \implies f\left(\frac{1}{x_k}\right) \ge -\frac{1}{\sqrt{k}}.$$

On the other hand,

$$c \ge f\left(\frac{1}{x_k}+x_k^2\right) = f\left(\frac{1}{x_k}\right)+f(x_k)^2 \ge -\frac{1}{\sqrt{k}}+\left(c-\frac{1}{k}\right)^2.$$

It follows that

$$\frac{1}{\sqrt{k}}-\frac{1}{k^2} \ge c\left(c-1-\frac{2}{k}\right),$$

which cannot hold for k sufficiently large.

Second solution. Assume that f exists and let n be the least integer such that $f(x) \le \frac{n}{4}$ for all x. Since $f(2) = 2$, we have $n \ge 8$. Let $f(x) > \frac{n-1}{4}$. Then $f(1/x) = f(x+1/x^2) - f(x) < 1/4$, so $f(1/x) > -1/2$. On the other hand, this implies $\left(\frac{n-1}{4}\right)^2 < f(x)^2 = f(1/x+x^2) - f(1/x) < \frac{n}{4}+\frac{1}{2}$, which is impossible when $n \ge 8$.

6. Let $y_i = x_{i+1} + \cdots + x_n$, $Y = \sum_{j=2}^n (j-1)x_j$, and $z_i = \frac{n(n-1)}{2}y_i - (n-i)Y$. Then

$$\frac{n(n-1)}{2}\sum_{i<j}x_ix_j - \left(\sum_{i=1}^{n-1}(n-i)x_i\right)Y = \frac{n(n-1)}{2}\sum_{i=1}^{n-1}x_iy_i - \sum_{i=1}^{n-1}(n-i)x_iY$$

$$= \sum_{i=1}^{n-1}x_iz_i,$$

so it remains to show that $\sum_{i=1}^{n-1}x_iz_i > 0$. Since $\sum_{i=1}^{n-1}y_i = Y$ and $\sum_{i=1}^{n-1}(n-i) = \frac{n(n-1)}{2}$, we have $\sum z_i = 0$. Note that $Y < \sum_{j=2}^n(j-1)x_n = \frac{n(n-1)}{2}x_n$, and consequently $z_{n-1} = \frac{n(n-1)}{2}x_n - Y > 0$. Furthermore, we have

$$\frac{z_{i+1}}{n-i-1} - \frac{z_i}{n-i} = \frac{n(n-1)}{2}\left(\frac{y_{i+1}}{n-i-1} - \frac{y_i}{n-i}\right) > 0,$$

which means that $\frac{z_1}{n-1} < \frac{z_2}{n-2} < \cdots < \frac{z_{n-1}}{1}$. Therefore there is a k for which $z_1,\ldots,z_k \leq 0$ and $z_{k+1},\ldots,z_{n-1} > 0$. But then $z_i(x_i - x_k) \geq 0$, i.e., $x_iz_i \geq x_kz_i$ for all i, so $\sum_{i=1}^{n-1}x_iz_i > \sum_{i=1}^{n-1}x_kz_i = 0$ as required.

Second solution. Set $X = \sum_{j=1}^{n-1}(n-j)x_j$ and $Y = \sum_{j=2}^n(j-1)x_j$. Since $4XY = (X+Y)^2 - (X-Y)^2$, the RHS of the inequality becomes

$$XY = \frac{1}{4}\left[(n-1)^2\left(\sum_{i=1}^n x_i\right)^2 - \left(\sum_{i=1}^n(2i-1-n)x_i\right)^2\right].$$

The LHS is $\frac{1}{4}\left((n-1)^2(\sum_{i=1}^n x_i)^2 - (n-1)\sum_{i<j}(x_j-x_i)^2\right)$. Since $\sum_{i=1}^n(2i-1-n)x_i = \sum_{i<j}(x_j-x_i)$ also holds, we must prove that

$$\left(\sum_{i<j}(x_j-x_i)\right)^2 > (n-1)\sum_{i<j}(x_j-x_i)^2. \tag{1}$$

Putting $x_{i+1} - x_i = d_i > 0$ (so, $x_j - x_i = d_i + d_{i+1} + \cdots + d_{j-1}$) and expanding the obtained expressions, we reduce this inequality to $\sum_k k^2(n-k)^2d_k^2 + 2\sum_{k<l}kl(n-k)(n-l)d_kd_l > \sum_k(n-1)k(n-k)d_k^2 + 2\sum_{k<l}(n-1)k(n-l)d_kd_l$, which is verified immediately by comparing coefficients.

Remark. An inequality significantly stronger than (1) in the second solution has appeared later, as *IMO 03-5*.

7. The result is trivial if O coincides with X or Y, so let us assume it does not. From $OB \cdot ON = OC \cdot OM = OX \cdot OY$ we deduce that $BCMN$ is a cyclic quadrilateral. Further, if O lies between X and Y, then $\angle MAD + \angle MND = \angle MAD + \angle MNB + \angle BND = \angle MAD + \angle MCA + \angle AMC = 180°$. Similarly, we also have $\angle MAD + \angle MND = 180°$ if O is not on the segment XY. Therefore $ADNM$ is cyclic. Now let AM and DN intersect at Z and let the line ZX intersect the two circles at Y_1

and Y_2. Then $ZX \cdot ZY_1 = ZM \cdot ZA = ZN \cdot ZD = ZX \cdot ZY_2$. Hence $Y_1 = Y_2 = Y$, implying that Z lies on XY.

Second solution. Let Z_1, Z_2 be the points in which AM, DN respectively meet XY, and $P = BC \cap XY$. Then, from $\triangle OPC \sim \triangle APZ_1$, we have $PZ_1 = \frac{PA \cdot PC}{PO} = \frac{PX^2}{PO}$ and analogously $PZ_2 = \frac{PX^2}{PO}$. Hence, we conclude that $Z_1 \equiv Z_2$.

8. Let A', B', C' be the points symmetric to A, B, C with respect to the midpoints of BC, CA, AB respectively. From the condition on X we have $XB^2 - XC^2 = AC^2 - AB^2 = A'B^2 - A'C^2$, and hence X must lie on the line through A' perpendicular to BC. Similarly, X lies on the line through B' perpendicular to CA. It follows that there is a unique position for X, namely the orthocenter of $\triangle A'B'C'$. It easily follows that this point X satisfies the original equations.

9. If EF is parallel to BC, $\triangle ABC$ must be isosceles and E, Y are symmetric to F, Z with respect to AD, so the result follows. Now suppose that EF meets BC at P. By Menelaus's theorem, $\frac{BP}{CP} = \frac{BF}{FA} \cdot \frac{AE}{EC} = \frac{BD}{DC}$ (since $BD = BF$, $CD = CE$, $AE = AF$). It follows that the point P depends only on D and not on A. In particular, the same point is obtained as the intersection of ZY with BC. Therefore $PE \cdot PF = PD^2 = PY \cdot PZ$, from which it follows that $EFZY$ is a cyclic quadrilateral.

Second solution. Since $CD = CY = CE$ and $BD = BZ = BF$, all angles of $EFZY$ can be calculated in terms of angles of ABC and $YZBC$. In fact, $\angle FEY = \frac{1}{2}(\angle A + \angle C + \angle BCY)$ and $\angle FZY = \frac{1}{2}(180° + \angle B + \angle BCY)$, which gives us $\angle FEY + \angle FZY = 180°$.

10. Let the two triangles be $X_1Y_1Z_1$, $X_2Y_2Z_2$, with $X_1 = BB_1 \cap CC_1$, $Y_1 = CC_1 \cap AA_1$, $Z_1 = AA_1 \cap BB_1$, $X_2 = BB_2 \cap CC_2$, $Y_2 = CC_2 \cap AA_2$, $Z_2 = AA_2 \cap BB_2$.

First, let us observe that $\angle ABB_2 = \angle ACC_1$ and $\angle ABB_1 = \angle ACC_2$. We now obtain that $\angle BZ_1A_1 = \angle BAA_1 + \angle ABB_1 = \angle BCC_2 + \angle C_2CA = \angle C$ and similarly $\angle AZ_2B_2 = \angle C$, $\angle AY_1C_1 = \angle CY_2A_2 = \angle B$. Also the triangles ABB_2 and ACC_1 are similar; hence $AC_1/AC = AB_2/AB$. From the law of sines we obtain

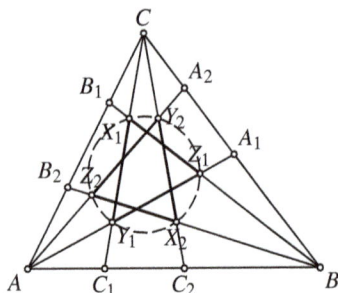

$$\frac{AZ_1}{\sin \angle ABZ_1} = \frac{AB}{\sin \angle AZ_1B} = \frac{AB}{\sin \angle C} = \frac{AC}{\sin \angle B} = \frac{AC}{\sin \angle AY_2C}$$

$$= \frac{AY_2}{\sin \angle ACY_2} \implies AZ_1 = AY_2.$$

Analogously, $BX_1 = BZ_2$ and $CY_1 = CX_2$. Furthermore, again from the sine formula,

$$\frac{AY_1}{\sin \angle AC_1Y_1} = \frac{AC_1}{\sin \angle AY_1C_1} = \frac{AC_1}{AC} \frac{AC}{\sin \angle B}$$
$$= \frac{AB_2}{AB} \frac{AB}{\sin \angle C} = \frac{AB_2}{\sin \angle AZ_2B_2} = \frac{AZ_2}{\sin \angle AB_2Z_2}.$$

Hence, $AY_1 = AZ_2$ and, analogously, $BZ_1 = BX_2$ and $CX_1 = CY_2$. We deduce that $Y_1Z_2 \parallel BC$ and $Z_2X_1 \parallel AC$, which gives us $\angle Y_1Z_2X_1 = 180° - \angle C = 180° - \angle Y_1Z_1X_1$. It follows that Z_2 lies on the circle circumscribed about $\triangle X_1Y_1Z_1$. Similarly, so do X_2 and Y_2.

Second solution. Let H be the orthocenter of $\triangle ABC$. Triangles AHB, BHC, CHA, ABC have the same circumradius R. Additionally,

$$\angle HAA_i = \angle HBB_i = \angle HCC_i = \theta \quad (i = 1,2).$$

Since $\angle HBX_1 = \angle HCX_1 = \theta$, BCX_1H is concyclic and therefore $HX_1 = 2R\sin\theta$. The same holds for $HY_1, HZ_1, HX_2, HY_2, HZ_2$. Hence X_i, Y_i, Z_i $(i = 1,2)$ lie on a circle centered at H.

11. Triangles BCD and EFA are equilateral, and hence BE is an axis of symmetry of $ABDE$. Let C', F' respectively be the points symmetric to C, F with respect to BE. The points G and H lie on the circumcircles of ABC' and DEF' respectively (because, for instance, $\angle AGB = 120° = 180° - \angle AC'B$); hence from Ptolemy's theorem we have $AG + GB = C'G$ and $DH + HE = HF'$. Therefore

$$AG + GB + GH + DH + HE = C'G + GH + HF' \geq C'F' = CF,$$

with equality if and only if G and H both lie on $C'F'$.

Remark. Since by Ptolemy's inequality $AG + GB \geq C'G$ and $DH + HE \geq HF'$, the result holds without the condition $\angle AGB = \angle DHE = 120°$.

12. Let O be the circumcenter and R the circumradius of $A_1A_2A_3A_4$. We have $OA_i^2 = (\overrightarrow{OG} + (\overrightarrow{OA_i} - \overrightarrow{OG}))^2 = OG^2 + GA_i^2 + 2\overrightarrow{OG} \cdot \overrightarrow{GA_i}$. Summing up these equalities for $i = 1,2,3,4$ and using that $\sum_{i=1}^{4} \overrightarrow{GA_i} = \overrightarrow{0}$, we obtain

$$\sum_{i=1}^{4} OA_i^2 = 4OG^2 + \sum_{i=1}^{4} GA_i^2 \iff \sum_{i=1}^{4} GA_i^2 = 4(R^2 - OG^2). \tag{1}$$

Now we have that the potential of G with respect to the sphere equals $GA_i \cdot GA_i' = R^2 - OG^2$. Plugging in these expressions for GA_i', we reduce the inequalities we must prove to

$$GA_1 \cdot GA_2 \cdot GA_3 \cdot GA_4 \leq (R^2 - OG^2)^2 \tag{2}$$

$$\text{and} \quad (R^2 - OG^2) \sum_{i=1}^{4} \frac{1}{GA_i} \geq \sum_{i=1}^{4} GA_i. \tag{3}$$

Inequality (2) immediately follows from (1) and the quadratic-geometric mean inequality for GA_i. From the Cauchy–Schwarz inequality we have $\sum_{i=1}^{4} GA_i^4 \geq$

$\frac{1}{4}\left(\sum_{i=1}^{4}GA_i\right)^2$ and $\left(\sum_{i=1}^{4}GA_i\right)\left(\sum_{i=1}^{4}\frac{1}{GA_i}\right) \geq 16$, hence the inequality (3) follows from (1) and from

$$\left(\sum_{i=1}^{4}GA_i^2\right)\left(\sum_{i=1}^{4}\frac{1}{GA_i}\right) \geq \frac{1}{4}\left(\sum_{i=1}^{4}GA_i\right)^2\left(\sum_{i=1}^{4}\frac{1}{GA_i}\right) \geq 4\sum_{i=1}^{4}GA_i.$$

13. If O lies on AC, then $ABCD$, $AKON$, and $OLCM$ are similar; hence $AC = AO + OC$ implies $\sqrt{S} = \sqrt{S_1} + \sqrt{S_2}$.

Assume that O does not lie on AC and that w.l.o.g. it lies inside triangle ADC. Let us denote by T_1, T_2 the areas of parallelograms $KBLO$, $NOMD$ respectively. Consider a line through O that intersects AD, DC, CB, BA respectively at X, Y, Z, W so that $OW/OX = OZ/OY$ (such a line exists by a continuity argument: the left side is smaller when $W = X = A$, but greater when $Y = Z = C$). The desired inequality is equivalent to $T_1 + T_2 \geq 2\sqrt{S_1 S_2}$. Since triangles WKO, OLZ, WBZ are similar and $WO + OZ = WZ$, we have $\sqrt{S_{WKO}} + \sqrt{S_{OLZ}} = \sqrt{S_{WBZ}} = \sqrt{S_{WKO} + S_{OLZ} + T_1}$, which implies $T_1 = 2\sqrt{S_{WKO}S_{OLZ}}$. Similarly, $T_2 = 2\sqrt{S_{XNO}S_{OMY}}$.

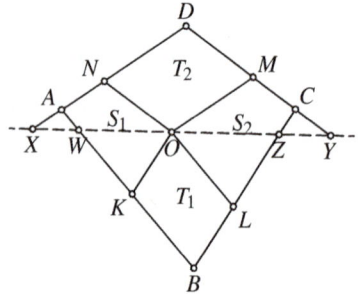

Since $OW/OZ = OX/OY$, we have $S_{WKO}/S_{XNO} = S_{OLZ}/S_{OMY}$. Therefore we obtain

$$T_1 + T_2 = 2\sqrt{S_{WKO}S_{OLZ}} + 2\sqrt{S_{XNO}S_{OMY}}$$
$$= 2\sqrt{(S_{WKO} + S_{XNO})(S_{OLZ} + S_{OMY})} \geq 2\sqrt{S_1 S_2}.$$

Second solution. Using an affine transformation of the plane one can transform any nondegenerate quadrilateral into a cyclic one, thereby preserving parallelness and ratios of areas. Thus we may assume w.l.o.g. that $ABCD$ is cyclic.

By a well-known formula, the area of a cyclic quadrilateral with sides a, b, c, d and semiperimeter p is given by

$$S = \sqrt{(p-a)(p-b)(p-c)(p-d)}.$$

Let us set $AK = a_1$, $KB = b_1$, $BL = a_2$, $LC = b_2$, $CM = a_3$, $MD = b_3$, $DN = a_4$, $NA = b_4$. Then the sides of quadrilateral $AKON$ are a_i, the sides of $CLOM$ are b_i, and the sides of $ABCD$ are $a_i + b_i$ ($i = 1, 2, 3, 4$). If p and q are the semiperimeters of $AKON$ and $CLOM$, and $x_i = p - a_i$, $y_i = q - b_i$, then we have $S_1 = \sqrt{x_1 x_2 x_3 x_4}$, $S_2 = \sqrt{y_1 y_2 y_3 y_4}$, and $S = \sqrt{(x_1 + y_1)(x_2 + y_2)(x_3 + y_3)(x_4 + y_4)}$. Thus we need to show that

$$\sqrt[4]{x_1 x_2 x_3 x_4} + \sqrt[4]{y_1 y_2 y_3 y_4} \leq \sqrt[4]{(x_1 + y_1)(x_2 + y_2)(x_3 + y_3)(x_4 + y_4)}.$$

By setting $y_i = t_i x_i$ we reduce this inequality to

$$1 + \sqrt[4]{t_1 t_2 t_3 t_4} \leq \sqrt[4]{(1+t_1)(1+t_2)(1+t_3)(1+t_4)}.$$

One way to prove the last inequality is to apply the simple inequality

$$1 + \sqrt{uv} \leq \sqrt{(1+u)(1+v)}$$

to $\sqrt{t_1 t_2}$, $\sqrt{t_3 t_4}$ and then to t_1, t_2 and t_3, t_4.

14. Let BB' cut CC' at P. Since $\angle B'BC = \angle B'CC'$, it follows that $\angle PBH = \angle PCH$. Let D and E be points such that $BPCD$ and $HPCE$ are parallelograms (consequently, so is $BHED$). Triangles BAC and $C'AB'$ are similar, from which we deduce that $\triangle B'H'C'$ and $\triangle BHC$ are similar, as well as $\triangle B'PC'$ and $\triangle BDC$. Hence $B'PC'H'$ and $BDCH$ are similar, from which we obtain $\angle H'PB' = \angle HDB$. Now $\angle CDE = \angle PBH = \angle PCH = \angle CHE$ implies that $HCED$ is a cyclic quadrilateral. Therefore $\angle BPH = \angle DCE = \angle DHE = \angle HDB = \angle H'PB'$; hence HH' also passes through P.

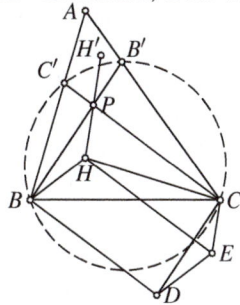

Second solution. Let us start with observations $\triangle HBC \sim \triangle H'B'C'$, $\angle PBH = \angle PCH$, and $\angle PB'H' = \angle PC'H'$.

By Ceva's theorem in trigonometric form applied to $\triangle BPC$ and the point H, we have $\frac{\sin \angle BPH}{\sin \angle HPC} = \frac{\sin \angle HBP}{\sin \angle HBC} \cdot \frac{\sin \angle HCB}{\sin \angle HCP} = \frac{\sin \angle HCB}{\sin \angle HBC}$. Similarly, Ceva's theorem for $\triangle B'PC'$ and point H' yields $\frac{\sin \angle B'PH'}{\sin \angle H'PC'} = \frac{\sin \angle H'C'B'}{\sin \angle H'B'C'}$. Thus it follows that

$$\frac{\sin \angle B'PH'}{\sin \angle H'PC'} = \frac{\sin \angle BPH}{\sin \angle HPC},$$

which finally implies that $\angle BPH = \angle B'PH'$.

15. We show by induction on k that there exists a positive integer a_k for which $a_k^2 \equiv -7 \pmod{2^k}$. The statement of the problem follows, since every $a_k + r2^k$ ($r = 0, 1, \ldots$) also satisfies this condition.

Note that for $k = 1, 2, 3$ one can take $a_k = 1$. Now suppose that $a_k^2 \equiv -7 \pmod{2^k}$ for some $k > 3$. Then either $a_k^2 \equiv -7 \pmod{2^{k+1}}$ or $a_k^2 \equiv 2^k - 7 \pmod{2^{k+1}}$. In the former case, take $a_{k+1} = a_k$. In the latter case, set $a_{k+1} = a_k + 2^{k-1}$. Then $a_{k+1}^2 = a_k^2 + 2^k a_k + 2^{2k-2} \equiv a_k^2 + 2^k \equiv -7 \pmod{2^{k+1}}$ because a_k is odd.

16. If A is odd, then every number in M_1 is of the form $x(x+A) + B \equiv B \pmod 2$, while numbers in M_2 are congruent to C modulo 2. Thus it is enough to take $C \equiv B + 1 \pmod 2$.

If A is even, then all numbers in M_1 have the form $\left(X + \frac{A}{2}\right)^2 + B - \frac{A^2}{4}$ and are congruent to $B - \frac{A^2}{4}$ or $B - \frac{A^2}{4} + 1$ modulo 4, while numbers in M_2 are congruent to C modulo 4. So one can choose any $C \equiv B - \frac{A^2}{4} + 2 \pmod 4$.

17. For $n = 4$, the vertices of a unit square $A_1A_2A_3A_4$ and $p_1 = p_2 = p_3 = p_4 = \frac{1}{6}$ satisfy the conditions. We claim that there are no solutions for $n = 5$ (and thus for any $n \geq 5$).

Suppose to the contrary that points A_i and p_i, $i = 1, \ldots, 5$, satisfy the conditions. Denote the area of $\triangle A_iA_jA_k$ by $S_{ijk} = p_i + p_j + p_k$, $1 \leq i < j < k \leq 5$. Observe that all the p_i's must be distinct. Indeed, if $p_4 = p_5$, then $S_{124} = S_{125}$ and $S_{234} = S_{235}$, which implies that A_4A_5 is parallel to A_1A_2 and A_2A_3, so A_1, A_2, A_3 are collinear, which is impossible. Also note that if $A_iA_jA_kA_l$ is convex, then $S_{ijk} + S_{ikl} = S_{ijl} + S_{jkl}$ gives $p_i + p_k = p_j + p_l$. Now consider the convex hull of A_1, A_2, A_3, A_4, A_5. There are three cases.

   (i) The convex hull is the pentagon $A_1A_2A_3A_4A_5$. We deduce that the quadrilaterals $A_1A_2A_3A_4$ and $A_1A_2A_3A_5$ are convex, so we have $p_1 + p_3 = p_2 + p_4$ and $p_1 + p_3 = p_2 + p_5$. Hence $p_4 = p_5$, a contradiction.

   (ii) The convex hull is w.l.o.g. the quadrilateral $A_1A_2A_3A_4$. Assume that A_5 lies within $A_1A_3A_4$. Then $A_1A_2A_3A_5$ is also convex, so as in (1) we get $p_4 = p_5$.

   (iii) The convex hull is w.l.o.g. the triangle $A_1A_2A_3$. Since $S_{124} + S_{134} + S_{234} = S_{125} + S_{135} + S_{235}$, we conclude that again $p_4 = p_5$.

18. Let $x = za$ and $y = zb$, where a and b are relatively prime. The given Diophantine equation becomes $a + zb^2 + z^2 = z^2ab$, so $a = zc$ for some $c \in \mathbb{Z}$. We obtain $c + b^2 + z = z^2cb$, or $c = \frac{b^2+z}{z^2b-1}$.

   (i) If $z = 1$, then $c = \frac{b^2+1}{b-1} = b + 1 + \frac{2}{b-1}$, so $b = 2$ or $b = 3$. These values yield two solutions: $(x, y) = (5, 2)$ and $(x, y) = (5, 3)$.

   (ii) If $z = 2$, then $16c = \frac{16b^2+32}{4b-1} = 4b + 1 + \frac{33}{4b-1}$, so $b = 1$ or $b = 3$. In this case $(x, y) = (4, 2)$ or $(x, y) = (4, 6)$.

   (iii) Let $z \geq 3$. First, we see that $z^2c = \frac{z^2b^2+z^3}{z^2b-1} = b + \frac{b+z^3}{z^2b-1}$. Thus $\frac{b+z^3}{z^2b-1}$ must be a positive integer, so $b + z^3 \geq z^2b - 1$, which implies $b \leq \frac{z^2-z+1}{z-1}$. It follows that $b \leq z$. But then $b^2 + z \leq z^2 + b < z^2b - 1$, with the last inequality because $(z^2 - 1)(b - 1) > 2$. Therefore $c = \frac{b^2+z}{z^2b-1} < 1$, a contradiction.

The only solutions for (x, y) are $(4, 2), (4, 6), (5, 2), (5, 3)$.

19. For each two people let n be the number of people exchanging greetings with both of them. To determine n in terms of k, we shall count in two ways the number of triples (A, B, C) of people such that A exchanged greetings with both B and C, but B and C mutually did not.

There are $12k$ possibilities for A, and for each A there are $(3k + 6)$ possibilities for B. Since there are n people who exchanged greetings with both A and B, there are $3k + 5 - n$ who did so with A but not with B. Thus the number of triples (A, B, C) is $12k(3k + 6)(3k + 5 - n)$. On the other hand, there are $12k$ possible choices of B, and $12k - 1 - (3k + 6) = 9k - 7$ possible choices of C; for every B, C, A can be chosen in n ways, so the number of considered triples equals $12kn(9k - 7)$.

Hence $(3k+6)(3k+5-n) = n(9k-7)$, i.e., $n = \frac{3(k+2)(3k+5)}{12k-1}$. This gives us that $\frac{4n}{3} = \frac{12k^2+44k+40}{12k-1} = k+4 - \frac{3k-44}{12k-1}$ is an integer too. It is directly verified that only $k = 3$ gives an integer value for n, namely $n = 6$.

Remark. The solution is complete under the assumption that such a k exists. We give an example of such a party with 36 persons, $k = 3$. Let the people sit in a 6×6 array $[P_{ij}]_{i,j=1}^6$, and suppose that two persons P_{ij}, P_{kl} exchanged greetings if and only if $i = k$ or $j = l$ or $i - j \equiv k - l \pmod{6}$. Thus each person exchanged greetings with exactly 15 others, and it is easily verified that this party satisfies the conditions.

20. We shall consider the set $M = \{0,1,\ldots,2p-1\}$ instead. Let $M_1 = \{0,1,\ldots,p-1\}$ and $M_2 = \{p,p+1,\ldots,2p-1\}$. We shall denote by $|A|$ and $\sigma(A)$ the number of elements and the sum of elements of the set A; also, let C_p be the family of all p-element subsets of M. Define the mapping $T : C_p \to C_p$ as $T(A) = \{x+1 \mid x \in A \cap M_1\} \cup \{A \cap M_2\}$, the addition being modulo p. There are exactly two fixed points of T: these are M_1 and M_2. Now if A is any subset from C_p distinct from M_1, M_2, and $k = |A \cap M_1|$ with $1 \le k \le p-1$, then for $i = 0,1,\ldots,p-1$, $\sigma(T^i(A)) = \sigma(A) + ik \pmod{p}$. Hence subsets $A, T(A), \ldots, T^{p-1}(A)$ are distinct, and exactly one of them has sum of elements divisible by p. Since $\sigma(M_1), \sigma(M_2)$ are divisible by p and $C_p \setminus \{M_1, M_2\}$ decomposes into families of the form $\{A, T(A), \ldots, T^{p-1}(A)\}$, we conclude that the required number is $\frac{1}{p}(|C_p| - 2) + 2 = \frac{1}{p}\left(\binom{2p}{p} - 2\right) + 2$.

Second solution. Let C_k be the family of all k-element subsets of $\{1,2,\ldots,2p\}$. Denote by M_k $(k = 1,2,\ldots,p)$ the family of p-element multisets with k distinct elements from $\{1,2,\ldots,2p\}$, exactly one of which appears more than once, that have sum of elements divisible by p. It is clear that every subset from C_k, $k < p$, can be complemented to a multiset from $M_k \cup M_{k+1}$ in exactly two ways, since the equation $(p-k)a \equiv 0 \pmod{p}$ has exactly two solutions in $\{1,2,\ldots,2p\}$. On the other hand, every multiset from M_k can be obtained by completing exactly one subset from C_k. Additionally, a multiset from M_k can be obtained from exactly one subset from C_{k-1} if $k < p$, and from exactly p subsets from C_{k-1} if $k = p$. Therefore $|M_k| + |M_{k+1}| = 2|C_k| = 2\binom{2p}{k}$ for $k = 1,2,\ldots,p-2$, and $|M_{p-1}| + p|M_p| = 2|C_{p-1}| = 2\binom{2p}{p-1}$. Since $M_1 = 2p$, it is not difficult to show using recursion that $|M_p| = \frac{1}{p}\left(\binom{2p}{p} - 2\right) + 2$.

Third solution. Let $\omega = \cos\frac{2\pi}{p} + i\sin\frac{2\pi}{p}$. We have $\prod_{i=1}^{2p}(x - \omega^i) = (x^p - 1)^2 = x^{2p} - 2x^p + 1$; hence comparing the coefficients at x^p, we obtain $\sum \omega^{i_1+\cdots+i_p} = \sum_{i=0}^{p-1} a_i\omega^i = 2$, where the first sum runs over all p-subsets $\{i_1,\ldots,i_p\}$ of the set $\{1,\ldots,2p\}$, and a_i is the number of such subsets for which $i_1 + \cdots + i_p \equiv i \pmod{p}$. Setting $q(x) = -2 + \sum_{i=0}^{p-1} a_ix^i$, we obtain $q(\omega^j) = 0$ for $j = 1,2,\ldots,p-1$. Hence $1 + x + \cdots + x^{p-1} \mid q(x)$, and since $\deg q = p-1$, we have $q(x) = -2 + \sum_{i=0}^{p-1} a_ix^i = c(1 + x + \cdots + x^{p-1})$ for some constant c. Thus $a_0 - 2 = a_1 = \cdots = a_{p-1}$, which together with $a_0 + \cdots + a_{p-1} = \binom{2p}{p}$ yields $a_0 = \frac{1}{p}\left(\binom{2p}{p} - 2\right) + 2$.

21. We shall show that there is no such n. Certainly, $n = 2$ does not work, so suppose $n \geq 3$. Let a, b be distinct elements of A_1, and c any integer greater than $-a$ and $-b$. We claim that $a + c, b + c$ belong to the same subsets. Suppose to the contrary that $a + c \in A_1$ and $b + c \in A_2$, and take arbitrary elements $x_i \in A_i$, $i = 3, \ldots, n$. The number $b + x_3 + \cdots + x_n$ is in A_2, so that $s = (a + c) + (b + x_3 + \cdots + x_n) + x_4 + \cdots + x_n$ must be in A_3. On the other hand, $a + x_3 + \cdots + x_n \in A_2$, so $s = (a + x_3 + \cdots + x_n) + (b + c) + x_4 + \cdots + x_n$ is in A_1, a contradiction. Similarly, if $a + c \in A_2$ and $b + c \in A_3$, then $s = a + (b + c) + x_4 + \cdots + x_n$ belongs to A_2, but also $s = b + (a + c) + x_4 + \cdots + x_n \in A_3$, which is impossible.
    For $i = 1, \ldots, n$ choose $x_i \in A_i$; set $s = x_1 + \cdots + x_n$ and $y_i = s - x_i$. Then $y_i \in A_i$. By what has been proved above, $2x_i = x_i + x_i$ belongs to the same subset as $x_i + y_i = s$ does. It follows that all numbers $2x_i$, $i = 1, \ldots, n$, are in the same subset. Since we can arbitrarily take x_i from each set A_i, it follows that all even numbers belong to the same set, say A_1. Similarly, $2x_i + 1 = (x_i + 1) + x_i$ is in the subset to which $(x_i + 1) + y_i = s + 1$ belongs for all $i = 1, \ldots, n$; hence all odd numbers greater than 1 are in the same subset, say A_2. By the above considerations, $3 - 2 = 1 \in A_2$ also. But then nothing remains in $A_3, \ldots, A_n$, a contradiction.

22. Let $u = \sqrt{2p} - \sqrt{x} - \sqrt{y}$ and $v = u(2\sqrt{2p} - u) = 2p - (\sqrt{2p} - u)^2 = 2p - x - y - \sqrt{4xy}$ for $x, y \in \mathbb{N}$, $x \leq y$. Obviously $u \geq 0$ if and only if $v \geq 0$, and u, v attain minimum positive values simultaneously. Note that $v \neq 0$. Otherwise $u = 0$ too, so $y = (\sqrt{2p} - \sqrt{x})^2 = 2p - x - 2\sqrt{2px}$, which implies that $2px$ is a square, and consequently x is divisible by $2p$, which is impossible.
    Now let z be the smallest integer greater than $\sqrt{4xy}$. We have $z^2 - 1 \geq 4xy$, $z \leq 2p - x - y$, and $z \leq p$ because $\sqrt{4xy} \leq (\sqrt{x} + \sqrt{y})^2 < 2p$. It follows that

$$v = 2p - x - y - \sqrt{4xy} \geq z - \sqrt{z^2 - 1} = \frac{1}{z + \sqrt{z^2 - 1}} \geq \frac{1}{p + \sqrt{p^2 - 1}}.$$

Equality holds if and only if $z = x + y = p$ and $4xy = p^2 - 1$, which is satisfied only when $x = \frac{p-1}{2}$ and $y = \frac{p+1}{2}$. Hence for these values of x, y, both u and v attain positive minima.

23. By putting $F(1) = 0$ and $F(361) = 1$, condition (c) becomes $F(F(n^{163})) = F(F(n))$ for $n \geq 2$. For $n = 2, 3, \ldots, 360$ let $F(n) = n$, and inductively define $F(n)$ for $n \geq 362$ as follows:

$$F(n) = \begin{cases} F(m), & \text{if } n = m^{163}, m \in \mathbb{N}; \\ \text{the least number not in } \{F(k) \mid k < n\}, & \text{otherwise.} \end{cases}$$

Obviously, (a) each nonnegative integer appears in the sequence because there are infinitely many numbers not of the form m^{163}, and (b) each positive integer appears infinitely often because $F(m^{163}) = F(m)$. Since $F(n^{163}) = F(n)$, (c) also holds.

Second solution. Another example of such a sequence is as follows: If $n = p_1^{\alpha_1} p_2^{\alpha_2} \cdots p_k^{\alpha_k}$ is the factorization of n into primes, we put $F(n) = \alpha_1 + \alpha_2 + $

$\cdots + \alpha_k$ and $F(1) = 0$. Conditions (a) and (b) are evidently satisfied for this F, while (c) follows from $F(F(n^{163})) = F(163F(n)) = F(F(n)) + 1$ (because 163 is a prime) and $F(F(361)) = F(F(19^2)) = F(2) = 1$.

24. The given condition is equivalent to $(2x_i - x_{i-1})(x_i x_{i-1} - 1) = 0$, so either $x_i = \frac{1}{2}x_{i-1}$ or $x_i = \frac{1}{x_{i-1}}$. We shall show by induction on n that for any $n \geq 0$, $x_n = 2^{k_n}x_0^{e_n}$ for some integer k_n, where $|k_n| \leq n$ and $e_n = (-1)^{n-k_n}$. Indeed, this is true for $n = 0$. If it holds for some n, then $x_{n+1} = \frac{1}{2}x_n = 2^{k_n-1}x_0^{e_n}$ (hence $k_{n+1} = k_n - 1$ and $e_{n+1} = e_n$) or $x_{n+1} = \frac{1}{x_n} = 2^{-k_n}x_0^{-e_n}$ (hence $k_{n+1} = -k_n$ and $e_{n+1} = -e_n$).
Thus $x_0 = x_{1995} = 2^{k_{1995}}x_0^{e_{1995}}$. Note that $e_{1995} = 1$ is impossible, since in that case k_{1995} would be odd, although it should equal 0. Therefore $e^{1995} = -1$, which gives $x_0^2 = 2^{k_{1995}} \leq 2^{1994}$, so the maximal value that x_0 can have is 2^{997}. This value is attained in the case $x_i = 2^{997-i}$ for $i = 0,\ldots,997$ and $x_i = 2^{i-998}$ for $i = 998,\ldots,1995$.

Second solution. First we show that there is an n, $0 \leq n \leq 1995$, such that $x_n = 1$. Suppose the contrary. Then each of x_n belongs to one of the intervals $I_{-i-1} = [2^{-i-1}, 2^{-i})$ or $I_i = (2^i, 2^{i+1}]$, where $i = 0, 1, 2, \ldots$. Let $x_n \in I_{i_n}$. Note that by the formula for x_n, i_n and i_{n-1} are of different parity. Hence i_0 and i_{1995} are also of different parity, contradicting $x_0 = x_{1995}$.
It follows that for some n, $x_n = 1$. Now if $n \leq 997$, then $x_0 \leq 2^{997}$, while if $n \geq 998$, we also have $x_0 = x_{1995} \leq 2^{997}$.

25. By the definition of $q(x)$, it divides x for all integers $x > 0$, so $f(x) = xp(x)/q(x)$ is a positive integer too. Let $\{p_0, p_1, p_2, \ldots\}$ be all prime numbers in increasing order. Since it easily follows by induction that all x_n's are square-free, we can assign to each of them a unique code according to which primes divide it: if p_m is the largest prime dividing x_n, the code corresponding to x_n will be $\ldots 0s_m s_{m-1}\ldots s_0$, with $s_i = 1$ if $p_i \mid x_n$ and $s_i = 0$ otherwise. Let us investigate how f acts on these codes. If the code of x_n ends with 0, then x_n is odd, so the code of $f(x_n) = x_{n+1}$ is obtained from that of x_n by replacing $s_0 = 0$ by $s_0 = 1$. Furthermore, if the code of x_n ends with $011\ldots1$, then the code of x_{n+1} ends with $100\ldots0$ instead. Thus if we consider the codes as binary numbers, f acts on them as an addition of 1. Hence the code of x_n is the binary representation of n and thus x_n uniquely determines n.
Specifically, if $x_n = 1995 = 3 \cdot 5 \cdot 7 \cdot 19$, then its code is 10001110 and corresponds to $n = 142$.

26. For $n = 1$ the result is trivial, since $x_1 = 1$. Suppose now that $n \geq 2$ and let $f_n(x) = x^n - \sum_{i=0}^{n-1} x^i$. Note that x_n is the unique positive real root of f_n, because $\frac{f_n(x)}{x^{n-1}} = x - 1 - \frac{1}{x} - \cdots - \frac{1}{x^{n-1}}$ is strictly increasing on $\mathbb{R}^+$.
Consider $g_n(x) = (x-1)f_n(x) = (x-2)x^n + 1$. Obviously $g_n(x)$ has no positive roots other than 1 and $x_n > 1$. Observe that $(1 - \frac{1}{2^n})^n > 1 - \frac{n}{2^n} \geq \frac{1}{2}$ for $n \geq 2$ (by Bernoulli's inequality). Since then

$$g_n\left(2 - \frac{1}{2^n}\right) = -\frac{1}{2^n}\left(2 - \frac{1}{2^n}\right)^n + 1 = 1 - \left(1 - \frac{1}{2^{n+1}}\right)^n > 0,$$

and

$$g_n\left(2-\frac{1}{2^{n-1}}\right)=-\frac{1}{2^{n-1}}\left(2-\frac{1}{2^{n-1}}\right)^n+1=1-2\left(1-\frac{1}{2^n}\right)^n<0,$$

we conclude that x_n is between $2-\frac{1}{2^{n-1}}$ and $2-\frac{1}{2^n}$, as required.

Remark. Moreover, $\lim_{n\to\infty}2^n(2-x_n)=1$.

27. Computing the first few values of $f(n)$, we observe the following pattern:

$$
\begin{array}{ll}
f(4k)=k,\ k\geq 3, & f(8)=3;\\
f(4k+1)=1,\ k\geq 4, & f(5)=f(13)=2;\\
f(4k+2)=k-3,\ k\geq 7, & f(2)=1, f(6)=f(10)=2,\\
 & f(14)=f(18)=3, f(26)=4;\\
f(4k+3)=2.
\end{array}
$$

We shall prove these statements simultaneously by induction on n, having verified them for $k\leq 7$.

(i) Let $n=4k$. Since $f(3)=f(7)=\cdots=f(4k-1)=2$, we have $f(4k)\geq k$.
But $f(n)\leq\max_{m<n}f(m)+1\leq(k-1)+1$, so $f(4k)=k$.

(ii) Let $n=4k+1$, $k\geq 7$. Since $f(4k)=k$ and $f(m)<k$ for $m<4k$, we deduce
that $f(4k+1)=1$.

(iii) Let $n=4k+2$, $k\geq 7$. Since $f(17)=f(21)=\cdots=f(4k+1)=1$, we obtain
$f(4k+2)\geq k-3$. On the other hand, if $f(4k+1)=f(4k+1-d)=1$, then
$d\geq 8$, and $4k+1-8(k-3)<0$. So $f(4k+2)=k-3$.

(iv) Let $n=4k+3$, $k\geq 7$. We have $f(4k+2)=k-3$ and $f(m)=k-3$ for
exactly one $m<4k+2$ (namely for $m=4k-12$); hence $f(4k+3)=2$.

Therefore, for example, $f(4n+8)=n+2$ for all n; hence we can take $a=4$ and
$b=8$.

28. Let $F(x)=f(x)-95$ for $x\geq 1$. Writing k for $m+95$, the given condition becomes

$$F(k+F(n))=F(k)+n, \qquad k\geq 96, n\geq 1. \tag{1}$$

Thus for $x,z\geq 96$ and an arbitrary y we have $F(x+y)+z=F(x+y+F(z))=$
$F(x+F(F(y)+z))=F(x)+F(y)+z$, and consequently $F(x+y)=F(x)+F(y)$
whenever $x\geq 96$. Moreover, since then $F(x+y)+F(96)=F(x+y+96)=$
$F(x)+F(y+96)=F(x)+F(y)+F(96)$ for any x,y, we obtain

$$F(x+y)=F(x)+F(y), \qquad x,y\in\mathbb{N}. \tag{2}$$

It follows by induction that $F(n)=nc$ for all n, where $F(1)=c$. Equation (1)
becomes $ck+c^2n=ck+n$, and yields $c=1$. Hence $F(n)=n$ and $f(n)=n+95$
for all n.

Finally, $\sum_{k=1}^{19}f(k)=96+97+\cdots+114=1995$.

Second solution. First we show that $f(n)>95$ for all n. If to the contrary $f(n)\leq$
95, we have $f(m)=n+f(m+95-f(n))$, so by induction $f(m)=kn+f(m+$
$k(95-f(n)))\geq kn$ for all k, which is impossible. Now for $m>95$ we have

$f(m+f(n)-95) = n+f(m)$, and again by induction $f(m+k(f(n)-95)) = kn+f(m)$ for all m,n,k. It follows that with n fixed,

$$(\forall m) \lim_{k\to\infty} \frac{f(m+k(f(n)-95))}{m+k(f(n)-95)} = \frac{n}{f(n)-95};$$

hence

$$\lim_{s\to\infty} \frac{f(s)}{s} = \frac{n}{f(n)-95}.$$

Hence $\frac{n}{f(n)-95}$ does not depend on n, i.e., $f(n) \equiv cn+95$ for some constant c. It is easily checked that only $c=1$ is possible.

4.37 Solutions to the Shortlisted Problems of IMO 1996

1. We have $a^5 + b^5 - a^2b^2(a+b) = (a^3 - b^3)(a^2 - b^2) \geq 0$, i.e. $a^5 + b^5 \geq a^2b^2(a+b)$. Hence

$$\frac{ab}{a^5 + b^5 + ab} \leq \frac{ab}{a^2b^2(a+b) + ab} = \frac{abc^2}{a^2b^2c^2(a+b) + abc^2} = \frac{c}{a+b+c}.$$

Now, the left side of the inequality to be proved does not exceed $\frac{c}{a+b+c} + \frac{a}{a+b+c} + \frac{b}{a+b+c} = 1$. Equality holds if and only if $a = b = c$.

2. Clearly $a_1 > 0$, and if $p \neq a_1$, we must have $a_n < 0$, $|a_n| > |a_1|$, and $p = -a_n$. But then for sufficiently large odd k, $-a_n^k = |a_n|^k > (n-1)|a_1|^k$, so that $a_1^k + \cdots + a_n^k \leq (n-1)|a_1|^k - |a_n|^k < 0$, a contradiction. Hence $p = a_1$.
   Now let $x > a_1$. From $a_1 + \cdots + a_n \geq 0$ we deduce $\sum_{j=2}^{n}(x - a_j) \leq (n-1)\left(x + \frac{a_1}{n-1}\right)$, so by the AM–GM inequality,

$$(x - a_2)\cdots(x - a_n) \leq \left(x + \frac{a_1}{n-1}\right)^{n-1} \leq x^{n-1} + x^{n-2}a_1 + \cdots + a_1^{n-1}. \quad (1)$$

The last inequality holds because $\binom{n-1}{r} \leq (n-1)^r$ for all $r \geq 0$. Multiplying (1) by $(x - a_1)$ yields the desired inequality.

3. Since $a_1 > 2$, it can be written as $a_1 = b + b^{-1}$ for some $b > 0$. Furthermore, $a_1^2 - 2 = b^2 + b^{-2}$ and hence $a_2 = (b^2 + b^{-2})(b + b^{-1})$. We prove that

$$a_n = (b + b^{-1})(b^2 + b^{-2})(b^4 + b^{-4})\cdots\left(b^{2^{n-1}} + b^{-2^{n-1}}\right)$$

by induction. Indeed, $\frac{a_{n+1}}{a_n} = \left(\frac{a_n}{a_{n-1}}\right)^2 - 2 = \left(b^{2^{n-1}} + b^{-2^{n-1}}\right)^2 - 2 = b^{2^n} + b^{-2^n}$. Now we have

$$\sum_{i=1}^{n}\frac{1}{a_i} = 1 + \frac{b}{b^2 + 1} + \frac{b^3}{(b^2 + 1)(b^4 + 1)} + \cdots$$
$$\cdots + \frac{b^{2^n - 1}}{(b^2 + 1)(b^4 + 1)\ldots(b^{2^n} + 1)}. \quad (1)$$

Note that $\frac{1}{2}(a + 2 - \sqrt{a^2 - 4}) = 1 + \frac{1}{b}$; hence we must prove that the right side in (1) is less than $\frac{1}{b}$. This follows from the fact that

$$\frac{b^{2^k}}{(b^2 + 1)(b^4 + 1)\cdots(b^{2^k} + 1)}$$
$$= \frac{1}{(b^2 + 1)(b^4 + 1)\cdots(b^{2^{k-1}} + 1)} - \frac{1}{(b^2 + 1)(b^4 + 1)\cdots(b^{2^k} + 1)};$$

hence the right side in (1) equals $\frac{1}{b}\left(1 - \frac{1}{(b^2+1)(b^4+1)\ldots(b^{2^n}+1)}\right)$, and this is clearly less than $1/b$.

4. Consider the function

$$f(x) = \frac{a_1}{x} + \frac{a_2}{x^2} + \cdots + \frac{a_n}{x^n}.$$

Since f is strictly decreasing from $+\infty$ to 0 on the interval $(0, +\infty)$, there exists exactly one $R > 0$ for which $f(R) = 1$. This R is also the only positive real root of the given polynomial.

Since $\ln x$ is a concave function on $(0, +\infty)$, Jensen's inequality gives us

$$\sum_{j=1}^{n} \frac{a_j}{A}\left(\ln \frac{A}{R^j}\right) \leq \ln\left(\sum_{j=1}^{n} \frac{a_j}{A} \cdot \frac{A}{R^j}\right) = \ln f(R) = 0.$$

Therefore $\sum_{j=1}^{n} a_j(\ln A - j\ln R) \leq 0$, which is equivalent to $A\ln A \leq B\ln R$, i.e., $A^A \leq R^B$.

5. Considering the polynomials $\pm P(\pm x)$ we may assume w.l.o.g. that $a, b \geq 0$. We have four cases:

    (1) $c \geq 0, d \geq 0$. Then $|a| + |b| + |c| + |d| = a + b + c + d = P(1) \leq 1$.
    (2) $c \geq 0, d < 0$. Then $|a| + |b| + |c| + |d| = a + b + c - d = P(1) - 2P(0) \leq 3$.
    (3) $c < 0, d \geq 0$. Then

$$|a| + |b| + |c| + |d| = a + b - c + d$$
$$= \frac{4}{3}P(1) - \frac{1}{3}P(-1) - \frac{8}{3}P(1/2) + \frac{8}{3}P(-1/2) \leq 7.$$

    (4) $c < 0, d < 0$. Then

$$|a| + |b| + |c| + |d| = a + b - c - d$$
$$= \frac{5}{3}P(1) - 4P(1/2) + \frac{4}{3}P(-1/2) \leq 7.$$

Remark. It can be shown that the maximum of 7 is attained only for $P(x) = \pm(4x^3 - 3x)$.

6. Let $f(x), g(x)$ be polynomials with integer coefficients such that

$$f(x)(x+1)^n + g(x)(x^n + 1) = k_0. \tag{1}$$

Write $n = 2^r m$ for m odd and note that $x^n + 1 = (x^{2^r} + 1)B(x)$, where $B(x) = x^{2^r(m-1)} - x^{2^r(m-2)} + \cdots - x^{2^r} + 1$. Moreover, $B(-1) = 1$; hence $B(x) - 1 = (x + 1)c(x)$ and thus

$$R(x)B(x) + 1 = (B(x) - 1)^n = (x+1)^n c(x)^n \tag{2}$$

for some polynomials $c(x)$ and $R(x)$.

The zeros of the polynomial $x^{2^r} + 1$ are ω_j, with $\omega_1 = \cos\frac{\pi}{2^r} + i\sin\frac{\pi}{2^r}$, and $\omega_j = \omega^{2j-1}$ for $1 \leq j \leq 2^r$. We have

$$(\omega_1 + 1)(\omega_2 + 1)\cdots(\omega_{2^r+1} + 1) = 2. \tag{3}$$

From (1) we also get $f(\omega_j)(\omega_j + 1)^n = k_0$ for $j = 1,2,\ldots,2^r$. Since $A = f(\omega_1)f(\omega_2)\cdots f(\omega_{2^r})$ is a symmetric polynomial in $\omega_1,\ldots,\omega_{2^r}$ with integer coefficients, A is an integer. Consequently, taking the product over $j = 1,2,\ldots,2^r$ and using (3) we deduce that $2^n A = k_0^{2^r}$ is divisible by $2^n = 2^{2^r m}$. Hence $2^m \mid k_0$. Furthermore, since $\omega_j + 1 = (\omega_1 + 1)p_j(\omega_1)$ for some polynomial p_j with integer coefficients, (3) gives $(\omega_1 + 1)^{2^r} p(\omega_1) = 2$, where $p(x) = p_2(x)\cdots p_{2^r}(x)$ has integer coefficients. But then the polynomial $(x + 1)^{2^r} p(x) - 2$ has a zero $x = \omega_1$, so it is divisible by its minimal polynomial $x^{2^r} + 1$. Therefore

$$(x+1)^{2^r} p(x) = 2 + (x^{2^r} + 1)q(x) \tag{4}$$

for some polynomial $q(x)$. Raising (4) to the mth power we get $(x+1)^n p(x)^n = 2^m + (x^{2^r} + 1)Q(x)$ for some polynomial $Q(x)$ with integer coefficients. Now using (2) we obtain

$$\begin{aligned}(x+1)^n c(x)^n (x^{2^r} + 1)Q(x) &= (x^{2^r} + 1)Q(x) + (x^{2^r} + 1)Q(x)B(x)R(x)\\ &= (x+1)^n p(x)^n - 2^m + (x^n + 1)Q(X)R(x).\end{aligned}$$

Therefore $(x+1)^n f(x) + (x^n + 1)g(x) = 2^m$ for some polynomials $f(x), g(x)$ with integer coefficients, and $k_0 = 2^m$.

7. We are given that $f(x+a+b) - f(x+a) = f(x+b) - f(x)$, where $a = 1/6$ and $b = 1/7$. Summing up these equations for $x, x+b,\ldots,x+6b$ we obtain $f(x+a+1) - f(x+a) = f(x+1) - f(x)$. Summing up the new equations for $x, x+a,\ldots,x+5a$ we obtain that

$$f(x+2) - f(x+1) = f(x+1) - f(x).$$

It follows by induction that $f(x+n) - f(x) = n[f(x+1) - f(x)]$. If $f(x+1) \neq f(x)$, then $f(x+n) - f(x)$ will exceed in absolute value an arbitrarily large number for a sufficiently large n, contradicting the assumption that f is bounded. Hence $f(x+1) = f(x)$ for all x.

8. Putting $m = n = 0$ we obtain $f(0) = 0$ and consequently $f(f(n)) = f(n)$ for all n. Thus the given functional equation is equivalent to

$$f(m + f(n)) = f(m) + f(n), \qquad f(0) = 0.$$

Clearly one solution is $(\forall x)\ f(x) = 0$. Suppose f is not the zero function. We observe that f has nonzero fixed points (for example, any $f(n)$ is a fixed point). Let a be the smallest nonzero fixed point of f. By induction, each ka ($k \in \mathbb{N}$) is a fixed point too. We claim that all fixed points of f are of this form. Indeed, suppose that $b = ka + i$ is a fixed point, where $i < a$. Then

$$b = f(b) = f(ka + i) = f(i + f(ka)) = f(i) + f(ka) = f(i) + ka;$$

hence $f(i) = i$. Hence $i = 0$.

Since the set of values of f is a set of its fixed points, it follows that for $i = 0, 1, \ldots, a-1$, $f(i) = an_i$ for some integers $n_i \geq 0$ with $n_0 = 0$.

Let $n = ka + i$ be any positive integer, $0 \leq i < a$. As before, the functional equation gives us

$$f(n) = f(ka+i) = f(i) + ka = (n_i + k)a.$$

Besides the zero function, this is the general solution of the given functional equation. To verify this, we plug in $m = ka + i$, $n = la + j$ and obtain

$$f(m + f(n)) = f(ka + i + f(la + j)) = f((k+l+n_j)a + i)$$
$$= (k + l + n_j + n_i)a = f(m) + f(n).$$

9. From the definition of $a(n)$ we obtain

$$a(n) - a([n/2]) = \begin{cases} 1 & \text{if } n \equiv 0 \text{ or } n \equiv 3 \pmod 4; \\ -1 & \text{if } n \equiv 1 \text{ or } n \equiv 2 \pmod 4. \end{cases}$$

Let $n = \overline{b_k b_{k-1} \ldots b_1 b_0}$ be the binary representation of n, where we assume $b_k = 1$. If we define $p(n)$ and $q(n)$ to be the number of indices $i = 0, 1, \ldots, k-1$ with $b_i = b_{i+1}$ and the number of $i = 0, 1, \ldots, k-1$ with $b_i \neq b_{i+1}$ respectively, we get

$$a(n) = p(n) - q(n). \tag{1}$$

(a) The maximum value of $a(n)$ for $n \leq 1996$ is 9 when $p(n) = 9$ and $q(n) = 0$, i.e., in the case $n = \overline{1111111111}_2 = 1023$.
   The minimum value is -10 and is attained when $p(n) = 0$ and $q(n) = 10$, i.e., only for $n = \overline{10101010101}_2 = 1365$.

(b) From (1) we have that $a(n) = 0$ is equivalent to $p(n) = q(n) = k/2$. Hence k must be even, and the $k/2$ indices i for which $b_i = b_{i+1}$ can be chosen in exactly $\binom{k}{k/2}$ ways. Thus the number of positive integers $n < 2^{11} = 2048$ with $a(n) = 0$ is equal to

$$\binom{0}{0} + \binom{2}{1} + \binom{4}{2} + \binom{6}{3} + \binom{8}{4} + \binom{10}{5} = 351.$$

But five of these numbers exceed 1996: these are

$$2002 = \overline{11111010010}_2, \quad 2004 = \overline{11111010100}_2,$$

$$2006 = \overline{11111010110}_2, \quad 2010 = \overline{11111011010}_2,$$

$$2026 = \overline{11111101010}_2.$$

Therefore there are 346 numbers $n \leq 1996$ for which $a(n) = 0$.

10. We first show that H is the common orthocenter of the triangles ABC and AQR.

Let G, G', H' be respectively the centroid of $\triangle ABC$, the centroid of $\triangle PBC$, and the orthocenter of $\triangle PBC$. Since the triangles ABC and PBC have a common circumcenter, from the properties of the Euler line we get $\overrightarrow{HH'} = 3\overrightarrow{GG'} = \overrightarrow{AP}$. But $\triangle AQR$ is exactly the image of $\triangle PBC$ under translation by $\overrightarrow{AP}$; hence the orthocenter of AQR coincides with H. (*Remark:* This can be shown by noting that $AHBQ$ is cyclic.)

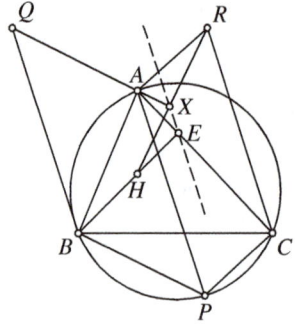

Now we have that $RH \perp AQ$; hence $\angle AXH = 90° = \angle AEH$. It follows that $AXEH$ is cyclic; hence

$$\angle EXQ = 180° - \angle AHE = 180° - \angle BCA = 180° - \angle BPA = \angle PAQ$$

(as oriented angles). Hence $EX \parallel AP$.

11. Let X, Y, Z respectively be the feet of the perpendiculars from P to BC, CA, AB. Examining the cyclic quadrilaterals $AZPY$, $BXPZ$, $CYPX$, one can easily see that $\angle XZY = \angle APB - \angle C$ and $XY = PC\sin\angle C$. The first relation gives that XYZ is isosceles with $XY = XZ$, so from the second relation $PB\sin\angle B = PC\sin\angle C$. Hence $AB/PB = AC/PC$. This implies that the bisectors BD and CD of $\angle ABP$ and $\angle ACP$ divide the segment AP in equal ratios; i.e., they concur with AP.

Second solution. Take that X, Y, Z are the points of intersection of AP, BP, CP with the circumscribed circle of ABC instead. We similarly obtain $XY = XZ$. If we write $AP \cdot PX = BP \cdot PY = CP \cdot PZ = k$, from the similarity of $\triangle APC$ and $\triangle ZPX$ we get

$$\frac{AC}{XZ} = \frac{AP}{PZ} = \frac{AP \cdot CP}{k},$$

i.e., $XZ = \frac{k \cdot AC \cdot BP}{AP \cdot BP \cdot CP}$. It follows again that $AC/AB = PC/PB$.

Third solution. Apply an inversion with center at A and radius r, and denote by $\overline{Q}$ the image of any point Q. Then the given condition becomes $\angle \overline{BCP} = \angle \overline{CBP}$, i.e., $\overline{BP} = \overline{PC}$. But

$$\overline{PB} = \frac{r^2}{AP \cdot AB} PB,$$

so $AC/AB = PC/PB$.

Remark. Moreover, it follows that the locus of P is an arc of the circle of Apollonius through C.

12. It is easy to see that P lies on the segment AC. Let E be the foot of the altitude BH and Y, Z the midpoints of AC, AB respectively. Draw the perpendicular HR to FP ($R \in FP$). Since Y is the circumcenter of $\triangle FCA$, we have $\angle FYA = 180° - 2\angle A$. Also, $OFPY$ is cyclic; hence $\angle OPF = \angle OYF = 2\angle A - 90°$. Next, $\triangle OZF$ and $\triangle HRF$ are similar, so $OZ/OF = HR/HF$.

This leads to $HR \cdot OF = HF \cdot OZ = \frac{1}{2}HF \cdot HC = \frac{1}{2}HE \cdot HB = HE \cdot OY$. This implies that $HR/HE = OY/OF$. Moreover, $\angle EHR = \angle FOY$; hence the triangles EHR and FOY are similar. Consequently $\angle HPC = \angle HRE = \angle OYF = 2\angle A - 90°$. We finally get $\angle FHP = \angle HPC + \angle HCP = \angle A$.

Second solution. As before, $\angle HFY = 90° - \angle A$, so it suffices to show that $HP \perp FY$. The points O, F, P, Y lie on a circle, say Ω_1 with center at the midpoint Q of OP. Furthermore, the points F, Y lie on the nine-point circle Ω of $\triangle ABC$ with center at the midpoint N of OH. The segment FY is the common chord of Ω_1 and Ω, from which we deduce that $NQ \perp FY$. However, $NQ \parallel HP$, and the result follows.

Third solution. Let H' be the point symmetric to H with respect to AB. Then H' lies on the circumcircle of ABC. Let the line FP meet the circumcircle at U, V and meet $H'B$ at P'. Since $OF \perp UV$, F is the midpoint of UV. By the butterfly theorem, F is also the midpoint of PP'. Therefore $\triangle H'FP' \cong FHP$; hence $\angle FHP = \angle FH'B = \angle A$.

Remark. It is possible to solve the problem using trigonometry. For example, $\frac{FZ}{ZO} = \frac{FK}{KP} = \frac{\sin(A-B)}{\cos C}$, where K is on CF with $PK \perp CF$. Then $\frac{CF}{KP} = \frac{\sin(A-B)}{\cos C} + \tan A$, from which one obtains formulas for KP and KH. Finally, we can calculate $\tan \angle FHP = \frac{KP}{KH} = \cdots = \tan A$.

Second remark. Here is what happens when $BC \leq CA$. If $\angle A > 45°$, then $\angle FHP = \angle A$. If $\angle A = 45°$, the point P escapes to infinity. If $\angle A < 45°$, the point P appears on the extension of AC over C, and $\angle FHP = 180° - \angle A$.

13. By the law of cosines applied to $\triangle CA_1B_1$, we obtain $A_1B_1^2 = A_1C^2 + B_1C^2 - A_1C \cdot B_1C \geq A_1C \cdot B_1C$. Analogously, $B_1C_1^2 \geq B_1A \cdot C_1A$ and $C_1A_1^2 \geq C_1B \cdot A_1B$, so that multiplying these inequalities yields

$$A_1B_1^2 \cdot B_1C_1^2 \cdot C_1A_1^2 \geq A_1B \cdot A_1C \cdot B_1A \cdot B_1C \cdot C_1A \cdot C_1B. \tag{1}$$

Now, the lines AA_1, BB_1, CC_1 concur, so by Ceva's theorem, $A_1B \cdot B_1C \cdot C_1A = AB_1 \cdot BC_1 \cdot CA_1$, which together with (1) gives the desired inequality. Equality holds if and only if $CA_1 = CB_1$, etc.

14. Let a, b, c, d, e, and f denote the lengths of the sides AB, BC, CD, DE, EF, and FA respectively. Note that $\angle A = \angle D$, $\angle B = \angle E$, and $\angle C = \angle F$. Draw the lines PQ and RS through A and D perpendicular to BC and EF

respectively ($P, R \in BC$, $Q, S \in EF$). Then $BF \geq PQ = RS$. Therefore $2BF \geq PQ + RS$, or

$$2BF \geq (a \sin B + f \sin C) + (c \sin C + d \sin B),$$

$$\text{and similarly,} \quad 2BD \geq (c \sin A + b \sin B) + (e \sin B + f \sin A), \qquad (1)$$

$$2DF \geq (e \sin C + d \sin A) + (a \sin A + b \sin C).$$

Next, we have the following formulas for the considered circumradii:

$$R_A = \frac{BF}{2 \sin A}, \qquad R_C = \frac{BD}{2 \sin C}, \qquad R_E = \frac{DF}{2 \sin E}.$$

It follows from (1) that

$$R_A + R_C + R_E \geq \frac{1}{4} a \left(\frac{\sin B}{\sin A} + \frac{\sin A}{\sin B} \right) + \frac{1}{4} b \left(\frac{\sin C}{\sin B} + \frac{\sin B}{\sin C} \right) + \cdots$$

$$\geq \frac{1}{2}(a + b + \cdots) = \frac{P}{2},$$

with equality if and only if $\angle A = \angle B = \angle C = 120°$ and $FB \perp BC$ etc., i.e., if and only if the hexagon is regular.

Second solution. Let us construct points A'', C'', E'' such that $ABA''F$, $CDC''B$, and $EFE''D$ are parallelograms. It follows that A'', C'', B are collinear and also C'', E'', B and E'', A'', F. Furthermore, denote by A' be the intersection of the perpendiculars through F and B to FA'' and BA'', respectively, and let C' and E' be analogously defined. Since $A'FA''B$ is cyclic with the diameter being $A'A''$ and since $\triangle FA''B \cong \triangle BAF$, it follows that $2R_A = A'A'' = x$. Similarly, $2R_C = C'C'' = y$ and $2R_E = E'E'' = z$. We also have $AB = AB = FA'' = y_a$, $A''B = z_a$, $CD = C''B = z_c$, $CB = C''D = x_c$, $EF = E''D = x_e$, and $ED = E''F = y_e$. The original inequality we must prove now becomes

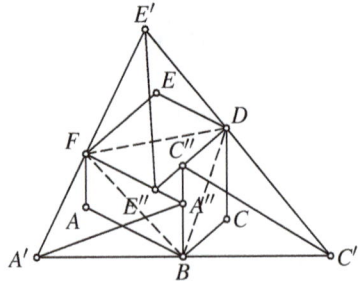

$$x + y + z \geq y_a + z_a + z_c + x_c + x_e + y_e. \qquad (1)$$

We now follow and generalize the standard proof of the Erdős–Mordell inequality (for the triangle $A'C'E'$), which is what (1) is equivalent to when $A'' = C'' = E''$.

We set $C'E' = a$, $A'E' = c$ and $A'C' = e$. Let A_1 be the point symmetric to A'' with respect to the bisector of $\angle E'A'C'$. Let F_1 and B_1 be the feet of the perpendiculars from A_1 to $A'C'$ and $A'E'$, respectively. In that case, $A_1F_1 = A''F = y_a$ and $A_1B_1 = A''B = z_a$. We have

$$ax = A'A_1 \cdot E'C' \geq 2S_{A'E'A_1C'} = 2S_{A'E'A_1} + 2S_{A'C'A_1}$$

$$= cz_a + ey_a.$$

Similarly, $cy \geq ex_c + az_c$ and $ez \geq ay_e + cx_e$. Thus

$$x + y + z \geq \frac{c}{a}z_a + \frac{a}{c}z_c + \frac{e}{c}x_c + \frac{c}{e}x_e + \frac{a}{e}y_e + \frac{e}{a}y_a$$

$$= \left(\frac{c}{a} + \frac{a}{c}\right)\left(\frac{z_a + z_c}{2}\right) + \left(\frac{c}{a} - \frac{a}{c}\right)\left(\frac{z_a - z_c}{2}\right) + \cdots . \tag{2}$$

Let us set $a_1 = \frac{x_c - x_e}{2}$, $c_1 = \frac{y_e - y_a}{2}$, $e_1 = \frac{z_a - z_c}{2}$. We note that $\triangle A''C''E'' \sim \triangle A'C'E'$ and hence $a_1/a = c_1/c = e_1/e = k$. Thus $\left(\frac{c}{a} - \frac{a}{c}\right)e_1 + \left(\frac{e}{c} - \frac{c}{e}\right)a_1 + \left(\frac{a}{e} - \frac{e}{a}\right)c_1 = k\left(\frac{ce}{a} - \frac{ae}{c} + \frac{ea}{c} - \frac{ca}{e} + \frac{ac}{e} - \frac{ec}{a}\right) = 0$. Equation (2) reduces to

$$x + y + z \geq \left(\frac{c}{a} + \frac{a}{c}\right)\left(\frac{z_a + z_c}{2}\right) + \left(\frac{e}{c} + \frac{c}{e}\right)\left(\frac{x_e + x_c}{2}\right)$$

$$+ \left(\frac{a}{e} + \frac{e}{a}\right)\left(\frac{y_a + y_e}{2}\right) .$$

Using $c/a + a/c$, $e/c + c/e$, $a/e + e/a \geq 2$ we finally get $x + y + z \geq y_a + z_a + z_c + x_c + x_e + y_e$.

Equality holds if and only if $a = c = e$ and $A'' = C'' = E'' =$ center of $\triangle A'C'E'$, i.e., if and only if $ABCDEF$ is regular.

Remark. From the second proof it is evident that the Erdős–Mordell inequality is a special case of the problem. If P_a, P_b, P_c are the feet of the perpendiculars from a point P inside $\triangle ABC$ to the sides BC, CA, AB, and $P_a PP_b P'_c, P_b PP_c P'_a, P_c PP_a P'_b$ parallelograms, we can apply the problem to the hexagon $P_a P'_c P_b P'_a P_c P'_b$ to prove the Erdős–Mordell inequality for $\triangle ABC$ and point P.

15. Denote by $ABCD$ and $EFGH$ the two rectangles, where $AB = a$, $BC = b$, $EF = c$, and $FG = d$. Obviously, the first rectangle can be placed within the second one with the angle α between AB and EF if and only if

$$a\cos\alpha + b\sin\alpha \leq c, \quad a\sin\alpha + b\cos\alpha \leq d. \tag{1}$$

Hence $ABCD$ can be placed within $EFGH$ if and only if there is an $\alpha \in [0, \pi/2]$ for which (1) holds.

The lines $l_1 (ax + by = c)$ and $l_2 (bx + ay = d)$ and the axes x and y bound a region $\mathcal{R}$. By (1), the desired placement of the rectangles is possible if and only if $\mathcal{R}$ contains some point $(\cos\alpha, \sin\alpha)$ of the unit circle centered at the origin $(0,0)$. This in turn holds if and only if the intersection point L of l_1 and l_2 lies outside the unit circle. It is easily computed that L has coordinates $\left(\frac{bd - ac}{b^2 - a^2}, \frac{bc - ad}{b^2 - a^2}\right)$. Now L being outside the unit circle is exactly equivalent to the inequality we want to prove.

Remark. If equality holds, there is exactly one way of placing. This happens, for example, when $(a,b) = (5,20)$ and $(c,d) = (16,19)$.

Second remark. This problem is essentially very similar to (SL89-2).

16. Let A_1 be the point of intersection of OA' and BC; similarly define B_1 and C_1. From the similarity of triangles OBA_1 and $OA'B$ we obtain $OA_1 \cdot OA' = R^2$. Now it is enough to show that $8OA_1 \cdot OB' \cdot OC' \le R^3$. Thus we must prove that

$$\lambda \mu v \le \frac{1}{8}, \quad \text{where} \quad \frac{OA_1}{OA} = \lambda, \ \frac{OB_1}{OB} = \mu, \ \frac{OC_1}{OC} = v. \tag{1}$$

On the other hand, we have

$$\frac{\lambda}{1+\lambda} + \frac{\mu}{1+\mu} + \frac{v}{1+v} = \frac{S_{OBC}}{S_{ABC}} + \frac{S_{AOC}}{S_{ABC}} + \frac{S_{ABO}}{S_{ABC}} = 1.$$

Simplifying this relation, we get

$$1 = \lambda\mu + \mu v + v\lambda + 2\lambda\mu v \ge 3(\lambda\mu v)^{2/3} + 2\lambda\mu v,$$

which cannot hold if $\lambda\mu v > \frac{1}{8}$. Hence $\lambda\mu v \le \frac{1}{8}$, with equality if and only if $\lambda = \mu = v = \frac{1}{2}$. This implies that O is the centroid of ABC, and consequently, that the triangle is equilateral.

Second solution. In the official solution, the inequality to be proved is transformed into

$$\cos(A - B)\cos(B - C)\cos(C - A) \ge 8\cos A\cos B\cos C.$$

Since $\frac{\cos(B-C)}{\cos A} = -\frac{\cos(B-C)}{\cos(B+C)} = \frac{\tan B\tan C+1}{\tan B\tan C-1}$, the last inequality becomes $(xy + 1)(yz+1)(zx+1) \ge 8(xy-1)(yz-1)(zx-1)$, where we write x, y, z for $\tan A$, $\tan B$, $\tan C$. Using the relation $x+y+z = xyz$, we can reduce this inequality to

$$(2x+y+z)(x+2y+z)(x+y+2z) \ge 8(x+y)(y+z)(z+x).$$

This follows from the AM–GM inequality: $2x + y + z = (x+y) + (x+z) \ge 2\sqrt{(x+y)(x+z)}$, etc.

17. Let the diagonals AC and BD meet in X. Either $\angle AXB$ or $\angle AXD$ is greater than or equal to $90°$, so we assume w.l.o.g. that $\angle AXB \ge 90°$. Let $\alpha, \beta, \alpha', \beta'$ denote $\angle CAB$, $\angle ABD$, $\angle BDC$, $\angle DCA$. These angles are all acute and satisfy $\alpha + \beta = \alpha' + \beta'$. Furthermore,

$$R_A = \frac{AD}{2\sin\beta}, \quad R_B = \frac{BC}{2\sin\alpha}, \quad R_C = \frac{BC}{2\sin\alpha'}, \quad R_D = \frac{AD}{2\sin\beta'}.$$

Let $\angle B + \angle D = 180°$. Then A, B, C, D are concyclic and trivially $R_A + R_C = R_B + R_D$.

Let $\angle B + \angle D > 180°$. Then D lies within the circumcircle of ABC, which implies that $\beta > \beta'$. Similarly $\alpha < \alpha'$, so we obtain $R_A < R_D$ and $R_C < R_B$. Thus $R_A + R_C < R_B + R_D$.

Let $\angle B + \angle D < 180°$. As in the previous case, we deduce that $R_A > R_D$ and $R_C > R_B$, so $R_A + R_C > R_B + R_D$.

18. We first prove the result in the simplest case. Given a 2-gon ABA and a point O, let a, b, c, h denote OA, OB, AB, and the distance of O from AB. Then $D = a + b$, $P = 2c$, and $H = 2h$, so we should show that

$$(a+b)^2 \geq 4h^2 + c^2. \tag{1}$$

Indeed, let l be the line through O parallel to AB, and D the point symmetric to B with respect to l. Then $(a+b)^2 = (OA + OB)^2 = (OA + OD)^2 \geq AD^2 = c^2 + 4h^2$. Now we pass to the general case. Let $A_1 A_2 \ldots A_n$ be the polygon $\mathscr{F}$ and denote by d_i, p_i, and h_i respectively OA_i, $A_i A_{i+1}$, and the distance of O from $A_i A_{i+1}$ (where $A_{n+1} = A_1$). By the case proved above, we have for each i, $d_i + d_{i+1} \geq \sqrt{4h_i^2 + p_i^2}$. Summing these inequalities for $i = 1, \ldots, n$ and squaring, we obtain

$$4D^2 \geq \left(\Sigma_{i=1}^n \sqrt{4h_i^2 + p_i^2} \right)^2.$$

It remains only to prove that

$$\sum_{i=1}^n \sqrt{4h_i^2 + p_i^2} \geq \sqrt{\sum_{i=1}^n (4h_i^2 + p_i^2)} = \sqrt{4H^2 + D^2}.$$

But this follows immediately from the Minkowski inequality.

Equality holds if and only if it holds in (1) and in the Minkowski inequality, i.e., if and only if $d_1 = \cdots = d_n$ and $h_1/p_1 = \cdots = h_n/p_n$. This means that $\mathscr{F}$ is inscribed in a circle with center at O and $p_1 = \cdots = p_n$, so $\mathscr{F}$ is a regular polygon and O its center.

19. It is easy to check that after 4 steps we will have all a, b, c, d even. Thus $|ab - cd|, |ac - bd|, |ad - bc|$ remain divisible by 4, and clearly are not prime. The answer is *no*.

Second solution. After one step we have $a + b + c + d = 0$. Then $ac - bd = ac + b(a+b+c) = (a+b)(b+c)$ etc., so

$$|ab - cd| \cdot |ac - bd| \cdot |ad - bc| = (a+b)^2 (a+c)^2 (b+c)^2.$$

However, the product of three primes cannot be a square, hence the answer is *no*.

20. Let $15a + 16b = x^2$ and $16a - 15b = y^2$, where $x, y \in \mathbb{N}$. Then we obtain

$$x^4 + y^4 = (15a + 16b)^2 + (16a - 15b)^2 = (15^2 + 16^2)(a^2 + b^2) = 481(a^2 + b^2).$$

In particular, $481 = 13 \cdot 37 \mid x^4 + y^4$. We have the following lemma.

Lemma. Suppose that $p \mid x^4 + y^4$, where $x, y \in \mathbb{Z}$ and p is an odd prime, where $p \not\equiv 1 \pmod{8}$. Then $p \mid x$ and $p \mid y$.

Proof. Since $p \mid x^8 - y^8$ and by Fermat's theorem $p \mid x^{p-1} - y^{p-1}$, we deduce that $p \mid x^d - y^d$, where $d = (p-1, 8)$. But $d \neq 8$, so $d \mid 4$. Thus $p \mid x^4 - y^4$, which implies that $p \mid 2y^4$, i.e., $p \mid y$ and $p \mid x$.

In particular, we can conclude that $13 \mid x, y$ and $37 \mid x, y$. Hence x and y are divisible by 481. Thus each of them is at least 481.

On the other hand, $x = y = 481$ is possible. It is sufficient to take $a = 31 \cdot 481$ and $b = 481$.

Second solution. Note that $15x^2 + 16y^2 = 481a^2$. It can be directly verified that the divisibility of $15x^2 + 16y^2$ by 13 and by 37 implies that both x and y are divisible by both primes. Thus $481 \mid x, y$.

21. (a) It clearly suffices to show that for every integer c there exists a quadratic sequence with $a_0 = 0$ and $a_n = c$, i.e., that c can be expressed as $\pm 1^2 \pm 2^2 \pm \cdots \pm n^2$. Since

$$(n+1)^2 - (n+2)^2 - (n+3)^2 + (n+4)^2 = 4,$$

we observe that if our claim is true for c, then it is also true for $c \pm 4$. Thus it remains only to prove the claim for $c = 0, 1, 2, 3$. But one immediately finds $1 = 1^2$, $2 = -1^2 - 2^2 - 3^2 + 4^2$, and $3 = -1^2 + 2^2$, while the case $c = 0$ is trivial.

(b) We have $a_0 = 0$ and $a_n = 1996$. Since $a_n \leq 1^2 + 2^2 + \cdots + n^2 = \frac{1}{6}n(n+1)(2n+1)$, we get $a_{17} \leq 1785$, so $n \geq 18$. On the other hand, a_{18} is of the same parity as $1^2 + 2^2 + \cdots + 18^2 = 2109$, so it cannot be equal to 1996. Therefore we must have $n \geq 19$. To construct a required sequence with $n = 19$, we note that $1^2 + 2^2 + \cdots + 19^2 = 2470 = 1996 + 2 \cdot 237$; hence it is enough to write 237 as a sum of distinct squares. Since $237 = 14^2 + 5^2 + 4^2$, we finally obtain

$$1996 = 1^2 + 2^2 + 3^2 - 4^2 - 5^2 + 6^2 + \cdots + 13^2 - 14^2 + 15^2 + \cdots + 19^2.$$

22. Let $a, b \in \mathbb{N}$ satisfy the given equation. It is not possible that $a = b$ (since it leads to $a^2 + 2 = 2a$), so we assume w.l.o.g. that $a > b$. Next, for $a > b = 1$ the equation becomes $a^2 = 2a$, and one obtains a solution $(a, b) = (2, 1)$.

Let $b > 1$. If $\left\lceil \frac{a^2}{b} \right\rceil = \alpha$ and $\left\lceil \frac{b^2}{a} \right\rceil = \beta$, then we trivially have $ab \geq \alpha\beta$. Since also $\frac{a^2 + b^2}{ab} \geq 2$, we obtain $\alpha + \beta \geq \alpha\beta + 2$, or equivalently $(\alpha - 1)(\beta - 1) \leq -1$. But $\alpha \geq 1$, and therefore $\beta = 0$. It follows that $a > b^2$, i.e., $a = b^2 + c$ for some $c > 0$. Now the given equation becomes $b^3 + 2bc + \left\lceil \frac{c^2}{b} \right\rceil = \left\lceil \frac{b^4 + 2b^2c + b^2 + c^2}{b^3 + bc} \right\rceil + b^3 + bc$, which reduces to

$$(c-1)b + \left\lceil \frac{c^2}{b} \right\rceil = \left\lceil \frac{b^2(c+1) + c^2}{b^3 + bc} \right\rceil. \tag{1}$$

If $c = 1$, then (1) always holds, since both sides are 0. We obtain a family of solutions $(a, b) = (n, n^2 + 1)$ or $(a, b) = (n^2 + 1, n)$. Note that the solution $(1, 2)$ found earlier is obtained for $n = 1$.

If $c > 1$, then (1) implies that $\frac{b^2(c+1) + c^2}{b^3 + bc} \geq (c-1)b$. This simplifies to

$$c^2(b^2 - 1) + b^2(c(b^2 - 2) - (b^2 + 1)) \leq 0. \tag{2}$$

Since $c \geq 2$ and $b^2 - 2 \geq 0$, the only possibility is $b = 2$. But then (2) becomes $3c^2 + 8c - 20 \leq 0$, which does not hold for $c \geq 2$.

Hence the only solutions are $(n, n^2 + 1)$ and $(n^2 + 1, n)$, $n \in \mathbb{N}$.

23. We first observe that the given functional equation is equivalent to

$$4f\left(\frac{(3m+1)(3n+1)-1}{3}\right) + 1 = (4f(m)+1)(4f(n)+1).$$

This gives us the idea of introducing a function $g : 3\mathbb{N}_0 + 1 \to 4\mathbb{N}_0 + 1$ defined as $g(x) = 4f\left(\frac{x-1}{3}\right) + 1$. By the above equality, g will be multiplicative, i.e.,

$$g(xy) = g(x)g(y) \qquad \text{for all } x, y \in 3\mathbb{N}_0 + 1.$$

Conversely, any multiplicative bijection g from $3\mathbb{N}_0 + 1$ onto $4\mathbb{N}_0 + 1$ gives us a function f with the required property: $f(x) = \frac{g(3x+1)-1}{4}$.

It remains to give an example of such a function g. Let P_1, P_2, Q_1, Q_2 be the sets of primes of the forms $3k+1$, $3k+2$, $4k+1$, and $4k+3$, respectively. It is well known that these sets are infinite. Take any bijection h from $P_1 \cup P_2$ onto $Q_1 \cup Q_2$ that maps P_1 bijectively onto Q_1 and P_2 bijectively onto Q_2. Now define g as follows: $g(1) = 1$, and for $n = p_1 p_2 \cdots p_m$ (p_is need not be different) define $g(n) = h(p_1)h(p_2)\cdots h(p_m)$. Note that g is well-defined. Indeed, among the p_is an even number are of the form $3k+2$, and consequently an even number of $h(p_i)$s are of the form $4k+3$. Hence the product of the $h(p_i)$s is of the form $4k+1$. Also, it is obvious that g is multiplicative. Thus, the defined g satisfies all the required properties.

24. We shall work on the array of lattice points defined by $\mathscr{A} = \{(x,y) \in \mathbb{Z}^2 \mid 0 \leq x \leq 19, \ 0 \leq y \leq 11\}$. Our task is to move from $(0,0)$ to $(19,0)$ via the points of $\mathscr{A}$ so that each move has the form $(x,y) \to (x+a, y+b)$, where $a, b \in \mathbb{Z}$ and $a^2 + b^2 = r$.

(a) If r is even, then $a + b$ is even whenever $a^2 + b^2 = r$ $(a, b \in \mathbb{Z})$. Thus the parity of $x + y$ does not change after each move, so we cannot reach $(19,0)$ from $(0,0)$.

If $3 \mid r$, then both a and b are divisible by 3, so if a point (x,y) can be reached from $(0,0)$, we must have $3 \mid x$. Since $3 \nmid 19$, we cannot get to $(19,0)$.

(b) We have $r = 73 = 8^2 + 3^2$, so each move is either $(x,y) \to (x \pm 8, y \pm 3)$ or $(x,y) \to (x \pm 3, y \pm 8)$. One possible solution is shown in Fig. 1.

(c) We have $97 = 9^2 + 4^2$. Let us partition $\mathscr{A}$ as $\mathscr{B} \cup \mathscr{C}$, where $\mathscr{B} = \{(x,y) \in \mathscr{A} \mid 4 \leq y \leq 7\}$. It is easily seen that moves of the type $(x,y) \to (x \pm 9, y \pm 4)$ always take us from the set $\mathscr{B}$ to $\mathscr{C}$ and vice versa, while the moves $(x,y) \to (x \pm 4, y \pm 9)$ always take us from $\mathscr{C}$ to $\mathscr{C}$. Furthermore, each move of the type $(x,y) \to (x \pm 9, y \pm 4)$ changes the parity of x, so to get from $(0,0)$ to $(19,0)$ we must have an odd number of such moves. On the other hand, with an odd number of such moves, starting from $\mathscr{C}$ we can end up only in $\mathscr{B}$, although the point $(19,0)$ is not in $\mathscr{B}$. Hence, the answer is no.

Remark. Part (c) can also be solved by examining all cells that can be reached from $(0,0)$. All these cells are marked in Fig. 2.

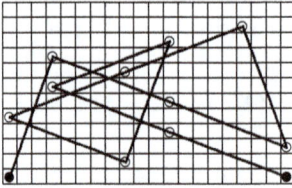   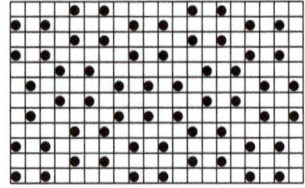

Fig. 1   Fig. 2

25. Let the vertices in the bottom row be assigned an arbitrary coloring, and suppose that some two adjacent vertices receive the same color. The number of such colorings equals $2^n - 2$. It is easy to see that then the colors of the remaining vertices get fixed uniquely in order to satisfy the requirement. So in this case there are $2^n - 2$ possible colorings.

Next, suppose that the vertices in the bottom row are colored alternately red and blue. There are two such colorings. In this case, the same must hold for every row, and thus we get 2^n possible colorings.

It follows that the total number of considered colorings is $(2^n - 2) + 2^n = 2^{n+1} - 2$.

26. Denote the required maximum size by $M_k(m,n)$. If $m < \frac{n(n+1)}{2}$, then trivially $M = k$, so from now on we assume that $m \geq \frac{n(n+1)}{2}$.

First we give a lower bound for M. Let $r = r_k(m,n)$ be the largest integer such that $r + (r+1) + \cdots + (r+n-1) \leq m$. This is equivalent to $nr \leq m - \frac{n(n-1)}{2} \leq n(r+1)$, so $r = \left[\frac{m}{n} - \frac{n-1}{2}\right]$. Clearly no n elements from $\{r+1, r+2, \ldots, k\}$ add up to m, so

$$M \geq k - r_k(m,n) = k - \left[\frac{m}{n} - \frac{n-1}{2}\right]. \qquad (1)$$

We claim that M is actually equal to $k - r_k(m,n)$. To show this, we shall prove by induction on n that if no n elements of a set $S \subseteq \{1, 2, \ldots, k\}$ add up to m, then $|S| \leq k - r_k(m,n)$.

For $n = 2$ the claim is true, because then for each $i = 1, \ldots, r_k(m,2) = \left[\frac{m-1}{2}\right]$ at least one of i and $m - i$ must be excluded from S. Now let us assume that $n > 2$ and that the result holds for $n - 1$. Suppose that $S \subseteq \{1, 2, \ldots, k\}$ does not contain n distinct elements with the sum m, and let x be the smallest element of S. We may assume that $x \leq r_k(m,n)$, because otherwise the statement is clear. Consider the set $S' = \{y - x \mid y \in S, y \neq x\}$. Then S' is a subset of $\{1, 2, \ldots, k-x\}$ no $n-1$ elements of which have the sum $m - nx$. Also, it is easily checked that $n - 1 \leq m - nx - 1 \leq k - x$, so we may apply the induction hypothesis, which yields that

$$|S| \leq 1 + k - x - r_k(m - nx, n - 1) = k - \left[\frac{m-x}{n-1} - \frac{n}{2}\right]. \qquad (2)$$

On the other hand, $\left(\frac{m-x}{n-1} - \frac{n}{2}\right) - r_k(m,n) = \frac{m-nx-\frac{n(n-1)}{2}}{n(n-1)} \geq 0$ because $x \leq r_k(m,n)$; hence (2) implies $|S| \leq k - r_k(m,n)$ as claimed.

27. Suppose that such sets of points $\mathscr{A}, \mathscr{B}$ exist.

    First, we observe that there exist five points A,B,C,D,E in $\mathscr{A}$ such that their convex hull does not contain any other point of $\mathscr{A}$. Indeed, take any point $A \in \mathscr{A}$. Since any two points of $\mathscr{A}$ are at distance at least 1, the number of points $X \in \mathscr{A}$ with $XA \leq r$ is finite for every $r > 0$. Thus it is enough to choose four points B,C,D,E of $\mathscr{A}$ that are closest to A. Now consider the convex hull $\mathscr{C}$ of A,B,C,D,E.

    Suppose that $\mathscr{C}$ is a pentagon, say $ABCDE$. Then each of the disjoint triangles ABC, ACD, ADE contains a point of $\mathscr{B}$. Denote these points by P, Q, R. Then $\triangle PQR$ contains some point $F \in \mathscr{A}$, so F is inside $ABCDE$, a contradiction.

    Suppose that $\mathscr{C}$ is a quadrilateral, say $ABCD$, with E lying within $ABCD$. Then the triangles ABE, BCE, CDE, DAE contain some points P, Q, R, S of $\mathscr{B}$ that form two disjoint triangles. It follows that there are two points of $\mathscr{A}$ inside $ABCD$, which is a contradiction.

    Finally, suppose that $\mathscr{C}$ is a triangle with two points of $\mathscr{A}$ inside. Then $\mathscr{C}$ is the union of five disjoint triangles with vertices in $\mathscr{A}$, so there are at least five points of $\mathscr{B}$ inside $\mathscr{C}$. These five points make at least three disjoint triangles containing three points of $\mathscr{A}$. This is again a contradiction.

    It follows that no such sets $\mathscr{A}, \mathscr{B}$ exist.

28. Note that w.l.o.g., we can assume that p and q are coprime. Indeed, otherwise it suffices to consider the problem in which all x_i's and p, q are divided by $\gcd(p,q)$.

    Let k, l be the number of indices i with $x_{i+1} - x_i = p$ and the number of those i with $x_{i+1} - x_i = -q$ ($0 \leq i < n$). From $x_0 = x_n = 0$ we get $kp = lq$, so for some integer $t > 1$, $k = qt$, $l = pt$, and $n = (p+q)t$.

    Consider the sequence $y_i = x_{i+p+q} - x_i$, $i = 0, \ldots, n-p-q$. We claim that at least one of the y_i's equals zero. We begin by noting that each y_i is of the form $up - vq$, where $u+v = p+q$; therefore $y_i = (u+v)p - v(p+q) = (p-v)(p+q)$ is always divisible by $p+q$. Moreover, $y_{i+1} - y_i = (x_{i+p+q+1} - x_{i+p+q}) - (x_{i+1} - x_i)$ is 0 or $\pm(p+q)$. We conclude that if no y_i is 0 then all y_i's are of the same sign. But this is in contradiction with the relation $y_0 + y_{p+q} + \cdots + y_{n-p-q} = x_n - x_0 = 0$. Consequently some y_i is zero, as claimed.

    Second solution. As before we assume $(p,q) = 1$. Let us define a sequence of points $A_i(y_i, z_i)$ ($i = 0, 1, \ldots, n$) in $\mathbb{N}_0^2$ inductively as follows. Set $A_0 = (0,0)$ and define (y_{i+1}, z_{i+1}) as $(y_i, z_i + 1)$ if $x_{i+1} = x_i + p$ and $(y_i + 1, z_i)$ otherwise. The points A_i form a trajectory L in $\mathbb{N}_0^2$ continuously moving upwards and rightwards by steps of length 1. Clearly, $x_i = pz_i - qy_i$ for all i. Since $x_n = 0$, it follows that $(z_n, y_n) = (kq, kp)$, $k \in \mathbb{N}$. Since $y_n + z_n = n > p+q$, it follows that $k > 1$. We observe that $x_i = x_j$ if and only if $A_i A_j \parallel A_0 A_n$. We shall show that such i, j with $i < j$ and $(i,j) \neq (0,n)$ must exist.

If L meets A_0A_n in an interior point, then our statement trivially holds. From now on we assume the opposite. Let P_{ij} be the rectangle with sides parallel to the coordinate axes and with vertices at (ip, jq) and $((i+1)p, (j+1)q)$. Let L_{ij} be the part of the trajectory L lying inside P_{ij}. We may assume w.l.o.g. that the endpoints of L_{00} lie on the vertical sides of P_{00}. Then there obviously exists $d \in \{1, \ldots, k-1\}$ such that the endpoints of L_{dd} lie on the horizontal sides of P_{dd}. Consider the translate L'_{dd} of L_{dd} for the vector $-d(p, q)$. The endpoints of L'_{dd} lie on the vertical sides of P_{00}. Hence L_{00} and L'_{dd} have some point $X \neq A_0$ in common. The translate Y of point X for the vector $d(p, q)$ belongs to L and satisfies $XY \parallel A_0A_n$.

29. Let the squares be indexed serially by the integers: $\ldots, -1, 0, 1, 2, \ldots$. When a bean is moved from i to $i+1$ or from $i+1$ to i for the first time, we may assign the index i to it. Thereafter, whenever some bean is moved in the opposite direction, we shall assume that it is exactly the one marked by i, and so on. Thus, each pair of neighboring squares has a bean stuck between it, and since the number of beans is finite, there are only finitely pairs of neighboring squares, and thus finitely many squares on which moves are made. Thus we may assume w.l.o.g. that all moves occur between 0 and $l \in \mathbb{N}$ and that all beans exist at all times within $[0, l]$.

Defining b_i to be the number of beans in the ith cell ($i \in \mathbb{Z}$) and b the total number of beans, we define the semi-invariant $S = \sum_{i \in \mathbb{Z}} i^2 b_i$. Since all moves occur above 0, the semi-invariant S increases by 2 with each move, and since we always have $S < b \cdot l^2$, it follows that the number of moves must be finite.

We now prove the uniqueness of the final configuration and the number of moves for some initial configuration $\{b_i\}$. Let $x_i \geq 0$ be the number of moves made in the ith cell ($i \in \mathbb{Z}$) during the game. Since the game is finite, only finitely many of x_i's are nonzero. Also, the number of beans in cell i, denoted as e_i, at the end is

$$(\forall i \in \mathbb{Z}) \ e_i = b_i + x_{i-1} + x_{i+1} - 2x_i \in \{0, 1\} . \tag{1}$$

Thus it is enough to show that given $b_i \geq 0$, the sequence $\{x_i\}_{i \in \mathbb{Z}}$ of nonnegative integers satisfying (1) is unique.

Suppose the assertion is false, i.e., that there exists at least one sequence $b_i \geq 0$ for which there exist distinct sequences $\{x_i\}$ and $\{x'_i\}$ satisfying (1). We may choose such a $\{b_i\}$ for which $\min\{\sum_{i \in \mathbb{Z}} x_i, \sum_{i \in \mathbb{Z}} x'_i\}$ is minimal (since $\sum_{i \in \mathbb{Z}} x_i$ is always finite). We choose any index j such that $b_j > 1$. Such an index j exists, since otherwise the game is over. Then one must make at least one move in the jth cell, which implies that $x_j, x'_j \geq 1$. However, then the sequences $\{x_i\}$ and $\{x'_i\}$ with x_j and x'_j decreased by 1 also satisfy (1) for a sequence $\{b_i\}$ where b_{j-1}, b_j, b_{j+1} is replaced with $b_{j-1} + 1, b_j - 2, b_{j+1} + 1$. This contradicts the assumption of minimal $\min\{\sum_{i \in \mathbb{Z}} x_i, \sum_{i \in \mathbb{Z}} x'_i\}$ for the initial $\{b_i\}$.

30. For convenience, we shall write $f^2, fg, \ldots$ for the functions $f \circ f, f \circ g, \ldots$. We need two lemmas.

Lemma 1. If $f(x) \in S$ and $g(x) \in T$, then $x \in S \cap T$.

Proof. The given condition means that $f^3(x) = g^2 f(x)$ and $gfg(x) = fg^2(x)$.
Since $x \in S \cup T = U$, we have two cases:

$x \in S$. Then $f^2(x) = g^2(x)$, which also implies $f^3(x) = fg^2(x)$. Therefore $gfg(x) = fg^2(x) = f^3(x) = g^2 f(x)$, and since g is a bijection, we obtain $fg(x) = gf(x)$, i.e., $x \in T$.

$x \in T$. Then $fg(x) = gf(x)$, so $g^2 f(x) = gfg(x)$. It follows that $f^3(x) = g^2 f(x) = gfg(x) = fg^2(x)$, and since f is a bijection, we obtain $x \in S$.

Hence $x \in S \cap T$ in both cases. Similarly, $f(x) \in T$ and $g(x) \in S$ again imply $x \in S \cap T$.

Lemma 2. $f(S \cap T) = g(S \cap T) = S \cap T$.

Proof. By symmetry, it is enough to prove $f(S \cap T) = S \cap T$, or in other words that $f^{-1}(S \cap T) = S \cap T$. Since $S \cap T$ is finite, this is equivalent to $f(S \cap T) \subseteq S \cap T$.

Let $f(x) \in S \cap T$. Then if $g(x) \in S$ (since $f(x) \in T$), Lemma 1 gives $x \in S \cap T$; similarly, if $g(x) \in T$, then by Lemma 1, $x \in S \cap T$.

Now we return to the problem. Assume that $f(x) \in S$. If $g(x) \notin S$, then $g(x) \in T$, so from Lemma 1 we deduce that $x \in S \cap T$. Then Lemma 2 claims that $g(x) \in S \cap T$ too, a contradiction. Analogously, from $g(x) \in S$ we are led to $f(x) \in S$. This finishes the proof.

4.38 Solutions to the Shortlisted Problems of IMO 1997

1. Let ABC be the given triangle, with $\angle B = 90°$ and $AB = m$, $BC = n$. For an arbitrary polygon $\mathscr{P}$ we denote by $w(\mathscr{P})$ and $b(\mathscr{P})$ respectively the total areas of the white and black parts of $\mathscr{P}$.

   (a) Let D be the fourth vertex of the rectangle $ABCD$. When m and n are of the same parity, the coloring of the rectangle $ABCD$ is centrally symmetric with respect to the midpoint of AC. It follows that $w(ABC) = \frac{1}{2}w(ABCD)$ and $b(ABC) = \frac{1}{2}b(ABCD)$; thus $f(m,n) = \frac{1}{2}|w(ABCD) - b(ABCD)|$. Hence $f(m,n)$ equals $\frac{1}{2}$ if m and n are both odd, and 0 otherwise.

   (b) The result when m,n are of the same parity follows from (a). Suppose that $m > n$, where m and n are of different parity. Choose a point E on AB such that $AE = 1$. Since by (a) $|w(EBC) - b(EBC)| = f(m-1,n) \leq \frac{1}{2}$, we have $f(m,n) \leq \frac{1}{2} + |w(EAC) - b(EAC)| \leq \frac{1}{2} + S(EAC) = \frac{1}{2} + \frac{n-1}{2} = \frac{n}{2}$. Therefore $f(m,n) \leq \frac{1}{2}\min(m,n)$.

   (c) Let us calculate $f(m,n)$ for $m = 2k+1$, $n = 2k$, $k \in \mathbb{N}$. With E defined as in (b), we have $BE = BC = 2k$. If the square at B is w.l.o.g. white, CE passes only through black squares. The white part of $\triangle EAC$ then consists of $2k$ similar triangles with areas $\frac{1}{2}\frac{i}{2k}\frac{i}{2k+1} = \frac{i^2}{4k(2k+1)}$, where $i = 1,2,\ldots,2k$. The total white area of EAC is

$$\frac{1}{4k(2k+1)}(1^2 + 2^2 + \cdots + (2k)^2) = \frac{4k+1}{12}.$$

Therefore the black area is $(8k-1)/12$, and $f(2k+1,2k) = (2k-1)/6$, which is not bounded.

2. For any sequence $X = (x_1, x_2, \ldots, x_n)$ let us define

$$\overline{X} = (1, 2, \ldots, x_1, 1, 2, \ldots, x_2, \ldots, 1, 2, \ldots, x_n).$$

Also, for any two sequences A, B we denote their concatenation by AB. It clearly holds that $\overline{AB} = \overline{A}\,\overline{B}$. The sequences $R_1, R_2, \ldots$ are given by $R_1 = (1)$ and $R_n = \overline{R_{n-1}}(n)$ for $n > 1$.

We consider the family of sequences Q_{ni} for $n, i \in \mathbb{N}$, $i \leq n$, defined as follows:

$$Q_{n1} = (1), \quad Q_{nn} = (n), \quad \text{and} \quad Q_{ni} = \overline{Q_{n-1,i-1}}Q_{n-1,i} \quad \text{if } 1 < i < n.$$

These sequences form a Pascal-like triangle, as shown in the picture below:

Q_{1i}:					1				
Q_{2i}:				1		2			
Q_{3i}:			1		12		3		
Q_{4i}:		1		112		123		4	
Q_{5i}:	1		1112		112123		1234		5

We claim that R_n is in fact exactly $Q_{n1}Q_{n2}\ldots Q_{nn}$. Before proving this, we observe that $Q_{ni} = \overline{Q_{n-1,i}}$. This follows by induction, since $Q_{ni} = \overline{Q_{n-1,i-1}}Q_{n-1,i} =$

$\overline{Q_{n-2,i-1}}\,\overline{Q_{n-2,i}} = \overline{Q_{n-1,i}}$ for $n \geq 3$, $i \geq 2$ (the cases $i = 1$ and $n = 1, 2$ are trivial). Now $R_1 = Q_{11}$ and

$$R_n = \overline{R_{n-1}}(n) = \overline{Q_{n-1,1} \cdots Q_{n-1,n-1}}(n) = Q_{n,1} \cdots Q_{n,n-1} Q_{n,n}$$

for $n \geq 2$, which justifies our claim by induction.

Now we know enough about the sequence R_n to return to the question of the problem. We use induction on n once again. The result is obvious for $n = 1$ and $n = 2$. Given any $n \geq 3$, consider the kth elements of R_n from the left, say u, and from the right, say v. Assume that u is a member of Q_{nj}, and consequently that v is a member of $Q_{n,n+1-j}$. Then u and v come from symmetric positions of R_{n-1} (either from $Q_{n-1,j}, Q_{n-1,n-j}$, or from $Q_{n-1,j-1}, Q_{n-1,n+1-j}$), and by the inductive hypothesis exactly one of them is 1.

3. (a) For $n = 4$, consider a convex quadrilateral $ABCD$ in which $AB = BC = AC = BD$ and $AD = DC$, and take the vectors $\overrightarrow{AB}, \overrightarrow{BC}, \overrightarrow{CD}, \overrightarrow{DA}$. For $n = 5$, take the vectors $\overrightarrow{AB}, \overrightarrow{BC}, \overrightarrow{CD}, \overrightarrow{DE}, \overrightarrow{EA}$ for any regular pentagon $ABCDE$.

   (b) Let us draw the vectors of V as originated from the same point O. Consider any maximal subset $B \subset V$, and denote by u the sum of all vectors from B. If l is the line through O perpendicular to u, then B contains exactly those vectors from V that lie on the same side of l as u does, and no others. Indeed, if any $v \notin B$ lies on the same side of l, then $|u + v| \geq |u|$; similarly, if some $v \in B$ lies on the other side of l, then $|u - v| \geq |u|$.

   Therefore every maximal subset is determined by some line l as the set of vectors lying on the same side of l. It is obvious that in this way we get at most $2n$ sets.

4. (a) Suppose that an $n \times n$ coveralls matrix A exists for some $n > 1$. Let $x \in \{1, 2, \ldots, 2n - 1\}$ be a fixed number that does not appear on the fixed diagonal of A. Such an element must exist, since the diagonal can contain at most n different numbers. Let us call the union of the ith row and the ith column the ith cross. There are n crosses, and each of them contains exactly one x. On the other hand, each entry x of A is contained in exactly two crosses. Hence n must be even. However, 1997 is an odd number; hence no coveralls matrix exists for $n = 1997$.

   (b) For $n = 2$, $A_2 = \begin{bmatrix} 1 & 2 \\ 3 & 1 \end{bmatrix}$ is a coveralls matrix. For $n = 4$, one such matrix is, for example,

   $$A_4 = \begin{bmatrix} 1 & 2 & 5 & 6 \\ 3 & 1 & 7 & 5 \\ 4 & 6 & 1 & 2 \\ 7 & 4 & 3 & 1 \end{bmatrix}.$$

   This construction can be generalized. Suppose that we are given an $n \times n$ coveralls matrix A_n. Let B_n be the matrix obtained from A_n by adding $2n$ to each entry, and C_n the matrix obtained from B_n by replacing each diagonal entry (equal to $2n + 1$ by induction) with $2n$. Then the matrix

$$A_{2n} = \begin{bmatrix} A_n & B_n \\ C_n & A_n \end{bmatrix}$$

is coveralls. To show this, suppose that $i \le n$ (the case $i > n$ is similar). The ith cross is composed of the ith cross of A_n, the ith row of B_n, and the ith column of C_n. The ith cross of A_i covers $1, 2, \ldots, 2n - 1$. The ith row of B_n covers all numbers of the form $2n + j$, where j is covered by the ith row of A_n (including $j = 1$). Similarly, the ith column of C_n covers $2n$ and all numbers of the form $2n + k$, where $k > 1$ is covered by the ith column of A_n. Thus we see that all numbers are accounted for in the ith cross of A_{2n}, and hence A_{2n} is a desired coveralls matrix. It follows that we can find a coveralls matrix whenever n is a power of 2.

Second solution for part b. We construct a coveralls matrix explicitly for $n = 2^k$. We consider the coordinates/cells of the matrix elements modulo n throughout the solution. We define the i-diagonal ($0 \le i < n$) to be the set of cells of the form $(j, j + i)$, for all j. We note that each cross contains exactly one cell from the 0-diagonal (the main diagonal) and two cells from each i-diagonal. For two cells within an i diagonal, x and y, we define x and y to be related if there exists a cross containing both x and y. Evidently, for every cell x not on the 0-diagonal there are exactly two other cells related to it. The relation thus breaks up each i-diagonal ($i > 0$) into cycles of length larger than 1. Due to the diagonal translational symmetry (modulo n), all the cycles within a given i-diagonal must be of equal length and thus of an even length, since $n = 2^k$.

The construction of a coveralls matrix is now obvious. We select a number, say 1, to place on all the cells of the 0-diagonal. We pair up the remaining numbers and assign each pair to an i-diagonal, say $(2i, 2i + 1)$. Going along each cycle within the i-diagonal we alternately assign values of $2i$ and $2i + 1$. Since the cycle has an even length, a cell will be related only to a cell of a different number, and hence each cross will contain both $2i$ and $2i + 1$.

5. We shall prove first the 2-dimensional analogue:

    Lemma. Given an equilateral triangle ABC and two points M, N on the sides AB and AC respectively, there exists a triangle with sides CM, BN, MN.

    Proof. Consider a regular tetrahedron $ABCD$. Since $CM = DM$ and $BN = DN$, one such triangle is DMN.

    Now, to solve the problem for a regular tetrahedron $ABCD$, we consider a 4-dimensional polytope $ABCDE$ whose faces $ABCD$, $ABCE$, $ABDE$, $ACDE$, $BCDE$ are regular tetrahedra. We don't know what it looks like, but it yields a desired triangle: for $M \in ABC$ and $N \in ADC$, we have $DM = EM$ and $BN = EN$; hence the desired triangle is EMN.

    Remark. A solution that avoids embedding in $\mathbb{R}^4$ is possible, but no longer so short.

6. (a) One solution is

$$x = 2^{n^2} 3^{n+1}, \qquad y = 2^{n^2 - n} 3^n, \qquad z = 2^{n^2 - 2n + 2} 3^{n-1}.$$

(b) Suppose w.l.o.g. that $\gcd(c,a) = 1$. We look for a solution of the form

$$x = p^m, \qquad y = p^n, \qquad z = qp^r, \qquad p,q,m,n,r \in \mathbb{N}.$$

Then $x^a + y^b = p^{ma} + p^{nb}$ and $z^c = q^c p^{rc}$, and we see that it is enough to assume $ma - 1 = nb = rc$ (there are infinitely many such triples (m,n,r)) and $q^c = p + 1$.

7. Let us set $AC = a$, $CE = b$, $EA = c$. Applying Ptolemy's inequality for the quadrilateral $ACEF$ we get

$$AC \cdot EF + CE \cdot AF \geq AE \cdot CF.$$

Since $EF = AF$, this implies $\frac{FA}{FC} \geq \frac{c}{a+b}$. Similarly $\frac{BC}{BE} \geq \frac{a}{b+c}$ and $\frac{DE}{DA} \geq \frac{b}{c+a}$.
Now,

$$\frac{BC}{BE} + \frac{DE}{DA} + \frac{FA}{FC} \geq \frac{a}{b+c} + \frac{b}{c+a} + \frac{c}{a+b}.$$

Hence it is enough to prove that

$$\frac{a}{b+c} + \frac{b}{c+a} + \frac{c}{a+b} \geq \frac{3}{2}. \tag{1}$$

If we now substitute $x = b+c$, $y = c+a$, $z = a+b$ and $S = a+b+c$ the inequality (1) becomes equivalent to $S(1/x + 1/y + 1/y) - 3 \geq 3/2$ which follows immediately form $1/x + 1/y + 1/z \geq 9/(x+y+z) = 9/(2S)$.
Equality occurs if it holds in Ptolemy's inequalities and also $a = b = c$. The former happens if and only if the hexagon is cyclic. Hence the only case of equality is when $ABCDEF$ is regular.

8. (a) Denote by b and c the perpendicular bisectors of AB and AC respectively. If w.l.o.g. b and AD do not intersect (are parallel), then $\angle BCD = \angle BAD = 90°$, a contradiction. Hence V, W are well-defined. Now, $\angle DWB = 2\angle DAB$ and $\angle DVC = 2\angle DAC$ as oriented angles, and therefore $\angle(WB, VC) = 2(\angle DVC - \angle DWB) = 2\angle BAC = 2\angle BCD$ is not equal to 0. Consequently CV and BW meet at some T with $\angle BTC = 2\angle BAC$.

   (b) Let B' be the second point of intersection of BW with Γ. Clearly $AD = BB'$. But we also have $\angle BTC = 2\angle BAC = 2\angle BB'C$, which implies that $CT = TB'$. It follows that $AD = BB' = |BT \pm TB'| = |BT \pm CT|$.

Remark. This problem is also solved easily using trigonometry.

9. For $i = 1,2,3$ (all indices in this problem will be modulo 3) we denote by O_i the center of C_i and by M_i the midpoint of the arc $A_{i+1}A_{i+2}$ that does not contain A_i. First we have that $O_{i+1}O_{i+2}$ is the perpendicular bisector of IB_i, and thus it contains the circumcenter R_i of A_iB_iI. Additionally, it is easy to show

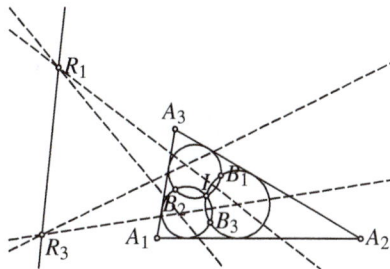

that $T_{i+1}A_i = T_{i+1}I$ and $T_{i+2}A_i = T_{i+2}I$ which implies that R_i lies on the line $T_{i+1}T_{i+2}$. Therefore $R_i = O_{i+1}O_{i+2} \cap T_{i+1}T_{i+2}$. Now, the lines T_1O_1, T_2O_2, T_3O_3 are concurrent at I. By Desargues's theorem, the points of intersection of $O_{i+1}O_{i+2}$ and $T_{i+1}T_{i+2}$, i.e., the R_i's, lie on a line for $i = 1, 2, 3$.

Second solution. The centers of three circles passing through the same point I and not touching each other are collinear if and only if they have another common point. Hence it is enough to show that the circles A_iB_iI have a common point other than I.

Now apply inversion at center I and with an arbitrary power. We shall denote by X' the image of X under this inversion. In our case, the image of the circle C_i is the line $B'_{i+1}B'_{i+2}$ while the image of the line $A_{i+1}A_{i+2}$ is the circle $IA'_{i+1}A'_{i+2}$ that is tangent to $B'_iB'_{i+2}$, and $B'_iB'_{i+2}$. These three circles have equal radii, so their centers P_1, P_2, P_3 form a triangle also homothetic to $\triangle B'_1B'_2B'_3$. Consequently, points A'_1, A'_2, A'_3, that are the reflections of I across the sides of $P_1P_2P_3$, are vertices of a triangle also homothetic to $B'_1B'_2B'_3$. It follows that $A'_1B'_1, A'_2B'_2, A'_3B'_3$ are concurrent at some point J', i.e., that the circles A_iB_iI all pass through J.

10. Suppose that $k \geq 4$. Consider any polynomial $F(x)$ with integer coefficients such that $0 \leq F(x) \leq k$ for $x = 0, 1, \ldots, k+1$. Since $F(k+1) - F(0)$ is divisible by $k+1$, we must have $F(k+1) = F(0)$. Hence

$$F(x) - F(0) = x(x - k - 1)Q(x)$$

for some polynomial $Q(x)$ with integer coefficients. In particular, $F(x) - F(0)$ is divisible by $x(k+1-x) > k+1$ for every $x = 2, 3, \ldots, k-1$, so $F(x) = F(0)$ must hold for any $x = 2, 3, \ldots, k-1$. It follows that

$$F(x) - F(0) = x(x-2)(x-3)\cdots(x-k+1)(x-k-1)R(x)$$

for some polynomial $R(x)$ with integer coefficients. Thus $k \geq |F(1) - F(0)| = k(k-2)!|R(1)|$, although $k(k-2)! > k$ for $k \geq 4$. In this case we have $F(1) = F(0)$ and similarly $F(k) = F(0)$. Hence, the statement is true for $k \geq 4$. It is easy to find counterexamples for $k \leq 3$. These are, for example,

$$F(x) = \begin{cases} x(2-x) & \text{for } k = 1, \\ x(3-x) & \text{for } k = 2, \\ x(2-x)^2(4-x) & \text{for } k = 3. \end{cases}$$

11. All real roots of $P(x)$ (if any) are negative: say $-a_1, -a_2, \ldots, -a_k$. Then $P(x)$ can be factored as

$$P(x) = C(x+a_1)\cdots(x+a_k)(x^2 - b_1x + c_1)\cdots(x^2 - b_mx + c_m), \qquad (1)$$

where $x^2 - b_ix + c_i$ are quadratic polynomials without real roots.
Since the product of polynomials with positive coefficients is again a polynomial with positive coefficients, it will be sufficient to prove the result for each of the factors in (1). The case of $x + a_j$ is trivial. It remains only to prove the claim for every polynomial $x^2 - bx + c$ with $b^2 < 4c$.

From the binomial formula, we have for any $n \in \mathbb{N}$,

$$(1+x)^n(x^2 - bx + c) = \sum_{i=0}^{n+2}\left[\binom{n}{i-2} - b\binom{n}{i-1} + c\binom{n}{i}\right]x^i = \sum_{i=0}^{n+2} C_i x^i,$$

where

$$C_i = \frac{n!\left((b+c+1)i^2 - ((b+2c)n + (2b+3c+1))i + c(n^2 + 3n + 2)\right)x^i}{i!(n-i+2)!}.$$

The coefficients C_i of x^i appear in the form of a quadratic polynomial in i depending on n. We claim that for large enough n this polynomial has negative discriminant, and is thus positive for every i. Indeed, this discriminant equals
$D = ((b+2c)n + (2b+3c+1))^2 - 4(b+c+1)c(n^2 + 3n + 2) = (b^2 - 4c)n^2 - 2Un + V$, where $U = 2b^2 + bc + b - 4c$ and $V = (2b+c+1)^2 - 4c$, and since $b^2 - 4c < 0$, for large n it clearly holds that $D < 0$.

12. *Lemma.* For any polynomial P of degree at most n, the following equality holds:

$$\sum_{i=0}^{n+1}(-1)^i\binom{n+1}{i}P(i) = 0.$$

Proof. See (SL81-13).

Suppose to the contrary that the degree of f is at most $p - 2$. Then it follows from the lemma that

$$0 = \sum_{i=0}^{p-1}(-1)^i\binom{p-1}{i}f(i) \equiv \sum_{i=0}^{p-1}f(i) \pmod{p},$$

since $\binom{p-1}{i} = \frac{(p-1)(p-2)\cdots(p-i)}{i!} \equiv (-1)^i \pmod{p}$. But this is clearly impossible if $f(i)$ equals 0 or 1 modulo p and $f(0) = 0$, $f(1) = 1$.

Remark. In proving the essential relation $\sum_{i=0}^{p-1}f(i) \equiv 0 \pmod{p}$, it is clearly enough to show that $S_k = 1^k + 2^k + \cdots + (p-1)^k$ is divisible by p for every $k \le p - 2$. This can be shown in two other ways.

(1) By induction. Assume that $S_0 \equiv \cdots \equiv S_{k-1} \pmod{p}$. By the binomial formula we have

$$0 \equiv \sum_{n=0}^{p-1}[(n+1)^{k+1} - n^{k+1}] \equiv (k+1)S_k + \sum_{i=0}^{k-1}\binom{k+1}{i}S_i \pmod{p},$$

and the inductive step follows.

(2) Using the primitive root g modulo p. Then

$$S_k \equiv 1 + g^k + \cdots + g^{k(p-2)} = \frac{g^{k(p-1)} - 1}{g^k - 1} \equiv 0 \pmod{p}.$$

13. Denote $A(r)$ and $B(r)$ by $A(n,r)$ and $B(n,r)$ respectively.

The numbers $A(n,r)$ can be found directly: one can choose r girls and r boys in $\binom{n}{r}^2$ ways, and pair them in $r!$ ways. Hence

$$A(n,r) = \binom{n}{r}^2 \cdot r! = \frac{n!^2}{(n-r)!^2 r!}.$$

Now we establish a recurrence relation between the $B(n,r)$'s. Let $n \geq 2$ and $2 \leq r \leq n$. There are two cases for a desired selection of r pairs of girls and boys:

   (i) One of the girls dancing is g_n. Then the other $r-1$ girls can choose their partners in $B(n-1, r-1)$ ways and g_n can choose any of the remaining $2n-r$ boys. Thus, the total number of choices in this case is $(2n-r)B(n-1, r-1)$.

   (ii) g_n is not dancing. Then there are exactly $B(n-1, r)$ possible choices.

Therefore, for every $n \geq 2$ it holds that

$$B(n,r) = (2n-r)B(n-1, r-1) + B(n-1, r) \quad \text{for } r = 2, \ldots, n.$$

Here we assume that $B(n,r) = 0$ for $r > n$, while $B(n,1) = 1 + 3 + \cdots + (2n-1) = n^2$.

It is directly verified that the numbers $A(n,r)$ satisfy the same initial conditions and recurrence relations, from which it follows that $A(n,r) = B(n,r)$ for all n and $r \leq n$.

14. We use the following nonstandard notation: $(1°)$ for $x, y \in \mathbb{N}$, $x \sim y$ means that x and y have the same prime divisors; $(2°)$ for a prime p and integers $r \geq 0$ and $x > 0$, $p^r \parallel x$ means that x is divisible by p^r, but not by p^{r+1}.

First, $b^m - 1 \sim b^n - 1$ is obviously equivalent to $b^m - 1 \sim \gcd(b^m - 1, b^n - 1) = b^d - 1$, where $d = \gcd(m, n)$. Setting $b^d = a$ and $m = kd$, we reduce the condition of the problem to $a^k - 1 \sim a - 1$. We are going to show that this implies that $a + 1$ is a power of 2. This will imply that d is odd (for even d, $a + 1 = b^d + 1$ cannot be divisible by 4), and consequently $b + 1$, as a divisor of $a + 1$, is also a power of 2. But before that, we need the following important lemma (Theorem 2.129).

Lemma. Let a, k be positive integers and p an odd prime. If $\alpha \geq 1$ and $\beta \geq 0$ are such that $p^\alpha \parallel a - 1$ and $p^\beta \parallel k$, then $p^{\alpha+\beta} \parallel a^k - 1$.

Proof. We use induction on β. If $\beta = 0$, then $\frac{a^k - 1}{a - 1} = a^{k-1} + \cdots + a + 1 \equiv k \pmod{p}$ (because $a \equiv 1$), and it is not divisible by p.

Suppose that the lemma is true for some $\beta \geq 0$, and let $k = p^{\beta+1}t$ where $p \nmid t$. By the induction hypothesis, $a^{k/p} = a^{p^\beta t} = mp^{\alpha+\beta} + 1$ for some m not divisible by p. Furthermore,

$$a^k - 1 = (mp^{\alpha+\beta} + 1)^p - 1$$
$$= (mp^{\alpha+\beta})^p + \cdots + \binom{p}{2}(mp^{\alpha+\beta})^2 + mp^{\alpha+\beta+1}.$$

Since $p \mid \binom{p}{2} = \frac{p(p-1)}{2}$, all summands except for the last one are divisible by $p^{\alpha+\beta+2}$. Hence $p^{\alpha+\beta+1} \parallel a^k - 1$, completing the induction.

Now let $a^k - 1 \sim a - 1$ for some $a, k > 1$. Suppose that p is an odd prime divisor of k, with $p^\beta \| k$. Then putting $X = a^{p^\beta - 1} + \cdots + a + 1$ we also have $(a - 1)X = a^{p^\beta} - 1 \sim a - 1$; hence each prime divisor q of X must also divide $a - 1$. But then $a^i \equiv 1 \pmod{q}$ for each $i \in \mathbb{N}_0$, which gives us $X \equiv p^\beta \pmod{q}$. Therefore $q \mid p^\beta$, i.e., $q = p$; hence X is a power of p.

On the other hand, since $p \mid a - 1$, we put $p^\alpha \| a - 1$. From the lemma we obtain $p^{\alpha + \beta} \| a^{p^\beta} - 1$, and deduce that $p^\beta \| X$. But X has no prime divisors other than p, so we must have $X = p^\beta$. This is clearly impossible, because $X > p^\beta$ for $a > 1$. Thus our assumption that k has an odd prime divisor leads to a contradiction: in other words, k must be a power of 2.

Now $a^k - 1 \sim a - 1$ implies $a - 1 \sim a^2 - 1 = (a - 1)(a + 1)$, and thus every prime divisor q of $a + 1$ must also divide $a - 1$. Consequently $q = 2$, so it follows that $a + 1$ is a power of 2. As we explained above, this gives that $b + 1$ is also a power of 2.

Remark. In fact, one can continue and show that k must be equal to 2. It is not possible for $a^4 - 1 \sim a^2 - 1$ to hold. Similarly, we must have $d = 1$. Therefore all possible triples (b, m, n) with $m > n$ are $(2^s - 1, 2, 1)$.

15. Let $a + bt, t = 0, 1, 2, \ldots$, be a given arithmetic progression that contains a square and a cube $(a, b > 0)$. We use induction on the progression step b to prove that the progression contains a sixth power.

    (i) $b = 1$: this case is trivial.

    (ii) $b = p^m$ for some prime p and $m > 0$. The case $p^m \mid a$ trivially reduces to the previous case, so let us have $p^m \nmid a$.

    Suppose that $\gcd(a, p) = 1$. If x, y are integers such that $x^2 \equiv y^3 \equiv a$ (here all the congruences will be mod p^m), then $x^6 \equiv a^3$ and $y^6 \equiv a^2$. Consider an integer y_1 such that $yy_1 \equiv 1$. It satisfies $a^2(xy_1)^6 \equiv x^6 y^6 y_1^6 \equiv x^6 \equiv a^3$, and consequently $(xy_1)^6 \equiv a$. Hence a sixth power exists in the progression.

    If $\gcd(a, p) > 1$, we can write $a = p^k c$, where $k < m$ and $p \nmid c$. Since the arithmetic progression $x_t = a + bt = p^k(c + p^{m-k}t)$ contains a square, k must be even; similarly, it contains a cube, so $3 \mid k$. It follows that $6 \mid k$. The progression $c + p^{m-k}t$ thus also contains a square and a cube; hence by the previous case it contains a sixth power and thus x_t does also.

    (iii) b is not a power of a prime, and thus can be expressed as $b = b_1 b_2$, where $b_1, b_2 > 1$ and $\gcd(b_1, b_2) = 1$. It is given that progressions $a + b_1 t$ and $a + b_2 t$ both contain a square and a cube, and therefore by the inductive hypothesis they both contain sixth powers: say z_1^6 and z_2^6, respectively. By the Chinese remainder theorem, there exists $z \in \mathbb{N}$ such that $z \equiv z_1 \pmod{b_1}$ and $z \equiv z_2 \pmod{b_2}$. But then z^6 belongs to both of the progressions $a + b_1 t$ and $a + b_2 t$. Hence z^6 is a member of the progression $a + bt$.

16. Let $d_a(X), d_b(X), d_c(X)$ denote the distances of a point X interior to $\triangle ABC$ from the lines BC, CA, AB respectively. We claim that $X \in PQ$ if and only if $d_a(X) + d_b(X) = d_c(X)$. Indeed, if $X \in PQ$ and $PX = kPQ$ then $d_a(X) = kd_a(Q), d_b(X) = (1 - k)d_b(P)$, and $d_c(X) = (1 - k)d_c(P) + kd_c(Q)$, and simple substitution yields

$d_a(X) + d_b(X) = d_c(X)$. The converse follows easily. In particular, $O \in PQ$ if and only if $d_a(O) + d_b(O) = d_c(O)$, i.e., $\cos \alpha + \cos \beta = \cos \gamma$.

We shall now show that $I \in DE$ if and only if $AE + BD = DE$. Let K be the point on the segment DE such that $AE = EK$. Then $\angle EKA = \frac{1}{2}\angle DEC = \frac{1}{2}\angle CBA = \angle IBA$; hence the points A, B, I, K are concyclic. The point I lies on DE if and only if $\angle BKD = \angle BAI = \frac{1}{2}\angle BAC = \frac{1}{2}\angle CDE = \angle DBK$, which is equivalent to $KD = BD$, i.e., to $AE + BD = DE$. But since $AE = AB\cos\alpha$, $BD = AB\cos\beta$, and $DE = AB\cos\gamma$, we have that $I \in DE \Leftrightarrow \cos\alpha + \cos\beta = \cos\gamma$. The conditions for $O \in PQ$ and $I \in DE$ are thus equivalent.

Second solution. We know that three points X, Y, Z are collinear if and only if for some $\lambda, \mu \in \mathbb{R}$ with sum 1, we have $\lambda \overrightarrow{CX} + \mu \overrightarrow{CY} = \overrightarrow{CZ}$. Specially, if $\overrightarrow{CX} = p\overrightarrow{CA}$ and $\overrightarrow{CY} = q\overrightarrow{CB}$ for some p, q, and $\overrightarrow{CZ} = k\overrightarrow{CA} + l\overrightarrow{CB}$, then Z lies on XY if and only if $kq + lp = pq$.

Using known relations in a triangle we directly obtain

$$\overrightarrow{CP} = \frac{\sin\beta}{\sin\beta + \sin\gamma}\overrightarrow{CB}, \qquad \overrightarrow{CQ} = \frac{\sin\alpha}{\sin\alpha + \sin\gamma}\overrightarrow{CA},$$

$$\overrightarrow{CO} = \frac{\sin 2\alpha \cdot \overrightarrow{CA} + \sin 2\beta \cdot \overrightarrow{CB}}{\sin 2\alpha + \sin 2\beta + \sin 2\gamma}; \qquad \overrightarrow{CD} = \frac{\tan\beta}{\tan\beta + \tan\gamma}\overrightarrow{CB},$$

$$\overrightarrow{CE} = \frac{\tan\beta}{\tan\beta + \tan\gamma}\overrightarrow{CA}, \qquad \overrightarrow{CI} = \frac{\sin\alpha \cdot \overrightarrow{CA} + \sin\beta \cdot \overrightarrow{CB}}{\sin\alpha + \sin\beta + \sin\gamma}.$$

Now by the above considerations we get that the conditions (1) P, Q, O are collinear and (2) D, E, I are collinear are both equivalent to $\cos\alpha + \cos\beta = \cos\gamma$.

17. We note first that x and y must be powers of the same positive integer. Indeed, if $x = p_1^{\alpha_1} \cdots p_k^{\alpha_k}$ and $y = p_1^{\beta_1} \cdots p_k^{\beta_k}$ (some of α_i and β_i may be 0, but not both for the same index i), then $x^{y^2} = y^x$ implies $\frac{\alpha_i}{\beta_i} = \frac{x}{y^2} = \frac{p}{q}$ for some $p, q > 0$ with $\gcd(p, q) = 1$, so for $a = p_1^{\alpha_1/p} \cdots p_k^{\alpha_k/p}$ we can take $x = a^p$ and $y = a^q$.

If $a = 1$, then $(x, y) = (1, 1)$ is the trivial solution. Let $a > 1$. The given equation becomes $a^{pa^{2q}} = a^{qa^p}$, which reduces to $pa^{2q} = qa^p$. Hence $p \neq q$, so we distinguish two cases:

(i) $p > q$. Then from $a^{2q} < a^p$ we deduce $p > 2q$. We can rewrite the equation as $p = a^{p-2q}q$, and putting $p = 2q + d, d > 0$, we obtain $d = q(a^d - 2)$. By induction, $2^d - 2 > d$ for each $d > 2$, so we must have $d \le 2$. For $d = 1$ we get $q = 1$ and $a = p = 3$, and therefore $(x, y) = (27, 3)$, which is indeed a solution. For $d = 2$ we get $q = 1$, $a = 2$, and $p = 4$, so $(x, y) = (16, 2)$, which is another solution.

(ii) $p < q$. As above, we get $q/p = a^{2q-p}$, and setting $d = 2q - p > 0$, this is transformed to $a^d = a^{(2a^d-1)p}$, or equivalently to $d = (2a^d - 1)p$. However, this equality cannot hold, because $2a^d - 1 > d$ for each $a \ge 2$, $d \ge 1$.

The only solutions are thus $(1, 1)$, $(16, 2)$, and $(27, 3)$.

18. By symmetry, assume that $AB > AC$. The point D lies between M and P as well as between Q and R, and if we show that $DM \cdot DP = DQ \cdot DR$, it will imply that M, P, Q, R lie on a circle.

Since the triangles ABC, AEF, AQR are similar, the points B, C, Q, R lie on a circle. Hence $DB \cdot DC = DQ \cdot DR$, and it remains to prove that

$$DB \cdot DC = DM \cdot DP.$$

However, the points B, C, E, F are concyclic, but so are the points E, F, D, M (they lie on the nine-point circle), and we obtain $PB \cdot PC = PE \cdot PF = PD \cdot PM$. Set $PB = x$ and $PC = y$. We have $PM = \frac{x+y}{2}$ and hence $PD = \frac{2xy}{x+y}$. It follows that $DB = PB - PD = \frac{x(x-y)}{x+y}$, $DC = \frac{y(x-y)}{x+y}$, and $DM = \frac{(x-y)^2}{2(x+y)}$, from which we immediately obtain $DB \cdot DC = DM \cdot DP = \frac{xy(x-y)^2}{(x+y)^2}$, as needed.

19. Using that $a_{n+1} = 0$ we can transform the desired inequality into

$$\sqrt{a_1 + a_2 + \cdots + a_{n+1}}$$
$$\leq \sqrt{1}\sqrt{a_1} + (\sqrt{2} - \sqrt{1})\sqrt{a_2} + \cdots + (\sqrt{n+1} - \sqrt{n})\sqrt{a_{n+1}}. \tag{1}$$

We shall prove by induction on n that (1) holds for any $a_1 \geq a_2 \geq \cdots \geq a_{n+1} \geq 0$, i.e., not only when $a_{n+1} = 0$. For $n = 0$ the inequality is obvious. For the inductive step from $n - 1$ to n, where $n \geq 1$, we need to prove the inequality

$$\sqrt{a_1 + \cdots + a_{n+1}} - \sqrt{a_1 + \cdots + a_n} \leq (\sqrt{n+1} - \sqrt{n})\sqrt{a_{n+1}}. \tag{2}$$

Putting $S = a_1 + a_2 + \cdots + a_n$, this simplifies to

$$\sqrt{S + a_{n+1}} - \sqrt{S} \leq \sqrt{na_{n+1} + a_{n+1}} - \sqrt{na_{n+1}}.$$

For $a_{n+1} = 0$ the inequality is obvious. For $a_{n+1} > 0$ we have that the function $f(x) = \sqrt{x + a_{n+1}} - \sqrt{x} = \frac{a_{n+1}}{\sqrt{x+a_{n+1}}+\sqrt{x}}$ is strictly decreasing on $\mathbb{R}^+$; hence (2) will follow if we show that $S \geq na_{n+1}$. However, the latter is true because $a_1, \ldots, a_n \geq a_{n+1}$.

Equality holds if and only if $a_1 = a_2 = \cdots = a_k$ and $a_{k+1} = \cdots = a_{n+1} = 0$ for some k.

Second solution. Setting $b_k = \sqrt{a_k} - \sqrt{a_{k+1}}$ for $k = 1, \ldots, n$ we have $a_i = (b_i + \cdots + b_n)^2$, so the desired inequality after squaring becomes

$$\sum_{k=1}^{n} k b_k^2 + 2 \sum_{1 \leq k < l \leq n} k b_k b_l \leq \sum_{k=1}^{n} k b_k^2 + 2 \sum_{1 \leq k < l \leq n} \sqrt{kl}\, b_k b_l,$$

which clearly holds.

20. To avoid dividing into cases regarding the position of the point X, we use oriented angles.

Let R be the foot of the perpendicular from X to BC. It is well known that the points P, Q, R lie on the corresponding Simson line. This line is a tangent to γ (i.e., the circle XDR) if and only if $\angle PRD = \angle RXD$. We have

$$\angle PRD = \angle PXB = 90^\circ - \angle XBA = 90^\circ - \angle XBC + \angle ABC$$
$$= 90^\circ - \angle DAC + \angle ABC$$

and

$$\angle RXD = 90^\circ - \angle ADB = 90^\circ + \angle BCA - \angle DAC;$$

hence $\angle PRD = \angle RXD$ if and only if $\angle ABC = \angle BCA$, i.e, $AB = AC$.

21. For any permutation $\pi = (y_1, y_2, \ldots, y_n)$ of $(x_1, x_2, \ldots, x_n)$, denote by $S(\pi)$ the sum $y_1 + 2y_2 + \cdots + ny_n$. Suppose, contrary to the claim, that $|S(\pi)| > \frac{n+1}{2}$ for any π.

Further, we note that if π' is obtained from π by interchanging two neighboring elements, say y_k and y_{k+1}, then $S(\pi)$ and $S(\pi')$ differ by $|y_k + y_{k+1}| \leq n+1$, and consequently they must be of the same sign.

Now consider the identity permutation $\pi_0 = (x_1, \ldots, x_n)$ and the reverse permutation $\overline{\pi_0} = (x_n, \ldots, x_1)$. There is a sequence of permutations $\pi_0, \pi_1, \ldots, \pi_m = \overline{\pi_0}$ such that for each i, π_{i+1} is obtained from π_i by interchanging two neighboring elements. Indeed, by successive interchanges we can put x_n in the first place, then x_{n-1} in the second place, etc. Hence all $S(\pi_0), \ldots, S(\pi_m)$ are of the same sign. However, since $|S(\pi_0) + S(\pi_m)| = (n+1)|x_1 + \cdots + x_n| = n+1$, this implies that one of $S(\pi_0)$ and $S(\overline{\pi_0})$ is smaller than $\frac{n+1}{2}$ in absolute value, contradicting the initial assumption.

22. (a) Suppose that f and g are such functions. From $g(f(x)) = x^3$ we have $f(x_1) \neq f(x_2)$ whenever $x_1 \neq x_2$. In particular, $f(-1), f(0)$, and $f(1)$ are three distinct numbers. However, since $f(x)^2 = f(g(f(x))) = f(x^3)$, each of the numbers $f(-1), f(0), f(1)$ is equal to its square, and so must be either 0 or 1. This contradiction shows that no such f, g exist.

   (b) The answer is *yes*. We begin with constructing functions $F, G : (1, \infty) \to (1, \infty)$ with the property $F(G(x)) = x^2$ and $G(F(x)) = x^4$ for $x > 1$. Define the functions φ, ψ by $F(2^{2^t}) = 2^{2^{\varphi(t)}}$ and $G(2^{2^t}) = 2^{2^{\psi(t)}}$. These functions determine F and G on the entire interval $(1, \infty)$, and satisfy $\varphi(\psi(t)) = t+1$ and $\psi(\varphi(t)) = t+2$. It is easy to find examples of φ and ψ: for example, $\varphi(t) = \frac{1}{2}t + 1$, $\psi(t) = 2t$. Thus we also arrive at an example for F, G:

$$F(x) = 2^{2^{\frac{1}{2}\log_2\log_2 x + 1}} = 2^2\sqrt{\log_2 x}, \qquad G(x) = 2^{2^{2\log_2\log_2 x}} = 2^{\log_2^2 x}.$$

It remains only to extend these functions to the whole of $\mathbb{R}$. This can be done as follows:

$$\widetilde{f}(x) = \begin{cases} F(x) & \text{for } x > 1, \\ 1/F(1/x) & \text{for } 0 < x < 1, \\ x & \text{for } x \in \{0,1\}; \end{cases} \qquad \widetilde{g}(x) = \begin{cases} G(x) & \text{for } x > 1, \\ 1/G(1/x) & \text{for } 0 < x < 1, \\ x & \text{for } x \in \{0,1\}; \end{cases}$$

and then $f(x) = \tilde{f}(|x|)$, $g(x) = \tilde{g}(|x|)$ for $x \in \mathbb{R}$.

It is directly verified that these functions have the required property.

23. Let K, L, M, and N be the projections of O onto the lines AB, BC, CD, and DA, and let $\alpha_1, \alpha_2, \alpha_3, \alpha_4, \beta_1, \beta_2, \beta_3, \beta_4$ denote the angles $OAB, OBC, OCD, ODA, OAD, OBA, OCB, ODC$, respectively.

We start with the following observation: Since NK is a chord of the circle with diameter OA, we have $OA \sin \angle A = NK = ON \cos \alpha_1 + OK \cos \beta_1$ (because $\angle ONK = \alpha_1$ and $\angle OKN = \beta_1$). Analogous equalities also hold: $OB \sin \angle B = KL = OK \cos \alpha_2 + OL \cos \beta_2$, $OC \sin \angle C = LM = OL \cos \alpha_3 + OM \cos \beta_3$ and $OD \sin \angle D = MN = OM \cos \alpha_4 + ON \cos \beta_4$. Now the condition in the problem can be restated as $NK + LM = KL + MN$ (i.e., $KLMN$ is circumscribed), i.e.,

$$OK(\cos \beta_1 - \cos \alpha_2) + OL(\cos \alpha_3 - \cos \beta_2)$$
$$+ OM(\cos \beta_3 - \cos \alpha_4) + ON(\cos \alpha_1 - \cos \beta_4) = 0. \tag{1}$$

To prove that $ABCD$ is cyclic, it suffices to show that $\alpha_1 = \beta_4$. Assume the contrary, and let w.l.o.g. $\alpha_1 > \beta_4$. Then point A lies inside the circle BCD, which is further equivalent to $\beta_1 > \alpha_2$. On the other hand, from $\alpha_1 + \beta_2 = \alpha_3 + \beta_4$ we deduce $\alpha_3 > \beta_2$, and similarly $\beta_3 > \alpha_4$. Therefore, since the cosine is strictly decreasing on $(0, \pi)$, the left side of (1) is strictly negative, yielding a contradiction.

24. There is a bijective correspondence between representations in the given form of $2k$ and $2k + 1$ for $k = 0, 1, \ldots$, since adding 1 to every representation of $2k$, we obtain a representation of $2k + 1$, and conversely, every representation of $2k + 1$ contains at least one 1, which can be removed. Hence, $f(2k + 1) = f(2k)$.

Consider all representations of $2k$. The number of those that contain at least one 1 equals $f(2k - 1) = f(2k - 2)$, while the number of those not containing a 1 equals $f(k)$ (the correspondence is given by division of summands by 2). Therefore

$$f(2k) = f(2k - 2) + f(k). \tag{1}$$

Summing these equalities over $k = 1, \ldots, n$, we obtain

$$f(2n) = f(0) + f(1) + \cdots + f(n). \tag{2}$$

We first prove the right-hand inequality. Since f is increasing, and $f(0) + f(1) = f(2)$, (2) yields $f(2n) \leq nf(n)$ for $n \geq 2$. Now $f(2^3) = f(0) + \cdots + f(4) = 10 < 2^{3^2/2}$, and one can easily conclude by induction that $f(2^{n+1}) \leq 2^n f(2^n) < 2^n \cdot 2^{n^2/2} < 2^{(n+1)^2/2}$ for each $n \geq 3$.

We now derive the lower estimate. It follows from (1) that $f(x + 2) - f(x)$ is increasing. Consequently, for each m and $k < m$ we have $f(2m + 2k) - f(2m) \geq f(2m + 2k - 2) - f(2m - 2) \geq \cdots \geq f(2m) - f(2m - 2k)$, so $f(2m + 2k) + f(2m - 2k) \geq 2f(2m)$. Adding all these inequalities for $k = 1, 2, \ldots, m$, we obtain $f(0) + f(2) + \cdots + f(4m) \geq (2m + 1)f(2m)$. But since $f(2) = f(3)$, $f(4) = f(5)$

etc., we also have $f(1) + f(3) + \cdots + f(4m-1) > (2m-1)f(2m)$, which together with the above inequality gives

$$f(8m) = f(0) + f(1) + \cdots + f(4m) > 4mf(2m). \tag{3}$$

Finally, we have that the inequality $f(2^n) > 2^{n^2/4}$ holds for $n = 2$ and $n = 3$, while for larger n we have by induction $f(2^n) > 2^{n-1}f(2^{n-2}) > 2^{n-1+(n-2)^2/4} = 2^{n^2/4}$. This completes the proof.

Remark. Despite the fact that the lower estimate is more difficult, it is much weaker than the upper estimate. It can be shown that $f(2^n)$ eventually (for large n) exceeds 2^{cn^2} for any $c < \frac{1}{2}$.

25. Let MR meet the circumcircle of triangle ABC again at a point X. We claim that X is the common point of the lines KP, LQ, MR. By symmetry, it will be enough to show that X lies on KP. It is easy to see that X and P lie on the same side of AB as K. Let $I_a = AK \cap BP$ be the excenter of $\triangle ABC$ corresponding to A. It is easy to calculate that $\angle AI_aB = \gamma/2$, from which we get $\angle RPB = \angle AI_aB = \angle MCB = \angle RXB$. Therefore R, B, P, X are concyclic. Now if P and K are on distinct sides

that $T_{i+1}A_i = T_{i+1}I$ and $T_{i+2}A_i = T_{i+2}I$ other case is similar), we have $\angle RXP = 180° - \angle RBP = 90° - \beta/2 = \angle MAK = 180° - \angle RXK$, from which it follows that K, X, P are collinear, as claimed.

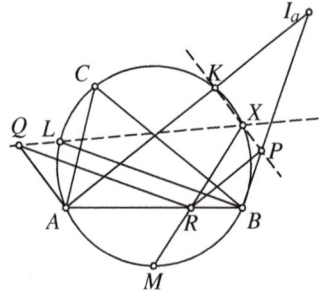

Remark. It is not essential for the statement of the problem that R be an internal point of AB. Work with cases can be avoided using oriented angles.

26. Let us first examine the case that all the inequalities in the problem are actually equalities. Then $a_{n-2} = a_{n-1} + a_n$, $a_{n-3} = 2a_{n-1} + a_n, \ldots, a_0 = F_n a_{n-1} + F_{n-1} a_n = 1$, where F_n is the nth Fibonacci number. Then it is easy to see (from $F_1 + F_2 + \cdots + F_k = F_{k+2}$) that $a_0 + \cdots + a_n = (F_{n+2} - 1)a_{n-1} + F_{n+1}a_n = \frac{F_{n+2}-1}{F_n} + \left(F_{n+1} - \frac{F_{n-1}(F_{n+2}-1)}{F_n}\right)a_n$. Since $\frac{F_{n-1}(F_{n+2}-1)}{F_n} \le F_{n+1}$, it follows that $a_0 + a_1 + \cdots + a_n \ge \frac{F_{n+2}-1}{F_n}$, with equality holding if and only if $a_n = 0$ and $a_{n-1} = \frac{1}{F_n}$.

We denote by M_n the required minimum in the general case. We shall prove by induction that $M_n = \frac{F_{n+2}-1}{F_n}$. For $M_1 = 1$ and $M_2 = 2$ it is easy to show that the formula holds; hence the inductive basis is true. Suppose that $n > 2$. The sequences $1, \frac{a_2}{a_1}, \ldots, \frac{a_n}{a_1}$ and $1, \frac{a_3}{a_2}, \ldots, \frac{a_n}{a_2}$ also satisfy the conditions of the problem. Hence we have

$$a_0 + \cdots + a_n = a_0 + a_1\left(1 + \frac{a_2}{a_1} + \cdots + \frac{a_n}{a_1}\right) \ge 1 + a_1 M_{n-1}$$

and

$$a_0 + \cdots + a_n = a_0 + a_1 + a_2 \left(1 + \frac{a_3}{a_2} + \cdots + \frac{a_n}{a_2}\right) \geq 1 + a_1 + a_2 M_{n-2}.$$

Multiplying the first inequality by $M_{n-2} - 1$ and the second one by M_{n-1}, adding the inequalities and using that $a_1 + a_2 \geq 1$, we obtain $(M_{n-1} + M_{n-2} + 1)(a_0 + \cdots + a_n) \geq M_{n-1}M_{n-2} + M_{n-1} + M_{n-2} + 1$, so

$$M_n \geq \frac{M_{n-1}M_{n-2} + M_{n-1} + M_{n-2} + 1}{M_{n-1} + M_{n-2} + 1}.$$

Since $M_{n-1} = \frac{F_{n+1}-1}{F_{n-1}}$ and $M_{n-2} = \frac{F_n - 1}{F_{n-2}}$, the above inequality easily yields $M_n \geq \frac{F_{n+2}-1}{F_n}$. However, we have shown above that equality can occur; hence $\frac{F_{n+2}-1}{F_n}$ is indeed the required minimum.

4.39 Solutions to the Shortlisted Problems of IMO 1998

1. We begin with the following observation: Suppose that P lies in $\triangle AEB$, where E is the intersection of AC and BD (the other cases are similar). Let M, N be the feet of the perpendiculars from P to AC and BD respectively. We have $S_{ABP} = S_{ABE} - S_{AEP} - S_{BEP} = \frac{1}{2}(AE \cdot BE - AE \cdot EN - BE \cdot EM) = \frac{1}{2}(AM \cdot BN - EM \cdot EN)$. Similarly, $S_{CDP} = \frac{1}{2}(CM \cdot DN - EM \cdot EN)$. Therefore, we obtain

$$S_{ABP} - S_{CDP} = \frac{AM \cdot BN - CM \cdot DN}{2}. \tag{1}$$

Now suppose that $ABCD$ is cyclic. Then P is the circumcenter of $ABCD$; hence M and N are the midpoints of AC and BD. Hence $AM = CM$ and $BN = DN$; thus (1) gives us $S_{ABP} = S_{CDP}$. On the other hand, suppose that $ABCD$ is not cyclic and let w.l.o.g. $PA = PB > PC = PD$. Then we must have $AM > CM$ and $BN > DN$, and consequently by (1), $S_{ABP} > S_{CDP}$. This proves the other implication.

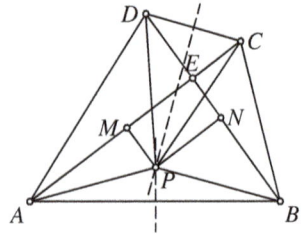

Second solution. Let F and G denote the midpoints of AB and CD, and assume that P is on the same side of FG as B and C. Since $PF \perp AB$, $PG \perp CD$, and $\angle FEB = \angle ABE$, $\angle GEC = \angle DCE$, a direct computation yields $\angle FPG = \angle FEG = 90° + \angle ABE + \angle DCE$.

Taking into account that $S_{ABP} = \frac{1}{2}AB \cdot FP = FE \cdot FP$, we note that $S_{ABP} = S_{CDP}$ is equivalent to $FE \cdot FP = GE \cdot GP$, i.e., to $FE/EG = GP/PF$. But this last is equivalent to triangles EFG and PGF being similar, which holds if and only if $EFPG$ is a parallelogram. This last is equivalent to $\angle EFP = \angle EGP$, or $2\angle ABE = 2\angle DCE$. Thus $S_{ABP} = S_{CDP}$ is equivalent to $ABCD$ being cyclic.

Remark. The problems also allows an analytic solution, for example putting the x and y axes along the diagonals AC and BD.

2. If AD and BC are parallel, then $ABCD$ is an isosceles trapezoid with $AB = CD$, so P is the midpoint of EF. Let M and N be the midpoints of AB and CD. Then $MN \parallel BC$, and the distance $d(E, MN)$ equals the distance $d(F, MN)$ because B and D are the same distance from MN and $EM/BM = FN/DN$. It follows that the midpoint P of EF lies on MN, and consequently $S_{APD} : S_{BPC} = AD : BC$.

If AD and BC are not parallel, then they meet at some point Q. It is plain that $\triangle QAB \sim \triangle QCD$, and since $AE/AB = CF/CD$, we also deduce that $\triangle QAE \sim \triangle QCF$. Therefore $\angle AQE = \angle CQF$. Further, from these similarities we obtain $QE/QF = QA/QC = AB/CD = PE/PF$, which in turn means that QP is the internal bisector of $\angle EQF$. But since $\angle AQE = \angle CQF$, this is also the internal bisector of $\angle AQB$. Hence P is at equal distances from AD and BC, so again $S_{APD} : S_{BPC} = AD : BC$.

Remark. The part $AB \parallel CD$ could also be regarded as a limiting case of the other part.

Second solution. Denote $\lambda = \frac{AE}{AB}$, $AB = a$, $BC = b$, $CD = c$, $DA = d$, $\angle DAB = \alpha$, $\angle ABC = \beta$. Since $d(P,AD) = \frac{c \cdot d(E,AD) + a \cdot d(F,AD)}{a+c}$, we have $S_{APD} = \frac{c S_{EAD} + a S_{FAD}}{a+c} = \frac{\lambda c S_{ABD} + (1-\lambda) a S_{ACD}}{a+c}$. Since $S_{ABD} = \frac{1}{2} ad \sin \alpha$ and $S_{ACD} = \frac{1}{2} cd \sin \beta$, we are led to $S_{APD} = \frac{acd}{a+c} [\lambda \sin \alpha + (1-\lambda) \sin \beta]$, and analogously $S_{BPC} = \frac{abc}{a+c} [\lambda \sin \alpha + (1-\lambda) \sin \beta]$. Thus we obtain $S_{APD} : S_{BPC} = d : b$.

3. *Lemma.* If U, W, V are three points on a line l in this order, and X a point in the plane with $XW \perp UV$, then $\angle UXV < 90°$ if and only if $XW^2 > UW \cdot VW$.

   Proof. Let $XW^2 > UW \cdot VW$, and let X_0 be a point on the segment XW such that $X_0 W^2 \geq UW \cdot VW$. Then $X_0 W / UW = VW / X_0 W$, so that triangles $X_0 WU$ and VWX_0 are similar. Thus $\angle UX_0 V = \angle UX_0 W + \angle WUX_0 = 90°$, which immediately implies that $\angle UXV < 90°$.

   Similarly, if $XW^2 \leq UW \cdot VW$, then $\angle UXV \geq 90°$.

   Since $BI \perp RS$, it will be enough by the lemma to show that $BI^2 > BR \cdot BS$. Note that $\triangle BKR \sim \triangle BSL$: in fact, we have $\angle KBR = \angle SBL = 90° - \beta/2$ and $\angle BKR = \angle AKM = \angle KLM = \angle BSL = 90° - \alpha/2$. In particular, we obtain $BR/BK = BL/BS = BK/BS$, so that $BR \cdot BS = BK^2 < BI^2$.

   Second solution. Let E, F be the midpoints of KM and LM respectively. The quadrilaterals $RBIE$ and $SBIF$ are inscribed in the circles with diameters IR and IS. Now we have $\angle RIS = \angle RMS + \angle IRM + \angle ISM = 90° - \beta/2 + \angle IBE + \angle IBF = 90° - \beta/2 + \angle EBF$.

   On the other hand, BE and BF are medians in $\triangle BKM$ and $\triangle BLM$ in which $BM > BK$ and $BM > BL$. We conclude that $\angle MBE < \frac{1}{2} \angle MBK$ and $\angle MBF < \frac{1}{2} \angle MBL$. Adding these two inequalities gives $\angle EBF < \beta/2$. Therefore $\angle RIS < 90°$.

   Remark. It can be shown (using vectors) that the statement remains true for an arbitrary line t passing through B.

4. Let K be the point on the ray BN with $\angle BCK = \angle BMA$. Since $\angle KBC = \angle ABM$, we get $\triangle BCK \sim \triangle BMA$. It follows that $BC/BM = BK/BA$, which implies that also $\triangle BAK \sim \triangle BMC$. The quadrilateral $ANCK$ is cyclic, because $\angle BKC = \angle BAM = \angle NAC$. Then by Ptolemy's theorem we obtain

$$AC \cdot BK = AC \cdot BN + AN \cdot CK + CN \cdot AK. \tag{1}$$

On the other hand, from the similarities noted above we get

$$CK = \frac{BC \cdot AM}{BM}, \quad AK = \frac{AB \cdot CM}{BM} \quad \text{and} \quad BK = \frac{AB \cdot BC}{BM}.$$

After substitution of these values, the equality (1) becomes

$$\frac{AB \cdot BC \cdot AC}{BM} = AC \cdot BN + \frac{BC \cdot AM \cdot AN}{BM} + \frac{AB \cdot CM \cdot CN}{BM},$$

which is exactly the equality we must prove multiplied by $\frac{AB \cdot BC \cdot CA}{BM}$.

5. Let G be the centroid of $\triangle ABC$ and $\mathcal{H}$ the homothety with center G and ratio $-\frac{1}{2}$. It is well-known that $\mathcal{H}$ maps H into O. For every other point X, let us denote by X' its image under $\mathcal{H}$. Also, let $A_2B_2C_2$ be the triangle in which A, B, C are the midpoints of B_2C_2, C_2A_2, and A_2B_2, respectively.

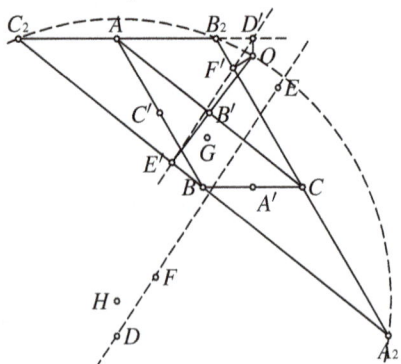

It is clear that A', B', and C' are the midpoints of BC, CA, and AB respectively. We also have that D' is the reflection of A' across $B'C'$. Thus D' must lie on B_2C_2 and $A'D' \perp B_2C_2$. However, it also holds that OA' and B_2C_2 are orthogonal, so we conclude that O, D', A' are collinear and D' is the projection of O on B_2C_2. Analogously, E', F' are the projections of O on C_2A_2 and A_2B_2.

Now we apply Simson's theorem. It claims that D', E', F' are collinear (which is equivalent to D, E, F being collinear) if and only if O lies on the circumcircle of $A_2B_2C_2$. However, this circumcircle is centered at H with radius $2R$, so the last condition is equivalent to $HO = 2R$.

6. Let P be the point such that $\triangle CDP$ and $\triangle CBA$ are similar and equally oriented. Since then $\angle DCP = \angle BCA$ and $\frac{BC}{CA} = \frac{DC}{CP}$, it follows that $\angle ACP = \angle BCD$ and $\frac{AC}{CP} = \frac{BC}{CD}$, so $\triangle ACP \sim \triangle BCD$. In particular, $\frac{BC}{CA} = \frac{DB}{PA}$.

Furthermore, by the conditions of the problem we have $\angle EDP = 360° - \angle B - \angle D = \angle F$ and $\frac{PD}{DE} = \frac{PD}{CD} \cdot \frac{CD}{DE} = \frac{AB}{BC} \cdot \frac{CD}{DE} = \frac{AF}{FE}$. Therefore $\triangle EDP \sim \triangle EFA$ as well, so that similarly as above we conclude that $\triangle AEP \sim \triangle FED$ and consequently $\frac{AE}{EF} = \frac{PA}{FD}$.

Finally, $\frac{BC}{CA} \cdot \frac{AE}{EF} \cdot \frac{FD}{DB} = \frac{DB}{PA} \cdot \frac{PA}{FD} \cdot \frac{FD}{DB} = 1$.

Second solution. Let a, b, c, d, e, f be the complex coordinates of A, B, C, D, E, F, respectively. The condition of the problem implies that $\frac{a-b}{b-c} \cdot \frac{c-d}{d-e} \cdot \frac{e-f}{f-a} = -1$. On the other hand, since $(a-b)(c-d)(e-f) + (b-c)(d-e)(f-a) = (b-c)(a-e)(f-d) + (c-a)(e-f)(d-b)$ holds identically, we immediately deduce that $\frac{b-c}{c-a} \cdot \frac{a-e}{e-f} \cdot \frac{f-d}{d-b} = -1$. Taking absolute values gives $\frac{BC}{CA} \cdot \frac{AE}{EF} \cdot \frac{FD}{DB} = 1$.

7. We shall use the following result.

Lemma. In a triangle ABC with $BC = a$, $CA = b$, and $AB = c$,

(a) $\angle C = 2\angle B$ if and only if $c^2 = b^2 + ab$;

(b) $\angle C + 180° = 2\angle B$ if and only if $c^2 = b^2 - ab$.

Proof.

(a) Take a point D on the extension of BC over C such that $CD = b$. The condition $\angle C = 2\angle B$ is equivalent to $\angle ADC = \frac{1}{2}\angle C = \angle B$, and thus to $AD = AB = c$. This is further equivalent to triangles CAD and ABD being similar, so $CA/AD = AB/BD$, i.e., $c^2 = b(a+b)$.

(b) Take a point E on the ray CB such that $CE = b$. As above, $\angle C + 180° = 2\angle B$ if and only if $\triangle CAE \sim \triangle ABE$, which is equivalent to $EB/BA = EA/AC$, or $c^2 = b(b-a)$.

Let F, G be points on the ray CB such that $CF = \frac{1}{3}a$ and $CG = \frac{4}{3}a$. Set $BC = a$, $CA = b$, $AB = c$, $EC = b_1$, and $EB = c_1$. By the lemma it follows that $c^2 = b^2 + ab$. Also $b_1 = AG$ and $c_1 = AF$, so Stewart's theorem gives us $c_1^2 = \frac{2}{3}b^2 + \frac{1}{3}c^2 - \frac{2}{9}a^2 = b^2 + \frac{1}{3}ab - \frac{2}{9}a^2$ and $b_1^2 = -\frac{1}{3}b^2 + \frac{4}{3}c^2 + \frac{4}{9}a^2 = b^2 + \frac{4}{3}ab + \frac{4}{9}a^2$. It follows that $b_1 = \frac{2}{3}a + b$ and $c_1^2 = b_1^2 - (ab + \frac{2}{3}a^2) = b_1^2 - ab_1$. The statement of the problem follows immediately from the lemma.

8. Let M be the point of intersection of AE and BC, and let N be the point on ω diametrically opposite A.

Since $\angle B < \angle C$, points N and B are on the same side of AE. Furthermore, $\angle NAE = \angle BAX = 90° - \angle ABE$; hence the triangles NAE and BAX are similar. Consequently, $\triangle BAY$ and $\triangle NAM$ are also similar, since M is the midpoint of AE. Thus $\angle ANZ = \angle ABZ = \angle ABY = \angle ANM$, implying that N, M, Z are collinear. Now we have $\angle ZMD = 90° - \angle ZMA = \angle EAZ = \angle ZED$ (the last equality because ED is tangent to ω); hence $ZMED$ is a cyclic quadrilateral. It follows that $\angle ZDM = \angle ZEA = \angle ZAD$, which is enough to conclude that MD is tangent to the circumcircle of AZD.

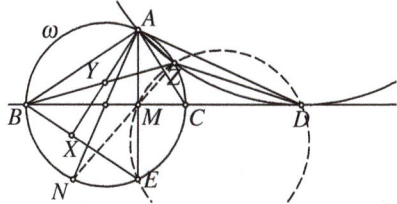

Remark. The statement remains valid if $\angle B \geq \angle C$.

9. Set $a_{n+1} = 1 - (a_1 + \cdots + a_n)$. Then $a_{n+1} > 0$, and the desired inequality becomes

$$\frac{a_1 a_2 \cdots a_{n+1}}{(1-a_1)(1-a_2)\cdots(1-a_{n+1})} \leq \frac{1}{n^{n+1}}.$$

To prove it, we observe that

$$1 - a_i = a_1 + \cdots + a_{i-1} + a_{i+1} + \cdots + a_{n+1} \geq n\sqrt[n]{a_1 \cdots a_{i-1} a_{i+1} \cdots a_{n+1}}.$$

Multiplying these inequalities for $i = 1, 2, \ldots, n+1$, we get exactly the inequality we need.

10. We shall first prove the inequality for n of the form 2^k, $k = 0, 1, 2, \ldots$. The case $k = 0$ is clear. For $k = 1$, we have

$$\frac{1}{r_1 + 1} + \frac{1}{r_2 + 1} - \frac{2}{\sqrt{r_1 r_2} + 1} = \frac{(\sqrt{r_1 r_2} - 1)(\sqrt{r_1} - \sqrt{r_2})^2}{(r_1 + 1)(r_2 + 1)(\sqrt{r_1 r_2} + 1)} \geq 0.$$

For the inductive step it suffices to show that the claim for k and 2 implies that for $k + 1$. Indeed,

$$\sum_{i=1}^{2^{k+1}} \frac{1}{r_i + 1} \geq \frac{2^k}{\sqrt[2^k]{r_1 r_2 \cdots r_{2^k}} + 1} + \frac{2^k}{\sqrt[2^k]{r_{2^k+1} r_{2^k+2} \cdots r_{2^{k+1}}} + 1}$$

$$\geq \frac{2^{k+1}}{\sqrt[2^{k+1}]{r_1 r_2 \cdots r_{2^{k+1}}} + 1}, \tag{1}$$

and the induction is complete.

We now show that if the statement holds for 2^k, then it holds for every $n < 2^k$ as well. Put $r_{n+1} = r_{n+2} = \cdots = r_{2^k} = \sqrt[n]{r_1 r_2 \ldots r_n}$. Then (1) becomes

$$\frac{1}{r_1 + 1} + \cdots + \frac{1}{r_n + 1} + \frac{2^k - n}{\sqrt[n]{r_1 \cdots r_n} + 1} \geq \frac{2^k}{\sqrt[n]{r_1 \cdots r_n} + 1}.$$

This proves the claim.

Second solution. Define $r_i = e^{x_i}$, where $x_i > 0$. The function $f(x) = \frac{1}{1+e^x}$ is convex for $x > 0$: indeed, $f''(x) = \frac{e^x(e^x - 1)}{(e^x + 1)^3} > 0$. Thus by Jensen's inequality applied to $f(x_1), \ldots, f(x_n)$, we get $\frac{1}{r_1 + 1} + \cdots + \frac{1}{r_n + 1} \geq \frac{n}{\sqrt[n]{r_1 \cdots r_n} + 1}$.

11. The given inequality is equivalent to $x^3(x+1) + y^3(y+1) + z^3(z+1) \geq \frac{3}{4}(x+1)(y+1)(z+1)$. By the A-G mean inequality, it will be enough to prove a stronger inequality:

$$x^4 + x^3 + y^4 + y^3 + z^4 + z^3 \geq \frac{1}{4}[(x+1)^3 + (y+1)^3 + (z+1)^3]. \tag{1}$$

If we set $S_k = x^k + y^k + z^k$, (1) takes the form $S_4 + S_3 \geq \frac{1}{4}S_3 + \frac{3}{4}S_2 + \frac{3}{4}S_1 + \frac{3}{4}$. Note that by the A-G mean inequality, $S_1 = x + y + z \geq 3$. Thus it suffices to prove the following:

> If $S_1 \geq 3$ and $m > n$ are positive integers, then $S_m \geq S_n$.

This can be shown in many ways. For example, by Hölder's inequality,

$$(x^m + y^m + z^m)^{n/m}(1 + 1 + 1)^{(m-n)/m} \geq x^n + y^n + z^n.$$

(Another way is using the Chebyshev inequality: if $x \geq y \geq z$ then $x^{k-1} \geq y^{k-1} \geq z^{k-1}$; hence $S_k = x \cdot x^{k-1} + y \cdot y^{k-1} + z \cdot z^{k-1} \geq \frac{1}{3}S_1 S_{k-1}$, and the claim follows by induction.)

Second solution. Assume that $x \geq y \geq z$. Then also $\frac{1}{(y+1)(z+1)} \geq \frac{1}{(x+1)(z+1)} \geq \frac{1}{(x+1)(y+1)}$. Hence Chebyshev's inequality gives that

$$\frac{x^3}{(1+y)(1+z)} + \frac{y^3}{(1+x)(1+z)} + \frac{z^3}{(1+x)(1+y)}$$
$$\geq \frac{1}{3}\frac{(x^3 + y^3 + z^3) \cdot (3 + x + y + z)}{(1+x)(1+y)(1+z)}.$$

Now if we put $x + y + z = 3S$, we have $x^3 + y^3 + z^3 \geq 3S$ and $(1+x)(1+y)(1+z) \leq (1+a)^3$ by the A-G mean inequality. Thus the needed inequality reduces to $\frac{6S^3}{(1+S)^3} \geq \frac{3}{4}$, which is obviously true because $S \geq 1$.

Remark. Both these solutions use only that $x + y + z \geq 3$.

12. The assertion is clear for $n = 0$. We shall prove the general case by induction on n. Suppose that $c(m,i) = c(m,m-i)$ for all i and $m \leq n$. Then by the induction hypothesis and the recurrence formula we have $c(n+1,k) = 2^k c(n,k) + c(n,k-1)$ and $c(n+1,n+1-k) = 2^{n+1-k}c(n,n+1-k) + c(n,n-k) = 2^{n+1-k}c(n,k-1) + c(n,k)$. Thus it remains only to show that

$$(2^k - 1)c(n,k) = (2^{n+1-k} - 1)c(n,k-1).$$

We prove this also by induction on n. By the induction hypothesis,

$$c(n-1,k) = \frac{2^{n-k}-1}{2^k-1}c(n-1,k-1)$$

and

$$c(n-1,k-2) = \frac{2^{k-1}-1}{2^{n+1-k}-1}c(n-1,k-1).$$

Using these formulas and the recurrence formula we obtain $(2^k - 1)c(n,k) - (2^{n+1-k} - 1)c(n,k-1) = (2^{2k} - 2^k)c(n-1,k) - (2^n - 3 \cdot 2^{k-1} + 1)c(n-1,k-1) - (2^{n+1-k} - 1)c(n-1,k-2) = (2^n - 2^k)c(n-1,k-1) - (2^n - 3 \cdot 2^{k-1} + 1)c(n-1,k-1) - (2^{k-1} - 1)c(n-1,k-1) = 0$. This completes the proof.

Second solution. The given recurrence formula resembles that of binomial coefficients, so it is natural to search for an explicit formula of the form $c(n,k) = \frac{F(n)}{F(k)F(n-k)}$, where $F(m) = f(1)f(2)\cdots f(m)$ (with $F(0) = 1$) and f is a certain function from the natural numbers to the real numbers. If there is such an f, then $c(n,k) = c(n,n-k)$ follows immediately.

After substitution of the above relation, the recurrence equivalently reduces to $f(n+1) = 2^k f(n-k+1) + f(k)$. It is easy to see that $f(m) = 2^m - 1$ satisfies this relation.

Remark. If we introduce the polynomial $P_n(x) = \sum_{k=0}^{n} c(n,k)x^k$, the recurrence relation gives $P_0(x) = 1$ and $P_{n+1}(x) = xP_n(x) + P_n(2x)$. As a consequence of the problem, all polynomials in this sequence are symmetric, i.e., $P_n(x) = x^n P_n(x^{-1})$.

13. Denote by $\mathscr{F}$ the set of functions considered. Let $f \in \mathscr{F}$, and let $f(1) = a$. Putting $n = 1$ and $m = 1$ we obtain $f(f(z)) = a^2 z$ and $f(az^2) = f(z)^2$ for all $z \in \mathbb{N}$. These equations, together with the original one, imply $f(x)^2 f(y)^2 = f(x)^2 f(ay^2) = f(x^2 f(f(ay^2))) = f(x^2 a^3 y^2) = f(a(axy)^2) = f(axy)^2$, which implies $f(axy) = f(x)f(y)$ for all $x,y \in \mathbb{N}$. Thus $f(ax) = af(x)$, and we conclude that

$$af(xy) = f(x)f(y) \qquad \text{for all } x,y \in \mathbb{N}. \tag{1}$$

We now prove that $f(x)$ is divisible by a for each $x \in \mathbb{N}$. In fact, we inductively get that $f(x)^k = a^{k-1}f(x^k)$ is divisible by a^{k-1} for every k. If p^α and p^β are the exact powers of a prime p that divide $f(x)$ and a respectively, we deduce that $k\alpha \geq (k-1)\beta$ for all k, so we must have $\alpha \geq \beta$ for any p. Therefore $a \mid f(x)$. Now we consider the function on natural numbers $g(x) = f(x)/a$. The above relations imply

$$g(1) = 1, \quad g(xy) = g(x)g(y), \quad g(g(x)) = x \quad \text{for all } x, y \in \mathbb{N}. \quad (2)$$

Since $g \in \mathscr{F}$ and $g(x) \leq f(x)$ for all x, we may restrict attention to the functions g only.

Clearly g is bijective. We observe that g maps a prime to a prime. Assume to the contrary that $g(p) = uv, u, v > 1$. Then $g(uv) = p$, so either $g(u) = 1$ or $g(v) = 1$. Thus either $g(1) = u$ or $g(1) = v$, which is impossible.

We return to the problem of determining the least possible value of $g(1998)$. Since $g(1998) = g(2 \cdot 3^3 \cdot 37) = g(2) \cdot g(3)^3 \cdot g(37)$, and $g(2)$, $g(3)$, $g(37)$ are distinct primes, $g(1998)$ is not smaller than $2^3 \cdot 3 \cdot 5 = 120$. On the other hand, the value of 120 is attained for any function g satisfying (2) and $g(2) = 3$, $g(3) = 2$, $g(5) = 37$, $g(37) = 5$. Hence the answer is 120.

14. If $x^2 y + x + y$ is divisible by $xy^2 + y + 7$, then so is the number $y(x^2 y + x + y) - x(xy^2 + y + 7) = y^2 - 7x$.

If $y^2 - 7x \geq 0$, then since $y^2 - 7x < xy^2 + y + 7$, it follows that $y^2 - 7x = 0$. Hence $(x, y) = (7t^2, 7t)$ for some $t \in \mathbb{N}$. It is easy to check that these pairs really are solutions.

If $y^2 - 7x < 0$, then $7x - y^2 > 0$ is divisible by $xy^2 + y + 7$. But then $xy^2 + y + 7 \leq 7x - y^2 < 7x$, from which we obtain $y \leq 2$. For $y = 1$, we are led to $x + 8 \mid 7x - 1$, and hence $x + 8 \mid 7(x + 8) - (7x - 1) = 57$. Thus the only possibilities are $x = 11$ and $x = 49$, and the obtained pairs $(11, 1), (49, 1)$ are indeed solutions. For $y = 2$, we have $4x + 9 \mid 7x - 4$, so that $7(4x + 9) - 4(7x - 4) = 79$ is divisible by $4x + 9$. We do not get any new solutions in this case.

Therefore all required pairs (x, y) are $(7t^2, 7t)$ $(t \in \mathbb{N})$, $(11, 1)$, and $(49, 1)$.

15. The condition is obviously satisfied if $a = 0$ or $b = 0$ or $a = b$ or a, b are both integers. We claim that these are the only solutions.

Suppose that a, b belong to none of the above categories. The quotient $a/b = \lfloor a \rfloor / \lfloor b \rfloor$ is a nonzero rational number: let $a/b = p/q$, where p and q are coprime nonzero integers.

Suppose that $p \notin \{-1, 1\}$. Then p divides $\lfloor an \rfloor$ for all n, so in particular p divides $\lfloor a \rfloor$ and thus $a = kp + \varepsilon$ for some $k \in \mathbb{N}$ and $0 \leq \varepsilon < 1$. Note that $\varepsilon \neq 0$, since otherwise $b = kq$ would also be an integer. It follows that there exists an $n \in \mathbb{N}$ such that $1 \leq n\varepsilon < 2$. But then $\lfloor na \rfloor = \lfloor knp + n\varepsilon \rfloor = knp + 1$ is not divisible by p, a contradiction. Similarly, $q \notin \{-1, 1\}$ is not possible. Therefore we must have $p, q = \pm 1$, and since $a \neq b$, the only possibility is $b = -a$. However, this leads to $\lfloor -a \rfloor = -\lfloor a \rfloor$, which is not valid if a is not an integer.

16. Let S be a set of integers such that for no four distinct elements $a, b, c, d \in S$, it holds that $20 \mid a + b - c - d$. It is easily seen that there cannot exist distinct elements a, b, c, d with $a \equiv b$ and $c \equiv d$ (mod 20). Consequently, if the elements of S give k different residues modulo 20, then S itself has at most $k + 2$ elements. Next, consider these k elements of S with different residues modulo 20. They give $\frac{k(k-1)}{2}$ different sums of two elements. For $k \geq 7$ there are at least 21 such sums, and two of them, say $a + b$ and $c + d$, are equal modulo 20; it is easy to

see that a, b, c, d are distinct. It follows that k cannot exceed 6, and consequently S has at most 8 elements.

An example of a set S with 8 elements is $\{0, 20, 40, 1, 2, 4, 7, 12\}$. Hence the answer is $n = 9$.

17. Initially, we determine that the first few values for a_n are 1, 3, 4, 7, 10, 12, 13, 16, 19, 21, 22, 25. Since these are exactly the numbers of the forms $3k+1$ and $9k+3$, we conjecture that this is the general pattern. In fact, it is easy to see that the equation $x + y = 3z$ has no solution in the set $K = \{3k+1, 9k+3 \mid k \in \mathbb{N}\}$. We shall prove that the sequence $\{a_n\}$ is actually this set ordered increasingly. Suppose $a_n > 25$ is the first member of the sequence not belonging to K. We have several cases:

  (i) $a_n = 3r+2, r \in \mathbb{N}$. By the assumption, one of $r+1, r+2, r+3$ is of the form $3k+1$ (and smaller than a_n), and therefore is a member a_i of the sequence. Then $3a_i$ equals $a_n + 1$, $a_n + 4$, or $a_n + 7$, which is a contradiction because $1, 4, 7$ are in the sequence.

  (ii) $a_n = 9r, r \in \mathbb{N}$. Then $a_n + a_2 = 3(3r+1)$, although $3r+1$ is in the sequence, a contradiction.

  (iii) $a_n = 9r+6, r \in \mathbb{N}$. Then one of the numbers $3r+3$, $3r+6$, $3r+9$ is a member a_j of the sequence, and thus $3a_j$ is equal to $a_n + 3$, $a_n + 12$, or $a_n + 21$, where $3, 12, 21$ are members of the sequence, again a contradiction.

  Once we have revealed the structure of the sequence, it is easy to compute a_{1998}. We have $1998 = 4 \cdot 499 + 2$, which implies $a_{1998} = 9 \cdot 499 + a_2 = 4494$.

18. We claim that, if $2^n - 1$ divides $m^2 + 9$ for some $m \in \mathbb{N}$, then n must be a power of 2. Suppose otherwise that n has an odd divisor $d > 1$. Then $2^d - 1 \mid 2^n - 1$ is also a divisor of $m^2 + 9 = m^2 + 3^2$. However, $2^d - 1$ has some prime divisor p of the form $4k - 1$, and by a well-known fact, p divides both m and 3. Hence $p = 3$ divides $2^d - 1$, which is impossible, because for d odd, $2^d \equiv 2 \pmod{3}$. Hence $n = 2^r$ for some $r \in \mathbb{N}$.

Now let $n = 2^r$. We prove the existence of m by induction on r. The case $r = 1$ is trivial. Now for any $r > 1$ note that

$$2^{2^r} - 1 = (2^{2^{r-1}} - 1)(2^{2^{r-1}} + 1).$$

The induction hypothesis claims that there exists an m_1 such that $2^{2^{r-1}} - 1 \mid m_1^2 + 9$. We also observe that $2^{2^{r-1}} + 1 \mid m_2^2 + 9$ for simple $m_2 = 3 \cdot 2^{2^{r-2}}$. By the Chinese remainder theorem, there is an $m \in \mathbb{N}$ that satisfies $m \equiv m_1 \pmod{2^{2^{r-1}} - 1}$ and $m \equiv m_2 \pmod{2^{2^{r-1}} + 1}$. It is easy to see that this $m^2 + 9$ will be divisible by both $2^{2^{r-1}} - 1$ and $2^{2^{r-1}} + 1$, i.e., that $2^{2^r} - 1 \mid m^2 + 9$. This completes the induction.

19. For $n = p_1^{\alpha_1} p_2^{\alpha_2} \cdots p_r^{\alpha_r}$, where p_i are distinct primes and α_i natural numbers, we have $\tau(n) = (\alpha_1 + 1) \cdots (\alpha_r + 1)$ and $\tau(n^2) = (2\alpha_1 + 1) \ldots (2\alpha_r + 1)$. Putting $k_i = \alpha_i + 1$, the problem reduces to determining all natural values of m that can be represented as

$$m = \frac{2k_1 - 1}{k_1} \cdot \frac{2k_2 - 1}{k_2} \cdots \frac{2k_r - 1}{k_r}. \tag{1}$$

Since the numerator $\tau(n^2)$ is odd, m must be odd too. We claim that every odd m has a representation of the form (1). The proof will be done by induction. This is clear for $m = 1$. Now for every $m = 2k - 1$ with k odd the result follows easily, since $m = \frac{2k-1}{k} \cdot k$, and k can be written as (1). We cannot do the same if k is even; however, in the case $m = 4k - 1$ with k odd, we can write it as $m = \frac{12k-3}{6k-1} \cdot \frac{6k-1}{3k} \cdot k$, and this works.

In general, suppose that $m = 2^t k - 1$, with k odd. Following the same pattern, we can write m as

$$m = \frac{2^t(2^t-1)k - (2^t-1)}{2^{t-1}(2^t-1)k - (2^{t-1}-1)} \cdots \frac{4(2^t-1)k - 3}{2(2^t-1)k - 1} \cdot \frac{2(2^t-1)k - 1}{(2^t-1)k} \cdot k.$$

The induction is finished. Hence m can be represented as $\frac{\tau(n^2)}{\tau(n)}$ if and only if it is odd.

20. We first consider the special case $n = 3^r$. Then the simplest choice $\frac{10^n - 1}{9} = 11\ldots1$ (n digits) works. This can be shown by induction: it is true for $r = 1$, while the inductive step follows from

$$10^{3^r} - 1 = \left(10^{3^{r-1}} - 1\right)\left(10^{2 \cdot 3^{r-1}} + 10^{3^{r-1}} + 1\right),$$

because the second factor is divisible by 3.

In the general case, let $k \geq n/2$ be a positive integer and $a_1, \ldots, a_{n-k}$ be nonzero digits. We have

$$A = (10^k - 1)\overline{a_1 a_2 \ldots a_{n-k}}$$
$$= a_1 a_2 \ldots a_{n-k-1} a'_{n-k} \underbrace{99\ldots99}_{2k-n} b_1 b_2 \ldots b_{n-k-1} b'_{n-k},$$

where $a'_{n-k} = a_{n-k} - 1$, $b_i = 9 - a_i$, and $b'_{n-k} = 9 - a'_{n-k}$. The sum of digits of A equals $9k$ independently of the choice of digits $a_1, \ldots, a_{n-k}$. Thus we need only choose $k \geq \frac{n}{2}$ and digits $a_1, \ldots, a_{n-k-1} \notin \{0,9\}$ and $a_{n-k} \in \{0,1\}$ in order for the conditions to be fulfilled. Let us choose

$$k = \begin{cases} 3^r, & \text{if } 3^r < n \leq 2 \cdot 3^r \text{ for some } r \in \mathbb{Z}, \\ 2 \cdot 3^r, & \text{if } 2 \cdot 3^r < n \leq 3^{r+1} \text{ for some } r \in \mathbb{Z}; \end{cases}$$

and $\overline{a_1 a_2 \ldots a_{n-k}} = \overline{22\ldots2}$. The number

$$A = \overline{\underbrace{22\ldots2}_{n-k-1} 1 \underbrace{99\ldots99}_{2k-n} \underbrace{77\ldots78}_{n-k-1}}$$

thus obtained is divisible by $2 \cdot (10^k - 1)$, which is, as explained above, divisible by $18 \cdot 3^r$. Finally, the sum of digits of A is either $9 \cdot 3^r$ or $18 \cdot 3^r$; thus A has the desired properties.

21. Such a sequence is obviously strictly increasing. We note that it must be unique. Indeed, given $a_0, a_1, \ldots, a_{n-1}$, then a_n is the least positive integer not of the form $a_i + 2a_j + 4a_k$, $i, j, k < n$.

We easily get that the first few a_n's are $0, 1, 8, 9, 64, 65, 72, 73, \ldots$. Let $\{c_n\}$ be the increasing sequence of all positive integers that consist of zeros and ones in base 8, i.e., those of the form $t_0 + 2^3 t_1 + \cdots + 2^{3q} t_q$ where $t_i \in \{0, 1\}$. We claim that $a_n = c_n$. To prove this, it is enough to show that each $m \in \mathbb{N}$ can be uniquely written as $c_i + 2c_j + 4c_k$. If $m = t_0 + 2t_1 + \cdots + 2^r t_r$ ($t_i \in \{0, 1\}$), then $m = c_i + 2c_j + 2^2 c_k$ is obviously possible if and only if $c_i = t_0 + 2^3 t_3 + 2^6 t_6 + \cdots$, $c_j = t_1 + 2^3 t_4 + \ldots$, and $c_k = t_2 + 2^3 t_5 + \cdots$.

Hence for $n = s_0 + 2s_1 + \cdots + 2^r s_r$ we have $a_n = s_0 + 8s_1 + \cdots + 8^r s_r$. In particular, $1998 = 2 + 2^2 + 2^3 + 2^6 + 2^7 + 2^8 + 2^9 + 2^{10}$, so

$$a_{1998} = 8 + 8^2 + 8^3 + 8^6 + 8^7 + 8^8 + 8^9 + 8^{10} = 1227096648.$$

Second solution. Define $f(x) = x^{a_0} + x^{a_1} + \cdots$. Then the assumed property of $\{a_n\}$ gives

$$f(x)f(x^2)f(x^4) = \sum_{i,j,k} x^{a_i + 2a_j + 4a_k} = \sum_n x^n = \frac{1}{1-x}.$$

We also get as a consequence $f(x^2)f(x^4)f(x^8) = \frac{1}{1-x^2}$, which gives $f(x) = (1 + x)f(x^8)$. Continuing this, we obtain

$$f(x) = (1 + x)(1 + x^8)(1 + x^{8^2}) \cdots.$$

Hence the a_n's are integers that have only 0's and 1's in base 8.

22. We can obviously change each x into $\lfloor x \rfloor$ or $\lceil x \rceil$ so that the column sums remain unchanged. However, this does not necessarily match the row sums as well, so let us consider the sum S of the absolute values of the changes in the row sums. It is easily seen that S is even, and we want it to be 0.

A row may have a higher or lower sum than desired. Let us mark a cell by $-$ if its entry x was changed to $\lfloor x \rfloor$, and by $+$ if it was changed to $\lceil x \rceil$ instead. We call a row R_2 *accessible* from a row R_1 if there is a column C such that $C \cap R_1$ is marked $+$ and $C \cap R_2$ is marked $-$. Note that a column containing a $+$ must contain a $-$ as well, because column sums are unchanged. Hence from each row with a higher sum we can access another row.

Assume that the row sum in R_1 is higher. If $R_1, R_2, \ldots, R_k$ is a sequence of rows such that R_{i+1} is accessible from R_i via some column C_i and such that the row sum in R_k is lower, then by changing the signs in $C_i \cap R_i$ and $C_i \cap R_{i+1}$ ($i = 1, 2, \ldots, k-1$) we decrease S by 2, leaving column sums unchanged. We claim that such a sequence of rows always exists.

Let $\mathscr{R}$ be the union of all rows that are accessible from R_1, directly or indirectly; let $\overline{\mathscr{R}}$ be the union of the remaining rows. We show that for any column C, the sum in $\mathscr{R} \cap C$ is not higher. If $\mathscr{R} \cap C$ contains no $+$'s, then this is clear. If $\mathscr{R} \cap C$

contains a +, since the rows of $\overline{\mathscr{R}}$ are not accessible, the set $\overline{\mathscr{R}} \cap C$ contains no $-$'s. It follows that the sum in $\overline{\mathscr{R}} \cap C$ is not lower, and since column sums are unchanged, we again come to the same conclusion. Thus the total sum in $\mathscr{R}$ is not higher. Therefore, there is a row in $\mathscr{R}$ with too low a sum, justifying our claim.

23. (a) If n is even, then every odd integer is unattainable. Assume that $n \geq 9$ is odd. Let a be obtained by addition from some b, and b from c by multiplication. Then a is $2c + 2$, $2c + n$, $nc + 2$, or $nc + n$, and is in every case congruent to $2c + 2$ modulo $n - 2$. In particular, if $a \equiv -2 \pmod{n-2}$, then also $b \equiv -4$ and $c \equiv -2 \pmod{n-2}$.

Now consider any $a = kn(n-2) - 2$, where k is odd. If it is attainable, but not divisible by 2 or n, it must have been obtained by addition. Thus all predecessors of a are congruent to either -2 or $-4 \pmod{n-2}$, and none of them equals 1, a contradiction.

  (b) Call an attainable number *addy* if the last operation is addition, and *multy* if the last operation is multiplication. We prove the following claims by simultaneous induction on k:

(1) $n = 6k$ is both addy and multy;
(2) $n = 6k + 1$ is addy for $k \geq 2$;
(3) $n = 6k + 2$ is addy for $k \geq 1$;
(4) $n = 6k + 3$ is addy;
(5) $n = 6k + 4$ is multy for $k \geq 1$;
(6) $n = 6k + 5$ is addy.

The cases $k \leq 1$ are easily verified. For $k \geq 2$, suppose all six statements hold up to $k - 1$.

Since $3k$ is addy, $6k$ is multy.

Next, $6k - 2$ is multy, so both $6k = (6k - 2) + 2$ and $6k + 1 = (6k - 2) + 3$ are addy.

Since $6k$ is multy, both $6k + 2$ and $6k + 3$ are addy.

Number $6k + 4 = 2 \cdot (3k + 2)$ is multy, because $3k + 2$ is addy (being either $6l + 2$ or $6l + 5$).

Finally, we have $6k + 5 = 3 \cdot (2k + 1) + 2$. Since $2k + 1$ is $6l + 1$, $6l + 3$, or $6l + 5$, it is addy except for 7. Hence $6k + 5$ is addy except possibly for 23. But $23 = ((1 \cdot 2 + 2) \cdot 2 + 2) \cdot 2 + 3$ is also addy.

This completes the induction. Now 1 is given and $2 = 1 \cdot 2$, $4 = 1 + 3$. It is easily checked that 7 is not attainable, and hence it is the only unattainable number.

24. Let $f(n)$ be the minimum number of moves needed to monotonize *any* permutation of n distinct numbers. Let us be given a permutation π of $\{1, 2, \ldots, n\}$, and let k be the first element of π. In $f(n-1)$ moves, we can transform π to either

$$(k, 1, 2, \ldots, k - 1, k + 1, \ldots, n) \text{ or } (k, n, n - 1, \ldots, k + 1, k - 1, \ldots, 1).$$

Now the former can be changed to $(k, k - 1, \ldots, 2, 1, k + 1, \ldots, n)$, which is then monotonized in the next move. Similarly, the latter also can be monotonized in two moves. It follows that $f(n) \leq f(n-1)+2$. Thus we shall be done if we show that $f(5) \leq 4$.

First we note that $f(3) = 1$. Consider a permutation of $\{1,2,3,4\}$. If either 1 or 4 is the first or the last element, we need one move to monotonize the other three elements, and at most one more to monotonize the whole permutation. Of the remaining four permutations, $(2,1,4,3)$ and $(3,4,1,2)$ can also be monotonized in two moves. The permutations $(2,4,1,3)$ and $(3,1,4,2)$ require 3 moves, but by this we can choose whether to change them into $(1,2,3,4)$ or $(4,3,2,1)$.

We now consider a permutation of $\{1,2,3,4,5\}$. If either 1 or 5 is in the first or last position, we can monotonize the rest in 3 moves, but in such a way that the whole permutation can be monotonized in the next move. If this is not the case, then either 1 or 5 is in the second or fourth position. Then we simply switch it to the outside in one move and continue as in the former case. Hence $f(5) = 4$, as desired.

25. We use induction on n. For $n = 3$, we have a single two-element subset $\{i,j\}$ that is split by (i,k,j) (where k is the third element of U). Assume that the result holds for some $n \geq 3$, and consider a family $\mathscr{F}$ of $n-1$ proper subsets of $U = \{1,2,\ldots,n+1\}$, each with at least 2 elements.

To continue the induction, we need an element $a \in U$ that is contained in all n-element subsets of $\mathscr{F}$, but in at most one of the two-element subsets. We claim that such an a exists. Let $\mathscr{F}$ contain k n-element subsets and m 2-element subsets $(k+m \leq n-1)$. The intersection of the n-element subsets contains exactly $n+1-k \geq m+2$ elements. On the other hand, at most m elements belong to more than one 2-element subset, which justifies our claim.

Now let A be the 2-element subset that contains a, if it exists; otherwise, let A be any subset from $\mathscr{F}$ containing a. Excluding a from all the subsets from $\mathscr{F} \setminus \{A\}$, we get at most $n-2$ subsets of $U \setminus \{a\}$ with at least 2 and at most $n-1$ elements. By the inductive hypothesis, we can arrange $U \setminus \{a\}$ so that we split all the subsets of $\mathscr{F}$ except A. It remains to place a, and we shall make a desired arrangement if we put it anywhere away from A.

26. Put $n = 2r+1$. Since each of the $\binom{n}{2}$ pairs of judges agrees on at most two candidates, the total number of agreements is at most $k\binom{n}{2}$. On the other hand, if the ith candidate is passed by x_i judges and failed by $n-x_i$ judges, then the number of agreements on this candidate equals

$$\binom{x_i}{2} + \binom{n-x_i}{2} = \frac{x_i^2 + (n-x_i)^2 - n}{2} \geq \frac{r^2 + (n-r)^2 - n}{2} = \frac{(n-1)^2}{4}.$$

Therefore the total number of agreements is at least $\frac{m(n-1)^2}{4}$, which implies that

$$k\binom{n}{2} \geq \frac{m(n-1)^2}{4}, \qquad \text{hence} \qquad \frac{k}{m} \geq \frac{n-1}{2n}.$$

Remark. The obtained inequality is sharp. Indeed, if $m = \binom{2r+1}{r}$ and each candidate is passed by a different subset of r judges, we get equality. A similar example shows that the result is not valid for even n. In that case the weaker estimate $\frac{k}{m} \geq \frac{n-2}{2n-2}$ holds.

27. Since this is essentially a graph problem, we call the points and segments vertices and edges of the graph. We first prove that the task is impossible if $k \leq 4$.

Cases $k \leq 2$ are trivial. If $k = 3$, then among the edges from a vertex A there are two of the same color, say AB and AC, so we don't have all the three colors among the edges joining A, B, C.

Now let $k = 4$, and assume that there is a desired coloring. Consider the edges incident with a vertex A. At least three of them have the same color, say blue. Suppose that four of them, AB, AC, AD, AE, are blue. There is a blue edge, say BC, among the ones joining B, C, D, E. Then four of the edges joining A, B, C, D are blue, and we cannot complete the coloring. So, exactly three edges from A are blue: AB, AC, AD. Also, of the edges connecting any three of the 6 vertices other than A, B, C, D, one is blue (because the edges joining them with A are not so). By a classical result, there is a blue triangle EFG with vertices among these six. Now one of EB, EC, ED must be blue as well, because none of BC, BD, CD is. Let it be EB. Then four of the edges joining B, E, F, G are blue, which is impossible.

For $k = 5$ the task is possible. Label the vertices $0, 1, \ldots, 9$. For each color, we divide the vertices into four groups and paint in this color every edge joining two from the same group, as shown below. Then among any 5 vertices, 2 must belong to the same group, and the edge connecting them has the considered color.

yellow:	01 12 20	36 69 93	57	48
red:	23 34 42	58 81 15	79	60
blue:	45 56 64	70 03 37	91	82
green:	67 78 86	92 25 59	13	04
orange:	89 90 08	14 47 71	35	26.

A desired coloring can be made for $k \geq 6$ as well. Paint the edge ij in the $(i+j)$th color for $i < j \leq 8$, and in the $2i$th color if $j = 9$ (the addition being modulo 9). We can ignore the edges painted with the extra colors. Then the edges of one color appear as five disjoint segments, so that any complete k-graph for $k \geq 5$ contains one of them.

28. Let A be the number of markers with white side up, and B the number of pairs of markers whose squares share a side.

We claim that $A + B$ does not change its parity as the game progresses. Suppose that in some move we remove a marker that has exactly k neighbors, among them r with white side up ($0 \leq r \leq k \leq 4$). Of course, this marker has its black side up. When it is removed, the r white markers get black side up, while the $k - r$ black ones become white. Thus A changes by $k - 2r$. As for B, it decreases by k. It follows that A decreases by $2r$ and preserves its parity, as claimed.

Initially, $A = mn - 1$ and $B = m(n-1) + n(m-1)$; hence

$$A + B = 3mn - m - n - 1.$$

If we succeed in removing all the markers, we end up with $A + B = 0$. Hence $3mn - m - n - 1 = (m-1)(n-1) + 2(mn-1)$ must be even, or equivalently at least one of m and n is odd.

On the other hand, the game can be finished successfully if m or n is odd. Assume that m is odd. As shown in the picture, we can arrive at the position (1) in m moves; with $\frac{m+1}{2}$ moves we reduce it to the position $(1\frac{1}{2})$, and with the next $\frac{m-1}{2}$ moves to the position (2). We continue until we empty all the columns.

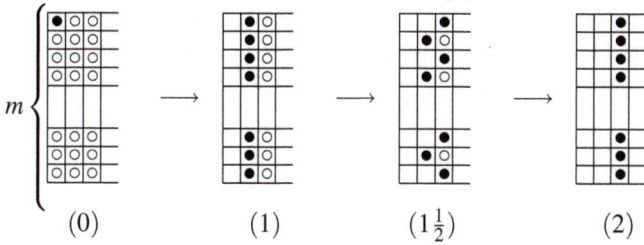

$$(0) \qquad (1) \qquad (1\tfrac{1}{2}) \qquad (2)$$

4.40 Solutions to the Shortlisted Problems of IMO 1999

1. Obviously $(1, p)$ (where p is an arbitrary prime) and $(2, 2)$ are solutions and the only solutions to the problem for $x < 3$ or $p < 3$.

   Let us now assume $x, p \geq 3$. Since p is odd, $(p-1)^x + 1$ is odd, and hence x is odd. Let q be the smallest prime divisor of x, which also must be odd. We have $q \mid x \mid x^{p-1} \mid (p-1)^x + 1 \Rightarrow (p-1)^x \equiv -1 \pmod{q}$. Also from Fermat's little theorem $(p-1)^{q-1} \equiv 1 \pmod{q}$. Since $q-1$ and x are coprime, there exist integers α, β such that $x\alpha = (q-1)\beta + 1$. We also note that α must be odd. We now have $p - 1 \equiv (p-1)^{(q-1)\beta+1} \equiv (p-1)^{x\alpha} \equiv -1 \pmod{q}$ and hence $q \mid p \Rightarrow q = p$. Since x is odd, $p \mid x$, and $x \leq 2p$, it follows $x = p$ for all $x, p \geq 3$. Thus

   $$p^{p-1} \mid (p-1)^x + 1 = p^2 \cdot \left(p^{p-2} - \binom{p}{1} p^{p-1} + \cdots - \binom{p}{p-2} + 1 \right).$$

   Since the expression in parenthesis is not divisible by p, it follows that $p^{p-1} \mid p^2$ and hence $p \leq 3$. One can easily verify that $(3, 3)$ is a valid solution.

   We have shown that the only solutions are $(1, p)$, $(2, 2)$, and $(3, 3)$, where p is an arbitrary prime.

2. We first prove that every rational number in the interval $(1, 2)$ can be represented in the form $\frac{a^3+b^3}{a^3+d^3}$. Taking b, d such that $b \neq d$ and $a = b + d$, we get $a^2 - ab + b^2 = a^2 - ad + d^2$ and

   $$\frac{a^3 + b^3}{a^3 + d^3} = \frac{(a+b)(a^2 - ab + b^2)}{(a+d)(a^2 - ad + d^2)} = \frac{a+b}{a+d}.$$

   For a given rational number $1 < m/n < 2$ we can select $a = m+n$ and $b = 2m - n$ such that along with $d = a - b$ we have $\frac{a+b}{a+d} = \frac{m}{n}$. This completes the proof of the first statement.

   For m/n outside of the interval we can easily select a rational number p/q such that $\sqrt[3]{\frac{n}{m}} < \frac{p}{q} < \sqrt[3]{\frac{2n}{m}}$. In other words $1 < \frac{p^3 m}{q^3 n} < 2$. We now proceed to obtain a, b and d for $\frac{p^3 m}{q^3 n}$ as before, and we finally have

   $$\frac{p^3 m}{q^3 n} = \frac{a^3 + b^3}{a^3 + d^3} \Rightarrow \frac{m}{n} = \frac{(aq)^3 + (bq)^3}{(ap)^3 + (dp)^3}.$$

   Thus we have shown that all positive rational numbers can be expressed in the form $\frac{a^3+b^3}{c^3+d^3}$.

3. We first prove the following lemma.

   Lemma. For $d, c \in \mathbb{N}$ and $d^2 \mid c^2 + 1$ there exists $b \in \mathbb{N}$ such that $d^2(d^2 + 1) \mid b^2 + 1$.

   Proof. It is enough to set $b = c + d^2 c - d^3 = c + d^2(c - d)$.

Using the lemma it suffices to find increasing sequences d_n and c_n such that $c_n - d_n$ is an increasing sequence and $d_n^2 \mid c_n^2 + 1$. We then obtain the desired sequences a_n and b_n from $a_n = d_n^2$ and $b_n = c_n + d_n^2(c_n - d_n)$. It is easy to check that $d_n = 2^{2n} + 1$ and $c_n = 2^{nd_n}$ satisfy the required conditions. Hence we have demonstrated the existence of increasing sequences a_n and b_n such that $a_n(a_n + 1) \mid b_n^2 + 1$.

Remark. There are many solutions to this problem. For example, it is sufficient to prove that the Pell-type equation $5a_n(a_n + 1) = b_n^2 + 1$ has an infinity of solutions in positive integers. Alternatively, one can show that $a_n(a_n + 1)$ can be represented as a sum of two coprime squares for infinitely many a_n, which implies the existence of b_n.

4. (a) The fundamental period of p is the smallest integer $d(p)$ such that $p \mid 10^{d(p)} - 1$.

     Let s be an arbitrary prime and set $N_s = 10^{2s} + 10^s + 1$. In that case $N_s \equiv 3 \pmod 9$. Let $p_s \neq 37$ be a prime dividing $N_s/3$. Clearly $p_s \neq 3$. We claim that such a prime exists and that $3 \mid d(p_s)$. The prime p_s exists, since otherwise N_s could be written in the form $N_s = 3 \cdot 37^k \equiv 3 \pmod 4$, while on the other hand for $s > 1$ we have $N_s \equiv 1 \pmod 4$.

     Now we prove $3 \mid d(p_s)$. We have $p_s \mid N_s \mid 10^{3s} - 1$ and hence $d(p_s) \mid 3s$. We cannot have $d(p_s) \mid s$, for otherwise $p_s \mid 10^s - 1 \Rightarrow p_s \mid (10^{2s} + 10^s + 1, 10^s - 1) = 3$; and we cannot have $d(p_s) \mid 3$, for otherwise $p_s \mid 10^3 - 1 = 999 = 3^3 \cdot 37$, both of which contradict $p_s \neq 3, 37$. It follows that $d(p_s) = 3s$. Hence for every prime s there exists a prime p_s such that $d(p_s) = 3s$. It follows that the cardinality of S is infinite.

     (b) Let $r = r(s)$ be the fundamental period of $p \in S$. Then $p \mid 10^{3r} - 1$, $p \nmid 10^r - 1 \Rightarrow p \mid 10^{2r} + 10^r + 1$. Let $x_j = \frac{10^{j-1}}{p}$ and $y_j = \{x_j\} = 0.a_j a_{j+1} a_{j+2} \ldots$. Then $a_j < 10 y_j$, and hence
     $$f(k, p) = a_k + a_{k+r} + a_{k+2r} < 10(y_k + y_{k+r} + y_{k+2r}).$$

     We note that $x_k + x_{k+s(p)} + x_{k+2s(p)} = \frac{10^{k-1} N_p}{p}$ is an integer, from which it follows that $y_k + y_{k+s(p)} + y_{k+2s(p)} \in \mathbb{N}$. Hence $y_k + y_{k+s(p)} + y_{k+2s(p)} \leq 2$. It follows that $f(k, p) < 20$. We note that $f(2, 7) = 4 + 8 + 7 = 19$. Hence 19 is the greatest possible value of $f(k, p)$.

5. Since one can arbitrarily add zeros at the end of m, which increases divisibility by 2 and 5 to an arbitrary exponent, it suffices to assume $2, 5 \nmid n$. If $(n, 10) = 1$, there exists an integer $w \geq 2$ such that $10^w \equiv 1 \pmod n$. We also note that $10^{iw} \equiv 1 \pmod n$ and $10^{jw+1} \equiv 10 \pmod n$ for all integers i and j. Let us assume that m is of the form $m = \sum_{i=1}^u 10^{iw} + \sum_{j=1}^v 10^{jw+1}$ for integers $u, v \geq 0$ (where if u or v is 0, the corresponding sum is 0). Obviously, the sum of the digits of m is equal to $u + v$, and also $m \equiv u + 10v \pmod n$. Hence our problem reduces to finding integers $u, v \geq 0$ such that $u + v = k$ and $n \mid u + 10v = k + 9v$. Since $(n, 9) = 1$, it follows that there exists some v_0 such that $0 \leq v_0 < n \leq k$

and $9v_0 \equiv -k \pmod{n} \Rightarrow n \mid k + 9v_0$. Taking this v_0 and setting $u_0 = k - v_0$ we obtain the desired parameters for defining m.

6. Let N be the smallest integer greater than M. We take the difference of the numbers in the progression to be of the form $10^m + 1$, $m \in \mathbb{N}$. Hence we can take $a_n = a_0 + n(10^m + 1) = \overline{b_s b_{s-1} \ldots b_0}$ where a_0 is the initial term in the progression and $\overline{b_s b_{s-1} \ldots b_0}$ is the decimal representation of a_n. Since $2m$ is the smallest integer x such that $10^x \equiv 1 \pmod{10^m + 1}$, it follows that $10^k \equiv 10^l \pmod{10^m + 1} \Leftrightarrow k \equiv l \pmod{2m}$. Hence

$$a_0 \equiv a_n = \overline{b_s b_{s-1} \ldots b_0} \equiv \sum_{i=0}^{2m-1} c_i 10^i \pmod{10^m + 1},$$

where $c_i = b_i + b_{2m+i} + b_{4m+i} + \cdots \geq 0$ for $i = 0, 1, \ldots, 2m - 1$ (these c_i also depend on n). We note that $\sum_{i=0}^{2m-1} c_i 10^i$ is invariant modulo $10^m + 1$ for all n and that $\sum_{i=0}^{2m-1} c_i = \sum_{j=0}^{s} b_j$ for a given n. Hence we must choose a_0 and m such that a_0 is not congruent to any number of the form $\sum_{i=0}^{2m-1} c_i 10^i$, where $c_0 + c_1 + \cdots + c_{2m-1} \leq N$ $(c_0, c_1, \ldots, c_{2m-1} \geq 0)$.

The number of ways to select the nonnegative integers $c_0, c_1, \ldots, c_{2m-1}$ such that $c_0 + c_1 + \cdots + c_{2m-1} \leq N$ is equal to the number of strictly increasing sequences $0 \leq c_0 < c_0 + c_1 + 1 < c_0 + c_1 + c_2 + 2 + \cdots < c_0 + c_1 + \cdots + c_{2m-1} + 2m - 1 \leq N + 2m - 1$, which is equal to the number of $2m$-element subsets of $\{0, 1, 2, \ldots, N + 2m - 1\}$, which is $\binom{N+2m}{N}$. For sufficiently large m we have $\binom{N+2m}{N} < 10^m$, and hence in this case one can select a_0 such that a_0 is not congruent to $\sum_{i=0}^{2m-1} c_i 10^i$ modulo $10^m + 1$ for any set of integers $c_0, c_1, \ldots, c_{2m-1}$ such that $c_0 + c_1 + \cdots + c_{2m-1} \leq N$. Thus we have found the desired arithmetic progression.

7. We use the following simple lemma.

   Lemma. Suppose that M is the interior point of a convex quadrilateral $ABCD$. Then it follows that $MA + MB < AD + DC + CB$.

   Proof. We repeatedly make use of the triangle inequality. The line AM, in addition to A, intersects the quadrilateral in a second point N. In that case $AM + MB < AN + NB < AD + DC + CB$.

   We now apply this lemma in the following way. Let D, E, and F be median points of BC, AC, and AB. Any point M in the interior of $\triangle ABC$ is contained in at least two of the three convex quadrilaterals $ABDE$, $BCEF$, and $CAFD$. Let us assume without loss of generality that M is in the interior of $BCEF$ and $CAFD$. In that case we apply the lemma to obtain $AM + CM < AF + FD + DC$ and $BM + CM < CE + EF + FB$ to obtain

$$CM + AM + BM + CM < AF + FD + DC + CE + EF + FB$$
$$= AB + AC + BC$$

from which the required conclusion immediately follows.

8. Let A, B, C, and D be inverses of four of the five points, with the fifth point being the pole of the inversion. A separator through the pole transforms into a line containing two of the remaining four points such that the remaining two points are on opposite sides of the line. A separator not containing the pole transforms into a circle through three of the points with the fourth point in its interior. Let K be the convex hull of A, B, C, and D. We observe two cases:

   (i) K is a quadrilateral, for example $ABCD$. In that case the four separators are the two diagonals and two circles ABC and ADC if $\angle A + \angle C < 180°$, or BAD and BCD otherwise. The remaining six viable circles and lines are clearly not separators.

   (ii) K is a triangle, for example ABC with D in its interior. In that case the separators are lines DA, DB, DC and the circle ABC. No other lines and circles qualify.

We have thus shown that any set of five points satisfying the stated conditions will have exactly four separators.

9. Let r_{PQ} denote a reflection about the planar bisector of PQ with $P, Q \in S$. Let G be the centroid of S. From $r_{PQ}(S) = S$ it follows that $r_{PQ}(G) = G$. Hence G belongs to the perpendicular bisector of PQ and thus $GP = GQ$. Consequently the whole of S lies on a sphere Σ centered at G. We note the following two cases:

   (a) S is a subset of a plane π. In this case S is included in a circle k, G being its center. Hence its n points form a convex polygon $A_1 A_2 \ldots A_n$. When applying $r_{A_i A_{i+2}}$ for some $0 < i < n - 1$ the point A_{i+1} transforms into some point of S lying on the same side of $A_i A_{i+1}$, which has to be A_{i+1} itself. It thus follows that $A_i A_{i+1} = A_{i+1} A_{i+2}$ for all $0 < i < n - 1$ and hence $A_1 A_2 \ldots A_n$ is a regular n-gon.

   (b) The points in S are not coplanar. It follows that S is a polyhedron P inscribed in a sphere Σ centered at G. By applying the previous case to the faces of the polyhedron, it follows that all faces are regular n-gons.

   Let us take an arbitrary vertex V and let VV_1, VV_2 and VV_3 be three consecutive edges stemming from V (V, V_1, V_2, and V_3 defining two adjacent faces of P). We now look at $r_{V_1 V_3}$. Since this transformation leaves the half-planes $[V_1 V_3, V_2$ and $[V_1 V_3, V$ invariant and since V_2 and V are the only points of P on the respective half-planes, it follows that $r_{V_1 V_3}$ leaves V and V_2 invariant. This transform also swaps V_1 and V_3. Hence, the face determined by $VV_1 V_2$ is transformed by $r_{V_1 V_3}$ into the face $VV_3 V_2$, and thus the two faces sharing VV_2 are congruent. We conclude that all faces are congruent and similarly that vertices are endpoints of the same number of edges; hence P is a regular polyhedron.

   Finally, we have to rule out S being vertices of a cube, a dodecahedron, or an icosahedron. In all of these cases if we select two diametrically opposite points P and Q, then $S \setminus \{P, Q\}$ is not symmetric with respect to the bisector of PQ, which prevents r_{PQ} from being an invariant transformation of S.

It thus follows that the only viable finite completely symmetric sets are vertices of regular n-gons, the tetrahedron, and the octahedron. It is not explicitly asked for, but it is easy to verify that all of these are indeed completely symmetric.

Remark. On the IMO, a simpler version of this problem was adopted, adding the condition that S belongs to a plane and thus eliminating the need for the second case altogether.

10. We use the following lemma.

> *Lemma.* Let ABC be a triangle and $X \in AB$ such that $\overrightarrow{AX} : \overrightarrow{XB} = m : n$. Then $(m+n)\cot\angle CXB = n\cot A - m\cot B$ and $m\cot\angle ACX = (n+m)\cot C + n\cot A$.
>
> *Proof.* Let CD be the altitude from C and h its length. Then using oriented segments we have $AX = AD + DX = h\cot A - h\cot\angle CXB$ and $BX = BD + DX = h\cot B + h\cot\angle CXB$. The first formula in the lemma now follows from $n \cdot AX = m \cdot BX$. The second formula immediately follows from the first part applied to the triangle ACX and the point $X' \in AC$ such that $XX' \parallel BC$.

Let us set $\cot A = x$, $\cot B = y$, and $\cot C = z$. Applying the second formula in the lemma to $\triangle ABC$ and the point X, we obtain $4\cot\angle ACX = 9z + 5x$. Applying the first formula in the lemma to $\triangle CXZ$ and the point Y and using $\angle XYZ = 45°$ and $\cot\angle CXZ = -y$, we obtain $3\cot\angle XYZ = \cot\angle ACX - 2\cot\angle CXZ = \frac{9z+5x}{4} + 2y \Rightarrow 5x + 8y + 9z = 12$.

We now use the well-known relation for cotangents of a triangle $xy + yz + xz = 1$ to get $9 = 9(x+y)z + 9xy = (x+y)(12 - 5x - 8z) + 9xy = 9 \Rightarrow (4y + x - 3)^2 + 9(x-1)^2 = 0 \Rightarrow x = 1$, $y = \frac{1}{2}$, $z = \frac{1}{3}$. It follows that x, y, and z have fixed values, and hence all triangles T in Σ are similar, with their smallest angle A having cotangent 1 and thus being equal to $\angle A = 45°$.

11. Let $\Omega(I, r)$ be the incircle of $\triangle ABC$. Let D, E, and F denote the points where Ω touches BC, AC, and AB, respectively. Let P, Q, and R denote the midpoints of EF, DF, and DE respectively. We prove that Ω_a passes through Q and R. Since $\triangle IQD \sim \triangle IDB$ and $\triangle IRD \sim \triangle IDC$, we obtain $IQ \cdot IB = IR \cdot IC = r^2$. We conclude that B, C, Q, and R lie on a single circle Γ_a. Moreover, since the power of I with respect to Γ_a is r^2, it follows for a tangent IX from I to Γ_a that X lies on Ω and hence Ω is perpendicular to Γ_a. From the uniqueness of Ω_a it follows that $\Omega_a = \Gamma_a$. Thus Ω_a contains Q and R. Similarly Ω_b contains P and R and Ω_c contains P and Q. Hence, $A' = P$, $B' = Q$ and $C' = R$. Therefore the radius of the circumcircle of $\triangle A'B'C'$ is half the radius of Ω.

12. We first introduce the following lemmas.

> *Lemma 1.* Let ABC be a triangle, I its inenter and I_a the center of the excircle touching BC. Let A' be the center of the arc $\overset{\frown}{BC}$ of the circumcircle not containing A. Then $A'B = A'C = A'I = A'I_a$.
>
> *Proof.* The result follows from a straightforward calculation of the relevant angles.

Lemma 2. Let two circles k_1 and k_2 meet each other at points X and Y and touch a circle k internally in points M and N, respectively. Let A be one of the intersections of the line XY with k. Let AM and AN intersect k_1 and k_2 respectively at C and E. Then CE is a common tangent of k_1 and k_2.

Proof. Since $AC \cdot AM = AX \cdot AY = AE \cdot AN$, the points M, N, E, C lie on a circle. Let MN meet k_1 again at Z. If M' is any point on the common tangent at M, then $\angle MCZ = \angle M'MZ = \angle M'MN = \angle MAN$ (as oriented angles), implying that $CZ \parallel AN$. It follows that $\angle ACE = \angle ANM = \angle CZM$. Hence CE is tangent to k_1 and analogously to k_2.

In the main problem, let us define E and F respectively as intersections of NA and NB with Ω_2. Then applying Lemma 2 we get that CE and DF are the common tangents of Ω_1 and Ω_2.

If the circles have the same radii, the result trivially holds. Otherwise, let G be the intersection of CE and DF. Let O_1 and O_2 be the centers of Ω_1 and Ω_2. Since $O_1 D = O_1 C$ and $\angle O_1 DG = \angle O_1 CG = 90°$, it follows that O_1 is the midpoint of the shorter arc of the circumcircle of $\triangle CDG$. The center O_2 is located on the bisector of $\angle CGD$, since Ω_2 touches both GC and GD. However, t also sits on Ω_1, and using Lemma 1 we obtain that O_2 is either at the incenter or at the excenter of $\triangle CDG$ opposite G. Hence, Ω_2 is either the incircle or the excircle of CDG and thus in both cases touches CD.

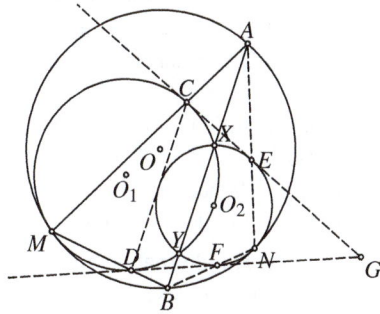

Second solution. Let O be the center of Γ, and r, r_1, r_2 the radii of $\Gamma, \Gamma_1, \Gamma_2$. It suffices to show that the distance $d(O_2, CD)$ is equal to r_2.

The homothety with center M and ratio r/r_1 takes Γ_1, C, D into Γ, A, B, respectively; hence $CD \parallel AB$ and $d(C, AB) = \frac{r-r_1}{r} d(M, AB)$. Let $O_1 O_2$ meet XY at R. Then $d(O_2, CD) = O_2 R + \frac{r-r_1}{r} d(M, AB)$, i.e.,

$$d(O_2, CD) = O_2 R + \frac{r-r_1}{r}[O_1 O_2 - O_2 R + r_1 \cos \angle OO_1 O_2], \qquad (1)$$

since O, O_1, and M are collinear. We have $O_1 X = O_1 O_2 = r_1$, $OO_1 = r - r_1$, $OO_2 = r - r_2$, and $O_2 X = r_2$. Using the cosine law in the triangles $OO_1 O_2$ and $XO_1 O_2$, we obtain that $\cos \angle OO_1 O_2 = \frac{2r_1^2 - 2rr_1 + 2rr_2 - r_2^2}{2r_1(r-r_1)}$ and $O_2 R = \frac{r_2^2}{2r_1}$. Substituting these values in (1) we get $d(O_2, CD) = r_2$.

13. Let us construct a convex quadrilateral $PQRS$ and an interior point T such that $\triangle PTQ \cong \triangle AMB$, $\triangle QTR \sim \triangle AMD$, and $\triangle PTS \sim \triangle CMD$. We then have $TS = \frac{MD \cdot PT}{MC} = MD$ and $\frac{TR}{TS} = \frac{TR \cdot TQ \cdot TP}{TQ \cdot TP \cdot TS} = \frac{MD \cdot MB \cdot MC}{MA \cdot MA \cdot MD} = \frac{MB}{MC}$ (using $MA = MC$). We also have $\angle STR = \angle BMC$ and therefore $\triangle RTS \sim \triangle BMC$. Now the relations between angles become

$$\angle TPS + \angle TQR = \angle PTQ \quad \text{and} \quad \angle TPQ + \angle TSR = \angle PTS,$$

implying that $PQ \parallel RS$ and $QR \parallel PS$. Hence $PQRS$ is a parallelogram and hence $AB = PQ = RS$ and $QR = PS$. It follows that $\frac{BC}{MC} = \frac{RS}{TS} = \frac{AB}{MD} \Rightarrow AB \cdot CM = BC \cdot MD$ and $\frac{AD \cdot BM}{AM} = \frac{AD \cdot QT}{AM} = QR = PS = \frac{CD \cdot TS}{MD} = CD \Rightarrow BM \cdot AD = MA \cdot CD$.

14. We first introduce the same lemma as in problem 12 and state it here without proof.

Lemma. Let ABC be a triangle and I the center of its incircle. Let M be the center of the arc $\widehat{BC}$ of the circumcircle not containing A. Then $MB = MC = MI$. Let the circle XO_1O_2 intersect the circle Ω again at point T. Let M and N be respectively the midpoints of arcs $\widehat{BC}$ and $\widehat{AC}$, and let P be the intersection of Ω and the line through C parallel to MN. Then the lemma gives $MP = NC = NI = NO_1$ and $NP = MC = MI = MO_2$. Since O_1 and O_2 lie on XN and XM respectively, we have $\angle NTM = \angle NXM = \angle O_1XO_2 = \angle O_1TO_2$ and hence $\angle NTO_1 = \angle MTO_2$. Moreover, $\angle TNO_1 = \angle TNX = \angle TMO_2$, from which it follows that $\triangle O_1NT \sim \triangle O_2MT$. Thus $\frac{NT}{MP} = \frac{NT}{NO_1} = \frac{MT}{MO_2} = \frac{MT}{NP} \Rightarrow MP \cdot MT = NP \cdot NT \Rightarrow S_{MPT} = S_{NPT}$. It follows that TP bisects the segment MN, and hence it passes through I. We conclude that T belongs to the line PI and does not depend on X.

Remark. An alternative approach is to apply an inversion at point C. Points O_1 and O_2 become excenters of $\triangle AXC$ and $\triangle BXC$, and T becomes the projection of I_c onto AB.

15. For all $x_i = 0$ any C will do, so we may assume the contrary. Since the equation is symmetric and homogeneous, we may assume $\Sigma_i x_i = 1$. The equation now becomes $F(x_1, x_2, \ldots, x_n) = \Sigma_{i<j} x_i x_j (x_i^2 + x_j^2) = \Sigma_i x_i^2 \Sigma_{j \neq i} x_j = \Sigma_i x_i^3 (1 - x_i) = \Sigma_i f(x_i) \leq C$, where we define $f(x) = x^3 - x^4$. We note that for $x, y \geq 0$ and $x + y \leq 2/3$,

$$f(x+y) + f(0) - f(x) - f(y) = 3xy(x+y)\left(\frac{2}{3} - x - y\right) \geq 0. \tag{1}$$

We note that if at least three elements of $\{x_1, x_2, \ldots, x_n\}$ are nonzero the condition of (1) always holds for the two smallest ones. Hence, applying (1) repeatedly, we obtain $F(x_1, x_2, \ldots, x_n) \leq F(a, 1 - a, 0, \ldots, 0) = \frac{1}{2}(2a(1 - a))(1 - 2a(1 - a)) \leq \frac{1}{8} = F\left(\frac{1}{2}, \frac{1}{2}, 0, \ldots, 0\right)$. Thus we have $C = \frac{1}{8}$ (for all n), and equality holds only when two x_i are equal and the remaining ones are 0.

Second solution. Let $M = x_1^2 + x_2^2 + \cdots + x_n^2$. Using $ab \leq (a + 2b)^2/8$ we have

$$\sum_{1 \leq i < j \leq n} x_i x_j (x_i^2 + x_j^2) \leq M \sum_{i<j} x_i x_j$$

$$\leq \frac{1}{8}\left(M + 2\sum_{i<j} x_i x_j\right)^2 = \frac{1}{8}\left(\sum_{i=1}^{n} x_i\right)^4.$$

Equality holds if and only if $M = 2\sum_{i<j} x_i x_j$ and $x_i x_j(x_i^2 + x_j^2) = M x_i x_j$ for all $i < j$, which holds if and only if $n - 2$ of the x_i are zero and the remaining two are equal.

Remark. Problems (SL90-26) and (SL91-27) are very similar.

16. Let $C(A)$ denote the characteristic of an arrangement A. We shall prove that max $C(A) = \frac{n+1}{n}$.

Let us prove first $C(A) \leq \frac{n+1}{n}$ for all A. Among elements $\{n^2 - n, n^2 - n + 1, \dots, n^2\}$, by the pigeonhole principle, in at least one row and at least one column there exist two elements, and hence one pair in the same row or column that is not $(n^2 - n, n^2)$. Hence

$$C(A) \leq \max\left\{\frac{n^2}{n^2 - n + 1}, \frac{n^2 - 1}{n^2 - n}\right\} = \frac{n^2 - 1}{n^2 - n} = \frac{n+1}{n}.$$

We now consider the following arrangement:

$$a_{ij} = \begin{cases} i + n(j - i - 1) & \text{if } i < j, \\ i + n(n - i + j - 1) & \text{if } i \geq j. \end{cases}$$

We claim that $C(a) = \frac{n+1}{n}$. Indeed, in this arrangement no two numbers in the same row or column differ by less than $n - 1$, and in addition, n^2 and $n^2 - n + 1$ are in different rows and columns, and hence

$$C(A) \geq \frac{n^2 - 1}{n^2 - n} = \frac{n+1}{n}.$$

17. A game is determined by the ordering $t_1, \dots, t_N$ of the $N = \binom{n}{2}$ transpositions (i, j) of the set $\{1, 2, \dots, n\}$. The game is nice if the permutation $P = t_N t_{N-1} \dots t_1$ has no fixed point, and tiresome if P is the identity (denoted by I). Recall that every permutation can be written as a composition of disjoint cycles.

We claim that there exists a nice game if and only if $n \neq 3$.

For $n = 2$, $P_2 = t_1 = (1, 2)$ is obviously nice. For $n = 3$ each game has the form $P = (b, c)(a, c)(a, b) = (a, c)$ for an appropriate notation of the players, which cannot be nice. Now for $n \geq 4$ we define $P_n = (1, 2)\,(1, 3)\,(2, 3)\,\cdots\,(1, n)\,(2, n)\,\cdots\,(n - 1, n)$. We obtain inductively that $P_n = P_{n-1}(1, n, n - 1, \dots, 2) = (1, n)(2, n - 1)\cdots(i, n + 1 - i)\cdots$ is nice for all even n.

Also, if $n = 2k + 1$ is odd, then $Q_n = P_{n-1}(1, n)(2, n)\cdots(k, n)(n - 1, n)(n - 2, n)\cdots(k + 1, n)$ maps i to $n + 1 - i$ for $i \leq k$, to $n - 1 - i$ for $k + 1 \leq i \leq 2k - 1$, and to $3k + 1 - i$ if $i \in \{2k, 2k + 1\}$. Hence Q_n is nice. This justifies our claim.

Now we prove that a tiresome game exists if and only if $n \equiv 0, 1 \pmod{4}$. Evidently every transposition changes the sign of the permutation. Thus the sign of P is $(-1)^{\binom{n}{2}}$ and for P to be the identity we must have $2 \mid \binom{n}{2} \Rightarrow n \equiv 0, 1 \pmod{4}$.

Let us now construct tiresome games for the allowed n. For $n = 4k$ we divide the girls into groups of 4. In each group we perform the following game: $(3, 4)(1, 3)(2, 4)(2, 3)(1, 4)(1, 2) = I$. On the other hand, among two different groups (call them $\{1, 2, 3, 4\}$ and $\{5, 6, 7, 8\}$) we perform

$$(4,7)(3,7)(4,6)(1,6)(2,8)(3,8)(2,7)(2,6)$$
$$(4,5)(4,8)(1,7)(1,8)(3,5)(3,6)(2,5)(1,5) = I.$$

For $n = 4k+1$ we divide into groups of four as before, with one girl remaining. Every time a group (denoted $\{1,2,3,4\}$) is to play a game the remaining girl (denoted 5) joins in, and they play

$$(3,5)(3,4)(4,5)(1,3)(2,4)(2,3)(1,4)(1,5)(1,2)(2,5) = I.$$

This completes the proof.

18. Define $f(x,y) = x^2 - xy + y^2$. Let us assume that three such sets A, B, and C do exist and that w.l.o.g. 1, b, and c ($c > b$) are respectively their smallest elements.
    Lemma 1. Numbers x, y, and $x+y$ cannot belong to three different sets.
    Proof. The number $f(x,x+y) = f(y,x+y)$ must belong to both the set containing y and the set containing x, a contradiction.
    Lemma 2. The subset C contains a multiple of b. Moreover, if kb is the smallest such multiple, then $(k-1)b \in B$ and $(k-1)b+1, kb+1 \in A$.
    Proof. Let r be the residue of c modulo b. If $r = 0$, the first statement automatically holds. Let $0 < r < b$. In that case $r \in A$, and $c - r$ is then not in B according to Lemma 1. Hence $c - r \in A$ and since $b \mid c - r$, it follows that $b \mid f(c-r,b) \in C$, thus proving the first statement. It follows immediately from Lemma 1 that $(k-1)b \in B$. Now by Lemma 1, $(k-1)b+1 = kb - (b-1)$ must be in A; similarly, $kb+1 = [(k-1)b+1]+b \in A$ as well.
    Let us show by induction that $(nk-1)b+1, nkb+1 \in A$ for all integers n. The inductive basis has been shown in Lemma 2. Assuming that $[(n-1)k-1]b+1 \in A$ and $(n-1)kb+1 \in A$, we get that $(nk-1)b+1 = ((n-1)kb+1)+(k-1)b = [((n-1)k-1)b+1]+kb$ belongs to A and $nkb+1 = ((nk-1)b+1)+b = ((n-1)kb+1)+kb \Rightarrow nkb+1 \in A$. This finishes the inductive step. In particular, $f(kb,kb+1) = (kb+1)kb+1 \in A$. However, since $kb \in C$, $kb+1 \in A$, it follows that $f(kb,kb+1) \in B$, which is a contradiction.

19. Let $A = \{f(x) \mid x \in \mathbb{R}\}$ and $f(0) = c$. Plugging in $x = y = 0$ we get $f(-c) = f(c)+c-1$, hence $c \neq 0$. If $x \in A$, then taking $x = f(y)$ in the original functional equation we get $f(x) = \frac{c+1}{2} - \frac{x^2}{2}$ for all $x \in A$.
    We now show that $A - A = \{x_1 - x_2 \mid x_1,x_2 \in A\} = \mathbb{R}$. Indeed, plugging in $y = 0$ into the original equation gives us $f(x-c)-f(x) = cx+f(c)-1$, an expression that evidently spans all the real numbers. Thus, each x can be represented as $x = x_1 - x_2$, where $x_1,x_2 \in A$. Plugging $x = x_1$ and $f(y) = x_2$ into the original equation gives us

$$f(x) = f(x_1 - x_2) = f(x_1)+x_1x_2+f(x_2)-1 = c - \frac{x_1^2+x_2^2}{2}+x_1x_2 = c - \frac{x^2}{2}.$$

Hence we must have $c = \frac{c+1}{2}$, which gives us $c = 1$. Thus $f(x) = 1 - \frac{x^2}{2}$ for all $x \in \mathbb{R}$. It is easily checked that this function satisfies the original functional equation.

20. We first introduce some useful notation. An arrangement around the circle will be denoted by $x = \{x_1, x_2, \ldots, x_n\}$, where the elements are arranged clockwise and x_1 is fixed to be the smallest number. We will call an arrangement *balanced* if $x_1 \le x_n \le x_2 \le x_{n-1} \le x_3 \le x_{n-2} \le \cdots$ (the string of inequalities continues until all the elements are accounted for). We will denote the permutation of $x = \{x_1, x_2, \ldots, x_n\}$ in ascending order by $x' = \{x_1', x_2', \ldots, x_n'\}$. We will let $f_i(x) = \{f_i(x)_1, f_i(x)_2, \ldots, f_i(x)_{n-1}\}$ denote the arrangement after one iteration of the algorithm where x_i was the deleted element.

  Lemma 1. If an arrangement x is balanced, then $f_1(x)$ is also balanced.

  Proof. In one iteration we have $\{x_1, \ldots, x_n\} \to \{x_n + x_2, x_2 + x_3, \ldots, x_{n-1} + x_n\}$. Since $x_n \le x_2 \le x_{n-1} \le x_3 \le x_{n-2} \le \cdots$, it follows that $x_n + x_2 \le x_n + x_{n-1} \le x_2 + x_3 \le x_{n-1} + x_{n-2} \le \cdots$, which means that $f_1(x)$ is balanced.

  We will first show by induction that $S_{\max}$ can be reached by using the balanced initial arrangement $\{a_1, a_3, a_5, \ldots, a_6, a_4, a_2\}$ and repeatedly deleting the smallest member. For $n = 3$ we have $S_3 = a_2 + a_3$, in accordance with the formula. Assuming that the formula holds for a given n, we note that for an arrangement $x = \{a_1, a_3, a_5, \ldots, a_6, a_4, a_2\}$ the arrangement $f_1(x)$ is also balanced. We now apply the induction hypothesis and use that $\binom{n-2}{i} + \binom{n-2}{i-1} = \binom{n-1}{i}$:

$$S(x) = S(f_1(x))$$
$$= \sum_{k=2}^{n-1} \binom{n-2}{\lfloor k/2 \rfloor - 1}(a_k + a_{k+2}) + \binom{n-2}{\lfloor n/2 \rfloor - 1}(a_n + a_{n+1}) = S_{\max}.$$

  We now prove that every other arrangement yields a smaller value. We shall write $\{x_1, \ldots, x_n\} \le \{y_1, \ldots, y_n\}$ whenever $x_n' + x_{n-1}' + \cdots + x_i' \le y_n' + y_{n-1}' + \cdots + y_i'$ holds for all $1 \le i \le n$.

  Lemma 2. Let x be an arbitrary arrangement and y a balanced arrangement, both of n elements, such that $x \le y$. Then it follows that $f_i(x) \le f_1(y)$, for all i.

  Proof. For any $1 \le j \le n-1$ there exists k_j such that $f_i(x)_j = x_{k_j} + x_{k_j+1}$ (assuming $k_j + 1 = 1$ if $k_j = n-1$). Then we have

$$f_i(x)_{n-1} + \cdots + f_i(x)_{n-j} = (x_{k_1} + x_{k_1+1}) + \cdots + (x_{k_j} + x_{k_j+1})$$
$$\le 2x_n' + \cdots + 2x_{n-i+1}' + x_{n-i}' + x_{n-i-1}'$$
$$= f_1(y)_{n-1} + \cdots + f_1(y)_{n-j}$$

  for all j, and hence $f_i(x) \le f_1(y)$.

  An immediate consequence of Lemma 2 is $f^{n-2}(x) \le f_1^{n-2}(y)$, implying $S = f^{n-2}(x)_1 + f^{n-2}(x)_2 \le f_1^{n-2}(y)_1 + f_1^{n-2}(y)_2 = S_{\max}(y)$. Thus the proof is finished.

21. Let us call $f(n, s)$ the number of paths from $(0, 0)$ to (n, n) that contain exactly s steps. Evidently, for all n we have $f(n, 1) = f(2, 2) = 1$, in accordance with the formula. Let us thus assume inductively for a given $n > 2$ that for all s we have $f(n, s) = \frac{1}{s}\binom{n-1}{s-1}\binom{n}{s-1}$. We shall prove that the given formula holds also for all $f(n+1, s)$, where $s \ge 2$.

We say that an $(n+1,s)$- or $(n+1,s+1)$-path is *related* to a given (n,s)-path if it is obtained from the given path by inserting a step EN between two moves or at the beginning or the end of the path. We note that by inserting the step between two moves that form a step one obtains an $(n+1,s)$-path; in all other cases one obtains an $(n+1,s+1)$-path. For each (n,s)-path there are exactly $2n+1-s$ related $(n+1,s+1)$-paths, and for each $(n,s+1)$-path there are $s+1$ related $(n+1,s+1)$-paths. Also, each $(n+1,s+1)$-path is related to exactly $s+1$ different (n,s)- or $(n,s+1)$-paths. Thus:

$$(s+1)f(n+1,s+1) = (2n+1-s)f(n,s) + (s+1)f(n,s+1)$$
$$= \frac{2n+1-s}{s}\binom{n-1}{s-1}\binom{n}{s-1} + \binom{n-1}{s}\binom{n}{s}$$
$$= \binom{n}{s}\binom{n+1}{s},$$

i.e., $f(n+1,s+1) = \frac{1}{s+1}\binom{n}{s}\binom{n+1}{s}$. This completes the proof.

22. (a) Color the first, third, and fifth row red, and the remaining squares white. There is total n pieces and $3n$ red squares. Since each piece can cover at most three red squares, it follows that each piece colors exactly three red squares. Then it follows that the two white squares it covers must be on the same row; otherwise, the piece has to cover at least three. Hence, each white row can be partitioned into pairs of squares belonging to the same piece. Thus it follows that the number of white squares in a row, which is n, must be even.

(b) Let a_k denote the number of different tilings of a $5 \times 2k$ rectangle. Let b_k be the number of tilings that cannot be partitioned into two smaller tilings along a vertical line (without cutting any pieces). It is easy to see that $a_1 = b_1 = 2$, $b_2 = 2$, $a_2 = 6 = 2 \cdot 3$, $b_3 = 4$, and subsequently, by induction, $b_{3k} \geq 4$, $b_{3k+1} \geq 2$, and $b_{3k+2} \geq 2$. We also have $a_k = b_k + \sum_{i=1}^{k-1} b_i a_{k-i}$. For $k \geq 3$ we now have inductively

$$a_k > 2 + \sum_{i=1}^{k-1} 2a_{k-i} \geq 2 \cdot 3^{k-1} + 2a_{k-1} \geq 2 \cdot 3^k.$$

23. Let $r(m)$ denote the rest period before the mth catch, $t(m)$ the number of minutes before the mth catch, and $f(n)$ as the number of flies caught in n minutes. We have $r(1) = 1$, $r(2m) = r(m)$, and $r(2m+1) = f(m) + 1$. We then have by induction that $r(m)$ is the number of ones in the binary representation of m. We also have $t(m) = \sum_{i=1}^{m} r(i)$ and $f(t(m)) = m$. From the recursive relations for r we easily derive $t(2m+1) = 2t(m) + m + 1$ and consequently $t(2m) = 2t(m) + m - r(m)$. We then have, by induction on p, $t(2^p m) = 2^p t(m) + p \cdot m \cdot 2^{p-1} - (2^p - 1)r(m)$.

(a) We must find the smallest number m such that $r(m+1) = 9$. The smallest number with nine binary digits is $\overline{111111111}_2 = 511$; hence the required m is 510.

(b) We must calculate $t(98)$. Using the recursive formulas we have $t(98) = 2t(49) + 49 - r(49)$, $t(49) = 2t(24) + 25$, and $t(24) = 8t(3) + 36 - 7r(3)$. Since we have $t(3) = 4$, $r(3) = 2$ and $r(49) = r(110001_2) = 3$, it follows $t(24) = 54 \Rightarrow t(49) = 133 \Rightarrow t(98) = 312$.

(c) We must find m_c such that $t(m_c) \leq 1999 < t(m_c + 1)$. One can estimate where this occurs using $t(2^p(2^q - 1)) = (p+q)2^{p+q-1} - p\,2^{p-1} - q\,2^p + q$, which follows from the recursive relations. It suffices to note that $t(462) = 1993$ and $t(463) = 2000$; hence $m_c = 462$.

24. Let $S = \{0, 1, \ldots, N^2 - 1\}$ be the group of residues (with respect to addition modulo N^2) and A an n-element subset. We will use $|X|$ to denote the number of elements of a subset X of S, and $\overline{X}$ to refer to the complement of X in S. For $i \in S$ we also define $A_i = \{a + i \mid a \in A\}$. Our task is to select $0 \leq i_1 < \cdots < i_N \leq N^2 - 1$ such that $\left| \bigcup_{j=1}^N A_{i_j} \right| \geq \frac{1}{2}|S|$. Each $x \in S$ appears in exactly N sets A_i. We have

$$\sum_{i_1 < \cdots < i_N} \left| \bigcap_{j=1}^N \overline{A}_{i_j} \right| = \sum_{i_1 < \cdots < i_N} |\{x \in S \mid x \notin A_{i_1}, \ldots, A_{i_N}\}|$$

$$= \sum_{x \in S} |\{i_1 < \cdots < i_N | x \notin A_{i_1}, \ldots, A_{i_N}\}|$$

$$= \sum_{x \in S} \binom{N^2 - N}{N} = \binom{N^2 - N}{N}|S|.$$

Hence

$$\sum_{i_1 < \cdots < i_N} \left| \bigcup_{j=1}^N A_{i_j} \right| = \sum_{i_1 < \cdots < i_N} \left(|S| - \left| \bigcap_{j=1}^N \overline{A}_{i_j} \right| \right)$$

$$= \left(\binom{N^2}{N} - \binom{N^2 - N}{N} \right)|S|.$$

Thus, by the pigeonhole principle, one can choose $i_1 < \cdots < i_N$ such that $\left| \bigcup_{j=1}^N A_{i_j} \right| \geq \left(1 - \binom{N^2 - N}{N} / \binom{N^2}{N} \right)|S|$. Since

$$\binom{N^2}{N} / \binom{N^2 - N}{N} \geq \left(\frac{N^2}{N^2 - N} \right)^N = \left(1 + \frac{1}{N-1} \right)^N > e > 2,$$

it follows that $\left| \bigcup_{j=1}^N A_{i_j} \right| \geq \frac{1}{2}|S|$; hence the chosen $i_1 < \cdots < i_N$ are indeed the elements of B that satisfy the conditions of the problem.

25. Let $n = 2k$. Color the cells neighboring the edge of the board black. Then color the cells neighboring the black cells white. Then in alternation color the still uncolored cells neighboring the white or black cells on the boundary the opposite color and repeat until all cells are colored.

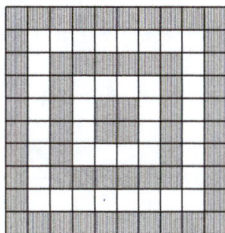

We call the cells colored the same color in each such iteration a "frame." In the color scheme described, each cell (white or black) neighbors exactly two black cells. The number of black cells is $2k(k+1)$, and hence we need to mark at least $k(k+1)$ cells.

On the other hand, going along each black-colored frame, we can alternately mark two consecutive cells and then not mark two consecutive cells. Every cell on the black frame will have one marked neighbor. One can arrange these sequences on two consecutive black frames such that each cell in the white frame in between has exactly one neighbor. Hence, starting from a sequence on the largest frame we obtain a marking that contains exactly half of all the black cells, i.e., $k(k+1)$ and neighbors every cell.

It follows that the desired minimal number of markings is $k(k+1)$.

Remark. For $n = 4k - 1$ and $n = 4k + 1$ one can perform similar markings to obtain minimal numbers $4k^2 - 1$ and $(2k+1)^2$, respectively.

26. We denote colors by capital initial letters. Let us suppose that there exists a coloring $f : \mathbb{Z} \to \{R,G,B,Y\}$ such that for any $a \in \mathbb{Z}$ we have $f\{a, a+x, a+y, a+x+y\} = \{R,G,B,Y\}$. We now define a coloring of an integer lattice $g : \mathbb{Z} \times \mathbb{Z} \to \{R,G,B,Y\}$ by the rule $g(i, j) = f(xi + yj)$. It follows that every unit square in g must have its vertices colored by four different colors.

If there is a row or column with period 2, then applying the condition to adjacent unit squares, we get (by induction) that all rows or columns, respectively, have period 2.

On the other hand, taking a row to be not of period 2, i.e., containing a sequence of three distinct colors, for example GRY, we get that the next row must contain in these columns YBG, and the following GRY, and so on. It would follow that a column in this case must have period 2. A similar conclusion holds if we start with an aperiodic column. Hence either all rows or all columns must have period 2.

Let us assume w.l.o.g. that all rows have a period of 2. Assuming w.l.o.g. $\{g(0,0), g(1,0)\} = \{G,B\}$, we get that the even rows are painted with $\{G,B\}$ and odd with $\{Y,R\}$. Since x is odd, it follows that $g(y,0)$ and $g(0,x)$ are of different color. However, since $g(y,0) = f(xy) = g(0,x)$, this is a contradiction. Hence the statement of the problem holds.

27. Denote $A = \{0,1,2\}$ and $B = \{0,1,3\}$. Let $f_T(x) = \sum_{a \in T} x^a$. Then define $F_T(x) = f_T(x) f_T(x^2) \cdots f_T(x^{p-1})$. We can write $F_T(x) = \sum_{i=0}^{p(p-1)} a_i x^i$, where a_i is the number of ways to select an array $\{x_1, \ldots, x_{p-1}\}$ where $x_i \in T$ for all i and $x_1 + 2x_2 + \cdots + (p-1)x_{p-1} = i$. Let $w = \cos(2\pi/p) + i\sin(2\pi/p)$, a pth root of unity. Noting that

$$1 + w^j + w^{2j} + \cdots + w^{(p-1)j} = \begin{cases} p, \ p \mid j, \\ 0, \ p \nmid j, \end{cases}$$

it follows that $F_T(1) + F_T(w) + \cdots + F_T(w^{p-1}) = pE(T)$.

Since $|A| = |B| = 3$, it follows that $F_A(1) = F_B(1) = 3^{p-1}$. We also have for $p \nmid i, j$ that $F_T(w^i) = F_T(w)$. Finally, we have

$$F_A(w) = \prod_{i=1}^{p-1}(1 + w^i + w^{2i}) = \prod_{i=1}^{p-1}\frac{1 - w^{3i}}{1 - w^i} = 1.$$

Hence, combining these results, we obtain

$$E(A) = \frac{3^{p-1} + p - 1}{p} \quad \text{and} \quad E(B) = \frac{3^{p-1} + (p-1)F_B(w)}{p}.$$

It remains to demonstrate that $F_B(w) \geq 1$ for all p and that equality holds only for $p = 5$. Since $E(B)$ is an integer, it follows that $F_B(w)$ is an integer and $F_B(w) \equiv 1 \pmod{p}$. Since $f_B(w^{p-i}) = \overline{f_B(w^i)}$, it follows that $F_B(w) = |f_B(w)|^2 |f_B(w^2)|^2 \cdots \left| f_B\left(w^{(p-1)/2}\right) \right|^2 > 0$. Hence $F_B(w) \geq 1$.

It remains to show that $F_B(w) = 1$ if and only if $p = 5$. We have the formula $(x - w)(x - w^2)\cdots(x - w^{p-1}) = x^{p-1} + x^{p-2} + \cdots + x + 1 = \frac{x^p-1}{x-1}$. Let $f_B(x) = x^3 + x + 1 = (x - \lambda)(x - \mu)(x - v)$, where λ, μ, and v are the three zeros of the polynomial $f_B(x)$. It follows that

$$F_B(w) = \left(\frac{\lambda^p - 1}{\lambda - 1}\right)\left(\frac{\mu^p - 1}{\mu - 1}\right)\left(\frac{v^p - 1}{v - 1}\right) = -\frac{1}{3}(\lambda^p - 1)(\mu^p - 1)(v^p - 1),$$

since $(\lambda - 1)(\mu - 1)(v - 1) = -f_B(1) = -3$. We also have $\lambda + \mu + v = 0$, $\lambda\mu v = -1$, $\lambda\mu + \lambda v + \mu v = 1$, and $\lambda^2 + \mu^2 + v^2 = (\lambda + \mu + v)^2 - 2(\lambda\mu + \lambda v + \mu v) = -2$. By induction (using that $(\lambda^r + \mu^r + v^r) + (\lambda^{r-2} + \mu^{r-2} + v^{r-2}) + (\lambda^{r-3} + \mu^{r-3} + v^{r-3}) = 0$), it follows that $\lambda^r + \mu^r + v^r$ is an integer for all $r \in \mathbb{N}$. Let us assume $F_B(x) = 1$. It follows that $(\lambda^p - 1)(\mu^p - 1)(v^p - 1) = -3$. Hence λ^p, μ^p, v^p are roots of the polynomial $p(x) = x^3 - qx^2 + (1+q)x + 1$, where $q = \lambda^p + \mu^p + v^p$. Since $f_B(x)$ is an increasing function in real numbers, it follows that it has only one real root (w.l.o.g.) λ, the other two roots being complex conjugates. From $f_B(-1) < 0 < f_B(-1/2)$ it follows that $-1 < \lambda < -1/2$. It also follows that λ^p is the x coordinate of the intersection of functions $y = x^3 + x + 1$ and $y = q(x^2 - x)$. Since $\lambda < \lambda^p < 0$, it follows that $q > 0$; otherwise, $q(x^2 - x)$ intersects $x^3 + x + 1$ at a value smaller than λ. Additionally, as p increases, λ^p approaches 0, and hence q must increase.

For $p = 5$ we have $1 + w + w^3 = -w^2(1 + w^2)$ and hence $G(w) = \prod_{i=1}^{p-1}(1 + w^{2j}) = 1$. For a zero of $f_B(x)$ we have $x^5 = -x^3 - x^2 = -x^2 + x + 1$ and hence $q = \lambda^5 + \mu^5 + v^5 = -(\lambda^2 + \mu^2 + v^2) + (\lambda + \mu + v) + 3 = 5$.

For $p > 5$ we also have $q \geq 6$. Assuming again $F_B(x) = 1$ and defining $p(x)$ as before, we have $p(-1) < 0$, $p(0) > 0$, $p(2) < 0$, and $p(x) > 0$ for a sufficiently large $x > 2$. It follows that $p(x)$ must have three distinct real roots. However, since $\mu^p, v^p \in \mathbb{R} \Rightarrow v^p = \overline{\mu^p} = \mu^p$, it follows that $p(x)$ has at most two real roots, which is a contradiction. Hence, it follows that $F_B(x) > 1$ for $p > 5$ and thus $E(A) \leq E(B)$, where equality holds only for $p = 5$.

4.41 Solutions to the Shortlisted Problems of IMO 2000

1. In order for the trick to work, whenever $x + y = z + t$ and the cards x, y are placed in different boxes, either z, t are in these boxes as well or they are both in the remaining box.

   Case 1. The cards $i, i+1, i+2$ are in different boxes for some i. Since $i + (i+3) = (i+1) + (i+2)$, the cards i and $i+3$ must be in the same box; moreover, $i-1$ must be in the same box as $i+2$, etc. Hence the cards $1, 4, 7, \ldots, 100$ are placed in one box, the cards $2, 5, \ldots, 98$ are in the second, while $3, 6, \ldots, 99$ are in the third box. The number of different arrangements of the cards is 6 in this case.

   Case 2. No three successive cards are all placed in different boxes. Suppose that 1 is in the blue box, and denote by w and r the smallest numbers on cards lying in the white and red boxes; assume w.l.o.g. that $w < r$. The card $w+1$ is obviously not red, from which it follows that $r > w+1$. Now suppose that $r < 100$. Since $w + r = (w-1) + (r+1)$, $r+1$ must be in the blue box. But then $(r+1) + w = r + (w+1)$ implies that $w+1$ must be red, which is a contradiction. Hence the red box contains only the card 100. Since $99 + w = 100 + (w-1)$, we deduce that the card 99 is in the white box. Moreover, if any of the cards k, $2 \le k \le 99$, were in the blue box, then since $k + 99 = (k-1) + 100$, the card $k-1$ should be in the red box, which is impossible. Hence the blue box contains only the card 1, whereas the cards $2, 3, \ldots, 99$ are all in the white box.

   In general, one box contains 1, another box only 100, while the remaining contains all the other cards. There are exactly 6 such arrangements, and the trick works in each of them.

   Therefore the answer is 12.

2. Since the volume of each brick is 12, the side of any such cube must be divisible by 6.

   Suppose that a cube of side $n = 6k$ can be built using $\frac{n^3}{12} = 18k^3$ bricks. Set a coordinate system in which the cube is given as $[0, n] \times [0, n] \times [0, n]$ and color in black each unit cube $[2p, 2p+1] \times [2q, 2q+1] \times [2r, 2r+1]$. There are exactly $\frac{n^3}{9} = 27k^3$ black cubes. Each brick covers either one or three black cubes, which is in any case an odd number. It follows that the total number of black cubes must be even, which implies that k is even. Hence $12 \mid n$.

   On the other hand, two bricks can be fitted together to give a $2 \times 3 \times 4$ box. Using such boxes one can easily build a cube of side 12, and consequently any cube of side divisible by 12.

3. Clearly $m(S)$ is the number of pairs of point and triangle $(P_t, P_i P_j P_k)$ such that P_t lies inside the circle $P_i P_j P_k$. Consider any four-element set $S_{ijkl} = \{P_i, P_j, P_k, P_l\}$. If the convex hull of S_{ijkl} is the triangle $P_i P_j P_k$, then we have $a_i = a_j = a_k = 0$, $a_l = 1$. Suppose that the convex hull is the quadrilateral $P_i P_j P_k P_l$. Since this quadrilateral is not cyclic, we may suppose that $\angle P_i + \angle P_k < 180° < \angle P_j + \angle P_l$.

In this case $a_i = a_k = 0$ and $a_j = a_l = 1$. Therefore $m(S_{ijkl})$ is 2 if P_i, P_j, P_k, P_l are vertices of a convex quadrilateral, and 1 otherwise. There are $\binom{n}{4}$ four-element subsets S_{ijkl}. If $a(S)$ is the number of such subsets whose points determine a convex quadrilateral, we have $m(S) = 2a(S) + (\binom{n}{4} - a(S)) = \binom{n}{4} + a(S) \leq 2\binom{n}{4}$. Equality holds if and only if every four distinct points of S determine a convex quadrilateral, i.e. if and only if the points of S determine a convex polygon. Hence $f(n) = 2\binom{n}{4}$ has the desired property.

4. By a *good placement* of pawns we mean the placement in which there is no block of k adjacent unoccupied squares in a row or column.

We can make a good placement as follows: Label the rows and columns with $0, 1, \ldots, n-1$ and place a pawn on a square (i, j) if and only if k divides $i + j + 1$. This is obviously a good placement in which the pawns are placed on three lines with k, $2n - 2k$, and $2n - 3k$ squares, which adds up to $4n - 4k$ pawns in total.

Now we shall prove that a good placement must contain at least $4n - 4k$ pawns. Suppose we have a good placement of m pawns. Partition the board into nine rectangular regions as shown in the picture. Let $a, b, \ldots, h$ be the numbers of pawns in the rectangles $A, B, \ldots, H$ respectively. Note that each row that passes through A, B, and C either contains a pawn inside B, or

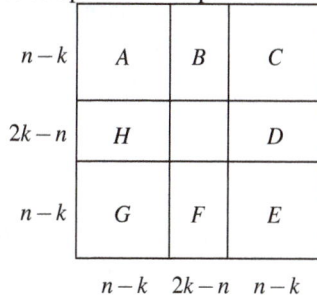

	$n-k$	$2k-n$	$n-k$
$n-k$	A	B	C
$2k-n$	H		D
$n-k$	G	F	E

contains a pawn in both A and C. It follows that $a + c + 2b \geq 2(n - k)$. We similarly obtain that $c + e + 2d$, $e + g + 2f$, and $g + a + 2h$ are all at least $2(n-k)$. Adding and dividing by 2 yields $a + b + \cdots + h \geq 4(n - k)$, which proves the statement.

5. We say that a vertex of a nice region is convex if the angle of the region at that vertex equals $90°$; otherwise (if the angle is $270°$), we say that a vertex is concave.

For a simple broken line C contained in the boundary of a nice region R we call the pair (R, C) a *boundary pair*. Such a pair is called *outer* if the region R is inside the broken line C, and *inner* otherwise. Let $\mathcal{B}_i, \mathcal{B}_o$ be the sets of inner and outer boundary pairs of nice regions respectively, and let $\mathcal{B} = \mathcal{B}_i \cup \mathcal{B}_o$. For a boundary pair $b = (R, C)$ denote by c_b and v_b respectively the number of convex and concave vertices of R that belong to C. We have the following facts:

(1) Each vertex of a rectangle corresponds to one concave angle of a nice region and vice versa. This correspondence is bijective, so $\sum_{b \in \mathcal{B}} v_b = 4n$.

(2) For a boundary pair $b = (R, C)$ the sum of angles of R that are on C equals $(c_b + v_b - 2)180°$ if b is outer, and $(c_b + v_b + 2)180°$ if b is inner. On the other hand the sum of angles is obviously equal to $c_b \cdot 90° + v_b \cdot 270°$. It immediately follows that $c_b - v_b = \begin{cases} 4 & \text{if } b \in \mathcal{B}_o, \\ -4 & \text{if } b \in \mathcal{B}_i. \end{cases}$

(3) Since every vertex of a rectangle appears in exactly two boundary pairs and each boundary pair contains at least one vertex of a rectangle, the number K of boundary pairs is less than or equal to $8n$.

(4) The set $\mathscr{B}_i$ is nonempty, because every boundary of the infinite region is inner.

Consequently, the sum of the numbers of the vertices of all nice regions is equal to

$$\sum_{b \in \mathscr{B}} (c_b + v_b) = \sum_{b \in \mathscr{B}} (2v_b + (c_b - v_b)) \le 2 \cdot 4n + 4(K-1) - 4 \le 40n - 8.$$

6. Every integer z has a unique representation $z = px + qy$, where $x, y \in \mathbb{Z}$, $0 \le x \le q - 1$. Consider the region T in the xy-plane defined by the last inequality and $px + qy \ge 0$. There is a bijective correspondence between lattice points of this region and nonnegative integers given by $(x, y) \mapsto z = px + qy$. Let us mark all lattice points of T whose corresponding integers belong to S and color in black the unit squares whose left-bottom vertices are at marked points. Due to the condition for S, this coloring has the property that all points lying on the right or above a colored point are colored as well. In particular, since the point $(0,0)$ is colored, all points above or on the line $y = 0$ are colored. What we need is the number of such colorings of T.

The border of the colored subregion C of T determines a path from $(0,0)$ to $(q, -p)$ consisting of consecutive unit moves either to the right or downwards. There are $\binom{p+q}{p}$ such paths in total. We must find the number of such paths not going below the line $l : px + qy = 0$.

Consider any path $\gamma = A_0 A_1 \ldots A_{p+q}$ from $A_0 = (0,0)$ to $A_{p+q} = (q, -p)$. We shall see the path γ as a sequence $G_1 G_2 \ldots G_{p+q}$ of moves to the right (R) or downwards (D) with exactly p D's and q R's.

Two paths are said to be *equivalent* if one is obtained from the other by a circular shift of the corresponding sequence $G_1 G_2 \ldots G_{p+q}$. We note that all the $p + q$ circular shifts of a path are distinct. Indeed, $G_1 \ldots G_{p+q} \equiv G_{i+1} \ldots G_{i+p+q}$ would imply $G_1 = G_{i+1} = G_{2i+1} = \cdots$ (where $G_{j+p+q} = G_j$), so $G_1 = \cdots = G_{p+q}$, which is impossible. Hence each equivalence class contains exactly $p + q$ paths. Let l_i, $0 \le i < p + q$, be the line through A_i that is parallel to the line l. Since $\gcd(p, q) = 1$, all these lines are distinct.

Let l_m be the unique lowest line among the l_is. The path $G_{m+1} G_{m+2} \ldots G_{m+p+q}$ must be above the line l. Every other cyclic shift gives rise to a path having at least one vertex below the line l. Thus each equivalence class contains exactly one path above the line l, so the number of such paths is equal to $\frac{1}{p+q}\binom{p+q}{p}$.

Therefore the answer is $\frac{1}{p+q}\binom{p+q}{p}$.

7. Note that at most one of the three factors is negative: for example, $a - 1 + \frac{1}{b}, b - 1 + \frac{1}{c} < 0$ implies $b, \frac{1}{b} < 1$, which is false. If exactly one factor is negative, the product is negative. Now assume that they are all positive. Elementary computation gives $\left(a - 1 + \frac{1}{b}\right)\left(b - 1 + \frac{1}{c}\right) = ab - a + \frac{a}{c} - b + 1 - \frac{1}{c} + 1 - \frac{1}{b} + \frac{1}{bc}$. Using $ab = \frac{1}{c}$ and $\frac{1}{bc} = a$ we obtain

$$\left(a-1+\frac{1}{b}\right)\left(b-1+\frac{1}{c}\right) = \frac{a}{c}-b-\frac{1}{b}+2\leq\frac{a}{c},$$

since $b+\frac{1}{b}\geq 2$. Similarly we obtain

$$\left(b-1+\frac{1}{c}\right)\left(c-1+\frac{1}{a}\right)\leq\frac{b}{a} \quad\text{and}\quad \left(c-1+\frac{1}{a}\right)\left(a-1+\frac{1}{b}\right)\leq\frac{c}{b}.$$

The desired inequality follows by multiplying the previous three inequalities. Equality holds if and only if $a=b=c=1$.

8. We note that $\{ta\}$ lies in $\left(\frac{1}{3},\frac{2}{3}\right]$ if and only if there is an integer k such that $k+\frac{1}{3}<ta\leq k+\frac{2}{3}$, i.e., if and only if $t\in I_k = \left(\frac{k+1/3}{a},\frac{k+2/3}{a}\right]$ for some k. Similarly, t should belong to the sets $J_m = \left(\frac{m+1/3}{b},\frac{m+2/3}{b}\right]$ and $K_n = \left(\frac{n+1/3}{c},\frac{n+2/3}{c}\right]$ for some m,n. We have to show that $I_k\cap J_m\cap K_n$ is nonempty for some integers k,m,n.

The intervals K_n are separated by a distance $\frac{2}{3c}$, and since $\frac{2}{3c}<\frac{1}{3b}$, each of the intervals J_m intersects at least one of the K_n's. Hence it is enough to prove that $J_m\subset I_k$ for some k,m.

Let u_m and v_m be the left and right endpoints of J_m. Since $av_m = au_m+\frac{a}{3b}<au_m+\frac{1}{6}$, it will suffice to show that there is an integer m such that the fractional part of au_m lies in $\left[\frac{1}{3},\frac{1}{2}\right]$.

Let $a=d\alpha$, $b=d\beta$, $\gcd(\alpha,\beta)=1$. Setting $m=d\mu$, we obtain that $au_m = a\frac{m+1/3}{b} = \frac{\alpha m}{d\beta}+\frac{\alpha}{3\beta} = \frac{\alpha\mu}{\beta}+\frac{\alpha}{3\beta}$. Since $\alpha\mu$ gives all possible residues modulo β, every term of the arithmetic progression $\frac{j}{\beta}+\frac{\alpha}{3\beta}$ ($j\in\mathbb{Z}$) has its fractional part equal to the fractional part of some au_m. Now for $\beta\geq 6$ the progression step is $\frac{1}{\beta}\leq\frac{1}{6}$, so at least one of the au_m has its fractional part in $[1/3,1/2]$. If otherwise $\beta\leq 5$, the only irreducible fractions $\frac{\alpha}{\beta}$ that satisfy $2\alpha<\beta$ are $\frac{1}{3},\frac{1}{4},\frac{1}{5},\frac{2}{5}$; hence one can take m to be $1,1,2,3$ respectively. This justifies our claim.

9. Let us first solve the problem under the assumption that $g(\alpha)=0$ for some α. Setting $y=\alpha$ in the given equation yields $g(x) = (\alpha+1)f(x) - xf(\alpha)$. Then the given equation becomes $f(x+g(y)) = (\alpha+1-y)f(x)+(f(y)-f(\alpha))x$, so setting $y=\alpha+1$, we get $f(x+n) = mx$, where $n=g(\alpha+1)$ and $m=f(\alpha+1)-f(\alpha)$. Hence f is a linear function, and consequently g is also linear. If we now substitute $f(x)=ax+b$ and $g(x)=cx+d$ in the given equation and compare the coefficients, we easily find that

$$f(x) = \frac{cx-c^2}{1+c} \quad\text{and}\quad g(x) = cx-c^2, \quad c\in\mathbb{R}\setminus\{-1\}.$$

Now we prove the existence of α such that $g(\alpha)=0$. If $f(0)=0$, then putting $y=0$ in the given equation, we obtain $f(x+g(0)) = g(x)$, so we can take $\alpha = -g(0)$.

Now assume that $f(0)=b\neq 0$. By replacing x by $g(x)$ in the given equation we obtain $f(g(x)+g(y)) = g(x)f(y)-yf(g(x))+g(g(x))$, and analogously,

$f(g(x)+g(y)) = g(y)f(x)-xf(g(y))+g(g(y))$. The given functional equation for $x = 0$ gives $f(g(y)) = a - by$, where $a = g(0)$. In particular, g is injective and f is surjective, so there exists $c \in \mathbb{R}$ such that $f(c) = 0$. Now the above two relations yield

$$g(x)f(y) - ay + g(g(x)) = g(y)f(x) - ax + g(g(y)). \tag{1}$$

Plugging $y = c$ in (1), we get $g(g(x)) = g(c)f(x) - ax + g(g(c)) + ac = kf(x) - ax + d$. Now (1) becomes $g(x)f(y) + kf(x) = g(y)f(x) + kf(y)$. For $y = 0$ we have $g(x)b + kf(x) = af(x) + kb$, whence

$$g(x) = \frac{a-k}{b}f(x) + k.$$

Note that $g(0) = a \neq k = g(c)$, since g is injective. From the surjectivity of f it follows that g is surjective as well, so it takes the value 0.

10. Clearly $F(0) = 0$ by (i). Moreover, it follows by induction from (i) that $F(2^n) = f_{n+1}$, where f_n denotes the nth Fibonacci number. In general, if $n = \varepsilon_k 2^k + \varepsilon_{k-1} 2^{k-1} + \cdots + \varepsilon_1 \cdot 2 + \varepsilon_0$ (where $\varepsilon_i \in \{0,1\}$), it is straightforward to verify that

$$F(n) = \varepsilon_k f_{k+1} + \varepsilon_{k-1} f_k + \cdots + \varepsilon_1 f_2 + \varepsilon_0 f_1. \tag{1}$$

We observe that if the binary representation of n contains no two adjacent ones, then $F(3n) = F(4n)$. Indeed, if $n = \varepsilon_{k_r} 2^{k_r} + \cdots + \varepsilon_{k_0} 2^{k_0}$, where $k_{i+1} - k_i \geq 2$ for all i, then $3n = \varepsilon_{k_r}(2^{k_r+1} + 2^{k_r}) + \cdots + \varepsilon_{k_0}(2^{k_0+1} + 2^{k_0})$. According to this, in computing $F(3n)$ each f_{i+1} in (1) is replaced by $f_{i+1} + f_{i+2} = f_{i+3}$, leading to the value of $F(4n)$.

We shall prove the converse: $F(3n) \leq F(4n)$ holds for all $n \geq 0$, with equality if and only if the binary representation of n contains no two adjacent ones.

We prove by induction on $m \geq 1$ that this holds for all n satisfying $0 \leq n < 2^m$. The verification for the early values of m is direct. Assume that it is true for a certain m and let $2^m \leq n \leq 2^{n+1}$. If $n = 2^m + p$, $0 \leq p < 2^m$, then (1) implies $F(4n) = F(2^{m+2} + 4p) = f_{m+3} + F(4p)$. Now we distinguish three cases:

(i) If $3p < 2^m$, then the binary representation of $3p$ does not carry into that of $3 \cdot 2^m$. Then it follows from (1) and the induction hypothesis that

$$F(3n) = F(3 \cdot 2^m) + F(3p) = f_{m+3} + F(3p) \leq f_{m+3} + F(4p) = F(4n).$$

Equality holds if and only if $F(3p) = F(4p)$, i.e. p has no two adjacent binary ones.

(ii) If $2^m \leq 3p < 2^{m+1}$, then the binary representation of $3p$ carries 1 into that of $3 \cdot 2^m$. Thus $F(3n) = f_{m+3} + (F(3p) - f_{m+1}) = f_{m+2} + F(3p) < f_{m+3} + F(4p) = F(4n)$.

(iii) If $2^{m+1} \leq p < 3 \cdot 2^m$, then the binary representation of $3p$ caries 10 into that of $3 \cdot 2^m$, which implies

$$F(3n) = f_{m+3} + f_{m+1} + (F(3p) - f_{m+2}) = 2f_{m+1} + F(3p) < F(4n).$$

It remains to compute the number of integers in $[0, 2^m)$ with no two adjacent binary 1's. Denote their number by u_m. Among them there are u_{m-1} less than 2^{m-1} and u_{m-2} in the segment $[2^{m-1}, 2^m)$. Hence $u_m = u_{m-1} + u_{m-2}$ for $m \geq 3$. Since $u_1 = 2 = f_3$, $u_2 = 3 = f_4$, we conclude that $u_m = f_{m+2} = F(2^{m+1})$.

11. We claim that for $\lambda \geq \frac{1}{n-1}$ we can take all fleas as far to the right as we want. In every turn we choose the leftmost flea and let it jump over the rightmost one. Let d and δ denote the maximal and the minimal distances between two fleas at some moment. Clearly, $d \geq (n-1)\delta$. After the leftmost flea jumps over the rightmost one, the minimal distance does not decrease, because $\lambda d \geq \delta$. However, the position of the leftmost flea moved to the right by at least δ, and consequently we can move the fleas arbitrarily far to the right after a finite number of moves. Suppose now that $\lambda < \frac{1}{n-1}$. Under this assumption we shall prove that there is a number M that cannot be reached by any flea. Let us assign to each flea the coordinate on the real axis on which it is settled. Denote by s_k the sum of all the numbers in the kth step, and by w_k the coordinate of the rightmost flea. Clearly, $s_k \leq n w_k$. We claim that the sequence w_k is bounded.

In the $(k+1)$th move let a flea A jump over B, landing at C, and let a, b, c be their respective coordinates. We have $s_{k+1} - s_k = c - a$. Then by the given rule, $\lambda(b - a) = c - b = s_{k+1} - s_k + a - b$, which implies $s_{k+1} - s_k = (1 + \lambda)(b - a) = \frac{1+\lambda}{\lambda}(c - b)$. Hence $s_{k+1} - s_k \geq \frac{1+\lambda}{\lambda}(w_{k+1} - w_k)$. Summing up these inequalities for $k = 0, \ldots, n-1$ yields $s_n - s_0 \geq \frac{1+\lambda}{\lambda}(w_n - w_0)$. Now using $s_n \leq n w_n$ we conclude that

$$\left(\frac{1+\lambda}{\lambda} - n\right) w_n \leq \frac{1+\lambda}{\lambda} w_0 - s_0.$$

Since $\frac{1+\lambda}{\lambda} - n > 0$, this proves the result.

12. Since $D(A) = D(B)$, we can define $f(i) > g(i) \geq 0$ that satisfy $b_i - b_{i-1} = a_{f(i)} - a_{g(i)}$ for all i.
The number $b_{i+1} - b_{i-1} \in D(B) = D(A)$ can be written in the form $a_u - a_v$, $u > v \geq 0$. Then $b_{i+1} - b_{i-1} = b_{i+1} - b_i + b_i - b_{i-1}$ implies $a_{f(i+1)} + a_{f(i)} + a_v = a_{g(i+1)} + a_{g(i)} + a_u$, so the B_3 property of A implies that $(f(i+1), f(i), v)$ and $(g(i+1), g(i), u)$ coincide up to a permutation. It follows that either $f(i+1) = g(i)$ or $f(i) = g(i+1)$. Hence if we define $R = \{i \in \mathbb{N}_0 \mid f(i+1) = g(i)\}$ and $S = \{i \in \mathbb{N}_0 \mid f(i) = g(i+1)\}$; it follows that $R \cup S = \mathbb{N}_0$.
Lemma. If $i \in R$, then also $i + 1 \in R$.
Proof. Suppose to the contrary that $i \in R$ and $i + 1 \in S$, i.e., $g(i) = f(i+1) = g(i+2)$. There are integers x and y such that $b_{i+2} - b_{i-1} = a_x - a_y$. Then $a_x - a_y = a_{f(i+2)} - a_{g(i+2)} + a_{f(i+1)} - a_{g(i+1)} + a_{f(i)} - a_{g(i)} = a_{f(i+2)} + a_{f(i)} - a_{g(i+1)} - a_{g(i)}$, so by the B_3 property, $(x, g(i+1), g(i))$ and $(y, f(i+2), f(i))$ coincide up to a permutation. But this is impossible, since $f(i+2), f(i) > g(i+2) = g(i) = f(i+1) > g(i+1)$. This proves the lemma. Therefore if $i \in R \neq \emptyset$, then it follows that every $j > i$ belongs to R. Consequently $g(i) = f(i+1) > g(i+1) = f(i+2) > g(i+2) = f(i+3) > \cdots$ is an infinite decreasing sequence of nonnegative integers, which is impossible. Hence $S =$

$\mathbb{N}_0$, i.e.,

$$b_{i+1} - b_i = a_{f(i+1)} - a_{f(i)} \qquad \text{for all } i \in \mathbb{N}_0.$$

Thus $f(0) = g(1) < f(1) < f(2) < \cdots$, implying $f(i) \geq i$. On the other hand, for any i there exist j, k such that $a_{f(i)} - a_i = b_j - b_k = a_{f(j)} - a_{f(k)}$, so by the B_3 property $i \in \{f(i), f(k)\}$ is a value of f. Hence we must have $f(i) = i$ for all i, which finally gives $A = B$.

13. One can easily find n-independent polynomials for $n = 0, 1$. For example, $P_0(x) = 2000x^{2000} + \cdots + 2x^2 + x + 0$ is 0-independent (for $Q \in M(P_0)$ it suffices to exchange the coefficient 0 of Q with the last term), and $P_1(x) = 2000x^{2000} + \cdots + 2x^2 + x - (1 + 2 + \cdots + 2000)$ is 1-independent (since any $Q \in M(P_1)$ vanishes at $x = 1$). Let us show that no n-independent polynomials exist for $n \notin \{0, 1\}$.

Consider separately the case $n = -1$. For any set T we denote by $S(T)$ the sum of elements of T. Suppose that $P(x) = a_{2000}x^{2000} + \cdots + a_1 x + a_0$ is -1-independent. Since $P(-1) = (a_0 + a_2 + \cdots + a_{2000}) - (a_1 + a_3 + \cdots + a_{1999})$, this means that for any subset E of the set $C = \{a_0, a_1, \ldots, a_{2000}\}$ having 1000 or 1001 elements there exist elements $e \in E$ and $f \in C \setminus E$ such that $S(E \cup \{f\} \setminus \{e\}) = \frac{1}{2}S(C)$, or equivalently that $S(E) - \frac{1}{2}S(C) = e - f$. We may assume w.l.o.g. that $a_0 < a_1 < \cdots < a_{2000}$.

Suppose that E is a 1000-element subset of C containing b_0, b_1 but not b_{1999}, b_{2000}. By the -1-independence of P there exist $e \in E$ and $f \in C \setminus E$ such that $S(E) - \frac{1}{2}S(C) = e - f$. The same must hold for the set $E' = E \cup \{b_{1999}, b_{2000}\} \setminus \{b_0, b_1\}$, so for some $e' \in E'$ and $f' \in C \setminus E'$ we have $S(E') - \frac{1}{2}S(C) = e' - f'$. It follows that $b_{1999} + b_{2000} - b_0 - b_1 = S(E') - S(E) = e + e' - f - f'$. Therefore the transposition $e \leftrightarrow f$ must involve at least one of the elements $b_0, b_1, b_{1999}, b_{2000}$.

There are 7994 possible transpositions involving one of these four elements. On the other hand, by (SL93-12) the subsets E of C containing b_0, b_1 but not b_{1999}, b_{2000} give at least $998 \cdot 999 + 1$ distinct sums of elements, far exceeding 7994. This is a contradiction.

For the case $|n| \geq 2$ we need the following lemma.

Lemma. Let $n \geq 2$ be a natural number and $P(x) = a_m x^m + \cdots + a_1 x + a_0$ a polynomial with distinct coefficients. Then the set $\{Q(n) \mid Q \in M(P)\}$ contains at least 2^m elements.

Proof. We shall use induction on m. The statement is easily verified for $m = 1$. Assume w.l.o.g. that $a_m < \cdots < a_1 < a_0$. Consider two polynomials Q_k and Q_{k+1} of the form

$$Q_k(x) = a_m x^m + \cdots + a_k x^k + a_0 x^{k-1} + b_{k-1} x^{k-2} + \cdots + b_1,$$
$$Q_{k+1}(x) = a_m x^m + \cdots + a_{k+1} x^{k+1} + a_0 x^k + c_k x^{k-1} + \cdots + c_1,$$

where $(b_{k-1}, \ldots, b_1)$ and $(c_k, \ldots, c_1)$ are respectively permutations of the sets $\{a_{k-1}, \ldots, a_1\}$ and $\{a_k, \ldots, a_1\}$. We claim that $Q_{k+1}(n) \geq Q_k(n)$. Indeed, since $a_0 - c_k \leq a_0 - a_k$ and $b_j - c_j < a_0 - a_k$ for $1 \leq j \leq n-1$, we have $Q_{k+1}(n) - Q_k(x) = (a_0 - a_k)n^k - (a_0 - c_k)n^{k-1} - (b_{k-1} - c_{k-1})n^{k-2} - \cdots - (b_1 - c_1) \geq (a_0 - a_k)(n^k - n^{k-1} - \cdots - n - 1) > 0$. Furthermore, by the induction hypothesis the polynomials of the form $Q_k(x)$ take at least 2^{k-2} values at $x = n$. Hence the total number of values of $Q(n)$ for $Q \in M(P)$ is at least $1 + 1 + 2 + 2^2 + \cdots + 2^{m-1} = 2^m$.

Now we return to the main result. Suppose that $P(x) = a_{2000}x^{2000} + a_{1999}x^{1999} + a_0$ is an n-independent polynomial. Since $P_2(x) = a_{2000}x^{2000} + a_{1998}x^{1998} + \cdots + a_2x^2 + a_0$ is a polynomial in $t = x^2$ of degree 1000, by the lemma it takes at least 2^{1000} distinct values at $x = n$. Hence $\{Q(n) \mid Q \in M(P)\}$ contains at least 2^{1000} elements. On the other hand, interchanging the coefficients b_i and b_j in a polynomial $Q(x) = b_{2000}x^{2000} + \cdots + b_0$ modifies the value of Q at $x = n$ by $(b_i - b_j)(n^i - n^j) = (a_k - a_l)(n^i - n^j)$ for some k, l. Hence there are fewer than 2001^4 possible modifications of the value at n. Since $2001^4 < 2^{1000}$, we have arrived at a contradiction.

14. The given condition is obviously equivalent to $a^2 \equiv 1 \pmod{n}$ for all integers a coprime to n. Let $n = p_1^{\alpha_1} p_2^{\alpha_2} \cdots p_k^{\alpha_k}$ be the factorization of n into primes. Since by the Chinese remainder theorem the numbers coprime to n can give any remainder modulo $p_i^{\alpha_i}$ except 0, our condition is equivalent to $a^2 \equiv 1 \pmod{p_i^{\alpha_i}}$ for all i and integers a coprime to p_i.

Now if $p_i \geq 3$, we have $2^2 \equiv 1 \pmod{p_i^{\alpha_i}}$, so $p_i = 3$ and $\alpha_i = 2$. If $p_j = 2$, then $3^2 \equiv 1 \pmod{2^{\alpha_j}}$ implies $\alpha_j \leq 3$. Hence n is a divisor of $2^3 \cdot 3 = 24$. Conversely, each $n \mid 24$ has the desired property.

15. Let $n = p_1^{\alpha_1} p_2^{\alpha_2} \cdots p_k^{\alpha_k}$ be the factorization of n onto primes ($p_1 < p_2 < \cdots < p_k$). Since $4n$ is a perfect cube, we deduce that $p_1 = 2$ and $\alpha_1 = 3\beta_1 + 1$, $\alpha_2 = 3\beta_2, \ldots, \alpha_k = 3\beta_k$ for some integers $\beta_i \geq 0$. Using $d(n) = (\alpha_1 + 1) \cdot (\alpha_2 + 1) \cdots (\alpha_k + 1)$ we can rewrite the equation $d(n)^3 = 4n$ as

$$(3\beta_1 + 2) \cdot (3\beta_2 + 1) \cdots (3\beta_k + 1) = 2^{\beta_1 + 1} p_2^{\beta_2} \cdots p_k^{\beta_k}.$$

Since $d(n)$ is not divisible by 3, it follows that $p_i \geq 5$ for $i \geq 2$. Thus the above equation is equivalent to

$$\frac{3\beta_1 + 2}{2^{\beta_1 + 1}} = \frac{p_2^{\beta_2}}{3\beta_2 + 1} \cdots \frac{p_k^{\beta_k}}{3\beta_k + 1}. \tag{1}$$

For $i \geq 2$ we have $p_i^{\beta_i} \geq (1 + 4)^{\beta_i} \geq 1 + 4\beta_i$; hence (1) implies that $\frac{3\beta_1 + 2}{2^{\beta_1 + 1}} \geq 1$, which leads to $\beta_1 \leq 2$.

For $\beta_1 = 0$ or $\beta_1 = 2$ we have that $\frac{3\beta_1 + 2}{2^{\beta_1 + 1}} = 1$, and therefore $\beta_2 = \cdots = \beta_k = 0$. This yields the solutions $n = 2$ and $n = 2^7 = 128$.

For $\beta_1 = 1$ the left-hand side of (1) equals $\frac{5}{4}$. On the other hand, if $p_i > 5$ or $\beta_i > 1$, then $\frac{p_i^{\beta_i}}{3\beta_i + 1} > \frac{5}{4}$, which is impossible. We conclude that $p_2 = 5$ and $k = 2$, so $n = 2000$.

Hence the solutions for n are $2, 128$, and 2000.

16. More generally, we will prove by induction on k that for each $k \in \mathbb{N}$ there exists $n_k \in \mathbb{N}$ that has exactly k distinct prime divisors such that $n_k \mid 2^{n_k} + 1$ and $3 \mid n_k$. For $k = 1$, $n_1 = 3$ satisfies the given conditions. Now assume that $k \geq 1$ and $n_k = 3^\alpha m$ where $3 \nmid m$, so that m has exactly $k - 1$ prime divisors. Then the number $3n_k = 3^{\alpha+1}m$ has exactly k prime divisors and $2^{3n_k} + 1 = (2^{n_k} + 1)(2^{2n_k} - 2^{n_k} + 1)$ is divisible by $3n_k$, since $3 \mid 2^{2n_k} - 2^{n_k} + 1$. We shall find a prime p not dividing n_k such that $n_{k+1} = 3pn_k$. It is enough to find p such that $p \mid 2^{3n_k} + 1$ and $p \nmid 2^{n_k} + 1$. Moreover, we shall show that for every integer $a > 2$ there exists a prime number p that divides $a^3 + 1 = (a+1)(a^2 - a + 1)$ but not $a + 1$. To prove this we observe that $\gcd(a^2 - a + 1, a + 1) = \gcd(3, a + 1)$. Now if $3 \nmid a + 1$, we can simply take $p = 3$; otherwise, if $a = 3b - 1$, then $a^2 - a + 1 = 9b^2 - 9b + 3$ is not divisible by 3^2; hence we can take for p any prime divisor of $\frac{a^2 - a + 1}{3}$.

17. Trivially all triples $(a, 1, n)$ and $(1, m, n)$ are solutions. Assume now that $a > 1$ and $m > 1$.
   If m is even, then $a^m + 1 \equiv (-1)^m + 1 \equiv 2 \pmod{a+1}$, which implies that $a^m + 1 = 2^t$. In particular, a is odd. But this is impossible, since $2 < a^m + 1 = (a^{m/2})^2 + 1 \equiv 2 \pmod 4$. Hence m is odd.
   Let p be an arbitrary prime divisor of m and $m = pm_1$. Then $a^m + 1 \mid (a+1)^n \mid (a^{m_1} + 1)^n$, so $b^p + 1 \mid (b+1)^n$ for $b = a^{m_1}$. It follows that

$$P = \frac{b^p + 1}{b + 1} = b^{p-1} - b^{p-2} + \cdots + 1 \mid (b+1)^n.$$

   Since $P \equiv p \pmod{b+1}$, we deduce that P has no prime divisors other than p; hence P is a power of p and $p \mid b + 1$. Let $b = kp - 1$, $k \in \mathbb{N}$. Then by the binomial formula we have $b^i = (kp - 1)^i \equiv (-1)^{i+1}(ikp - 1) \pmod{p^2}$, and therefore $P \equiv -kp((p-1) + (p-2) + \cdots + 1) + p \equiv p \pmod{p^2}$. We conclude that $P \leq p$. But we also have $P \geq b^{p-1} - b^{p-2} \geq b^{p-2} > p$ for $p > 3$, so we must have $P = p = 3$ and $b = 2$. Since $b = a^{m_1}$, we obtain $a = 2$ and $m = 3$. The triple $(2, 3, n)$ is indeed a solution if $n \geq 2$.
   Hence the set of solutions is $\{(a, 1, n), (1, m, n) \mid a, m, n \in \mathbb{N}\} \cup \{(2, 3, n) \mid n \geq 2\}$.
   Remark. This problem is very similar to (SL97-14).

18. The area of the triangle is $S = pr = p^2/n$ and $S = \sqrt{p(p-a)(p-b)(p-c)}$. It follows that $p^3 = n^2(p-a)(p-b)(p-c)$, which by putting $x = p - a$, $y = p - b$, and $z = p - c$ transforms into

$$(x + y + z)^3 = n^2 xyz. \tag{1}$$

   We will be done if we show that (1) has a solution in positive integers for infinitely many natural numbers n. Let us assume that $z = k(x + y)$ for an integer $k > 0$. Then (1) becomes $(k+1)^3(x+y)^2 = kn^2xy$. Further, by setting $n = 3(k+1)$ this equation reduces to

$$(k+1)(x+y)^2 = 9kxy. \tag{2}$$

Set $t = x/y$. Then (2) has solutions in positive integers if and only if $(k+1)(t+1)^2 = 9kt$ has a rational solution, i.e., if and only if its discriminant $D = k(5k-4)$ is a perfect square. Setting $k = u^2$, we are led to show that $5u^2 - 4 = v^2$ has infinitely many integer solutions. But this is a classic Pell-type equation, whose solution is every Fibonacci number $u = F_{2i+1}$. This completes the proof.

19. Suppose that a natural number N satisfies $N = a_1^2 + \cdots + a_k^2$, $2N = b_1^2 + \cdots + b_l^2$, where a_i, b_j are natural numbers such that none of the ratios a_i/a_j, b_i/b_j, a_i/b_j, b_j/a_i is a power of 2.

We claim that every natural number $n > \sum_{i=0}^{4N-2}(2iN+1)^2$ can be represented as a sum of distinct squares. Suppose $n = 4qN + r$, $0 \le r < 4N$. Then

$$n = 4Ns + \sum_{i=0}^{r-1}(2iN+1)^2$$

for some positive integer s, so it is enough to show that $4Ns$ is a sum of distinct even squares. Let $s = \sum_{c=1}^{C} 2^{2u_c} + \sum_{d=1}^{D} 2^{2v_d+1}$ be the binary expansion of s. Then

$$4Ns = \sum_{c=1}^{C}\sum_{i=1}^{k}(2^{u_c+1}a_i)^2 + \sum_{d=1}^{D}\sum_{j=1}^{l}(2^{u_d+1}b_j)^2,$$

where all the summands are distinct by the condition on a_i, b_j.

It remains to choose an appropriate N: for example $N = 29$, because $29 = 5^2 + 2^2$ and $58 = 7^2 + 3^2$.

Second solution. It can be directly checked that every odd integer $67 < n \le 211$ can be represented as a sum of distinct squares. For any $n > 211$ we can choose an integer m such that $m^2 > \frac{n}{2}$ and $n - m^2$ is odd and greater than 67, and therefore by the induction hypothesis can be written as a sum of distinct squares. Hence n is also a sum of distinct squares.

20. Denote by k_1, k_2 the given circles and by k_3 the circle through A, B, C, D. We shall consider the case that k_3 is inside k_1 and k_2, since the other case is analogous.

Let AC and AD meet k_1 at points P and R, and BC and BD meet k_2 at Q and S respectively. We claim that PQ and RS are the common tangents to k_1 and k_2, and therefore P, Q, R, S are the desired points. The circles k_1 and k_3 are tangent to each other, so we have $DC \parallel RP$. Since

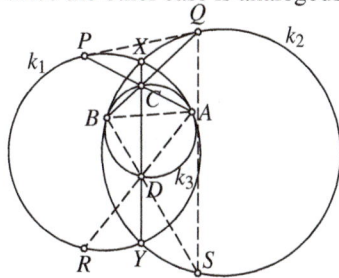

$$AC \cdot CP = XC \cdot CY = BC \cdot CQ,$$

the quadrilateral $ABQP$ must be cyclic, implying that the angles $\angle APQ$, $\angle ABQ$, $\angle ADC$, $\angle ARP$ are all equal. It follows that PQ is tangent to k_1. Similarly, PQ is tangent to k_2.

21. Let K be the intersection point of the lines MN and AB.

Since $KA^2 = KM \cdot KN = KB^2$, it follows that K is the midpoint of the segment AB, and consequently M is the midpoint of AB. Thus it will be enough to show that $EM \perp PQ$, or equivalently that $EM \perp AB$. However, since AB is tangent to the circle G_1 we have $\angle BAM = \angle ACM = \angle EAB$, and similarly $\angle ABM = \angle EBA$. This implies that the triangles EAB and MAB are congruent. Hence E and M are symmetric with respect to AB; hence $EM \perp AB$.

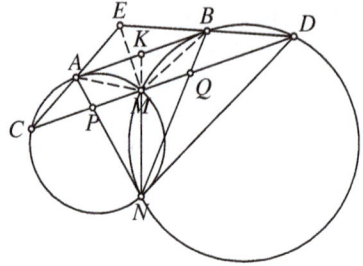

Remark. The proposer has suggested an alternative version of the problem: to prove that EN bisects the angle CND. This can be proved by noting that $EANB$ is cyclic.

22. Let L be the point symmetric to H with respect to BC. It is well known that L lies on the circumcircle k of $\triangle ABC$. Let D be the intersection point of OL and BC. We similarly define E and F. Then

$$OD + DH = OD + DL = OL = OE + EH = OF + FH.$$

We shall prove that AD, BE, and CF are concurrent. Let line AO meet BC at D'. It is easy to see that $\angle OD'D = \angle ODD'$; hence the perpendicular bisector of BC bisects DD' as well. Hence $BD = CD'$. If we define E' and F' analogously, we have $CE = AE'$ and $AF = BF'$. Since the lines AD', BE', CF' meet at O, it follows that:

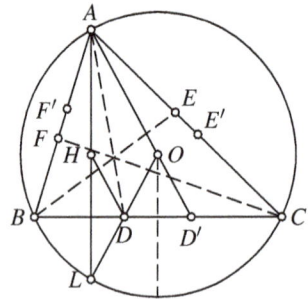

$$\frac{BD}{DC} \cdot \frac{CE}{EA} \cdot \frac{AF}{FB} = \frac{BD'}{D'C} \cdot \frac{CE'}{E'A} \cdot \frac{AF'}{F'B} = 1.$$

This proves our claim by Ceva's theorem.

23. First, suppose that there are numbers (b_i, c_i) assigned to the vertices of the polygon such that

$$A_iA_j = b_jc_i - b_ic_j \quad \text{for all } i, j \text{ with } 1 \le i \le j \le n. \tag{1}$$

In order to show that the polygon is cyclic, it is enough to prove that A_1, A_2, A_3, A_i lie on a circle for each i, $4 \le i \le n$, or equivalently, by Ptolemy's theorem, that $A_1A_2 \cdot A_3A_i + A_2A_3 \cdot A_iA_1 = A_1A_3 \cdot A_2A_i$. But this is straightforward with regard to (1).

Now suppose that $A_1A_2 \ldots A_n$ is a cyclic quadrilateral. By Ptolemy's theorem we have $A_iA_j = A_2A_j \cdot \frac{A_1A_i}{A_1A_2} - A_2A_i \cdot \frac{A_1A_j}{A_1A_2}$ for all i, j. This suggests taking $b_1 =$

$-A_1A_2$, $b_i = A_2A_i$ for $i \geq 2$ and $c_i = \frac{A_1A_i}{A_1A_2}$ for all i. Indeed, using Ptolemy's theorem, one easily verifies (1).

24. Since $\angle ABT = 180° - \gamma$ and $\angle ACT = 180° - \beta$, the law of sines gives $\frac{BP}{PC} = \frac{S_{ABT}}{S_{ACT}} = \frac{AB \cdot BT \cdot \sin\gamma}{AB \cdot BT \cdot \sin\beta} = \frac{AB\sin\gamma}{AC\sin\beta} = \frac{c^2}{b^2}$, which implies $BP = \frac{c^2a}{b^2+c^2}$. Denote by M and N the feet of perpendiculars from P and Q on AB. We have $\cot\angle ABQ = \frac{BN}{NQ} = $

$\frac{2BN}{PM} = \frac{BA+BM}{BP\sin\beta} = \frac{c+BP\cos\beta}{BP\sin\beta} = \frac{b^2+c^2+ac\cos\beta}{ca\sin\beta} = \frac{2(b^2+c^2)+a^2+c^2-b^2}{2ca\sin\beta} = \frac{a^2+b^2+3c^2}{4S_{ABC}} = $
$2\cot\alpha + 2\cot\beta + \cot\gamma$. Similarly, $\cot\angle BAS = 2\cot\alpha + 2\cot\beta + \cot\gamma$; hence $\angle ABQ = \angle BAS$.

Now put $p = \cot\alpha$ and $q = \cot\beta$. Since $p+q \geq 0$, the A-G mean inequality gives us $\cot\angle ABQ = 2p + 2q + \frac{1-pq}{p+q} \geq 2p + 2q + \frac{1-(p+q)^2/4}{p+q} = \frac{7}{4}(p+q) + \frac{1}{p+q} \geq 2\sqrt{\frac{7}{4}} = \sqrt{7}$. Hence $\angle ABQ \leq \arctan\frac{1}{\sqrt{7}}$. Equality holds if and only if $\cot\alpha = \cot\beta = \frac{1}{\sqrt{7}}$, i.e., when $a:b:c = 1:1:\frac{1}{\sqrt{2}}$.

25. By the condition of the problem, $\triangle ADX$ and $\triangle BCX$ are similar. Then there exist points Y' and Z' on the perpendicular bisector of AB such that $\triangle AY'Z'$ is similar and oriented the same as $\triangle ADX$, and $\triangle BY'Z'$ is (being congruent to $\triangle AY'Z'$) similar and oriented the same as $\triangle BCX$. Since then $AD/AY' = AX/AZ'$ and $\angle DAY' = \angle XAZ'$, $\triangle ADY'$ and $\triangle AXZ'$ are also similar, implying $\frac{AD}{AX} = \frac{DY'}{XZ'}$. Analogously, $\frac{BC}{BX} = \frac{CY'}{XZ'}$. It follows from $\frac{AD}{AX} = \frac{BC}{BX}$ that $CY' = DY'$, which means that Y' lies on the perpendicular bisector of CD. Hence $Y' \equiv Y$.
Now $\angle AYB = 2\angle AYZ' = 2\angle ADX$, as desired.

26. The problem can be reformulated in the following way: *Given a set S of ten points in the plane such that the distances between them are all distinct, for each point $P \in S$ we mark the point $Q \in S \setminus \{P\}$ nearest to P. Find the least possible number of marked points.*
Observe that each point $A \in S$ is the nearest to at most five other points. Indeed, for any six points $P_1,\ldots,P_6$ one of the angles P_iAP_j is at most 60°, in which case P_iP_j is smaller than one of the distances AP_i, AP_j. It follows that at least two points are marked.
Now suppose that exactly two points, say A and B, are marked. Then AB is the minimal distance of the points from S, so by the previous observation the rest of the set S splits into two subsets of four points according to whether the nearest point is A or B. Let these subsets be $\{A_1,A_2,A_3,A_4\}$ and $\{B_1,B_2,B_3,B_4\}$ respectively. Assume that the points are labelled so that the angles A_iAA_{i+1} are successively adjacent as well as the angles B_iBB_{i+1}, and that A_1,B_1 lie on one side of AB, and A_4,B_4 lie on the other side. Since all the angles A_iAA_{i+1} and B_iBB_{i+1} are greater than 60°, it follows that

$$\angle A_1AB + \angle BAA_4 + \angle B_1BA + \angle ABB_4 < 360°.$$

Therefore $\angle A_1AB + \angle B_1BA < 180°$ or $\angle A_4AB + \angle B_4BA < 180°$. Without loss of generality, let us assume the first inequality.

On the other hand, note that the quadrilateral ABB_1A_1 is convex because A_1 and B_1 are on different sides of the perpendicular bisector of AB. From $A_1B_1 > A_1A$ and $BB_1 > AB$ we obtain $\angle A_1AB_1 > \angle A_1B_1A$ and $\angle BAB_1 > \angle AB_1B$. Adding these relations yields $\angle A_1AB > \angle A_1B_1B$. Similarly, $\angle B_1BA > \angle B_1A_1A$. Adding these two inequalities, we get

$$180° > \angle A_1AB + \angle B_1BA > \angle A_1B_1B + \angle B_1A_1A;$$

hence the sum of the angles of the quadrilateral ABB_1A_1 is less than $360°$, which is a contradiction. Thus at least 3 points are marked.

An example of a configuration in which exactly 3 gangsters are killed is shown below.

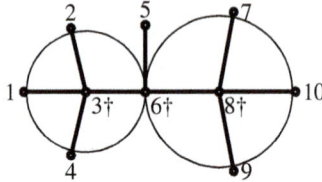

27. Denote by $\alpha_1, \alpha_2, \alpha_3$ the angles of $\triangle A_1A_2A_3$ at vertices A_1, A_2, A_3 respectively. Let T_1, T_2, T_3 be the points symmetric to L_1, L_2, L_3 with respect to A_1I, A_2I, and A_3I respectively. We claim that $T_1T_2T_3$ is the desired triangle.

Denote by S_1 and R_1 the points symmetric to K_1 and K_3 with respect to L_1L_3. It is enough to show that T_1 and T_3 lie on the line R_1S_1. To prove this, we shall prove that $\angle K_1S_1T_1 = \angle K'K_1S_1$ for a point K' on the line K_1K_3 such that K_3 and K' lie on different sides of K_1. We show first that $S_1 \in A_1I$. Let X be the point of intersection of lines A_1I and L_1L_3. We see from the triangle A_1L_3X that $\angle L_1XI = \alpha_3/2 = \angle L_1A_3I$, which implies that L_1XA_3I is cyclic. We now have $\angle A_1XA_3 = 90° = \angle A_1K_1A_3$; hence $A_1K_1XA_3$ is also cyclic. This fact now implies that $\angle K_1XI = \angle K_1A_3A_1 = \alpha_3 = 2\angle L_1XI$. Therefore X_1L_1 bisects the angle K_1X_1I. Hence $S_1 \in XI$ as claimed. Now we have $\angle K_1S_1T_1 = \angle K_1S_1L_1 + 2\angle L_1S_1X = \angle S_1K_1L_1 + 2\angle L_1K_1X$. It remains to prove that the line K_1X bisects $\angle A_3K_1K'$. From the cyclic quadrilateral $A_1K_1XA_3$ we see that $\angle XK_1A_3 = \alpha_1/2$. Since $A_1K_3K_1A_3$ is cyclic, we also have $\angle K'K_1A_3 = \alpha_1 = 2\angle XK_1A_3$, which proves the claim.

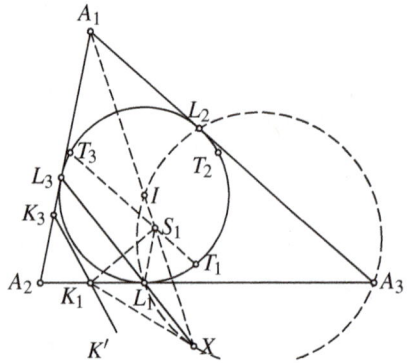

4.42 Solutions to the Shortlisted Problems of IMO 2001

1. First, let us show that there is at most one such function. Suppose that f_1 and f_2 are two such functions, and consider $g = f_1 - f_2$. Then g is zero on the boundary and satisfies

$$g(p,q,r) = \frac{1}{6}[g(p+1,q-1,r) + \cdots + g(p,q-1,r+1)],$$

i.e., $g(p,q,r)$ is equal to the average of the values of g at six points $(p+1,q-1,r),\ldots$ that lie in the plane π given by $x+y+z = p+q+r$. Suppose that (p,q,r) is the point at which g attains its maximum in absolute value on $\pi \cap T$. The averaging property of g implies that the values of g at $(p+1,q-1,r)$ etc. are all equal to $g(p,q,r)$. Repeating this argument we obtain that g is constant on the whole of $\pi \cap T$, and hence it equals 0 everywhere. Therefore $f_1 \equiv f_2$.
   It remains to guess f. It is natural to try $\overline{f}(p,q,r) = pqr$ first: it satisfies $\overline{f}(p,q,r) = \frac{1}{6}[\overline{f}(p+1,q-1,r) + \cdots + \overline{f}(p,q-1,r+1)] + \frac{p+q+r}{3}$. Thus we simply take

$$f(p,q,r) = \frac{3}{p+q+r}\overline{f}(p,q,r) = \frac{3pqr}{p+q+r}$$

and directly check that it satisfies the required property. Hence this is the unique solution.

2. It follows from Bernoulli's inequality that for each $n \in \mathbb{N}$, $\left(1+\frac{1}{n}\right)^n \geq 2$, or $\sqrt[n]{2} \leq 1+\frac{1}{n}$. Consequently, it will be enough to show that $1+a_n > \left(1+\frac{1}{n}\right)a_{n-1}$. Assume the opposite. Then there exists N such that for each $n \geq N$,

$$1+a_n \leq \left(1+\frac{1}{n}\right)a_{n-1}, \quad \text{i.e.,} \quad \frac{1}{n+1} + \frac{a_n}{n+1} \leq \frac{a_{n-1}}{n}.$$

Summing for $n = N,\ldots,m$ yields $\frac{a_m}{m+1} \leq \frac{a_{N-1}}{N} - \left(\frac{1}{N+1} + \cdots + \frac{1}{m+1}\right)$. However, it is well known that the sum $\frac{1}{N+1} + \cdots + \frac{1}{m+1}$ can be arbitrarily large for m large enough, so that $\frac{a_m}{m+1}$ is eventually negative. This contradiction yields the result.

Second solution. Suppose that $1+a_n \leq \sqrt[n]{2}a_{n-1}$ for all $n \geq N$. Set $b_n = 2^{-(1+1/2+\cdots+1/n)}$ and multiply both sides of the above inequality to obtain $b_n + b_n a_n \leq b_{n-1}a_{n-1}$. Thus

$$b_N a_N > b_N a_N - b_n a_n \geq b_N + b_{N+1} + \cdots + b_n.$$

However, it can be shown that $\sum_{n>N} b_N$ diverges: in fact, since $1 + \frac{1}{2} + \cdots + \frac{1}{n} < 1 + \ln n$, we have $b_n > 2^{-1-\ln n} = \frac{1}{2}n^{-\ln 2} > \frac{1}{2n}$, and we already know that $\sum_{n>N} \frac{1}{2n}$ diverges.

Remark. As can be seen from both solutions, the value 2 in the problem can be increased to e.

3. By the arithmetic–quadratic mean inequality, it suffices to prove that

$$\frac{x_1^2}{(1+x_1^2)^2}+\frac{x_2^2}{(1+x_1^2+x_2^2)^2}+\cdots+\frac{x_n^2}{(1+x_1^2+\cdots+x_n^2)^2} < 1.$$

Observe that for $k \geq 2$ the following holds:

$$\frac{x_k^2}{(1+x_1^2+\cdots+x_k^2)^2} \leq \frac{x_k^2}{(1+\cdots+x_{k-1}^2)(1+\cdots+x_k^2)}$$

$$= \frac{1}{1+x_1^2+\cdots+x_{k-1}^2} - \frac{1}{1+x_1^2+\cdots+x_k^2}.$$

For $k = 1$ we have $\frac{x_1^2}{(1+x_1)^2} \leq 1 - \frac{1}{1+x_1^2}$. Summing these inequalities, we obtain

$$\frac{x_1^2}{(1+x_1^2)^2}+\cdots+\frac{x_n^2}{(1+x_1^2+\cdots+x_n^2)^2} \leq 1 - \frac{1}{1+x_1^2+\cdots+x_n^2} < 1.$$

Second solution. Let $a_n(k) = \sup\left(\frac{x_1}{k^2+x_1^2}+\cdots+\frac{x_n}{k^2+x_1^2+\cdots+x_n^2}\right)$ and $a_n = a_n(1)$. We must show that $a_n < \sqrt{n}$. Replacing x_i by kx_i shows that $a_n(k) = a_n/k$. Hence

$$a_n = \sup_{x_1}\left(\frac{x_1}{1+x_1^2}+\frac{a_{n-1}}{\sqrt{1+x_1^2}}\right) = \sup_\theta(\sin\theta\cos\theta+a_{n-1}\cos\theta), \qquad (1)$$

where $\tan\theta = x_1$. The above supremum can be computed explicitly:

$$a_n = \frac{1}{8\sqrt{2}}\left(3a_{n-1}+\sqrt{a_{n-1}^2+8}\right)\sqrt{4-a_{n-1}^2+a_{n-1}\sqrt{a_{n-1}^2+8}}.$$

However, the required inequality is weaker and can be proved more easily: if $a_{n-1} < \sqrt{n-1}$, then by (1) $a_n < \sin\theta + \sqrt{n-1}\cos\theta = \sqrt{n}\sin(\theta+\alpha) \leq \sqrt{n}$, for $\alpha \in (0, \pi/2)$ with $\tan\alpha = \sqrt{n}$.

4. Let $(*)$ denote the given functional equation. Substituting $y = 1$ we get $f(x)^2 = xf(x)f(1)$. If $f(1) = 0$, then $f(x) = 0$ for all x, which is the trivial solution. Suppose $f(1) = C \neq 0$. Let $G = \{y \in \mathbb{R} \mid f(y) \neq 0\}$. Then

$$f(x) = \begin{cases} Cx & \text{if } x \in G, \\ 0 & \text{otherwise.} \end{cases} \qquad (1)$$

We must determine the structure of G so that the function defined by (1) satisfies $(*)$.

  (i) Clearly $1 \in G$, because $f(1) \neq 0$.
  (ii) If $x \in G$, $y \notin G$, then by $(*)$ it holds $f(xy)f(x) = 0$, so $xy \notin G$.
  (iii) If $x, y \in G$, then $x/y \in G$ (otherwise by (ii), $y(x/y) = x \notin G$).

(iv) If $x, y \in G$, then by (iii) we have $x^{-1} \in G$, so $xy = y/x^{-1} \in G$.

Hence G is a set that contains 1, does not contain 0, and is closed under multiplication and division. Conversely, it is easy to verify that every such G in (1) gives a function satisfying $(*)$.

5. Let $a_1, a_2, \ldots, a_n$ satisfy the conditions of the problem. Then $a_k > a_{k-1}$, and hence $a_k \geq 2$ for $k = 1, \ldots, n$. The inequality $(a_{k+1} - 1)a_{k-1} \geq a_k^2(a_k - 1)$ can be rewritten as

$$\frac{a_{k-1}}{a_k} + \frac{a_k}{a_{k+1} - 1} \leq \frac{a_{k-1}}{a_k - 1}.$$

Summing these inequalities for $k = i+1, \ldots, n-1$ and using the obvious inequality $\frac{a_{n-1}}{a_n} < \frac{a_{n-1}}{a_{n-1}}$, we obtain $\frac{a_i}{a_{i+1}} + \cdots + \frac{a_{n-1}}{a_n} < \frac{a_i}{a_{i+1}-1}$. Therefore

$$\frac{a_i}{a_{i+1}} \leq \frac{99}{100} - \frac{a_0}{a_1} - \cdots - \frac{a_{i-1}}{a_i} < \frac{a_i}{a_{i+1} - 1} \qquad \text{for } i = 1, 2, \ldots, n-1. \qquad (1)$$

Consequently, given $a_0, a_1, \ldots, a_i$, there is at most one possibility for a_{i+1}. In our case, (1) yields $a_1 = 2$, $a_2 = 5$, $a_3 = 56$, $a_4 = 280^2 = 78400$. These values satisfy the conditions of the problem, so that this is a unique solution.

6. We shall determine a constant $k > 0$ such that

$$\frac{a}{\sqrt{a^2 + 8bc}} \geq \frac{a^k}{a^k + b^k + c^k} \qquad \text{for all } a, b, c > 0. \qquad (1)$$

This inequality is equivalent to $(a^k + b^k + c^k)^2 \geq a^{2k-2}(a^2 + 8bc)$, which further reduces to

$$(a^k + b^k + c^k)^2 - a^{2k} \geq 8a^{2k-2}bc.$$

On the other hand, the AM–GM inequality yields

$$(a^k + b^k + c^k)^2 - a^{2k} = (b^k + c^k)(2a^k + b^k + c^k) \geq 8a^{k/2}b^{3k/4}c^{3k/4},$$

and therefore $k = 4/3$ is a good choice. Now we have

$$\frac{a}{\sqrt{a^2 + 8bc}} + \frac{b}{\sqrt{b^2 + 8ca}} + \frac{c}{\sqrt{c^2 + 8ab}}$$

$$\geq \frac{a^{4/3}}{a^{4/3} + b^{4/3} + c^{4/3}} + \frac{b^{4/3}}{a^{4/3} + b^{4/3} + c^{4/3}} + \frac{c^{4/3}}{a^{4/3} + b^{4/3} + c^{4/3}} = 1.$$

Second solution. The numbers $x = \frac{a}{\sqrt{a^2 + 8bc}}$, $y = \frac{b}{\sqrt{b^2 + 8ca}}$ and $z = \frac{c}{\sqrt{c^2 + 8ab}}$ satisfy

$$f(x, y, z) = \left(\frac{1}{x^2} - 1\right)\left(\frac{1}{y^2} - 1\right)\left(\frac{1}{z^2} - 1\right) = 8^3.$$

Our task is to prove $x + y + z \geq 1$.

Since f is decreasing on each of the variables x, y, z, this is the same as proving that $x, y, z > 0$, $x + y + z = 1$ implies $f(x, y, z) \geq 8^3$. However, since $\frac{1}{x^2} - 1 = \frac{(x+y+z)^2 - x^2}{x^2} = \frac{(2x+y+z)(y+z)}{x^2}$, the inequality $f(x, y, z) \geq 8^3$ becomes

$$\frac{(2x+y+z)(x+2y+z)(x+y+2z)(y+z)(z+x)(x+y)}{x^2y^2z^2} \geq 8^3,$$

which follows immediately by the AM–GM inequality.

Third solution. We shall prove a more general fact: the inequality $\dfrac{a}{\sqrt{a^2+kbc}}+$

$\dfrac{b}{\sqrt{b^2+kca}}+\dfrac{c}{\sqrt{c^2+kab}} \geq \dfrac{3}{\sqrt{1+k}}$ is true for all $a,b,c > 0$ if and only if $k \geq 8$.

Firstly suppose that $k \geq 8$. Setting $x = bc/a^2$, $y = ca/b^2$, $z = ab/c^2$, we reduce the desired inequality to

$$F(x,y,z) = f(x) + f(y) + f(z) \geq \frac{3}{\sqrt{1+k}}, \quad \text{where } f(t) = \frac{1}{\sqrt{1+kt}}, \quad (2)$$

for $x,y,z > 0$ such that $xyz = 1$. We shall prove (2) using the method of Lagrange multipliers.

The boundary of the set $D = \{(x,y,z) \in \mathbb{R}^3_+ \mid xyz = 1\}$ consists of points (x,y,z) with one of x,y,z being 0 and another one being $+\infty$. If w.l.o.g. $x = 0$, then $F(x,y,z) \geq f(x) = 1 \geq 3/\sqrt{1+k}$.

Suppose now that (x,y,z) is a point of local minimum of F on D. There exists $\lambda \in \mathbb{R}$ such that (x,y,z) is stationary point of the function $F(x,y,z) + \lambda xyz$. Then (x,y,z,λ) is a solution to the system $f'(x) + \lambda yz = f'(y) + \lambda xz = f'(z) + \lambda xy = 0$, $xyz = 1$. Eliminating λ gives us

$$xf'(x) = yf'(y) = zf'(z), \quad xyz = 1. \quad (3)$$

The function $tf'(t) = \frac{-kt}{2(1+kt)^{3/2}}$ decreases on the interval $(0,2/k]$ and increases on $[2/k,+\infty)$ because $(tf'(t))' = \frac{k(kt-2)}{4(1+kt)^{5/2}}$. It follows that two of the numbers x,y,z are equal. If $x = y = z$, then $(1,1,1)$ is the only solution to (3). Suppose that $x = y \neq z$. Since $(yf'(y))^2 - (zf'(z))^2 = \frac{k^2(z-y)(k^3y^2z^2-3kyz-y-z)}{4(1+ky)^3(1+kz)^3}$, (3) gives us $y^2z = 1$ and $k^3y^2z^2 - 3kyz - y - z = 0$. Eliminating z we obtain an equation in y, $k^3/y^2 - 3k/y - y - 1/y^2 = 0$, whose only real solution is $y = k - 1$. Thus $(k-1, k-1, 1/(k-1)^2)$ and the cyclic permutations are the only solutions to (3) with x,y,z being not all equal. Since $F(k-1, k-1, 1/(k-1)^2) = (k+1)/\sqrt{k^2-k+1} > F(1,1,1) = 1$, the inequality (2) follows.

For $0 < k < 8$ we have that

$$\frac{a}{\sqrt{a^2+kbc}} + \frac{b}{\sqrt{b^2+kca}} + \frac{c}{\sqrt{c^2+kab}}$$
$$> \frac{a}{\sqrt{a^2+8bc}} + \frac{b}{\sqrt{b^2+8ca}} + \frac{c}{\sqrt{c^2+8ab}}$$
$$\geq 1.$$

If we fix c and let a,b tend to 0, the first two summands will tend to 0 while the third will tend to 1. Hence the inequality cannot be improved.

7. It is evident that arranging of A in increasing order does not diminish m. Thus we can assume that A is nondecreasing. Assume w.l.o.g. that $a_1 = 1$, and let b_i be the number of elements of A that are equal to i $(1 \le i \le n = a_{2001})$. Then we have $b_1 + b_2 + \cdots + b_n = 2001$ and

$$m = b_1 b_2 b_3 + b_2 b_3 b_4 + \cdots + b_{n-2} b_{n-1} b_n. \tag{1}$$

Now if b_i, b_j $(i < j)$ are two largest b's, we deduce from (1) and the AM–GM inequality that $m \le b_i b_j (b_1 + \cdots + b_{i-1} + b_{i+1} + \cdots + b_{j-1} + b_{j+1} + b_n) \le \left(\frac{2001}{3}\right)^3 = 667^3$ $(b_1 b_2 b_3 \le b_1 b_i b_j$, etc.). The value 667^3 is attained for $b_1 = b_2 = b_3 = 667$ (i.e., $a_1 = \cdots = a_{667} = 1$, $a_{668} = \cdots = a_{1334} = 2$, $a_{1335} = \cdots = a_{2001} = 3$). Hence the maximum of m is 667^3.

8. Suppose to the contrary that all the $S(a)$'s are different modulo $n!$. Then the sum of $S(a)$'s over all permutations a satisfies $\sum_a S(a) \equiv 0 + 1 + \cdots + (n! - 1) = \frac{(n!-1)n!}{2} \equiv \frac{n!}{2} \pmod{n!}$. On the other hand, the coefficient of c_i in $\sum_a S(a)$ is equal to $(n-1)!(1 + 2 + \cdots + n) = \frac{n+1}{2} n!$ for all i, from which we obtain

$$\sum_a S(a) \equiv \frac{n+1}{2} (c_1 + \cdots + c_n) n! \equiv 0 \pmod{n!}$$

for odd n. This is a contradiction.

9. Consider one such party. The result is trivially true if there is only one 3-clique, so suppose there exist at least two 3-cliques C_1 and C_2. We distinguish two cases:
   (i) $C_1 = \{a, b, c\}$ and $C_2 = \{a, d, e\}$ for some distinct people a, b, c, d, e. If the departure of a destroys all 3-cliques, then we are done. Otherwise, there is a third 3-clique C_3, which has a person in common with each of C_1, C_2 and does not include a: say, $C_3 = \{b, d, f\}$ for some f. We thus obtain another 3-clique $C_4 = \{a, b, d\}$, which has two persons in common with C_3, and the case (ii) is applied.
   (ii) $C_1 = \{a, b, c\}$ and $C_2 = \{a, b, d\}$ for distinct people a, b, c, d. If the departure of a, b leaves no 3-clique, then we are done. Otherwise, for some e there is a clique $\{c, d, e\}$.
   We claim that then the departure of c, d breaks all 3-cliques. Suppose the opposite, that a 3-clique C remains. Since C shares a person with each of the 3-cliques $\{c, d, a\}, \{c, d, b\}, \{c, d, e\}$, it must be $C = \{a, b, e\}$. However, then $\{a, b, c, d, e\}$ is a 5-clique, which is assumed to be impossible.

10. For convenience let us write $a = 1776$, $b = 2001$, $0 < a < b$. There are two types of historic sets:

    (1) $\{x, x + a, x + a + b\}$ and   (2) $\{x, x + b, x + a + b\}$.

    We construct a sequence of historic sets $H_1, H_2, H_3, \ldots$ inductively as follows:
    (i) $H_1 = \{0, a, a + b\}$, and
    (ii) Let y_n be the least nonnegative integer not occurring in $U_n = H_1 \cap \cdots \cap H_n$. We take H_{n+1} to be $\{y_n, y_n + a, y_n + a + b\}$ if $y_n + a \notin U_n$, and $\{y_n, y_n + b, y_n + a + b\}$ otherwise.

It remains to show that this construction never fails. Suppose that it failed at the construction of H_{n+1}. The element $y_n + a + b$ is not contained in U_n, since by the construction the smallest elements of $H_1, \ldots, H_n$ are all less than y_n. Hence the reason for the failure must be the fact that both $y_n + a$ and $y_n + b$ are covered by U_n. Further, $y_n + b$ must have been the largest element of its set H_k, so the smallest element of H_k equals $y_n - a$. But since y_n is not covered, we conclude that H_k is of type (2). This is a contradiction, because y_n was free, so by the algorithm we had to choose for H_k the set of type (1) (that is, $\{y_n - a, y_n, y_n + b\}$) first.

11. Let $(x_0, x_1, \ldots, x_n)$ be any such sequence: its terms are clearly nonnegative integers. Also, $x_0 = 0$ yields a contradiction, so $x_0 > 0$. Let m be the number of positive terms among $x_1, \ldots, x_n$. Since x_i counts the terms equal to i, the sum $x_1 + \cdots + x_n$ counts the total number of positive terms in the sequence, which is known to be $m + 1$. Therefore among $x_1, \ldots, x_n$ exactly $m - 1$ terms are equal to 1, one is equal to 2, and the others are 0. Only x_0 can exceed 2, and consequently at most one of $x_3, x_4, \ldots$ can be positive. It follows that $m \leq 3$.
    (i) $m = 1$: Then $x_2 = 2$ (since $x_1 = 2$ is impossible), so $x_0 = 2$. The resulting sequence is $(2, 0, 2, 0)$.
    (ii) $m = 2$: Either $x_1 = 2$ or $x_2 = 2$. These cases yield $(1, 2, 1, 0)$ and $(2, 1, 2, 0, 0)$ respectively.
    (iii) $m = 3$: This means that $x_k > 0$ for some $k > 2$. Hence $x_0 = k$ and $x_k = 1$. Further, $x_1 = 1$ is impossible, so $x_1 = 2$ and $x_2 = 1$; there are no more positive terms in the sequence. The resulting sequence is $(p, 2, 1, \underbrace{0, \ldots, 0}_{p-3}, 1, 0, 0, 0)$.

    Remark. The same problem occurred as (LL83-15), proposed by Canada.

12. For each balanced sequence $a = (a_1, a_2, \ldots, a_{2n})$ denote by $f(a)$ the sum of j's for which $a_j = 1$ (for example, $f(100101) = 1 + 4 + 6 = 11$). Partition the $\binom{2n}{n}$ balanced sequences into $n + 1$ classes according to the residue of f modulo $n + 1$. Now take S to be a class of minimum size: obviously $|S| \leq \frac{1}{n+1}\binom{2n}{n}$. We claim that every balanced sequence a is either a member of S or a neighbor of a member of S. We consider two cases.
    (i) Let a_1 be 1. It is easy to see that moving this 1 just to the right of the kth 0, we obtain a neighboring balanced sequence b with $f(b) = f(a) + k$. Thus if $a \notin S$, taking a suitable $k \in \{1, 2, \ldots, n\}$ we can achieve that $b \in S$.
    (ii) Let a_1 be 0. Taking this 0 just to the right of the kth 1 gives a neighbor b with $f(b) = f(a) - k$, and the conclusion is similar to that of (i).
    This justifies our claim.

13. At any moment, let p_i be the number of pebbles in the ith column, $i = 1, 2, \ldots$. The final configuration has obvious properties $p_1 \geq p_2 \geq \cdots$ and $p_{i+1} \in \{p_i, p_i - 1\}$. We claim that $p_{i+1} = p_i > 0$ is possible for at most one i.
    Assume the opposite. Then the final configuration has the property that for some r and $s > r$ we have $p_{r+1} = p_r$, $p_{s+1} = p_s > 0$ and $p_{r+k} = p_{r+1} - k + 1$ for all $k = 1, \ldots, s - r$. Consider the earliest configuration, say C, with this property. What

was the last move before C? The only possibilities are moving a pebble either from the rth or from the sth column; however, in both cases the configuration preceding this last move had the same property, contradicting the assumption that C is the earliest.

Therefore the final configuration looks as follows: $p_1 = a \in \mathbb{N}$, and for some r, p_i equals $a - (i-1)$ if $i \leq r$, and $a - (i-2)$ otherwise. It is easy to determine a, r: since $n = p_1 + p_2 + \cdots = \frac{(a+1)(a+2)}{2} - r$, we get $\frac{a(a+1)}{2} \leq n < \frac{(a+1)(a+2)}{2}$, from which we uniquely find a and then r as well.

The final configuration for $n = 13$:

14. We say that a problem is *difficult for boys* if at most two boys solved it, and *difficult for girls* if at most two girls solved it.

    Let us estimate the number of pairs boy-girl both of whom solved some problem difficult for boys. Consider any girl. By the condition (ii), among the six problems she solved, at least one was solved by at least 3 boys, and hence at most 5 were difficult for boys. Since each of these problems was solved by at most 2 boys and there are 21 girls, the considered number of pairs does not exceed $5 \cdot 2 \cdot 21 = 210$.

    Similarly, there are at most 210 pairs boy-girl both of whom solved some problem difficult for girls. On the other hand, there are $21^2 > 2 \cdot 210$ pairs boy-girl, and each of them solved one problem in common. Thus some problems were difficult neither for girls nor for boys, as claimed.

    Remark. The statement can be generalized: if $2(m-1)(n-1) + 1$ boys and as many girls participated, and nobody solved more than m problems, then some problem was solved by at least n boys and n girls.

15. Let $MNPQ$ be the square inscribed in $\triangle ABC$ with $M \in AB, N \in AC, P, Q \in BC$, and let AA_1 meet MN, PQ at K, X respectively. Put $MK = PX = m$, $NK = QX = n$, and $MN = d$. Then

    $$\frac{BX}{XC} = \frac{m}{n} = \frac{BX+m}{XC+n} = \frac{BP}{CQ} = \frac{d\cot\beta + d}{d\cot\gamma + d} = \frac{\cot\beta + 1}{\cot\gamma + 1}.$$

    Similarly, if BB_1 and CC_1 meet AC and BC at Y, Z respectively then $\frac{CY}{YA} = \frac{\cot\gamma + 1}{\cot\alpha + 1}$ and $\frac{AZ}{ZB} = \frac{\cot\alpha + 1}{\cot\beta + 1}$. Therefore $\frac{BX}{XC} \frac{CY}{YA} \frac{AZ}{ZB} = 1$, so by Ceva's theorem, AX, BY, CZ have a common point.

    Second solution. Let A_2 be the center of the square constructed over BC outside $\triangle ABC$. Since this square and the inscribed square corresponding to the side BC are homothetic, A, A_1, and A_2 are collinear. Points B_2, C_2 are analogously defined. Denote the angles $BAA_2, A_2AC, CBB_2, B_2BA, ACC_2, C_2CB$ by $\alpha_1, \alpha_2, \beta_1, \beta_2, \gamma_1, \gamma_2$. By the law of sines we have

$$\frac{\sin\alpha_1}{\sin\alpha_2} = \frac{\sin(\beta+45°)}{\sin(\gamma+45°)}, \quad \frac{\sin\beta_1}{\sin\beta_2} = \frac{\sin(\gamma+45°)}{\sin(\alpha+45°)}, \quad \frac{\sin\gamma_1}{\sin\gamma_2} = \frac{\sin(\alpha+45°)}{\sin(\beta+45°)}.$$

Since the product of these ratios is 1, by the trigonometric Ceva's theorem AA_2, BB_2, CC_2 are concurrent.

16. Since $\angle OCP = 90° - \angle A$, we are led to showing that $\angle OCP > \angle COP$, i.e., $OP > CP$. By the triangle inequality it suffices to prove $CP < \frac{1}{2}CO$.
Let $CO = R$. The law of sines yields

$$CP = AC\cos\gamma = 2R\sin\beta\cos\gamma < 2R\sin\beta\cos(\beta+30°).$$

Finally, we have

$$2\sin\beta\cos(\beta+30°) = \sin(2\beta+30°) - \sin 30° \le \frac{1}{2},$$

which completes the proof.

17. Let us investigate a more general problem, in which G is any point of the plane such that AG, BG, CG are sides of a triangle.
Let F be the point in the plane such that $BC : CF : FB = AG : BG : CG$ and F, A lie on different sides of BC. Then by Ptolemy's inequality, on $BPCF$ we have
$AG \cdot AP + BG \cdot BP + CG \cdot CP = AG \cdot AP + \frac{AG}{BC}(CF \cdot BP + BF \cdot CP) \ge AG \cdot AP + \frac{AG}{BC}BC \cdot PF$. Hence

$$AG \cdot AP + BG \cdot BP + CG \cdot CP \ge AG \cdot AF, \tag{1}$$

where equality holds if and only if P lies on the segment AF and on the circle BCF. Now we return to the case of G the centroid of $\triangle ABC$.
We claim that F is then the point $\widehat{G}$ in which the line AG meets again the circumcircle of $\triangle BGC$. Indeed, if M is the midpoint of AB, by the law of sines we have $\frac{BC}{CG} = \frac{\sin\angle B\widehat{G}C}{\sin\angle CB\widehat{G}} = \frac{\sin\angle BGM}{\sin\angle AGM} = \frac{AG}{BG}$, and similarly $\frac{BC}{BG} = \frac{AG}{CG}$. Thus (1) implies

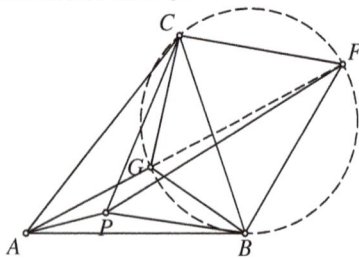

$$AG \cdot AP + BG \cdot BP + CG \cdot CP \ge AG \cdot A\widehat{G}.$$

It is easily seen from the above considerations that equality holds if and only if $P \equiv G$, and then the (minimum) value of $AG \cdot AP + BG \cdot BP + CG \cdot CP$ equals

$$AG^2 + BG^2 + CG^2 = \frac{a^2 + b^2 + c^2}{3}.$$

Second solution. Notice that $AG \cdot AP \ge \overrightarrow{AG} \cdot \overrightarrow{AP} = \overrightarrow{AG} \cdot (\overrightarrow{AG} + \overrightarrow{GP})$. Summing this inequality with analogous inequalities for $BG \cdot BP$ and $CG \cdot CP$ gives us
$AG \cdot AP + BG \cdot BP + CG \cdot CP \ge AG^2 + BG^2 + CG^2 + (\overrightarrow{AG} + \overrightarrow{BG} + \overrightarrow{CG}) \cdot \overrightarrow{GP} = AG^2 + BG^2 + CG^2 = \frac{a^2+b^2+c^2}{3}$. Equality holds if and only if $P \equiv Q$.

18. Let α_1, β_1, γ_1, α_2, β_2, γ_2 denote the angles $\angle MAB$, $\angle MBC$, $\angle MCA$, $\angle MAC$, $\angle MBA$, $\angle MCB$ respectively. Then

$$\frac{MB' \cdot MC'}{MA^2} = \sin\alpha_1 \sin\alpha_2, \quad \frac{MC' \cdot MA'}{MB^2} = \sin\beta_1 \sin\beta_2, \quad \frac{MA' \cdot MB'}{MC^2} = \sin\gamma_1 \sin\gamma_2;$$

hence $p(M)^2 = \sin\alpha_1 \sin\alpha_2 \sin\beta_1 \sin\beta_2 \sin\gamma_1 \sin\gamma_2$. Since

$$\sin\alpha_1 \sin\alpha_2 = \frac{1}{2}(\cos(\alpha_1 - \alpha_2) - \cos(\alpha_1 + \alpha_2)) \leq \frac{1}{2}(1 - \cos\alpha) = \sin^2\frac{\alpha}{2},$$

we conclude that

$$p(M) \leq \sin\frac{\alpha}{2} \sin\frac{\beta}{2} \sin\frac{\gamma}{2}.$$

Equality occurs when $\alpha_1 = \alpha_2$, $\beta_1 = \beta_2$, and $\gamma_1 = \gamma_2$, that is, when M is the incenter of $\triangle ABC$.

It is well known that $\mu(ABC) = \sin\frac{\alpha}{2} \sin\frac{\beta}{2} \sin\frac{\gamma}{2}$ is maximal when $\triangle ABC$ is equilateral (it follows, for example, from Jensen's inequality applied to $\ln\sin x$). Hence $\max\mu(ABC) = \frac{1}{8}$.

19. The hexagon $AEBFCD$ is obviously convex. Therefore $\angle AEB + \angle BFC + \angle CDA = 360°$. Using this relation we obtain that the circles ω_1, ω_2, ω_3 with centers at D, E, F and radii DA, EB, FC respectively all pass through a common point O. Indeed, if $\omega_1 \cap \omega_2 = \{O\}$, then $\angle AOB = 180° - \angle AEB/2$ and $\angle BOC = 180° - \angle BFC/2$. We now have $\angle COA = 180° - \angle CDA/2$ as well, i.e., $O \in \omega_3$. The point O is the reflection of A with

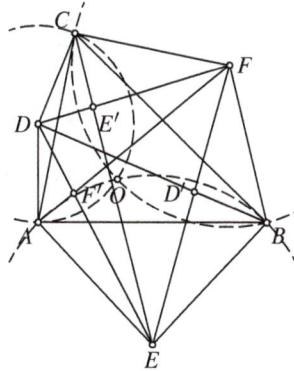

respect to DE. Similarly, it is also the reflection of B with respect to EF, and that of C with respect to FD. Hence

$$\frac{DB}{DD'} = 1 + \frac{D'B}{DD'} = 1 + \frac{S_{EBF}}{S_{EDF}} = 1 + \frac{S_{OEF}}{S_{DEF}}.$$

Analogously $\frac{EC}{EE'} = 1 + \frac{S_{ODE}}{S_{DEF}}$ and $\frac{FA}{FF'} = 1 + \frac{S_{ODE}}{S_{DEF}}$. Adding these relations gives us

$$\frac{DB}{DD'} + \frac{EC}{EE'} + \frac{FA}{FF'} = 3 + \frac{S_{OEF} + S_{ODF} + S_{ODE}}{S_{DEF}} = 4.$$

20. By Ceva's theorem, we can choose real numbers x, y, z such that

$$\frac{\overrightarrow{BD}}{\overrightarrow{DC}} = \frac{z}{y}, \quad \frac{\overrightarrow{CE}}{\overrightarrow{EA}} = \frac{x}{z}, \quad \text{and} \quad \frac{\overrightarrow{AF}}{\overrightarrow{FB}} = \frac{y}{x}.$$

The point P lies outside the triangle ABC if and only if x, y, z are not all of the same sign. In what follows, S_X will denote the signed area of a figure X.

Let us assume that the area S_{ABC} of $\triangle ABC$ is 1. Since $S_{PBC} : S_{PCA} : S_{PAB} = x : y : z$ and $S_{PBD} : S_{PDC} = z : y$, it follows that $S_{PBD} = \frac{z}{y+z} \frac{x}{x+y+z}$. Hence $S_{PBD} = \frac{1}{y(y+z)} \frac{xyz}{x+y+z}$, $S_{PCE} = \frac{1}{z(z+x)} \frac{xyz}{x+y+z}$, $S_{PAF} = \frac{1}{x(x+y)} \frac{xyz}{x+y+z}$. By the condition of the problem we have $|S_{PBD}| = |S_{PCE}| = |S_{PAF}|$, or

$$|x(x+y)| = |y(y+z)| = |z(z+x)|.$$

Obviously x, y, z are nonzero, so that we can put w.l.o.g. $z = 1$. At least two of the numbers $x(x+y), y(y+1), 1(1+x)$ are equal, so we can assume that $x(x+y) = y(y+1)$. We distinguish two cases:

(i) $x(x+y) = y(y+1) = 1+x$. Then $x = y^2 + y - 1$, from which we obtain $(y^2 + y - 1)(y^2 + 2y - 1) = y(y+1)$. Simplification gives $y^4 + 3y^3 - y^2 - 4y + 1 = 0$, or

$$(y-1)(y^3 + 4y^2 + 3y - 1) = 0.$$

If $y = 1$, then also $z = x = 1$, so P is the centroid of $\triangle ABC$, which is not an exterior point. Hence $y^3 + 4y^2 + 3y - 1 = 0$. Now the signed area of each of the triangles PBD, PCE, PAF equals

$$S_{PAF} = \frac{yz}{(x+y)(x+y+z)}$$
$$= \frac{y}{(y^2 + 2y - 1)(y^2 + 2y)} = \frac{1}{y^3 + 4y^2 + 3y - 2} = -1.$$

It is easy to check that not both of x, y are positive, implying that P is indeed outside $\triangle ABC$. This is the desired result.

(ii) $x(x+y) = y(y+1) = -1 - x$. In this case we are led to

$$f(y) = y^4 + 3y^3 + y^2 - 2y + 1 = 0.$$

We claim that this equation has no real solutions. In fact, assume that y_0 is a real root of $f(y)$. We must have $y_0 < 0$, and hence $u = -y_0 > 0$ satisfies $3u^3 - u^4 = (u+1)^2$. On the other hand,

$$3u^3 - u^4 = u^3(3-u) = 4u \left(\frac{u}{2}\right)\left(\frac{u}{2}\right)(3-u)$$
$$\leq 4u \left(\frac{u/2 + u/2 + 3 - u}{3}\right)^3 = 4u$$
$$\leq (u+1)^2,$$

where at least one of the inequalities is strict, a contradiction.

Remark. The official solution was incomplete, missing the case (ii).

21. We denote by $p(XYZ)$ the perimeter of a triangle XYZ.

If O is the circumcenter of $\triangle ABC$, then A_1, B_1, C_1 are the midpoints of the corresponding sides of the triangle, and hence $p(A_1B_1C_1) = p(AB_1C_1) = p(A_1BC_1) = p(A_1B_1C)$.

Conversely, suppose that $p(A_1B_1C_1) \geq p(AB_1C_1), p(A_1BC_1), p(A_1B_1C)$. Let α_1, α_2, β_1, β_2, γ_1, γ_2 denote the angles $\angle B_1A_1C$, $\angle C_1A_1B$, $\angle C_1B_1A$, $\angle A_1B_1C$, $\angle A_1C_1B$, $\angle B_1C_1A$.

Suppose that $\gamma_1, \beta_2 \geq \alpha$. If A_2 is the fourth vertex of the parallelogram $B_1AC_1A_2$, then these conditions imply that A_1 is in the interior or on the border of $\triangle B_1C_1A_2$, and therefore $p(A_1B_1C_1) \leq p(A_2B_1C_1) = p(AB_1C_1)$. Moreover, if one of the inequalities $\gamma_1 \geq \alpha$, $\beta_2 \geq \alpha$ is strict, then $p(A_1B_1C_1)$ is strictly less than $p(AB_1C_1)$, contrary to the assumption. Hence

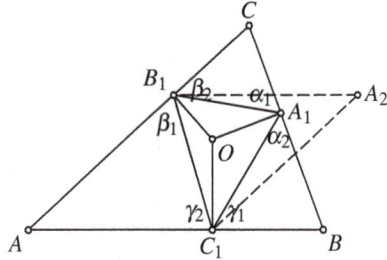

$$\beta_2 \geq \alpha \implies \gamma_1 \leq \alpha,$$
$$\gamma_2 \geq \beta \implies \alpha_1 \leq \beta, \qquad (1)$$
$$\alpha_2 \geq \gamma \implies \beta_1 \leq \gamma,$$

the last two inequalities being obtained analogously to the first one. Because of the symmetry, there is no loss of generality in assuming that $\gamma_1 \leq \alpha$. Then since $\gamma_1 + \alpha_2 = 180° - \beta = \alpha + \gamma$, it follows that $\alpha_2 \geq \gamma$. From (1) we deduce $\beta_1 \leq \gamma$, which further implies $\gamma_2 \geq \beta$. Similarly, this leads to $\alpha_1 \leq \beta$ and $\beta_2 \geq \alpha$. To sum up,

$$\gamma_1 \leq \alpha \leq \beta_2, \qquad \alpha_1 \leq \beta \leq \gamma_2, \qquad \beta_1 \leq \gamma \leq \alpha_2.$$

Since OA_1BC_1 and OB_1CA_1 are cyclic, we have $\angle A_1OB = \gamma_1$ and $\angle A_1OC = \beta_2$. Hence $BO : CO = \cos\beta_2 : \cos\gamma_1$, hence $BO \leq CO$. Analogously, $CO \leq AO$ and $AO \leq BO$. Therefore $AO = BO = CO$, i.e., O is the circumcenter of ABC.

22. Let S and T respectively be the points on the extensions of AB and AQ over B and Q such that $BS = BP$ and $QT = QB$. It is given that $AS = AB + BP = AQ + QB = AT$. Since $\angle PAS = \angle PAT$, the triangles APS and APT are congruent, from which we deduce that $\angle ATP = \angle ASP = \beta/2 = \angle QBP$. Hence $\angle QTP = \angle QBP$.

If P does not lie on BT, then the last equality implies that $\triangle QBP$ and $\triangle QTP$ are congruent, so P lies on the internal bisector of $\angle BQT$. But P also lies on the internal bisector of $\angle QAB$; consequently, P is an excenter of $\triangle QAB$, thus lying on the internal bisector of $\angle QBS$ as well. It follows that $\angle PBQ = \beta/2 = \angle PBS = 180° - \beta$, so $\beta = 120°$, which is impossible.

Therefore $P \in BT$, which means that $T \equiv C$. Now from $QC = QB$ we conclude that $120° - \beta = \gamma = \beta/2$, i.e., $\beta = 80°$ and $\gamma = 40°$.

23. For each positive integer x, define $\alpha(x) = x/10^r$ if r is the positive integer satisfying $10^r \leq x < 10^{r+1}$. Observe that if $\alpha(x)\alpha(y) < 10$ for some $x, y \in \mathbb{N}$, then

$\alpha(xy) = \alpha(x)\alpha(y)$. If, as usual, $[t]$ means the integer part of t, then $[\alpha(x)]$ is actually the leftmost digit of x.

Now suppose that n is a positive integer such that $k \le \alpha((n+k)!) < k+1$ for $k = 1,2,\ldots,9$. We have

$$1 < \alpha(n+k) = \frac{\alpha((n+k)!)}{\alpha((n+k-1)!)} < \frac{k+1}{k-1} \le 3 \quad \text{for } 2 \le k \le 9,$$

from which we obtain $\alpha(n+k+1) > \alpha(n+k)$ (the opposite can hold only if $\alpha(n+k) \ge 9$). Therefore

$$1 < \alpha(n+2) < \cdots < \alpha(n+9) \le \frac{5}{4}.$$

On the other hand, this implies that $\alpha((n+4)!) = \alpha((n+1)!)\alpha(n+2)\alpha(n+3)\alpha(n+4) < (5/4)^3\alpha((n+1)!) < 4$, contradicting the assumption that the leftmost digit of $(n+4)!$ is 4.

24. We shall find the general solution to the system. Squaring both sides of the first equation and subtracting twice the second equation we obtain $(x-y)^2 = z^2 + u^2$. Thus $(z, u, x-y)$ is a Pythagorean triple. Then it is well known that there are positive integers t, a, b such that $z = t(a^2 - b^2)$, $u = 2tab$ (or vice versa), and $x - y = t(a^2 + b^2)$. Using that $x + y = z + u$ we come to the general solution:

$$x = t(a^2 + ab), \quad y = t(ab - b^2); \quad z = t(a^2 - b^2), \quad u = 2tab.$$

Putting $a/b = k$ we obtain

$$\frac{x}{y} = \frac{k^2 + k}{k-1} = 3 + (k-1) + \frac{2}{k-1} \ge 3 + 2\sqrt{2},$$

with equality for $k - 1 = \sqrt{2}$. On the other hand, k can be arbitrarily close to $1 + \sqrt{2}$, and so x/y can be arbitrarily close to $3 + 2\sqrt{2}$. Hence $m = 3 + 2\sqrt{2}$.

Remark. There are several other techniques for solving the given system. The exact lower bound of m itself can be obtained as follows: by the system $\left(\frac{x}{y}\right)^2 - 6\frac{x}{y} + 1 = \left(\frac{z-u}{y}\right)^2 \ge 0$, so $x/y \ge 3 + 2\sqrt{2}$.

25. Define $b_n = |a_{n+1} - a_n|$ for $n \ge 1$. From the equalities $a_{n+1} = b_{n-1} + b_{n-2}$, from $a_n = b_{n-2} + b_{n-3}$ we obtain $b_n = |b_{n-1} - b_{n-3}|$. From this relation we deduce that $b_m \le \max(b_n, b_{n+1}, b_{n+2})$ for all $m \ge n$, and consequently b_n is bounded.

   Lemma. If $\max(b_n, b_{n+1}, b_{n+2}) = M \ge 2$, then $\max(b_{n+6}, b_{n+7}, b_{n+8}) \le M - 1$.

   Proof. Assume the opposite. Suppose that $b_j = M$, $j \in \{n, n+1, n+2\}$, and let $b_{j+1} = x$ and $b_{j+2} = y$. Thus $b_{j+3} = M - y$. If $x, y, M - y$ are all less than M, then the contradiction is immediate. The remaining cases are these:

   (i) $x = M$. Then the sequence has the form $M, M, y, M - y, y, y, \ldots$, and since $\max(y, M - y, y) = M$, we must have $y = 0$ or $y = M$.

(ii) $y = M$. Then the sequence has the form $M, x, M, 0, x, M - x, \ldots$, and since $\max(0, x, M - x) = M$, we must have $x = 0$ or $x = M$.

(iii) $y = 0$. Then the sequence is $M, x, 0, M, M - x, M - x, x, \ldots$, and since $\max(M - x, x, x) = M$, we have $x = 0$ or $x = M$.

In every case M divides both x and y. From the recurrence formula M also divides b_i for every $i < j$. However, $b_2 = 12^{12} - 11^{11}$ and $b_4 = 11^{11}$ are relatively prime, a contradiction.

From $\max(b_1, b_2, b_3) \le 13^{13}$ and the lemma we deduce inductively that $b_n \le 1$ for all $n \ge 6 \cdot 13^{13} - 5$. Hence $a_n = b_{n-2} + b_{n-3}$ takes only the values 0, 1, 2 for $n \ge 6 \cdot 13^{13} - 2$. In particular, $a_{14^{14}}$ is 0, 1, or 2. On the other hand, the sequence a_n modulo 2 is as follows: $1, 0, 1, 0, 0, 1, 1; 1, 0, 1, 0, \ldots$; and therefore it is periodic with period 7. Finally, $14^{14} \equiv 0$ modulo 7, from which we obtain $a_{14^{14}} \equiv a_7 \equiv 1 \pmod{2}$. Therefore $a_{14^{14}} = 1$.

26. Let C be the set of those $a \in \{1, 2, \ldots, p - 1\}$ for which $a^{p-1} \equiv 1 \pmod{p^2}$. At first, we observe that $a, p - a$ do not both belong to C, regardless of the value of a. Indeed, by the binomial formula,

$$(p - a)^{p-1} - a^{p-1} \equiv -(p - 1)p\, a^{p-2} \not\equiv 0 \pmod{p^2}.$$

As a consequence we deduce that $|C| \le \frac{p-1}{2}$. Further, we observe that $p - k \in C \Leftrightarrow k \equiv k(p - k)^{p-1} \pmod{p^2}$, i.e.,

$$p - k \in C \Leftrightarrow k \equiv k(k^{p-1} - (p - 1)p\, k^{p-2}) \equiv k^p + p \pmod{p^2}. \qquad (1)$$

Now assume the contrary to the claim, that for every $a = 1, \ldots, p - 2$ one of $a, a + 1$ is in C. In this case it is not possible that $a, a + 1$ are both in C, for then $p - a, p - a - 1 \notin C$. Thus, since $1 \in C$, we inductively obtain that $2, 4, \ldots, p - 1 \notin C$ and $1, 3, 5, \ldots, p - 2 \in C$. In particular, $p - 2, p - 4 \in C$, which is by (1) equivalent to $2 \equiv 2^p + p$ and $4 \equiv 4^p + p \pmod{p^2}$.

However, squaring the former equality and subtracting the latter, we obtain $2^{p+1}p \equiv p \pmod{p^2}$, or $4 \equiv 1 \pmod{p}$, which is a contradiction unless $p = 3$. This finishes the proof.

27. The given equality is equivalent to $a^2 - ac + c^2 = b^2 + bd + d^2$. Hence $(ab + cd)(ad + bc) = ac(b^2 + bd + d^2) + bd(a^2 - ac + c^2)$, or equivalently,

$$(ab + cd)(ad + bc) = (ac + bd)(a^2 - ac + c^2). \qquad (1)$$

Now suppose that $ab + cd$ is prime. It follows from $a > b > c > d$ that

$$ab + cd > ac + bd > ad + bc; \qquad (2)$$

hence $ac + bd$ is relatively prime with $ab + cd$. But then (1) implies that $ac + bd$ divides $ad + bc$, which is impossible by (2).

Remark. Alternatively, (1) could be obtained by applying the law of cosines and Ptolemy's theorem on a quadrilateral $XYZT$ with $XY = a$, $YZ = c$, $ZT = b$, $TX = d$ and $\angle Y = 60°$, $\angle T = 120°$.

28. Yes. The desired result is an immediate consequence of the following fact applied on $p = 101$.

Lemma. For any odd prime number p, there exist p nonnegative integers less than $2p^2$ with all pairwise sums mutually distinct.

Proof. We claim that the numbers $a_n = 2np + (n^2)$ have the desired property, where (x) denotes the remainder of x upon division by p.

Suppose that $a_k + a_l = a_m + a_n$. Assume that $k \leq l$ and $m \leq n$. By the construction of a_i, we have $2p(k+l) \leq a_k + a_l < 2p(k+l+1)$. Hence we must have $k + l = m + n$, and therefore also $(k^2) + (l^2) = (m^2) + (n^2)$. Thus

$$k + l \equiv m + n \quad \text{and} \quad k^2 + l^2 \equiv m^2 + n^2 \pmod{p}.$$

But then it holds that $(k-l)^2 = 2(k^2 + l^2) - (k+l)^2 \equiv (m-n)^2 \pmod{p}$, so $k - l \equiv \pm(m-n)$, which leads to $\{k, l\} = \{m, n\}$. This proves the lemma.

4.43 Solutions to the Shortlisted Problems of IMO 2002

1. Consider the given equation modulo 9. Since each cube is congruent to either $-1, 0$ or 1, whereas $2002^{2002} \equiv 4^{2002} = 4 \cdot 64^{667} \equiv 4 \pmod 9$, hence $t \geq 4$. On the other hand,

$$2002^{2002} = 2002 \cdot (2002^{667})^3 = (10^3 + 10^3 + 1^3 + 1^3)(2002^{667})^3$$

is a solution with $t = 4$. Hence the answer is 4.

2. Set $S = d_1 d_2 + \cdots + d_{k-1} d_k$. Since $d_i/n = 1/d_{k+1-i}$, we have $\frac{S}{n^2} = \frac{1}{d_k d_{k-1}} + \cdots + \frac{1}{d_2 d_1}$. Hence

$$\frac{1}{d_2 d_1} \leq \frac{S}{n^2} \leq \left(\frac{1}{d_{k-1}} - \frac{1}{d_k}\right) + \cdots + \left(\frac{1}{d_1} - \frac{1}{d_2}\right) = 1 - \frac{1}{d_k} < 1,$$

or (since $d_1 = 1$) $1 < \frac{n^2}{S} \leq d_2$. This shows that $S < n^2$.
Also, if S is a divisor of n^2, then n^2/S is a nontrivial divisor of n^2 not exceeding d_2. But d_2 is obviously the least prime divisor of n (and also of n^2), so we must have $n^2/S = d_2$, which holds if and only if n is prime.

3. We observe that if a, b are coprime odd numbers, then $\gcd(2^a + 1, 2^b + 1) = 3$. In fact, this g.c.d. divides $\gcd(2^{2a} - 1, 2^{2b} - 1) = 2^{\gcd(2a,2b)} - 1 = 2^2 - 1 = 3$, while 3 obviously divides both $2^a + 1$ and $2^b + 1$. In particular, if $3 \nmid b$, then $3^2 \nmid 2^b + 1$, so $2^a + 1$ and $(2^b + 1)/3$ are coprime; consequently $2^{ab} + 1$ (being divisible by $2^a + 1, 2^b + 1$) is divisible by $\frac{(2^a+1)(2^b+1)}{3}$.
Now we prove the desired result by induction on n. For $n = 1$, $2^{p_1} + 1$ is divisible by 3 and exceeds 3^2, so it has at least 4 divisors. Assume that $2^a + 1 = 2^{p_1 \cdots p_{n-1}} + 1$ has at least 4^{n-1} divisors and consider $N = 2^{ab} + 1 = 2^{p_1 \cdots p_n} + 1$ (where $b = p_n$). As above, $2^a + 1$ and $\frac{2^b+1}{3}$ are coprime, and thus $Q = (2^a + 1)(2^b + 1)/3$ has at least $2 \cdot 4^{n-1}$ divisors. Moreover, N is divisible by Q and is greater than Q^2 (indeed, $N > 2^{ab} > 2^{2a}2^{2b} > Q^2$ if $a, b \geq 5$). Then N has at least twice as many divisors as Q (because for every $d \mid Q$ both d and N/d are divisors of N), which counts up to 4^n divisors, as required.

Remark. With some knowledge of cyclotomic polynomials, one can show that $2^{p_1 \cdots p_n} + 1$ has at least $2^{2^{n-1}}$ divisors, far exceeding 4^n.

4. For $a = b = c = 1$ we obtain $m = 12$. We claim that the given equation has infinitely many solutions in positive integers a, b, c for this value of m.
After multiplication by $abc(a+b+c)$ the equation $\frac{1}{a} + \frac{1}{b} + \frac{1}{c} + \frac{1}{abc} - \frac{12}{a+b+c} = 0$ becomes

$$a^2(b+c) + b^2(c+a) + c^2(a+b) + a + b + c - 9abc = 0. \tag{1}$$

Suppose that (a, b, c) is one such solution with $a \leq b \leq c$. Regarding (1) as a quadratic equation in a, we see by Vieta's formula that $\left(b, c, \frac{bc+1}{a}\right)$ also satisfies (1).

Define $(a_n)_{n=0}^\infty$ by $a_0 = a_1 = a_2 = 1$ and $a_{n+1} = \frac{a_n a_{n-1}+1}{a_{n-2}}$ for each $n > 1$. We show that all a_n's are integers. This procedure is fairly standard. The above relation for n and $n - 1$ gives $a_{n+1}a_{n-2} = a_n a_{n-1} + 1$ and $a_{n-1}a_{n-2} + 1 = a_n a_{n-3}$, so that adding yields $a_{n-2}(a_{n-1}+a_{n+1}) = a_n(a_{n-1}+a_{n-3})$. Therefore $\frac{a_{n+1}+a_{n-1}}{a_n} = \frac{a_{n-1}+a_{n-3}}{a_{n-2}} = \cdots$, from which it follows that

$$\frac{a_{n+1}+a_{n-1}}{a_n} = \begin{cases} \frac{a_2+a_0}{a_1} = 2 \text{ for } n \text{ odd}; \\ \frac{a_3+a_1}{a_2} = 3 \text{ for } n \text{ even}. \end{cases}$$

It is now an immediate consequence that every a_n is integral. Also, the above consideration implies that (a_{n-1}, a_n, a_{n+1}) is a solution of (1) for each $n \geq 1$. Since a_n is strictly increasing for $n \geq 2$, this gives an infinity of solutions in integers.

Remark. There are infinitely many values of $m \in \mathbb{N}$ for which the given equation has at least one solution in integers, and each of those values admits an infinity of solutions.

5. Consider all possible sums $c_1 a_1 + c_2 a_2 + \cdots + c_n a_n$, where each c_i is an integer with $0 \leq c_i < m$. There are m^n such sums, and if any two of them give the same remainder modulo m^n, say $\sum c_i a_i \equiv \sum d_i a_i \pmod{m^n}$, then $\sum (c_i - d_i) a_i$ is divisible by m^n, and since $|c_i - d_i| < m$, we are done. We claim that two such sums must exist.

Suppose to the contrary that the sums $\sum_i c_i a_i$ ($0 \leq c_i < m$) give all the different remainders modulo m^n. Consider the polynomial

$$P(x) = \sum x^{c_1 a_1 + \cdots + c_n a_n},$$

where the sum is taken over all $(c_1, \ldots, c_n)$ with $0 \leq c_i < m$. If ξ is a primitive m^nth root of unity, then by the assumption we have

$$P(\xi) = 1 + \xi + \cdots + \xi^{m^n - 1} = 0.$$

On the other hand, $P(x)$ can be factored as

$$P(x) = \prod_{i=1}^n (1 + x^{a_i} + \cdots + x^{(m-1)a_i}) = \prod_{i=1}^n \frac{1 - x^{m a_i}}{1 - x^{a_i}},$$

so that none of its factors is zero at $x = \xi$ because $m a_i$ is not divisible by m^n. This is obviously a contradiction.

Remark. The example $a_i = m^{i-1}$ for $i = 1, \ldots, n$ shows that the condition that no a_i is a multiple of m^{n-1} cannot be removed.

6. Suppose that (m, n) is such a pair. Assume that division of the polynomial $F(x) = x^m + x - 1$ by $G(x) = x^n + x^2 - 1$ gives the quotient $Q(x)$ and remainder $R(x)$. Since $\deg R(x) < \deg G(x)$, for x large enough $|R(x)| < |G(x)|$; however, $R(x)$ is divisible by $G(x)$ for infinitely many integers x, so it is equal to zero infinitely often. Hence $R \equiv 0$, and thus $F(x)$ is exactly divisible by $G(x)$.

The polynomial $G(x)$ has a root α in the interval $(0,1)$, because $G(0) = -1$ and $G(1) = 1$. Then also $F(\alpha) = 0$, so that

$$\alpha^m + \alpha = \alpha^n + \alpha^2 = 1.$$

If $m \geq 2n$, then $1 - \alpha = \alpha^m \leq (\alpha^n)^2 = (1 - \alpha^2)^2$, which is equivalent to $\alpha(\alpha - 1)(\alpha^2 + \alpha - 1) \geq 0$. But this last is not possible, because $\alpha^2 + \alpha - 1 > \alpha^m + \alpha - 1 = 0$; hence $m < 2n$.

Now we have $F(x)/G(x) = x^{m-n} - (x^{m-n+2} - x^{m-n} - x + 1)/G(x)$, so $H(x) = x^{m-n+2} - x^{m-n} - x + 1$ is also divisible by $G(x)$; but $\deg H(x) = m - n + 2 \leq n + 1 = \deg G(x) + 1$, from which we deduce that either $H(x) = G(x)$ or $H(x) = (x - a)G(x)$ for some $a \in \mathbb{Z}$. The former case is impossible. In the latter case we must have $m = 2n - 1$, and thus $H(x) = x^{n+1} - x^{n-1} - x + 1$; on the other hand, putting $x = 1$ gives $a = 1$, so $H(x) = (x - 1)(x^n + x^2 - 1) = x^{n+1} - x^n + x^3 - x^2 - x + 1$. This is possible only if $n = 3$ and $m = 5$.

Remark. It is an old (though difficult) result that the polynomial $x^n \pm x^k \pm 1$ is either irreducible or equals $x^2 \pm x + 1$ times an irreducible factor.

7. To avoid working with cases, we use oriented angles modulo $180°$. Let K be the circumcenter of $\triangle BCD$, and X any point on the common tangent to the circles at D. Since the tangents at the ends of a chord are equally inclined to the chord, we have $\angle BAC = \angle ABD + \angle BDC + \angle DCA = \angle BDX + \angle BDC + \angle XDC = 2\angle BDC = \angle BKC$. It follows that B, C, A, K are concyclic, as required.

8. Construct equilateral triangles ACP and ABQ outside the triangle ABC. Since $\angle APC + \angle AFC = 60° + 120° = 180°$, the points A, C, F, P lie on a circle; hence $\angle AFP = \angle ACP = 60° = \angle AFD$, so D lies on the segment FP; similarly, E lies on FQ. Further, note that

$$\frac{FP}{FD} = 1 + \frac{DP}{FD} = 1 + \frac{S_{APC}}{S_{AFC}} \geq 4$$

with equality if F is the midpoint of the smaller arc AC: hence $FD \leq \frac{1}{4}FP$ and $FE \leq \frac{1}{4}FQ$. Then by the law of cosines,

$$\begin{aligned} DE &= \sqrt{FD^2 + FE^2 + FD \cdot FE} \\ &\leq \frac{1}{4}\sqrt{FP^2 + FQ^2 + FP \cdot FQ} = \frac{1}{4}PQ \leq AP + AQ = AB + AC. \end{aligned}$$

Equality holds if and only if $\triangle ABC$ is equilateral.

9. Since $\angle BCA = \frac{1}{2}\angle BOA = \angle BOD$, the lines CA and OD are parallel, so that $ODAI$ is a parallelogram. It follows that $AI = OD = OE = AE = AF$. Hence

$$\angle IFE = \angle IFA - \angle EFA = \angle AIF - \angle ECA = \angle AIF - \angle ACF = \angle CFI.$$

Also, from $AE = AF$ we get that CI bisects $\angle ECF$. Therefore I is the incenter of $\triangle CEF$.

10. Let O be the circumcenter of A_1A_2C, and O_1, O_2 the centers of S_1, S_2 respectively.

First, from

$$\angle A_1QA_2 = 180° - \angle PA_1Q - \angle QA_2P = \frac{1}{2}(360° - \angle PO_1Q - \angle QO_2P)$$
$$= \angle O_1QO_2$$

we obtain $\angle A_1QA_2 = \angle B_1QB_2 = \angle O_1QO_2$.
We now have $\angle A_1QA_2 = \angle B_1QP +$
$\angle PQB_2 = \angle CA_1P + \angle CA_2P = 180° -$
$\angle A_1CA_2$, from which we conclude that
Q lies on the circumcircle of $\triangle A_1A_2C$.
Hence $OA_1 = OQ$. However, we also
have $O_1A_2 = O_1Q$. Consequently, O,
O_1 both lie on the perpendicular bisector of A_1Q, so $OO_1 \perp A_1Q$. Similarly,
$OO_2 \perp A_2Q$, leading to

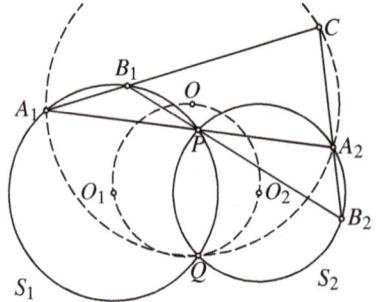

$$\angle O_2OO_1 = 180° - \angle A_1QA_2 = 180° - \angle O_1QO_2.$$

Hence, O lies on the circle through O_1, O_2, Q, which is fixed.

11. When S is the set of vertices of a regular pentagon, then it is easily verified that
$\frac{M(S)}{m(S)} = \frac{1+\sqrt{5}}{2} = \alpha$. We claim that this is the best possible. Let A, B, C, D, E be
five arbitrary points, and assume that $\triangle ABC$ has the area $M(S)$. We claim that
some triangle has area less than or equal to $M(S)/\alpha$.
Construct a larger triangle $A'B'C'$ with $C \in A'B' \parallel AB, A \in B'C' \parallel BC, B \in C'A' \parallel$
CA. The point D, as well as E, must lie on the same side of $B'C'$ as BC, for
otherwise $\triangle DBC$ would have greater area than $\triangle ABC$. A similar result holds
for the other edges, and therefore D, E lie inside the triangle $A'B'C'$ or on its
boundary. Moreover, at least one of the triangles $A'BC, AB'C, ABC'$, say ABC',
contains neither D nor E. Hence we can assume that D, E are contained inside
the quadrilateral $A'B'AB$.
An affine linear transformation does not change the ratios between areas. Thus if
we apply such an affine transformation mapping A, B, C into the vertices $ABMCN$
of a regular pentagon, we won't change $M(S)/m(S)$. If now D or E lies inside
$ABMCN$, then we are done. Suppose that both D and E are inside the triangles
CMA', CNB'. Then $CD, CE \leq CM$ (because $CM = CN = CA' = CB'$) and $\angle DCE$
is either less than or equal to $36°$ or greater than or equal to $108°$, from which
we obtain that the area of $\triangle CDE$ cannot exceed the area of $\triangle CMN = M(S)/\alpha$.
This completes the proof.

12. Let $l(MN)$ denote the length of the shorter arc MN of a given circle.
    Lemma. Let PR, QS be two chords of a circle k of radius r that meet each other
    at a point X, and let $\angle PXQ = \angle RXS = 2\alpha$. Then $l(PQ) + l(RS) = 4\alpha r$.

Proof. Let O be the center of the circle. Then $l(PQ) + l(RS) = \angle POQ \cdot r + \angle ROS \cdot r = 2(\angle QSP + \angle RPS)r = 2\angle QXP \cdot r = 4\alpha r$.

Consider a circle k, sufficiently large, whose interior contains all the given circles. For any two circles C_i, C_j, let their exterior common tangents PR, QS ($P, Q, R, S \in k$) form an angle 2α. Then $O_i O_j = \frac{2}{\sin \alpha}$, so $\alpha > \sin \alpha = \frac{2}{O_i O_j}$. By the lemma we have $l(PQ) + l(RS) = 4\alpha r \geq \frac{8r}{O_i O_j}$, and hence

$$\frac{1}{O_i O_j} \leq \frac{l(PQ) + l(RS)}{8r}. \tag{1}$$

Now sum all these inequalities for $i < j$. The result will follow if we show that every point of the circle k belongs to at most $n - 1$ arcs such as PQ, RS. Indeed, that will imply that the sum of all the arcs is at most $2(n - 1)\pi r$, hence from (1) we conclude that $\sum \frac{1}{O_i O_j} \leq \frac{(n-1)\pi}{4}$.

Consider an arbitrary point T of k. We prove by induction (the basis $n = 1$ is trivial) that the number of pairs of circles that are simultaneously intercepted by a ray from T is at most $n - 1$. Let Tu be a ray touching k at T. If we let this ray rotate around T, it will at some moment intercept a pair of circles for the first time, say C_1, C_2. At some further moment the interception with one of these circles, say C_1, is lost and never established again. Thus the pair (C_1, C_2) is the only pair containing C_1 that is intercepted by some ray from T. On the other hand, by the inductive hypothesis the number of such pairs not containing C_1 does not exceed $n - 2$, justifying our claim.

13. Let k be the circle through B, C that is tangent to the circle Ω at point N'. We must prove that K, M, N' are collinear. Since the statement is trivial for $AB = AC$, we may assume that $AC > AB$. As usual, $R, r, \alpha, \beta, \gamma$ denote the circumradius and the inradius and the angles of $\triangle ABC$, respectively.

We have $\tan \angle BKM = DM/DK$. Straightforward calculation gives $DM = \frac{1}{2}AD = R \sin \beta \sin \gamma$ and $DK = \frac{DC - DB}{2} - \frac{KC - KB}{2} = R \sin(\beta - \gamma) - R(\sin \beta - \sin \gamma) = 4R \sin \frac{\beta - \gamma}{2} \sin \frac{\beta}{2} \sin \frac{\gamma}{2}$, so we obtain

$$\tan \angle BKM = \frac{\sin \beta \sin \gamma}{4 \sin \frac{\beta - \gamma}{2} \sin \frac{\beta}{2} \sin \frac{\gamma}{2}} = \frac{\cos \frac{\beta}{2} \cos \frac{\gamma}{2}}{\sin \frac{\beta - \gamma}{2}}.$$

To calculate the angle BKN', we apply the inversion ψ with center at K and power $BK \cdot CK$. For each object X, we denote by $\widehat{X}$ its image under ψ. The incircle Ω maps to a line $\widehat{\Omega}$ parallel to $\widehat{BC}$, at distance $\frac{BK \cdot CK}{2r}$ from $\widehat{BC}$. Thus the point $\widehat{N'}$ is the projection of the midpoint $\widehat{U}$ of $\widehat{BC}$ onto $\widehat{\Omega}$. Hence

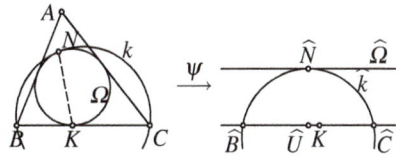

$$\tan \angle BKN' = \tan \angle \widehat{B}K\widehat{N'} = \frac{\widehat{U}\widehat{N'}}{\widehat{U}K} = \frac{BK \cdot CK}{r(CK - BK)}.$$

Again, one easily checks that $KB \cdot KC = bc \sin^2 \frac{\alpha}{2}$ and $r = 4R \sin \frac{\alpha}{2} \cdot \sin \frac{\beta}{2} \cdot \sin \frac{\gamma}{2}$, which implies

$$\tan \angle BKN' = \frac{bc \sin^2 \frac{\alpha}{2}}{r(b-c)}$$

$$= \frac{4R^2 \sin \beta \sin \gamma \sin^2 \frac{\alpha}{2}}{4R \sin \frac{\alpha}{2} \sin \frac{\beta}{2} \sin \frac{\gamma}{2} \cdot 2R(\sin \beta - \sin \gamma)} = \frac{\cos \frac{\beta}{2} \cos \frac{\gamma}{2}}{\sin \frac{\beta - \gamma}{2}}.$$

Hence $\angle BKM = \angle BKN'$, which implies that K, M, N' are indeed collinear; thus $N' \equiv N$.

14. Let G be the other point of intersection of the line FK with the arc BAD.

Since $BN/NC = DK/KB$ and $\angle CEB = \angle BGD$ the triangles CEB and BGD are similar. Thus $BN/NE = DK/KG = FK/KB$.

From $BN = MK$ and $BK = MN$ it follows that $MN/NE = FK/KM$. But we also have that $\angle MNE = 90° + \angle MNB = 90° + \angle MKB = \angle FKM$, and hence $\triangle MNE \sim \triangle FKM$.

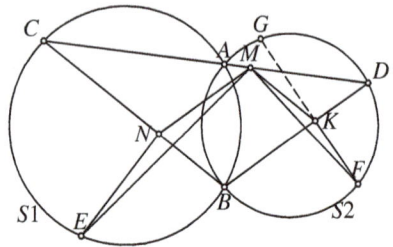

Now $\angle EMF = \angle NMK - \angle NME - \angle KMF = \angle NMK - \angle NME - \angle NEM = \angle NMK - 90° + \angle BNM = 90°$ as claimed.

15. Notice that f is surjective. Indeed, setting $y = -f(x)$ gives $f(f(-f(x)) - x) = f(0) - 2x$, where the right-hand expression can take any real value. In particular, there exists x_0 for which $f(x_0) = 0$. Now setting $x = x_0$ in the functional equation yields $f(y) = 2x_0 + f(f(y) - x_0)$, so we obtain

$$f(z) = z - x_0 \quad \text{for } z = f(y) - x_0.$$

Since f is surjective, z takes all real values. Hence for all z, $f(z) = z + c$ for some constant c, and this is indeed a solution.

16. For $n \geq 2$, let $(k_1, k_2, \ldots, k_n)$ be the permutation of $\{1, 2, \ldots, n\}$ with $a_{k_1} \leq a_{k_2} \leq \cdots \leq a_{k_n}$. Then from the condition of the problem, using the Cauchy–Schwarz inequality, we obtain

$$c \geq a_{k_n} - a_{k_1} = |a_{k_n} - a_{k_{n-1}}| + \cdots + |a_{k_3} - a_{k_2}| + |a_{k_2} - a_{k_1}|$$

$$\geq \frac{1}{k_1 + k_2} + \frac{1}{k_2 + k_3} + \cdots + \frac{1}{k_{n-1} + k_n}$$

$$\geq \frac{(n-1)^2}{(k_1 + k_2) + (k_2 + k_3) + \cdots + (k_{n-1} + k_n)}$$

$$= \frac{(n-1)^2}{2(k_1 + k_2 + \cdots + k_n) - k_1 - k_n} \geq \frac{(n-1)^2}{n^2 + n - 3} \geq \frac{n-1}{n+2}.$$

Therefore $c \geq 1 - \frac{3}{n+2}$ for every positive integer n. But if $c < 1$, this inequality is obviously false for all $n > \frac{3}{1-c} - 2$. We conclude that $c \geq 1$.

Remark. The least value of c is not greater than $2\ln 2$. An example of a sequence $\{a_n\}$ with $0 \leq a_n \leq 2\ln 2$ can be constructed inductively as follows: Given $a_1, a_2, \ldots, a_{n-1}$, then a_n can be any number from $[0, 2\ln 2]$ that does not belong to any of the intervals $\left(a_i - \frac{1}{i+n}, a_i + \frac{1}{i+n}\right)$ $(i = 1, 2, \ldots, n-1)$, and the total length of these intervals is always less than or equal to

$$\frac{2}{n+1} + \frac{2}{n+2} + \cdots + \frac{2}{2n-1} < 2\ln 2.$$

17. Let x, y be distinct integers satisfying $xP(x) = yP(y)$; this is equivalent to $a(x^4 - y^4) + b(x^3 - y^3) + c(x^2 - y^2) + d(x - y) = 0$. Dividing by $x - y$ we obtain

$$a(x^3 + x^2 y + xy^2 + y^3) + b(x^2 + xy + y^2) + c(x + y) + d = 0.$$

Putting $x + y = p$, $x^2 + y^2 = q$ leads to $x^2 + xy + y^2 = \frac{p^2 + q}{2}$, so the above equality becomes

$$apq + \frac{b}{2}(p^2 + q) + cp + d = 0, \quad \text{i.e.} \quad (2ap + b)q = -(bp^2 + 2cp + 2d).$$

Since $q \geq p^2/2$, it follows that $p^2|2ap + b| \leq 2|bp^2 + 2cp + 2d|$, which is possible only for finitely many values of p, although there are infinitely many pairs (x, y) with $xP(x) = yP(y)$. Hence there exists p such that $xP(x) = (p - x)P(p - x)$ for infinitely many x, and therefore for all x.
If $p \neq 0$, then p is a root of $P(x)$. If $p = 0$, the above relation gives $P(x) = -P(-x)$. This forces $b = d = 0$, so $P(x) = x(ax^2 + c)$. Thus 0 is a root of $P(x)$.

18. Putting $x = z = 0$ and $t = y$ into the given equation gives $4f(0)f(y) = 2f(0)$ for all y. If $f(0) \neq 0$, then we deduce $f(y) = \frac{1}{2}$, i.e., f is identically equal to $\frac{1}{2}$. Now we suppose that $f(0) = 0$. Setting $z = t = 0$ we obtain

$$f(xy) = f(x)f(y) \quad \text{for all } x, y \in \mathbb{R}. \tag{1}$$

Thus if $f(y) = 0$ for some $y \neq 0$, then f is identically zero. So, assume $f(y) \neq 0$ whenever $y \neq 0$.
Next, we observe that f is strictly increasing on the set of positive reals. Actually, it follows from (1) that $f(x) = f(\sqrt{x})^2 \geq 0$ for all $x \geq 0$, so that the given equation for $t = x$ and $z = y$ yields $f(x^2 + y^2) = (f(x) + f(y))^2 \geq f(x^2)$ for all $x, y \geq 0$. Using (1) it is easy to get $f(1) = 1$. Now plugging $t = y$ into the given equation, we are led to

$$2[f(x) + f(z)] = f(x - z) + f(x + z) \quad \text{for all } x, z. \tag{2}$$

In particular, $f(z) = f(-z)$. Further, it is easy to get by induction from (2) that $f(nx) = n^2 f(x)$ for all integers n (and consequently for all rational numbers as well). Therefore $f(q) = f(-q) = q^2$ for all $q \in \mathbb{Q}$. But f is increasing for $x > 0$, so we must have $f(x) = x^2$ for all x.
It is easy to verify that $f(x) = 0$, $f(x) = \frac{1}{2}$ and $f(x) = x^2$ are indeed solutions.

19. Write $m = [\sqrt[3]{n}]$. To simplify the calculation, we shall assume that $[b] = 1$. Then
$a = \sqrt[3]{n}$, $b = \frac{1}{\sqrt[3]{n}-m} = \frac{1}{n-m^3}\left(m^2 + m\sqrt[3]{n} + \sqrt[3]{n^2}\right)$, $c = \frac{1}{b-1} = u + v\sqrt[3]{n} + w\sqrt[3]{n^2}$
for certain rational numbers u, v, w. Obviously, integers r, s, t with $ra + sb + tc = 0$
exist if (and only if) $u = m^2 w$, i.e., if $(b-1)(m^2 w + v\sqrt[3]{n} + w\sqrt[3]{n^2}) = 1$ for some
rational v, w.

When the last equality is expanded and simplified, comparing the coefficients at
$1, \sqrt[3]{n}, \sqrt[3]{n^2}$ one obtains

$$
\begin{aligned}
1: & & v + ((m^2 + m^3 - n)m^2 + m)w &= n - m^3, \\
\sqrt[3]{n}: & (m^2 + m^3 - n)v + & (m^3 + n)w &= 0, \qquad (1) \\
\sqrt[3]{n^2}: & mv + & (2m^2 + m^3 - n)w &= 0.
\end{aligned}
$$

In order for the system (1) to have a solution v, w, we must have $(2m^2 + m^3 - n)(m^2 + m^3 - n) = m(m^3 + n)$. This quadratic equation has solutions $n = m^3$ and $n = m^3 + 3m^2 + m$. The former is not possible, but the latter gives $a - [a] > \frac{1}{2}$, so $[b] = 1$, and the system (1) in v, w is solvable. Hence every number $n = m^3 + 3m^2 + m$, $m \in \mathbb{N}$, satisfies the condition of the problem.

20. Assume to the contrary that $\frac{1}{b_1} + \cdots + \frac{1}{b_n} > 1$. Certainly $n \geq 2$ and A is infinite. Define $f_i : A \to A$ as $f_i(x) = b_i x + c_i$ for each i. By condition (ii), $f_i(x) = f_j(y)$ implies $i = j$ and $x = y$; iterating this argument, we deduce that $f_{i_1}(\ldots f_{i_m}(x)\ldots) = f_{j_1}(\ldots f_{j_m}(x)\ldots)$ implies $i_1 = j_1, \ldots, i_m = j_m$ and $x = y$.
As an illustration, we shall consider the case $b_1 = b_2 = b_3 = 2$ first. If a is large enough, then for any $i_1, \ldots, i_m \in \{1, 2, 3\}$ we have $f_{i_1} \circ \cdots \circ f_{i_m}(a) \leq 2.1^m a$. However, we obtain 3^m values in this way, so they cannot be all distinct if m is sufficiently large, a contradiction.
In the general case, let real numbers $d_i > b_i$, $i = 1, 2 \ldots, n$, be chosen such that $\frac{1}{d_1} + \cdots + \frac{1}{d_n} > 1$: for a large enough, $f_i(x) < d_i a$ for each $x \geq a$. Also, let $k_i > 0$ be arbitrary rational numbers with sum 1; denote by N_0 the least common multiple of their denominators.
Let N be a fixed multiple of N_0, so that each $k_i N$ is an integer. Consider all combinations $f_{i_1} \circ \cdots \circ f_{i_N}$ of N functions, among which each f_i appears exactly $k_i N$ times. There are $F_N = \frac{N!}{(k_1 N)! \cdots (k_n N)!}$ such combinations, so they give F_N distinct values when applied to a. On the other hand, $f_{i_1} \circ \cdots \circ f_{i_N}(a) \leq (d_1^{k_1} \cdots d_n^{k_n})^N a$. Therefore

$$
(d_1^{k_1} \cdots d_n^{k_n})^N a \geq F_N \qquad \text{for all } N, N_0 \mid N. \qquad (1)
$$

It remains to find a lower estimate for F_N. In fact, it is straightforward to verify that F_{N+N_0}/F_N tends to Q^{N_0}, where $Q = 1/\left(k_1^{k_1} \cdots k_n^{k_n}\right)$. Consequently, for every $q < Q$ there exists $p > 0$ such that $F_N > pq^N$. Then (1) implies that

$$
\left(\frac{d_1^{k_1} \cdots d_n^{k_n}}{q}\right)^N > \frac{p}{a} \quad \text{for every multiple } N \text{ of } N_0,
$$

and hence $d_1^{k_1}\cdots d_n^{k_n}/q \geq 1$. This must hold for every $q < Q$, and so we have $d_1^{k_1}\cdots d_n^{k_n} \geq Q$, i.e.,

$$(k_1 d_1)^{k_1}\cdots(k_n d_n)^{k_n} \geq 1.$$

However, if we choose $k_1,\ldots,k_n$ such that $k_1 d_1 = \cdots = k_n d_n = u$, then we must have $u \geq 1$. Therefore $\frac{1}{d_1}+\cdots+\frac{1}{d_n} \leq k_1+\cdots+k_n = 1$, a contradiction.

21. Let a_i be the number of blue points with x-coordinate i, and b_i the number of blue points with y-coordinate i. Our task is to show that $a_0 a_1 \cdots a_{n-1} = b_0 b_1 \cdots b_{n-1}$. Moreover, we claim that $a_0,\ldots,a_{n-1}$ is a permutation of $b_0,\ldots,b_{n-1}$, and to show this we use induction on the number of red points.

The result is trivial if all the points are blue. So, choose a red point (x,y) with $x+y$ maximal: clearly $a_x = b_y = n-x-y-1$. If we change this point to blue, a_x and b_y will decrease by 1. Then by the induction hypothesis, $a_0,\ldots,a_{n-1}$ with a_x decreased by 1 is a permutation of $b_0,\ldots,b_{n-1}$ with b_y decreased by 1. However, $a_x = b_y$, and the claim follows.

Remark. One can also use induction on n: it is not more difficult.

22. Write $n = 2k+1$. Consider the black squares at an odd height: there are $(k+1)^2$ of them in total and no two can be covered by one tromino. Thus, we always need at least $(k+1)^2$ trominoes, which cover $3(k+1)^2$ squares in total. However, $3(k+1)^2$ is greater than n^2 for $n = 1,3,5$, so we must have $n \geq 7$.

The case $n = 7$ admits such a covering, as shown in Figure 1. For $n > 7$ this is possible as well: it follows by induction from Figure 2.

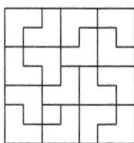

Fig. 1

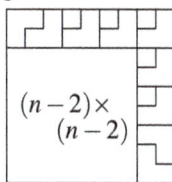

Fig. 2

23. We claim that there are $n!$ full sequences. To show this, we construct a bijection with the set of permutations of $\{1,2,\ldots,n\}$.

Consider a full sequence $(a_1, a_2,\ldots, a_n)$, and let m be the greatest of the numbers $a_1,\ldots,a_n$. Let S_k, $1 \leq k \leq m$, be the set of those indices i for which $a_i = k$. Then $S_1,\ldots S_m$ are nonempty and form a partition of the set $\{1,2,\ldots,n\}$. Now we write down the elements of S_1 in descending order, then the elements of S_2 in descending order and so on. This maps the full sequence to a permutation of $\{1,2,\ldots,n\}$. Moreover, this map is reversible, since each permutation uniquely breaks apart into decreasing sequences $S_1', S_2',\ldots, S_m'$, so that $\max S_i' > \min S_{i-1}'$. Therefore the full sequences are in bijection with the permutations of $\{1,2,\ldots,n\}$.

Second solution. Let there be given a full sequence of length n. Removing from it the first occurrence of the highest number, we obtain a full sequence of length $n-1$. On the other hand, each full sequence of length $n-1$ can be obtained from exactly n full sequences of length n. Therefore, if x_n is the number of full sequences of length n, we deduce $x_n = n x_{n-1}$.

24. Two moves are not sufficient. Indeed, the answer to each move is an even number between 0 and 54, so the answer takes at most 28 distinct values. Consequently, two moves give at most $28^2 = 784$ distinct outcomes, which is less than $10^3 = 1000$.

We now show that three moves are sufficient. With the first move $(0,0,0)$, we get the reply $2(x+y+z)$, so we now know the value of $s = x+y+z$. Now there are several cases:

(i) $s \leq 9$. Then we ask $(9,0,0)$ as the second move and get $(9-x-y)+(9-x-z)+(y+z) = 18-2x$, so we come to know x. Asking $(0,9,0)$ we obtain y, which is enough, since $z = s-x-y$.

(ii) $10 \leq s \leq 17$. In this case the second move is $(9,s-9,0)$. The answer is $z + (9-x) + |x+z-9| = 2k$, where $k = z$ if $x+z \geq 9$, or $k = 9-x$ if $x+z < 9$. In both cases we have $z \leq k \leq y+z \leq s$.

Let $s-k \leq 9$. Then in the third move we ask $(s-k,0,k)$ and obtain $|z-k| + |k-y-z| + y$, which is actually $(k-z)+(y+z-k)+y = 2y$. Thus we also find out $x+z$, and thus deduce whether k is z or $9-x$. Consequently we determine both x and z.

Let $s-k > 9$. In this case, the third move is $(9,s-k-9,k)$. The answer is $|s-k-x-y| + |s-9-y-z| + |k+9-z-x| = (k-z)+(9-x)+(9-x+k-z) = 18+2k-2(x+z)$, from which we find out again whether k is z or $9-x$. Now we are easily done.

(iii) $18 \leq s \leq 27$. Then as in the first case, asking $(0,9,9)$ and $(9,0,9)$ we obtain x and y.

25. Assume to the contrary that no set of size less than r meets all sets in $\mathscr{F}$.

Consider any set A of size less than r that is contained in infinitely many sets of $\mathscr{F}$. By the assumption, A is disjoint from some set $B \in \mathscr{F}$. Then of the infinitely many sets that contain A, each must meet B, so some element b of B belongs to infinitely many of them. But then the set $A \cup \{b\}$ is contained in infinitely many sets of $\mathscr{F}$ as well.

Such a set A exists: for example, the empty set. Now taking for A the largest such set we come to a contradiction.

26. Write $n = 2m$. We shall define a directed graph G with vertices $1,\ldots,m$ and edges labelled $1,2,\ldots,2m$ in such a way that the edges issuing from i are labelled $2i-1$ and $2i$, and those entering it are labelled i and $i+m$. What we need is an Euler circuit in G, namely a closed path that passes each edge exactly once. Indeed, if x_i is the ith edge in such a circuit, then x_i enters some vertex j and x_{i+1} leaves it, so $x_i \equiv j \pmod{m}$ and $x_{i+1} = 2j-1$ or $2j$. Hence $2x_i \equiv 2j$ and $x_{i+1} \equiv 2x_i$ or $2x_i - 1 \pmod{n}$, as required.

The graph G is connected: by induction on k there is a path from 1 to k, since 1 is connected to j with $2j = k$ or $2j-1 = k$, and there is an edge from j to k. Also, the in-degree and out-degree of each vertex of G are equal (to 2), and thus by a known result, G contains an Euler circuit.

27. For a graph G on 120 vertices (i.e., people at the party), write $q(G)$ for the number of weak quartets in G. Our solution will consist of three parts.

First, we prove that some graph G with maximal $q(G)$ breaks up as a disjoint union of complete graphs. This will follow if we show that any two adjacent vertices x, y have the same neighbors (apart from themselves). Let G_x be the graph obtained from G by "copying" x to y (i.e., for each $z \neq x, y$, we add the edge zy if zx is an edge, and delete zy if zx is not an edge). Similarly G_y is the graph obtained from G by copying y to x. We claim that $2q(G) \leq q(G_x) + q(G_y)$. Indeed, the number of weak quartets containing neither x nor y is the same in G, G_x, and G_y, while the number of those containing both x and y is not less in G_x and G_y than in G. Also, the number containing exactly one of x and y in G_x is at least twice the number in G containing x but not y, while the number containing exactly one of x and y in G_y is at least twice the number in G containing y but not x. This justifies our claim by adding. It follows that for an extremal graph G we must have $q(G) = q(G_x) = q(G_y)$. Repeating the copying operation pair by pair (y to x, then their common neighbor z to both x, y, etc.) we eventually obtain an extremal graph consisting of disjoint complete graphs.

Second, suppose the complete graphs in G have sizes $a_1, a_2, \ldots, a_n$. Then

$$q(G) = \sum_{i=1}^{n} \binom{a_i}{2} \sum_{\substack{j<k \\ j,k \neq i}} a_j a_k.$$

If we fix all the a_i except two, say p, q, then $p + q = s$ is fixed, and for some constants C_i, $q(G) = C_1 + C_2 pq + C_3 \left(\binom{p}{2} + \binom{q}{2} \right) + C_4 \left(q\binom{p}{2} + p\binom{q}{2} \right) = A + Bpq$, where A and B depend only on s. Hence the maximum of $q(G)$ is attained if $|p - q| \leq 1$ or $pq = 0$. Thus if $q(G)$ is maximal, any two nonzero a_i's differ by at most 1.

Finally, if G consists of n disjoint complete graphs, then $q(G)$ cannot exceed the value obtained if $a_1 = \cdots = a_n$ (not necessarily integral), which equals

$$Q_n = \frac{120^2}{n} \binom{120/n}{2} \binom{n-1}{2} = 30 \cdot 120^2 \frac{(n-1)(n-2)(120-n)}{n^3}.$$

It is easy to check that Q_n takes its maximum when $n = 5$ and $a_1 = \cdots = a_5 = 24$, and that this maximum equals $15 \cdot 23 \cdot 24^3 = 4769280$.

4.44 Solutions to the Shortlisted Problems of IMO 2003

1. Consider the points $O(0,0,0)$, $P(a_{11},a_{21},a_{31})$, $Q(a_{12},a_{22},a_{32})$, $R(a_{13},a_{23}, a_{33})$ in three-dimensional Euclidean space. It is enough to find a point $U(u_1,u_2,u_3)$ in the interior of the triangle PQR whose coordinates are all positive, all negative, or all zero (indeed, then we have $\overrightarrow{OU} = c_1\overrightarrow{OP} + c_2\overrightarrow{OQ} + c_3\overrightarrow{OR}$ for some $c_1, c_2, c_3 > 0$ with $c_1 + c_2 + c_3 = 1$).

   Let $P'(a_{11},a_{21},0)$, $Q'(a_{12},a_{22},0)$, and $R'(a_{13},a_{23},0)$ be the projections of P, Q, and R onto the Oxy plane. We see that P', Q', R' lie in the fourth, second, and third quadrants, respectively. We have the following two cases:

   (i) O is in the exterior of $\triangle P'Q'R'$.

   Set $S' = OR' \cap P'Q'$ and let S be the point of the segment PQ that projects to S'. The point S has its z coordinate negative (because the z coordinates of P and Q are negative). Thus any point of the segment SR sufficiently close to S has all coordinates negative.

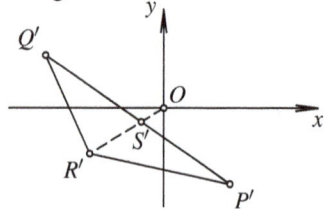

   (ii) O is in the interior or on the boundary of $\triangle P'Q'R'$.

   Let T be the point in the plane PQR whose projection is O. If $T = O$, then all coordinates of T are zero, and we are done. Otherwise O is interior to $\triangle P'Q'R'$. Suppose that the z coordinate of T is positive (negative). Since x and y coordinates of T are equal to 0, there is a point U inside PQR close to T with both x and y coordinates positive (respectively negative), and this point U has all coordinates of the same sign.

2. We can rewrite (ii) as $-(f(a) - 1)(f(b) - 1) = f(-(a-1)(b-1) + 1) - 1$. So putting $g(x) = f(x+1) - 1$, this equation becomes $-g(a-1)g(b-1) = g(-(a-1)(b-1))$ for $a < 1 < b$. Hence

$$-g(x)g(y) = g(-xy) \text{ for } x < 0 < y,$$

$$\text{and } g \text{ is nondecreasing with } g(-1) = -1, g(0) = 0. \tag{1}$$

   Conversely, if g satisfies (1), then f is a solution of our problem.

   Setting $y = 1$ in (1) gives $-g(-x)g(1) = g(x)$ for each $x > 0$, and therefore (1) reduces to $g(1)g(yz) = g(y)g(z)$ for all $y, z > 0$. We have two cases:

   1° $g(1) = 0$. By (1) we have $g(z) = 0$ for all $z > 0$. Then any nondecreasing function $g : \mathbb{R} \to \mathbb{R}$ with $g(-1) = -1$ and $g(z) = 0$ for $z \geq 0$ satisfies (1) and gives a solution: f is nondecreasing, $f(0) = 0$ and $f(x) = 1$ for every $x \geq 1$

   2° $g(1) \neq 0$. Then the function $h(x) = \frac{g(x)}{g(1)}$ is nondecreasing and satisfies $h(0) = 0$, $h(1) = 1$, and $h(xy) = h(x)h(y)$. Fix $a > 0$, and let $h(a) = b = a^k$ for some $k \in \mathbb{R}$. It follows by induction that $h(a^q) = h(a)^q = (a^q)^k$ for every rational number q. But h is nondecreasing, so $k \geq 0$, and since the set $\{a^q \mid q \in \mathbb{Q}\}$ is dense in $\mathbb{R}^+$, we conclude that $h(x) = x^k$ for every

$x > 0$. Finally, putting $g(1) = c$, we obtain $g(x) = cx^k$ for all $x > 0$. Then $g(-x) = -x^k$ for all $x > 0$. This g obviously satisfies (1). Hence

$$f(x) = \begin{cases} c(x-1)^k, & \text{if } x > 1; \\ 1, & \text{if } x = 1; \\ 1-(1-x)^k, & \text{if } x < 1, \end{cases} \quad \text{where } c > 0 \text{ and } k \geq 0.$$

3. (a) Given any sequence c_n (in particular, such that C_n converges), we shall construct a_n and b_n such that A_n and B_n diverge.
   First, choose n_1 such that $n_1 c_1 > 1$ and set $a_1 = a_2 = \cdots = a_{n_1} = c_1$: this uniquely determines $b_2 = c_2, \ldots, b_{n_1} = c_{n_1}$. Next, choose n_2 such that $(n_2 - n_1)c_{n_1+1} > 1$ and set $b_{n_1+1} = \cdots = b_{n_2} = c_{n_1+1}$; again $a_{n_1+1}, \ldots, a_{n_2}$ is hereby determined. Then choose n_3 with $(n_3 - n_2)c_{n_2+1} > 1$ and set $a_{n_2+1} = \cdots = a_{n_3} = c_{n_2+1}$, and so on. It is plain that in this way we construct decreasing sequences a_n, b_n such that $\sum a_n$ and $\sum b_n$ diverge, since they contain an infinity of subsums that exceed 1; on the other hand, $c_n = \min\{a_n, b_n\}$ and C_n is convergent.
   (b) The answer changes in this situation. Suppose to the contrary that there is such a pair of sequences (a_n) and (b_n). There are infinitely many indices i such that $c_i = b_i$ (otherwise all but finitely many terms of the sequence (c_n) would be equal to the terms of the sequence (a_n), which has an unbounded sum). Thus for any $n_0 \in \mathbb{N}$ there is $j \geq 2n_0$ such that $c_j = b_j$. Then we have

$$\sum_{k=n_0}^{j} c_k \geq \sum_{k=n_0}^{j} c_j = (j - n_0)\frac{1}{j} \geq \frac{1}{2}.$$

   Hence the sequence (C_n) is unbounded, a contradiction.

4. (a) By the Cauchy–Schwarz inequality we have

$$\left(\sum_{i,j=1}^{n} (i-j)^2\right) \left(\sum_{i,j=1}^{n} (x_i - x_j)^2\right) \geq \left(\sum_{i,j=1}^{n} |i-j| \cdot |x_i - x_j|\right)^2. \quad (1)$$

On the other hand, it is easy to prove (for example by induction) that

$$\sum_{i,j=1}^{n} (i-j)^2 = (2n-2) \cdot 1^2 + (2n-4) \cdot 2^2 + \cdots + 2 \cdot (n-1)^2 = \frac{n^2(n^2-1)}{6}$$

and that

$$\sum_{i,j=1}^{n} |i-j| \cdot |x_i - x_j| = \frac{n}{2} \sum_{i,j=1}^{n} |x_i - x_j|.$$

Thus the inequality (1) becomes

$$\frac{n^2(n^2-1)}{6} \left(\sum_{i,j=1}^{n} (x_i - x_j)^2\right) \geq \frac{n^2}{4} \left(\sum_{i,j=1}^{n} |x_i - x_j|\right)^2,$$

which is equivalent to the required one.

(b) Equality holds if and only if it holds in (1), i.e., if and only if there is $\lambda \in \mathbb{R}$ such that $|x_i - x_j| = \lambda |i - j|$ for all i, j. This is equivalent to (x_i) being an arithmetic sequence.

5. Placing $x = y = z = 1$ in (i) leads to $4f(1) = f(1)^3$, so by the condition $f(1) > 0$ we get $f(1) = 2$. Also, setting $x = y = t$, $z = 1/t$ yields $f(t) = f(1/t)$; hence $f(t) \geq f(1) = 2$ for each t. Thus $f(t) = c + c^{-1}$ for some $c = c(t) \geq 1$. Now setting $(x, y, z) = (ts, \frac{t}{s}, \frac{s}{t})$ in (i) gives

$$f(t)f(s) = f(ts) + f\left(\frac{t}{s}\right).$$

In particular, $s = t$ yields $f(t^2) = f(t)^2 - 2$, so if $f(t) = c + c^{-1}$, then $f(t^2) = c^2 + c^{-2}$. A simple induction shows that $f(t^n) = c^n + c^{-n}$ for each $n \in \mathbb{N}$. Consequently, $f(t^{n/m}) = c^{n/m} + c^{-n/m}$. Now if $t > 1$ is fixed and $\lambda \in \mathbb{R}$ is such that $c = t^\lambda$, we obtain $f(t^q) = t^{\lambda q} + t^{-\lambda q}$, i.e.,

$$f(x) = x^\lambda + x^{-\lambda} \quad \text{for} \quad x \in T = \{t^q \mid q \in \mathbb{Q}\}. \tag{1}$$

But since the set T is dense in $\mathbb{R}^+$ and f is monotone on $(0, 1]$ and $[1, \infty)$, it follows that (1) holds for all $x > 0$. It is directly verified that this function satisfies (i) and (ii).

6. Set $X = \max\{x_1, \ldots, x_n\}$ and $Y = \max\{y_1, \ldots, y_n\}$. By replacing x_i by $x_i' = \frac{x_i}{X}$, y_i by $y_i' = \frac{y_i}{Y}$ and z_i by $z_i' = \frac{z_i}{\sqrt{XY}}$, we may assume that $X = Y = 1$. It is sufficient to prove that

$$M + z_2 + \cdots + z_{2n} \geq x_1 + \cdots + x_n + y_1 + \cdots + y_n, \tag{1}$$

because this implies the result by the A-G mean inequality.

To prove (1) it is enough to prove that for any r, *the number of terms greater than r on the left-hand side of* (1) *is at least that number on the right-hand side of* (1).

If $r \geq 1$, then there are no terms on the right-hand side greater than r. Suppose that $r < 1$ and consider the sets $A = \{i \mid 1 \leq i \leq n, x_i > r\}$ and $B = \{i \mid 1 \leq i \leq n, y_i > r\}$. Set $a = |A|$ and $b = |B|$. If $x_i > r$ and $y_j > r$, then $z_{i+j} \geq \sqrt{x_i y_j} > r$; hence

$$C = \{k \mid 2 \leq k \leq 2n, z_k > r\} \supseteq A + B = \{\alpha + \beta \mid \alpha \in A, \beta \in B\}.$$

It is easy to verify that $|A + B| \geq |A| + |B| - 1$. It follows that the number of z_k's greater than r is $\geq a + b - 1$. But in that case $M > r$, implying that at least $a + b$ elements of the left-hand side of (1) are greater than r, which completes the proof.

7. Consider the set $D = \{x - y \mid x, y \in A\}$. Obviously, the number of elements of the set D is less than or equal to $101 \cdot 100 + 1$. The sets $A + t_i$ and $A + t_j$ are disjoint if and only if $t_i - t_j \notin D$. Now we shall choose inductively 100 elements $t_1, \ldots, t_{100}$. Let t_1 be any element of the set $S \setminus D$ (such an element exists, since the number of elements of S is greater than the number of elements of D). Suppose now that we have chosen k ($k \leq 99$) elements $t_1, \ldots, t_k$ from D such that the difference of

any two of the chosen elements does not belong to D. We can select t_{k+1} to be an element of S that does not belong to any of the sets $t_1 + D, t_2 + D, \ldots, t_k + D$ (this is possible to do, since each of the previous sets has at most $101 \cdot 100 + 1$ elements; hence their union has at most $99(101 \cdot 100 + 1) = 999999 < 1000000$ elements).

8. Let S be the disk with the smallest radius, say s, and O the center of that disk. Divide the plane into 7 regions: one bounded by disk s and 6 regions $T_1, \ldots, T_6$ shown in the figure.

Any of the disks different from S, say D_k, has its center in one of the seven regions. If its center is inside S then D_k contains point O. Hence the number of disks different from S having their centers in S is at most 2002.

Consider a disk D_k that intersects S and whose center is in the region T_i. Let P_i be the point such that OP_i bisects the region T_i and $OP_i = s\sqrt{3}$. We claim that D_k contains P_i. Divide the region T_i by a line l_i through P_i perpendicular to OP_i into two regions U_i and V_i, where O and U_i are on the same side of l_i. Let K be the center of D_k. Consider two cases:

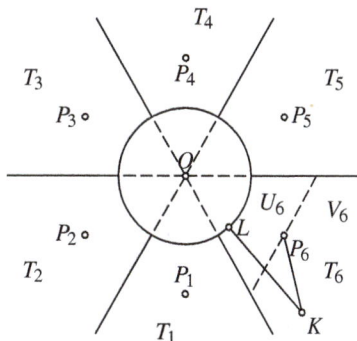

(i) $K \in U_i$. Since the disk with the center P_i and radius s contains U_i, we see that $KP_i \leq s$. Hence D_k contains P_i.

(ii) $K \in V_i$. Denote by L the intersection point of the segment KO with the circle s. We want to prove that $KL > KP_i$. It is enough to prove that $\angle KP_iL > \angle KLP_i$. However, it is obvious that $\angle LP_iO \leq 30°$ and $\angle LOP_i \leq 30°$, hence $\angle KLP_i \leq 60°$, while $\angle NP_iL = 90° - \angle LP_iO \geq 60°$. This implies $\angle KP_iL \geq \angle NP_iL \geq 60° \geq \angle KLP_i$ (N is the point on the edge of T_i as shown in the figure). Our claim is thus proved.

Now we see that the number of disks with centers in T_i that intersect S is less than or equal to 2003, and the total number of disks that intersect S is not greater than $2002 + 6 \cdot 2003 = 7 \cdot 2003 - 1$.

9. Suppose that k of the angles of an n-gon are right. Since the other $n - k$ angles are less than $360°$ and the sum of the angles is $(n-2)180°$, we have the inequality $k \cdot 90° + (n-k)360° > (n-2)180°$, which is equivalent to $k < \frac{2n+4}{3}$. Since n and k are integers, it follows that $k \leq \left[\frac{2n}{3}\right] + 1$.

If $n = 5$, then $\left[\frac{2n}{3}\right] + 1 = 4$, but if a pentagon has four right angles, the other angle is equal to $180°$, which is impossible. Hence for $n = 5$, $k \leq 3$. It is easy to construct a pentagon with 3 right angles, e.g., as in the picture below.

Now we shall show by induction that for $n \geq 6$ there is an n-gon with $\left[\frac{2n}{3}\right] + 1$ internal right angles. For $n = 6, 7, 8$ examples are presented in the picture. Assume that there is a $(n-3)$gon with $\left[\frac{2(n-3)}{3}\right] + 1 = \left[\frac{2n}{3}\right] - 1$ internal right angles. Then one of the internal angles, say $\angle BAC$, is not convex. Interchange the vertex A with four new vertices A_1, A_2, A_3, A_4 as shown in the picture such that $\angle BA_1A_2 = \angle A_3A_4C = 90°$.

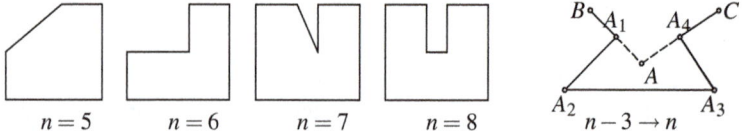

$n = 5$ $n = 6$ $n = 7$ $n = 8$ $n-3 \rightarrow n$

10. Denote by b_{ij} the entries of the matrix B. Suppose the contrary, i.e., that there is a pair (i_0, j_0) such that $a_{i_0,j_0} \neq b_{i_0,j_0}$. We may assume without loss of generality that $a_{i_0,j_0} = 0$ and $b_{i_0,j_0} = 1$.

Since the sums of elements in the i_0th rows of the matrices A and B are equal, there is some j_1 for which $a_{i_0,j_1} = 1$ and $b_{i_0,j_1} = 0$. Similarly, from the fact that the sums in the j_1th columns of the matrices A and B are equal, we conclude that there exists i_1 such that $a_{i_1,j_1} = 0$ and $b_{i_1,j_1} = 1$. Continuing this procedure, we construct two sequences i_k, j_k such that $a_{i_k,j_k} = 0$, $b_{i_k,j_k} = 1$, $a_{i_k,j_{k+1}} = 1$, $b_{i_k,j_{k+1}} = 0$. Since the set of the pairs (i_k, j_k) is finite, there are two different numbers t, s such that $(i_t, j_t) = (i_s, j_s)$. From the given condition we have that $x_{i_k} + y_{i_k} < 0$ and $x_{i_{k+1}} + y_{i_{k+1}} \geq 0$. But $j_t = j_s$, and hence $0 \leq \sum_{k=s}^{t-1}(x_{i_k} + y_{j_{k+1}}) = \sum_{k=s}^{t-1}(x_{i_k} + y_{j_k}) < 0$, a contradiction.

11. (a) By the pigeonhole principle there are two different integers $x_1, x_2, x_1 > x_2$, such that $|\{x_1\sqrt{3}\} - \{x_2\sqrt{3}\}| < 0.001$. Set $a = x_1 - x_2$. Consider the equilateral triangle with vertices $(0,0), (2a,0), (a, a\sqrt{3})$. The points $(0,0)$ and $(2a,0)$ are lattice points, and we claim that the point $(a, a\sqrt{3})$ is at distance less than 0.001 from a lattice point. Indeed, since $0.001 > |\{x_1\sqrt{3}\} - \{x_2\sqrt{3}\}| = |a\sqrt{3} - ([x_1\sqrt{3}] - [x_2\sqrt{3}])|$, we see that the distance between the points $(a, a\sqrt{3})$ and $(a, [x_1\sqrt{3}] - [x_2\sqrt{3}])$ is less than 0.001, and the point $(a, [x_1\sqrt{3}] - [x_2\sqrt{3}])$ is with integer coefficients.

(b) Suppose that $P'Q'R'$ is an equilateral triangle with side length $l \leq 96$ such that each of its vertices P', Q', R' lies in a disk of radius 0.001 centered at a lattice point. Denote by P, Q, R the centers of these disks. Then we have $l - 0.002 \leq PQ, QR, RP \leq l + 0.002$. Since PQR is not an equilateral triangle, two of its sides are different, say $PQ \neq QR$. On the other hand, PQ^2, QR^2 are integers, so we have $1 \leq |PQ^2 - QR^2| = (PQ + QR)|PQ - QR| \leq 0.004(PQ + QR) \leq (2l + 0.004) \cdot 0.004 \leq 2 \cdot 96.002 \cdot 0.004 < 1$, which is a contradiction.

12. Denote by $\overline{a_{k-1}a_{k-2}\ldots a_0}$ the decimal representation of a number whose digits are $a_{k-1}, \ldots, a_0$. We will use the following well-known fact:

$$\overline{a_{k-1}a_{k-2}\ldots a_0} \equiv i \pmod{11} \iff \sum_{l=0}^{k-1}(-1)^l a_l \equiv i \pmod{11}.$$

Let m be a positive integer. Define A as the set of integers n $(0 \leq n < 10^{2m})$ whose right $2m - 1$ digits can be so permuted to yield an integer divisible by 11, and B as the set of integers n $(0 \leq n < 10^{2m-1})$ whose digits can be permuted so that the resulting is an integer divisible by 11.

Suppose that $a = \overline{a_{2m-1} \ldots a_0} \in A$. Then it satisfies

$$\sum_{l=0}^{2m-1} (-1)^l a_l \equiv 0 \ (\mathrm{mod}\ 11). \tag{1}$$

The $2m$-tuple $(a_{2m-1}, \ldots, a_0)$ satisfies (1) if and only if the $2m$-tuple $(ka_{2m-1} + l, \ldots, ka_0 + l)$ satisfies (1), where $k, l \in \mathbb{Z}$, $11 \nmid k$.

Since $a_0 + 1 \not\equiv 0 \ (\mathrm{mod}\ 11)$, we can choose k from the set $\{1, \ldots, 10\}$ such that $(a_0 + 1)k \equiv 1 \ (\mathrm{mod}\ 11)$. Thus there is a permutation of the $2m$-tuple $((a_{2m-1} + 1)k - 1, \ldots, (a_1 + 1)k - 1, 0)$ satisfying (1). Interchanging odd and even positions if necessary, we may assume that this permutation keeps the 0 at the last position. Since $(a_i + 1)k$ is not divisible by 11 for any i, there is a unique $b_i \in \{0, 1, \ldots, 9\}$ such that $b_i \equiv (a_i + 1)k - 1 \ (\mathrm{mod}\ 11)$. Hence the number $\overline{b_{2m-1} \ldots b_1} \in B$.

Thus for fixed $a_0 \in \{0, 1, 2, \ldots, 9\}$, to each $a \in A$ such that the last digit of a is a_0 we associate a unique $b \in B$. Conversely, having $a_0 \in \{0, 1, 2, \ldots, 9\}$ fixed, from any number $\overline{b_{2m-1} \ldots b_1} \in B$ we can reconstruct $\overline{a_{2m-1} \ldots a_1 a_0} \in A$. Hence $|A| = 10|B|$, i.e., $f(2m) = 10f(2m-1)$.

13. Denote by K and L the intersections of the bisectors of $\angle ABC$ and $\angle ADC$ with the line AC, respectively. Since $AB : BC = AK : KC$ and $AD : DC = AL : LC$, we have to prove that

$$PQ = QR \iff \frac{AB}{BC} = \frac{AD}{DC}. \tag{1}$$

Since the quadrilaterals $AQDR$ and $QPCD$ are cyclic, we deduce that $\angle RDQ = \angle BAC$ and $\angle QDP = \angle ACB$. By the law of sines it follows that

$$\frac{AB}{BC} = \frac{\sin \angle ACB}{\sin \angle BAC}$$

and that $QR = AD \sin \angle RDQ$, $QP = CD \sin \angle QDP$. Now we have

$$\frac{AB}{BC} = \frac{\sin \angle ACB}{\sin \angle BAC} = \frac{\sin \angle QDP}{\sin \angle RDQ} = \frac{AD \cdot QP}{QR \cdot CD}.$$

The statement (1) follows directly.

14. Denote by R the intersection point of the bisector of $\angle AQC$ and the line AC. From $\triangle ACQ$ we get

$$\frac{AR}{RC} = \frac{AQ}{QC} = \frac{\sin \angle QCA}{\sin \angle QAC}.$$

By the sine version of Ceva's theorem we have $\frac{\sin \angle APB}{\sin \angle BPC} \cdot \frac{\sin \angle QAC}{\sin \angle PAQ} \cdot \frac{\sin \angle QCP}{\sin \angle QCA} = 1$, which is equivalent to

$$\frac{\sin \angle APB}{\sin \angle BPC} = \left(\frac{\sin \angle QCA}{\sin \angle QAC}\right)^2$$

because $\angle QCA = \angle PAQ$ and $\angle QAC = \angle QCP$. Denote by $S(XYZ)$ the area of a triangle XYZ. Then

$$\frac{\sin \angle APB}{\sin \angle BPC} = \frac{AP \cdot BP \cdot \sin \angle APB}{BP \cdot CP \cdot \sin \angle BPC} = \frac{S(\triangle ABP)}{S(\triangle BCP)} = \frac{AB}{BC},$$

which implies that $\left(\frac{AR}{RC}\right)^2 = \frac{AB}{BC}$. Hence R does not depend on Γ.

15. From the given equality we see that $0 = (BP^2 + PE^2) - (CP^2 + PF^2) = BF^2 - CE^2$, so $BF = CE = x$ for some x. Similarly, there are y and z such that $CD = AF = y$ and $BD = AE = z$. It is easy to verify that D, E, and F must lie on the segments BC, CA, AB.

   Denote by a, b, c the length of the segments BC, CA, AB. It follows that $a = z + y$, $b = z + x$, $c = x + y$, so D, E, F are the points where the excircles touch the sides of $\triangle ABC$. Hence P, D, and I_A are collinear and

$$\angle PI_AC = \angle DI_AC = 90° - \frac{180° - \angle ACB}{2} = \frac{\angle ACB}{2}.$$

   In the same way we obtain that $\angle PI_BC = \frac{\angle ACB}{2}$ and $PI_B = PI_A$. Analogously, we get $PI_C = PI_B$, which implies that P is the circumcenter of the triangle $I_AI_BI_C$.

16. Apply an inversion with center at P and radius r; let $\hat{X}$ denote the image of X. The circles $\Gamma_1, \Gamma_2, \Gamma_3, \Gamma_4$ are transformed into lines $\hat{\Gamma}_1, \hat{\Gamma}_2, \hat{\Gamma}_3, \hat{\Gamma}_4$, where $\hat{\Gamma}_1 \parallel \hat{\Gamma}_3$ and $\hat{\Gamma}_2 \parallel \hat{\Gamma}_4$, and therefore $\hat{A}\hat{B}\hat{C}\hat{D}$ is a parallelogram.
   Further, we have $AB = \frac{r^2}{PA \cdot PB}\hat{A}\hat{B}$, $BC = \frac{r^2}{PB \cdot PC}\hat{B}\hat{C}$, $CD = \frac{r^2}{PC \cdot PD}\hat{C}\hat{D}$, $DA = \frac{r^2}{PD \cdot PA}\hat{D}\hat{A}$ and $PB = \frac{r^2}{P\hat{B}}$, $PD = \frac{r^2}{P\hat{D}}$. The equality to be proven becomes

$$\frac{P\hat{D}^2}{P\hat{B}^2} \cdot \frac{\hat{A}\hat{B} \cdot \hat{B}\hat{C}}{\hat{A}\hat{D} \cdot \hat{D}\hat{C}} = \frac{P\hat{D}^2}{P\hat{B}^2},$$

   which holds because $\hat{A}\hat{B} = \hat{C}\hat{D}$ and $\hat{B}\hat{C} = \hat{D}\hat{A}$.

17. The triangles PDE and CFG are homothetic; hence lines FD, GE, and CP intersect at one point. Let Q be the intersection point of the line CP and the circumcircle of $\triangle ABC$. The required statement will follow if we show that Q lies on the lines GE and FD. Since $\angle CFG = \angle CBA = \angle CQA$, the quadrilateral $AQPF$ is cyclic. Analogously, $BQPG$ is cyclic. However, the isosceles trapezoid $BDPG$ is also cyclic; it

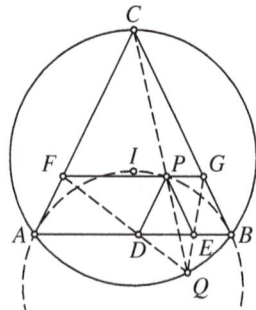

follows that B,Q,D,P,G lie on a circle. Therefore we get

$$\angle PQF = \angle PAC, \quad \angle PQD = \angle PBA. \tag{1}$$

Since I is the incenter of $\triangle ABC$, we have $\angle CAI = \frac{1}{2}\angle CAB = \frac{1}{2}\angle CBA = \angle IBA$; hence CA is the tangent at A to the circumcircle of $\triangle ABI$. This implies that $\angle PAC = \angle PBA$, and it follows from (1) that $\angle PQF = \angle PQD$, i.e., that F,D,Q are also collinear. Similarly, G,E,Q are collinear and the claim is thus proved.

18. Let $ABCDEF$ be the given hexagon. We shall use the following lemma.

   Lemma. If $\angle XZY \geq 60°$ and if M is the midpoint of XY, then $MZ \leq \frac{\sqrt{3}}{2}XY$, with equality if and only if $\triangle XYZ$ is equilateral.

   Proof. Let Z' be the point such that $\triangle XYZ'$ is equilateral. Then Z is inside the circle circumscribed about $\triangle XYZ'$. Consequently $MZ \leq MZ' = \frac{\sqrt{3}}{2}XY$, with equality if and only if $Z = Z'$.

   Set $AD \cap BE = P$, $BE \cap CF = Q$, and $CF \cap AD = R$. Suppose $\angle APB = \angle DPE > 60°$, and let K,L be the midpoints of the segments AB and DE respectively. Then by the lemma,

$$\frac{\sqrt{3}}{2}(AB+DE) = KL \leq PK + PL < \frac{\sqrt{3}}{2}(AB+DE),$$

   which is impossible. Hence $\angle APB \leq 60°$ and similarly $\angle BQC \leq 60°$, $\angle CRD \leq 60°$. But the sum of the angles APB, BQC, CRD is $180°$, from which we conclude that these angles are all equal to $60°$, and moreover that the triangles APB, BQC, CRD are equilateral. Thus $\angle ABC = \angle ABP + \angle QBC = 120°$, and in the same way all angles of the hexagon are equal to $120°$.

19. Let D, E, F be the midpoints of BC, CA, AB, respectively. We construct smaller semicircles $\Gamma_d, \Gamma_e, \Gamma_f$ inside $\triangle ABC$ with centers D,E,F and radii $d = \frac{s-a}{2}$, $e = \frac{s-b}{2}$, $f = \frac{s-c}{2}$ respectively. Since $DE = d + e$, $DF = d + f$, and $EF = e + f$, we deduce that Γ_d, Γ_e, and Γ_f touch each other at the points D_1, E_1, F_1 of tangency of the incircle γ of $\triangle DEF$ with its sides ($D_1 \in EF$, etc.). Consider the circle Γ_g with center O and radius g that lies inside $\triangle DEF$ and tangents $\Gamma_d, \Gamma_e, \Gamma_f$.

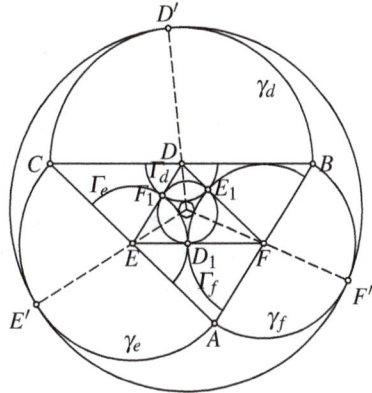

Now let OD, OE, OF meet the semicircles $\Gamma_d, \Gamma_e, \Gamma_f$ at D', E', F' respectively. We have $OD' = OD + DD' = g + d + \frac{a}{2} = g + \frac{s}{2}$ and similarly $OE' = OF' = g + \frac{s}{2}$. It follows that the circle with center O and radius $g + \frac{s}{2}$ touches all three semicircles, and consequently $t = g + \frac{s}{2} > \frac{s}{2}$. Now set the coordinate system such that we have the points $D_1(0,0)$, $E(-e,0)$, $F(f,0)$ and such that the y coordinate of D is positive.

Apply the inversion with center D_1 and unit radius. This inversion maps the circles Γ_e and Γ_f to the lines $\widehat{\Gamma}_e\left[x = -\frac{1}{2e}\right]$ and $\widehat{\Gamma}_e\left[x = \frac{1}{2f}\right]$ respectively, and the circle γ goes to the line $\widehat{\gamma}\left[y = \frac{1}{r}\right]$. The images $\widehat{\Gamma}_d$ and $\widehat{\Gamma}_g$ of Γ_d, Γ_g are the circles that touch the lines $\widehat{\Gamma}_e$ and $\widehat{\Gamma}_f$. Since $\widehat{\Gamma}_d, \widehat{\Gamma}_g$ are perpendicular to γ, they have radii equal to $R = \frac{1}{4e} + \frac{1}{4f}$ and centers at $\left(-\frac{1}{4e} + \frac{1}{4f}, \frac{1}{r}\right)$ and $\left(-\frac{1}{4e} + \frac{1}{4f}, \frac{1}{r} + 2R\right)$ respectively. Let p and P be the distances from $D_1(0,0)$ to the centers of Γ_g and $\widehat{\Gamma}_g$ respectively. We have that $P^2 = \left(\frac{1}{4e} - \frac{1}{4f}\right)^2 + \left(\frac{1}{r} + 2R\right)^2$, and that the circles Γ_g and $\widehat{\Gamma}_g$ are homothetic with center of homothety D_1; hence $p/P = g/R$. On the other hand, $\widehat{\Gamma}_g$ is the image of Γ_g under inversion; hence the product of the tangents from D_1 to these two circles is equal to 1. In other words, we obtain $\sqrt{p^2 - g^2} \cdot \sqrt{P^2 - R^2} = 1$. Using the relation $p/P = g/R$ we get $g = \frac{R}{P^2 - R^2}$. The inequality we have to prove is equivalent to $(4 + 2\sqrt{3})g \leq r$. This can be proved as follows:

$$r - (4 + 2\sqrt{3})g = \frac{r(P^2 - R^2 - (4 + 2\sqrt{3})R/r)}{P^2 - R^2}$$

$$= \frac{r\left(\left(\frac{1}{r} + 2R\right)^2 + \left(\frac{1}{4e} - \frac{1}{4f}\right)^2 - R^2 - (4 + 2\sqrt{3})\frac{R}{r}\right)}{P^2 - R^2}$$

$$= \frac{r}{P^2 - R^2}\left(\left(R\sqrt{3} - \frac{1}{r}\right)^2 + \left(\frac{1}{4e} - \frac{1}{4f}\right)^2\right) \geq 0.$$

Remark. One can obtain a symmetric formula for g:

$$\frac{1}{2g} = \frac{1}{s - a} + \frac{1}{s - b} + \frac{1}{s - c} + \frac{2}{r}.$$

20. Let r_i be the remainder when x_i is divided by m. Since there are at most m^m types of m-consecutive blocks in the sequence (r_i), some type will repeat at least twice. Then since the entire sequence is determined by one m-consecutive block, the entire sequence will be periodic.

The formula works both forward and backward; hence using the rule $x_i = x_{i+m} - \sum_{j=1}^{m-1} x_{i+j}$ we can define $x_{-1}, x_{-2}, \ldots$. Thus we obtain that

$$(r_{-m}, \ldots, r_{-1}) = (0, 0, \ldots, 0, 1).$$

Hence there are $m - 1$ consecutive terms in the sequence (x_i) that are divisible by m.

If there were m consecutive terms in the sequence (x_i) divisible by m, then by the recurrence relation all the terms of (x_i) would be divisible by m, which is impossible.

21. Let a be a positive integer for which $d(a) = a^2$. Suppose that a has $n + 1$ digits, $n \geq 0$. Denote by s the last digit of a and by f the first digit of c. Then $a = \overline{* \ldots * s}$,

where $*$ stands for a digit that is not important to us at the moment. We have $\overline{*\ldots*s^2} = a^2 = d = \overline{*\ldots*f}$ and $b^2 = \overline{s*\ldots*^2} = c = \overline{f*\ldots*}$.

We cannot have $s = 0$, since otherwise c would have at most $2n$ digits, while a^2 has either $2n + 1$ or $2n + 2$ digits. The following table gives all possibilities for s and f:

s	1	2	3	4	5	6	7	8	9
$f = $ last digit of $\overline{*\ldots*s^2}$	1	4	9	6	5	6	9	4	1
$f = $ first digit of $\overline{s*\ldots*^2}$	1,2,3	4 – 8	9,1	1,2	2,3	3,4	4,5,6	6,7,8	8,9

We obtain from the table that $s \in \{1,2,3\}$ and $f = s^2$, and consequently $c = b^2$ and d have exactly $2n + 1$ digits each. Put $a = 10x + s$, where $x < 10^n$. Then $b = 10^n s + x$, $c = 10^{2n} s^2 + 2 \cdot 10^n sx + x^2$, and $d = 2 \cdot 10^{n+1} sx + 10x^2 + s^2$, so from $d = a^2$ it follows that $x = 2s \cdot \frac{10^n - 1}{9}$. Thus $a = \underbrace{6\ldots63}_{n}$, $a = \underbrace{4\ldots42}_{n}$ or $a = \underbrace{2\ldots21}_{n}$.

For $n \geq 1$ we see that a cannot be $a = 6\ldots63$ or $a = 4\ldots42$ (otherwise a^2 would have $2n + 2$ digits). Therefore a equals 1, 2, 3 or $\underbrace{2\ldots21}_{n}$ for $n \geq 0$. It is easy to verify that these numbers have the required property.

22. Let a and b be positive integers for which $\frac{a^2}{2ab^2 - b^3 + 1} = k$ is a positive integer. Since $k > 0$, it follows that $2ab^2 \geq b^3$, so $2a \geq b$. If $2a > b$, then from $2ab^2 - b^3 + 1 > 0$ we see that $a^2 > b^2(2a - b) + 1 > b^2$, i.e. $a > b$. Therefore, if $a \leq b$, then $a = b/2$.

We can rewrite the given equation as a quadratic equation in a, $a^2 - 2kb^2 a + k(b^3 - 1) = 0$, which has two solutions, say a_1 and a_2, one of which is in $\mathbb{N}_0$. From $a_1 + a_2 = 2kb^2$ and $a_1 a_2 = k(b^3 - 1)$ it follows that the other solution is also in $\mathbb{N}_0$. Suppose w.l.o.g. that $a_1 \geq a_2$. Then $a_1 \geq kb^2$ and

$$0 \leq a_2 = \frac{k(b^3 - 1)}{a_1} \leq \frac{k(b^3 - 1)}{kb^2} < b.$$

By the above considerations we have either $a_2 = 0$ or $a_2 = b/2$. If $a_2 = 0$, then $b^3 - 1 = 0$ and hence $a_1 = 2k$, $b = 1$. If $a_2 = b/2$, then $b = 2t$ for some t, and $k = b^2/4$, $a_1 = b^4/2 - b/2$. Therefore the only solutions are

$$(a,b) \in \{(2t,1), (t,2t), (8t^4 - t, 2t) \mid t \in \mathbb{N}\}.$$

It is easy to show that all of these pairs satisfy the given condition.

23. Assume that $b \geq 6$ has the required property. Consider the sequence $y_n = (b - 1)x_n$. From the definition of x_n we easily find that $y_n = b^{2n} + b^{n+1} + 3b - 5$. Then $y_n y_{n+1} = (b - 1)^2 x_n x_{n+1}$ is a perfect square for all $n > M$. Also, straightforward calculation implies

$$\left(b^{2n+1} + \frac{b^{n+2} + b^{n+1}}{2} - b^3\right)^2 < y_n y_{n+1} < \left(b^{2n+1} + \frac{b^{n+2} + b^{n+1}}{2} + b^3\right)^2.$$

Hence for every $n > M$ there is an integer a_n such that $|a_n| < b^3$ and

$$y_n y_{n+1} = \left(b^{2n} + b^{n+1} + 3b - 5\right)\left(b^{2n+2} + b^{n+2} + 3b - 5\right)$$
$$= \left(b^{2n+1} + \frac{b^{n+1}(b+1)}{2} + a_n\right)^2. \tag{1}$$

Now considering this equation modulo b^n we obtain $(3b-5)^2 \equiv a_n^2$, so that assuming that $n > 3$ we get $a_n = \pm(3b-5)$.

If $a_n = 3b - 5$, then substituting in (1) yields $\frac{1}{4}b^{2n}(b^4 - 14b^3 + 45b^2 - 52b + 20) = 0$, with the unique positive integer solution $b = 10$. Also, if $a_n = -3b+5$, we similarly obtain $\frac{1}{4}b^{2n}(b^4 - 14b^3 - 3b^2 + 28b + 20) - 2b^{n+1}(3b^2 - 2b - 5) = 0$ for each n, which is impossible.

For $b = 10$ it is easy to show that $x_n = \left(\frac{10^n + 5}{3}\right)^2$ for all n. This proves the statement.

Second solution. In problems of this type, computing $z_n = \sqrt{x_n}$ asymptotically usually works.

From $\lim_{n \to \infty} \frac{b^{2n}}{(b-1)x_n} = 1$ we infer that $\lim_{n \to \infty} \frac{b^n}{z_n} = \sqrt{b-1}$. Furthermore, from $(bz_n + z_{n+1})(bz_n - z_{n+1}) = b^2 x_n - x_{n+1} = b^{n+2} + 3b^2 - 2b - 5$ we obtain

$$\lim_{n \to \infty}(bz_n - z_{n+1}) = \frac{b\sqrt{b-1}}{2}.$$

Since the z_n's are integers for all $n \geq M$, we conclude that $bz_n - z_{n+1} = \frac{b\sqrt{b-1}}{2}$ for all n sufficiently large. Hence $b - 1$ is a perfect square, and moreover b divides $2z_{n+1}$ for all large n. It follows that $b \mid 10$; hence the only possibility is $b = 10$.

24. Suppose that $m = u + v + w$ where u, v, w are good integers whose product is a perfect square of an odd integer. Since uvw is an odd perfect square, we have that $uvw \equiv 1 \pmod 4$. Thus either two or none of the numbers u, v, w are congruent to 3 modulo 4. In both cases $u + v + w \equiv 3 \pmod 4$. Hence $m \equiv 3 \pmod 4$.

Now we shall prove the converse: every $m \equiv 3 \pmod 4$ has infinitely many representations of the desired type. Let $m = 4k + 3$. We shall represent m in the form

$$4k + 3 = xy + yz + zx, \quad \text{for } x, y, z \text{ odd.} \tag{1}$$

The product of the summands is an odd square. Set $x = 1 + 2l$ and $y = 1 - 2l$. In order to satisfy (1), z must satisfy $z = 2l^2 + 2k + 1$. The summands xy, yz, zx are distinct except for finitely many l, so it remains only to prove that for infinitely many integers l, $|xy|$, $|yz|$, and $|zx|$ are not perfect squares. First, observe that $|xy| = 4l^2 - 1$ is not a perfect square for any $l \neq 0$.

Let $p, q > m$ be fixed different prime numbers. The system of congruences $1 + 2l \equiv p \pmod{p^2}$ and $1 - 2l \equiv q \pmod{q^2}$ has infinitely many solutions l by the Chinese remainder theorem. For any such l, the number $z = 2l^2 + 2k + 1$ is divisible by neither p nor q, and hence $|xz|$ (respectively $|yz|$) is divisible by p, but not by p^2 (respectively by q, but not by q^2). Thus xz and yz are also good numbers.

25. Suppose that for every prime q, there exists an n for which $n^p \equiv p \pmod{q}$. Assume that $q = kp + 1$. By Fermat's theorem we deduce that $p^k \equiv n^{kp} = n^{q-1} \equiv 1 \pmod{q}$, so $q \mid p^k - 1$.

It is known that any prime q such that $q \mid \frac{p^p - 1}{p - 1}$ must satisfy $q \equiv 1 \pmod{p}$. Indeed, from $q \mid p^{q-1} - 1$ it follows that $q \mid p^{\gcd(p, q-1)} - 1$; but $q \nmid p - 1$ because $\frac{p^p - 1}{p - 1} \equiv 1 \pmod{p - 1}$, so $\gcd(p, q - 1) \neq 1$. Hence $\gcd(p, q - 1) = p$.

Now suppose q is any prime divisor of $\frac{p^p - 1}{p - 1}$. Then $q \mid \gcd(p^k - 1, p^p - 1) = p^{\gcd(p,k)} - 1$, which implies that $\gcd(p, k) > 1$, so $p \mid k$. Consequently $q \equiv 1 \pmod{p^2}$. However, the number $\frac{p^p - 1}{p - 1} = p^{p-1} + \cdots + p + 1$ must have at least one prime divisor that is not congruent to 1 modulo p^2. Thus we arrived at a contradiction.

Remark. Taking $q \equiv 1 \pmod{p}$ is natural, because for every other q, n^p takes all possible residues modulo q (including p too). Indeed, if $p \nmid q - 1$, then there is an $r \in \mathbb{N}$ satisfying $pr \equiv 1 \pmod{q - 1}$; hence for any a the congruence $n^p \equiv a \pmod{q}$ has the solution $n \equiv a^r \pmod{q}$.

The statement of the problem itself is a special case of the Chebotarev's theorem.

26. Define the sequence x_k of positive reals by $a_k = \cosh x_k$ (cosh is the hyperbolic cosine defined by $\cosh t = \frac{e^t + e^{-t}}{2}$). Since $\cosh(2x_k) = 2a_k^2 - 1 = \cosh x_{k+1}$, it follows that $x_{k+1} = 2x_k$ and thus $x_k = \lambda \cdot 2^k$ for some $\lambda > 0$. From the condition $a_0 = 2$ we obtain $\lambda = \log(2 + \sqrt{3})$. Therefore

$$a_n = \frac{(2 + \sqrt{3})^{2^n} + (2 - \sqrt{3})^{2^n}}{2}.$$

Let p be a prime number such that $p \mid a_n$. We distinguish the following two cases:

(i) There exists an $m \in \mathbb{Z}$ such that $m^2 \equiv 3 \pmod{p}$. Then we have

$$(2 + m)^{2^n} + (2 - m)^{2^n} \equiv 0 \pmod{p}. \tag{1}$$

Since $(2 + m)(2 - m) = 4 - m^2 \equiv 1 \pmod{p}$, multiplying both sides of (1) by $(2 + m)^{2^n}$ gives $(2 + m)^{2^{n+1}} \equiv -1 \pmod{p}$. It follows that the multiplicative order of $(2 + m)$ modulo p is 2^{n+2}, or $2^{n+2} \mid p - 1$, which implies that $2^{n+3} \mid (p - 1)(p + 1) = p^2 - 1$.

(ii) $m^2 \equiv 3 \pmod{p}$ has no integer solutions. We will work in the algebraic extension $\mathbb{Z}_p(\sqrt{3})$ of the field $\mathbb{Z}_p$. In this field $\sqrt{3}$ plays the role of m, so as in the previous case we obtain $(2 + \sqrt{3})^{2^{n+1}} = -1$; i.e., the order of $2 + \sqrt{3}$ in the multiplicative group $\mathbb{Z}_p(\sqrt{3})^*$ is 2^{n+2}. We cannot finish the proof as in the previous case: in fact, we would conclude only that 2^{n+2} divides the order $p^2 - 1$ of the group. However, it will be enough to find a $u \in \mathbb{Z}_p(\sqrt{3})$ such that $u^2 = 2 + \sqrt{3}$, since then the order of u is equal to 2^{n+3}. Note that $(1 + \sqrt{3})^2 = 2(2 + \sqrt{3})$. Thus it is sufficient to prove that $\frac{1}{2}$ is a perfect square in $\mathbb{Z}_p(\sqrt{3})$. But we know that in this field $a_n = 0 = 2a_{n-1}^2 - 1$, and hence $2a_{n-1}^2 = 1$ which implies $\frac{1}{2} = a_{n-1}^2$. This completes the proof.

27. Let $p_1, p_2, \ldots, p_r$ be distinct primes, where $r = p - 1$. Consider the sets $B_i = \{p_i, p_i^{p+1}, \ldots, p_i^{(r-1)p+1}\}$ and $B = \bigcup_{i=1}^r B_i$. Then B has $(p-1)^2$ elements and satisfies (i) and (ii).

Now suppose that $|A| \geq r^2 + 1$ and that A satisfies (i) and (ii). Let $\{t_1, \ldots, t_{r^2+1}\}$ be distinct elements of A, where $t_j = p_1^{\alpha_{j_1}} \cdot p_2^{\alpha_{j_2}} \cdots p_r^{\alpha_{j_r}}$. We shall show that the product of some elements of A is a perfect pth power, i.e., that there exist $\tau_j \in \{0, 1\}$ $(1 \leq j \leq r^2 + 1)$, not all equal to 0, such that $T = t_1^{\tau_1} \cdot t_2^{\tau_2} \cdots t_{r^2+1}^{\tau_{r^2+1}}$ is a pth power. This is equivalent to the condition that

$$\sum_{j=1}^{r^2+1} \alpha_{ij} \tau_j \equiv 0 \pmod{p}$$

holds for all $i = 1, \ldots, r$.

By Fermat's theorem it is sufficient to find integers $x_1, \ldots, x_{r^2+1}$, not all zero, such that the relation

$$\sum_{j=1}^{r^2+1} \alpha_{ij} x_j^r \equiv 0 \pmod{p}$$

is satisfied for all $i \in \{1, \ldots, r\}$. Set $F_i = \sum_{j=1}^{r^2+1} \alpha_{ij} x_j^r$. We want to find $x_1, \ldots, x_r$ such that $F_1 \equiv F_2 \equiv \cdots \equiv F_r \equiv 0 \pmod{p}$, which is by Fermat's theorem equivalent to

$$F(x_1, \ldots, x_r) = F_1^r + F_2^r + \cdots + F_r^r \equiv 0 \pmod{p}. \tag{1}$$

Of course, one solution of (1) is $(0, \ldots, 0)$: we are not satisfied with it because it generates the empty subset of A, but it tells us that (1) has at least one solution. We shall prove that the number of solutions of (1) is divisible by p, which will imply the existence of a nontrivial solution and thus complete the proof. To do this, consider the sum $\sum F(x_1, \ldots, x_{r^2+1})^r$ taken over all elements $(x_1, \ldots, x_{r^2+1})$ in the vector space $\mathbb{Z}_p^{r^2+1}$. Our statement is equivalent to

$$\sum F(x_1, \ldots, x_{r^2+1})^r \equiv 0 \pmod{p}. \tag{2}$$

Since the degree of F^r is r^2, in each monomial in F^r at least one of the variables is missing. Consider any of these monomials, say $b x_{i_1}^{a_1} x_{i_2}^{a_2} \cdots x_{i_k}^{a_k}$. Then the sum $\sum b x_{i_1}^{a_1} x_{i_2}^{a_2} \cdots x_{i_k}^{a_k}$, taken over the set of all vectors $(x_1, \ldots, x_{r^2+1}) \in \mathbb{Z}_p^{r^2+1}$, is equal to

$$p^{r^2+1-u} \cdot \sum_{(x_{i_1}, \ldots, x_{i_k}) \in \mathbb{Z}_p^k} b x_{i_1}^{a_1} x_{i_2}^{a_2} \cdots x_{i_k}^{a_k},$$

which is divisible by p, so that (2) is proved. Thus the answer is $(p-1)^2$.

4.45 Solutions to the Shortlisted Problems of IMO 2004

1. By symmetry, it is enough to prove that $t_1 + t_2 > t_3$. We have

$$\left(\sum_{i=1}^{n} t_i \right) \left(\sum_{i=1}^{n} \frac{1}{t_i} \right) = n^2 + \sum_{i<j} \left(\frac{t_i}{t_j} + \frac{t_j}{t_i} - 2 \right). \tag{1}$$

All the summands on the RHS are positive, and therefore the RHS is not smaller than $n^2 + T$, where $T = (t_1/t_3 + t_3/t_1 - 2) + (t_2/t_3 + t_3/t_2 - 2)$. We note that T is increasing as a function in t_3 for $t_3 \geq \max\{t_1, t_2\}$. If $t_1 + t_2 = t_3$, then $T = (t_1 + t_2)(1/t_1 + 1/t_2) - 1 \geq 3$ by the Cauchy–Schwarz inequality. Hence, if $t_1 + t_2 \leq t_3$, we have $T \geq 1$, and consequently the RHS in (1) is greater than or equal to $n^2 + 1$, a contradiction.

Remark. It can be proved, for example using Lagrange multipliers, that if $n^2 + 1$ in the problem is replaced by $(n + \sqrt{10} - 3)^2$, then the statement remains true. This estimate is the best possible.

2. We claim that the sequence $\{a_n\}$ must be unbounded.
   The condition of the sequence is equivalent to $a_n > 0$ and $a_{n+1} = a_n + a_{n-1}$ or $a_n - a_{n-1}$. In particular, if $a_n < a_{n-1}$, then $a_{n+1} > \max\{a_n, a_{n-1}\}$.
   Let us remove all a_n such that $a_n < a_{n-1}$. The obtained sequence $(b_m)_{m \in \mathbb{N}}$ is strictly increasing. Thus the statement of the problem will follow if we prove that $b_{m+1} - b_m \geq b_m - b_{m-1}$ for all $m \geq 2$.
   Let $b_{m+1} = a_{n+2}$ for some n. Then $a_{n+2} > a_{n+1}$. We distinguish two cases:
   (i) If $a_{n+1} > a_n$, we have $b_m = a_{n+1}$ and $b_{m-1} \geq a_{n-1}$ (since b_{m-1} is either a_{n-1} or a_n). Then $b_{m+1} - b_m = a_{n+2} - a_{n+1} = a_n = a_{n+1} - a_{n-1} = b_m - a_{n-1} \geq b_m - b_{m-1}$.
   (ii) If $a_{n+1} < a_n$, we have $b_m = a_n$ and $b_{m-1} \geq a_{n-1}$. Consequently, $b_{m+1} - b_m = a_{n+2} - a_n = a_{n+1} = a_n - a_{n-1} = b_m - a_{n-1} \geq b_m - b_{m-1}$.

3. The answer is *yes*. Every rational number $x > 0$ can be uniquely expressed as a continued fraction of the form $a_0 + 1/(a_1 + 1/(a_2 + 1/(\cdots + 1/a_n)))$ (where $a_0 \in \mathbb{N}_0, a_1, \ldots, a_n \in \mathbb{N}$). Then we write $x = [a_0; a_1, a_2, \ldots, a_n]$. Since n depends only on x, the function $s(x) = (-1)^n$ is well-defined. For $x < 0$ we define $s(x) = -s(-x)$, and set $s(0) = 1$. We claim that this $s(x)$ satisfies the requirements of the problem.
   The equality $s(x)s(y) = -1$ trivially holds if $x + y = 0$.
   Suppose that $xy = 1$. We may assume w.l.o.g. that $x > y > 0$. Then $x > 1$, so if $x = [a_0; a_1, a_2, \ldots, a_n]$, then $a_0 \geq 1$ and $y = 0 + 1/x = [0; a_0, a_1, a_2, \ldots, a_n]$. It follows that $s(x) = (-1)^n$, $s(y) = (-1)^{n+1}$, and hence $s(x)s(y) = -1$.
   Finally, suppose that $x + y = 1$. We consider two cases:
   (i) Let $x, y > 0$. We may assume w.l.o.g. that $x > 1/2$. Then there exist natural numbers $a_2, \ldots, a_n$ such that $x = [0; 1, a_2, \ldots, a_n] = 1/(1 + 1/t)$, where $t = [a_2, \ldots, a_n]$. Since $y = 1 - x = 1/(1 + t) = [0; 1 + a_2, a_3, \ldots, a_n]$, we have $s(x) = (-1)^n$ and $s(y) = (-1)^{n-1}$, giving us $s(x)s(y) = -1$.

(ii) Let $x > 0 > y$. If $a_0, \ldots, a_n \in \mathbb{N}$ are such that $-y = [a_0; a_1, \ldots, a_n]$, then $x = [1 + a_0; a_1, \ldots, a_n]$. Thus $s(y) = -s(-y) = -(-1)^n$ and $s(x) = (-1)^n$, so again $s(x)s(y) = -1$.

4. Let $P(x) = a_0 + a_1 x + \cdots + a_n x^n$. For every real number x the triple $(a, b, c) = (6x, 3x, -2x)$ satisfies the condition $ab + bc + ca = 0$. Then the condition on P gives us $P(3x) + P(5x) + P(-8x) = 2P(7x)$ for all x, implying that for all $i = 0, 1, 2, \ldots, n$ the following equality holds:

$$\left(3^i + 5^i + (-8)^i - 2 \cdot 7^i\right) a_i = 0.$$

Suppose that $a_i \neq 0$. Then $K(i) = 3^i + 5^i + (-8)^i - 2 \cdot 7^i = 0$. But $K(i)$ is negative for i odd and positive for $i = 0$ or $i \geq 6$ even. Only for $i = 2$ and $i = 4$ do we have $K(i) = 0$. It follows that $P(x) = a_2 x^2 + a_4 x^4$ for some real numbers a_2, a_4. It is easily verified that all such $P(x)$ satisfy the required condition.

5. By the general mean inequality ($M_1 \leq M_3$), the LHS of the inequality to be proved does not exceed

$$E = \frac{3}{\sqrt[3]{3}} \sqrt[3]{\frac{1}{a} + \frac{1}{b} + \frac{1}{c}} + 6(a + b + c).$$

From $ab + bc + ca = 1$ we obtain that $3abc(a + b + c) = 3(ab \cdot ac + ab \cdot bc + ac \cdot bc) \leq (ab + ac + bc)^2 = 1$; hence $6(a + b + c) \leq \frac{2}{abc}$. Since $\frac{1}{a} + \frac{1}{b} + \frac{1}{c} = \frac{ab + bc + ca}{abc} = \frac{1}{abc}$, it follows that

$$E \leq \frac{3}{\sqrt[3]{3}} \sqrt[3]{\frac{3}{abc}} \leq \frac{1}{abc},$$

where the last inequality follows from the AM–GM inequality $1 = ab + bc + ca \geq 3\sqrt[3]{(abc)^2}$, i.e., $abc \leq 1/(3\sqrt{3})$. The desired inequality now follows. Equality holds if and only if $a = b = c = 1/\sqrt{3}$.

6. Let us make the substitution $z = x + y$, $t = xy$. Given $z, t \in \mathbb{R}$, x, y are real if and only if $4t \leq z^2$. Define $g(x) = 2(f(x) - x)$. Now the given functional equation transforms into

$$f\left(z^2 + g(t)\right) = (f(z))^2 \quad \text{for all } t, z \in \mathbb{R} \text{ with } z^2 \geq 4t. \tag{1}$$

Let us set $c = g(0) = 2f(0)$. Substituting $t = 0$ into (1) gives us

$$f(z^2 + c) = (f(z))^2 \quad \text{for all } z \in \mathbb{R}. \tag{2}$$

If $c < 0$, then taking z such that $z^2 + c = 0$, we obtain from (2) that $f(z)^2 = c/2$, which is impossible; hence $c \geq 0$. We also observe that

$$x > c \quad \text{implies} \quad f(x) \geq 0. \tag{3}$$

If g is a constant function, we easily find that $c = 0$ and therefore $f(x) = x$, which is indeed a solution.

Suppose g is nonconstant, and let $a, b \in \mathbb{R}$ be such that $g(a) - g(b) = d > 0$. For some sufficiently large K and each $u, v \geq K$ with $v^2 - u^2 = d$ the equality $u^2 + g(a) = v^2 + g(b)$ by (1) and (3) implies $f(u) = f(v)$. This further leads to $g(u) - g(v) = 2(v - u) = \frac{d}{u + \sqrt{u^2 + d}}$. Therefore every value from some suitably chosen segment $[\delta, 2\delta]$ can be expressed as $g(u) - g(v)$, with u and v bounded from above by some M.

Consider any x, y with $y > x \geq 2\sqrt{M}$ and $\delta < y^2 - x^2 < 2\delta$. By the above considerations, there exist $u, v \leq M$ such that $g(u) - g(v) = y^2 - x^2$, i.e., $x^2 + g(u) = y^2 + g(v)$. Since $x^2 \geq 4u$ and $y^2 \geq 4v$, (1) leads to $f(x)^2 = f(y)^2$. Moreover, if we assume w.l.o.g. that $4M \geq c^2$, we conclude from (3) that $f(x) = f(y)$. Since this holds for any $x, y \geq 2\sqrt{M}$ with $y^2 - x^2 \in [\delta, 2\delta]$, it follows that $f(x)$ is eventually constant, say $f(x) = k$ for $x \geq N = 2\sqrt{M}$. Setting $x > N$ in (2) we obtain $k^2 = k$, so $k = 0$ or $k = 1$.

By (2) we have $f(-z) = \pm f(z)$, and thus $|f(z)| \leq 1$ for all $z \leq -N$. Hence $g(u) = 2f(u) - 2u \geq -2 - 2u$ for $u \leq -N$, which implies that g is unbounded. Hence for each z there exists t such that $z^2 + g(t) > N$, and consequently $f(z)^2 = f(z^2 + g(t)) = k = k^2$. Therefore $f(z) = \pm k$ for each z.

If $k = 0$, then $f(x) \equiv 0$, which is clearly a solution. Assume $k = 1$. Then $c = 2f(0) = 2$ (because $c \geq 0$), which together with (3) implies $f(x) = 1$ for all $x \geq 2$. Suppose that $f(t) = -1$ for some $t < 2$. Then $t - g(t) = 3t + 2 > 4t$. If also $t - g(t) \geq 0$, then for some $z \in \mathbb{R}$ we have $z^2 = t - g(t) > 4t$, which by (1) leads to $f(z)^2 = f(z^2 + g(t)) = f(t) = -1$, which is impossible. Hence $t - g(t) < 0$, giving us $t < -2/3$. On the other hand, if X is any subset of $(-\infty, -2/3)$, the function f defined by $f(x) = -1$ for $x \in X$ and $f(x) = 1$ satisfies the requirements of the problem.

To sum up, the solutions are $f(x) = x$, $f(x) = 0$ and all functions of the form

$$f(x) = \begin{cases} 1, & x \notin X, \\ -1, & x \in X, \end{cases}$$

where $X \subset (-\infty, -2/3)$.

7. Let us set $c_k = A_{k-1}/A_k$ for $k = 1, 2, \ldots, n$, where we define $A_0 = 0$. We observe that $a_k/A_k = (kA_k - (k-1)A_{k-1})/A_k = k - (k-1)c_k$. Now we can write the LHS of the inequality to be proved in terms of c_k, as follows:

$$\sqrt[n]{\frac{G_n}{A_n}} = \sqrt[n^2]{c_2 c_3^2 \cdots c_n^{n-1}} \quad \text{and} \quad \frac{g_n}{G_n} = \sqrt[n]{\prod_{k=1}^{n}(k - (k-1)c_k)}.$$

By the $AM - GM$ inequality we have

$$n\sqrt[n^2]{1^{n(n+1)/2} c_2 c_3^2 \cdots c_n^{n-1}} \leq \frac{1}{n}\left(\frac{n(n+1)}{2} + \sum_{k=2}^{n}(k-1)c_k\right)$$

$$= \frac{n+1}{2} + \frac{1}{n}\sum_{k=1}^{n}(k-1)c_k. \tag{1}$$

Also by the AM–GM inequality, we have

$$\sqrt[n]{\prod_{k=1}^{n}(k-(k-1)c_k)} \leq \frac{n+1}{2} - \frac{1}{n}\sum_{k=1}^{n}(k-1)c_k. \qquad (2)$$

Adding (1) and (2), we obtain the desired inequality. Equality holds if and only if $a_1 = a_2 = \cdots = a_n$.

8. Let us write $n = 10001$. Denote by $\mathscr{T}$ the set of ordered triples $(a, C, \mathscr{S})$, where a is a student, C a club, and $\mathscr{S}$ a society such that $a \in C$ and $C \in \mathscr{S}$. We shall count $|\mathscr{T}|$ in two different ways.

Fix a student a and a society $\mathscr{S}$. By (ii), there is a unique club C such that $(a, C, \mathscr{S}) \in \mathscr{T}$. Since the ordered pair $(a, \mathscr{S})$ can be chosen in nk ways, we have that $|\mathscr{T}| = nk$.

Now fix a club C. By (iii), C is in exactly $(|C| - 1)/2$ societies, so there are $|C|(|C| - 1)/2$ triples from $\mathscr{T}$ with second coordinate C. If $\mathscr{C}$ is the set of all clubs, we obtain $|\mathscr{T}| = \sum_{C \in \mathscr{C}} \frac{|C|(|C|-1)}{2}$. But we also conclude from (i) that

$$\sum_{C \in \mathscr{C}} \frac{|C|(|C|-1)}{2} = \frac{n(n-1)}{2}.$$

Therefore $n(n-1)/2 = nk$, i.e., $k = (n-1)/2 = 5000$.

On the other hand, for $k = (n-1)/2$ there is a desired configuration with only one club C that contains all students and k identical societies with only one element (the club C). It is easy to verify that (i)–(iii) hold.

9. Obviously we must have $2 \leq k \leq n$. We shall prove that the possible values for k and n are $2 \leq k \leq n \leq 3$ and $3 \leq k \leq n$. Denote all colors and circles by $1, \ldots, n$. Let $F(i, j)$ be the set of colors of the common points of circles i and j.

Suppose that $k = 2 < n$. Consider the ordered pairs (i, j) such that color j appears on the circle i. Since $k = 2$, clearly there are exactly $2n$ such pairs. On the other hand, each of the n colors appears on at least two circles, so there are at least $2n$ pairs (i, j), and equality holds only if each color appears on exactly 2 circles. But then at most two points receive each of the n colors and there are $n(n-1)$ points, implying that $n(n-1) = 2n$, i.e., $n = 3$. It is easy to find examples for $k = 2$ and $n = 2$ or 3.

Next, let $k = 3$. An example for $n = 3$ is given by $F(i, j) = \{i, j\}$ for each $1 \leq i < j \leq 3$. Assume $n \geq 4$. Then an example is given by $F(1, 2) = \{1, 2\}$, $F(i, i+1) = \{i\}$ for $i = 2, \ldots, n-2$, $F(n-1, n) = \{n-2, n-1\}$ and $F(i, j) = n$ for all other $i, j > i$.

We now prove by induction on k that a desired coloring exists for each $n \geq k \geq 3$. Let there be given n circles. By the inductive hypothesis, circles $1, 2, \ldots, n-1$ can be colored in $n-1$ colors, k of which appear on each circle, such that color i appears on circle i. Then we set $F(i, n) = \{i, n\}$ for $i = 1, \ldots, k$ and $F(i, n) = \{n\}$ for $i > n$. We thus obtain a coloring of the n circles in n colors, such that $k+1$ colors (including color i) appear on each circle i.

10. The least number of edges of such a graph is n.

We note that deleting edge AB of a 4-cycle $ABCD$ from a connected and non-bipartite graph G yields a connected and nonbipartite graph, say H. Indeed, the connectedness is obvious; also, if H were bipartite with partition of the set of vertices into P_1 and P_2, then w.l.o.g. $A, C \in P_1$ and $B, D \in P_2$, so $G = H \cup \{AB\}$ would also be bipartite with the same partition, a contradiction.

Any graph that can be obtained from the complete n-graph in the described way is connected and has at least one cycle (otherwise it would be bipartite); hence it must have at least n edges.

Now consider a complete graph with vertices $V_1, V_2, \ldots, V_n$. Let us remove every edge $V_i V_j$ with $3 \leq i < j < n$ from the cycle $V_2 V_i V_j V_n$. Then for $i = 3, \ldots, n-1$ we remove edges $V_2 V_i$ and $V_i V_n$ from the cycles $V_1 V_i V_2 V_n$ and $V_1 V_i V_n V_2$ respectively, thus obtaining a graph with exactly n edges: $V_1 V_i$ ($i = 2, \ldots, n$) and $V_2 V_n$.

11. Consider the matrix $A = (a_{ij})_{i,j=1}^n$ such that a_{ij} is equal to 1 if $i, j \leq n/2$, -1 if $i, j > n/2$, and 0 otherwise. This matrix satisfies the conditions from the problem and all row sums and column sums are equal to $\pm n/2$. Hence $C \geq n/2$.

Let us show that $C = n/2$. Assume to the contrary that there is a matrix $B = (b_{ij})_{i,j=1}^n$ all of whose row sums and column sums are either greater than $n/2$ or smaller than $-n/2$. We may assume w.l.o.g. that at least $n/2$ row sums are positive and, permuting rows if necessary, that the first $n/2$ rows have positive sums. The sum of entries in the $n/2 \times n$ submatrix B' consisting of first $n/2$ rows is greater than $n^2/4$, and since each column of B' has sum at most $n/2$, it follows that more than $n/2$ column sums of B', and therefore also of B, are positive. Again, suppose w.l.o.g. that the first $n/2$ column sums are positive. Thus the sums R^+ and C^+ of entries in the first $n/2$ rows and in the first $n/2$ columns respectively are greater than $n^2/4$. Now the sum of all entries of B can be written as

$$\sum a_{ij} = R^+ + C^+ + \sum_{\substack{i > n/2 \\ j > n/2}} a_{ij} - \sum_{\substack{i \leq n/2 \\ j \leq n/2}} a_{ij} > \frac{n^2}{2} - \frac{n^2}{4} - \frac{n^2}{4} = 0,$$

a contradiction. Hence $C = n/2$, as claimed.

12. We say that a number $n \in \{1, 2, \ldots, N\}$ is *winning* if the player who is on turn has a winning strategy, and *losing* otherwise. The game is of type A if and only if 1 is a losing number.

Let us define $n_0 = N$, $n_{i+1} = [n_i/2]$ for $i = 0, 1, \ldots$ and let k be such that $n_k = 1$. Consider the sets $A_i = \{n_{i+1} + 1, \ldots, n_i\}$. We call a set A_i *all-winning* if all numbers from A_i are winning, *even-winning* if even numbers are winning and odd are losing, and *odd-winning* if odd numbers are winning and even are losing.

  (i) Suppose A_i is even-winning and consider A_{i+1}. Multiplying any number from A_{i+1} by 2 yields an even number from A_i, which is a losing number. Thus $x \in A_{i+1}$ is winning if and only if $x + 1$ is losing, i.e., if and only if it is even. Hence A_{i+1} is also even-winning.

  (ii) Suppose A_i is odd-winning. Then each $k \in A_{i+1}$ is winning, since $2k$ is losing. Hence A_{i+1} is all-winning.

(iii) Suppose A_i is all-winning. Multiplying $x \in A_{i+1}$ by two is then a losing move, so x is winning if and only if $x+1$ is losing. Since n_{i+1} is losing, A_{i+1} is odd-winning if n_{i+1} is even and even-winning otherwise.

We observe that A_0 is even-winning if N is odd and odd-winning otherwise. Also, if some A_i is even-winning, then all $A_{i+1}, A_{i+2}, \ldots$ are even-winning and thus 1 is losing; i.e., the game is of type A. The game is of type B if and only if the sets $A_0, A_1, \ldots$ are alternately odd-winning and all-winning with A_0 odd-winning, which is equivalent to $N = n_0, n_2, n_4, \ldots$ all being even. Thus N is of type B if and only if all digits at the odd positions in the binary representation of N are zeros.

Since $2004 = \overline{11111010100}$ in the binary system, 2004 is of type A. The least $N > 2004$ that is of type B is $\overline{100000000000} = 2^{11} = 2048$. Thus the answer to part (b) is 2048.

13. Since $X_i, Y_i, i = 1, \ldots, 2004$, are 4008 distinct subsets of the set $S_n = \{1, 2, \ldots, n\}$, it follows that $2^n \geq 4008$, i.e. $n \geq 12$.

Suppose $n = 12$. Let $\mathscr{X} = \{X_1, \ldots, X_{2004}\}$, $\mathscr{Y} = \{Y_1, \ldots, Y_{2004}\}$, $\mathscr{A} = \mathscr{X} \cup \mathscr{Y}$. Exactly $2^{12} - 4008 = 88$ subsets of S_n do not occur in $\mathscr{A}$.

Since each row intersects each column, we have $X_i \cap Y_j \neq \emptyset$ for all i, j. Suppose $|X_i|, |Y_j| \leq 3$ for some indices i, j. Since then $|X_i \cup Y_j| \leq 5$, any of at least $2^7 > 88$ subsets of $S_n \setminus (X_i \cap Y_j)$ can occur in neither $\mathscr{X}$ nor $\mathscr{Y}$, which is impossible. Hence either in $\mathscr{X}$ or in $\mathscr{Y}$ all subsets are of size at least 4. Suppose w.l.o.g. that $k = |X_l| = \min_i |X_i| \geq 4$. There are

$$n_k = \binom{12-k}{0} + \binom{12-k}{1} + \cdots + \binom{12-k}{k-1}$$

subsets of $S \setminus X_l$ with fewer than k elements, and none of them can be either in $\mathscr{X}$ (because $|X_l|$ is minimal in $\mathscr{X}$) or in $\mathscr{Y}$. Hence we must have $n_k \leq 88$. Since $n_4 = 93$ and $n_5 = 99$, it follows that $k \geq 6$. But then none of the $\binom{12}{0} + \cdots + \binom{12}{5} = 1586$ subsets of S_n is in $\mathscr{X}$, hence at least $1586 - 88 = 1498$ of them are in $\mathscr{Y}$. The 1498 complements of these subsets also do not occur in $\mathscr{X}$, which adds to 3084 subsets of S_n not occurring in $\mathscr{X}$. This is clearly a contradiction.

Now we construct a golden matrix for $n = 13$. Let

$$A_1 = \begin{bmatrix} 1 & 1 \\ 2 & 3 \end{bmatrix} \quad \text{and} \quad A_m = \begin{bmatrix} A_{m-1} & A_{m-1} \\ A_{m-1} & B_{m-1} \end{bmatrix} \quad \text{for } m = 2, 3, \ldots,$$

where B_{m-1} is the $2^{m-1} \times 2^{m-1}$ matrix with all entries equal to $m+2$. It can be easily proved by induction that each of the matrices A_m is golden. Moreover, every upper-left square submatrix of A_m of size greater than 2^{m-1} is also golden. Since $2^{10} < 2004 < 2^{11}$, we thus obtain a golden matrix of size 2004 with entries in S_{13}.

14. Suppose that an $m \times n$ rectangle can be covered by "hooks". For any hook H there is a unique hook K that covers its "inside" square. Then also H covers the inside square of K, so the set of hooks can be partitioned into pairs of type

$\{H, K\}$, each of which forms one of the following two figures consisting of 12 squares:

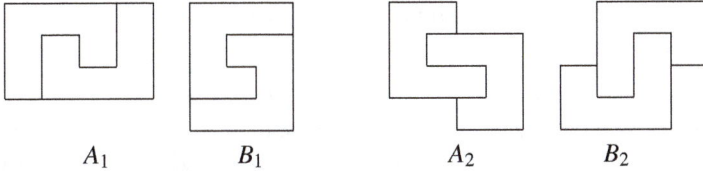

A_1 B_1 A_2 B_2

Thus the $m \times n$ rectangle is covered by these tiles. It immediately follows that $12 \mid mn$.

Suppose one of m, n is divisible by 4. Let w.l.o.g. $4 \mid m$. If $3 \mid n$, one can easily cover the rectangle by 3×4 rectangles and therefore by hooks. Also, if $12 \mid m$ and $n \notin \{1, 2, 5\}$, then there exist $k, l \in \mathbb{N}_0$ such that $n = 3k + 4l$, and thus the rectangle $m \times n$ can be partitioned into 3×12 and 4×12 rectangles all of which can be covered by hooks. If $12 \mid m$ and $n = 1, 2,$ or 5, then it is easy to see that covering by hooks is not possible.

Now suppose that $4 \nmid m$ and $4 \nmid n$. Then m, n are even and the number of tiles is odd. Assume that the total number of tiles of types A_1 and B_1 is odd (otherwise the total number of tiles of types A_2 and B_2 is odd, which is analogous). If we color in black all columns whose indices are divisible by 4, we see that each tile of type A_1 or B_1 covers three black squares, which yields an odd number in total. Hence the total number of black squares covered by the tiles of types A_2 and B_2 must be odd. This is impossible, since each such tile covers two or four black squares.

15. Denote by $V_1, \ldots, V_n$ the vertices of a graph G and by E the set of its edges. For each $i = 1, \ldots, n$, let A_i be the set of vertices connected to V_i by an edge, G_i the subgraph of G whose set of vertices is A_i, and E_i the set of edges of G_i. Also, let $v_i, e_i,$ and $t_i = f(G_i)$ be the numbers of vertices, edges, and triangles in G_i respectively.

The numbers of tetrahedra and triangles one of whose vertices is V_i are respectively equal to t_i and e_i. Hence

$$\sum_{i=1}^{n} v_i = 2|E|, \quad \sum_{i=1}^{n} e_i = 3f(G) \quad \text{and} \quad \sum_{i=1}^{n} t_i = 4g(G).$$

Since $e_i \leq v_i(v_i - 1)/2 \leq v_i^2/2$ and $e_i \leq |E|$, we obtain $e_i^2 \leq v_i^2|E|/2$, i.e., $e_i \leq v_i\sqrt{|E|/2}$. Summing over all i yields $3f(G) \leq 2|E|\sqrt{|E|/2}$, or equivalently $f(G)^2 \leq 2|E|^3/9$. Since this relation holds for each graph G_i, it follows that

$$t_i = f(G_i) = f(G_i)^{1/3}f(G_i)^{2/3} \leq \left(\frac{2}{9}\right)^{1/3} f(G)^{1/3}e_i.$$

Summing the last inequality for $i = 1, \ldots, n$ gives us

$$4g(G) \leq 3\left(\frac{2}{9}\right)^{1/3} f(G)^{1/3} \cdot f(G), \quad \text{i.e.,} \quad g(G)^3 \leq \frac{3}{32}f(G)^4.$$

The constant $c = 3/32$ is the best possible. Indeed, in a complete graph C_n it holds that $g(K_n)^3/f(K_n)^4 = \binom{n}{4}^3 \binom{n}{3}^{-4} \to \frac{3}{32}$ as $n \to \infty$.

Remark. Let N_k be the number of complete k-subgraphs in a finite graph G. Continuing inductively, one can prove that $N_{k+1}^k \le \frac{k!}{(k+1)^k} N_k^{k+1}$.

16. Note that $\triangle ANM \sim \triangle ABC$ and consequently $AM \ne AN$. Since $OM = ON$, it follows that OR is a perpendicular bisector of MN. Thus, R is the common point of the median of MN and the bisector of $\angle MAN$. Then it follows from a well-known fact that R lies on the circumcircle of $\triangle AMN$.

    Let K be the intersection of AR and BC. We then have $\angle MRA = \angle MNA = \angle ABK$ and $\angle NRA = \angle NMA = \angle ACK$, from which we conclude that $RMBK$ and $RNCK$ are cyclic. Thus K is the desired intersection of the circumcircles of $\triangle BMR$ and $\triangle CNR$ and it indeed lies on BC.

17. Let H be the reflection of G about AB $(GH \parallel \ell)$. Let M be the intersection of AB and ℓ. Since $\angle FEA = \angle FMA = 90°$, it follows that $AEMF$ is cyclic and hence $\angle DFE = \angle BAE = \angle DEF$. The last equality holds because DE is tangent to Γ. It follows that $DE = DF$ and hence $DF^2 = DE^2 = DC \cdot DA$ (the power of D with re-

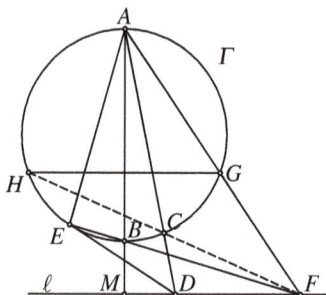

    spect to Γ). It then follows that $\angle DCF = \angle DFA = \angle HGA = \angle HCA$. Thus it follows that H lies on CF as desired.

18. It is important to note that since $\beta < \gamma$, $\angle ADC = 90° - \gamma + \beta$ is acute. It is elementary that $\angle CAO = 90° - \beta$. Let X and Y respectively be the intersections of FE and GH with AD. We trivially get $X \in EF \perp AD$ and $\triangle AGH \cong \triangle ACB$. Consequently, $\angle GAY = \angle OAB = 90° - \gamma = 90° - \angle AGY$. Hence, $GH \perp AD$ and thus $GH \parallel FE$. That $EFGH$ is a rectangle is now equivalent to $FX = GY$ and $EX = HY$.

    We have that $GY = AG \sin \gamma = AC \sin \gamma$ and $FX = AF \sin \gamma$ (since $\angle AFX = \gamma$). Thus,

    $$FX = GY \Leftrightarrow CF = AF = AC \Leftrightarrow \angle AFC = 60° \Leftrightarrow \angle ADC = 30°.$$

    Since $\angle ADC = 180° - \angle DCA - \angle DAC = 180° - \gamma - (90° - \beta)$, it immediately follows that $FX = GY \Leftrightarrow \gamma - \beta = 60°$. We similarly obtain $EX = HY \Leftrightarrow \gamma - \beta = 60°$, proving the statement of the problem.

19. Assume first that the points A, B, C, D are concyclic. Let the lines BP and DP meet the circumcircle of $ABCD$ again at E and F, respectively. Then it follows from the given conditions that $\widehat{AB} = \widehat{CF}$ and $\widehat{AD} = \widehat{CE}$; hence $BF \parallel AC$ and $DE \parallel AC$. Therefore $BFED$ and $BFAC$ are isosceles trapezoids and thus $P = BE \cap DF$ lies on the common bisector of segments BF, ED, AC. Hence $AP = CP$.

Assume in turn that $AP = CP$. Let P w.l.o.g. lie in the triangles ACD and BCD. Let BP and DP meet AC at K and L, respectively. The points A and C are isogonal conjugates with respect to $\triangle BDP$, which implies that $\angle APK = \angle CPL$. Since $AP = CP$, we infer that K and L are symmetric with respect to the perpendicular bisector p of AC. Let E be the reflection of D in p. Then E lies on the line BP, and the triangles APD and CPE are congruent. Thus $\angle BDC = \angle ADP = \angle BEC$, which means that the points B, C, E, D are concyclic. Moreover, A, C, E, D are also concyclic. Hence, $ABCD$ is a cyclic quadrilateral.

20. We first establish the following lemma.

    Lemma. Let $ABCD$ be an isosceles trapezoid with bases AB and CD. The diagonals AC and BD intersect at S. Let M be the midpoint of BC, and let the bisector of the angle BSC intersect BC at N. Then $\angle AMD = \angle AND$.

    Proof. It suffices to show that the points A, D, M, N are concyclic. The statement is trivial for $AD \parallel BC$. Let us now assume that AD and BC meet at X, and let $XA = XB = a$, $XC = XD = b$. Since SN is the bisector of $\angle CSB$, we have

$$\frac{a - XN}{XN - b} = \frac{BN}{CN} = \frac{BS}{CS} = \frac{AB}{CD} = \frac{a}{b},$$

    and an easy computation yields $XN = \frac{2ab}{a+b}$. We also have $XM = \frac{a+b}{2}$; hence $XM \cdot XN = XA \cdot XD$. Therefore A, D, M, N are concyclic, as needed.

Denote by C_i the midpoint of the side A_iA_{i+1}, $i = 1, \ldots, n-1$.

By definition $C_1 = B_1$ and $C_{n-1} = B_{n-1}$. Since $A_1A_iA_{i+1}A_n$ is an isosceles trapezoid with $A_1A_i \parallel A_{i+1}A_n$ for $i = 2, \ldots, n-2$, it follows from the lemma that $\angle A_1B_iA_n = \angle A_1C_iA_n$ for all i. The sum in consideration thus equals $\angle A_1C_1A_n + \angle A_1C_2A_n + \cdots + \angle A_1C_{n-1}A_n$. Moreover, the triangles $A_1C_iA_n$ and $A_{n+2-i}C_1A_{n+1-i}$ are congruent (a rotation about the center of the n-gon carries the first one to the second), and consequently $\angle A_1C_iA_n = \angle A_{n+2-i}C_1A_{n+1-i}$ for $i = 2, \ldots, n-1$. Hence

$$\Sigma = \angle A_1C_1A_n + \angle A_nC_1A_{n-1} + \cdots + \angle A_3C_1A_2 = \angle A_1C_1A_2 = 180°.$$

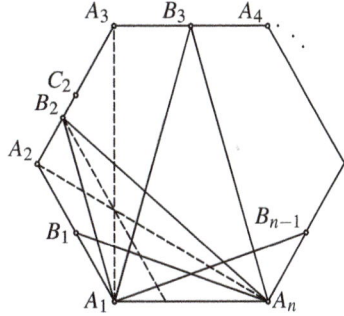

21. Let ABC be the triangle of maximum area S contained in $\mathscr{P}$ (it exists because of compactness of $\mathscr{P}$). Draw parallels to BC, CA, AB through A, B, C, respectively, and denote the triangle thus obtained by $A_1B_1C_1$ ($A \in B_1C_1$, etc.). Since each triangle with vertices in $\mathscr{P}$ has area at most S, the entire polygon $\mathscr{P}$ is contained in $A_1B_1C_1$.

Next, draw lines of support of $\mathscr{P}$ parallel to BC, CA, AB and not intersecting the triangle ABC. They determine a convex hexagon $U_aV_aU_bV_bU_cV_c$ containing $\mathscr{P}$, with $V_b, U_c \in B_1C_1$, $V_c, U_a \in C_1A_1$, $V_a, U_b \in A_1B_1$. Each of the line segments U_aV_a, U_bV_b, U_cV_c contains points of $\mathscr{P}$. Choose such points A_0, B_0, C_0 on

U_aV_a, U_bV_b, U_cV_c, respectively. The convex hexagon $AC_0BA_0CB_0$ is contained in $\mathscr{P}$, because the latter is convex. We prove that $AC_0BA_0CB_0$ has area at least 3/4 the area of $\mathscr{P}$.

Let x, y, z denote the areas of triangles U_aBC, U_bCA, and U_cAB. Then $S_1 = S_{AC_0BA_0CB_0} = S + x + y + z$. On the other hand, the triangle $A_1U_aV_a$ is similar to $\triangle A_1BC$ with similitude $\tau = (S-x)/S$, and hence its area is $\tau^2 S = (S-x)^2/S$. Thus the area of quadrilateral U_aV_aCB is $S - (S-x)^2/S = 2z - z^2/S$. Analogous formulas hold for quadrilaterals U_bV_bAC and U_cV_cBA. Therefore

$$S_{\mathscr{P}} \leq S_{U_aV_aU_bV_bU_cV_c} = S + S_{U_aV_aCB} + S_{U_bV_bAC} + S_{U_cV_cBA}$$
$$= S + 2(x+y+z) - \frac{x^2+y^2+z^2}{S}$$
$$\leq S + 2(x+y+z) - \frac{(x+y+z)^2}{3S}.$$

Now $4S_1 - 3S_{\mathscr{P}} \geq = S - 2(x+y+z) + (x+y+z)^2/S = (S-x-y-z)^2/S \geq 0$; i.e., $S_1 \geq 3S_{\mathscr{P}}/4$, as claimed.

22. The proof uses the following observation:

> **Lemma.** In a triangle ABC, let K, L be the midpoints of the sides AC, AB, respectively, and let the incircle of the triangle touch BC, CA at D, E, respectively. Then the lines KL and DE intersect on the bisector of the angle ABC.
>
> *Proof.* Let the bisector ℓ_b of $\angle ABC$ meet DE at T. One can assume that $AB \neq BC$, or else $T \equiv K \in KL$. Note that the incenter I of $\triangle ABC$ is between B and T, and also $T \neq E$. From the triangles BDT and DEC we obtain $\angle ITD = \alpha/2 = \angle IAE$, which implies that A, I, T, E are concyclic. Then $\angle ATB = \angle AEI = 90°$. Thus L is the circumcenter of $\triangle ATB$ from which $\angle LTB = \angle LBT = \angle TBC \Rightarrow LT \parallel BC \Rightarrow T \in KL$, which is what we were supposed to prove.

Let the incircles of $\triangle ABX$ and $\triangle ACX$ touch BX at D and F, respectively, and let them touch AX at E and G, respectively. Clearly, DE and FG are parallel. If the line PQ intersects BX and AX at M and N, respectively, then $MD^2 = MP \cdot MQ = MF^2$, i.e., $MD = MF$ and analogously $NE = NG$. It follows that PQ is parallel to DE and FG and equidistant from them.

The midpoints of AB, AC, and AX lie on the same line m, parallel to BC. Applying the lemma to $\triangle ABX$, we conclude that DE passes through the common point U of m and the bisector of $\angle ABX$. Analogously, FG passes through the common point V of m and the bisector of $\angle ACX$. Therefore PQ passes through the midpoint W of the line segment UV. Since U, V do not depend on X, neither does W.

23. To start with, note that point N is uniquely determined by the imposed properties. Indeed, $f(X) = AX/BX$ is a monotone function on both arcs AB of the circumcircle of $\triangle ABM$. Denote by P and Q respectively the second points of intersection of the line EF with the circumcircles of $\triangle ABE$ and $\triangle ABF$.

The problem is equivalent to showing that $N \in PQ$. In fact, we shall prove that N coincides with the midpoint $\overline{N}$ of segment PQ.

The quadrilaterals $APBE$, $AQBF$, and $ABCD$ are cyclic. This implies that $\angle APE = \angle ABE$ therefore $\angle APQ = \angle ADC$. We can also conclude that $\angle AQP = \angle AQF = \angle ABF = \angle ACD$. It follows that the triangles APQ and ADC are similar, and consequently the same holds for the triangles $\triangle A\overline{N}P$ and $\triangle AMD$. Analogously $\triangle B\overline{N}P \sim \triangle BMC$. Therefore $A\overline{N}/AM = PQ/DC = B\overline{N}/BM$, i.e.,

$A\overline{N}/B\overline{N} = AM/BM$. Moreover, $\angle A\overline{N}B = \angle A\overline{N}P + \angle P\overline{N}B = \angle AMD + \angle BMC = 180° - \angle AMB$, which means that point $\overline{N}$ lies on the circumcircle of $\triangle AMB$. By the uniqueness of N, we conclude that $\overline{N} \equiv N$, which completes the solution.

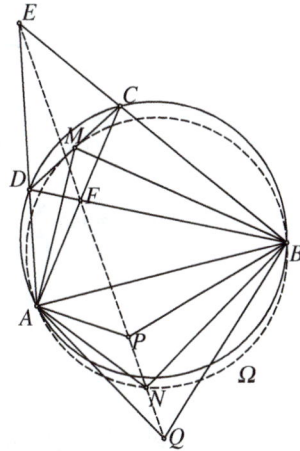

24. Setting $m = an$ we reduce the given equation to $m/\tau(m) = a$.

Let us show that for $a = p^{p-1}$ the above equation has no solutions in $\mathbb{N}$ if $p > 3$ is a prime. Assume to the contrary that $m \in \mathbb{N}$ is such that $m = p^{p-1}\tau(m)$. Then $p^{p-1} \mid m$, so we may set $m = p^{\alpha}k$, where $\alpha, k \in \mathbb{N}$, $\alpha \geq p-1$, and $p \nmid k$. Let $k = p_1^{\alpha_1} \cdots p_r^{\alpha_r}$ be the decomposition of k into primes. Then $\tau(k) = (\alpha_1 + 1) \cdots (\alpha_r + 1)$ and $\tau(m) = (\alpha + 1)\tau(k)$. Our equation becomes

$$p^{\alpha-p+1}k = (\alpha+1)\tau(k). \tag{1}$$

We observe that $\alpha \neq p - 1$: otherwise the RHS would be divisible by p and the LHS would not be so. It follows that $\alpha \geq p$, which also easily implies that $p^{\alpha-p+1} \geq \frac{p}{p+1}(\alpha+1)$.

Furthermore, since $\alpha+1$ cannot be divisible by $p^{\alpha-p+1}$ for any $\alpha \geq p$, it follows that $p \mid \tau(k)$. Thus if $p \mid \tau(k)$, then at least one $\alpha_i + 1$ is divisible by p and consequently $\alpha_i \geq p - 1$ for some i. Hence $k \geq \frac{p_i^{\alpha_i}}{\alpha_i+1}\tau(k) \geq \frac{2^{p-1}}{p}\tau(k)$. But then we have

$$p^{\alpha-p+1}k \geq \frac{p}{p+1}(\alpha+1) \cdot \frac{2^{p-1}}{p}\tau(k) > (\alpha+1)\tau(k),$$

contradicting (1). Therefore (1) has no solutions in $\mathbb{N}$.

Remark. There are many other values of a for which the considered equation has no solutions in $\mathbb{N}$: for example, $a = 6p$ for a prime $p \geq 5$.

25. Let n be a natural number. For each $k = 1, 2, \ldots, n$, the number (k, n) is a divisor of n. Consider any divisor d of n. If $(k, n) = n/d$, then $k = nl/d$ for some $l \in \mathbb{N}$, and $(k, n) = (l, d)n/d$, which implies that l is coprime to d and $l \leq d$. It follows that (k, n) is equal to n/d for exactly $\varphi(d)$ natural numbers $k \leq n$. Therefore

$$\psi(n) = \sum_{k=1}^{n}(k,n) = \sum_{d|n}\varphi(d)\frac{n}{d} = n\sum_{d|n}\frac{\varphi(d)}{d}. \tag{1}$$

(a) Let n,m be coprime. Then each divisor f of mn can be uniquely expressed as $f = de$, where $d \mid n$ and $e \mid m$. We now have by (1)

$$\psi(mn) = mn\sum_{f|mn}\frac{\varphi(f)}{f} = mn\sum_{d|n,\,e|m}\frac{\varphi(de)}{de}$$

$$= mn\sum_{d|n,\,e|m}\frac{\varphi(d)}{d}\frac{\varphi(e)}{e} = \left(n\sum_{d|n}\frac{\varphi(d)}{d}\right)\left(m\sum_{e|m}\frac{\varphi(e)}{e}\right)$$

$$= \psi(m)\psi(n).$$

(b) Let $n = p^k$, where p is a prime and k a positive integer. According to (1),

$$\frac{\psi(n)}{n} = \sum_{i=0}^{k}\frac{\varphi(p^i)}{p^i} = 1 + \frac{k(p-1)}{p}.$$

Setting $p = 2$ and $k = 2(a-1)$ we obtain $\psi(n) = an$ for $n = 2^{2(a-1)}$.

(c) We note that $\psi(p^p) = p^{p+1}$ if p is a prime. Hence, if a has an odd prime factor p and $a_1 = a/p$, then $x = p^p 2^{2a_1-2}$ is a solution of $\psi(x) = ax$ different from $x = 2^{2a-2}$.

Now assume that $a = 2^k$ for some $k \in \mathbb{N}$. Suppose $x = 2^\alpha y$ is a positive integer such that $\psi(x) = 2^k x$. Then $2^{\alpha+k}y = \psi(x) = \psi(2^\alpha)\psi(y) = (\alpha + 2)2^{\alpha-1}\psi(y)$, i.e., $2^{k+1}y = (\alpha+2)\psi(y)$. We notice that for each odd y, $\psi(y)$ is (by definition) the sum of an odd number of odd summands and therefore odd. It follows that $\psi(y) \mid y$. On the other hand, $\psi(y) > y$ for $y > 1$, so we must have $y = 1$. Consequently $\alpha = 2^{k+1} - 2 = 2a - 2$, giving us the unique solution $x = 2^{2a-2}$.

Thus $\psi(x) = ax$ has a unique solution if and only if a is a power of 2.

26. For $m = n = 1$ we obtain that $f(1)^2 + f(1)$ divides $(1^2 + 1)^2 = 4$, from which we find that $f(1) = 1$.
Next, we show that $f(p-1) = p-1$ for each prime p. By the hypothesis for $m = 1$ and $n = p-1$, $f(p-1)+1$ divides p^2, so $f(p-1)$ equals either $p-1$ or $p^2 - 1$. If $f(p-1) = p^2 - 1$, then $f(1) + f(p-1)^2 = p^4 - 2p^2 + 2$ divides $(1 + (p-1)^2)^2 < p^4 - 2p^2 + 2$, giving a contradiction. Hence $f(p-1) = p-1$.
Let us now consider an arbitrary $n \in \mathbb{N}$. By the hypothesis for $m = p-1$, $A = f(n) + (p-1)^2$ divides $(n + (p-1)^2)^2 \equiv (n - f(n))^2 \pmod{A}$, and hence A divides $(n - f(n))^2$ for any prime p. Taking p large enough, we can obtain A to be greater than $(n - f(n))^2$, which implies that $(n - f(n))^2 = 0$, i.e., $f(n) = n$ for every n.

27. Set $a = 1$ and assume that $b \in \mathbb{N}$ is such that $b^2 \equiv b + 1 \pmod{m}$. An easy induction gives us $x_n \equiv b^n \pmod{m}$ for all $n \in \mathbb{N}_0$. Moreover, b is obviously coprime to m, and hence each x_n is coprime to m.

It remains to show the existence of b. The congruence $b^2 \equiv b+1 \pmod{m}$ is equivalent to $(2b-1)^2 \equiv 5 \pmod{m}$. Taking $2b-1 \equiv 2k$, i.e., $b \equiv 2k^2 + k - 2 \pmod{m}$, does the job.

Remark. A desired b exists whenever 5 is a quadratic residue modulo m, in particular, when m is a prime of the form $10k \pm 1$.

28. If n is divisible by 20, then every multiple of n has two last digits even and hence it is not alternate. We shall show that any other n has an alternate multiple.
    (i) Let n be coprime to 10. For each k there exists a number

    $$A_k(n) = \overline{10\ldots010\ldots01\ldots0\ldots01} = \frac{10^{mk}-1}{10^k-1}$$

    ($m \in \mathbb{N}$) that is divisible by n (by Euler's theorem, choose $m = \varphi[n(10^k - 1)]$). In particular, $A_2(n)$ is alternate.
    (ii) Let $n = 2 \cdot 5^r \cdot n_1$, where $r \geq 1$ and $(n_1, 10) = 1$. We shall show by induction that, for each k, there exists an alternative k-digit odd number M_k that is divisible by 5^k. Choosing the number $10A_{2r}(n_1)M_{2r}$ will then solve this case, since it is clearly alternate and divisible by n.
    We can trivially choose $M_1 = 5$. Let there be given an alternate r-digit multiple M_r of 5^r, and let $c \in \{0,1,2,3,4\}$ be such that $M_r/5^r \equiv -c \cdot 2^r \pmod{5}$. Then the $(r+1)$digit numbers $M_r + c \cdot 10^r$ and $M_r + (5+c) \cdot 10^r$ are respectively equal to $5^r(M_r/5^r + 2^r \cdot c)$ and $5^r(M_r/5^r + 2^r \cdot c + 5 \cdot 2^r)$, and hence they are divisible by 5^{r+1} and exactly one of them is alternate: we set it to be M_{r+1}.
    (iii) Let $n = 2^r \cdot n_1$, where $r \geq 1$ and $(n_1, 10) = 1$. We show that there exists an alternate $2r$-digit number N_r that is divisible by 2^{2r+1}. Choosing the number $A_{2r}(n_1)N_r$ will then solve this case.
    We choose $N_1 = 16$, and given N_r, we can prove that one of $N_r + m \cdot 10^{2r}$, for $m \in \{10, 12, 14, 16\}$, is divisible by 2^{2r+3} and therefore suitable for N_{r+1}. Indeed, for $N_r = 2^{2r+1}d$ we have $N_r + m \cdot 10^{2r} = 2^{2r+1}(d + 5^r m/2)$ and $d + 5^r m/2 \equiv 0 \pmod{4}$ has a solution $m/2 \in \{5,6,7,8\}$ for each d and r.

Remark. The idea is essentially the same as in (SL94-24).

29. Let $S_n = \{x \in \mathbb{N} \mid x \leq n,\ n \mid x^2 - 1\}$. It is easy to check that $P_n \equiv 1 \pmod{n}$ for $n = 2$ and $P_n \equiv -1 \pmod{n}$ for $n \in \{3,4\}$, so from now on we assume $n > 4$.
    We note that if $x \in S_n$, then also $n - x \in S_n$ and $(x,n) = 1$. Thus S_n splits into pairs $\{x, n-x\}$, where $x \in S_n$ and $x \leq n/2$. In each of these pairs the product of elements gives remainder -1 upon division by n. Therefore $P_n \equiv (-1)^m$, where S_n has $2m$ elements. It remains to find the parity of m.
    Suppose first that $n > 4$ is divisible by 4. Whenever $x \in S_n$, the numbers $|n/2 - x|$, $n - x, n - |n/2 - x|$ also belong to S_n (indeed, $n \mid (n/2 - x)^2 - 1 = n^2/4 - nx + x^2 - 1$ because $n \mid n^2/4$, etc.). In this way the set S_n splits into four-element subsets $\{x, n/2 - x, n/2 + x, n - x\}$, where $x \in S_n$ and $x < n/4$ (elements of these subsets are different for $x \neq n/4$, and $n/4$ doesn't belong to S_n for $n > 4$). Therefore $m = |S_n|/2$ is even and $P_n \equiv 1 \pmod{m}$.

Now let n be odd. If $n \mid x^2 - 1 = (x-1)(x+1)$, then there exist natural numbers a, b such that $ab = n$, $a \mid x - 1$, $b \mid x + 1$. Obviously a and b are coprime. Conversely, given any odd $a, b \in \mathbb{N}$ such that $(a, b) = 1$ and $ab = n$, by the Chinese remainder theorem there exists $x \in \{1, 2, \ldots, n-1\}$ such that $a \mid x - 1$ and $b \mid x + 1$. This gives a bijection between all ordered pairs (a, b) with $ab = n$ and $(a, b) = 1$ and the elements of S_n. Now if $n = p_1^{\alpha_1} \cdots p_k^{\alpha_k}$ is the decomposition of n into primes, the number of pairs (a, b) is equal to 2^k (since for every i, either $p_i^{\alpha_i} \mid a$ or $p_i^{\alpha_i} \mid b$), and hence $m = 2^{k-1}$. Thus $P_n \equiv -1 \pmod{n}$ if n is a power of an odd prime, and $P_n \equiv 1$ otherwise.

Finally, let n be even but not divisible by 4. Then $x \in S_n$ if and only if x or $n - x$ belongs to $S_{n/2}$ and x is odd. Since $n/2$ is odd, for each $x \in S_{n/2}$ either x or $x + n/2$ belongs to S_n, and by the case of n odd we have $S_n \equiv \pm 1 \pmod{n/2}$, depending on whether or not $n/2$ is a power of a prime. Since S_n is odd, it follows that $P_n \equiv -1 \pmod{n}$ if $n/2$ is a power of a prime, and $P_n \equiv 1$ otherwise.

Second solution. Obviously S_n is closed under multiplication modulo n. This implies that S_n with multiplication modulo n is a subgroup of $\mathbb{Z}_n$, and therefore there exist elements $a_1 = -1, a_2, \ldots, a_k \in S_n$ that generate S_n. In other words, since the a_i are of order two, S_n consists of products $\prod_{i \in A} a_i$, where A runs over all subsets of $\{1, 2, \ldots, k\}$. Thus S_n has 2^k elements, and the product of these elements equals $P_n \equiv (a_1 a_2 \cdots a_k)^{2^{k-1}} \pmod{n}$. Since $a_i^2 \equiv 1 \pmod{n}$, it follows that $P_n \equiv 1$ if $k \geq 2$, i.e., if $|S_n| > 2$. Otherwise $P_n \equiv -1 \pmod{n}$.

We note that $|S_n| > 2$ is equivalent to the existence of $a \in S_n$ with $1 < a < n - 1$. It is easy to find that such an a exists if and only if neither of n, $n/2$ is a power of an odd prime.

30. We shall denote by k the given circle with diameter p^n.

Let A, B be lattice points (i.e., points with integer coordinates). We shall denote by $\mu(AB)$ the exponent of the highest power of p that divides the integer AB^2. We observe that if S is the area of a triangle ABC where A, B, C are lattice points, then $2S$ is an integer. According to Heron's formula and the formula for the circumradius, a triangle ABC whose circumcenter has diameter p^n satisfies

$$2AB^2 BC^2 + 2BC^2 CA^2 + 2CA^2 AB^2 - AB^4 - BC^4 - CA^4 = 16S^2 \tag{1}$$

and
$$AB^2 \cdot BC^2 \cdot CA^2 = (2S)^2 p^{2n}. \tag{2}$$

Lemma 1. Let A, B, and C be lattice points on k. If none of AB^2, BC^2, CA^2 is divisible by p^{n+1}, then $\mu(AB), \mu(BC), \mu(CA)$ are $0, n, n$ in some order.

Proof. Let $\kappa = \min\{\mu(AB), \mu(BC), \mu(CA)\}$. By (1), $(2S)^2$ is divisible by $p^{2\kappa}$. Together with (2), this gives us $\mu(AB) + \mu(BC) + \mu(CA) = 2\kappa + 2n$. On the other hand, if none of AB^2, BC^2, CA^2 is divisible by p^{n+1}, then $\mu(AB) + \mu(BC) + \mu(CA) \leq \kappa + 2n$. Therefore $\kappa = 0$ and the remaining two of $\mu(AB), \mu(BC), \mu(CA)$ are equal to n.

Lemma 2. Among every four lattice points on k, there exist two, say M, N, such that $\mu(MN) \geq n + 1$.

Proof. Assume that this doesn't hold for some points A, B, C, D on k. By Lemma 1, μ for some of the segments $AB, AC, \ldots, CD$ is 0, say $\mu(AC) = 0$. It easily follows by Lemma 1 that then $\mu(BD) = 0$ and $\mu(AB) = \mu(BC) = \mu(CD) = \mu(DA) = n$. Let $a, b, c, d, e, f \in \mathbb{N}$ be such that $AB^2 = p^n a$, $BC^2 = p^n b$, $CD^2 = p^n c$, $DA^2 = p^n d$, $AC^2 = e$, $BD^2 = f$. By Ptolemy's theorem we have $\sqrt{ef} = p^n \left(\sqrt{ac} + \sqrt{bd} \right)$. Taking squares, we get that

$\frac{ef}{p^{2n}} = \left(\sqrt{ac} + \sqrt{bd} \right)^2 = ac + bd + 2\sqrt{abcd}$ is rational and hence an integer. It follows that ef is divisible by p^{2n}, a contradiction.

Now we consider eight lattice points $A_1, A_2, \ldots, A_8$ on k. We color each segment $A_i A_j$ red if $\mu(A_i A_j) > n$ and black otherwise, and thus obtain a graph G. The degree of a point X will be the number of red segments with an endpoint in X. We distinguish three cases:

(i) There is a point, say A_8, whose degree is at most 1. We may suppose w.l.o.g. that $A_8 A_7$ is red and $A_8 A_1, \ldots, A_8 A_6$ black. By a well-known fact, the segments joining vertices $A_1, A_2, \ldots, A_6$ determine either a red triangle, in which case there is nothing to prove, or a black triangle, say $A_1 A_2 A_3$. But in the latter case the four points A_1, A_2, A_3, A_8 do not determine any red segment, a contradiction to Lemma 2.

(ii) All points have degree 2. Then the set of red segments partitions into cycles. If one of these cycles has length 3, then the proof is complete. If all the cycles have length at least 4, then we have two possibilities: two 4-cycles, say $A_1 A_2 A_3 A_4$ and $A_5 A_6 A_7 A_8$, or one 8-cycle, $A_1 A_2 \ldots A_8$. In both cases, the four points A_1, A_3, A_5, A_7 do not determine any red segment, a contradiction.

(iii) There is a point of degree at least 3, say A_1. Suppose that $A_1 A_2$, $A_1 A_3$, and $A_1 A_4$ are red. We claim that A_2, A_3, A_4 determine at least one red segment, which will complete the solution. If not, by Lemma 1, $\mu(A_2 A_3)$, $\mu(A_3 A_4)$, $\mu(A_4 A_2)$ are $n, n, 0$ in some order. Assuming w.l.o.g. that $\mu(A_2 A_3) = 0$, denote by S the area of triangle $A_1 A_2 A_3$. Now by formula (1), $2S$ is not divisible by p. On the other hand, since $\mu(A_1 A_2) \geq n + 1$ and $\mu(A_1 A_3) \geq n + 1$, it follows from (2) that $2S$ is divisible by p, a contradiction.

4.46 Solutions to the Shortlisted Problems of IMO 2005

1. Clearly, $p(x)$ has to be of the form $p(x) = x^2 + ax \pm 1$, where a is an integer. For $a = \pm 1$ and $a = 0$, polynomial p has the required property: it suffices to take $q = 1$ and $q = x + 1$, respectively.

   Suppose now that $|a| \geq 2$. Then $p(x)$ has two real roots, say x_1, x_2, which are also roots of $p(x)q(x) = x^n + a_{n-1}x^{n-1} + \cdots + a_0$, $a_i = \pm 1$. Thus

   $$1 = \left| \frac{a_{n-1}}{x_i} + \cdots + \frac{a_0}{x_i^n} \right| \leq \frac{1}{|x_i|} + \cdots + \frac{1}{|x_i|^n} < \frac{1}{|x_i| - 1},$$

   which implies $|x_1|, |x_2| < 2$. This immediately rules out the case $|a| \geq 3$ and the polynomials $p(x) = x^2 \pm 2x - 1$. The remaining two polynomials $x^2 \pm 2x + 1$ satisfy the condition for $q(x) = x \mp 1$.

   Therefore, the polynomials $p(x)$ with the desired property are $x^2 \pm x \pm 1$, $x^2 \pm 1$, and $x^2 \pm 2x + 1$.

2. Given $y > 0$, consider the function $\varphi(x) = x + yf(x)$, $x > 0$. This function is injective: indeed, if $\varphi(x_1) = \varphi(x_2)$, then $f(x_1)f(y) = f(\varphi(x_1)) = f(\varphi(x_2)) = f(x_2)f(y)$, so $f(x_1) = f(x_2)$, so $x_1 = x_2$ by the definition of φ. Now if $x_1 > x_2$ and $f(x_1) < f(x_2)$, we have $\varphi(x_1) = \varphi(x_2)$ for $y = \frac{x_1 - x_2}{f(x_2) - f(x_1)} > 0$, which is impossible; hence f is nondecreasing. The functional equation now yields $f(x)f(y) = 2f(x + yf(x)) \geq 2f(x)$ and consequently $f(y) \geq 2$ for $y > 0$. Therefore

   $$f(x + yf(x)) = f(xy) = f(y + xf(y)) \geq f(2x)$$

   holds for arbitrarily small $y > 0$, implying that f is constant on the interval $(x, 2x]$ for each $x > 0$. But then f is constant on the union of all intervals $(x, 2x]$ over all $x > 0$, that is, on all of $\mathbb{R}^+$. Now the functional equation gives us $f(x) = 2$ for all x, which is clearly a solution.

   Second Solution. In the same way as above we prove that f is nondecreasing, and hence its discontinuity set is at most countable. We can extend f to $\mathbb{R} \cup \{0\}$ by defining $f(0) = \inf_x f(x) = \lim_{x \to 0} f(x)$, and the new function f is continuous at 0 as well. If x is a point of continuity of f we have $f(x)f(0) = \lim_{y \to 0} f(x)f(y) = \lim_{y \to 0} 2f(x + yf(x)) = 2f(x)$, hence $f(0) = 2$. Now, if f is continuous at $2y$, then $2f(y) = \lim_{x \to 0} f(x)f(y) = \lim_{x \to 0} 2f(x + yf(x)) = 2f(2y)$. Thus $f(y) = f(2y)$, for all but countably many values of y. Being nondecreasing f is a constant; hence $f(x) = 2$.

3. Assume without loss of generality that $p \geq q \geq r \geq s$. We have

   $$(pq + rs) + (pr + qs) + (ps + qr) = \frac{(p + q + r + s)^2 - p^2 - q^2 - r^2 - s^2}{2} = 30.$$

   It is easy to see that $pq + rs \geq pr + qs \geq ps + qr$, which gives us $pq + rs \geq 10$. Now setting $p + q = x$, we obtain $x^2 + (9 - x)^2 = (p + q)^2 + (r + s)^2 = 21 + 2(pq + rs) \geq 41$, which is equivalent to $(x - 4)(x - 5) \geq 0$. Since $x = p + q \geq r + s$, we conclude that $x \geq 5$. Thus

$$25 \le p^2 + q^2 + 2pq = 21 - (r^2 + s^2) + 2pq \le 21 + 2(pq - rs),$$

or $pq - rs \ge 2$, as desired.

Remark. The quadruple $(p,q,r,s) = (3,2,2,2)$ shows that the estimate 2 is the best possible.

4. Setting $y = 0$ yields $(f(0) + 1)(f(x) - 1) = 0$, and since $f(x) = 1$ for all x is impossible, we get $f(0) = -1$. Now plugging in $x = 1$ and $y = -1$ gives us $f(1) = 1$ or $f(-1) = 0$. In the first case setting $x = 1$ in the functional equation yields $f(y+1) = 2y + 1$, i.e., $f(x) = 2x - 1$, which is one solution.
Suppose now that $f(1) = a \ne 1$ and $f(-1) = 0$. Plugging $(x,y) = (z,1)$ and $(x,y) = (-z,-1)$ in the functional equation yields

$$f(z+1) = (1-a)f(z) + 2z + 1$$
$$f(-z-1) = f(z) + 2z + 1.$$

It follows that $f(z+1) = (1-a)f(-z-1) + a(2z+1)$, i.e. $f(x) = (1-a)f(-x) + a(2x-1)$. Analogously, $f(-x) = (1-a)f(x) + a(-2x-1)$, which together with the previous equation yields

$$(a^2 - 2a)f(x) = -2a^2x - (a^2 - 2a).$$

Now $a = 2$ is clearly impossible. For $a \notin \{0,2\}$ we get $f(x) = \frac{-2ax}{a-2} - 1$. This function satisfies the requirements only for $a = -2$, giving the solution $f(x) = -x - 1$. In the remaining case, when $a = 0$, we have $f(x) = f(-x)$. Setting $y = z$ and $y = -z$ in the functional equation and subtracting yields $f(2z) = 4z^2 - 1$, so $f(x) = x^2 - 1$, which satisfies the equation.
Thus the solutions are $f(x) = 2x - 1$, $f(x) = -x - 1$, and $f(x) = x^2 - 1$.

5. The desired inequality is equivalent to

$$\frac{x^2 + y^2 + z^2}{x^5 + y^2 + z^2} + \frac{x^2 + y^2 + z^2}{y^5 + z^2 + x^2} + \frac{x^2 + y^2 + z^2}{z^5 + x^2 + y^2} \le 3. \qquad (1)$$

By the Cauchy inequality we have $(x^5 + y^2 + z^2)(yz + y^2 + z^2) \ge (x^{5/2}(yz)^{1/2} + y^2 + z^2)^2 \ge (x^2 + y^2 + z^2)^2$ and therefore

$$\frac{x^2 + y^2 + z^2}{x^5 + y^2 + z^2} \le \frac{yz + y^2 + z^2}{x^2 + y^2 + z^2}.$$

We get analogous inequalities for the other two summands in (1). Summing these yields

$$\frac{x^2 + y^2 + z^2}{x^5 + y^2 + z^2} + \frac{x^2 + y^2 + z^2}{y^5 + z^2 + x^2} + \frac{x^2 + y^2 + z^2}{z^5 + x^2 + y^2} \le 2 + \frac{xy + yz + zx}{x^2 + y^2 + z^2},$$

which together with the well-known inequality $x^2 + y^2 + z^2 \ge xy + yz + zx$ gives us the result.

Second solution. Multiplying both sides by the common denominator and using notation in Chapter 2 (*Muirhead's inequality*), we get

$$T_{5,5,5} + 4T_{7,5,0} + T_{5,2,2} + T_{9,0,0} \geq T_{5,5,2} + T_{6,0,0} + 2T_{5,4,0} + 2T_{4,2,0} + T_{2,2,2}.$$

By Schur's and Muirhead's inequalities we have that $T_{9,0,0} + T_{5,2,2} \geq 2T_{7,2,0} \geq 2T_{7,1,1}$. Since $xyz \geq 1$ we have that $T_{7,1,1} \geq T_{6,0,0}$. Therefore

$$T_{9,0,0} + T_{5,2,2} \geq 2T_{6,0,0} \geq T_{6,0,0} + T_{4,2,0}. \tag{2}$$

Moreover, Muirhead's inequality combined with $xyz \geq 1$ gives us $T_{7,5,0} \geq T_{5,5,2}$, $2T_{7,5,0} \geq 2T_{6,5,1} \geq 2T_{5,4,0}$, $T_{7,5,0} \geq T_{6,4,2} \geq T_{4,2,0}$, and $T_{5,5,5} \geq T_{2,2,2}$. Adding these four inequalities to (2) yields the desired result.

6. A room will be called *economic* if some of its lamps are on and some are off. Two lamps sharing a switch will be called *twins*. The twin of a lamp l will be denoted by $\bar{l}$.

   Suppose we have arrived at a state with the minimum possible number of un-economic rooms, and that this number is strictly positive. Let us choose any uneconomic room, say R_0, and a lamp l_0 in it. Let $\bar{l}_0$ be in a room R_1. Switch-ing l_0, we make R_0 economic; therefore, since the number of uneconomic rooms cannot be decreased, this change must make room R_1 uneconomic. Now choose a lamp l_1 in R_1 having the twin $\bar{l}_1$ in a room R_2. Switching l_1 makes R_1 economic, and thus must make R_2 uneconomic. Continuing in this manner we obtain a se-quence $l_0, l_1, \dots$ of lamps with l_i in a room R_i and $\bar{l}_i \neq l_{i+1}$ in R_{i+1} for all i. The lamps $l_0, l_1, \dots$ are switched in this order. This sequence has the property that switching l_i and $\bar{l}_i$ makes room R_i economic and room R_{i+1} uneconomic.

   Let $R_m = R_k$ with $m > k$ be the first repetition in the sequence (R_i). Let us stop switching the lamps at l_{m-1}. The room R_k was uneconomic prior to switching l_k. Thereafter, lamps l_k and $\bar{l}_{m-1}$ have been switched in R_k, but since these two lamps are distinct (indeed, their twins $\bar{l}_k$ and l_{m-1} are distinct), the room R_k is now economic, as well as all the rooms $R_0, R_1, \dots, R_{m-1}$. This decreases the number of uneconomic rooms, contradicting our assumption.

7. Let v be the number of video winners. One easily finds that for $v = 1$ and $v = 2$, the number n of customers is at least $2k + 3$ and $3k + 5$ respectively. We prove by induction on v that if $n \geq k + 1$, then $n \geq (k+2)(v+1) - 1$.

   Without loss of generality, we can assume that the total number n of customers is minimum possible for given $v > 0$. Consider a person P who was convinced by nobody but himself. Then P must have won a video; otherwise, P could be removed from the group without decreasing the number of video winners. Let Q and R be the two persons convinced by P. We denote by $\mathscr{C}$ the set of persons influenced by P through Q to buy a sombrero, including Q, and by $\mathscr{D}$ the set of all other customers excluding P. Let x be the number of video winners in $\mathscr{C}$. Then there are $v - x - 1$ video winners in $\mathscr{D}$. We have $|\mathscr{C}| \geq (k+2)(x+1) - 1$, by the induction hypothesis if $x > 0$ and because P is a winner if $x = 0$. Similarly, $|\mathscr{D}| \geq (k+2)(v-x) - 1$. Thus $n \geq 1 + (k+2)(x+1) - 1 + (k+2)(v-x) - 1$, i.e., $n \geq (k+2)(v+1) - 1$.

8. Suppose that a two-sided $m \times n$ board T is considered, where exactly k of the squares are transparent. A transparent square is colored only on one side (then it looks the same from the other side), while a nontransparent one needs to be colored on both sides, not necessarily in the same color.

Let $C = C(T)$ be the set of colorings of the board in which there exist two black paths from the left edge to the right edge, one on top and one underneath, not intersecting at any transparent square. If $k = 0$ then $|C| = N^2$. We prove by induction on k that $2^k |C| \le N^2$. This will imply the statement of the problem, since $|C| = M$ for $k = mn$.

Let q be a fixed transparent square. Consider any coloring B in C: If q is converted into a nontransparent square, a new board T' with $k - 1$ transparent squares is obtained, so by the induction hypothesis $2^{k-1} |C(T')| \le N^2$. Since B contains two black paths at most one of which passes through q, coloring q in either color on the other side will result in a coloring in C'; hence $|C(T')| \ge 2|C(T)|$, implying $2^k |C(T)| \le N^2$ and finishing the induction.

Second solution. By a *path* we shall mean a black path from the left edge to the right edge. Let $\mathscr{A}$ denote the set of pairs of $m \times n$ boards each of which has a path. Let $\mathscr{B}$ denote the set of pairs of boards such that the first board has two nonintersecting paths. Obviously, $|\mathscr{A}| = N^2$ and $|\mathscr{B}| = 2^{mn} M$. To prove $|\mathscr{A}| \ge |\mathscr{B}|$, we will construct an injection $f : \mathscr{B} \to \mathscr{A}$.

Among paths on a given board we define path x to be *lower* than y if the set of squares "under" x is a subset of the squares under y. This relation is a relation of incomplete order. However, for each board with at least one path there exists a lowest path (comparing two intersecting paths, we can always take the "lower branch" on each nonintersecting segment). Now, for a given element of $\mathscr{B}$, we "swap" the lowest path and all squares underneath on the first board with the corresponding points on the other board. This swapping operation is the desired injection f. Indeed, since the first board still contains the highest path (which didn't intersect the lowest one), the new configuration belongs to $\mathscr{A}$. On the other hand, this configuration uniquely determines the lowest path on the original element of $\mathscr{B}$; hence no two different elements of $\mathscr{B}$ can go to the same element of $\mathscr{A}$. This completes the proof.

9. Let $[XY]$ denote the label of segment XY, where X and Y are vertices of the polygon. Consider any segment MN with the maximum label $[MN] = r$. By condition (ii), for any $P_i \ne M, N$, exactly one of P_iM and P_iN is labeled by r. Thus the set of all vertices of the n-gon splits into two complementary groups: $\mathscr{A} = \{P_i \mid [P_iM] = r\}$ and $\mathscr{B} = \{P_i \mid [P_iN] = r\}$. We claim that a segment XY is labelled by r if and only if it joins two points from different groups. Assume without loss of generality that $X \in \mathscr{A}$. If $Y \in \mathscr{A}$, then $[XM] = [YM] = r$, so $[XY] < r$. If $Y \in \mathscr{B}$, then $[XM] = r$ and $[YM] < r$, so $[XY] = r$ by (ii), as we claimed.

We conclude that a labeling satisfying (ii) is uniquely determined by groups $\mathscr{A}$ and $\mathscr{B}$ and labelings satisfying (ii) within A and B.

(a) We prove by induction on n that the greatest possible value of r is $n-1$. The degenerate cases $n = 1, 2$ are trivial. If $n \geq 3$, the number of different labels of segments joining vertices in $\mathscr{A}$ (resp. $\mathscr{B}$) does not exceed $|\mathscr{A}| - 1$ (resp. $|\mathscr{B}| - 1$), while all segments joining a vertex in $\mathscr{A}$ and a vertex in $\mathscr{B}$ are labeled by r. Therefore $r \leq (|\mathscr{A}| - 1) + (|\mathscr{B}| - 1) + 1 = n - 1$. Equality is achieved if all the mentioned labels are different.

(b) Let a_n be the number of labelings with $r = n - 1$. We prove by induction that $a_n = \frac{n!(n-1)!}{2^{n-1}}$. This is trivial for $n = 1$, so let $n \geq 2$. If $|\mathscr{A}| = k$ is fixed, the groups $\mathscr{A}$ and $\mathscr{B}$ can be chosen in $\binom{n}{k}$ ways. The set of labels used within $\mathscr{A}$ can be selected among $1, 2, \ldots, n - 2$ in $\binom{n-2}{k-1}$ ways. Now the segments within groups $\mathscr{A}$ and $\mathscr{B}$ can be labeled so as to satisfy (ii) in a_k and a_{n-k} ways, respectively. In this way, every labeling has been counted twice, since choosing $\mathscr{A}$ is equivalent to choosing $\mathscr{B}$. It follows that

$$
\begin{aligned}
a_n &= \frac{1}{2} \sum_{k=1}^{n-1} \binom{n}{k} \binom{n-2}{k-1} a_k a_{n-k} \\
&= \frac{n!(n-1)!}{2(n-1)} \sum_{k=1}^{n-1} \frac{a_k}{k!(k-1)!} \cdot \frac{a_{n-k}}{(n-k)!(n-k-1)!} \\
&= \frac{n!(n-1)!}{2(n-1)} \sum_{k=1}^{n-1} \frac{1}{2^{k-1}} \cdot \frac{1}{2^{n-k-1}} = \frac{n!(n-1)!}{2^{n-1}}.
\end{aligned}
$$

10. Denote by L the leftmost and by R the rightmost marker. To start with, note that the parity of the number of black-side-up markers remains unchanged. Hence, if only two markers remain, these markers must have the same color up.

We shall show by induction on n that the game can be successfully finished if and only if $n \equiv 0$ or $n \equiv 2 \pmod 3$, and that the upper sides of L and R will be black in the first case and white in the second case.

The statement is clear for $n = 2, 3$. Assume that we have finished the game for some n, and denote by k the position of the marker X (counting from the left) that was last removed. Having finished the game, we have also finished the subgames with the k markers from L to X and with the $n - k + 1$ markers from X to R (inclusive). Thereby, before X was removed, the upper side of L had been black if $k \equiv 0$ and white if $k \equiv 2 \pmod 3$, while the upper side of R had been black if $n - k + 1 \equiv 0$ and white if $n - k + 1 \equiv 2 \pmod 3$. Markers L and R were reversed upon the removal of X. Therefore, in the final position, L and R are white if and only if $k \equiv n - k + 1 \equiv 0$, which yields $n \equiv 2 \pmod 3$, and black if and only if $k \equiv n - k + 1 \equiv 2$, which yields $n \equiv 0 \pmod 3$.

On the other hand, a game with n markers can be reduced to a game with $n - 3$ markers by removing the second, fourth, and third markers in this order. This finishes the induction.

Second solution. An invariant can be defined as follows. To each white marker with k black markers to its left we assign the number $(-1)^k$. Let S be the sum of the assigned numbers. Then it is easy to verify that the remainder of S modulo

3 remains unchanged throughout the game: For example, when a white marker with two white neighbors and k black markers to its left is removed, S decreases by $3(-1)^t$.

Initially, $S = n$. In the final position with two markers remaining, S equals 0 if the two markers are black and 2 if these are white (note that, as before, the two markers must be of the same color). Thus $n \equiv 0$ or 2 (mod 3).

Conversely, a game with n markers is reduced to $n - 3$ markers as in the first solution.

11. Assume that there were n contestants, a_i of whom solved exactly i problems, where $a_0 + \cdots + a_5 = n$. Let us count the number N of pairs (C,P), where contestant C solved the pair of problems P. Each of the 15 pairs of problems was solved by at least $\frac{2n+1}{5}$ contestants, implying $N \geq 15 \cdot \frac{2n+1}{5} = 6n + 3$. On the other hand, a_i students solved $\frac{i(i-1)}{2}$ pairs; hence

$$6n + 3 \leq N \leq a_2 + 3a_3 + 6a_4 + 10a_5 = 6n + 4a_5 - (3a_3 + 5a_2 + 6a_1 + 6a_0).$$

Consequently $a_5 \geq 1$. Assume that $a_5 = 1$. Then we must have $N = 6n + 4$, which is possible only if 14 of the pairs of problems were solved by exactly $\frac{2n+1}{5}$ students and the remaining one by $\frac{2n+1}{5} + 1$ students, and all students but the winner solved 4 problems.

The problem t not solved by the winner will be called *tough* and the pair of problems solved by $\frac{2n+1}{5} + 1$ students *special*.

Let us count the number M_p of pairs (C,P) for which P contains a fixed problem p. Let b_p be the number of contestants who solved p. Then $M_t = 3b_t$ (each of the b_t students solved three pairs of problems containing t), and $M_p = 3b_p + 1$ for $p \neq t$ (the winner solved four such pairs). On the other hand, each of the five pairs containing p was solved by $\frac{2n+1}{5}$ or $\frac{2n+1}{5} + 1$ students, so $M_p = 2n + 2$ if the special pair contains p, and $M_p = 2n + 1$ otherwise.

Now since $M_t = 3b_t = 2n + 1$ or $2n + 2$, we have $2n + 1 \equiv 0$ or 2 (mod 3). But if $p \neq t$ is a problem not contained in the special pair, we have $M_p = 3b_p + 1 = 2n + 1$; hence $2n + 1 \equiv 1$ (mod 3), which is a contradiction.

12. Suppose that there exist desired permutations σ and τ for some sequence $a_1, \ldots, a_n$. Given a sequence (b_i) with sum divisible by n that differs modulo n from (a_i) in only two positions, say i_1 and i_2, we show how to construct desired permutations σ' and τ' for sequence (b_i). In this way, starting from an arbitrary sequence (a_i) for which σ and τ exist, we can construct desired permutations for any other sequence with sum divisible by n. All congruences below are modulo n.

We know that $\sigma(i) + \tau(i) \equiv b_i$ for all $i \neq i_1, i_2$. We construct the sequence $i_1, i_2, i_3, \ldots$ as follows: for each $k \geq 2$, i_{k+1} is the unique index such that

$$\sigma(i_{k-1}) + \tau(i_{k+1}) \equiv b_{i_k}. \tag{1}$$

Let $i_p = i_q$ be the repetition in the sequence with the smallest q. We claim that $p = 1$ or $p = 2$. Assume to the contrary that $p > 2$. Summing (1) for $k = p, p+1,$

..., $q-1$ and taking the equalities $\sigma(i_k) + \tau(i_k) = b_{i_k}$ for $i_k \neq i_1, i_2$ into account, we obtain $\sigma(i_{p-1}) + \sigma(i_p) + \tau(i_{q-1}) + \tau(i_q) \equiv b_p + b_{q-1}$. Since $i_q = i_p$, it follows that $\sigma(i_{p-1}) + \tau(i_{q-1}) \equiv b_{q-1}$ and therefore $i_{p-1} = i_{q-1}$, a contradiction. Thus $p = 1$ or $p = 2$ as claimed.

Now we define the following permutations:

$$\sigma'(i_k) = \sigma(i_{k-1}) \text{ for } k = 2,3,\ldots,q-1 \text{ and } \sigma'(i_1) = \sigma(i_{q-1}),$$

$$\tau'(i_k) = \tau(i_{k+1}) \text{ for } k = 2,3,\ldots,q-1 \text{ and } \tau'(i_1) = \begin{cases} \tau(i_2) & \text{if } p = 1, \\ \tau(i_1) & \text{if } p = 2; \end{cases}$$

$$\sigma'(i) = \sigma(i) \text{ and } \tau'(i) = \tau(i) \quad \text{for } i \notin \{i_1,\ldots,i_{q-1}\}.$$

Permutations σ' and τ' have the desired property. Indeed, $\sigma'(i) + \tau'(i) = b_i$ obviously holds for all $i \neq i_1$, but then it must also hold for $i = i_1$.

13. For every green diagonal d, let C_d denote the number of green–red intersection points on d. The task is to find the maximum possible value of the sum $\sum_d C_d$ over all green diagonals.

Let d_i and d_j be two green diagonals and let the part of polygon M lying between d_i and d_j have m vertices. There are at most $n - m - 1$ red diagonals intersecting both d_i and d_j, while each of the remaining $m - 2$ diagonals meets at most one of d_i, d_j. It follows that

$$C_{d_i} + C_{d_j} \leq 2(n - m - 1) + (m - 2) = 2n - m - 4. \tag{1}$$

We now arrange the green diagonals in a sequence $d_1, d_2, \ldots, d_{n-3}$ as follows. It is easily seen that there are two green diagonals d_1 and d_2 that divide M into two triangles and an $(n-2)$-gon; then there are two green diagonals d_3 and d_4 that divide the $(n-2)$-gon into two triangles and an $(n-4)$-gon, and so on. We continue this procedure until we end up with a triangle or a quadrilateral. Now, the part of M between d_{2k-1} and d_{2k} has at least $n - 2k$ vertices for $1 \leq k \leq r$, where $n - 3 = 2r + e$, $e \in \{0,1\}$; hence, by (1), $C_{d_{2k-1}} + C_{d_{2k}} \leq n + 2k - 4$. Moreover, $C_{d_{n-3}} \leq n - 3$. Summing yields

$$C_{d_1} + C_{d_2} + \cdots + C_{d_{n-3}} \leq \sum_{k=1}^{r} (n + 2k - 4) + e(n - 3)$$

$$= 3r^2 + e(3r + 1) = \left\lceil \frac{3}{4}(n-3)^2 \right\rceil.$$

This value is attained in the following example. Let $A_1 A_2 \ldots A_n$ be the n-gon M and let $l = \left\lceil \frac{n}{2} \right\rceil + 1$. The diagonals $A_1 A_i$, $i = 3,\ldots,l$, and $A_l A_j$, $j = l+2,\ldots,n$ are colored green, whereas the diagonals $A_2 A_i$, $i = l+1,\ldots,n$, and $A_{l+1} A_j$, $j = 3,\ldots,l-1$ are colored red.

Thus the answer is $\left\lceil \frac{3}{4}(n-3)^2 \right\rceil$.

14. Let F be the point of tangency of the incircle with AC and let M and N be the respective points of tangency of AB and BC with the corresponding excircles. If I is the incenter and I_a and P respectively the center and the tangency point with ray AC of the excircle corresponding to A, we have $\frac{AI}{IL} = \frac{AI}{IF} = \frac{AI_a}{I_a P} = \frac{AI_a}{I_a N}$, which

implies that $\triangle AIL \sim \triangle AI_aN$. Thus L lies on AN, and analogously K lies on CM. Define $x = AF$ and $y = CF$. Since $BD = BE$, $AD = BM = x$, and $CE = BN = y$, the condition $AB + BC = 3AC$ gives us $DM = y$ and $EN = x$. The triangles CLN and MKA are congruent since their altitudes KD and LE satisfy $DK = EL$, $DM = CE$, and $AD = EN$. Thus $\angle AKM = \angle CLN$, implying that $ACKL$ is cyclic.

15. Let P be the fourth vertex of the rhombus $C_2A_1A_2P$. Since $\triangle C_2PC_1$ is equilateral, we easily conclude that $B_1B_2C_1P$ is also a rhombus. Thus $\triangle PB_1A_2$ is equilateral and $\angle(C_2A_1, C_1B_2) = \angle A_2PB_1 = 60°$. It easily follows that $\triangle AC_1B_2 \cong \triangle BA_1C_2$ and consequently $AC_1 = BA_1$; similarly, $BA_1 = CB_1$. Therefore triangle $A_1B_1C_1$ is equilateral. Now it follows from $B_1B_2 = B_2C_1$ that A_1B_2 bisects $\angle C_1A_1B_1$. Similarly, B_1C_2 and C_1A_2 bisect $\angle A_1B_1C_1$ and $\angle B_1C_1A_1$; hence A_1B_2, B_1C_2, C_1A_2 meet at the incenter of $A_1B_1C_1$, i.e. at the center of ABC.

16. Since $\angle ADL = \angle KBA = 180° - \frac{1}{2}\angle BCD$ and $\angle ALD = \frac{1}{2}\angle AYD = \angle KAB$, triangles ABK and LDA are similar. Thus $\frac{BK}{BC} = \frac{BK}{AD} = \frac{AB}{DL} = \frac{DC}{DL}$, which together with $\angle LDC = \angle CBK$ gives us $\triangle LDC \sim \triangle CBK$. Therefore $\angle KCL = 360° - \angle BCD - (\angle LCD + \angle KCB) = 360° - \angle BCD - (\angle CKB + \angle KCB) = 180° - \angle CBK$, which is constant.

17. To start with, we note that points B, E, C are the images of D, F, A respectively under the rotation around point O for the angle $\omega = \angle DOB$, where O is the intersection of the perpendicular bisectors of AC and BD. Then $OE = OF$ and $\angle OFE = \angle OAC = 90 - \frac{\omega}{2}$; hence the points A, F, R, O are on a circle and $\angle ORP = 180° - \angle OFA$. Analogously, the points B, E, Q, O are on a circle and $\angle OQP = 180° - \angle OEB = \angle OEC = \angle OFA$. This shows that $\angle ORP = 180° - \angle OQP$, i.e. the point O lies on the circumcircle of $\triangle PQR$, thus being the desired point.

18. Let O and O_1 be the circumcenters of triangles ABC and ADE, respectively. It is enough to show that $HM \parallel OO_1$. Let AA' be the diameter of the circumcircle of ABC. We note that if B_1 is the foot of the altitude from B, then HE bisects $\angle CHB_1$. Since the triangles COM and CHB_1 are similar (indeed, $\angle CHB = \angle COM = \angle A$), we have $\frac{CE}{EB_1} = \frac{CH}{HB_1} = \frac{CO}{OM} = \frac{2CO}{AH} = \frac{A'A}{AH}$. Thus, if Q is the intersection point of the bisector of $\angle A'AH$ with HA', we obtain $\frac{CE}{EB_1} = \frac{A'Q}{QH}$, which together with $A'C \perp AC$ and $HB_1 \perp AC$ gives us $QE \perp AC$. Analogously, $QD \perp AB$. Therefore AQ is a diameter of the circumcircle of $\triangle ADE$ and O_1 is the midpoint of AQ. It follows that OO_1 is the line passing through the midpoints of AQ and AA'; hence $OO_1 \parallel HM$.

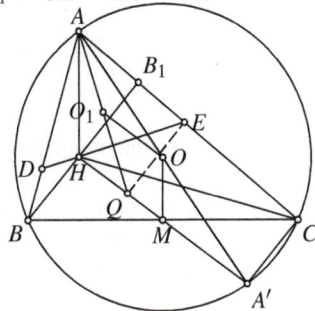

Second solution. We again prove that $OO_1 \parallel HM$. Since $AA' = 2AO$, it suffices to prove $AQ = 2AO_1$.

Elementary calculations of angles give us $\angle ADE = \angle AED = 90° - \frac{\alpha}{2}$. Applying the law of sines to $\triangle DAH$ and $\triangle EAH$ we now have $DE = DH + EH = \frac{AH\cos\beta}{\cos\frac{\alpha}{2}} + \frac{AH\cos\gamma}{\cos\frac{\alpha}{2}}$. Since $AH = 2OM = 2R\cos\alpha$, we obtain

$$AO_1 = \frac{DE}{2\sin\alpha} = \frac{AH(\cos\beta + \cos\gamma)}{2\sin\alpha\cos\frac{\alpha}{2}} = \frac{2R\cos\alpha\sin\frac{\alpha}{2}\cos(\frac{\beta-\gamma}{2})}{\sin\alpha\cos\frac{\alpha}{2}}.$$

We now calculate AQ. Let N be the intersection of AQ with the circumcircle. Since $\angle NAO = \frac{\beta-\gamma}{2}$, we have $AN = 2R\cos(\frac{\beta-\gamma}{2})$. Noting that $\triangle QAH \sim \triangle QNM$ (and that $MN = R - OM$), we have

$$AQ = \frac{AN\cdot AH}{MN + AH} = \frac{2R\cos(\frac{\beta-\gamma}{2})\cdot 2\cos\alpha}{1 + \cos\alpha} = \frac{2R\cos(\frac{\beta-\gamma}{2})\cos\alpha}{\cos^2\frac{\alpha}{2}} = 2AO_1.$$

19. We denote by D, E, F the points of tangency of the incircle with BC, CA, AB, respectively, by I the incenter, and by Y' the intersection of AX and LY. Since EF is the polar line to the point A with respect to the incircle, it meets AL at point R such that $A, R; K, L$ are conjugate, i.e., $\frac{KR}{RL} = \frac{KA}{AL}$. Then $\frac{KX}{LY'} = \frac{KA}{AL} = \frac{KR}{RL} = \frac{KX}{L\overline{Y}}$ and therefore $LY = L\overline{Y}$, where $\overline{Y}$ is the intersection of XR and LY. Thus showing that $LY = LY'$

(which is the same as showing that $PM = MQ$, i.e., $CP = QB$) is equivalent to showing that XY contains R. Since $XKYL$ is an inscribed trapezoid, it is enough to show that R lies on its axis of symmetry, that is, DI.

Since AM is the median, the triangles ARB and ARC have equal areas, and since $\angle(RF, AB) = \angle(RE, AC)$ we have that $1 = \frac{S_{\triangle ABR}}{S_{\triangle ACR}} = \frac{(AB\cdot FR)}{(AC\cdot ER)}$. Hence $\frac{AB}{AC} = \frac{ER}{FR}$.

Let I' be the point of intersection of the line through F parallel to IE with the line IR. Then $\frac{FI'}{EI} = \frac{FR}{RE} = \frac{AC}{AB}$ and $\angle I'FI = \angle BAC$ (angles with orthogonal rays). Thus the triangles ABC and FII' are similar, implying that $\angle FII' = \angle ABC$. Since $\angle FID = 180° - \angle ABC$, it follows that R, I, and D are collinear.

20. We shall prove the inequalities $p(ABC) \geq 2p(DEF)$ and $p(PQR) \geq \frac{1}{2}p(DEF)$. The statement of the problem will immediately follow.

Let D_b and D_c be the reflections of D in AB and AC, and let A_1, B_1, C_1 be the midpoints of BC, CA, AB, respectively. It is easy to see that D_b, F, E, D_c are collinear. Hence $p(DEF) = D_bF + FE + ED_c = D_bD_c \leq D_bC_1 + C_1B_1 + B_1D_c = \frac{1}{2}(AB + BC + CA) = \frac{1}{2}p(ABC)$.

To prove the second inequality we observe that P, Q, and R are the points of tangency of the excircles with the sides of $\triangle DEF$. Let $FQ = ER = x$, $DR = FP = y$, and $DQ = EP = z$, and let $\delta, \varepsilon, \varphi$ be the angles of $\triangle DEF$ at D, E, F, respectively. Let Q' and R' be the projections of Q and R onto EF, respectively. Then $QR \geq Q'R' = EF - FQ' - R'E = EF - x(\cos\varphi + \cos\varepsilon)$. Summing this with the analogous inequalities for FD and DE, we obtain

$$p(PQR) \geq p(DEF) - x(\cos\varphi + \cos\varepsilon) - y(\cos\delta + \cos\varphi) - z(\cos\delta + \cos\varepsilon).$$

Assuming without loss of generality that $x \leq y \leq z$, we also have $DE \leq FD \leq FE$ and consequently $\cos\varphi + \cos\varepsilon \geq \cos\delta + \cos\varphi \geq \cos\delta + \cos\varepsilon$. Now Chebyshev's inequality gives us $p(PQR) \geq p(DEF) - \frac{2}{3}(x+y+z)(\cos\varepsilon + \cos\varphi + \cos\delta) \geq p(DEF) - (x+y+z) = \frac{1}{2}p(DEF)$, where we used $x+y+z = \frac{1}{2}p(DEF)$ and the fact that the sum of the cosines of the angles in a triangle does not exceed $\frac{3}{2}$. This finishes the proof.

21. We will show that 1 is the only such number. It is sufficient to prove that for every prime number p there exists some a_m such that $p \mid a_m$. For $p = 2, 3$ we have $p \mid a_2 = 48$. Assume now that $p > 3$. Applying Fermat's theorem, we have

$$6a_{p-2} = 3 \cdot 2^{p-1} + 2 \cdot 3^{p-1} + 6^{p-1} - 6 \equiv 3 + 2 + 1 - 6 = 0 \pmod{p}.$$

Hence $p \mid a_{p-2}$, i.e. $\gcd(p, a_{p-2}) = p > 1$. This completes the proof.

22. It immediately follows from the condition of the problem that all the terms of the sequence are distinct. We also note that $|a_i - a_n| \leq n - 1$ for all integers i, n where $i < n$, because if $d = |a_i - a_n| \geq n$ then $\{a_1, \ldots, a_d\}$ contains two elements congruent to each other modulo d, which is a contradiction. It easily follows by induction that for every $n \in \mathbb{N}$ the set $\{a_1, \ldots, a_n\}$ consists of consecutive integers. Thus, if we assumed that some integer k did not appear in the sequence $a_1, a_2, \ldots$, the same would have to hold for all integers either larger or smaller than k, which contradicts the condition that infinitely many positive and negative integers appear in the sequence. Thus, the sequence contains all integers.

23. Let us consider the polynomial

$$P(x) = (x+a)(x+b)(x+c) - (x-d)(x-e)(x-f) = Sx^2 + Qx + R,$$

where $Q = ab + bc + ca - de - ef - fd$ and $R = abc + def$.

Since $S \mid Q, R$, it follows that $S \mid P(x)$ for every $x \in \mathbb{Z}$. Hence, $S \mid P(d) = (d+a)(d+b)(d+c)$. Since $S > d+a, d+b, d+c$ and thus cannot divide any of them, it follows that S must be composite.

24. We will show that n has the desired property if and only if it is prime.

For $n = 2$ we can take only $a = 1$. For $n > 2$ and even, $4 \mid n!$, but $a^n + 1 \equiv 1, 2 \pmod 4$, which is impossible. Now we assume that n is odd. Obviously $(n! - 1)^n + 1 \equiv (-1)^n + 1 = 0 \pmod{n!}$. If n is composite and d its prime divisor, then $\left(\frac{n!}{d} - 1\right)^n + 1 = \sum_{k=1}^n \binom{n}{k} \frac{n!^k}{d^k}$, where each summand is divisible by $n!$ because $d^2 \mid n!$; therefore $n!$ divides $\left(\frac{n!}{d} - 1\right)^n + 1$. Thus, all composite numbers are ruled out.

It remains to show that if n is an odd prime and $n! \mid a^n + 1$, then $n! \mid a + 1$, and therefore $a = n! - 1$ is the only relevant value for which $n! \mid a^n + 1$. Consider any prime number $p \leq n$. If $p \mid \frac{a^n+1}{a+1}$, we have $p \mid (-a)^n - 1$ and by Fermat's theorem $p \mid (-a)^{p-1} - 1$. Therefore $p \mid (-a)^{(n,p-1)} - 1 = -a - 1$, i.e. $a \equiv -1 \pmod{p}$. But then $\frac{a^n+1}{a+1} = a^{n-1} - a^{n-2} + \cdots - a + 1 \equiv n \pmod{p}$, implying that $p = n$. It follows that $\frac{a^n+1}{a+1}$ is coprime to $(n-1)!$ and consequently $(n-1)!$ divides $a+1$. Moreover, the above consideration shows that n must divide $a+1$. Thus $n! \mid a+1$ as claimed. This finishes our proof.

25. We will use the abbreviation HD to denote a "highly divisible integer." Let $n = 2^{\alpha_2(n)} 3^{\alpha_3(n)} \cdots p^{\alpha_p(n)}$ be the factorization of n into primes. We have $d(n) = (\alpha_2(n) + 1) \cdots (\alpha_p(n) + 1)$. We start with the following two lemmas.

Lemma 1. If n is an HD and p, q primes with $p^k < q^l$ ($k, l \in \mathbb{N}$), then

$$k\alpha_q(n) \leq l\alpha_p(n) + (k+1)(l-1).$$

Proof. The inequality is trivial if $\alpha_q(n) < l$. Suppose that $\alpha_q(n) \geq l$. Then np^k/q^l is an integer less than q, and $d(np^k/q^l) < d(n)$, which is equivalent to $(\alpha_q(n) + 1)(\alpha_p(n) + 1) > (\alpha_q(n) - l + 1)(\alpha_p(n) + k + 1)$ implying the desired inequality.

Lemma 2. For each p and k there exist only finitely many HD's n such that $\alpha_p(n) \leq k$.

Proof. It follows from Lemma 1 that if n is an HD with $\alpha_p(n) \leq k$, then $\alpha_q(n)$ is bounded for each prime q and $\alpha_q(n) = 0$ for $q > p^{k+1}$. Therefore there are only finitely many possibilities for n.

We are now ready to prove both parts of the problem.

(a) Suppose that there are infinitely many pairs (a, b) of consecutive HD's with $a \mid b$. Since $d(2a) > d(a)$, we must have $b = 2a$. In particular, $d(s) \leq d(a)$ for all $s < 2a$. All but finitely many HD's a are divisible by 2 and by 3^7. Then $d(8a/9) < d(a)$ and $d(3a/2) < d(a)$ yield

$$(\alpha_2(a) + 4)(\alpha_3(a) - 1) < (\alpha_2(a) + 1)(\alpha_3(a) + 1) \Rightarrow 3\alpha_3(a) - 5 < 2\alpha_2(a),$$
$$\alpha_2(a)(\alpha_3(a) + 2) \leq (\alpha_2(a) + 1)(\alpha_3(a) + 1) \Rightarrow \alpha_2(a) \leq \alpha_3(a) + 1.$$

We now have $3\alpha_3(a) - 5 < 2\alpha_2(a) \leq 2\alpha_3(a) + 2 \Rightarrow \alpha_3(a) < 7$, which is a contradiction.

(b) Assume for a given prime p and positive integer k that n is the smallest HD with $\alpha_p \geq k$. We show that $\frac{n}{p}$ is also an HD. Assume the opposite, i.e., that there exists an HD $m < \frac{n}{p}$ such that $d(m) \geq d(\frac{n}{p})$. By assumption, m must also satisfy $\alpha_p(m) + 1 \leq \alpha_p(n)$. Then

$$d(mp) = d(m)\frac{\alpha_p(m) + 2}{\alpha_p(m) + 1} \geq d\left(\frac{n}{p}\right)\frac{\alpha_p(n) + 1}{\alpha_p(n)} = d(n),$$

contradicting the initial assumption that n is an HD (since $mp < n$). This proves that $\frac{n}{p}$ is an HD. Since this is true for every positive integer k, the proof is complete.

26. Assuming $b \neq a$, it trivially follows that $b > a$. Let $p > b$ be a prime number and let $n = (a+1)(p-1)+1$. We note that $n \equiv 1 \pmod{p-1}$ and $n \equiv -a \pmod{p}$. It follows that $r^n = r \cdot (r^{p-1})^{a+1} \equiv r \pmod{p}$ for every integer r. We now have $a^n + n \equiv a - a = 0 \pmod{p}$. Thus, $a^n + n$ is divisible by p, and hence by the condition of the problem $b^n + n$ is also divisible by p. However, we also have $b^n + n \equiv b - a \pmod{p}$, i.e., $p \mid b - a$, which contradicts $p > b$. Hence, it must follow that $b = a$. We note that $b = a$ trivially fulfills the conditions of the problem for all $a \in \mathbb{N}$.

27. Let p be a prime and $k < p$ an even number. We note that $(p-k)!(k-1)! \equiv (-1)^{k-1}(p-k)!(p-k+1)\cdots(p-1) = (-1)^{k-1}(p-1)! \equiv 1 \pmod{p}$ by Wilson's theorem. Therefore

$$(k-1)!^n P((p-k)!) = \sum_{i=0}^n a_i [(k-1)!]^{n-i}[(p-k)!(k-1)!]^i$$
$$\equiv \sum_{i=0}^n a_i [(k-1)!]^{n-i} = S((k-1)!) \pmod{p},$$

where $S(x) = a_n + a_{n-1}x + \cdots + a_0 x^n$. Hence $p \mid P((p-k)!)$ if and only if $p \mid S((k-1)!)$. Note that $S((k-1)!)$ depends only on k. Let $k > 2a_n + 1$. Then, $s = (k-1)!/a_n$ is an integer that is divisible by all primes smaller than k. Hence $S((k-1)!) = a_n b_k$ for some $b_k \equiv 1 \pmod{s}$. It follows that b_k is divisible only by primes larger than k. For large enough k we have $|b_k| > 1$. Thus for every prime divisor p of b_k we have $p \mid P((p-k)!)$.

It remains to select a large enough k for which $|P((p-k)!)| > p$. We take $k = (q-1)!$, where q is a large prime. All the numbers $k+i$ for $i = 1, 2, \ldots, q-1$ are composite (by Wilson's theorem, $q \mid k+1$). Thus $p = k+q+r$, for some $r \geq 0$. We now have $|P((p-k)!)| = |P((q+r)!)| > (q+r)! > (q-1)!+q+r = p$, for large enough q, since $n = \deg P \geq 2$. This completes the proof.

Remark. The above solution actually also works for all linear polynomials P other than $P(x) = x + a_0$. Nevertheless, these particular cases are easily handled. If $|a_0| > 1$, then $P(m!)$ is composite for $m > |a_0|$, whereas $P(x) = x+1$ and $P(x) = x-1$ are both composite for, say, $x = 5!$. Thus the condition $n \geq 2$ was redundant.

4.47 Solutions to the Shortlisted Problems of IMO 2006

1. If $a_0 \geq 0$ then $a_i \geq 0$ for each i and $[a_{i+1}] \leq a_{i+1} = [a_i]\{a_i\} < [a_i]$ unless $[a_i] = 0$. Eventually 0 appears in the sequence $[a_i]$ and all subsequent a_k's are 0. Now suppose that $a_0 < 0$; then all $a_i \leq 0$. Suppose that the sequence never reaches 0. Then $[a_i] \leq -1$ and so $1 + [a_{i+1}] > a_{i+1} = [a_i]\{a_i\} > [a_i]$, so the sequence $[a_i]$ is nondecreasing and hence must be constant from some term on: $[a_i] = c < 0$ for $i \geq n$. The defining formula becomes $a_{i+1} = c\{a_i\} = c(a_i - c)$, which is equivalent to $b_{i+1} = cb_i$, where $b_i = a_i - \frac{c^2}{c-1}$. Since (b_i) is bounded, we must have either $c = -1$, in which case $a_{i+1} = -a_i - 1$ and hence $a_{i+2} = a_i$, or $b_i = 0$ and thus $a_i = \frac{c^2}{c-1}$ for all $i \geq n$.

2. We use induction on n. We have $a_1 = 1/2$; assume that $n \geq 1$ and $a_1, \ldots, a_n > 0$. The formula gives us $(n+1)\sum_{k=1}^{m} \frac{a_k}{m-k+1} = 1$. Writing this equation for n and $n+1$ and subtracting yields

$$(n+2)a_{n+1} = \sum_{k=1}^{n} \left(\frac{n+1}{n-k+1} - \frac{n+2}{n-k+2} \right) a_k,$$

which is positive, as is the coefficient at each a_k.

Remark. Using techniques from complex analysis such as contour integrals, one can obtain the following formula for $n \geq 1$:

$$a_n = \int_1^\infty \frac{dx}{x^n(\pi^2 + \ln^2(x-1))} > 0.$$

3. We know that $c_n = \frac{\phi^{n-1} - \psi^{n-1}}{\phi - \psi}$, where $\phi = \frac{1+\sqrt{5}}{2}$ and $\psi = \frac{1-\sqrt{5}}{2}$ are the roots of $t^2 - t - 1$. Since $c_{n-1}/c_n \to -\psi$, taking $\alpha = \psi$ and $\beta = 1$ is a natural choice. For every finite set $J \subseteq \mathbb{N}$ we have

$$-1 = \sum_{n=0}^{\infty} \psi^{2n+1} < \psi x + y = \sum_{j \in J} \psi^{j-1} < \sum_{n=0}^{\infty} \psi^{2n} = \phi.$$

Thus $m = -1$ and $M = \phi$ is an appropriate choice. We now prove that this choice has the desired properties by showing that for any $x, y \in \mathbb{N}$ with $-1 < K = x\psi + y < \phi$, there is a finite set $J \subset \mathbb{N}$ such that $K = \sum_{j \in J} \psi^j$. Given such K, there are sequences $i_1 \leq \cdots \leq i_k$ with $\psi^{i_1} + \cdots + \psi^{i_k} = K$ (one such sequence consists of y zeros and x ones). Consider all such sequences of minimum length n. Since $\psi^m + \psi^{m+1} = \psi^{m+2}$, these sequences contain no two consecutive integers. Order such sequences as follows: If $i_k = j_k$ for $1 \leq k \leq t$ and $i_t < j_t$, then $(i_r) \prec (j_r)$. Consider the smallest sequence $(i_r)_{r=1}^n$ in this ordering. We claim that its terms are distinct. Since $2\psi^2 = 1 + \psi^3$, replacing two equal terms m, m by $m-2, m+1$ for $m \geq 2$ would yield a smaller sequence, so only 0 or 1 can repeat among the i_r. But $i_t = i_{t+1} = 0$ implies $\sum_r \psi^{i_r} > 2 + \sum_{k=0}^{\infty} \psi^{2k+3} = \phi$, while $i_t = i_{t+1} = 1$ similarly implies $\sum_r \psi^{i_r} < -1$, so both cases are impossible, proving our claim. Thus $J = \{i_1, \ldots, i_n\}$ is a required set.

4. Since $\frac{ab}{a+b} = \frac{1}{4}\left(a+b - \frac{(a-b)^2}{a+b}\right)$, the left hand side of the desired inequality equals

$$A = \sum_{i<j} \frac{a_i a_j}{a_i + a_j} = \frac{n-1}{4}\sum_k a_k - \frac{1}{4}\sum_{i<j} \frac{(a_i - a_j)^2}{a_i + a_j}.$$

The righthand side of the inequality is equal to

$$B = \frac{n}{2}\frac{\sum a_i a_j}{\sum a_k} = \frac{n-1}{4}\sum_k a_k - \frac{1}{4}\sum_{i<j} \frac{(a_i - a_j)^2}{\sum a_k}.$$

Now $A \le B$ follows from the trivial inequality $\sum \frac{(a_i - a_j)^2}{a_i + a_j} \ge \sum \frac{(a_i - a_j)^2}{\sum a_k}$.

5. Let $x = \sqrt{b} + \sqrt{c} - \sqrt{a}$, $y = \sqrt{c} + \sqrt{a} - \sqrt{b}$, and $z = \sqrt{a} + \sqrt{b} - \sqrt{c}$. All of these numbers are positive because a, b, c are sides of a triangle. Then $b + c - a = x^2 - \frac{1}{2}(x-y)(x-z)$ and

$$\frac{\sqrt{b+c-a}}{\sqrt{b}+\sqrt{c}-\sqrt{a}} = \sqrt{1 - \frac{(x-y)(y-z)}{2x^2}} \le 1 - \frac{(x-y)(x-z)}{4x^2}.$$

Now it is enough to prove that

$$x^{-2}(x-y)(x-z) + y^{-2}(y-z)(y-x) + z^{-2}(z-x)(z-y) \ge 0,$$

which directly follows from Schur's inequality.

6. Assume, without loss of generality, that $a \ge b \ge c$. The lefthand side of the inequality equals $L = (a-b)(b-c)(a-c)(a+b+c)$. From $(a-b)(b-c) \le \frac{1}{4}(a-c)^2$ we get $L \le \frac{1}{4}(a-c)^3|a+b+c|$. The inequality $(a-c)^2 \le 2(a-b)^2 + 2(b-c)$ implies $(a-c)^2 \le \frac{2}{3}[(a-b)^2 + (b-c)^2 + (a-c)^2]$. Therefore

$$L \le \frac{\sqrt{2}}{2}\left(\frac{(a-b)^2 + (b-c)^2 + (a-c)^2}{3}\right)^{3/2}(a+b+c).$$

Finally, the mean inequality gives us

$$L \le \frac{\sqrt{2}}{2}\left(\frac{(a-b)^2 + (b-c)^2 + (a-c)^2 + (a+b+c)^2}{4}\right)^2$$

$$= \frac{9\sqrt{2}}{32}(a^2 + b^2 + c^2)^2.$$

Equality is attained if and only if $a - b = b - c$ and $(a-b)^2 + (b-c)^2 + (a-c)^2 = 3(a+b+c)^2$, which leads to $a = \left(1 + \frac{3}{\sqrt{2}}\right)b$ and $c = \left(1 - \frac{3}{\sqrt{2}}\right)b$. Thus $M = \frac{9\sqrt{2}}{32}$.

Second solution. We have $L = |(a-b)(b-c)(c-a)(a+b+c)|$. Without loss of generality, assume that $a + b + c = 1$ (the case $a + b + c = 0$ is trivial). The monic cubic polynomial with roots $a - b$, $b - c$, and $c - a$ is of the form

$$P(x) = x^3 + qx + r, \quad q = \frac{1}{2} - \frac{3}{2}(a^2 + b^2 + c^2), \quad r = -(a-b)(b-c)(c-a).$$

Then $M^2 = \max r^2 / \left(\frac{1-2q}{3}\right)^4$. Since $P(x)$ has three real roots, its discriminant $(q/3)^3 + (r/2)^2$ must be positive, so $r^2 \geq -\frac{4}{27}q^3$. Thus $M^2 \leq f(q) = -\frac{4}{27}q^3 / \left(\frac{1-2q}{3}\right)^4$. The function f attains its maximum $3^4/2^9$ at $q = -3/2$, so $M \leq \frac{9\sqrt{2}}{32}$. The case of equality is easily computed.

Third solution. Assume that $a^2 + b^2 + c^2 = 1$ and write $u = (a+b+c)/\sqrt{3}$, $v = (a + \varepsilon b + \varepsilon^2 c)/\sqrt{3}$, $w = (a + \varepsilon^2 b + \varepsilon c)/\sqrt{3}$, where $\varepsilon = e^{2\pi i/3}$. Then analogous formulas hold for a, b, c in terms of u, v, w, from which one directly obtains $|u|^2 + |v|^2 + |w|^2 = a^2 + b^2 + c^2 = 1$ and

$$a + b + c = \sqrt{3}u, \quad |a-b| = |v - \varepsilon w|, \quad |a-c| = |v - \varepsilon^2 w|, \quad |b-c| = |v - w|.$$

Thus $L = \sqrt{3}|u||v^3 - w^3| \leq \sqrt{3}|u|(|v|^3 + |w|^3) \leq \sqrt{\frac{3}{2}|u|^2(1 - |u|^2)^3} \leq \frac{9\sqrt{2}}{32}$. It is easy to trace back a, b, c to the equality case.

7. (a) We show that for $n = 2^k$ all lamps will be switched on in $n-1$ steps and off in n steps. For $k = 1$ the statement is true. Suppose it holds for some k and let $n = 2^{k+1}$; define $L = \{L_1, \ldots, L_{2^k}\}$ and $R = \{L_{2^k+1}, \ldots, L_{2^{k+1}}\}$. The first $2^k - 1$ steps are performed without any influence on or from the lamps from R; thus after $2^k - 1$ steps the lamps in L are on and those from R are off. After the 2^kth step, L_{2^k} and L_{2^k+1} are on and the other lamps are off. Notice that from now on, L and R will be symmetric (i.e., L_i and $L_{2^{k+1}-i}$ will have the same state) and will never influence each other. Since R starts with only the leftmost lamp on, in 2^k steps all its lamps will be off. The same will happen to L. There are $2^k + 2^k = 2^{k+1}$ steps in total.

    (b) We claim that for $n = 2^k + 1$ the lamps cannot be switched off. After the first step, only L_1 and L_2 are on. According to (a), after $2^k - 1$ steps all lamps but L_n will be on, so after the 2^kth step all lamps will be off except for L_{n-1} and L_n. Since this position is symmetric to the one after the first step, the procedure will never end.

8. We call a triangle *odd* if it has two odd sides. To any odd isosceles triangle $A_iA_jA_k$ we assign a pair of sides of the 2006-gon. We may assume that $k - j = j - i > 0$ is odd. A side of the 2006-gon is said to *belong* to triangle $A_iA_jA_k$ if it lies on the polygonal line $A_iA_{i+1}\ldots A_k$. At least one of the odd number of sides $A_iA_{i+1}, \ldots, A_{j-1}A_j$ and at least one of the sides $A_jA_{j+1}, \ldots, A_{k-1}A_k$ do not belong to any other odd isosceles triangle; assign those two sides to $\triangle A_iA_jA_k$. This ensures that every two assigned pairs are disjoint; therefore there are at most 1003 odd isosceles triangles.

    An example with 1003 odd isosceles triangles can be attained when the diagonals $A_{2k}A_{2k+2}$ are drawn for $k = 0, \ldots, 1002$, where $A_0 = A_{2006}$.

9. The number $c(P)$ of points inside P is equal to $n - a(P) - b(P)$, where $n = |S|$. Writing $y = 1 - x$, the considered sum becomes

$$\sum_P x^{a(P)} y^{b(P)} (x+y)^{c(P)} = \sum_P \sum_{i=0}^{c(P)} \binom{c(P)}{i} x^{a(P)+i} y^{b(P)+c(P)-i}$$

$$= \sum_P \sum_{k=a(P)}^{a(P)+c(P)} \binom{c(P)}{k-a(P)} x^k y^{n-k}.$$

Here the coefficient at $x^k y^{n-k}$ is the sum $\sum_P \binom{c(P)}{k-a(P)}$, which equals the number of pairs (P,Z) of a convex polygon P and a k-element subset Z of S whose convex hull is P, and is thus equal to $\binom{n}{k}$. Now the required statement immediately follows.

10. Denote by $S_{\mathscr{A}}(R)$ the number of strawberries of arrangement $\mathscr{A}$ inside rectangle R. We write $\mathscr{A} \le \mathscr{B}$ if for every rectangle Q containing the top left corner O we have $S_{\mathscr{B}}(Q) \ge S_{\mathscr{A}}(Q)$. In this ordering, every switch transforms an arrangement to a larger one. Since the number of arrangements is finite, it is enough to prove that whenever $\mathscr{A} < \mathscr{B}$ there is a switch taking $\mathscr{A}$ to $\mathscr{C}$ with $\mathscr{C} \le \mathscr{B}$. Consider the highest row t of the cake that differs in $\mathscr{A}$ and $\mathscr{B}$; let X and Y be the positions of the strawberries in t in $\mathscr{A}$ and $\mathscr{B}$ respectively. Clearly Y is to the left from X and the strawberry of $\mathscr{A}$ in the column of Y is below Y. Now consider the highest strawberry X' of $\mathscr{A}$ below t whose column is between X and Y (including Y). Let s be the row of X'. Now switch X, X' to the other two vertices Z, Z' of the corresponding rectangle, obtaining an arrangement $\mathscr{C}$. We claim that $\mathscr{C} \le \mathscr{B}$. It is enough to verify that $S_{\mathscr{C}}(Q) \le S_{\mathscr{B}}(Q)$ for those rectangles $Q = OMNP$ with N lying inside $XZX'Z'$. Let $Q' = OMN_1 P_1$ be the smallest rectangle containing X. Our choice of s ensures that $S_{\mathscr{C}}(Q) = S_{\mathscr{A}}(Q') \ge S_{\mathscr{B}}(Q') \ge S_{\mathscr{B}}(Q)$, as claimed.

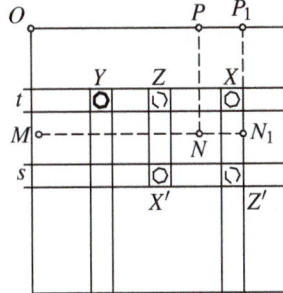

11. Let q be the largest integer such that $2^q \mid n$. We prove that an (n,k)-tournament exists if and only if $k < 2^q$.

The first l rounds of an (n,k)-tournament form an (n,l)-tournament. Thus it is enough to show that an $(n, 2^q - 1)$-tournament exists and an $(n, 2^q)$-tournament does not.

If $n = 2^q$, we can label the contestants and rounds by elements of the additive group $\mathbb{Z}_2^q$. If contestants x and $x + j$ meet in the round labeled j, it is easy to verify the conditions. If $n = 2^q p$, we can divide the contestants into p disjoint groups of 2^q and perform a $(2^q, 2^q - 1)$-tournament in each, thus obtaining an $(n, 2^q - 1)$-tournament.

For the other direction let $\mathscr{G}_i$ be the graph of players with edges between any two players who met in the first i rounds. We claim that the size of each connected component of $\mathscr{G}_i$ is a power of 2. For $i = 1$ this is obvious; assume that it holds for i. Suppose that the components C and D merge in the $(i+1)$th round. Then some $c \in C$ and $d \in D$ meet in this round. Moreover, each player in C meets a player in D. Indeed, for every $c' \in C$ there is a path $c = c_0, c_1, \ldots, c_k = c'$ with $c_j c_{j+1} \in \mathcal{G}_i$; then if d_j is the opponent of c_j in the $(i+1)$th round, condition (ii) shows that each $d_j d_{j+1}$ belongs to $\mathcal{G}_i$, so $d_k \in D$. Analogously, all players in D meet players in C, so $|C| = |D|$, proving our claim. Now if there are 2^q rounds, every component has size at least $2^q + 1$ and is thus divisible by 2^{q+1}, which is impossible if $2^{q+1} \nmid n$.

12. Let U and D be the sets of upward and downward unit triangles, respectively. Two triangles are *neighbors* if they form a diamond. For $A \subseteq D$, denote by $F(A)$ the set of neighbors of the elements of A.

    If a holey triangle can be tiled with diamonds, in every upward triangle of side l there are l^2 elements of D, so there must be at least as many elements of U and at most l holes.

    Now we pass to the other direction. It is enough to show the condition (ii) of the marriage theorem: For every set $X \subset D$ we have $|F(X)| \geq |X|$. Assume the contrary, that $|F(X)| < |X|$ for some set X. Note that any two elements of D with a common neighbor must share a vertex; this means that we can focus on connected sets X. Consider an upward triangle of side 3. It contains three elements of D; if two of them are in X, adding the third one to X increases $F(X)$ by at most 1, so $|F(X)| < |X|$ still holds. Continuing this procedure, we will end up with a set X forming an upward subtriangle of T and satisfying $|F(X)| < |X|$, which contradicts the conditions of the problem. This contradiction proves that $|F(X)| \geq |X|$ for every set X, and an application of the Hall's marriage theorem establishes the result.

13. Consider a polyhedron $\mathscr{P}$ with v vertices, e edges, and f faces. Consider the map σ to the unit sphere S taking each vertex, edge, or face x of $\mathscr{P}$ to the set of outward unit normal vectors (i.e., points on S) to the support planes of $\mathscr{P}$ containing x. Thus σ maps faces to points on S, edges to shorter arcs of big circles connecting some pairs of these points, and vertices to spherical regions formed by these arcs. These points, arcs, and regions on S form a "spherical polyhedron" $\mathscr{G}$.

    We now translate the conditions of the problem into the language of $\mathscr{G}$. Denote by $\bar{x}$ the image of x through reflection with the center in the center of S. No edge of $\mathscr{P}$ being parallel to another edge or face means that the big circle of any edge e of $\mathscr{G}$ does not contain any vertex V nonincident to e. Also note that vertices A and B of $\mathscr{P}$ are antipodal if and only if $\sigma(A)$ and $\overline{\sigma(B)}$ intersect, and that the midpoints of edges a and b are antipodal if and only if $\sigma(a)$ and $\overline{\sigma(b)}$ intersect. Consider the union $\mathscr{F}$ of $\mathscr{G}$ and $\overline{\mathscr{G}}$. The faces of $\mathscr{F}$ are the intersections of faces of $\mathscr{G}$ and $\overline{\mathscr{G}}$, so their number equals $2A$. Similarly, the edges of $\mathscr{G}$ and $\overline{\mathscr{G}}$ have $2B$ intersections, so $\mathscr{F}$ has $2e + 4B$ edges and $2f + 2B$ vertices. Now Euler's theorem for $\mathscr{F}$ gives us $2e + 4B + 2 = 2A + 2f + 2B$, and therefore $A - B = e - f + 1$.

14. The condition of the problem implies that $\angle PBC + \angle PCB = 90° - \alpha/2$, i.e., $\angle BPC = 90° + \alpha/2 = \angle BIC$. Thus P lies on the circumcircle ω of $\triangle BCI$. It is

well known that the center M of ω is the second intersection of AI with the circumcircle of $\triangle ABC$. Therefore $AP \geq AM - MP = AM - MI = AI$, with equality if and only if $P \equiv I$.

15. The relation $AK/KB = DL/LC$ implies that AD, BC, and KL have a common point O. Moreover, since $\angle APB = 180° - \angle ABC$ and $\angle DQC = 180° - \angle BCD$, line BC is tangent to the circles APB and CQD. These two circles are homothetic with respect to O, so if OP meets circle APB again at P', we have $\angle PQC = \angle PP'B = \angle PBC$, showing that P, Q, B, C lie on a circle.

16. Let the diagonals AC and BD meet at Q and AD and CE meet at R. The quadrilaterals $ABCD$ and $ACDE$ are similar, so $AQ/QC = AR/RD$. Now if AP meets CD at M, Ceva's theorem gives us $\frac{CM}{MD} = \frac{CQ}{QA} \cdot \frac{AR}{RD} = 1$.

17. Let M be the point on AC such that $JM \parallel KL$. It is enough to prove that $AM = 2AL$.

    From $\angle BDA = \alpha$ we obtain that $\angle JDM = 90° - \frac{\alpha}{2} = \angle KLA = \angle JMD$; hence $JM = JD$, and the tangency point of the incircle of $\triangle BCD$ with CD is the midpoint T of segment MD. Therefore, $DM = 2DT = BD + CD - BC = AB - BC + CD$, which gives us

$$AM = AD + DM = AC + AB - BC = 2AL.$$

18. Assume that A_1B_1 and CJ intersect at K. Then JK is parallel and equal to C_1D and $DC_1/C_1J = JK/JB_1 = JB_1/JC = C_1J/JC$, so the right triangles DC_1J and C_1JC are similar; hence $C_1C \perp DJ$. Thus E belongs to CC_1. The points A_1, B_1, and E lie on the circle with diameter CJ. Therefore $\angle DBA_1 = \angle A_1CJ = \angle A_1ED$, implying that BEA_1D is cyclic; hence $\angle A_1EB = 90°$. Likewise, $ADEB_1$ is cyclic because $\angle EB_1A = \angle EJC = \angle EDC_1$, so $\angle AEB_1 = 90°$.

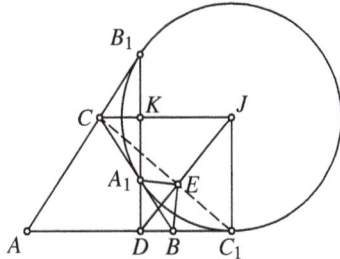

    Second solution. The segments JA_1, JB_1, JC_1 are tangent to the circles with diameters A_1B, AB_1, C_1D. Since $JA_1^2 = JB_1^2 = JC_1^2 = JD \cdot JE$, E lies on the first two circles (with diameters A_1B and AB_1), so $\angle AEB_1 = \angle A_1EB = 90°$.

19. The homothety with center E mapping ω_1 to ω maps D to B, so D lies on BE; analogously, D lies on AF. Let AE and BF meet at point C. The lines BE and AF are altitudes of triangle ABC, so D is the orthocenter and C lies on t. Let the line through D parallel to AB meet AC at M. The centers O_1 and O_2 are the midpoints of DM and DN respectively.

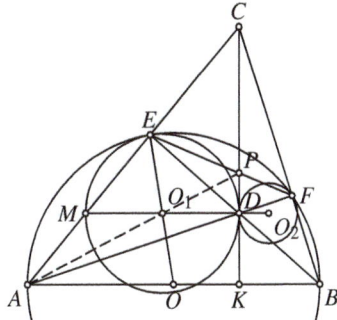

We have thus reduced the problem to a classical triangle geometry problem: If CD and EF intersect at P, we should prove that points A, O_1 and P are collinear (analogously, so are B, O_2, P). By Menelaus's theorem for triangle CDM, this is equivalent to $\frac{CA}{AM} = \frac{CP}{PD}$, which is again equivalent to $\frac{CK}{KD} = \frac{CP}{PD}$ (because $DM \parallel AB$), where K is the foot of the altitude from C to AB. The last equality immediately follows from the fact that the pairs C, D; P, K are harmonically adjoint.

20. Let I be the incenter of $\triangle ABC$. It is well known that $T_a T_c$ and $T_a T_b$ are the perpendicular bisectors of the segments BI and CI respectively. Let $T_a T_b$ meet AC at P and ω_b at U, and let $T_a T_c$ meet AB at Q and ω_c at V. We have $\angle BIQ = \angle IBQ = \angle IBC$, so $IQ \parallel BC$; similarly $IP \parallel BC$. Hence PQ is the line through I parallel to BC.

    The homothety from T_b mapping ω_b to the circumcircle ω of ABC maps the tangent t to ω_b at U to the tangent to ω at T_a that is parallel to BC. It follows that $t \parallel BC$. Let t meet AC at X. Since $XU = XM_b$ and $\angle PUM_b = 90°$, X is the midpoint of PM_b. Similarly, the tangent to ω_c at V meets QM_c at its midpoint Y. But since $XY \parallel PQ \parallel M_b M_c$, points U, X, Y, V are collinear, so t coincides with the common tangent p_a. Thus p_a runs midway between I and $M_b M_c$. Analogous conclusions hold for p_b and p_c, so these three lines form a triangle homothetic to the triangle $M_a M_b M_c$ from center I in ratio $\frac{1}{2}$, which is therefore similar to the triangle ABC in ratio $\frac{1}{4}$.

21. The following proposition is easy to prove:

    Lemma. For an arbitrary point X inside a convex quadrilateral $ABCD$, the circumcircles of triangles ADX and BCX are tangent at X if and only if $\angle ADX + \angle BCX = \angle AXB$.

    Let Q be the second intersection point of the circles ABP and CDP (we assume $Q \neq P$; the opposite case is similarly handled). It follows from the conditions of the problem that Q lies inside quadrilateral $ABCD$ (since $\angle BCP + \angle BAP < 180°$, C is outside the circumcircle of APB; the same holds for D). If Q is inside $\triangle APD$ (the other case is similar), we have $\angle BQC = \angle BQP + \angle PQC = \angle BAP + \angle CDP \leq 90°$. Similarly, $\angle AQD \leq 90°$. Moreover, $\angle ADQ + \angle BCQ = \angle ADP + \angle BCP = \angle APB = \angle AQB$ implies that circles ADQ and BCQ are tangent at Q. Therefore the interiors of the semicircles with diameters AD and BC are disjoint, and if M, N are the midpoints of AD and BC respectively, we have $2MN \geq AD + BC$. On the other hand, $2MN \leq AB + CD$ because $\overrightarrow{BA} + \overrightarrow{CD} = 2\overrightarrow{MN}$, and the statement of the problem immediately follows.

22. We work with oriented angles modulo $180°$. For two lines a, b we denote by $\angle(l, m)$ the angle of counterclockwise rotation transforming a to b; also, by $\angle ABC$ we mean $\angle(BA, BC)$.

    It is well known that the circles AB_1C_1, BC_1A_1, and CA_1B_1 have a common point, say P. Let O be the circumcenter of ABC. Set $\angle PB_1C = \angle PC_1A = \angle PA_1B = \varphi$. Let A_2P, B_2P, C_2P meet the circle ABC again at A_4, B_4, C_4, respectively. Since

$\angle A_4A_2A = \angle PA_2A = \angle PC_1A = \varphi$ and thus $\angle A_4OA = 2\varphi$ etc., $\triangle ABC$ is the image of $\triangle A_4B_4C_4$ under rotation $\mathscr{R}$ about O by 2φ.

Therefore $\angle(AB_4, PC_1) = \angle B_4AB + \angle AC_1P = \varphi - \varphi = 0$, so $AB_4 \parallel PC_1$. Let PC_1 intersect A_4B_4 at C_5; define A_5, B_5 analogously. Then $\angle B_4C_5P = \angle A_4B_4A = \varphi$, so $AB_4C_5C_1$ is an isosceles trapezoid with $BC_3 = AC_1 = B_4C_5$. Similarly, $AC_3 = A_4C_5$, so C_3 is the image of C_5 under $\mathscr{R}$; similar statements hold for A_5, B_5. Thus $\triangle A_3B_3C_3 \cong \triangle A_5B_5C_5$. It remains to show that $\triangle A_5B_5C_5 \sim \triangle A_2B_2C_2$.

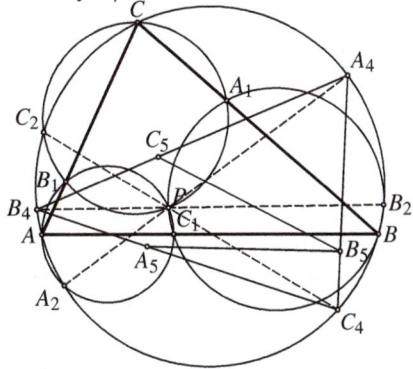

We have seen that $\angle A_4B_5P = \angle B_4C_5P$, which implies that P lies on the circle $A_4B_5C_5$. Analogously, P lies on the circle $C_4A_5B_5$. Therefore

$$\angle A_2B_2C_2 = \angle A_2B_2B_4 + \angle B_4B_2C_2 = \angle A_2A_4B_4 + \angle B_4C_4C_2$$
$$= \angle PA_4C_5 + \angle A_5C_4P = \angle PB_5C_5 + \angle A_5B_5P = \angle A_5B_5C_5,$$

and similarly for the other angles, which is what we wanted.

23. Let S_i be the area assigned to side A_iA_{i+1} of polygon $\mathscr{P} = A_1 \ldots A_n$ of area S. We start with the following auxiliary statement.

   Lemma. At least one of the areas $S_1, \ldots, S_n$ is not smaller than $2S/n$.

   Proof. It suffices to prove the statement for even n. The case of odd n will then follow immediately from this case applied to the degenerate $2n$-gon $A_1A_1' \ldots A_nA_n'$, where A_i' is the midpoint of A_iA_{i+1}.

   Let $n = 2m$. For $i = 1, 2, \ldots, m$, denote by T_i the area of the region $\mathscr{P}_i$ inside the polygon bounded by the diagonals A_iA_{m+i}, $A_{i+1}A_{m+i+1}$ and the sides A_iA_{i+1}, $A_{m+i}A_{m+i+1}$. We observe that the regions $\mathscr{P}_i$ cover the entire polygon. Indeed, let X be an arbitrary point inside the polygon, to the left (without loss of generality) of the ray A_1A_{m+1}. Then X is to the right of the ray $A_{m+1}A_1$, so there is a k such that X is to the left of ray A_kA_{k+m} and to the right of ray $A_{k+1}A_{k+m+1}$, i.e., $X \in \mathscr{P}_k$. It follows that $T_1 + \cdots + T_m \geq S$; hence at least one T_i is not smaller than $2S/n$, say $T_1 \geq 2S/n$.

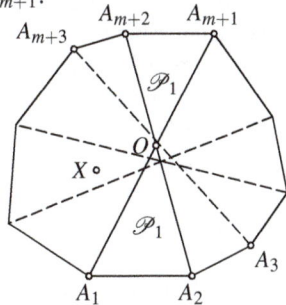

   Let O be the intersection point of A_1A_{m+1} and A_2A_{m+2}, and let us assume without loss of generality that $S_{A_1A_2O} \geq S_{A_{m+1}A_{m+2}O}$ and $A_1O \geq OA_{m+1}$. Then required result now follows from

$$S_1 \geq S_{A_1A_2A_{m+2}} = S_{A_1A_2O} + S_{A_1A_{m+2}O} \geq S_{A_1A_2O} + S_{A_{m+1}A_{m+2}O} = T_1 \geq \frac{2S}{n}.$$

If, contrary to the assertion, $\sum \frac{S_i}{S} < 2$, we can choose rational numbers $q_i = 2m_i/N$ with $N = m_1 + \cdots + m_n$ such that $q_i > S_i/S$. However, considering the given polygon as a degenerate N-gon obtained by division of side A_iA_{i+1} into m_i equal parts for each i and applying the lemma, we obtain $S_i/m_i \geq 2S/N$, i.e., $S_i/S \geq q_i$ for some i, a contradiction.

Equality holds if and only if $\mathscr{P}$ is centrally symmetric.

Second solution. We say that vertex V is assigned to side a of a convex (possibly degenerate) polygon $\mathscr{P}$ if the triangle determined by a and V has the maximum area S_a among the triangles with side a contained in $\mathscr{P}$. Define $\sigma(\mathscr{P}) = \sum_a S_a$ and $\delta(\mathscr{P}) = \sigma(\mathscr{P}) - 2[\mathscr{P}]$. We use induction on the number n of pairwise non-parallel sides of $\mathscr{P}$ to show that $\delta(\mathscr{P}) \geq 0$ for every polygon $\mathscr{P}$. This is obviously true for $n = 2$, so let $n \geq 3$.

There exist two adjacent sides AB and BC whose respective assigned vertices U and V are distinct. Let the lines through U and V parallel to AB and BC respectively intersect at point X. Assume, without loss of generality, that there are no sides of $\mathscr{P}$ lying on UX and VX. Call the sides and vertices of $\mathscr{P}$ lying within the triangle UVX *passive* (excluding vertices U and V). It is easy to see that no passive vertex is assigned to any side of $\mathscr{P}$ and that vertex B is assigned to every passive side. Now replace all passive vertices of $\mathscr{P}$ by X, obtaining a polygon $\mathscr{P}'$. Vertex B is assigned to sides UX and VX of $\mathscr{P}'$. Therefore the sum of areas assigned to passive sides increases by the area S of the part of quadrilateral $BUXV$ lying outside $\mathscr{P}$; the other assigned areas do not change. Thus σ increases by S. On the other hand, the area of the polygon also increases by S, so δ must decrease by S.

Note that the change from $\mathscr{P}$ to $\mathscr{P}'$ decreases the number of nonparallel sides. Thus by the inductive hypothesis we have $\delta(\mathscr{P}) \geq \delta(\mathscr{P}') \geq 0$.

Third solution. To each convex n-gon $\mathscr{P} = A_1A_2\ldots A_n$ we assign a centrally symmetric $2n$-gon $\mathscr{Q}$, called the *associate* of $\mathscr{P}$, as follows. Attach the $2n$ vectors $\pm \overrightarrow{A_iA_{i+1}}$ at a common origin and label them $b_1, \ldots, b_{2n}$ counterclockwise so that $b_{n+i} = -b_i$ for $1 \leq i \leq n$. Then take $\mathscr{Q}$ to be the polygon $B_1B_2\ldots B_{2n}$ with $\overrightarrow{B_iB_{i+1}} = b_i$. Denote by a_i the side of $\mathscr{P}$ corresponding to b_i ($i = 1, \ldots, n$). The distance between the parallel sides B_iB_{i+1} and $B_{n+i}B_{n+i+1}$ of $\mathscr{Q}$ equals twice the maximum height of $\mathscr{P}$ to the side a_i. Thus, if O is the center of $\mathscr{Q}$, the area of $\triangle B_iB_{i+1}O$ ($i = 1, \ldots, n$) is exactly the area S_i assigned to side a_i of $\mathscr{P}$; therefore $[\mathscr{Q}] = 2\sum S_i$. It remains to show that $d(\mathscr{P}) = [\mathscr{Q}] - 4[\mathscr{P}] \geq 0$.

(i) Suppose that $\mathscr{P}$ has two parallel sides a_i and a_j, where $a_j \geq a_i$, and remove from it the parallelogram D determined by a_i and a part of side a_j. We obtain a polygon $\mathscr{P}'$ with a smaller number of nonparallel sides. Then the associate of $\mathscr{P}'$ is obtained from $\mathscr{Q}$ by removing a parallelogram similar to D in ratio 2 (and with area four times that of D); thus $d(\mathscr{P}') = d(\mathscr{P})$.

(ii) Suppose that there is a side b_i ($i \leq n$) of $\mathscr{Q}$ such that the sum of the angles at its endpoints is greater than $180°$. Extend the pairs of sides adjacent to b_i and b_{n+i} to their intersections U and V, thus enlarging $\mathscr{Q}$ by two congruent triangles to a polygon $\mathscr{Q}'$. Then $\mathscr{Q}'$ is the associate of the polygon $\mathscr{P}'$

obtained from $\mathscr{P}$ by attaching a triangle congruent to $B_i B_{i+1} U$ to the side a_i. Therefore $d(\mathscr{P}')$ equals $d(\mathscr{P})$ minus twice the area of the attached triangle.

By repeatedly performing the operations (i) and (ii) to polygon $\mathscr{P}$ we will eventually reduce it to a parallelogram E, thereby decreasing the value of d. Since $d(E) = 0$, it follows that $d(\mathscr{P}) \geq 0$.

Remark. Polygon $\mathscr{Q}$ is the Minkowski sum of $\mathscr{P}$ and a polygon centrally symmetric to $\mathscr{P}$. Thus the inequality $[\mathscr{Q}] \geq 4[\mathscr{P}]$ is a direct consequence of the Brunn–Minkowski inequality.

24. Obviously $x \geq 0$. For $x = 0$ the only solutions are $(0, \pm 2)$. Now let (x, y) be a solution with $x > 0$. Without loss of generality, assume that $y > 0$. The equation rewritten as $2^x(1 + 2^{x+1}) = (y-1)(y+1)$ shows that one of the factors $y \pm 1$ is divisible by 2 but not by 4 and the other by 2^{x-1} but not by 2^x; hence $x \geq 3$. Thus $y = 2^{x-1}m + \varepsilon$, where m is odd and $\varepsilon = \pm 1$. Plugging this in the original equation and simplifying yields

$$2^{x-2}(m^2 - 8) = 1 - \varepsilon m. \qquad (1)$$

Since $m = 1$ is obviously impossible, we have $m \geq 3$ and hence $\varepsilon = -1$. Now (1) gives us $2(m^2 - 8) \leq 1 + m$, implying $m = 3$, which leads to $x = 4$ and $y = 23$. Thus all solutions are $(0, \pm 2)$ and $(4, \pm 23)$.

25. If x is rational, its digits repeat periodically starting at some point. If n is the length of the period of x, the sequence $2, 2^2, 2^3, \ldots$ is eventually periodic modulo n, so the corresponding digits of x (i.e., the digits of y) also make an eventually periodic sequence, implying that y is rational.

26. Consider $g(n) = \left[\frac{n}{1}\right] + \left[\frac{n}{2}\right] + \cdots + \left[\frac{n}{n}\right] = nf(n)$ and define $g(0) = 0$. Since for any k the difference $\left[\frac{n}{k}\right] - \left[\frac{n-1}{k}\right]$ equals 1 if k divides n and 0 otherwise, we obtain that $g(n) = g(n-1) + d(n)$, where $d(n)$ is the number of positive divisors of n. Thus $g(n) = d(1) + d(2) + \cdots + d(n)$ and $f(n)$ is the arithmetic mean of the numbers $d(1), \ldots, d(n)$. Therefore, (a) and (b) will follow if we show that each of $d(n+1) > f(n)$ and $d(n+1) < f(n)$ holds infinitely often. But $d(n+1) < f(n)$ holds whenever $n+1$ is prime, and $d(n+1) > f(n)$ holds whenever $d(n+1) > d(1), \ldots, d(n)$ (which clearly holds for infinitely many n).

27. We first show that every fixed point x of Q is in fact a fixed point of $P \circ P$. Consider the sequence given by $x_0 = x$ and $x_{i+1} = P(x_i)$ for $i \geq 0$. Assume $x_k = x_0$. We know that $u - v$ divides $P(u) - P(v)$ for every two distinct integers u and v. In particular,

$$d_i = x_{i+1} - x_i \mid P(x_{i+1}) - P(x_i) = x_{i+2} - x_{i+1} = d_{i+1}$$

for all i, which together with $d_k = d_0$ implies $|d_0| = |d_1| = \cdots = |d_k|$. Suppose that $d_1 = d_0 = d \neq 0$. Then $d_2 = d$ (otherwise $x_3 = x_1$ and x_0 will never occur in the sequence again). Similarly, $d_3 = d$ etc., and hence $x_i = x_0 + id \neq x_0$ for all i, a contradiction. It follows that $d_1 = -d_0$, so $x_2 = x_0$ as claimed. Thus we can assume that $Q = P \circ P$.

If every integer t with $P(P(t)) = t$ also satisfies $P(t) = t$, the number of solutions is clearly at most $\deg P = n$. Suppose that $P(t_1) = t_2$, $P(t_2) = t_1$, $P(t_3) = t_4$, and $P(t_4) = t_3$, where $t_1 \neq t_{2,3,4}$ (but not necessarily $t_3 \neq t_4$). Since $t_1 - t_3$ divides $t_2 - t_4$ and vice versa, we conclude that $t_1 - t_3 = \pm(t_2 - t_4)$. Assume that $t_1 - t_3 = t_2 - t_4$, i.e. $t_1 - t_2 = t_3 - t_4 = u \neq 0$. Since the relation $t_1 - t_4 = \pm(t_2 - t_3)$ similarly holds, we obtain $t_1 - t_3 + u = \pm(t_1 - t_3 - u)$ which is impossible. Therefore, we must have $t_1 - t_3 = t_4 - t_2$, which gives us $P(t_1) + t_1 = P(t_3) + t_3 = c$ for some c. It follows that all integral solutions t of the equation $P(P(t)) = t$ satisfy $P(t) + t = c$, and hence their number does not exceed n.

28. Every prime divisor p of $\frac{x^7-1}{x-1} = x^6 + \cdots + x + 1$ is congruent to 0 or 1 modulo 7. Indeed, if $p \mid x - 1$, then $\frac{x^7-1}{x-1} \equiv 1 + \cdots + 1 \equiv 7 \pmod{p}$, so $p = 7$; otherwise the order of x modulo p is 7 and hence $p \equiv 1 \pmod 7$. Therefore every positive divisor d of $\frac{x^7-1}{x-1}$ satisfies $d \equiv 0$ or 1 $\pmod 7$.

Now suppose (x,y) is a solution of the given equation. Since $y - 1$ and $y^4 + y^3 + y^2 + y + 1$ divide $\frac{x^7-1}{x-1} = y^5 - 1$, we have $y \equiv 1$ or 2 and $y^4 + y^3 + y^2 + y + 1 \equiv 0$ or 1 $\pmod 7$. However, $y \equiv 1$ or 2 implies that $y^4 + y^3 + y^2 + y + 1 \equiv 5$ or 3 $\pmod 7$, which is impossible.

29. All representations of n in the form $ax + by$ $(x,y \in \mathbb{Z})$ are given by $(x,y) = (x_0 + bt, y_0 - at)$, where x_0, y_0 are fixed and $t \in \mathbb{Z}$ is arbitrary. The following lemma enables us to determine $w(n)$.

Lemma. The equality $w(ax + by) = |x| + |y|$ holds if and only if one of the following conditions holds:
   (i) $\frac{a-b}{2} < y \leq \frac{a+b}{2}$ and $x \geq y - \frac{a+b}{2}$;
   (ii) $-\frac{a-b}{2} \leq y \leq \frac{a-b}{2}$ and $x \in \mathbb{Z}$;
   (iii) $-\frac{a+b}{2} \leq y < -\frac{a-b}{2}$ and $x \leq y + \frac{a+b}{2}$.

Proof. Without loss of generality, assume that $y \geq 0$. We have $w(ax + by) = |x| + y$ if and only if $|x + b| + |y - a| \geq |x| + y$ and $|x - b| + (y + a) \geq |x| + y$, where the latter is obviously true and the former clearly implies $y < a$. Then the former inequality becomes $|x + b| - |x| \geq 2y - a$. We distinguish three cases: if $y \leq \frac{a-b}{2}$, then $2y - a \leq b$ and the previous inequality always holds; for $\frac{a-b}{2} < y \leq \frac{a+b}{2}$, it holds if and only if $x \geq y - \frac{a+b}{2}$; and for $y > \frac{a+b}{2}$, it never holds.

Now let $n = ax + by$ be a local champion with $w(n) = |x| + |y|$. As in the lemma, we distinguish three cases:

   (i) $\frac{a-b}{2} < y \leq \frac{a+b}{2}$. Then $x + 1 \geq y - \frac{a+b}{2}$ by the lemma, so $w(n+a) = |x+1| + y$ (because $n + a = a(x+1) + by$). Since $w(n+a) \leq w(n)$, we must have $x < 0$. Likewise, $w(n-a)$ equals either $|x-1| + y = w(n) + 1$ or $|x+b-1| + a - y$. The condition $w(n - a) \leq w(n)$ leads to $x \leq y - \frac{a+b-1}{2}$; hence $x = y - \lceil \frac{a+b}{2} \rceil$ and $w(n) = \lceil \frac{a+b}{2} \rceil$. Now $w(n - b) = -x + y - 1 = w(n) - 1$ and $w(n + b) = (x + b) + (a - 1 - y) = a + b - 1 - \lceil \frac{a+b}{2} \rceil \leq w(n)$, so n is a local champion. Conversely, every $n = ax + by$ with $\frac{a-b}{2} < y \leq \frac{a+b}{2}$ and $x = y - \lceil \frac{a+b}{2} \rceil$ is a local champion. Thus we obtain $b-1$ local champions, which are all distinct.

(ii) $|y| \le \frac{a-b}{2}$. Now we conclude from the lemma that $w(n-a) = |x-1|+|y|$ and $w(n+a) = |x+1|+|y|$, and at least one of these two values exceeds $w(n) = |x|+|y|$. Thus n is not a local champion.

(iii) $-\frac{a+b}{2} \le y < -\frac{a-b}{2}$. By taking x,y to $-x,-y$ this case is reduced to case (i), so we again have $b-1$ local champions $n = ax+by$ with $x = y + [\frac{a+b}{2}]$.

It is easy to check that the sets of local champions from cases (i) and (iii) coincide if a and b are both odd (so we have $b-1$ local champions in total), and are otherwise disjoint (then we have $2(b-1)$ local champions).

30. We shall show by induction on n that there exists an arbitrarily large m satisfying $2^m \equiv -m \pmod{n}$. The case $n = 1$ is trivial; assume that $n > 1$.

Recall that the sequence of powers of 2 modulo n is eventually periodic with the period dividing $\varphi(n)$; thus $2^x \equiv 2^y$ whenever $x \equiv y \pmod{\varphi(n)}$ and x and y are large enough. Let us consider m of the form $m \equiv -2^k \pmod{n\varphi(n)}$. Then the congruence $2^m \equiv -m \pmod{n}$ is equivalent to $2^m \equiv 2^k \pmod{n}$, and this holds whenever $-2^k \equiv m \equiv k \pmod{\varphi(n)}$ and m,k are large enough. But the existence of m and k is guartanteed by the inductive hypothesis for $\varphi(n)$, so the induction is complete.

4.48 Solutions to the Shortlisted Problems of IMO 2007

1. (a) Assume that $d = d_m$ for some index m, and let k and l be the indices such that $k \leq m \leq l$ such that $d_m = a_k - a_l$. Then $d_m = a_k - a_l \leq (a_k - x_k) + (x_l - a_l)$, and hence $a_k - x_k \geq d/2$ or $x_l - a_l \geq d/2$. The claim follows immediately.

   (b) Let $M_i = \max\{a_j : 1 \leq j \leq i\}$ and $m_i = \min\{a_j : i \leq j \leq n\}$. Set $x_i = \frac{m_i + M_i}{2}$. Clearly, $m_i \leq a_i \leq M_i$ and both (m_i) and (M_i) are nondecreasing. Furthermore, $-\frac{d_i}{2} = \frac{m_i - M_i}{2} = x_i - M_i \leq x_i - a_i$. Similarly $x_i - a_i \leq \frac{d_i}{2}$; hence $\max\{|x_i - a_i| : 1 \leq i \leq n\} \leq \max\left\{\frac{d_i}{2}, 1 \leq i \leq n\right\}$. Thus, the equality holds in (1) for the sequence $\{x_i\}$.

2. Placing $n = 1$ we get $f(m+1) \geq f(m) + f(f(1)) - 1 \geq f(m)$; hence the function is nondecreasing. Let n_0 be the smallest integer such that $f(n_0) > 1$. If $f(n) = n + k$ for some $k, n \geq 1$, then placing $m = 1$ gives that $f(f(n)) = f(n+k) \geq f(k) + f(f(n)) - 1$, which implies $f(k) = 1$. We immediately get $k < n_0$. Choose maximal k_0 such that there exists $n \in \mathbb{N}$ for which $f(n) = n + k_0$. Then we have $2n + k_0 \geq f(2n) \geq f(n) + f(f(n)) - 1 = n + k_0 + f(n+k_0) - 1 \geq n + k_0 + f(n) - 1 = 2n + (2k_0 - 1)$; hence $2k_0 - 1 \leq k_0$, or $k_0 \leq 1$. Therefore $f(n) \leq n + 1$ and $f(2007) \leq 2008$.

   Now we will prove that $f(2007)$ can be any of the numbers $1, 2, \ldots, 2008$. Define the functions

   $$f_j(n) = \begin{cases} 1, & n \leq 2007 - j, \\ n + j - 2007, & \text{otherwise} \end{cases}, \quad \text{for } j \leq 2007, \text{ and}$$

   $$f_{2008}(n) = \begin{cases} n, & 2007 \nmid n, \\ n + 1, & 2007 \mid n. \end{cases}$$

   It is easy to verify that the functions f_j satisfy the conditions of the problem for $j = 1, 2, \ldots, 2008$.

3. The inequality $\frac{1+t^2}{1+t^4} < \frac{1}{t}$ holds for all $t \in (0, 1)$ because it is equivalent to $0 < t^4 - t^3 - t + 1 = (1 - t)(1 - t^3)$. Applying it to $t = x^k$ and summing over $k = 1, \ldots, n$; we get $\sum_{k=1}^{n} \frac{1+x^{2k}}{1+x^{4k}} < \sum_{k=1}^{n} \frac{1}{x^k} = \frac{x^n - 1}{x^n(x-1)} = \frac{y^n}{x^n(1-x)}$. Writing the same relation for y and multiplying by this one gives the desired inequality.

4. Notice that $f(x) > x$ for all x. Indeed, $f(x + f(y)) \neq f(x+y)$ and if $f(y) < y$ for some y, setting $x = y - f(y)$ yields a contradiction.

   Now we will prove that $f(x) - x$ is injective. If we assume that $f(x) - x = f(y) - y$ for some $x \neq y$, we will have $x + f(y) = y + f(x)$, hence $f(x+y) + f(y) = f(x+y) + f(x)$, implying $f(x) = f(y)$, which is impossible. From the functional equation we conclude that $f(f(x) + f(y)) - (f(x) + f(y)) = f(x+y)$, hence $f(x) + f(y) = f(x') + f(y')$ whenever $x + y = x' + y'$. In particular, we have $f(x) + f(y) = 2f(\frac{x+y}{2})$.

   Our next goal is to prove that f is injective. If $f(x) = f(x + h)$ for some $h > 0$ then $f(x) + f(x + 2h) = 2f(x + h) = 2f(x)$; hence $f(x) = f(x + 2h)$, and by induction $f(x+nh) = f(x)$. Therefore, $0 < f(x+nh) - (x+nh) = f(x) - x - nh$ for every n, which is impossible.

We now have $f(f(x) + f(y)) = f(f(x) + y) + f(y) = 2f(\frac{f(x)}{2} + y)$ and by symmetry $f(f(x) + f(y)) = 2f(\frac{f(y)}{2} + x)$. Hence $\frac{f(x)}{2} + y = \frac{f(y)}{2} + x$, and thus $f(x) - 2x = c$ for some $c \in \mathbb{R}$. The functional equation forces $c = 0$. It is easy to verify that $f(x) = 2x$ satisfies the given relation.

5. Defining $a(0)$ to be 0, the relations in the problem continue to hold. It follows by induction that $a(n_1 + n_2 + \cdots + n_k) \leq 2a(n_1) + 2^2 a(n_2) + \cdots + 2^k a(n_k)$. We also have $a(n_1 + n_2 + \cdots + n_{2^i}) \leq 2a(n_1 + \cdots + n_{2^{i-1}}) + 2a(n_{2^{i-1}+1} + \cdots + n_{2^i}) \leq \cdots \leq 2^i(a(n_1) + \cdots + a(n_{2^i}))$. For integer $k \in [2^i, 2^{i+1})$ we have

$$a(n_1 + \cdots + n_k) = a(n_1 + \cdots + n_k + \underbrace{0 + \cdots + 0}_{2^{i+1} - k})$$
$$\leq 2^{i+1}(a(n_1) + \cdots + a(n_k) + (2^{i+1} - k)a(0))$$
$$\leq 2k(a(n_1) + \cdots + a(n_k)).$$

Assume now that $N \in \mathbb{N}$ is given and let $N = \sum_{i=0}^{K} b_i 2^i$ be its binary representation ($K \in \mathbb{N}$, $b_i \in \{0, 1\}$). For each increasing sequence $(\tau_n)_{n \in \mathbb{N}}$ of integers we have

$$a(N) = a\left(\sum_n \sum_{i=\tau_{n-1}}^{\max\{\tau_n, K\}} b_i 2^i\right) \leq \sum_n 2^n a\left(\sum_{i=\tau_{n-1}}^{\max\{\tau_n, K\}} b_i 2^i\right)$$
$$\leq \sum_n 2^n \cdot 2(\tau_n - \tau_{n-1} + 1) \sum_{i=\tau_{n-1}}^{\max\{\tau_n, K\}} a(2^i)$$
$$\leq \sum_n (\tau_n - \tau_{n-1} + 1)^2 \frac{2^{n+1}}{(\tau_{n-1} + 1)^c} \leq \sum_n \frac{2^{n+1} \left(\frac{\tau_n + 1}{\tau_{n-1} + 1}\right)^2}{(\tau_{n-1} + 1)^{c-2}}.$$

Choosing $\tau_n = 2^{\alpha n} - 1$, we get $a(N) \leq 2^{2\alpha} \cdot 2^2 \sum_n 2^{n-1-\alpha(c-2)(n-1)}$. Thus, choosing any $\alpha > \frac{1}{c-2}$ would give us the sequence τ_n for which the last series is bounded, which proves the required statement.

6. Using the Cauchy–Schwarz inequality we can bound the left-hand side in the following way: $\frac{1}{3}[a_1(a_{100}^2 + 2a_1 a_2) + a_2(a_1^2 + 2a_2 a_3) + \cdots + a_{100}(a_{99}^2 + 2a_{100}a_1)] \leq \frac{1}{3}(a_1^2 + \cdots + a_{100}^2)^{1/2} \cdot (\sum_{k=1}^{100}(a_k^2 + 2a_{k+1}a_{k+2})^2)^{1/2}$ (the indices are modulo 100). It suffices to show that

$$\sum_{k=1}^{100} (a_k^2 + 2a_{k+1}a_{k+2})^2 \leq 2.$$

Each term of the last sum can be seen as $a_k^4 + 4a_{k+1}^2 a_{k+2}^2 + 4a_k^2(a_{k+1} \cdot a_{k+2}) \leq (a_k^4 + 2a_k^2 a_{k+1}^2 + 2a_k^2 a_{k+2}^2) + 4a_{k+1}^2 a_{k+2}^2$. The required inequality now follows from $\sum_{k=1}^{100}(a_k^4 + 2a_k^2 a_{k+1}^2 + 2a_k^2 a_{k+2}^2) \leq (a_1^2 + \cdots + a_{100}^2)^2 = 1$, and $\sum_{k=1}^{100} a_k^2 a_{k+1}^2 \leq (a_1^2 + a_3^2 + \cdots + a_{99}^2) \cdot (a_2^2 + a_4^2 + \cdots + a_{100}^2) \leq \frac{1}{4}(a_1^2 + \cdots + a_{100}^2)^2 = \frac{1}{4}$.

7. The union of the planes $x = i$, $y = i$, and $z = i$ for $1 \leq i \leq n$ contains S and doesn't contain 0. Assume now that there exists a collection $\{a_i x + b_i y + c_i z + d_i = 0 : 1 \leq i \leq N\}$ of $N < 3n$ planes with the described properties. Consider the polynomial $P(x,y,z) = \prod_{i=1}^{N}(a_i x + b_i y + c_i z + d_i)$.

Let $\delta_0 = 1$, and choose the numbers $\delta_1, \ldots, \delta_n$ such that $\sum_{i=0}^{n} \delta_i i^m = 0$ for $m = 0, 1, 2, \ldots, n-1$ (here we assume that $0^0 = 1$). The choice of these numbers is possible because the given linear system in $(\delta_1, \ldots, \delta_n)$ has the Vandermonde determinant.

Let $S = \sum_{i=0}^{n} \sum_{j=0}^{n} \sum_{k=0}^{n} \delta_i \delta_j \delta_k P(i,j,k)$. By the construction of P we know that $P(0,0,0) \neq 0$ and $P(i,j,k) = 0$ for all other choices of $i,j,k \in \{0,1,\ldots,n\}$. Therefore $S = \delta_0^3 P(0,0,0)$. On the other hand, expanding P as $P(x,y,z) = \sum_{\alpha+\beta+\gamma \leq N} P_{\alpha,\beta,\gamma} x^\alpha y^\beta z^\gamma$ we get

$$S = \sum_{i=0}^{n} \sum_{j=0}^{n} \sum_{k=0}^{n} \delta_i \delta_j \delta_k \sum_{\alpha+\beta+\gamma \leq N} P_{\alpha,\beta,\gamma} i^\alpha j^\beta k^\gamma$$

$$= \sum_{\alpha+\beta+\gamma \leq N} P_{\alpha,\beta,\gamma} \left(\sum_{i=0}^{n} \delta_i i^\alpha \right) \left(\sum_{j=0}^{n} \delta_j j^\beta \right) \left(\sum_{k=0}^{n} \delta_k k^\gamma \right) = 0,$$

because for every choice of α, β, γ at least one of them is less than n, making the corresponding sum in the last expression equal to 0. This is a contradiction; hence the required number of planes is $3n$.

8. Let $S_k^m = a_k + a_{k+1} + \cdots + a_m$. Since $S_1^n < S_{n+1}^{2n} < \cdots < S_{n^2+1}^{n^2+n}$ and since each of these $n+1$ numbers belongs to $\{0,1,\ldots,n\}$, we have that $S_{in+1}^{(i+1)n} = i$. We immediately get $a_1 = a_2 = \cdots = a_n = 0$ and $a_{n^2+1} = a_{n^2+2} = \cdots = a_{n^2+n} = 1$. For every $0 \leq k \leq n$, consider the sequence $l_k = \left(S_{k+1}^{k+n}, S_{k+n+1}^{k+2n}, \ldots, S_{k+n^2-n+1}^{k+n^2} \right)$. The sequence is strictly increasing and its elements are from the set $\{0,1,2,\ldots,n\}$. Let m be the number that doesn't appear in l_k. Define U_k to be the total sum of elements of l_k. Since $a_1 + \cdots + a_{n^2+n} = S_1^n + S_{n+1}^{2n} + \cdots + S_{n^2+1}^{n^2+n} = \frac{n(n+1)}{2} = S_1^k + U_k + S_{k+n^2+1}^{n^2+n} = U_k + n - k$, we get $m = n - k$. Therefore

$$S_{k+sn+1}^{k+(s+1)n} = \begin{cases} s, & \text{if } s < n-k, \\ s+1, & \text{if } s \geq n-k. \end{cases}$$

Using the previous equality we conclude

$$S_{k+sn+1}^{k+(s+1)n} = S_{k-1+sn+1}^{k-1+(s+1)n} + a_{k+(s+1)n} - a_{k-1+sn+1}$$

$$= S_{k+sn}^{k-1+(s+1)n} + a_{k+(s+1)n} - a_{k+sn},$$

hence for $s+k < n$ we have $S_{k+sn+1}^{k+(s+1)n} = S_{k+sn}^{k-1+(s+1)n} = s$, and for $s+k \geq n+1$ we have $S_{k+sn+1}^{k+(s+1)n} = S_{k+sn}^{k-1+(s+1)n} = s+1$. Hence $a_{k+(s+1)n} = a_{k+sn}$ if either $s+k < n$ or $s+k \geq n+1$. If $s+k = n$, then $S_{k+sn}^{k-1+(s+1)n} = s$, while $S_{k+sn+1}k + (s+1)n = $

$s+1$. Hence $a_{k+(s+1)n} = 1$ and $a_{k+sn} = 0$. Now by induction we can easily get that for $1 \leq u \leq n$ and $0 \leq v \leq n$,

$$a_{u+vn} = \begin{cases} 0, & \text{if } u+v \leq n, \\ 1, & \text{if } u+v > n. \end{cases}$$

It is easy to verify that the above sequence satisfies the required properties.

9. Assume the contrary. Consider the minimal such dissection of the square $ABCD$ (i.e., the dissection with the smallest number of rectangles). No two rectangles in this minimal dissection can share an edge. Let $AMNP$ be the rectangle containing the vertex A, and let $UBVW$ be the rectangle containing B. Assume that $MN \leq BV$. Let $MXYZ$ be another rectangle containing the point M (this one could be the same as $UBVW$). We can have either $MN > MZ$ or $MZ > MN$. In the first case the rectangle containing the point Z would have to touch the side CD (it can't touch BC because $WU \geq NM > MZ$). The line MN doesn't intersect any of the interiors of the rectangles. Contradiction.
If $MZ > MN$, consider the rectangle containing the point N. It can't touch AD because it can't share the entire side with $AMNP$. Hence it has to touch CD, and again, MN would be a line that doesn't intersect any of the interiors. Contradiction.

10. Let $T = \{(x,y,z) \in S \times S \times S : x+y+z \text{ is divisible by } n\}$. For any pair $(x,y) \in S \times S$ there exists unique $z \in S$ such that $(x,y,z) \in T$; hence $|T| = n^2$. Let $M \subseteq T$ be the set of those triples that have all elements of the same color. Denote by R and B the sets of red and blue numbers and assume that the number r of red numbers is not less than $n/2$. Consider the following function $F : T \setminus M \to R \times B$: If $(x,y,z) \in T \setminus M$, then $F(x,y,z)$ is defined to be one of the pairs (x,y), (y,z), (z,x) that belongs to $R \times B$ (there exists exactly one such pair). For each element $(p,q) \in R \times B$ there is unique $s \in S$ for which $n \mid p+q+s$. Then $F(p,q,s) = F(s,p,q) = F(q,s,p) = (p,q)$. Hence $|T \setminus M| = 3|R \times B| = 3r(n-r)$ and $|T| = n^2 - 3r(n-r) = n^2 - 3rn + 3r^2$.
It remains to solve $n^2 - 3nr + 3r^2 = 2007$ in the set $\mathbb{N} \times \mathbb{N}$. First of all, $n = 3k$ for some $k \in \mathbb{N}$. Therefore $9k^2 - 9kr + 3r^2 = 2007$ and we see that $3 \mid r$. Let $r = 3s$. The equation becomes $k^2 - 3kr + 3r^2 = 223$. From our assumption $r \geq n/2$ we get $223 = k^2 - 3kr + 3r^2 = (k-r)(k-2r) + r^2 \leq r^2$. Furthermore, $4 \cdot 223 = (2k-3r)^2 + 3r^2 \geq 3r^2 \geq 3 \cdot 223$. Hence $r \in \{15, 16, 17\}$. For $r = 15$ and $r = 16$, $4 \cdot 223 - 3r^2$ is not a perfect square, and for $r = 17$ we get $(2k-3r)^2 = 25$, hence $2k - 3 \cdot 17 = \pm 5$. Both $k = 28$ and $k = 23$ lead to solutions $(n,r) = (84,51)$ and $(n,r) = (69,51)$.

11. Denote by $a_{k,1}, a_{k,2}, \ldots, a_{k,n}$ the elements of A_k, and let $Q_k = \sum_{i=1}^{n} a_{k,i}^2$. Assume the contrary, that $|a_{k,i}| < n/2$ for all k,i. This means that the number of elements in $\bigcup_{k \in \mathbb{N}} A_k$ is finite. Hence there are different $p,q \in \mathbb{N}$ such that $A_p = A_q$. For any $I \subseteq \{1, 2, \ldots, n\}$, define $S_k(I) = \sum_{i \in I} a_{k,i}$. Let (I_k, J_k) be the partition that was chosen in constructing A_{k+1} from A_k. We have

$$Q_{k+1} - Q_k = \sum_{i \in I_k} ((a_{k,i}+1)^2 - (a_{k,i})^2) + \sum_{j \in J_k} ((a_{k,j}-1)^2 - (a_{k,j})^2)$$

$$= n + 2\left(S_k(I_k) - S_k(J_k)\right) = n - 2\min_{I,J}\{|S_k(I) - S_k(J)|\},$$

where the last minimum is taken over all partitions (I, J) of the set $\{1, 2, \ldots, n\}$. However, for each k, it is easy to inductively build a partition (I', J') for which $|S_k(I') - S_k(J')| < n/2$. Take I_0 and J_0 to be empty sets, and assume we made the partition I_l, J_l of $\{1, \ldots, l\}$ in such a way that $|S_k(I_l) - S_j(I_l)| < n/2$. Now, take

$$(I_{l+1}, J_{l+1}) = \begin{cases} (I_l \cup \{l+1\}, J_l), & \text{if } S(I_l) \le S(J_l), \\ (I_l, J_l \cup \{l+1\}), & \text{if } S(I_l) > S(J_l). \end{cases}$$

Therefore $Q_{k+1} - Q_k > n - 2\frac{n}{2} = 0$ and Q_k is increasing. This contradicts the previously established fact that $A_p = A_q$ for some $p \ne q$.

12. Assume that $a > b$, $a = a_1 d$, $b = b_1 d$, $(a_1, b_1) = 1$. There exist $p, q \in \mathbb{Z}$ such that $pa + qb = d$. We may assume that the pairs (S_n, S_{n+a}) and (S_n, S_{n+b}) are of different colors, since otherwise the statement would follow immediately. By induction we get that $(S_n, S_{n+ua+vb})$ are of the same color if and only if $u + v$ is even. From $ab_1 = ba_1$ we conclude that $S_{ab_1} = S_{ba_1}$, which means that a_1 and b_1 must be of the same parity; hence they are both odd and $a_1 \ge 3$. Furthermore, $pa_1 + qb_1 = 1$ gives $2 \nmid p+q$, which implies that the strips $S_{n+d} = S_{n+pa+qb}$ and S_n are of different colors. Now S_n and S_{n+2d} have the same color.

Consider the rectangle $MNPQ$ such that $MQ = NP = a$, $MN = PQ = b$ and the difference between the x coordinates of M and Q (and consequently N and P) is $2d$. It suffices to show that we can choose a rectangle in such a way that M and N are of the same color. Simple calculation shows that the difference of the x coordinates of M and N is $\tau = \frac{\sqrt{a_1^2-4}}{a_1}b \notin \mathbb{Q}$. Let s be one of the longest single-colored (say red) segments on the x-axis. The translation s' of s to the left by τ has noninteger endpoints, hence it can't be single-colored. Thus, there is a red point on s'; choose this point to be N. Other points are now easily determined.

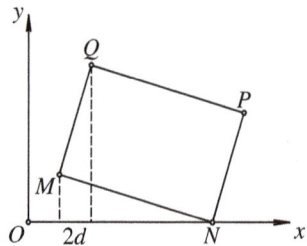

Remark. We used the fact $a_1 > 2$, which is implied by $a \ne b$. The statement doesn't hold for squares (a counterexample is the unit square).

13. Consider one of the cliques of maximal size $2n$ and put its members in a room X. Call these students Π-*students*. Put the others in the room Y. Let $d(X)$ and $d(Y)$ be the maximal sizes of cliques in X and Y at a given moment. If a student moves from X to Y, then $d(X) - d(Y)$ decreases by 1 or 2. Repeating this procedure, we can make this difference 0 or -1.

Assume that it is -1 and $d(X) = l$, $d(Y) = l+1$. If the room Y contains a Π-student that doesn't belong to a Y-clique of size $l+1$, after moving that student to X we will manage to have $d(X) = d(Y)$. Therefore assume that all $2n - l$ Π-students in Y belong to all Y-cliques of size $l+1$. Each such clique has to contain $2(l-n)+1 \geq 1$ non-Π-students. Take an arbitrary clique in Y of size $l+1$ and move a non-Π-student from it to X. Repeat this procedure as long as there are $l+1$ cliques in Y. We claim that $d(X)$ remains l after each such move. If not, consider an $(l+1)$-clique in X. All its members would know all of $2n - l$ Π-students in Y. Together with them they would form a $2n+1$ clique, which is impossible. Hence we will end up with the configuration with $d(X) = d(Y) = l$.

14. Assume that $A_1, \ldots, A_k$ are disjoint m-element subsets of $\{1, \ldots, n\}$. Let

$$\mathscr{S} = \{S \subseteq \{1, \ldots, n\} : S \cap A_i \neq \emptyset \text{ for all } i\} \quad \text{and}$$
$$\mathscr{T} = \{T \subseteq \{1, \ldots, n\} : T \supseteq A_i \text{ for some } i, \text{ but } T \cap A_j = \emptyset \text{ for some } j\}.$$

For each $A \in \mathscr{S}$ and $B \in \mathscr{T}$ we have $A \cap B \neq \emptyset$. It suffices to prove:

Lemma. For $k = k(m) = \left\lceil 2^m \cdot \log \frac{3+\sqrt{5}}{2} \right\rceil$ and $n = mk$ we have

$$\lim_{m \to \infty} \frac{|\mathscr{S}|}{2^n} = \lim_{m \to \infty} \frac{|\mathscr{T}|}{2^n} = \frac{3 - \sqrt{5}}{2}.$$

Proof. For simplicity set $\rho = \log \frac{3+\sqrt{5}}{2}$. For every i, each set in $\mathscr{S}$ must contain one of $2^m - 1$ nonempty subsets of A_i. Hence $|\mathscr{S}| = (2^m - 1)^k$. Using the substitution $r = 2^m$, we get

$$\lim_{m \to \infty} \frac{|\mathscr{S}|}{2^{km}} = \lim_{m \to \infty} \frac{(2^m - 1)^k}{2^{mk}} = \lim_{r \to \infty} \frac{(r-1)^{\rho r}}{r^{\rho r}}$$
$$= \lim_{r \to \infty} \left(1 - \frac{1}{r}\right)^{\rho r} = e^{-\rho} = \frac{3 - \sqrt{5}}{2}.$$

In order to calculate $|\mathscr{T}|$, notice that $\mathscr{T} = \mathscr{U} \setminus (\mathscr{U} \cap \mathscr{S})$ where $\mathscr{U} = \{T \subseteq \{1, \ldots, n\} : T \supseteq A_i \text{ for some } i\}$. Obviously, $|\mathscr{U}^c| = (2^m - 1)^k$, which gives us $|\mathscr{U}| = 2^{mk} - (2^m - 1)^k$. Furthermore, $|\mathscr{U} \cap \mathscr{S}| = |\mathscr{S}| - |\mathscr{S} \setminus \mathscr{U}|$. From $\mathscr{S} \setminus \mathscr{U} = \{T \subseteq \{1, \ldots, n\} : T \cap A_i \neq \emptyset \text{ and } T \cap A_i \neq A_i \text{ for all } i\}$ we get $|\mathscr{S} \setminus \mathscr{U}| = (2^m - 2)^k$. Using the substitution $r = 2^m$, we get

$$\lim_{m \to \infty} \frac{|\mathscr{T}|}{2^n} = \lim_{m \to \infty} \frac{2^{km} - 2(2^m - 1)^k + (2^m - 2)^k}{2^{mk}}$$
$$= 1 - 2e^{-\rho} + \lim_{r \to \infty} \frac{(r-2)^{\rho r}}{r^{\rho r}} = 1 - 2e^{-\rho} + \lim_{r \to \infty} \left(1 - \frac{2}{r}\right)^{\rho r}$$
$$= 1 - 2e^{-\rho} + e^{-2\rho} = e^{-\rho} = \frac{3 - \sqrt{5}}{2}.$$

15. For each vertex V of P consider all good triangles whose one vertex is V. All the vertices of these triangles belong to the unit circle centered at V. Label them counterclockwise as $V_1, \ldots, V_i$. Denote by $f(V)$ and $l(V)$ the first and the last of

these vertices $(f(V) = V_1, l(V) = V_i)$. Denote by $T_f(V)$ the good triangle with vertices V and $f(V)$. $T_l(V)$ is defined analogously. We call $T_f(V)$ and $T_l(V)$ the triangles *associated* with V (they might be the same). Let α be the total number of associated triangles, and t the total number of good triangles. It is enough to prove that each good triangle is associated with at least three vertices. Indeed this would imply $3t \leq \alpha \leq 2n$. It suffices to show that for an arbitrary good triangle ABC oriented counterclockwise we have $A = l(B)$ or $A = f(C)$. Assume that $A \neq l(B)$ and $A \neq f(C)$. Then $l(B)$ and $f(C)$ would belong to the half-plane $[BC, A$. Define $A' = l(B)$ if $\angle ABl(B) \leq 60°$. If this angle is bigger than 60°, define A' to be the third vertex of $T_l(B)$. We define A'' similarly. We have $\angle ABA' < 60°$, $\angle ACA'' < 60°$, hence A belongs to the interior of the rectangle $A'BCA''$ and P can't be convex. This concludes the proof of our claim.

Remark. It is easy to refine the proof to show that $t \leq [\frac{2}{3}(n-1)]$. This result is sharp and the example of a $(3k+1)$-gon with $2k$ good triangles is not hard to construct: rotate a rhombus $ABCD$ ($AB = BC = DA = 1$) around A by small angles k times.

16. Let O be the circumcenter of the triangle ABC. We know that $\angle CPK = \angle CQL$, and hence $\frac{S_{RPK}}{S_{RQL}} = \frac{RP \cdot PK}{RQ \cdot QL}$. Since $\triangle PKC \sim \triangle QLC$, we have $\frac{PK}{QL} = \frac{PC}{QC}$. Since ROC is isosceles and $\angle OPR = \angle OQC$, we get $\triangle ROQ \cong \triangle COP$ and $RQ = PC$. This finally implies $\frac{S_{RPK}}{S_{RQL}} = 1$.

17. Let Y be the midpoint of BT. Then $MY \| CT$ and $TY \perp XY$; hence T, Y, M, X belong to a circle. Thus $\angle MTB - \angle CTM = \angle MXY - \angle YMT = \angle MXY - \angle TXY = \angle MXY - \angle YXB = \angle MXB = \angle BAM$.

18. Let X be the point on the line PQ such that $XC \| AQ$. Then $XC : AQ = CP : PA = BC : AD$, which implies $\triangle BCX \sim \triangle DAQ$. Hence $\angle DAQ = \angle BCX$ and $\angle BXC = \angle DQA = \angle BQC$. Therefore B, C, Q, X belong to a circle, which implies that $\angle BCX = \angle BQX = \angle BQP$.

19. Let K and L be the midpoints of FC and CG respectively. Then $EK \perp CD$ and $EL \perp BC$ hence KL is the Simson's line of the triangle BCD and intersects BD at point M such that $EM \perp BD$. We also have that $KL \| l$ and KL bisects the side CA of $\triangle ACG$. Hence KL passes through the intersection of the diagonals of $ABCD$, and thus KL bisects BD. Therefore $DM = MB$ and DEB is isosceles. From $\triangle DEB \sim \triangle KEL$ we get that $EK = EL$; hence $CF = CG$. Thus $\angle DAF = \angle FGC = \angle GFC = \angle FAB$.

20. Denote by A_0, B_0, and C_0 the given intersection points. Applying Pascal's theorem to the points $APCC'BA'$ gives us that $B_0 \in C_1A_1$. Similarly, we get $C_0 \in A_1B_1$ and $A_0 \in B_1C_1$. From $B_0A_1 \| AB_1$ we get $\frac{C_0B_0}{C_0A} = \frac{C_0A_1}{C_0B_1}$. Since $BA_1 \| B_1A_0$, we get $\frac{C_0A_1}{C_0B_1} = \frac{C_0B}{C_0A_0}$. Hence $\frac{C_0B_0}{C_0A} = \frac{C_0B}{C_0A_0}$ or $C_0B_0 \cdot C_0A_0 = C_0A \cdot C_0B$. Therefore $S_{A_0B_0C_0} = S_{ABC_0}$. However, $S_{ABC_0} = S_{ABB_1}$ (because $A_1B_1 \| AB$); hence $S_{A_0B_0C_0} = S_{ABB_1} = \frac{1}{2} S_{ABC}$.

21. Let us prove that $S_1 \geq S$, i.e., that $k \leq 1$.

Lemma. If X', Y', Z' are the points on the sides YZ, ZX, XY of $\triangle XYZ$, then $S_{X'Y'Z'} \geq \min\{S_{XY'Z'}, S_{YZ'X'}, S_{ZX'Y'}\}$.

Proof. Denote by X_1, Y_1, Z_1 the midpoints of YZ, ZX, XY. If two of X', Y', Z' belong to one of the triangles XY_1Z_1, YZ_1X_1, ZX_1Y_1, then the statement follows immediately. Indeed, if Y', $Z' \in XY_1Z_1$, then $Y'Z'$ intersects the altitude from X to YZ at the point Q inside $\triangle XY_1Z_1$, which forces $d(X, Z'Y') \leq d(X, Z_1Y_1) = \frac{1}{2}d(X, YZ) \leq d(X', Y'Z')$. Assume now, without loss of generality, that $X' \in X_1Z$, $Y' \in Y_1X$, $Z' \in Z_1Y$. Then $d(Z', X'Y') > d(Z_1, X'Y')$, hence $S_{X'Y'Z'} > S_{X'Y'Z_1}$. Similarly $S_{X'Y'Z_1} > S_{X'Y_1Z_1}$. Since $S_{X'Y_1Z_1} = S_{X_1Y_1Z_1}$ we have $S_{X'Y'Z'} > S_{X_1Y_1Z_1} = \frac{1}{4}S_{XYZ}$, hence $S_{X'Y'Z'} > \min\{S_{XY'Z'}, S_{YZ'X'}, S_{ZX'Y'}\}$.

If $S_{A_1B_1C_1} \geq \min\{S_{A_1BB_1}, S_{B_1CC_1}\}$ and $S_{A_1C_1D_1} \geq \min\{S_{C_1DD_1}, S_{D_1AA_1}\}$ then the problem is trivial. The same holds for $S_{A_1B_1D_1}$ and $S_{B_1C_1D_1}$.
Without loss of generality assume that $S_{A_1B_1C_1} < \min\{S_{A_1BB_1}, S_{B_1CC_1}\}$ and $S_{A_1B_1D_1} < \min\{S_{A_1BB_1}, S_{AA_1D_1}\}$. Assume also that $S_{A_1B_1C_1} \leq S_{A_1B_1D_1}$. Then the line C_1D_1 intersects the ray $(BC$ at some point U. The lines AB and CD can't intersect at a point that is on the same side of A_1C_1 as B_1. Otherwise, any line through C_1 that intersects $(BA$ and $(BC$, say at M and N, would force $S_{C_1B_1N} > S_{A_1B_1C_1}$ and $S_{A_1BB_1} > S_{A_1B_1C_1}$. The lemma would imply that $S_{MA_1C_1} \leq S_{A_1B_1C_1}$. This is impossible since we can make $S_{MA_1C_1}$ arbitrarily large. Therefore $C_1D_1 \cap (BA = V$. Applying the lemma to $\triangle VBU$, we get $S_{A_1B_1C_1} \geq S_{VA_1C_1} > S_{A_1C_1D_1}$, a contradiction.
To show that the constant $k = 1$ is the best possible, consider the cases close to the degenerate one in which ACD is a triangle, D_1, C_1 the midpoints of AD and CD, and $B = A_1 = B_1$ the midpoint of AC.

22. We will prove that $\angle KID = \frac{\beta - \gamma}{2}$. Let AA' be the diameter of the circumcircle k of $\triangle ABC$. Denote by M the intersection of AI with k.

Since $A'M \perp AM$, we have that K, M, and A' are collinear. Let A_1, B_1, and X be the feet of perpendiculars from I to BC, CA, and AD. Then $\triangle XIB_1 \sim \triangle A'MB$, since the corresponding angles are equal (this is easy to verify). Since $MB = MI = MC$, we conclude that $IM : KM = BM : MA' = IB_1 : IX = XD : IX$; hence $\triangle KIM \sim \triangle IDX$. Therefore $\angle KID = \angle XIM - (\angle XID + \angle KIM) = \angle XIM - 90° = 180° - \angle AA'M - 90° = \angle MAA' = \frac{\beta - \gamma}{2}$.
Assume now that $IE = IF$. Since $\beta > \gamma$, we have that A_1 belongs to the segment FC and hence $\angle C = \angle DIA_1 + \angle EIB_1 = \angle DIF + 2\angle FIA_1$. This is equivalent to $2\angle FIA_1 = \gamma - \frac{\beta - \gamma}{2}$; thus $\beta < 3\gamma$.

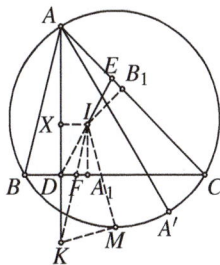

23. Let J be the center of circle k tangent to the lines AB, DA, and BC respectively. Denote by a and b the incircles of $\triangle ADP$ and $\triangle BCP$.

First we will prove that $F \in IJ$. Observe that A is the center of homothethy that maps a to k; K is the center of negative homothethy that maps a to ω; and denote by $\hat{F}$ the center of negative homothethy that maps ω to k. By the consequence of Desargues's theorem we have that A, K, and $\hat{F}$ are collinear. Similarly. we prove that $\hat{F} \in BL$, and therefore $F = \hat{F}$, and

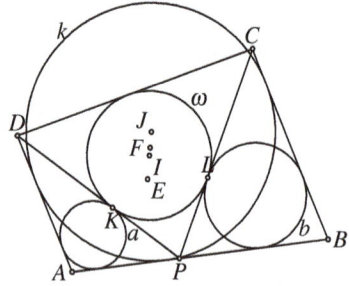

$F \in IJ$. Now we will prove that $E \in IJ$. Denote by X and Y the centers of inversions that map a and b to ω. Comparing the lengths of the tangents from A, B, C, D to the circles k and a, we get that $AP + DC = AD + PC$. Hence there exists a circle d inscribed in $APCD$. Let X be the center of homothethy that maps a to ω. Using the same consequence of Desargues's theorem, we see that A, C, and X are collinear. Consider the circles a, ω, and k again. We use again that A is the center of homothethy that maps a to k. Notice that X is the center of homothethy that maps a to ω. Therefore XA contains the center $\hat{E}$ of homothethy with positive coefficient that maps ω to k. Similarly $\hat{E} \in BX$; hence $\hat{E} = E$ and $E \in IJ$.

24. $k^4 + n^2$ must be even because $7^k - 3^n$ is even. If both k and n are odd, then $k^4 + n^2 \equiv 2 \pmod 4$ while $7^k - 3^n \equiv 7 - 3 \equiv 0 \pmod 4$. Assume that $k = 2k'$ and $n = 2n'$. Then $7^k - 3^n = \frac{7^{k'} - 3^{n'}}{2} \cdot 2(7^{k'} + 3^{n'})$ and $2(7^{k'} + 3^{n'})$ must be a divisor of $2(8k'^4 + 2n'^2)$; hence $7^{k'} + 3^{n'} \leq 8k'^2 + 2n'^2$. It is easy to prove by induction that $7^{k'} > 8k'^4$ for $k' \geq 4$ and $3^{n'} > 2n'^2$ for $n' \geq 1$. Therefore $k' \in \{1, 2, 3\}$.

For $k' = 1$ we must have $7 + 3^{n'} \mid 8 + 2n'^2$. An easy induction gives $7 + 3^{n'} > 8 + 2n'^2$ for $n' \geq 3$. Hence $n' \leq 2$. For $n' = 1$ we get $(k, n) = (2, 2)$, which doesn't satisfy the given conditions; $n' = 2$ implies $(k, n) = (2, 4)$, which is a solution. Assume now that $k' = 2$. Then $|7^k - 3^n| = |7^2 - 3^{n'}| \cdot (7^2 + 3^{n'}) \geq 22 \cdot (49 + 3^{n'}) > 4^4 + 4n'^2$. This contradiction proves that $k' \neq 2$.

If we assume that $k' = 3$, then $|7^k - 3^n| = |7^3 - 3^{n'}| \cdot (7^3 + 3^{n'}) = |343 - 3^{n'}| \cdot (7^3 + 3^{n'}) \geq 100 \cdot (7^3 + 3^{n'}) > 6^4 + 4n'^2$. This is again a contradiction.

Thus $(k, n) = (2, 4)$ is the only solution.

25. Assume that $b = p_1^{\alpha_1} \cdots p_l^{\alpha_l}$, where $p_1, \ldots, p_l$ are prime numbers. Since $b - a_{b^2}^n$ is divisible by b^2 we get that $p_i^{\alpha_i} \mid a_{b^2}^n$ but $p_i^{\alpha_i + 1} \nmid a_{b^2}^n$ for each i. This implies that $n \mid \alpha_i$ for each i; hence b is a complete nth power.

26. A set Z of integers will be called *good* if $47 \nmid a - b + c - d + e$ for any $a, b, c, d, e \in Z$. Notice that the set $G = \{-9, -7, -5, -3, -1, 1, 3, 5, 7, 9\}$ is good. For each integer $k \in \{1, 2, \ldots, 46\}$ the set $G_k = \{x \in X \mid \exists g \in G : kx \equiv g \pmod{47}\}$ is good as well. Indeed, if $a_i \in G_k$ ($1 \leq i \leq 5$) are such that $47 \mid a_1 - a_2 + a_3 - a_4 + a_5$, then $47 \mid ka_1 - ka_2 + ka_3 - ka_4 + ka_5$. There are elements $b_i \in G$ for which $ka_i \equiv b_i \pmod{47}$ which is impossible. Each element of x is contained in exactly 10

of the sets G_k; hence $10|X| = \sum_{i=1}^{46} |A_k|$, and therefore $|A_k| > 2173 > 2007$ for at least one k.

27. The difference of the two binomial coefficients can be written as

$$D = \binom{2^{k+1}}{2^k} - \binom{2^k}{2^{k-1}} = \frac{(2^{k+1})!}{(2^k)! \cdot (2^k)!} - \frac{1}{(2^k)!} \cdot \left(\frac{(2^k)!}{(2^{k-1})!} \right)^2$$

$$= \frac{2^{2^k}}{(2^k)!} \cdot (2^{k+1} - 1)!! - \frac{2^{2^{k-1} \cdot 2}}{(2^k)!} \cdot ((2^k - 1)!!)^2$$

$$= \frac{2^{2^k} \cdot (2^k - 1)!!}{(2^k)!} \cdot P(2^k),$$

for $P(x) = (x+1)(x+3) \cdots (x + 2^k - 1) - (x-1) \cdot (x-3) \cdots (x - 2^k + 1)$. The exponent of 2 in $(2^k)!$ is equal to $2^k - 1$; hence the exponent of 2 in D is bigger by 1 than the exponent of 2 in $P(2^k)$. Since $P(-x) = -P(x)$, we get $P(x) = \sum_{i=1}^{2^{k-1}} c_i x^{2i-1}$. Being the coefficient near x, c_1 satisfies

$$c_1 = 2 \cdot (2^k - 1)!! \cdot \sum_{i=1}^{2^{k-1}} \frac{1}{2i-1} = (2^k - 1)!! \cdot \sum_{i=1}^{2^{k-1}} \left(\frac{1}{2i-1} + \frac{1}{2^k - 2i + 1} \right)$$

$$= 2^k \cdot \sum_{i=1}^{2^{k-1}} \frac{(2^k - 1)!!}{(2i-1)(2^k - 2i + 1)}.$$

Let a_i be the solution of $a_i \cdot (2i - 1) \equiv 1 \pmod{2^k}$. Then

$$\sum_{i=1}^{2^{k-1}} \frac{(2^k - 1)!!}{(2i-1)(2^k - 2i + 1)} \equiv -\sum_{i=1}^{2^{k-1}} (2^k - 1)!! \cdot a_i^2$$

$$= -(2^k - 1)!! \sum_{i=1}^{2^{k-1}} (2i - 1)^2$$

$$= -(2^k - 1)!! \frac{2^{k-1}(2^k + 1)(2^k - 1)}{3}$$

$$\equiv 2^{k-1} \pmod{2^k}.$$

The exponent of 2 in c_1 has to be $2k - 1$. Now $P(2^k) = c_1 \cdot 2^k + 2^{3k} Q(2^k)$ for some polynomial Q. Clearly, $2^{3k-1} \mid P(2^k)$ but $2^{3k} \nmid P(2^k)$. Finally, the exponent of 2 in D is equal to $3k$.

28. Fix a prime number p. Let $d \in \mathbb{N}$ be the smallest number for which $p \mid f(d)$ (it exists because f is surjective). By induction we have $p \mid f(kd)$ for every $k \in \mathbb{N}$. If $p \mid f(x)$ but $d \nmid x$, from the minimality of d we conclude that $x = kd + r$, where $r \in \{1, 2, \ldots, d - 1\}$. Now we have $p \mid f(kd) + f(r)$ hence $p \mid f(r)$, which is impossible. Therefore $d \mid x \Leftrightarrow p \mid f(x)$.

If $x \equiv y \pmod{d}$, let $D > x$ be some number such that $d \mid D$. Then $d \mid (y - x + D)$ hence $p \mid f(y - x + D)$, and $p \mid f(y) + f(D - x)$. Since $p \mid f(x + D - x)$ we get $p \mid f(x) + f(D - x)$. We have obtained $x \equiv y \pmod{d} \Rightarrow f(x) \equiv f(y) \pmod{p}$.

Now assume that $f(x) \equiv f(y)$ (mod p). Taking the same D as above and assuming that $y > x$, we get $0 \equiv f(x) + f(D-x) \equiv f(y) + f(D-x) \equiv f(y+D-x) \equiv f(y-x) + f(D) \equiv f(y-x)$ (mod p). This implies that $d \mid y-x$, and thus $x \equiv y$ (mod d) $\Leftrightarrow f(x) \equiv f(y)$ (mod p).

Now we know that $f(1), \ldots, f(d)$ have different residues modulo p hence $p \geq d$. Since f is surjective there are numbers $x_1, \ldots, x_p$ such that $f(x_1) = 1, \ldots, f(x_p) = p$. They all give different residues modulo p hence $x_1, \ldots, x_p$ must give p distinct residues modulo d, implying $p = d$.

Now we have $p|x \Leftrightarrow p|f(x)$ and $x \equiv y$ (mod p) $\Leftrightarrow f(x) \equiv f(y)$ (mod p) for every $x,y \in \mathbb{N}$ and every prime number p. Since no prime divides 1, we must have $f(1) = 1$. We will prove by induction that $f(n) = n$. Assume that $f(k) = k$ for every $k < n$. If $f(n) > n$, then $f(n) - n + 1 \geq 2$ will have a prime factor p. This is impossible because $f(n) \equiv n - 1 = f(n-1)$ (mod p), and hence $n \equiv n - 1$ (mod p). If $f(n) < n$, let p be a prime factor of $n - f(n) + 1 \geq 2$. Now we have $n \equiv f(n) - 1$ (mod p) and $f(n) \equiv f(f(n) - 1) = f(n) - 1$ (mod p), contradiction. Thus $f(n) = n$ is the only possible solution. It is easy to verify that this f satisfies the given conditions.

29. The statement will follow from the following lemma applied to $x = k$ and $y = 2n$.

    Lemma. Given two positive integers x and y, the number $4xy - 1$ divides the number $(4x^2 - 1)^2$ if and only if $x = y$.

    Proof. If $x = y$ it is obvious that $4xy - 1 \mid (4x^2 - 1)^2$. Assume that there is a pair (x,y) of two distinct positive integers such that $4xy - 1 \mid (4x^2 - 1)$. Choose such a pair for which $2x + y$ is minimal. From $(4y^2 - 1)^2 \equiv (4y^2 - (4xy)^2)^2 \equiv 16y^2(4x^2 - 1)^2 \equiv 0$ (mod $4xy - 1$) we get that (y,x) is such pair as well; hence $2y + x > 2x + y$ and $y > x$.

    Assume that $(4x^2 - 1)^2 = k \cdot (4xy - 1)$. Multiplying $4xy - 1 \equiv -1$ (mod $4x$) by k, we get $(4x^2 - 1)^2 \equiv -k$ (mod $4x$); hence $k = 4xl - 1$ for some positive integer l. However, this means that $4xl - 1 \mid (4x^2 - 1)^2$, and since $y > x$, we must have $l < x$, implying $2l + y < 2x + y$, and this contradicts the minimality of (x,y).

    Remark: Using the same method one can prove the following more general theorem: If $k > 1$ is an integer, then $kab - 1 \mid (ka^2 - 1)^2 \Leftrightarrow a = b$.

30. Denote by $f_i(n)$ the remainder when $v_{p_i}(n)$ is divided by d. Let $f(n) = (f_1(n), \ldots, f_k(n))$. Consider the sequence of integers n_j defined inductively as $n_1 = 1$ and $n_{j+1} = (p_1 \cdots p_k)^{n_j}$. Let us first prove that $v_p(r + lp^m) = v_p(r) + v_p(lp^m)$ for $r < p^m$. This follows from $(r + lp^m)! = (lp^m)! \cdot (lp^m + 1) \cdots (lp^m + r)$ and for each $i < p^m$, the exponent of p in $lp^m + i$ is equal to the exponent of p in i. If $j_1 < j_2 < \cdots < j_u$, the exponent of p_i in each of $n_{j_2}, \ldots, n_{j_u}$ is bigger than n_{j_1}; hence $f_i(n_{j_1} + n_{j_2} + \cdots + n_{j_u}) \equiv f_i(n_{j_1}) + f_i(n_{j_2} + \cdots + n_{j_u})$ (mod d). Continuing by induction, we get $f_i(n_{j_1} + \cdots + n_{j_u}) \equiv f_i(n_{j_1}) + \cdots + f_i(n_{j_u})$ (mod d). Since the range of f has at most $(d+1)^k$ elements, we see that there is an infinite subsequence of n_i on which f is constant. Then for any d elements $n_{l_1}, \ldots, n_{l_d}$ of this subsequence we have $f(n_{l_1} + \cdots + n_{l_d}) \equiv f(n_{l_1}) + \cdots + f(n_{l_d}) \equiv df(n_{l_1}) \equiv (0, 0, \ldots, 0)$ (mod d).

4.49 Solutions to the Shortlisted Problems of IMO 2008

1. For $x = y = z = w$ the functional equation gives $f(x)^2 = f(x^2)$ for all $x \in \mathbb{R}_+$. In particular, $f(1) = 1$. Setting $\sqrt{w}, \sqrt{x}, \sqrt{y}, \sqrt{z}$ in the equation yields

$$\frac{f(w) + f(x)}{f(y) + f(z)} = \frac{w + x}{y + z}, \quad \text{whenever } wx = yz.$$

Choosing $z = 1$, we get $w = y/x$ and $f(y/x) + f(x) = \left(\frac{y}{x} + x\right) \cdot \frac{f(y)+1}{y+1}$. Now if we place $y = x^2$, we get $f(x) = x \cdot \frac{f(x^2)+1}{x^2+1}$, which is equivalent to $(f(x) - x)(f(x) - \frac{1}{x}) = 0$. Assume that there are $x, w \in \mathbb{R}_+ \setminus \{1\}$ such that $f(x) = x$ and $f(w) = \frac{1}{w}$. Choosing $y = z = \sqrt{wx}$ and placing in the equation implies $\frac{1}{w} + x = (w + x) \cdot \frac{f(\sqrt{wx})}{\sqrt{wx}}$. If $f(\sqrt{wx}) = \sqrt{wx}$ then we have $w = 1$. Otherwise, if $f(\sqrt{wx}) = 1/\sqrt{wx}$, we get $x = 1$, contrary to our assumption. Therefore we either have $f(x) = x$ for all $x \in \mathbb{R}_+$ or $f(x) = \frac{1}{x}$ for all $x \in \mathbb{R}_+$. It is easy to verify that both functions satisfy the original equation.

2. (a) Substituting $a = \frac{x}{x-1}, b = \frac{y}{y-1}, c = \frac{z}{z-1}$, the inequality becomes equivalent to $a^2 + b^2 + c^2 \geq 1$, while the constraint becomes $a + b + c = ab + bc + ca + 1$. The last equation is equivalent to $2(a + b + c) = (a + b + c)^2 - (a^2 + b^2 + c^2) + 2 = (a + b + c)^2 + 1 - [(a^2 + b^2 + c^2) - 1]$, or $[(a + b + c) - 1]^2 = (a^2 + b^2 + c^2) - 1$, which immediately implies that $a^2 + b^2 + c^2 \geq 1$.

   (b) Equality holds if and only if $a + b + c = 1$ and $ab + bc + ca = 0$. Expressing $c = 1 - a - b$ yields $-ab + a + b - a^2 - b^2 = 0$. It suffices to prove that there are infinitely many rational numbers a for which the quadratic equation $b^2 + (a - 1)b + a(a - 1) = 0$ has a rational solution b. This equation has a rational solution if and only if its discriminant $(a - 1)^2 - 4a(a - 1) = (1 - a)(1 + 3a)$ is the square of a rational number. We want to find infinitely many rational numbers $\frac{p}{q}$ such that $\left(1 - \frac{p}{q}\right)\left(1 + 3\frac{p}{q}\right)$ is the square of a rational number, which is equivalent to $(q - p)(q + 3p)$ being the square of an integer. However, for each $m, n \in \mathbb{N}$ the system $q - p = (2m + 1)^2$ and $q + 3p = (2n + 1)^2$ has a solution $p = n^2 + n - m^2 - m$, $q = n^2 + n + 3m^2 + 3m + 1$. Keeping m fixed and increasing n would guarantee that we are getting infinitely many different fractions $a = \frac{p}{q}$.

3. (a) Assume that (f, g) is a Spanish couple on $\mathbb{N}$. If $g(a) > g(b)$, then $a > b$ ($a \leq b$ would yield $g(a) \leq g(b)$). Let us introduce the notation

$$g^k(x) = \underbrace{g(g(\cdots g(x) \cdots))}_{k}.$$

If we assume that $g(x) < x$ for some $x \in \mathbb{N}$ we get $g(g(x)) < g(x) < x$, and by induction $(g^k(x))_{k=1}^{\infty}$ is an infinite decreasing sequence from $\mathbb{N}$ which is impossible. Hence $g(x) \geq x$ for all $x \in \mathbb{N}$. The same holds for f. If for some $x \in \mathbb{N}$ we had $f(x) \leq g(x)$, then $g(f(x)) \leq g(g(x)) \leq f(g(g(x)))$, which contradicts (ii). Therefore $f(x) > g(x)$ for all $x \in \mathbb{N}$. From $g(f(x)) > f(g^2(x)) > g(g^2(x))$ we conclude that $f(x) > g^2(x)$, and now easy induction implies $f(x) > g^n(x)$ for all $n \in \mathbb{N}$. This is impossible if $(g^n(x))_{n=1}^{\infty}$ is an infinite increasing sequence. Hence $g(x) = x$ for all x. This can't be true either, because of (ii). This proves that there is no Spanish couple on $\mathbb{N}$.

(b) The functions $f\left(a - \frac{1}{b}\right) = 3a - \frac{1}{b}$ and $g(a,b) = a - \frac{1}{a+b}$ form a Spanish couple on the given set S.

4. Using the given properties we get $f(2) = f(2^0 + 0) = f(2^0 - 3) - f(0) = -1$, $f(3) = f(2+1) = f(2-2) - f(1) = 2$, $f(4) = -2$. For every $i \in \{1,2,3\}$ we have $f(2^n - i) = f(2^{n-1} + 2^{n-1} - i) = f(2^{n-1} - t(2^{n-1} - i)) - f(2^{n-1} - i)$. If $2 \mid n$ then $2^{n-1} - i \equiv 2 - i \equiv -(i+1) \pmod 3$, and if $2 \nmid n$ then $2^{n-1} - i \equiv -(i-1) \pmod 3$. Set $a_k^i = f(2^{2k} - i)$ and $b_k^i = f(2^{2k+1} - i)$. From the previous calculation we get $a_k^i = b_{k-1}^{i+1} - b_{k-1}^i$ and $b_k^i = a_k^{i-1} - a_k^i$. This further implies that $a_k^i = 2a_{k-1}^i - a_{k-1}^{i-1} - a_{k-1}^{i+1}$. By induction we now obtain $a_k^1 = 2 \cdot 3^{k-1}$, $a_k^2 = a_k^3 = -3^{k-1}$. This further gives $b_k^1 = a_k^3 - a_k^1 = -3^k$, $b_k^2 = 3^k$, $b_k^3 = 0$. Moreover, for each $n \in \mathbb{N}$ and each $i \in \{1,2,3\}$: $f(2^n - i) > 0$ if and only if $3 \mid 2^n - i$. If $3 \mid 2^n - i$, then $f(2^n - i) \geq 3^{n/2}$. In addition, $|f(2^n - i)| \leq 2 \cdot 3^{(n-2)/2}$. Assume that $p \geq 1$ is an integer. Let $\alpha_1, \ldots, \alpha_l$ be positive integers such that $3p = 2^{\alpha_1} + \cdots + 2^{\alpha_l}$. In order to prove $f(3p) \geq 0$ let us start with

$$f(3p) = f\left(2^{\alpha_1} - t\left(2^{\alpha_2} + \cdots + 2^{\alpha_l}\right)\right) - f\left(2^{\alpha_2} - t\left(2^{\alpha_3} + \cdots + 2^{\alpha_l}\right)\right)$$
$$+ f\left(2^{\alpha_3} + \cdots + 2^{\alpha_l}\right).$$

Since $x + y \equiv x - t(y) \pmod 3$ the first term on the right-hand side is $\geq 3^{\alpha_1/2}$, while $f\left(2^{\alpha_2} - t(2^{\alpha_3} + \cdots + 2^{\alpha_l})\right) \leq 0$. It suffices to establish $f(2^{\alpha_3} + \cdots + 2^{\alpha_l}) \leq 3^{\alpha_1/2}$. Let us prove the following: $|f(m)| \leq 3^{n/2}$ for all m, n with $m < 2^n$. The last statement is true for $n = 1, 2$. Assume that the statement holds for some n and assume that $m < 2^{n+1}$. If $m < 2^n$ we are done. Otherwise, let $m = 2^n + j$, for some $0 \leq j < 2^n$. Then $|f(m)| = |f(2^n - t(j)) - f(j)| \leq |f(2^n - t(j))| + |f(j)|$. Since $|f(2^n - t(j))| \leq 2 \cdot 3^{(n-2)/2}$ and $|f(j)| \leq 3^{n/2}$ we get $|f(m)| \leq 3^{(n+1)/2}$.

5. Consider the quantity

$$S = 3\left(\frac{a}{b} + \frac{b}{c} + \frac{c}{d} + \frac{d}{a}\right) + \left(\frac{b}{a} + \frac{c}{b} + \frac{d}{c} + \frac{a}{d}\right).$$

Using the inequality between the arithmetic and geometric means, we get

$$S = \left(2\frac{a}{b} + \frac{b}{c} + \frac{a}{d}\right) + \left(2\frac{b}{c} + \frac{c}{d} + \frac{b}{a}\right) + \left(2\frac{c}{d} + \frac{d}{a} + \frac{c}{b}\right) + \left(2\frac{d}{a} + \frac{a}{b} + \frac{d}{c}\right)$$
$$\geq 4\sqrt[4]{\frac{a^3}{bcd}} + 4\sqrt[4]{\frac{b^3}{cda}} + 4\sqrt[4]{\frac{c^3}{dab}} + 4\sqrt[4]{\frac{d^3}{abc}} = 4(a+b+c+d)$$
$$> 3\left(\frac{a}{b} + \frac{b}{c} + \frac{c}{d} + \frac{d}{a}\right) + (a+b+c+d).$$

The required inequality now follows immediately.

6. Assume the contrary, that f is onto. There exists a sequence of numbers a_n such that $f(a_n) = n$.

If $f(u) = f(v)$ then $f(u+1/n) = f(a_n + 1/f(u)) = f(a_n + 1/f(v)) = f(v+1/n)$ for all $n \in \mathbb{N}$, and by induction $f(u+m/n) = f(v+m/n)$ for all $m, n \in \mathbb{N}$. Applying this to $u = a_1$, $v = a_1 + 1/f(a_1 - 1)$, and $n = f(a_1 - 1)$, we get $1 = f(a_1) = f(a_1 + 1/n) = f(a_1 + 2/n) = \cdots = f(a_1 + 1)$. This further implies that $f(a_1 + q + 1) = f(a_1 + q)$ for all rational numbers q.

For fixed y we have

$$\Gamma(y) = \left\{ f\left(y + \frac{1}{n}\right) : n \in \mathbb{N} \right\} = \left\{ f\left(y + \frac{1}{f(x)}\right) : x \in \mathbb{R} \right\}$$
$$= \left\{ f\left(x + \frac{1}{f(y)}\right) : x \in \mathbb{R} \right\} = \mathbb{N}.$$

Particularly, $\Gamma(a_1) = \mathbb{N}$, so we could assume that $a_2, a_3, \ldots$ are chosen from $\Gamma(a_1)$. Hence for each $n \geq 2$ there exists $k_n \in \mathbb{N}$ such that $a_n = a_1 + 1/k_n$. Now we have $f(a_1 + 1/n) = f(a_1 + 1/f(a_n)) = f(a_n + 1) = f(a_1 + \frac{1}{k_n} + 1) = f(a_1 + 1/k_n) = f(a_n) = n$.

Since $\Gamma(a_1 + \frac{1}{3}) = \mathbb{N}$, there exists d such that $f(a_1 + \frac{1}{3} + \frac{1}{d}) = 1$. Assume that $\frac{1}{3} + \frac{1}{d} = \frac{p}{q}$ for relatively prime numbers p and q. Since $\frac{1}{3} + \frac{1}{d} \neq 1$, we have $q > 1$. Let k be an integer such that $kp \equiv 1 \pmod{q}$. Then $1 = f(a_1 + p/q) = f(a_1 + kp/q) = f(a_1 + 1/q) = q$ which is a contradiction.

7. The left-hand side can be rewritten as

$$L = \frac{(a-c)^2}{a+b+c} + \frac{(b-d)^2}{b+c+d}$$
$$+ (a-c)(b-d) \cdot \left(\frac{2d+b}{(b+c+d)(d+a+b)} - \frac{2c+a}{(a+b+c)(c+d+a)} \right).$$

In order to prove that $L \geq 0$, it suffices to establish the following inequality:

$$(a-c)(b-d) \left(\frac{2d+b}{(b+c+d)(d+a+b)} - \frac{2c+a}{(a+b+c)(c+d+a)} \right)$$
$$\geq \frac{-2|(a-c)(b-d)|}{\sqrt{a+b+c} \cdot \sqrt{b+c+d}}.$$

If $a = c$ or $b = d$, the inequality is obvious. Assume that $a > c$ and $b > d$. Our goal is to prove that

$$\left| \frac{2d+b}{(b+c+d)(d+a+b)} - \frac{2c+a}{(a+b+c)(c+d+a)} \right| \leq \frac{2}{\sqrt{a+b+c} \cdot \sqrt{b+c+d}}.$$

Both fractions on the left-hand side are positive hence it is enough to prove that each of them is smaller than the right-hand side. These two inequalities are analogous, so let us prove the first one. After squaring both sides, cross-multiplying, and subtracting, we get

$$-(2d+b)^2(a+b+c)+4(d+a+b)^2(b+c+d)$$
$$= -(2d+b)^2(a+b+c)+[(2d+b)+(2a+b)]^2(b+c+d)$$
$$= -(2d+b)^2\cdot a+(2d+b)^2\cdot d+(2a+b)^2(b+c+d)$$
$$+2\cdot(2d+b)(2a+b)(b+c+d)$$
$$> -(2d+b)^2\cdot a+(2d+b)(2a+b)(2b+2c+2d)$$
$$> -(2d+b)^2\cdot a+(2d+b)^2(2a+b)>0.$$

Equality holds if and only if $a=c$ and $b=d$.

8. The largest such n is equal to 6. Six boxes can be placed in a plane as shown in the picture.

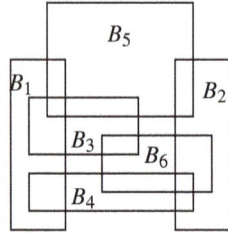

Let us now prove that $n \leq 6$. Denote by X_i and Y_i the projections of the box B_i to the lines Ox and Oy respectively. If $i \neq j \pm 1 \pmod{n}$ then $X_i \cap X_j \neq \emptyset$ and $Y_i \cap Y_j \neq \emptyset$. If $i \equiv j \pm 1 \pmod{n}$, then $X_i \cap X_j = \emptyset$ or $Y_i \cap Y_j = \emptyset$.

We will now prove that there are at most 3 values for i such that $X_i \cap X_{i+1} = \emptyset$. Assume that $X_1 = [a_1, b_1], \ldots, X_n = [a_n, b_n]$. Without loss of generality we may assume that $a_1 = \max\{a_1, a_2, \ldots, a_n\}$. If $b_i < a_1$ for some $i \in \{2, 3, \ldots, n\}$, then $X_i \cap X_1 = \emptyset$, hence $i \in \{2, n\}$. Therefore $a_1 \in X_3 \cap X_4 \cap \cdots \cap X_{n-1}$ and $X_2 \cap X_3$, $X_n \cap X_{n-1}$, $X_1 \cap X_2$, and $X_1 \cap X_n$ are the only possible intersections that could be empty. We will prove that not all of these sets can be empty. Assume the contrary. Then $a_3 \in (b_2, a_1)$, $b_n \in (a_3, a_1)$, and $a_{n-1} \in (b_n, a_1)$. This implies that $a_{n-1} > b_2$ and $X_{n-1} \cap X_2 = \emptyset$, a contradiction.

In a similar way we prove that at most three of the intersections $Y_1 \cap Y_2, \ldots, Y_n \cap Y_1$ are empty. However, there are n empty sets among the intersections $X_1 \cap X_2$, $X_2 \cap X_3, \ldots, X_n \cap X_1, Y_1 \cap Y_2, Y_2 \cap Y_3, \ldots, Y_n \cap Y_1$, yielding $n \leq 3+3 = 6$.

9. Let us call a permutation *nice* if it satisfies the stated property. We want to calculate the number x_n of nice permutations. For $n = 1, 2, 3$ every permutation is nice hence $x_n = n!$ for $n \leq 3$.

Assume now that $n \geq 4$. From $(n-1) \mid 2(a_1 + \cdots + a_{n-1}) = 2[(1 + \cdots + n) - a_n] = n(n+1) - 2a_n = (n+2)(n-1) - 2(a_n - 1)$ we conclude that $(n-1) \mid 2(a_n - 1)$. If n is even we immediately conclude that $a_n = n$ or $a_n = 1$.

Let us prove that $a_n \in \{1, n\}$ for odd n. Assume the contrary. Then $n - 1 = 2(a_n - 1)$, i.e., $a_n = \frac{n+1}{2}$. Then $n - 2 \mid 2(a_1 + \cdots + a_{n-2}) = n(n+1) - 2a_n - 2a_{n-1} = n(n+1) - (n+1) - 2a_{n-1} = (n-2)(n+2) + 3 - 2a_{n-1}$ which gives $n - 2 \mid 2a_{n-1} - 3$. Since $\frac{2a_{n-1}-3}{n-2} \leq \frac{2n-3}{n-2} = 1 + \frac{1}{n-2} < 2$ we get $n - 2 = 2a_{n-1} - 3$. This implies that $a_{n-1} = \frac{n+1}{2} = a_n$, which is a contradiction.

Therefore, for $n \geq 4$ we must have $a_n \in \{1, n\}$. There are x_{n-1} nice permutations for $a_n = n$, and for $a_n = 1$, the problem reduces to counting the nice permutations of the set $\{2, 3, \ldots, n\}$ satisfying the given property. However, since

$2(a_1 + \cdots + a_k) = 2k + 2((a_1 - 1) + \cdots (a_k - 1))$, we get $k \mid 2(a_1 + \cdots + a_k)$ if and only if $k \mid 2[(a_1 - 1) + \cdots + (a_k - 1)]$. This provides a bijection between the nice permutations of $\{2, 3, \ldots, n\}$ and the nice permutations of $\{1, 2, \ldots, n-1\}$. Thus we have $x_n = 2x_{n-1}$ for $n \geq 4$, which implies $x_n = 2^{n-3} \cdot x_3 = 6 \cdot 2^{n-3}$ for $n \geq 4$.

10. Since the area of a triangle ABC is equal to $\frac{1}{2}|\overrightarrow{AB} \times \overrightarrow{AC}|$, we have that $(0,0)$ and (a,b) are k-friends if and only if there exists a point (x,y) such that $ay - bx = \pm 2k$. According to Euclidian algorithm such integers x and y will exist if and only if $\gcd(a,b) \mid 2k$. Similarly, (a,b) and (c,d) are k-friends if and only if $\gcd(c-a, d-b) \mid 2k$.
    Assume that there exists a k-clique S of size $n^2 + 1$ for some $n \geq 1$. Then there are two elements $(a,b), (c,d) \in S$ such that $a \equiv c \pmod{n}$ and $b \equiv d \pmod{n}$. This implies $n \mid \gcd(a-c, b-d) \mid 2k$, or equivalently, $n \mid 2k$.
    Therefore, for a k-clique of size 200 to exist, we must have $n \mid 2k$ for all $n \in \{1, 2, \ldots, 14\}$. Therefore $k \geq 4 \cdot 9 \cdot 5 \cdot 7 \cdot 11 \cdot 13 = 180180$.
    It is easy to see that all lattice points from the square $[0, 14]^2$ are 180180-friends.

11. The number of sequences in which the lamp i is switched (on or off) exactly α_i times $(i = 1, 2, \ldots, 2n)$ is equal to $\frac{k!}{\alpha_1! \cdot \alpha_2! \cdots \alpha_{2n}!}$. Therefore

$$M = k! \cdot \sum \left\{ \frac{1}{\alpha_1! \cdots \alpha_n!} : \alpha_1 + \cdots + \alpha_n = k, 2 \nmid \alpha_1, \ldots, 2 \nmid \alpha_n \right\}. \quad (1)$$

Similarly, we get

$$N = k! \cdot \sum \frac{1}{\alpha_1! \cdots \alpha_n! \cdot \beta_1! \cdots \beta_n!},$$

where the summation is over all possible $\alpha_1, \ldots, \alpha_n, \beta_1, \ldots, \beta_n$ that satisfy $\alpha_1 + \cdots + \alpha_n + \beta_1 + \cdots + \beta_n = k$, $2 \nmid \alpha_1, \ldots, 2 \nmid \alpha_n, 2 \mid \beta_1, \ldots, 2 \mid \beta_n$. We see that the sum in (1) is equal to the coefficient of X^k in the expansion

$$f(X) = \left(X + \frac{X^3}{3!} + \frac{X^5}{5!} + \cdots \right)^n = \sinh^n(X),$$

while the sum in (2) is equal to the coefficient of X^k in the expansion

$$g(X) = \left(X + \frac{X^3}{3!} + \frac{X^5}{5!} + \cdots \right)^n \cdot \left(1 + \frac{X^2}{2!} + \frac{X^4}{4!} + \cdots \right)^n$$

$$= \sinh^n(X) \cdot \cosh^n(X) = \frac{1}{2^n} \sinh^n(2X).$$

Assume that $\sinh^n(X) = \sum_{i=0}^{\infty} a_i X^i$ for some real numbers $a_1, a_2, \ldots$. Then we have $\sinh^n(2X) = \sum_{i=0}^{\infty} a_i (2X)^i = \sum_{i=0}^{\infty} a_i \cdot 2^i \cdot X^i$. Therefore

$$\frac{N}{M} = \frac{k! \cdot a_k}{\frac{1}{2^n} \cdot k! \cdot a_k \cdot 2^k} = 2^{n-k}.$$

12. Let m be the average of all elements from S, i.e. $m = \frac{1}{k+l}(x_1 + x_2 + \cdots + x_{k+l})$.
Define $\Gamma(A) = \frac{1}{k}\sum_{x \in A} x$. Clearly, set A is nice if and only if $|\Gamma(A) - m| \leq \frac{1}{2k}$.
For each permutation $\pi = (\pi_1, \ldots, \pi_{k+l})$ of S consider the sets $A_1^\pi, A_2^\pi, \ldots, A_{k+l}^\pi$
defined as $A_i^\pi = \{\pi_i, \pi_{i+1}, \ldots, \pi_{i+k-1}\}$ (indices are modulo $k+l$). We will prove
that at least two of the sets A_i^π are nice. Let us paint the sets A_i^π red and green in
the following way: A_i^π is green if and only if $\Gamma(A_i^\pi) \geq m$. Notice that $\Gamma(A_i^\pi) - \Gamma(A_{i+1}^\pi) = \frac{1}{k}(\pi_i - \pi_{i+k})$ is of absolute value $\leq \frac{1}{k}$. Therefore $|\Gamma(A_i^\pi) - m| \leq \frac{1}{2k}$
or $|\Gamma(A_{i+1}^\pi) - m| \leq \frac{1}{2k}$. Hence whenever two consecutive sets in the sequence
$A_1^\pi, \ldots, A_{k+l}^\pi$ are of different colors, one of them must be nice. If there are ≥ 2
sets of each of the colors, it is obvious that at least two of the sets will be nice.
Assume that there is only one red set and that it is the only nice set. Without
loss of generality assume that A_1^π is red. Then $m(k+l) = \Gamma(A_1^\pi) + \Gamma(A_2^\pi) + \cdots + \Gamma(A_{k+l}^\pi) > (m - \frac{1}{2k}) + m + \frac{1}{2k} + \cdots + m + \frac{1}{2k} \geq (k+l)m$, which is a contradiction.
Now we can prove the required statement. To each of the $(k+l)!$ permutations
of S we assign at least two nice sets. Each set is counted $(k+l) \cdot k! \cdot l!$ times, so
there are at least $\frac{2(k+l)!}{k+l} \cdot \frac{1}{k! \cdot l!}$ nice sets.

13. We will prove a stronger result, that for each k we have

$$\sum_{i=1}^{k} (|S_i| - (n+1)) \leq (2n+1) \cdot 2^{n-1}. \tag{1}$$

The desired statement follows from (1) because if $|S_i| \geq 2n+2$ for each i, then
$(2n+1) \cdot 2^{n-1} \geq 2^n \cdot (n+1)$ which is impossible.
We will use induction on n to prove (1). First, for $n = 1$ and any subsets
$S_1, S_2, \ldots, S_k$ of $\{1, 2, 3, 4\}$ we want to prove $|S_1| + \cdots + |S_k| \leq 2^0 \cdot 3 + k(1+1) = 3 + 2k$. It suffices to verify this only when $|S_i| \geq 3$ for each i. If there is one set
with four elements, then $k = 1$ and the inequality is satisfied. If all sets have
cardinality 3, then $k \leq 3$ ($\{1,3,4\}$ and $\{2,3,4\}$ can't both be among the chosen
sets), hence $3k \leq 6 + 2k$.
Assume now that the statement is true for $n - 1$. Let us divide the subsets
$S_1, \ldots, S_k$ of $\{1, 2, \ldots, 2^{n+1}\}$ into two families: $\mathscr{A} = \{A_1, \ldots, A_l\}$, those with all
elements greater than 2^n, and $\mathscr{B} = \{B_1, \ldots, B_{k-l}\}$, the remaining subsets. Using
the induction hypothesis we obtain $\sum_{i=1}^{l}(|A_i| - (n+1)) \leq (2n-1) \cdot 2^{n-2} - l$. Let
us denote by α_i the smallest element of $B_i \cap \{2^n + 1, \ldots, 2^{n+1}\}$ if it exists. Let
$H_i = B_i \cap \{1, \ldots, 2^n\}$ and $G_i = B_i \cap \{2^n + 1, \ldots, 2^{n+1}\} \setminus \{\alpha_i\}$. If $i < j$ we claim
that $G_i \cap G_j = \emptyset$. If not, considering $z \in G_i \cap G_j$ and taking $y = \alpha_i$, $x \in H_j$, we
get a contradiction since $x < y < z$ and $x, z \in B_j$, $y, z \in B_i$.
Sets G_i are all disjoint, and the induction hypothesis holds for sets H_i. Hence
$\sum_{i=1}^{k-l}(|B_i| - (n+1)) = \sum_{i=1}^{k-l}(|H_i| - n) + \sum_{i=1}^{k-l}|G_i| \leq (2n-1)2^{n-2} + 2^n$. Thus

$$\sum_{i=1}^{k} (|S_i| - (n+1)) \leq (2n-1) \cdot 2^{n-2} \cdot 2 - l + 2^n \leq (2n+1)2^{n-1}.$$

14. Let A', B', C' be the midpoints of BC, CA, AB, respectively, and let A'', B'', C''
be the midpoints of HA, HB, HC respectively. Let O be the circumcenter of

$\triangle ABC$ and R its circumradius. The Pythagorean theorem implies $OA_1^2 = OA'^2 + A'A_1^2 = OA'^2 + A'H^2$. Since $HA'OA''$ is a parallelogram we have that $OA'^2 + A'H^2 = \frac{1}{2}(OH^2 + A'A''^2)$. However, since $A''A'OA$ is a parallelogram, we have that $A'A'' = OA = R$. Thus $OA_1^2 = \frac{1}{2}(R^2 + OH^2)$. Similar relations for OA_2, OB_1, OB_2, OC_1, OC_2 imply that the points A_1, A_2, B_1, B_2, C_1, C_2 lie on a circle with center O.

15. Assume that the distribution of the points is such that $ABEF$ is a convex quadrilateral and C belongs to the segment BE (other cases are analogous).

    Let $JF = p$, $FI = q$, $IK = r$. Then $KE = q + r$. Let us further define $DI = s$, $IC = t$, $JA = x$, $AB = y$. Since $ABEF$ is cyclic, we have $JA \cdot JB = JF \cdot JE$, i.e., $x(x + y) = p(p + 2q + 2r)$. From $CD \| JB$ we have $\frac{s}{x} = \frac{q}{p}$ and $\frac{t}{x+y} = \frac{q+2r}{p+2q+2r}$. The last three equalities imply that $st = q(q + 2r)$. The quadrilateral $ABKI$ is cyclic if and only if $x(x + y) = (p + q)(p + q + r)$. And $JCKD$ is cyclic if and only if $(p + q)r = st$. We want to prove that

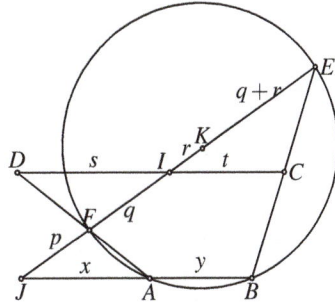

$$x(x + y) = (p + q)(p + q + r) \iff (p + q)r = st.$$

Using the equalities we already have, we can eliminate x, $x + y$, s, and t from the previous equivalence. Hence it suffices to prove that

$$\left[\frac{p(p + 2q + 2r)}{(p + q)(p + q + r)} = 1 \iff \frac{(p + q)r}{q(q + 2r)} = 1\right]$$
$$\iff [p(p + 2q + 2r) = (p + q)(p + q + r) \iff (p + q)r = q(q + 2r)].$$

The last equivalence becomes obvious once we multiply out all the terms.

16. Let us first consider the case $EQ \neq EP$. Assume that $EQ < EP$ and denote by A' and D' the intersections of EA and ED with the circumcircle of $APQD$. Then $\angle PAA' = \angle PAD' + \angle D'AA' = \angle PAD' + \angle D'DA'$, while $\angle QDD' = \angle QDA' + \angle A'DD'$ hence $\angle QDA' = \angle PAD'$. This means that $A'Q = PD'$ and $A'D' \| QP$. Therefore $\angle DEQ = \angle DD'A' = \angle DAA'$; hence QE is a tangent to the circumcircle of $\triangle DAE$. Let M be the intersection of AD and PQ. Then $ME^2 = MD \cdot MA$. Since $APQD$ is cyclic we have that $MD \cdot MA = MQ \cdot MP$; hence $ME^2 = MQ \cdot MP$. Assume that BC intersects PQ at a point N. Then $NE^2 = NQ \cdot NP$, and since there is a unique point X on the line PQ for which $XE^2 = XQ \cdot XP$, we conclude that $M \equiv N$. Now from $MD \cdot MA = ME^2 = MC \cdot MB$ we get that $ABCD$ is cyclic.
    If $EQ = EP$ then it is easy to prove that the perpendicular bisectors of AD, BC, PQ coincide hence $ABCD$ is an isosceles trapezoid, hence it is cyclic.

17. Let M be the intersection point of QF and PE. We need to prove that $\angle QMP = \angle BAC$. Since $\angle MQP = \angle QAB$ (QB is a tangent to the circle around $\triangle QFA$),

it is enough to prove that $\angle QAB + \angle BAC = \angle QMP + \angle MQP$, or equivalently, $\angle QAE = \angle EPC$. Therefore we need to prove that $AQPE$ is a cyclic quadrilateral. From $BQ^2 = BF \cdot BA = BP^2$ we get $BP = BQ$. Adding $BF \cdot BA = BP^2$ to $AF \cdot AB = AE \cdot AC$ (which holds since $BCEF$ is cyclic) we get $AB^2 = AE \cdot AC + BP^2$. From the Pythagorean theorem we have $AB^2 = AE^2 + BE^2 = AE^2 + BC^2 - CE^2$, from which we get $BC^2 - CE^2 = AE \cdot EC + BP^2$. This implies that $BC^2 - BP^2 = CE^2 + AE \cdot EC$, or equivalently

$$CE \cdot (CE + AE) = (BC + BP)(BC - BP) = CQ \cdot CP.$$

Thus $CE \cdot CA = CP \cdot CQ$ and $QPEA$ is cyclic.

18. We will use induction on k. The statement is valid for $k = 0$, since there is at least one point P for which (OP) doesn't intersect any of the lines from L. Assume that the statement holds for $k - 1$. Consider the point O and the line (or one of the lines if there are more) whose distance from O is the smallest. Denote this line by l. This line contains $n - 1$ points from I. We will first prove that there are at least $k + 1$ red points on l. We start by noticing that there exists a point $P \in l \cap I$ such that (OP) doesn't intersect any of the lines from L. Point P divides the line l in two rays; assume that one of them contains the points $P_1, P_2, \ldots, P_u \in I$, while the other ray contains the points $Q_1, \ldots, Q_{n-2-u} \in I$. Assume that P_i's are sorted according to their distance from P, and the same holds for Q_i's. Consider the open segments (OP_i) and (OP_{i+1}). Each line not containing any of P_i and P_{i+1} must intersect either both or none of these segments. The line passing through P_i (other than l) could intersect (OP_{i+1}), and a similar fact holds for the line passing through P_{i+1}. Hence the numbers of intersections of (OP_i) and (OP_{i+1}) with lines from L differ by at most 1. Therefore $P_1, P_2, \ldots, P_{\min\{k,u\}}$ are all red. A similar result holds for Q_i's hence there are at least $k + 1$ red points on l. If we remove l together with $n - 1$ points on it, the remaining configuration allows us to apply the induction hypothesis. There are at least $\frac{1}{2}k \cdot (k+1)$ points G from $I \setminus \{l\}$ for which (OG) intersects at most $k - 1$ lines from $L \setminus \{l\}$. Therefore there are at least $\frac{1}{2}k \cdot (k+1) + k + 1 = \frac{1}{2}(k+1)(k+2)$ red points.

19. Assume first that there exists a point P inside $ABCD$ with the described property. Let K, L, M, N be the feet of perpendiculars from P to AB, BC, CD, and DA respectively. We have $\angle KNM = \angle KNP + \angle PNM = \angle KAP + \angle PDM = 90°$ and similarly $\angle NKL = \angle KLM = \angle LMN = 90°$; hence $KLMN$ is a rectangle. Denote by W, X, Y, Z the feet of perpendiculars from P to KL, LM, MN, and NK. From $\triangle PLX \sim \triangle PCM$ we get that $CM = \frac{XL}{PX} \cdot PM = PM \cdot \frac{PW}{PX}$. Similarly $DM = PM \cdot \frac{PZ}{PY}$, $CL = PL \cdot \frac{PY}{PX}$, $BL = PL \cdot \frac{PZ}{PW}$. Notice that

$$CM : DM = \frac{PW \cdot PY}{PZ \cdot PX} = CL : BL;$$

hence $BD \| ML$. Similarly $AC \| LK$; hence $AC \perp BD$.

Conversely, assume that $ABCD$ is a convex quadrilateral for which $AC \perp BD$. Let P' be any point in the plane and consider a triangle $M'P'L'$ for which $M'L' \| BD$,

$P'M' \perp CD$, and $P'L' \perp CB$. Let K' be the point for which $P'K' \perp AB$ and $K'L' \| AC$. Let N' be the point such that $K'L'M'N'$ is a rectangle. Consider the four lines α_k, α_l, α_m, α_n through K', L', M', N' perpendicular to $P'K'$, $P'L'$, $P'M'$, and $P'N'$ respectively. Let $A' = \alpha_k \cap \alpha_n$, $B' = \alpha_k \cap \alpha_l$, $C' = \alpha_l \cap \alpha_m$, and $D' = \alpha_m \cap \alpha_n$. Using the previously established result we have $A'C' \| K'L'$ and $B'D' \| M'L'$. We also have $C'D' \| CD$, $B'C' \| BC$, $A'B' \| AB$ hence $\triangle DCB \sim \triangle D'C'B'$ and $\triangle ABC \sim \triangle A'B'C'$. Thus there exists a homothety that takes $A'B'C'D'$ to $ABCD$, and this homothethy will map P' into the point P with the required properties.

20. Let M, N, P, Q be the points of tangency of ω with AB, BC, CD, and DA, respectively. We have that $AB + AD = AB + AQ - QD = AB + AM - DP = BM - CP + CD = BN - CN + CD = BC + CD$. Denote by X and Y the points of tangency of ω_1 and ω_2 with AC. Then we have $AB = AX + BC - CX$ and $AD = AY + CD - CY$.

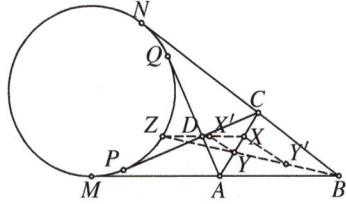

Together with $AB + AD = BC + CD$, this yields $AX - CX = CY - AY$. Since $AX + CX = CY + AY$, we conclude that $AX = CY$; hence Y is the point of tangency of AC and the excircle ω_B of $\triangle ABC$ that corresponds to B. Similarly, the excircle ω_D corresponding to D of $\triangle ADC$ passes through X.

Consider the homothety that maps ω_B to ω. Denote by Z the image of Y under this homothety. Then Z belongs to the tangent of ω that is parallel to AC. Therefore Z is the image of X under the homothety with center D that maps ω_D to ω. Denote by X' and Y' the intersections of DX and BY with ω_2 and ω_1 respectively. Circles ω_1 and ω_B are homothetic with center B; hence Y' the image of Y under this homothety. Moreover, Y' belongs to the tangent of ω_1 that is parallel to AC. This implies that XY' is a diameter of ω_1. Similarly, $X'Y$ is a diameter of ω_2. This implies that $X'Y \| XY'$, which means that $\triangle ZX'Y \sim \triangle ZXY'$ and Z is the center of the homothety that maps ω_2 to ω_1. This finishes the proof of the required statement.

21. Assume the contrary. If two of the numbers are the same, then so are all three of them. Let us therefore assume that all of a, b, c are different. The given conditions imply that

$$\frac{a^n - b^n}{a - b} \cdot \frac{b^n - c^n}{b - c} \cdot \frac{c^n - a^n}{c - a} = -p^3,$$

which immediately implies that some of the numbers a, b, c have to be negative. Moreover, n can't be odd since otherwise each of the fractions would be positive. Assume first that p is odd. Since $2 \nmid \frac{a^n - b^n}{a - b} = a^{n-1} + a^{n-2}b + \cdots + b^{n-1}$, the numbers a and b have to be of different parity. Similarly, $2 \nmid b - c$ and $2 \nmid c - a$ which is not possible.

We are left with the case $p = 2$. Writing $n = 2m$, we derive $(a^m + b^m) \cdot (b^m + c^m) \cdot (c^m + a^m) \cdot \frac{a^m - b^m}{a - b} \cdot \frac{b^m - c^m}{b - c} \cdot \frac{c^m - a^m}{c - a} = -8$. This means that $a^m + b^m = \pm 2$, $a^m - b^m = \pm(a - b)$, and analogous equalities hold for the pairs (b, c) and (c, a).

If m is even, then $|a| = |b| = |c| = 1$, which means that at least two of a, b, c have to be the same.

If m is odd, then $\pm 2 = a^m + b^m$ is divisible by $a + b$. Since $a^m + b^m \equiv a + b \pmod{2}$, we conclude that $a + b = \pm 2$. Similarly $b + c = \pm 2$ and $c + a = \pm 2$. At least two of a, b, c have to be the same, which is a contradiction.

Remark. The statement of the problem remains valid if we replace the assumption that p is prime with the assumption $2 \nmid p$ or $p = 2$.

22. Assume the contrary. Without loss of generality we may assume that these numbers are relatively prime (otherwise we could divide them by their common divisor). We may also assume that $a_1 < a_2 < \cdots < a_n$. For each $i \in \{1, 2, \ldots, n-1\}$ there exists $j \in \{1, 2, \ldots, n-1\}$ such that $a_n + a_i \mid 3a_j$. This together with $a_n + a_i > a_j$ implies that $a_n + a_i$ is divisible by 3 for all i.

There exists $k \in \{1, 2\}$ such that $a_n \equiv k \pmod{3}$ and $a_i \equiv 3 - k \pmod{3}$ for all $i \neq n$. For each $i \in \{1, 2, \ldots, n-2\}$ there exists j such that $a_{n-1} + a_i \mid 3a_j$. Since $a_{n-1} + a_i$ is not divisible by 3, we must have $a_{n-1} + a_i \mid a_j$; hence $j = n$, and we conclude that $a_{n-1} + a_i \mid a_n$ for all $i \in \{1, 2, \ldots, n-2\}$. Let $l \in \{1, 2, \ldots, n\}$ be such an integer for which $a_n + a_{n-1} \mid 3a_l$. Adding the inequalities $a_n + a_{n-1} \leq 3a_l$ and $a_{n-1} + a_l \leq a_n$ gives that $a_{n-1} \leq a_l$; thus either $l = n$ or $l = n-1$.

In the first case, $u(a_{n-1} + a_n) = 3a_n$ for some $u \in \mathbb{N}$. We immediately see that $u < 3$ and $u > 1$. Hence $u = 2$ and $2a_{n-1} = a_n$. However, this is impossible, since for each $i \in \{1, 2, \ldots, n-2\}$ the number $a_{n-1} + a_i$ divides $a_n = 2a_{n-1}$.

On the other hand, if $a_{n-1} + a_n \mid 3a_{n-1}$ then there exists $v \in \mathbb{N}$ for which $v(a_{n-1} + a_n) = 3a_{n-1}$. If $v \geq 2$, then $2a_{n-1} + 2a_n \leq 3a_{n-1}$, which is impossible. Hence $v = 1$, and we get $a_n = 2a_{n-1}$. In the same way as in the previous case we get a contradiction.

23. We will use induction on n. Observe that $a_n \geq (a_{n+1}, a_n) > a_{n-1}$. Obviously, $a_0 \geq 1$, and $a_1 \geq a_0 + 1 \geq 2$. From $a_{k+1} - a_k \geq (a_{k+1}, a_k) \geq a_{k-1} + 1$ we get $a_2 \geq 4$ and $a_3 \geq 7$. It is impossible to have $a_3 = 7$, since $(a_3, a_2) > a_1 = 2$ would imply $a_2 = 7 = a_3$. Hence we have that the statement is satisfied for $n \in \{0, 1, 2, 3\}$. Assume that $n \geq 2$ and $a_i \geq 2^i$ for all $i \in \{0, 1, \ldots, n\}$. We need to prove that $a_{n+1} \geq 2^{n+1}$. Let us define $d_n = (a_{n+1}, a_n)$. We have $d_n > a_{n-1}$. Let $a_{n+1} = kd_n$ and $a_n = ld_n$. If $k \geq 4$ we are done, because $a_{n+1} \geq 4d_n > 4a_{n-1} \geq 4 \cdot 2^{n-1} = 2^{n+1}$. If $l \geq 3$, then $a_{n+1} > a_n$ implies $k \geq 4$. If $l = 1$, then $a_{n+1} \geq 2a_n \geq 2^{n+1}$. Hence the only remaining case to consider is $a_n = 3d_n$, $a_{n-1} = 2d_n$. Obviously, $d_{n-1} = (2d_n, a_{n-1}) > a_{n-2}$. If $a_{n-1} = d_{n-1}$, then from $a_{n-1} < d_n$ and $a_{n-1} \mid 2d_n$ we get $\frac{2d_n}{a_{n-1}} \geq 3$ and $d_n \geq \frac{3}{2}a_{n-1} \geq \frac{3}{2} \cdot 2^{n-1}$. Now $a_{n+1} = 3d_n \geq 9 \cdot 2^{n-2} > 2^{n+1}$. If $a_{n-1} \geq 3d_{n-1}$ then $d_n > a_{n-1} \geq 3d_{n-1}$. Since $d_{n-1} = (2d_n, a_{n-1})$, there exists $s \in \mathbb{N}$ such that $2d_n = sd_{n-1}$. This implies that $d_n > 3 \cdot \frac{2d_n}{s}$, which means that $s > 6$, or $s \geq 7$. Therefore $2d_n \geq 7d_{n-1} > 7 \cdot 2^{n-2}$ and $a_{n+1} = 3d_n > 2^{n+1}$.

It remains to consider the case $a_{n-1} = 2d_{n-1}$. From $(2d_n, 2d_{n-1}) = d_{n-1}$ we conclude that $d_n = \frac{d_{n-1}}{2}w$ for some odd integer $w \geq 3$. From $a_{n-1} < d_n$ we get $2d_{n-1} < d_n$; hence $w \geq 5$. If $w \geq 7$, then $a_{n+1} \geq 3 \cdot 7 \cdot \frac{d_{n-1}}{2} > 21 \cdot 2^{n-3} > 2^{n+1}$; hence it remains to consider the case $w = 5$. We now have $2^{n-3} \leq a_{n-3} <$

$d_{n-2} = (2d_{n-1}, a_{n-2})$. If $a_{n-2} \geq 2d_{n-2}$, then $2d_{n-1} \geq 3d_{n-2} > 3 \cdot 2^{n-3}$. Therefore $a_{n+1} = 3 \cdot \frac{5d_{n-1}}{2} \geq 45 \cdot 2^{n-4} > 2^{n+1}$. If $a_{n-2} = d_{n-2}$, then from $a_{n-2} < d_{n-1}$ we get again $2d_{n-1} \geq 3d_{n-2}$ and $a_{n+1} \geq 2^{n+1}$.

24. First we prove that the numbers $\binom{2^n-1}{k}$ are all odd. Let M be the largest integer for which 2^M divides $(2^n - 1)!$. Then $M = \sum_{i=1}^{n-1} \left[2^{n-i} - \frac{1}{2^i} \right] = \sum_{i=1}^{n-1} \left(2^{n-i} - 1 \right)$. The largest number N for which 2^N divides $k! \cdot (2^n - 1 - k)!$ satisfies

$$N = \sum_{i=1}^{n-1} \left(\left[\frac{k}{2^i} \right] + \left[2^{n-i} - \frac{k+1}{2^i} \right] \right).$$

Each summand on the right-hand side is equal to $2^{n-i} - 1$ (write $k = q_i \cdot 2^i + r_i$, for $0 \leq r_i < 2^i$). Hence $M = N$ and $\binom{2^n-1}{k}$ is odd.

Let us prove that the numbers $\binom{2^n-1}{k}$ give different remainders modulo 2^n. This is valid for $n = 1$. Assume that this holds for some $n > 1$. We claim that the sets $A_i = \left\{ \binom{2^{n+1}-1}{2i}, \binom{2^{n+1}-1}{2i+1} \right\}$ and $B_i = \left\{ \binom{2^n-1}{i}, 2^{n+1} - \binom{2^n-1}{i} \right\}$ are the same modulo 2^{n+1} for each $i = 0, 1, \dots, 2^{n-1} - 1$. We also claim that that all numbers from $\bigcup_{i=0}^{2^{n-1}-1} B_i$ are different modulo 2^{n+1}. These two claims will imply the desired result. Let us show that $\binom{2^{n+1}-1}{2i} \equiv -\binom{2^{n+1}-1}{2i+1} \pmod{2^{n+1}}$ and that one of these two numbers is congruent to $\binom{2^n-1}{i}$. The first congruence follows from

$$\binom{2^{n+1}-1}{2i} = \binom{2^{n+1}}{2i+1} - \binom{2^{n+1}-1}{2i+1}$$

$$= \frac{2^{n+1}}{2i+1} \binom{2^{n+1}-1}{2i} - \binom{2^{n+1}-1}{2i+1}$$

$$\equiv -\binom{2^{n+1}-1}{2i+1} \pmod{2^{n+1}},$$

while the second is true because

$$\binom{2^{n+1}-1}{2i} = \prod_{k=0}^{i-1} \frac{2^{n+1} - (2k+1)}{2k+1} \cdot \prod_{k=1}^{i} \frac{2^{n+1} - 2k}{2k}$$

$$= \prod_{k=0}^{\left[\frac{i-1}{2}\right]} \frac{2^{n+1} - (2k+1)}{2k+1} \cdot \prod_{k=1}^{i} \frac{2^n - k}{k}$$

$$\equiv (-1)^i \cdot \binom{2^n-1}{i} \pmod{2^{n+1}}.$$

It remains to show that $\bigcup_{i=0}^{2^{n-1}-1} B_i$ have all elements different modulo 2^{n+1}. The induction hypothesis implies that $\binom{2^n-1}{i}$ has a different remainder from $\binom{2^n-1}{j}$ for $i \neq j$. The same holds for $2^{n+1} - \binom{2^n-1}{i}$ and $2^{n+1} - \binom{2^n-1}{j}$. From $\binom{2^n-1}{2k} + \binom{2^n-1}{2k+1} = 2^n$ we have that $\binom{2^n-1}{i} \equiv 2^{n+1} - \binom{2^n-1}{j} \equiv 0 \pmod{2^{n+1}}$ if and only if there exists k such that $\{i, j\} = \{2k, 2k+1\}$ for some k. However, in that case $\binom{2^n-1}{i} + \binom{2^n-1}{j} = \frac{2^n}{2k+1} \binom{2^n-1}{2k} \neq 0 \pmod{2^{n+1}}$.

25. If p is a prime number, then $d(f(p))$ has p divisors, and must be a power of a prime. Hence $f(p) = q^{p-1}$ for some prime number q. Let us show that $q = p$. Consider first the case $p > 2$. From $f(2p) \mid (2-1) \cdot p^{2p-1} \cdot f(2)$ and $f(2p) \mid (p-1) \cdot 2^{2p-1} \cdot f(p) = (p-1) \cdot 2^{2p-1} \cdot q^{p-1}$ we conclude that $f(2p) \mid (p^{2p-1} \cdot f(2), (p-1) \cdot 2^{2p-1} \cdot q^{p-1}) = (f(2), (p-1) \cdot 2^{2p-1} \cdot q^{p-1})$. Since $f(2p)$ has $2p$ divisors and $f(2)$ is prime, this is a contradiction. We also have $f(2) = 2$. Indeed, this follows from $f(6) \mid 3^{6-1} \cdot f(2)$, $f(6) \mid 2 \cdot 2^{6-1} \cdot 3^{3-1}$, and $d(f(6)) = 6$.
    Assume now that $x = p_1^{a_1} \cdots p_n^{a_n}$ is a prime factorization of x with $p_1 < \cdots < p_n$. Let $f(x) = q_1^{b_1} \cdots q_m^{b_m}$. From $d(f(x)) = p_1^{a_1} \cdots p_n^{a_1} = (b_1+1) \cdots (b_m+1)$ we conclude that $b_i \geq p_1 - 1$ for all i. The relation $f(x) \mid (p_1-1) \cdot (p_1^{a_1-1} \cdots p_n^{a_n})^{x-1} \cdot f(p_1)$ yields $q_1, \ldots, q_m \in \{p_1, \ldots, p_n\}$. Hence for each prime p and each $a \in \mathbb{N}$ there is $b \in \mathbb{N}$ such that $f(p^a) = p^b$. From $p^a = b+1$ we get $f(p^a) = p^{p^a-1}$.
    Now assume that $x \in \mathbb{N}$. There are integers $a_1, \ldots, a_n, b_1, \ldots, b_n$ such that $x = p_1^{a_1} \cdots p_n^{a_n}$ and $f(x) = p_1^{b_1} \cdots p_n^{b_n}$. For each $i \in \{1, \ldots, n\}$ we have $f(x) \mid (p_i^{a_i} - 1) \cdot (x/p_i^{a_i})^{x-1} \cdot p_i^{p_i^{a_i}-1}$; hence $p_i^{b_i} \mid p_i^{p_i^{a_i}-1}$, which implies $b_i + 1 \leq p_i^{a_i}$. Multiplying this for $i = 1, \ldots, n$ we get $d(f(x)) = (b_1+1) \cdots (b_n+1) \leq p_1^{a_1} \cdots p_n^{a_n} = x$. Since $d(f(x)) = x$, we must have $b_i = p_i^{a_i} - 1$ for all i and $f(x) = p_1^{p_1^{a_1}-1} \cdots p_n^{p_n^{a_n}-1}$. It is easy to verify that the function f defined by the previous relation satisfies the required conditions.

26. If p is any prime number of the form $p \equiv 1 \pmod 4$; we know that $\left(\frac{-1}{p}\right) = 1$, and there are exactly two numbers $n, m \in \{0, 1, 2, \ldots, p-1\}$ whose square is congruent to -1 modulo p. Since the sum of these two numbers is equal to p, one of them is smaller than $p/2$. Assuming that $n < p/2$; let us set $k = p - 2n$. It suffices to prove that there exist infinitely many prime numbers p for which $k > \sqrt{2n}$. From $p \mid n^2 + 1 = \frac{p^2 - 2pk + k^2}{4} + 1$ we conclude that $p \mid k^2 + 4$. This implies that $k^2 \geq p - 4$. It suffices to prove that $p - 4 > 2n$, i.e., $4 < p - 2n = k$ for infinitely many values of p. However, this will be satisfied, since $k \geq \sqrt{p-4} > 4$ for $p > 20$, and there are infinitely many prime numbers greater than 20 that are congruent to 1 modulo 4.

4.50 Solutions to the Shortlisted Problems of IMO 2009

1. Notice that b_{2009}, r_{2009}, and w_{2009} always form a triangle. In order to prove this fact it suffices to show that the largest of these three numbers (say w_{2009}) is less than the sum of the other two. Since there are b_i and r_j such that w_{2009}, b_i, and r_j form a triangle, we get $w_{2009} < b_i + r_j \leq b_{2009} + r_{2009}$. This proves that $k \geq 1$. We will now provide an example in which only one triple (b_i, r_i, w_i) forms a triangle. Let us set $w_i = 2i$, $r_i = i$, for $i = 1, 2, \ldots, 2009$; and let us define $b_i = i$ for $i \in \{1, 2, \ldots, 2008\}$, and $b_i = 2i$ for $i = 2009$. We will now form triangles $T_1, \ldots, T_{2009}$ such that each has one blue, one red, and one white side. For $j = 1, 2, \ldots, 2008$ we can define the triangle T_j to be the one with sides w_j, r_{j+1}, and b_j because $2j < j + j + 1$. The sides of T_{2009} can be $w_{2009} = 4018$, $r_1 = 1$, and $b_{2009} = 4018$. The conditions of the problem are clearly satisfied.

2. The given conditions on a, b, and c imply that $abc(a + b + c) = ab + bc + ca$. Applying the inequality $(X + Y + Z)^2 \geq 3(XY + YZ + ZX)$ to $X = ab$, $Y = bc$, $Z = ca$ gives us $(ab + bc + ca)^2 \geq 3abc(a + b + c) = 3(ab + bc + ca)$, which means that $ab + bc + ca \geq 3$.
   From $2a + b + c \geq 2\sqrt{(a+b)(a+c)}$ and two analogous inequalities we deduce

$$\frac{1}{(2a+b+c)^2} + \frac{1}{(2b+c+a)^2} + \frac{1}{(2c+a+b)^2} \leq \frac{2(a+b+c)}{4(a+b)(b+c)(c+a)}$$

$$\leq \frac{9(a+b+c)}{16(ab+bc+ca)(a+b+c)}.$$

The last inequality follows from $9(a+b)(b+c)(c+a) = 9(a^2b + ab^2 + b^2c + bc^2 + c^2a + ca^2) + 18abc = 8(ab + bc + ca)(a + b + c) + (a^2b + ab^2 + b^2c + bc^2 + c^2a + ca^2 - 6abc) \geq 8(ab + bc + ca)(a + b + c)$.
The required inequality now follows from $ab + bc + ca \geq 3$. Equality holds if and only if $a = b = c = 1$.

3. Assume that f satisfies the given requirements. Let us prove that $f(1) = 1$. Assume that $K = f(1) - 1 > 0$. Let m be the minium of f, and b any number for which $f(b) = m$. Since 1, $m = f(b)$, and $f(b + f(1) - 1) = f(b + K)$ form a triangle we must have $f(b + K) < 1 + f(b)$. The minimality of m implies $f(b + K) = m$, and by induction $f(b + nK) = m$ for all $n \in \mathbb{N}$. There exists a triangle with sides $b + nK$, $f(1)$, and $f(m)$; hence $b + nK < f(1) + f(m)$ for each n. This contradiction implies $f(1) = 1$.
   The numbers a, $1 = f(1)$, and $f(f(a))$ form a triangle for every a. Therefore $a - 1 < f(f(a)) < a + 1$; hence $f(f(a)) = a$, and f is a bijection. We now have that $f(a)$, $f(b)$, and $f(b + a - 1)$ determine a triangle for all $a, b \in \mathbb{N}$.
   Let $z = f(2)$. Clearly, $z > 1$. Since $f(z)$, $f(z)$, and $f(2z - 1)$ form a triangle, we get $f(2z - 1) < f(z) + f(z) = 2f(f(2)) = 4$. This implies that $f(2z - 1) \in \{1, 2, 3\}$. Since f is a bijection and $f(1) = 1$, $f(z) = 2$, we must have $f(2z - 1) = 3$. Let us prove that $f(k) = (k - 1)z - k + 2$ for all $k \in \mathbb{N}$.
   The statement is true for 1 and 2. Assume that it holds for all of $1, 2, \ldots, k$. Since $f((k - 1)z - k + 2)$, $f(z)$, and $f(kz - k + 1)$ form a triangle, we have $f(kz - k +$

1) $\leq k+1$. The function f is injective; hence $f(kz-k+1) \neq i$ unless $kz-k+1 = (i-1)z-i+2$, i.e., $k+1 = i$. Therefore $f(kz-k+1) = k+1$, or $f(k+1) = kz-k+1$, and the induction is complete. Furthermore, f is increasing.
If $z > 2$ then $2 = f(z) > f(2) = z$, a contradiction. Thus $z = 2$ and $f(k) = 2(k-1)-k+2 = k$. It is easy to verify that $f(k) = k$ satisfies the given condition.

4. We first prove that for $a, b > 0$ the following inequality holds:

$$\sqrt{\frac{a^2+b^2}{2}} + \sqrt{ab} \leq a+b. \tag{1}$$

After dividing both sides by $\sqrt{ab}$ and substituting $\sqrt{\frac{a}{b}} = x$ it becomes

$$\frac{1}{\sqrt{2}}\sqrt{x^2+\frac{1}{x^2}} + 1 \leq \left(x+\frac{1}{x}\right).$$

Taking squares transforms this inequality into $\left(x+\frac{1}{x}-2\right)^2 \geq 0$, which obviously holds. Equality occurs if and only if $x = 1$, or $a = b$.
Rewriting (1) in the form $\sqrt{\frac{a^2+b^2}{a+b}} + \sqrt{2}\sqrt{\frac{ab}{a+b}} \leq \sqrt{2}\sqrt{a+b}$ and summing it with the two analogous inequalities for the pairs (b,c) and (c,a), we obtain

$$\sqrt{\frac{a^2+b^2}{a+b}} + \sqrt{\frac{b^2+c^2}{b+c}} + \sqrt{\frac{c^2+a^2}{c+a}} + \sqrt{2}\left(\sqrt{\frac{ab}{a+b}} + \sqrt{\frac{bc}{b+c}} + \sqrt{\frac{ca}{c+a}}\right)$$
$$\leq \sqrt{2}\left(\sqrt{a+b} + \sqrt{b+c} + \sqrt{c+a}\right).$$

It suffices to prove that $\sqrt{\frac{ab}{a+b}} + \sqrt{\frac{bc}{b+c}} + \sqrt{\frac{ca}{c+a}} \geq \frac{3}{\sqrt{2}}$. Applying the mean inequality $M_1 \geq M_{-2}$ to the sequence $\sqrt{\frac{ab}{a+b}}, \sqrt{\frac{bc}{b+c}}, \sqrt{\frac{ca}{c+a}}$ gives us

$$\sqrt{\frac{ab}{a+b}} + \sqrt{\frac{bc}{b+c}} + \sqrt{\frac{ca}{c+a}} \geq 3\sqrt{\frac{3}{\frac{a+b}{ab} + \frac{b+c}{bc} + \frac{c+a}{ca}}}$$
$$= 3 \cdot \sqrt{\frac{3abc}{2(ab+bc+ca)}} \geq \frac{3}{\sqrt{2}}.$$

This completes the proof of the required inequality. Equality holds if and only if $a = b = c$.

5. Assume the contrary, that $f(x-f(y)) \leq yf(x)+x$ for all $x, y \in \mathbb{R}$. Setting $y = 0$, $x = z+f(0)$ gives $f(z) \leq z+f(0)$. For $x = f(y)$ we get $f(0) \leq yf(f(y))+f(y) \leq yf(f(y))+y+f(0)$; hence we can conclude that $y(f(f(y))+1) \geq 0$. This means that if $y > 0$, then $f(f(y)) \geq -1$, and if $y < 0$, then $f(f(y)) \leq -1$.
If $f(x) > 0$ for some x, then each $y < x-f(0)$ must satisfy $f(y) - f(0) \leq y < x-f(0)$, therefore $f(y) < x$. From this we conclude that $-1 \leq f(f(x-f(y))) \leq$

$f(x-f(y))+f(0) \le yf(x)+x+f(0)$ and $y \ge \frac{-1-x-f(0)}{f(x)}$. We have concluded that each $y < x - f(0)$ must be greater than or equal to $\frac{-1-x-f(0)}{f(x)}$, which is impossible. Therefore $f(x) \le 0$ for all x. In particular, for all real x we have $f(x) \le x + f(0) \le x$. For any $z > 0$ we now have $f(-1) = f[(f(z)-1)-f(z)] \le zf(f(z)-1)+f(z)-1 \le z(f(z)-1)+f(z)-1 = (z+1)(f(z)-1) \le -z-1$. Thus each $z > 0$ satisfies $z \le -f(-1) - 1$, which is a contradiction.

6. Assume that $s_{s_n} = an+b$ and $s_{s_n+1} = cn+d$ for some $a,b,c,d \in \mathbb{Z}$. If $m > n > 0$ are two integers, then $s_m - s_n = (s_{n+1}-s_n)+(s_{n+2}-s_{n+1})+\cdots+(s_m-s_{m-1}) \ge m - n$ because the sequence s_n is increasing. Hence $s_{n+1} - s_n \le s_{s_{n+1}} - s_{s_n} = a$. Denote by m and M the minimal and maximal values of $s_{n+1} - s_n$ for $n \in \mathbb{N}$. Our goal is to prove that $M = m$. Assume the contrary, $m < M$. If $s_{k+1} - s_k = m$ for some $k \in \mathbb{N}$, we get $a = s_{s_{k+1}} - s_{s_k} = (s_{s_{k+1}} - s_{s_k}) + (s_{s_k+2} - s_{s_k+1}) + \cdots + (s_{s_{k+1}} - s_{s_{k+1}-1}) \le m \cdot M$. Similarly, if $s_{l+1} - s_l = M$ for some $l \in \mathbb{N}$, we get that $a = s_{s_{l+1}} - s_{s_l} \ge m \cdot M$. In particular, these two inequalities imply that

$$M \cdot m = a,$$
$$s_{s_k+1} - s_{s_k} = M, \quad \text{whenever } s_{k+1} - s_k = m, \text{ and}$$
$$s_{s_l+1} - s_{s_l} = m, \quad \text{whenver } s_{l+1} - s_l = M.$$

Take any $k \in \mathbb{N}$ such that $s_{k+1} - s_k = m$. Then $M = s_{s_k+1} - s_{s_k} = ck+d-(ak+b) = (c-a)k+d-b$. Furthermore, we have $m = s_{s_{s_k}+1} - s_{s_{s_k}} = (c-a)s_k+d-b$. Repeating the same argument yields $M = (c-a)s_{s_k}+d-b$. Consequently the equation $(c-a)x+d-b = M$ has two solutions, $x = k$ and $x = s_{s_k}$, which yields $s_k = k$. Since $s_k = s_1 + (s_2 - s_1) + \cdots + (s_k - s_{k-1}) \ge k$, we conclude that $s_i = i$ for $i = 1,2,\ldots,k$. Thus $M = s_{s_k+1} - s_{s_k} = s_{k+1} - s_k = m$, a contradiction.

7. Substituting $x = 0$ in the given relation gives $f(0) = f(yf(0))$ for all y. Therefore $f(0) = 0$, because otherwise for each $z \in \mathbb{R}$ we could take $y = z/f(0)$ to get $f(z) = f(0)$, meaning that f is constant (which is obviously impossible). We now have $f(xf(x)) = f(xf(x+0)) = f(0f(x))+x^2 = x^2$ and $0 = f(xf(x-x)) = f(-xf(x))+x^2$, implying $f(-xf(x)) = -x^2$. Hence f is onto. If $f(z) = 0$ for some $z \ne 0$ we would have $0 = f(zf(z)) = z^2$, a contradiction. Assume that $f(x) = f(y)$ for some $x,y \in \mathbb{R}$. Then $x^2 = f(xf(x)) = f(xf(y)) = f((y-x)f(x))+x^2$, giving that $f((y-x)f(x)) = 0$; hence $f(x) = 0$ or $x-y = 0$. Both cases now yield $x = y$; therefore f is one-to-one. Now we will prove that $f(-x) = -f(x)$ for all $x \in \mathbb{R}$. Assume that $x \ne 0$ (the other case is trivial). If $f(x) > 0$, there exists z such that $f(x) = z^2$. Since f is injective and $f(zf(z)) = z^2$, we conclude that $x = zf(z)$; hence $f(-x) = f(-zf(z)) = -z^2 = -f(x)$. The case $f(x) < 0$ is similar.
On the other hand, we have

$$f(yf(x)) = -x^2 + f(xf(x+y)) = -x^2 + (x+y)^2 - [(x+y)^2 + f(-xf(x+y))]$$
$$= y^2 + 2xy - f((x+y)f(y)) = 2xy + [(-y)^2 + f((x+y)f(-y))]$$
$$= 2xy + f(-yf(x)),$$

which implies that $f(xf(y)) = xy$. Analogously, $f(yf(x)) = xy$; hence $xf(y) = yf(x)$. Hence $f(x) = cx$ for some $c \in \mathbb{R}$. The equation $f(xf(x)) = x^2$ implies that $c \in \{-1, 1\}$. Clearly, both $f(x) = x$ and $f(x) = -x$ satisfy the given conditions.

8. (a) If we denote gold cards by 1, and black by 0, the entire sequence of cards corresponds to a number in binary representation. After each of the moves, the number decreases; hence the game has to end.

   (b) We will show that the second player wins game no matter how the players play. Consider the cards whose position (counted from the right) is divisible by 50. There is a total of 40 such cards, and in each move exactly one of these cards is turned over. In the beginning, all 40 of these cards are 1, and in the end all 40 are 0; hence the second player must win.

9. Assume that $(a_i, b_i, c_i)_{i=1}^N$ satisfy the conditions of the problem. Then

$$\sum_{i=1}^{N} a_i \geq \frac{N(N-1)}{2},$$

and similarly the two analogous inequalities hold for the sequences (b_i) and (c_i). Hence $3N(N-1)/2 \leq \sum_{i=1}^{N}(a_i + b_i + c_i) = nN$, which implies that $N \leq \left[\frac{2n}{3}\right] + 1$. To prove that there are sequences of length $\left[\frac{2n}{3}\right] + 1$ with the given properties, let us consider the following cases:

   1° $n = 3k$ for some $k \in \mathbb{N}$. We can take $(a_i, b_i, c_i) = (i-1, k+i-1, 2k-2i+2)$ for $i = 1, 2, \ldots, k+1$, and $(a_i, b_i, c_i) = (3k-i+2, 2k-i+1, 2(i-k)-3)$ for $i = k+2, \ldots, 2k+1$.

   2° $n = 3k+1$ for some $k \in \mathbb{N}$. Take $(a_i, b_i, c_i) = (i-1, k+i-1, 2k-2i+3)$ for $i = 1, 2, \ldots, k+1$, and $(a_i, b_i, c_i) = (3k-i+2, 2k-i+1, 2(i-k)-2)$ for $i = k+2, \ldots, 2k+1$.

   3° $n = 3k-1$ for some $k \in \mathbb{N}$. Define $(a_i, b_i, c_i) = (i-1, k+i-1, 2k-2i+1)$ for $i = 1, 2, \ldots, k$, and $(a_i, b_i, c_i) = (3k-i+1, 2k-i, 2i-2k-2)$ for $i = k+1, \ldots, 2k$.

10. For a binary sequence $(\varepsilon)_{n-1} = (\varepsilon_i)_{i=1}^{n-1}$, let us define $f(u, v, (\varepsilon)_{n-1}) = c_n$ where the sequence $(c_i)_{i=1}^{n}$ is defined as: $c_0 = u$, $c_1 = v$, and

$$c_{i+1} = \begin{cases} 2c_{i-1} + 3c_i, & \text{if } \varepsilon_i = 0, \\ 3c_{i-1} + c_i, & \text{if } \varepsilon_i = 1, \end{cases} \quad \text{for } i = 1, \ldots, n-1.$$

The given sequences (a_n) and (b_n) can be now rewritten as $a_n = f(1, 7, (\varepsilon)_{n-1})$, $b_n = f(1, 7, (\overline{\varepsilon})_{n-1})$ where $(\overline{\varepsilon})_{i=1}^{n-1}$ is defined as $\overline{\varepsilon}_i = \varepsilon_{n-i}$. Using induction on n we will prove that $f(1, 7, (\varepsilon)_{n-1}) = f(1, 7, (\overline{\varepsilon})_{n-1})$. This is straightforward to verify for $n = 2$ and $n = 3$, so assume that $n > 3$ and that the statement is true for all binary sequences $(\varepsilon)_k$ of length smaller than n. Notice that $f(\alpha u_1 + \beta u_2, \alpha v_1 + \beta v_2, (\varepsilon)_m) = \alpha f(u_1, v_1, (\varepsilon)_m) + \beta f(u_2, v_2, (\varepsilon)_m)$ (this easily follows by induction on m). Assuming that $\varepsilon_n = 0$, we obtain

$$f(1,7,(\varepsilon)_n) = 2f(1,7,(\varepsilon)_{n-2}) + 3f(1,7,(\varepsilon)_{n-1})$$
$$= 2f(1,7,\overline{(\varepsilon)_{n-2}}) + 3f(1,7,\overline{(\varepsilon)_{n-1}})$$
$$= 2f(1,7,\overline{(\varepsilon)_{n-2}}) + 3f(7,f(1,7,(\varepsilon_{n-1})_1),\overline{(\varepsilon)_{n-2}})$$
$$= f(23,14 + 3f(1,7,(\varepsilon_{n-1})_1),\overline{(\varepsilon)_{n-2}}).$$

Using $14 + 3f(1,7,(\varepsilon_{n-1})_1) = f(1,7,(0,\varepsilon_{n-1})_2)$ and $23 = f(1,7,(0)_1)$, we get

$$f(23,14 + 3f(1,7,(\varepsilon_{n-1})_1),\overline{(\varepsilon)_{n-2}})$$
$$= f(f(1,7,(0)_1),f(1,7,(0,\varepsilon_{n-1})_2),\overline{(\varepsilon)_{n-2}})$$
$$= f(1,7,\overline{(\varepsilon)_n}).$$

To finish the proof, it remains to see that for $\varepsilon_n = 1$ we have

$$f(1,7,(\varepsilon)_n) = 3f(1,7,\overline{(\varepsilon)_{n-2}}) + f(1,7,\overline{(\varepsilon)_{n-1}})$$
$$= f(f(1,7,(1)_1),f(1,7,(1,\varepsilon_{n-1})_2),\overline{(\varepsilon)_{n-2}})$$
$$= f(1,7,\overline{(\overline{\varepsilon})_n}).$$

11. Denote by (i,j) the cells of the table, and assume that the diagonal cells (i,i) form separate rectangles. We will prove by induction that it is possible to have a perimeter of $p_m = (m+1)2^{m+2}$. The case $m = 0$ is obvious, and assume that the statement holds for some $m \geq 0$. Divide the $2^{m+1} \times 2^{m+1}$ board into four equal boards. Each of the two off-diagonal squares has perimeter $4 \cdot 2^m$, while the other two can be partitioned into rectangles of total perimeter p_m each. The total perimeter is therefore equal to $2 \cdot 4 \cdot 2^m + 2p_m = (m+2)2^{m+3}$.
Let us now prove the other direction, that the total perimeter P satisfies $P \geq (m+1)2^{m+2}$. Assume that the table is partitioned into n rectangles in the described way. Denote by R_i the set of those rectangles that contain at least one square from the ith row. Similarly, let C_i be the set of rectangles that contain the squares from the ith column. Clearly the intersection $C_i \cap R_i$ contains only the diagonal square (i,i). We certainly have

$$P = 2\left(\sum_{i=1}^{2^m} |R_i| + \sum_{i=1}^{2^m} |C_i|\right).$$

Let $\mathscr{F}$ be the collection of all subsets of rectangles in the partition. Since there are n rectangles, we have $|\mathscr{F}| = 2^n$. Let $\mathscr{F}_i$ denote the collection of those subsets S that satisfy $(R_i \setminus (i,i)) \subseteq S$ and $C_i \cap S \subseteq \{(i,i)\}$. Since $\mathscr{F}_1, \mathscr{F}_2, \ldots, \mathscr{F}_{2^m}$ are disjoint sets and $|\mathscr{F}_i| = 2^{n-|C_i|-|R_i|+2}$, using Jensen's inequality applied to $f(x) = 2^{-x}$, we obtain

$$2^n \geq 2^{n+2}\sum_{i=1}^{2^m} 2^{-|C_i|-|R_i|} \geq 2^{n+2} \cdot 2^m \cdot 2^{-\frac{1}{2^m}\Sigma(|C_i|+|R_i|)} = 2^{n+m+2-\frac{1}{2^m}\cdot\frac{P}{2}}.$$

This yields $\frac{P}{2} \geq (m+1) \cdot 2^{m+1}$, which is the relation we wanted to prove.

12. We will show that Cinderella can always make sure that after each of her moves:

   (i) The total amount of water in all buckets is less than $3/2$; and

   (ii) The amount of water in each pair of nonadjacent buckets is smaller than 1.

The condition (ii) ensures that the stepmother won't be able to make a bucket overflow. Both (i) and (ii) hold in the beginning of the game. Assume that they are satisfied after the kth round, and let us prove that in round $k+1$, Cinderella can make them both hold again. Denote the buckets by $1, \dots, 5$ (counterclockwise), and let x_i be the amount of water in bucket i. After the $(k+1)$st move of the stepmother we have $x_1 + \cdots + x_5 < \frac{5}{2}$ and $x_i + x_{i+2} < 2$ for each i (summation of indices is modulo 5). It is impossible that $x_i + x_{i+2} \geq 1$ for each i; hence we may assume that $x_2 + x_5 < 1$. If $x_1 + x_2 + x_5 < \frac{3}{2}$, it would be safe for Cinderella to empty buckets 3 and 4. Assume therefore that $x_1 + x_2 + x_5 \geq \frac{3}{2}$. Hence $x_1 > \frac{1}{2}$. If both $x_2 + x_4 \geq 1$ and $x_3 + x_5 \geq 1$, hold then we must have $x_1 \leq \frac{1}{2}$, a contradiction. Assume therefore that $x_2 + x_4 < 1$. If $x_2 + x_3 + x_4 < \frac{3}{2}$ she could empty 1 and 5, so assume that $x_2 + x_3 + x_4 \geq \frac{3}{2}$. This gives $x_3 > \frac{1}{2}$, $x_2 + \frac{5}{2} > (x_1 + x_2 + x_5) + (x_2 + x_3 + x_4) \geq 6$, which yields $x_2 > \frac{1}{2}$. Therefore at least one of $x_1 + x_4 < 1$ and $x_3 + x_5 < 1$ holds, say the first one. Thus, if Cinderella empties buckets 2 and 3, condition (ii) will be satisfied. Condition (i) will hold as well because $x_2 + x_3 > \frac{1}{2} + \frac{1}{2} > 1$.

13. Let $k \geq 3$. We will prove that the longest cyclic route in a $(4k-1) \times (4k-1)$ board has length $4 \cdot [(2k-1)^2 - 1]$. Let us label the cells 1, 2, 3, 4 using the pattern from Figure 1, so that the top-left corner is labeled 1.

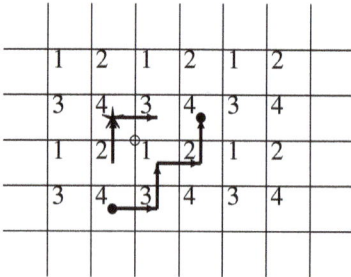

Figure 1

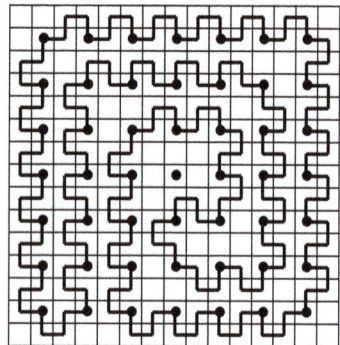

Figure 2

Any four consecutive moves land on squares with different labels. Therefore, each cyclic path has equal numbers of squares of each label. The label 4 appears exactly $(2k-1)^2$ times, but we will prove now that the limp rook can't visit all of them. Assume the contrary, that it is possible for a cyclic route to pass through all the squares labeled 4. Let us paint all squares labeled 4 alternately black and white so that the top left square is black. We see that the number of black squares is bigger by 1 than the number of white ones. Therefore, a cyclic route has two consecutive black squares. Assume that these squares are those denoted by •.

Without loss of generality we may assume that the part of route is as shown in
Figure 1. Since the route always visits $4,3,1,2$ in that order, immediately before
visiting $\star$, the rook has to land on the cell labeled 2 that is exactly below $\star$. The
rook has to leave $\star$ by visiting 3 exactly to the right of it. Each point of the two-
dimensional plane must be all the time either to the left or to the right of the rook
when it is passing next to it. However, this is not the case with the point marked
by $\circ$. A contradiction. Therefore the rook can visit at most $4 \cdot ((2k-1)^2 - 1)$
squares in a cyclic route. Figure 2 shows how it is possible to recursively make a
route that omits only the central square and visits all the other squares labeled 4.

14. We will prove the statement by induction. The case $n = 1$ is trivial, so let us
    assume that $n > 1$ and that the statement holds for $1, 2, \ldots, n-1$. Assume that
    $a_1 < \cdots < a_n$. Let $m \in M$ be the smallest element. Consider the following cases:
    1° $m < a_n$: If $a_n \notin M$, then if the grasshopper makes the first jump of size
        a_n the problem gets reduced to the sequence $a_1, \ldots, a_{n-1}$ and the set $M \setminus$
        $\{m\}$, which immediately follows by induction. Let us assume that $a_n \in M$.
        Consider the following $n-1$ pairs: $(a_1, a_1 + a_n), \ldots, (a_{n-1}, a_{n-1} + a_n)$. All
        numbers from these pairs belong to the $(n-2)$-element set $M \setminus \{a_n\}$; hence
        one of these pairs, say $(a_k, a_k + a_n)$, has both of its members outside of M.
        If the first two jumps of the grasshopper are a_k, and $a_k + a_n$, it has jumped
        over at least two members of M: m and a_n. There are at most $n-3$ more
        elements of M to jump over, and $n-2$ more jumps, so the claim follows by
        induction.
    2° $m \geq a_n$: By the induction hypothesis the grasshopper can start from the
        point $s = a_1 + \cdots + a_n$, make $n-1$ jumps of sizes $a_1, \ldots, a_{n-1}$ to the left,
        and avoid all the points of $M \setminus \{m\}$. If it misses the point m as well, then
        we are done (first make a jump of size a_n and reverse the previously made
        jumps). Suppose that after making the jump a_k the grasshopper landed at
        site m. If it changes the jump a_k to the jump a_n, it will miss the site m and
        all subsequent jumps will lend outside of M because m is the leftmost point.

15. For each $i = 0, 1, \ldots, 9$ denote by N_i the set of all finite strings whose terms are
    from $\{i, i+1, \ldots, 9\}$ (the empty string ϕ belongs to each of N_i). Define functions
    $m_i : N_i \to \mathbb{N}$ recursively: For each $x \in N_9$ we set $m_9(x) = 1 +$ the number of digits
    of x (set $m_9(\phi) = 1$). Once $m_9, \ldots, m_{i+1}$ are defined we construct m_i as follows:
    Write each $x \in N_i$ in the form $x = \overline{x_0 i x_1 i \ldots x_{t-1} i x_t}$ where $x_0, \ldots, x_t \in N_{i+1}$. We
    can now choose
    $$m_i(x) = \sum_{s=0}^{t} 4^{m_{i+1}(x_s)}.$$

    Let us prove that $m = m_0$ satisfies $m(h(n)) < m(n)$ whenever $n \neq \phi$. If the last
    digit of n is 0, then $n = \overline{l_0 0 l_1 \ldots 0 l_t 0}$ for some $l_0, \ldots, l_t \in N_1$ and $m(n) - m(h(n)) =$
    $4^{m_1(\phi)} > 0$. Assume that $n = \overline{Ler(d+1)}$ for $9 \geq d \geq e \geq 0$ and $r \in N_{d+1}$. If $L \in N_i$
    for some $i < e$ then $L = \overline{l_0 i l_1 i \ldots i l_t}$ with $l_0, \ldots, l_t \in N_{i+1}$. Set $n' = \overline{l_t er(d+1)}$.
    Then $h(n') = \overline{l_t erdrd}$ and $m_i(n) - m_i(h(n)) = 4^{m_{i+1}(n')} - 4^{m_{i+1}(h(n'))}$. To prove
    that $m(n) > m(h(n))$ it suffices to prove that $m_{i+1}(n') > m_{i+1}(h(n'))$. Repeating

this argument, we reduce our problem to the case $i = e$. There are $l_0, \ldots, l_t$, $r \in N_{d+1}$ such that $n = \overline{l_0 e \ldots e l_t er(d+1)}$. For $e = d$ we need to prove that $0 < m_d(n) - m_d(h(n)) = 4^{m_{d+1}(\overline{r(d+1)})} - 2 \cdot 4^{m_{d+1}(r)}$. This inequality suffices even in the case $e < d$ because $r \in N_{d+1}$ implies

$$m_{e+1}(\overline{r(d+1)}) = 4^{\cdot^{\cdot^{\cdot^{4^{m_d(\overline{r(d+1)})}}}}} \quad, \quad \text{and} \quad m_{e+1}(\overline{rdrd}) = 4^{\cdot^{\cdot^{\cdot^{4^{m_d(\overline{rdrd})}}}}} \quad;$$

where 4 appears $d - e - 1$ times in the exponents. If $d + 1 = 9$, then $m_d(n) - m_d(h(n)) = 4^{k+1} - 2 \cdot 4^k > 0$, where k is the number of digits of r. If $d < 8$, then we can write $r = \overline{r_0(d+1) \cdots (d+1) r_s}$ for some $r_0, \ldots, r_s \in N_{d+2}$. We get $m_{d+1}(r) = 4^{m_{d+2}(r_0)} + \cdots + 4^{m_{d+2}(r_s)}$ and $m_{d+1}(\overline{r(d+1)}) = m_{d+1}(r) \cdot 4^{4^{m_{d+2}(\phi)}}$. Hence $m_d(n) - m_d(h(n)) = 4^{m_{d+1}(r)} (4^{4^{m_{d+2}(\phi)}} - 2) > 0$.

Therefore the function m is positive and decreasing on the sequence n, $h(n)$, $h(h(n))$, $\ldots$, forcing this sequence to eventually become equal to ϕ. The only way this can occur is if the last terms of the sequence are $1, 00, 0, \phi$.

Remark. The number of operations required can be huge. For example, $n = 5$ will reduce to 1 after approximately $0.828753 \cdot 2^{2^{127}}$ iterations.

16. Denote by I and L the incenters of $\triangle ABC$ and $\triangle BDA$. From $\angle ALI = \angle LBA + \angle LAB = 45°$ we see that $AL \| EK$. Let L' be the intersection of DK and BI. From $\angle DL'I = \angle BID - \angle IDK = \angle A/4 = \angle LAI$ we conclude that A, L, D, and L' belong to a circle. Hence $\angle LAL' = 180° - \angle LDL' = 90°$. Now consider $\triangle AKL'$. The segment KE is the altitude from K and $\angle KAE = \angle KL'E$. We will now prove that this is equivalent to E being the orthocenter or $\triangle AKL'$ being isosceles (with $KA = KL'$). Denote by P and Q the intersections of $L'E$ and AE with AK and KL' respectively. If $KA = KL'$, then the statement is obvious. If this is not the case, then PQ intersects AL' at some point M. Then KE is the polar line of the point M with respect to the circumcircle k of $PQL'A$, and since $MA \perp KE$, we conclude that MA contains the center of k. Then we must have $\angle APL' = \angle AQL' = 90°$. If E is the orthocenter of $\triangle KL'A$ then from $\triangle ABP$ we conclude that $3\angle A/4 + 45° - \angle A/4 = 90°$ which yields $\angle A = 90°$. If $KA = KL'$, this together with $KA = AL = AL'$ implies that $\triangle AKL'$ is equilateral and $\angle A/4 = \angle KL'E = 60° - \angle LL'A = 15°$. This means that $\angle A = 60°$.

It is easy to verify that $\angle A \in \{60°, 90°\}$ implies $\angle BEK = 45°$.

17. From $MK \| AB$ and $ML \| AC$ we get $\angle KML = \angle BAC$. Also, $\angle AQP = \angle KMQ = \angle MLK$, because of the assumption that PQ is a tangent to Γ. Therefore $\triangle AQP \sim \triangle MLK$; hence

$$\frac{AQ}{AP} = \frac{ML}{MK} = \frac{PC}{QB},$$

and $AQ \cdot QB = AP \cdot PC$. The quantity on the left-hand side of the last equality represents the power of the point Q with respect to the circumcircle of $\triangle ABC$ and it is equal to $OA^2 - OQ^2$. Similarly, $AP \cdot PC = OA^2 - OP^2$. Thus $OA^2 - OP^2 = OA^2 - OQ^2$, implying $OP = OQ$.

18. Consider the excircle k_a corresponding to the vertex A. Let X be the point of tangency of the incircle k and BC, and X_a the point of tangency of k_a and BC. Similarly, let us denote by Z_a and Y_a the points of tangency of k_a with AB and AC. Then we have $ZZ_a = ZB + BZ_a = BX + BX_a = BX + CX = BC = ZS$. We used the fact that $BX_a = CX$. On the other hand, $CY_a = CX_a = BX = BZ = CS$. Denote by s the degenerate circle with center S and radius 0. Points Z and C belong to the radical axis of the circles s and k_a. Similarly, we prove that B and Y belong to the radical axis of the circles k_a and r, where r is the circle with center R and radius 0. Thus G is the radical center of r, s, and k_a; hence $GS = GR$.

19. Denote by M and N respectively the points symmetric to E with respect to G and H. Notice that $\angle FAM = \angle FAC + \angle BEC = \angle EBC + \angle CEB = \angle FCE$. From $\triangle FDC \sim \triangle FBA$ we have $\frac{FA}{FC} = \frac{AB}{CD}$, and from $\triangle AEB \sim \triangle DEC$ we have $\frac{AB}{DC} = \frac{BE}{CE}$. Therefore

$$\frac{FA}{FC} = \frac{BE}{CE} = \frac{AM}{CE};$$

hence $\triangle FAM \sim \triangle FCE$ and $\angle AFM = \angle EFC$. Similarly we prove that $\triangle FDN \sim \triangle FBE$ and $\angle AFN = \angle EFC$. This implies that F, N, and M are collinear. Since H and G are the midpoints of EN and EM, it suffices to show that FE is the tangent to the circumcircle of $\triangle NEM$. From $\triangle FDN \sim \triangle FBE$ we have $\frac{FD}{FB} = \frac{FN}{FE}$, while the similarity $\triangle FAM \sim \triangle FCE$ implies that $\frac{FC}{FA} = \frac{FE}{FM}$. Using $\triangle FDC \sim \triangle FBA$ again, we finally obtain $\frac{FD}{FB} = \frac{FC}{FA}$. Thus $\frac{FE}{FM} = \frac{FN}{FE}$ and FE is tangent to the circumcircle of $\triangle MEN$.

20. Let A and B be two vertices of the polygon P for which $S_{\triangle AOB}$ is maximal. Let C and D be the points symmetric to A and B with respect to O. Let $WXYZ$ be the parallelogram such that A, B, C, and D are the midpoints of XY, YZ, ZW, and WX, and $WX \| OA$, $XY \| OB$. The polygon P is contained inside $WXYZ$. Let U, V, M, and N be the intersections of P with XZ and WY such that the order of points on lines is $W - M - N - Y$ and $X - U - V - Z$. There are two parallel lines u and v through U and V such that P is within the strip between u and v; similarly, there are two parallel lines $n \ni N$ and $m \ni M$ such that P is within the strip between these two lines. The lines u, v, n, and m determine another parallelogram $EFGH$. We will prove that $S_{EFGH} \leq \sqrt{2} S_P$ or $S_{WXYZ} \leq \sqrt{2} S_P$.

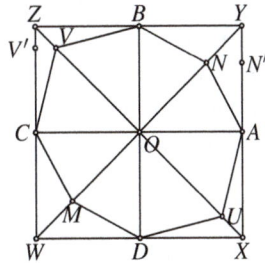

By performing affine transformations to the plane, the ratios of the areas of the figures don't change, and we can choose the transformations in such a way that $WXYZ$ maps to a square. Then $EFGH$ maps to a rectangle, and P maps to another convex polygon. We may thus assume that $WXYZ$ was a square to start with, and $EFGH$ was a rectangle. Let V' be the projection of V to WZ, and let N' be the projection of N to YX. Set $a = OA$, $x = ZV'$, and $y = YN'$. Then $S_P \geq$

$S_{ANBVCMDU} = 4S_{\triangle COV} + 4S_{\triangle OAN} = 2a(a-x) + 2a(a-y) = 2a(2a-(x+y))$.
We also have $S_{WXYZ} = 4a^2$ and $S_{EFGH} = 4OV \cdot ON = 8(a-x)(a-y)$. If we
assume that $S_{WXYZ} > \sqrt{2}S_P$ and $S_{EFGH} > \sqrt{2}S_P$, multiplying these two inequalities gives us $32a^2(a-x)(a-y) > 8a^2(2a-(x+y))^2$, which after simplification
becomes $4xy > (x+y)^2$, or equivalently $(x-y)^2 < 0$, a contradiction.

21. Assume that the perpendicular from E_1 to CD and the perpendicular from E_2 to
    AB intersect H_1H_2 at S_1 and S_2 respectively. We will prove that $H_1S_1 : H_1H_2 =$
    $H_1S_2 : H_1H_2$, which will imply $S_1 = S_2$. Let $M_1 = S_1E_1 \cap PH_1$ and $M_2 = S_2E_2 \cap$
    PH_2. It suffices to establish the relation $H_1M_1 : H_1P = PM_2 : H_2P$.

Let us denote by N_1 and N_2 the midpoints of PH_1 and PH_2. Without loss of
generality, assume that M_1 is between
P and N_1, and consequently N_2 is between P and M_2. Our goal is to prove
that $\frac{H_1M_1}{H_1N_1} = \frac{PM_2}{PN_2}$, or after subtracting 1
from both sides:

$$\frac{M_1N_1}{H_1N_1} = \frac{M_2N_2}{PN_2}. \qquad (1)$$

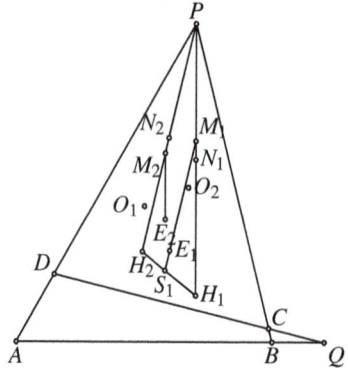

From $E_1M_1 \| PH_2$ and $E_2M_2 \| PH_1$ we
conclude that $\angle N_2M_2E_2 = 180° -$
$\angle E_1M_1N_1$. Observe that $E_1N_1 \| PO_1$
hence $\angle M_1E_1N_1 = \angle O_1PH_2 = \angle DPH_2 - \angle APO_1$. From $\triangle DCP$ we obtain the
equality $\angle DPH_2 = 90° - \angle CDP$ and from $\triangle ABP$ we have $\angle APO_1 = 90° -$
$\angle ABP$. Therefore $\angle M_1E_1N_1 = \angle ABP - \angle CDP$. In a similar way we prove that
$\angle M_2E_2N_2 = \angle O_2PH_1 = \angle ABP - \angle CDP$ which gives us $\angle M_1E_1N_1 = \angle M_2E_2N_2$.
Applying the law of sines to $\triangle E_2N_2M_2$ and $\triangle E_1N_1M_1$ we get $\frac{M_1N_1}{E_1N_1} = \frac{M_2N_2}{E_2N_2}$.
Hence in order to prove (1) we need to verify that $\frac{E_1N_1}{H_1N_1} = \frac{E_2N_2}{PN_2}$, or equivalently, $\frac{PO_1}{PH_1} = \frac{PO_2}{PH_2}$. From $PH_1 = 2O_1X$, where X is the midpoint of AB, we
see that $\frac{PO_1}{PH_1} = \frac{AO_1}{2O_1X} = \frac{1}{2\cos \angle AO_1X} = \frac{1}{2\cos \angle APB}$. Analogously we prove that
$\frac{PO_2}{PH_2} = \frac{1}{2\cos \angle CPD}$ and this completes the proof of the required statement.

22. Let us denote by α, β, and γ the angles of $\triangle ABC$. Then we calculate $\angle BIX =$
    $\angle XIC = \frac{1}{2}\angle BIC = 45° + \frac{\alpha}{4}$, and two similar formulas hold for $\angle ZIX$ and $\angle ZIY$.
    If we denote by P and Q the feet of perpendiculars from I to CX and CY; then
    $IP = IQ$. Since $\angle PIX = \angle PIC - \angle XIC = (90° - \angle XCI) - 45° - \frac{\alpha}{4} = 45° - \frac{\alpha}{4} -$
    $\frac{\gamma}{4} = \frac{\beta}{4}$, and similarly $\angle QIY = \frac{\alpha}{4}$, we deduce that $IX/\cos\frac{\beta}{4} = IY/\cos\frac{\alpha}{4}$. If we
    denote the previous quantity by ρ, we get $IX = \rho\cos\frac{\alpha}{4}$, $IY = \rho\cos\frac{\beta}{4}$, and anal-
    ogously, $IZ = \rho\cos\frac{\gamma}{4}$. Applying the law of cosines to $\triangle ZIX$ gives us that $ZX^2 -$
    $ZI^2 = IX^2 - 2IX \cdot ZI \cdot \cos\left(90° + \frac{\alpha+\gamma}{4}\right) = \rho^2\left(\cos^2\frac{\alpha}{4} + 2\cos\frac{\alpha}{4}\cos\frac{\gamma}{4}\sin\frac{\alpha+\gamma}{4}\right)$. Us-
    ing the analogous relation for $ZY^2 - ZI^2$ and the assumption $ZX = ZY$, we get

$$0 = \cos^2\frac{\alpha}{4} - \cos^2\frac{\beta}{4} + 2\cos\frac{\alpha}{4}\sin\frac{\alpha}{4}\cos^2\frac{\gamma}{4} - 2\cos\frac{\beta}{4}\sin\frac{\beta}{4}\cos^2\frac{\gamma}{4}$$

$$+2\cos^2\frac{\alpha}{4}\sin\frac{\gamma}{4}\cos\frac{\gamma}{4} - 2\cos^2\frac{\beta}{4}\sin\frac{\gamma}{4}\cos\frac{\gamma}{4}$$

$$= \left(\cos^2\frac{\alpha}{4} - \cos^2\frac{\beta}{4}\right)\cdot\left(1 - \sin\frac{\gamma}{2}\right) + \cos^2\frac{\gamma}{4}\left(\sin\frac{\alpha}{2} - \sin\frac{\beta}{2}\right).$$

We now use the formulas $\cos^2 X - \cos^2 Y = \frac{1}{2}(\cos(2X) - \cos(2Y)) = -\sin(X + Y)\sin(X - Y)$, and $\sin X - \sin Y = 2\sin\frac{X-Y}{2}\cos\frac{X+Y}{2}$ to obtain

$$0 = \sin\frac{\alpha - \beta}{4}\left(2\cos^2\frac{\gamma}{4}\cos\frac{\alpha+\beta}{4} - \left(1 - \sin\frac{\gamma}{2}\right)\sin\frac{\alpha+\beta}{4}\right). \qquad (1)$$

Let E be the expression from the last parenthesis. Since $\frac{\alpha+\beta}{4} = 45° - \frac{\gamma}{4}$, then

$$\sqrt{2}E = \left(1 + \cos\frac{\gamma}{2}\right)\left(\cos\frac{\gamma}{4} + \sin\frac{\gamma}{4}\right) - \left(1 - \sin\frac{\gamma}{2}\right)\left(\cos\frac{\gamma}{4} - \sin\frac{\gamma}{4}\right)$$

$$= 2\sin\frac{\gamma}{4} + \cos\frac{\gamma}{2}\cos\frac{\gamma}{4} - \sin\frac{\gamma}{2}\sin\frac{\gamma}{4} + \cos\frac{\gamma}{2}\sin\frac{\gamma}{4} + \sin\frac{\gamma}{2}\cos\frac{\gamma}{4}.$$

Hence $\sqrt{2}E = 2\sin\frac{\gamma}{4} + \cos\frac{3\gamma}{4} + \sin\frac{3\gamma}{4} = 2\sin\frac{\gamma}{4} + \sqrt{2}\sin\left(45° + \frac{3\gamma}{4}\right)$. Clearly, the last quantity is positive, since the sine is a positive function on $(0, 180°)$. Thus from (1) we get $\alpha = \beta$. Similarly, we prove that $\beta = \gamma$ and $\triangle ABC$ is equilateral.

23. Denote by k_1, k_2, and k_3 the incircles of $\triangle ABM$, $\triangle MNC$, and $\triangle ADN$, respectively. Let R, S, T be the points of tangency of k_1 with AB, BM, and MA; U, V, W the tangency points of k_3 with ND, DA, and AN; and P and Q the points of tangency of tangents from C to k_3 and k_1 different from CD and CB, respectively. Assume that the configuration of the points is as in the picture. From $CD + AB = CB + DA$ we get $CU + UD + AR + RB = DV + VA + BS + SC$, which together with $DU = DV$, $AR = AW + WT = AV + WT$, and $BR = BS$ implies that $CU + WT = CS = CQ$. On the other hand, $CU + WT = CP + WT \geq CP + PQ$ because $PQ \leq WT$, and equality holds if and only if PQ is a common tangent of the circles k_1 and k_3. We conclude that $CQ \geq CP + PQ$. The triangle inequality yields $CQ = CP + PQ$ hence C, P, and Q are collinear.

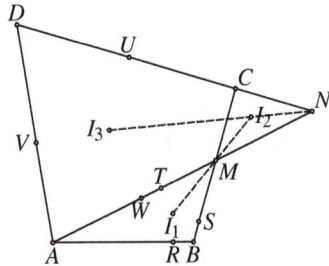

We now have that $\angle I_3 C I_1 = \frac{1}{2}\angle DCB = 90° - \angle I_2 CM = \angle I_3 I_2 M$; hence C belongs to the circle circumscribed about $\triangle I_1 I_2 I_3$. The Simson's line corresponding to C bisects the segment CH, where H is the orthocenter of $\triangle I_1 I_2 I_3$. It remains to notice that the line g is the image of the Simson's line under the homothety with center C and coefficient 2. Indeed, the reflections of C with respect to $I_1 I_2$ and $I_2 I_3$ belong to g because $I_1 I_2$ and $I_2 I_3$ are the bisectors of $\angle CMN$ and $\angle CNM$.

24. Assume the contrary, that $a_i a_{i+1} \equiv a_i \pmod{n}$, for $i = 1, 2, \ldots, k$ (summation of indices is modulo k). The case $k = 2$ is trivial, since we have $a_1 a_2 \equiv a_1$ and $a_1 a_2 \equiv a_2$, which yields the immediate contradiction $a_1 \equiv a_2$. Here and in the sequel, all of the congruences are modulo n. Assume that $1 < i < k$. Multiplying the congruence $a_i a_{i+1} \equiv a_i$ by $a_1 \cdots a_{i-1}$ we get $a_1 \cdots a_{i+1} \equiv a_1 \cdots a_i$. By induction we get that $a_1 \cdots a_k \equiv a_1$. Since everything is cyclic, in an analogous way we obtain $a_1 \cdots a_k \equiv a_2$, which yields $a_1 \equiv a_2$, and this is a contradiction.

25. (a) To each $n \in \mathbb{N}$ we can associate a sequence $(s_1, \ldots, s_{50})$ of numbers from $\{0, 1\}$ such that
$$s_i = \begin{cases} 0, & \text{if } n + i \text{ is balanced,} \\ 1, & \text{if } n + i \text{ is not balanced.} \end{cases}$$

Since there are at most 2^{50} such sequences corresponding to natural numbers, we see that there are $a, b \in \mathbb{N}$ that correspond to the same sequence. For such a choice of a and b the number $P(i)$ is balanced for each $i \in \{1, 2, \ldots, 50\}$.

(b) Assume that $a < b$ and that $P(n) = (n + a)(n + b)$ is balanced for each $n \in \mathbb{N}$. For each $k > a$, consider the number $x = k(b - a) - a$. Then $P(x) = (b - a)^2 k(k + 1)$ is balanced, which means that k and $k + 1$ are *equibalanced* (if one is balanced then so is the other) whenever $k > a$. Then all numbers greater than a are equibalanced, which can't be true, since squares are balanced but primes are not.

26. Assume the contrary: There are finitely many prime numbers $p_1, p_2, \ldots, p_m$ such that no other prime can be a divisor of $f(n)$ for $n \in \mathbb{N}$. Assume that $\alpha_1, \ldots, \alpha_m$ are nonnegative integers for which $f(1) = p_1^{\alpha_1} \cdots p_m^{\alpha_m}$. For any sequence $\beta = (\beta_1, \ldots, \beta_m)$ satisfying $\beta_1 > \alpha_1, \ldots, \beta_m > \alpha_m$, consider the number $a_\beta = p_1^{\beta_1} \cdots p_m^{\beta_m}$. Assume that $f(a_\beta + 1) = p_1^{\gamma_1} \cdots p_m^{\gamma_m}$, for some $\gamma_1, \ldots, \gamma_m \in \mathbb{N}_0$. Since $a_\beta \mid f(a_\beta + 1) - f(1)$, we can conclude that $\gamma_1 = \alpha_1, \ldots, \gamma_m = \alpha_m$; hence $f(a_\beta + 1) = f(1)$. If n is a positive integer for which $f(n) \neq f(1)$, then $a_\beta + 1 - n \mid f(a_\beta + 1) - f(n) = f(1) - f(n)$. This relation has to hold for every sequence β satisfying $\beta_1 > \alpha_1, \ldots, \beta_n > \alpha_1$, which is impossible.

27. We will prove that $n \leq 4$ by showing that there is no sequence of length 5, and that there is a sequence of length 4, satisfying the conditions.
Assume that $a_1, \ldots, a_5$ is a sequence of length 5. If $2 \nmid a_k$ for some $k \leq 3$, from $a_{k+1}^2 + 1 = (a_k + 1)(a_{k+2} + 1)$ we see that $2 \nmid a_{k+1}$ as well. Notice that a_1 and a_2 are even. Indeed, if $2 \nmid a_k$ for $k \in \{1, 2\}$, then $2 \nmid a_{k+1}$ and $2 \nmid a_{k+2}$. We then have $a_{k+1}^2 + 1 \equiv 2 \pmod 4$, while $4 \mid (a_k + 1)(a_{k+2} + 1)$, which is a contradiction. Since a_1 and a_2 are even, we get from $a_2^2 + 1 = (a_1 + 1)(a_3 + 1)$ that a_3 is even as well. We now have $a_3 + 1 \mid a_2^2 + 1$ and $a_2 + 1 \mid a_3^2 + 1$. Let us prove that there are no two positive even integers x and y satisfying $x + 1 \mid y^2 + 1$ and $y + 1 \mid x^2 + 1$. Assume the contrary, that (x, y) is one such pair for which $x + y$ is minimal and $x \geq y$. Let $d = \gcd(x + 1, y + 1)$. From $d \mid x + 1$ one gets $d \mid x^2 - 1$. Since $d \mid y + 1 \mid x^2 + 1$, we derive that $d \mid (x^2 + 1) - (x^2 - 1) = 2$. Since $x + 1$ is odd, we see that $d = 1$. Therefore $x + 1 \mid y^2 + 1 + x^2 - 1 = x^2 + y^2$ and $y + 1 \mid x^2 + y^2$ imply

$(x+1)(y+1) \mid x^2+y^2$. There exists $m \in \mathbb{N}$ such that $m(x+1)(y+1) = x^2 + y^2$. Consider the quadratic polynomial $P(\lambda) = \lambda^2 - m(y+1)\lambda - m(y+1) + y^2$. Since $P(x) = 0$, there exists a positive integer x' such that $P(\lambda) = (\lambda - x)(\lambda - x')$. From $x + x' = m(y+1)$ and $xx' = y^2 - m(y+1)$ we get that x' is even and $y^2 + 1 = (x+1)(x'+1)$. We now must have $x' < y \le x$; hence (x',y) is another pair of even natural numbers such that $x' + 1 \mid y^2 + 1$ and $y + 1 \mid x'^2 + 1$, a contradiction.

One sequence of length 4 is $a_1 = 4$, $a_2 = 33$, $a_3 = 217$, $a_4 = 1384$.

28. Assume that there exist $T : \mathbb{Z} \to \mathbb{Z}$ and a polynomial P with integer coefficients such that $T^n(x) = x$ has exactly $P(n)$ solutions for each $n \in \mathbb{N}$. For $k \in \mathbb{N}$ denote by $B(k)$ the set of those x such that $T^k(x) = x$ but $T^l(x) \ne x$ for all $0 \le l < k$. Take any $x \in A(n) \cap B(k)$, and assume that $n = ak + b$, for $a \in \mathbb{N}_0$ and $0 \le b \le k - 1$. Then $x = T^n(x) = T^b(T^{ak}(x)) = T^b(x)$. We conclude that $b = 0$ and $k \mid n$. Hence $A(n) = \bigcup_{k \mid n} B(k)$ and moreover,

$$|A(n)| = \sum_{k \mid n} |B(k)|.$$

Assume now that $x \in B(n)$, and consider the sequence $\{T^i(x)\}_{i=0}^{n-1}$. Each $T^i(x)$ belongs to $A(n)$, since $T^i(x) = T^i(T^n(x)) = T^n(T^i(x))$. If $T^i(x) = T^{i+j}(x)$ for $0 \le i \le n-1$ and $0 \le j \le n-1$, then $x = T^n(x) = T^{n-i}(T^i(x)) = T^{n-i}(T^{i+j}(x)) = T^{n+j}(x) = T^j(T^n(x)) = T^j(x)$; which means that $j = 0$. Therefore, $T^{i_1}(x) \ne T^{i_2}(x)$ whenever $i_1 \ne i_2$ and $i_1, i_2 \in \{0, 1, \ldots, n-1\}$. In addition, each of $T^i(x)$ belongs to $B(n)$. This means that $B(n)$ partitions into sequences of n elements in each, and thus $n \mid B(n)$.

Let p be a prime number. We have $P(p) = |A(p)| = |B(1)| + |B(p)|$. If q is also prime, then $P(pq) = |B(1)| + |B(p)| + |B(q)| + |B(pq)|$ hence $P(pq) \equiv |B(1)| + |B(q)| \pmod{p}$. However, from $P(pq) \equiv P(0) \pmod{p}$ we get that $P(0) - |B(1)| - |B(q)|$ is divisible by p. If we fix q, this continues to hold for each prime p. Therefore $P(0) = |B(1)| + |B(q)| = P(q)$. However, this is now true for every prime q, and hence P must be constant, contrary to our assumptions.

29. For each $k \in \mathbb{N}$ there exist $q_k \in \mathbb{Z}$ and a polynomial $P_k(x)$ of degree $k - 1$ with integer coefficients such that $xP_k(x) = x^k + P_k(x-1) + q_k$. Indeed, the coefficients of $P_k(x) = c_{k-1}x^{k-1} + \cdots + c_0$ form a system of linear equations which we can explicitly solve (we can see that $c_{k-1} = 1$). The sequence $b_n = a_n - P_k(n)$ satisfies the recurrence relation $b_n = \frac{b_{n-1}}{n} - \frac{q_k}{n}$. Inductively we prove that $b_n = \frac{b_0}{n!} - \frac{q_k}{n!} \cdot \sum_{i=0}^{n-1} i!$, hence $a_n - P_k(n) = \frac{a_0 - P_k(0)}{n!} - \frac{q_k}{n!} \cdot \sum_{i=0}^{n-1} i!$. All of $a_n - P_k(n)$ are integers, and $|a_n - P_k(n)| \le \frac{|a_0 - P_k(0)|}{n!} + \frac{|q_k|}{n} + |q_k| \cdot \sum_{i=0}^{\infty} \frac{1}{i^2} \to 0$. Hence we conclude that $a_0 = P_k(0)$ and $q_k = 0$.

We will finish the proof by showing that q_k is even only when $k \equiv 2 \pmod{3}$. We start with the equality $\Gamma_k(x) = xP_k(x) - x^k - P_k(x-1) - q_k = 0$ and use the fact that $x(x+1)\Gamma_k(x) - \Gamma_{k+1}(x) - \Gamma_{k+2}(x) = 0$. After simplifying this becomes equivalent to $xT_k(x) = T_k(x-1) + 2xP_k(x-1) + 2xq_k - (q_k + q_{k+1} + q_{k+2})$, for $T_k(x) = x(x+1)P_k(x) - P_{k+1}(x) - P_{k+2}(x) - q_k x$. Therefore $xT_k(x) - T_k(x-1) \equiv$

$q_k + q_{k+1} + q_{k+2}$ (mod 2). For each polynomial f one of the following two identities hold: either $f(x) \equiv x$ (mod 2) for all $x \in \mathbb{Z}$, or $f(x) \equiv 0$ (mod 2) for all $x \in \mathbb{Z}$. Since $xT_k(x) - T_k(x-1)$ is a constant polynomial modulo 2, it must be 0, and $q_k + q_{k+1} + q_{k+2} = 0$. We can easily calculate $q_1 = -1$ and $q_2 = 0$, and now by induction it is straightforward to establish $q_k \equiv 0$ (mod 2) if and only if $k \equiv 2$ (mod 3).

30. We will prove a stronger statement by assuming that a and b are perfect squares. Assume that each of $\rho_n = \sqrt{(a^n - 1)(b^n - 1)}$ is an integer. Consider the Taylor representation of $f(x) = (1-x)^{1/2} = \sum_{k=0}^{\infty} \alpha_k x^k$ for $x \in (-1,1)$ (the sequence α_k is fixed here). There exist real numbers $(c_{k,l})_{k,l \geq 0}$ such that

$$g(x,y) = (1-x)^{\frac{1}{2}}(1-y)^{\frac{1}{2}} = \sum_{k,l \geq 0} c_{k,l} x^k y^l, \quad \text{for all } x,y \in (-1,1). \tag{1}$$

Therefore $\rho_n = \sum_{k,l \geq 0} c_{k,l} \left(\sqrt{ab}/(a^k b^l) \right)^n$. Take $k_0, l_0 \in \mathbb{N}$ for which $a^{k_0} > \sqrt{ab}$ and $b^{l_0} > \sqrt{ab}$. Consider the polynomial $P(x) = \prod_{k=0}^{k_0} \prod_{l=0}^{l_0} (a^k b^l x - \sqrt{ab})$. There are $d_0, \ldots, d_{k_0 l_0} \in \mathbb{Z}$ such that $P(x) = \sum_{i=0}^{k_0 l_0} d_i x^i$. For each $n \geq 0$ define

$$\sigma_n = \sum_{i=0}^{k_0 l_0} d_i \rho_{n+i} = \sum_{k,l \geq 0} \left(\frac{\sqrt{ab}}{a^k b^l} \right)^n c_{k,l} P\left(\frac{\sqrt{ab}}{a^k b^l} \right) = \sum_{k>k_0 \text{ or } l>l_0} \gamma_{k,l} \left(\frac{\sqrt{ab}}{a^k b^l} \right)^n,$$

where $\gamma_{k,l} = c_{k,l} P\left(\sqrt{ab}/(a^k b^l) \right)$. The series for σ_n is absolutely convergent because it is a finite linear combination of absolutely convergent series in ρ_{n+i}. Since $\sqrt{ab}/(a^k b^l) < \max\{1/a, 1/b\} \leq 1/2$ then $\sigma_n < \frac{1}{2}\sigma_{n-1}$. This means that $\lim_{n \to \infty} \sigma_n = 0$, and since all of σ_n are integers there exists N such that $\sigma_n = 0$ for $n \geq N$. For $n \geq N$ we have $\sum_{i=0}^{k_0 l_0} d_i \rho_{n+i} = 0$.

Assume first that $a^k \neq b^l$ for each pair (k,l) of positive integers. Solving the system of recurrence equations for $(\rho_n)_{n \geq N}$ we obtain constants $e_{k,l}$ for $0 \leq k \leq k_0$, $0 \leq l \leq l_0$ such that $\rho_n = \sum_{k=0}^{k_0} \sum_{l=0}^{l_0} e_{k,l} \left(\sqrt{ab}/(a^k b^l) \right)^n$. This together with (1) implies that if $(x,y) = (1/a^n, 1/b^n)$ for some $n \geq N$, then $\sum_{k=0}^{\infty} \sum_{l=0}^{\infty} c_{k,l} x^k y^l = \sum_{k=0}^{k_0} \sum_{l=0}^{l_0} e_{k,l} x^k y^l$. These two are Taylor series for $g(x,y)$ and have to be the same. Hence $c_{k,l} = e_{k,l}$ if $k \leq k_0$ or $l \leq l_0$. If either $k > k_0$ or $l > l_0$, then $c_{k,l} = 0$. We conclude that $g(x,y)$ has a finite Taylor expansion around 0, which is impossible.

If there were k and l such that $a^k = b^l$, there would exist an integer p such that $a = p^l$, and $b = p^k$. In a similar way we get a contradiction by proving the finiteness of the Taylor expansion of $(1-x^l)^{1/2}(1-x^k)^{1/2}$.

A

Notation and Abbreviations

A.1 Notation

We assume familiarity with standard elementary notation of set theory, algebra, logic, geometry (including vectors), analysis, number theory (including divisibility and congruences), and combinatorics. We use this notation liberally.

We assume familiarity with the basic elements of the game of chess (the movement of pieces and the coloring of the board).

The following is notation that deserves additional clarification.

- $\mathcal{B}(A,B,C)$, $A-B-C$: indicates the relation of *betweenness*, i.e., that B is between A and C (this automatically means that A,B,C are different collinear points).

- $A = l_1 \cap l_2$: indicates that A is the intersection point of the lines l_1 and l_2.

- AB: line through A and B, segment AB, length of segment AB (depending on context).

- $[AB$: ray starting in A and containing B.

- $(AB$: ray starting in A and containing B, but without the point A.

- (AB): open interval AB, set of points between A and B.

- $[AB]$: closed interval AB, segment AB, $(AB) \cup \{A,B\}$.

- $(AB]$: semiopen interval AB, closed at B and open at A, $(AB) \cup \{B\}$.
  The same bracket notation is applied to real numbers, e.g., $[a,b) = \{x \mid a \leq x < b\}$.

- ABC: plane determined by points A,B,C, triangle ABC ($\triangle ABC$) (depending on context).

- $[AB,C$: half-plane consisting of line AB and all points in the plane on the same side of AB as C.

- $(AB,C$: $[AB,C$ without the line AB.

- $\langle \vec{a}, \vec{b} \rangle$, $\vec{a} \cdot \vec{b}$: scalar product of $\vec{a}$ and $\vec{b}$.

- $a, b, c, \alpha, \beta, \gamma$: the respective sides and angles of triangle ABC (unless otherwise indicated).

- $k(O, r)$: circle k with center O and radius r.

- $d(A, p)$: distance from point A to line p.

- $S_{A_1 A_2 \ldots A_n}$, $[A_1 A_2 \ldots A_n]$: area of n-gon $A_1 A_2 \ldots A_n$ (special case for $n = 3$, S_{ABC}: area of $\triangle ABC$).

- $\mathbb{N}$, $\mathbb{Z}$, $\mathbb{Q}$, $\mathbb{R}$, $\mathbb{C}$: the sets of natural, integer, rational, real, complex numbers (respectively).

- $\mathbb{Z}_n$: the ring of residues modulo n, $n \in \mathbb{N}$.

- $\mathbb{Z}_p$: the field of residues modulo p, p being prime.

- $\mathbb{Z}[x]$, $\mathbb{R}[x]$: the rings of polynomials in x with integer and real coefficients respectively.

- R^*: the set of nonzero elements of a ring R.

- $R[\alpha]$, $R(\alpha)$, where α is a root of a quadratic polynomial in $R[x]$: $\{a + b\alpha \mid a, b \in R\}$.

- X_0: $X \cup \{0\}$ for X such that $0 \notin X$.

- X^+, X^-, $aX + b$, $aX + bY$: $\{x \mid x \in X, x > 0\}$, $\{x \mid x \in X, x < 0\}$, $\{ax + b \mid x \in X\}$, $\{ax + by \mid x \in X, y \in Y\}$ (respectively) for $X, Y \subseteq \mathbb{R}$, $a, b \in \mathbb{R}$.

- $[x]$, $\lfloor x \rfloor$: the greatest integer smaller than or equal to x.

- $\lceil x \rceil$: the smallest integer greater than or equal to x.

The following is notation simultaneously used in different concepts (depending on context).

- $|AB|$, $|x|$, $|S|$: the distance between two points AB, the absolute value of the number x, the number of elements of the set S (respectively).

- (x, y), (m, n), (a, b): (ordered) pair x and y, the greatest common divisor of integers m and n, the open interval between real numbers a and b (respectively).

A.2 Abbreviations

We tried to avoid using nonstandard notation and abbreviations as much as possible. However, one nonstandard abbreviation stood out as particularly convenient:

- w.l.o.g.: without loss of generality.

Other abbreviations include:

- RHS: right-hand side (of a given equation).

- LHS: left-hand side (of a given equation).

- QM, AM, GM, HM: the quadratic mean, the arithmetic mean, the geometric mean, the harmonic mean (respectively).

- gcd, lcm: greatest common divisor, least common multiple (respectively).

- i.e.: in other words.

- e.g.: for example.

B

Codes of the Countries of Origin

ARG	Argentina	HRV	Croatia	POL	Poland
ARM	Armenia	HUN	Hungary	POR	Portugal
AUS	Australia	IDN	Indonesia	PRI	Puerto Rico
AUT	Austria	IND	India	PRK	Korea, North
BEL	Belgium	IRL	Ireland	ROU	Romania
BGR	Bulgaria	IRN	Iran	RUS	Russia
BLR	Belarus	ISL	Iceland	SAF	South Africa
BRA	Brazil	ISR	Israel	SCG	Serbia and
CAN	Canada	ITA	Italy		Montenegro
CHN	China	JPN	Japan	SGP	Singapore
COL	Colombia	KAZ	Kazakhstan	SRB	Serbia
CUB	Cuba	KOR	Korea, South	SVK	Slovakia
CYP	Cyprus	KWT	Kuwait	SVN	Slovenia
CZE	Czech Republic	LTU	Lithuania	SWE	Sweden
CZS	Czechoslovakia	LUX	Luxembourg	THA	Thailand
ESP	Spain	LVA	Latvia	TUN	Tunisia
EST	Estonia	MAR	Morocco	TUR	Turkey
FIN	Finland	MEX	Mexico	TWN	Taiwan
FRA	France	MKD	Macedonia	UKR	Ukraine
FRG	Germany, FR	MNG	Mongolia	UNK	United Kingdom
GDR	Germany, DR	NLD	Netherlands	USA	United States
GEO	Georgia	NOR	Norway	USS	Soviet Union
GER	Germany	NZL	New Zealand	UZB	Uzbekistan
HEL	Greece	PER	Peru	VNM	Vietnam
HKG	Hong Kong	PHI	Philippines	YUG	Yugoslavia

C

Authors of Problems

1959-02 C. Ionescu-Tiu, ROU – IMO2
1959-06 Cezar Cosnita, ROU – IMO6

1960-03 G.D. Simionescu, ROU – IMO3

1961-06 G.D. Simionescu, ROU – IMO6

1962-04 Cezar Cosnita, ROU – IMO4

1963-05 Wolfgang Engel, GDR – IMO5

1964-05 G.D. Simionescu, ROU – IMO5

1965-05 G.D. Simionescu, ROU – IMO5

1967-04 Tullio Viola, ITA – IMO4

1968-06 David Monk, UNK – IMO6

1969-01 Wolfgang Engel, GDR – IMO1
1969-05 Abish Mekei, MNG – IMO5

1970-02 D. Batinetzu, ROU – IMO2
1970-06 D. Gerll, A. Warusfel, FRA
1970-10 Åke Samuelsson, SWE – IMO6

1973-05 Georges Glaeser, FRA
1973-10 Åke Samuelsson, SWE – IMO6
1973-11 Åke Samuelsson, SWE – IMO3
1973-13 Đorđe Dugošija, YUG
1973-14 Đorđe Dugošija, YUG – IMO4

1974-01 Murray Klamkin, USA – IMO1
1974-06 Ciprian Borcea, ROU – IMO3
1974-08 Jan van de Craats, NLD – IMO5
1974-10 Matti Lehtinen, FIN – IMO2

1975-08 Jan van de Craats, NLD – IMO3
1975-10 David Monk, UNK – IMO6
1975-14 Vladimir Janković, YUG

1975-10 R. Lyness,
     D. Monk, UNK – IMO6
1976-05 Jan van de Craats, NLD – IMO5
1976-06 Jan van de Craats, NLD – IMO3
1976-09 Matti Lehtinen, FIN – IMO2
1976-10 Murray Klamkin, USA – IMO4

1977-01 Ivan Prodanov, BGR – IMO6
1977-03 Hermann Frasch, FRG – IMO5
1977-04 Arthur Engel, FRG
1977-05 Arthur Engel, FRG
1977-06 Helmut Bausch, GDR
1977-07 David Monk, UNK – IMO4
1977-10 Jan van de Craats, NLD – IMO3
1977-12 Jan van de Craats, NLD – IMO1
1977-15 P. Đức Chính, VNM – IMO2

1978-10 Jan van de Craats, NLD – IMO6
1978-12 Murray Klamkin, USA – IMO4
1978-13 Murray Klamkin, USA – IMO2
1978-15 Dragoslav Ljubić, YUG
1978-16 Dragoslav Ljubić, YUG

1979-04 N. Hadzhiivanov,
     N. Nenov, BGR – IMO2
1979-07 Arthur Engel, FRG – IMO1
1979-08 Arthur Engel, FRG
1979-09 Arthur Engel, FRG – IMO6
1979-12 Hans-Dietrich Gronau, GDR

1979-15 Joe Gillis, ISR – IMO5
1979-22 N. Vasilyev,
    I.F. Sharygin, USS – IMO3
1979-25 Murray Klamkin, USA – IMO4
1979-26 Milan Božić, YUG

1981-05 Rafael Marino, COL
1981-07 Juha Oikkonen, FIN – IMO6
1981-08 Arthur Engel, FRG – IMO2
1981-09 Arthur Engel, FRG
1981-12 Jan van de Craats, NLD – IMO3
1981-15 David Monk, UNK – IMO1
1981-19 Vladimir Janković, YUG

1981-01 David Monk, UNK – IMO1
1982-03 A.N. Grishkov, USS – IMO3
1982-05 Jan van de Craats, NLD – IMO5
1982-06 V. Như Cương, VNM – IMO6
1982-13 Jan van de Craats, NLD – IMO2
1982-16 David Monk, UNK – IMO4

1983-08 Juan Ochoa, ESP
1983-09 Murray Klamkin, USA – IMO6
1983-12 David Monk, UNK – IMO1
1983-13 Lucien Kieffer, LUX
1983-14 Marcin Kuczma, POL – IMO5
1983-16 Konrad Engel, GDR
1983-17 Hans-Dietrich Gronau, GDR
1983-18 Arthur Engel, FRG – IMO3
1983-19 Titu Andreescu, ROU
1983-23 Igor F. Sharygin, USS – IMO2

1984-04 T. Dashdorj, MNG – IMO5
1984-05 M. Stoll,
    B. Haible, FRG – IMO1
1984-07 Horst Sewerin, FRG
1984-08 Ioan Tomescu, ROU – IMO3
1984-12 Aiko Tiggelaar, NLD – IMO2
1984-14 L. Panaitopol, ROU – IMO4
1984-15 Lucien Kieffer, LUX
1984-17 Arthur Engel, FRG

1985-01 R. Gonchigdorj, MNG – IMO4
1985-03 Jan van de Craats, NLD – IMO3
1985-04 George Szekeres, AUS – IMO2
1985-17 Åke Samuelsson, SWE – IMO6

1985-20 Frank Budden, UNK – IMO1
1985-22 Igor F. Sharygin, USS – IMO5

1986-01 David Monk, UNK – IMO5
1986-05 Arthur Engel, FRG – IMO1
1986-06 Johannes Notenboom, NLD
1986-08 Cecil Rousseau, USA
1986-09 Konrad Engel, GDR – IMO6
1986-12 Elias Wegert, GDR – IMO3
1986-13 Arthur Engel, FRG
1986-15 Johannes Notenboom, NLD
1986-16 Sven Sigurðsson, ISL – IMO4
1986-17 G. Chang, D. Qi, CHN – IMO2

1987-06 Dimitris Kontogiannis, HEL
1987-13 H.-D. Gronau, GDR – IMO5
1987-14 Arthur Engel, FRG
1987-15 Arthur Engel, FRG – IMO3
1987-16 Horst Sewerin, FRG – IMO1
1987-20 Vsevolod F. Lev, USS – IMO6
1987-21 I.A. Kushnir, USS – IMO2
1987-22 N. Minh Đức, VNM – IMO4

1988-09 Stephan Beck, FRG – IMO6
1988-12 Dimitris Kontogiannis, HEL
1988-13 D. Kontogiannis, HEL – IMO5
1988-16 Finbarr Holland, IRL – IMO4
1988-18 Lucien Kieffer, LUX – IMO1
1988-26 David Monk, UNK – IMO3

1989-01 Geoffrey Bailey, AUS – IMO2
1989-05 Joaquín Valderrama, COL
1989-10 Theodoros Bolis, HEL
1989-13 Eggert Briem, ISL – IMO4
1989-14 S.A. Shirali, IND
1989-15 Fergus Gaines, IRL
1989-20 Harm Derksen, NLD – IMO3
1989-21 Johannes Notenboom, NLD
1989-22 Jose Marasigan, PHI – IMO1
1989-23 Marcin Kuczma, POL – IMO6
1989-25 Myung-Hwan Kim, KOR
1989-30 Bernt Lindström, SWE – IMO5

1990-03 Pavol Černek, CZS – IMO2
1990-06 Hagen von Eitzen, FRG – IMO5
1990-07 Theodoros Bolis, HEL
1990-11 C.R. Pranesachar, IND – IMO1

1990-12 Fergus Gaines, IRL
1990-14 Toshio Seimiya, JPN
1990-16 Harm Derksen, NLD – IMO6
1990-17 Johannes Notenboom, NLD
1990-18 Istvan Beck, NOR
1990-23 L. Panaitopol, ROU – IMO3
1990-25 Albert Erkip, TUR – IMO4

1991-02 Toshio Seimiya, JPN
1991-04 Johan Yebbou, FRA – IMO5
1991-06 A. Skopenkov, USS – IMO1
1991-08 Harm Derksen, NLD
1991-10 Cecil Rousseau, USA – IMO4
1991-12 Chengzhang Li, CHN – IMO3
1991-16 L. Panaitopol, ROU – IMO2
1991-19 Tom Laffey, IRL
1991-20 Tom Laffey, IRL
1991-23 C.R. Pranesachar, IND
1991-24 C.R. Pranesachar, IND
1991-28 Harm Derksen, NLD – IMO6

1992-04 Chengzhang Li, CHN – IMO3
1992-05 Germán Rincón, COL
1992-06 B.J. Venkatachala, IND – IMO2
1992-07 S.A. Shirali, IND
1992-08 C.R. Pranesachar, IND
1992-10 Stefano Mortola, ITA – IMO5
1992-13 Achim Zulauf, NZL – IMO1
1992-14 Marcin Kuczma, POL
1992-18 Robin Pemantle, USA
1992-19 Tom Laffey, IRL
1992-20 Johan Yebbou, FRA – IMO4
1992-21 Tony Gardiner, UNK – IMO6

1993-03 Francisco Bellot Rosado, ESP
1993-04 Francisco Bellot Rosado, ESP
1993-05 Kerkko Luosto, FIN – IMO3
1993-06 Elias Wegert, GER – IMO5
1993-08 S.A. Shirali, IND
1993-09 V.S. Joshi, IND
1993-10 C.S. Yogananda, IND
1993-11 Tom Laffey, IRL – IMO1
1993-12 Fergus Gaines, IRL
1993-13 Tom Laffey, IRL
1993-15 D. Dimovski, MKD – IMO4
1993-17 N.G. de Bruijn, NLD – IMO6

1993-21 Chirstopher Bradley, UNK
1993-22 David Monk, UNK – IMO2
1993-23 David Monk, UNK
1993-24 Titu Andreescu, USA

1994-01 Titu Andreescu, USA
1994-02 V. Lafforgue, FRA – IMO1
1994-03 David Monk, UNK – IMO5
1994-05 Marcin Kuczma, POL
1994-06 Vadym Radchenko, UKR
1994-07 David Berenstein, COL
1994-09 H. Nestra, R. Palm, EST
1994-14 Vyacheslav Yasinskiy, UKR
1994-16 H. Lausch,
          G. Tonoyan, AUS/ARM – IMO2
1994-19 Hans Lausch, AUS – IMO4
1994-20 Kerkko Luosto, FIN – IMO6
1994-22 Gabriel Istrate, ROU – IMO3
1994-24 David Monk, UNK

1995-01 Nazar Agakhanov, RUS – IMO2
1995-03 Vadym Radchenko, UKR
1995-04 Titu Andreescu, USA
1995-05 Vadym Radchenko, UKR
1995-06 Toru Yasuda, JPN
1995-07 B. Mihailov, BGR – IMO1
1995-08 Arthur Engel, GER
1995-10 Vyacheslav Yasinskiy, UKR
1995-11 A. McNaughton, NZL – IMO5
1995-12 Titu Andreescu, USA
1995-14 Germán Rincón, COL
1995-16 Alexander S. Golovanov, RUS
1995-17 Jaromír Šimša, CZE – IMO3
1995-19 Tom Laffey, IRL
1995-20 Marcin Kuczma, POL – IMO6
1995-22 Stephan Beck, GER
1995-23 Igor Mitelman, UKR
1995-24 Marcin Kuczma, POL – IMO4
1995-25 Marcin Kuczma, POL
1995-28 B.J. Venkatachala, IND

1996-01 Peter Anastasov, SVN
1996-02 F. Gaines, T. Laffey, IRL
1996-03 Michalis Lambrou, HEL
1996-06 Tom Laffey, IRL
1996-08 M. Becheanu, ROU – IMO3

1996-09 Marcin Kuczma, POL
1996-10 David Monk, UNK
1996-11 J.P. Grossman, CAN – IMO2
1996-12 David Monk, UNK
1996-13 Titu Andreescu, USA
1996-14 N.M. Sedrakyan, ARM – IMO5
1996-16 Christopher Bradley, UNK
1996-18 Vyacheslav Yasinskiy, UKR
1996-19 Igor Mitelman, UKR
1996-20 A.S. Golovanov, RUS – IMO4
1996-21 Emil Kolev, BGR
1996-24 Kerkko Luosto, FIN – IMO1
1996-25 Vadym Radchenko, UKR
1996-26 T. Andreescu, K. Kedlaya, USA
1996-28 Laurent Rosaz, FRA – IMO6
1996-30 F. Gaines, T. Laffey, IRL

1997-01 Igor Voronovich, BLR – IMO1
1997-03 Arthur Engel, GER
1997-04 E.S. Mahmoodian,
       M. Mahdian, IRN – IMO4
1997-05 Mircea Becheanu, ROU
1997-06 F. Gaines, T. Laffey, IRL
1997-07 Valentina Kirichenko, RUS
1997-08 David Monk, UNK – IMO2
1997-09 T. Andreescu, K. Kedlaya, USA
1997-11 N.G. de Bruijn, NLD
1997-12 Roberto Dvornicich, ITA
1997-13 C.R. Pranesachar, IND
1997-14 B.J. Venkatachala, IND
1997-15 Alexander S. Golovanov, RUS
1997-16 Igor Voronovich, BLR
1997-17 Petr Kaňovský, CZE – IMO5
1997-18 David Monk, UNK
1997-19 Finbarr Holland, IRL
1997-20 Kevin Hutchinson, IRL
1997-21 A. Kachurovskiy,
       O.Bogopolskiy, RUS – IMO3
1997-22 Vadym Radchenko, UKR
1997-23 Christopher Bradley, UNK
1997-24 G. Alkauskas, LTU – IMO6
1997-25 Marcin Kuczma, POL
1997-26 Stefano Mortola, ITA

1998-01 Charles Leytem, LUX – IMO1
1998-02 Waldemar Pompe, POL

1998-03 V. Yasinskiy, UKR – IMO5
1998-06 Waldemar Pompe, POL
1998-07 David Monk, UNK
1998-08 Sambuddha Roy, IND
1998-12 Marcin Kuczma, POL
1998-13 O. Mushkarov,
       N. Nikolov, BGR – IMO6
1998-14 David Monk, UNK – IMO4
1998-16 Vadym Radchenko, UKR
1998-17 David Monk, UNK
1998-19 Igor Voronovich, BLR – IMO3
1998-21 Murray Klamkin, CAN
1998-22 Vadym Radchenko, UKR
1998-25 Arkadii Slinko, NZL
1998-26 Ravi B. Bapat, IND – IMO2

1999-01 Liang-Ju Chu, TWN – IMO4
1999-02 Nairi M. Sedrakyan, ARM
1999-07 Nairi M. Sedrakyan, ARM
1999-08 Shin Hitotsumatsu, JPN
1999-09 Jan Villemson, EST – IMO1
1999-10 David Monk, UNK
1999-12 P. Kozhevnikov, RUS – IMO5
1999-15 Marcin Kuczma, POL – IMO2
1999-19 Tetsuya Ando, JPN – IMO6
1999-21 C.R. Pranesachar, IND
1999-22 Andy Liu, jury, CAN
1999-23 David Monk, UNK
1999-24 Ben Green, UNK
1999-25 Ye. Barabanov,
       I.Voronovich, BLR – IMO3
1999-26 Mansur Boase, UNK

2000-01 Sándor Dobos, HUN – IMO4
2000-02 Roberto Dvornicich, ITA
2000-03 Federico Ardila, COL
2000-07 Titu Andreescu, USA – IMO2
2000-08 I. Leader, P. Shiu, UNK
2000-10 David Monk, UNK
2000-12 Gordon Lessells, IRL
2000-16 Valeriy Senderov, RUS – IMO5
2000-21 Sergey Berlov, RUS – IMO1
2000-22 C.R. Pranesachar, IND
2000-24 David Monk, UNK
2000-27 L. Emelyanov,
       T. Emelyanova, RUS – IMO6

2001-01 B. Rajarama Bhat, IND
2001-02 Marcin Kuczma, POL
2001-04 Juozas Juvencijus Mačys, LTU
2001-05 O. Mushkarov,
    N. Nikolov, BGR
2001-06 Hojoo Lee, KOR – IMO2
2001-07 Federico Ardila, COL
2001-08 Bill Sands, CAN – IMO4
2001-10 Michael Albert, NZL
2001-12 Bill Sands, CAN
2001-14 Christian Bey, GER – IMO3
2001-15 Vyacheslav Yasinskiy, UKR
2001-16 Hojoo Lee, KOR – IMO1
2001-17 Christopher Bradley, UNK
2001-19 Sotiris Louridas, HEL
2001-20 C.R. Pranesachar, IND
2001-22 Shay Gueron, ISR – IMO5
2001-23 Australian PSC, AUS
2001-24 Sandor Ortegón, COL
2001-25 Kevin Buzzard, UNK
2001-27 A. Ivanov, BGR – IMO6
2001-28 F. Petrov, D. Đukić, RUS

2002-02 Mihai Manea, ROU – IMO4
2002-03 G. Bayarmagnai, MNG
2002-04 Stephan Beck, GER
2002-06 L. Panaitopol, ROU – IMO3
2002-08 Hojoo Lee, KOR
2002-09 Hojoo Lee, KOR – IMO2
2002-11 Angelo Di Pasquale, AUS
2002-12 V. Yasinskiy, UKR – IMO6
2002-16 Dušan Đukić, SCG
2002-17 Marcin Kuczma, POL
2002-18 B.J. Venkatachala, IND – IMO5
2002-19 C.R. Pranesachar, IND
2002-20 Omid Naghshineh, IRN
2002-21 Federico Ardila, COL – IMO1
2002-23 Federico Ardila, COL
2002-24 Emil Kolev, BGR
2002-26 Marcin Kuczma, POL
2002-27 Michael Albert, NZL

2003-01 Kiran Kedlaya, USA
2003-02 A. Di Pasquale,
    D. Mathews, AUS
2003-04 Finbarr Holland, IRL – IMO5

2003-05 Hojoo Lee, KOR
2003-06 Reid Barton, USA
2003-07 C.G. Moreira, BRA – IMO1
2003-09 Juozas Juvencijus Mačys, LTU
2003-12 Dirk Laurie, SAF
2003-13 Matti Lehtinen, FIN – IMO4
2003-15 C.R. Pranesachar, IND
2003-17 Hojoo Lee, KOR
2003-18 Waldemar Pompe, POL – IMO3
2003-19 Dirk Laurie, SAF
2003-20 Marcin Kuczma, POL
2003-21 Zoran Šunić, USA
2003-22 A. Ivanov, BGR – IMO2
2003-23 Laurenţiu Panaitopol, ROU
2003-24 Hojoo Lee, KOR
2003-25 Johan Yebbou, FRA – IMO6

2004-01 Hojoo Lee, KOR – IMO4
2004-02 Mihai Bălună, ROU
2004-03 Dan Brown, CAN
2004-04 Hojoo Lee, KOR – IMO2
2004-07 Finbarr Holland, IRL
2004-08 Guihua Gong, PRI
2004-09 Horst Sewerin, GER
2004-10 Norman Do, AUS
2004-11 Marcin Kuczma, POL
2004-12 A. Slinko, S. Marshall, NZL
2004-14 J. Villemson,
    M. Pettai, EST – IMO3
2004-15 Marcin Kuczma, POL
2004-16 D. Şerbănescu,
    V. Vornicu, ROU – IMO1
2004-18 Hojoo Lee, KOR
2004-19 Waldemar Pompe, POL – IMO5
2004-20 Dušan Đukić, SCG
2004-21 B. Green, E. Crane, UNK
2004-23 Dušan Đukić, SCG
2004-26 Mohsen Jamali, IRN
2004-27 Jarosław Wróblewski, POL
2004-28 M. Jamali,
    A. Morabi, IRN – IMO6
2004-29 John Murray, IRL
2004-30 Alexander Ivanov, BGR

2005-02 Nikolai Nikolov, BGR
2005-04 B.J. Venkatachala, IND

2005-05 Hojoo Lee, KOR – IMO3
2005-06 Australian PSC, AUS
2005-09 Federico Ardila, COL
2005-10 Dušan Đukić, SCG
2005-11 R. Gologan,
        D. Schwarz, ROU – IMO6
2005-12 R. Liu, Z. Feng, USA
2005-13 Alexander Ivanov, BGR
2005-14 Dimitris Kontogiannis, HEL
2005-15 Bogdan Enescu, ROU – IMO1
2005-16 Vyacheslav Yasinskiy, UKR
2005-17 Waldemar Pompe, POL – IMO5
2005-20 Hojoo Lee, KOR
2005-21 Mariusz Skałba, POL – IMO4
2005-22 N.G. de Bruijn, NLD – IMO2
2005-24 Carlos Caicedo, COL
2005-26 Mohsen Jamali, IRN

2006-01 Härmel Nestra, EST
2006-02 Mariusz Skałba, POL
2006-04 Dušan Đukić, SRB
2006-05 Hojoo Lee, KOR
2006-06 Finbarr Holland, IRL – IMO3
2006-08 Dušan Đukić, SRB – IMO2
2006-09 Federico Ardila, COL
2006-12 Federico Ardila, COL
2006-13 Kei Irie, JPN
2006-14 Hojoo Lee, KOR – IMO1
2006-15 Vyacheslav Yasinskiy, UKR
2006-16 Zuming Feng, USA
2006-18 Dimitris Kontogiannis, HEL
2006-20 Tomáš Jurík, SVK
2006-21 Waldemar Pompe, POL
2006-23 Dušan Đukić, SRB – IMO6
2006-24 Zuming Feng, USA – IMO4
2006-25 J.P. Grossman, CAN
2006-26 Johan Meyer, SAF
2006-27 Dan Schwarz, ROU – IMO5
2006-29 Zoran Šunić, USA
2006-30 Juhan Aru, EST

2007-01 Michael Albert, NZL – IMO1
2007-02 Nikolai Nikolov, BGR
2007-03 Juhan Aru, EST
2007-05 Vjekoslav Kovač, HRV
2007-06 Waldemar Pompe, POL

2007-07 G. Woeginger, NLD – IMO6
2007-08 Dušan Đukić, SRB
2007-09 Kei Irie, JPN
2007-10 Gerhard Woeginger, NLD
2007-11 Omid Hatami, IRN
2007-12 R. Gologan, D. Schwarz, ROU
2007-13 Vasiliy Astakhov, RUS – IMO3
2007-14 Gerhard Woeginger, AUT
2007-15 Vyacheslav Yasinskiy, UKR
2007-16 Marek Pechal, CZE – IMO4
2007-17 Farzin Barekat, CAN
2007-18 Vyacheslav Yasinskiy, UKR
2007-19 Charles Leytem, LUX – IMO2
2007-20 Christopher Bradley, UNK
2007-21 Z. Feng, O. Golberg, USA
2007-22 Davoud Vakili, IRN
2007-23 Waldemar Pompe, POL
2007-24 Stephan Wagner, AUT
2007-25 Dan Brown, CAN
2007-26 Gerhard Woeginger, NLD
2007-27 Jerzy Browkin, POL
2007-28 M. Jamali,
        N. Ahmadi Pour Anari, IRN
2007-29 K. Buzzard,
        E. Crane, UNK – IMO5
2007-30 N.V. Tejaswi, IND

2008-01 Hojoo Lee, KOR – IMO4
2008-02 Walther Janous, AUT – IMO2
2008-05 Pavel Novotný, SVK
2008-06 Žymantas Darbėnas, LTU
2008-09 Vidan Govedarica, SRB
2008-10 Jorge Tipe, PER
2008-11 B. Le Floch,
        I. Smilga, FRA – IMO5
2008-14 A. Gavrilyuk, RUS – IMO1
2008-15 Charles Leytem, LUX
2008-16 John Cuya, PER
2008-19 Dušan Đukić, SRB
2008-20 V. Shmarov, RUS – IMO6
2008-21 Angelo Di Pasquale, AUS
2008-24 Dušan Đukić, SRB
2008-26 K. Česnavičius, LTU – IMO3

2009-01 Michal Rolínek, CZE
2009-03 Bruno Le Floch, FRA – IMO5

2009-06 Gabriel Carroll, USA – IMO3
2009-07 Japanese PSC, JPN
2009-08 Michael Albert, NZL
2009-14 D. Khramtsov, RUS – IMO6
2009-15 Gerhard Woeginger, AUT
2009-16 H. Lee,
       P. Vandendriessche,
       J. Vonk, BEL – IMO4

2009-17 Sergei Berlov, RUS – IMO2
2009-19 David Monk, UNK
2009-21 Eugene Bilopitov, UKR
2009-24 Ross Atkins, AUS – IMO1
2009-25 Jorge Tipe, PER
2009-28 József Pelikán, HUN
2009-29 Okan Tekman, TUR

References

1. M. Aassila, *300 Défis Mathématiques*, Ellipses, 2001.
2. M. Aigner, G.M. Ziegler, *Proofs from THE BOOK*, Springer; 4th edition, 2009.
3. G.L. Alexanderson, L.F. Klosinski, L.C. Larson, *The William Lowell Putnam Mathematical Competition, Problems and Solutions: 1965–1984*, The Mathematical Association of America, 1985.
4. N. Altshiller-Court, *College Geometry: An Introduction to the Modern Geometry of the Triangle and the Circle*, Dover Publications, 2007.
5. T. Andreescu, D. Andrica, *360 Problems for Mathematical Contests*, GIL Publishing House, Zalău, 2003.
6. T. Andreescu, D. Andrica, *An Introduction to Diophantine Equations*, GIL Publishing House, 2002.
7. T. Andreescu, D. Andrica, *Complex Numbers from A to ... Z*, Birkhäuser, Boston, 2005.
8. T. Andreescu, D. Andrica, Z. Feng, *104 Number Theory Problems*, Birkhauser, Boston, 2006.
9. T. Andreescu, V. Cartoaje, G. Dospinescu, M. Lascu, *Old and New Inequalities*, GIL Publishing House, 2004.
10. T. Adnreescu, G. Dospinescu, *Problems from the Book*, XYZ Press, 2008.
11. T. Andreescu, B. Enescu, *Mathematical Treasures*, Birkhäuser, Boston, 2003.
12. T. Andreescu, Z. Feng, *102 Combinatorial Problems*, Birkhäuser Boston, 2002.
13. T. Andreescu, Z. Feng, *103 Trigonometry Problems: From the Training of the USA IMO Team*, Birkhäuser Boston, 2004.
14. T. Andreescu, Z. Feng, *Mathematical Olympiads 1998–1999, Problems and Solutions from Around the World*, The Mathematical Association of America, 2000.
15. T. Andreescu, Z. Feng, *Mathematical Olympiads 1999–2000, Problems and Solutions from Around the World*, The Mathematical Association of America, 2002.
16. T. Andreescu, Z. Feng, *Mathematical Olympiads 2000–2001, Problems and Solutions from Around the World*, The Mathematical Association of America, 2003.
17. T. Andreescu, Z. Feng, *A Path to Combinatorics for Undergraduates: Counting Strategies*, Birkhauser, Boston, 2003.
18. T. Andreescu, R. Gelca, *Mathematical Olympiad Challenges*, Birkhäuser, Boston, 2000.
19. T. Andreescu, K. Kedlaya, P. Zeitz, *Mathematical Contests 1995-1996, Olympiads Problems and Solutions from Around the World*, American Mathematics Competitions, 1997.
20. T. Andreescu, K. Kedlaya, *Mathematical Contests 1996-1997, Olympiads Problems and Solutions from Around the World*, American Mathematics Competitions, 1998.

21. T. Andreescu, K. Kedlaya, *Mathematical Contests 1997-1998, Olympiads Problems and Solutions from Around the World*, American Mathematics Competitions, 1999.
22. T. Andreescu, O. Mushkarov, L. Stoyanov, *Geometric Problems on Maxima and Minima*, Birkhauser Boston, 2005.
23. M. Arsenović, V. Dragović, *Functional Equations (in Serbian)*, Mathematical Society of Serbia, Beograd, 1999.
24. M. Ašić et al., *International Mathematical Olympiads (in Serbian)*, Mathematical Society of Serbia, Beograd, 1986.
25. M. Ašić et al., *60 Problems for XIX IMO (in Serbian)*, Society of Mathematicians, Physicists, and Astronomers, Beograd, 1979.
26. A. Baker, *A Concise Introduction to the Theory of Numbers*, Cambridge University Press, Cambridge, 1984.
27. E.J. Barbeau, *Polynomials*, Springer, 2003.
28. E.J. Barbeau, M.S. Klamkin, W.O.J. Moser, *Five Hundred Mathematical Challenges*, The Mathematical Association of America, 1995.
29. E.J. Barbeau, *Pell's Equation*, Springer-Verlag, 2003.
30. M. Becheanu, *International Mathematical Olympiads 1959–2000. Problems. Solutions. Results*, Academic Distribution Center, Freeland, USA, 2001.
31. E.L. Berlekamp, J.H. Conway, R.K. Guy, *Winning Ways for Your Mathematical Plays*, Volumes 1-4, AK Peters, Ltd., 2nd edition, 2001 – 2004.
32. G. Berzsenyi, S.B. Maurer, *The Contest Problem Book V*, The Mathematical Association of America, 1997.
33. R. Brualdi, *Introductory Combinatorics*, 4th edition, Prentice-Hall, 2004.
34. C.J. Bradley, *Challenges in Geometry : for Mathematical Olympians Past and Present*, Oxford University Press, 2005.
35. P.S. Bullen, D.S. Mitrinović , M. Vasić, *Means and Their Inequalities*, Springer, 1989.
36. C. Chuan-Chong, K. Khee-Meng, *Principles and Techniques in Combinatorics*, World Scientific Publishing Company, 1992.
37. H.S.M. Coxeter, *Introduction to Geometry*, John Willey & Sons, New York, 1969
38. H.S.M. Coxeter, S.L. Greitzer, *Geometry Revisited*, Random House, New York, 1967.
39. I. Cuculescu, *International Mathematical Olympiads for Students (in Romanian)*, Editura Tehnica, Bucharest, 1984.
40. A. Engel, *Problem Solving Strategies*, Springer, 1999.
41. D. Fomin, A. Kirichenko, *Leningrad Mathematical Olympiads 1987–1991*, MathPro Press, 1994.
42. A.A. Fomin, G.M. Kuznetsova, *International Mathematical Olympiads (in Russian)*, Drofa, Moskva, 1998.
43. A. Gardiner, *The Mathematical Olympiad Handook*, Oxford, 1997.
44. R. Gelca, T. Andreescu, *Putnam and Beyond*, Springer 2007.
45. A.M. Gleason, R.E. Greenwood, L.M. Kelly, *The William Lowell Putnam Mathematical Competition, Problems and Solutions: 1938–1964*, The Mathematical Association of America, 1980.
46. R.K. Guy, *Unsolved Problems in Number Theory*, Springer, 3rd edition, 2004.
47. S.L. Greitzer, *International Mathematical Olympiads 1959–1977*, M.A.A., Washington, D. C., 1978.
48. R.L. Graham, D.E. Knuth, O. Patashnik, *Concrete Mathematics*, 2nd Edition, Addison-Wesley, 1989.
49. L-S. Hahn, *Complex Numbers & Geometry*, New York, 1960.
50. L. Hahn, *New Mexico Mathematics Contest Problem Book*, University of New Mexico Press, 2005.

51. G.H. Hardy, J.E. Littlewood, G. Pólya, *Inequalities*, Cambridge University Press; 2nd edition, 1998.
52. G.H. Hardy, E.M. Wright, *An Introduction to the Theory of Numbers*, Oxford University Press; 5th edition, 1980.
53. R. Honsberger, *From Erdos to Kiev: Problems of Olympiad Caliber*, MAA, 1996.
54. R. Honsberger, *In Polya's Footsteps: Miscelaneous Problems and Essays*, MAA, 1997.
55. D. Hu, X. Tao, *Mathematical Olympiad – the Problems Proposed to the 31st IMO (in Chinese)*, Peking University Press, 1991.
56. V. Janković, Z. Kadelburg, P. Mladenović, *International and Balkan Mathematical Olympiads 1984–1995 (in Serbian)*, Mathematical Society of Serbia, Beograd, 1996.
57. V. Janković, V. Mićić, *IX & XIX International Mathematical Olympiads*, Mathematical Society of Serbia, Beograd, 1997.
58. R.A. Johnson, *Advanced Euclidean Geometry*, Dover Publications, 1960.
59. Z. Kadelburg, D. Djukić, M. Lukić, I. Matić, *Inequalities (in Serbian)*, Mathematical Society of Serbia, Beograd, 2003.
60. N.D. Kazarinoff, *Geometric Inequalities*, Mathematical Association of America (MAA), 1975.
61. K.S. Kedlaya, B. Poonen, R. Vakil, *The William Lowell Putnam Mathematical Competition 1985–2000 Problems, Solutions and Commentary*, The Mathematical Association of America, 2002.
62. A.P. Kiselev (author), A. Givental (editor), *Kiselev's Geometry / Book I Planimetry*, Sumizdat, 2006.
63. A.P. Kiselev (author), A. Givental (editor), *Kiselev's Geometry / Book II Stereometry*, Sumizdat, 2008.
64. M.S. Klamkin, *International Mathematical Olympiads 1979–1985 and Forty Supplementary Problems*, M.A.A., Washington, D.C., 1986.
65. M.S. Klamkin, *International Mathematical Olympiads 1979–1986*, M.A.A., Washington, D.C., 1988.
66. M.S. Klamkin, *USA Mathematical Olympiads 1972–1986*, M.A.A., Washington, D.C., 1988.
67. M.E. Kuczma, *144 Problems of the Austrian-Polish Mathematics Competition 1978–1993*, The Academic Distribution Center, Freeland, Maryland, 1994.
68. J. Kurshak, *1Hungarian Problem Book I*, MAA, 1967.
69. J. Kurshak, *1Hungarian Problem Book II*, MAA, 1967.
70. A. Kupetov, A. Rubanov, *Problems in Geometry*, MIR, Moscow, 1975.
71. S. Lando, *Lectures on Generating Functions*, AMS, 2003.
72. H.-H. Langmann, *30th International Mathematical Olympiad, Braunschweig 1989*, Bildung und Begabung e.V., Bonn 2, 1990.
73. L.C. Larson, *Problem Solving Through Problems*, Springer, 1983.
74. H. Lausch, C. Bosch-Giral *Asian Pacific Mathematics Olympiads 1989–2000*, AMT, Canberra, 2000.
75. H. Lausch, P. Taylor, *Australian Mathematical Olympiads 1979–1995*, AMT, Canberra 1997.
76. P.K. Hung, *Secrets in Inequalities*, GIL Publishing House, 2007.
77. A. Liu, *Chinese Mathematical Competitions and Olympiads 1981–1993*, AMT, Canberra 1998.
78. A. Liu, *Hungarian Problem Book III*, MAA, 2001.
79. E. Lozansky, C. Rousseau, *Winning Solutions*, Springer-Verlag, New York, 1996.
80. V. Mićić, Z. Kadelburg, D. Djukić, *Introduction to Number Theory (in Serbian)*, 4th edition, Mathematical Society of Serbia, Beograd, 2004.

81. D.S. Mitrinović , J. Pečarić, A.M. Fink, *Classical and New Inequalities in Analysis*, Springer, 1992.
82. D.S. Mitrinović, J.E. Pečarić, V. Volenec, *Recent Advances in Geometric Inequalities*, Kluwer Academic Publishers, 1989.
83. P. Mladenović, *Combinatorics*, 3rd edition, Mathematical Society of Serbia, Beograd, 2001.
84. P.S. Modenov, *Problems in Geometry*, MIR, Moscow, 1981.
85. P.S. Modenov, A.S. Parhomenko, *Geometric Transformations*, Academic Press, New York, 1965.
86. L. Moisotte, *1850 exercices de mathémathique*, Bordas, Paris, 1978.
87. L.J. Mordell, *Diophantine Equations*, Academic Press, London and New York, 1969.
88. E.A. Morozova, I.S. Petrakov, V.A. Skvortsov, *International Mathematical Olympiads (in Russian)*, Prosveshchenie, Moscow, 1976.
89. I. Nagell, *Introduction to Number Theory*, John Wiley & Sons, Inc., New York, Stockholm, 1951.
90. I. Niven, H.S. Zuckerman, H.L. Montgomery, *An Introduction to the Theory of Numbers*, John Wiley and Sons, Inc., 1991.
91. G. Polya, *How to Solve It: A New Aspect of Mathematical Method*, Princeton University Press, 2004.
92. C.R. Pranesachar, Shailesh A. Shirali, B.J. Venkatachala, C.S. Yogananda, *Mathematical Challenges from Olympiads*, Interline Publishing Pvt. Ltd., Bangalore, 1995.
93. V.V. Prasolov, *Problems of Plane Geometry*, Volumes 1 and 2, Nauka, Moscow, 1986.
94. V.V. Prasolov, V.M. Tikhomirov *Geometry*, Volumes 1 and 2, American Mathematical Society, 2001.
95. A.S. Posamentier, C.T. Salking, *Challenging Problems in Algebra*, Dover Books in Mathematics, 1996.
96. A.S. Posamentier, C.T. Salking, *Challenging Problems in Geometry*, Dover Books in Mathematics, 1996.
97. I. Reiman, J. Pataki, A. Stipsitz, *International Mathematical Olympiad: 1959–1999*, Anthem Press, London, 2002.
98. I.F. Sharygin, *Problems in Plane Geometry*, Imported Pubn, 1988.
99. W. Sierpinski, *Elementary Theory of Numbers*, Polski Academic Nauk, Warsaw, 1964.
100. W. Sierpinski, *250 Problems in Elementary Number Theory*, American Elsevier Publishing Company, Inc., New York, PWN, Warsaw, 1970.
101. A.M. Slinko, *USSR Mathematical Olympiads 1989–1992*, AMT, Canberra, 1998.
102. C.G. Small, *Functional Equations and How to Solve Them*, Springer 2006.
103. Z. Stankova, T. Rike, *A Decade of the Berkeley Math Circle*, American Mathematical Society, 2008.
104. R.P. Stanley, *Enumerative Combinatorics*, Volumes 1 and 2, Cambridge University Press; New Ed edition, 2001.
105. D. Stevanović, M. Milošević, V. Baltić, *Discrete Mathematics: Problem Book in Elementary Combinatorics and Graph Theory (in Serbian)*, Mathematical Society of Serbia, Beograd, 2004.
106. A.M. Storozhev, *International Mathematical Tournament of the Towns, Book 5: 1997–2002*, AMT Publishing, 2006.
107. P.J. Taylor, *International Mathematical Tournament of the Towns, Book 1: 1980–1984*, AMT Publishing, 1993.
108. P.J. Taylor, *International Mathematical Tournament of the Towns, Book 2: 1984–1989*, AMT Publishing, 2003.

109. P.J. Taylor, *International Mathematical Tournament of the Towns, Book 3: 1989–1993*, AMT Publishing, 1994.

110. P.J. Taylor, A.M. Storozhev, *International Mathematical Tournament of the Towns, Book 4: 1993–1997*, AMT Publishing, 1998.

111. J. Tattersall, *Elementary Number Theory in Nine Chapters*, 2nd edition, Cambridge University Press, 2005.

112. I. Tomescu, R.A. Melter, *Problems in Combinatorics and Graph Theory*, John Wiley & Sons, 1985.

113. I. Tomescu et al., *Balkan Mathematical Olympiads 1984-1994 (in Romanian)*, Gil, Zalău, 1996.

114. R. Vakil, *A Mathematical Mosaic: Patterns and Problem Solving*, 2nd edition, MAA, 2007.

115. J.H. van Lint, R.M. Wilson, *A Course in Combinatorics*, second edition, Cambridge University Press, 2001.

116. I.M. Vinogradov, *Elements of Number Theory*, Dover Publications, 2003.

117. I.M. Vinogradov, *The Method of Trigonometrical Sums in the Theory of Numbers*, Dover Books in Mathematics, 2004.

118. H.S. Wilf, *Generatingfunctionology*, Academic Press, Inc.; 3rd edition, 2006.

119. A.M. Yaglom, I.M. Yaglom, *Challenging Mathematical Problems with Elementary Solutions*, Dover Publications, 1987.

120. I.M. Yaglom, *Geometric Transformations*, Vols. I, II, III, The Mathematical Association of America (MAA), 1962, 1968, 1973.

121. P. Zeitz, *The Art and Craft of Problem Solving*, Wiley; International Student edition, 2006.